# Direct and Inverse Methods in Radar Polarimetry

Part 1



This Advanced Research Workshop was co-sponsored by:
NATO, Scientific Affairs Division
DLR, Institute for Radio Frequency Technology
U.S.-ARO, European Branch Office/Geosciences Division
U.S. Offices of Naval Research & Naval Technology
TST, Telefunken System Technik
Carl Zeiss, Electronics R&D Center
UEN, High Frequency Engineering Laboratories
UIC/EECS, Communications & Sensing Laboratory

# NATO ASI Series

**Advanced Science Institutes Series**

*A Series presenting the results of activities sponsored by the NATO Science Committee, which aims at the dissemination of advanced scientific and technological knowledge, with a view to strengthening links between scientific communities.*

The Series is published by an international board of publishers in conjunction with the NATO Scientific Affairs Division

| | |
|---|---|
| **A Life Sciences**<br>**B Physics** | Plenum Publishing Corporation<br>London and New York |
| **C Mathematical and Physical Sciences**<br>**D Behavioural and Social Sciences**<br>**E Applied Sciences** | Kluwer Academic Publishers<br>Dordrecht, Boston and London |
| **F Computer and Systems Sciences**<br>**G Ecological Sciences**<br>**H Cell Biology**<br>**I Global Environmental Change** | Springer-Verlag<br>Berlin, Heidelberg, New York, London, Paris and Tokyo |

**NATO-PCO-DATA BASE**

The electronic index to the NATO ASI Series provides full bibliographical references (with keywords and/or abstracts) to more than 30000 contributions from international scientists published in all sections of the NATO ASI Series.
Access to the NATO-PCO-DATA BASE is possible in two ways:

– via online FILE 128 (NATO-PCO-DATA BASE) hosted by ESRIN, Via Galileo Galilei, I-00044 Frascati, Italy.

– via CD-ROM "NATO-PCO-DATA BASE" with user-friendly retrieval software in English, French and German (© WTV GmbH and DATAWARE Technologies Inc. 1989).

The CD-ROM can be ordered through any member of the Board of Publishers or through NATO-PCO, Overijse, Belgium.

**Series C: Mathematical and Physical Sciences - Vol. 350 Part 1**

# Direct and Inverse Methods in Radar Polarimetry

Part 1

edited by


Wolfgang-M. Boerner
Communications & Sensing Laboratory,
EECS Department,
University of Illinois at Chicago, U.S.A.

Hans Brand
High-Frequency Engineering
Laboratories, University Erlangen-
Nürnberg, Germany

Leonard A. Cram
Thorn EMI, Electronics Ltd., U.K.

William A. Holm
GTRI-RAIL/MAL,
Atlanta, GA,
U.S.A.

David E. Stein
Westinghouse Electric Corporation,
Baltimore, MD,
U.S.A.

Werner Wiesbeck
IHE, University of Karlsruhe,
Karlsruhe, Germany

Wolfgang Keydel
DLR, Institute for Radio-Frequency
Technology, Oberpfaffenhofen, Germany

Dino Giuli
Radar Research Laboratory,
Department of Electronics Engineering,
University of Florence, Italy

Dag T. Gjessing
ESTP, Royal Norwegian Council of
Industrial & Scientific Research, Kjeller,
Norway

Frédéric A. Molinet
Société Mothesim,
Le Plessis-Robinson,
France



Springer-Science+Business Media, B.V.

Proceedings of the NATO Advanced Research Workshop on
Direct and Inverse Methods in Radar Polarimetry
Bad Windsheim, Franconia, Germany
September 18–24, 1988

**Library of Congress Cataloging-in-Publication Data**

NATO Advanced Research Workshop on Direct and Inverse Methods in Radar Polarimetry (1988 : Bad Windsheim, Germany)
Direct and inverse methods in radar polarimetry / edited by Wolfgang-M. Boerner ... [et al.].
p. cm. -- (NATO ASI series. Series C, Mathematical and physical sciences ; vol. 350)
"Proceedings of the NATO Advanced Research Workshop on Direct and Inverse Methods in Radar Polarimetry, Bad Windsheim, Franconia, F.R.G., September 18-24, 1988"--T.p. verso.
"Published in cooperation with the NATO Scientific Affairs Division."
Includes index.
ISBN 978-94-010-9245-6 ISBN 978-94-010-9243-2 (eBook)
DOI 10.1007/978-94-010-9243-2
1 : alk. paper). -- ISBN 0-7923-1497-2 (pt. 2 : alk. paper)
1. Radar--Congresses. 2. Polarimetry--Congresses. 3. Electromagnetic waves--Polarization--Congresses. I. Boerner, Wolfgang M., 1937- . II. Title. III. Series: NATO ASI series. Series C, Mathematical and physical sciences ; no. 350.
TK6573.N28 1988
621.3848--dc20 91-41267
CIP

ISBN 978-94-010-9245-6

*Printed on acid-free paper*

**TABLE OF CONTENTS (Part 1)** v

## ADDRESS OF THE HOST NATION

Besides amplitude, frequency and phase, the knowledge of the polarization behavior is essential for the understanding of the scattering mechanism and the interaction between electromagnetic waves and respective targets. The need of polarization information utilization for military and civilian reconnaissance and remote sensing is steadily increasing. However, the total information content of 'polarization' is, up to now, not known exactly and the respective methods and techniques for polarization measurement, polarimetric systems calibration, data evaluation and data interpretation are in many aspects not yet sufficiently developed.

Therefore, this workshop on direct and inverse methods in radar polarimetry has been facilitated in order to assess and promote the state of the art of high-resolution radar polarimetry. Especially, the use of polarimetry for target-clutter separation and target- and clutter-classification and imaging as well should be main topics for workshop deliberations. For this purpose, more than 100 experts came together in Bad Windsheim concentrating on direct and inverse methods exclusively related to radar polarimetry which has recently become an indispensible tool in modern electromagnetic sensor technology in the civil and military sectors as well and in the different fields of remote sensing and radar meteorology.

If so many experts which represent the key persons in this important scientific field from countries of the eastern and western world meet for a workshop, then the scientific community will expect relevant results. Those results have been brought together in these proceedings. A more general text and source book on this subject matter does not exist. It seems an impossible task for one man to write a suitable book because of the great variety of subjects and the wide range of applications; even if a single expert or a group had to produce a book, it would certainly have removed them from active research for several years. Therefore, in preparing the proceedings of this workshop, several review papers have been introduced, which together with more specific papers, should provide a good overview on the subject matter. Thus, may the resulting proceedings serve as a source and reference book where none existed before.

At this place, I want to express my gratitude to Professor Wolfgang-M. Boerner. He was the conceiver, the organizer and the promoter of this workshop. He solicited the necessary funds, his enthusiasm encouraged leading scientists to participate and to contribute to the workshop and to these proceedings; he established the program, organized the sessions and selected the cultural program, and edited this textbook. Wolfgang-M. Boerner was one of the first who realized the importance of polarimetry for reconnaissance and remote sensing and he always acted as a forerunner in this field. Therefore, he was the driver and the motor, all in one, of this workshop. We all owe him the debt of gratitude.

Wolfgang Keydel, Director
Institute of Radio Frequency Technology
German Space & Aerospace Research Establishment
DLR-Oberpfaffenhofen, FR Germany

## OPENING REMARKS

It was my honor and pleasure to attend the Inverse Methods in Electromagnetic Imaging (IMEI) NATO Advanced Research Workshop in 1983 and the Direct and Inverse Methods in Radar Polarimetry (DIMRP) NATO Advanced Research Workshop in 1988. Both workshops were held in late September at the Kur-und Kongresshotel Residenz which is adjacent to the Kurpark in Bad Windsheim, FRG.

After attending the NATO-ARW-IMEI in 1983, it was my opinion that the workshop was one of the best organized and most enlightening technical meetings I had attended in many years. In retrospect, I must say that the 1988 NATO-ARW-DIMRP may have been an even greater success. The director, Dr. Wolfgang-M. Boerner, assembled more than 100 renowned scientists and engineers from 14 different countries including China and Poland. All were active participants in the workshop as speaker, session chairman, working group chairman/member/reporter, moderator or critical observer.

The primary objective of the 1988 NATO-ARW-DIMRP was to provide a forum for free and efficient interchange of radar polarimetric theory, techniques, applications and research on an international basis. The objective was met through the presentation of some 72 technical papers by polarimetry experts from countries in Europe, North America and Asia.

The Director and his Codirectors facilitated the international interaction and technical exchange by organizing the workshop participants into six working groups. Each working group was assigned to address a specific unresolved issue relating to radar polarimetry and to provide recommendations for potential future research projects which will require active interactions among engineering scientists of all NATO-member countries. The Reporter for each working group presented a ten-minute summary of his group's activities, conclusions and recommendations. The final reports are included in these Proceedings of the NATO-ARW-DIMRP'88.

After inspection of the headings and the dedication address you will observe that our workshop has been dedicated to one of the pioneering radar polarimetrists, Dr. J. Richard Huynen, who so strongly advanced theory and concepts next to Prof. Edward Morton Kennaugh to whom the first workshop was dedicated. It gives me special pleasure to congratulate Rick on behalf of all participants of this NATO Advanced Research Workshop of 1988 and to present him with this deserved honor.

The workshop was a huge success in its own right. When coupled with the beauty of the Franconian region along with the social and cultural program arranged by the Director, the conference was truly outstanding. I am certain that all workshop participants agree with my assessment of the 1988 NATO-ARW-DIMRP and that they also look forward to participating in future NATO-ARW meetings at the Kur-und Kongresshotel Residenz in Bad Windsheim.

Jerry L. Eaves
Associate Director,
Radar and Instrumentation Laboratory
Georgia Tech. Research Institute
Georgia Institute of Technology
Atlanta, GA USA

JEAN RICHARD HUYNEN

## IN DEDICATION TO

**SENIOR RADAR POLARIMETRIST: JEAN RICHARD HUYNEN**
(*1920 October 8, Batavia, Nederlands Oost Indie)

Among the engineering scientists who have most decisively contributed toward forefront advances for

**"The Development of Polarimetric Radar Techniques and Radar Target Phenomenology which have Advanced Theory, Metrology, and Applications of Radar Target Detection, Classification and Identification,"**

Dr. J. Richard Huynen stands out, next to the late Professor Edward M. Kennaugh, as one of the towering pioneers.

Rick developed the Phenomenological Theory of Radar Targets in the course of his long professional life during the years of about 1949 to 1977, based on the earlier work of Sinclair, Rumsey, Deschamps, Gent, Graves, and Copeland, which culminated in his D.Sc. dissertation delivered at the Delft University of Technology under the promotion of famed Professor Dr. Ir. Johannes P. Schouten on 1970 December 16. By mere coincidence, Ir. André J. Poelman sat in on Rick's defense and, without delay, immediately thereafter went about and built the first truly dual polarization coherent radar system, culminating in the design of the first high power rapid electronic polarization switch, in order to verify Rick's visionary concepts of the target characteristic equations expressed in terms of his **Polarization Fork.** In the meantime, this monograph has been studied world-wide; and so have many of his subsequent treatises been discussed and assessed most critically, in the West and the East alike, and those have become the fruit of thought and inspiration for many pioneering advancements in polarimetric radar technology such as those by André J. Poelman, Vladimir A. Potekhin, Jerry L. Eaves, Anatolij Ivanovich Kozlov, Donald J.R. Stock, and many others.

In particular, with the recent resurgence of polarimetric radar device technology his basic work was immediately in demand and has become instrumental in advancing polarimetric radar theory, metrology, signal/image processing, and technology from every possible angle of approach as is also documented in these Proceedings. Although not all his visionary concepts are accepted today, it is with great pleasure to have him still contribute to the field so actively and it is also with the sincerest gratitude from all of us to him for his congenial invigorating stimulations and continued strive for advancing radar polarimetry that we are dedicating these Proceedings of the NATO-ARW on Direct and Inverse Methods in Radar Polarimetry (1988) to him.

On 1990 December 23, Rick suffered an aneurysm and had to undergo a nine-hour craniotomy, from which he recovered rather well. We wish him well and we hope we all continue to enjoy our interactions with this vivid and witty nature for a long time to come.

On behalf of all participants,

Wolfgang-M. Boerner
University of Illinois Senior Scholar

**NATO-ARW-DIMRP'88**
Group of Participants
present during break of
FINAL WORKING GROUP SESSION
KuK Hotel Residenz, Bad Windsheim
Friday, September 23, 1988, 14:30
- Exit to the Kurpark -
"Auf der Brücke"

By courtesy of
Photohaus Karl Drumm
Bad Windsheim

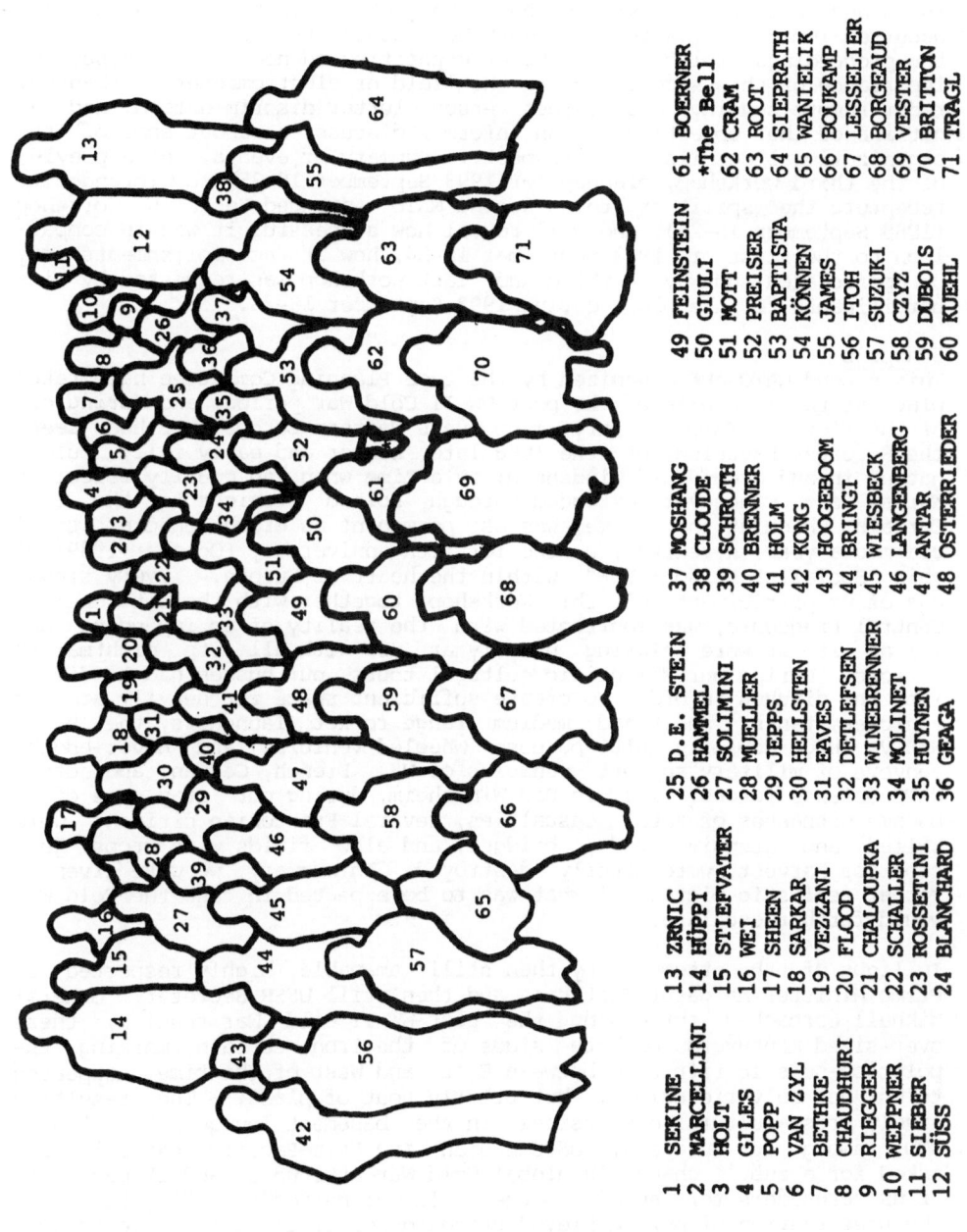

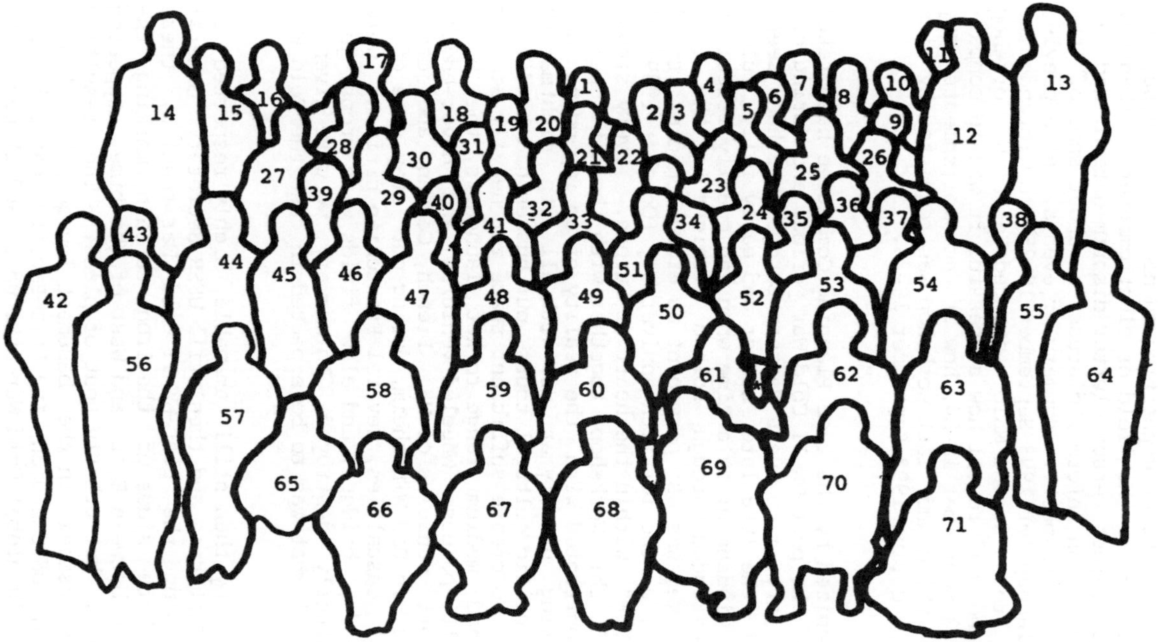

| | | | | | |
|---|---|---|---|---|---|
| 1 SEKINE | 13 ZRNIC | 25 D.E. STEIN | 37 MOSHANG | 49 FEINSTEIN | 61 BOERNER |
| 2 MARCELLINI | 14 HÜPPI | 26 HAMMEL | 38 CLOUDE | 50 GIULI | *The Bell |
| 3 HOLT | 15 STIEFVATER | 27 SOLIMINI | 39 SCHROTH | 51 MOTT | 62 CRAM |
| 4 GILES | 16 WEI | 28 MUELLER | 40 BRENNER | 52 PREISER | 63 ROOT |
| 5 POPP | 17 SHEEN | 29 JEPPS | 41 HOLM | 53 BAPTISTA | 64 SIEPRATH |
| 6 VAN ZYL | 18 SARKAR | 30 HELLSTEN | 42 KONG | 54 KÖNNEN | 65 WANIELIK |
| 7 BETHKE | 19 VEZZANI | 31 EAVES | 43 HOOGEBOOM | 55 JAMES | 66 BOUKAMP |
| 8 CHAUDHURI | 20 FLOOD | 32 DETLEFSEN | 44 BRINGI | 56 ITOH | 67 LESSELIER |
| 9 RIEGGER | 21 CHALOUPKA | 33 WINEBRENNER | 45 WIESBECK | 57 SUZUKI | 68 BORGEAUD |
| 10 WEPPNER | 22 SCHALLER | 34 MOLINET | 46 LANGENBERG | 58 CZYZ | 69 VESTER |
| 11 SIEBER | 23 ROSSETINI | 35 HUYNEN | 47 ANTAR | 59 DUBOIS | 70 BRITTON |
| 12 SÜSS | 24 BLANCHARD | 36 GEAGA | 48 OSTERRIEDER | 60 KUEHL | 71 TRAGL |

## DIRECTOR'S FOREWORD

This foreword deals exclusively with the planning, organization, and execution of the Workshop's scientific as well as cultural programs. It is opened with a synopsis on how the global political changes that occurred immediately after the Workshop caused the delay in producing the proceedings, followed by a brief exposition on need, timeliness, and importance of this second ARW in the field of electromagnetic imaging, radar remote sensing, and target versus clutter discrimination; and an outline of the objectives. An informal discussion about some of the organizational details, a retrospective summary of events, and a preview of the third workshop, planned for 1993 September 19-25, is intended to recapture the spirit of this second NATO Advanced Research Workshop (1988 September 18-24), and will reveal how successful it was in comparison to the first of 1983 September 18-24, how its accomplishments may be appreciated and why a third and last workshop was requested by its participants to take place during 1993 September 19-25.

Synopsis

This second NATO-ARW organized by the same Planning Committee has taken place at the last peak of the post-WW-II Cold War period, and during one of the fiercest NATO military manoeuvers Western Europe ever had seen, the Reforger Exercises of the 1988 late summer and early fall, culminating in and around Bad Windsheim; at a time when an equally ferocious Warsaw Pact manoeuver extended through Poland straight to the Iron Curtain, which at that time, was as permanent as Erich Honecker still claimed a year thereafter, on the fortieth anniversary (October 1989) of this "Socialist Prison: DDR", within the heart of Europe. Every single one of us participants of this Workshop, together with the citizens of Central Franconia, was confronted with the reality of an impending WW-III as if it were glaring nakedly at us with all its nightmarish horrors. During our scientific/cultural tours, our busses had to divert into the ditches in order to create sufficient space for heavily armored vehicles, tanks, short and medium range rocket launchers, the bulky HMMWVs (High Mobility Multi-purpose Wheeled Vehicles) and never-ending convoys of military support vehicles of US, French, German, and other NATO defense forces. Yes, in Bad Windsheim, during our stay some of us became witnesses of fatal casualties, several Franconian citizens were killed, and extensive roads, bridges, and also fields with crops just ripe for harvest, were utterly destroyed. In summary, we were given a rather realistic picture of what was to be expected in case the Cold War would further escalate.

In light of the attempts by then still amenable, highly respected UK Prime Minister Margaret Thatcher, and then still USSR Secretary General Mikhail Gorbachev for easing the post-WW-II Cold War tension, these over-sized maneuvers on both sides of the Iron Curtain, marking the politico-seismic rift zone between East and West of the time, appeared to us not only ridiculous, but utterly out of place. The resulting lively cross-cultural discussions in the basement lounge after "DAY'S END," among our European, US/Canadian and Asia-Pacific participants, asked for a subtle change in global Cold War attitudes, and it made many of us "crusaders for an end to the Cold War period" and "fighters for the preservation of our achieved standards of civilization, religions, and cultures - developed over milleniums of human hardship, suffering and persistent struggle," which are being threatened by the nihilistic "USSR Socialist" ideology.

Thus by the end of the workshop and thereafter, some of us got heavily involved in bilateral East-West rapprochements which gathered daily in momentum culminating in repeated time- and energy-absorbing visits to Eastern block countries and in the visit of USSR and other Eastern block experts to the "Capitalist West" and especially to UIC-EECS/CSL. As a result of all these many international research travels and the time-consuming post-event reporting, the completion of the manuscript of the Proceedings of this NATO-ARW-DIMRP'88 got delayed successively more, for which I request, as editor of the Proceedings, my sincerest pardon.

Yet, in retrospect, let me share with you some of the very mindboggling observations and my deep felt conclusions. As was stated incorrectly in the concluding remarks of the Workshop Discussion Group Final Session [IX-7] by Dr. Tapan K. Sarkar et al., the Workshop Director "sometimes extends the cultural aspects beyond the limit of analyticity"! Namely, after what we have now repeatedly experienced and seen existing in true reality during our visits to the Eastern block countries including Poland, East Germany (DDR), Czechoslovakia, the USSR and PR China, where civilization and with it the regional cultural and religious bases were annihilated, there just cannot be any exaggeration - and, for what other global goals are we struggling but to preserve our global environmental and cultural basis of existence and to carefully expand on it. Therefore, I consider it absolutely necessary and certainly worthwhile to instill the high values of defending mankind's past achievements irrespective of its cultural or religious origins also, and especially during our NATO Advanced Research Workshops. Thus, may all of you view the execution of these important scientific-cultural events in the light of "global planetary environmental and cultural defense" to which our series of workshops is intended to be contributing profoundly.

Rationale

In adherence to the workshop programme and schedule objectives worked out for the first NATO-ARW for planning and execution of 1983 Sept. 18-24 in this series, also for the second workshop of 1988 Sept. 18-24 a rather ambitious program with a densely packed schedule was prepared. Essentially, the same organizational committee was retained with the addition of Dr. Ernst Lüneburg replacing retired Dr. Martin Vogel of DLR, Oberpfaffenhofen; and of Professors Dino Giuli, Werner Wiesbeck, Major David E. Stein and of Dr. Frédéric A. Molinet with whom several well functioning short course lecture series in France, Italy, FRG and the USA were conducted since 1983.

The response to our proposal for a second workshop was overwhelming, and we obtained immediate requests for participation in this planned or similar future ARWs by more than 200 renowned scientists in closely related fields. However, upon request by the NATO-ARW-PANEL we had reduced the size of the second ARW for 1988 Sept. 18-24 to only sixty invitees of NATO-member countries so that experts from as many NATO-member countries as feasible could join, in addition to several experts from Austral-Asia-Pacific Rim allied countries.

All invitees approached guaranteed participation for the duration of five complete working days. The lecture and time schedules were so arranged that optimal interaction among the various participants was possible. There were no simultaneous lectures scheduled. The scheduling of the six workshop groups was so arranged as to allow for active interaction among groups, and the exchange of ideas among the groups was strongly promoted. The chairperson of each workshop discussion group was given special goals to meet and was requested to inform participants about detailed program objectives well in advance of the workshop and by July 1988. The main purpose of this Advanced Research Workshop was to provide a FORUM for international experts to expand interaction and to enlarge the scope of activities for promoting and advancing this rapidly growing, new interdisciplinary field of "Direct and Inverse Methods in Polarimetric Radar Theory, Scattering, Metrology, Calibration, Data Processing and its Applications". (It was, however not intended to be misused for family gatherings or for having a paid-for vacation round-trip of Central Europe.) The organizing committee also desired to ensure that participants from as many NATO-member as well as allied countries from the Far East could join in this event for the express purpose of closing the communicative gap between the allied free countries of Europe, America and the Asia Pacific Rim. Also, a first successful attempt was made to include participants from neighboring East-block Poland, Czechoslovakia, Hungary, and from P.R. China.

Need, Timeliness and Importance

This was the second NATO-ARW of the organizing committee in the field of electromagnetic imaging, radar remote sensing, and target versus clutter discrimination. During the first 1983 ARW, inverse methods in electromagnetic imaging were assessed primarily for solving radar scattering problems including mathematical and numerical inversion techniques, signal and image processing with the specific emphasis on the vector (polarization) nature of electromagnetic fields. The resulting two-volume Proceedings are highly praised in the international literature, and stimulating letters from previous workshop participants and many other readers — from East and West — encouraged us to organize this second ARW on a more specific topic. During this second ARW we decided to concentrate on direct and inverse methods exclusively related to radar polarimetry.

High resolution radar polarimetry has most recently become an indispensable tool in modern electromagnetic sensor technology, both in the civil and military sectors, as well as in remote sensing and radar meteorology. From the outset, we wished to emphasize that by incorporating coherent polarimetric phase information into radar signal and image processing, one can anticipate a breakthrough which is at least comparable to that brought about by the advent of holography and computer-assisted tomography and its application to Synthetic Aperture Radar (SAR) and Inverse Synthetic Aperture Radar (ISAR). It will also play an essential role in developing Ultra-Wideband Impulsive Radar (UWIR) theory, metrology and technology urgently required for local and global environmental surveillance of terrestrial and planetary atmospheres and crusts and for the instantaneous detection, ranging, discrimination, and specification of pollutants threatening such environments. Instead of providing an overview of the lecture topics presented during the workshop which are appended in the second volume of these Proceedings, with a summary of its papers presented in the Opening Paper [O-1] of Volume One; in the following, the functioning of the workshop events will be summarized.

Execution of the NATO-ARW-DIMRP'88 in the KuK Congress Hotel Residenz, Bad Windsheim, FRG

The structure of the workshop schedule is very similar to that of the NATO-ARW-IMEI'83 as described in the Director's Foreword of Volume One of its Proceedings. We again met in the idyllic Franconian spa and retreat center of Bad Windsheim, slightly removed from the tourist centers in Rothenburg ob der Tauber and Nürnberg an der Pegnitz. The facilities of both the Congress Hotel and the surrounding Kurpark had been highly enriched since the last workshop, and an exciting hiking and bicycle day-tour program in the surrounding Frankengau, Steigerwald and Frankenhöhe was added.

Again, it was a prime objective of the Workshop Organizing Committee to provide a forum for internationally renowned key experts to expand the interaction, collaboration and scope of their activities in pursuit of this rapidly growing new radar and remote sensing discipline of "Radar Polarimetry." The available guest rooms of the KuK Hotel Residenz and of closely surrounding hotels and pensions within the Kurpark were filled to the brim with a total of more than 160 participants attending some of the key sessions. We were able to assemble all participants under one conference roof so as to allow optimum direct interaction and the ultimate use of a day's working hours (from 7:00 to 22:00 hours). Similar to the first workshop, early (6:00) daily wake-up calls for all participants were arranged for all hotels, have now been accepted as a routine, and this time no loud complaints were made.

We were able to schedule no overlapping sessions, and in total, seventy-five papers were presented. As a consequence, the duration for oral presentation and subsequent discussions was regretfully limited. Therefore, the speakers were strictly advised to use the allotted time optimally (3 min: problem identification; 10 min: succinct summary; 3 min: highlighting relevant contributions; 4 min: discussion). This procedure was well accepted and this time the KuK-Center bell was rarely used for "Call to Order". The main purpose of the presentations was to identify important topics for the workshop Discussion Group activities which, at times, were extended into the evening. In retrospect, everyone will agree that this approach not only worked, but as so many stated in their enthusiastic, positive thank-you notes, it so contributed to optimum information exchange, stimulation of new ideas, to the planning of additional research retreats on highly specific topics, the initiation of long-lasting interaction and cross-continental collaboration among participants who only knew of each other by names or publications, and will now desire to meet regularly in the future. In conclusion, the Organizing Committee wishes to extend their sincere thanks to all those speakers and session chairmen who demonstrated harmonious collaboration for the excellence of paper content, as well as paper presentation and open-minded question handling.

Cultural/Scientific Events

In enriching the overall program, cultural events were scheduled for the evenings on a daily basis in addition to the two main scientific-cultural events of Wednesday, September 21 and Saturday, September 24, 1988. Furthermore, deliberately on late call, a rather functional spouses program was intelligently improvised again with the dear assistance of Ms. Ursula Allmendinger, Verkehrsamt, Bad Windsheim; by Mrs. Monika Kuehl, UEN, Erlangen; by Ms. Jutta Brockhoff of the KuK Hotel Residenz; by Mr. Johann Schmidt of the Wilhelm Thürauf Reisebüro; and by Mrs. Anita Zierlein, a very knowledgeable multi-lingual tour guide from Rothenburg-/oT, and participating spouses, whose inspiration made the cultural spouses shopping, walking and biking tours a memorable event for the accompanients and also for some of the scientists.

The entire scientific/cultural program was planned in detail by the Workshop Director during his tenure as a Senior US Scientist awardee of the Alexander von Humboldt foundation during the winter, spring, and early summer of 1987 (every single bus stop for visits of cultural treasures, scientific displays, meals, shopping and sightseeing walks was painstakenly pre-determined and timed as described below). He was greatly assisted in this function by Prof. Dr.-Ing. Hans Brand, then the dean of Engineering at UEN, his knowledgeable senior assistant, Dr.-Ing. Gerd Schaller of UEN, and their secretary, Mrs. Monika Kuehl; by Dr.-Ing. Siegfried Osterrieder, FHS, Ravenburg; by Dr. Wolfgang Keydel, the director of DLR NE-HF at Oberpfaffenhofen and members of his radar electronics, optics, and remote sensing divisions: Dr. Helmut Süß, Dr.-Ing. Arno Schroth, Dr. Martin Vogel and Dr. Ernst Lüneburg; by Dr. Rudolf Großkopf, Director of the Carl Zeiss Electronics R&D Division, Oberkochen, FRG and by Dr. Wolf-Dieter Teuchert, Dipl. Phys. Hans Werner Flack and Dr. Joachim Heppner, experts of his infra-red and mm-wave sensors divisions; and by Dr. Gerhard Boucke, director of the TST, Radar Electronics Division in Ulm an der Donau and Dr. Gerd Wanielik, Dr. Klaus Solbach, Dr. Werner Sieprath and Dipl.-Ing. Jörg Schroth of his R&D sections. The cultural evening programs were planned for in advance with the assistance of Ms. Jutta Brockhoff, the Congress Center managess, Mr. Ralph M. Erlenbach, the hotel manager, and Ms. Ursula Allmendinger, who all went out of their way to help, to advise, and assist us in any way they could. Therefore, we wish to use this occasion to express, on behalf of all the participants of the second ARW, our admiration and gratitude for the thorough and delicate assistance we received during our stay in this pleasant and well developing spa at Bad Windsheim, not only from its friendly citizens during our walks, shopping sprees and visits of the splendid Franconian inns, but, in particular, to the staff and management of the KuK Hotel Residenz. It was especially delightful to meet some faces known to us already from the first ARW during 1983, specifically 'Frau' Eva-Maria Schorr, the head receptionist, and her friendly assistants; 'Herr' Joachim Stein, the head concierge (bell captain); and 'Herr' Gotthard Reiter, the chief cook with his delicately selected team of friendly, attentive waitresses. The cuisine of the KuK Hotel was superb and in comparison to the first workshop, we found that decisive changes had also occurred in the Franconian diet and menu toward lighter meals and plentiful salad bars.

It was a pleasure to be their guests because the excellent service we received again, and the peaceful atmosphere at the elegantly enlarged restaurant enabled us to work hard, yet at the same time, to eat well and to relax in the tastefully selected exercise and swimming facilities of the center. Then, in keeping up with the tradition, after "DAY'S END", a daily increasing number of us disappeared downstairs into the hotel lounge, listening to a well chosen musical trio playing Franconian, Bavarian, and Czech tunes. The enlarged KuK Hotel Residenz with its varied recreational facilities and the adjacent Kurpark and newly added Herb and Botanic Gardens, provided an ideal stimulating, yet tranquil environment for another work-laden Advanced Research Workshop in Bad Windsheim, lying as it did before, and more so since the unification of East and West Germany into the Federal Republic of Germany of today, in the historical heart of the modern German culture, Franconia.

In the following, some of the highlights of our scientific/cultural events will be summarized with the intention of concluding the description of our NATO-ARW-DIMRP'88 which, again, was a successful, well-rounded scientific as well as cultural experience opening for all of us the cultural border region between the medieval and modern states of Franken and Württemberg/Schwaben through Rothenburg, Dinkelsbühl, Heidenheim/Brenz, Oberkochen, Ulm, Augsburg, and Oberpfaffenhofen near München, where the workshop came to an end on Saturday evening, 1988 September 24.

<u>Sunday Evening Opening Session (September 18, 1988): Preview of Daily Events</u>

Whereas during the NATO-ARW-IMEI'83 the scientific/cultural tours were directed toward the industrial Regnitz/Main regions of Central Franconia (Nürnberg, Erlangen, Bamberg, and Würzburg: Germanisches Museum, Siemens Medical Research Division, Schloß Banz, Mainfränkisches Museum in the Marienburg and the Roccocco Residenz) during the NATO-ARW-DIMRP '88 we selected the Western inter-Franconian-Swabian (Württemberg) industrial corridor (Rothenburg, Dinkelsbühl, Heidenheim/Brenz, Okerkochen: Carl Zeiss, Ulm: TST), traveling along the venerable Romantic Road, Romantische Straße, a commercial trade route dating back to before the Romans. A preview of these events was given after Dr. Günther P. Können presented a most invigorating, lively lecture on "Polarization in Nature" [O-2], with a clear explanation of atmospheric optic polarimetric effects which we observe daily and thoughtlessly take for granted.

<u>Monday Evening Event (September 19, 1988)</u>

Following established tradition, a formal reception by the First Mayor, Mr. Otmar Schaller, and the Second Mayor, Mr. Theodor Michel, at the Historischer Rathaussaal in the City Hall in downtown Bad Windsheim marked the first social event. This time, Mr. Michel, himself a decendant of Huguenotic families, introduced us to the early remodernization period of Franconia at the end of the Seventeenth Century, after the plague had eliminated or almost erased the original population of Central Franconia. It was mainly the protestant Salzburger emigrants from Austria and the protestant Huguenotic immigrants from France who contributed to the restoration of the area; and, the townships of Schwabach, Erlangen, Bad Windsheim and Ansbach had a sizeable French immigrant population of which the Reformed Churches, like the one in Erlangen, are still existing reminders which was of special interest to our French participants. It was a pleasant evening with all of us enjoying the hospitable and cozy downtown inns, and the excellent local Bürgerbräu beer.

Tuesday Evening Event (September 20, 1988)

A very special event was the organ recital on the newly restored pipe organ of the baroque/renaissance church, St. Kilian, in downtown Bad Windsheim. The local organist, Cantor Bernd Uhlmann, took extra efforts to provide tutorial introductions to the functioning, operations, and exposition of this supreme master instrument of European music. He played selected works of D. Buxtehude, J. Pachhâvel, J.S. Bach and C. Frank in order to demonstrate the versatility and beauty of the pipe organ. Some of us stayed until late into the evening enjoying Cantor Uhlman's mastery. Certainly, such an organ recital at St. Kilian ought to become a highly welcomed event by all participants of future international events of this kind in Bad Windsheim.

Wednesday, Scientific/Cultural Tour (September 21, 1988)

By now, all of the participants of this NATO-ARW-DIMRP'88 were well aware that the NATO Reforger exercises were reaching its climax, with low flying aircraft, booming even during the middle of the night, screeching helicopters from the nearby base at Illesheim and tanks roaring throughout the night. It was like in the war movies. During this Wednesday bus travel to Oberkochen and Ulm, we got into the midst of it: the "blues" chasing the "reds", tanks attacking straight across the highways in front of us ...! To many non-German participants, and especially US and Canadian Americans, who barely ever see any military vehicles on their roads and certainly not rummaging through harvest-ripe fields, this was a true eye-opener on what the "Cold War" is all about, well fitting with the execution of a NATO-sponsored research workshop. It again enticed many of us to look toward some means of putting an end to these non-sensical global war games of destruction. More to the day's events, it put us into the right frame of mind for exploring the superb scientific programs on "night-vision" laser-optical, infrared, millimeterwave and radar remote sensing and battlefield surveillance technology, which then was demonstrated to us at the Carl Zeiss and TST R&D Centers.

We split into two groups, one bus heading for a visit to TST at Ulm, and the other to Carl Zeiss at Oberkochen along the recently completed new autobahn section, Crailsheim-Ulm. Here, we wish to extend a special note of thanks to Carl Zeiss, Electronics R&D Center Director Dr. Rudolf Großkopf and his co-workers, Dr. Wolf-Dieter Teuchert, Dipl-Phys. Hans-Werner Flack and Dr. Joachim Heppner for introducing us to their HI-TECH forefront R&D on thermal sight and night vision tactical surveillance. All of us were very delighted by the pleasing and thought-provoking presentations, the mastery of display technology for which Carl Zeiss was and still is very famed for, but so also were we deeply impressed by the excellence of R&D technology developed here in selected military technologies. Similarly, the other group was highly impressed by the equally outstanding hospitality of the TST Radio & Radar Systems Division at Ulm, where we were welcomed by its Director, Dr. Gerhard Boucke receiving indepth expositions by Dr. Werner Sieprath on seeker guidance; by Dr. Klaus Solbach on the mostly inhouse developed near-future field antenna measurement range; by Dipl.-Ing. Jörg Schroth in mm-wave device technology in which TST has become one of the European leaders; and by Mr. Karl Fischer, who provided the overview and had arranged the details for our visit. Again, we thank all of the R&D staff, who so well prepared a superb introduction to two of NATO's leading European HI-TECH R&D&E centers, and on behalf of all participants, we extend our thanks for receiving the workshop maps filled with selected meaningful notes on Europe 92, on NATO R&D facilities in Europe, and on FR (West) German research.

After being severely delayed on our bus ride back due to the heavy activity of the Reforger exercises of the day, we turned into "Die Schranne", a newly refurbished meeting hall in a famed medieval "spelt & wheat" storage elevator at the Weinmarkt, in the heart of Dinkelsbühl ("Spelt hillock"), a renowned medieval Reichsstadt (independent empirical state-city), similar, if not more scenic than nearby Rothenburg. We toured the medieval city hall and the famed "high-gothic" St. Georgen Cathedral before being given a "night-watchman's" stroll with trumpeter through the night-lit narrow alleys of the center town to the Spital by Frau Lucas and Herr Wagner, where the "Dinkelsbühl Trio" arranged for an unforgettable evening, playing skits of medieval days. This outing which almost got us bogged down amidst advancing tanks, swerving "HMMWVs" and those never-ending convoys of military trucks, then after all ended on a very pleasant note, and close before midnight, we turned in to the KuK Residenz for a well-deserved rest, though yet with the distant sound of roaring tanks in our ears — all through the night!

Thursday, Late Afternoon/Early Evening (September 22, 1988)

Since our last visit to the Franconian Open Air Museum ('83 September 20), the number of reconstructed original villages, farms, etc. had expanded considerably. It was another enjoyful leisurely stroll at a beautiful sunny afternoon in the mellow Franconian fall atmosphere which remains as the permanent highlight of our memories of this eventful week.

However, the Workshop Director was very disgusted by the unexpected rowdy behavior of some "Kulturbanausen as ever long those come" and especially the rude noise of some lesser-experienced catfish-fresser(s), almost ruining the fine traditional carp dinner prepared diligently according to an original medieval culinary recipe. Such "rude pranks" will not be tolerated in the future and the willful destruction of the workshop director's scientific camera and the removal of the "instrument-to-order" was uncalled for! Yet, after some extra coaxing, we were able to have the "Bad Windsheimer Sänger" rejoin us in spite of the unweildly rude noise just before they packed up, and to regain the festive enjoyful mood which prevailed earlier on during the afternoon. Our deep-felt gratitude is extended to Professor Horst Steinmetz, who, with his capable artists of medieval Franconian musical culture, Georg Föster, Heiner Böe, Fritz Eckardt and Georg Eggermeier introduced us to folk and art songs of the Fifteenth/Sixteenth Century, selecting a set of enjoyful originals created within the vicinity of Rothenburg, Nürnberg, and Bad Windsheim, not duplicating what was presented to us by them during their 1983 presentation.

Friday Evening Workshop Division (September 23, 1988)

After our most experienced, senior participants of NATO-ARWs, Len Cram and Dag Gjessing, had collected the working discussion group reports, marking the end of the lecture and discussion group programs; we assembled once more in the newly enlarged main festivities hall of the Congress Hotel. After honoring some of our distinguished foreign guests as Dr. Zbigniew H. Czyż from Poland, Professors Tsutomu Suzuki, Matsuo Sekine and Kiyohiko Itoh from Japan, Dr. Walter A. Flood and Mr. Jerry L. Eaves from the USA, the Workshop was dedicated to one of our still highly active pioneers, Dr. Jean Richard Huynen. Thereafter, the Workshop director was presented with a new "dingle bell" with the "sound" of the true instrument frozen into an ice block until the next workshop during 1993, September 18-24.

In the final presentation of the evening, Dr. Ing. Siegfried Osterrieder provided a preview of the highlights of Saturday's bus travel along the "Romantische Straße" with the anticipated stop-overs at Rothenburg, Heidenheim/Brenz and Ulm/Donau; and Dr.-Ing. Arno Schroth delivered a very pleasing preview of the visit to the German Space and Aerospace Research Establishment (DLR) at Oberpfaffenhofen.

<u>Saturday Cultural/Scientific Close-Down Tour (September 24, 1988)</u>
Also during our last day of travel, we got deeply into the Reforger exercises with an unexpected battlefield formation of Medium Range and ICBM launching vehicles forcing us to take major detours into Rothenburg. Therefore, our stop-over in Rothenburg was shortened considerably and another in Heidenheim/Brenz cancelled altogether so that we were able to enjoy one of Europe's largest gothic cathedrals of the Twelfth Century, the Ulmer Dom, with the highest existing gothic steeple added later on. Many of us, next to awing the grandeur of the architecture, of the mighty organ pipes, historical wall paintings, etc., enjoyed the ongoing organ recital on the newly restored mighty pipe organ, one of the most grandious of its kind.

<u>Visit of the German Space and Aerospace Research Establishment (Deutsche Forschungs-und Versuchs-Anstalt für Luft und Raumfahrt) DLR-Oberpfaffenhofen, Oberbayern, FRG: Saturday Afternoon</u>
After traversing the Danube and escaping the Reforger exercises, we soon reached Fürstenfeldbruck and then Oberpfaffenhofen to the West of München, where we arrived in time for the scientific program at DLR. We were welcomed by the principal hosts of our workshop, the Director General of this rapidly expanding Space and Aerospace Research and Test Center, Dr. Heinz Häberle, who along with division directors Dr. Wolfgang Keydel (radio/radar systems) and Dr. Manfred Reinhardt (meteorology) introduced us to their very excellent optic, infrared, mm/micro-wave radar-meteorologic measurement, data-processing and image restoration facilities. DLR, until recently known as DFVLR, is the main non-profit German space and aerospace research and test facility with a rich history of scientific accomplishments dating back to 1937, when the Institute for Air Traffic Radio Communications and Radio Navigation FFO (Flugfunk Forschungs-Institut Oberpfaffenhofen) was founded by famed Prof. Dr. Max Dieckmann, then the director of the near-by "Research and Test Station Gräfelfing for Atmospheric Electricity and Wireless Communications: DVG (Drahtlos-telegraphische und Luft-elektrische Versuchsstation Gräfelfing, Obb.)", which he founded as a post-doctoral fellow in 1908.

Major German developments of air traffic control, navigation and especially weather forecasting by means of electromagnetic waves were here developed which became the basis of wireless communications during WWI and of radar navigation during WW-II. After WW-II, the entire research instrumentation facilities and the precious scientific library were transferred to the Wright-Patterson AFB, were never returned, and have become a part of the dust-covered inventory of its Air Museum near Dayton, Ohio. Research at Oberpfaffenhofen came to a virtual standstill until the founding of the Federal Republic of Germany in 1948/49; and after the establishment of its Ministry of Defense (Verteidigungsministerium) in 1955, the Institute for Wireless Radio Navigation and Microwave Technology/FFM (Institut für Flugfunk and Mikrowellen) was founded under Prof. Günter Ulbricht. In 1978 it was renamed the INSTITUT FÜR HOCHFREQUENZTECHNIK under its current director Dr. Wolfgang Keydel, a renowned expert in radar remote sensing, formerly at AEG-TELEFUNKEN, Ulm/Donau, now renamed Telefunken System Technik (TST).

Among the European NATO Research, Development & Engineering Centers in electromagnetic sensing and imaging covering the entire non-invasive spectral range, it ranks among the very top at a level as RSRE, Great Malvern, UK; ONERA, Chatillon-Bagneux and CELAR, Bruz, FR; TNO, Scheveningen, NL; and NTNF, Kjeller, NO. Thus, it was indeed a great treat to be invited there and view the superb research instrumentation facilities on a splendid fall afternoon with the characteristic Bavarian blue sky dotted with the fluffy "baroque clouds" above us.

We were sub-divided into four rotating groups after the introductory lectures (one hour) to view (i) the Wideband Polarimetric Target RCS Radar Instrumentation facility (800 MHz - to 45 GHz), [VI-4]), (ii) the m-to-mm-wave and infrared radiometric and thermal sensing & imaging division, (iii) the optical & infrared air & space sensors division, (iv) the fleet of Dornier research aircrafts and its ground-based meteorologic instrumentation facilities. Although all expositions were truly exceptionally planned by Dr. Wolfgang Keydel and Dr. Manfred Reinhardt and their staff, the absolute highlight was the introduction to one of the most celebrated new instrumentation additions, the Wolkenradar Poldirad (C-band Dual Polarization Doppler Radar), designed and planned at DFVLR-OPH in collaboration with Electronic Enterprice Corporation of Enterprice, Alabama, USA. Our special thanks are extended to Dr. Arno Schroth, Dr. Herbert Meischner, Dr. Helmut Schuster, Dr. Karl Tragl, Dr. Madhu Chandra and Dr. Gottlieb Schnabl for the very pleasing and informative scientific introduction [VIII-1], [II-9] to the Poldirad, and to Drs. Bernd Röde, Ulrich Fuchs and Reinhard Hammel for demonstrating the impressive Wideband Polarimetric Target RCS Radar Instrumentation Facility. To all of the DLR research staff we would like to extend our sincerest thanks for this exciting afternoon, as well as to the truly excellent support staff, Frau Margareth Malchow, Fräulein Gabriele Bierl, and Herrn Reiner Weppner for hosting us before we departed for München-Hauptbahnhof, where the Workshop came to a full-filled end.

Post-Workshop Engagements: The End of the Post-WW-II Cold War, the Unification of Germany, and the Break-up of the USSR

As a result of the realities of an escalating Cold War, so vividly demonstrated to all of us by the NATO Reforger exercises, the Workshop

Director considered it his prime duty to work toward the de-escalation of these nonsensical all-destructive cold-war-games. The opportunity arose concurrently and he was invited repeated times to Eastern Europe (1988-90), the People's Republic of China (1988/89) and then also to the USSR (1988/89/90/91). In all of these Eastern block countries, we observed a very distinct double-class system, with the Communist Party cadre plus the scientific and technologic establishment replacing the Czarist and similar nobility and living fully separated from the very great majority (85%) of its citizens. Ironically, counter to the original spirit of the Communist's "peasant, laborer and soldier's paradise", the common people are misused more or less as slaves of the system. It is not only a few million Soviet and Chinese citizens that were annihilated, mainly during Stalin's earlier era (1922-1933) and the destructive period of the Red guards, but tens of millions were erased in what amounts to ethnic genocide versus minority groups. More so, the once blooming regional cultures and civilizations as that of the Buryatian and the Evink of Sacred Lake Baikal region were destroyed; synagogues, churches, temples, and spiritual sites totally burned down; and it all replaced by some nihilistic empty Communist ideology with very little respect for individual rights and none what-so-ever for the preservation of the natural environment and original cultural resources and values.

Certainly, not only the Europeans, but the entire world owes a great debt to the United States, its brave statesmen and citizens, for defending the free world; fully justifying the realities of Cold War as exercised in the Reforger manoeuvers we all experienced so vividly.

However, for many explainable reasons, the Eastern Communist systems are breaking apart very rapidly, and due to the courageous struggles of all Eastern European countries, due to the wisdom of President Mikhail Gorbachev and his co-strugglers within the USSR and that of our NATO statesmen, are we witnessing how the global political Cold War structure is rapidly disintegrating and a rapprochement between East and West is and can now be taking place.

Although we must always be aware of the true serious threat any "nihilistic ideology" still poses to us, we must understand that the East block citizens were and still are imprisoned, have been and are still treated like slaves — and this for about forty-five years, and for more than seventy-five years for those living within today's USSR borders. Soon the "iron curtains" surrounding the East Block will deterioriate completely leading to an explosive mass migration which could reach tens of millions originating from within the current USSR alone by the turn of the century. Unless we accept that such "migration tsunamies" will totally inundate not only North-Central Europe and Austral-Asia, but also America and especially Canada and the U.S., NATO will have to change its approach toward its recent WW-II Cold War enemies totally and completely.

Namely, in order to dampen and reduce mass migration, we must reach out and provide generously expertise not only in rebuilding trade, commerce, business, banking, sanitary facilities (in most part non-existent!!) communal government, etc., but we must at the same time, provide solid education in regional cultural, linguistic and religious aspects because all of these "suppressed" ethnic groups suffered from the complete destruction of their natural habitat, their once generous natural resources, of their civilizations, religions, and cultural bases during their recent episode of "communist imprisonment". Thus, we will have to not only rebuild their economic and industrial bases from scratch, but also develop strong global environmental planetary defense measures, so their entire life's spiritual and religious needs will be slowly regenerated and redeveloped.

As regards the series of these NATO-ARWs on the advancement of "Direct and Inverse Methods in Electromagnetic Sensing and Imaging," it has now been clearly established world-wide that there exists a great need for developing acoustic, electromagnetic, and seismic sensing and imaging methods for the "instantaneous localization, ranging, detection, specification and identification of environmental pollutants" of any kind and source, resulting in the request for another NATO-ARW.

Therefore, in light of all of these changes, our third NATO-ARW in this series will be dealing with "ultra-wideband polarimetric impulsive radar sensing and imaging" in "wide-area electromagnetic surveillance of our terrestrial atmosphere and crust". In order for us to advance this scientifically very demanding technology rapidly, we will attempt to request participation of a large number of experts residing in regions of the former "Eastern Block" and have them join us during 1993 September 19-25 at the KuK Congress Hotel Residenz at Bad Windsheim, FRG.

Concluding Remarks

In meeting so many diverse tasks simultaneously, and in guaranteeing a full success for the future distribution of the proceedings for this Workshop, our gratitude foremost is extended to our prime supporters, the NATO Scientific Affairs Division, ARW Programmes; and, here, very special thanks are extended to the late Dr. Mario DiLullo, under whose directorship the planning for this second ARW was initiated, and to Dr. Craig Sinclair who continued with the encouraging and always enthusiastic approach until he was replaced after his retirement recently by Dr. Giovanni A. Venturi, who also showed deep understanding for our approach. Similarly, we wish to thank Dr. Walter A. Flood, US Army Research Office, Dr. Karl H. Steinbach, US Army RDSG (UK) and Dr. Wolfgang Keydel, DLR NE-HF, for their extensive financial and moral support. Last but not least, we wish to thank both Dr. Tilo and Mrs. Barbara Kester of the NATO Publication Coordination Office and our publishers Mrs. Nel M. Pols-van der Heijden and Mrs. Nel de Boer of Kluwer Academic Publishers/D. Reidel, for their continual encouragement and advice.

The preparation of these Proceedings was again strongly facilitated by the superb managerial and secretarial capabilities of Mr. Richard W. Foster and also by Major David E. Stein and Dr. Ernst Lüneburg who assisted us in many extra hours in the completion of this enormous task.

Again, my understanding wife, Eileen Annette and our children, deserve my deep appreciation for their patient tolerance of the many pre- and post-workshop travels, the many long nights and lost weekends of extra hours, often far from home, during the entire pre-to-post workshop engagement for the past five (5) years.

Wolfgang-M. Boerner
The Workshop Director

## ORGANIZING COMMITTEE

### PLANNING COMMITTEE

DIRECTOR: Professor Wolfgang-M. Boerner
CSL-EECS, University of Illinois at Chicago
Chicago, IL 60680 USA

CO-DIRECTORS:

Professor Hans Brand
University Erlangen-Nürnberg
Erlangen, FR GERMANY

Professor Dino Giuli
University of Florence
Florence, ITALY

Professor Dag T. Gjessing
ESTP, Royal Norwegian Council for Industrial and Scientific Research, Kjeller, NORWAY

Professor Werner Wiesbeck
IHE, University of Karlsruhe
Karlsruhe, FR Germany

Mr. Leonard A. Cram
THORN EMI, Radar Electronics
Wells/Somerset, UK

Dr. Wolfgang Keydel
DLR, Institute for Radio Frequency Technology
Oberpfaffenhofen, FR GERMANY

Dr. Frédéric A. Molinet
Société Mothesim
Le Plessis-Robinson, FRANCE

Dr. William A. Holm
GTRI-RAIL/MAL, Georgia Institute of Technology, Atlanta, GA, USA

### NATO-ADVISORS

Drs. Craig Sinclair/Giovanni Venturi
ARW/ASI Programmes, Scientific Affairs Division
NATO HQ, Brussels, BELGIUM

Dr. Tilo & Barbara Kester
NATO-Publications Coordination Office
Overijse, BELGIUM

### TECHNICAL COMMITTEE AND EDITORIAL BOARD

Professor Wolfgang-M. Boerner
Professor Hans Brand
Dr.-Ing. Gerd Schaller, UEN
Dr. Ernst Lüneburg, DLR
Professor Dino Giuli
Mr. Leonard A. Cram
Dr. Wolfgang Keydel
Professor Dag T. Gjessing
Dr. Frédéric A. Molinet
Major David E. Stein, USAFR

### COORDINATORS OF SCIENTIFIC/CULTURAL TOURS

Dr. Helmut Süß, DLR
Dr.-Ing. Siegfried Osterrieder, UEN
Dr.-Ing. Gerd Schaller, UEN
Dr. Rudolf Großkopf, CZ
Dr. Gerhard Boucke, TST
Professor Wolfgang-M. Boerner

### HOSTESSES OF THE SPOUSES/CULTURAL PROGRAMMES

Mrs. Monika Kuehl, UEN
Mrs. Jutta Brockhoff, KuK
Mrs. Ursula Allmendinger, B.Wh.
Mrs. Anita Zierlein, Tourist Guide

### PLANNING AND RESOURCES MANAGEMENT

Professor Wolfgang-M. Boerner
Dr. Wolfgang Keydel
Mr. Reiner Weppner
Mrs. Monika Kuehl

### WORKSHOP PROGRAM PROCEEDINGS DEVELOPMENT AND PREPARATION

Mr. Richard W. Foster, UIC
Mrs. Julie A. Furlong, UIC
Ms. Deborah A. Foster, UIC
Ms. Mirian R. Mailey, UIC

# HISTORICAL DEVELOPMENT OF RADAR POLARIMETRY, INCENTIVES FOR THIS WORKSHOP, AND OVERVIEW OF CONTRIBUTIONS TO THESE PROCEEDINGS

**Wolfgang-Martin Boerner**
**University of Illinois at Chicago**
**UIC-EECS/CSL, 840 W. Taylor St, SEL-4210**
**CHICAGO, ILLINOIS/USA 60680-4348 (M/C 154)**

**ABSTRACT:** A succinct overview of historical events in the development of polarimetry, in optics and in radar, is followed by a critical assessment of unresolved problems and a listing of recent workshop activities, providing the incentives for this workshop. Based on this critical assessment, the outlay of the Proceedings Topics is developed and a succinct summary of the papers is presented.

## I. INTRODUCTION

Radar Polarimetry, i.e., the utilization of complete electromagnetic vector wave information, has become an indispensible tool in modern electromagnetic sensor and imaging technology, both in the civil and military sectors, and increasingly more in environmental remote sensing of the terrestrial and planetary crusts. From the outset, we emphasize that by incorporating coherent polarimetric phase and amplitude information into radar signal and image processing, one can anticipate and already is witnessing a breakthrough that is comparable to that brought about by the advent of holography and computer assisted (Radon projection) tomography and its applications to Synthetic Aperture Radar (SAR) and Inverse Synthetic Aperture Radar (ISAR).

### 1.1 Radar Polarimetry

In early RADAR (RAdio Detection And Ranging) only amplitude information of the electromagnetic wave at a suitable frequency was utilized which, since its inception by Hülsmeyer in 1904 [1] and its implementation in the late twenties [2], has become a key element in civil and military operations on land, at sea, and in the air [3]. Then, some forty to fifty years later, it was possible to build wide-band radar systems which, in addition to frequency and amplitude, also utilize relative and absolute phase information for resolving physical features of scatterers and the background topology [4]. The increased resolution capability has provided the means of extending the original RADAR concept of radio detection and ranging to include capabilities for high resolution mapping, profiling, and imaging unrelated to either detection or ranging [5]. However, in order to further improve on high resolution techniques for carrying out traditional radar tasks of search, track, and weapon control in increasingly more difficult surveillance environments with



*W.-M. Boerner et al. (eds.), Direct and Inverse Methods in Radar Polarimetry, Part 1*, 1–32.

the simultaneous increase in target camouflaging capabilities; in addition to amplitude, frequency, relative and absolute phase also complete coherent polarization information must be incorporated into the target versus background clutter image contrast enhancement algorithms [6]. This problem is still one of the most formidable known under electromagnetic vector inverse scattering [7], requiring extensive resources for developing the underlying theory, metrology and computer numerical tools of vector signal/image processing [8,9].

## 1.2 Direct and Inverse Methods

The problem of determining the scattered electric and magnetic fields everywhere, given the exact shape and distribution of the material parameters ($\varepsilon$, $\mu$, $\sigma$), expressed in terms of a coordinate system placed in the center of the object for a given incident field, is known as the direct electromagnetic scattering problem [6]. Inversely, the problem of determining the size, the shape, and electromagnetic properties of an unknown scatterer, given the incident field and the measured scattered electromagnetic fields, is defined to be the electromagnetic inverse scattering problem [7]. The case in which the amplitude, phase, and, in addition, polarization are known and relevant, is known as the electromagnetic vector inverse scattering problem [8], which is the problem we are dealing with in this NATO - Advanced Research Workshop on "Direct and Inverse Methods in Radar Polarimetry (DIMRP)".

For the purpose of describing this new discipline of high resolution polarimetric radar imaging, it was found convenient to introduce a mathematical representation of the imaging process, which was considered in depth during the NATO-ARW-IMEI'83 [7] serving to highlight the roles of direct and inverse theories, image quality and image reconstruction techniques. Here, imaging can be described as the generation of the image field, **I**, from an object field, **O**, by means of appropriate electromagnetic and computational system, **L**,:

$$\mathbf{I} = \mathbf{L} \cdot \mathbf{O}$$

Given **O** and **L** and determining **I** defines the direct problem explicitly, whereas if **O** is not known as in remote sensing, nondestructive evaluation, medical imaging, deep sounding, geo-physical exploration, and similar applications; it must be estimated by indirect or inverse methods, $\mathbf{L}^{-1}$, from data contained in the "image fields", **I**. This procedure is conveniently written as

$$\mathbf{O} = \mathbf{L}^{-1} \cdot \mathbf{I}$$

although we may not be able to define $\mathbf{L}^{-1}$ mathematically. Inverse problems of this type may have inherent instabilities so that small variations in the image data, **I**, can cause large errors in the reconstructed object, **O**. Here, image quality is of specific interest, and it is measured by the "closeness" of **I** to **O** in some appropriate mathematical presentation, i.e.,

$$\| \mathbf{I} - \mathbf{O} \| < \varepsilon \quad \text{alternatively,} \quad \textbf{distance}\,\{\mathbf{I},\mathbf{O}\} < \varepsilon \;.$$

The optimal recovery of the image is related closely to the "illposedness of solution" and is treated by a variety of methods [6-8]: constraints on the solution for O, use of a priori knowledge of O, relaxation of the error limit $\varepsilon$ on image quality, and the addition of more data such as "polarization descriptors" in the process of image collection, as defined below.

### 1.3 High Resolution Polarimetric Radar Imagery

In high resolution polarimetric radar imaging, it is the objective to utilize the complete vector nature of electromagnetic waves, i.e., in addition to amplitude, frequency and phase, also polarization state information of the transmitted and received waves is incorporated into signal (bin-by-bin) and image (pixel-by-pixel) processing, requiring a 2x2 Sinclair matrix [S] and/or a 4×4 Mueller matrix [M] description of the scatterer for the coherent and partially polarized cases, respectively. Although there still exist some "grey areas" in both theory and techniques of radar polarimetry, in recent years considerable progress was made in theory, device technology and algorithm development for broadband polarimetric vector signal and tensorial image processing [9]. These dramatic advances will have a definite impact on the merging military but also on the civil technologies of the 1990s, especially as regards the design and manufacture of high precision polarimetric antennas, multi-channel polarimetric transceiver systems, real-time polarimetric wideband signal and image processing modules including neural networking, etc., which, in a next step, need to be developed for polarimetric ultra-wideband impulsive radar imaging methods [10]. Therefore, in keeping abreast with the dynamic advances, here an up-to-date state-of-the-art assessment is attempted concluding with the identification of viable new high-resolution polarimetric radar techniques which are to address the rapidly changing needs of civil and military remote sensing operations of the future in dealing with "extremely low RCS" targets well "camouflaged" into increasingly more complex background environments.

Next, a brief historical review with pertinent references is provided, followed by an overview of the papers collected in these proceedings.

## II. HISTORICAL REVIEW ON THE UTILIZATION OF POLARIZED LIGHT IN NATURE AND THE DEVELOPMENT OF RADAR POLARIMETRY

In an effort to clear up some of the still remaining open questions, misconceptions and utilities of the polarization state transformation capabilities of the electromagnetic wave, it is useful to first present a brief assessment of polarimetric bionics, i.e., the intrinsic facility of "polarization perception" by insects, birds, fish and mammals, and then to provide a historical development of the theory of polarization and the fundamentals of radar polarimetry. Based on such a critical historical assessment, the appropriate formalism for handling basic polarimetric radar concepts may then readily be developed providing clarification, qualification and identification of still unresolved problems.

## 2.1 Visual Observations of Polarized Light by Insects, Fish, Mammals and Humans

Insects (bees, ants, hornets, wasps, water fleas, fruit flies), cray fish, eels, various bottom feeding fish, migrating birds, and also some mammals are able to distinguish between polarized light and unpolarized light as easily as we can distinguish colors (Können [11]). When making use of this ability to navigate using polarimetric parameter manipulations, these creatures sometimes perceive the orientation and ellipticity of light at very low (10%) degrees of polarization, and use these inputs for controlling dynamic flight motion even under the severest storm conditions. On the other hand, humans are almost "polarization-blind" and, generally, have to use a polarization filter to determine the polarization of light. Nonetheless, when light has an extremely high degree of polarization (e.g., $\geqslant$ 60%), most humans (and especially those living in regions with almost permanent foggy conditions), can still perceive polarization with the naked eye when properly trained to observe the so-called "Haidinger brush" which was discovered by the geologist and natural scientist Wilhelm Ritter von Haidinger (1795-1871) as early as 1844 [12]. It turns out that in a plane emitting polarized light, we may see a tiny yellowish figure appear, which we do not perceive in unpolarized light. This happens especially for shorter wavelengths.

The orientation of this so-called "Haidinger brush" [12] depends on the direction of vibration of light and "co-rotates" with it if this plane is rotating. If one wishes to visualize the outward appearance of the 'brush', the best way is to compare it with "Brewster's brush" which one can observe by looking through certain minerals (pleochroic minerals: glaucophane, cordierite, etc.) and we refer to the excellent recent books of Können (1980, 1985 [11]) on this subject and to his recent invited address [0-2] during this NATO-ARW-DIMRP'88.

We note that the cornea and retina of the eye of different species of birds, fish, eels, and mammals are not polarization isotropical [11-18] and may even change their polarimetric properties drastically during the species' life cycle. For example, the color sensitivity of the eel's eye is optimal towards the red during infancy and shifts to the blue-green before the eel departs on its cross-oceanic journey and, at the same time, the cornea, passing purely linear polarization states during infancy, changes to passing purely circularly polarized light during maturity. On the other hand, the cornea of grass feeding animals like cattle, horse, sheep, goats, hare, etc., mainly pass 'horizontal' linearly polarized light with strong suppression of 'vertically' polarized light. Furthermore, various sections of the cornea, for example, of migrating birds, possess different polarimetric properties over different view angle domains. This is also true of many fish and of the human eyes. We are just in the process of discovering, let alone understanding some of the curious polarimetric optical properties of the eye's cornea and retina, and extensive future trans-disciplinary research in this area is required in order to discover the intrinsic reasons why "Mother Nature" makes use of them [13,15,18].

In 1949, Karl von Frisch discovered [14] that many insects, such as bees, possess the ability to orient themselves using the distribution of polarization of the daylight to find their way, as well as to land on moving platforms (leaves, twigs, stems, blades of grass, etc.) under the most adverse weather conditions although direct access to sunlight is obscured. Although very considerable additional research is conducted in this field of polarimetric bionics [15,18:T.H. Waterman], there also exist very pertinent open questions regarding the dynamic polarization state handling of the bee's eyes and its feedback system [14].

It is not only the cornea and the retina of the eyes that possess polarimetric anisotropic properties, but also the body surface of some species of insects, beetles, fish, moths, butterflies, etc., possess polarimetric "camouflaging capabilities". For example, some beetles and moths are able to camouflage themselves so well against the background that they seem completely invisible when viewed with the eye of other species. For example, the beetle species of the rose chafer (cetonia aurata) [11] reflects incoming unpolarized light as almost purely circularly polarized light; and a serious search for other species displaying similar polarimetric effects is ongoing. In most cases, this camouflaging operation is a strongly polarimetric phenomenon and with the use of a naturally built-in polarization filter, optimized not only at circular or linear polarizations, complete contrast enhancement against the background is obtained. This indicates that not only is the body surface polarimetrically agile and adaptive in the sense of an "electromagnetically smart skin" but also that the cornea of the eye of these creatures is too and that there may exist an interrelated adaptive feedback control loop [15] which yet needs to be explored.

We have introduced this section on the visual observation of polarized light in Nature because from extended physiologic behavioral studies of the pertinent species of insects, fish, birds and mammals, we may be able to discern new and improved design and vector (polarization) signal processing approaches for future polarimetric radar and lidar systems, particularly in obtaining a better understanding on how and why polarization anisotropy is used in Nature.

## 2.2 Historical Development

The history of polarimetric radar can be traced from the first recorded analytic descriptions of polarized light [16,17,18,19,20]. In the following, we shall first discuss the discovery of the properties of polarized light in nature (Können, 1985 [11]) and then of the discoveries of the phenomenon of polar electromagnetic energy (Gehrels, 1974 [18]) spanning the entire electromagnetic spectrum, such as, the m-to-sub-mm-wavelength regions, leading to the emergence of the new discipline of polarimetric radar and polarimetric lidar technology.

### 2.2.1 Historical Discovery of Polarization

First, let us recall that the Vikings, using "the findlings stone (sunstone)", produced from some dichroic mineral like cordierite, were able to navigate in the absence of direct sunlight. Thus, the Vikings during the eleventh (11th) century when navigating in the Baltic Sea and Arctic waters prior to traversing the Atlantic, may have been one of the first human beings to utilize polarimetric effects as a tool.

The discovery of the phenomenon of polar electromagnetic energy dates back as early as 1669, when the first known quantitative work on the subject was published by Erasmus Bartolinus (1625-1695). It contained his observations of objects viewed through a calcite crystal "doubled" and of an incident light ray splitting into ordinary and extraordinary rays [17].

Bartolinus was followed by Christian Huygens (1629-1695) who, observing the intrinsic difference of the two calcite rays, contributed most significantly to the field of optics by proposing the wave nature of light (although incorrectly proposing them to be of longitudinal wave character) and discovering polarized light (1677). E. Louis Malus (1755-1821) proved Newton's suggestions that polarization is an intrinsic property of light and not something added by a crystal (1808). The next significant contribution to this field was added by Augustine Fresnel (1788-1827), who proposed that light could be considered as a transverse wave. His reflection formulas, deduced from experimentation, are still in use today and have been rederived using rigorous electromagnetic theory by Paul Drude in the late nineteenth century (1890). The last of the early pioneers was Sir David Brewster (1781-1868), who by extending the work of Malus, discovered the relationship between the polarizing angle and the relative refractive power of dielectric materials (1815). The transition to the formulation of a rigorous electromagnetic theory was paved by Michael Faraday (1791-1867) with his postulations of the physical laws of electromagnetism in 1832, which was soon followed by his discovery in 1845 of the rotation of the polarization plane in magnetic fields. In 1852, George Gabriel Stokes (1819-1903) laid the foundations of future mathematical theories to describe polarized, unpolarized and partially polarized streams of light by introducing his four parameters now known as "Stokes parameters".

In 1873, James Clerk Maxwell (1831-1879) the foremost, "principator electromagneticus", succeeded in providing a rigorous formulation of Faraday's postulates which led to the formulation of the diffraction theories by Helmholtz (1881) and Kirchoff (1883) and a strict mathematical treatment was then provided in 1896 by Arnold Sommerfeld (1868-1951) though earlier attempted in 1881 by Lord Rayleigh, (1842-1919). A significant contribution to the understanding of polarized light was made in 1892 by Henri Poincaré, who showed that all possible states of polarization could be represented by points on the Riemann(1872) sphere, the latitude and longitude of each point defining the eccentricity and inclination of the polarization orientation angle of the polarization ellipse. Based on these discoveries, during 1886-1888, Heinrich Hertz (1857-1894) demonstrated the application of the electromagnetic theory as it applies to lower frequencies such as radio waves including its polarization state dependence. This marks the advent of modern applications of electromagnetic waves leading to radio wave communication, object detection and ranging, which was implemented first during 1896 to 1904 by Christian Hülsmeyer, now the acknowledged true father of radar [1]. The work of most of these scientists are available in a translated form collected by William Swindel, [17], and are also portrayed in Born and Wolf [20], Shurcliff [16], in Gehrels [18], Egan [19], Können [11],

Skolnik [3] and Pritchard [1], in which various chronological event tables and major references are included.

### 2.2.2 Historical Development of Radar Polarimetry

The use of radio-to-microwave frequencies for ship and air-craft detection and the design of the first radars were accomplished concurrently in Europe and America, commencing with Hülsmeyer's demonstrations (1902-1904), re-explored during the twenties and thirties (see Pritchard [1]), and further advanced before and during World War II (1937-1945) by all of the competing factions [1,21, 9,3] (also see 0-5 for references).

Very important early basics on the properties of partially polarized waves were discovered by Norbert Wiener (1927-1929) on harmonic analysis in quantum mechanics showing that the coherency matrix is a linear combination of the Pauli spin matrices with the Stokes parameters. These studies by Wiener and Pauli had direct influence on the later work of R. Clark Jones (1941-1944) [22], under the guidance of Professor Hans Mueller at the Massachusetts Institute of Technology, Cambridge, MA [23]. Earlier on, during 1924-29, Wolfgang Pauli introduced the concept of the spinor in quantum mechanics which has proven to be also an ideally successful tool in the proper description of polarimetric radar problems. Then extensive polarimetric wave propagation analyses were carried out at MIT leading to the so-called "Jones calculus" as so well described in Azzam and Bashara [23], which now has also become of great value to forward scattering radar and lidar analyses, for example, in polarimetric radar meteorology and oceanography as well as for microwave propagation path estimation [24].

One of the first extensive fundamental studies on radar polarization was initiated by George Sinclair [25] in 1946 at the Ohio State University Antenna (later called Electro Science) Laboratory. He showed that a radar target acts as a polarization transformer and he expressed the properties of a coherent radar target by the 2x2 coherent scattering matrix, now commonly denoted as the Sinclair matrix [S]. These studies were further pursued by Victor Rumsey (1949-1951) and particularly by Edward Morton Kennaugh (1948-1954, 1952) [26]. Other basic studies [27] were conducted by Booker (1950), Deschamps (1951) [27], Kales (1951), Bohnert (1951), and independently in the UK by Gent (1954) [28]; and here we refer particularly to the series of papers in the Proc. IRE, May 1951, of which the paper by Georges A. Deschamps [27] is still widely in use. Based on these studies, in 1952, Kennaugh (1922-1983) introduced a new approach to radar theory [26] and developed the "Optimal Target Polarization Concept" for the reciprocal, monostatic relative phase case which was immediately of direct interest to meteorological radar studies (circular polarization rain clutter rejection or cancellation) [7]. There were a number of other isolated studies initiated, for example, the GIT-Project A235 (July 1955) for the purpose of using polarization to distinguish between targets and clutter, which was well reviewed and summarized by Root [29]. The decade of the fifties closed without any real recognition for the need of polarimetric decoy discrimination, and theory and techniques of polarimetric radar still remained highly under-developed. However, an extensive amount of measurements on the relative phase scattering matrix were made in the late fifties with the guidance of J. Richard Huynen [30] then at the Lockheed Aircraft Corporation, Palo Alto, CA, who exploited Kennaugh's optimal target polarization concept (May 1960) and developed, during the sixties, his own approach to radar target phenomenology which culminated in his remarkable and visionary dissertation monograph (1970) [31]. We also note here that Copeland (1960 [32]), under the guidance of Kennaugh, developed the first practical scheme for classification and identification of radar targets which, later on utilized by Huynen (1960-1970), was based purely on radar polarimetric concepts for symmetric, reciprocal targets [30,31].

Except for the pioneering polarimetric radar developments by Jerry L. Eaves [33] and co-workers of GIT-EES, who developed the concept of intra-pulse polarization agile radar (IPPAR) pulse switching and by André J. Poelman, who introduced the polarmetric multi-notch filter (PMNF) concept [34], only very few relevant studies on advancing radar polarimetry were carried out in the West during the sixties and early seventies. Whereas, we note that in the Russian literature ([35] to [42]: Kanareykin et al., 1965, 1968; Kozlov, 1979; Shupyatsky and Morgunov, 1968; Varshanchuk and Kobak, 1971; Zhivotovski, 1973, 1978; Bogorodsky et. al., 1980; Radimov and Popovski, 1984) the potential applicability of radar polarimetry to target/clutter analysis was recognized early on; and, immediately after the appearance of Huynen's dissertation [31], polarimetric radar investigations were pursued with new vigor in the USSR [O-3, O-4].

No real progress was made in advancing the fundamentals of radar polarimetry until the early eighties when a renewed effort was re-initiated in 1978 at UIC-EECS/CSL (Chicago, IL) [43], to assess, most critically, the previous works of Kennaugh and Huynen [6] which were essentially unknown or forgotten by most workers in the field, and especially in the US and NATO Europe [29,9,7,44]. We generalized the target characteristic operator concept to the general bistatic case giving special consideration to polarization scattering matrix measurements in any orthogonal elliptic basis. Whereas Kennaugh [26] and Huynen [31] were primarily concerned with determining the characteristic polarization states for the coherent monostatic case, at UIC also optimization procedures for the bistatic case [45] and for the 4×4 Mueller matrix [M] for the degenerate coherent, for the partially polarized and for the partially coherent cases were achieved, as summarized in [46], concurrent with the investigation of Van Zyl [47] at the CAL-TECH, Jet Propulsion Laboratory. However, we must draw attention to the ongoing lively discussions on Huynen's still controversial "target decomposition concepts," based on the Mueller matrix formulation for the partially polarized case [31], which will require extensive future analysis and verification studies before those may be accepted for implementation in practical systems [7].

## 2.3 Recent Post-Workshop Retreats, Workshops and Pertinent Symposia Sessions

In the meantime, the potentials of polarimetric radar target phenomenology have been recognized worldwide as is evident: (i) from the large number of recent workshops, retreats and coordinated topical/special symposia sessions on **"Direct and Inverse Methods in Radar Polarimetry"**, and on **"Polarimetric Radar/Scatterometer/Synthetic Aperture Radar Data Processing and Applications"**; (ii) from the large body of new books on the subject matter such as for example by Mott [48], Ulaby, Fung and Moore [49], Ulaby and Elachi [50] and Kong [51]; and (iii) from the increasing number of international measurement campaigns and closely related experimenters and data processing retreats such as the Joint NASA/ESA/ISAS Polarimetric Radar/Scatterometer/SAR Calibration Workshop at DLR-Oberpfaffenhofen FRG, rescheduled due to the Mid-East crisis from 1991 January 25-30 to 1991 October 14-19. We are confident to state that these increased activities were, in major parts, stimulated by (i) the IEEE Ant & Prop, Special Issue March 1981 (Vol AP-29) on "Inverse Methods in Electromagnetics" [8], (ii) the three Polarimetric Technology Workshops of MICOM, Redstone Arsenal/Huntsville, AL planned, organized and executed primarily by L.W. Root and coworkers [9,29]; and (iii) the first NATO-ARW-IMEI'83 [7] as well as by our NATO-ARW-DIMRP'88. We refer the reader to the resulting proceedings, and also to those of other pertinent workshops such as CCG/DLR-86(87-89) [52], at Oberpfaffenhofen, FRG; PRWL 89 in Leningrad, USSR [53, 57]; JIPR'90 [54, 59] at Nantes, FR and also the proceedings of the recent IEEE/AP-S and IEEE/GSRS-S International Symposia plus URSI North American Radio Science Meetings, the European Microwave Conferences and of the Asia-Pacific Microwave Conferences since about 1986.

As regards the pure and applied mathematical treatments of Inverse Scattering Theory, we refer to the annually reoccuring workshops RCP 264 (Rencontre Interdisciplinaire Problèmes Inverses) organized every year in late November/early December by Professor Pierre Célestin Sabatier at USTL, Montpellier, France [54]. In cyclic order the emphasis of these exciting, research stimulating get-togethers focuses every year on different application areas and scientific issues of inverse problems.

Here, another two specific recent applied workshops deserve the attention of the readers; and their main scientific events are briefly summarized:

### 2.4 1989, May 13–May 26 Leningrad, USSR Meeting on "The Advancement of Inverse Methods in High Resolution Polarimetric Radar Imaging and Its Application to Hostile Object Discrimination Against And the Remote Sensing of Terrerstrial and Planetary Atmospheres" [53].

Upon the initial assistance of Dr. Zbigniew H. Czyż, Warsaw, Poland, contact with polarimetric radar experts via Prof. Pyotr Yakobovich Ufimtsev, USSR Academy of Sciences, Institute of Radioengineering and Electronics, Moscow, USSR, a first get-together was arranged by Dr. Gennady N. Gromov of the All-Union R&D Institute for Radio & Radar Systems, Leningrad, USSR during 1989, May 13-26. It was a series of very lively presentations also by the famed radar polarimetrists Prof. Vladimir A. Potekhin, Prof. Anatolij I. Kozlov, Prof. Radij V. Ostrovityanov, Dr. Dimitrij B. Kanareykin, Dr. Victor N. Tatarinov, Prof. Arkadij B. Shupyatsky, Dr. Levitan A. Zhivotovsky, Prof. Boris Sh. Lande, Dr. Valdimir D. Stepanenko, Dr. Aleksander V. Ryzhkov, Prof. Aleksander I. Logvin, Dr. Konstantin G. Tskhakaya, and other famed polarimetric radar meteorologists and applications specialists from within the USSR. It was very apparent that radar polarimetry within the USSR had been advanced very strongly especially since the appearance of Dr. J. Richard Huynen's dissertation monograph in 1970 [31] (which is referenced already in 1972 in various Russian papers). Unfortunately, no commercially available translations of most of their monographs and books exist; and, therefore, the Workshop director requested from them the preparation of two review papers for these Proceedings **[O-3 and O-4]**. Furthermore, as a result of this Leningrad May 1989 PRWL meeting [53], and during a brief stop-over of 1989 November 10-11 in Moscow [55], a reciprocating visit of several radar polarimetrists of the USSR to the 1990 May 05-11 IEEE/ AP-S & MTT-S plus URSI-NARSM at Dallas, TX [56] and the 1990 May 14-17 SPIE Polarimetry Conference [24] at Huntsville, AL was negotiated. It resulted in the visit of Profs. Anatolij I. Kozlov and Aleksander I. Logvin of the Moscow University of Aeronautical & Space Science & Technology; of Prof. Arkadij B. Shupyatsky of the USSR Central Aerologic Observatory, Dolgoprudny; of Dr. Aleksander V. Ryzhkov and Dr. Sergey Yu. Matrosov of the Main Geophysical Observatory, Leningrad; and of Profs. Levitan A. Zhivotovsky and Radiy V. Ostrovityanov from the Leningrad Universities of Radioengineering LETI (named after Lenin: V.I.Ul'yanov) and Naval Electronics, respectively. Eighteen (18) major papers were presented by our guests during their two-weeks first-time stay in the USA of which several were re-edited and expanded for inclusion in these proceedings (**[II-8]**, **[IV-3]**, **[IV-10]**, **[VI-9]**, **[VIII-10]**). This rapidly intensifying interaction was further advanced by the recent participation in the Joint US-USSR Global Environmental Preservation Program on "Save the Terrestrial Large Lakes: Baikal/Michigan Environmental Theatre Festival with Ecological Research Retreats", Ulan-Ude 1990 Aug 18 - Sept 04 [57] during which joint polarimetric radar research and POL-SAR measurement campaigns (USA:JPL - FRG:DLR - USSR:ASREI - JAP:ISAS/NASDA) were initiated in coordination with Dr. Namzhil B. Chimitdorzhiyev, USSR Academy of Sciences, Siberian Division, Buryat Region, USSR [58]. Members of this Buryatian boreal environmental remote sensing research center were revisiting the USA during 1990 December for preparing the 1992 July event in coordination with the URSI-NARSM and IEEE-AP-S Symposium in Chicago, IL/USA, 1992, July 18-25.

As is clearly stated in the resulting travel reports [53-57], it is this bilateral USSR-NATO interaction which is desperately required to develop the badly needed advanced HI-TECH in **"Ultra-wideband Polarimetric Radar Sensing & Imaging of the Terrestrial and Planetary Atmospheres and Crusts"** for the **"Instantaneous Localization, Ranging, Specification and Identification of Pollutants, its Sources and Transmitters"**. It is the very definite objective to deal with these timely and important **"global environmental planetary defense"** issues henceforth and especially during the third NATO-ARW in 1993, Sept. 19-25.

### 2.5 International Workshop on Radar Polarimetry: Wideband Polarimetric Doppler Radar Sensing & Imaging, JIPR, L'IRESTE, University of Nantes, Atlanpole, La Chantrerie, France, 1990 March 19-22 [54,59]

During 1990 March 19-22, the First French International Workshop on Radar Polarimetry; JIPR'90, was held at L'IRESTE, Université de Nantes, La Chantrerie - Nantes, France, organized by Professor Joseph Saillard

[54], the director of its High Frequencies Systems & Signatures Laboratory (SZHF), in coordination with Drs. Pierre Gaudon and Denis Raguin of the Center for Defense Electronics (CELAR), Division of Electromagnetic Signatures of Targets and Environmental Clutter (ASRE) at Bruz near Rennes, West France; and Dr. Frédéric A. Molinet, president of Le Société Mothesim at Le Plessis-Robinson, France, who organized the first short course lecture series on the subject matter in France during 1983 to 1988 [59]. JIPR'90 is the first of a periodically reoccurring French workshop on radar polarimetry with about a two-to-three year interval, similar to some of the workshops organized by the (FR) German Carl Cranz Gesellschaft (CCG) in Oberpfaffenhofen, Mannheim and Hamburg, FRG [52]. It was sponsored mainly by DRET/DGA, the French Department of Defense, the University of Nantes, and the French radar industry, especially the Thomson Radar Application Center (LCTAR) of Vélizy - Villacoublay, Central France, with the generous assistance of Dr. François Le Chevalier.

The workshop was well attended by about 145 radar and electromagnetic sensing experts from eight countries with the majority from France representing the NATO radar defense R&D centers. The lecturers were selected from experts of various scientific and engineering disciplines in radar polarimetry, most of whom had met before during the NATO-ARW-IMEI'83 [7] and NATO-ARW-DIMRP'88. The workshop was separated into three full-day programs, with presentations on basic direct and inverse scattering theories and fundamentals on coherent and partially coherent treatments of radar polarimetry on the first; basic issues regarding dual polarization radar systems, conceptualization, metrology, calibration, and vector signal processing during the second; and applications to military and civil radar target detection, discrimination of useful target signal versus dynamic background clutter and noise, the introduction of the polarimetric matched signal/image filter designs and its applications to POL-RAD, POL-SCAT and POL-SAR signal/image interpretation on the third and last day. The introduction and concluding lectures of the entire program were presented by the author of this report, who was chosen because of his function as "le porte parole international du radar doppler polarimétrique et l'expert mondial de la science mathématique-physique inverse electromagnétique". The main theme was to stimulate increased enthusiasm and understanding for the rapid advancement of the underlying basic sciences and HI-TECH advanced technologies for **"wideband polarimetric Doppler radar sensing & imaging for the INSTANTANEOUS DETECTION, LOCALIZATION, IMAGING, SPECIFICATION, and IDENTIFICATION of SOURCES & OBJECTS OF MAN-MADE & NATURAL, MILITARY, ENVIRONMENTAL, SOCIO-POL/ECONOMIC POLLUTANTS"** — One of the most demanding and still unresolved problems of our times.

The entire workshop program was executed within the new spacious modern glass complex of IRESTE, Université Nantes at Atlanpole, La Chantrerie, providing a most pleasant atmosphere for stimulating research discussions among the thirteen lecturers, the eight systems research experts from CELAR, and the one hundred and twenty attendees from eight countries. The pre-workshop event was a cleared visit for radar polarimetrists from NATO-member countries to CELAR on 1990 March 19 [59], providing an excellent introduction to ongoing French defense activities in low observable and RCS theory, modeling and metrology, stressing the need for advancing wideband polarimetric doppler radar theory, metrology and techniques very rapidly. The research discussions were carried over to the culminating social event, a river boat dinner sojourn on the River Erdre along its pastoral shorelines with many remarkable castles extending the beauty of the famed Loire Valley, a true cradle of occidental culture, leading to the culmination of the French social and cultural style of life dominating most of Europe, including Czarist Russia before WW-I.

As summarized in the post-workshop report [59] and in the well done Workshop Proceedings [54], the lecturers were renowned radar polarimetrists including, among others, J.R. Huynen, S.R. Cloude, F.Le Chevalier, G. Wanielik, D. Giuli, F.A. Molinet, Th. Le Toan, O. Crop, E. Pottier, and J. Saillard. Of specific interest is the interesting dissertation of Eric Pottier [60], resulting from these interactions, and especially with J. Richard Huynen.

It was pleasing to witness that the "Spirit of Europe 92" was carrying the lively atmosphere and that a relaxed attitude toward integrating Eastern European invitees such as Zbigniew H. Czyż, Vladimir A. Potekhin, Victor N. Tatarinov, Boris Sh. Lande, Radij V. Ostrovityanov, Dimitrij B. Kanareykin, Anatolij I. Kozlov, Arkadij B. Shupyatsky, et al into similar forthcoming events was highly welcomed.

## III. OUTLAY OF PROCEEDINGS TOPICS WITH SUMMARIES

The specific objectives of this NATO-ARW are to advance direct and inverse methods exclusively related to radar polarimetry including every single presentation as described in the Director's Forward. Although considerable R&D efforts had been expanded during the past decades, and primarily since the NATO-ARW-IMEI'83, there still exist many "grey areas" in both theory and techniques of radar polarimetry which were chosen for consideration during this NATO-ARW-DIMRP'88, covering the meter-to-submillimeter wavelength, infrared and also optical regions of the electromagnetic spectrum. The emphasis was placed on the basic principles of electromagnetic wave interrogation with natural and/or man-made media and objects, the optimal selection of illumination and sensing, the optimal recovery of useful target signal, mathematical and data processing methods, and representative applications.

The complete proceedings of this NATO-ARW-DIMRP'88 are collected in two parts. A perusal of the Table of Contents reveals the diversity of topics presented, ranging from mathematical expositions, metrological approaches, experimental results to computer-numerical methods of vector signal/image processing and its applications. For such a rapidly developing research area, it is necessary to classify the progress reported here approximately according to topics, of which we identified basically eight and collected about ninety-two papers in separate topical sections plus an introductory overview and a working discussion group activities reporting section. This arrangement, in ten sections - with Topics **O - IV** contained in **Part One**, and Topics **V - IX** in **Part Two** - will enable the reader to become acquainted with the current research trends.

- PART ONE -

## 3.0 Historical Trends and Overviews: Topic O

**Reviews of pertinent direct and inverse methods used in Radar Polarimetry with historical developments, current trends in both Western and Eastern countries are assessed for the identification of the current state-of-the-art.**

First, an attempt is made of providing an up-to-date review and assessment of direct and inverse methods in radar polarimetry together with those of recent events culminating in the overview for these proceedings **[O-1]**. It is followed by the "Keynote Address" on"Polarization in Nature" by Günther P. Können **[O-2]**, the famed Polar (Arctic & Antarctic) Atmospheric 'Polarimetrist'; by two special post-workshop invitations to our radar 'polarimetrists' from the USSR, each providing one state-of-the-art-review on fundamental developments **[O-3]** and on applications of radar polarimetry to meteorology, oceanography and planetary remote sensing **[O-4]**, by a succint review on meteorologic weather radar polarimetry in North America **[0-5]**, and concluded with a comparison of fundamental approaches to radar polarimetry **[0-6]**.

## 3.1 Basic Polarimetric Radar Theory: Topic I

**Although the polarization properties of electromagnetic waves are well understood [20,23,48], various unresolved questions, both in theory and application, are addressed for the coherent, partially polarized and partially coherent cases. Fundamental hitherto unknown issues, such as polarization sensitivity of SEM (Singularity Expansion Method) and EEM (Eigenmode Expansion Method) and time-domain polarization descriptors are identified.**

There definitely exists a great number of truly deserving invitees based on their past and recent contributions to radar polarimetry; and we were very delighted to have a few of them join us, for instance Professor Harold Mott, one of the most recent book authors on the subject. He was given the topic of introducing the basic definitions of polarization in the opening paper [I-1], which is followed by that on basic equations in radar polarimetry of Boerner et al **[I- 2]**. Our Eastern-European Dr. Zbigniew H. Czyż struggled with the formulation of alternative fundamental approaches **[I-3]** and the newly appointed Prof. Shane R. Cloude with the uniqueness of target decomposition theorems in radar polarimetry **[I-4]**, first introduced by Dr. J. Richard Huynen to whom these Proceedings are dedicated **[I-7]**. His unique discovery and formulation of the **"Polarization Fork"** concept, derived from E.M. Kennaugh's original work, to which the first NATO- ARW-IMEI'83 [7] was dedicated, is readdressed in the companion papers **[I-5]** and **[I-6]** comparing two entirely different approaches converging, as they should, in the same results.

Because the workshop director is of the opinion that the broadband partially polarized case can still not be broached satisfactorily; instead of further pursuing the controversial subject of **"Huynen's decomposition theories"** (see papers **[IV-2]** and **[III-12]** of the previous Workshop Proceedings [7], and **[II-5]** of these Proceedings), here the question of wideband polarimetric radar vector signal analysis is being considered by Sarkar et al, first introducing the T-Pulse concept [I-8], based on Kennaugh's original Kill or **"Kennaugh"-Pulse** concept [8], and then by Baum, who is analysing the polarization dependence of the SEM and EEM Scattering matrix in wideband radar vector signal processing [I-9]. This topical Section (I) is concluded in a novel approach by Chamberlain [I-10] of dealing with "transient polarization" aspects, a topic of growing vital importance to the rapid advancement of "ultrawideband polarimetric impulsive radar vector signal processing" [61], to be investigated in depth during our forthcoming third workshop NATO-ARW-WPDR'93 [IX-8].

### 3.2 Polarimetric Scattering Theories: Topic II

**Analyses of geometric and physical optics inverse scattering theories and their relevance to polarimetric radar target phenomenology are still ongoing and will have to be further advanced for a long time to come. Specific emphasis is placed during this NATO-ARW-DIMRP'88 on the extensions of the Kennaugh-Cosgriff transient ramp response methods to the slightly bistatic case and its applications in polarimetric target phenomenology and remote sensing.**

The topic of polarimetric scattering theories, briefly broached in [I-8] to [I-10] from a low-frequency wideband interpretation, in this topical Section (II) is exclusively handled from the Geometric Optics (GO) and Physical Optics (PO) viewpoints as introduced in the well done overview by Dr. Alon Schatzberg [II-1]. Here, the underlying concept of polarization correction of backscattering from conducting smooth convex surfaces is assessed which is based on Kennaugh's original work, reproduced here in the commemorative paper [II-3], and the research endeavors of the workshop director and his students in extending the results to the bistatic case as well [II-2]. Chaloupka, who together with his able doctoral students also contributed profoundly to these investigations, is expanding on both the PO/GO and GTD (Geometric Theory of Diffraction) inversion via polarimetric linear prediction [II-4], and the related paper by Chaudhuri, Foo and Boerner [II-5] is displaying some interesting features of Huynen's Mueller matrix parameters in applying the polarization corrected PO results derived in [II-2] for the monostatic and slightly bistatic cases. Huynen presents a conjecture on how to approach the "torsional parameters" in his Mueller matrix decomposition theory in [II- 6] which yet requires a satisfactory mathematical proof and physical verification. In a very industrious post-workshop invitation, Dr. Volker Stein puts together on what hitherto has been accomplished in polarimetric scattering theory and how it may be applied in numerical modeling [II-7]. Pertinent contributions by the polarimetric target/clutter modelers Kozlov and Logvin of the Moscow University of Aviation & Aerospace Science & Technology are summarized in [II-8], whereas Dr. Karl Tragl develops an elegant extension of these polarimetric scattering theories to the determination of optimal polarization states for reciprocal random targets applicable to meteorologic radar backscatter description [II-9]. We refer here also to his excellent dissertation [62] dealing with a co-variance matrix optimization procedure. A polarimetric model for multipath imaging and scrutinization, based on polarimetric GO/PO scattering theory extensions by Chaudhuri and Boerner **[II-10]**, rounds off this topical section **(II)** providing input for the analysis of polarimetric multistatic broadband inverse scattering approaches which are to be assessed during the forthcoming NATO-ARW-WPDR'93 **[IX-8]**.

### 3.3 Polarimetric Metrology & Systems Calibration: Topic III

**The design of the optimal polarimetric radar system still requires resolution of several questions pertaining to basic polarimetric radar metrology including proper standardization of measurement procedures for scattering matrix measurements, polarization state transformation, dual polarization antenna measurements, etc. Also, the questions of designing the most suitable set of calibration targets as well as defining proper pre/in/post-flight calibration methods are still not resolved as is discovered — in parts — in this section.**

Since the test-bed design attempts of the first dual polarization radars by Kennaugh [26], Huynen [30], Gent [28], van Etten [63], Nathanson [64], Poelman [65], Eaves [33], Root [29] and many more; persistent drives by Lloyd W. Root, particularly in recent years, brought about a subtle change in device technology as reported primarily in three recent MICOM Polarimetric Radar Technology Workshops at the (Wernherr von Braun) Rocket Auditorium within Redstone Arsenal [9,29]. Associated with these mindboggling advances are the innovative new approaches to polarimetric metrology and polarimetric systems calibration which elevated radar polarimetry to a mature engineering science discipline in remote sensing and high resolution imaging during the 'eighties'. Professor Werner Wiesbeck and his many able, enthusiastic doctoral students delivered a fine introduction in **[III-1]** to the subject matter which is followed by another brilliant contribution of Dr. Shane R. Cloude **[III-2]** all dealing with m-to-sub-mm wavelength analyses of polarimetric scattering matrix measurements, of which a systematic measurement error evaluation scheme is provided in **[III-3]**. Interspersed are the papers by Prof. Rasheed M.A. Azzam and Prof. Walter K. Kahn and students on efficient metrology and calibration in the optical **[III-4]** and the microwave **[III-5]** spectral regions, followed by a meditative analysis on reciprocity by Feinstein **[III-6]**, a topic that stirred up a lot of heated commotion during the various Workshop discussion group activities (see **[IX-4]** & **[IX-5]**). The "factotum non comparis" of our workshop, the highly esteemed, dynamic Lloyd W. Root, contributed a humorous and witty summary on his innovative pioneering efforts of developing the concept of "Active Polarimetric Calibrators" and of demonstrating practical calibration results **[III-7]** which since have been copied ruthlessly by many without mentioning the originator's name. The last three papers of this topical section **(III)** are dealing with various aspects of system metrology and calibration of POL-RAD **[III-8]**, **POL-SAR** **[III-9]** and combined **[III-10]** systems aspects. Every single paper of **(Topical Section III)** contributed profoundly to the lively discussions of the entire working discussing group efforts **W-A** to **W-E** and also to **W-F** in that obvious discrepancies in the usage of nomenclature, symbols, concepts and standards become blatantly apparent as highlighted in every report of **[IX-1]** to **[IX-7]**, and to be reassessed during the third NATO-ARW-WPDR'93 **[IX-8]**.

### 3.4 Polarimetric Vector Signal Processing: Target versus Clutter Discrimination: Topic IV

**Provided that the scattering matrices are known on a POL-RAD bin-by-bin, POL-SAR pixel-by-pixel basis, novel robust signal and image processing tools are feasible and are being introduced in order to display the rapid advancements made.**

Closely related with the topic of metrology and calibration (III) is that of "Polarimetric Vector Signal Processing" in polarimetric sensing and imaging with direct application to optimal target versus clutter discrimination, being the topic of this section (IV). The first paper is by the author of the now famed review on "Polarization Diversity in Radar of 1986" [44], Prof. Dino Giuli and his senior assistant of the University of Florence, Italy providing an up-to-date overview on Polarimetric Signal Processing Techniques [IV-1]. Here, we add that after this workshop came to a close, Prof. Dino Giuli has introduced another very important contribution to realizing "instantaneous recovery" of the 2x2 Sinclair scattering matrix parameters by using an orthogonal ambiguity function approach for simultaneous dual polarization radar measurements [VI-10] as will be discussed in detail during a special session of the forthcoming PIERS' 91, and more so during NATO-ARW-WPDR'93. Similarly, the next three papers by Wanielik [IV-2], Zhivotovskij [IV-3] and Sekine [IV-4] deal with novel multi-dimensional concepts of dual polarization radar signal processing which deserve our full attention including Sekine's polarimetric Weibull radar clutter modeling approaches (see Topic IX-1, reference [48]). Following the exposition of Preiser on "Adaptive Radar Polarimetry" [IV-5], which strongly reflects on earlier studies by Compton [67] and Fujimoto et al [68] on the subject matter; Wanielik, in another well done presentation, introduces his very ambitious and complex novel approach to a "Polarimetric CFAR-detector"[IV-6], stimulating extensive future investigations on Poelman's recent true original contribution of the "Multi-notch Logic-product Polarization Filter" approach for enhancing automated instantaneous target detection in severe dynamic background clutter [34,65]. It was most sincerely regretted that Poelman was prohibited from taking part in this NATO-ARW-DIMRP'88 due to internal restructuring of his home base in order to present another innovative contribution [69] which was finally released for inclusion in these Proceedings under [VI-6] . Now, we strongly request his participation in the forthcoming NATO-ARW-WPDR'93 addressing primarily the advancement of "Optimal Polarimetric Multi-Static Wideband Low RCS Target Detection in Severe Dynamic Background Clutter". The next two papers deal with the assessment of in-house polarimetric discrimination algorithm developments at GTRI-RAIL by Holm [IV-7] and at Selenia by Farina et al [IV-8]. A very interesting paper was contributed by Dr. Pax Samuel P. Wei of Boeing Aerospace, Kent Space Center dealing with scattering matrix analyses in a non-reciprocal magneto-ionic medium [IV-9], followed by another exposition of Logvin and Kozlov on optimal processing of polarization of higher order harmonics generated by electromagnetic wave interaction with non-linear materials [IV-10].

In concluding this Topical Section (IV), we need to refer to the two review papers [O-3] and [O-4], in which other recent very outstanding contributions of radar "polarimetrists" from the USSR are cited, and especially those of Potekhin, Kanareykin, and Tatarinov deserve our fullest attention as will become evident during NATO-ARW-WPDR'93, and here we refer to Section **V** of this introductory paper **[O-1]** in which the events **92-2** and **92-3** next to **93-1** are most pertinent.

**- PART TWO -**

### 3.5 Vector (Polarization) Diffraction Tomography & Environmental Sensing: Topic V

**Imaging in inhomogeneous media requires the generalization of straight-line projection tomography to diffraction tomography by satisfying at least the Rytov-Born approximations. In case the true depolarizing terms ($\nabla\varepsilon/\varepsilon$) and/or ($\nabla\sigma/\sigma$) can no longer be neglected in the inhomogeneous wave equation, a further extension to vector diffraction tomography is required. Applications to the detection and imaging of concealed and buried objects in strongly inhomogeneous media are discussed.**

The first two papers are presented by Prof. Karl J. Langenberg **[V-1]** and Prof. Pierre Célestin Sabatier **[V-2]**, who recently together with G.T. Herman and H.K.Tuy contributed the outstanding tutorial research text on **"Basic Methods of Tomography and Inverse Problems"** [70], here, however focusing on the polarimetric aspects of the multi-dimensional problem in inhomogeneous propagation space; a problem which was also discussed violently during the Working Discussion Group **W-C** activities **[IX-3]**. More practical aspects of microwave and ultrasonic imaging are considered by the expert team of SUPERLEC **[V- 3]**, a preliminary analysis of "de-polarizing effects" in vector (polarization) diffraction tomography is attempted in **[V-4]**, and more recently further advanced by the workshop director and students in [71]. The remaining six papers of this topical section all deal with the implementation of polarization ependent tomographic principles, where in [**V-5**] Prof. Jürgen Detlefsen of the famed TU München, Microwave Research Laboratory demonstrates practical results of a 94 GHz imaging radar for autonomous vehicle operation. Dr. Robert H. Giles et al consider the related problems of submillimeter wavelength modeling of dielectric materials in polarimetric interferometric imaging radar approaches **[V-6]**. In **[V-7]** diffraction limited polarimetric backscatter by rough terrain surfaces is discussed. The next three papers deal with polarimetric sensing and imaging of "volumetric scattering scenarios" such as on underground radar imaging by famed Professor Tsutomo Suzuki and students of the Electro-Communications University at Chofu-Shi, Tokyo/Japan **[V-8]**; polarimetric microwave remote sensing of surface and volumetric subterranean terrain scattering by Prof. Jin-Au Kong and former students **[V-9]**; and, is concluded with a comparative assessment on the dynamic modelling of volumetric sea icescatter and its dependence on geophysical ground truth parameters by Dr. Dale P. Winebrenner and his professor, Leung Tsang, and colleagues of the renowned University of Washington, Applied Physics Laboratory, Oceanographic Imaging Division, in Seattle, WA/USA **[V-10]**.

In concluding this topical section **(V)**, we need to refer to the proceedings of other related workshops of the URSI Int'l Commission F, Wave Propagation Symposia and of those of PIERS 87/89/91 in relation to applying some of these important topics to polarimetric remote sensing [51] utilizing POL-RAD/SCAT and POL-SAR/ISAR measurement systems as was further pursued during the working discussion group activities **W-B [IX-2], W-C [IX-3], W-D [IX-4]** and in **W-E [IX-5]** as regards the development of proper standards, and in **W-F [IX-6]** as relates to current and future inter-NATO country joint measurement campaigns.

### 3.6 Polarimetric Radar (POL-RAD) Systems Design and Operation: Topic VI

**In order to deploy polarimetric radar systems in practice, one requires polarimetric channel isolation of about 35 dB or more, sidelobe reduction of about 30 dB or better, etc. In addition, we require that the polarimetric antenna state switching is reciprocal and the radiation patterns are polarization state isotropic. Some of the recent advances on how to achieve these and similar goals and how to assemble state-of-the-art dual polarization radar systems are discussed.**

The papers of this topical section describe a variety of recent polarimetric test-bed instrumentation radar systems, presenting the state-of-the-art of the late 1980s, and with the current rapid advances made in device technology, these soon may seem to look like "pre-historic, large beasts" of "unwieldy collossal radar systems". Nevertheless, for the purpose of integrating practical aspects with metrology and calibration and also with theory and signal processing, the inclusion of this section was considered most vital. The workshop director considers the integrated multi-level theory-metrology-information-technology approach an absolute necessity for advancing this complex trans-disciplinary field; and, theoreticians are especially invited to not only peruse the papers of this section but to have a true thorough look at them in order to be able to focus on the inherent technologic complexities and current limitations of realizing some of the more "utopic polarimetric algorithm" implementations.

The description of the RADC POL-RAD System, **[VI-1]**, so well presented by Ken Stiefvater and well defended during the working discussion group activities **W-D [IX-4], W-E [IX-5]** and **W-F [IX-6]** is about the top of the current-state-of-the-polarimetric-technology. Similarly, the RSRE/THORN-EMI **[VI-2]**, the UKAN [VI-3], the DLR [VI-4], the RAT-SCAT **[VI-5]** and the SHAPE-TC **[VI-6]** systems descriptions well illustrate the true progress made in polarimetric radar system technology as well as in the professional operation and calibration of such systems. The paper on integrated micro-strip polarimetric patch antenna devices [VI-7] by one of its true pioneers, Prof. Kiyohiko Itoh, is applicable to direction-finding and these polarimetric integrated, fast-switching antenna approaches will soon allow the implementation of real-time polarimetric doppler radar processing and operation in practice, which are so urgently required in severe storm and tornado prediction [81]. Whereas, the well received papers on "polarimetry-dependent angulometry" by renowned polarimetric scatterometrists Prof. Dag T. Gjessing and Dr. Fan-Nian Kong **[VI-8]** and its relevance to glint-suppression by famed polarimetric radiometrist Prof. Radiy V. Ostrovityanov [VI-9] demonstrate the true capabilities of circumnavigating the inherent limitations of the classical non-polarimetric radars, and the venerable ground/ship-based monopolarization radars in particular; the new concept of applying general ambiguity function methods to the simultaneous transmission and reception of two orthogonally polarized signals is presented by Professor Dino Giuli and co-worker in [VI-10]. This latter new concept may open up new avenues of designing polarimetric Doppler radar systems for the analysis of target vibrational modes, etc., which certainly will become a subject of intense future R&D activities.

Whereas the polarimetric systems described during the workshop were mainly CW and rather narrow-band polarimetric radar systems, it is anticipated that we will be dealing with the description of wide-band (20% bandwidth), broad-band (50% bandwidth), and truly ultra-wide band (many octaves, e.g. 300 KHz to - 300 GHz desired non-carrier-frequency impulses) polarimetric (simultaneous orthogonal channel transmit-receive processor) systems during the forthcoming NATO-ARW-WPDR'93.

### 3.7 Polarimetric Synthetic Aperture Radar (POL-SAR) and Inverse SAR (POL-ISAR) Systems: Topic VII

**Most recently, POLarimetric Synthetic-Aperture-Radar (POL-SAR) systems and its inverse configurations (POL-ISAR) have become available and thus allow one to deploy the novel concept of the Polarimetric Matched Image Filter (PMIF) which is based on purely polarimetric (scattering matrix) optimization procedures. The rapid development of these intriguing new target enhancement versus background clutter rejection methods is discussed and verified with the use of POL-SAR data sets recently made available by NASA-JPL.**

At the time of our first NATO-ARW-IMEI'83, the near future availability of truly complete Polarimetric SAR systems was still considered a dream of the far-removed future. However, with the many spectacular advances made at CAL-TECH/JPL, Pasadena by Drs. Walter E. Brown, Jr., David Held and in particular by Dr. Fuk Li; at Loral (formerly Goodyear Aerospace Systems Corp.) Lichfield,AZ by Russell Blair, and Drs. Paul W. Goetz and Phil Murray; at ERIM, Ann Arbor by William E. Brown, Dale A. Ausherman, Roger J. Sullivan, Robert A. Shachman and Robert F. Rawson, such POL-SAR systems have now — within a short decade — become a "household item" in global terrestrial and planetary remote sensing. Thus, it is appropriate to start with two papers of CAL/TECH/JPL, introducing here Dr. Jakob J. Van Zyl, who deserves high praise and credit for refocusing the CAL-TECH/JPL efforts onto the proper path [VII-1] by de-emphasizing the brute-force Mueller matrix data reduction procedures explained in [VII-2]. This set of two papers is followed by two papers of the ERIM POL-SAR image processing team, which under its team leader Dr. Ivan J. LaHaie also contributed most profoundly to POL-SAR "in vitro/in vito" scatter image analysis, where [VII-3] deals with statistical classification of POL-SAR images, [VII-4] with its applications to target characterization and discrimination, and [VII-5] with computer simulation of polarimetric radar and laser imagery. In **VII-6**, Krogager investigates various decomposition schemes of the Sinclair matrix with specific applications to high resolution radar target imaging. From the very inception of Polarimetric SAR systems design and implementation, the workshop director's laboratory effort was strongly focused on the complete utilization of polarization information dating back to 1974 in polarization-dependent microwave holographic image formation [72,73] and to 1978 in polarimetric microwave tomographic image formulation [43]. Especially, we were the first to draw attention to the span-invariant [74] and its use in achieving speckle reduction [75]. Whereas, in **[VII-7]** a well-done computer-animated image interpretation for POL-SAR parameters was contribed by Geaga unfortunately placing too much emphasis on "computer image cosmetics" instead of clearly establishing what complete polarimetric utilization in POL-SAR image analysis can truly achieve, this indeed is accomplished in the well-orchestrated post-workshop contribution of the workshop director and students **[VII-8]**. In **[VII-8]** various POL-SAR speckle reduction and discrimination optimizers are compared with those by others [76,77], and the further advancements of all of these efforts need to be followed up carefully! The next paper by Hoogeboom **[VII-9]** surveys the efforts of the Dutch L-band POL-SAR design, implementation and operation; and, we refer here to the forthcoming joint NASA/ESA/ISAS event of POL-SAR Systems Calibration, DLR-Oberpfaffenhofen, rescheduled due to the Mid-East Petroleum crisis from 1991 January 26-28 to 1991 October 9-18, where other European (POL-SAR: DK) and Japanese (POL-SAR: ISAS) systems will be scrutinized on their pre/in-flight/post operational calibration capabilities. This section **(VII)** is then concluded with an innovative POL-ISAR target detection study by Profs. Andrew J. Blanchard and Adrian K. Fung and co-workers of the famed Wave Scattering Laboratory at the University of Texas at Arlington **[VII- 10]**. All of these papers provided plenty fruits of thought for the activities of working discussion group **W-D [IX-4]** and also **W-E** [IX-5] and **W-F [IX-6]** as relates to the development of standards and calibration procedures.

### 3.8 Polarimetric Radar Meteorology and Oceanography: Topic VIII

**Polarimetric radar systems implementation are especially well suited for the analysis of meteorologic and oceanographic scatter for which the term ($\nabla\varepsilon/\varepsilon$) shows marked local changes. Recent outstanding advances in these fields are highlighted. Various polarimetric doppler radar systems facilities are compared for illustrating the inherent inter/transdisciplinary complexities of these fields.**

Two of the principle areas of application of polarimetric radar, scatterometry, SAR and ISAR systems are radar meteorology and wide-area ocean surface and coastal region air-borne and space surveillance. At the time of execution of this NATO-ARW-DIMRP'88, polarimetric radar meteorology was much further advanced in comparison to POL-SAR wide-area ocean surveillance which, however, has caught up very rapidly in the meantime [78]. Based on the principle interests at that time of the prime institutional workshop supporter, the DLR at Oberpfaffenhofen, major and special emphasis during our workshop was placed on assessing the progress made in Polarimetric Radar Meteorology (also refer to **[O-4]**), starting off with the descriptions of systems performance, calibration and measurement capabilities of some of the very best test-bed instrumentation Polarimetric Doppler Radar Systems of the time: **DLR [VIII-1]**, CHILL **[VIII-2]**, DUT **[VIII-3]**, FIR **[VIII-4]**, NRC **[VIII-5]** with the RADC and RAT-SCAT systems already described in **[VI-1]** and **[VI-5]**.

A thorough up-to-date polarimetric radar assessment of our current understanding on the propagation through rain is presented by **Dr. Anthony R. Holt** in **[VIII-6]**. Here, we refer to the comparative analysis provided by Seliga and Humphries and Metcalf [79] and by Bringi and Hendry [80], respectively, on the history of radar meteorology and on the development of POL-DOP-RAD systems for radar meteorologic applications in the Western hemisphere as presented in [81], whereas, those accomplished - during the same period within the Soviet block - are succinctly summarized by Kanareikin et al in the invited review **[O-4]**. The next two papers by Dr. Dusan S. Zrnic and colleagues **[VIII-7]** and by Prof. Viswanathan N. Bringi and students **[VIII-8]** are outstanding expositions on the utility of polarimetric Doppler radar measurements of weather radar and convective storms. The last two papers provide a simplified backscatter model of fluctuating distributed rain backscatter **[VIII-9]** by Agrawal and Boerner, and of rough sea surface scatter by Lande **[VIII-10]**. Much emphasis will be placed on multi-band POL-SAR (P/L/S/C/X/K/V/W) and UWB-POL-DOP-RAD analyses of sea-surface and also meteorologic scatter during the forthcoming third NATO-ARW-WPDR'93. Here, however, we emphasize that contrary to what was initially anticipated by some non-believers in POL-RAD/SAR applications [Atlas, 81], polarimetry will also play a major role in air/space-borne POL-SAR global terrestrial and planetary meteorologic and oceanographic remote sensing [82]. Authors and speakers of these topical papers **(VIII)** were truly most instrumental in guaranteeing the success of the working discussion group activities which is evident from the perusal of names of contributors of all of the resulting reports.

### 3.9 Final Reporting of Working Discussion Group Activities: Topic IX

The workshop discussion group activities are intended to provide recommendations for potential future research projects. The six basic topics (**W-A** to **W-F**) chosen define issues for which immediate answers are required. Details on the Working Discussion Group Topics and Activities are provided in the Background Section **[IX-0]**, summarized in the Concluding Remarks **[IX-7]**, and a preview on a resulting third NATO-ARW-WPDR'93 is given in **[IX-8]**. These reports (**[IX-1]** to **[IX-6]**), which are collected at the end of **Part Two**, provide an assessment of the progress made after execution of the first NATO-ARW-IMEI'83, and they present the integral status and the important future emphasis of these polarimetric radar techniques to electromagnetic sensing and imaging. We encourage the readers of these Proceedings to peruse first the final reports of the working discussion group activities of topical Section **IX**, before embarking into the critical assessment of individual papers presented in Topical Sections **I** to **VIII** within Parts One and Two of these

Proceedings.

## IV. RELATION TO OTHER NATO ADVANCED STUDY INSTITUTES (ASI) AND ADVANCED RESEARCH WORKSHOPS (ARW)

Over the past decades, NATO-ASI/ARWs have provided a continuing forum for fundamental results and latest developments also in the topical fields of "Signal & Image Processing", "Pattern Recognition and Target Identification", "Non-destructive Material Testing and Non-Invasive Radiologic Diagnosis", "Inverse Problems and Profile Reconstruction", etc. Whereas, continual up-dates on all newly appearing NATO-ASI/ARW Proceedings are provided in the Quarterly NATO Newsletter of the NATO-SAD, Brussels and of the NATO Publications Coordination Office, here several recent ASIs and ARWs pertinent to this series of NATO-ARWs: (IMEI'83, DIMRP'88, WPDR'93) are listed below with reference to the pertinent proceedings given in the List of References:

Physics and Engineering of Medical Imaging [83]

Industrial Robotic Vision [84]

Vision & Image Understanding [85]

Pictorial Data Analysis [86]

Diagnostic Imaging in Medicine [87]

Atmospheric Effects on Radar Target Identification and Imaging [88]

Surveillance of Environmental Pollution and Resources by Electromagnetic Waves [89]

Theoretical Methods for Determining the Interaction of Electromagnetic Waves with Structures [90]

Pattern Recognition Theory and Applications [91]

Remote Sensing Applications in Marine Science and Technology [92]

Optical Metrology [93]

The Application of Laser Light Scattering to the Study of Biological Motion [94]

Nonlinear Phenomena at Phase Transitions and Instabilities [95]

Image Sequence Processing and Dynamic Scene Analysis [96]

Aspects of Signal Processing with Emphasis on Underwater Acoustics[97]

New Directions in Signal Processing in Communications and Control [98]

EM Modelling & Measurements for Analysis & Synthesis Problems [99]

Signal Processing and Pattern Recognition in Non-destructive Evaluation of Materials [100]

Pattern Recognition and Signal Processing [101]

The Impact of Processing Techniques on Communications [102]

Fast Electrical and Optical Measurements [103]

Modern Topics in Microwave Propagation and Air-Sea Interaction [104]

ELF/VLF Radio Wave Propagation [105]

Pattern Recognition Theory and Applications [106]

Advanced Physical Oceanographic Numerical Modelling [107]

Large-Scale Transport Processes in Ocean and Atmospheres [108]

Remote Sensing Applications in Meteorology and Climatology [109]

Electromagnetic Coupling in the Polar Clefts and Caps [110]

Microwave Remote Sensing for Oceanographic and Marine Weather-Forecast Models [111]

In conclusion, this NATO Advanced Research Workshop has demonstrated both the basic unity of direct & inverse methods in radar polarimetry and the diversity of their applications to high resolution imaging and remote sensing. The Proceedings of this workshop represent, however, only one sampling of the development of this active new area of science and technology. Yet, these Proceedings will provide workers with sufficient reference material and suggestions for new research topics to further their own continuing research development. The following rounding-off section gives a preview of related forthcoming events for which requests for active participation are invited, and the resulting proceedings should be studied carefully.

## V. PREVIEW OF NEAR-FUTURE NATO AND OTHER INTERNATIONAL ARW ACTIVITIES FOR THE ADVANCEMENT OF DIRECT AND INVERSE METHODS IN ULTRA-WIDEBAND POLARIMETRIC RADAR TECHNOLOGY

The momentum gained during this rather time-consuming task of assembling these Proceedings of the NATO-ARW-DIMRP'88, and subsequent research workshops, retreats, and symposia during 1989 [53] and 1990 [54,59], resulted in the planning of Special Sessions on "Direct and Inverse Methods in Radar Polarimetry," and additional workshops for 1991, 1992, 1993 and already thereafter. Here, we alert the readers' attention to the forthcoming proceedings and research reports of some of these events.

**<u>1991</u>**

**91-1** January 27-31: NASA/ESA/ISAS POL-SAR Calibration Workshop, DLR-Oberpfaffenhofen, FRG: Seven (7) sessions on POL-RAD/SAR Calibration (due to Mid-East Petroleum crisis cancelled and now rescheduled for 1991 October 9-18);

**91-2** April 15-18: Seventh International Conference on Antennas & Propagation (ICAP), University of York, UK: Two (2) special sessions on radar polarimetry plus workshop (April 18-20);

**91-3** May 20-24: MIKON'90, Ninth Polish Microwave Conference, Rydzyna-Poznan, Poland: Two (2) special sessions on Radar Polarimetry plus workshop (May 23-25); follow-up workshop/short-course lecture series considered either during week before or week after, at the Technical University Dresden, FRG(0);

**91-4** June 24-28: International IEEE/AP-S & URSI Radio Science Meeting, The University of Western Ontario, London, Ontario/-Canada: Two (2) special sessions on radar polarimetry;

91-5 July 1-5: PIERS'91, MIT, Cambridge, MA: Various Special Sessions on "Direct and Inverse Methods in Radar Polarimetry" (W-M. Boerner);

91-6 September 17-20: Second Italian International Conference on Electromagnetics and Aerospace Applications, "ICEAA 91", Politechnico di Torino, Torino, Italia: Two (2) special sessions on fundamental polarimetric radar theory;

<u>1992</u>

92-1 March 20-22: Second French International Workshop on Radar Polarimetry: JIRP'92, IRESTE, University of Nantes, Nantes, France;

92-2 June 28-30: The International URSI-F Microwave Signatures Conference, Igls/Innsbruck, Austria (Dr. Wolfgang Keydel and Prof. H Rott, Co-directors): Several sessions on "Polarimetric Signatures, Techniques and Systems", (W-M. Boerner, W. Wiesbeck and J.J. van Zyl, conveners);

92-3 July 18-25: International IEEE/AP-S & URSI Radio Science Meeting, Hyatt Regency Hotel, Chicago, IL (W-M. Boerner, Vice Chair): Two (2) sessions on Radar Polarimetry;

92-4 August 17-20: URSI International Commission-B, Electromagnetic Wave Workshop, Sidney Convention Center, Sidney, Australia: Two (2) sessions on Radar Polarimetry (W-M. Boerner, Convener);

92-5 August 24-26: Fourth (biennial) Asia-Pacific Microwave Conference-APMC'92, Adelaide Convention Centre, Adelaide, S.A., Australia. Two (2) Special Sessions on Radar Polarimetry - Theory/Metrology/Systems/Technology/Applications (W-M. Boerner and B. Haywood, conveners);

92-6 September 1-18: First USSR Conference on "Polarimetric Radar Remote Sensing and Surveillance of the Terrestrial and Planetary Environments", USSR Academy of Sciences, Polar & Antarctic Division, Leningrad (92 Sept. 1-7) and Siberian Division, Lake Baikail, Ulan-Ude, Buryat Region, USSR (1992 Sept. 10-16) with A.B. Shupyatkij, A.I. Kozlov, N.B. Chimotdorzhiev and W-M. Boerner, conveners);

92-7 September 22-25: ISAP'92, Fifth International Japanese Symposium on Antennas and Propagation, University of Hakkaido, Sapporo, Japan: Two (2) sessions on Radar Polarimetry plus 3-day workshop (Y. Yamaguchi, M. Tanaka, H.J. Eom and W-M. Boerner).

<u>1993</u>

93-1 **September 19-25: NATO-ARW-WPDR'93: Wideband Polarimetric Doppler Radar Sensing, KuK Hotel Residenz, Bad Windsheim, FRG (although, currently proposed for 1991, September 18-24; we expect this NATO-ARW-WPDR to take place during 1993, September 19-25; and not during 1992, because of scheduling conflicts): W-M. Boerner, Director and Editor of Proceedings.**

In concluding this overview section, it should be observed that a strong international interaction program on "Radar Polarimetry: Theory-Metrology Instrumentation - Data Processing - Applications" is now developing as a result of the previous NATO-ARW-IMEI'83, our NATO-ARW-DIMRP'88, PIERS'89, JIPR'90, and in particular, of the recent USA-USSR interactions, as summarized in the Technical Research Travel Interaction Reports [54-59].

## VI. ACKNOWLEDGEMENTS

The preparation of this proceedings overview was supported, in parts, by an Extension Grant No. NATO-SA.5-2-04(ARW.923/86)620, of the NATO Scientific Affairs Division, by a Special Contract No. US Army (ARO) DAAL-03-89-K-0075, of the US Army Research Office for the completion of these Workshop proceedings, and substantially by the University of Illinois Foundation. In addition, without the contributions from an Alexander von Humboldt Foundation, US Senior Scientist Award, the very time-consuming diligent preparations for this NATO-ARW-DIMRP'88, would not have been made possible; and, this indirect assistance is gratefully acknowledged.

The manuscript typing was skillfully carried out by Richard W. Foster, Mirian R. Mailey, and Jasmin Chao-Hui Tuan, and sincere thanks are also extended to Drs. Ernst Lüneburg, Dusan S. Zrnic, and especially, David E. Stein for proof-reading the manuscript.

## VII. REFERENCES

[1] D. Pritchard, The Radar War, Wellingborough/North Hampshire, England: Patrick Stephens Ltd., Thorsons Publ. Group, 1989

[2] R.S. Berkowitz, Modern Radar, New York: John Wiley & Sons, 1965

[3] M.I. Skolnik, Ed., Radar Handbook, New York, NY: McGraw-Hill, 1970 (also see: M.I. Skolnik, Introduction to Radar Systems, 2nd ed., New York: McGraw-Hill, 1980)

[4] D.L. Mensa, High Resolution Radar Imaging, Dedham, MA: Artech House, Inc, 1981 (also see:ibid, High Resolution Radar Cross-Section Imaging, Boston: Artech House, Inc., 1991)

[5] D.R. Wehner, High Resolution Radar, Norwood, MA:Artech House, 1987

[6] W-M. Boerner, "Polarization Utilization in Electromagnetic Inverse Scattering", Cpt. 7 in Inverse Scattering Problems in Optics, Vol. 2 (Ed. H.P. BALTES), Topics in Current Physics, Vol. 20, Heidelberg: Springer Verlag, July 1980, pp. 237-305.

[7] W-M. Boerner, et al (eds), Inverse Methods in Electromagnetic Imaging, NATO-ARW-IMEI, 1983 Sept 18-24, Bad Windsheim, FRG, NATO-ASI Series C, Math & Phys. Sci., Vol. 143, Dordrecht, NL: D. Reidel Publ. Co., 1985

[8] W-M. Boerner, A.K. Jordan and I.W. Kay (Guest Editors), "Introduction to the Special Issue on Inverse Methods in Electromagnetics", in IEEE Trans. Ant. & Propag., AP-29, March 1981, pp 185-189

[9] L-W. Root and B.L. Matkin, Chairmen/Editors, Proceedings, (Third) Polarimetric (Technology) Workshop, Redstone Arsenal, AL, 1988 August 16-18, Chicago, IL: GACIAC/IITRI, 1990

[10] J.W. Battles and L.W. Root, eds., Handbook of Polarimetric Technology, Chicago, IL: GACIAC/IITRI, 1991

[11] G.P. Können, Polarized Light in Nature,(English Translation), Cambridge, UK:Cambridge University Press,1985; also see:ibid,Gepolariseerd Licht in de Natuur, Zutphen, NL:B.V.W.J. Thieme & Cie, 1980

[12] W. Ritter von Haidinger, "Über das direkte Erkennen des polarisierten Lichts und der Lage der Polarisationsebene", Pockendorff's Annalen, Vol. 63, pp. 29-39 , 1844, (also see: ibid, Über den Pleochroismus des Amethysts", Pockendorff's Annalen, Vol. 70, pp. 531-544, 1847: Discovery of circular dichroism)

[13] R.F. Hochheimer and H.A. Kues, "Retinal polarization effects", Applied Optics, Vol. 21, No. 21, pp. 3811-3818, Nov. 1982 (also see:B.F. Hochheimer, Polarized light retinal photography of a monkey eye, Vision Research (Pergamon Press) Vol. 18, pp. 19-23)

[14] Karl von Frisch, Bees, Their Vision, Chemical Senses and Language, Heidelberg: Springer Verlag 1971 (Translation of German Version: Tanzsprache und Orientierung der Bienen, Heidelberg: Springer, 1965)

[15] A.W. Snyder and R. Menzel, eds., Photoreceptor Optics, Proceedings of a Workshop at Darmstadt, FRG, October, 1974, Heidelberg: Springer Verlag 1975 (see Section D: Polarisation Sensitivity and Dichroism: T.H. Waterman, The Optics of polarization sensitivity, pp. 339-371)

[16] W.A. Shurcliff, Polarized Light, Cambridge, MA: Harvard Press,1962

[17] W. Swindell, ed., Polarized Light, Stroudsburg, PA: Halsted Press, 1975

[18] T. Gehrels, ed., Planets, Stars and Nebulae Studied with Photopolarimetry, Tucson, AZ: The University of Arizona Press (Extensive List of Important References), 1974 (see: T.H. Waterman, Polarimeters in Animals, pp. 472-494)

[19] W.G. Egan, Photometry and Polarization in (Optical) Remote Sensing (.185 to 12μm), New York: Elsevier, 1985

[20] M. Born and E. Wolf, Principles of Optics, 3rd ed., New York: Pergamon Press, 1964

[21] W.E. Knowles Middleton, Radar Development in Canada: The Radio Branch of the National Research Council of Canada; 1939-1946, Waterloo, Ontario/Canada: Wilfrid Laurier University Press, 1981

[22] R.C. Jones, "A New Calculus for the Treatment of Optical Systems", J. Optical Society of America: Part I (Description and Discussion the Calculus: JOSA 31, pp 488-493, 1941), Part III (The Söhncke Theory of Optical Activity: JOSA 31, pp500-503, 1941), Part IV (Interpretation: JOSA 32, pp 486-493, 1942), Part IV (More General Formulation, and Description of Another Calculus, JOSA 37, pp 107-110, 1947)

[23] R.M.M. Azzam and N.M. Bashara, Ellipsometry and Polarized Light, Amsterdam: North Holland Publ. Co., 1977

[24] R.A. Chipman and J.W. Morris, eds., Proceedings, SPIE Int'l. Workshop on Polarimetry: Radar Infrared, Visible, Ultraviolet, and X-Ray, 1990 May 15-17, Huntsville, AL, SPIE-PROC.-1317, 1990

[25] G. Sinclair, The Transmission and Reception of Elliptically Polarized Radar Waves, Proc. IRE, Vol 38, pp. 148-151, Feb 1950

[26] E.M. Kennaugh, Polarization Properties of Radar Reflectors, M.Sc. Thesis, Dept. of Electr. Eng., The Ohio State University, Columbus, OH 43212, 1952 (also see: E.M. Kennaugh and D.L. Moffatt, Transient Current Density Waveforms on a Perfectly Conducting Sphere, Memorial Paper in [7, 0.1, pp 1-31] containing references to all of Kennaugh's pertinent reports, 1985)

[27] G.A. Deschamps, "Part 2: Geometrical Representation of the Polarization State of a Plane Electromagnetic Wave", Proc IRE Vol 39, pp 540-544, 1951 (also see papers by Booker (Pt 1), Kales (Pt 3), Rumsey (Pt 4), and of Bohnert (Pt 5), same issue)

[28] H. Gent, Elliptically polarized waves and their reflections from radar targets: A theoretical analysis, Telecommunications Research Establishment, Cheltenham, England, UK: TRE-MEMO 584, March 1954

[29] L.W. Root, Chairman & Editor, Workshop on "Polarimetric Radar Technology", Redstone Arsenal, AL, 1981, June 25-26, Chicago, IL: GACIAC, IITRI, 1982

[30] J.R. Huynen, "Study on Ballistic-Missile Sorting Based on Radar Cross-Section Data", Special Report No. 4, "Radar Target Sorting Based on Polarization Signature Analysis", Palo Alto, CA, Lockheed Aircraft Corp., Missiles and Space Division, Rept. LMSD-288216, (Contract No. AF19(604)5550), May 1960

[31] J.R. Huynen, Phenomenological Theory of Radar Targets, Ph.D. Dissertation, Technical University Delft, The Netherlands, 1970 (revised: 1987/available from author; also see:ibid, Chapter 11 in "Electromagnetic Scattering", P.L.E. Uslenghi, ed., New York: Academic Press, 1978)

[32] J.D. Copeland, "Radar Target Classification by Polarization Properties", Proc. IRE, Vol. 48, pp. 1290-1296, July 1960

[33] J.L. Eaves and E.K. Reedy, Principles of Modern Radar, New York, NY: Van Nostrand Reinhold Co., 1987

[34] A.J. Poelman and J.R.F. Guy, "Polarization Information Utilization in Primary Radar: An Introduction & Update to Activities at Shape Technical Center", Proc. NATO-ARW on IMEI, Bad Windsheim, FR. Germany, Sept. 18-24, 1983, Session RP.5, Section III: Paper No. III.2, pp. 521-572.

[35] V.V. Bogorodsky, D.B. Kanareykin and A.E. Kozlov, Polarization of the Scattered Radio Radiation of the Earth Covers, Leningrad: Gidrometeorizdat, (in Russian) 1981

[36] M.M. Gorshkov, Ellipsometry, Moscow: Sovetskoye Radio Press, (in Russian), 1974

[37] D.B. Kanareykin, N.F. Pavlov, and U.A. Potekhin, The Polarization of Radar Signals, Moscow: Sovyetskoye Radio, Chap. 1-10 (in

Russian), 1966, (English Translation of Chpts. 10-12: Radar Polarization Effects, CCM Inf. Corp., G. Collier and McMillan, 900 Third Ave., New York, N.Y. 10023)

[38] D.B. Kanareykin, V.A. Potekhin, and M.F. Shisikin, Maritime (Marine) Polarimetry, Leningrad: Sudostroyenie, 1968

[39] A.L. Kozlov, "Radar Contrast of Two Objects", Izvestiya Vuz., Radioelektronika, Vol. 22, No. 7, July 1979, pp. 63-67

[40] A.G. Shupyatskij and S.P. Morganov, "The Application of Polarization Methods to Radar Studies of Clouds and Precipitation", AFCRL, L.G. Hanscom Field MA, Techn. Rept. AFCRL-68-0483, Sept. 1968 (Translation from Russian)

[41] M.L. Varshanchuk and V.O. Kobak, "Cross-correlation of Orthogonally Polarized Components of the Electromagnetic Field Scattered by an Extended Object", Radio Eng. & Electr. Phys. 16, 1971, pp 201-205.

[42a] L.A. Zhivotovskiy, "Optimum Polarization of Radar Signals", Radio Eng. and Electronic Phys., 18(4), 1973, pp. 630-632.

[42b] L.A. Zhivotovskiy, "Calculation of the Depolarization Properties of Radar Targets", Radio Eng. 31, Part 2, pp. 46-48, 1973

[42c] L.A. Zhivotovskiy, "Dependence of the Power Flux Density of Echoe Signals on Radar Signal Polarization", Radio Eng. 31, Part 2, pp. 49-53, 1976.

[43] W-M. Boerner, "Polarization Utilization in Microwave Imaging, Radar Target Mapping, and Electromagnetic Inverse Scattering", A State-of-the-Art Review, Comm. Lab. Rept. No. 78-3, Communications Lab., Inf. Eng. Dept., SEO-1124, UICC, Chicago, Oct. 1978

[44] D. Giuli, "Polarization Diversity in Radar", Proc. IEEE, Vol. 74(2), pp. 245-269, Feb. 1986

[45a] M. Davidovitz and W-M. Boerner, "Extension of Kennaugh's Optimal Polarization Concept to the Asymmetric Matrix Case", IEEE Trans. A&P, Vol. AP-34(4), pp. 569-574, Apr. 1986.

[45b] M. Davidovitz and W-M. Boerner, "Reduction of Bistatic Scattering Matrix Measurements for Inversely Symmetric Radar Targets", IEEE Trans. A&P, Vol. AP-31, pp. 237-242, March 1983

[46] W-M. Boerner, W-L. Yan, A-Q. Xi, "Basic Principles of Radar Polarimetry", Chapter 2 in: J.W. Battles and L.W. Root, eds., Handbook of Radar Polarimetry, Chicago, IL: GACIAC/IITRI, 1991

[47] J.J. Van Zyl, On the Importance of Polarization in Radar Scattering Problems, Ph.D. Dissertation, California Institute of Technology, Pasadena, CA, January 1986

[48] H. Mott, Polarization in Antennas and Radar, New York: John Wiley & Sons, 1986 (second greatly up-dated edition in preparation)

[49] F.T. Ulaby, R.K. Moore, and A.K. Fung, Microwave Remote Sensing: Active and Passive, Vol. I(81), Vol. II(82), Vol III(85), Addison-Wesley, Reading, MA, 1981

[50] F.T. Ulaby and C. Elachi, eds., Radar Polarimetry for Geoscience Applications, Norwood, MA: Artech House, Inc., 1990

[51] J.A. Kong, ed., Polarimetric Remote Sensing, PIER-3, New York: Elsevier, 1990

[52] Carl Cranz Gesellschaft e.V. (CCG), Short Course Lecture Series, Sensor Technology: W-M. Boerner, Advanced Radar Polarimetry and Its Applications, Lecture No.S-2.19, Oberpfaffenhofen, FRG: 1986, February 2-6

[53] W-M. Boerner, International Travel Report on the 1989 May 13 to May 26 USSR Travel, "The Advancement of Inverse Methods in High Resolution Polarimetric Radar Imaging and Its Applications to Hostile Object Discrimination Against/And the Remote Sensing of Terrestrial and Planetary Environments", UIC-EECS/CSL Report No. 89-05-13, 1990 July 15.

[54] J. Saillard, ed., Journées Internationales de la Polarimétrie Radar, Recueil des Communications, IRESTE, Nantes, 1990 March 20-22 (French/English)

[55] W-M. Boerner, International Travel Report on 1989 November FRG-(USSR)-JAP Travel, UIC-EECS/CSL Rept No. 1989 11-03, (1989)

[56] W-M. Boerner, Report on 1990 May 5 to May 18, Visit of Six Polarimetric Radar Experts from the USSR (Technical Report & Budget, including summaries of events by D.S. Zrnic and J.R. Huynen), 1990 May 21

[57] W-M. Boerner, Final Report on a Cultural & Research Interaction Travel to the Buryat Autonomous Soviet Republic in South-Eastern Siberia, USSR of 1990 August 17 to September 04, UIC-EECS/CSL Technical Report, 1990 December 03/26

[58] N.B. Chimitdorzhiyev, P.N. Dagurov and Yu. L. Lomukhin, Propagation, Scattering and Diffraction of Microwaves, Novosibirsk: USSR Acad. Sci., Nauka Press, 1987 (in Russian)

[59] W-M. Boerner, International Travel Report on 1990 March 17 to 25, NATO-Europe Research Travel of W-M. Boerner for Advancing Interaction on "Wideband Polarimetric Doppler Radar Target Detection, Imaging & Identification & Environmental Sensing", UIC-EECS/CSL Technical Report, 1990 March 30

[60] Eric Pottier, Contribution de la Polarimetrie dans la discrimination de cibles radar - Application á l'Imagerie Electro-magnétique Haute Resolution, Ph.D. thesis (IRESTE), l'Université de Rennes I, 13 Dec. 1990 (in French)

[61] J.D. Taylor, editor, Handbook of Ultra-wideband Impulsive Radar Technology, New York: Academic Press, 1991

[62] K. Tragl, Polarimetrische Radarbeobachtungen von zeit-veränderlichen Zufallszielen, Dr.-Ing. Dissertation, Universität Kaiserlautern, 1990 Oct. 26 (in German)

[63] P. van Etten, Polarization Radar Method and System, US Patent No. 4053882, 1977 Oct. 11 (RADC/OCTS, Griffiss AFB, NY. 13441-5700, Attn: Kathy Paige: +[1](315)330-2087)

[64] F.E. Nathanson, Radar Design Principles, McGraw Hill Book Co., New York, N.Y., 1969

[65a] A.J. Poelman, "Virtual Polarization Adaptation, a Method of Increasing the Detection Capability of a Radar System Through Polarization-Vector Processing", IEE Proc., Part F, Vol. 128, No. 5, pp. 261-270, Oct. 1981

[65b] A.J. Poelman, "Polarization Vector Translation in Radar Systems", IEE Proc., Part F, Vol. 130, No. 2, pp. 161-165, March 1983

[66] J.L. Eaves, "Intrapulse Agile Radar System (IPAR)", Dept. of the Army, Washington, D.C., Patent Appl., Rept. PAT-APPL-6-359 646, March 19, 1982

[67] R.T. Compton, Jr., Adaptive Antennas, Concepts and Performances, Englewood Cliffs, NJ: Prentice Hall, 1988

[68] K. Fujimoto, A. Henderson, K. Hirasawa and J.R. James, Small Antennas, New York: John Wiley, 1987

[69] A.J. Poelman and C.J. Hilgers, "The effectiveness of multi-notch logic-product polarisation filters in radar for countering rain clutter", IEE Proceedings, Part F, Vol. 138, in print, 1991

[70] P.C. Sabatier, ed., Tomography and Inverse Problems, Bristol, UK: Adam Hilger, Modern Physics Series, 1987

[71] Y. Yamaguchi, M. Mochida, T. Abe and W-M. Boerner, "Effect of View Angle Variations in Vector Diffraction Tomography", Japan Journal for Electronics and Communications Engineering (JEICE Trans Com), Subm. Dec. 1990, Vol. 71, in print, Spring 1991

[72] W-M. Boerner, "Identification of Stationary and Moving Scatterers in Clutter", Annual Research Report 1974, DRB Grant No. 3880-08 (UG), Oct. 1974

[73] M.L.A. Gassend and W-M. Boerner, "Polarization Holography with Local Reference Beam Obtained by Optical Frequency Separation", Appl. Phys. (Springer), 13, pp. 71-79, Jan. 1977

[74] W-M. Boerner, C-Y. Chan, "Inverse Methods in Electromagnetic Imaging", Chapt 2 in Proc. on Electromagnetic Dosimetric Imagery, L.E. Larsen and J.H. Jacobi, Eds., Walter Reed Army Institute of Research, Washington, D.C., June 3-4, 1980, Mack Printing Co., Easton PA, 1981.

[75] W-M. Boerner, "Polarization Microwave Holography: An Extension of Scalar to Vector Holography" (INVITED), 1980 International Optics Computing Conference, SPIE's Techn. Symposium East, Washington, DC, April 9, 1980, Session 3B, SPIE Symposium Report 231, Paper No. 231-23, pp. 188-198, 1980

[76] L.M. Novak, M.B. Sechtin, M.C. Burl, "Algorithms for Optimal Processing of Polarimetric Radar Data", Lexington, MA: MIT-LL, Proj. Rpt. No. TT-73, 1989 Nov. 6

[77] L.M. Novak, M.B. Sechtin, M.J. Candullo, "Studies of Target Detection Algorithms That Use Polarimetric Radar Data", IEEE Trans. AES, Vol. 25, No. 2, 1989 (March)

[78] IEEE Transactions on Geoscience and Remote Sensing, Special Issue on "SAR Ocean Applications", IEEE Trans. GSRS Vol. 29, No. 1, January 1991

[79] T.A. Seliga, R.G. Humphries and J.I. Metcalf, "Polarization Diversity in Radar Meteorlogy: Early Developments", in: D. Atlas, ed., Radar in Meteorology, Boston, MA: American Meteorological Society, pp. 109-114, 1990

[80] V.N. Bringi and A. Hendry, "Technology of Polarization Diversity Radar for Meteorology", in: D. Atlas, ed., Radar in Meteorology, Boston, MA: American Meteorological Society, pp. 153-190, 1990

[81] D. Atlas, Radar in Meteorology: Battan Memorial and 40th Anniversary Radar Meteorology Conference, Boston, MA: American Meteorological Society, 1990

[82] J.S. Theon and N. Fugono, Tropical Rainfall Measurements, Hampton, VA: A. Deepak Publ., 1988

[83] R. Guzzardi, ed., Physics and Engineering of Medical Imaging, NATO Advanced Study Institute, 23 Sept.-5 Oct. 1984; Maratea, Italy

[84] A. Dosterlinck, ed., Industrial Robotic Vision, NATO Adanced Study Institute, 6-17 August 1984; Leuven, Belgium

[85] A. Borsellino, ed., Vision & Image Understanding, NATO Advanced Study Institute, 1-12 July, 1984; Erice, Italy

[86] R.M. Haralick, ed., Pictorial Data Analysis, Proc. of ASI, Springer Verlag, 1983

[87] R.C. Reba, D.J. Goodenough, eds., Diagnostic Imaging in Medicine, Proc. of ASI, M. Nijhoff Publishers, 1983

[88] H. Jeske, ed., Atmospheric Effects on Radar Target Identification and Imaging, Proc. of ASI, D. Reidel Publishing Co., 1979

[89] T. Lund, ed., Surveillance of Environmental Pollution and Resources by Electromagnetic Waves, Proc. of ASI, D. Reidel Publishing Co., 1976

[90] J.F. Skwirzynski, ed., Theoretical Methods for Determining the Interaction of Electromagnetic Waves with Structures, Proc. of ASI, M. Nijhoff Publishers, 1981

[91] J. Kittler, K. Fu, L. Pau, eds., Pattern Recognition Theory and Applications, Proc. of ASI, D. Reidel Publishing Co., 1982

[92] A.P. Cracknell, ed., Remote Sensing Applications in Marine Science and Technology, Proc. of ASI, D. Reidel Publishing Co., 1983

[93] O.D.D. Soares, ed., Optical Metrology, NATO Advanced Study Institute, 16-27 July, 1984, Viana do Castelo, Portugal

[94] J.C. Earnshaw, ed., The Application of Laser Light Scattering to the Study of Biological Motion, NATO Advanced Study Institute, Plenum Publishing Co., 1983

[95] T. Riste, ed., Nonlinear Phenomena at Phase Transitions and Instabilities, NATO Advanced Study Institute, Plenum Press, 1982

[96] T.S. Huang, ed., Image Sequence Processing and Dynamic Scene Analysis, NATO-ASI, Series F2, Heidelberg: Springer, 1983

[97] G. Tacconi, ed., Aspects of Signal Processing with Emphasis on Underwater Acoustics (Parts I & II), NATO-ASI Series C-33, Dordrecht: Reidel, 1977

[98] J.K. Skwirzynski, ed., New Directions in Signal Processing in Communication and Control, NATO-ASI Series E-12, Dordrecht: Martinus Nijhoff, 1975

[99] J.K. Skwirzynski and B. de Neumann, eds., EM Modelling & Measurements for Analysis & for Synthesis Problems, NATO-ASI Applied Sciences Series E-199, 1991

[100] C.H. Chen, ed., Signal Processing and Pattern Recognition in Non-destructive Evaluation of Materials, NATO-ASI Series F-44, 1988

[101] C.H. Chen, ed., Pattern Recognition and Signal Processing, Proc. of ASI, Paris, France, June/July 1978

[102] J.K. Skwirzynski, ed., The Impact of Processing Techniques on Communications, Proceedings of the NATO Advanced Study Institute, Chateau de Bonus, France, July 11-22, 1983

[103] J.E. Thompson and L.H. Luessen, eds., Fast Electrical and Optical Measurements, Proceedings of the NATO Advanced Study Institute, Il Ciocco, Castelvecchio, Pascoli, Italy, July 10-24, 1983

[104] A. Zancla, ed., Modern Topics in Microwave Propagation and Air-Sea Interaction, Proceedings of the NATO ASI, Sorrento, Italy, June 5-14, 1973

[105] J.A. Holtet, ed., ELF/VLF Radio Wave Propagation, Proceedings of the NATO ASI, Spåtind, Norway, April 17-27, 1974

[106] J. Kittler, K.S. Fu and L.F. Pau, eds., Pattern Recognition Theory and Applications, Proceedings of the NATO ASI, Oxford, England, March 29-April 10, 1981

[107] J.J. O'Brien, ed., Advanced Physical Oceanographic Numerical Modelling, Proceedings of the NATO Advanced Study Institute, Banyuls-sur-Mer, France, June 2-15, 1985

[108] J. Willebrand and D.L.T. Anderson, eds., Large-Scale Transport Processes in Oceans and Atmosphere, Proceedings of the NATO Advanced Study Institute, Les Houches, France, February 11-22, 1985

[109] R.A. Vaughan, ed., Remote Sensing Applications in Meteorology and Climatology, Proceedings of the NATO Advanced Study Institute, Dundee, Scotland, August 17-September 5, 1986

[110] P.E. Sandholt and A. Egeland, eds., Electromagnetic Coupling in the Polar Clefts and Caps, Proceedings of the NATO Advanced Research Workshop, Lillehammer, Norway, September 20-24, 1988

[111] R.A. Vaughn, ed., Microwave Remote Sensing for Oceanographic and Marine Weather-Forecast Models, Proceedings of NATO Advanced Study Institute, Dundee, Scotland, August 14-September 3, 1988

**POLARIZATION IN NATURE**
(INVITED)

**GÜNTHER P. KÖNNEN**
Royal Netherlands Meteorological Institute
P.O. Box 201, Wilhelminalaan 10
NL-3730 AE De Bilt,
The Netherlands

**ABSTRACT:** In the open, we are surrounded not only by a mass of colour, but also by much polarization. The latter normally remains invisible to us, but with the aid of a simple polarizing sheet, like those in Polaroid sunglasses, one suddenly becomes aware how much polarization there is. In this article, a number of observations are described and the regularities in the natural polarization pattern are discussed.

## Polarization

Light is characterized by three properties: intensity, colour, and polarization. The state of the light around us is thus completely defined only if these three characteristics are all known. Natural light displays not only strong variations in its intensity and colour, but also in its linear polarization. In other words, much of the light around us has a preferential plane of vibration, and the strength of this tendency (degree of polarization) and the position of this preferential plane (direction of polarization) may vary from place to place. Unfortunately, however, such variations remain almost entirely hidden from us, since the human eye is barely able to distinguish between unpolarized and polarized light: we are, so to say, "polarization-blind". Many animals share this handicap with us, but not all. Bees, for instance, can observe the polarization of light as easily as we can see colours, so that this particular aspect of light is vividly visible to them. If such animals were smart enough, they could use the additional information gained from their polarization observations to their advantage.

This happens, indeed, to be the case: in 1949 K. von Frisch showed that bees are able to infer the position of the Sun from the polarization pattern of the blue sky and use this to orient themselves. Since then, many other insects are found to do the same thing.

## The Polarized World

With the aid of a simple polarizing filter, we can easily overcome our polarization-blindness. Such filters are easily accessible, since they are present in common Polaroid sunglasses. A polarizing filter



*W.-M. Boerner et al. (eds.), Direct and Inverse Methods in Radar Polarimetry, Part 1*, 33–44.

transmits only one direction of polarization. If this transmission direction corresponds to the polarization direction of light, it is transmitted by the filter; if the directions are perpendicular to each other, the light is weakened. The stronger the degree of polarization, the stronger the weakening.

Inspecting the world with a polarizing filter rotating before the eye, one suddenly becomes aware how much and how varied this polarization is in Nature: some objects change their appearance considerably during the rotation of the filter, others hardly at all, and the direction of polarization may vary from object to object. A typical example of strongly polarized light is the blue sky, particularly at about 90° from the Sun. Fig. 1 shows the sky observed with a polarizing filter directed in such a way that it maximally transmits its light; its appearance does not differ very much from that viewed with the naked eye. In Fig. 2, however, the filter has been rotated by a quarter of a turn and the aspect has changed drastically: in this case the contrast with the clouds is even entirely reversed. Watching a clear sky with a filter in the latter position, one perceives an impressing dark band at 90° from the Sun marking the area of maximal polarization. D.F.J. Arago was the first to report this remarkable polarization; his observation dates back to 1809.

As mentioned, some animals do see polarization clearly and even make use of it. But that does not mean that they see such a dark band in the sky as we see with the filter: they basically see "something different". Fortunately, it does not require very much imagination to know what they are actually seeing, since humans are still able to catch a feeble glimpse of the real appearance of the polarized world. This capability was recognized by K. Haidinger in 1844. The fact is, that in polarized light, we see in the centre of the retina a yellowish structure, which is absent in unpolarized light. Usually, however, some practice is needed to see it. The procedure is as follows: Take a filter for the eye to produce strongly polarized light, rotate it slowly, look to infinity, and relax. After a few seconds the yellow structure which co-rotates with the filter becomes visible. Once achieved, it becomes progressively easier to see this so-called Haidinger brush and it then may even reveal the polarization of the blue sky to our naked eyes.

Nevertheless, our sensitivity to polarization is only weak and we will restrict ourselves further to observations by filters, giving far more spectacular results, among them the "dark band" in the sky already mentioned. We have to realize, however, that the existence of this band is just a consequence of the specific properties of the filter, which is translating polarization shades into intensity shades for us. Through a filter, one thus perceives a very remarkable world, in which real intensity shades are intermixed with shades resulting from the translation of polarization into intensity. As the eye responds logarithmically to intensity, the most striking differences between Nature viewed with and without a filter occur when the filter is in a position close to maximal extinction. Moreover, in this position, the image is very sensitive to a small rotation of the filter. One may say that the filter then shows us at its best its translation of the "polarized world".

It is amazing that such a simple device as a polarizing filter suffices to give parts of our world such a dramatically different appearance. This is the consequence of two facts: our almost complete polarization blindness and the large amount of polarization around us. Since relatively few people seemed to be familiar with the polarized world and no guide for it existed, I recently wrote a book on the subject [1]. In this article I describe some major effects in a historical perspective and discuss the regularities in the natural polarization pattern.

## Gloss, Reflections, and Umov's Rule

Although the blue sky is a nice example of strong polarization in Nature, it is certainly not the best known one which is provided by the gloss of smooth surfaces like water or window panes. Its polarization was discovered in 1808 by E. Malus, when he was looking at glass through a calcite crystal. In such a crystal, light is split into two rays of opposite polarization. This effect had already been observed by C. Huygens in 1690, although he did not know the reason of the intrinsic difference of the two rays. Malus found that the two images produced by calcite (sometimes called "drunkard's-glass") were extinguished alternately when he rotated the crystal. This discovery marks the start of the exploration of polarized light in Nature.

Figs. 3 and 4 show the polarization of the gloss at a water surface of a pond. Again, the appearance changes drastically: when the gloss is maximally extinguished by the filter, even plants at the bottom of the pond become visible as the glare is eliminated. Apparently, at the angle at which we are observing the water surface in this picture, the polarization is almost complete. If one looks at a steeper or a more grazing angle, the polarization is less: this also can be seen in the pictures. Indeed, as was pointed out by D. Brewster in 1812, there is one particular angle of reflection at which unpolarized light is converted completely into polarized light. For most natural substances this Brewster angle of reflection is about 50°. As a consequence, the completely polarized image of the Sun, reflected at this Brewster angle, is at an angular distance of about 80° from the real Sun.

The direction of polarization of gloss is parallel to the reflecting surface. Polaroid sunglasses make use of this, the sheet polarizers being set in the frames with their transmission direction vertically. Consequently, they reduce the glare from horizontal smooth surfaces without greatly affecting the appearance of other objects. Of course, for vertical surfaces like window panes, such sunglasses do not confer this advantage: on the contrary, they make the situation more unfavourable.

Many objects in Nature are not smooth enough to have a gloss: they scatter light in all directions. But coloured, diffusely reflected light is also polarized, although the degree of polarization is by no means as strong as that of a glossy surface. Interestingly, the degree of polarization of rough objects is governed by a simple rule, given by N. Umov in 1905: the darker the surface, the stronger the polarization. Hence, the light reflected from concrete is less polarized than that from dark asphalt. Likewise, the diffusely reflected light of wet bricks is more strongly polarized than that from dry bricks. An interesting consequence of this rule is that colours of objects in Nature are more pronounced if the intensity is maximally extinguished with a polarizer. The direction of polarization of diffusely reflected light is perpendicular to the plane of scattering (that is, the plane containing the Sun, the reflecting particle and the observer); this holds also for the blue sky. Again, maximum polarization occurs if the reflecting object is at about 90° from the Sun. Thus, the polarization pattern of scattered light (the blue sky), diffusely reflected light, and light reflected by smooth objects are similar, differing only in the degree of polarization [1].

By far the greater part of the light encountered in Nature is scattered or (diffusely) reflected sunlight: consequently, the polarization pattern in Nature is very uniform. In this respect, the polarization pattern differs markedly from the colour pattern around us.

## A Uniform Polarization Pattern

The simplicity of this overall polarization pattern (direction of polarization perpendicular to the scattering plane, maximum polarization at about 90° from the Sun) suggests that the explanation of it is simple too, and this is, indeed, the case. The reason is that light waves are transverse. This means that if they progress in the z-direction, their vibrations take place in the x-y plane only. If such a wave conceals a small scatterer, the latter starts to oscillate also and to emit light. But since oscillations in the z-direction are lacking, light emitted perpendicularly to this z-direction vibrates only in one direction: for example, light radiating in the x-direction has left vibrations only in the y-direction. Thus, this light should be completely polarized; light scattered in another angle with the z-axis partially polarized; and backward or forward scattered light unpolarized. Hence, the overall polarization pattern is caused simply by the anisotropic structure of light waves, which always lacks one of the three vibration modes. In practice, however, the maximal polarization is often weaker than this qualitative description suggests. The anisotropy of waves also explains immediately why the direction of polarization in the case of scattering by small particles, or in case of reflection, is perpendicular to the plane of scattering. This implies, that in the event the Sun is obscured but its resulting polarization pattern is still present in the sky, the solar position can be easily inferred from polarization observations. This is in fact what bees do. Interestingly, some ten years ago, the archaeologist T. Ramskou found indications that our Viking ancestors had adopted the same technique, using a cordierite crystal as polarizing filter. Such a navigation tool would have been very useful, as it works even under a cloud deck or in fog if their vertical dimensions are not too large.

## Exotic Sky Phenomena

Not all the light around us emerges from direct reflection or from scattering by very small particles. Sometimes remarkable structures appear in the sky; among them the rainbow is a familiar example. These phenomena are caused by scattering of light by larger objects: in the case of the rainbow these are water drops. Such a phenomenon may have a striking appearance and display its own specific polarization. This polarization usually differs markedly from the general polarization pattern outlined above. For these reasons, such exotic phenomena are among the most delightful objects to be observed in Nature. Their complete description, however, could easily fill a full chapter in a book [1,7-10] and will not be attempted here. In this article we restrict ourselves to a brief discussion of three typical examples of this group: the rainbow, the 22° ice-crystal halo, and the glory.

The polarization of the rainbow was first observed by J.B. Biot in 1811. It was found to be extremely strong, as can be seen from Figs. 5-6. Indeed, with a polarizing filter the rainbow can be completely "rotated away". The direction of polarization is perpendicular to the scattering plane, thus tangent to the bow. As the bow appears at about 140° from the Sun, this strong polarization manifests itself at a very unusual place in the sky, hence disturbing the usual polarization pattern completely. The polarization is caused by the fact that the rainbow-generating light-path through the drop consists of the sequence: refraction-internal reflection-refraction, and that the reflection happens to occur very close to the Brewster angle. The rainbow is without doubt one of the most spectacular examples of polarized light in Nature.

The second exotic phenomenon is the group of the 22° ice crystal haloes, which often appear in cirrus clouds. A prominent member of this group is the parhelion, which shows up at 22° at the left or the right of the Sun. Such a halo is shown in Fig. 7. It is caused by refraction of sunlight by tiny, oriented hexagonal ice crystals, which act as prisms. Unlike the rainbow, a parhelion cannot be extinguished with a polarizing filter. In 1977, however, I found that something else happens: it shifts its position to and fro as the filter rotates. The distance at which the shift occurs corresponds to about one quarter of the apparent diameter of the Moon. As the angular width of the halo itself is much larger, this means that its inner, red edge is completely polarized. So, again, strong polarization is produced at an unusual place in the sky, albeit only in a small angular range. The direction of polarization of the halo-edge is in the plane of scattering, hence opposite to what one normally encounters. Although this halo-edge polarization has been known for only about ten years, it is very easy to observe with a filter. It arises from the birefringence of ice, which causes the halo to consist of two completely polarized components, which appear at a slightly different angular distance from the Sun.

The last example of our list of exotic phenomena is the glory, which is a kind of aureole around our shadow (Fig. 8). It is often visible from aircraft if the shadow point falls on clouds of small water drops. Normally, hardly any polarization is to be expected near one's shadow point, but the glory is an exception. Its polarization, first reported by J. Kiessling in 1884, is found to display to remarkable structure: its direction at the coloured ring is in the plane of scattering, but that of the bright area within this ring, close to the shadow point, is perpendicular to this. As the glory is concentrated in a small solid angle, this polarization causes a regular structure to appear in the glory if viewed through a filter. Obviously, this structure co-rotates with the filter. The strongest polarization appears when the glory is relatively large, which means that the generating drops are small. The mechanism causing the glory polarization will not be discussed here, as it is rather complex and outside the scope of this paper. Suffice it to say that the first step toward a correct explanation was given in 1947 by H.C. van de Hulst, and that the complete description by H.M. Nussenzveig dates from 1977.

From this discussion the following conclusion may be drawn. Although it is true that there is a large variation in the polarization, there is also clearly a system in it. All cases display a cylindrical symmetry in the polarization with respect to the primary light source. This means on the one hand that the magnitude of the polarization depends only on the angular distance to the Sun, and on the other hand, that the direction of polarization is either in the plane of scattering or perpendicular to it, but not in between. Apparently, this cylindrical symmetry is a basic feature of our polarized world if dominated by single scattering or single reflection.

## Hierarchy and Exceptions

However, this symmetry is not valid for all light in Nature, for part of it reaches us via more than one object. Its polarization then differs from that of singly scattered or singly reflected light.

For simplicity we may divide light in Nature into three groups. Group 1 consists of the direct light sources; group 2 of light reaching us via one object; and group 3 of light reaching us via more than one object. The main characteristics of the light and its polarization of the subsequent groups are the following.

The direct light sources, group 1, are at night the most conspicuous part of our world; our eyes are inevitably drawn to them. As the light emerges from particles at high temperatures in chaotic motion, it is usually unpolarized. Examples of this group are stars and artificial lights. The Sun also belongs in this group, but as its intensity is so dazzling we take care not to look straight at it. So during the day the most striking feature is not the light source itself, but the many objects around us which it illuminates. These provide light belonging to the next groups.

Group 2, light that reaches us via one object, is the next brightest after the sources. As indicated above, it basically has a cylindrical symmetrical polarization with respect to the original light source, in which the direction of polarization is predominantly perpendicular to the scattering plane.

Group 3, light that reaches us via more than one object, is the least intense contribution. In its complicated route to our eyes its polarization may have changed drastically, so that the polarization pattern bears no obvious relationship to the position of the last object with respect to the original source. Nevertheless, strong polarization can be present if the light path is favourable. In this case, during its light path, polarized light may even become circularly polarized. This has to be observed with other types of filters than those suitable for linearly polarized light. Obviously circularly polarized light is rare in Nature.

Although the polarization of most light in the groups fits this description, there are still some exceptions. For there is light which definitely must be included within groups 1 or 2, but yet shows a polarization belonging to the next group. Indeed, there are some strongly polarized light sources in Nature. One example, the glow of clean, incandescent metals was discovered by D.J.F. Arago in 1824; a second example, discovered by V.A. Dombrovski in 1954, is the synchrotron radiation from the crab nebula. But perhaps the most outstanding example of exceptional polarization concerns a certain family of beetles, to which, for example, the rose chafer (**Cetonia aurata**) belongs. These beetles display a metallic gloss, which turns out to be completely circularly polarized! This unique feature was discovered in 1911 by A.A. Michelson. Although liquid crystals were later found to have the same property, no further examples of this particular characteristic outside this family of beetles have been found to exist in Nature. Its biological function, if any, is completely obscure [11].

## Some Applications of Optical Polarization

As said at the beginning of this article, polarization is as important a property as colour in light. But since the state of polarization is determined by other factors, it often contains information which cannot easily be inferred from spectral analysis. Today very sensitive polarimeters exist, which have been successfully applied in, for example, astronomy; also interplanetary spacecrafts are often equipped with sensitive polarimeters. A further important use is in the quality control of materials: from the way they alter the state of polarization of light passing through, information can be obtained about their crystalline structure while weaknesses (mechanical tensions) in them can be detected. In daily life we are sometimes directly confronted with this latter application when we drive a car: the coloured spots which often appear in the windscreen when we wear Polaroid sunglasses are caused by this very effect. Applications of polarized light like the ones mentioned are very widespread in research, technology, and industry. It is all the more surprising that so few people seem to be aware that in the open field there is so much polarized light, in which gloss, the blue sky, and the rainbow are only just a few of many spectacular examples.

## Acknowledgements

This paper is an updated version of a review paper recently published by the author on "Viewing our world with polarizing glasses" in Endeavour, New Series, Vol.10, No.3, pp.121-124, 1986.

The bibliography, below, includes key references for further reading. References [1] to [6] are primarily focussed on polarization and scattering; references [7] - [10] on rainbows, halos and other atmospheric optical phenomena.

## Bibliography

[1] G.P. Können, Polarized Light in Nature, London, U.K., Cambridge University Press, 1985 (original Dutch version: G.P. Können, Gepolariseerd Licht in de Natuur, Zutphen, Nederland: B.V.W.J. Thieme & Cie, 1980).

[2] W.A. Shurcliff and Ballard, S.S., Princeton, NJ/U.S.A., Polarized Light, New York: Van Nostrand, 1964.

[3] W.A. Shurcliff, Polarized Light, Production and Use, Cambridge, MA U.S.A., Harvard University Press, 1966.

[4] T. Gehrels, (Ed.), Planets, Stars and Nebulae Studied with Photopolarimetry, Tucson, AZ/U.S.A., University of Arizona Press, 1974.

[5] H.C. van de Hulst, Light Scattering by Small Particles, New York: John Wiley & Sons, 1955.

[6] C.F. Bohren and D.R. Huffman, Absorption and Scattering of Light by Small Particles, New York: John Wiley & Sons, 1983.

[7] M. Minnaert, Light and Colour in Open Air, London, U.K., Dover, 1968.

[8] R.G. Greenler, Rainbows, Halos and Glories, Cambridge University Press, London, U.K., 1980.

[9] W. Tape, Halos: Celestial manifestation of crystalline bearty, in preparation.

[10] F. Linke and F. Möller, eds., Physik der Atmosphäre I, in Handbuch der Physik, Band 8, Berlin: Borntraeger, 1961.

Figs. 1 and 2 The blue sky at 90° from the Sun viewed through a polarizing filter in two positions. Above, the filter transmits the light maximally; the aspect of the sky is similar to that viewed with the naked eye. Below, the polarized light is maximally extinguished, causing a drastic change in its appearance and, in this case, also a contrast reversal with respects to the weaker polarized clouds.

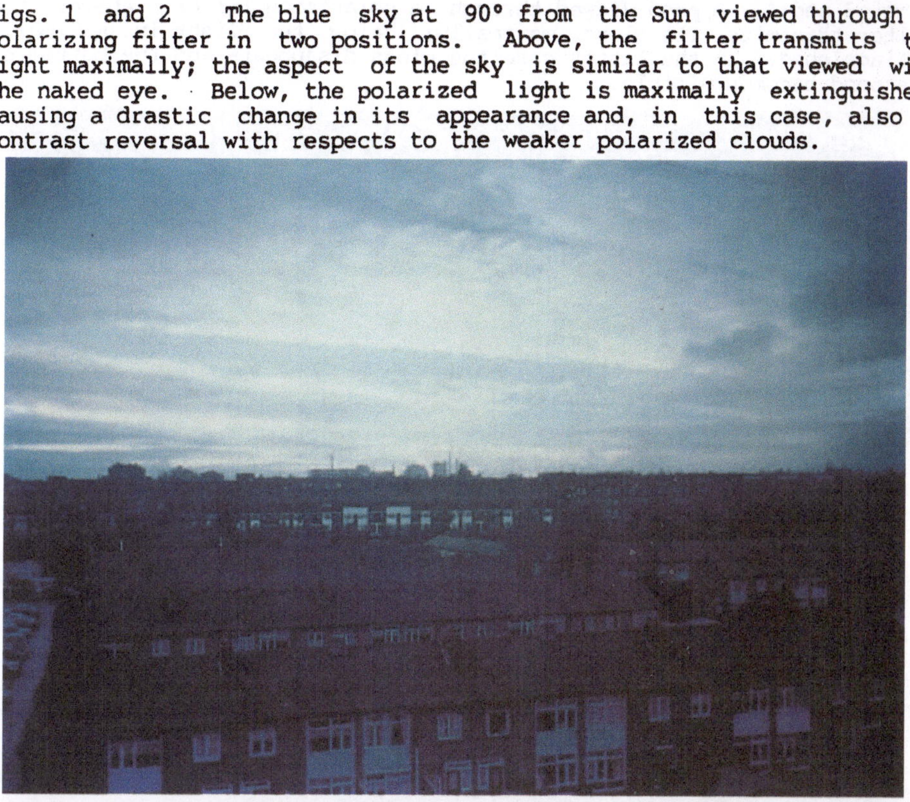

Figs. 3 and 4 A pond viewed through a polarizing filter. Above, the filter transmits its gloss maximally; below, the light is maximally extinguished. In the latter case the gloss disappears and the bottom of the pond becomes visible.

Figs. 5 and 6 A rainbow viewed through a polarizing filter. Above, its light is maximally transmitted and the rainbow is markedly brighter than when viewed with the naked eye. Below, the polarized light is maximally extinguished with the filter and the bow disappears completely.

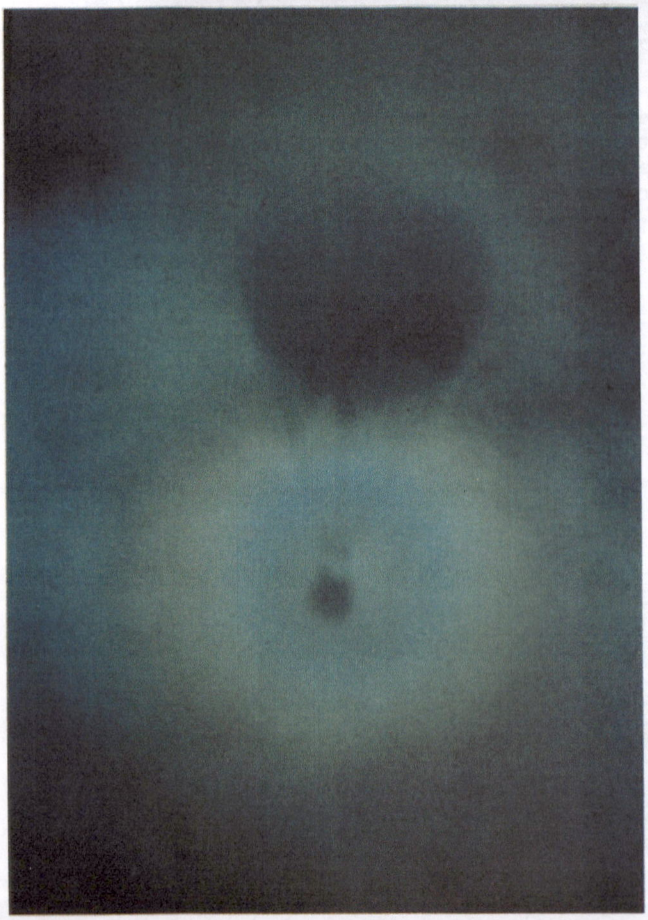

Fig. 7 A parhelion, member of the 22° halo group. Its position in the sky shifts by .1° to and fro if viewed through a polarizing filter which is rotated; the inner red edge of the halo being completely polarized.

Fig. 8 The glory, around the shadow of the basket of a balloon. Despite it being located near the shadow point, it still produces strong polarization. (Photograph by A.F.G. Kip)

# REVIEW OF PAST AND CURRENT RESEARCH IN THE USSR ON THE FUNDAMENTALS AND BASICS OF RADAR POLARIMETRY AND HIGH RESOLUTION RADAR IMAGING

(INVITED)

Anatolij I. Kozlov(1), Aleksander I. Logvin(1)
and Lev A. Zhivotovsky(2)

(1) Moscow Civil Aviation Engineering Institute
MIIGA, Kronshtadsky Boulevard 20
USSR 125-493 MOSCOW

(2) LENIN, (V.I. Ulyanov) Electronic Technical Institute
LETI, Popova Street 5
USSR 197-022 LENINGRAD

**ABSTRACT:** The development of polarimetric radar research studies within the USSR, initiated during the late 1940's, is reviewed. Major monographs, research texts and books and some of the outstanding pioneering research papers are cited. It is demonstrated that polarimetric radar theory, metrology, systems technology, calibration and data processing are well developed in the USSR.

## 1. INTRODUCTION

In the USSR, research on polarization properties of electromagnetic waves was initiated during the late 1940's; a certain part of this research had matured by the mid-1960's. The results obtained by this time were summed up in two monographs by Kanareykin, Pavlov and Potekhin, "Radar Signals Polarization" [1] and by Kanareykin, Potekhin and Schischkin, "Marine Polarimetry" [2], published in 1966 [1] and 1968, respectively. These two publications stimulated greater interest of Soviet researchers in theoretical and practical problems with regard to a wider use of incorporating information on the polarization state of electromagnetic waves for increasing the informative capabilities of various kinds of radar systems.

The above mentioned monographs considered different aspects associated with the use of polarization properties of reflected and scattered electromagnetic waves. They contain descriptions of polarization of plane electromagnetic waves, fully or partially polarized; the identification of basic statistic parameters of the polarization structure of electromagnetic waves; the development of practical methods of measurement of electromagnetic wave polarization parameters; and polarization properties of radar targets. Special attention was paid to the practical side of polarization discrimination. Here, the problems of eliminating the interfering reflections by exploiting the polarization differences in signals reflected from useful and interfering targets were considered. The monographs also analyzed possibilities of the use of polarization



*W.-M. Boerner et al. (eds.), Direct and Inverse Methods in Radar Polarimetry, Part 1*, 45–59.

properties of reflected signals for studying the environment, for solving many problems in meteorology and in oceanography, and in related fields of geo-electromagnetic remote sensing and deep sounding.

Further investigations in radar wave polarization were conducted in several basic directions, which are being worked out intensively nowadays. These topics include:

- development of the general theory of radar wave polarization;
- investigation of the problem of polarization discrimination and the design of the corresponding sensor devices;
- creation of special algorithms for receiver-processor agility on the basis of classical theories of detection, discrimination, filtering, and evaluation of the vector signal parameters;
- investigation of polarization modulation;
- study of polarization phenomena in the systems of passive radiolocation;
- polarimetric radiometry and polarimetric scatterometry; and,
- utilization of polarization analysis of scattered and reflected radio waves for studying the environment — for solving problems in meteorology, hydrology, oceanography, glaciology, geophysical exploration, etc.

Now, we shall give a retrospective analysis of each of the above mentioned directions from the point-of-view of summarizing some of the results obtained hitherto and currently by Soviet scientists.

## 2. OVERALL REVIEW

We will refer only to monographs and basic articles (containing extensive lists of pertinent references) since it is physically impossible to make a list of (or to mention) all the relevant Soviet studies in this field. Besides, we do not dwell on the question of originality and priority; such should be tackled by historians of modern science.

It should be emphasized that the separation of the content of this manuscript into seven parts is only a convenient means of presenting the material.

### A. General Theory of Radio Wave Polarization

A great contribution toward the development of the theory of radio wave polarization was made recently as summarized in the two monographs by Bogorodsky, Kanareykin and Kozlov, "The Polarization and Earth Surface Radio Radiation Properties" [3], and Potekhin and Tatarinov, "The Theory of Coherence of Electromagnetic Fields" [4], published in 1981 and 1978, respectively.

Monograph [4] gives a physical basis of the electromagnetic wave polarization process, and it may serve as a methodologic basis for advancing the overall investigations in this direction. Monograph [3] considers a wide range of problems associated with general theoretical concepts of radio wave polarization, polarization parameters of stable and fluctuating targets, antenna systems, and soil surfaces. It considers problems of polarization discrimination.

As a whole, monograph [3] is a logical development of ideas and directions which were given in monographs [1, 2] and to some extent it sums up computer numerical polarization investigations conducted in the USSR during the late 1970's. Publication of the following two basic monographs: Pozdniak and Melititsky, "The Introduction of the Statistical Theory of Radio Wave Polarization" [5], and Bass and Fuks, "Electromagnetic Wave Scattering from Statistically Rough Surfaces" [6], published in 1972 and 1974, respectively, was an impelling force for further development of investigations in the field of statistic radio polarimetry within the USSR. Several chapters of these monographs considered in detail various statistic parameters of polarized signals.

Monograph [6] deals with the radio wave scattering over a rough surface and aims at solving problems of newly conceived methods of remote radioprobing and sensing of the environment.

Monograph [5] describes probable models of partially polarized radio waves, statistic characteristics of polarization parameters of partially polarized radio waves, and it contains a detailed description of statistic characteristics of polarization factors of reception.

Monograph [5] also gives a system of the main concepts of detection and discrimination of polarized signals when there exists normal partially polarized interference. Monographs [5,6] led to further numerical studies culminating in publications such as monographs [7-9]. These investigations, with the aid of the Poincaré sphere and of the 4-dimensional polarization sphere introduced by Zhivotovskij [10, 11], greatly assist in solving the problem of addition and decomposition of partially polarized plane waves.

Much work has been done to develop geometric methods of presentation and analysis of partially polarized waves on planar maps and three/four-dimensional configurations [8-11].

To analyse the field in space, Stokes parameters were verified, the concept of the informative Stokes space was introduced, and its dimensions — three versus four-dimensional polarization sphere — were determined [10,11].

## B. Polarization Theory of Radar Targets

The polarization properties of stable and fluctuating targets and their combination were considered in monograph [1]. The monograph describes the scattering matrix in different polarization bases and provides the formulation of the characteristic polarization states of stable radar targets. These problems were further developed in monograph [3], and by Kozlov [7], in which the statistic concept of polarization properties of radar targets becomes much clearer.

A considerable contribution to the development of radar target polarization theory was made in monographs [12, 13]. The most important contribution of the monograph by Ostrovityanov and Basalov,"Statistical Theory of Distributed Radar Targets", [12] is the fact that this monograph, for the first time, considers and systematizes investigations associated with the concept of "a prolonged target with distributed scattering centers," i.e., a radar target, which in real applications cannot be considered as a point target. Monograph [12] also considers the influence of polarization on the statistics of the signal-to-noise ratio of prolonged targets, determines statistic characteristics of the signal-to-noise ratio of prolonged targets at different polarizations of the measurement antenna, and gives examples of calculating characteristics of the signal-to-noise ratio of prolonged targets for different polarization properties of the polarimetric antenna measurement system. This monograph [12] has become of vital importance to the advancement of polarimetric glint analysis both in the East and in the West (see **VI-9**).

A monograph on "Radiolocation Characteristics of Aircraft" by Varganov, Zinovjev, and Astanin [13] gives an analysis of descriptive methods, measurement and formulation of radar target characteristics. Within the framework of this analysis the polarization characteristics of radar targets are described on the basis of scattering matrix applications. The analysis considers a "one-position scattering matrix" (the case of monostatic radio-location without obligatory mutual antenna characteristics), the matrix of two-position scattering (bistatic radiolocation), and of local scattering characteristics, i.e., multistatic polarimetric radar target analyses.

Monograph [13] also investigates problems of determination of probable characterisitcs of the reflected signal with the aid of the given statistic radar characteristics of aircraft. It also describes the presentation of scattering characteristics of fluctuating objects in different polarization bases; and, it states that the form of the law of distribution of properties of statistic scattering matrices in any polarization transformation basis remains the same, establishing the basis for polarimetric doppler radar analysis by fluctuating targets.

The monographs by Zhivotovskij [14, 15] describe the development of constructive methods of presentation and analysis of depolarizing properties of fluctuating radar targets. They show that the field of signals reflected from a fluctuating radar target may be presented as a sum of 4 non-coherent polarized waves and a non-polarized wave, thus making it possible to determine the radar target "polarization portrait" — a spheric quadrangle on Poincaré's sphere, which in some characteristic polarization basis assumes a canonic appearance.

It is shown that fluctuating radar targets with a symmetric matrix of scattering can be compared with a random vector in 3-dimensional unique space with decomposition into three non-coherent orthogonal components. In this case the task of matrix decomposition of radar target scattering is solved in the most natural way [14,15].

Monographs [16, 17, 18] consider some problems of matrix scattering having some independent meaning. Monograph [16] investigates the interrelation between the statistic scattering matrix, the Stokes reflection matrix, and the Mueller matrix. In particular, monograph [18] offers a design approach to maximize the efficiency of polarimetric radar 'vector signal receiving' and 'front-end processing' devices for the optimal detection of a minimally depolarized signal. Monograph [17] considers certain peculiarities and tendencies associated with the description of fluctuating target properties on the basis of a four-dimensional complex vector concept, leading to the co-variance matrix of the order of 4×4 of which the non-diagonal elements are the complex correlation factors, and the diagonal ones are the root-mean-square values.

### C. Polarization Discrimination

Polarization discrimination problems are closely associated with the radar target polarization characteristics. The polarization discrimination concept, however, is more general than the polarization differentiation of two radar targets. The important aspect of this approach is that the polarization discrimination can be associated both with the communication systems (systems with the repeated usage of frequency) and with radio-navigation (i.e., radio direction finders), etc.

The characteristic polarization qualities of some radar targets in terms of the resulting effect of the polarization dependence for symmetric point-wise, surface-distributed and/or body-distributed scattering centers have been considered in [1].

Monograph [3] looks at the problems of the polarization-compensation method of hum suppression, of the possibility of resolution between the two radio signals in an affine basis, of the full polarization scanning mode, of the selection of a radar target by the polarization scanning method, of the synthesis of radar targets, and of principles of synthesizing the target with some given polarization characteristics.

The most successful results are achieved just with solving the problems of the two radio signals resolved in an affine basis; and, second in describing the possibilities of the full polarization scanning mode. The criteria that the signals of the two targets are divided between the channels is the condition of coincidence of unit vectors of an affine basis with the polarization state of the polarized components of the received waves.

The full polarization scanning mode is treated in monograph [13] in terms of the theoretical analysis, which shows that this mode makes it possible to emit all types of the traveling wave polarizations during the definite time interval (with some "loci of the form of a helix" on the Poincaré sphere). At the same time, monograph [3] deals with the problems of technical realization of the full polarization scanning mode on the basis of obtaining - at high frequencies - the two quasi-coherent oscillations, having a small but controlled frequency shift. The application of such oscillations provides both the full polarization scanning mode and a smooth change in scanning frequencies within a wide range (See monographs [19, 20] for more details).

Soviet scientists paid much attention to the problem of synthesis and analysis of polarization discrimination algorithms for different classes of reflecting objects. Monographs [3, 5, 13] deal with these problems. In addition, one can find more detailed consideration of these problems also in monographs [21-23] and monographs [24-26]. These studies consider first the increase of the power contrast between useful signals and distributed signals; and secondly, pattern recognition objectives. The studies by Zhivotovskij [21, 22] determine the optimal polarization of radar system reception by involving Poincaré sphere manipulations. The concept of statistical control theory and Poincaré sphere manipulations discussed in Rodimov and Popovsky [24], in Gusav, Filatov and Sobolev [25] and in Bogorodsky and Kozlov [26], helped on the one hand to formulate the assessment of potential efficiency algorithms in polarization discrimination and, on the other hand, to determine the optimal selector's invariances for definite changes in a priori information.

To analyze the problems of accomplishing polarization discrimination, the studies [5, 27, 28] have assisted in introducing and developing the concept of the polarization reception factor. Moreover, in monograph [5] one can find information about the density of probability of the polarization reception factor and its statistical characteristics in circular and linear bases presented on the Poincaré sphere and on suitable map projections.

Pozdniak [29] offers the concept of the signal polarization discrimination factor against the background of interference; and its probable distribution function is investigated and is obtained.

## D. The Development of the Polarization Signal Reception Algorithms

Soviet scientists have contributed much to the general theory of polarimetry in terms of the development of optimal reception algorithms. The questions of detection, of polarization signal parameters discrimination, of filtering and of analysis have been considered. One of the first studies in this field was provided by Kiselev in [30], in which the integral equation determining the structure of the optimal receiver has been obtained. This study as well as [31] solved the problem of detection of the elliptically polarized signal. The system of the optimal reception of polarization-modulated signals has been considered in [25], in which the main concepts have been introduced and the necessary polarization discrimination ratios have been defined. In addition, the problems of optimization in the reception of polarized signals against the hum of interference have been determined and the problems of potential noise immunity in the reception systems have been investigated. Further studies associated with the detection and discrimination of polarized signals, when normal partially polarized interference may be assumed, are offered in [5]. Monograph [5] considers the matrix of the determined signals when normal partially polarized, correlated interferences are assumed, leading to: (i) the detection of polarized determined signals; (ii) the detection of polarized signals with random initial phase; and (iii) the detection of polarized signals with random amplitude and phase after deteriorating detection of polarized signals (when normal partially polarized interferences are assumed) - which is divided into amplitude and phase detection methods. The difference in polarized signals has been analysed on the basis of polarization parameters such as the polarization factor, that is the modulus of the ratio between the orthogonal components of the polarized signals.

Monograph [24] looks deeply into the problem of assessment of parameters and filtering of polarized signals. This study investigates the polarimetry of partially polarized signals and interference, and also the problems of polarization time-dependent signals and interference processing. There the assessment of parameters of the coherence matrix of partially polarized signals is given, the linear and non-linear assessment algorithms of polarized signals and interference parameters are obtained, and also the adaptive assessment algorithms of polarized signals and interferences are determined. The synthesis of control algorithms for polarization parameters of aerial systems, together with the analysis of control algorithms for partially polarized signals, and interference (with optimal weight factor vector assessment), are considered.

The latter algorithms have the universal character, however, that they first refer to communication systems and that they don't take into account the specifics of the polarized radar signals which include the detection criteria, the information parameters choice, and the transmitted signal character [24].

The polarized radar signals' filtering has been solved on the basis of non-linear Markov optimal filtering methods by Logvin [32, 34]. Monograph [32] gives the filtering algorithm of linear polarized signals with a random polarization parameter, in particular, the "polarization surface" slope angle to the chosen mathematical coordination basis. The optimal receiver includes the system of information extraction and the system of tracking the change of the angle of the polarization surface orientation of the electromagnetic wave. Monograph [33] concentrates on the algorithms of optimal filtering, when random parameters of polarized radar signals are the geometric parameters of the polarization ellipse; and, at last, monograph [34] gives the structure of the optimal receiver of the polarized signal when amplitude and phase of orthogonal components of the received elliptically polarized signal are taken as random polarized parameters. In this case, the optimal receiver contains the system of the extraction of information parameters, two devices "PLL" (Phase Locked Loop) and two "AGC" (Automatic Gain Control) with the crossed connections, conditioned by the correlation between orthogonal components of the interference. Such a receiver constantly implements the matching of polarization characteristics of the electromagnetic waves and of the aerial system, which makes it possible to expand the use far beyond the power potential of radar systems.

The attention is focused on the problems of the optimal processing of polarized signals, where numerical processing methods, robust algorithms of reception, non-gaussian distribution of polarization parameters, etc., are being introduced, developed and optimized.

### E. Polarization Modulation

Monograph [25] reviews the problems of polarization modulation. The theoretical analysis is carried out by means of a double complex surface, which is the basis for the elliptically polarized electromagnetic waves. The signals with continuous and discrete polarization modulations are considered.

The modulation of the angle of ellipticity, the angles of orientation of polarization surfaces and their combination are described in the study [25]. Spectral images of polarized modulated signals are analysed. The principles of polarization transformers and modulators are given. At the same time the methods of reception of polarization-modulated signals are described in detail. In parallel, the influences of additive multistatic signal and noise interference on polarization-modulated signals are evaluated.

Monograph [35] gives the relation for the design of noise immunity in a receiver, in case the polarization modulation is accomplished on the basis of the angle of ellipticity, as proposed in [25].

Monograph [36] analyses the spectral content of linearly polarized signals with the rotating surface of polarization for the selection and recognition of radar targets. The surface frequency rotation harmonic in the received signal spectrum is used as an information parameter. This approach offers the amplitudes and phases on spectral components at frequencies multiple to the frequency of polarization modulation for different polarization parameters of a radar target. The authors used the orientation of the target coordinate system itself, the difference of arguments of scattering target matrix values and the anisotropy factor. The phases of spectral components were counted off relatively to the phases of definite supporting signals. It was demonstrated that if one or two target polarization characteristics are known a priori, it is possible to obtain single valued relations of target spectral components and target polarization characteristics, i.e., for the purpose of meteorology and oceanography.

Monograph [37] deals with the polarization state manipulation (linear polarization versus circular polarization states). By means of the polarizationally manipulated signal, one is able to evaluate the degree of scattering target anisotropy. The study provides some illustrative examples of the evaluation method.

The information given above [35-37] proves that polarizationally modulated or polarizationally manipulated signals can be applied in radiolocation for the distant probing of targets even at extremely low frequencies within the ULF/ELF/VLF ranges. For micro- and millimeter-waves, this problem is considered below.

## F. Polarization Phenomena in Thermal-Microwave Imaging Radar Scatterometry, and in Radiometry

Recently there appeared a large number of scientific Soviet publications associated with the utilization of polarization properties of the thermal radio radiation of different objects. The thermo-microwave imaging radars help to solve the problems of distant probing for environmental studies. The authors considered it reasonable to devote a special part of the review to the problem of polarimetry in radiometry and scatterometry because very considerable advances are accomplished by

Soviet experts in this important field.

The investigations in the field of polarimetric radiometry in the USSR were initiated in the sixties; and, already in the early seventies witnessed the application of polarization utilization of thermal radio radiation components, when the measurements of this kind were accomplished from the earth-orbiting satellite Meteor. The outcome of these investigations is reflected in [3]. This study considers both the polarization of scattering and radio radiation as well as the sensitivity of a target's reradiation to the polarization state of illumination. Monograph [3] indicates that thermal radio radiation is partially polarized; that's why the full discription of its properites can be fulfilled by means of a coherency matrix analysis. However, this idea has some serious limitations, because the matrix elements only contain the characteristics of the received field and the matrix isn't expressed through the radio brightness temperature, and furthermore, it doesn't contain the information about basic parameters of the target which caused the reradiation. In this context it was appropriate to find a new way of presenting the coherence matrix in another formulation which would avoid the above drawbacks. This task can be solved by introducing a loss matrix. In general, a loss matrix consists of some integral coefficients characterizing the body's response to it being exposed to an electromagnetic wave. This matrix is analogous to the Graves power matrix and coherence intensity matrix. The loss matrix is a Hermitian matrix, 2x2, which testifies to the fact the the radiating capacity of an arbitrary object is defined by four quantities. Invariants of the loss matrix are its determinant and trace. In radiometry a non-dimensional parameter is often used which is called a polarization factor. It is equal to the relation of the difference of extreme values of brightness temperatures on the orthogonal polarizations to their sum. Hence, the radiobrightness temperature of objects is independent of polarization type if the object is polarization-isotropical.

As a result of the transformations the loss matrix can be expressed through radiobrightness temperatures. The elements of such a matrix are: radiobrightness temperature taken in the summation channel; radiobrightness temperature taken in one of the channels; radiobrightness temperature taken in the orthogonal channel; radiobrightness temperature taken in the first channel provided the aerial is turned 45° in space.

By using the loss matrix elements, in monograph [3] a relationship between radiobrightness and Stokes parameters has been derived and the task of defining the thermo-microwave radar contrast of two objects has been solved.

Polarimetry and thermo-microwave imaging radar concepts were developed in monograph [26], which considers the physical principles of creating polarization-dependent radiation by multi-layered heterogenous media.

Polarization of microwave radio radiation of the earth's surface and its statistic characteristics are treated taking into account the pecularities of the earth's surface relief. Great attention is paid in Bogorodsky and Kozlov [26] to the problems of using polarization analysis of the earth's surface microwave radio radiation in defining the inherent electro-physical and geometric parameters when there exist intermediate absorbing layers and when such layers are missing.

The informative character of polarization measurements of thermal microwave radiation of the earth's surface from one or more directions is considered also in [26].

The information on the temperature, roughness and dielectric permittivity of the earth's surface, obtained on the basis of polarization analysis, is given. In a similar way the polarization analysis of thermal microwave radiation to estimate integral atmospheric absorption, for defining electro-physical and geometric parameters of the earth's surface through its vegetation, is carried out.

The possibilities of using the polarization analysis of the earth's surface microwave radiation to facilitate finding weak-contrast structures on the earth's surface are shown. Theoretical statements are illustrated by the results of estimating the parameters of rough and flat surfaces, of probed sea ice floes and to estimate their age.

It should be pointed out that in [26] the problems of optimum reception of the earth's surface microwave radiation are dealt with. On the basis of the Markov theory of non-linear filtering, the algorithms of processing thermal microwave-radiation signals with Gaussian and non-Gaussian distributions of radiation polarization parameters have been obtained.

The assessment of noise immunity of earth surface microwave radiation reception for different receiving devices is given and the adaptive filtering of radiometric signals in the background of Gaussian interference is described.

Thus, monograph [26] provides a generalization and the summarization of all the research done in the USSR on polarization associated with the sensing of the environment by thermal-microwave imaging radar methods. All the directions of research mentioned above are being further advanced and aimed, above all, at accumulating experimental results and their further interpretation of truly polarimetric ultra-wideband measurements.

## G. Polarization Analysis of Scattered and Reflected Radiowaves for Environmental Remote Sensing Studies in Meteorology, Oceanography, Glaciology, etc.

The use of the polarization analysis for obtaining data on meteo-particles (hydro-meteors), on rain intensity, rain drop dimension and orientation, in particular, was suggested for the first time in monograph [1]. In monograph [2], the authors suggested using the polarization analysis of scattered microwaves for studying sea surfaces. Then due to the extensive development in the USSR of remote sensing and probing systems, polarization methods of solving environmental problems are, therefore, widely used and with increasing frequency. It is impossible to list here even only a very minute part of Soviet publications where the principles of polarization analysis are used to study the sea surface and the inland ice, to study the atmosphere and hydrometeor-formations, to study the surfaces of natural outer space objects (meteorites) and our Earth's and the solar system's planets (Moon, Mars, Venus, etc.), to solve agricultural problems, etc.

That's why mention will be made, at least, of some Soviet studies which play a central role in introducing microwave polarimetric methods in modern planetary and extra-terrestrial remote sensing and probing systems of the future.

In studying the properties of ice surfaces, changes in the polarization condition of scattered electromagnetic waves make it possible to solve different problems of glaciology. For instance, the estimation of ice salinity and voluble content of moisture can be achieved. The development of micro-polarimetric methods for studying sea and inland ice began in monograph [38], where it was shown experimentally in what way the polarization of radar signals changes in vertical icefield probing. Similar results were obtained for shelf ice and later for sea ice probing of arctic and sub-arctic ice floes. Interesting results were obtained in the field of ice microwave polarimetry in [39]. The Stokes parameter measurements were carried out in icefields and the functions of these parameters were processsed and correlated, which made it possible to make proper basic assumptions for icefield models in the form of a doubly diffracting plate with proper linear polarizations. The polarimetric analysis of an icefield makes it possible to define its stress condition, to find the directions of main tensions and the modulus of their differences. Defining the tension and stress conditions of ice is a very important practical task in glaciology and the stress tensor analysis of ice, and it is only with the help of microwave polarimetric remote probing that this task can be and is being solved [38].

A lot of research has been done by Soviet scientists to study sea surfaces by polarimetric sensing and probing methods. The relationship between scattered field characteristics for different types of polarization of incident and scattered waves and virtually observed parameters and sea state fluctuational characteristics was studied. Representing the statistical polarization matrix of the scattered field with the elements defined by geometrical shape characteristics and electro-physical parameters of sea surfaces was considered in [38-40]. The abnormally high level of the backscattered horizontal polarization field intensity as compared with the vertically polarized component of monostatic and bistatic rough sea surface scattering, was investigated and accounted for. All these questions are of a general theoretical character and are of interest, not only to a particular sea surface analyses, but also for a more general use with appropriate models applied to extra-terrestrial planetary rough surface sensing and probing. The studies reported in monographs [34] to [42] belong to those topics, wherein a great variety of these complicated, mixed surface with voluminous underburden scattering problems are treated.

## 3. CONCLUSIONS

The scientists, engineers and environmentalists of the Central Geophysical Observatory of Leningrad, named after V.I. Voyeykov and founded upon the advice of Alexander von Humboldt in 1834 (thus becoming one of, or indeed, the first Geophysical Observatories anywhere), and the Central Aerological Observatory of Dolgoprudny/Metropolitan Moscow carry out research in the field of meteorology using radio polarimetric analyses. Similar research is being done by different organizations engaged in studying plants, forests, agricultural areas and other kinds of earth surface environments such as at the USSR Academy of Sciences, Siberian Division, Buryat Region, Baikal Basin Remote Sensing Center of Ulan Ude [43], where ultra-wideband polarimetric sensing techniques for environmental remote sensing of sub-arctic and boreal regions are being developed. The number of Soviet researchers being interested in a wider use of radio polarimetric methods for solving remote sensing probing tasks is growing very rapidly; and, especially, in extra-terrestrial planetary remote sensing and probing. New generation miniaturized microwave/electro-optic devices which can emit and receive different types of polarized signals and process them instantaneously with the help of electro-optic micro-computers open up growing prospects for real-time polarization utilization in modern terrestrial and extra-terrestrial planetary remote sensing sciences and engineering as well as in ground, sea and air traffic control.

Let the joint trans-continental efforts of scientists from different countries aimed at solving cognitive tasks be developed in every direction possible for the ever improving surveillance of our fragile terrestrial environment in the true sense of "global environmental planetary defense" for contributing toward the preservation of our precious natural terrestrial as well as planetary ecospheres.

## 4. ACKNOWLEGDEMENTS

The authors wish to express their sincere gratitude to Professor Boernerov, Wolfgang-Martinovich, the workshop director and chief editor of these proceedings for his timeless efforts for coming to visit us in Moscow, Leningrad and at Baikal Lake, and in assisting us, most generously, for visiting the USA in various post-workshop events during May 1990 - to - January 1991. It is our sincere desire to further expand on this first very lively interaction which also provided the opportunity for us to meet with J. Richard Huynen, one of the true pioneers of radar polarimetry, whose thought-provoking dissertation monograph was known to us since the early 1970's.

## 5. REFERENCES: (all monographs, books and publications are available mainly in Russian: Several journal publications* are also available in English)

1. Kanareykin D.B., Pavlov N.F., Potekhin V.A., Radar signal polarization, Moscow: M. Sov. Radio, (1966); **English translation:** D.B. Kanareykin, et. al., Radar Polarization Effects, CCM Information Corp., Subsidary of Crowell Collier and Macmillan, Inc., No. 8409-0162, JPRS: 46, 624, 1968 Oct. 9

2. Kanareykin D.B., Potekhin V.A., Schischkin M.F., Marine (maritime) polarimetry, Leningrad: Sudostroyenie, (1968); **English translation:** D.B. Kanareykin, et. al., Radio Polarimetry at Sea, JPRS 49, 302, 1969 Nov. 24

3. Bogorodsky V.V., Kanareykin D.B., Kozlov A.I., The polarization and earth surface radio radiation properties, Leningrad: Gidrometioizdat, (1981)

4. Potekhin V.A., Tatarinov V.N., The theory of coherence of electromagnetic fields, Moscow: M.Sov. Radio, (1978)

5. Pozdniak S.I., Melititsky V.A., The introduction of the statistical theory of radio waves polarization, Moscow: M. Sov. Radio, (1974)

6. Bass T.G., Fuks I.M., Electromagnetic wave scattering from statistically rough surfaces, Moscow: M. Nauka, (1972); English translation

7. Kozlov A.I., 'The properties of statistical parameters of the scattering radar target matrix', Izvestija VUZov Radioelectronic, Vol. 22, No. 1, pp. 14-18, (1972)

8. Meletitsky V.A., Mosionshik A.I., 'Probability model of non-gaussian, periodically non-stationary radio signals', Radiotechnika i Elect- ronica, No. 4, pp. 747-754, (1987)*

9. Pozdniak S.I., Mits U.K., 'Coherence matrix and Stokes parameters of partially polarized waves in three-dimensional space', Radiotechnika, No. 4 pp. 80-82, (1987)*

10. Zhivotovskij L.A., 'Generalized Stokes parameters equidistant ellipses of polarization in space', Radiotechnica i Electronica, No. 11, pp. 2111-2115, (1984)*

11. Zhivotovskij L.A., 'The polarization sphere modification (to four-dimension) for the representation of partially polarized electromagnetic waves', Radiotechnica i Electronica, Vol. 30, No. 8, pp. 1497-1504, (1985)*

12. Ostrovitjanov R.V., Basalov F.A., Stretched (distributed) targets radiolocation statistical theory, Moscow: Radio i svjaz (1982)* **In English:** ibid, Statistical Theory of Extended (Distributed) Targets, Dedham, MA: Artech House, 364 pages, 1985

13. Varganov M.E., Zinovjev U.S., Astanin L.U., Edited by Tucshkov L.T. Radiolocation characteristics of aircraft, M. Radio i svjaz,(1985)

14. Zhivotovskij L.A., 'Decomposition and fluctuating radar targets analysis in aerial space', Radiotechnika, Vol. **33**, No. 6, pp. 1186-1191, (1988)*

15. Zhivotovskij L.A., 'Proper bases and polarization pictures of stable and fluctuating radar targets', Radioengineering and Electronics, Vol. **33**, No. 2, pp. 326-330, (1988)*

16. Poliansky V.A., Kanareykin D.B., 'On the relationship between the statistic scattering matrix and Mueller matrix', Radioengineering and Electronics, No. 11 pp. 2407-2410, (1974)*

17. Kozlov A.S., Demidov G.M., 'Some properties of the covariance scattering matrix', Radioengineering and Electronics, Vol. **21**, No. 11, (1976)*

18. Krasnov O.A., 'The polarization structure of an electromagnetic wave scattered by a stable radar target', Electromagnetic waves scattering, Taganrog, Vol. **6**, pp. 68–71, (1986)

19. Demidov G.M., Kozlov A.I., Ustinivitch V.B., 'On polarizational selection of reflected signals', Radioengineering and Electronics, Vol. **20**, No. 5, pp. 1099–1100, (1975)*

20. Demidov G.M., Kozlov A.I., Kransnitsky G.A., 'Polarization processing signal aerial devices', Radioelectronics (Higher School's information), Vol. **21**, No. 8, pp. 122–124, (1978)

21. Zhivotovskij L.A., 'Optimum polarization of receiving antennas', Radioengineering and Electronics, Vol. **17**, No. 11, pp. 2427–2428, (1972)*

22. Zhivotovskij L.A., 'The Poincaré sphere and optimum selection', Radioelectronics (Technical Universities' information), Vol. **16**, No. 12, pp. 48–53, (1973)

23. Kozlov A.I., 'Radar contrast of two objects', Radioelectronics (Higher School's information), Vol. **22**, No. 7, pp. 63–67, (1979)

24. Rodimov A.Z., Popovsky V.V., The statistic theory of signals and noise polarization-time processing, Moscow: M. Radio and Communication, (1984); English translation:

25. Gusav K.G., Filatov A.D., Sobolev A.Z., Polarization modulation, Moscow: M. Soviet Radio, (1974)

26. Bogorodsky V.V., Kozlov A.I., Microwave radiometry of earth surface Leningrad: Hydrometeoizdat, (1985)

27. Maksimov M.V., Editor, Radio noise protection, Moscow: M. Soviet Radio, 1976.

28. Kostryukov A.M., Gusev K.S., 'The assessment of fluctuational polarization interference suppression by the polarization selection method', Radioelectronics (Higher School's information), 1973, Vol. 16, No. 1, pp. 73–78.

29. Pozdniak S.I. 'The distribution of polarization discrimination of signals against the interference background', Radiotechnika i electronica, 1989. Vol. 34, No. 4, pp. 880–882.*

30. Kiselev A.Z., 'Optimal reception of elliptically polarized signals, when random polarized noise is available', Radiotechnika, 1972, Vol. 27, No. 6, pp. 6–10.*

31. Pozdniak S.I., Radzievsky V.G., Trifonov A.L., 'The analysis of optimal reception of elliptically polarized signal', Radiotechnika 1972, Vol. 27, No. 6, pp. 6-10.*

32. Logvin A.I., 'Non-linear filtering of polarized radar signals', Radiotechnika, 1983, No. 12, pp. 32-34.*

33. Logvin A.I., 'Non-linear filtering of elliptically polarized impulse signals', Izvestia VUZov Radiotechnika, 1985, Vol. 28, No. 3, pp. 72-74.*

34. Logvin A.I., 'Non-linear filtering of radar signals with random polarization parameters of electromagnetic waves', Radiotechnica, 1985, No. 6, pp. 56-58.*

35. Krasnov O.A., Uzlenkov A.V., 'Noise immunity of reception of radio signals on angle of ellipticity', Kharkov, 1986, Deprukopis VINITI No. 164-D88.

36. Badulin N.N., Gulko V.L., 'Spectral characteristics of electro-signals at radiation RLC polarization modulation', Izv. VUZov Radioelectronica, 1988, Vol. 31, No. 4, pp. 74-76.

37. Tatarinov V.N., Lukjanov S.P., Masalov E.V., 'Rejector comb filtering of polarizationally manipulated radar signals', Izv VUZov Radioelectronica 1989, Vol. 32, No. 5, pp. 3-7.

38. Bogorodsky V.V., Grekov G.V., Fjodorov B.A., Polarization changes of radar signals at vertical probing of icefields,* T.F., 1976, Vol. 46,2.

39. Nikitin S.A., Menschikov V.A., Vesnin A.V., Selin G.A., Tatarinov V.N., 'The investigation of Altai icefields by impulse radiation and wideband SHF - ellipsometric methods', Trudi AANII, 1985, No. 395, pp. 68-80.

40. Eshenko S.D., Dande B. Sh., 'On the problem of radar imaging of sea surfaces', Radiotechnica i Electronica, 1972, Vol. 17, No. 8.*

41. Zuikov B.A., Kulekin G.P., Lutsenko V.I., 'The specification of scattering SHF radiation of the sea at small angles of gliding (glint)', Izvestia VUZov Radiofizica, 1981, Vol. 24, No. 7, pp. 831-839.

42. Melnitshuk U.V., Chernikov A.A., 'About a back scattering matrix of centimeter radio waves of sea excited surfaces', Trudi CAO, 1975, issue 121, pp. 58-70.

43. N.B. Chimitdorzhiev, P.N. Dagurov and Yu. L. Lomcekhin, Polarization-dependent propagation, scattering and diffraction of microwaves, Novosibirsk: USSR Acad. Sci., Nauka Press, 1987

# APPLYING THE POLARIZATION SELECTION TECHNIQUES TO METEOROLOGIC AND OCEANOGRAPHIC RADAR REMOTE SENSING (REVIEW OF SOVIET STUDIES)

(INVITED)

**Dimitrij B. Kanareykin, Boris Sh. Lande, Yurij A. Melnik, Aleksander V. Ryzhkov, Vladimir D. Stepanenko, Sergeij Yu. Matrosov and Arkadij B. Shupyatsky**
**Voeikov Main Geophysical Observatory,**
**Karbyshev Str. 7,**
**USSR - 194 018, Leningrad**

**ABSTRACT:** We survey 30 years of Soviet studies in applying polarization techniques to radar systems for meteorologic and oceanographic remote sensing. Principal results are formulated in radar identification of hail by means of polarimeters which permit, besides the radar reflectivity, to measure the depolarization ratio in the linear basis. Next, the effectiveness of different techniques of polarization modulation aimed at retrieving the necessary physical information is analyzed. Particular attention is paid to recent results in measuring the differential reflectivity of clouds and precipitation, and also to synchronous measurements of differential reflectivity and of the depolarization ratio in a linear basis. Certain theoretical conclusions on classification of various meteorological targets are presented in terms of invariant polarization parameters. This approach points to potential capabilities of polarization selection of meteorologic targets by means of visual graphical presentation on the Poincaré sphere and of its appropriate map projections.

## 1. REVIEW OF THEORETICAL POLARIMETRIC REMOTE SENSING STUDIES OF METEOROLOGIC AND OCEANOGRAPHIC BACKSCATTER

The first Soviet studies of the polarization structure of radar signals reflected from hydrometeorologic targets date back to the late 1950s and to the early 1960s. These were pioneered by A.B. Shupyatsky, D.B. Kanareykin, V.D. Stepanenko, N.F. Pavlov, V.A. Potekhin, and others. One of the first theoretical publications in this field was an article by A.B. Shupyatsky [1], in which its author calculated the power of both the co-polar and cross-polar components of the reflected signal as well as the value of the depolarization ratio in an arbitrary polarization basis. These calculations were applied to the case of UHF scattering by an ensemble of non-spherical Rayleigh particles, randomly oriented in space. Calculations covered the cases of both liquid water droplets and ice crystals. Monographs by D.B. Kanareykin, N.F. Pavlov and V.A. Potekhin: "Polarization of Radar Signals" [2] and by D.B. Kanareykin, V.A. Potekhin and N.F. Schischkin: "Marine Polarimetry" [3], published in 1966 and 1969, respectively, were decisive in stimulating active application of radar polarimetry techniques to environmental remote sensing within the USSR. These monographs were among the first studies to extend the fundamental approaches of polarization theory, which



*W.-M. Boerner et al. (eds.), Direct and Inverse Methods in Radar Polarimetry, Part 1*, 61–83.

before that were mainly applied to optics, to the UHF electromagnetic range and to multiple radar channels - particularly to the problem of radar remote sensing of the environment. These pioneering monographs summarized the data from previous studies, adding original results obtained by their authors themselves; and, treated the subjects of the mathematical presentation of the polarization state of electromagnetic waves, and described various polarization state transformation devices.

First, field studies in radar polarimetry were carried out using meteorologic radars with either two antennas, providing for simultaneous reception of two orthogonal components of the echo-signal, so that microwaves of varying polarization were independently emitted, or just one, when the target was successively irradiated by pulses of various different polarizations (linear, circular, elliptical) [4-6].

The authors of study [7] experimentally retrieved the distribution of liquid and ice crystal hydrometeors in cumulo-nimbus (Cb) clouds in the mid-USSR from the polarization analysis of the reflected radar signals. They concluded that the distribution of hydrometeors of various phases within the cloud is characterized by a high inhomogeneity and strong variability. Often the zero isotherm boundary in the Cb clouds does not separate liquid droplets from ice crystals. Zones of supercooled water droplets at negative temperatures are found in the upper parts of such clouds. This study of 1966 also attempted to make a classification of meteorologic targets of various phase compositions by the values of their depolarization ratio in the linear basis, LDR (presently the common term is the "linear depolarization ratio"). It was demonstrated that at a probability of at least 83% clouds and precipitation may be considered liquid if their LDRs in the X-band are below -18 dB. At LDRs above -9 dB they would quite probably be crystalline. Finally, when the LDRs are in the $-9$ dB $>$ LDR $> -18$ dB range both clouds and precipitation would be of a mixed phase. Such a classification is supported by the results from direct aircraft microphysical sampling inside the target clouds [7]. Values agreeing with the above stated results were also obtained in [5], which gave the following typical ranges for the depolarization ratio: rain: (-16.5)-(-19.5) dB; wet snow: -4 dB; dry snow: -2 dB.

N.F. Pavlov and V.D. Stepanenko [6] studied the amplitude distribution and correlation functions for the echo-signals from precipitating rain and dry snow, using as their database the pulse-by-pulse recordings from a radar with controlled polarization state in both its transmission and reception channels. It was demonstrated that the amplitude distribution is Rayleigh and that the correlation functions are determined by relative motions of the scatterers, and not by their changing spatial orientation.

As for studying radar characteristics of the surface extended natural objects, we refer to works by S.B. Zubkovich. In monograph [8], he classified and described in minute detail the scattering properties of various types of the underlying surface, including water, and accounted for their polarization properties and characteristics. The classical two-dimensional model of microwave scattering by the rough sea surface was introduced by S.Ya. Braude, F.G. Bass, I.M. Fuchs, A.I. Kalmyukov, et al [9-11]. When studying the polarization matrix of microwave scattering by the sea surface certain anomalous phenomena are encountered, so that some experimental results disagree with the above model. This is why the studies by Yu.V. Melnichuk and A.A. Chernikov [12,13], in which the intensity of the scattered radiation in the basic polarization states for both the emitted and received radiation - the vertical, the horizontal, and the crossed - were examined, are worth a special notice. They demonstrated the dynamics of the growing intensity of scattered radiation in its horizontally polarized wave component, which accompanies higher wind speed, comparable to a similar increase in the intensity of the vertically polarized radiation for rougher seas. These studies also treated the spectral parameters of the scattered field at various polarizations of both the incident and the reflected waves. Attempts to explain the experimentally found polarization effects were undertaken by researchers of the Kharkov school. These explanations were used to account for various inhomogeneities (including foam and spray) above the sea surface. Note, in particular, studies by A.I. Kalmyukov and his co-authors [10], who explained particular features of the anomalous increase in the intensity of horizontally polarized fields at heavy sea roughness as the effect of capsizing sea waves' crests. N.P. Krasyuk and B.Sh. Lande [14,15] accounted for hydrometeors hitting the sea surface (i.e., scattering from columns of liquid). They also obtained the relationships which explained changes in the spectrum width for signals scattered from the sea surface; these were explained as modulation of fine capillary waves by the major macro-scale ones. Earlier, this modulation effect was studied in certain works by A.D. Rosenberg, summarized and updated in [16]. All the experimentally observed effects in the scattered radar emission are adequately presented by the equivalent dipole model; this was demonstrated by B.Sh. Lande with co-authors in a series of their works. For example, they combined in [17] the two-scale model of scattering with another one - of scattering from eddy tropospheric inhomogeneities. Such studies made it possible to relate radiophysical models to newly constructed phenomenological models of the polarization state of radar signals scattered by the rough sea surface. The latter model best fits the needs of microwave instrumentation design and construction; those have been worked out by D.B. Kanareykin with a group of his co-authors [2,3,18].

This short review rounds up the subject of scattering by surface objects such as sea and sea ice surfaces, so that further expositions may deal exclusively with polarization effects in meteorological radar sounding.

## 2. POLARIZATION STUDIES OF HAIL AND SEVERE STORMS

Note that initially, when radar polarimetry started to develop as a branch of radar meteorology, researchers mostly used the linear depolarization ratio (LDR) as their basic polarization parameter to characterize meteorological targets. In particular, the technique of detecting hail zones, conceived and designed by A.B. Shupyatsky and co-workers at the Central Aerological Observatory, Dolgoprudny/Metropolitan Moscow, and tested for several years at the Moldavian field sites [19-21], was based on a joint analysis of the depolarization ratio and the traditional parameter of radar reflectivity, Z. It was found that hail-bearing clouds feature zones of radar reflectivity above the threshold value of

32 dBZ, their LDRs simultaneously exceeding (-10)-(-12.5) dB.

Zones of higher depolarization are generated in such hail clouds 10 minutes or more before the hail event itself. Therefore, this important coherent polarimetric radar study [20] claimed that the composite data on cloud structures including their depolarization profiles, radar reflectivity, vertical development, and shape, plus temporal trends of these characteristics make possible nowcasting of hail at approximately 10 minute notice at a probability level of approximately 95%. This technique found its implementation in a special polarization device aimed for the All-Union-Soviet network meteorological radars MRL-1, MRL-2 and MRL-5 [22]. This device, operating in the 3.2cm range permits transmitting linear (horizontal and vertical) and circularly polarized microwaves and fast switching of polarization to reception. The device is build around a ferrite circulator and a high-speed p-i-n diode switch. Its high speed provides for pulse-to-pulse reception agility of both orthogonal components of the reflected signal using the same receiver.

When discussing the task of polarimetric identification of hail cells, study [23] should be mentioned, which suggested the so-called amplitude differential technique of identifying hail and rain shower cloud cells. To implement this technique in a ROZ-1 radar, a microwave polarizer was designed, capable of transmitting linearly polarized microwaves at 45° vector tilt, and of receiving both orthogonal components of the reflected wave with increased mutual isolation. Using this polarizer, the Cb clouds were studied in the Odessa region during the spring-summer seasons of 1980-1981. Hail cells were identified from average differences between the power reflected at horizontal and vertical polarizations. Recall that the initially emitted wave had its vector tilted to 45°.

The classical scheme of a radar polarimeter with two receiving channels, making possible the simultaneous processing of both orthogonal polarization components of the reflected signals, poses harsh demands on the degree of isolation of both channels, which are not always easy to satisfy in practice. Besides, the existing network weather radars are single-channel systems (for fixed wavelength range). Constructing a two-channel polarimeter on the basis of a single-channel weather radar presents considerable technological difficulties. Therefore, lately, an approach using a single-channel weather radar, with polarization modulated in both its transmission and reception channels, finds more and more popularity within the meteorological radar community of the USSR.

## 3. POLARIZATION MODULATION TECHNIQUES

### 3.1 Theoretical Foundations

Let us set forth some definitions and notations, needed for further discussion.

The polarization structure of the radar signal scattered by an ensemble of hydrometeors is described by the classical scattering matrix [S], which may be presented as a sum of scattering matrices for single hydrometeor elements, $[S^{(l)}]$, multiplied by appropriate phase coefficients, which correspond to various distances of these hydrometeors from the receiver:

$$[S] = \sum_{l} [S^{(l)}] e^{i\phi_l} \tag{1}$$

Modeling these hydrometeors as rotational ellipsoids, the scattering matrix $[S^l]$ for a single hydrometeor (l) will have the following form in the horizontal-vertical polarization basis:

$$[S^{(l)}] = \begin{bmatrix} S_{hh}^{(l)} & S_{vh}^{(l)} \\ S_{hv}^{(l)} & S_{vv}^{(l)} \end{bmatrix} = \tag{2}$$

$$= \begin{bmatrix} \left(s_a^{(l)} - s_b^{(l)}\right) \sin^2\delta^{(l)} \sin^2\theta^{(l)} + s_b^{(l)} & \left(s_a^{(l)} - s_b^{(l)}\right) \sin^2\delta^{(l)} \sin\theta^{(l)} \cos\theta^{(l)} \\ \left(s_a^{(l)} - s_b^{(l)}\right) \sin^2\delta^{(l)} \sin\theta^{(l)} \cos\theta^{(l)} & \left(s_a^{(l)} - s_b^{(l)}\right) \sin^2\delta^{(l)} \cos^2\theta^{(l)} + s_b^{(l)} \end{bmatrix}$$

Here index "h" corresponds to horizontal polarization, while "v" - to vertical. The value $s_a^{(l)}$ in (2) is a complex scattering amplitude, obtained when our ellipsoid is irradiated by a linearly polarized wave, its $\vec{E}_o$ vector coinciding in its direction with the ellipsoid rotation axis, and $s_b^{(l)}$ - the respective amplitude for the same ellipsoid irradiated by a wave, linearly polarized in the direction orthogonal to the ellipsoid axis. Angles $\delta^{(l)}$ and $\theta^{(l)}$ determine the orientation of that axis, $\vec{N}^{(l)}$. Note, that $\delta^{(l)}$ is the angle between the vector $\vec{N}^{(l)}$ and that of the direction of wave propagation $\vec{K}$ - of electromagnetic wave propagation, while $\theta^{(l)}$ - the angle between the vertical and the projection of vector $\vec{N}^{(l)}$ upon a plane normal to the direction of wave propagation (traditionally called canting angle) (see Fig. 1).

The particular form of the scattering matrix depends on the polarization basis chosen. For example, if the latter is circular, the scattering matrix for a single hydrometeor will have the form:

$$\left[S_c^{(l)}\right] = \frac{1}{2} \begin{bmatrix} \left(s_a^{(l)} - s_b^{(l)}\right) \sin^2\delta^{(l)} e^{-i2\theta^{(l)}} & \left(s_a^{(l)} - s_b^{(l)}\right) \sin^2\delta^{(l)} + 2s_b^{(l)} \\ \left(s_a^{(l)} - s_b^{(l)}\right) \sin^2\delta^{(l)} + 2s_b^{(l)} & \left(s_a^{(l)} - s_b^{(l)}\right) \sin^2\delta^{(l)} e^{i2\theta^{(l)}} \end{bmatrix} \tag{3}$$

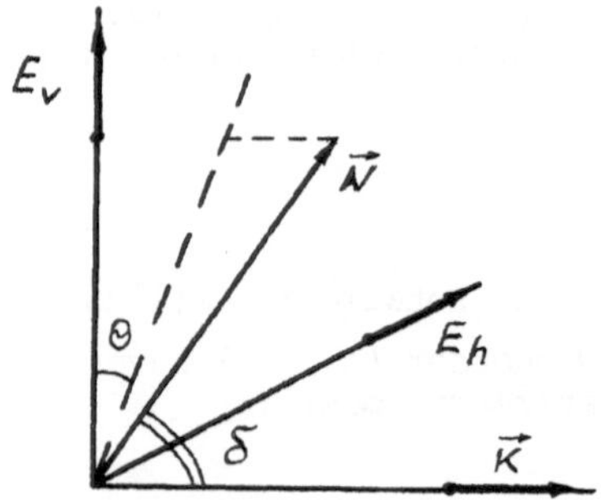


Fig. 1: Orientation of hydrometeor rotation axis, $\vec{N}$, relative to the wave phase front and the direction of wave propagation, $\vec{K}$

It is well known (see example [2,3,18]) that any radar target may be characterized by a certain set of invariant polarization parameters (i.e., those independent of the choice of a polarization basis). In case radar targets fluctuate (meteorological targets do fall into such a category), the eigenvalues of the Graves matrix $<[G]> = <[S]^+[S]>$ may serve as such invariants. Here $[S]^+$ is the Hermitian conjugated matrix for [S], and the symbol of < > denotes temporal averaging. Other polarization invariants are the parameters describing the distribution of co-pol and cross-pol nulls on the Poincaré sphere, as discussed in more detail below (see Section 4). The eigenvalues are determined from the relationship [18]:

$$\lambda_{1,2} = \frac{1}{2}\,\mathrm{Sp}\,<[G]>\left\{ 1 \pm \left[ 1 - \frac{4\det<[G]>}{(\mathrm{Sp}<[G]>)^2} \right]^{1/2} \right\} \tag{4}$$

Using (1), (2) and (4), one may easily demonstrate that for scattering hydrometeors:

$$\begin{aligned} \lambda_1 &= < |(s_a - s_b)\sin^2\delta + s_b|^2 > \\ \lambda_2 &= < |s_b|^2 > \end{aligned} \tag{5}$$

Now the sign < > in relationship (5) denotes averaging over the ensemble of hydrometeors. One of the invariants of the radar target, most convenient for practical use, is its anisotropy coefficient, as determined from the relationship:

$$q = \frac{\lambda_1 - \lambda_2}{\lambda_1 + \lambda_2} \tag{6}$$

If particles do not display strong anisotropy, the relationship

$|(s_a - s_b)/s_b| << 1$ is usually satisfied. In that case, relationships (2) and (3) easily yield that the parameter q of (6) may, to a first approximation, be estimated by:

$$q = \frac{\mathrm{Re}\ \langle s_b^* (s_a - s_b)\sin^2\delta\rangle}{\langle |s_b|^2\rangle} \tag{7}$$

Since the parameter $|(s_a - s_b)/s_b|$ is minor, the latter would be equal to the ratio of co-polar returns in the circular and linear bases, respectively.

### 3.2 Measurements of the anisotropy coefficient: q

Following the above formulation, the authors of [24] assessed the anisotropy coefficient from the difference between the averaged circularly and linearly polarized signals at the output of the logarithmic receiver. Radiation was polarized by manipulating a quarter-wave rotating phase plate built into the transmitting UHF tract of the MRL-2 weather radar. The phase plate was constructed so as to periodically, every few seconds, switch the polarization of the emitted signal from linear to circular. Results of field observations, carried out in [24], are presented as vertical profiles of the anisotropy coefficient, q, for precipitating stratiform clouds and convective clouds. According to [24], the range of the q variability was (-11)-(-20) dB. The highest value of q (-11 dB) was most frequently observed around the zero isotherm. According to [24], the anisotropy coefficient noticeably increases for heavy precipitation, which is a natural consequence of stronger deformations of raindrops of larger size. Strictly speaking (see expression (3)), the value of the anisotropy coefficient is determined not only by the shape and the dielectric properties of the reflecting particles, but by the distribution of hydrometeors by their orientation angle, $\delta$; this fact should be accounted for when interpreting the experimental results. One of the approaches to retrieving the anisotropy coefficient suggests to assess it from the amplitude of the fourth harmonic of the modulation frequency for the signal scattered by hydrometeor formations, when the phase plate in the radar waveguide is quickly rotated [24].

### 3.3 Retrieval of the Mueller matrix

Another approach to technical realization of the polarization modulation in a single-channel radar consists in using a rotating polarizing grating, which is set in front of the feed horn of a weather radar antenna [25-27]. Measuring the reflected signal power at different polarizations of the emitted wave, which are determined by the rotation angle of the polarizing grating, one might, in principle, retrieve all the 16 elements of the 4×4 statistical scattering matrix - the Mueller matrix. In general, if polarization of the transmitting antenna coincides with that of the receiving antenna, the measured power P of the reflected signal would be determined by:

$$P = \vec{R}^T[M]\vec{R} \tag{8}$$

Here $\vec{R}$ is the Stokes vector-parameter for the transmitting and the receiving antennas, [M] - the Mueller matrix for a standard radar target. A more detailed expansion of the relationship (8) looks as follows:

$$P = J^2m_{11} + V^2m_{22} + Q^2m_{33} + U^2m_{44} + 2JVm_{12} + 2JQm_{13} + 2JUm_{14} + 2VQm_{23} + \\ + 2VUm_{24} + 2QUm_{34} \qquad (9)$$

The elements J, V, Q, U in (9) are but the elements of the Stokes vector-parameter, and the $m_{kl}$ - the Mueller matrix elements. In case of backscattering, the Mueller matrix is symmetrical and is described by nine independent parameters. Theoretically, to determine nine unknown elements of the Mueller matrix, one has to solve a system of nine linear equations of structure (9) for nine different polarization states of the emitted wave. It is important to note, that at least one of these nine polarizations should differ from linear, or the system would not have a unique solution and that for reducing inherent measurement errors, computational errors, and uncertainies (ill-posedness), the nine polarization states ought to be distributed equally over the entire polarization sphere. V.K. Zavirukha [25,26] performed experiments for a set of different polarizations characterized by various ellipticity angles, α, and ellipse orientation angles, β. Notations chosen for α and β are illustrated by Fig. 2, where the respective polarization ellipse and the point, corresponding to it, on the Poincaré sphere are presented.

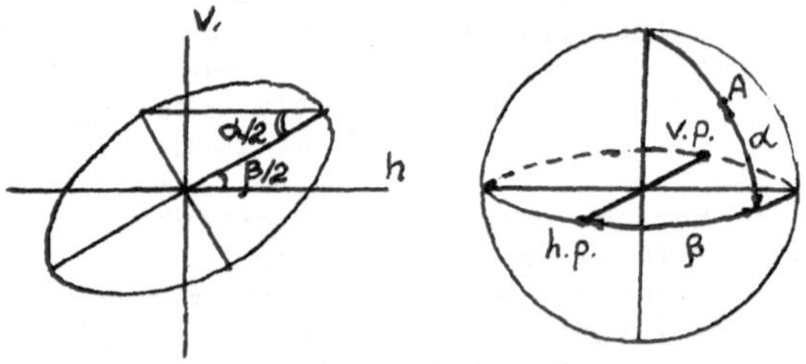


Fig. 2: Presentation of a polarization state on the Poincaré sphere

The solutions of the above system may become unstable if the input data contain certain errors, accumulated when measuring the reflected signal power. Thus, one is forced to resort to some regularization procedure (similar to those applied to ill-posed inverse problems). As yet this question remains open. By now the polarization modulation technique based on the use of the rotating grating has been tested at a wavelength λ = 3.2 cm (MRL-2 and MRL-5 radars) and also in the 10 cm channel of the MRL-5 radar. Application of the polarization grating appeals for its technological simplicity and does not entail any interference with the waveguide of the radar. However, one should remember certain flaws of polarization gratings. It was found that when rotating the grating, the extinction it produces and the shape of the antenna beam may noticeably vary. If polarization modulation is used for sounding meteorologic targets of complex structures, one can hardly account too strictly for variations in the width of the main lobe of the antenna pattern. Another drawback of the technique is the large time needed to complete the measurement cycle. However, the analysis time might be considerably reduced if we stick to reduced sets of most informative emitted wave polarizations and measured polarization parameters, and by using complete, dual polarization radar systems with fast polarization state switching capabilities (the adaptively chosen polarization states ideally covering the entire polarization sphere).

### 3.4 The differential reflectivity technique

The simplest polarization modulation is a horizontal to vertical polarization switch of the transmitted/returned signals every other period, followed by computation of the ratio of the averaged reflected power in both polarizations, i.e., of the so-called differential reflectivity:

$$Z_{DR} = 10 \lg P_H/P_V = 10 \lg P_H - 10 \lg P_V \qquad (10)$$

The differential reflectivity technique, first introduced by A.B. Shupyatsky [1], has lately found wide application both in the West and in the Soviet Union. Speaking about its merits, one should mention its simplicity, its lenient demands on the quality of the antenna-feeder system, as compared to other techniques of polarization selection (including the channel polarization isolation, antenna cross-polarization, etc.), and its lack of energy potential losses. (When assessing the depolarization ratio in either linear or circular bases, $L_{DR}$ and $C_{DR}$, one of the orthogonal components is 15 to 20 dB weaker than the other, which makes operation with weak signals impossible). First, Soviet measurements of the differential reflectivity of clouds and precipitation were performed by A.V. Kochin and V.R. Megalinski [28,29]. When technically implementing the polarimeter, particular attention was given in these investigations to the special design of the feed horn, which provides for the identity of the antenna patterns in both orthogonal planes. The feed horn best fitting the demands of measuring $Z_{DR}$ is the scalar one constructed as a horn with broken generant. Besides symmetrizing the antenna patterns in the orthogonal planes of any polarization, this horn also makes possible a distribution of the field at the antenna aperture which is close to rectangular, thus, increasing the effectiveness of its use and hence, its gain. The designed Cassegrain antenna made possible the antenna pattern identical to -15 dB in its principal lobe in the two orthogonal planes. In accordance with the results from [28], the vertical profiles of differential reflectivity of Cumulus congestus (Cu cong) clouds (up to 5-6 km height) usually have three maxima: in the upper part of the cloud, around its zero isotherm, and close to the ground in the precipitation zone. Higher $Z_{DR}$ at cloud tops coincide with the zone of decreased radar reflectivity, apparently associated with the presence of crystal particles, which strongly deviate in their shape from spheres and possess low dielectric permittivity. The surface maximum in $Z_{DR}$ suggests an increase in the average droplet size due to coagulation growth of precipitating particles. As for the "melting layer", according to [30] the maximum of differential reflectivity surface, where $Z_{DR}$ reaches about 1-3 dB, is found at the bottom of the layer of higher radar reflectivity; this maximum is about 100 m thick. Immediately above this layer the value of $Z_{DR}$ drops practically to zero. A.V. Kochin suggested the following physical interpretation of this $Z_{DR}$ maximum observed at the altitudes of the "melting layer". When falling through this layer, the crystals first to melt are the smallest ones. The velocity of their fall sharply increases then, so that these newly born droplets start to sediment on larger crystals and fluffs below, which go on falling at considerably slower velocities. This results in a sharp increase in size of the largest crystals. At the final melting stage such particles are in excess of 4-5 mm in diameter and disintegrate and turn into proper droplets. However, prior to this disintegration, they strongly contribute to the radar reflectivity, and, due to significant non-sphericity also cause noticeable differential reflectivity (polarization state transformation effects).

Systematic studies of differential reflectivity of various cloud types and precipitation in the X-band have been lately carried out at the Voeikov Main Geophysical Observatory, Leningrad [31] using an automatized radar instrumentation facility built around the MRL-2 weather radar. It includes a polarization device, a pre-processing system for radar data, and a computer with colour display. The principal element of the polarization device is a ferrite horizontal-to-vertical polarization switch operating at radar pulse repetition frequencies, i.e., 500-1000 Hz, and having a switching time below 250 $\mu$s. The ferrite switch uses the Faraday effect, i.e., that of rotation of the polarization plane of the linearly polarized wave in a 'gyrotropic medium', when forced by external electromagnetic fields of varying intensity and direction. Data on the radar reflectivity, Z, and on differential reflectivity, $Z_{DR}$, are processed at a stepping increment of 250 m; the maximum dwell time is about 1 s. However, it should be emphasized that when using 'gyrotropic ferrite magnetic' switches, asymmetries are inflicted on the two dual polarization channels (1,2) of the transceiver, i.e., $S_{12} \neq S_{21}$, and great care in turn must be taken in adjusting for these enforced errors by proper polarimetric calibration of the transceiver system.

Study [32] was dedicated to assessing the statistical accuracy of the differential reflectivity retrievals from theoretical calculations and to testing the validity of such calculations against the pulse-to-pulse recordings of the reflected signals, about 8 s each. It was demonstrated that in case polarization is switched every odd period (for transmission and reception simultaneously), the statistical accuracy of determining $Z_{DR}$ starts to significantly depend on the value of the odd period correlation coefficient of the reflected signal; in its turn this coefficient depends on the wavelength, the pulse frequency, and the RMS width of the Doppler spectrum of the received signal. At a fixed wavelength the $Z_{DR}$ measurement accuracy improves for longer wavelengths, higher pulse repetition frequencies, and narrower spectra. Calculations show that upon transition from non-correlated to highly correlated pulses (correlation coefficient $\rho = 0.9$), the $Z_{DR}$ error decreases by approximately a factor of four (for a fixed dwell time). At a fixed error level this would be equivalent to spending 16 times less time on signal accumulation for the strongly correlated pulses. Another conclusion is that in the C- and S-bands the RMS deviations of $Z_{DR}$ (SD ($Z_{DR}$)) of about 0.2 dB would be provided by the dwell time of 0.1 – 0.2 s if the pulse repetition frequency is 1000 Hz, and the spectrum width is 1 m/s. At the wavelength of 3.2 cm and the pulse frequency of 500 Hz, (standard values for non-coherent weather radars) the same accuracy of assessing $Z_{DR}$ is reached for an order of magnitude longer dwell times. Analysis of the pulse-to-pulse signal recordings basically supports such theoretical estimates, except in cases when the process itself is non-stationary.

The statistics of routine cloud and precipitation $Z_{DR}$ cross-sections, produced by varying the elevation angle of the antenna, testifies to assessible differences between the spatial distributions of this parameter in stratiform and convective clouds. This difference agrees with the differring phase composition of these two cloud types. In the overwhelming majority of cases the zero isotherm in stratiform clouds distinctly separates the zones of positive $Z_{DR}$ (below that isotherm) and those where $Z_{DR}$ is close to zero (i.e., above the zero isotherm). In other words, the liquid and solid phases in those clouds are clearly separated. A minor local maximum of $Z_{DR}$ is often observed in the area of the melting layer. As for convective clouds, particularly those at their early developmental stage, the respective vertical distributions of $Z_{DR}$ might be quite complicated. Zones of positive $Z_{DR}$ are frequently observed above the zero isotherm, while negative values of $Z_{DR}$ are also encountered below it. Presumably, such a patchy distribution is associated with the local distribution of the updrafts and downdrafts inside the cloud. Note, the possibility of identifying the zones of ground clutter contamination when $Z_{DR}$ is measured against the background of radar returns form meteorologic targets. Anomalously high negative values of $Z_{DR}$ are typical for such clutter. The typical vertical distributions of $Z_{DR}$ for Ns and Cb clouds are presented in Fig. 3.

Since the value of differential reflectivity is determined both by the shape of the reflecting particles and by their spatial orientation, it is impossible, strictly speaking, to differentiate between the liquid and solid phases in meteorologic targets by their $Z_{DR}$s alone. Indeed, fine spherical droplets and chaotically oriented non-spherical crystals would yield approximately equal differential reflectivities, both close to zero. Following the suggestion by A.B. Shupyatsky such an ambiguity may be resolved by measuring a second, additional polarization parameter - the depolarization ratio in the linear vertical-horizontal basis. Indeed, a possibility was first demonstrated in [33] of identifying the character of the reflecting cylindrical or disk-shaped particles from

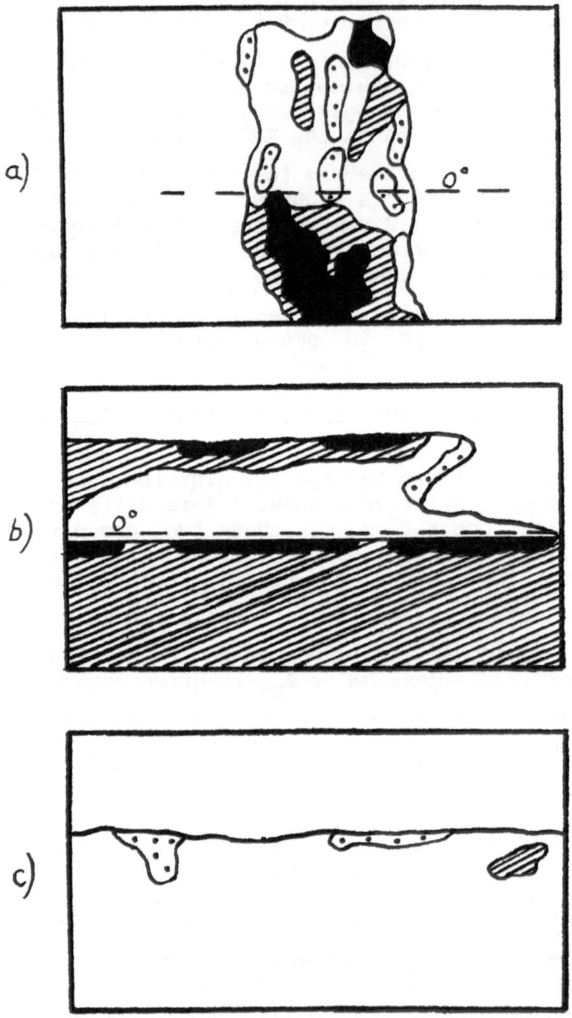


Fig.3. Cloud differential reflectivity cross-sections vs. elevation angle.
a) Cumulonimbus clouds;
b) Nimbostratus clouds;
c) Stratiform clouds, snow.

$Z_{DR} < 0$ dB    $0.5 < Z_{DR} \leqslant 1$ dB

$0 \leqslant Z_{DR} \leqslant 0.5$ dB    $Z_{DR} > 1$ dB.

measurements of such parameters. Moreover, having the respective distribution versus the elevation angle, their typical orientation may be determined . The results from GATE polarization measurements from aboard a research vessel in the ocean demonstrated that such polarization parameters are related to certain characteristics of the ocean surface roughness, i.e., the direction and height of sea waves [34,35]. Operating additionally in the bistatic reception mode, the roughness parameter may be enhanced so that further measured data, collected from sounding clouds and precipitation, would yield the size of those large reflecting particles which exceed some preset levels [36,37].

We now address the problem of relating the depolarization and the differential reflectivity of radar signals to the microstructure of reflecting particles. Theoretically this relationship was treated in detail in [38], while [39] was specifically dedicated to investigating certain polydispersive distributions and composite cloud forms. Values of the above parameters were computed as functions of particle canting angle $\theta$ (see Fig. 1) and of their shape, approximated as either elongated (prolate spheroidal) or pressed (oblate spheroidal) ellipsoids. Calculations were based on computational relationships for the mutually orthogonal components, earlier derived in [1] and then reproduced in the now popular Soviet monographs on polarization [2] and radar meteorology [4]. Applicability of such calculations of polarization components to large particles in the centimeter radar wave-length range was demonstrated for rain in [26], and for hail - in studies by R. Petrov (Institute of Hydrology and Meteorology, Sofia, Bulgaria).

Calculations demonstrated the values of differential reflectivity and depolarization to depend strongly on particle orientation. Moreover, in various ranges of canting angle these dependencies are different for both polarization characteristics. Simultaneous retrieval of the values of depolarization and of differential reflectivity makes possible identification of both spherical and non-spherical particles of arbitrary orientation at higher certainty. In case the canting angle of the scattering particles is close to either zero or 90°, the respective differential reflectivity would differ from zero, while depolarization would be zero; an inverse situation is also true: random particles' orientation leads to zero differential reflectivity, while the depolarization component would be strongly displayed. Consequently, the difference from zero of at least one of the above polarization components would mean that non-spherical particles are present in the scattering volume; thus, capabilities of the polarization selection techniques in identifying the phase of cloud particles, including hail, are increased.

Besides the purely qualitative aspects, simultaneous measurements of differential reflectivity and depolarization open the possibility of determining both the shape and the proper orientation of the reflecting particles. By itself, the particles' orientation results from interaction of aerodynamic, gravitational, and in certain cases, electric forces in the atmosphere. The latter factor is, in turn, an important predictor in studies of pre-thunderstorm and thunderstorm conditions in clouds. Theoretically, the possibility of determining the particle's orientation from the state of circular polarization was demonstrated as early as 1961 [35,41]. Using this technique, the difference in particle orientation in shower and thunderstorm clouds was discovered [42].

Applying the differential reflectivity technique, a difference between the fluctuation spectra of that polarization parameter for shower and thunderstorm clouds was also found. Later stepwise changes in the values of differential reflectivity (caused by sudden shape changes from prolate to oblate spheroidal shape of the hydrometeors, etc.), which accompany lightning discharges in clouds, were identified with high certainty.

The particle's shape is definitely related to its size. Therefore, assuming a certain particle size distribution, one may retrieve various particle size distribution moments, parameters of their distribution, and furthermore, one may retrieve cloud water content and precipitation intensity for the given droplet fall velocity [43]. Employing the polarization parameters based on completely coherent dual orthogonal polarization meteorological concepts makes it possible to increase precipitation measurement accuracy rather effectively, as compared to conventional single fixed antenna polarization approaches.

Polarization measurement techniques acquire particular importance for cloud modification studies aimed at preventing hail damage and at increasing the amount of precipitation. Essentially such modification amounts to the control of the microphysical parameters of clouds. Using a polarization diversity radar for differential reflectivity measurements [44] makes it possible not only to detect with high certainty the presence of hail in clouds, but also to lower the radar reflectivity threshold of such detection (from about 33-35 dBZ for depolarization alone [20] to 27-28 dBZ when adding the differential reflectivity observations).

The same cited study [44] presents the typical vertical profiles of reflectivity and depolarization, and the associated profiles of the reflecting particle's shape, orientation, and size for the weakly and strongly precipitating clouds and for those with hail. One may see considerable differences in these clouds' microstructure.

Employment of the polarization techniques for the retrieval of the cloud microstructure may also emcompass applied problems - in particular, those of remote sensing of cloud zones dangerous for aircraft navigation, including icing, high electric activity, wind shear, down burst regions, extreme localized temperature changes, and other dangerous phenomena [45].

Besides studying hydrometeors, several authors addressed the problem of polarization properties of signals reflected from "clean air", their possible sources being bird flocks, insect formations, plant seeds and pollen, and eddy inhomogeneities of the refractive index [46,47]. In this respect the authors of study [48] undertook laboratory studies of polarization properties of separate meteorologic and ornithologic targets using an echoless camera. In [48] experimental results were obtained on the radar cross-sections of artificial hail and certain bird species at various polarizations. As demonstrated by A.A. Chernikov in

[47], retrieval of polarization characterisitics of the reflected signals is an effective means of differentiating the "clear air" targets into two classes: eddy dielectric inhomogeneities of air and foreign bodies. Lack of any depolarization in the reflected signal proves that the source of such a reflection is the dielectric inhomogeneity of air, while the appearance of a depolarized component testifies to generation of the reflections by some non-spherical targets of certain dielectric permittivity, strongly differing from unity such as caused by insects, plant seeds, coarse aerosols, etc. In their work, A.A. Chernikov and A.B. Shupyatsky [46] measured depolarization ratios for the point-type reflected signals from clear skies at $\lambda = 3.2$ cm. Experiments were executed during warm days of weak cloudiness and well developed convection. Reflections from birds, identified by means of an optical system, were excluded from analysis. Measurements demonstrated the most probable value of the depolarized component to reach (-12)-(-13) dB relative to the co-polarized one. Since dielectric inhomogeneities of air cannot provide such depolarization, this result should be envisaged as a proof that point type signals are produced by insects, plant seeds, pollen and other foreign bodies in the atmosphere including those introduced by man-induced pollution of the atmosphere.

## 4. OPTIMAL POLARIZATION PROCESSING OF RADAR SIGNALS: FORMULATION OF THE PROBLEM

Polarization identification of radar targets is based on their capability to alter the polarization state of the incident wave in the course of reflection. These properties should be described by certain invariant polarization parameters, i.e., independent of the polarization basis of the incident radiation. To assess the degree of effectiveness in identification of stable targets, it is convenient and expedient to use their COPOL null and XPOL null polarization characteristics and associated distribution functions.

Recall that a radar target's XPOL null polarization is by definition that polarization of the incident wave which lacks a cross-polarized component in the reflected signal (i.e., that for which its coefficients $S_{12}$, $S_{21}$, of the scattering matrix [S] are zero). Two mutually orthogonal polarizations would satisfy this condition; they are described by two diametrally antipodal points (XPOL nulls) on the Poincaré sphere. Polarization properties of the antenna are also known to be presented by a certain single point on the Poincaré sphere, whereas for a point target two such points are required (one copol null and one cross-pol null by definition of Huynen's polarization fork). If this point coincides with one of the XPOL nulls of the target, the signal reflected by such a target would be completely received by such an antenna.

The COPOL null polarization state of the given stable target is such a polarization of the incident wave that its reflected wave is orthogonal to the polarization of the transmitting antenna and cannot be received by it. A pair of COPOL null polarizations is, in general, mutually non-orthogonal. These are presented by a COPOL null pair symmetrical to the orthogonal XPOL null pair on the Poincaré sphere. The XPOL and COPOL nulls reside on one and the same main circle of the sphere. The radii connecting these points with the sphere center form the so-called polarization fork (the Huynen polarization fork). The position of this fork and hence polarization properties of the target are completely determined by one XPOL null and one COPOL null.

However, the informative modulation of polarization during reflection from a multiple target is a random process. The radar signal of randomly modulated polarization is traditionally treated as a partly polarized oscillation - in other words, a superposition of polarized and non-polarized waves. The notions of XPOL null polarization and COPOL null polarization lose their meaning for such oscillations, since it has no basis in which the power of either the co-polar or cross-polar component of the signal would turn to zero.

However, the approach we follow on the problem of polarization identification of targets suggests that we use the same invariant polarization characteristics which were used for stable targets, i.e., the COPOL/XPOL null polarization concepts, to analyze fluctuating signals from statistical targets. This approach appears to be possible since in actual observational conditions during the time interval while the multiple target is irradiated by an incident radar pulse, this target may be considered to be of a certain frozen configuration. At those wavelengths which are used in meteorologic radar sounding even the most rapidly fluctuating signals from hydrometeors noticeably vary in a few milliseconds or even tens of milliseconds. This would mean that for at least one pulse repetition period the fluctuating object may be considered a stable radar target and be characterized by the above invariant polarization characteristics. The instant COPOL/XPOL null polarizations may then be presented on the Poincaré sphere by points which randomly scatter across the sphere, but remain within certain areas (clustering zones), characterized by two-dimensional probability distributions.

Thus, polarization properties of a multiple target are completely characterized by two two-dimensional probability density distributions of the coordinates of XPOL $W_x(\alpha_x,\beta_x)$ and COPOL $W_c(\alpha_c,\beta_c)$ nulls on the Poincaré sphere. If such distributions are known for various hydrometeoric targets, the task of determining identification procedures for them and of assessing its effectiveness is solved by classical techniques of image recognition.

It may be seen that the principally new element in such an approach to optimal polarization processing of radar signals is the search for the above probability distribution densities, $W_x(\alpha_x,\beta_x)$ and $W_c(\alpha_c,\beta_c)$ for the observed objects. Further treatment is dedicated to calculating such polarization invariants for an ensemble of hydrometeors, modelled as a combination of rotational ellipsoids. To determine the function $W_x(\alpha_x,\beta_x)$, we use scattering matrices (1), (2), presented in the linear polarization basis, together with the well known relationships for coordinates of the instant XPOL null polarization basis on the Poincaré sphere $(\alpha_x,\beta_x)$ [3]:

$$\operatorname{tg} \beta_x = \frac{2 \operatorname{Re} (S_{hh}^* S_{vh} + S_{vh}^* S_{vv})}{|S_{hh}|^2 - |S_{vv}|^2} \tag{11}$$

$$\operatorname{tg} \alpha_x = \frac{2 \operatorname{Im} (S_{hh}^* S_{vh} + S_{vh}^* S_{vv})}{|S_{hh}|^2 + |S_{vv}|^2 + 2|S_{vh}|^2} \cos\beta_x \tag{12}$$

Analytic expressions for the distribution functions $W_x(\beta_x)$, $W_x(\alpha_x)$ were obtained in [49]; we omit them here because of their bulky form.

The first important conclusion from the analysis of the distribution $W_x(\alpha_x)$ is in that the XPOL nulls occupy two areas on the Poincaré sphere, symmetric with respect to $\beta = 0$ and $\beta = \pi$; they protract along the equator. The stretch of these areas is proportional to the width of angular distriubtions, characterized by the RMS angular deviation $\delta$. For a liquid droplet cloud at $\delta = 20°$, this distribution area $W_x(\beta_x)$ reaches about 50° at the level of 0.5; and, for plate crystals at $\delta = 40°$, it is about 100°. In a particular case of a homogeneous distribution of particle orientations the XPOL null area becomes homogeneously distributed along the equator or the sphere. If all the particles have identical orientations, the distribution $W_x(\beta_x)$ turns into a $\delta$-function.

To determine the COPOL nulls, it is helpful to use the well-known expression [2]:

$$p(\alpha_x, \beta_x) = \pm i\sqrt{S_{11}/S_{22}} \tag{13}$$

Here, $S_{11}$, $S_{22}$ are the co-polarized coefficients of the scattering matrix, similar to (2), however, taken in a different new basis of XPOL null polarizations, rotated by an angle of $\beta_x$ with respect to the initial one. With the help of expression (13), one determines the conditional distribution function for the phasor of the COPOL null polarization $W_c(p(\alpha_x, \beta_x))$. Then the transition to a non-conditional distribution is determined as:

$$W_c(p) = \int W_c(p(\alpha_x, \beta_x)) W_x(\alpha_x, \beta_x) d\alpha_x \, d\beta_x \tag{14}$$

The sought distribution $W_c(\alpha_c, \beta_c)$ is obtained form this expression, since the coordinates of COPOL nulls are related to the argument $\kappa$ and the module of the phasor $p = \operatorname{tg}\gamma \exp(i\kappa)$ by the relationship:

$$\sin\alpha_c = \sin 2\gamma \sin\kappa, \quad \operatorname{tg}\beta_c = \operatorname{tg} 2\gamma \cos\kappa \tag{15}$$

Examples of the COPOL null distributions, presented as cross-sections of the distribution $W_c(\alpha_c, \beta_c)$ at the level of 0.5, projected on the equatorial plane of the Poincaré sphere, are presented in Fig. 4. [50].

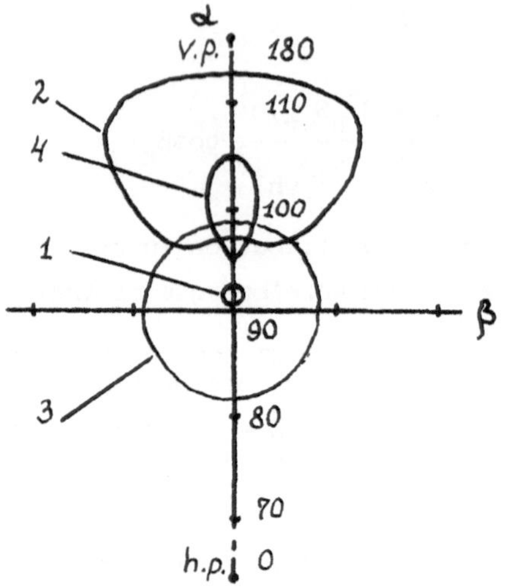


Fig. 4: Null polarization areas for hydrometeors of various types on a cylindrical projection of the Poincaré sphere. 1 - Water droplets; 2-crystal plates, $\gamma_o = 15°$; $s_a/s_b = 0.3$; $\delta = 20°$; 3 - same as 2, completely random orientation, $s_a/s_b = 0.3$; 4 - crystal needles (randomly oriented in the horizontal), $\gamma_o = 15°$; $s_a/s_b = 2$

COPOL nulls for liquid water clouds cluster within 1° of the sphere poles. For ice plates (flattened-oblate-spheroids) such a cluster and its distance from the poles increase for higher anisotropy of particles and lower antenna elevations ($\gamma_o$). For example, when the ratio $s_a/s_b$ reduces from 0.6 to 0.3, and $\gamma_o$ - from 75° to 15°, the center of COPOL null cluster shifts from 1° to 25° from the sphere pole, while its meridianal size increases from 1.4° to 15°. The respective cluster in Fig. 4 is shown for $s_a/s_b$ = 0.3, and $\gamma_o$ = 15° with curve 2 plotted for RMS angular deviation of particles $\delta$ = 10°, and curve 3 - for the completely random case.

For needle crystals (elongated rotational (prolate) ellipsoids), randomly oriented in the horizontal, the COPOL (and XPOL) null cluster size is determined by the ratio $s_a/s_b$ and the antenna elevation, $\gamma_o$. Reduction of the latter from 75° to 15° shifts the center of the COPOL null cluster meridianally within 88-100°, the cluster width narrows form 16° to 9°. Curve 4, Fig. 4, shows such a cluster for $\gamma_o$ = 15° and $s_a/s_b$ = 2.

It may be seen from these examples that the principal difference between polarization properties of different cloud types is displayed in the position and size of their COPOL null clusters. While for the liquid phase, it approaches the δ-function, for crystal clouds the cluster angular size reaches several tens of degrees. Leaving apart the last stage of the solution - decision making and assessing its results, we note by way of an example that identification of targets presented by curves 1 and 2 (or 4) in Fig. 4 is, in principle, completely unambiguous (if we disregard the "tails" of distribution functions).

Thus, the relative position and the degree of overlap of COPOL/XPOL null clusters make possible conclusions on the potential effectiveness of polarization techniques in identifying clouds of various phase composition, and a rational choice of the polarization basis; they also make possible foundations of optimal and sub-optimal algorithms for polarization processing of radar signals which will strongly advance micro-cloud hydrometeoric metrology in the near future.

## 5. CONCLUSIONS

The Soviet radar meteorologists have come to appreciate the virtues of complete utilization of coherent polarization information in various fields of meteorologic research and, increasingly more also, in meteorologic forecasting, especially of severe storms, hail, cyclones, and the like. Both polarimetric theory and metrology have been well advanced during the past thirty years within the USSR; and further rapid advances in polarimetric Doppler radar meteorology are currently being achieved at various Soviet meteorologic radar observatories.

Here, we would like to propose and suggest strong future collaboration in carrying out joint large scale measurement campaigns with experts from, for example, the DLR-POLDIRAD meteorologic instrumentation radar facility at Oberpfaffenhofen, FRG; with NOAA-ERL-NSSC at Norman, OK/USA; with CRL at Koganei-shi/Tokyo, Japan; etc., utilizing simultaneously ground-based, airborne and space borne polarimetric radar systems operated by various international trans-continental research centers. The Soviet polarimetric meteorologic radar research community is ready for this challenge and is eager for initiating such international large scale meteorologic measurement campaigns at the earliest possible occasion.

## 6. ACKNOWLEDGEMENTS

The authors are indebted to the workshop director for his kind post-workshop invitation of this review on Soviet polarimetric radar meteorologic studies for inclusion in these proceedings. We also thank Dr. Wolfgang-M. Boerner for his diligent and generous assistance for several of us to visit the USA and Western Europe during 1990 May to 1991 January; and here we are extending our sincere gratitude to the administrators and polarimetric radar meteorologists of NOAA-ERL-WPL & NSSL, and especially to Drs. Dusan S. Zrnic and Richard J. Doviak of NOAA-ERL-NSSL/OU-CIMMS at Norman, OK.

## 7. REFERENCES


[1] A.B. Shupyatsky, 'Microwave scattering by non-spherical particles', Trudy CAO, Vol. 30, pp. 39-52, 1959

[2] D.B. Kanareykin, N.F. Pavlov, V.A. Potekhin, Polarization of Radar Signals, Moscow, Sovetskoye Radio, 440 p., 1966

[3] D.B. Kanareykin, V.A. Potekhin, I.F. Schischkin, Marine Polarimetry, Leningrad, Sudostroyenie, 1969

[4] V.D. Stepanenko, Microwave Sounding in Meteorology, Leningrad, Gidrometeoizdat, 1979

[5] N.F. Pavlov, 'Extinction of Precipitation Returns for Linear Polarization Radars', In: Proc. III All-Union Conf. Radar Meteorology, pp. 115-119, 1968

[6] N.F. Pavlov, V.D. Stepanenko, 'Techniques for and Results from Experimental Studies of Polarization Properties of Meteorologic Radar Targets', In: Proc. III All-Union Conf. Radar Meteorology, pp. 87-95, 1968

[7] V.E. Minervin, A.B. Shupyatsky, 'Radar Studies of Cloud Phase Composition', Izvestia USSR Ac.Sci.Physics Atm. Oc., Vol. 11, No. 9, pp. 933-944, 1966

[8] Yu.A. Melnik, ed., Radar Technologies in Earth Studies, Moscow, Sovyetskoye Radio, 1980

[9] F.G. Bass, I.M. Fuchs, Wave Scattering by Statistically Rough Surfaces, Moscow, 276 p., 1972

[10] A.I. Kalmyukov, A.S. Kurekhin, Yu.A. Lementa, I.E. Ostrovsky, V.V. Pustovoitenko, 'Particularities of Microwave Scattering by Capsizing Sea Waves', Izvestia VUZov Radiofizika XIX, Vol. 9, pp. 1315-1321, 1976

[11] S.Ya. Braude, ed., Microwave Oceanographic Studies of Sea Roughness, Kiev, Ac.Sci.Ukraine SSR, 371 p., 1962

[12] Yu.V. Melnichuk, A.A. Chernikov, 'Spectra of Radar Signals from the Sea Surface for Various Polarization States of Radar Soundings and Returns', Izvestia USSR Ac.Sci. Physics Atm. Oc., Vol. 7, No. 1, pp. 28-39, 1971

[13] Yu.V. Melnichuk, A.A. Chernikov, 'On the Backscattering Matrix for the Centimeter Range Microwaves Reflected from Rough Sea Surface', Trudy CAO, Vol. 121, pp. 58-70, 1975

[14] N.P. Krasyuk, V.I. Rosenberg, eds., Ship Radars and Meteorology, Leningrad, Sudosroyenie.q, 190 p., 1970

[15] N.P. Krasyuk, V.L. Koblov, eds., Tropospheric and Surface Effects in Radar Operation, Moscow Radio i Svyaz.,1988

[16] O.Yu. Lavrova, A.D. Rosenberg, eds., 'Experimental Studies of Ripple Modulation by Long Waves', Preprint 1428, Instit. Space Studies, USSR, Ac.Sci. (58 p.), 1988

[17] B.Sh. Lande, N.N. Putyashev, 'Dipole Model of the "Sea Surface - Low Contrast Inhomogeneous Overlaying Medium"', In: Scattering and Diffraction of Microwave Signals and Their Information Content, Higher School Coll. Works., Leningrad, SZPI, pp. 16-19, 1984

[18] V.V. Bogorodsky, D.B. Kanareykin, A.I. Kozlovz, Polarization of Scattered and Emitted Microwave Radiation from the Earth Surface, Leningrad, Gidrometeoizdat, 1981

[19] L.A. Dinevich, A.B. Shupyatsky, 'Polarization States of Hail Clouds in Molfavia', Trudy CAO, Vol. 126, pp. 14-124, 1976

[20] A.B. Shupyatsky, L.A. Dinevich, R.P. Tychina, 'Remote Sensing of Hail in Clouds from Polarization Properties of Radar Returns', In: All-Union Conf. Radar Meteorology, Moscow, Gidrometeoizdat, pp. 101-103, 1981

[21] A.B. Shupyatsky, L.A. Dinevich, R.P. Tychina, 'Remote Sensing of Hail in Clouds from Polarization Properties of Radar Returns', Trudy CAO, Vol. 121, pp. 18-27, 1975

[22] A.I. Korotov, A.V. Kochin, A.E. Myachkov, Yu.I. Rudenko, A.B. Shupyatsky, 'Automated Radar Facility for Studying Clouds and Precipitation', Abstracts, Technical Mean for State System of Control of Natural Environment, Obninsk, Part II, 293 p., 1981

[23] V.Kh. Korban, N.F. Pavlov, V.D. Stepanenko, 'Radar Studies of Rain and Hail Clouds by Amplitude Differential Techniques', In: Radar Meteorology: Proc. Socialist Countries Methodological Center Radar Meteorology, Leningrad, pp. 79-86, 1987

[24] N.N. Badulin, A.P. Batsula, E.B. Kulsheneva, S.P. Lukyanov, E.V. Masalov, V.N. Tatarinov, 'Experimental Studies of Anisotropy of Radar Signals Scattered by Clouds and Precipitation', Izvestia USSR Ac.Sci. Physics Atm. Oc., Vol. 20, No. 6, pp. 505-510, 1984

[25] V.K. Zavirukha, 'On the Possibility of Complete Retrieval of Polarization Characteristics of Meteorologic Targets from Single Channel Radar Observations', Trudy GGO, Vol. 451, pp. 1201-1208, 1982

[26] V.K. Zavirukha, 'On Optimal Technique of Determining Polarization Properties of Meteorological Targets', Trudy GGO, Vol. 440, pp. 21-24, 1985

[27] M.Yu. Pashkevich, O.S. Bogomolov, Kh.M. Lalov, Yu.P. Streltsov, 'Some Results from Assessing Physical Parameters of Thunderstorm-Hail Processes from Polarization Measurements', Abstracts - All Union Conf. Wea. Modification, Kiev, 121, 1987

[28] A.V. Kochin, V.R. Megalinski, 'Polarization Studies of Cu Cong Clouds', Proc. VI All-Union Conf. Radar Meteorology, Leningrad, Gidrometeoizdat, pp. 240-242, 1984

[29] A.V. Kochin, V.R. Megalinski, 'Application of Scalar Feed Horns to Weather Radar Antennas', Abstracts, VII All-Union Conf. Radar Meteorology, 81, 1986

[30] A.V. Kochin, 'Studies of the "Melting Layer" by a Polarization Radar', Abstracts VII All-Union Conf. Radar Meteorology, 94, 1986

[31] S.M. Vaksenberg, V.B. Zhuravlev, Yu.A. Melnik, D.B. Kanareykin, A.V. Ryzhkov, 'Radar Complex for Measuring Differential Reflectivity of Clouds and Precipitation', Trudy GGO, Vol. 526, pp. 132-135, 1988

[32] V.B. Zhuravlev, N.A. Rybakova, A.V. Ryzhkov, 'On Statistical Accuracy of Determining Differential Reflectivity', Trudy GGO, Vol. 526, pp. 23-29, 1988

[33] V.E. Minervin, A.B. Shupyatsky, 'Polarization Properties of Radar Signals from an Atmospheric Cloud of Artificial Particles', Trudy CAO, Vol. 102, pp. 89-101, 1971

[34] A.B. Shupyatsky, 'Relationship between Polarization Properties of Radar Signals and Certain Parameters of the Sea Surface', Reprints All-Union Workshop Remote Techs. Measure. Sea Parameters, Moscow, Gidrometeoizdat, 1975

[35] A.B. Shupyatsky, S.P. Morgunov, 'Application of Elliptically Polarized Microwaves to Cloud and Precipitation Studies', Doklady USSR Ac. Sci., Vol. 140, Vo. 3, pp. 591-594, 1961

[36] A.B. Shupyatsky, A.I. Korotov, R.S. Pastushkov, 'Radar Studies of Evolution and Microstructure of Cloudiness in the Eastern Atlantic', Proc. Interagency Expedition TROPEX-74, Vol. 1, Atmosphere, pp. 508-514, 1976

[37] K.G. Tskhakaya, A.B. Shupyatsky, 'Application of Radar Signals Depolarization to Studies of Cloud and Precipitation Microstructure in Bistatic Regime', Izvestia USSR Ac.Sci. Physics Atm. Ocean., Vol. 11, pp. 1227-1230, 1974

[38] G.I. Morgunova, A.B. Shupyatsky, 'On the Relationship Between Depolarization, Differential Reflectivity of Radar Signals and Microstructure of the Reflecting Particles', Abstracts VII All-Union Conf. Radar Meteorology, Moscow, 103, 1986

[39] G.K. Zagorin, N.I. Kozhevnikova, A.B. Shupyatsky, 'Retrieval of the Crystal and Mixed Phase Clouds Microstructure from Polarization Characteristics of Radar Returns', Trudy CAO, Vol. 168, pp. 37-49, 1989

[40] S.Yu. Matrosov, 'On Applicability of the Rayleigh Approximation to Calculating Differential Reflectivity of Liquid Precipitation', Trudy GGO, Vol. 526, pp. 40-43, 1989

[41] Yu.M. Gershenzon, A.B. Shupyatsky, 'Scattering of Elliptically Polarized Microwaves by Atmospheric Non-precipitating particles', Trudy CAO, Vol. 36, 1961

[42] R.P. Tychina, 'Polarization Studies of Hydrometeor Orientation', Proc. V. All-Union Conf. Radar Meteorology, Moscow, Gidrometeoizdat, pp. 101-102, 1981

[43] A.V. Kochin, 'Retrieval of Rain Droplet Size Distribution by Differential Reflectivity Technique', Abstracts VII All-Union Conf. Radar Meteorology, Moscow, 105, 1986

[44] L.A. Dinevich, I.P. Kapitalchuk, D.K. Shtivelman, A.B. Shupyatsky, 'Retrieval of Cloud and Precipitation Microphysical Characteristics Using a Double-Polarization Radar', Collected Sci. Works Moldavian Weather Modif. Serv., Shtiintsa, Kishinev, pp. 5-18, 1989

[45] A.B. Shupyatsky, F.I. Yanovsky, 'Application of Polarization and Statistical Properties of Radar Signals to Identification of Cloud Zones Dangerous for Aircraft Navigation', Abstracts, All-Union Tech. Sci. Conf. "Statistical Techniques in Theory of Receival and Transformation of Informational Signals", KIIGA, Kiev, 68 p., 1988

[46] A.A. Chernikov, A.B. Shupyatsky, 'Polarization Properties of Radar Returns from "Clear Air"', Izvestia USSR Ac. Sci. Physics Atm. Oc., Vol. 2, pp. 136-143, 1967

[47] A.A. Chernikov, Radar Returns from "Clean Air", Leningrad, Gidrometeoizdat, 1979

[48] V.K. Zavirukha, V.A. Sarychev, V.D. Stepanenko, Yu.N. Shehepkin, Trudy GGO, Vol. 395, pp. 40-46, 1977

[49] A.V. Ryzhkov, 'Invariant Polarization Characteristics of Meteorological Radar Targets', Trudy GGO, Vol. 490, pp. 11-16, 1985

[50] Yu.A. Melnik, A.V. Ryzhkov, 'Presentation of Meteorologic Radar Targets of Various Phase Composition on the Poincaré sphere', Trudy GGO, Vol. 490, pp. 17-20, 1985

# METEOROLOGIC RADAR POLARIMETRY IN NORTH AMERICA DURING 1950-1991

Eugene A. Mueller and Venkatachalam Chandrasekar
Colorado State University
Fort Collins, CO 80523

**ABSTRACT:** A brief summary of early polarimetric radar development pertinent to radar meteorology in North America over the past forty years (1950-1991) is presented. Rather than providing an in-depth assessment, only important events leading to major advances are succinctly reviewed by providing the pertinent references.

## I. INTRODUCTION

This paper is intended to outline some of the important polarimetric research that took place in North America during the first forty years of applying polarization diversity radars to the problems of meteorology (1950-1991)[1]. There was obviously polarimetric research in other fields of radar applications which influenced the development of the meteorologic radar applications and only the more important contributions in these areas are mentioned. There was also considerable important early work accomplished by researchers in countries other than in North America and this work is reported in accompanying papers [0-1, 0-3, 0-4, IV-1, VI-6, and VIII-1] of these proceedings [2]. It is also true that much of the work has been classified [VI-1, VI-7, VI-5] and has not yet appeared in the open literature and thus is not part of this review.

Furthermore, this is not an exhaustive reference of works that have been accomplished, but instead represents to the authors some chronologic developments that have led to the present widespread usage of polarimetric parameters in the areas of cloud physics, severe weather identification, and rainfall measurement in North America, and we here refer in particular to the Proceedings of the recent Battan Memorial and 40th Anniversary Workshop 1990 on "Radar in Meteorology" [1]. In these areas there has been significant development of polarimetric techniques which emphasizes the scientific need for a modern fully polarization-diverse coherent instrumentation radar system as described, for example, in various papers of these proceedings [VI-1, VIII-1]. Unfortunately, an advanced completely coherent, fast switching polarization-diverse Meteorologic Radar System is not available so far in North America for the Scientific Community at large.



*W.-M. Boerner et al. (eds.), Direct and Inverse Methods in Radar Polarimetry, Part 1, 85–97.*

## II. HISTORIC DEVELOPMENT OF POLARIMETRIC RADAR METEOROLOGY

In order to provide relevant references of the chronologic development of radar, the prehistory of polarimetric radar (circa 1886-1950) is first perused, and then followed by succinct summaries on the subsequent subperiods of its recent developmental phases.

### 2.1 Early Development of Radar: 1886-1950

The use of electromagnetic waves for remote object "radio-location" (ranging) dates back to the fundamental experiment of Heinrich Hertz (1886-1888) [2]; and, the first true radar ranging experiments for ship and major obstacle detection commenced with Hulsmeyer's demonstration (1902-1904) resulting in the first patent as reviewed in Pritchard [3]. The use of radio-to-microwave frequencies for ship and aircraft detection and the design of the first radars were accomplished concurrently in Europe, America and also in Asia mainly during the Twenties and Thirties, and rapidly advanced before, during, and immediately after World War II (1937-1945) by all the competing factions. For the interested reader, we refer to various most recent historic reviews on the subject matter and especially to Guerlac [4], Burns [5], Swords [6], Price [7], Pritchard [3], Zahl [8], Knowles-Middleton [9] and Allison [10] just to mention a few. Recently, Skolnik [11] assessed the historic radar literature, highly praising Guerlac's and Burn's expositions [4,5]; and here we also refer to Skolnik's Radar Handbook [12] for additional references.

Although some mention is made about applications of radar to meteorology and to polarization diversity in earlier studies (1900-1955) [3,4,5], such polarimetric radar systems were truly developed well after World War II (circa 1950 to 1990), and in this paper major emphasis will be given to the development of polarization diversity radars for meteorological applications during 1950 to 1991 mainly in North America. Separate historic treatises do not exist on the development of polarimetric radar, and we refer to isolated monographs contained in [1] and in the Proceedings of the three Polarimetric Radar Technology Workshops at the Missile Laboratory of Redstone Arsenal, Al, organized by Lloyd W. Root (and coworkers), [13,14,15] as summarized in the Polarimetric Technology Handbook [16]. Also we refer to the Proceedings of the two NATO Advanced Research Workshops, the NATO-ARW-IMEI'83 [17] and the recent NATO-ARW-DIMRP'88 [2], organized, directed and edited by Wolfgang-M. Boerner, one of the international crusaders for polarimetric radar theory, metrology and technology [18].

#### a) Analytic Studies

Some of the earliest theoretical work in polarization was that of Kerker and Hitschfeld [19]. In this work the primary emphasis was on the depolarization of the electromagnetic wave by non-spherical targets. The prior work of Gans [20] was the basis of their investigations. A paper by Atlas, Kerker and Hitschfeld [21] containing an extension of this work, as applied to radar scattering by meteorologic targets, was published in 1953. These were all theoretical investigations of the expected value of the depolarization of linear electromagnetic waves. The work of Gans [20] was extended by Stevenson [22] to a higher degree of approximation. Stevenson's work was used by Mathur and Mueller [23,24] to better determine the effect of polarization on the co-polar returns of prolate and oblate water spheres, as well as the depolarization effects of tilted oblate and prolate spheroids.

**b) Experimental Studies**

The earliest experimental work in polarization studies of meteorologic phenomena was conducted at the Massachusetts Institute of Technology in the mid 1950's. This work was reported on by Newell, Geotis, Stone and Fleisher [25,26]. In these investigations they describe a cancellation as a comparison between horizontal linear polarization return and circular polarization return. They found that the effects of the shapes of scatterers were significant in snow and in the bright band, but rather surprisingly not so in rain. The present authors believe that the rain in which their data was taken was either quite light or the sensitivity of their instrumentation was not sufficient to note the values of differential reflectivities that are presently measured by modern polarization systems even for moderate rainfall. There was considerable interest in the determination of the particle makeup of the bright band in this era and there were a number of papers devoted to depolarization effects and the interpretation of these effects in the determination of the makeup of the bright band [27,28,29]. These studies used the cancellation ratio as defined by Newell. At the time the composition of the particles responsible for the scattering enhancement in the bright band was not clearly understood and the added information of the shape was responsible for the understanding of this phenomena. A paper by Newell [30] demonstrated the use of this early definition of the cancellation ratio to the observation of hail. In this particular case the hail was very irregular in nature and produced a very low value of the cancellation ratio of around 7 db.

In a remarkable set of reports by Kennaugh in the early 50's, many of the fundamental theories of radar polarization were determined and also, in principle, experimentally verified [31,32]. These included the ideas of optimum co- and cross-polarization nulls and maxima of isolated and also distributed radar targers, theoretic studies of polarization properties of both simulated and natural rain, and the procedure of investigating the properties of rain by use of multiple transmitted and received polarizations. It is unfortunate that much of this excellent work has been overlooked because of the lack of wide distributions of these reports. Recently many of these reports have been re-edited by Moffatt [32] and there are additional papers published in [2,18], to which the reader is directed.

## 2.3 The decades of the 1960's to early 1970's

The latter part of the 1950's and the early part of the 1960's saw a diminishing of interest in the meteorologic usage of polarization. This may have been largely a result of what appeared to be concentration of research in other areas such as the usage of Doppler radar and more sophisticated radar signal processing [2]. Interest in polarization studies was revived by work in Canada by McCormick [33] who developed a radar system that was capable of simultaneously recording both the co-polar and cross-polar returns from a circularly polarized transmitter. This permitted a measurement of the circular depolarization ratio which was related to the detection of hail by Barge, Humphries and McCormick along with others [34,35,36].

McCormick, Hendry and their colleagues at NRC developed extensive theoretic formulations and associated measurements to study precipitation using a polarized radar systems at S-band using a circular polarization basis [37-41]. Based on their research they concluded that raindrops tended to fall with their symmetry axis vertical and the degree of correlation between the main orthogonal components was high for rain. One of the main hindrances to circular polarization observations is the effect of the propagation medium on the measurements. Under circular polarization basis the polarization state incident on the scatterer is perturbed by the differential propagation phase in the intervening precipitation medium [42,32]. Furthermore, the principal co-polarization nulls for rain cluster at circular polarization states (optimal rain clutter suppression [32]). This resulted in a shift towards using horizontal and vertical linear polarization transmit states.

## III. THE DECADE OF THE 1970's AND 1980's

Using the fact that raindrop oblateness is related to size, and that drops are highly oriented, Seliga and Bringi [43] showed on theoretic grounds that the median volume diameter could be related to the differential reflectivity $Z_{DR}$. They also proposed a scheme whereby the $Z_{DR}$ could be measured by switching the transmitted polarization state between horizontal (H) and vertical (V) states on a pulse-to-pulse basis and receiving the copolar signal via the same receiver and processer chain. This method was not susceptible to the propagation effect and soon became very popular (see Appendix). The fact that $Z_{DR}$ could be measured in rainfall and in the range predicted by theory was first shown conclusively using data from the Illinois State Water Survey's CHILL Radar [44], which used sequential block switching between H and V states rather than pulse-to-pulsing switching. The first fast polarization switching scheme to switch the polarization state of the transmitted wave on a pulse-to-pulse basis was implemented using a rotating waveguide switch at the Rutherford Appleton Laboratory in the United Kingdom [45]. Subsequently many radars in the U.S. implemented the fast switching scheme suggested by Seliga and Bringi [43] to measure $Z_{DR}$. The first radar to implement pulse to pulse fast switching measurements was the CHILL radar and the first measurements were taken during CCOPE (Cooperative Convective Precipitation Experiment) conducted in the vicinity of Miles City, Montana during the summer of 1981 [28-46]. Subsequently the CP-2 radar and the CIMMARRON radar were modified to measure $Z_{DR}$ with a fast polarization switch in the years 1983 and 1986, respectively. Since the first implementation of the fast polarization switch in CHILL and CP-2 several experiments were conducted which led to the discovery that $Z_{DR}$ can be used to detect hail [47]. In 1984 the CP-2 radar was upgraded to measure the linear depolarization ratio ($L_{DR}$) at X-band. Most of the radars that were modified to make polarization measurements were already capable of making Doppler measurements, and this resulted in the simultaneous measurement of differential phase ($\phi_{DP}$) in addition to $Z_{DR}$. Mueller's [48] algorithm to estimate $\phi_{DP}$ is currently used as the standard to estimate $\phi_{DP}$. Time series data from the CIMARRON radar at NSSL (National Severe Storms Laboratory) were used to obtain estimates of the $\phi_{DP}$ simultaneously with $Z_{DR}$ [49]. Mueller's algorithm has been successfully implemented in other radars like CHILL and CP-2. As a parallel development the dual polarized CHILL radar was made a national facility by the National Science Foundation starting 1985. Currently there are approximately half a dozen active polarimetric radars in North America that are used for meteorologic applications and we refer to the table in the appendix comparing some twenty-one polarimetric radar systems currently in use for meteorologic radar studies worldwide (repeated from [50]). From the ones mentioned in the above text a partial list of polarimetric radars that are currently being used in North America, as given in this table, are the S-band polarimetric radar at the Alberta Research Council, the S-band polarimetric radar at AFGL (Air Force Geophysics Laboratory), Spandor located at the NASA Wallops Island Research facility in Virginia, the X-band polarimetric radar operated by the NOAA(National Oceanic and Atmospheric Administration) and the X-band polarimetric radar at New Mexico Institute of Mining and Technology. The reader is referred to the paper by Bringi and Hendry [50] for further details as provided in [1].

## IV. RECENT THEORETIC AND EXPERIMENTAL DEVELOPMENTS

The continued advancement of polarimetric radar technology must utilize the more broadly based theoretic work dealing with the target scattering matrix and the coherency matrix [51,52,53]. These theoretic developments are very important for polarimetric radar measurements and interpretation. The relationship between the coherency matrix and the properties of precipitation particles was described by McCormick and Hendry [37]. The cross-correlation between orthogonally polarized components was emphasized by Poelman [55,56] who has advanced the basic studies of Huynen [56]. Optimal polarization and their representation by the Huynen Fork [57] on the Poincare Sphere, have been suggested for meteorologic applications by Boerner and El-Arini [58], Nespor [59], Agrawal and Boerner [60], and more recently also by McCormick [61]. The extensive development in polarimetric radar was also accompanied by microphysical modelling efforts to interpret the polarimetric data. The reader is referred to the review article by Jameson and Johnson [62] for details as regards cloud microphysics.

## V. SUMMARY AND CONCLUSIONS

There have been many theoretic investigations by Kennaugh, Huynen, Boerner and others in the usage of scattering matrices and the various other partial polarization matrix representations of electromagnetic waves which have not been well presented in this paper due to the current limitations of these papers in the application to existing meteorological radars. The absence of practical applications of these concepts is regrettable. In defense of the lack of the practical demonstration of some of these more advanced polarimetric measurements is the lack within the United States of a fully polarization diverse Doppler radar for general usage among investigators. This lack of available equipment has been due in large measure to the lack of funding for polarimetry in radar investigations of the atmosphere and in oceanography.

At present the only polarization radar facility available for general usage by researchers in the meteorological field is the venerable CHILL system which is funded by the National Science Foundation and now operated by Colorado State University at Fort Collins, Colorado.

There have been extensive investigations of polarization theory and measurements by a number of investigators around the world. In particular the recent work done by the Deutsche (Forschungs-und Versuchsanstalt fur) Luft and Raumfahrt (DLR) should be mentioned as their radar was specially designed for work in meteorology [63]. There is a discussion on the field of polarization diversity in the recent American Meteorological Society publication "Radar in Meteorology" by Seliga, Humphries and Metcalf. The reader interested in a more complete synopsis of the field is encouraged to refer to this article [64].

## ACKNOWLEDGEMENTS

In preparing this historic review we relied upon input from many of our expert colleagues, and especially by Professors V. N. Bringi and Thomas A. Seliga as well as by Drs. Robert G. Humphries and Dusan S. Zrnic who all contributed profoundly to the chapters on polarimetric radar meteorology [50,69,70] in the recent AMS publication "Radar in Meteorology". In addition, we have received constructive advice and corrections from Professor Wolfgang-Martin Boerner of the University of Illinois at Chicago. This paper was skillfully typed by Ms. Betty Rinker.

We also would like to acknowledge the support of the National Science Foundation under Grant Number ATM-9019596 and Cooperative Agreement Number ATM-8919080.

## REFERENCES

[1] D. Atlas, (1990) Radar in Meteorology: Battan Memorial and 40th Anniversary Radar Meteorology Conference, Boston: Am. Meteor. Soc.

[2] W-M. Boerner, et al., (1991): eds. Direct and Inverse Methods in Radar Polarimetry, Proc. NATO-ARW-DIMRP'88, Bad Windshiem FRG, Sept. 18-24, NATO ASI, Mathematics and Physical Science, Vol. C., Dordrecht, NL: D. Reidel Publ. Co.

[3] Dr. Pritchard, (1989): The Radar War, Wellingborough/North Hampshire, England. Patrick Stephens Ltd., Thorsons Publ. Group.

[4] H. E. Guerlac, (1987): Radar in World War II, in "The History of Modern Physics 1800-1950," The American Institute of Physics, New York: Tomash Publishers.

[5] R. W. Burns, (1988): Radar Development to 1945, London, UK; IEE Publ./Series No. 2 (Radar Sonar Navigation & Avionics: ISBN 0 86341 1398).

[6] S. S. Swords, (1985): Technical History of the Beginnings of Radar, London, UK: IEE Publications (ISBN 0863410219).

[7] A. Price, (1985): Instruments of Darkness, The History of Electronic Warfare, New York: Charles Scribners's Sons Publ.

[8] H. A. Zahl, (1987): Radar Spelled Backwards, Vantage Press.

[9] W. E. Knowles Middleton, (1981) Radar Development in Canada: The radio branch of the National Research Council of Canada: 1939-1946; Waterloo, Ontario Canada: Wilfred Laurier University Press.

[10] D. K. Allison, (1981): New Eye for the Navy: The origin of radar at the US Navy Naval Research Laboratory, Washington, DC: NRL Publ. (Rept. 8466) (Sept. 29).

[11] M. I. Skolnik, (1987): Detecting Radar's Development: Radar in World War II, Reviews, The Scientist, October 19, p. 22.

[12] M. I. Skolnik, (1990): ed., Radar Handbook, 2nd edition, New York, NY: McGraw-Hill.

[13] L. W. Root, Chairman, (1981): First Workshop on "Polarimetric Radar Technology," 1980, June 25-26, U.S. Army Missile Command, Redstone Arsenal, AL, Proceedings Chicago, IL: GACIAC-11TRI, Dept. PR-81-02, Feb. 1982.

[14] L. W. Root, Chairman, (1983): Second Workshop on "Polarimetric Radar Technology," 1983, May 3-5, U.S. Army Missile Command, Redstone Arsenal, Proceedings, Chicago, IL: GACIA-11tri, PR-83-01, August.

[15] L. W. Root and B. L. Matkins, Co-Chairmen/Editors, (1990): Proceedings (Third) Polarimetric (Technology) Workshop, U.S. Army Missile Laboratory, Redstone Arsenal, AL, 1988, Aug. 16-18, Chicago, IL GACIAC-IITRI, May.

[16] J. W. Battles and L. W. Root, eds., (1991): Polarimetric Technology Handbook, Chicago, IL: GACIAC-TRI, July.

[17] W-M. Boerner, et al., eds. (1985): Inverse Methods in Electromagnetic Imaging, Proc. NATO ARW-1MEI'88, FRG, 1983, Sppt. 18-24, NATO ASI Series C, Math & Phys. Sci., Vol. C-143, Dordrecht, Holland: D. Reidel Publ. Co.

[18] W-M. Boerner, (1980): Polarization Utilization in Electromagnetic Inverse Scattering, Chapt. 7, in Inverse Scattering Problems in Optics, Edit. by H. P. Baltes, Vol. 2, Topics in Current Physics, Vol. 20, Heidelberg: Springer Verlag, July, pp. 237-305.

[19] M. Kerker and W. Hitschfeld, (1951): Effects of Particle Shape and Secondary Scattering on Microwave Reflections from Clouds and Precipitation, Scientific Report MW-1, Stormy Weather Group, McGill University, Montreal Canada.

[20] R. Gans, (1912): Uber die Form ultramikroskopischer Goldteilchen, Ann. Phys. 37, pp. 881-900.

[21] D. Atlas, M. Kerker and W. Hitschfeld, (1953): Scattering and Attenuation by Non-spherical Atmospheric particles, J. Atmos. and Terest. Phys., V. 3, pp. 108-119.

[22] A.F. Stevenson, (1953): Electromagnetic Scattering by an Ellipsoid in the Third Approximation, J. Appl. Phys., V. 23, pp. 102-108.

[23] P. M. Mathur and E.A. Mueller, (1955): Radar Backscattering from Non-Spherical Scatterers, Rept. of Investigation No. 28, Illinois State Water Survey Division, University of Illinois, Urbana, Illinois.

[24] P.M. Mathur and E. A. Mueller, (1956): Radar Backscattering Cross Sections for Non-Spherical Targets, IRE Transactions on Antennas and Propagation V. AP-4, pp. 51-53.

[25] R.E. Newell, S. G. Geotis, M.L. Stone, and A. Fleisher, (1955): How Round are Raindrops? Proc. Fifth Weather Radar Conf., Fort Monmouth, New Jersey, Amer. Meteor. Soc., pp. 261-268.

[26] R.E. Newell, S.G. Geotis and A. Fleisher, (1957): The Shape of Rain and Snow at Microwavelengths, Mass. Inst. of Tech., Weather Radar Res. Rep. No. 28, 103 pp.

[27] I.M. Hunter, (1954): Polarization of Radar Echoes from Meteorological Precipitation, Nature V. 173, pp. 165-166.

[28] I. C. Brown and N. P. Robinson, (1952): Cross-polarization of the Radar Melting-Band, Nature, V. 173, pp. 165-166.

[29] R. Wexler, (1955): An Evaluation of the Physical Effects in the Melting Layer, Proc. Fifth Weather Radar Conf., Fort Monmouth, New Jersey, Amer. Meteor. Soc., pp. 329-334.

[30] R.E. Newell (1958): A Comparison of Solid Hydrometeor Shape at Wavelengths of 3 cm and 5000Å, Proceedings of Seventh Weather Radar Conference, Miami, Florida, American Meteorological Society, pp. B1-B6.

[31] E. M. Kennaugh, (1949-1954): Effects of Type of Polarization on Echo Characteristics, Reports 389-1 to 389-24, Antenna Laboratory, Ohio State University, Columbus Ohio.

[32] E. M. Kennaugh, (1985): Polarization Properties of Radar Reflectors, M. Sc. Thesis (1952), Department of Electrical Engineering, The Ohio State University, Columbus OH 43212: Together with two bound volumes of pertinent parametric radar reports (1949-1954/1956-1958: Rep. 381-1 to 394-24), edited by Prof. David L. Moffatt, OSU-ESL, Columbus, OH.

[33] G.C. McCormick, (1968): An Antenna for Obtaining Polarization Related Data with the Alberta Hail Radar, Proc. of the 13th Radar Meteor. Conf., Montreal Canada, American Meteor. Soc., pp. 225-230.

[34] B.L. Barge, (1972): Hail Detection with a Polarization Diverse Radar, Sci. Rep. MW-71, Stormy Weather Group, McGill Univ., Montreal, Quebec, Canada

[35] G.C. McCormick and A. Hendry, (1970): The Study of Precipitation Backscatter at 1.8 cm with a Polarization Diverse Radar, Preprints 14th Radar Meteor. Conf., Tucson, Arizona, Amer. Meteor. Soc.225-230.

[36] R.G. Humphries, (1974): Depolarization Effects at 3 Ghz Due to Precipitation, Sci. Rep. MW-82, Stormy Weather Group McGill Univ., Montreal, Quebec, Canada.

[37] G. C. McCormick and A. Hendry, (1975): Principles for the Radar Determination of the Polarization Properties of Precipitation, Radio Sci., 10, 421-434. (Chapters 14, 19a)

[38] G. C. McCormick and A. Hendry, (1979): Techniques for the Determination of the Polarization Properties of Precipitation, Radio Sci., 14, 1027-1040 (Chapters 14, 19a).

[39] G. C. McCormick, (1981): Polarization Errors in a Two Channel System, Radio Sci., 16, 67-75 (Chapter 19a).

[40] G. C. McCormick and A. Hendry, (1974): Polarization Properties of Transmission Through Precipitation Over a Communication Link., J. Rech. Atmos., 8, 175-187. (Chapters 19a, 21a).

[41] A. Hendry and G. C. McCormick, (1974): Polarization Properties of Precipitation Particles Related to Storm Structure. J. Rech. Atmos., 8, 189-200 (Chapter 23a).

[42] T. A. Seliga and V. N. Bringi, (1978): Differential Reflectivity and Differential Phase Shift: Applications in Radar Meteorology, Radio Sci., 13, 271-275. (Chapter 14).

[43] T. A. Seliga and V. N. Bringi, (1976): Potential Use of Radar Differential Reflectivity Measurements at Orthogonal Polarizations for Measuring Precipitation, J. Appl. Meteor., 15, 69-76 (Chapters 12, 14, 19a, 19b, 23a, 29a, 29b).

[44] T. A. Seliga, V. N. Bringi and H. H. Al-Khatib, (1979): Differential Reflectivity Measurements in Rain: First Experiments, IEEE Trans. Geosci. Electron, 17, 240-244 (Chapters 14, 19a).

[45] M.P.M. Hall, S. M. Cherry, J.W.F. Goddard and G. R. Kennedy, (1980): Raindrop Sizes and Rainfall Rate Measured by Dual-Polarization Radar. Nature, 285, 195-198.

[46] V. N. Bringi, T. A. Seliga and W. A. Cooper, (1984): Analysis of Aircraft Hydrometeor Spectra and Differential Reflectivity During the Cooperative Convective Precipitation Experiment, Radio Science, 19(1), 157-167.

[47] V. N. Bringi, T. A. Seliga and K. Aydin, (1984): Hail Detection with a Differential Reflectivity Radar, Science, 225, 1145-1174.

[48] E. A. Mueller, (1984): Calculation Procedure for Differential Propagation Phase Shift. Preprints 22nd Radar Meteorology Conf., Zurich, AMS, 397-399.

[49] M. Sachidananda and D. S. Zrnic, (1986): Differential Propagation Phase Shift and Rainfall Rate Estimation, Radio Sci., 21, 235-247.

[50] V. N. Bringi and A. Hendry, (1990): Technology of Polarization Diversity Radars for Meteorology, Radar in Meteorology, Chap. 19a, AMS, Ed. David Atlas.

[51] W-M. Boerner, (1981): Use of the Optimal Polarization Concept in Electromagnetic Imaging of Hydrometeor Distributions, Proc. Workshop on Precipitation Measurement from Space, NASA Gooddard, Eds. D. Atlas and O. W. Thiele.

[52] D. Giuli, (1986): Polarization Diversity in Radar, Proc. IEEE, Vol. 74(2), pp. 245-269, Feb.

[53] W-M. Boerner, W.-L. Yan and A.-Q. Xi, (1990): Basic Equations of Radar Polarimetry and its Solutions: Proceedings of SPIE, Workshop on Polarimetry: Radar, Infrared, Visible, UV and X-Ray, 1990 Vol. 1317, pp. 16-79.

[54] H. Mott, (1986): Polarization in Antennas and Radar, John Wiley, & Sons, Inc., Englewood Cliffs, N.Y. (2nd ed. in preparation).

[55] A. J. Poelman, (1976): Cross-Correlation of Orthogonally Polarized Backscatter Components, IEEE Trans. Aerosp. Electron. Syst., AES-12, 674-681.

[56] A. J. Poelman and C. J. Hilgers, (1991): The Effectiveness of Multi-notch Logic-Product Polarization Filters in Radar for Countering Rain Clutter, Proc. IEEE, Part F, Vol. 138.

[57] J. R. Huynen, (1965): Measurement of the Target Scattering Matrix, Proc. IEEE, 53(8), 936-946.

[58] W-M. Boerner and M. B. El-Arini, (1981): Utilization of the Optimal Polarization Concept in Radar Meteorology, Preprints, 20th Radar Meteorology Conf., Boston, AMS, 656-665.

[59] J. D. Nespor, (1983): Theory and Design of a Dual-Polarization Radar for Meteorological Studies. M.S. Thesis, Univ. of Illinois at Chicago.

[60] A. P. Agrawal and W-M. Boerner, (1990): Description of a Monostatic Polarimetric Radar Model for Fluctuating Distributed Scatters, Papers V1-3 in Proc. of the Third Polarimetric Radar Technology Workshop, U.S. Army Missiles Command, Redstone Arsenal, AL, 1988 IITRI/GACIAC, Chicago, Vol. 1, part 2, pp. 143-178.

[61] G. C. McCormick, (1990): The Theory of Polarization Diversity Systems, Journal of Electromagnetic Waves and Applications, JEWA Vol. 4 (No. 8), pp. 707-725.

[62] A. R. Jameson and D. B. Johnson, (1990): Cloud Microphysics and Radar, Radar in Meteorology, Chap. 23a, AMS, Ed. David Atlas.

[63] A. Schroth, K. Tragl, E. Luneburg, M. Chandra, (1992): Polarimetric Signal Processing of Random Target Observations, European Transactions on Telecommunications, Vol. 2, in print.

[64] T. A. Seliga, R. G. Humphries, and J. I. Metcalf, (1990): Polarization Diversity in Radar Meteorology: Early Developments, D. Atlas Editor, Amer. Meter. Soc., Boston Mass.

## Appendix: Characteristics of 21 Polarization Diversity Radars

| AGENCY | RUTHERFORD APPLETON LABORTORY | NCAR CP-2 S-BAND | NCAR CP-2 X-BAND | NRC of CANADA | NRC of CANADA | DFVLR | CHILL S-BAND |
|---|---|---|---|---|---|---|---|
| GEOGRAPHIC LOCATION OF RADAR | Chilbolton, U.K. | Boulder, Colorado | Boulder, Colorado | Ottawa, Ontario | Ottawa, Ontario | Oberpfaffenhofen, F.R.G. | Urbana, Illinois |
| POLARIZATION TYPE | $Z_{DR}$, LDR | $Z_{DR}$ | LDR | CDR, LDR (φ) | CDR, LDR (φ) | CDR, LDR (φ) $Z_{DR}$ (φ) | $Z_{DR}$ |
| WAVELENGTH, cm | 9.75 | 10.7 | 3.2 | 3.1 | 1.8 | 5.3 | 10.7 |
| PEAK POWER, kW | 560 | 1200 | 200 | 100 | 40 | 500 | 1000 |
| PULSE LENGTH, μs | 0.5 | 0.15-1.2 | 1.0 | 0.8 | 0.4 | 0.5, 1, 2 | 0.25, 0.5, 1 |
| PRF, $s^{-1}$ | 610 | ≤1700 | ≤1000 | 1000 | 2000 | 490, 1200, 2400 | 400-1250 |
| ANTENNA TYPE | Centre Fed Paraboloid | Centre Fed Paraboloid | Two Cassegrainian Paraboloids | Offset Fed Paraboloid | Offset Fed Paraboloid | Offset Fed Paraboloid | Centre Fed Paraboloid |
| ANTENNA SIZE, m | 25 | 8.5 | 2.4 | 2.7 | 1.6 | 5 | 8.5 |
| FEED TYPE | Scalar Feed, Turnstile Junction | Potter Horn | Rectangular Waveguide | Potter Horn, Turnstile Junction | Potter Horn, Turnstile Junction | Corrugated Horn, OMT | Potter Horn |
| BEAMWIDTH, DEGREES | 0.25 | 0.93 | 0.94 | 0.9 | 0.89 | 1 | 0.96 |
| AZ. COVERAGE,DEGREES | 360 | 360 | 360 | 330 | 330 | 360 | 360 |
| ELEV. COVERAGE,DEGREES | 90 | 90 | 90 | 93 | 92 | 92 | 90 |
| POLARIZATIONS RADIATED | LIN-V or LIN-H | LIN-V or LIN-H | LIN-H | LHC, RHC Any Linear or Elliptical | LHC, RHC Any Linear or Elliptical | Any Elliptical, 3° Increments in Direction | LIN-V or LIN-H |
| SIMULTANEOUS 2-CHANNEL RECEPTION | N | N | Y | Y | Y | Y | N |
| POLARIZATIONS RECEIVED | Copolar or Orthogonal to Transmit | Copolar to Transmit | LIN-H & LIN-V | Copolar and Orthogonal to Transmit | Copolar and Orthogonal to Transmit | Copolar and Orthogonal to Transmit | Copolar to Transmit |
| MAX. SIDELOBE LEVEL, dB | -20 | -21 | ≈ -30 | <-30 | <-30 | <-32 | <-20 |
| MAX. ANTENNA LINEAR X-POL LOBES, dB | ≈ -32 | -21 | -36 | <-29 | <-28 | <-28 | ≈ -20 |
| ANTENNA ICR, dB (CIRCULAR POLARIZATION) | | | | ≈ 40 | ≈ 40 | | |
| PHASE & CORRELATION CAPABILITY | N | N | N | Y | Y | Y | N |
| POLARIZATON CONTROL METHOD | Mechanical, (Transmit); PIN Diode, (Receive) | Ferrite Switch | N.A. | Ferrite Switches (Transmit Path) | Manually Operated Polarizer | Ferrite Switch | Ferrite Switch |
| POLARIZATION CONTROL PERIOD OR RATE | Pulse by pulse | Pulse by pulse | N.A. | 1.2 s | <10 s | Pulse by pulse | Pulse by pulse |
| CHANNEL-CHANNEL ISOLATION (EXCLUDING ANTENNA), dB | ≈ 40 | >30 | >30 | >52 | >52 | ≈ 30 | 25 |
| DOPPLER CAPABILITY | N | Y | N | N | N | Y | Y |
| NUMBER OF RANGE GATES | 256 | 512 | 512 | 12 | 12 | 452 | 1024-4096 |
| RANGE RESOLUTION, m | 300 | >30 Typical 150 | >30 Typical 150 | 150 | 150 | 75 | 37 1/2, 75 & 150 |
| REAL-TIME VECTOR SIGNAL PROCESSING CAPABILITY | N.A. | N.A. | N.A. | N | N | Y | N.A. |
| POLARIZATION QUANTITIES MEASURED | $Z_{DR}$, LDR | $Z_{DR}$ | LDR | CDR, LDR, Cross-Corr. (Mag. & Phase) | CDR, LDR, Cross-Corr. (Mag. & Phase) | Scatt. Matrix, Complex Cross Correlation, Optimal Polarizations | $Z_{DR}$ |

Adapted from [50]

Table 1 of 3

Appendix: Characteristics of 21 Polarization Diversity Radars

| AGENCY | ALBERTA RESEARCH COUNCIL | VPI & SU | SHAPE TECHNICAL CENTRE | NSSL | AFGL | ESA & FGJ/IAS | SPANDAR |
|---|---|---|---|---|---|---|---|
| GEOGRAPHIC LOCATION OF RADAR | Penhold Alberta | Blacksburg Virginia | The Hague Netherlands | Cimarron, Oklahoma | Sudbury, Mass. | Graz, Austria | Wallops Is. Virginia |
| POLARIZATION TYPE | CDR, LDR | $Z_{DR}$ ($\phi$) | General Elliptical | $Z_{DR}$ | $Z_{DR}$ | $Z_{DR}$, Two Magnetrons | $Z_{DR}$, with Freq. Agility |
| WAVELENGTH, cm | 10.4 | 10.7 | 3.1 | 10.7 | 10 | 5.31-5.35 Freq. Agile | 10.56 |
| PEAK POWER, kW | 500 | 250 | 100 | 500 | 1000 | 250 | 1000 |
| PULSE LENGTH, μs | 1.75 | 0.8 | 0.4-2.4 | 1 | 1 | 2 | 0.5, 1.0 |
| PRF, $s^{-1}$ | 480 | 480 | 1-10,000 | 1300 | 100-1302 | 1 kHz Each Magnetron | 960 |
| ANTENNA TYPE | Centre Fed Paraboloid | Paraboloid | Cassegrainian | Centre Fed Paraboloid | Centre Fed Paraboloid | Centre Fed Paraboloid | Paraboloid |
| ANTENNA SIZE, m | 6.7 | 4 | 1.8 | 9.1 | 7.32 | 3.66 | 18.3 |
| FEED TYPE | Potter Horn, Turnstile Junction | Rotating Horn | Diagonal Horn | Potter Horn | Potter Horn | Circular Horn, Polarization Combiner | Circular Horn |
| BEAMWIDTH, DEGREES | 1.15 | 2.5 | 1.2 | 0.88 H 0.92 V | 1 | 1 | 0.4 |
| AZ. COVERAGE, DEGREES | 360 | 360 | 360 | 360 | 360 | ±365 | 360 |
| ELEV. COVERAGE, DEGREES | 21 | 90 | -3 to 65 | 90 | 90 | -10 to 95 | 90 |
| POLARIZATIONS RADIATED | LHC, RHC LIN | LIN, Stepped 22 1/2°/ Pulse Incrementable in 1/2° Steps | 28 Discrete Linear & 137 RH &LH Elliptical | LIN-V or LIN-H | LIN-V LIN-H RHC or LHC | LIN-V and LIN-H | LIN-V or LIN-H |
| SIMULTANEOUS 2-CHANNEL RECEPTION | Y | N | Y | N | N | Y | N |
| POLARIZATIONS RECEIVED | Copolar and Orthogonal to Transmit Polarization | Copolar to Transmit | Copolar and Orthogonal to Transmit Polarization | Copolar to Transmit | Copolar to Transmit | Copolar and Orthogonal To Transmit | Copolar to Transmit |
| MAX. SIDELOBE LEVEL, dB | -23 AZ, -30 EL | ≈ -20 | <-25 | <-24.5 | -27 | -28 | ≈ -20 |
| MAX. ANTENNA LINEAR X-POL LOBES, dB | | ≈ -20 | | <-25 | | -35 | |
| ANTENNA ICR, dB (CIRCULAR POLARIZATION) | ≈ 37 | | | | | | N.A. |
| PHASE & CORRELATION CAPABILITY | Y | N | Y | N | N | N | N |
| POLARIZATON CONTROL METHOD | Motor Driven Polarizer | Continuously Rotating Polarizer | Digitally Cont. Ferrite Phase Shifter | Ferrite Switch | Ferrite & Mechanical Switches | Separate H & V Transmitters | Mechanical Switch |
| POLARIZATION CONTROL PERIOD OR RATE | 30 s | Stepped 22 1/2° in 2.08 ms | Variable, up to 10 kHz Rate | Pulse by Pulse | Pulse by Pulse | Pulse to Pulse | 0.7 s |
| CHANNEL-CHANNEL ISOLATION (EXCLUDING ANTENNA), dB | >40 | | ≈ 22 dB X-Pol Isolation | ≈ 27 | >40 | >60 | ≈ 30 |
| DOPPLER CAPABILITY | N | N | Y | Y | Y | N | Y |
| NUMBER OF RANGE GATES | 148 | 255 | 8, 16, 32 | 768 | 1000 | 255 Max. | 1-450 |
| RANGE RESOLUTION, m | 1000 | 125/255 | Pulse Width$^{-1}$ | 150-255 | 150 | 300 | 75, 150 |
| REAL-TIME VECTOR SIGNAL PROCESSING CAPABILITY | N | N.A. | Y | N.A. | N | N | |
| POLARIZATION QUANTITIES MEASURED | CDR, LDR, Cross Correlation (Mag. & Phase) | $Z_{DR}$ in 4 Planes | Scattering Matrix, Complex Cross-Corr. | $Z_{DR}$ | $Z_{DR}$ | $Z_{DR}$, LDR with Pulse to Pulse Freq. Shift | $Z_{DR}$ Doppler |

Adapted from [50]

Table 2 of 3

Appendix: Characteristics of 21 Polarization Diversity Radars

| AGENCY | NOAA ERL/WPL | NOAA ERL/WPL | LABORATOIRE D'AEROLOGIE (RABELAIS) | CNR/ INSTITUTO DI ELECTRONICA | NEW MEXICO TECH | PWRI (DND) OF JAPAN | DELFT UNIV. OF TECH. |
|---|---|---|---|---|---|---|---|
| GEOGRAPHIC LOCATION OF RADAR | Boulder, Colorado | Boulder, Colorado | Lannemezan, France | Italy | Socorro, N.M. | Toyama, Japan | Delft, The Netherlands |
| POLARIZATION TYPE | CDR, LDR ($\phi$) | CDR, LDR | $Z_{DR}$, LDR CDR | $Z_{DR}$, LDR | $Z_{DR}$, LDR | $Z_{DR}$ | $Z_{DR}(\phi)$, LDR ($\phi$) |
| WAVELENGTH, cm | 0.87 | 3.22 | 0.86 | 5.5 | 3 | 5.68 | 8.9 |
| PEAK POWER, kW | 160 | 40 | 70 | 500 | 20-25 | 75 | 0.1 kW (avg.) |
| PULSE LENGTH, $\mu s$ | 0.25 | 0.5-2.0 | 0.3 | 0.5, 1.5, 3 | 1/3-4/3 | 0.5 (Doppler) 2.0 (Pol.) | 625, 1250 |
| PRF, $s^{-1}$ | Variable 2000 Avg. | Variable 1250 Avg. | 3125/2688 | 1200; 500, 250 | 1, 2, 4, 8, kHz | 1120 (Doppler) 280 (Pol.) | FM-CW |
| ANTENNA TYPE | Two Collimated Cassegrainian Paraboloids | Cassegrainian | Two Centre-fed Paraboloids | Offset Fed Paraboloid | Cassegrainian Paraboloid | Centre Fed Paraboloid | Two Centre Fed Paraboloids |
| ANTENNA SIZE, m | 1.2 | 3.1 | 1.4 | 4.6 | 3.7 | 2 | 4.2, 2.1 |
| FEED TYPE | Corrugated Horn | Corrugated Horn | Circular Horn | Corrugated Horn with Integral OMT | Corrugated Horn | Rear Feed | Corrugated Feed |
| BEAMWIDTH, DEGREES | 0.5 | 0.8 | 0.4 | 0.92 Az 1.02 El | 0.8 | 1.8 | 1.4 |
| AZ. COVERAGE, DEGREES | 360 | 360 | 360 | 360 | 360 | 360 | |
| ELEV. COVERAGE, DEGREES | 170 | 90 | 90 | -2 to 90 | 360 | 90 | |
| POLARIZATIONS RADIATED | LHC, RHC, LIN-V, LIN-H | RHC, LHC, LIN-V, LIN-H | LIN-V, LIN-H RHC, LHC | LIN-V or LIN-H | LIN-V or LIN-H | LIN-V or LIN-H | Any Linear |
| SIMULTANEOUS 2-CHANNEL RECEPTION | Y | Y | Y | Y | Y | N | N |
| POLARIZATIONS RECEIVED | RHC/LHC or Vert./Horiz. | RHC/LHC or Vert./Horiz. | Copolar & Orthogonal, Linear & Circular | Copolar and Orthogonal | Copolar and Orthogonal to Transmit | Copolar to Transmit | Any Linear Independent of Transmit Polarization |
| MAX. SIDELOBE LEVEL, dB | <-20 | ≈ -20 | ≈ -24 | < -32 | ≈ -30 | <-23 | <-23 |
| MAX. ANTENNA LINEAR X-POL LOBES, dB | ≈ -20 | | -37 | <-30 | | <-29 | <-20 |
| ANTENNA ICR, dB (CIRCULAR POLARIZATION) | ≈ 23 | ≈ 23 | 18 | | | | |
| PHASE & CORRELATION CAPABILITY | Y | Y | N | Y | Y | N | |
| POLARIZATON CONTROL METHOD | Motor Driven Polarizer | | Ferrite Switch | Ferrite Switch | Pin Diode Switch | Ferrite Switch | Mechanical |
| POLARIZATION CONTROL PERIOD OR RATE | 0.25 s | | 3125 Hz Pulse to Pulse | Pulse by Pulse | Pulse by Pulse | 1/9 Pulses (32.1 ms) | 10 ms |
| CHANNEL-CHANNEL ISOLATION (EXCLUDING ANTENNA), dB | >30 | >40 | >35 | >25 | ≈ 55 | >35 | >30 |
| DOPPLER CAPABILITY | Y | Y | Y | Y | Y | Y | Y |
| NUMBER OF RANGE GATES | 328 | 328 | 256 | 1024 Max | ≈ 375 | 80 | 64, 128 |
| RANGE RESOLUTION, m | 37.5 N, N=1,2,3, etc. | 37.5 N, N=1,2,3, etc. | 75 | 75 | 50, 100 150, 200 | 500 | 75 Typical |
| REAL-TIME VECTOR SIGNAL PROCESSING CAPABILITY | N | N | N | N | N | N | Y |
| POLARIZATION QUANTITIES MEASURED | CDR, LDR, Cross Correlation (Mag. & Phase) | CDR, LDR, Cross Correlation (Mag. & Phase) | $Z_{DR}$, CDR, LDR | $Z_{DR}$, LDR | Cross-Spectra, Coherence, $Z_{DR}$, LDR vs Doppler Freq. | $Z_{DR}$ | Relative Scattering Matrix |

Adapted from [50]

Table 3 of 3

# COMPARISON OF FUNDAMENTAL APPROACHES TO RADAR POLARIMETRY

Zbigniew H. Czyż
Telecommunications Research Institute
Poligonowa 30
00-991 Warsaw
Poland

ABSTRACT. Comparison of different definitions of polarization state of EM waves in their amplitude and power representations leading to different transmission equations and transformation rules has been presented. Special attention is given to the problem of transformation of non-depolarizing bistatic radar scattering and Mueller propagation matrices by choosing their characteristic polarization bases. Inversion, rotation and Lorentz transformation of polarization vectors are considered.

## 1. INTRODUCTION

There are many approaches to radar polarimetry possible. For instance, you may apply amplitude or power approaches depending on whether you will use Jones, u , or Stokes, P , polarization vectors. Also, you may describe these vectors in terms of the so called analytical, $2\delta$ , $2\gamma$ , $2\varepsilon$ , or geometrical, $2\tau$ , $2\psi$, $2\chi$ parameters expressing angular coordinates of the polarization point on the Poincaré sphere. The first choice is more convenient for change-of-basis transformations, whereas, the second is appropriate for representations in linear or circular polarization bases. To describe polarization properties of scattering objects, e.g. in power approach, you may use the Mueller [1] or radar power scattering matrices [2, 3] , target correlation [4], or 4x4 matrices transforming the so called "null vectors" [5] which are complex and preserve information about the phase.

All these approaches can be applied simultaneously without causing any confusion as long as only one direction of propagation, say along positive direction of Oz axes of all Oxyz coordinate systems involved, is considered. This was always the case during the early studies of electromagnetic polarimetry when opticians examined propagation of polychromatic light waves passing through transparent media. But the same physical process can be considered as propagation or scattering depending on whether the Oz axis, along which the outgoing wave is travelling, has been directed to or from the observer receiving the wave.



W.-M. Boerner et al. (eds.), Direct and Inverse Methods in Radar Polarimetry, Part 1, 99–116.

Radar engineers were dealing with the problem of scattering the mono-or quasimonochromatic waves and used antennas as measuring instruments. So they had to redevelop the whole theory. As a consequence, different approaches, from that point of view also, appeared in the literature. It seems advisable to compare them, and to suggest some unified formulations.

The differences begin in definitions of fundamental polarimetric quantities such as state of polarization /SP/ or target operator /TO/. It is obvious that, in order to define a vector or an operator, it is not enough to present their components but also indicating rules for their change-of-basis transformations is required. And these rules cannot depend on reversal of the coordinate system. Nevertheless, because of the need to compare the SP of two waves propagating in opposite directions, or of two antennas operating in a direct transmission system, two different definitions of the SP are still possible. The two approaches related to these definitions will be called "polarization ellipse" /PE/ and "polarization helix" /PH/ approach.

## 2. POLARIZATION ELLIPSE AND POLARIZATION HELIX APPROACHES

### 2.1. Amplitude definition of the SP in the PE approach

The first approach, PE, relates the SP to the unit complex amplitude of the wave, independently of its direction of propagation [6]. According to such definition, if two waves travelling in opposite directions have the same SP expressed by the unit Jones vector $u_p$, with P denoting the polarization point on the Poincaré sphere, then these waves can be presented as

$$E^{\pm} = u_p \; e^{j(\omega t \mp kz)} \tag{1}$$

Because of the one-to-one correspondence between the complex amplitude $u_p$ and the polarization ellipse, the name of the approach seems to be evident.

The transformation of the $u_p$ components under: $1^o$ - reversal of the coordinate system Oxyz by rotation about, say, vertical Oy axis by $180^o$, and $2^o$ - change of orthogonal polarization basis, can be presented for both directions of propagation by the succesive equations

$$u_p^o = C^o u_p \tag{2}$$

$$u_p' = \widetilde{C}^* u_p \tag{3}$$

with unitary transformation matrices: of reversal

$$C^{o} = \begin{bmatrix} -1 & 0 \\ 0 & 1 \end{bmatrix} \tag{4}$$

and of change-of-basis

$$C^{o} = \begin{bmatrix} a & -b^{*} \\ b & a^{*} \end{bmatrix} ; \quad a\,a^{*} + b\,b^{*} = 1 \tag{5}$$

where a and b are e.g. Cayley-Klein rotation parameters

$$\begin{aligned} a &= \cos\gamma \;\; e^{-j\,(\delta+\varepsilon)} \\ b &= \sin\gamma \;\; e^{+j\,(\delta-\varepsilon)} \end{aligned} \tag{6}$$

## 2.2. Amplitude definition of the SP in the PH approach

The idea of the second approach comes from consideration of the equation of the received voltage $V_r$ at the output of the receiving antenna of polarization given by the unit Jones vector $u^P$, matched to the polarization of the incoming wave of complex amplitude $u^{P*}$:

$$V_r = \tilde{u}^P \, u^{P*} = 1 \tag{7}$$

Following identification of the SP of the receiving antenna with the SP of the wave the antenna would radiate on transmit, two equations for the oppositely propagating waves of the same polarization instead of one equation(1) should now be written:

$$E^{+} = u^{P} \, e^{\,j(\omega t - kz)}$$

and (8a)

$$E^{-} = u^{P*} e^{\,j(\omega t + kz)} \tag{8b}$$

In these equations the upper index P, instead of the lower P in the previous approach, is thought to be a mean of distinguishing between the two vectors $u_p$ and $u^p$. They represent the same SP and also the same complex amplitude of the $E^+$ waves in (1) and (8a). But in (8b) $u^p$ denotes the SP vector which is complex conajugate in relation to the complex amplitude $u^{p*}$ of the wave $E^-$. That way the SP vector has been defined in the PH approach as different entity than complex amplitude.

The name of the approach becomes evident when considering real parts of the two equations (8a) and (8b) as two same helices which are moving, without rolling, along the Oz axis in both directions. For the time t = 0 they are coinciding:

$$\mathrm{Re}\,(u^{P} e^{-jkz}) = \mathrm{Re}\,(u^{P*} e^{jkz}) \tag{9}$$

thus representing not only the same polarization but also the same /spatial/ phase. That is why the $u^{P}$ SP will also be called the polarization and phase /PP/ vector.

The reversal and change-of-basis transformations should take the form

$$u^{Po} = C^{o}\, u^{P*} \tag{10}$$

and

$$u^{P,} = \widetilde{C}^{*}\, u^{P} \tag{11}$$

The difference in transformations (10) and (2) will soon become clear when considering the direct /one-way/ transmission equations.

## 2.3. Comparison of amplitude direct transmission equations for the two approaches: PE and PM

Following transformations (3) and (11) for the case when the receiving antenna is not necessarily matched to the incoming wave, we obtain for the two approaches,

$$\text{PE:} \qquad V_r = \widetilde{u}_R\, u_S = \widetilde{u}'_R\, \widetilde{C}\, C\, u'_S \tag{12}$$

$$\text{PM:} \qquad V_r = \widetilde{u}^{R}\, u^{S*} = \widetilde{u}^{R,}\, u^{S,*} \tag{13}$$

where for $u^{R} = u_R$ and $u^{S*} = u_S$ we have

$$u^{S,*} = \widetilde{C}\, C\, u'_S \tag{14}$$

Thos evident result has been shown e.g. in [6]. It is worth noticing that the hermitian product of two PP vectors appeared in (13) and preserved its form after change-of-basis.

The incoming wave can also be considered as radiated by the antenna previously oriented along the positive Oz direction and then reversed to allow for transmission.

Following the reversal and change-of-basis transformations similar forms will be obtained as (12) and (13) for the two approaches,

$$\text{PE:} \qquad V_r = \tilde{u}_R\, u_T^o = \tilde{u}'_R\, \tilde{C}\, C\, u_T^{o\prime} = \tilde{u}'_R\, \tilde{C}\, C^o\, C\, u'_T \tag{15}$$

$$\text{PH:} \qquad V_r = \widetilde{u^R}\, u^{To*} = \widetilde{u^{R\prime}}\, u^{To*\prime} = \widetilde{u^{R\prime}}\, \tilde{C}\, C^o\, C\, u^{T\prime} \tag{16}$$

The additional last expressions in (15) and (16) confirm obeying the reciprocity condition, thus justifying postulated transformations. Moreover, the change-of-basis for $C^o$

$$C^{o\prime} = \tilde{C}\, C^o\, C \tag{17}$$

identical for both approaches PE and PH can be observed.

## 2.4. Comparison of amplitude equations for transmission through the intervenient medium for two approaches: PE and PH

In both approaches the target operator can be represented by the scattering, A, or propagation, $A^o$, matrices related to each other in the linear polarization basis by the equation

$$A^o = C^o A \tag{18}$$

Complex amplitudes of the outgoing wave and corresponding transmission equations then are -

$$\text{PE, scattering:} \quad A\, u_T = u_S\,, \qquad V_r = \tilde{u}_R\, A\, u_T \tag{19}$$

$$\text{PE, Propagation:} \quad A^o\, u_T = u_S^o\,, \qquad V_r = \widetilde{u_R^o}\, A^o\, u_T \tag{20}$$

$$\text{PH, scattering:} \quad A\, u^T = u^{S*}\,, \qquad V_r = \widetilde{u^R}\, A\, u^T \tag{21}$$

$$\text{PH, propagation:} \quad A^o\, u^T = u^{So}\,, \qquad V_r = \widetilde{U^{Ro*}}\, A^o\, u^T \tag{22}$$

After change-of-basis, the amplitude matrices and transmission equations are -

$$\text{PE, scattering:} \quad A'_E = \widetilde{C^*}\, A\, C\,, \qquad V_r = \tilde{u}'_R\, \tilde{C}\, C\, A'_E\, u'_T \tag{23}$$

$$\text{PE, propagation:} \quad A^{o\prime} = \widetilde{C^*}\, A^o\, C, \qquad V_r = \widetilde{u_R^{o\prime}}\, \tilde{C}\, C\, A^{o\prime}\, u'_T \tag{24}$$

PH, scattering: $A' = \tilde{C}\, A\, C$ , $V_r = \widetilde{u^{R}}' A'\, u^{T}'$ (25)

PH, propagation: $A^{o}{}' = \tilde{C}^{*} A^{o} C$ , $V_r = \widetilde{u^{Ro}}'^{*} A^{o}{}' u^{T}{}'$ (26)

As can be seen from the above, scattering matrices are being transformed differently in the two approaches. Therefore A, after change-of-basis in PE, has been denoted by $A_E'$ .

## 2.5. PE and PH approaches in the power /Stokes/ representation

Using 4 x 4 unitary matrix

$$U = \frac{1}{\sqrt{2}} \begin{bmatrix} 1 & 1 & 0 & 0 \\ 0 & 0 & 1 & -j \\ 0 & 0 & 1 & j \\ 1 & -1 & 0 & 0 \end{bmatrix} \tag{27}$$

and Kronecker multiplication rules for matrices:

and
$$\begin{aligned} (AB) \otimes (CD) &= (A \otimes C)(B \otimes D) \\ \widetilde{(A \otimes B)\, C} &= \tilde{C}\, (\tilde{A} \otimes \tilde{B}) \end{aligned} \tag{28}$$

amplitude transmission equations can be transformed easily into power form. So, the received power, $P_r$ , can be obtained as Kronecker product /for details - see [3]/:

$$P_r = V_r \otimes V_r^{*} \tag{29}$$

The Stokes four-vector, representing the SP of a wave or an antenna:

$$P = \tilde{U}^{*} ( u \otimes u^{*} ) \tag{30}$$

the same for both approaches: PE and PH, can be transformed by the three different Stokes matrices

PE, scattering: $F_E = \tilde{U}^{*} (A \otimes A^{*})\, U$ (31)

PE and PH, propagation: $F^{o} = \tilde{U}^{*} (A^{o} \otimes A^{o*})\, U$ (32)

PH, scattering: $F = \tilde{U}\, (A \otimes A^{*})\, U$ (33)

And again, the difference in the transformation rules for scattering matrices in the PE and PH approaches schould be noticed.

The transmission equations in the power representation have the forms -

PE: $P_r = (\widetilde{P}_R\, D^V) F_E\, P_T = (\widetilde{P}'_R\, D^{V'}) F'_E\, P'_T = (\widetilde{P}_R\, D^o) F^o\, P_T$ (34)

PH: $P_r = \widetilde{P^R}\, F\, P^T = \widetilde{P^{R'}}\, F'\, P^{T'} = \widetilde{P^{Ro'}}\, F^o\, P^T$ (35)

with the following transformation rules applied, identical for the two approaches, PE and PH:

$$P^o = D^o\, P, \qquad D^o = \widetilde{U}\,(C^o \times C^o)\, U = \begin{bmatrix} 1 & & & \\ & 1 & & \\ & & -1 & \\ & & & 1 \end{bmatrix} \qquad (36)$$

and

$$P' = D\, P\,, \qquad D = \widetilde{U^*}\,(C \times C^*)\, U \qquad (37)$$

and with

$$D^V = \widetilde{U}\, U = \begin{bmatrix} 1 & & & \\ & 1 & & \\ & & 1 & \\ & & & -1 \end{bmatrix}, \quad D^{V'} = \widetilde{D}\, D^V\, D \qquad (38)$$

It should be observed that in the PE approach, in equations (34), the receiving vectors, in brackets, are different from the transmitting ones, even if they represent the same antennas, what is not the case for the scattering transmission equations in the PH approach /compare the two first equations (35) /.
Transformation rules for Stokes matrices are:

$$F^o = D^o\, F\,, \qquad F^o_E = D^o\, F_E \qquad (39)$$

and

$$F' = \widetilde{D}\, F\, D \qquad (40)$$

for both approaches, PE and PH

The following names can be proposed for Stokes matrices (31)-(33):
$F_E$ - Mueller scattering matrix,

$F^o$ - Mueller propagation /or optical/ matrix, and
$F$ - radar scattering matrix

Because $F_E$ and $F^o$ have common property - their determinants are nonnegative, both will be called "of propagation type", whereas F matrix, having nonpositive determinant, will be called "of scattering type".

## 2.6. Examples of other power representations /in the PH approach/

The Cloude's correlation matrix [4], $T_C$, can be obtained by succesively denoting:

$$A = \begin{bmatrix} A_2 & A_3 \\ A_4 & A_1 \end{bmatrix}, \qquad v = \begin{bmatrix} A_2 \\ A_3 \\ A_4 \\ A_1 \end{bmatrix}, \qquad k = \widetilde{U^*}\, v \tag{41}$$

and finally, the mean value:

$$T_C = \langle k \quad \widetilde{k^*} \rangle \tag{42}$$

The received power expression

$$P_r = \widetilde{V_r}\; V_r^* = \widetilde{S}\; T_C\; S^* \tag{43}$$

with

$$S = \widetilde{U}\,(u^R \otimes u^T), \qquad \widetilde{V_r} = \widetilde{S}\; k \tag{44}$$

suggests that $T_C$ can be treated as an extension of the Graves power scattering matrix concept to the case of partially depolarizing target operators, for $T_C$ and S statistically independent.

$T_C$ is fully equivalent to the mean Stokes radar scattering matrix

$$F = U \langle A \otimes A^* \rangle U \tag{45}$$

The two equivalent matrices in the parametric forms are shown beneath for comparison:

$$T_C = \begin{bmatrix} A_o+A & C-jD & H+jG & I-jJ \\ C+jD & B_o+B & E+jF & K-jL \\ H-jG & E-jF & B_o-B & M+jN \\ I+jJ & K+jL & M-jN & A_o-A \end{bmatrix}, \quad F = \begin{bmatrix} A_o+B_o & C+N & H+L & F+I \\ C-N & A+B & E+J & G+K \\ H-L & E-J & A-B & D+M \\ F-I & G-K & D-M & -A_o+B_o \end{bmatrix} \tag{46}$$

The $T_C$ coherency matrix, as positively semidefinite, can serve as a best test of physical realizability of related Stokes radar scattering matrix for depolarizing target.

In order to present another example of power representation, the orthogonality/reversal transformation will be introduced, e.g. in the PH approach:

$$C^x\, u = u^{x\,*} \tag{47}$$

with

$$C^x = \begin{bmatrix} 0 & -1 \\ 1 & 0 \end{bmatrix} \tag{48}$$

and

$$u = \begin{bmatrix} a \\ b \end{bmatrix} \tag{49}$$

with a and b given by (6). Using this transformation, the "null four-vector, Q, will be defined of the form

$$Q = \widetilde{U}^{*} (u \otimes u^{x\,*}) = \frac{1}{\sqrt{2}} \begin{bmatrix} 0 \\ \cdots \\ q \end{bmatrix} \tag{50}$$

with complex

$$q = \begin{bmatrix} -2ab \\ a^2 - b^2 \\ j(a^2 + b^2) \end{bmatrix} \tag{51}$$

and the matrix

$$T_q = \widetilde{U}_3 < A \otimes (C^x\, A\, \widetilde{C}^x) > U_3 \tag{52}$$

with

$$U_3 = \frac{1}{\sqrt{2}} \begin{bmatrix} 1 & 0 & 0 \\ 0 & 1 & -j \\ 0 & 1 & j \\ -1 & 0 & 0 \end{bmatrix} \tag{53}$$

representing the partially depolarizing scatterer with its power and phase transformation properties expressed by the transmission equation

$$< \lambda^2 \quad q^{S*} > = T_q \quad q^T \tag{54}$$

Similar equation, in the PE approach, has been derived by Wanielik [5].

## 3. PROPAGATION AND SCATTERING APPROACHES

### 3.1. Representation of the orthonormal null-phase polarization basis on the Poincare sphere

Convenient representation of the original linear, orthogonal, $(1_x, 1_y)$ polarization basis is shown in Fig 1.
As can be seen, both real SP versors are tangent to the Poincare sphere at the point corresponding to the horizontal and vertical polarizations, respectively, and both are directed along the equator being parallel shifted of one another by $180^o$. Any orthonormal polarization basis, elliplic in general, can be obtained by proper rotation of this

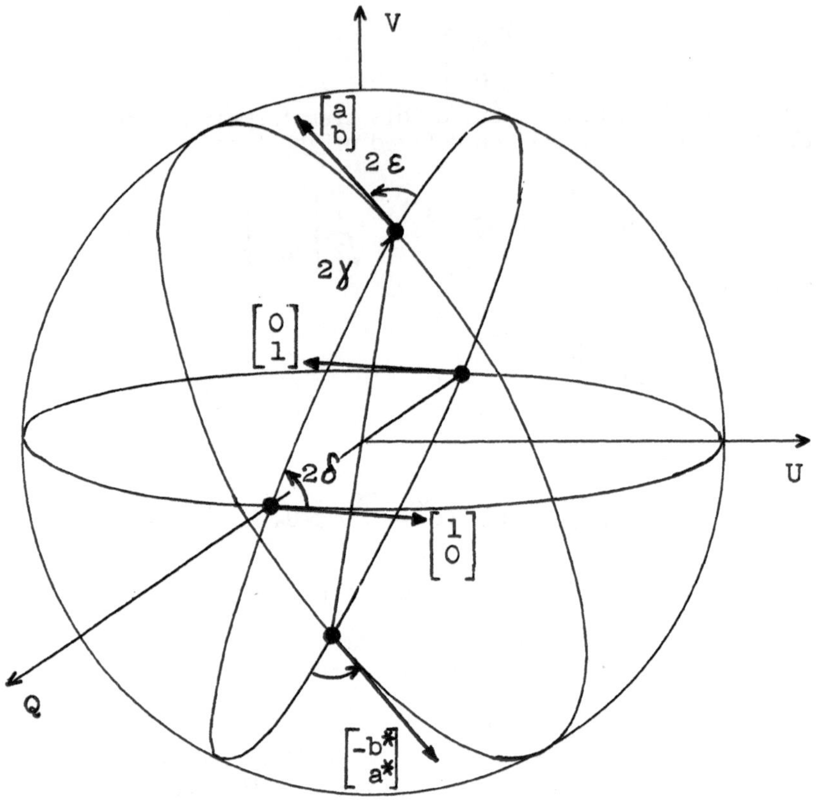


Fig.1. Orthonormal linear polarization bases

original basis. Analytically such rotation may be performed by applying unimodular unitary matrix C, as defined in (5), with Cayley-Klein parameters (6) in terms of three "analytical" Euler angles, or alternatively, with Huynen's parametres [7]

$$a = (\cos\psi \ \cos\tau \ -j \ \sin\psi \ \sin\tau\ )e^{j\chi}$$
$$b = (\sin\psi \ \cos\tau \ +j \ \cos\psi \ \sin\tau\ )e^{j\chi} \qquad (55)$$

in terms of three "geometrical" Euler angles and, eventually with Wanielik's parametres [5]

$$a = \cos\phi - j\, n_1 \sin\phi$$
$$b = (n_3 - j\, n_2) \sin\phi \qquad (55a)$$

in terms of one angle of rotation, $2\phi$, and oriented axis of right-hand rotation n. Mutual dependences between, for example, the first and third representations of the C matrix are worth noticing:

$$\cos\phi = \cos\gamma \cos(\delta + \varepsilon)$$
$$n_1 \sin\phi = \cos\gamma \sin(\delta + \varepsilon) \qquad (55b)$$
$$n_2 \sin\phi = -\sin\gamma \sin(\delta - \varepsilon)$$
$$n_3 \sin\phi = \sin\gamma \cos(\delta - \varepsilon)$$

Both vectors of the rotated basis

$$u = C\begin{bmatrix}1\\0\end{bmatrix} = \begin{bmatrix}a\\b\end{bmatrix} \quad \text{and} \quad u^{x} = C\begin{bmatrix}0\\1\end{bmatrix} = \begin{bmatrix}-b^{*}\\a^{*}\end{bmatrix} \qquad (56)$$

can be represented on the Poincare sphere by tangent Argand phasors [3], or by real parts of proper three-vectors (51). Dividing q into the real and imaginary parts:

$$q = q_R + j\, q_I, \quad |q_R| = |q_I| = 1 \qquad (57)$$

we get real vector

$$q_R = \begin{bmatrix}-\sin 2\gamma \cos 2\varepsilon\\ \cos 2\gamma \cos 2\delta \cos 2\varepsilon - \sin 2\delta \sin 2\varepsilon\\ \cos 2\gamma \sin 2\delta \cos 2\varepsilon + \cos 2\delta \sin 2\varepsilon\end{bmatrix} \qquad (58)$$

tangent to the Poincare sphere of unit radius at the point

$$p = q_R \times q_I = \sqrt{2}\,\widetilde{U_3^{*}}\,(u \times u^{*}) = \begin{bmatrix}\cos 2\gamma\\ \sin 2\gamma \cos 2\delta\\ \sin 2\gamma \sin 2\delta\end{bmatrix} \qquad (59)$$

and coinciding exactly with the Argand phasor described in [3].

## 3.2. Propagation and scattering matrices in their characteristic coordinate systems /CCS/

By proper rotating the original orthogonal polarization basis to the position of the so called characteristic basis, together with the orthogonal coordinate system of Stokes parameters, we arrive at the CCS of the scatterer. In the CCS scattering matrices have simple forms and simple expressions describe special polarization points of the scatterer on the Poincare sphere.

The procedure of finding the $u^K$ and $u^L$ PP vectors of the characteristic basis for the amplitude scattering matrix A in the PH approach is well known and has been described e.g. in [3]. According to that procedure we find the change-of-basis matrix

$$C^K = \begin{bmatrix} u^K & u^L \end{bmatrix} \tag{60}$$

for which the scattering matrix has the new form

$$A^K = \widetilde{C}^K \, A \, C^K \, e^{-j\mu^K} = \begin{bmatrix} A_2^K & B_1^K + jB_2^K \\ -B_1^K - jB_2^K & A_1^K \end{bmatrix} \tag{61}$$

with $A_2^K \geq A_1^K$ positive real, and $B_1^K$ and $B_2^K$ real. The corresponding form of Stokes radar scattering matrix is

$$F^K = \widetilde{U} \, ( A \times A^* ) \, U = \begin{bmatrix} a_1 & b_1 & b_3 & b_5 \\ b_1 & a_2 & b_4 & b_6 \\ -b_3 & -b_4 & a_3 & 0 \\ -b_5 & -b_6 & 0 & a_4 \end{bmatrix} \tag{62}$$

with

$$a_2 = a_1 + a_3 + a_4 \tag{63}$$

and where the upper indeces K were omitted for simplicity.
Having the above result we can find the $u^{K^o}$ and $u^{L^o}$ PP vectors of the characteristic basis for the propagation matrix $A^o$ by inspection its transformation equation which, according to (26), can be written in the form

$$A^{K^o} = \widetilde{C}^{K^o *} \, A^o \, C^{K^o} \, e^{-j\mu^{K^o}} \tag{64}$$

with, initially, unknown matrix

$$C^{K^o} = \begin{bmatrix} u^{K^o} & u^{L^o} \end{bmatrix} \tag{65}$$

To this equation the orthogonality/reversal transformation will be applied, making use of special properties of the $C^x$ matrix:

$$C^x \, \widetilde{C}^x = \begin{bmatrix} 1 & 0 \\ 0 & 1 \end{bmatrix} \tag{66}$$

and

$$\widetilde{C}^{*}\, C^{x} = C^{x}\, \widetilde{C} \tag{67}$$

which are valid for any orthogonal polarization basis, because

$$\widetilde{C}\; C^{x}\; C \;=\; C^{x} \tag{68}$$

The procedure will be as follows. Using succesively (66) and (67) we get:

$$A^{oK^{o}} = \widetilde{C^{K^{o}}}^{*}(C^{x}\;\; \widetilde{C^{x}})\; A^{o}\; C^{K^{o}}\;\; e^{-j\mu^{K^{o}}} \tag{69}$$

$$= C^{x}\; \widetilde{C^{K^{o}}}(\widetilde{C^{x}}\; A^{o})\; C^{K^{o}}\;\; e^{-j\mu^{K^{o}}} \tag{70}$$

$$= C^{x}\; (\widetilde{C^{x}}\; A^{o})^{K^{o}} \tag{71}$$

$$= \begin{bmatrix} B_1^{K^{o}} + jB_2^{K^{o}} & -A_1^{K^{o}} \\ A_2^{K^{o}} & B_1^{K^{o}} + jB_2^{K^{o}} \end{bmatrix} \tag{72}$$

In the course of the above procedure the propagation matrix $A^{o}$ has been changed into scattering matrix $(C^{x}\; A^{o})$ which has been changed back into propagation resultant matrix. The form (72) of the propagation matrix in its characteristic polarization basis has been shown at the end.

The same procedure in power domain would involve the Stokes orthogonality transformation performed by the matrix

$$D^{x} = \widetilde{U}\,(C^{x} \times C^{x})\; U = \begin{bmatrix} 1 & & & \\ & -1 & & \\ & & -1 & \\ & & & -1 \end{bmatrix} \tag{73}$$

and the Stokes equivalent of equation (71) is

$$F^{oK^{o}} = D^{x}\; (D^{x}\; F^{o})^{K^{o}} \tag{74}$$

resulting in the following form of the Stokes propagation matrix in its CCS:

$$F^{oK^{o}} = \begin{bmatrix} a_1 & b_1 & b_3 & b_5 \\ -b_1 & -a_2 & -b_4 & -b_6 \\ b_3 & b_4 & -a_3 & 0 \\ b_5 & b_6 & 0 & -a_4 \end{bmatrix} \tag{75}$$

where the upper indeces $K^o$ were omitted for simplicity.

In order to enable the direct computation of the $A^{oK^o}$ matrix elements the following explicit formulae will be given at the end of the Section:

$$A_2^{K^o} = |A_2'| , \quad A_1^{K^o} = |A_1'| \quad \text{/always } A_2^{K^o} \geq A_1^{K^o}\text{/,} \tag{76}$$

$$B_1^{K^o} + j\, B_2^{K^o} = A_3'\, e^{-j\mu^{K^o}}$$

where for

$$A^o = \begin{bmatrix} A_2^o & A_3^o \\ A_4^o & A_1^o \end{bmatrix} \tag{77}$$

we have:

$$\begin{aligned} A_2' &= \left[A_4^o + (A_1^o - A_2^o)\varrho - A_3^o \varrho^2\right] / N \\ A_1' &= \left[A_4^o \varrho^{*2} - (A_1^o - A_2^o)\varrho^* - A_3^o\right] / N \\ A_3' &= \left[-A_4^o \varrho^* + A_1^o + A_2^o \varrho\varrho^* - A_3^o \varrho\right] / N \end{aligned} \tag{78}$$

$$N = 1 + \varrho\varrho^*$$

$$\mu^{K^o} = \frac{1}{2}\,(\arg A_2' + \arg A_1') \tag{79}$$

and characteristic polarization ratio:

$$\varrho = \varrho^{K^o} = (R_1 \mp \sqrt{R_1^2 + R_2^2}) / R_2 \tag{80}$$

where

$$\begin{aligned} R_1 &= A_4^o A_4^{o*} - A_3^o A_3^{o*} \\ R_2 &= A_3^o (A_1^{o*} - A_2^{o*}) - A_4^{o*}(A_1^o - A_2^o) \end{aligned}$$

The three Euler angles of rotation the $(1_x, 1_y)$ basis to the characteristic position are:

$$\begin{aligned} 2\delta &= \arg \varrho \\ 2\gamma &= 2 \tan^{-1} |\varrho| \\ 2\varepsilon &= \frac{1}{2}(\arg A_2' - \arg A_1') - 2\delta \end{aligned} \tag{81}$$

Of course, after exchange of elements

$$\tilde{C}^x A^o = \begin{bmatrix} A_4^o & A_1^o \\ -A_2^o & -A_3^o \end{bmatrix} \longrightarrow \begin{bmatrix} A_2 & A_3 \\ A_4 & A_1 \end{bmatrix} = A \tag{82}$$

the same formulae (78) - (81) can serve to compute the transformation of the scattering matrix to the form in its characteristic basis.

Elements of $A^K$ matrix (61) insted of $A^{K^o}$ of (72) will then be obtained using formulae (76).

### 3.3. Normalization of TO matrices and target geometrical models

As a next step towards presentation for targets their pure polarization properties, the normalization of their TO amplitude and power matrices will be proposed.

The effective crossection $\sigma_o$ will be suggested as a measure of "target magnitude". For pure target states / following definition by Holm and Barnes [8] /:

$$\sigma_o = \text{Span } A + 2 \,|\det A| \tag{83}$$

and for mixed target states also

$$\sigma_o = 2\,(a_1 + a_o) \tag{84}$$

with

$$a_o = a_1^2 - b_0^2 \tag{85}$$

and

$$b_o^2 = b_1^2 + b_3^3 + b_5^2 \tag{86}$$

where all as' and bs' are elements of averaged Stokes matrices of the type (62) or (75), not necessarily in the CCS.

It is well known [3] that the geometrical model of polarization properties of a target can be presented as Poincare sphere in its CCS with an inversion point situated in some allowed regions inside the sphere and responsible for transformation of incident polarization and power. The radius of the sphere can be taken

$$r_o = \sqrt{\sigma_o}\,/2 = 1 \tag{87}$$

thus giving the value $\sigma_{on} = 4$, and proposed normalization formulae:

$$A_n = \frac{2}{\sqrt{\sigma_o}}\,A \tag{88}$$

and

$$F_n = \frac{4}{\sigma_o}\,F$$

It is worth observing that $\sigma_o/4$ plays the same role for normalized TO matrices as determinat for unimodular transformation matrices, because the unimodular transformation matrices considered as TO matrices in the PH approach have their $\sigma_o/4 = 1$.
It is evident, because they have inversion points I in the center O of the Poincaré sphere model. For $r_o = 1$, the distance (OI) is

$$(OI) = b_o/2 \tag{89}$$

So, for $b_o = 0$, there is: $a_o = a_1$ , and $\sigma_o = 4$

### 3.4. Comparison of geometrical models of propagation and scattering matrices

Like in Sec. 3.2., considerations will be limited to the PH approach only because in the PE approach matrices describing either propagation or scattering are both of the propagation type, as has been observed at the end of Sec. 2.5.

The main feature of the geometrical model of the scattering matrix is that it presents the transformation of polarization as inversion and rotation of the Poincare sphere. For the propagation matrix we have to consider additionally the orthogonality, known also as the pairity transformation, which is an inversion through the center of the sphere. Observe, please, that owing to such small change in the model, all formulae for coordinates of all special polarization points for the scattering matrix are still valid also in the case of propagation, only the meaning of these points is changing accordingly. For instance, the COPOL NULLS, or Kennaugh's zero polarizations, now become XPOL NULLS, or eigenpolarizations, and vice versa. On the contrary, maximal polarizations /i.e. polarizations of maximum scattered power/ for bistatic scattering still remain maximal for propagation. Some other interesting facts can also be predicted when comparing matrices of the two types. For scattering, contrary to propagation, zero polarizations can always be found, while the eigenpolarizations require fulfillment of well known conditions for their existence. For propagation eigenpolarizations always exist. The free space property of preserving all propagated polarizations correspond to the property of some possible scatterers that change all incident polarizations for orthogonal.

Many other examples for different properties of propagation and scattering matrices can be given, but there is one especially worth mentioning. The pure inversion Stokes scattering matrix differs from Lorentz transformation matrix by the $D^x$ factor only.
Namely,

$$F^{LOR} = D^x F^{INV} \tag{90}$$

This agrees with the known fact that

$$F^{SCATT} = \widetilde{D^{ROT}} F^{INV} \tag{91}$$

i.e. scattering can be considered as inversion followed by rotation /or vice versa/. If we add the $D^x$ transformation to the above scattering operator, we get:

$$\begin{aligned} F^{PROP} &= D^x \widetilde{D^{ROT}} F^{INV} \\ &= \widetilde{D^{ROT}} D^x F^{INV} \end{aligned}$$

$$= \widetilde{D^{ROT}} F^{LOR} \qquad (92)$$

This explains the fact that the propagation matrix can be presented as Lorentz transformation and rotation.
The Lorentz transformation of polarization has been presented in [5] by the use of ellipsoidal shell of constant scattered power. This can be compared with Kennaugh's inversion followed by orthogonality transformation, as shown in Fig.2.

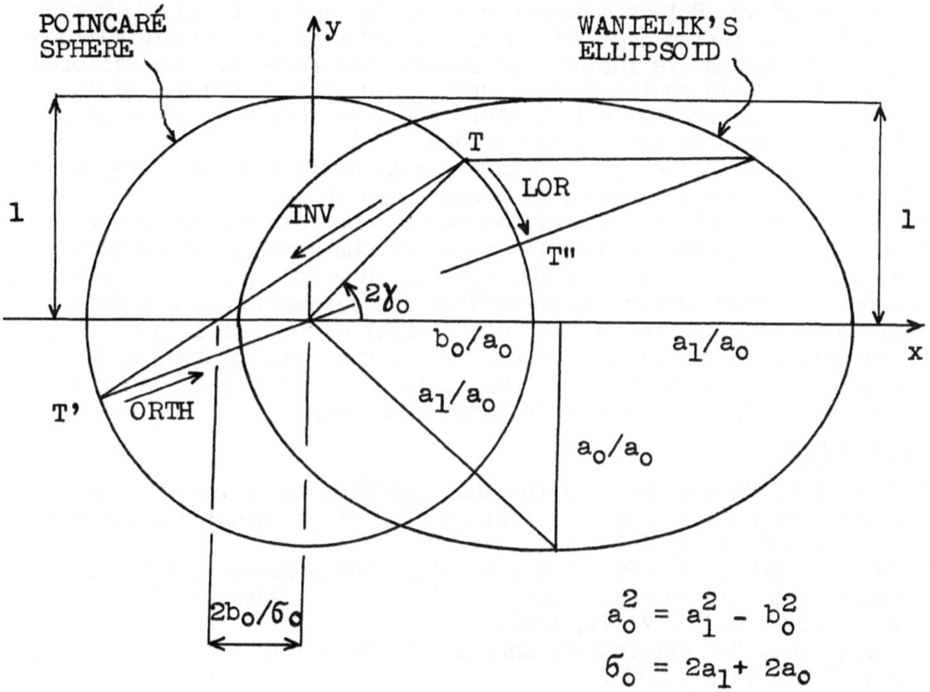


Fig.2 Lorentz transformation geometrical models

The equation of the shell in terms of TO geometrical model parameters is

$$\left(\frac{y}{a_0}\right)^2 + \left(\frac{x - b_0}{a_1}\right)^2 = 1 \qquad (93)$$

and the equation joining the two models /also the value of outgoing power /:

$$(I \quad T)^2 = (O \; W) = a_1 + b_0 \cos 2\gamma_0 \qquad (94)$$

T"is the outgoing polarization for the model without rotation.

## 4. CONCLUDING REMARKS

Two main fundamental approaches to radar polarimetry have been discussed. The first one is based on the PE definition of the SP and uses the propagation type of TO matrices for both propagation and scattering. The second approach is based on the PH definition of the SP and uses the two types: of propagation and scattering TO matrices accordingly.

The propagation type is characterized by Lorentz and rotation transformations of the Poincare sphere geometrical model of the TO matrix, whereasthe scattering type, by inversion and rotation transformations.

Both approaches are formally equivalent and correct. Main difference between them is in analysing scattering phenomena by the use of two different transformation rules. There are also different advantages and disadvantages of the two approaches.

The PE approach, restricting considerations to one only type of TO matrices, prevents the reader from confusion in interpretation of measuring data published in the literature, whereas the PH approach offers simpler transmission formulae after change-of-polarization-basis and enables one to consider the same physical process as propagation or scattering depending on whether the outgoing wave has to be examined as passing through the intervenient medium and illuminating the target, or as being scattered by the target and received by the observer. Which one of the two fundamental approaches will be finally accepted will depend on our scientific community.

## REFERENCES

1. Jones,R.C.,"A new calculus for the treatment of systems,V. A more general formulation and description of another calculus", J.Opt.Soc. Am., Vol.37, No.2, 107 - 112, 1947.
2. Boerner, W-M., El-Arini, M.B., et al., "Polarization dependence in electromagnetic inverse problems", IEEE Trans. Antennas Propagat., Vol.AP-29, No.2, 262-271, 1981.
3. Czyż, Z.H., "An alternative approach to foundations of radar polarimetry", in this volume.
4. Cloude, S.R., "Group theory and polarisation algebra", Optik, Vol.75, No.1, 26-36, 1986.
5. Wanielik, G., "Polarization radar signal description: modelling and processing of polarimetric pulse radar data", in this volume.
6. Kostinski, A.B., Boerner, W-M., "On foundations of radar polarimetry" IEEE Trans, Antennas Propagat., Vol.AP-34, No.12, 1395-1404, 1986; Mieras, H., "Comments on" On foundations of radar polarimetry"" ibid, 1470-1471; Kostinski, A.B., Boerner, W-M., "Authors reply", ibid., 1971-1973.
7. Huynen, J.R., "Target decomposition theorem for real target polarimetric data analysis", International Conference on Radar-Versailles, April 24-28, 1989.
8. Holm, W.A., Barnes, R.M. "On radar polarization mixed target state decomposition techniques", IEEE National Radar Conference, University of Michigan, Ann Arbor, MI, April 20-21, 1988.

# DEFINITIONS OF POLARIZATION IN RADAR

Harold Mott
The University of Alabama
Electrical Engineering Department
Tuscaloosa, Alabama, U.S.A

ABSTRACT. A review of definitions and methods used in the polarization-dependent analysis of radar is presented. Careful attention is given to coordinate systems used in the definitions of polarization descriptors. Deficiencies in the IEEE Standards for antennas are pointed out.

## 1. INTRODUCTION

In the literature on polarization of electromagnetic waves there is a substantial degree of inconsistency and vagueness in the definition and usage of terms. This is not an unexpected situation since polarization terminology has been developed by optics and physics researchers on one hand and somewhat independently by those interested in antennas and lower frequency wave propagation. Some of the differences are trivial; physicists call left-handed the same wave that antenna engineers call right-handed. Some differences are more serious, however. Engineers normally use time variation $\exp(j\omega t)$ and those in optics use $\exp(-i\omega t)$. The same time-invariant equation for the electric field leads to two physically different time-varying fields, depending on the time variation assumed. Even coordinate systems can lead to misunderstandings. The rectangular components of a wave backscattered from a target can be given either in the coordinate system used for the incident wave or in a reversed system. In a final example, it is commonly noted that if the polarizations of transmitting and receiving antennas are plotted on the Poincaré sphere, the angular separation of the points is a measure of polarization efficiency. It is less common to see the qualification that one polarization is, in fact, not the polarization but the receiving polarization.

It may well be that the best approach to clarifying the nomenclature and removing the inconsistencies is to present an overall view of polarization descriptors and methods of treating polarization problems, while being careful to define coordinate systems, time variation, polarization and receiving polarization, etc. That approach is adopted here. The discussion, it is hoped, is a consistent, clear



*W.-M. Boerner et al. (eds.), Direct and Inverse Methods in Radar Polarimetry, Part 1*, 117–153.

introduction to polarization theory, including transmission between antennas and scattering from a target, with consideration of completely polarized and partially polarized waves.

## 2. THE POLARIZATION OF WAVES

### 2.1. The Plane Wave and Coordinate Systems

The electric field vector of a single-frequency plane wave traveling in the z direction of Fig. 1 can be written as

$$\overline{E} = E_x \overline{u}_x + E_y \overline{u}_y \tag{1}$$

where $E_x$ and $E_y$ are complex rectangular components. In (1), a time and distance variation $\exp[j(\omega t-kz)]$, where

$$k = 2\pi/\lambda \tag{2}$$

is assumed and suppressed for convenience. The coordinate system is chosen so that

$$\overline{u}_x \times \overline{u}_y = \overline{u}_z \tag{3}$$

Fig. 1 Coordinate Systems for Wave Propagation

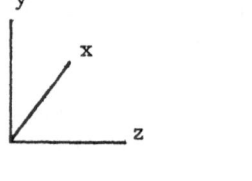

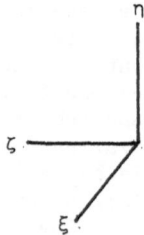

In Fig. 1 is also shown a reversed coordinate system appropriate for a plane wave traveling in the -z, or $\zeta$, direction. Like the x,y,z system, the $\xi$, $\eta$, $\zeta$ system is right handed.

An important parameter in the study of wave polarization is the linear polarization ratio P, to be discussed in greater detail in a later section but defined here as the ratio of the complex field components of $\overline{E}$,

$$P = E_y/E_x \tag{4}$$

Examination of Fig. 1 shows that a rotation of the coordinate system by $\pi/2$ around the z axis will change the value of P for a given wave, so the definition of P is incomplete. A more informative definition of P is therefore

$$P = \frac{E_{\text{vertical}}}{E_{\text{horizontal}}} \tag{5}$$

The unit vectors must be chosen so that

$$\overline{u}_{\text{horizontal}} \times \overline{u}_{\text{vertical}} = \overline{u}_{\text{propagation direction}} \tag{6}$$

Near the earth's surface, "horizontal" may be loosely defined as parallel to the surface. More precisely, if a line is drawn from the coordinate origin to the earth's center, the horizontal axis of the coordinate system lies in a plane perpendicular to the line. The direction of propagation of the wave may be arbitrary and the vertical axis is perpendicular to the horizontal axis and the direction of propagation and not necessarily perpendicular to the earth's surface. Far from the earth's surface the coordinate system is essentially arbitrary.

The radiated field of an antenna is normally defined by a spherical coordinate system as in Fig. 2. The radiated wave, having only $E_\theta$ and $E_\phi$ components, travels radially outward. If the xy plane is parallel to the earth's surface, then $E_\phi$ is the horizontal wave component and $-E_\theta$ the vertical. The appropriate definition of linear polarization ratio is then

$$P = - E_\theta/E_\phi \tag{7}$$

If the antenna coordinate system has an xy plane not parallel to the earth's surface, then neither $E_\theta$ nor $E_\phi$ is in general horizontal (parallel to the earth's surface) and P cannot be defined by vertical and horizontal components. Equation (7) still holds, however, and "horizontal" should be taken as parallel to the xy plane.

## 2.2. The Polarization Ellipse

In (1) the wave components may be written as

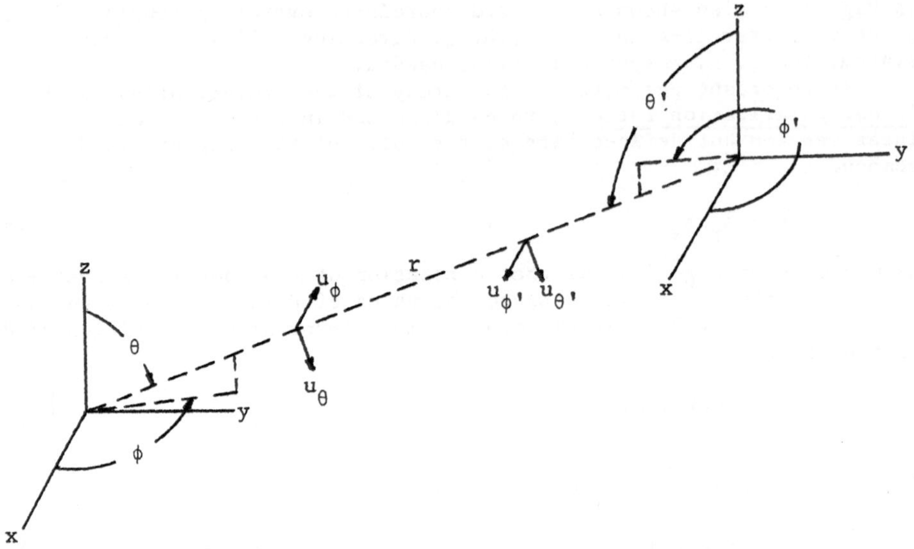


Fig. 2 Spherical Coordinate Systems for Antennas

$$E_x = |E_x| e^{j\phi_x} \quad (a) \qquad E_y = |E_y| e^{j\phi_y} \quad (b) \tag{8}$$

Then the time-varying electric field associated with $\overline{E}$ is

$$\overline{E}_t = \mathrm{Re}(\overline{E}\, e^{j(\omega t - kz)})$$

$$= \overline{u}_x\, |E_x| \cos(\omega t - kz + \phi_x) + \overline{u}_y\, |E_y| \cos(\omega t - kz + \phi_y) \tag{9}$$

It is a straightforward procedure to manipulate the components of the time-varying field vector into the form

$$\frac{E_{tx}^2}{|E_x|^2} - 2\,\frac{E_{tx}}{|E_x|}\,\frac{E_{ty}}{|E_y|}\cos\phi + \frac{E_{ty}^2}{|E_y|^2} = \sin^2\phi \tag{10}$$

where

$$\phi = \phi_y - \phi_x \tag{11}$$

Equation (10) represents an ellipse, as shown by Fig. 3. If auxiliary angles are defined by

$$\tan\alpha = \frac{|E_y|}{|E_x|} \qquad 0 \leq \alpha \leq \pi/2 \tag{12}$$

$$\sin 2\delta = \sin 2\alpha \sin\phi \qquad -\pi/4 \leq \delta \leq \pi/4 \tag{13}$$

then the tilt angle, measured from the positive x axis, and axial ratio of the ellipse are given by

$$\tan 2\tau = \tan 2\alpha \cos\phi \qquad 0 \leq \tau \leq \pi \tag{14}$$

$$R = m/n = \pm\cot\delta \qquad R \geq 0 \tag{15}$$

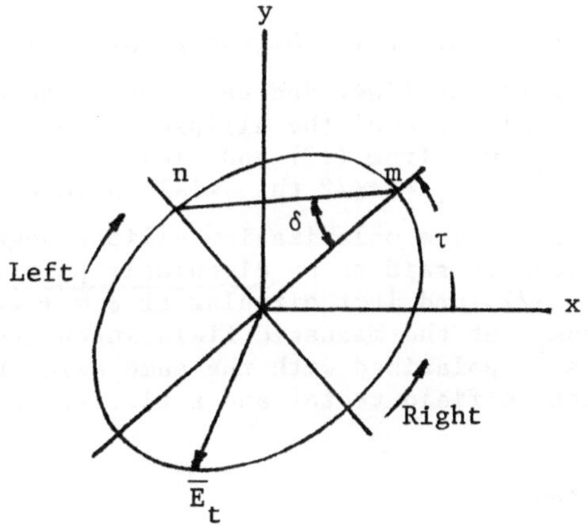


Fig. 3 The Polarization Ellipse

As time increases, $\overline{E}_t$ in Fig. 3 rotates clockwise or counterclockwise. IEEE Standard 145-1983 (Standard Definitions of Terms for

Antennas) specifies that the plane in which the ellipse lies be viewed looking in the direction of propagation. If the tip of the vector $E_t$ rotates in a clockwise direction as time increases, the sense of rotation is called right-handed. Counterclockwise rotation is called left-handed. The sense of rotation is right handed for $\delta < 0$ and left handed for $\delta > 0$. In optics and much of physics, usage is reversed from IEEE custom.

There is not a common usage in the literature of polarization ellipse descriptors. The axial ratio is often written as the ratio of minor to major axes, the inverse of (15). The shape of the ellipse may also be given by its ellipticity, the ratio of the difference of major and minor axes to the major axis

$$\text{Ellipticity} = (m-n)/m \tag{16}$$

Unfortunately, ellipticity is sometimes taken as synonymous with axial ratio.

A useful descriptor of ellipse shape is the ellipticity angle, $\delta$. It is defined by (15) and shown in the polarization ellipse of Fig. 3, with limits given by (13). Then right circular, linear, and left circular polarizations are given respectively by $\delta = -\pi/4$, 0, and $+\pi/4$. Here again, confusion exists; the ellipticity angle is sometimes called the ellipticity.

## 2.3. Linear and Circular Polarization

If $|E_x| = 0$, or $|E_y| = 0$, or $\phi = 0$, the polarization ellipse degenerates to a straight line, and the wave is said to be linearly polarized. The axial ratio of the ellipse is infinite, and the tilt angle may still be found from (12) and (14).

If $|E_x| = |E_y|$ and $\phi = \pm\pi/2$ the axial ratio as given by (15) becomes equal to one. The polarization ellipse degenerates to a circle, and the wave is said to be circularly polarized - right circular if $\phi = -\pi/2$, and left circular if $\phi = +\pi/2$.

It is obvious that the magnetic field intensity vector of the wave is also elliptically polarized with the same axial ratio and rotation sense as the electric field vector and a tilt angle which differs by $\pi/2$.

## 2.4. Time Variation

In this discussion, a time variation of $\exp(j\omega t)$ is used, which is standard among those with engineering backgrounds. In optics and physics, however, the use of $\exp(-i\omega t)$ has a strong tradition. If the positive exponent is chosen, the Maxwell curl equation

$$\nabla\text{x}\ \overline{E}_t(\overline{r},t) = -\frac{\partial\ \overline{B}_t(\overline{r},t)}{\partial t} \tag{17}$$

reduces to the time invariant form

$$\nabla x\ \overline{E}(\overline{r}) = -\ j\omega\overline{B}(\overline{r}) \tag{18}$$

and if the negative time exponent is used, it becomes

$$\nabla x\ \overline{E}(\overline{r}) = +\ i\omega\overline{B}(\overline{r}) \tag{19}$$

The other Maxwell curl equation is similarly altered.

Consider the plane wave traveling in the z direction

$$\overline{E} = \overline{u}_x + j\ \overline{u}_y \tag{20}$$

If time and distance variation exp[j(ωt-kz)] is assumed, the associated time-varying field is

$$\overline{E}_t = \overline{u}_x \cos(\omega t - kz) + \overline{u}_y \cos(\omega t - kz + \pi/2) \tag{21}$$

If one looks in the direction of wave travel, it is seen that the field vector rotates counterclockwise as time increases and is, by IEEE Standard 145 definition, left circular.

On the other hand, if exp[-i(ωt-kz)] is used, the time-varying field is

$$\overline{E}_t' = \overline{u}_x \cos(\omega t - kz) + \overline{u}_y \cos(\omega t - kz - \pi/2) \tag{22}$$

This wave appears to the same observer to rotate clockwise, and is right circular.

This difference between engineering and optics practices is more significant than mere nomenclature, such as the discrepancy between "right" and "left" handedness. If positive time exponent is assumed, the same (time-invariant) field equation leads to a wave with one rotation sense, and a negative time exponent gives a wave of opposite sense. It is unlikely that the notational difference can be reconciled soon. Until it is, a reasonable notation is to use j for the positive exponent and i for the negative.

## 2.5. Circular Wave Components

The electric field intensity has been written in terms of rectangular components by (1). It may also be written as the vector sum of two circular wave components, one rotating in a left handed sense and the other with a right handed sense. Consider the complex vectors

$$\overline{u}_L = \frac{1}{\sqrt{2}} (\overline{u}_x + j\ \overline{u}_y) \tag{23}$$

$$\overline{u}_R = \frac{1}{\sqrt{2}} (\overline{u}_x - j\ \overline{u}_y) \tag{24}$$

Like the basis set $\overline{u}_x$ and $\overline{u}_y$, the vectors $\overline{u}_L$ and $\overline{u}_R$ are unit vectors, since

$$\bar{u}_L \cdot \bar{u}_L^* = \bar{u}_R \cdot \bar{u}_R^* = 1 \qquad (25)$$

and orthogonal, since

$$\bar{u}_L \cdot \bar{u}_R^* = \bar{u}_L^* \cdot \bar{u}_R = 0 \qquad (26)$$

If multiplied by the phase term exp(-jkz), the vector $\bar{u}_L$ represents a left-circularly polarized wave propagating in the z direction. The real time-varying field associated with it is

$$\mathrm{Re}(\bar{u}_L e^{j\omega t} e^{-jkz}) = \frac{1}{\sqrt{2}}\,\bar{u}_x \cos(\omega t-kz) + \frac{1}{\sqrt{2}}\,\bar{u}_y \cos(\omega t-kz+\pi/2) \qquad (27)$$

which is clearly a left-circular wave ($|E_x|=|E_y|$, $\phi = +\pi/2$). Similarly, $\bar{u}_R$ can represent a right circular wave.

A plane wave that can be represented by the rectangular components of (1) can clearly be represented also by the sum of left and right circular waves, since the complex vectors $\bar{u}_L$ and $\bar{u}_R$ are defined in terms of the rectangular vectors. The plane wave field can be written as

$$\bar{E} = E_x\,\bar{u}_x + E_y\,\bar{u}_y = E_L\,\bar{u}_L + E_R\,\bar{u}_R \qquad (28)$$

Substituting (16) and (17) into (21) and equating coefficients yields

$$E_x = \frac{1}{\sqrt{2}}\,(E_L + E_R) \qquad (29) \qquad\qquad E_L = \frac{1}{\sqrt{2}}\,(E_x - jE_y) \qquad (31)$$

$$E_y = \frac{j}{\sqrt{2}}\,(E_L - E_R) \qquad (30) \qquad\qquad E_R = \frac{1}{\sqrt{2}}\,(E_x + jE_y) \qquad (32)$$

It has been seen here that the plane harmonic wave can be considered to be composed of orthogonal rectangular components or orthogonal circular components. It will be seen in a later section that it may also be considered the sum of two orthogonal, elliptically polarized waves with the polarization of one of the components chosen arbitrarily.

## 2.6 The Polarization State and Polarization Ratios

In many instances, the absolute phase of the wave is neither known nor of interest. In addition, the magnitude of the field vector does not affect polarization efficiency and other quantities of interest. All of the necessary polarization information is carried by the vector <u>polarization state</u> of the wave. It is the normalized electric field vector with the absolute phase removed, and may be written as

$$\bar{E}_N = \begin{bmatrix} \cos\alpha \\ \sin\alpha \quad e^{j\phi} \end{bmatrix} \tag{33}$$

where $\phi$ is the phase angle by which $E_y$ leads $E_x$, and $\alpha$ is defined by (12). The polarization state of (33) is expressed in an xy or linear polarization base.

The polarization information is also contained in a scalar, the linear polarization ratio, defined by

$$P = \frac{E_y}{E_x} = \frac{|E_y|}{|E_x|} \; e^{j\phi} \tag{34}$$

The qualifier "linear" is sometimes dropped, and the parameter is called the "polarization ratio" if the context is such that ambiguity can be avoided. It will be seen that from P the polarization ellipse parameters can be obtained.

Another ratio that contains all polarization information is the ratio of the left and right circular wave components.

$$Q = \frac{E_L}{E_R} = \frac{|E_L|}{|E_R|} \; e^{-j\theta} \tag{35}$$

The IEEE Standard 149-1979 (Test Procedures for Antennas) definition of the circular polarization ratio is the inverse

$$E_R/E_L$$

of the definition given here. The form used here is that of Rumsey.

Since $\bar{E}_N$, P, and Q all describe the wave polarization they are not independent. Appropriate substitution shows these relations:

$$P = \tan\alpha \quad e^{j\phi} \quad \text{(a)} \qquad E_N = \frac{1}{\sqrt{1+|P|^2}} \begin{bmatrix} 1 \\ P \end{bmatrix} \quad \text{(b)}$$

$$P = j\,\frac{Q-1}{Q+1} \quad \text{(c)} \qquad Q = \frac{1-jP}{1+jP} \quad \text{(d)} \tag{36}$$

Rumsey has shown that jP and Q are related by the same bilinear transforms used for the Smith chart, and use can be made of the commercially available Smith chart in describing polarization.

## 2.7. Polarization Ellipse Characteristics in Terms of P and Q

The time-varying electric field may be written as

$$\bar{E}_t = \frac{1}{2}\,[E_R(Q\bar{u}_L + \bar{u}_R)e^{j(\omega t-kz)} + E_R^*(Q^*\bar{u}_L^* + \bar{u}_R^*)e^{-j(\omega t-kz)}] \tag{37}$$

Using

$$\beta = \text{ang}\ (E_L) \tag{38}$$

$$\beta + \theta = \text{ang}\ (E_R) \tag{39}$$

and multiplying $\overline{E}_t$ by its conjugate gives

$$2|\overline{E}_t|^2/|E_R|^2 = 1 + |Q|^2 + 2|Q|\cos(2\omega t + 2\beta - 2kz + \theta) \tag{40}$$

The axial ratio of the polarization ellipse is the ratio of maximum to minimum values of $|\overline{E}_t|$ and is readily seen from (40) to be

$$R = \left|\frac{1 + |Q|}{1 - |Q|}\right| \tag{41}$$

The angle made by $\overline{E}_t$ with the x axis is readily found by substituting (23) and (24) into (37) to find the rectangular components of $\overline{E}_t$. This angle is the tilt angle when $|\overline{E}_t|$ is maximum, which from (40) occurs when

$$\omega t + \beta - kz = -\theta/2 \tag{42}$$

Then tilt angle $\tau$ is given by

$$\tau = \tan^{-1}(E_{ty}/E_{tx})\Big|_{\omega t + \beta - kz = -\theta/2} = \theta/2,\ \pi + \theta/2 \tag{43}$$

The first form is used for positive $\theta$ and the second for negative $\theta$ so the tilt angle is in the range 0-$\pi$.

Rotation sense is immediately found from $|Q|(= |E_L|/|E_R|)$, with a value >1 giving left elliptic rotation.

The use of (35) and (36d) in (41) and (43) gives the ellipse axial ratio and tilt angle in terms of P,

$$R = \left|\frac{|1+jP| + |1-jP|}{|1+jP| - |1-jP|}\right| \tag{44}$$

$$e^{-j2\tau} = \frac{(1-jP)/(1+jP)}{|(1-jP)/(1+jP)|} \tag{45}$$

Rotation sense is left handed for $0 < \phi < \pi$, and right handed otherwise.

## 2.8. Stokes Parameters and the Poincaré Sphere

The Stokes parameters may be used to describe amplitude and polarization of a wave. For monochromatic waves they are:

$$G_0 = |E_x|^2 + |E_y|^2 \qquad \text{(a)}$$

$$G_1 = |E_x|^2 - |E_y|^2 \qquad \text{(b)}$$

(46)

$$G_2 = 2|E_x|\ |E_y| \cos \phi \qquad (c)$$

$$G_3 = 2|E_x|\ |E_y| \sin \phi \qquad (d)$$

where $\phi$ is the phase lead of $E_y$ over $E_x$. It is readily seen that

$$G_0{}^2 = G_1{}^2 + G_2{}^2 + G_3{}^2 \qquad (47)$$

The Stokes parameters are readily related to the terms used previously to describe the polarization ellipse. From (12) and (13) we note that the auxiliary angle $\delta$ used to find the axial ratio can be related to the Stokes parameters by

$$\sin 2\delta = \sin 2\alpha \sin \phi = \frac{2|E_x|\ |E_y|}{|E_x|^2+|E_y|^2} \sin \phi = \frac{G_3}{G_0} \qquad (48)$$

The tilt angle $\tau$ can be found from (12) and (14) as

$$\tan 2\tau = \tan 2\alpha \cos \phi = \frac{2 \tan \alpha \ \cos \phi}{1 - \tan^2\alpha}$$

$$= \frac{2\ |E_x|\ |E_y| \cos \phi}{|E_x|^2 - |E_y|^2} = \frac{G_2}{G_1} \qquad (49)$$

Equations (46) through (49) can be used to write the Stokes parameters in terms of the angles defined previously.

$$G_1 = G_0 \cos 2\delta \cos 2\tau \qquad (a)$$

$$G_2 = G_0 \cos 2\delta \sin 2\tau \qquad (b) \qquad (50)$$

$$G_3 = G_0 \sin 2\delta \qquad (c)$$

The equations suggest a geometric interpretation of the Stokes parameters. $G_1$, $G_2$, and $G_3$ are the Cartesian coordinates of a point on a sphere of radius $G_0$, as shown in Fig. 4. Angles $2\delta$ and $2\tau$ are the latitude and azimuth angles measured to the point. To every polarization state there corresponds one point on the sphere, called the Poincaré sphere, and vice versa.

A few special points on the Poincaré sphere are of interest. For a left circular wave, $|E_x| = |E_y|$, $\phi = \pi/2$, and $2\delta = \pi/2$. From either (46) or (50) it may be seen that $G_1 = G_2 = 0$, $G_3 = G_0$, and the point representing left circular polarization is the "north pole," the + z axis, of the sphere. In a similar way, it can be determined that the south pole represents right circular polarization, all points representing linear polarization lie on the equator, all left elliptic points are in the northern hemisphere, and all right elliptic points are in the southern hemisphere.

It has been shown that points on the Poincaré sphere can be mapped by a stereographic projection into finite regions of planes representing the complex polarization ratios P or Q.

2.9. Orthogonal Elliptical Components of a Wave

It was seen in a previous section that the electric field can be written as the sum of rectangular or circular components

$$\overline{E} = E_x \overline{u}_x + E_y \overline{u}_y = E_L \overline{u}_L + E_R \overline{u}_R \tag{28}$$

where the vectors have unit length and form orthogonal sets. It will now be seen that $\overline{E}$ is the sum of two orthogonally polarized elliptical waves.

Let

$$\overline{E} = \overline{E}_1 + \overline{E}_2 = E_1 \overline{u}_1 + E_2 \overline{u}_2 \tag{51}$$

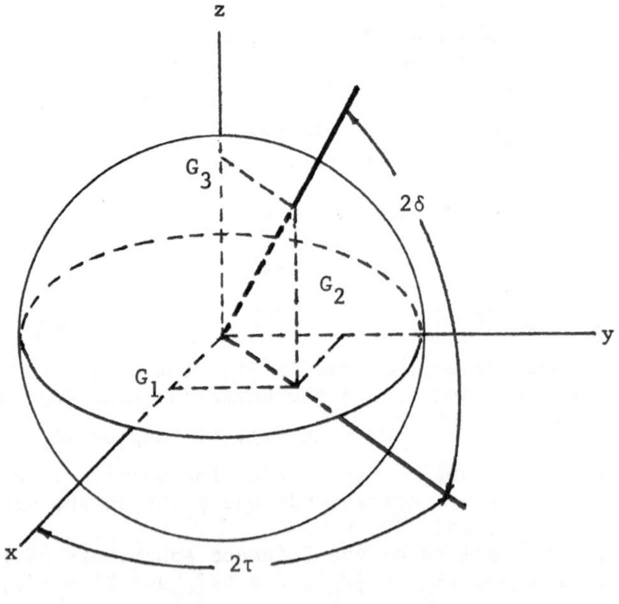


Fig. 4 The Poincaré Sphere

where $\bar{u}_1$ and $\bar{u}_2$ are orthogonal unit vectors. If one of the components, say $\bar{E}_1$, is to have a specified polarization, then $\bar{u}_1$ may be written as

$$\bar{u}_1 = u_{1x}\,\bar{u}_x + u_{1y}\,\bar{u}_y = u_{1x}\,(\bar{u}_x + P_1\,\bar{u}_y) \tag{52}$$

The phase of $\bar{E}_1$ may be included in the multiplier $E_1$ and $u_{1x}$ taken as real. Doing so and noting that $\bar{u}_1$ has unit length leads to

$$\bar{u}_1 = \frac{\bar{u}_x + P_1\,\bar{u}_y}{(1 + |P_1|^2)^{\frac{1}{2}}} \tag{53}$$

In the same way

$$\bar{u}_2 = \frac{\bar{u}_x + P_2\,\bar{u}_y}{(1+|P_2|^2)^{\frac{1}{2}}} \tag{54}$$

The orthogonality condition

$$\bar{u}_1^{\,*} \cdot \bar{u}_2 = 0 \tag{55}$$

leads quickly to

$$1 + P_1^{\,*}\,P_2 = 0 \tag{56}$$

or

$$P_2 = -\,1/P_1^{\,*} \tag{57}$$

At this point only the formal concept of orthogonality of the basis vectors has been used. In a later section it will be seen that if a receiving antenna has a polarization that causes it to receive the wave component $\bar{E}_1$ without loss, then it cannot receive component $\bar{E}_2$ at all.

## 3. POLARIZATION MATCHING OF ANTENNAS

### 3.1. Effective Length of an Antenna

The effective length of an antenna is defined in terms of the field it radiates by

$$\bar{E}^t(r,\theta,\phi) = \frac{j\,Z_o\,I}{2\lambda r}\,e^{-jkr}\,\bar{h}(\theta,\phi) \tag{58}$$

where $\bar{h}$ is the effective length, $\bar{E}^t$ the radiated field, $Z_o$ the characteristic impedance of free space, I a current at arbitrarily chosen terminals, and $\theta$ and $\phi$ polar and azimuth angles. This definition is a generalization of the equation of the field of a short dipole. IEEE Standard 145 defines effective length only for a linear antenna, which is a regrettable oversight since it is a useful concept for other antennas. To represent a general field, $\bar{h}$ must be complex.

The polarization state of an antenna is, by definition, the same as the polarization state of the wave that it transmits when fed by a single-frequency wave. The antenna polarization state is then

$$\bar{h}_N = \begin{bmatrix} \cos\alpha \\ \sin\alpha \quad e^{j\phi} \end{bmatrix} = \bar{E}_N^t \tag{59}$$

where

$$\tan\alpha = \frac{|h_y|}{|h_x|} \tag{60}$$

and $\phi$ is the phase lead of $h_y$ over $h_x$. The antenna polarization ratio is also, by definition, the same as the polarization ratio of the radiated wave.

If an antenna with effective length $\bar{h}$ located at the origin of the $(r,\theta,\phi)$ coordinates of Fig. 2 receives a wave with electric field $\bar{E}^i$ from an antenna at the origin of the $(r,\theta',\phi')$ coordinates, it may be shown by reciprocity that the open circuit voltage induced in the receiving antenna is

$$V = \bar{E}^i \cdot \bar{h} \tag{61}$$

It should be noted that both $\bar{E}^i$ and $\bar{h}$ are measured in the same coordinate system, the $(r,\theta,\phi)$ system of the receiving antenna.

### 3.2. Maximum Received Power

The power to the receiving antenna, neglecting load impedance, is

$$W = VV^* = |\bar{E}^i \cdot \bar{h}|^2 = |h_\theta E_\theta^i + h_\phi E_\phi^i|^2 \tag{62}$$

If $\bar{E}^i$ is varied, subject to the constraint that the incident power density is constant, or

$$\bar{E}^i \cdot \bar{E}^{i*} = \text{constant} \tag{63}$$

the maximum received power is found to be

$$W_m = |\bar{h}|^2 \, |\bar{E}^i|^2 \tag{64}$$

which occurs when

$$\frac{h_\phi}{h_\theta} = \left(\frac{E_\phi^i}{E_\theta^i}\right)^* \tag{65}$$

A polarization efficiency may then be defined by

$$\rho = \frac{|\bar{E}^i \cdot \bar{h}|^2}{|\bar{E}^i|^2 |\bar{h}|^2} \tag{66}$$

It is worth repeating that $\bar{E}^i$ and $\bar{h}$ are measured in the same coordinate system.

The polarization efficiency may be written in terms of the linear polarization ratios of the two antennas. Consider antenna #1 to be receiving. Its polarization ratio, $P_1$, is defined in terms of the field it transmits, $\bar{E}^t$, which from (58) is directly related to its effective length, so that

$$P_1 = -\frac{E^t_\theta}{E^t_\phi} = -\frac{h_\theta}{h_\phi} \tag{67}$$

The polarization ratio of the transmitting antenna, #2, is defined in the $r,\theta',\phi'$ coordinate system

$$P_2 = -\frac{E^i_{\theta'}}{E^i_{\phi'}} \tag{68}$$

But from Fig. 2, $E^i_{\theta'} = E^i_\theta$, and $E^i_{\phi'} = -E^i_\phi$, so that

$$P_2 = +\frac{E^i_\theta}{E^i_\phi} \tag{69}$$

If (68) and (69) are substituted into (66) the polarization efficiency becomes, in terms of the linear polarization ratios of the two antennas,

$$\rho = \frac{(1-P_1P_2)(1-P_1{}^*P_2{}^*)}{(1+P_1P_1{}^*)(1+P_2P_2{}^*)} \tag{70}$$

The efficiency $\rho$ is a real quantity in the range 0-1.

The polarization efficiency can be written in terms of the other polarization descriptors:

Circular polarization ratio

$$\rho = \frac{(1+Q_1Q_2)(1+Q_1{}^*Q_2{}^*)}{(1+Q_1Q_1{}^*)(1+Q_2Q_2{}^*)} \tag{71}$$

Ellipse parameters, antennas same sense

$$\rho = \frac{(R_1R_2+1)^2 + (R_1+R_2)^2 + (R_1{}^2-1)(R_2{}^2-1)\cos 2(\tau_1+\tau_2)}{2(R_1{}^2+1)(R_2{}^2+1)} \tag{72}$$

Ellipse parameters, antennas opposite sense

$$\rho = \frac{(R_1R_2-1)^2 + (R_1-R_2)^2 + (R_1{}^2-1)(R_2{}^2-1)\cos 2(\tau_1+\tau_2)}{2(R_1{}^2+1)(R_2{}^2+1)} \tag{73}$$

Stokes parameters

$$\rho = \frac{1}{2}\left(1 + \frac{G_1{}^a}{G_0{}^a}\frac{G_1{}^b}{G_0{}^b} - \frac{G_2{}^a}{G_0{}^a}\frac{G_2{}^b}{G_0{}^b} + \frac{G_3{}^a}{G_0{}^a}\frac{G_3{}^b}{G_0{}^b}\right) \tag{74}$$

It is stressed that, in all equations, the polarization descriptors of each antenna are defined in a right handed coordinate system appropriate for that antenna as a transmitting antenna, with the r axis pointing outward. Thus the coordinate systems used to define $P_1$ and $P_2$, for example, are reversed, as in Fig. 1. Note that because of the subscripted Stokes parameters, the two antennas are differentiated as a and b.

### 3.3. Polarization-Matched Antennas

In (70), if the polarization efficiency is set equal to one, $P_1$ and $P_2$ are found to be related by

$$P_1 = - P_2{}^* \tag{75}$$

Antennas with polarization ratios satisfying (75) are polarization matched, with no polarization loss when they are used in a communication pair.

The receiving polarization of an antenna is by definition the polarization of a plane wave, incident from a given direction, that gives maximum available power at the antenna terminals. Thus an antenna with polarization ratio P has a receiving polarization ratio $-P^*$. In the convention used here, the polarization ratio is determined in coordinates appropriate to the antenna as a transmitter.

Substitution of (36d) in (75) shows that polarization matched antennas have circular polarization ratios related by

$$Q_1 = Q_2{}^* \tag{76}$$

It follows from (41) that the axial ratios of the polarization ellipses are equal

$$R_1 = R_2 \tag{77}$$

and it is clear from (76) that the rotation senses are the same. From (76)

$$\theta_1 = -\theta_2 \tag{78}$$

and from

$$\tau = \theta/2 \quad , \quad \pi + \theta/2 \tag{43}$$

it follows that

$$\tau_2 = \pi + \theta_2/2 = \pi - \theta_1/2 = \pi - \tau_1 \tag{79}$$

where the first form of (43) is chosen for $\tau_1$ and the second for $\tau_2$ in order to keep $\tau_2$ in the range 0-$\pi$. It may be seen from the values of axial ratio and tilt angle that the polarization ellipses of matched antennas coincide when viewed from one position in space. Both ellipses have the same sense, but when viewed from one position the field vectors radiated by the two antennas appear to rotate in opposite directions.

When points corresponding to polarizations of two antennas in a transmit-receive configuration are plotted on the Poincaré sphere an elegant result is obtained. Antenna a is considered to transmit and the Stokes parameters $G_1^a$, $G_2^a$, $G_3^a$ are plotted on the sphere. Antenna b is the receiving antenna and its receiving polarization, $-P_b^*$, is plotted. But if $G_1^b$, $G_2^b$, $G_3^b$ correspond to $P_b$, the corresponding parameters for $-P_b^*$ are $G_1^b$, $-G_2^b$, $G_2^b$. If two rays are drawn from the Poincare sphere center to these plotted polarization points, the angle between the rays is given by

$$\cos \beta = \frac{G_1^a}{G_0^a}\frac{G_1^b}{G_0^b} - \frac{G_2^a}{G_0^a}\frac{G_2^b}{G_0^b} + \frac{G_3^a}{G_0^a}\frac{G_3^b}{G_0^b} \tag{80}$$

from which it follows that

$$\cos^2 \frac{\beta}{2} = \frac{1}{2}\left(1 + \frac{G_1^a}{G_0^a}\frac{G_1^b}{G_0^b} - \frac{G_2^a}{G_0^a}\frac{G_2^b}{G_0^b} + \frac{G_3^a}{G_0^a}\frac{G_3^b}{G_0^b}\right) \tag{81}$$

The right side of this equation is equal to the polarization efficiency given by (74). Thus

$$\rho = \cos^2 \frac{\beta}{2} \tag{82}$$

It follows that coincident points on the Poincaré sphere correspond to the matched polarization condition.

## 3.4. Cross-Polarized Antennas

If the polarization efficiency of (70) is zero, then

$$P_1 = \frac{1}{P_2} \tag{83}$$

$$Q_1 = -\frac{1}{Q_2} \tag{84}$$

and the antennas are cross polarized. Axial ratios for the two antennas are the same, and the senses of rotation are opposite. From (84)

$$\theta_1 = -\theta_2 + \pi \tag{85}$$

and from (43)

$$\tau_2 = \pi/2 - \tau_1 \tag{86}$$

The polarization ellipses of waves from the two antennas have the same axial ratio, but the major axis of one ellipse coincides with the minor axis of the other. When viewed from one point in space the field vectors radiated by the two antennas rotate in the same direction.

From (82) it is clear that if the polarization of one antenna and the receiving polarization of the other antenna are plotted on the Poincaré sphere, they will be at opposite points on a sphere diameter.

## 3.5. Orthogonal Wave Components Reexamined

In a previous section it was seen that a wave can be considered composed of two orthogonal elliptical waves with polarization ratios $P_1$ and $-1/P_1^*$, but a physical interpretation of orthogonality was not given. It will be seen now that if a receiving antenna has a polarization that causes it to receive one of the elliptical wave components without loss, it cannot receive the other component at all.

Let the orthogonal components of the transmitted waves have polarization ratios $P_1$ and $-1/P_1^*$. The polarization ratio of the receiving antenna is $P_2$. If it is matched to receive polarization $P_1$ without loss, so that $\rho=1$, from (70),

$$P_2 = -P_1^* \tag{87}$$

Consider next the match between $P_2$ and the transmitted component, $-1/P_1^*$. Substitution in (70), using $-1/P_1^*$ instead of $P_1$, and $P_2 = -P_1^*$ gives

$$\rho = 0 \tag{88}$$

# 4. SCATTERING BY A TARGET

## 4.1. The Scattering Matrix

Coordinate systems that may be used to describe scattering from a target are shown in Fig. 5.

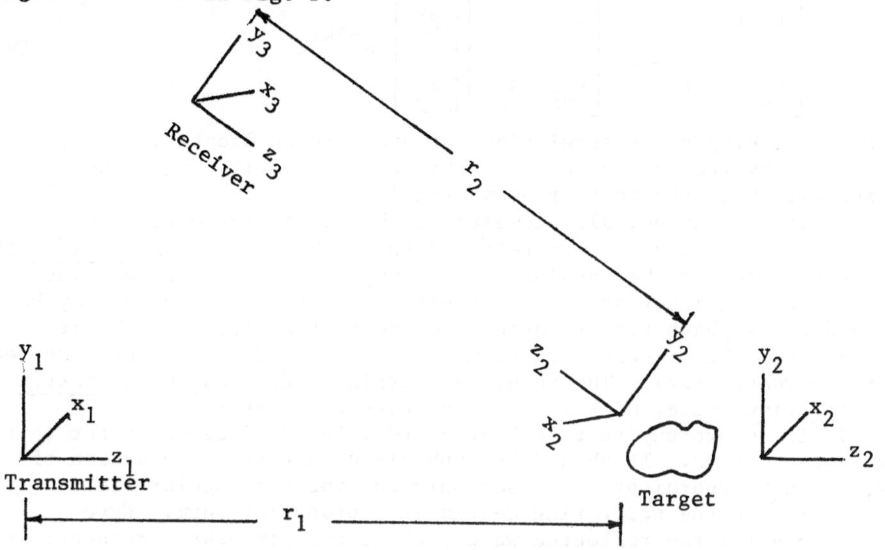


Fig. 5 Coordinate Systems for Scattering

The field incident on the target from a transmitting antenna of effective length $\bar{h}$ is

$$\begin{bmatrix} E_{x1}^{i} \\ E_{y1}^{i} \end{bmatrix} = j\,\frac{Z_o I}{2\lambda r_1}\begin{bmatrix} h_{x1} \\ h_{y1} \end{bmatrix} \tag{89}$$

The incident wave causes a scattered wave at the receiver that may be written either in $x_2y_2$ or $x_3y_3$ coordinates. The scattered and incident waves are related by a matrix, thus

$$\begin{bmatrix} E_{x2}^{r} \\ E_{y2}^{r} \end{bmatrix} = \frac{1}{\sqrt{4\pi}\, r_2}\begin{bmatrix} L_{x2x1} & L_{x2y1} \\ L_{y2x1} & L_{y2y1} \end{bmatrix}\begin{bmatrix} E_{x1}^{i} \\ E_{y1}^{i} \end{bmatrix} e^{-jkr_2} \tag{90}$$

or by

$$\begin{bmatrix} E_{x3}^{r} \\ E_{y3}^{r} \end{bmatrix} = \frac{1}{\sqrt{4\pi}\, r_2}\begin{bmatrix} S_{x3x1} & S_{x3y1} \\ S_{y3x1} & S_{y3y1} \end{bmatrix}\begin{bmatrix} E_{x1}^{i} \\ E_{y1}^{i} \end{bmatrix} e^{-jkr_2} \tag{91}$$

Use of the second form, in $x_1x_1$ and $x_3y_3$ coordinates, is standard practice in radar. The matrix [S] is the target scattering matrix or simply the scattering matrix. Even for bistatic scattering, as shown, if the coordinate systems are kept in mind the numeric subscripts can be omitted and the scattered field written as

$$\begin{bmatrix} E_x^r \\ E_y^r \end{bmatrix} = \frac{1}{\sqrt{4\pi}\ r_2} \begin{bmatrix} S_{xx} & S_{xy} \\ S_{yx} & S_{yy} \end{bmatrix} \begin{bmatrix} E_x^i \\ E_y^i \end{bmatrix} e^{-jkr_2} \tag{92}$$

If the receiving and transmitting antennas are co-located, the $x_1y_1z_1$ and $x_3y_3z_3$ systems coincide. For this situation, it may be shown by reciprocity that the scattering matrix is symmetric.

As it is written, all elements of the scattering matrix are complex, and it is sometimes referred to as the scattering matrix with absolute phase, or the absolute scattering matrix. If the absolute phase is not of interest, a common situation, the exponential may be dropped and a phase term removed from the matrix elements. If it is removed from the off-diagonal terms, then two elements can be made real (for backscattering). The resulting matrix is the scattering matrix with relative phase, or the relative scattering matrix.

It is noteworthy that IEEE Standard 145-1983 does not refer to the scattering matrix. It should be emphasized that the combination of $x_1y_1$ and $x_3y_3$ coordinates is used only for the rectangular field components and the scattering matrix in rectangular form. Wave descriptors for the reflected wave such as the circular components, the polarization ratios, and the Stokes parameters are given in $x_2y_2z_2$ coordinates. Target descriptors such as the scattering matrix in circular components and the Mueller matrix use the two systems $x_1y_1z_1$ and $x_2y_2z_2$. Transmission by the target is obviously best treated by $x_1y_1z_1$ and $x_2y_2z_2$, which coincide if the wave direction is unchanged.

## 4.2. Back-Scattering Efficiency

A monostatic radar with normalized transmitting antenna effective length $\bar{h}_1$ and receiving length $\bar{h}_2$ has an open-circuit induced voltage, when scattered wave $\bar{E}^r$ is received from the target, of

$$V = \bar{h}_2 \cdot \bar{E}^r = \frac{j\ Z_o\ I}{\sqrt{4\pi}\ (2\lambda r^2)} [h_1]^T\ [S]\ [h_2] \tag{93}$$

The magnitude of V may be maximized by the choice of receiver effective length, with

$$\frac{h_{2y}}{h_{2x}} = \frac{E_y^r}{E_x^r}^* \tag{94}$$

If normalized effective length vectors (polarization states) are defined in new coordinates by

$$[h_1] = [R][H_1] \quad (a) \qquad [h_2] = [R][H_2] \quad (b) \tag{95}$$

where [R] is unitary, and $\overline{H}_1$ and $\overline{H}_2$ are of unit length, V becomes

$$V = \frac{j\, Z_o\, I_J}{\sqrt{4\pi}\,(2\lambda r^2)} \; [H_2]^T\, [R]^T\, [S]\, [R]\, [H_1] \tag{96}$$

If [S] is symmetric, then

$$[D] = [R]^T\, [S]\, [R] \tag{97}$$

is diagonal, with complex elements in general, or

$$[D] = \begin{bmatrix} D_a & o \\ o & D_b \end{bmatrix} \tag{98}$$

with this substitution the received voltage magnitude is

$$|V| = \frac{Z_o\, I}{\sqrt{4\pi}\,(2\lambda r^2)} \, |H_{1a}\, H_{2a}\, D_a + H_{1b}\, H_{2b}\, D_b| \tag{99}$$

The voltage magnitude may be maximized by the choice of angles for the effective lengths, giving

$$|V|_m = |H_{1a}|\, |H_{2a}|\, |D_a| + |H_{1b}|\, |H_{2b}|\, |D_b| \tag{100}$$

Now if $|D_a| > |D_b|$, then $|V|_m$ has its greatest value if

$$|H_{1a}| = |H_{2a}| = 1 \quad (a) \qquad |H_{1b}| = |H_{2b}| = 0 \quad (b)$$

and if $|D_a| < |D_b|$ when (101)

$$|H_{1a}| = |H_{2a}| = 0 \quad (c) \qquad |H_{1b}| = |H_{2b}| = 1 \quad (d)$$

The angle constraint leading to (100), and the length requirement of (101) are satisfied if

$$\overline{H}_1 = \overline{H}_2 \tag{102}$$

and the normalized effective lengths of the transmitting and receiving antennas are equal

$$\overline{h}_1 = \overline{h}_2 \tag{103}$$

This is an important result; the received power backscattered from a target can be a maximum if one antenna, with the proper polarization state, is used both for transmitting and receiving.

Equation (94) for the optimum receiving antenna may be written as

$$\alpha\, \overline{h}_2^{*} = \overline{E}^{r} \tag{104}$$

where $\alpha$ is an arbitrary constant, and combined with

$$\overline{E}^r = \frac{j\, Z_o\, I}{\sqrt{4\pi}\, (2\lambda r^2)} \tag{105}$$

to give

$$[S]\,[h_1] = \gamma\, \overline{h}_2^* \tag{106}$$

where $\gamma$ is arbitrary.

Now, since maximum power can be received for transmitting and receiving antennas the same, the subscripts can be dropped, and the optimum monostatic radar antenna satisfies

$$[S]\,[h_{opt}] = \gamma\, \overline{h}_{opt}^* \tag{107}$$

This equation can be manipulated to give

$$([S]^*[S] - |\gamma^2|\,[I])\,[h_{opt}] = [0] \tag{108}$$

Since $[S]^*[S]$ is Hermitian, the eigenvalues $|\gamma|^2$ are real, given by

$$|\gamma_1|^2,\ |\gamma_2|^2 = \frac{B}{2} \pm \frac{1}{2}\sqrt{B^2 - 4C} \tag{109}$$

where

$$B = |S_{xx}|^2 + 2|S_{xy}|^2 + |S_{yy}|^2 \qquad \text{(a)}$$
$$C = |S_{xx}|S_{yy}|^2 + |S_{xy}|^4 - 2\mathrm{Re}(S_{xx}\, S_{xy}^{*2}\, S_{yy}) \qquad \text{(b)} \tag{110}$$

The optimum $\overline{h}$ corresponding to these eigenvalues is found from

$$[S]^*[S][h_{opt}] = |\gamma^2|\,[h_{opt}] \tag{111}$$

and leads to optimum polarization ratios for the radar antenna (using the relative scattering matrix with $S_{xy}$ real)

$$P_1 = -S_{xy}\,\frac{S_{xx} + S_{yy}^*}{S_{xy}^2 + |S_{yy}|^2 - |\gamma_1|^2} \qquad \text{(a)}$$
$$P_2 = -S_{xy}\,\frac{S_{xx} + S_{yy}^*}{S_{xy}^2 + |S_{yy}|^2 - |\gamma_2|^2} \qquad \text{(b)} \tag{112}$$

The received voltage with optimum polarization is given by (83), appropriately modified,

$$V_{opt} = \frac{j\, Z_o\, I}{\sqrt{4\pi}\, (2\lambda r^2)}\,[h_{opt}]^T\,[S]\,[h_{opt}] \tag{113}$$

and with the use of (106) may be changed to

$$V_{opt} = \frac{j\, Z_o\, I\, \gamma_1}{\sqrt{4\pi}\,(2\lambda r^2)} [h_{opt}]^T [h_{opt}]^* \tag{114}$$

where the larger of the eigenvalues, $|\gamma_1|^2$, is used to give maximum received voltage. This maximum voltage may be used to normalize the received voltage for any polarization, and the ratio defines a backscatter polarization efficiency

$$\rho_s = \frac{|[h]^T[S][h]|^2}{|\gamma_1|^2} \tag{115}$$

where normalized effective lengths (antenna polarization states) are used.

The antenna polarization state that gives a zero polarization efficiency in (115) is readily found. If $\rho_s$ is set to zero, two antenna polarization ratios are readily found to be

$$P_3 = \frac{R - S_{xy}}{S_{yy}} \quad \text{(a)} \qquad P_4 = -\frac{R + S_{xy}}{S_{yy}} \quad \text{(b)} \tag{116}$$

where

$$R = \sqrt{S_{xy}^2 - S_{xx}\, S_{yy}} \tag{117}$$

and the relative scattering matrix is used.

In a monostatic radar, the power received by the same antenna that transmits is called the co-polarized power. Then the polarization states corresponding to $\rho_s = 0$ are called the co-polarized power null states, or succinctly, the co-pol nulls. The states corresponding to $\rho_s = 1$ are the co-polarized power maxima, or the co-pol maxima. The power that is received by an antenna orthogonal to the transmitting antenna is the cross-polarized power. The transmitting antenna polarization that causes this power to be zero is a cross-polarized null state, frequently called a cross-pol null, or an X-pol null. It may be shown that the X-pol nulls coincide with the co-pol maxima.

## 4.3. Scattering Matrix for Circular Wave Components

It is often convenient to use the scattering parameters in left and right circular component form so that the scattered fields are given by

$$\begin{bmatrix} E_R^r \\ E_L^r \end{bmatrix} = \frac{1}{\sqrt{4\pi}\, r_2} \begin{bmatrix} S_{RR} & S_{RL} \\ S_{LR} & S_{LL} \end{bmatrix} \begin{bmatrix} E_R^i \\ E_L^i \end{bmatrix} e^{-jkr_2} \tag{118}$$

The elements of the circular scattering matrix may be found in terms of the Cartesian matrix if $E_R$ and $E_L$ are replaced by their equivalents from (31) and (32). This procedure focuses on a difference between the rectangular and circular scattering matrices. The rectangular component fields are measured in one coordinate system for both incident and reflected waves, but it is inappropriate to do that for circular components. For the incident field, therefore, (31) and (32) are used directly for $E_R$ and $E_L$, but for the reflected wave a right handed coordinate system with unit vectors $\bar{u}_x$, $-\bar{u}_y$, $-\bar{u}_z$ is utilized to define the reflected circular components. For the reflected wave, $E_y$ is replaced by $-E_y$, and (118) becomes

$$\begin{bmatrix} E_x^r - j\,E_y^r \\ E_x^r + j\,E_y^r \end{bmatrix} = \frac{1}{\sqrt{4\pi}\,r_2} \begin{bmatrix} S_{RR} & S_{RL} \\ S_{LR} & S_{LL} \end{bmatrix} \begin{bmatrix} E_x^i + j\,E_y^i \\ E_x^i - j\,E_y^i \end{bmatrix} e^{-jkr_2} \qquad (119)$$

When this equation is compared to (92) the circular matrix elements are found to be

$$S_{RR} = \frac{1}{2}(S_{xx} - j\,S_{xy} - j\,S_{yx} - S_{yy}) \qquad (a)$$

$$S_{RL} = \frac{1}{2}(S_{xx} + j\,S_{xy} - j\,S_{yx} + S_{yy}) \qquad (b)$$

$$S_{LR} = \frac{1}{2}(S_{xx} - j\,S_{xy} + j\,S_{yx} + S_{yy}) \qquad (c) \qquad (120)$$

$$S_{LL} = \frac{1}{2}(S_{xx} + j\,S_{xy} + j\,S_{yx} - S_{yy}) \qquad (d)$$

which may be simplified further if it is recognized that $S_{xy} = S_{yx}$ for backscattering.

### 4.4. Polarization Ratio of the Reflected Wave

The polarization ratios of incident and reflected waves are

$$P^i = \frac{E_y^i}{E_x^i} \quad (a) \qquad\qquad P^r = -\frac{E_y^r}{E_x^r} \quad (b) \qquad (121)$$

where again it is noted that $P^r$ must be defined in a reversed coordinate system. Substitution into (92) gives

$$\begin{bmatrix} 1 \\ P^r \end{bmatrix} = \frac{1}{\sqrt{4\pi}\,r} \left[\frac{E_x^i}{E_x^r}\right] \begin{bmatrix} S_{xx} & S_{xy} \\ S_{yx} & S_{yy} \end{bmatrix} \begin{bmatrix} 1 \\ P^i \end{bmatrix} \qquad (122)$$

which is readily solved for the polarization ratio of the reflected wave,

$$P^r = - \frac{S_{yy} P^i + S_{yx}}{S_{xy} P^i + S_{xx}} \tag{123}$$

## 4.5 Scattering Cross Section and Radar Cross Section

In much of the literature the phrases scattering cross section and radar cross section are not satisfactorily distinguished. IEEE Standard 145-1983 uses total radiation intensity of the scattered wave, independent of polarization, to define scattering cross section. The Standard uses radar cross section to be "that portion of the scattering cross section corresponding to a specified polarization component of the scattered wave." Both cross sections depend on the polarization of the incident wave; the radar cross section also depends on the polarization of the receiving antenna, whereas the scattering cross section does not.

## 4.6. Depolarization and Repolarization

A completely polarized wave reflected from a complex moving target has a degree of polarization less than unity. It would be appropriate to say that the wave is depolarized, were it not that depolarization is widely used for another phenomenon, the change in polarization state when a completely polarized wave is reflected as a completely polarized wave with a different polarization ratio. It is unfortunate that "depolarization" is used to describe a change of state while the degree of polarization is unchanged, since the prefix "de" means to do the opposite of, to reverse, or to remove from, and none of these happens to the polarization of the reflected wave. The prefix "re" has an origin involving "back" or "again", as in "restore". In some words, such as "reclassify" and "rename", it clearly means to alter the classification or alter the name. In this second sense it is applicable to the change of polarization state of a wave.

It is recommended that a change in wave state that decreases its degree of polarization without changing the polarization ratio of the polarized part of the wave be called depolarization. Using the second sense of the prefix "re", meaning to alter, a change in wave state that changes the polarization ratio of the wave without changing its degree of polarization should be called repolarization. Then the completely polarized wave reflected from a stationary target is repolarized and one reflected from a moving target is both repolarized and depolarized.

# 5. PARTIAL POLARIZATION

## 5.1. Description of Partially Polarized Waves

A field component of a partially-polarized wave can be represented by the analytic signal of Gabor.

$$E(t) = a(t)\, e^{j[\omega t + \phi(t)]} \tag{124}$$

where a(t) and $\phi(t)$ vary slowly enough that the real part of E(t) approximates a cosine. Such a wave is called quasi-monochromatic. The fields of a wave traveling in the z direction are

$$E_x(t) = a_x(t)\, e^{j[\omega t + \phi_x(t)]} \quad \text{(a)}$$
$$E_y(t) = a_y(t)\, e^{j[\omega t + \phi_y(t)]} \quad \text{(b)} \tag{125}$$

and we may write the component at angle $\theta$ to the x axis as

$$E(t,\theta) = E_x(t) \cos\theta + E_y(t) \sin\theta \tag{126}$$

The time average of this wave is of interest and it will be written as

$$I = \langle E(t,\theta)\, E^*(t,\theta)\rangle = \lim_{T\to\infty} \frac{1}{2T} \int_{-T}^{T} E_T\, E_T^*\, dt \tag{127}$$

where $E_T = E(\theta,t)$ in the integration range and zero outside.

The time average of the wave of (128) is

$$I = \langle E_x(t)\, E_x^*(t)\rangle \cos^2\theta + \langle E_y(t)\, E_y^*(t)\rangle \sin^2\theta$$
$$+ [\langle E_x(t)\, E_y^*(t)\rangle + \langle E_x^*(t)\, E_y(t)\rangle] \sin\theta \cos\theta \tag{128}$$

It is desirable because of these time averages to define a coherency matrix for the wave as

$$[J] = \frac{1}{2\,Z_o} \begin{bmatrix} \langle E_x(t)\, E_x^*(t)\rangle & \langle E_x(t)\, E_y^*(t)\rangle \\ \langle E_x^*(t)\, E_y(t)\rangle & \langle E_y(t)\, E_y^*(t)\rangle \end{bmatrix} \tag{129}$$

which, using (125), may also be written as

$$[J] = \frac{1}{2\,Z_o} \begin{bmatrix} \langle a_x^2\rangle & \langle a_x\, a_y\, e^{j(\phi_x-\phi_y)}\rangle \\ \langle a_x\, a_y\, e^{j(\phi_y-\phi_x)}\rangle & \langle a_y^2\rangle \end{bmatrix} \tag{130}$$

If the amplitude and phase functions in (125) vary so slowly that the time derivatives in the Maxwell equations can be replaced by $j\omega$, the time-average Poynting vector is equal to the trace of the coherency matrix

$$U = \langle a_x^2\rangle + \langle a_y^2\rangle \tag{131}$$

The off-diagonal terms of the coherency matrix may be normalized to give

$$\mu_{xy} = \frac{J_{xy}}{\sqrt{J_{xx}}\,\sqrt{J_{yy}}} \tag{132}$$

For unpolarized waves, the time average I of (128) must be independent of angle θ, and also independent of a fixed phase change in one of the field components. This requires that

$$J_{xx} = J_{yy} \qquad J_{xy} = J_{yx} = 0 \qquad \mu_{xy} = 0 \tag{133}$$

The coherency matrix becomes

$$[J] = \frac{S}{2} \begin{bmatrix} 1 & 0 \\ 0 & 1 \end{bmatrix} \tag{134}$$

A monochromatic wave is completely polarized. Then

$$[J] = \frac{1}{2\,Z_o} \begin{bmatrix} a_x^2 & a_x\,a_y\,e^{-j\phi} \\ a_x\,a_y\,e^{j\phi} & a_y^2 \end{bmatrix} \tag{135}$$

where

$$\phi = \phi_y - \phi_x \tag{136}$$

It follows that

$$|\mu_{xy}| = 1 \tag{137}$$

and this may be used as a definition of <u>complete coherence</u> and <u>complete polarization</u>. A non-monochromatic wave may also be completely polarized if $a_x$, $a_y$, $\phi_x$, $\phi_y$ depend on time in such a way that the amplitude ratio and the phase difference are time-independent.

$$\frac{a_y(t)}{a_x(t)} = C_1 \ \text{(a)} \qquad \phi = \phi_y(t) - \phi_x(t) = C_2 \ \text{(b)} \tag{138}$$

With these constraints the coherency matrix becomes

$$[J] = \frac{1}{2\,Z_o} \begin{bmatrix} \langle a_x^2\rangle & C_1\langle a_x^2\rangle e^{-jC_2} \\ C_1\langle a_x^2\rangle e^{jC_2} & C_1^2\langle a_x^2\rangle \end{bmatrix} \tag{139}$$

from which

$$\mu_{xy} = e^{-jC_2} \tag{140}$$

The wave is completely polarized.

The coherency matrix may be written as

$$[J] = [J^{(1)}] + [J^{(2)}] \tag{141}$$

where

$$[J^{(1)}] = \begin{bmatrix} A & 0 \\ 0 & A \end{bmatrix} \qquad [J^{(2)}] = \begin{bmatrix} B & D \\ D^* & C \end{bmatrix} \tag{142}$$

with A, B, C real and positive, and BC-DD* = 0. The first matrix is clearly that of an unpolarized wave, and since $|\mu_{xy}| = 1$ the second matrix is for a completely polarized wave. A, B, C, and D are readily found from the elements of [J].

The Poynting vector magnitudes of the total wave and the polarized part are

$$U_t = J_{xx} + J_{yy} \qquad \text{(a)}$$
$$U_p = B+C = [(J_{xx}+J_{yy})^2 - ||J||]^{\frac{1}{2}} \qquad \text{(b)} \tag{143}$$

and their ratio is defined as the <u>degree</u> <u>of</u> <u>polarization</u>,

$$R_p = \frac{U_p}{U_t} = \left(1 - \frac{4||J||}{(J_{xx}+J_{yy})^2}\right)^{\frac{1}{2}} \tag{144}$$

which has a range 0-1.

The Stokes parameters were originally introduced to describe partially polarized light, for which they are

$$G_0 = \langle a_x^2 \rangle + \langle a_y^2 \rangle \qquad \text{(a)}$$
$$G_1 = \langle a_x^2 \rangle - \langle a_y^2 \rangle \qquad \text{(b)}$$
$$G_2 = 2\langle a_x\, a_y \cos\phi \rangle \qquad \text{(c)}$$
$$G_3 = 2\langle a_x\, a_y \sin\phi \rangle \qquad \text{(d)} \tag{145}$$

If the wave component amplitude and phase are time independent, these equations reduce to (46). It is obvious that these parameters are related to the coherency matrix elements of (130). The relationship is

$$G_0 = 2\, Z_o\, (J_{xx} + J_{yy}) \qquad \text{(a)}$$
$$G_1 = 2\, Z_o\, (J_{xx} - J_{yy}) \qquad \text{(b)}$$
$$G_2 = 2\, Z_o\, (J_{xy} + J_{yx}) \qquad \text{(c)}$$
$$G_3 = 2\, j\, Z_o\, (J_{xy} - J_{yx}) \qquad \text{(d)} \tag{146}$$

With appropriate substitutions in (144) it is straightforward to show that the degree of polarization of a wave is

$$R_p = \frac{(G_1^2 + G_2^2 + G_3^2)^{\frac{1}{2}}}{S_0} \tag{147}$$

If a target has a scattering matrix the coherency matrix of a wave reflected from it can be expressed in terms of the scattering matrix and the coherency matrix of the incident wave. Let the incident and reflected wave coherency matrices be $[J^i]$ and $[J^r]$ respectively, where

$$[J] = \frac{1}{2\, Z_o} <[E][E]^{T*}> \tag{148}$$

Using the relation

$$[E^r] = \frac{1}{\sqrt{4\pi}\ r^2} \quad [L][E^i] \tag{149}$$

to write the field at the receiver at distance r from the target, the coherency matrix of the reflected wave can be expressed as

$$[J^r] = \frac{1}{4\pi\ r^2} \ \frac{1}{2\ Z_o} <[L][E^i][E^i]^{T*}[L]^{T*}> \tag{150}$$

which becomes, if [L] is time invariant

$$[J^r] = \frac{1}{4\pi\ r^2} \quad [L][J^i][L]^{T*} \tag{151}$$

It is appropriate to define $[J^i]$ in the $x_1y_1z_1$ coordinates of Fig. 5 and $[J^r]$ in $x_2y_2z_2$ coordinates. Then matrix [L] must be defined using the two systems, as discussed previously, and differs from the common form of the scattering matrix.

## 5.2 Reception of Partially Polarized Waves

The open circuit voltage induced by an incident field $\overline{E}$ in an antenna of effective length $\overline{h}$ is

$$V = \overline{E} \cdot \overline{h} \tag{152}$$

whether $\overline{E}$ is completely polarized or not. The power to a matched load is then

$$W = \frac{<V\ V^*>}{8\ R_a} \tag{153}$$

where $R_a$ is the antenna resistance. It is related to the antenna effective length and effective receiving area by

$$R_a = \frac{Z_o \; \overline{h} \cdot \overline{h}^*}{4 \; A_e} \tag{154}$$

Then

$$W = \frac{A_e}{2 \; Z_o \; \overline{h} \cdot \overline{h}^*} < (\overline{E} \cdot \overline{h})(\overline{E} \cdot \overline{h})^* > \tag{155}$$

If it is noted that time averaging is unnecessary for the receiving antenna, and if the elements of the coherency matrix are used for the incident wave, the received power becomes

$$W = \frac{A_e}{\overline{h} \cdot \overline{h}} \left(|h_x|^2 \; J_{xx} + h_x \; h_y^* \; J_{xy} + h_x^* \; h_y \; J_{yx} + |h_y|^2 \; J_{yy}\right) \tag{156}$$

It was seen previously that a partially polarized wave is the sum of completely polarized and unpolarized waves. In (156), if the matrix elements of (142) are substituted, the received power becomes

$$\begin{aligned} W &= W' + W'' \\ &= A_e \; A + \frac{A_e}{\overline{h} \cdot \overline{h}^*} \left(|h_x|^2 \; B + h_x \; h_y^* \; D + h_x^* \; h_y \; D^* + |h_y|^2 \; C\right) \end{aligned} \tag{157}$$

The first term, which may be put into the form

$$W' = A_e \; A = \frac{1}{2} \; A_e \; U_t \; (1 - R_p) \tag{158}$$

represents the power received from the unpolarized portion of the wave. $U_t$ is the power density of the total wave. It should be noted that if the wave is unpolarized ($R_p = 0$) the maximum power is one half the power that could be extracted from a completely polarized wave matched to the receiving antenna.

The second term in (157), which represents the power that can be extracted from the polarized part of the wave, can be maximized by choosing

$$\frac{h_y}{h_x} = \frac{C}{D*} = \frac{D}{B} \tag{159}$$

and is given by

$$W''_m = A_e \; U_p = A_e \; U_t \; R_p \tag{160}$$

which is of course the effective area of the receiving antenna multiplied by the power density of the polarized part of the wave. Then the maximum total received power is

$$W_m = W' + W''_m = \frac{1}{2} A_e U_t (1+R_p) \tag{161}$$

## 5.3. The Stokes Vector and Received Power

A widely used method for treating partially-coherent scattering uses the Stokes vector and the Mueller matrix. It will be introduced initially with monochromatic waves and extended to the quasi-monochromatic case. The Stokes vector is comprised of the Stokes parameters, thus

$$[G] = \text{col}[G_0 \;\; G_1 \;\; G_2 \;\; G_3] \tag{162}$$

Some writers use the elements in a different order, but the one shown is common. In analogy to the polarization ratio, the Stokes vector of an antenna is related to the Stokes vector of the wave it transmits.

Consider two antennas with a receiving and b transmitting. The Stokes vector for antenna a is given in a coordinate system appropriate to a as a transmitter, e.g. the xyz system of Fig. 1, even though it is receiving. Likewise, the Stokes vectors for antenna b and its radiated wave are given in coordinates appropriate to it as a transmitter, the $\xi\eta\zeta$ system of Fig. 1.

The Stokes vector of the wave incident from antenna b is

$$[G^i] = \begin{bmatrix} <|E_\xi|^2 + |E_\eta|^2> \\ <|E_\xi|^2 - |E_\eta|^2> \\ 2 \text{ Re } <E_\xi^* E_\eta> \\ 2 \text{ Im } <E_\xi^* E_\eta> \end{bmatrix} \tag{163}$$

The Stokes vector of antenna a is related to the Stokes vector of its radiated wave, which in turn is proportional to the effective length of a. It is therefore appropriate to define the Stokes vector of an antenna as

$$[G_A] = \begin{bmatrix} |h_x|^2 + |h_y|^2 \\ |h_x|^2 - |h_y|^2 \\ 2 \text{ Re } (h_x^* h_y) \\ 2 \text{ Im } (h_x^* h_y) \end{bmatrix} \tag{164}$$

Note that for the antenna time averaging is unnecessary.

It is reasonable to ask if a product of $[G^i]$ and $[G_A]$ will give the received power. In the given form, it will not, but if a receiving Stokes vector of the antenna is defined by replacing x,y by $\xi,\eta$ and changing the sign of $G_3$ in the antenna Stokes vector, thus

$$[G_R] = \begin{bmatrix} |h_\xi|^2 + |h_\eta|^2 \\ |h_\xi|^2 - |h_\eta|^2 \\ 2\ \mathrm{Re}\ (h_\xi^* h_\eta) \\ -2\ \mathrm{Im}\ (h_\xi^* h_\eta) \end{bmatrix} \tag{165}$$

then the power received is readily seen to be

$$W = \frac{1}{16\ R_a} [G^i]^T [G_R] \tag{166}$$

5.4. The Mueller Matrix

If a partially-polarized wave is scattered by a target it remains partially-polarized. In this case, the scattering matrix formulation of (92) loses much of its usefulness. The Stokes vectors retain their validity, however.

If the Stokes vector of the wave incident on the target of Fig. 5 is $\overline{G}^i$, the Stokes vector at the receiving antenna at distance $r_2$ from the target is

$$[G^r(x_2,y_2,z_2)] = \frac{1}{4\pi\ r_2^2} [M][G^i(x_1,y_1,z_1)] \tag{167}$$

where [M] is a 4 x 4 matrix called the Mueller matrix. The functional notation is employed to stress that $\overline{G}^r$ and $\overline{G}^i$ are defined in different coordinate systems. If the incident wave is completely polarized and the target immobile, the reflected wave is also completely polarized. If the incident wave is partially-polarized, so will be the reflected wave, and the Mueller matrix elements may be determined at the mean frequency of the quasi-monochromatic wave [20]. Even with a completely polarized incident wave, target motion will cause the reflected wave to be partially-polarized. Note that (167) is not restricted to the monostatic scattering case.

Consider the backscatter problem with the same antenna used to transmit and receive. Let its Stokes vector be $[G_A]$. The field incident on the target is

$$G^i = \frac{Z_o^{\ 2}\ I^2}{4\lambda^2 r^2} [G_A] \tag{168}$$

and the Stokes vector of the reflected wave at the receiving antenna is

$$[G^r] = \frac{Z_o^{\ 2}\ I^2}{16\pi\lambda^2 r^4} [M][G_A] \tag{169}$$

If this is substituted into (166), noting that the refected wave of (169) is the incident wave of (166) the received power in the monostatic radar is

$$W = \frac{Z_o^2 I^2}{256\pi R_a \lambda^2 r^4} [G_A]^T [M]^T [G_A] \tag{170}$$

5.5. The Coherency Matrix and Received Power

In Section 5.3 a Stokes vector for an antenna was defined, using effective length components instead of field intensities. If, in analogous manner, a coherency matrix for an antenna is defined as

$$[J_A] = \begin{bmatrix} |h_x|^2 & h_x h_y^* \\ h_x^* h_y & |h_y|^2 \end{bmatrix} \tag{171}$$

then the power received from an incoming wave whose coherency matrix is known can be determined. Recall that it was necessary in Section 5.3 to use a receiving Stokes vector. The coherency matrix corresponding to the receiving Stokes vector of an antenna is

$$[J_R] = \begin{bmatrix} |h_\xi|^2 & h_\xi^* h_\eta \\ h_\xi h_\eta^* & |h_\eta|^2 \end{bmatrix} \tag{172}$$

By definition, this is the receiving coherency matrix of the antenna.

The incoming wave to the receiving antenna has coherency matrix

$$[J^i] = \frac{1}{2 Z_o} \begin{bmatrix} <|E_\xi|^2> & <E_\xi E_\eta^*> \\ <E_\xi^* E_\eta> & <|E_\eta|^2> \end{bmatrix} \tag{173}$$

The product of these matrices is a matrix whose trace is

$$\mathrm{Tr}([J^i][J_R]) = \frac{1}{2 Z_o} = <|\bar{E} \cdot \bar{h}|^2> \tag{174}$$

The received power is then

$$W = \frac{1}{4 R_a} \mathrm{Tr}\,([J^i][J_R]) \tag{175}$$

5.6 The Power Scattering Matrix

If the incident wave on a radar target is completely polarized and the scattered wave is depolarized by target motion, the partially-polarized scattered wave can be treated by use of the Graves' polarization power scattering matrix, or, as it is commonly called, the power scattering matrix. It is introduced here by relating it to the scattering matrix, but its existence does not depend on that of the scattering matrix.

For a monostatic radar, the power density at the receiving antenna is

$$U = \frac{1}{2\,Z_o}\,[E^r]^{T*}\,[E^r] = \frac{1}{8\pi Z_o r^2}\,[E^i]^{T*}\,[S]^{T*}\,[S]\,[E^i] \qquad (176)$$

The matrix

$$[\sigma] = [S]^{T*}\,[S] = \begin{bmatrix} |S_{xx}|^2 + |S_{yx}|^2 \\ S_{xx}\,S_{xy}^* + S_{yx}\,S_{yy}^* \end{bmatrix} \begin{bmatrix} S_{xx}^*\,S_{xy} + S_{yx}^*\,S_{yy} \\ |S_{xy}|^2 + |S_{yy}|^2 \end{bmatrix} \qquad (177)$$

is the power scattering matrix of the target. It simplifies somewhat if the relative scattering matrix, with $S_{xy}$ real, is used.

The power scattering matrix can be decomposed into

$$[\sigma] = [\sigma_H] + [\sigma_V] \qquad (178)$$

where $[\sigma_H]$ determines the power that can be received by a horizontal (x directed) antenna and $[\sigma_V]$ the power that can be received by a vertical (y directed) antenna. The split is achieved by noting that the voltage measured by an x-directed receiving antenna is

$$V_H = \frac{h_x}{\sqrt{4\pi}\;r}\,(S_{xx}\,E_x^i + S_{xy}\,E_y^i) \qquad (179)$$

and that measured by a y-directed antenna is

$$V_V = \frac{h_y}{\sqrt{4\pi}\;r}\,(S_{yx}\,E_x^i + S_{yy}\,E_y^i) \qquad (180)$$

Comparison of received powers shows that the appropriate decomposition of $[\sigma]$ is

$$[\sigma_H] = \begin{bmatrix} |S_{xx}|^2 & S_{xx}^*\,S_{xy} \\ S_{xx}\,S_{xy}^* & |S_{xy}|^2 \end{bmatrix} \;(a) \qquad [\sigma_V] = \begin{bmatrix} |S_{yx}|^2 & S_{yx}^*\,S_{yy} \\ S_{yx}\,S_{yy}^* & |S_{yy}|^2 \end{bmatrix} \;(b) \qquad (181)$$

If the target does not have a scattering matrix, the scattered power density can be found in terms of the Mueller matrix and the incident Stokes vector. Comparison then shown that the power scattering matrix can be written, if the incident wave is completely polarized, as

$$[\sigma] = \begin{bmatrix} M_{11} + M_{12} & M_{13} - jM_{14} \\ M_{13} + jM_{14} & M_{11} - M_{12} \end{bmatrix} \qquad (182)$$

## 6. CONCLUSION

In this overview of polarization theory, careful attention was paid to the appropriate coordinate systems for defining polarization parameters for two antennas in a transmit-receive configuration and for the reception of a wave reflected from a target. A distinction was made between the polarization and receiving polarization of an antenna. A Stokes vector and a receiving Stokes vector (and coherency matrices) for an antenna were defined, and the coordinate systems used in the definitions were specified.

The effective length of an antenna was given a prominent place in the discussion, and care was taken that equations be dimensionally correct.

Some inconsistencies in the literature definitions of polarization terms were noted, and a few inconsistencies and deficiencies in IEEE Standards 145-1983 and 149-1979 were pointed out.

## 7. BIBLIOGRAPHY


1. M. J. Beran and G. B. Parrent, Jr., Theory of Partial Coherence, Prentice-Hall, Englewood Cliffs, NJ, 1964.

2. P. Beckmann, "Optimum Polarization for Polarization Discrimination," Proc. IEEE, Vol. 56, No. 10, pp. 1755-1756, October 1968.

3. P. Beckmann and A. Spizzichino, The Scattering of Electromagnetic Waves from Rough Surfaces, Pergamon Press, New York, 1963.

4. P. Beckmann, The Depolarization of Electromagnetic Waves, The Golem Press, Boulder, 1968.

5. M. Born and E. Wolf, Principles of Optics, 3rd ed., Pergamon Press, Oxford, 1965.

6. S. R. Cloude, "Polarimetric Techniques in Radar Signal Processing," Microwave Journal, pp. 119-127, July 1983.

7. R. E. Collin and F. J. Zucker, Antenna Theory, Part 1, McGraw-Hill, New York, NY, 1969.

8. J. R. Copeland, "Radar Target Classification by Polarization Properties," Proc. IRE, Vol. 48, No. 7, pp. 1290-1296, July 1960.

9. D. Gabor, "Theory of Communication," Journal of the Institution of Electrical Engineers, Vol. 93, Part III, pp. 429-457, 1946.

10. C. D. Graves, "Radar Polarization Power Scattering Matrix," Proc. IRE, Vol. 44, No. 2, pp. 248-252, February 1956.

11. T. G. Hickman, et al., "Polarization Measurements," Chapter 10 in J. S. Hollis, et al., Microwave Antenna Measurements, Scientific-Atlanta, GA, 1970.

12. J. R. Huynen, "Phenomenological Theory of Radar Targets," Drukkerij Bronder-Offset, Rotterdam, 1970.

13. E. M. Kennaugh, "Effects of Type of Polarization on Echo Characteristics," Reports 389-1 to 389-24, Antenna Laboratory, The Ohio State University, Columbus, OH, 1949-1954.

14. G. H. Knittel, "The Polarization Sphere as a Graphical Aid in Determining the Polarization of an Antenna by Amplitude Measurements Only," IEEE Trans. on Antennas and Propagation, Vol. AP-15, No. 2, pp. 217-221, March 1967.

15. H. C. Ko, "On the Reception of Quasi-Monochromatic, Partially Polarized Radio Waves," Proc. IRE, Vol. 50, No. 9, pp. 1950-1957, September 1962.

16. H. C. Ko, "The Interaction of Radio Antennas with Statistical Radiation," Notes for Short Course in Antennas, Ohio State University, 1965.

17. A. B. Kostinski and W-M. Boerner, "On Foundations of Radar Polarimetry," IEEE Trans. on Antennas and Propagation, Vol. AP-34, No. 12, pp. 1395-1404, December 1986.

18. H. Mott, Polarization in Antennas and Radar, Wiley-Interscience, New York, NY, 1986.

19. E. L. O'Neill, Introduction to Statistical Optics, Addison-Wesley, Reading, MA, 1963.

20. C. H. Papas, Theory of Electromagnetic Wave Propagation, McGraw-Hill, New York, 1965.

21. R. Rubin, "Antenna Measurements," Chapter 34 in H. Jasik, ed., Antenna Engineering Handbook, McGraw-Hill, New York, 1961.

22. G. T. Ruck, D. E. Barrick, W. D. Stuart, and C. K. Krichbaum, Radar Cross Section Handbook, Plenum Press, New York, 1970.

23. V. H. Rumsey, G. A. Deschamps, M. L. Kales, and J. I. Bohnert, "Techniques for Handling Elliptically Polarized Waves with Special Reference to Antennas," Proc. IRE, Vol. 39, No. 5, pp. 533-552, May 1951.

24. G. Sinclair, "The Transmission and Reception of Elliptically Polarized Waves," Proc. IRE, Vol. 38, No. 2, pp. 148-151, February 1950.

25. G. Sinclair, "Modification of the Radar Range Equation for Arbitrary Targets and Arbitrary Polarization," Report #302-19, The Ohio State University Research Foundation, September 25, 1948.

26. J. A. Stratton, Electromagnetic Theory, McGraw-Hill, New York, 1941.

27. Definitions of Terms for Antennas, IEEE Standard 145-1983.

28. Test Procedures for Antennas, IEEE Standard 149-1979.

# BASIC CONCEPTS OF RADAR POLARIMETRY

Wolfgang-M. Boerner, Wei-Ling Yan, An-Qing Xi,
and Yoshio Yamaguchi
University of Illinois at Chicago, UIC-EECS/CL,
840 W. Taylor St., SEL-4210, M/C 154
Chicago, IL 60680-4348

**ABSTRACT:** A comprehensive overview of the basic principles of radar polarimetry is presented. The relevant fundamental field equations are first provided in order to introduce the polarization state formulations of electromagnetic waves in the frequency domain, including the Jones and the Stokes vector formalism and its presentation on the Poincaré sphere and on relevant map projections. In a next step, the scattering matrices [S] and [M] are given together with change of polarization bases transformation operators, where upon the optimal (characteristic) polarization states are determined for the coherent and partially coherent cases, respectively. This chapter is concluded with a set of simple examples.

## 1. INTRODUCTION

Radar polarimetry, i.e., utilization of complete electromagnetic vector wave information, has become an indispensible tool in modern electromagnetic sensor technology, both in the civil and the military sectors, and increasingly more in environmental remote sensing of the terrestrial and planetary atmospheres and crusts. From the outset, we emphasize that by incorporating coherent polarimetric phase and amplitude information into radar signal and image processing, one can anticipate and already is witnessing a breakthrough which is at least comparable to that brought about by the advent of holography and computer assisted (Radon projection) tomography and its applications to Synthetic Aperture Radar (SAR) and Inverse Synthetic Aperture Radar (ISAR).

In early RADAR (RAdio Detection And Ranging) only amplitude information of the electromagnetic wave at a suitable frequency was utilized which, since its inception at the beginning of this century, has become a key element in civil and military operations on land, at sea, and in the air. Then, some forty to fifty years later, it was possible to build wide-band radar systems which, in addition to frequency and amplitude, also utilize relative and absolute phase information for resolving physical features of scatterers and the background environment (vehicles, ships, aircraft, space objects, terrestrial and planetary surface structures). The increased resolution capability has provided the means of extending the original RADAR concept of radio detection and ranging to include capabilities for high resolution mapping, profiling, and



*W.-M. Boerner et al. (eds.), Direct and Inverse Methods in Radar Polarimetry, Part 1*, 155–245.

imaging unrelated to either detection and mapping. However, in order to further improve high resolution techniques for carrying out traditional radar tasks of search, track, and weapon control in increasingly difficult surveillance environments with increasing simultaneous target camouflaging capabilities; in addition to amplitude, frequency, relative and absolute target phase also complete coherent polarization information must be incorporated into the target versus background clutter image contrast enhancement algorithms. Here, the electromagnetic inverse problem is considered. This is one of the most formidable problems, requiring extensive resources for developing the underlying theory, metrology, and computer numerical tools of vector signal/image processing.

The problem of determining the scattered electric and magnetic fields everywhere, given the exact shape and distribution of material parameters ($\varepsilon, \mu, \sigma$), expressed in terms of a coordinate system placed in the center of the object for a given incident field, is known as the direct electromagnetic scattering problem. Inversely, the problem of determining the size, the shape, and electromagnetic properties of an unknown scatterer, given the incident field and the measured scattered electromagnetic fields, is defined to be the electromagnetic inverse scattering problem. The case in which the amplitude, phase, and, in addition, **polarization** are known and **relevant**, is known as the electromagnetic **vector inverse scattering problem**. The general mathematico-physico-electronic processing science of INVERSE METHODS or PROFILING is still in a rather embryonic phase of development, although such inverse methods have been developed to investigate a vast range of problems of the physical sciences, where the characteristics of a medium are estimated from experimental data in a given situation.

In high resolution polarimetric radar imaging, it is the objective to utilize the complete vector nature of eletromagnetic waves, i.e., in addition to amplitude, frequency and phase, also polarization state information of the transmitted and received waves is incorporated into signal and image (pixel-by-pixel) processing, requiring a 2x2 Sinclair matrix [S] and/or a 4x4 Mueller matrix [M] description of the scatterer for the coherent and partially polarized cases, respectively. Although there still exist some "grey areas" in both theory and techniques of radar polarimetry, in recent years considerable progress was made in theory, device technology and algorithm development for broadband polarimetric vector signal and image processing. These dramatic advances will have a definite impact on the merging military technologies of the 1990s, especially as regards the design and manufacture of high precision polarimetric antennas, multi-channel polarimetric transceiver systems, real-time polarimetric wideband signal and image processing modules including neural networking, etc., which, in a next step needs to be developed for polarimetric ultra-wideband impulsive radar imaging methods.

Whereas, in the introduction paper by Mott on the 'Definitions of Polarization in Radar', the basic formalism for handling polarization descriptors in radar polarimetry is presented, in the following a cohesive succinct summary of the basic principles for the determination of the optimal polarization states is developed. Although some overlap may exist, the approaches of the two papers are rather distinct and different.

## 2. FUNDAMENTAL FIELD EQUATIONS

In order to assess the presentation of the polarization state of electromagnetic plane waves, in this section the relevant fundamental field equations are briefly summarized, and we refer to other texts for details [58]. Essentially, the basic nomenclature, terminology and basic assumptions of analyzing basic polarization state properties of monochromatic and quasimonochromatic plane waves are here introduced.

### 2.1 Maxwell's equations in the time domain

At every point in space $\vec{r}$ the electromagnetic field vectors at any instant of time t are subject to Maxwell's equations

$$\nabla \times \vec{E}(\vec{r},t) + \frac{\partial}{\partial t} \vec{B}(\vec{r},t) = 0 \tag{2.1}$$

$$\nabla \times \vec{H}(\vec{r},t) - \frac{\partial}{\partial t} \vec{D}(\vec{r},t) = \vec{J}(\vec{r},t) \tag{2.2}$$

where $\vec{E}$ and $\vec{H}$ denote the electric and magnetic fields, and $\vec{B}$ and $\vec{D}$ the magnetic induction and dielectric displacement densities, respectively. The divergence of the field intensities leads to

$$\nabla \cdot \vec{B} = 0 \tag{2.3}$$

$$\nabla \cdot \vec{J}(\vec{r},t) + \frac{\partial}{\partial t} \rho (\vec{r},t) = 0 \tag{2.4}$$

$$\nabla \cdot \vec{D} = \rho \tag{2.5}$$

where $\vec{J}$ and $\rho$ represent the current and charge densities, and (2.4) defines the continuity of current equation. The macroscopic properties of the medium, in which the fields reside, are expressed in terms of the constitutive parameters $\varepsilon$, $\mu$, and $\sigma$, being the dielectric permittivity $\varepsilon$, the magnetic permiability $\mu$, and the electric conductivity $\sigma$, respectively, so that

$$\vec{D} = \varepsilon \vec{E} \tag{2.6}$$

$$\vec{H} = \frac{1}{\mu} \vec{B} \tag{2.7}$$

$$\vec{J} = \sigma \vec{E} \tag{2.8}$$

where it is assumed here that the medium is isotropic.

## 2.2 The vector wave equations in the time domain

Introducing the vector differential operator identity

$$\nabla \times (\nabla \times \vec{A}) = \nabla (\nabla \cdot \vec{A}) - \nabla \cdot \nabla \vec{A}$$

and operating with the rotation operator (curl: $\nabla$ x) on Eqs. 2.1 and 2.2 and by taking Eqs. 2.3 to 2.4 into consideration, we obtain the vector wave equations for $\vec{E}(\vec{r}, t)$ and $\vec{H}(\vec{r}, t)$

$$\nabla^2\vec{E} - \sigma\mu \frac{\partial}{\partial t} \vec{E} - \varepsilon\mu \frac{\partial^2}{\partial t^2} \vec{E} = \mu \frac{\partial}{\partial t} \vec{J}_S + \frac{1}{\varepsilon} \nabla\rho \tag{2.9}$$

$$\nabla^2\vec{H} - \sigma\mu \frac{\partial}{\partial t} \vec{H} - \varepsilon\mu \frac{\partial^2}{\partial t^2} \vec{H} = - \nabla \times \vec{J}_S \tag{2.10}$$

where the current density vector $\vec{J}$ was decomposed into the conduction current $\vec{J}_C = \sigma\vec{E}$ and the source current $\vec{J}_S$ density terms, i.e., $\vec{J} = \sigma\vec{E} + \vec{J}_S$.

In general, it would be desirable to determine a time-dependent polarization state presentation satisfying Eq. 2.9 and Eq. 2.10 for which, however, hitherto, no simple useful presentation was found to exist. Instead, we will be treating plane wave propagation in a source-free region for which we assume harmonic excitation which allows a frequency domain treatment of the wave equation for which a unique monochromatic presentation of the polarization state does exist.

## 2.3 The vector wave equation in the frequency domain: the Helmholtz equation.

In the following and throughout the remainder of this text, we shall adopt the harmonic time phasor definition according to the IEEE Standard 149-1979 (Standard Test Procedures For Antenna Measurements) [31]

$$\vec{A}(\vec{r},t) = \vec{A}(\vec{r}) \exp(j\omega t) \tag{2.11}$$

so that for the source-free region ($\rho = 0$, $\vec{J}_S = 0$), Eqs 2.9 and 2.10 become

$$\nabla^2\vec{E}(\vec{r}) - j\omega\mu(\sigma+j\omega\varepsilon)\vec{E}(\vec{r}) = 0$$

$$\nabla^2\vec{H}(\vec{r}) - j\omega\mu(\sigma+j\omega\varepsilon)\vec{H}(\vec{r}) = 0$$

and by defining the complex propagation constant $\tilde{k}$ so that we obtain the Helmoltz equations

$$\nabla^2\vec{E}(\vec{r}) + \tilde{k}^2\vec{E}(\vec{r}) = 0 \tag{2.12a}$$

$$\nabla^2\vec{H}(\vec{r}) + \tilde{k}^2\vec{H}(\vec{r}) = 0 \tag{2.12b}$$

where $\tilde{k}^2 = -j\omega\mu(\sigma + j\omega\varepsilon) = \omega^2\varepsilon\mu - j\omega\mu\sigma$, the space-time solution to Eq. 2.12 is then given with $\vec{k} = \tilde{k}\,\hat{a}_r$ for a wave traveling in $\hat{a}_r$ direction by

$$\vec{E}_{\pm}(\vec{r},t) = \vec{E}_0 \exp j(\omega t \mp \vec{k}\cdot\vec{r}) \tag{2.13a}$$

$$\vec{H}_{\pm}(\vec{r},t) = \vec{H}_0 \exp j(\omega t \mp \vec{k}\cdot\vec{r}) \tag{2.13b}$$

According to the IEEE definition, the phasor $\exp[+j(\omega t - \vec{k}\cdot\vec{r})]$ corresponds to a wave traveling in positive $\vec{r}$ direction (+) for advancing time, and the phasor $\exp[+j(wt + \vec{k}\cdot\vec{r})]$ to a wave traveling in negative $\vec{r}$ direction (−), respectively, as for example in Stratton [58].

It ought to be emphasized that great care must be taken in first verifying which phasor notation is adopted when dealing with the analysis of polarization state transformation properties. For example, in the standard physics literature the "complex-conjugated notations" of $\exp[+i(\vec{k}\cdot\vec{r} - \omega t)]$, and $\exp[i(\vec{k}\cdot\vec{r} + \omega t)]$, are adopted for waves traveling in positive and negative directions for advancing time, respectively, as for example in Born and Wolf [9].

Note, in an isotropic, lossless, homogeneous medium ($\varepsilon$, $\mu$) which is the case considered throughout the remainder of this chapter, the propagation constant becomes

$$k = \omega\sqrt{\varepsilon\mu} = \omega/v = 2\pi/\lambda \tag{2.14}$$

with $v$ denoting the speed of electromagnetic propagation and $\lambda$ representing the unbounded space medium wavelength.

## 2.4 The monochromatic transverse electromagnetic (TEM) traveling wave

Because, in the following, we are dealing only with TEM wave presentations in unbounded space, we will assume that the waves are traveling along rectilinear straight paths, where the z-variable is chosen to be aligned with the travel direction ($\hat{a}_r = \pm\hat{a}_z$) and the x-y plane presents the transverse plane in which the $\vec{E}_{\pm}(z,t)$ and $\vec{H}_{\pm}(z,t)$ field vectors reside, and, according to the Poynting's theorem those vectors are related via the instantaneous Poynting vector $\vec{S}(z,t)$ as

$$\vec{S}_{\pm}(z,t) = \vec{E}_{\pm}(z,t) \times \vec{H}_{\pm}(z,t) \tag{2.15}$$

$$= \pm\,\hat{a}_z \frac{E_0^{\,2}}{\eta} \exp[+2j(\omega t \mp kz)]$$

where the medium intrinsic impedance $\eta$ for the lossy case is given by

$$\eta = \sqrt{\frac{j\omega\mu}{\sigma+j\omega\varepsilon}} \tag{2.16}$$

and $\vec{H}_{\pm}(\vec{r},t)$ is related to $\vec{E}_{\pm}(\vec{r},t)$, as shown in Fig. 2.1 as

$$\vec{H}_{\pm}(\vec{r},t) = \pm \frac{1}{\eta} \hat{a}_z \times \vec{E}_{\pm}(r,t) = \pm \frac{1}{\eta} \hat{a}_z \times \vec{E}_{0\pm} \exp j(\omega t \mp \vec{k}\cdot\vec{r})$$

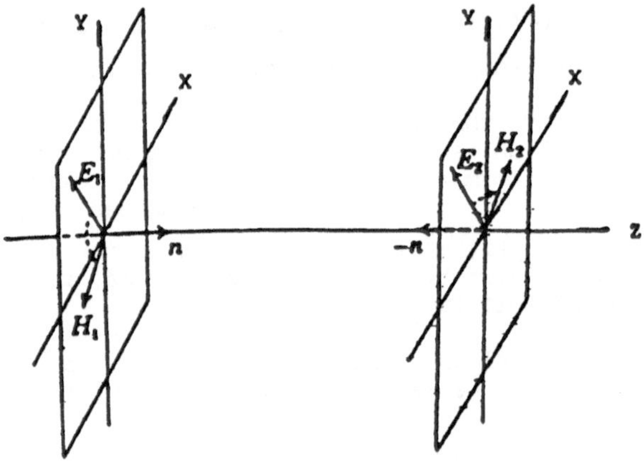


Fig. 2.1 Relative Directions of the Electric and Magnetic Vectors

For further details we refer the reader to standard texts (e.g., J.A. Stratton, Electromagnetic Theory, McGraws Hill, New York, 1941[58]). In all of the following derivations only the properties of the electric field vector will be considered.

## 3. THE REPRESENTATIONS OF THE POLARIZATION STATE

Polarization is a property of single-frequency electromagnetic radiation describing the shape and orientation of the locus of the extremity of the field vectors as a function of time [31]. In general, the electric vector of a harmonic plane wave traces an ellipse in the transverse plane with time [46]. In this section, we introduce several representations and descriptive parameters, commonly used to describe wave polarization, namely, the size A of the ellipse, the ellipticity angle $\varepsilon$, the tilt (orientation) angle $\tau$, the polarization transformation ratio $\rho$, the relative phase $\phi = \phi_y - \phi_x$ (phase difference between the two orthogonal components of the electric field of the wave), and the Stokes parameters $g_i$, and we illustrate how these are related to each other. The Poincaré sphere is also introduced and utilized to summarize this section.

### 3.1 The general representation

A single monochromatic, uniform, TEM (transverse electromagnetic) traveling plane wave can be decomposed into three orthogonal components, which are linear in simple harmonic oscillations. In the right-handed cartesian coordinate system, the electric vector of this plane wave may be expressed in terms of the x, y, and z components. If the wave is traveling in the positive z direction, then the real instantaneous electric field is written as

$$\vec{\varepsilon}(z,t) = \begin{bmatrix} \varepsilon_x(z,t) \\ \varepsilon_y(z,t) \\ \varepsilon_z(z,t) \end{bmatrix} = \begin{bmatrix} |E_x|\cos(\omega t-kz+\phi_x) \\ |E_y|\cos(\omega t-kz+\phi_y) \\ 0 \end{bmatrix} . \tag{3.1}$$

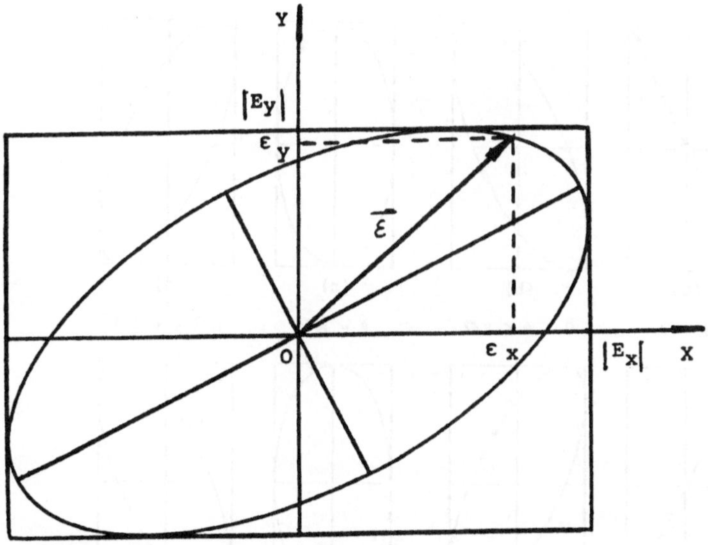


Fig. 3.1 Superposition of $\varepsilon_x$ and $\varepsilon_y$ Yields the Polarization Ellipse

The non-zero components are $\varepsilon_x(z,t)$ and $\varepsilon_y(z,t)$, where $|E_x|$ and $|E_y|$ represent the magnitudes of these components, and $\phi_x$ and $\phi_y$ represent the respective phases of these oscillations. It is easy to show [46] that at a fixed value of $\vec{z}$, on the transverse plane, the electric vector $\varepsilon$ of a harmonic plane wave rotates as a function of time, with the tip of the vector describing an ellipse. For simplicity, consider the plane of z = 0, Eq. 3.1 becomes

$$\vec{\varepsilon}(t) = \begin{bmatrix} \varepsilon_x(t) \\ \\ \varepsilon_y(t) \end{bmatrix} = \begin{bmatrix} |E_x|\cos(\omega t+\phi_x) \\ \\ |E_y|\cos(\omega t+\phi_y) \end{bmatrix} \quad . \tag{3.2}$$

These two independent, linear, simple harmonic vibrations $\varepsilon_x(t)$ and $\varepsilon_y(t)$, which are along two orthogonal directions x and y, compose an elliptical vibration as shown in Fig. 3.1.

The locus of the extremity of the electric field vectors is, in general, an ellipse that may degenerate into a segment of a straight line or into a circle. Correspondingly, the polarization is called elliptical, linear, or circular [31]. The shape and the orientation of the polarization ellipse depend on the relative phase $\phi = \phi_y - \phi_x$. Fig. 3.2 displays different kinds of polarization ellipses for different relative phases $\phi$.

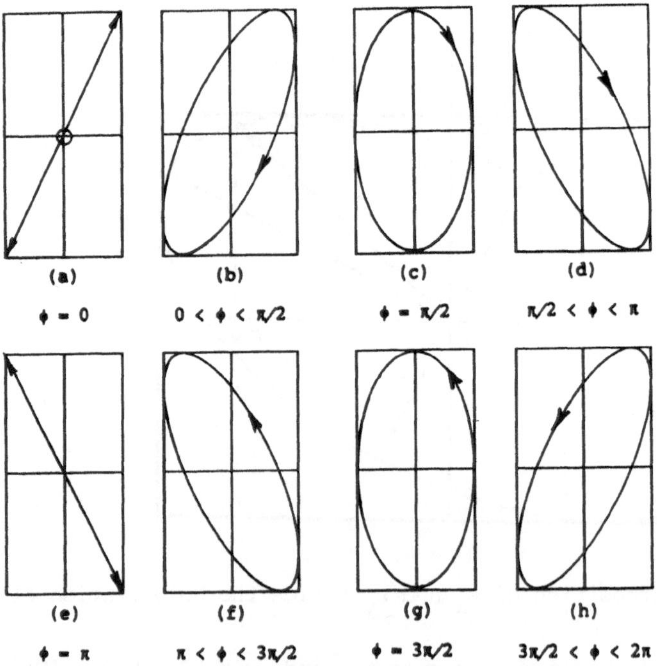


Fig. 3.2 General Polarization States: (a) and (e) are linear polarizations; (b), (c) and (d) are left-handed polarizations; (f), (g) and (h) are right-handed polarizations.

When $\phi_x$ and $\phi_y$ differ by an integer number of $2\pi$, the two components are in phase; the wave is said to be linearly polarized and the straight line locus traverses the first and third quadrants (Fig. 3-2(a)). When

$\phi_x$ and $\phi_y$ differ by an odd integer multiple of $\pi$, the components are 180° out of phase, and the straight line locus traverses the second and fourth quadrants (Fig. 3-2(e)). When $\phi_y$ leads $\phi_x$ by an angle $\phi$ which is less than $\pi$, the two components form an elliptic locus which traverses counterclockwise when looking down along +z-direction, and it defines left-handed polarization as shown in Figs. 3-2(b), (c) and (d). When $\phi_y$ lags $\phi_x$ by an angle $\phi$ less than $\pi$, it defines right-handed polarization as shown in Figs. 3-2(f), (g) and (h).

### 3.2 The polarization ellipse

Because the locus of the extremity of the field vectors is, in general, an ellipse, we can use the geometric parameters of the ellipse, $\tau$, $\varepsilon$, and A, to represent the polarization state of the plane wave [30]. These geometric parameters are shown in Fig. 3.3.

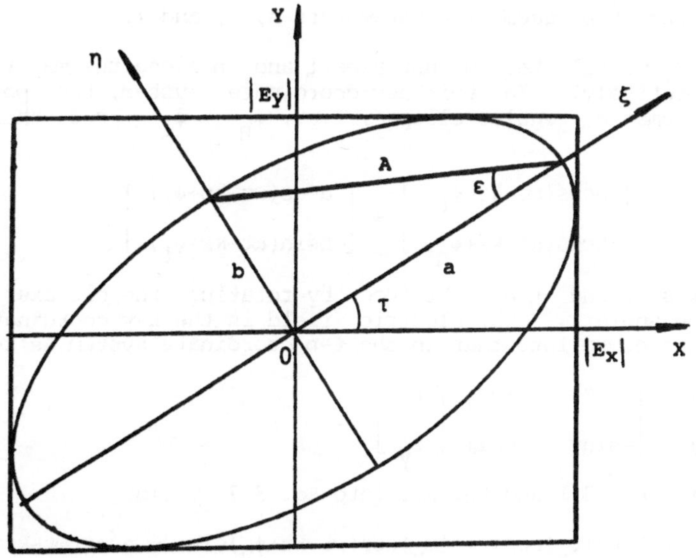


Fig. 3.3 The Three Parameters That Define the Ellipse of Polarization are the Orientation (or Tilt) Angle $\tau$, the Ellipcity Angle $\varepsilon$ and the Size A.

They are defined as follows: The orientation (or tilt) angle $\tau$ is the angle between the major axis of the ellipse and the positive x-axis. The ellipticity angle $\varepsilon$ is defined by

$$\varepsilon = \tan^{-1}(b/a) \tag{3.3}$$

where b is the length of the semi-minor axis of the ellipse and a is the length of its semi-major axis. The size of the ellipse is

$$A = \sqrt{a^2 + b^2} \quad . \tag{3.4}$$

The polarization of the plane wave can be determined in terms of the geometric parameters [30],

$$\vec{\varepsilon}(A,\phi_x,\tau,\varepsilon) = A \begin{bmatrix} \cos\tau & -\sin\tau \\ \sin\tau & \cos\tau \end{bmatrix} \begin{bmatrix} \cos\varepsilon \\ i\sin\varepsilon \end{bmatrix} \cos\phi_x \quad . \tag{3.5}$$

In the next step, we will find the relationship between the parameters in the two equivalent methods of specifying the elliptical polarization state of a monochromatic wave. The first method describes the elliptic vibration of the electric vector as a superposition in terms of $|E_x|$, $|E_y|$, $\phi_x$, and $\phi_y$. And the second method specifies the elliptic polarization by the three geometric parameters A, $\varepsilon$, and $\tau$.

In Fig. 3.3, we selected the new axes $\xi$ and $\eta$ along the major and minor axes, respectively. In this new coordinate system, the polarization ellipse becomes a regular ellipse, i.e., $\phi_\eta - \phi_\xi = \pi/2$, thus Eq. 3.1 becomes

$$\begin{bmatrix} \varepsilon_\xi \\ \varepsilon_\eta \end{bmatrix} = \begin{bmatrix} a\cos(\omega t-kz+\phi_\xi) \\ b\cos(\omega t-kz+\phi_\eta) \end{bmatrix} = \begin{bmatrix} a\cos(\omega t-kz+\phi_\xi) \\ -b\sin(\omega t-kz+\phi_\xi) \end{bmatrix} \quad . \tag{3.6}$$

The new axes $\xi$ and $\eta$ are obtained by rotating the old axes x and y through the angle $\tau$. The electric field in the x-y coordinate system can be transformed into that in the $\xi$-$\eta$ coordinate system as follows,

$$\begin{bmatrix} \varepsilon_\xi \\ \varepsilon_\eta \end{bmatrix} = \begin{bmatrix} \cos\tau & \sin\tau \\ -\sin\tau & \cos\tau \end{bmatrix} \begin{bmatrix} \varepsilon_x \\ \varepsilon_y \end{bmatrix} \quad . \tag{3.7}$$

Substituting Eq. 3.1 and Eq. 3.6 into Eq. 3.7, yields

$$|E_x|\cos(\omega t-kz+\phi_x)\cos\tau + |E_y|\cos(\omega t-kz+\phi_y)\sin\tau = a\cos(\omega t-kz+\phi_\xi)$$

$$-|E_x|\cos(\omega t-kz+\phi_x)\sin\tau + |E_y|\cos(\omega t-kz+\phi_y)\cos\tau = -b\sin(\omega t-kz+\phi_\xi) \quad .$$

Expanding cos and sin terms, and equating the coefficients of $\cos(\omega t-kz)$ and $\sin(\omega t-kz)$, yields:

$$\begin{aligned} |E_x|\cos\phi_x\cos\tau + |E_y|\cos\phi_y\sin\tau &= a\cos\phi_\xi \\ |E_x|\sin\phi_x\cos\tau + |E_y|\sin\phi_y\sin\tau &= a\sin\phi_\xi \\ -|E_x|\cos\phi_x\sin\tau + |E_y|\cos\phi_y\cos\tau &= -b\sin\phi_\xi \\ |E_x|\sin\phi_x\sin\tau - |E_y|\sin\phi_y\cos\tau &= -b\cos\phi_\xi \quad . \end{aligned} \tag{3.8}$$

Through simple algebra, the following relationship can be obtained:

$$a^2 + b^2 = |E_x|^2 + |E_y|^2 \tag{3.9}$$

$$\frac{b}{a} = \frac{|E_x|\cos\phi_x\sin\tau - |E_y|\cos\phi_y\cos\tau}{|E_x|\sin\phi_x\cos\tau + |E_y|\sin\phi_y\sin\tau}$$

and

$$\frac{b}{a} = \frac{-|E_x|\sin\phi_x\sin\tau + |E_y|\sin\phi_y\cos\tau}{|E_x|\cos\phi_x\cos\tau + |E_y|\cos\phi_y\sin\tau} \quad .$$

After equating the above two equations, we find

$$(|E_x|^2 - |E_y|^2)\,\frac{1}{2}\,\sin2\tau = |E_x|\,|E_y|\cos(\phi_y - \phi_x)\cos2\tau \quad ,$$

so that with $\phi = \phi_y - \phi_x$ ,

$$\tan2\tau = \frac{2|E_x|\,|E_y|\cos\phi}{|E_x|^2 - |E_y|^2} \quad . \tag{3.10}$$

Also from Eq. 3.8, we find

$$ab\sin^2\phi_\xi + ab\cos^2\phi_\xi = |E_x|\,|E_y|\sin\phi_y\cos\phi_x - |E_x|\,|E_y|\sin\phi_x\cos\phi_y \quad ,$$

thus

$$ab = |E_x|\,|E_y|\sin(\phi_y - \phi_x) = |E_x|\,|E_y|\sin\phi \quad . \tag{3.11}$$

According to the definition of the ellipticity angle $\varepsilon$ of Eq. 3.3, we obtain

$$\sin2\varepsilon = \frac{2\tan\varepsilon}{1 + \tan^2\varepsilon} = \frac{2(b/a)}{1 + (b^2/a^2)} = \frac{2|E_x|\,|E_y|\sin\phi}{|E_x|^2 + |E_y|^2} \quad . \tag{3.12}$$

Eq. 3.10 and Eq. 3.12 give us the relationship between $\varepsilon$, $\tau$, $|E_x|$, $|E_y|$, and $\phi$. The polarization states in terms of $\varepsilon$ and $\tau$ are given in Table 3.1, following Section 3.4.

### 3.3 The Jones vector [2,15]

In the previous sections we described the electric field using the real measurable quantity $\vec{\varepsilon}$. Now, we introduce the complex electric field vector $\vec{E}$. For a monochromatic field whose variation in space is also sinusoidal, the instantaneous vector $\vec{\varepsilon}$ takes the form [2,15]

$$\vec{\varepsilon}(\vec{r},t) = \mathrm{Re}\{\vec{E}(\vec{r})e^{j\omega t}\} \tag{3.13}$$

where the complex vector $\vec{E}(\vec{r})$ depends on the position vector $\vec{r}$ and is given by

$$\vec{E}(\vec{r}) = \vec{E}e^{-j\vec{k}\cdot\vec{r}} \quad . \tag{3.14}$$

$\vec{E}$ is a complex-constant amplitude vector and is given by

$$\vec{E} = \begin{bmatrix} E_x \\ E_y \\ E_z \end{bmatrix} = \begin{bmatrix} |E_x|e^{j\phi_x} \\ |E_y|e^{j\phi_y} \\ |E_z|e^{j\phi_z} \end{bmatrix} \quad . \tag{3.15}$$

For a plane wave which propagates along the +z-axis, the complex electric field is given by

$$\vec{E}(z,t) = \begin{bmatrix} E_x(z,t) \\ E_y(z,t) \end{bmatrix} = \begin{bmatrix} |E_x|e^{j(\omega t-kz+\phi_x)} \\ |E_y|e^{j(\omega t-kz+\phi_y)} \end{bmatrix} \quad . \tag{3.16}$$

It is the same as Eq. 3.1, where $E_x(z,t)$ and $E_y(z,t)$ are the complex time invariant terms associated with each real, time-varying electric field component.

In a time harmonic (monochromatic) TEM wave field, each cartesian component of the electric vector varies sinusoidally with time, at all points in space. We can use the phasor notation to suppress the temporal term, thus Eq. 3.16 can be replaced by

$$\vec{E}(z) = \begin{bmatrix} |E_x|e^{j\phi_x} \\ |E_y|e^{j\phi_y} \end{bmatrix} e^{-jkz} \quad .$$

Also, the spatial term can be dropped because the electric field has the same phase at all points on a z = constant plane. If we consider the plane at z = 0, the above equation becomes

$$\vec{E}(0) = \begin{bmatrix} |E_x| e^{j\phi_x} \\ |E_y| e^{j\phi_y} \end{bmatrix} . \qquad (3.17)$$

This vector is called the "Jones Vector" of the wave. If we extract and disregard $\phi_x$, the absolute phase of $E_x$, then Eq. 3.17 becomes

$$\vec{E} = \begin{bmatrix} |E_x| \\ |E_y| e^{j\phi} \end{bmatrix} \qquad (3.18)$$

where $\phi = \phi_y - \phi_x$ denotes the relative phase.

### 3.4 The polarization ratio $\rho$

Any wave can be resolved into two orthogonal components (they may be two orthogonal linearly, circularly, or elliptically polarized components) [31] in the plane transverse to the propagation direction. So, we can choose any two orthogonal polarized components as the polarization basis. For an arbitrary polarization basis (AB) with unit vectors $\hat{A}$ and $\hat{B}$, one may define the polarization state

$$\vec{E}(AB) = E_A\hat{A} + E_B\hat{B} \qquad (3.19)$$

where the two components, $E_A$ and $E_B$, are complex numbers. The polarization ratio $\rho_{AB}$ in an arbitrary basis (AB) is also a complex number, and it may be defined as

$$\rho_{AB} = \frac{E_B}{E_A} = \frac{|E_B|}{|E_A|} e^{j(\phi_B-\phi_A)} = |\rho_{AB}| e^{j\phi_{AB}} \qquad (3.20)$$

where $|\rho_{AB}|$ is the ratio of magnitude of the two orthogonal components of the field $E_A$ and $E_B$, and $\phi_{AB}$ is the phase difference between $E_A$ and $E_B$. The complex polarization ratio $\rho_{AB}$ depends on the polarization basis (AB) and can be used to specify the polarization of an electromagnetic wave

$$\vec{E}(AB) = \begin{bmatrix} E_A \\ E_B \end{bmatrix} = |E_A| e^{j\phi_A} \begin{bmatrix} 1 \\ \rho_{AB} \end{bmatrix}$$

$$= |E_A| e^{j\phi_A} \frac{\sqrt{1 + \frac{E_B E_B^*}{E_A E_A^*}}}{\sqrt{1 + \frac{E_B E_B^*}{E_A E_A^*}}} \begin{bmatrix} 1 \\ \rho_{AB} \end{bmatrix}$$

$$= |E| e^{j\phi_A} \frac{1}{\sqrt{1 + \rho_{AB}\rho_{AB}^*}} \begin{bmatrix} 1 \\ \rho_{AB} \end{bmatrix} \tag{3.21}$$

where $|E| = \sqrt{E_A E_A^* + E_B E_B^*}$ is the amplitude of the wave $\vec{E}(AB)$. If we choose $|E| = 1$ and disregard the absolute phase $\phi_A$, the above representation becomes

$$\vec{E}(AB) = \frac{1}{\sqrt{1 + \rho_{AB}\rho_{AB}^*}} \begin{bmatrix} 1 \\ \rho_{AB} \end{bmatrix} . \tag{3.22}$$

The above representation of the polarization state using the polarization ratio $\rho_{AB}$ is very useful. For example, if we want to represent a left-handed circular (LHC) polarization state and a right-handed circular (RHC) polarization state in a linear basis (HV) using the polarization ratio. For a left-handed circular polarization, $|E_H| = |E_V|$, $\phi_{HV} = \phi_V - \phi_H = \pi/2$, and according to Eq. 3.20, the polarization ratio $\rho_{HV}$ of the LHC is j. One uses Eq. 3.22 with $\rho_{AB} = j$ and obtains for the left-handed circular polarization

$$\vec{E}(HV) = \frac{1}{\sqrt{1 + j(-j)}} \begin{bmatrix} 1 \\ j \end{bmatrix} = \frac{1}{\sqrt{2}} \begin{bmatrix} 1 \\ j \end{bmatrix} . \tag{3.23}$$

Similarly, the polarization ratio $\rho_{HV}$ of a right-handed circular polarization in the (HV) basis is $-j$ because the relative phase $\phi_{HV}$ is $-\pi/2$. The representation of RHC in the linear (HV) basis is

$$\vec{E}(HV) = \frac{1}{\sqrt{2}} \begin{bmatrix} 1 \\ -j \end{bmatrix} .$$

Note: The polarization ratio $\rho_{AB}$ is important in polarimetry. However, the value of the polarization ratio $\rho$ defined in a certain polarization basis is different from that defined in the other polarization basis even if the physical polarization state is the same (see Table 3.1).

### 3.4.1 The polarization ratio $\rho_{HV}$ in the linear basis (HV).

In the linear (HV) basis with unit vectors $\hat{H}$ and $\hat{V}$, a polarization state may be expressed as

$$\vec{E}(HV) = E_H\hat{H} + E_V\hat{V} \quad .$$

The polarization ratio $\rho_{HV}$, according to Eq. 3.20, can be described as

$$\rho_{HV} = \frac{E_V}{E_H} = \frac{|E_V|}{|E_H|} e^{j(\phi_V - \phi_H)} = \tan\gamma_{HV} e^{j\phi_{HV}} \quad . \tag{3.24}$$

where the angle $\gamma_{HV}$ is defined in Fig. 3.4, only in the (HV) basis, and

$$|E_H| = \sqrt{E_H^2 + E_V^2}\cos\gamma_{HV} \tag{3.25}$$

$$|E_V| = \sqrt{E_H^2 + E_V^2}\sin\gamma_{HV} \quad .$$

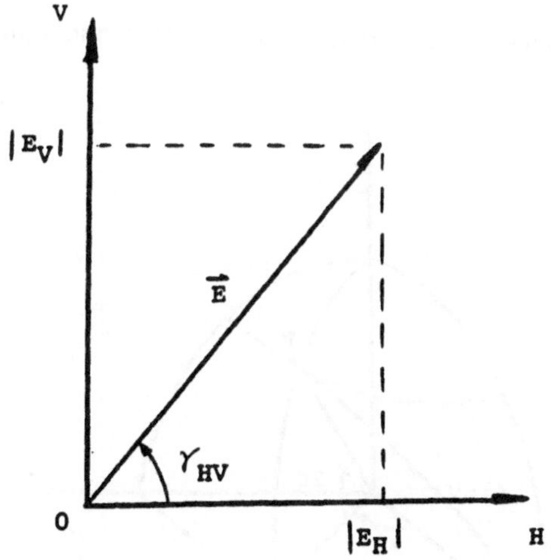


Fig. 3.4 The Parameter $\gamma_{HV}$

In a cartesian coordinate system, the +x-axis is commonly chosen as the horizon- tal basis and the +y-axis as the vertical basis. Substituting Eq. 3.25 into Eq. 3.1, we find

$$\vec{\varepsilon}(z,t) = \begin{bmatrix} \sqrt{E_x^2 + E_y^2}\cos\gamma\cos(\omega t - kz + \phi_x) \\ \sqrt{E_x^2 + E_y^2}\sin\gamma\cos(\omega t - kz + \phi_y) \end{bmatrix} =$$

$$= \sqrt{E_x^2 + E_y^2}\ \mathrm{Re}\left\{ \begin{bmatrix} \cos\gamma \\ \sin\gamma e^{j\phi} \end{bmatrix} e^{j(\omega t - kz + \phi_x)} \right\}$$

where $\phi = \phi_y - \phi_x$ is the relative phase. In the literature, $\phi_x$ is commonly denoted by $\alpha$, thus

$$\vec{\varepsilon}(z,t) = \sqrt{E_x^2 + E_Y^2}\ \mathrm{Re}\left\{ \begin{bmatrix} \cos\gamma \\ \sin\gamma e^{j\phi} \end{bmatrix} e^{j(\omega t - kz + \alpha)} \right\} . \tag{3.26}$$

The expression in the square bracket is a spinor which is independent of the time-space dependence of the traveling wave. The spinor parameters $(\gamma,\phi)$ are easy to be located on the Poincaré sphere and can be used to represent the polarization state of a plane wave.

In Fig. 3.5, the polarization state, described by the point P on the Poincaré sphere, can be expressed in terms of these two angles, where

$2\gamma_{HV}$ = the angle subtended by the great circle drawn from the point P on the equator measured from H toward V.

$\phi_{HV}$ = the angle between the great circle and the equator.

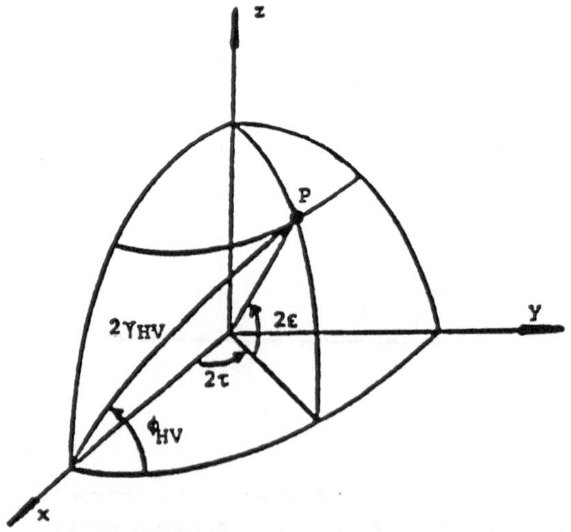


Fig. 3.5 Poincaré Sphere Showing Relation of Angles $\gamma_{HV}$, $\tau$, $\varepsilon$, and $\phi_{HV}$

After insertion of Eq. 3.25 into Eq. 3.12, the spinor parameters, $\gamma_{HV}$

and $\phi_{HV}$, can be related to the elliptic parameters, $\varepsilon$ and $\tau$, in the following manner,

$$\sin 2\varepsilon = 2\cos\gamma_{HV}\sin\gamma_{HV}\sin\phi_{HV} = \sin 2\gamma_{HV}\sin\phi_{HV} \quad . \tag{3.27}$$

Dividing the numerator and the denominator of Eq. 3.10 by $|E_x|^2$, we find

$$\tan 2\tau = \frac{2\tan\gamma_{HV}}{1 - \tan^2\gamma_{HV}} \cos\phi_{HV} = \tan 2\gamma_{HV}\cos\phi_{HV} \quad . \tag{3.28}$$

Eq. 3.27 and Eq. 3.28 describe the elllipticity angle $\varepsilon$ and the tilt angle $\tau$ in terms of the variables $\gamma_{HV}$ and $\phi_{HV}$. Also, we can derive an inverse pair which describes the $\gamma_{HV}$ and $\phi_{HV}$ in terms of $\varepsilon$ and $\tau$. Eq. 3.27 and Eq. 3.28 provide $\sin\phi_{HV}$ and $\cos\phi_{HV}$ expressions, thus

$$\frac{\sin^2 2\varepsilon}{\sin^2 2\gamma_{HV}} + \frac{\tan^2 2\tau}{\tan^2 2\gamma_{HV}} = 1 \quad ,$$

by multiplying both sides by $\sin^2 2\gamma_{HV}\cos^2 2\tau$ and moving $\sin^2 2\tau\cos^2 2\gamma_{HV}$ to the right side, and after some calculations, we obtain

$$\cos 2\gamma_{HV} = \cos 2\tau\cos 2\varepsilon \quad . \tag{3.29}$$

The ratio of Eq. 3.27 and Eq. 3.28 yields

$$\tan\phi_{HV} = \frac{\sin 2\varepsilon}{\tan 2\tau\cos 2\gamma_{HV}} = \frac{\tan 2\varepsilon}{\sin 2\tau} \quad . \tag{3.30}$$

It is convenient to describe the polarization state by either of the two sets of angles $(\gamma_{HV}, \phi_{HV})$ or $(\varepsilon, \tau)$ which describe a point on the Poincaré sphere. The complex polarization ratio $\rho_{HV}$ can be used to specify the polarization of an electromagnetic wave expressed in an (HV) basis. Some common polarization states expressed in terms of $\rho_{HV}$ are listed in Table 3.1, following Sect. 3.4.

### 3.4.2 Polarization ratio $\rho_{LR}$ in the cicular basis (LR).

In the circular basis (LR), we have two unit vectors $\hat{L}$ and $\hat{R}$. $\hat{L}$ stands for left-handed circular unit vector and $\hat{R}$ is the right-handed circular unit vector. Any polarization state of a plane wave can be expressed by

$$\vec{E}(LR) = E_L\hat{L} + E_R\hat{R}$$

where $E_L$ and $E_R$ are the L component and the R component, respectively.

A unit amplitude left-handed circular polarization has only the L component in the circular basis (LR). It can be expressed by

$$\vec{E}(LR) = 1 * \hat{L} + 0 * \hat{R} = \begin{bmatrix} 1 \\ 0 \end{bmatrix} .$$

The above representation of a unit LHC polarization in the circular basis (LR) is different from that in the linear basis (HV) of Eq. 3.23. Similarly, a unit amplitude right-handed circular polarization has only the R component in the circular basis (LR).

$$\vec{E}(LR) = 0 * \hat{L} + 1 * \hat{R} = \begin{bmatrix} 0 \\ 1 \end{bmatrix} .$$

Also, this representation of RHC polarization in the circular basis (LR) is different from that in the linear (HV) basis.

The polarization ratio $\rho_{LR}$, according to the definition of the polarization ratio, Eq. 3.20, is

$$\rho_{LR} = \frac{E_R}{E_L} = \frac{|E_R|}{|E_L|} e^{j(\phi_R - \phi_L)} = |\rho_{LR}| e^{j\phi_{LR}} = \tan\gamma_{LR} e^{j\phi_{LR}} \qquad (3.31)$$

where $|\rho_{LR}|$ is the ratio of magnitudes of the two orthogonal components $E_L$ and $E_R$, and $\phi_{LR}$ the phase difference. The angles $\gamma_{LR}$ and $\phi_{LR}$ are also easy to be found on the Poincaré sphere like the angles $\gamma_{HV}$ and $\phi_{HV}$ (see Fig. 3.8 of Section 3.6). Some common polarization states, expressed in terms of $\rho_{LR}$, are listed in Table 3.1.

### 3.4.3 Polarization ratio $\rho_{45°135°}$ in the Linear basis (45°135°)

In the linear (45°135°) basis with unit vectors in $\hat{45°}$ and $\hat{135°}$, a polarization state may be expressed as

$$\vec{E}_{(45°135°)} = E_{45°}\, \hat{45°} + E_{135°} \hat{135°} ,$$

where $E_{45°}$ and $E_{135°}$ are the 45° component and the 135° component, respectively. The polarization ratio $\rho_{45\ 135°}$, according to the definition of the polarization ratio of Eq. 3.20, can be written as

$$\rho_{45°135°} = \frac{E_{135°}}{E_{45°}} = \frac{|E_{135°}|}{|E_{45°}|} e^{j(\phi_{135°} - \phi_{45°})} = |\rho_{45°135°}|\, e^{j\phi_{45°135°}} =$$

$$= \tan \gamma_{45°135°}\, e^{j\phi_{45°135°}}$$

where $|\rho_{45°135°}|$ is the ratio of the two orthogonal component magnitudes $|E_{135°}|$ and $|E_{45°}|$, and $\phi_{45°135°}$ is the phase difference. The angles $\gamma_{45°135°}$ and $\phi_{45°135°}$ can be found on the Poincaré sphere in a later Section 3.6.2 .

**Table 3.1: Examples of polarization states expressed in terms of geometric parameters ($\epsilon$, $\tau$), polarization ratio $\rho$, and normalized Jones vector $\vec{E}$.**

| Polarization | $\epsilon$ | $\tau$ | (HV)basis | | (45°135°) | | (LR)basis | |
|---|---|---|---|---|---|---|---|---|
| | | | $\rho_{HV}$ | $\vec{E}$ | $\rho_{45°135°}$ | $\vec{E}$ | $\rho_{LR}$ | $\vec{E}$ |
| Linear Horizontal | 0 | 0 | 0 | $\begin{bmatrix} 1 \\ 0 \end{bmatrix}$ | -1 | $\frac{1}{\sqrt{2}}\begin{bmatrix} 1 \\ -1 \end{bmatrix}$ | 1 | $\frac{1}{\sqrt{2}}\begin{bmatrix} 1 \\ 1 \end{bmatrix}$ |
| Linear Vertical | 0 | $+\frac{\pi}{2}$ | $\infty$ | $\begin{bmatrix} 0 \\ 1 \end{bmatrix}$ | 1 | $\frac{1}{\sqrt{2}}\begin{bmatrix} 1 \\ 1 \end{bmatrix}$ | -1 | $\frac{1}{\sqrt{2}}\begin{bmatrix} -j \\ j \end{bmatrix}$ |
| 45°Linear | 0 | $\frac{\pi}{4}$ | 1 | $\frac{1}{\sqrt{2}}\begin{bmatrix} 1 \\ 1 \end{bmatrix}$ | 0 | $\begin{bmatrix} 1 \\ 0 \end{bmatrix}$ | j | $\frac{1}{2}\begin{bmatrix} 1 & -j \\ 1 & +j \end{bmatrix}$ |
| 135° Linear | 0 | $-\frac{\pi}{4}$ | -1 | $\frac{1}{\sqrt{2}}\begin{bmatrix} -1 \\ 1 \end{bmatrix}$ | $\infty$ | $\begin{bmatrix} 0 \\ 1 \end{bmatrix}$ | -j | $\frac{1}{2}\begin{bmatrix} -1 & -j \\ -1 & +j \end{bmatrix}$ |
| Left-Handed Circular | $\frac{\pi}{4}$ | | j | $\frac{1}{\sqrt{2}}\begin{bmatrix} 1 \\ j \end{bmatrix}$ | j | $\frac{1}{2}\begin{bmatrix} 1 & +j \\ -1 & +j \end{bmatrix}$ | 0 | $\begin{bmatrix} 1 \\ 0 \end{bmatrix}$ |
| Right-Handed Circular | $-\frac{\pi}{4}$ | | -j | $\frac{1}{\sqrt{2}}\begin{bmatrix} 1 \\ -j \end{bmatrix}$ | -j | $\frac{1}{2}\begin{bmatrix} 1 & -j \\ -1 & -j \end{bmatrix}$ | $\infty$ | $\begin{bmatrix} 0 \\ 1 \end{bmatrix}$ |

## 3.5 THE STOKES PARAMETERS [12]

The previous sections deal with completely polarized waves, i.e., waves for which $|E_A|$, $|E_B|$, and $\phi_{AB}$ are constants (or at least slowly varying functions of time). If we need to deal with partial polarization, it is convenient to use the Stokes parameters, introduced by Sir George Stokes in 1852 [44].

### 3.5.1 The Stokes vector for the completely polarized wave

First, let us consider the simple case. For a monochromatic wave, in terms of the linear (HV) basis, the four Stokes parameters are

$$g_0 = |E_H|^2 + |E_V|^2$$

$$g_1 = |E_H|^2 - |E_V|^2$$

$$g_2 = 2|E_H|\,|E_V|\cos\phi$$

$$g_3 = 2|E_H|\,|E_V|\sin\phi \quad , \tag{3.32}$$

where $|E_H|$, $|E_V|$, and $\phi$ are the magnitudes and the phase difference between the two orthogonal components $E_H$ and $E_V$, respectively. For a completely polarized wave, there are only three independent parameters, which are related in the following manner,

$$g_0^2 = g_1^2 + g_2^2 + g_3^2 \quad . \tag{3.33}$$

The Stokes parameters are sufficient to characterize the magnitude, phase, and polarization of a wave. It is possible to show that the Stokes parameter $g_0$ is always equal to the total power (density) of the wave, $g_1$ is equal to the power in the linear horizontal or vertical polarized components, $g_2$ is equal to the power in the linearly polarized components at tilt angles $\tau = 45°$ or $135°$, and $g_3$ is equal to the power in the left-handed or right-handed circular polarized components [44]. If any of the parameters $g_1$, $g_2$, or $g_3$ has a non-zero value, it indicates the presence of a polarized component in the plane wave. The Stokes parameters are not only related to $|E_H|$, $|E_V|$, and $\phi$ through Eq. 3.32, but also related to the geometric parameters A, $\varepsilon$, and $\tau$, used previously for describing the polarization ellipse. From Eq. 3.4, Eq. 3.9, and Eq. 3.32, we obtain

$$A^2 = |E_x|^2 + |E_y|^2 = g_0 \tag{3.34}$$

From Eq. 3.12 and Eq. 3.32, we obtain

$$\sin 2\varepsilon = \frac{g_3}{g_0} \quad ,$$

or $$g_3 = g_0 \sin 2\varepsilon = A^2 \sin 2\varepsilon \quad . \tag{3.35}$$

From Eq. 3.10 and Eq. 3.32, we obtain

$$\tan 2\tau = \frac{g_2}{g_1}$$

$$\text{or } g_2 = g_1 \tan 2\tau \tag{3.36}$$

By inserting Eq. 3.35 and Eq. 3.36 into Eq. 3.33, we find

$$g_1 = g_0 \cos 2\varepsilon \cos 2\tau = A^2 \cos 2\varepsilon \cos 2\tau \quad . \tag{3.37}$$

Substitution of this equation into Eq. 3.36, in turn, gives

$$g_2 = g_0 \cos 2\varepsilon \sin 2\tau = A^2 \cos 2\varepsilon \sin 2\tau \quad . \tag{3.38}$$

Eq. 3.34, Eq. 3.37, Eq. 3.38, and Eq. 3.35 form a set of expressions which describe the Stokes parameters in terms of the geometric parameters A, ε, and τ. Now, we can group the Stokes parameters in a 4 x 1 column vector called the "Stokes Vector"(in terms of the two sets of parameters given below in Eq 3.39),

$$\vec{g} = \begin{bmatrix} g_0 \\ g_1 \\ g_2 \\ g_3 \end{bmatrix} = \begin{bmatrix} |E_H|^2 + |E_V|^2 \\ |E_H|^2 - |E_V|^2 \\ 2|E_H||E_V|\cos\phi \\ 2|E_H||E_V|\sin\phi \end{bmatrix} = A^2 \begin{bmatrix} 1 \\ \cos 2\varepsilon \cos 2\tau \\ \cos 2\varepsilon \sin 2\tau \\ \sin 2\varepsilon \end{bmatrix} \quad . \tag{3.39}$$

### 3.5.2 The Stokes vector for the partially polarized wave

The partially polarized wave can be defined quantitatively by introducing a coherency matrix [J]. The polarization coherency matrix [J] results from multiplication of the electric field and its Hermitian conjugate, as defined by:

$$[J] = \langle \vec{E} \; \vec{E}^{+} \rangle = \begin{bmatrix} \langle E_H E_H^* \rangle & \langle E_H E_V^* \rangle \\ \langle E_V E_H^* \rangle & \langle E_V E_V^* \rangle \end{bmatrix} = \begin{bmatrix} J_{HH} & J_{HV} \\ J_{VH} & J_{VV} \end{bmatrix} \tag{3.40}$$

where $\langle\cdots\rangle = \lim_{T\to\infty} [\ \frac{1}{2T} \int_{-T}^{T} \langle\cdots\rangle \, dt\ ]$.

We can associate the Stokes Vector $\vec{g}$ with the coherency matrix [J]. From Eq. 3.32, we have

$$g_0 = |E_H|^2 + |E_V|^2 = \langle E_H E_H^*\rangle + \langle E_V E_V^*\rangle = J_{HH} + J_{VV} \qquad (3.41)$$

$$g_1 = |E_H|^2 - |E_V|^2 = \langle E_H E_H^*\rangle - \langle E_V E_V^*\rangle = J_{HH} - J_{VV}$$

$$g_2 = 2|E_H||E_V|\cos\phi = |E_H||E_V|(e^{i\phi} + e^{-j\phi}) = \langle E_H E_V^*\rangle + \langle E_V E_H^*\rangle = J_{HV} + J_{VH}$$

$$g_3 = 2|E_H||E_V|\sin\phi = -j|E_H||E_V|(e^{i\phi} - e^{-j\phi}) = +j\langle E_H E_V^*\rangle - j\langle E_V E_H^*\rangle = +jJ_{HV} - jJ_{VH}$$

A partially polarized wave may be regarded as the sum of a completely unpolarized wave and a completely polarized wave. Thus, we may write $\vec{g}$ for a partially polarized wave as follows,

$$\begin{bmatrix} g_0 \\ g_1 \\ g_2 \\ g_3 \end{bmatrix} = A^2 \begin{bmatrix} 1-p \\ 0 \\ 0 \\ 0 \end{bmatrix} + A^2 \begin{bmatrix} p \\ p\cos 2\varepsilon \cos 2\tau \\ p\cos 2\varepsilon \sin 2\tau \\ p\sin 2\varepsilon \end{bmatrix} \qquad (3.42)$$

where p is the degree of polarization which is defined as the ratio of completely polarized power to the total power and can be written as

$$p = \frac{\sqrt{g_1^2 + g_2^2 + g_3^2}}{g_0} \quad . \qquad (3.43)$$

Using Eq. 3.41 and Eq. 3.43, we can obtain an expression of the degree of polarization p in terms of the coherency matrix [J] as follow

$$p = \sqrt{1 - \frac{4\,(\det [J])}{(J_{HH} + J_{VV})^2}} \quad . \qquad (3.44)$$

Table 3.2 gives the Jones Vector $\vec{E}$, the coherency vector $\vec{J}$ (to be introduced in Section 4.4), and the Stokes vector $\vec{g}$ for special cases of purely monochromatic wave fields in specific states of polarization.

### 3.6 THE POINCARÉ SPHERE

The Poincaré sphere, shown in Fig. 3.6, is a useful graphical aid for the visualization of polarization effects. There is a one-to-one correspondence between all possible polarizations states and points on the Poincaré sphere.

Table 3.2 Jones Vector $\vec{E}$, Coherency Vector $\vec{J}$, and Stokes Vector $\vec{g}$ for Some States of Polarization.

| Polarization | $\vec{E}$ (HV) | $\vec{J}$ (HV) | $\vec{g}$ |
|---|---|---|---|
| Linear Horizontal | [ 1 0 ] | [ 1 0 0 0 ] | [ 1 1 0 0 ] |
| Linear Vertical | [ 0 1 ] | [ 0 0 0 1 ] | [ 1 -1 0 0 ] |
| 45° Linear | $\frac{1}{\sqrt{2}}$ [ 1 1 ] | $\frac{1}{2}$ [ 1 1 1 1 ] | [ 1 0 1 0 ] |
| 135° Linear | $\frac{1}{\sqrt{2}}$ [-1 1 ] | $\frac{1}{2}$ [ 1 -1 -1 1 ] | [ 1 0 -1 0 ] |
| Left-handed Circular | $\frac{1}{\sqrt{2}}$ [ 1 j ] | $\frac{1}{2}$ [ 1 -j j 1 ] | [ 1 0 0 1 ] |
| Right-handed Circular | $\frac{1}{\sqrt{2}}$ [ 1 -j ] | $\frac{1}{2}$ [ 1 j -j 1 ] | [ 1 0 0 -1 ] |

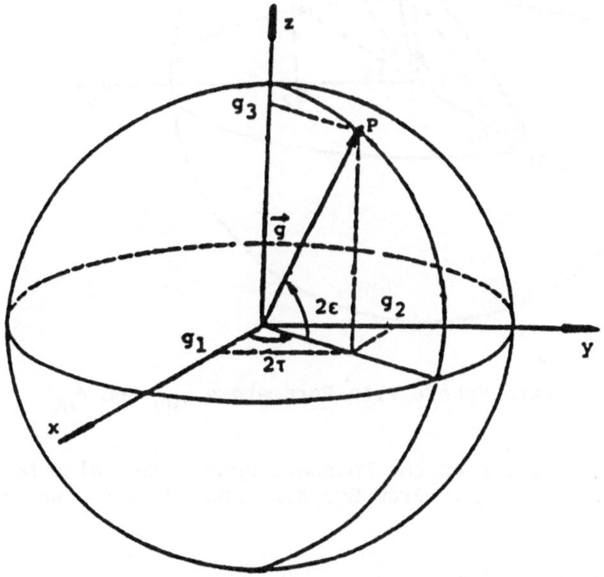


Fig. 3.6 Representation of a Polarization State on the Poincaré Sphere with Definition of Angle τ and ε

### 3.6.1 The polarization state on the Poincaré sphere described in terms of the sets of angles ($\tau$, $\varepsilon$) and ($\gamma_{HV}$, $\phi_{HV}$) for the HV basis

In the Poincaré sphere representation, the polarization state is described by a point P on the sphere, where the three cartesian coordinate components are $g_1$, $g_2$ and $g_3$ according to Eq. 3.39. So, for any state of a completely polarized wave, there corresponds one point $P(g_1,g_2,g_3)$ on the sphere of radius $g_0$, and vice versa. From Figure 3.6 and Eq. 3.38, it can be seen that the longitude and latitude of the point P are related to the geometric parameters of the polarization ellipses as follow,

$$\text{longitude} = 2\tau$$

$$\text{latitude} = 2\varepsilon$$

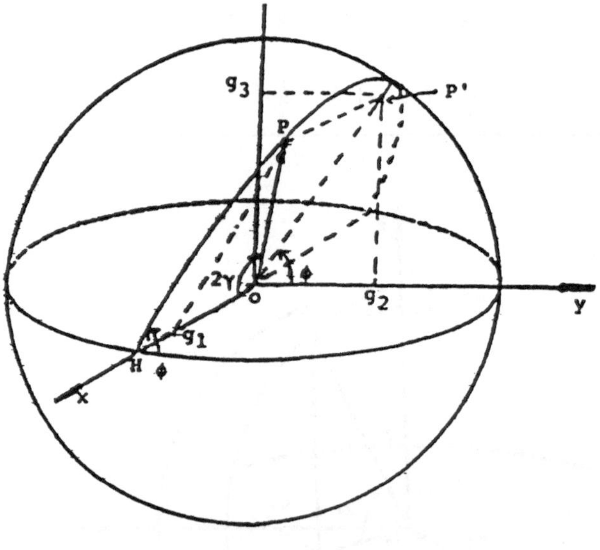


Fig. 3.7 The Poincaré Sphere with Parameters $\gamma_{HV}$ and $\phi_{HV}$

In addition, the point P on the Poincaré sphere can also be represented by the angles $\gamma_{HV}$ and $\phi_{HV}$. From Eq. 3.29 and Eq. 3.39, we find that

$$\frac{g_1}{g_0} = \cos 2\varepsilon \cos 2\tau = \cos 2\gamma_{HV} \quad ,$$

where cos $2\gamma_{HV}$ is the direction cosine of the Stokes vector $\vec{g}$ with respect to the x-axis, i.e., the angle $2\gamma_{HV}$ is the angle between $\vec{g}$ and the x-axis. So, we may define the arc length from the point H to the point P on the great circle as $_{HP} = g_o \ (2\gamma_{HV}) = 2\gamma_{HV}$ measured in the direction from H toward V. The angle $\phi_{HV}$ is the angle between the equator and the great circle with basis diameter HV through the point P, and it is equal to the angle between the XOY plane and the XOP plane. Drawing a projecting line from point P to the YOZ plane, the intersecting point P′ is on the XOP plane, so, $\phi_{HV} = \angle YOP'$. On the YOZ plane, we find that

$$\tan\phi_{HV} = \tan(\angle YOP') = g_3/g_2 \text{ , which satisfies Eq. 3.39 ,}$$

### 3.6.2 The polarization ratio in the Poincaré sphere presentation for different polarization bases

Also, it can be shown that a polarization state can be represented in different polarization bases. Any polarization basis consists of two unit vectors which are located at two corresponding antipodal points on the Poincaré sphere. Fig. 3.8 shows how the polarization state P on the Poincaré sphere can be represented in three polarization bases, (HV), (45°135°), and (LR). The complex polarization ratios are given by

$$\rho_{HV} = |\rho_{HV}| e^{j\phi_{HV}} = \begin{cases} \tan\gamma_{HV} \, e^{j\phi_{HV}} & 0 < \gamma_{HV} < \frac{\pi}{2} \\ -\tan\gamma_{HV} \, e^{j\phi_{HV}} & \frac{\pi}{2} < \gamma_{HV} < \pi \end{cases} \tag{3.45a}$$

$$\rho_{45°135°} = |\rho_{45°135°}| e^{j\phi_{45°135°}} = \begin{cases} \tan\gamma_{45°135°} e^{j\phi_{45°135°}} & 0 < \gamma_{45°135°} < \frac{\pi}{2} \\ -\tan\gamma_{45°135°} e^{j\phi_{45°135°}} & \frac{\pi}{2} < \gamma 45°135° < \pi \end{cases} \tag{3.45b}$$

$$\rho_{LR} = |\rho_{LR}| e^{j\phi_{LR}} = \begin{cases} \tan\gamma_{LR} e^{j\phi_{LR}} & 0 < \gamma_{LR} < \frac{\pi}{2} \\ -\tan\gamma_{LR} e^{j\phi_{LR}} & \frac{\pi}{2} < \gamma_{LR} < \pi \end{cases} \tag{3.45c}$$

where $\tan\gamma_{HV}$, $\tan\gamma_{45°135°}$ , and $\tan\gamma_{LR}$ are the ratios of the magnitudes of the corresponding orthogonal components, and $\phi_{HV}$, $\phi_{45°135°}$, and $\phi_{LR}$ are the phase differences between the corresponding orthogonal components.

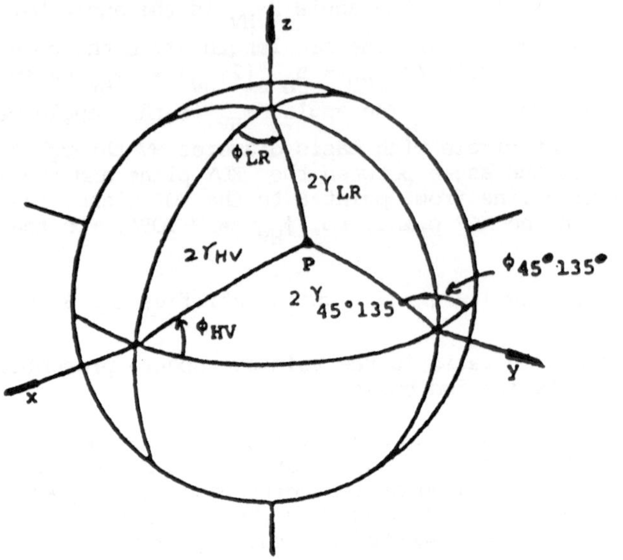


Fig. 3.8 The Polarization State P in the Different Polarization Bases

### 3.6.3 The relationship between the Stokes vector and the polarization ratio

**a) $\rho_{HV}$ – Presentation**

First, consider the polarization ratio $\rho_{HV}$ defined in the (HV) basis. Because $\cos 2\gamma_{HV}$ is the direction cosine of the Stokes vector $\vec{g}$ with respect to the x-axis, we find

$$\frac{g_1}{g_0} = \cos(2\gamma_{HV}) = \frac{1 - \tan^2\gamma_{HV}}{1 + \tan^2\gamma_{HV}} = \frac{1 - |\rho_{HV}|^2}{1 + |\rho_{HV}|^2} , \qquad (3.46a)$$

the straight forward solution for $|\rho_{HV}|$ is

$$|\rho_{HV}| = \sqrt{\frac{g_0 - g_1}{g_0 + g_1}} . \qquad (3.46b)$$

From Subsection 3.6.1, we find

$$\phi_{HV} = \angle YOP' = \tan^{-1}\left(\frac{g_3}{g_2}\right) . \qquad (3.46c)$$

Combining above two equations yields

$$\rho_{HV} = |\rho_{HV}|e^{j\phi_{HV}} = \sqrt{\frac{g_0 - g_1}{g_0 + g_1}}\, e^{j\tan^{-1}\left(\frac{g_3}{g_2}\right)} \tag{3.46d}$$

For a completely polarized wave, we may obtain the Stokes vector in terms of the polarization ratio $\rho_{HV}$ by applying

$$g_0 = \sqrt{g_1^2+g_2^2+g_3^2} = 1 , \tag{3.47a}$$

$$g_1 = \frac{1 - |\rho_{HV}|^2}{1 + |\rho_{HV}|^2} = \cos(2\gamma_{HV}) \tag{3.47b}$$

$$g_2 = \frac{2|\rho_{HV}|\cos(\phi_{HV})}{1 + |\rho_{HV}|^2} = \frac{2\tan(\gamma_{HV})\cos(\phi_{HV})}{1 + |\tan(\gamma_{HV})|^2} = \sin(2\gamma_{HV})\cos(\phi_{HV}) \tag{3.47c}$$

$$g_3 = \frac{2|\rho_{HV}|\, \sin(\phi_{HV})}{1 + |\rho_{HV}|^2} = \sin(2\gamma_{HV})\sin(\phi_{HV}) \tag{3.47d}$$

The sign of the three components of the Stokes vector $\vec{g}$ can be summarized in Table 3.3.

**b) $\rho_{45^\circ 135^\circ}$ – Presentation**

Secondly, consider the polarization ratio $\rho_{45^\circ 135^\circ}$ defined in the (45° 135°) basis. The cos $2\gamma_{45^\circ 135^\circ}$ is the direction cosine of the Stokes vector $\vec{g}$ with respect to the y-axis. So similarly, with

$$g_0 = 1,\ |\rho_{45^\circ 135^\circ}| = \sqrt{\frac{g_0 - g_2}{g_0 + g_2}}\ ,\ \phi_{45^\circ 135^\circ} = \tan^{-1}\left(-\frac{g_3}{g_1}\right) . \tag{3.48a,b,c}$$

Then the polarization ratio $\rho_{45^\circ 135^\circ}$ can be determined by the Stokes vector $\vec{g}$ as

$$\rho_{45^\circ 135^\circ} = \sqrt{\frac{g_0 - g_2}{g_0 + g_2}}\, e^{j\tan^{-1}\left(-\frac{g_3}{g_1}\right)} . \tag{3.48d}$$

Also, the Stokes vector $\vec{g}$ can be determined by the polarization ratio $\rho_{45^\circ 135^\circ}$ as follows:

$$g_0 = 1$$

Table 3.3. The sign of the $g_1$, $g_2$, and $g_3$ parameters in the HV basis

| $\phi_{HV}$ | $\gamma_{HV}$ | $g_1$ | $g_2$ | $g_3$ |
|---|---|---|---|---|
| $0 < \phi_{HV} < \frac{\pi}{2}$ | $0 < 2\gamma_{HV} < \frac{\pi}{2}$ | + | + | + |
| | $\frac{\pi}{2} < 2\gamma_{HV} < \pi$ | − | + | + |
| | $\pi < 2\gamma_{HV} < \frac{3\pi}{2}$ | − | − | − |
| | $\frac{3\pi}{2} < 2\gamma_{HV} < 2\pi$ | + | − | − |
| $-\frac{\pi}{2} < \phi_{HV} < 0$ | $0 < 2\gamma_{HV} < \frac{\pi}{2}$ | + | + | − |
| | $\frac{\pi}{2} < 2\gamma_{HV} < \pi$ | − | + | − |
| | $\pi < 2\gamma_{HV} < \frac{3\pi}{2}$ | − | − | + |
| | $\frac{3\pi}{2} < 2\gamma_{HV} < 2\pi$ | + | − | + |

$$g_1 = -\frac{2|\rho_{45°135°}|}{1+|\rho_{45°135}|^2}\cos(\phi_{45°135°}) = -\sin(2\gamma_{45°135°})\cos(\phi_{45°135°})$$

$$g_2 = \frac{1-|\rho_{45°135°}|^2}{1+|\rho_{45°135°}|^2} = \cos(2\gamma_{45°135°}) \tag{3.49}$$

$$g_3 = \frac{2|\rho_{45°135°}|}{1+|\rho_{45°135°}|^2}\sin(\phi_{45°135°}) = \sin(2\gamma_{45°135°})\sin(\phi_{45°135°})$$

**c) $\rho_{LR}$ – Presentation**

Finally, consider the polarization ratio $\rho_{LR}$ defined in the (LR) basis. Similarly, because the cos $2\gamma_{LR}$ is the direction cosine of the Stokes vector $\vec{g}$ with respect to the z-axis, the polarization ratio $\rho_{LR}$ can be determined by the Stokes vector $\vec{g}$ as:

$$\rho_{LR} = \sqrt{\frac{g_0 - g_3}{g_0 + g_3}}\; e^{j\tan^{-1}\left(\frac{g_2}{g_1}\right)} \quad . \tag{3.50}$$

Inversely,

$$\begin{aligned} g_0 &= 1 \\ g_1 &= \frac{2|\rho_{LR}|\cos\phi_{LR}}{1 + |\rho_{LR}|^2} = \sin(2\gamma_{LR})\cos\phi_{LR} \\ g_2 &= \frac{2|\rho_{LR}|\sin\phi_{LR}}{1 + |\rho_{LR}|^2} = \sin(2\gamma_{LR})\sin\phi_{LR} \\ g_3 &= \frac{1 - |\rho_{LR}|^2}{1 + |\rho_{LR}|^2} = \cos 2\gamma_{LR} \quad . \end{aligned} \tag{3.51}$$

**Table 3.4: Alternate Expressions for Normalized Stokes Vector Presentations on the Polarization Sphere**

| | $\varepsilon$ , $\tau$ | $\gamma_{HV}$ , $\phi_{HV}$ | $\gamma_{45°135°}$ , $\phi_{45°135°}$ | $\gamma_{LR}$ , $\phi_{LR}$ |
|---|---|---|---|---|
| $g_0$ | 1 | 1 | 1 | 1 |
| $g_1$ | $\cos 2\varepsilon\cos 2\tau$ | $\cos(2\gamma_{HV})$ | $-\sin(2\gamma_{45°135°})\cos(\phi_{45°135°})$ | $\sin(2\gamma_{LR})\cos(\phi_{LR})$ |
| $g_2$ | $\cos 2\varepsilon\sin 2\tau$ | $\sin(2\gamma_{HV})\cos(\phi_{HV})$ | $\cos(2\gamma_{45°135°})$ | $\sin(2\gamma_{LR})\sin(\phi_{LR})$ |
| $g_3$ | $\sin 2\varepsilon$ | $\sin(2\gamma_{HV})\sin(\phi_{HV})$ | $\sin(2\gamma_{45°135°})\sin(\phi_{45°135°})$ | $\cos(2\gamma_{LR})$ |

### 3.6.4. The Riemann transformation

So far, we have shown how to map a polarization state onto a unique point on the surface of the Poincaré Sphere using the magnitude and the phase of the complex polarization ratio $\rho$. Now, we will introduce another method which gives a relationship between the polarization ratio and its corresponding spherical coordinates on the Poincaré Sphere. First, we need to introduce an auxiliary complex parameter $u(\rho)$ which is defined by the Riemann transformation of the surface of the sphere onto the polar grid as follows,

$$u(\rho) = \frac{1 - j\rho}{1 + j\rho} \tag{3.52}$$

In the (HV) basis, $\rho_{HV} = \tan\gamma_{HV}\, e^{j\phi_{HV}} = \tan\gamma_{HV}\,(\cos\phi_{HV} + j\sin\phi_{HV})$

$$u = \frac{(1 + \tan\gamma_{HV}\,\sin\phi_{HV}) - j\tan\gamma_{HV}\,\cos\phi_{HV}}{(1 - \tan\gamma_{HV}\,\sin\phi_{HV}) + j\tan\gamma_{HV}\,\cos\phi_{HV}}$$

$$|u|^2 = \frac{1 + 2\tan\gamma_{HV}\,\sin\phi_{HV} + \tan^2\gamma_{HV}}{1 - 2\tan\gamma_{HV}\,\sin\phi_{HV} + \tan^2\gamma_{HV}}$$

$$\frac{|u|^2 - 1}{|u|^2 + 1} = \frac{2\tan\gamma_{HV}}{1 + \tan^2\gamma_{HV}}\sin\phi_{HV} = \sin 2\gamma_{HV}\,\sin\phi_{HV} \quad (= g_3)$$

according to Eq. 3.27 and Fig. 3.6, the polar angle $\Theta = \pi/2 - 2\varepsilon$ can be obtained from

$$\frac{|u|^2 - 1}{|u|^2 + 1} = \sin 2\varepsilon = \sin(\pi/2 - \Theta) = \cos\Theta$$

so that

$$\Theta = \cos^{-1}\left\{\frac{|u|^2 - 1}{|u|^2 + 1}\right\} \tag{3.53}$$

also, according to Eq. 3.28 and Fig. 3.6, the spherical azimuthal angle $\Phi = 2\tau$ can be obtained from

$$-\frac{\mathrm{Im}\{u\}}{\mathrm{Re}\{u\}} = \frac{2\tan\gamma_{HV}\cos\phi_{HV}}{1 - \tan^2\gamma_{HV}} = \tan 2\tau = \tan\Phi$$

so that the spherical azimuthal angle $\Phi$ becomes

$$\Phi = -\tan^{-1}\left(\frac{\mathrm{Im}\{u\}}{\mathrm{Re}\{u\}}\right) \tag{3.54}$$

### 3.7 MAP PROJECTION TECHNIQUES

The Poincaré Sphere is very useful in polarization analyses as described earlier. But since it is a sphere, at times it is hard to visualize where the point (corresponding to a polarization state) is located on the spherical surface. Especially, it is difficult to use computer plotting methods for the Poincaré Sphere polarization diagrams. Instead, map projection techniques can be applied directly to the polarimetric analysis in order to solve for the plotting problem. Through map projection, we obtain a two-dimensional representation of the Poincaré Polarization Sphere Surface. In computer plotting of polarimetric radar data, all the input coordinates are spherical coordinates:

Input Coordinates

Radius of the Sphere: R

Scale factor: S

Latitude: $2\varepsilon$, $2\varepsilon_0$ (the coordinate of the center)

Longitude: $2\tau$, $2\tau_0$ (the coordinate of the center)

$0 \leq \varepsilon < 45°$ upper semisphere

$-45° < \varepsilon \leq 0$ lower semisphere

### 3.7.1 The Mercator cylindrical conformal projection

The Mercator projection, introduced in 1569 by Gerardus Mercator, served as an aid to navigation from the age of ocean exploration to the age of space exploration which marked the beginning of modern map projections. This projection is a cylindrical projection.

The mathematical transformation of the projection process involves a transformation from spherical coordinates to a rectangular coordinate grid.

Output Coordinates: (rectangular coordinate)

x coordinate: x (horizontal axis)
y coordinate: y (vertical axis)

Let the x-coordinate be arbitrarily defined at the beginning of the derivation, the y-coordinate is determined by the condition of conformality.

Coordinate Transformation: [47]

$$\begin{pmatrix} x \\ y \end{pmatrix} = \begin{pmatrix} RS(2\tau - 2\tau_0) \\ RS \ln \tan(\pi/4 + \varepsilon) \end{pmatrix} \tag{3.55}$$

In Fig. 3.9, the X axis is along the equator, the y axis is along the central meridian, all other meridians are equally spaced straight lines, and the parallels are unequally spaced straight lines. The center of the map corresponds to ($2\tau_0 = 0°$, $2\varepsilon_0 = 0°$), which represents the hori-

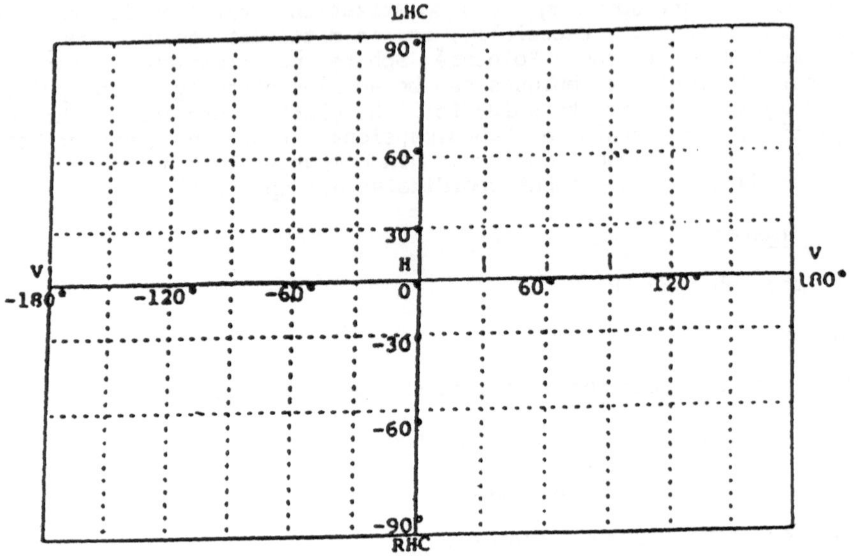


Fig. 3.9: The Mercator Cylindrical Conformal Projection

zontal polarization state. Because of this, the vertical polarization appears at the two extreme points on the right and left of the map. The equator, which is the x-axis of the projection, represents the linear polarization states.

The advantages: First, the meridians and parallels are simply vertical and horizontal. The second advantage is the readability.

From the nature of the map projection, each sufficiently narrow belt of latitude in any part of the Poincaré Sphere, except for the polar region, is enlarged or reduced to a desired scale, and represents the true form of the locality for ready use. This presentation was used extensively by Huynen [30].

Note that the scale factor increases continuously when the map extends beyond NL 60°. In which case the map presentation is strongly distorted, where the top horizontal line (90°) corresponds to LCP and the bottom line to RCP. Thus, this particular Mercator projection is not area-true.

### 3.7.2 The $\tau$-$\varepsilon$ plots [40]

There exists another more useful method of mapping a polarization state from the Poincaré Sphere onto a rectangular coordinate grid. In the coherent case, the Stokes parameters, $g_i$ of every polarization states can be uniquely calculated in terms of the ellipticity angle $\varepsilon$ and orientation (or tilt) angle $\tau$ according to Eq. 3.39. Let us define $\varepsilon$ in the range $[-\pi/4, \pi/4]$ and z in the range $[-\pi/2, \pi/2]$.

Output Coordinates: (Rectangular Coordinate)

x-coordinate: x (horizontal axis)
y-coordinate: y (vertical axis)

Coordinate Transformation:

$$\begin{pmatrix} x \\ y \end{pmatrix} = \begin{pmatrix} \tau \\ \varepsilon \end{pmatrix} \tag{3.56}$$

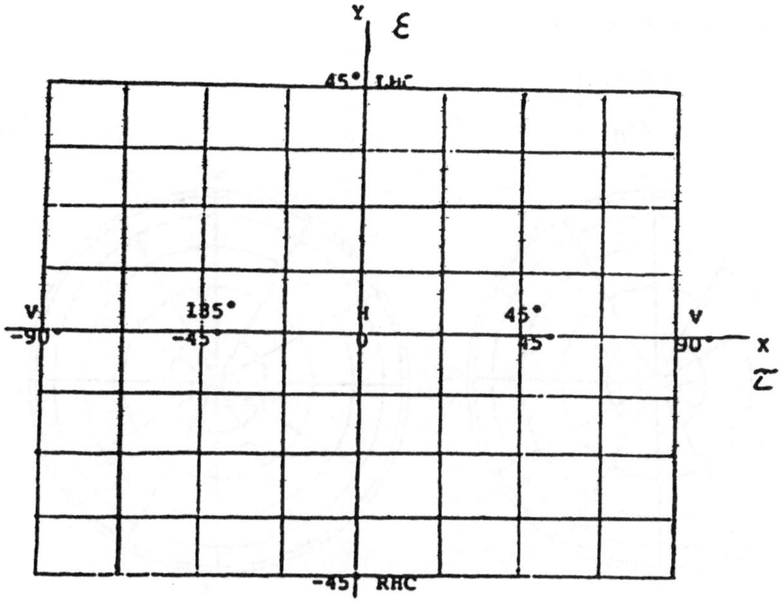


Fig. 3.10: Ellipticity - Tilt Plot

Any polarization state is determined by a point on a plane with coordinates $(\tau, \varepsilon)$, as in Fig. 3.10. Points on the x-axis ($\varepsilon = 0$) correspond to the linear Polarizations, and LHC and RHC are given by ($\varepsilon = \pi/4$) and ($\varepsilon = -\pi/4$) points. The conjugate, symmetric and orthogonal ellipses are given correspondingly by ($x \to x, y \to -y$), ($x \to -x, y \to -y$), and ($x \to x-\pi/2, y \to -y$). This presentation was used extensively in [1], [39], [41] and [64].

### 3.7.3 The Lambert azimuthal equal area polar projection

This map projection is suitable in the analysis of the clustering of multiple polarization states because of its area conformality. The mathematical transformation involves a transformation from the spherical coordinate to a polar coordinate system. Particularly, the projection process involves a parameter generator which is performed by solving a sinusoidal function (3.51).

<u>Output Coordinates:</u> (Polar Coordinates)

$\gamma$ coordinate: r(radius)
$\Theta$ coordinate: $\Theta$(azimuthal angle)

<u>Intermediate Parameter Expression:</u>

$$c = \cos^{-1}\{(\sin(2\varepsilon)\sin(2\varepsilon_0) + \cos(2\varepsilon_0)\cos(2\tau - 2\tau_0)\} \tag{3.57}$$

<u>Transformation Operation:</u> [47]

$$\begin{pmatrix} \gamma \\ \Theta \end{pmatrix} = \begin{pmatrix} 2\sin(c/2) \\ 2\tau - 2\tau_0 \end{pmatrix} \tag{3.58}$$

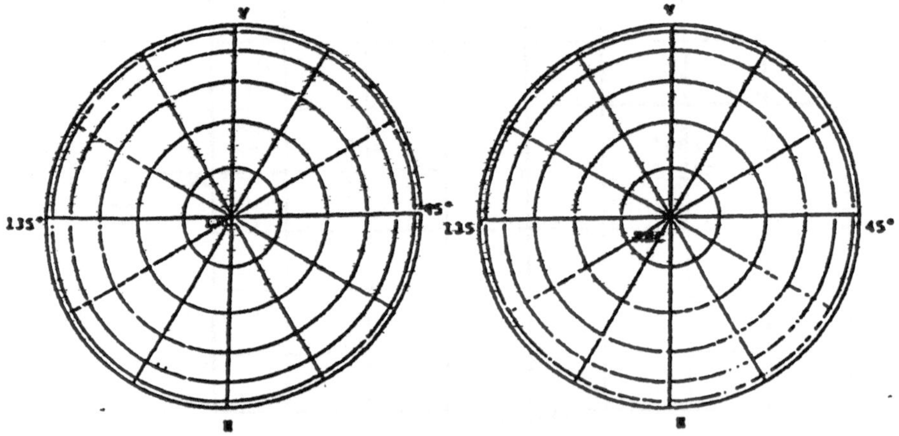


Fig. 3.11: The Lambert Azimuthal Equal Area Polar Projection

The Lambert equal area projection is directly derived from the conical projection surface, i.e., the projection surface is a cone. The equal longitude lines (Meridians) converge at the pole (the center of the map) and the equal latitude lines (parallels) are concentric circles with respect to the center of the map.

The prime property for this projection is that areas on the projected surface are proportional to the corresponding area on the sphere surface. In fact, it means the ratio of areas of any part of the map is equal to the ratio of areas of the entire map. In the analysis of the clustering of multiple polarization states, the mean value and the distribution of polarization nulls are truly visible on the map because of the true area representation. From its circular appearance, this map projection is suited to represent the mapping of the areas of considerable extent in all directions. For a clustering of polarization states near to the pole, that is, a group of near circular polarization states, this map projection is suggested. The disadvantage of this projection is that we require each one projection for the top and one for bottom semisphere, i.e., one for left-sensed and one for right-polarization states. This presentation was and is being used extensively by Poelman [48] and now also by Giuli [24].

### 3.7.4 The Aitoff-Hammer elliptical equal area projection

The Aitoff-Hammer projection is yet another equal area projection based on a spherical model.

The mathematical transformation of this projection process involves a transformation from a spherical system to a rectangular coordinate system. It consists of two steps. First, generating three parameters used in the transformation and second, making the transformation.

Output Coordinates: (Rectangular Coordinates)

x coordinate: x(longitude line)
y coordinate: y(latitude line)

Intermediate Parameter Expressions: $A_1$ , $A_2$

$$A_1 = \sqrt{2(1 - \cos 2\varepsilon \cos \frac{2\tau-2\tau_0}{2})} \tag{3.59}$$

$$A_2 = \tan^{-1}\left(\frac{\sin \frac{2\tau-2\tau_0}{2}}{\tan 2\varepsilon}\right) \tag{3.60}$$

Transformation Operations: [47]

$$\begin{pmatrix} x \\ y \end{pmatrix} = \begin{bmatrix} 2RS\ A_1 \sin(A_2) \\ RS\ A_1 \cos(A_2) \end{bmatrix} \tag{3.61}$$

Fig. 3.12 is generated by the map construction program developed for the Aitoff-Hammer projection with setting of the center of projection to be $\tau_0 = 0°$ and $\varepsilon_0 = 0°$ which is the location of the horizontal polarization state on the Poincaré Sphere. In the figure, the equator and the central meridian are the only straight lines, and the rest of the parallels and the meridians are curves. This equal area map gives a clear view of the entire sphere surface at one glance. It makes this projection useful in representing given data of unknown polarization states, and it was used extensively by Boerner et al [6,7] and first used in radar polarimetry by Raven, included in [6], for the analysis of the clustering of characteristic polarization states of clutter [6, 7].

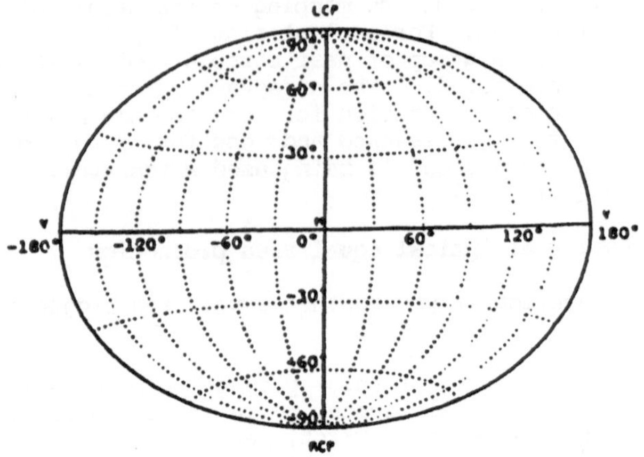


Fig. 3.12: The Aitoff-Hammer Elliptical Equal Area Projection

It is also very useful to represent distributions on the Poincaré Sphere surface as does the Lambert cylindrical equal area projection. Furthermore, because of its circular shape in regions near to the pole, this projection can give much more accurate representation for the near-pole regions, and, like the Lambert cylindrical equal area projection it is area-true, i.e., the ratio of areas of any part of the map is equal to the ratio of areas of the entire map. Note that the Aitoff-Hammer map projection is usually preferred to the use of the closely related Mollweide projection which is used extensively by Wanielik [61].

### 3.7.5. The Mollweide equal area projection

This projection, based on a spherical model of the Earth, is often used in world maps. The main problem in this projection is spacing the parallels so that the property of equivalence of area is maintained. The mathematical transformation of this projection process involves a transformation from a spherical system to a rectangular coordinate system.

<u>Output Coordinates</u>: (Rectangular Coordinates)

X-Coordinate: X (Horizontal axis)
Y-Coordinate: Y (Vertical axis)

**Coordinate Transformation [47]:**

$$\begin{bmatrix} X \\ Y \end{bmatrix} = \begin{bmatrix} \frac{\sqrt{2}}{90^\circ} (2\tau - 2\tau_0) \text{ RS } \cos(2\varepsilon) \\ \\ \sqrt{2} \text{ RS } \sin(2\varepsilon) \end{bmatrix} \tag{3.62}$$

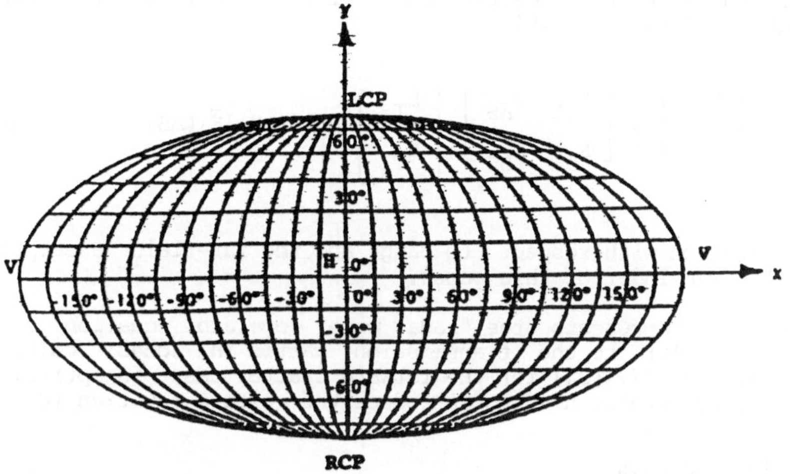


Fig. 3.13: The Mollweide Projection

A representative grid for this projection, with setting at the center of the projection to be $\tau = 0°$ and $\varepsilon = 0°$, appears in Fig. 3.13. The equator and parallels are horizontal straight lines. The polarization state on the equator is a linear polarization state. All of the meridians are ellipses except the central meridian which is a rectilinear ellipse, or a straight line prependicu- lar to the equator. The two extreme points at the central meridian represent the LHC and the RHC polarization states, respectively.

## 4. RADAR SCATTERING MATRICES [S] AND [M]

In radar polarimetry, one is concerned with the polarization dependence of a radar scattering process as seen at the receiving antenna terminal at a fixed point in space. Let us use $\vec{E}_T$ and $\vec{E}_R$ to represent transmitted and reflected monochromatic TEM traveling plane waves, respectively. In addition, we can use them to describe the polarization properties of transmitter and receiver. Then we use the antenna height $\vec{h}$ to describe the wave which the receiving antenna would radiate if it acted as a transmitter [17].

When an electromagnetic plane wave illuminates a target, the polarzation of the scattered wave $\vec{E}_R(AB)$, in general, is different from that of the incident wave $\vec{E}_T(AB)$ expressed in the general polarization basis (AB). This change in polarization states represents the characteristic properties of the target expressed by the coherent scattering matrix [S(AB)] also known as the Sinclair matrix [53]. At the target an incident wave $\vec{E}_T(AB)$ gives rise to a scattered wave $\vec{E}_R(AB)$ which at the receiver is related via the scattering matrix [S(AB)] for the polarization basis (AB), as presented in Mott [45]

$$\vec{E}_R(AB) = \begin{bmatrix} E_{R_A} \\ E_{R_B} \end{bmatrix} = \begin{bmatrix} S_{AA} & S_{AB} \\ S_{BA} & S_{BB} \end{bmatrix} \begin{bmatrix} E_{T_A} \\ E_{T_B} \end{bmatrix} = [S(AB)]\vec{E}_T(AB) \tag{4.1}$$

using normalization with respect to range and the far-field assumption throughout the remainder of this paper.

Whereas, Eq. 4.1 represents the "radar wave" operator equation which behaves very similar to the transmission matrix in Jones' calculus (Azzam and Bashara 1977 [2]) for the forward scatter case in optics, in radar polarimetry we also need to introduce the voltage equation (Collin 1985 [17])

$$V = \vec{h}^T\vec{E}_R = \vec{h}^T[S]\vec{E}_T \tag{4.2}$$

which mathematically behaves like a bilinear form when transformation from one basis to another occurs (Kostinski and Boerner 1986 [39]). The range normalized scattering matrix [S] is used to describe the properties of the radar target at a given direction (or fixed direction of illumination and reception) and the radar frequency, independent of the measurement system.

### 4.1 Coordinate system and polarization bases

The value of the elements $S_{ij}$ of the scattering matrix [S] depend on the chosen coordinate system and polarization basis. It is convenient to choose a fixed cartesian coordinate system (x,y,z) with the origin at the center inside of the scattering target as shown in Fig. 4.1, where the (x,y,z) direction are chosen to coincide with the invariant symmetry axes of the target. However, there are many choices of polarization bases because any two orthogonal unit vectors in the transverse plane can be chosen for a polarization basis. We would like to define a polarization basis uniquely determined by a general coordinate system [14,18,27] described in Fig. 4.1.

At any point $T(r_1,\theta_1,\phi_1)$ in the space, we consider a distinct orthogonal basis formed by three spherical unit vectors $-\hat{r}_1$, $-\hat{\theta}_1$, and $\hat{\phi}_1$. They are defined as

$$-\hat{r}_1 = -\sin\theta_1\cos\phi_1\hat{x} - \sin\theta_1\sin\phi_1\hat{y} - \cos\theta_1\hat{z}$$

$$-\hat{\theta}_1 = -\cos\theta_1\cos\phi_1\hat{x} - \cos\theta_1\sin\phi_1\hat{y} + \sin\theta_1\hat{z}$$

$$\hat{\phi}_1 = -\sin\phi_1\hat{x} + \cos\phi_1\hat{y}$$

These unit vectors define a right-handed vector triplet, i.e.,

$$(-\hat{r}_1)\times(-\hat{\theta}_1) = \hat{\phi}_1 \tag{4.3}$$

$$(-\hat{\theta}_1)\times\hat{\phi}_1 = -\hat{r}_1$$

$$\hat{\phi}_1\times(-\hat{r}_1) = -\hat{\theta}_1$$

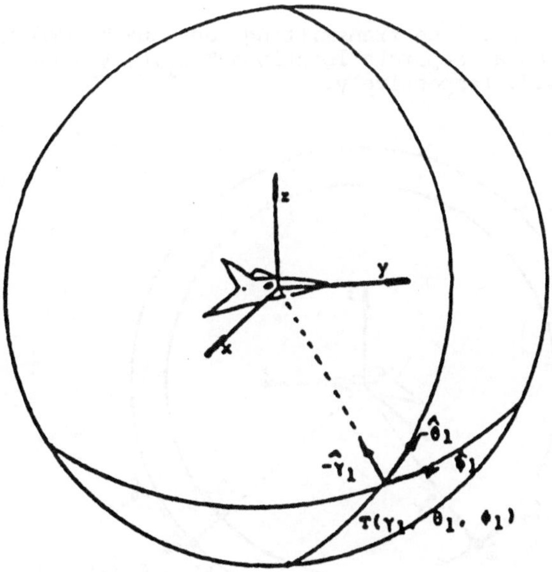


Fig. 4.1 Coordinate System and Polarization Basis for the Monostatic Case

For an electromagnetic plane wave traveling through the point T, or an antenna located at the point T, we can choose the direction of the unit vector $-\hat{r}_1$ as the plane wave traveling direction $\hat{k}_1 = -\hat{r}_1$, and then use the other two spherical unit vectors $-\hat{\theta}_1$ and $\hat{\phi}_1$, to form a polarization basis for this plane wave traveling in the $-\hat{r}_1$ direction.

For the monostatic (backscattering) case, the transmitting antenna T and the receiving antenna R are located at the same position $T(r_1,\theta_1,\phi_1)$. The polarization bases for transmission and reception are chosen to be the same, i.e., $-\hat{\theta}_1$ and $\hat{\phi}_1$. The polarization states of the transmitter and the receiver can be written as

$$\vec{E}_T = E_{T_{\theta_1}}(-\hat{\theta}_1) + E_{T_{\phi_1}}(\hat{\phi}_1) \tag{4.4}$$

$$\vec{E}_R = E_{R_{\theta_1}}(-\hat{\theta}_1) + E_{R_{\phi_1}}(\hat{\phi}_1)$$

$$\vec{h} = h_{\theta_1}(-\hat{\theta}_1) + h_{\phi_1}(\hat{\phi}_1)$$

For the bistatic case, the transmitting antenna T and the receiving antenna R are placed at separate locations $T(r_1,\theta_1,\phi_1)$ and $R(r_2,\theta_2,\phi_2)$, as shown in Fig. 4.2, respectively.

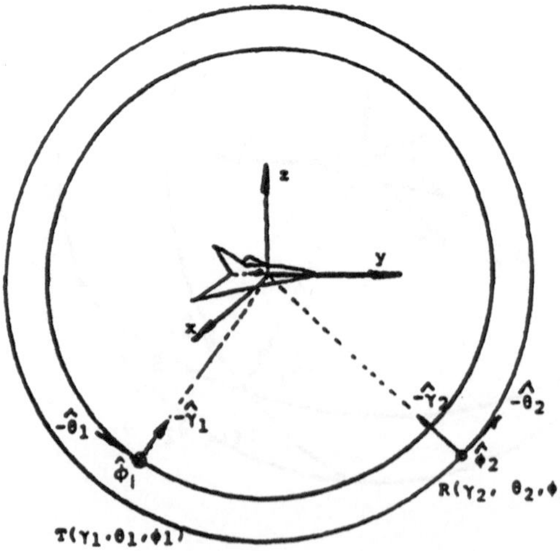


Fig. 4.2 Coordinate System and Polarization Basis for the Bistatic Case

The two sets of spherical unit vectors $(-\hat{\theta}_1, \hat{\phi}_1)$ and $(-\hat{\theta}_2, \hat{\phi}_2)$ form the transmitter polarization basis and the receiver polarization basis, respectively. Note,

$$-\hat{r}_2 = -\sin\theta_2\cos\phi_2\hat{x} - \sin\theta_2\sin\phi_2\hat{y} - \cos\theta_2\hat{z}$$

$$-\hat{\theta}_2 = -\cos\theta_2\cos\phi_2\hat{x} - \cos\theta_2\sin\phi_2\hat{y} + \sin\theta_2\hat{z}$$

$$\hat{\phi}_2 = -\sin\phi_2\hat{x} + \cos\phi_2\hat{y}$$

Thus the polarization states can be written as follows,

$$\vec{E}_T = E_{T_{\theta_1}}(-\hat{\theta}_1) + E_{T_{\phi_1}}(\hat{\phi}_1) \qquad (4.5)$$

$$\vec{E}_R = E_{R_{\theta_2}}(-\hat{\theta}_2) + E_{R_{\phi_2}}(\hat{\phi}_2)$$

$$\vec{h} = h_{\theta_2}(-\hat{\theta}_2) + h_{\phi_2}(\hat{\phi}_2)$$

where $E_{T_\theta}$, $E_{R_\theta}$ and $h_\theta$ are the $\theta$ components of $E_T$, $E_R$ and h, and $E_{T_\phi}$, $E_{R_\phi}$ and $h_\phi$ are the $\phi$ components of $E_T$, $E_R$ and h.

## 4.2 The scattering matrix for the bistatic case

From Eq. 4.5, we know the polarization of incident wave is

$$\vec{E}_T = E_{T_{\theta_1}}(-\hat{\theta}_1) + E_{T_{\phi_1}}(\hat{\phi}_1) = E_{T_A}\hat{A} + E_{T_B}\hat{B}$$

and the polarization of the scattered wave is

$$\vec{E}_R = E_{R_{\theta_2}}(-\hat{\theta}_2) + E_{R_{\phi_2}}(\hat{\phi}_2) = E_{R_A}\hat{A} + E_{R_B}\hat{B}$$

Now the bistatic scattering process according to Eq. 4.1, can be written as

$$\begin{bmatrix} E_{R_{\theta_2}} \\ E_{R_{\phi_2}} \end{bmatrix} = \begin{bmatrix} S_{\theta_2\theta_1} & S_{\theta_2\phi_1} \\ S_{\phi_2\theta_1} & S_{\phi_2\phi_1} \end{bmatrix} \begin{bmatrix} E_{T_{\theta_1}} \\ E_{T_{\phi_1}} \end{bmatrix} \qquad (4.6)$$

here $S_{\theta_2\phi_1}$ is a complex number, which is proportional to the received

$-\hat{\theta}_2$ component of the scattered wave, while the target is illuminated by $\hat{\phi}_1$ polarized transmitter only, i.e., $E_{R\theta_2} = S_{\theta_2\phi_1} E_{T\phi_1}$ with $E_{T\theta_1} = 0$. In general, $-\hat{\theta}_2$ and $\hat{\phi}_1$ are not orthogonal, thus, in the strict sense, the incident wave polarized in $\hat{\phi}_1$ direction and the scattered wave polarized in $-\hat{\theta}_2$ direction do not constitute a cross-polarized channel. But, in a broad sense, we may treat them as a cross-polarization channel; similarly, $\hat{\phi}_1$ and $\hat{\phi}_2$ are not parallel, and in the strict sense, they do not form a pair of co-polarized channels.

In general, the scattering matrix [S] is not symmetric. Thus, a total of eight parameters which depend on $\hat{\theta}_1$, $\hat{\theta}_2$, $\hat{\phi}_1$, and $\hat{\phi}_2$ are required to give the complete bi-static scattering matrix, however, if only the relative phases are of interest [14,18,27], the total number of parameters would be seven.

### 4.3 The scattering matrix in the backscattering case

In the monostatic (backscattering) case, most imaging radars use the same antenna for transmission and reception of the signals located at position $T(r_1,\theta_1,\phi_1)$. The incident wave $\vec{E}_I$ for this target is the electromagnetic traveling plane wave produced by the transmitting antenna $\vec{E}_T$. The scattered wave $\vec{E}_S$ from this target is the electromagnetic traveling plane wave reflected by the target $\vec{E}_R$.

From Eq. 4.4, we know that $\vec{E}_T$ and $\vec{E}_R$ have the same polarization basis ($-\hat{\theta}_1$ and $\hat{\phi}_1$). For convenience, let us call unit vector $-\hat{\theta}_1$ as horizontal unit vector $\hat{H}$, and unit vector $\hat{\phi}_1$ as vertical unit sector $\hat{V}$. In the basis (HV), one may define a complex 2x2 scattering matrix [S(HV)] according to Eq. 4.1 and Eq. 4.4 as

$$\begin{bmatrix} E_{S_H} \\ E_{S_V} \end{bmatrix} = \begin{bmatrix} S_{HH} & S_{HV} \\ S_{VH} & S_{VV} \end{bmatrix} \begin{bmatrix} E_{I_H} \\ E_{I_V} \end{bmatrix} \tag{4.7}$$

Here, $S_{HV}$ is also a complex number which is proportional to the received horizontally polarized component of the scattered signal (in amplitude and phase) when the target is only illuminated by a vertically polarized transmitter, i.e., $E_{S_H} = S_{HV}E_{I_V}$ with $E_{I_H} = 0$, and similarly for the other elements. From Eq. 4.6, we can see that the elements $S_{HH}$ and $S_{VV}$ produce the power return in the copolarized channel and the elements $S_{HV}$

and $S_{VH}$ produce the power return in the cross-polarized channel. If the role of the transmitting and the receiving antennas are interchanged, the reciprocity theorem requires that the scattering matrix be symmetrical, provided that the propagation medium is reciprocal, that is

$$S_{HV} = S_{VH} \tag{4.8}$$

There are only five independent parameters in the scattering matrix [S] if only the relative phases are of interest.

## 4.4 The Mueller matrix [M]

If scatterer properties change with time, or a number of independent incoherent scatterers are of interest, [M] must be used. The Mueller matrix [M] can be defined by

$$\vec{g}_S = [M]\vec{g}_I \tag{4.9}$$

Here, $\vec{g}_S$ is the scattered Stokes vector and $\vec{g}_I$ is the incident Stokes vector.

The Mueller matrix [M] is a 4×4 real matrix and its elements $M_{ij}$ are the same for measurements in any orthogonal polarization basis (AB), whereas the elements of [S(AB)] depend on the basis, i.e., $S_{ij} = S_{ij}(AB)$. For measurements in the (HV) basis the Mueller matrix [M] can be related to the scattering matrix [S(HV)] for the coherent case. In order to obtain the relationship, we introduce a coherency vector $\vec{J}(HV)$ which is formed by the elements of the coherency matrix [J] as

$$\vec{J}(HV) = \begin{bmatrix} E_H E_H^* \\ E_H E_V^* \\ E_V E_H^* \\ E_V E_V^* \end{bmatrix} = \begin{bmatrix} J_{HH} \\ J_{HV} \\ J_{VH} \\ J_{VV} \end{bmatrix} \tag{4.10}$$

According to Eq.3.41, it is

$$\begin{bmatrix} g_0 \\ g_1 \\ g_2 \\ g_3 \end{bmatrix} = \begin{bmatrix} J_{HH} + J_{VV} \\ J_{HH} - J_{VV} \\ J_{HV} + J_{VH} \\ +jJ_{HV} - jJ_{VH} \end{bmatrix}$$

If we define a Kronecker product matrix [A(HV)] as

$$[A(HV)] = \begin{bmatrix} 1 & 0 & 0 & 1 \\ 1 & 0 & 0 & -1 \\ 0 & 1 & 1 & 0 \\ 0 & +j & -j & 0 \end{bmatrix} \tag{4.11}$$

then $\vec{g} = [A(HV)]\vec{J}(HV)$ (4.12)

Now, by using the relationship between the incident and the scattered electric field of the wave, from Eq. 4.10 and Eq. 4.1

$$\vec{J}_S(HV) = \vec{E}_S(HV) \otimes \vec{E}_S^*(HV) = ([S(HV)]\vec{E}_I(HV)) \otimes ([S^*(HV)]\vec{E}_I^*(HV))$$

$$= \begin{bmatrix} S_{HH}E_{I_H} + S_{HV}E_{I_V} \\ S_{VH}E_{I_H} + S_{VV}E_{I_V} \end{bmatrix} \otimes \begin{bmatrix} S_{HH}^*E_{I_H}^* + S_{HV}^*E_{I_V}^* \\ S_{VH}^*E_{I_H}^* + S_{VV}^*E_{I_V}^* \end{bmatrix}$$

$$= \left( \begin{bmatrix} S_{HH} & S_{HV} \\ S_{VH} & S_{VV} \end{bmatrix} \otimes \begin{bmatrix} S_{HH}^* & S_{HV}^* \\ S_{VH}^* & S_{VV}^* \end{bmatrix} \right) \begin{bmatrix} J_{I_{HH}} \\ J_{I_{HV}} \\ J_{I_{VH}} \\ J_{I_{VV}} \end{bmatrix}$$

$$= ([S(HV)] \otimes [S(HV)]^*)\vec{J}_I(HV)$$

so that the scattered Stokes vector $\vec{g}_S$ can be written as

$$\vec{g}_S = [A(HV)]\vec{J}_S(HV) = [A(HV)]([S(HV)] \otimes [S^*(HV)])\vec{J}_I(HV)$$
$$= [A(HV)]([S(HV)] \otimes [S^*(HV)])[A(HV)]^{-1}\vec{g}_I(HV)$$

Comparing the above equation with Eq. 4.9, we obtain the relationship between the Mueller matrix [M] and the scattering matrix [S(HV)].

$$[M] = [A(HV)]([S(HV)] \otimes [S^*(HV)])[A(HV)]^{-1} \tag{4.13}$$

The elements of [M] can be given in terms of the elements of the scattering matrix [S(HV)] by

$$M_{11} = \tfrac{1}{2}(S_{HH}S_{HH}^{*}+S_{VH}S_{VH}^{*}+S_{HV}S_{HV}^{*}+S_{VV}S_{VV}^{*}) = \tfrac{1}{2}(|S_{HH}|^{2}+|S_{VH}|^{2}+|S_{HV}|^{2}+|S_{VV}|^{2})$$

$$M_{12} = \tfrac{1}{2}(S_{HH}S_{HH}^{*}+S_{VH}S_{VH}^{*}-S_{HV}S_{HV}^{*}-S_{VV}S_{VV}^{*}) = \tfrac{1}{2}(|S_{HH}|^{2}+|S_{VH}|^{2}-|S_{HV}|^{2}-|S_{VV}|^{2})$$

$$M_{13} = \tfrac{1}{2}(S_{HH}S_{HV}^{*}+S_{VH}S_{VV}^{*}+S_{HV}S_{HH}^{*}+S_{VV}S_{VH}^{*}) = \mathrm{Re}(S_{HH}S_{HV}^{*}) + \mathrm{Re}(S_{VH}S_{VV}^{*})$$

$$M_{14} = -(j/2)(S_{HH}S_{HV}^{*}+S_{VH}S_{VV}^{*}-S_{HV}S_{HH}^{*}-S_{VV}S_{VH}^{*}) = -\mathrm{Im}(S_{HH}^{*}S_{HV}) - \mathrm{Im}(S_{VH}^{*}S_{VV})$$

$$M_{21} = \tfrac{1}{2}(S_{HH}S_{HH}^{*}-S_{VH}S_{VH}^{*}+S_{HV}S_{HV}^{*}-S_{VV}S_{VV}^{*}) = \tfrac{1}{2}(|S_{HH}|^{2}-|S_{VH}|^{2}+|S_{HV}|^{2}-|S_{VV}|^{2})$$

$$M_{22} = \tfrac{1}{2}(S_{HH}S_{HH}^{*}-S_{VH}S_{VH}^{*}-S_{HV}S_{HV}^{*}+S_{VV}S_{VV}^{*}) = \tfrac{1}{2}(|S_{HH}|^{2}-|S_{VH}|^{2}-|S_{HV}|^{2}+|S_{VV}|^{2})$$

$$M_{23} = \tfrac{1}{2}(S_{HH}S_{HV}^{*}-S_{VH}S_{VV}^{*}+S_{HV}S_{HH}^{*}-S_{VV}S_{VH}^{*}) = \mathrm{Re}(S_{HH}S_{HV}^{*}) - \mathrm{Re}(S_{VH}S_{VV}^{*})$$

$$M_{24} = -(j/2)(S_{HH}S_{HV}^{*}-S_{VH}S_{VV}^{*}-S_{HV}S_{HH}^{*}+S_{VV}S_{VH}^{*}) = -\mathrm{Im}(S_{HH}^{*}S_{HV}) - \mathrm{Im}(S_{VV}^{*}S_{VH})$$

$$M_{31} = \tfrac{1}{2}(S_{HH}S_{VH}^{*}+S_{VH}S_{HH}^{*}+S_{HV}S_{VV}^{*}+S_{VV}S_{HV}^{*}) = \mathrm{Re}(S_{HH}S_{VH}^{*}) + \mathrm{Re}(S_{HV}S_{VV}^{*})$$

$$M_{32} = \tfrac{1}{2}(S_{HH}S_{VH}^{*}+S_{VH}S_{HH}^{*}-S_{HV}S_{VV}^{*}-S_{VV}S_{HV}^{*}) = \mathrm{Re}(S_{HH}S_{VH}^{*}) - \mathrm{Re}(S_{HV}S_{VV}^{*})$$

$$M_{33} = \tfrac{1}{2}(S_{HH}S_{VV}^{*}+S_{VH}S_{HV}^{*}+S_{HV}S_{VH}^{*}+S_{VV}S_{HH}^{*}) = \mathrm{Re}(S_{HH}S_{VV}^{*}) + \mathrm{Re}(S_{HV}S_{VH}^{*})$$

$$M_{34} = -(j/2)(S_{HH}S_{VV}^{*}+S_{VH}S_{HV}^{*}-S_{HV}S_{VH}^{*}-S_{VV}S_{HH}^{*}) = -\mathrm{Im}(S_{HH}^{*}S_{VV}) - \mathrm{Im}(S_{VH}^{*}S_{HV})$$

$$M_{41} = -(j/2)(S_{VH}S_{HH}^{*}-S_{HH}S_{VH}^{*}+S_{VV}S_{HV}^{*}-S_{HV}S_{VV}^{*}) = -\mathrm{Im}(S_{VH}^{*}S_{HH}) - \mathrm{Im}(S_{VV}^{*}S_{HV})$$

$$M_{42} = -(j/2)(S_{VH}S_{HH}^{*}-S_{HH}S_{VH}^{*}-S_{VV}S_{HV}^{*}+S_{HV}S_{VV}^{*}) = -\mathrm{Im}(S_{VH}^{*}S_{HH}) - \mathrm{Im}(S_{HV}^{*}S_{VV})$$

$$M_{43} = -(j/2)(S_{VH}S_{HV}^{*}-S_{HH}S_{VV}^{*}+S_{VV}S_{HH}^{*}-S_{HV}S_{VH}^{*}) = -\mathrm{Im}(S_{VH}^{*}S_{HV}) - \mathrm{Im}(S_{VV}^{*}S_{HH})$$

$$M_{44} = -\tfrac{1}{2}(S_{VH}S_{HV}^{*}-S_{HH}S_{VV}^{*}-S_{VV}S_{HH}^{*}+S_{HV}S_{VH}^{*}) = \mathrm{Re}(S_{HH}S_{VV}^{*}) - \mathrm{Re}(S_{HV}S_{VH}^{*}) \quad (4.14)$$

The transformation relationships of Eq. 4.14 are different from those for measurements made in another basis (AB). In general, the relationship between the Mueller matrix [M] and the scattering matrix [S(AB)] is

$$[M] = [A(AB)]\ \{[S(AB)\otimes[S(AB)]^{*}\}\ [A(AB)]^{-1} \quad (4.15)$$

The Kronecker product matrix [A(AB)] must be derived separately for each specific polarization basis (AB), where [A(LR)] for the circular basis (LR) is given in Section 5.5.

## 4.5 Alternative derivation of the Mueller matrix

There is another way of obtaining the Mueller matrix [M] from the scattering matrix [S(HV)] measured in the (HV) basis. In terms of the coherency matrix [J(HV)] of Eq. 3.40, the coherency matrix $[J_S(HV)]$ for the scattered field $\vec{E}_S(HV)$ can be written as

$$
\begin{aligned}
[J_S(HV)] &= \langle \vec{E}_S(HV)\ \vec{E}_S^{+}(HV) \rangle \\
&= \langle \{[S(HV)]\vec{E}_I(HV)\}\ \{[S(HV)]\vec{E}_I(HV)\}^{+} \rangle \\
&= \langle [S(HV)]\vec{E}_I(HV)\ \vec{E}_I^{+}[S(HV)]^{+} \rangle \\
&= \langle [S(HV)]\ [\vec{J}_I(HV)]\ [S(HV)]^{+} \rangle
\end{aligned}
\tag{4.16}
$$

where the subscript I denotes the incident field.

In addition, the polarization coherency matrix [J(HV)] can be expressed by the Stokes vector $\vec{g}$, according to Eqs. 3.40 and 3.41, as follows,

$$
[J(HV)] = \begin{bmatrix} J_{HH} & J_{HV} \\ J_{VH} & J_{VV} \end{bmatrix} = \frac{1}{2}\begin{bmatrix} g_0+g_1 & g_2-jg_3 \\ g_2+jg_3 & g_0-g_1 \end{bmatrix} = \frac{1}{2}\sum_{\nu=0}^{3}[\sigma_\nu]g_\nu \tag{4.17}
$$

where the $[\sigma_i]$'s are the Pauli spin matrices:

$$
[\sigma_0] = \begin{bmatrix} 1 & 0 \\ 0 & 1 \end{bmatrix},\quad [\sigma_1] = \begin{bmatrix} 1 & 0 \\ 0 & -1 \end{bmatrix},\quad [\sigma_2] = \begin{bmatrix} 0 & 1 \\ 1 & 0 \end{bmatrix},\quad [\sigma_3] = \begin{bmatrix} 0 & -j \\ j & 0 \end{bmatrix} \tag{4.18}
$$

Using the trace orthogonality of the Pauli matrices, we can obtain the components of the Stokes vector, $g_\mu$, by multipling $[\sigma_\mu]$ by Eq. 4.17 and taking the trace

$$
g_\mu = \frac{1}{2}\,\mathrm{Tr}\ \{[\sigma_\mu]\sum_{\nu=0}^{3}[\sigma_\nu]g_\nu\ \} = \mathrm{Tr}\ \{[\sigma_\mu][J(HV)]\ \}
$$

So, we can multiply both sides of Eq. 4.16 by $[\sigma_\mu]$, and take the trace, which yields

$$
g_{S\mu} = \mathrm{Tr}\ \{[\sigma_\mu]\ [J_S(HV)]\ \} = \mathrm{Tr}\ \{[\sigma_\mu][S(HV)][\vec{J}_I(HV)][S(HV)]^{+}\} =
$$

$$= \frac{1}{2} \operatorname{Tr} \{[\sigma_\mu][S(HV)] \sum_{\nu=0}^{3} [\sigma_\nu] \; g_{I_\nu} [S(HV)]^+\}$$

$$= \sum_{\nu=0}^{3} \{ \frac{1}{2} \operatorname{Tr} \{ [\sigma_\mu][S(HV)][\sigma_\nu][S(HV)]^+ \} \; g_{I_\nu} \}$$

according to the definition of the Mueller matrix [M] of Eq. 4.9,

$$M_{\mu\nu} = \frac{1}{2} \operatorname{Tr} \{ [\sigma_\mu] \; [S(HV)] \; [\sigma_\nu] \; [S(HV)]^+ \} \tag{4.19}$$

where it can be shown that Eq. 4.19 yields the elements $M_{\mu\nu}$ identical to those given in Eq. 4.14.

## 4.6 Distinction between different formulations of the Mueller matrix

There are different kinds of Stokes reflection matrices known as Mueller matrices in use. These matrices have different definitions and representations and must be carefully distinguished from one another. We briefly introduce them and give the relationships for them.

### 4.6.1 The modified Mueller matrix

In radiative transfer theory, it is more common to use the "modified Stokes vectors," used frequently in Ishimaru[33], and given here in our notation (see Eqs. 2.13 and 3.18) by:

$$\vec{I} = \begin{pmatrix} I_0 \\ I_1 \\ I_2 \\ I_3 \end{pmatrix} = \begin{pmatrix} \langle E_x\rangle^2 \\ \langle E_y\rangle^2 \\ 2\langle|E_x||E_y|\cos\phi\rangle \\ 2\langle|E_x||E_y|\sin\phi\rangle \end{pmatrix} = \begin{pmatrix} \langle E_x\rangle^2 \\ \langle E_y\rangle^2 \\ 2\mathrm{Re}\langle E_x E_y^*\rangle \\ 2\mathrm{Im}\langle E_x^* E_y\rangle \end{pmatrix} . \tag{4.20}$$

Then the modified Mueller matrix [Mm] can be defined for the general bistatic scattering process, similar to Eq. 4.9, by

$$\vec{I}_S = [Mm] \; \vec{I}_I \tag{4.21}$$

where $\vec{I}_I$ and $\vec{I}_S$ represent the modified Stokes vectors of the incident and the scattered waves. The modified Mueller matrix [Mm] has a relation with the scattering matrix [S] of Eq. 4.1, and in our notation becomes:

$$[Mm] = \begin{bmatrix} S_{xx}S_{xx}^* & S_{xy}S_{xy}^* & Re(S_{xx}S_{xy}^*) & -Im(S_{xx}S_{xy}^*) \\ S_{yx}S_{yx}^* & S_{yy}S_{yy}^* & Re(S_{yx}S_{yy}^*) & -Im(S_{yx}S_{yy}^*) \\ 2Re(S_{xx}S_{yx}^*) & 2Re(S_{xy}S_{yy}^*) & Re(S_{xx}S_{yy}^* + S_{xy}S_{yx}^*) & -Im(S_{xx}^*S_{yy} - S_{xy}^*S_{yx}) \\ 2Im(S_{xx}^*S_{yx}) & 2Im(S_{xy}^*S_{yy}) & -Im(S_{xx}S_{yy}^* + S_{xy}S_{yx}^*) & Re(S_{xx}S_{yy}^* - S_{xy}S_{yx}^*) \end{bmatrix} \quad (4.22)$$

which differs from that given in [33].

### 4.6.2 The Stokes reflection matrices

Some authors [30],[62] prefer to define the Stokes reflection matrix using the received power equation as

$$P = \vec{g}^T [\bar{M}_p] \vec{g} \quad , \qquad (4.23)$$

where $\vec{g}$ is the Stokes vector of the transmitter, and $[\bar{M}_p]$ is called the Stokes reflection matrix (also called Mueller matrix by some authors). Because the received power depends on the polarization state of the receiving antenna, the Stokes reflection matrix $[\bar{M}_p]$ is not unique to the power P, but uniquely determined in a specific channel. In fact, Eq. 4.23 gives the Stokes reflection matrix $[\bar{M}_p]$ in the co-polarization channel.

Next we only consider monostatic scattering and derive $[\bar{M}_p]$ in three cases, one is for the co-polarization channel without and with satisfying the matching condition and the other is for the cross-polarization channel.

#### 4.6.2a The co-polarization channel case

In the co-polarization channel, for the monostatic case, the polarization state of the effective antenna height $\vec{h}$ is the same as that of the transmitting antenna, i.e.,

$$\vec{h} = \vec{E}_T / ||\vec{E}_T|| = \vec{e}_T \qquad (4.24)$$

The voltage equation given by Eq. 4.2, becomes

$$V_c = \vec{E}_T^T [S] \vec{E}_T / ||\vec{E}_T|| = \vec{e}_T^T [S] \vec{E}_T \quad , \qquad (4.25)$$

from which the received power in the co-pol channel is determined by

$$P_c = |V_c|^2 = |\vec{E}_T^T [S] \vec{E}_T|^2 / ||\vec{E}_T||^2 = |\vec{e}_T^T [S] \vec{E}_T|^2$$

$$= \vec{g}^T \left([A]^{-1}\right)^T \left([S]\otimes[S]^*\right) [A]^{-1} \vec{g} \tag{4.26}$$

Comparing Eq. 4.23 and Eq. 4.20, yields the Stokes reflection matrix in the co-pol channel

$$[M_{pc}] = \left([A]^{-1}\right)^T \left([S] \otimes [S]^*\right)[A]^{-1} \tag{4.27}$$

The Stokes reflection matrix $[M_{pc}]$ can be related to the Mueller matrix $[M]$, given by Eq. 4.14 by the following relation:

$$[M_{pc}] = \frac{1}{2} \begin{pmatrix} 1 & 0 & 0 & 0 \\ 0 & 1 & 0 & 0 \\ 0 & 0 & 1 & 0 \\ 0 & 0 & 0 & -1 \end{pmatrix} [M] \tag{4.28}$$

**4.6.2b The X-polarization channel case**

In the cross-polarization channel for the monostatic case, the polarization state of the effective antenna height $\vec{h}$ is orthogonal to the polarization state of the transmitter, i.e.,

$$\vec{h} = \vec{E}_{T\perp} / \|\vec{E}_{T\perp}\| = \vec{e}_{T\perp} \tag{4.29}$$

where the subscript $\perp$ denotes the orthogonal channel. A normalized completely polarized transmitting wave $\vec{E}_T$ can be expressed by Eq. 3.21 using the polarization ratio $\rho$ as

$$\vec{e}_T = \begin{bmatrix} e_x \\ e_y \end{bmatrix} = \frac{e^{j\phi_x}}{\sqrt{1 + \rho\rho^*}} \begin{pmatrix} 1 \\ \rho \end{pmatrix} \tag{4.30}$$

and the orthogonal polarization state of the transmitting wave can be expressed by

$$\vec{e}_{T\perp} = \begin{bmatrix} e'_x \\ e'_y \end{bmatrix} = \frac{e^{j\phi'_y}}{\sqrt{1 + \rho\rho^*}} \begin{pmatrix} -\rho^* \\ 1 \end{pmatrix} \tag{4.31}$$

The voltage equation 4.2 in the x-pol channel becomes

$$V_x = \vec{e}_{T\perp}^{\,T} [S] \vec{E}_T = \frac{e^{j\phi_x} e^{j\phi'_y}}{1 + \rho\rho^*} (-\rho^* \quad 1)[S] \begin{bmatrix} 1 \\ \rho \end{bmatrix} \tag{4.32}$$

Then the recieved power through some simple algebra can be written as

$$P_x = |V_x|^2 = \vec{g}^T [M_{px}] \vec{g} \quad , \tag{4.33}$$

where

$$[M_{px}] =$$

$$([A]^{-1})^T \left[ \begin{pmatrix} 0 & 0 & 0 & 1 \\ 0 & 0 & 0 & 0 \\ 0 & 0 & 0 & 0 \\ 1 & 0 & 0 & 0 \end{pmatrix} ([S]\otimes[S]^*) - \begin{pmatrix} 0 & 0 & 0 & 0 \\ 0 & 1 & 0 & 0 \\ 0 & 0 & 1 & 0 \\ 0 & 0 & 0 & 0 \end{pmatrix} ([S]\otimes[S]^*) \right] [A]^{-1}$$

$$= \frac{1}{2} \begin{pmatrix} 1 & 0 & 0 & 0 \\ 0 & -1 & 0 & 0 \\ 0 & 0 & -1 & 0 \\ 0 & 0 & 0 & 1 \end{pmatrix} [M] \tag{4.34}$$

which is the Stokes reflection matrix $[M_{px}]$ defined in the cross-polarization channel.

#### 4.6.2c The matched two-antenna case

If there are two separate or joint antennas, one being the transmitting antenna which radiates the transmitted wave $\vec{E}_T$ as the incident wave on the target; and, the other being the receiving antenna; this optimum received voltage condition, in polarimetry, is given for the case in which the polarization state of the receiving antenna is matched to that of the scattered field $\vec{E}_S$. For example in [39], this case is treated which is also known as the "pseudo-eigenvalue" problem, i.e., for:

$$\vec{h} = \frac{\vec{E}_S^*}{||\vec{E}_S||} = \vec{e}_S^* \tag{4.35}$$

The optimum receiving voltage becomes

$$V_2 = \left( \frac{\vec{E}_S^*}{||\vec{E}_S||} \right)^T [S]\vec{E}_T = \frac{\vec{E}_T^+ [S]^+ [S]\vec{E}_T}{||[S]\vec{E}_T||} \quad \frac{\vec{E}_T^+[G]\bar{E}_T}{||[S]\bar{E}_T||} = \frac{\vec{E}_S \cdot \vec{E}_S}{||\vec{E}_S||} \tag{4.36}$$

so that the received power in the matched two antennas case can be derived as follows with the Graves power matrix [G] being defined in more detail in Section 6.1 [26].

$$P_2 = |V_2|^2 = |\vec{E}_S|^2 \tag{4.37}$$

Using the Stokes vector representation yields

$$\begin{aligned} P_2 &= g_{S0} = M_{00} + M_{01}g_{T1} + M_{02}g_{T2} + M_{03}g_{T3} \\ &= \vec{g}_T^{\,T}\,[M_2]\vec{g}_T \end{aligned} \tag{4.38}$$

where

$$[M_2] = \begin{pmatrix} 1 & 0 & 0 & 0 \\ 0 & 0 & 0 & 0 \\ 0 & 0 & 0 & 0 \\ 0 & 0 & 0 & 0 \end{pmatrix} [M] = [M_{pc}] + [M_{px}]$$

is the particular Stokes reflection matrix defined in the specific matched two-antenna case.

In concluding the introduction of these particular special cases, we need to alert the reader that there exist additional possible alternatives for the presentation of specific Stokes reflection matrices as for example for the forward scattering case, which need to be analyzed separately in each case.

## 5. CHANGE OF POLARIZATION BASIS

The change of polarization basis plays an important role in applications of radar polarimetry. All measurable quantities such as voltage, energy density, etc., must remain invariant and orthonormality of any two vectors must be preserved under the change of basis transformation.

### 5.1 The basis transformation matrix [T] and the vector transformation matrix [U]

In general, consider that there are two polarization bases, let the old basis be (XY), formed by two orthogonal unit vectors $\hat{x}$ and $\hat{y}$; and the new basis be (AB), formed by another two orthogonal unit vectors $\hat{a}$ and $\hat{b}$. We can express the unit vectors of the new basis (AB), $\hat{a}$ and $\hat{b}$, in terms of the unit vectors $\hat{x}$ and $\hat{y}$ as

$$\hat{a}(XY) = a_1\hat{x} + a_2\hat{y} \tag{5.1}$$

$$\hat{b}(XY) = b_1\hat{x} + b_2\hat{y}$$

and in matrix form as

$$\begin{bmatrix} \hat{a}(XY) \\ \hat{b}(XY) \end{bmatrix} = \begin{bmatrix} a_1 & a_2 \\ b_1 & b_2 \end{bmatrix} \begin{bmatrix} \hat{x} \\ \hat{y} \end{bmatrix} = [T] \begin{bmatrix} \hat{x} \\ \hat{y} \end{bmatrix} \tag{5.2}$$

Eq. 5.2 is the transformation from the old basis to the new basis, $[T(XY\to AB)]$ is the basis transformation matrix

$$[T(XY\to AB)] = \begin{bmatrix} a_1 & a_2 \\ b_1 & b_2 \end{bmatrix} \tag{5.3}$$

where $a_1$ and $a_2$ are the components of the unit vector $\hat{a}$ of the new basis (AB) presented in the old basis (XY), and $b_1$ and $b_2$ are the components of the $\hat{b}$ presented in the old basis (XY). It can be shown that the vector transformation matrix [U] is the transpose matrix of the bases transformation matrix [T], i.e.,

$$[U] = [T]^T \quad , \quad |[U]| = 1 \tag{5.4}$$

and

$$\vec{E}(XY) = [U(AB\to XY)]\,\vec{E}(AB) \tag{5.5}$$

where [U] is the vector transformation matrix. It is a unitary matrix, and transforms a vector $\vec{E}$ from the new basis (AB) to the old basis (XY). From Eqs. 5.4 and 5.3 we obtain

$$[U(AB\to XY)] = \begin{bmatrix} a_1 & b_1 \\ a_2 & b_2 \end{bmatrix} \tag{5.6}$$

which indicates that we can use the new basis unit vector $\hat{a}$ and $\hat{b}$ expressed in the old basis (XY) as the column to form the unitary vector transformation matrix $[U(AB\to XY)]$.

### 5.2 Unitary transformation matrix in terms of the polarization ratio

A vector $\vec{E}$ can be expressed in both bases. In the old basis (XY)

$$\vec{E}(XY) = \begin{bmatrix} E_x \\ E_y \end{bmatrix} = E_x\hat{x} + E_y\hat{y} \tag{5.7}$$

and in the new basis(AB) by

$$\vec{E}(AB) = \begin{bmatrix} E_a \\ E_b \end{bmatrix} = E_a\hat{a} + E_b\hat{b} \tag{5.8}$$

From Eq. 5.6, the unitary transformation matrix [U] is

$$[U(AB \rightarrow XY)] = ( \hat{a}(XY) \quad \hat{b}(XY) ) \tag{5.9}$$

where $\hat{a}(XY)$ and $\hat{b}(XY)$ are the column matrices which are the unit vectors of the new basis (AB) and expressed in terms of the old system (XY). From Eq. 5.5, the vector transformation is

$$\begin{bmatrix} E_x \\ E_y \end{bmatrix} = [U(AB \rightarrow XY)] \begin{bmatrix} E_a \\ E_b \end{bmatrix} \tag{5.10}$$

One can express the unitary transformation matrix [U(AB→XY)] in terms of the polarization ratio $\rho$. From Eq. 3.21, the unit vectors $\hat{a}$ and $\hat{b}$ of the new basis (AB) can be expressed in the old basis (XY) in terms of the polarization ratio $\rho$

$$\hat{a}(XY) = \begin{bmatrix} a_x \\ a_y \end{bmatrix} = \frac{e^{j\phi_{a_x}}}{\sqrt{1 + \rho\rho^*}} \begin{bmatrix} 1 \\ \rho \end{bmatrix} \tag{5.11}$$

$$\hat{b}(XY) = \begin{bmatrix} b_x \\ b_y \end{bmatrix} = \frac{e^{j\phi_{b_x}}}{\sqrt{1 + \rho'\rho'^*}} \begin{bmatrix} 1 \\ \rho' \end{bmatrix}$$

where

$$\rho = \frac{a_y}{a_x} = \frac{|a_y|}{|a_x|} e^{j\phi} \quad \text{and } \phi = \phi_{a_y} - \phi_{a_x}$$

$$\rho' = \frac{b_y}{b_x} = \frac{|b_y|}{|b_x|} e^{j\phi'} \quad \text{and } \phi' = \phi_{b_y} - \phi_{b_x}$$

To find the relationship between $\rho$ and $\rho'$, and because $\hat{a}$ and $\hat{b}$ are orthonormal, we find

$$|| \hat{a} || = || \hat{b} || = 1 \tag{5.12}$$

and

$$\hat{a} \cdot \hat{b} = 0 \tag{5.13}$$

The dot product of $\hat{a}$ with $\hat{b}$, yields

$$\hat{a}\cdot\hat{b} = \left( \frac{e^{j\phi_{a_x}}}{\sqrt{1 + \rho\rho^*}} \begin{bmatrix} 1 & \rho^* \end{bmatrix} \right) \left( \frac{e^{-j\phi_{b_x}}}{\sqrt{1 + \rho'\rho'^*}} \begin{bmatrix} 1 \\ \rho' \end{bmatrix} \right)$$

$$= \frac{e^{j(\phi_{a_x} - \phi_{b_x})}}{\sqrt{1 + \rho\rho^*}\ \sqrt{1+\rho'\rho'^*}} (1 + \rho^*\rho')$$

then comparing the result with Eq. 5.13, yields $1 + \rho^*\rho' = 0$, that means $\rho' = -1/\rho^*$, substituting this into Eq. 5.11, we find

$$\hat{b}(XY) = \frac{e^{j\phi_{b_x}}}{\sqrt{1 + (-1/\rho^*)(-1/\rho)}} \begin{bmatrix} 1 \\ -1/\rho^* \end{bmatrix}$$

$$= \frac{|\rho| e^{j\phi_{b_x}}}{\sqrt{1 + \rho\rho^*}} (-1/\rho^*) \begin{bmatrix} -\rho^* \\ 1 \end{bmatrix}$$

$$= \frac{|\rho| e^{j\phi_{b_x}}}{\sqrt{1 + \rho\rho^*}} \frac{-1}{|\rho| e^{-j\phi}} \begin{bmatrix} -\rho^* \\ 1 \end{bmatrix}$$

$$= \frac{e^{j(\phi_{b_x} + \phi)}}{\sqrt{1 + \rho\rho^*}} \begin{bmatrix} +\rho^* \\ -1 \end{bmatrix} \tag{5.14}$$

Using Eq. 5.11 and Eq. 5.14 to form unitary transformation matrix [U] by Eq. 5.9, yields

$$[U(AB\rightarrow XY)] = (\hat{a}(XY) \quad \hat{b}(XY)) = \frac{1}{\sqrt{1+\rho\rho^*}} \begin{bmatrix} e^{j\phi_{a_x}} & \rho^* e^{j(\phi_{b_x}+\phi)} \\ \rho e^{j\phi_{a_x}} & -e^{j(\phi_{b_x}+\phi)} \end{bmatrix} .$$

$$= \frac{1}{\sqrt{1+\rho\rho^*}} \begin{bmatrix} e^{j\psi_1} & \rho^* e^{j\psi_4} \\ \rho e^{j\psi_1} & -e^{j\psi_4} \end{bmatrix} \tag{5.15}$$

## 5.3 Transformation of a polarization state from linear to circular basis

Assuming an old basis (XY) is the linear basis (HV), where $\hat{H}$ is the horizontal unit vector, and $\hat{V}$ is the vertical unit vector; and a new basis (AB) is the circular basis (LR), where $\hat{L}$ is the left-handed circular unit vector, and $\hat{R}$ is the right-handed circular unit vector. From Eq. 3.23, expressed in the (HV) basis, $\hat{L}$ and $\hat{R}$ can be written as

$$\hat{L}(HV) = \frac{1}{\sqrt{2}} \begin{bmatrix} 1 \\ j \end{bmatrix} \tag{5.16a}$$

$$\hat{R}(HV) = \frac{1}{\sqrt{2}} \begin{bmatrix} 1 \\ -j \end{bmatrix} \tag{5.16b}$$

According to Eq. 5.9, the unitary transformation matrix $[U(LR\rightarrow HV)]$ then becomes

$$[U(LR\rightarrow HV)] = (\hat{L}(HV) \quad \hat{R}(HV)) = \frac{1}{\sqrt{2}} \begin{bmatrix} 1 & 1 \\ j & -j \end{bmatrix} \tag{5.16c}$$

If a polarization state is $\vec{E}(HV)$ in the linear basis (HV), and is $\vec{E}'(LR)$ in the circular basis (LR), then the polarization state transformation from Eq. 5.5, is

$$\vec{E}(HV) = [U(LR\rightarrow HV)]\vec{E}'(LR) \tag{5.17}$$

or

$$\vec{E}'(LR) = [U(LR\rightarrow HV)]^{-1}\vec{E}(HV) = [U(HV\rightarrow LR)]\vec{E}(HV) \tag{5.18}$$

Because $[U(LR\rightarrow HV)]$ is a unitary matrix, we find

$$[U(LR\rightarrow HV)]^{-1} = [U]^{*T} = \frac{1}{\sqrt{2}} \begin{bmatrix} 1 & -j \\ 1 & j \end{bmatrix} \tag{5.19}$$

As an example, we transform a left handed circular polarization state from a circular basis to a linear basis. We know that this polarization state is $\hat{L}(HV) = \frac{1}{\sqrt{2}}\begin{bmatrix}1\\j\end{bmatrix}$ in the linear basis (HV) and is $\hat{L}(LR) = \begin{bmatrix}1\\0\end{bmatrix}$ in the circular basis (LR). Applying Eq. 5.16 and Eq. 5.17, we obtain

$$\hat{L}(HV) = [U(LR\rightarrow HV)]\hat{L}(LR) = \frac{1}{\sqrt{2}}\begin{bmatrix}1 & 1\\ j & -j\end{bmatrix}\begin{bmatrix}1\\0\end{bmatrix} = \frac{1}{\sqrt{2}}\begin{bmatrix}1\\j\end{bmatrix}$$

Also, we can use Eq. 5.18 and Eq. 5.19 to transform this polarization state from a linear basis to the circular basis:

$$\hat{L}(LR) = [U(LR\rightarrow HV)]^{-1}\hat{L}(HV) = \left(\frac{1}{\sqrt{2}}\begin{bmatrix}1 & -j\\ 1 & j\end{bmatrix}\right)\left(\frac{1}{\sqrt{2}}\begin{bmatrix}1\\j\end{bmatrix}\right) = \frac{1}{2}\begin{bmatrix}2\\0\end{bmatrix} = \begin{bmatrix}1\\0\end{bmatrix}$$

In order to represent this transformation matrix [U] by the polarization ratio $\rho$, from Eq. 5.11 and above equatons, we obtain

$$\hat{L}(HV) = \frac{1}{\sqrt{1+\rho\rho^*}}\begin{bmatrix}1\\ \rho\end{bmatrix} = \frac{1}{\sqrt{2}}\begin{bmatrix}1\\j\end{bmatrix}$$

which gives $\rho = j$.

From Eq. 5.14, we obtain

$$\hat{R}(HV) = \frac{e^{j\phi}}{\sqrt{1+\rho\rho^*}}\begin{bmatrix}\rho^*\\ -1\end{bmatrix} = \frac{1}{\sqrt{2}}\begin{bmatrix}1\\ -j\end{bmatrix}$$

and we find from $e^{j\phi} = j$ that $\phi = \pi/2$. Substituting above values into Eq. 5.15, we find

$$[U(LR\rightarrow HV)] = \frac{1}{\sqrt{1+\rho\rho^*}}\begin{bmatrix}1 & +\rho^* e^{+j\pi/2}\\ \rho & -e^{+j\pi/2}\end{bmatrix} = \frac{1}{\sqrt{1+\rho\rho^*}}\begin{bmatrix}1 & j\rho^*\\ \rho & -j\end{bmatrix} \tag{5.20}$$

**5.4 Transformation of the scattering matrix [S] from the (HV) basis to a new basis (AB) for the monostatic case**

The received voltage according to Eq. 4.2, is

$$V = \vec{h}^T[S]\vec{E}_T = \vec{E}_R{}^T[S]\vec{E}_T \tag{5.21}$$

In the linear (HV) basis, the voltage is

$$V(HV) = E_R(HV)^T[S(HV)]E_T(HV) \tag{5.22}$$

In the new basis (AB), the voltage is

$$V'(AB) = E'_R(AB)^T[S'(AB)]E'_T(AB) \tag{5.23}$$

Appling the unitary transformation matrix [U] in Eq. 5.22, yields

$$\begin{aligned} V(HV) &= ([U]E'_R(AB))^T[S(HV)]([U]E'_T(AB)) \\ &= E'_R(AB)^T[U]^T[S(HV)][U]E'_T(AB) \\ &= E'_R(AB)^T([U]^T[S(HV)][U])E'_T(AB) \end{aligned}$$

Comparing this result with Eq. 5.23, and because the received voltage equation remains unchanged under a transformation of the basis, the transformation of the scattering matrix [S(AB)] becomes

$$[S'(AB)] = [U]^T[S(HV)][U] = \begin{bmatrix} S'_{AA} & S'_{AB} \\ S'_{AB} & S'_{BB} \end{bmatrix} \tag{5.24}$$

Substituting Eq. 5.20 into Eq. 5.24, yields

$$[S'(AB)] = \frac{1}{1+\rho\rho^*}\begin{bmatrix} 1 & \rho \\ j\rho^* & -j \end{bmatrix}\begin{bmatrix} S_{HH} & S_{HV} \\ S_{HV} & S_{VV} \end{bmatrix}\begin{bmatrix} 1 & j\rho^* \\ \rho & -j \end{bmatrix}$$

We can get the following elements of [S'(AB)] in terms of the elements of [S(HV)],

$$S'_{AA} = \frac{1}{1+\rho\rho^*}(S_{HH} + 2\rho S_{HV} + \rho^2 S_{VV}) \tag{5.25}$$

$$S'_{AB} = \frac{-j}{1+\rho\rho^*}(\rho S_{VV} + (1-\rho\rho^*)S_{HV} - \rho^* S_{HH})$$

$$S'_{BB} = \frac{-1}{1+\rho\rho^*}(\rho^{*2} S_{HH} - 2\rho^* S_{HV} + S_{VV})$$

In general, it can be shown that

$$\text{Span}\{[S'(AB)]\} = |S'_{AA}|^2 + |S'_{AB}|^2 + |S'_{BA}|^2 + |S'_{BB}|^2 \tag{5.26}$$

$$= |S_{HH}|^2 + |S_{HV}|^2 + |S_{VH}|^2 + |S_{VV}|^2$$

$$= \text{Span}\{[S(HV)]\}$$

Thus, the Span{[S]} is invariant under the change of basis transformation.

## 5.5 Derivation of the matrix [A(LR)] for measurements of $\vec{J}(LR)$ in the circular polarization basis (LR)

From Eq. 4.10, we know how the Stokes vector $\vec{g}$ can be represented by the coherency vector $\vec{J}(HV)$ and the Jones vector $\vec{E}(HV)$ in the linear basis (HV). Now we apply the polarization state basis transformation to obtain the Stokes vector $\vec{g}$ represented by the coherency vector $\vec{J}(LR)$ and the Jones vector $\vec{E}(LR)$ in the circular basis (LR), where

$$\vec{J}(LR) = \begin{bmatrix} J_{LL} \\ J_{LR} \\ J_{RL} \\ J_{RR} \end{bmatrix} = \begin{bmatrix} E_L E_L^* \\ E_L E_R^* \\ E_R E_L^* \\ E_R E_R^* \end{bmatrix} \tag{5.27}$$

Then applying the unitary transformation of Eqs. 5.18 and 5.17 to transform an arbitrary polarization state vector $\vec{E}(LR)$ to $\vec{E}(HV)$, we find that

$$\begin{bmatrix} E_H \\ E_V \end{bmatrix} = \frac{1}{\sqrt{2}} \begin{bmatrix} 1 & 1 \\ j & -j \end{bmatrix} \begin{bmatrix} E_L \\ E_R \end{bmatrix} = \frac{1}{\sqrt{2}} \begin{bmatrix} E_L + E_R \\ jE_L - jE_R \end{bmatrix}$$

and

$$\begin{bmatrix} E_H \\ E_V \end{bmatrix}^* = \frac{1}{\sqrt{2}} \begin{bmatrix} 1 & 1 \\ j & -j \end{bmatrix}^* \begin{bmatrix} E_L \\ E_R \end{bmatrix}^* = \frac{1}{\sqrt{2}} \begin{bmatrix} E_L^* + E_R^* \\ -jE_L^* + jE_R^* \end{bmatrix}$$

Substituting above equations into Eq. 4.10, we obtain

$$J_{HH} = E_H E_H^* = (1/2)(E_L E_L^* + E_R E_R^* + E_R E_L^* + E_L E_R^*)$$

$$J_{HV} = E_H E_V^* = (1/2)(-jE_L E_L^* + jE_R E_R^* - jE_R E_L^* + jE_L E_R^*)$$

$$J_{VH} = E_V E_H^* = (1/2)(jE_L E_L^* - jE_R R_R^* - jE_R E_L^* + jE_L E_R^*)$$

$$J_{VV} = E_V E_V^* = (1/2)(E_L E_L^* + E_R E_R^* - E_R E_L^* - E_L E_R^*)$$

Using Eqs. 3.41, 5.27, and above equations, we can derive the relationship between the Stokes vector $\vec{g}$ and the coherency vector $\vec{J}(LR)$ as

$$\vec{g} = \begin{bmatrix} J_{HH} + J_{VV} \\ J_{HH} - J_{VV} \\ J_{HV} + J_{VH} \\ jJ_{HV} - jJ_{VH} \end{bmatrix} = \begin{bmatrix} E_L E_L^* + E_R E_R^* \\ E_R E_L^* + E_L E_R^* \\ jE_L E_R^* - jE_R E_L^* \\ E_L E_L^* - E_R E_R^* \end{bmatrix} = \begin{bmatrix} J_{LL} + J_{RR} \\ J_{RL} + J_{LR} \\ jJ_{LR} - jJ_{RL} \\ J_{LL} - J_{RR} \end{bmatrix}$$

and according to Eq. 4.12, we find that

$$[A(LR)] = \begin{bmatrix} 1 & 0 & 0 & 1 \\ 0 & 1 & 1 & 0 \\ 0 & +j & -j & 0 \\ 1 & 0 & 0 & -1 \end{bmatrix} \qquad (5.28)$$

which differs from the expression of [A(HV)] given in Eq. 4.11; and care must be taken in establishing the proper relationship for deriving [M] from [J(LR)] versus [J(HV)], and [J(AB)] in general.

## 6. OPTIMAL POLARIZATION STATES

The "Optimal Polarization State" problem is to find such polarization states of the transmitted and received waves for a target of known scattering matrix [S], that the voltage developed across the receiving antenna terminals is maximized (or minimized). Mathematically, it is to find $\vec{E}_T$ and $\vec{h}$ so that the received voltage in the voltage Eq. 5.21 is maximized (or minimized) for a given [S].

### 6.1 Optimal polarization states for the completely polarized wave

First, let us consider the completely polarized wave case, i.e., both of the transmitted and the received waves are completely polarized.

#### 6.1.1 Optimal polarizations states for the monostatic case

There are two methods to solve this problem.

**Method 1.** Graves [26] and Kennaugh [37] defined the so-called "power scattering matrix [G]" as

$$[G] = [S]^+[S] \qquad (6.1)$$

As usual, the $[\,]^{+} = [*]^{T}$ denotes Hermitian conjugate. This problem becomes an eigenvalue problem.

$$[G]\vec{E}_{T_n} = \lambda_n \vec{E}_{T_n} \tag{6.2}$$

where n = 1,2,3,.... The eigenvectors $\vec{E}_{Tn}$ of [G] are "optimal" polarization states which yield the optimal backscattered power at the transmitter location T for the given target of known scattering matrix [S]. [G] is Hermitian for any [S]. The eigenvalues $\lambda_n$ of [G] are real which yield the power returned from the target at position T. It can be obtained by premultiplying $\vec{E}_T$ on both sides of Eq. 6.2.

$$\vec{E}_{T_n}^{+}([S]^{+}S)\vec{E}_{T_n} = ([S]\vec{E}_{T_n})^{+}([S]\vec{E}_{T_n}) = |\vec{E}_{R_n}|^{2} = \lambda_n$$

**Method 2.** Kennaugh's "Pseudo-Eigenvalue Equation"

Kennaugh (1952) [36] originated the concept of optimal polarization states in the monostatic case and demonstrated that the optimal polarization states for a given target satisfy the following nonlinear equation,

$$[S]X_n = \mu_n X_n^{*} \tag{6.3}$$

where n = 1,2,3,.... The eigenvalues $\mu_n$ are complex numbers. $X_n$ are optimal in the sense that the absolute values of the complex scalar obtained by the Euclidean inner product as the $|C_n| = |\langle X_n \cdot [S]X_n\rangle| = |\mu_n|$ are optimal. Kennaugh denoted the $C_n$ to be the backscatter voltages.

### 6.1.2 The three-step optimization approach for the general case

This method enables one to treat symmetric, asymmetric, monostatic and bistatic cases in an identical manner [37] as was first considered in [13].

**Step 1**

The total normalized (power flux density) W in the scattered wave is given by $\vec{E}_R^{+}\vec{E}_R$, where

$$W = \vec{E}_R^{+}\vec{E}_R = ([S]\vec{E}_T)^{+}[S]\vec{E}_T = \vec{E}_T^{+}[S]^{+}[S]\vec{E}_T = \vec{E}_T^{+}[G]\vec{E}_T \tag{6.4}$$

We need to find such an $\vec{E}_T$ for which $P_W$ is extremum for a given [S]. It translates into the following eigenvalue problem:

$$[G]\vec{E}_T = \lambda\vec{E}_T \tag{6.5}$$

The characteristic equation of Eq. 6.5, is

$$\{[G] - \lambda[I]\}\vec{E}_{T.OPT} = 0 \tag{6.6}$$

where [I] is the identity matrix. The explicit solution for the eigenvalues is given by a simple quadratic equation

$$\lambda^2 - (G_{11} + G_{22})\lambda + (G_{11}G_{22} - G_{12}G_{21}) = 0 \tag{6.7a}$$

where

$$\lambda_{1,2} = \frac{1}{2}\,[\mathrm{Tr}\{[G]\} \pm \sqrt{\mathrm{Tr}^2\{[G]\} - 4\det\{[G]\}}$$

The eigenvalues $\lambda_1$ and $\lambda_2$ are real because [G] is Hermitian which agrees with their physical interpretation as power. Substituting the eigenvalues $(\lambda_1,\lambda_2)$ into Eq. 6.6 and solving for the components of $\vec{E}_{T,OPT}$, the eigenvector $\vec{E}_{T,OPT}$ is the polarization state of a transmitter so that the power in the scattered wave is maximized. If the two eigenvalues $(\lambda_1,\lambda_2)$ are not equal, the two eigenvectors $(\vec{E}_{T,OPT})$ are orthogonal and the maximum is achieved by $\vec{E}_{T,OPT}$ corresponding to the largest eigenvalue (the minimum--for the smallest one). Furthermore, the sum of the two eigenvalues, i.e., the total energy, is an invariant, as it was shown before in Eq. 5.26

$$\lambda_1 + \lambda_2 = \mathrm{Tr}\{[G]\} = \mathrm{Span}\{[S]\} = \text{invariant} \tag{6.7b}$$

and so is the product of the two eigenvalues

$$\lambda_1\lambda_2 = \det\{[G]\} = (\det\{[S]\})(\det\{[S]^*\}) = \text{invariant} \quad . \tag{6.7c}$$

**Step 2**

Compute this scattered wave by using the known scattering matrix [S] and $\vec{E}_{T,OPT}$ from Eq. 6.6

$$\vec{E}_{R,OPT} = [S]\vec{E}_{T,OPT} \tag{6.8}$$

so that the scattered polarization state $\vec{E}_{R,OPT}$, associated with the optimal transmitting polarization state $\vec{E}_{T,OPT}$, is completely specified. In general, $\vec{E}_{R,OPT} \neq \vec{E}_{T,OPT}$, because the eigenvectors of [G] are not identical to those of [S], unless [S] is a normal matrix.

**Step 3**

In order to ensure polarization match, i.e., to receive all of the power contained in the scattered wave, one must adjust the receiver polarization state, which results in

$$\vec{h}_{OPT} = \frac{\vec{E}^{*}_{R,OPT}}{||\vec{E}_{R,OPT}||} = \frac{\{[S]\vec{E}_{T,OPT}\}^{*}}{||[S]\vec{E}_{T,OPT}||} \tag{6.9}$$

where $||\cdot||$ indicates the norm. Eq. 6.9 completes the optimization process. Note, by using the "three stage procedure", we can obtain the maximum polarization state for optimal reception, i.e., for satisfying the condition Eq. 4.35, only.

### 6.1.3. Characteristic Polarization state determination by the basis transformation procedures.

Let $\vec{E}_T$ and $\vec{E}_R$ represent the normalized vectors of the transmitting and receiving monochromatic traveling plane waves in the old basis(HV), and $\vec{E}'_T$ and $\vec{E}'_R$ in the new basis (AB), respectively. The vectors in the two bases are related as:

$$\vec{E}_T = [U]\,\vec{E}'_T \tag{6.10}$$

where [U] is the general unitary vector transformation matrix represented by

$$[U] = \frac{1}{\sqrt{1+\rho\rho^{*}}}\begin{bmatrix} e^{j\psi_1} & -\rho^{*}e^{j\psi_4} \\ \rho e^{j\psi_1} & e^{j\psi_4} \end{bmatrix} \tag{6.11}$$

For the coherent case, the total received power according to the voltage equation (4.2) can be expressed in the old and new bases, respectively, as

$$P = |V|^2 = |\vec{E}_R{}^T[S]\vec{E}_T|^2 \,/\, ||\vec{E}_R{}^T||^2 \tag{6.12a}$$

$$P = |V|^2 = |\vec{E}'^{T}_{R}[S']\vec{E}'|^2 \,/\, ||\vec{E}_R{}^T||^2 \,, \tag{6.12b}$$

so that for the monostatic backscattering case, the co-polarized power $P_{co}$ can be calculated for $\vec{E}_R = \vec{E}_T = \vec{E}$ and $\vec{E}_R{}' = \vec{E}_T{}' = \vec{E}'$ as

$$P_{co} = |V_{co}|^2 = |\vec{E}^T[S]\vec{E}|^2 \,/\, ||\vec{E}^T||^2 \tag{6.13a}$$

$$P_{co} = |V_{co}|^2 = |\vec{E}'^T[S]\vec{E}'|^2 \,/\, ||\vec{E}'^T||^2 \tag{6.13b}$$

Similarly, the cross-polarized power $P_x$ can be calculated for $\vec{E}_R = \vec{E}_{T\perp} = \vec{E}_\perp$ and $\vec{E}'_R = \vec{E}'_{T\perp} = \vec{E}'_\perp$ and with $(\cdot_\perp)$ denoting the orthogonal vector as

$$P_x = |V_x|^2 = |\vec{E}_\perp^T [S]\, \vec{E}|^2 \,/\, ||\vec{E}_\perp^T||^2 \tag{6.14a}$$

$$P_x = |V_x|^2 = |\vec{E}_\perp^T [S']\, \vec{E}'|^2 \,/\, ||\vec{E}_\perp^T||^2 \tag{6.14b}$$

In reference to the optimal polarization states, we refer to those polarization states which produce maximum and minimum returns in the co/cross-polarized channels. Following Kennaugh [37] and Huynen [30], it is shown in Agrawal and Boerner [1] and in Boerner and Xi [62] that there exist five pairs of optimal polarization states: (i) the cross-polarization null states (X-POL NULLS), (ii) the co-polarization null states (CO-POL NULLS), (iii) the cross-polarization maximum states (X-POL MAXS); (iv) the co-polarization maximum states (CO-POL MAXS)and (v) the cross-polarization saddle, turning point states (X-POL SADS), which may be derived as follows:

<u>X-POL NULL and CO-POL Maximum States</u>

It can be shown for the monostatie reciprocal case that the X-POL Nulls and the CO-POL Maxima are identical, which can be determined from Eq. 6.13 or Eq. 6.14., either in the old basis (HV) or in the new basis (AB). The simplest way to find the X-POL null expression, in the old basis (HV) is using Eq. 5.25 and setting $S'_{AB} = 0$, which gives:

$$\rho_{xn1,2} = \frac{-B \pm \sqrt{B^2 - 4AC}}{2A} \qquad \text{(in old basis)} \tag{6.15}$$

with $A = S_{HH}^* S_{HV} + S_{HV}^* S_{VV}$, $B = |S_{HH}|^2 - |S_{VV}|^2$, $C = -A^*$. Where $\rho_{xn1}\rho_{xn2}^* = -1$, satisfying the orthonormality condition. The corresponding X-POL NULL states in the new arbitrary,basis (AB) expressed in terms of the "new basis polarization ratios" $\rho$ (AB), can be found by determining the critical points of Eq. 6.14b [61] as:

$$\left.\begin{aligned} \rho'_{xn1} &= 0 \\ \rho'_{xn2} &= \infty \end{aligned}\right\} \qquad \text{(in the new basis)} \tag{6.16a}$$

Equivalently, if we use the new basis (AB) with the column vectors of the unitary transformation matrix as basis vectors, the scattering matrix becomes

$$[S_d] = \begin{bmatrix} \lambda_1 & 0 \\ 0 & \lambda_2 \end{bmatrix} \tag{6.17}$$

where

$$\lambda_1 = S'_{AA}(\rho_1) = \frac{1}{1 + \rho_1\rho_1^*} (S_{HH} + 2\rho_1 S_{HV} + \rho_1^2 S_{VV}) e^{2j\psi_1} \tag{6.18a}$$

$$\lambda_2 = S_{B}{}'_{B}(\rho_1) = \frac{1}{1 + \rho_1\rho_1^*} (\rho_1^{*2} S_{HH} - 2\rho_1^* S_{HV} + S_{VV}) e^{2j\psi_4} \tag{6.18b}$$

as expressed in Eq. 5.25 with $\rho$ replaced by $\rho_1$, or in the exponential representation

$$\lambda_1 = |\lambda_1| e^{j\delta_1} \text{ and } \lambda_2 = |\lambda_2| e^{j\delta_2}$$

By using the power expression in the new basis (AB), the maximum power becomes

$$P_{co1}(\rho_{xn1}) = |\lambda_1|^2$$

$$P_{co2}(\rho_{xn2}) = |\lambda_2|^2$$

<u>CO-POL Nulls, X-POL Maxima and X-POL SADDLES</u>

From inspection of Eq. 6.13a it follows that the CO-POL null states in the old basis, can be found from Eq. 5.25 by setting $S'_{AA} = 0$ as

$$\rho_{cn1,2} = \frac{- S_{HV} \pm \sqrt{S_{HV}^2 - S_{HH}S_{VV}}}{S_{VV}} \quad \text{(in the old basis)} \tag{6.19}$$

and for the new basis (AB) by finding the critical points of Eq. 6.13b, where

$$\rho'_{cn1,2} = \pm \left( - \frac{\lambda_1}{\lambda_2} \right)^{1/2} \quad \text{(in the new basis)} \tag{6.20}$$

The cross-polarization Maxima $\rho'_{xm1,2}$ and saddle points extrema $\rho'_{xs1,2}$ can also be found from (6.14b) by using the "critical point" method described in [62] as

$$\rho'_{xm1,2} = \pm j \left( \frac{\lambda_1\lambda_2^*}{\lambda_1^*\lambda_2} \right)^{1/4} \quad \text{(in the new basis)} \tag{6.21}$$

$$\rho'_{xs1,2} = \pm \left( \frac{\lambda_1\lambda_2^*}{\lambda_1^*\lambda_2} \right)^{1/4} \quad \text{(in the new basis)} \tag{6.22}$$

and the corresponding power returns to the receiver of the co/cross polarization channels then become

$$P_x(\rho_{x}'{}_{m1,2}) = 1/4(|\lambda_1| + |\lambda_2|)^2 \quad (6.23)$$

$$P_x(\rho'_{xs1,2}) = 1/4(|\lambda_1| - |\lambda_2|)^2 \quad (6.24)$$

$$P_c(\rho'_{xs1,2}) = 1/4(|\lambda_1| - |\lambda_2|)^2 \quad (6.25)$$

$$P_c(\rho'_{xs1,2}) = 1/4(|\lambda_1| + |\lambda_2|)^2 \quad (6.26)$$

Note that the following conditions are satisfied:

$$\rho'_{xm1}\ \rho'^{*}_{xm2} = -1 \quad (6.27)$$

$$\rho'_{xs1}\ \rho'^{*}_{xs2} = -1 \quad (6.28)$$

that means $\rho'_{xm1}$, $\rho'_{xm2}$ are orthogonal and so are $\rho'_{xs1}$ and $\rho'_{xs2}$.

## The Polarization Fork

In order to determine the relation among the five pairs of characteristic polarization states on the Poincaré sphere, resulting in the polarization Fork, we introduce the Huynen [30] geometric parameters.

$m$ – target size

$\tau$ – target orientation (or tilt) angle

$\varepsilon_m$ – ellipticity angle of the co-pol maximum (6.29)

$\nu$ – target skip angle

$\mu$ – target characteristic angle

$\phi_m$ – the phase of the polarization ratio of the co-pol max

$\gamma_m$ – the spinor parameter

defined by

$$\lambda_1 = m\, e^{j\delta_1} \quad (6.30a)$$

$$\lambda_2 = |\lambda_2| e^{j\delta_2} \quad (6.30b)$$

$$\nu = \frac{\delta_1 - \delta_2}{4} \tag{6.30c}$$

$$\sin 2\varepsilon_m = \sin 2\gamma_m \sin\phi_m \tag{6.30d}$$

$$\tan 2\tau_m = \tan 2\gamma_m \cos \phi_m \tag{6.30e}$$

$$\tan\mu = \sqrt{\left|\frac{\lambda_2}{\lambda_1}\right|} \tag{6.30f}$$

$$\phi_m = \arg\{\rho_1\} \tag{6.30g}$$

$$\tan\gamma_m = |\rho_1| \tag{6.30h}$$

The five pairs of characteristic polarization states can be written using the geometric parameters as:

$$\rho'_{xn1}(AB) = 0 = \tan(90°-90°)\exp\{arb\} = \rho'_{cm1}(AB) \tag{6.31a}$$

$$\rho'_{xn2}(AB) = \infty = \tan(90°- 0°)\exp\{arb\} = \rho'_{cm2}(AB) \tag{6.31b}$$

$$\rho_{cn1}(AB) = \tan(90° - \mu)\exp\{j(2\nu + \pi/2)\} \tag{6.31c}$$

$$\rho_{cn2}(AB) = \tan(90° + \mu)\exp\{j(2\nu + \pi/2)\} \tag{6.31d}$$

$$\rho'_{xm1}(AB) = \tan(90° - 45°)\exp\{j(2\nu + \pi/2)\} \tag{6.31e}$$

$$\rho'_{xm2}(AB) = \tan(90° + 45°)\exp\{j(2\nu + \pi/2)\} \tag{6.31f}$$

$$\rho'_{xs1}(AB) = \tan(90° - 45°)\exp\{j2\nu\} \tag{6.31g}$$

$$\rho'_{xs2}(AB) = \tan(90° + 45°)\exp\{j2\nu\} \tag{6.31h}$$

On the Poincaré sphere, we can first allocate the pair of antipodal points $(X_1, X_2)$ for the cross-pol nulls $(\rho'_{xn1}, \rho'_{xn2})$ because they are orthogonal and have arbitrary phase. Then, we may determine the positions for other polarization states according to the Poincaré sphere presentation rules [20], [19]:

If we present the polarization ratio as

$$\rho = \tan\gamma \exp\{j\,\phi\} = \tan(90^\circ - \eta)\exp\{j(2\nu+\beta)\} \tag{6.32}$$

then

(i) All points corresponding to polarization ratios with the same relative phase $\phi$ = constant lie on one and the same great circle with the diameter joining $X_1$ and $X_2$.

(ii) Choosing the radius $OX_2$ as the angle origin refered to $X_1X_2$, any point corresponding to (6.32) will be at an angle $2\eta$ with respect to the angle origin.

(iii) Any two sets of points of constant phase but with different $\beta$ will lie on two separate main circles rotated by an angle $\beta$ about the diameter $X_1X_2$ against each other.

Based on these rules, we can determine the interrelations of locations of $X_1, X_2$; $C_1(\rho'_{cn1})$, $C_2(\rho'_{cn2})$; $S_1(\rho'_{xm1})$, $S_2(\rho'_{xm2})$; $T_1(\rho'_{xs1})$, $T_2(\rho'_{xs2})$. Since $\rho'_{cn1}$, $\rho'_{cn2}$ $\rho'_{xm1}$ and $\rho'_{xm2}$ have the same phase $(2\nu + \pi/2)$ as $P'_{xn1,2}$ so that they lie on the same great circle. Further, according to the magnitudes, $\tan(90^\circ \pm \mu)$ for $\rho'_{cn1,2}$ and $\tan(90^\circ \pm 45^\circ)$ for $\rho'_{xm1,2}$, the angle $\angle\, C_1OX_2$ and $\angle\, C_2OX_2$ are equal to $2\mu$, the line joining $X_1$ and $X_2$ bi-sects the angle spanned by the co-pol nulls at the origin of the sphere, and $\rho'_{xm1}$, $\rho'_{xm2}$ are orthogonal with $\eta = \pm 45^\circ$, so that $S_1$, $S_2$ are a pair of antipodal points and $S_1S_2$ perpendicular to $X_1X_2$. The cross-polarization saddle point ratios $\rho'_{xs1,2}$ have the same phase as $\rho_{xn1,2}$ but a different phase of $\pi/2$ with $\rho'_{cn1,2}$, $\rho'_{xm1,2}$, so $T_1, T_2, X_1, X_2$ lie on another great circle rotated by 90 degree from the first great circle. The four pairs of points form a polarization Fork on the Poincaré sphere as shown in Fig. 6.1.

**The polarization ratio relation between the old basis and the new basis**

The relation between the polarization ratios in these two bases can be determined by using their corresponding field vectors, expressed as:

$$\vec{E}' = [U]\,\vec{E}$$

where [U] is the unitary vector transformation matrix as expressed in (6.11) and

$$\vec{E} = \frac{1}{\sqrt{1 + \rho\rho^*}} \begin{bmatrix} 1 \\ \rho \end{bmatrix} \tag{6.33a}$$

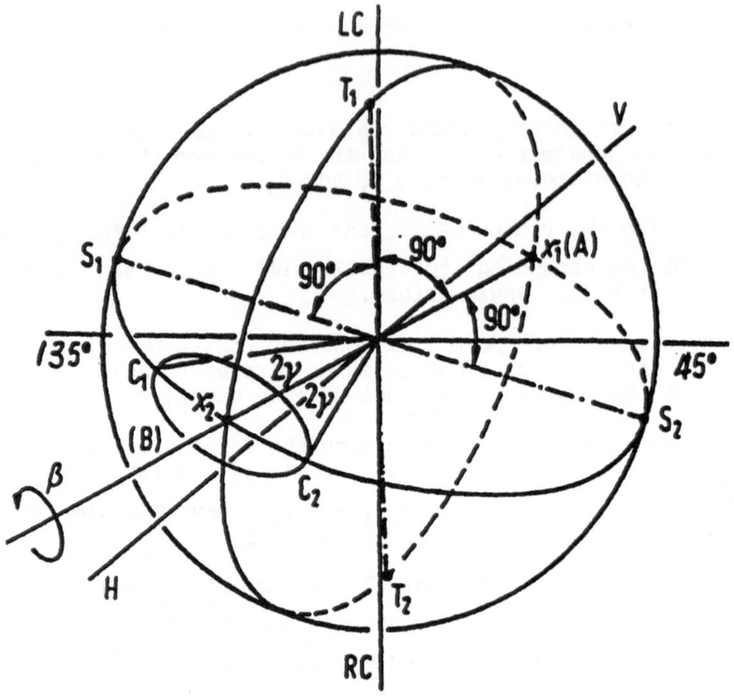


Fig. 6.1. Representation of the characteristic polarization states on the Poincaré sphere ($X_1$: cross-pol null and co-pol max; $X_2$: cross-pol null and co-pol extremum; $C_{1,2}$: co-pol nulls; $S_{1,2}$: cross-pol maxima; $T_{1,2}$: cross-pol saddlepoints; $\gamma$: target characteristic angle)

$$\vec{E}' = \frac{1}{\sqrt{1 + \rho'\rho'^*}} \begin{bmatrix} 1 \\ \rho' \end{bmatrix} \tag{6.33b}$$

with ′ representing the new basis. The matrix [U] can be determined uniquely if the polarization geometry or the scattering matrix [S] is given. To determine [U], three parameters $\rho$, $\psi_1$ and $\psi_4$ must be obtained. The $\rho$ is the polarization ratio for maximum power return to the cross-pol channel which is known from Eq. 6.15, and $\psi_1$, $\psi_4$ can be determined as:

$$\psi_1 = - \frac{\phi_2}{2} - \frac{\pi}{4} \tag{6.34a}$$

$$\psi_4 = \frac{\phi_2}{2} - \frac{\pi}{4} \tag{6.34b}$$

where $\phi_2$ is the phase of the polarization ratio for which the submaximum power is returned to the cross-pol channel. Next we need to show that the result of the polarization ratio formulation is equivalent to Huynen's polarization concept, and that we can obtain the standardized polarization fork presentation of Huynen straightforwardly as presented in Fig 6.2.

$\psi$ - target orientation or tilt angle
$\nu$ - target skip angle
$\varepsilon$ - target ellipticity angle
$\mu$ - target characteristic angle
$\rho$ - $\tan\alpha \exp j\delta$ - polarization ratio

In [30], Huynen represented the scattering matrix [S] using geometric parameters as:

$$[S] = [U^*(\psi,\varepsilon_m,\nu)]\; m \begin{bmatrix} 1 & 0 \\ 0 & \tan^2\mu \end{bmatrix} [U^{*T}(\psi,\varepsilon_m,\nu)] \exp(j\zeta) \tag{6.35}$$

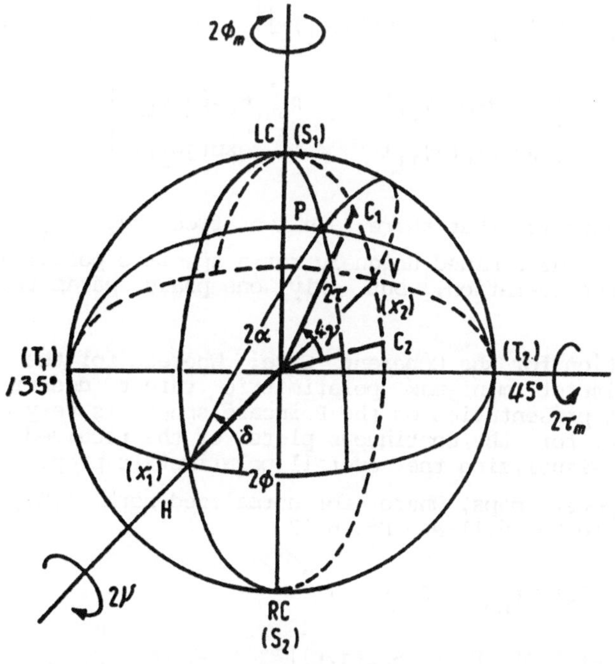


Fig. 6.2. The standarized "Polarization Fork of Huynen" with definition of Huynen's geometrical parameters

where $[U] = \exp\{\psi[J]\} \; \exp\{\varepsilon_m[K]\} \; \exp\{\nu[L]\}$ and $\zeta$ is called the absolute phase of the target [30]; it disappears with power measurements, and it may be altered arbitrarily by moving the target alonng the line-of-sight direction, leaving the targets attitude otherwise unaltered.

In the case of using the polarization ratio formulation, the scattering matrix [S] is expressed as:

$$[S] = [U(\rho_1)]^*[S_d]\;[U(\rho_1)]^{*T} \tag{6.36}$$

$$= \frac{1}{(1+\rho_1\rho_1^*)}\begin{bmatrix}\exp(-j\psi_1) & -\rho_1^*\exp(-j\psi_4)\\ \rho_1^*\exp(-\psi_1) & \exp(-j\psi_4)\end{bmatrix}\begin{bmatrix}m\exp(j2\nu) & 0\\ 0 & m\tan^2\mu\exp(-2j\nu)\end{bmatrix}$$

$$\cdot\begin{bmatrix}\exp(-j\psi_1) & \rho_1^*\exp(-\psi_1)\\ -\rho_1\exp(j\psi_4) & \exp(-j\psi_4)\end{bmatrix} \tag{6.37}$$

Working on (6.35), Huynen's expression can be transformed as to yield the same result (6.37) as:

$$[S] = \frac{1}{1+\rho_1\rho_1^*}\begin{bmatrix}\exp(-j\psi_{11}) & -\rho_1\exp(j\psi_{11})\\ -\rho_1\exp(-j\psi_{11}) & \exp(j\psi_{11})\end{bmatrix}\begin{bmatrix}m\exp(j2\nu) & 0\\ 0 & m\tan^2\mu\exp(-2j\nu)\end{bmatrix}$$

$$\cdot\begin{bmatrix}\exp(-\psi_{11}) & \rho_1^*\exp(-j\psi_{11})\\ -\rho_1\exp(j\psi_{11}) & \exp(j\psi_{11})\end{bmatrix} \tag{6.38}$$

The only difference is that there are two free phases $\psi_1$ and $\psi_4$ in (6.37) which can be determined uniquely for a specific polarization fork geometry under consideration, but only one phase parameter $\psi_{11}$ in (6.38).

**Graphic Presentation for the Coherent Case:** Whereas for the illustration of the discret mini-max polarimetric target description the polarization fork presentation on the Poincaré sphere is very useful, it is found expedient for the continuous plots of the received co/cross-polarized powers visualizing the overall polarimetric properties using planar spectral power maps, where for normalized norm $||\vec{E}||^2 = 1$, we obtain according to Eq. 6.11 and Eq. 6.12

$$\text{Co-Pol Power Plot: } P_{co} = |S_{AA}(\tau,\varepsilon)|^2 \tag{6.39}$$

$$\text{Cross-Pol Power Plot: } P_x = |S_{AB}(\tau,\varepsilon)|^2 \tag{6.40}$$

as well as the relative

Co-Pol Phase Diff.: $\Phi(\tau,\varepsilon) = |\phi_{AA}-\phi_{BB}|$ (6.41)

Cross-Pol Phase Diff.: $\chi(\tau,\varepsilon) = |\phi_{AA}-\phi_{AB}|$ (6.42)

expressed in terms of the general transceiver polarization states $(\Phi,\tau)$.

For one specific case, the scattering matrix together with the Poincaré sphere polarization fork description and the four spectral plots are presented in Fig. 6.3 following Agrawal and Boerner [1], Yamaguchi and Boerner [63], where many additional specific scattering matrix cases are treated in detail. Fig.6.3(a) describes the sysmmetric scattering matrix, Fig. 6.3(b) presents the polarization fork, Fig. 6.3(c,d) displays the co-/cross-polarization power spectral plots $|S_{AA}(\tau,\varepsilon)|^2$, $|S_{AB}(\tau,\varepsilon)|^2$, and Fig.6.3(e,f) displays the relative co/ cross-polarization phase plots $\Phi(\tau,\varepsilon)$, $\chi(\tau,\varepsilon)$, respectively [1].

Thus, in extension of previous results, it was found that there exist eight distinct characteristic polarization states for the symmetric matrix case, the three pairs of orthogonal pairs whose diameters are mutually orthogonal on the polarization sphere: the X-pol null pair (identical to co-pol max pair), the x-pol max pair and the x-pol saddle (turning point) pair. In addition, there exists a pair of co-pol nulls lying in the plane spanned by the x-pol null and the x-pol max pairs, the target characteristic plane with the line (diameter) joining the two x-pol nulls bisecting the angle between the two co-pol nulls on this target characteristic circle. As a result of these unique polarization fork properties, one can show that once the two co-pol nulls have been found, the entire polarization fork can be recovered; i.e., for the descritpion of a radar target we require the specification of two distinct points on the polarization sphere, whereas, only one for the description of a completely polarized wave. In particular, our polarization transformation ratio $\rho$ formulation is in complete agreement with Huynen's formulation and shows, given a measured matrix [S], that the Huynen target characteristic parameters m, $\phi_m$, $\nu$, $\gamma$, $\delta_m$ and $\alpha_m$, can be uniquely determined; or inversely, given these parameters the scattering matrix [S] can be uniquely reconstructed [7]. Hence, the resulting Huynen fork concept represents a unique example of a fundamental polarimetric radar inverse problem.

## 6.2. Optimal Polarization States Using Stokes Vector and the Stokes Reflection Matrix Formulism.

The received power does not only depend on the scattered wave, but also depends on the receiving antenna height. So, it follows that we should consider two cases for the coherent case, respectively. One is the one-antenna case for which the polarization state of the receiving antenna is fixed, and the other is the two-antenna case for which the polarization state of the receiver can be changed.

First, consider the one-antenna case. In this case, one antenna is used simultaneously both as the transmitter and the receiver and care must be

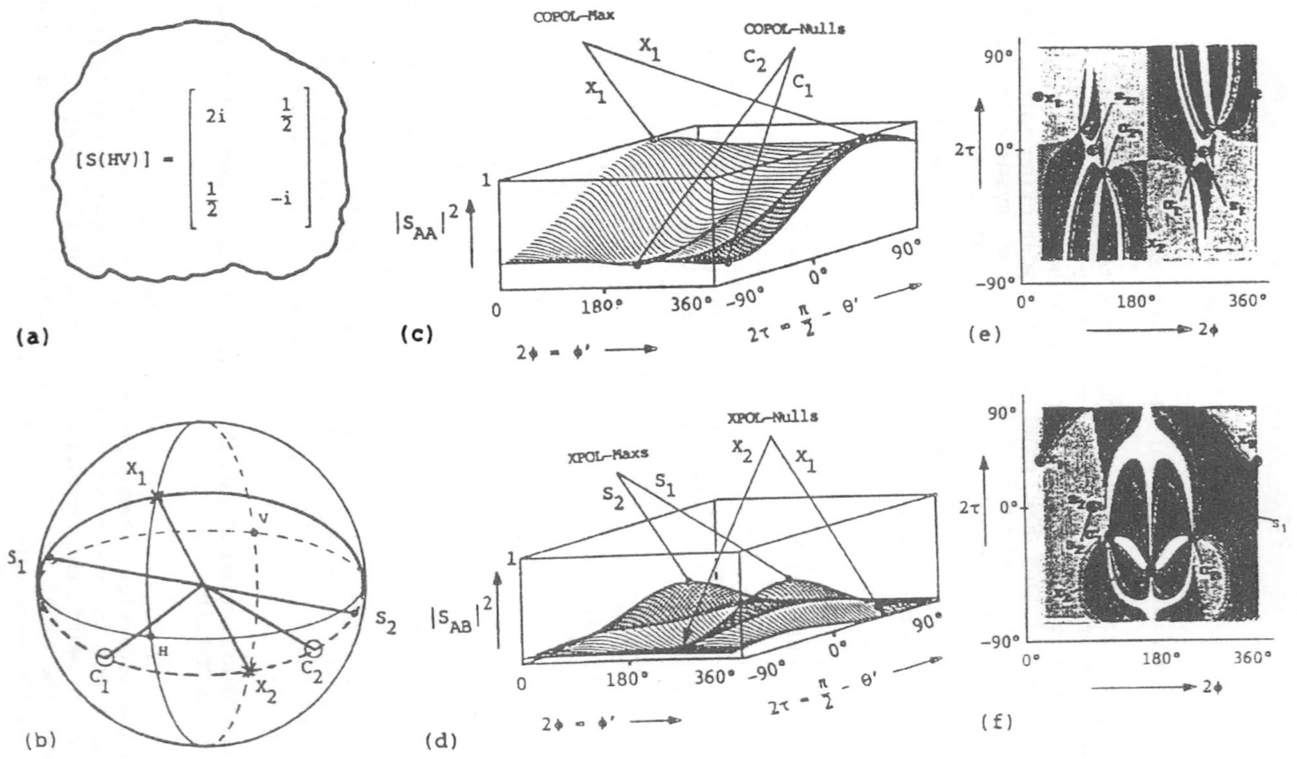


**Fig. 6.3: A General Example**
(a) Target Shape
(b) Polarization Fork
(c) Co-Polarization Spectrum
(d) Cross-Polarization Spectrum
(e) Relative Co-Polarization Phase $\Phi(\phi,\tau)$
(f) Relative Cross-Polarization Phase $X(\phi,\tau)$

taken in dealing separately with the received power in the co-polarized channel and the cross-polarized channel. The received power in the co-pol channel, according to Eq. 4.26 and Eq. 4.27, can be written as:

$$P_c = \vec{g}^T [M_{pc}] \vec{g} \tag{6.43}$$

where $[M_{pc}]$ denotes the Stokes reflection matrix in the co-pol channel, and the received power in the x-pol channel, according to Eq. 4.33, can be obtained as:

$$P_x = \vec{g}^T [M_{px}] \vec{g} \tag{6.44}$$

where $[M_{px}]$ denotes the Stokes reflection matrix in the x-pol channel. Secondly, consider the two-matched-antenna case. In the bistatic or monostatic separated two-antenna case, one antenna serves as the source of the transmitted wave, and the other as the receiver. The optimal received voltage condition in polarimetry is given for the case in which the polarization of the receiving antennea is matched to that of the scattered field $\vec{g}_S$. So, the received power in the matched two-antenna case, according to Eq. 4.38, can be written as:

$$P_2 = \vec{g}_T^T [M_2] \vec{g}_T \tag{6.45}$$

where $[M_2]$ is the Stokes reflection matrix defined in the matched two-antenna case.

The problem of a search for the optimal received power is to find $\vec{g}_T$ such that the received power P is maximized for the given [S]. For simplicity, we always choose the transmitting wave $\vec{g}_T$ as a normalized, completely polarized wave; i.e., the components of $\vec{g}_T$ satisfy the following constraint,

$$g_{T0} = \sqrt{g_{T1}^2 + g_{T2}^2 + g_{T3}^2} = 1$$

The received power P is a function of $g_{T1}$, $g_{T2}$, and $g_{T3}$ according to the previous sub-sections, which is subject to the above constraint. This requirement dictates the use of the method of Lagrangian multipliers to find the extremum of P.

Let our equation of constraint be given by

$$\Phi(g_{T1}, g_{T2}, g_{T3}) = \sqrt{g_{T1}^2 + g_{T2}^2 + g_{T3}^2} - 1 = 0 \quad , \tag{6.46}$$

then the Lagrangian multipliers method for finding the extreme value of the power $P\ (g_{T1}\ ,\ g_{T2}\ ,\ g_{T3})$ results in

$$\begin{cases} \dfrac{\partial P}{\partial g_{T1}} - \mu \dfrac{\partial \Phi}{\partial g_{T1}} = 0 \\[2ex] \dfrac{\partial P}{\partial g_{T2}} - \mu \dfrac{\partial \Phi}{\partial g_{T2}} = 0 \\[2ex] \dfrac{\partial P}{\partial g_{T3}} - \mu \dfrac{\partial \Phi}{\partial g_{T3}} = 0 \quad . \end{cases} \tag{6.47}$$

When Eq. 6.46 and Eq. 6.47 are satisfied, $dP(g_{T1}\ ,\ g_{T2}\ ,\ g_{T3}) = 0$, and P is an extremum.

**Example**

Let the scattering matrix [S] be given and in correspondence with Agrawal and Boerner [1], Xi and Boerner [62] by

$$[S] = \begin{bmatrix} 2j & \frac{1}{2} \\ \frac{1}{2} & -j \end{bmatrix} \tag{6.48}$$

for which the corresponding "degenerate" Mueller matrix [M] is obtained by solving Eq. 4.15 as

$$[M] = \begin{pmatrix} 2.75 & 1.5 & 0 & 1.5 \\ 1.5 & 2.25 & 0 & 0.5 \\ 0 & 0 & -1.75 & 0 \\ -1.5 & -0.5 & 0 & -2.25 \end{pmatrix} \tag{6.49}$$

Similarly, the Stokes reflection matrices $[M_{pc}]$, $[M_{px}]$, and $[M_2]$ are obtained by solving Eqs. 4.27, 4.34, and 4.38, respectively, as

$$[M_{pc}] = \begin{pmatrix} 1.375 & 0.75 & 0 & 0.75 \\ 0.75 & 1.125 & 0 & 0.25 \\ 0 & 0 & -0.875 & 0 \\ 0.75 & 0.25 & 0 & 1.125 \end{pmatrix} \tag{6.50}$$

$$[M_{px}] = \begin{bmatrix} 1.375 & 0.75 & 0 & 0.75 \\ -0.75 & -1.125 & 0 & -0.25 \\ 0 & 0 & 0.875 & 0 \\ -0.75 & -0.25 & 0 & -1.125 \end{bmatrix} \tag{6.51}$$

$$[M_2] = \begin{bmatrix} 2.75 & 1.5 & 0 & 1.5 \\ 0 & 0 & 0 & 0 \\ 0 & 0 & 0 & 0 \\ 0 & 0 & 0 & 0 \end{bmatrix} \tag{6.52}$$

First, consider the received power in the co-pol channel. By substituting Eq. 6.50 into Eq. 4.26, yields

$$P_c = 1.375+1.5\, g_{T1}+1.5\, g_{T3}+1.125\, g_{T1}^2+0.5g_{T1}g_{T3}-0.825\, g_{T2}^2+1.125\, g_{T3}^2 .$$

Subject to the constraint equation

$$\Phi = \sqrt{g_{T1}^2 + g_{T2}^2 + g_{T3}^2} - 1 = 0 \quad ,$$

for optimizing $P_c$, and applying the Lagrangian multiplier method, yields

$$\begin{cases} \dfrac{\partial P_c}{\partial g_{T1}} - \mu \dfrac{\partial \Phi}{\partial g_{T1}} = 1.5 + 2.25g_{T1} + 0.5g_{T3} - \mu g_{T1} = 0 \\[2ex] \dfrac{\partial P_c}{\partial g_{T2}} - \mu \dfrac{\partial \Phi}{\partial g_{T2}} = -1.75g_{T2} - \mu g_{T2} = 0 \\[2ex] \dfrac{\partial P_c}{\partial g_{T3}} - \mu \dfrac{\partial \Phi}{\partial g_{T3}} = 1.5 + 0.5g_{T1} + 2.25g_{T3} - \mu g_{T3} = 0 \end{cases}$$

Solving the above equations yields four solutions for the polarization state $\vec{g}_T$; where two solutions correspond to the two maxima and two correspond to the nulls, listed in Table 6.1 and shown in Fig. 6.3(b), respectively.

Second, for the received power in the x-pol channel, we obtain by substituting Eq. 6.51 into Eq. 4.33,

$$P_x = 1.375 - 1.125g_{T1}^2 - 0.5g_{T1}g_{T3} + 0.875g_{T2}^2 - 1.125g_{T3}^2$$

and similarly by applying the Lagrangian multiplier method

$$\begin{cases} \dfrac{\partial P_x}{\partial g_{T1}} - \mu \dfrac{\partial \Phi}{\partial g_{T1}} = -2.25\, g_{T1} - 0.5g_{T3} - \mu g_{T1} = 0 \\ \dfrac{\partial P_x}{\partial g_{T2}} - \mu \dfrac{\partial \Phi}{\partial g_{T2}} = 1.75g_{T2} - \mu g_{T2} = 0 \\ \dfrac{\partial P_x}{\partial g_{T3}} - \mu \dfrac{\partial \Phi}{\partial g_{T3}} = -0.5g_{T1} - 2.25\, g_{T3} - \mu g_{T3} = 0 \end{cases}$$

Solving the above equations yields the six polarization states $\vec{g}_T$ shown in Table 6.1 and Fig 6.3(b). These solutions correspond to the two cross-pol maxima, two cross-pol turning points and two cross-pol nulls, where the latter are identical to the co-pol max extrema.

To optimize the received power P for the matched two-antennas case, using Eqs. 6.52 and 4.38 yields

$$P_2 = 2.75 + 1.5g_{T1} + 1.5g_{T3}$$

$$\begin{cases} \dfrac{\partial P_2}{\partial g_{T1}} - \mu \dfrac{\partial \Phi}{\partial g_{T1}} = 1.5 - \mu g_{T1} = 0 \\ \dfrac{\partial P_2}{\partial g_{T2}} - \mu \dfrac{\partial \Phi}{\partial g_{T2}} = -\mu g_{T2} = 0 \\ \dfrac{\partial P_2}{\partial g_{T3}} - \mu \dfrac{\partial \Phi}{\partial g_{T3}} = 1.5 - \mu g_{T3} = 0 \end{cases}$$

Solving the above equations yields two solutions, $(1, 0.707,0,0.707)^T$ and $(1,-0.707,0,-0.707)^T$, which are the same as the co-pol max pair of Table 6.1.

### 6.3 Optimization procedures for the partially polarized case

We consider the case for which a time-dependent scatterer is illuminated by a monochromatic (completely polarized) wave $\vec{E}_T$, and the reflected wave $\vec{E}_R$, which is, in general, non-monochromatic and therefore, partial-

Table 6.1. Results for First Example

| | | $g_{T0}$ | $g_{T1}$ | $g_{T2}$ | $g_{T3}$ | $P_c$ | $P_x$ |
|---|---|---|---|---|---|---|---|
| co-pol | max | 1 | 0.707 | 0 | 0.707 | 4.87 | |
| | | 1 | -0.707 | 0 | -0.707 | 0.63 | |
| | null | 1 | -0.333 | 0.882 | -0.333 | 0 | |
| | | 1 | -0.333 | -0.882 | -0.333 | 0 | |
| x-pol | max | 1 | 0 | 1 | 0 | | 2.25 |
| | | 1 | 0 | -1 | 0 | | 2.25 |
| | turning point | 1 | 0.707 | 0 | -0.707 | | 0.5 |
| | | 1 | -0.707 | 0 | 0.707 | | 0.5 |
| | null | 1 | 0.707 | 0 | 0.707 | | 0 |
| | | 1 | -0.707 | 0 | -0.707 | | 0 |

ly polarized. From here on, we will be concerned with the partially polarized case, consequently, the Stokes vector and Mueller matrix formalism will be employed.

Let $\vec{g}_T$ be the Stokes vector of the transmitted wave $\vec{E}_T$,

$$\vec{g}_T = \begin{bmatrix} g_{T0} \\ g_{T1} \\ g_{T2} \\ g_{T3} \end{bmatrix} \tag{6.53}$$

Because the transmitted wave incident on the scatterer is assummed to be completely polarized, p = 1, we have

$g_{T0} = \sqrt{g_{T1}^2 + g_{T2}^2 + g_{T3}^2} = 1$ for the normalized input.

Let $\vec{g}_S$ stand for the Stokes vector of the reflected wave $\vec{E}_R$,

$$\vec{g}_S = \begin{bmatrix} g_{S0} \\ g_{S1} \\ g_{S2} \\ g_{S3} \end{bmatrix} \tag{6.54}$$

Then the scattering process according to Eq. 4.9, becomes

$$\vec{g}_S = \begin{bmatrix} g_{S0} \\ g_{S1} \\ g_{S2} \\ g_{S3} \end{bmatrix} = \begin{bmatrix} M_{00} & M_{01} & M_{02} & M_{03} \\ M_{10} & M_{11} & M_{12} & M_{13} \\ M_{20} & M_{21} & M_{22} & M_{23} \\ M_{30} & M_{31} & M_{32} & M_{33} \end{bmatrix} \begin{bmatrix} g_{T0} \\ g_{T1} \\ g_{T2} \\ g_{T3} \end{bmatrix} \tag{6.55}$$

As it is partially polarized, the degree of polarization p for the reflected wave is

$$p = \frac{\sqrt{g_{S1} + g_{S2}^{2} + g_{S3}^{2}}}{g_{S0}} \tag{6.56}$$

According to Eq. 3.42, the reflected wave $\vec{g}_S$ can be decomposed into its completely polarized component $\vec{g}_P$ and unpolarized component $\vec{g}_u$ as

$$\vec{g}_S = \vec{g}_P + \vec{g}_u = \begin{bmatrix} pg_{S0} \\ g_{S1} \\ g_{S2} \\ g_{S3} \end{bmatrix} + \begin{bmatrix} (1-p)gS_0 \\ 0 \\ 0 \\ 0 \end{bmatrix} \tag{6.57}$$

Since the energy density is given by the first element of the Stokes vector $\vec{g}_S$, we can write the following expression for the total available intensity (normalized),

$$P = pg_{S0} + \frac{1}{2}(1-p)g_{S0} = \frac{1}{2}(1+p)g_{S0} \tag{6.58}$$

where the first term represents adjustable intensity and the second term corresponds to noise-like intensity with 50% reception efficiency [40].

### 6.3.1 Determination of optimization criteria

There are four types of intensity terms that can be optimized:

$g_{S0}$ — Total intensity in the scattered wave before it reaches the receiver (GS); (6.59a)

$pg_{S0}$ — Completely polarized part of the intensity; (6.59b)

$(1-p)g_{S0}$ — Noise or the unpolarized part: Regardless of the receiver, one half of the unpolarized part is always accepted. Thus, $g_{S0}$ $(1-p)/2$ is always accepted; (6.59c)

$(1+p)g_{S0}/2$ — $pg_{S0} + (1-p)g_{S0}/2$ = maximum of the total receptable intensity (i.e., the sum of the matched polarized part and one half the unpolarized part). However, if the polarized part is mismatched (cancelled with the proper receiver tuning), the total received power is minimal and equal half the unpolarized part. (6.59d)

**Optimization of the adjustable intensity $pg_{S0}$**

The energy density $pg_{S0}$, contained in the completely polarized part $g_p$, is called the adjustable intensity because one may adjust the polarization state of the receiver to ensure the polarization match, i.e., to receive all of the power in the completely polarized part of the scattered wave. Kostinski, James and Boerner [41] obtained the optimal reception of the adjustable intensity $(P\ g_{so})^2$ by using the rotation transformation to diagnalize part of the Mueller matrix and then one applying the Lagranian multipliers method [57]. Actually, the step of the rotation transformation is not necessary and the reception of the adjustable intensity can be accomplished by applying the Lagrangian multiplier method directly as is shown next. We can rewrite the scattering process of Eq. 6.55 in index notation as

$$g_{Si} = \sum_{j=0}^{3} M_{ij}\, g_{Tj}$$

where j = 0, 1, 2, 3. The first term in Eq. 6.57, is a completely polarized wave which has the following property:

$$pg_{S0} = (g_{S1}^2 + g_{S2}^2 + g_{S3}^2)^{1/2} \quad . \tag{6.60}$$

So, in the index notation, the adjustable intensity can be expressed as

$$pg_{S0} = \left( \sum_{i=1}^{3} g_{Si}^2 \right)^{1/2} = \left[ \sum_{i=1}^{3} \left( \sum_{j=0}^{3} M_{ij} g_{Tj} \right)^2 \right]^{1/2} \tag{6.61}$$

where $g_{Ti}$'s are the elements of the Stokes vector of the transmitting wave. The adjustable intensity $pg_{S0}$ is a function of the $g_{Ti}$'s. Let the $g_{Ti}$'s be the variables, then the partial derivative of $(pg_{S0})^2$ with respect to $g_{Tk}$ can be derived as

$$\frac{\partial (pg_{S0})^2}{\partial g_{Tk}} = \sum_{i=1}^{3} \frac{\partial g_{Si}^2}{\partial g_{Tk}} = 2 \sum_{i=1}^{3} g_{Si} M_{ik} = 2 \sum_{i=1}^{3} \sum_{j=0}^{3} M_{ij} M_{ik} g_{Tj} \qquad (6.62)$$

For optimizing the adjustable intensity, we apply the method of Lagrangian multipliers, which yields

$$\begin{cases} \dfrac{\partial (pg_{S0})^2}{\partial g_{T1}} - \mu \dfrac{\partial \Phi}{\partial g_{T1}} = 2 \displaystyle\sum_{i=1}^{3} \sum_{j=0}^{3} M_{ij} M_{i1} g_{Tj} - \mu g_{T1} = 0 \\ \dfrac{\partial (pg_{S0})^2}{\partial g_{T2}} - \mu \dfrac{\partial \Phi}{\partial g_{T2}} = 2 \displaystyle\sum_{i=1}^{3} \sum_{j=0}^{3} M_{ij} M_{i2} g_{Tj} - \mu g_{T2} = 0 \\ \dfrac{\partial (pg_{S0})^2}{\partial g_{T3}} - \mu \dfrac{\partial \Phi}{\partial g_{T3}} = 2 \displaystyle\sum_{i=1}^{3} \sum_{j=0}^{3} M_{ij} M_{i3} g_{Tj} - \mu g_{T3} = 0 \end{cases} \qquad (6.63)$$

where $\Phi$ is the constraint equation of Eq. 6.46. Eq. 6.63 is a set of inhomogeneous linear equations in $g_{T1}(\mu)$, $g_{T2}(\mu)$, and $g_{T3}(\mu)$. Then the straightforward solutions for the $g_{Ti}(\mu)$ are three functions of $\mu$. Substituting $g_{Ti}(\mu)$, $(i = 1,2,3)$ into the constraint condition of Eq. 6.46 leads to a sixth-order polynomial equation of $\mu$. This equation is analytically solvable in $\mu$. For each $\mu$ value, we calculate $g_{T1}$, $g_{T2}$, $g_{T3}$, and $pg_{S0}$ according to the formula in Eq. 6.61. The largest (or smallest) intensity is used to choose the optimal intensity, the corresponding $\vec{g}_T$ is the optimal polarization state of the transmitted wave.

**Minimizing the noise-like energy density term: $(1-p)g_{SO}$**

An unpolarized wave can always be represented by an incoherent sum of any two orthogonal completely polarized waves of equal intensity [24], which leads to a 50% efficiency for the reception of the unpolarized wave. We would like to minimize the noise-like energy so that as much energy as possible may be received. The total energy density of the unpolarized part of the scattered wave is given by:

$$(1 - p)g_{S0} = g_{S0} - pg_{S0} = \sum_{j=0}^{3} M_{0j} g_{Tj} - \sqrt{\sum_{i=1}^{3} \left( \sum_{j=0}^{3} M_{ij} g_{Tj} \right)^2} \qquad (6.64)$$

There is no simple way of giving the analytic closed form solution for the minimum solusion. Also, the Lagrangian multiplier method becomes far too involved, and instead computer numerical analysis must be used.

**Maximizing the receivable intensity in the scattered wave: $1/2(1 + p)g_{so}$**

The total receivable energy density consists of two parts: 100% reception efficiency for the completely polarized part of the scattered wave and 50% reception efficiency for the unpolarized part. We may write the following expression for the total receivable intensity:

$$1/2(1 + p)g_{S0} = pg_{S0} + 1/2(1 - p)g_{S0}$$

$$= \frac{1}{2}\sum_{j=0}^{3} M_{0j}\, g_{Tj} + \frac{1}{2}\sqrt{\sum_{i=1}^{3}\left(\sum_{j=0}^{3} M_{ij}\, g_{Tj}\right)^2} \qquad (6.65)$$

Also, this equation can only be solved using numerical analysis.

**Numerical Examples.**

Consider the following Mueller matrix, which corresponds to experimental data taken for a combined collimator-radiometer system [23].

$$M = \begin{bmatrix} 0.7599 & -0.0623 & 0.0295 & 0.1185 \\ -0.0573 & 0.4687 & -0.1811 & -0.1863 \\ 0.0384 & -0.1714 & 0.5394 & 0.0282 \\ 0.1240 & -0.2168 & -0.0120 & 0.6608 \end{bmatrix} \qquad (6.66)$$

By applying the method of Lagrangian multipliers for optimizing the adjustable intensity of Eq. 6.63, a set of inhomogeneous linear equations is obtained as follows,

$$\begin{pmatrix} 0.2961 - \mu & -0.1747 & -0.2354 \\ -0.1747 & 0.3293 - \mu & 0.0410 \\ -0.2354 & 0.0410 & 0.4722 - \mu \end{pmatrix}\begin{pmatrix} g_{T1} \\ g_{T2} \\ g_{T3} \end{pmatrix} = \begin{pmatrix} 0.0603 \\ -0.0296 \\ -0.0937 \end{pmatrix} \qquad (6.67)$$

Solving the linear equations results in three solutions $g_{T1}(\mu)$, $g_{T2}(\mu)$, and $g_{T3}(\mu)$ which are functions of $\mu$. Substituting the three solutions into the constraint equation of Eq. 6.46 leads to a sixth-order polynomial equation in $\mu$,

$$\mu^6 - 2.1842\mu^5 + 1.7813\mu^4 - 0.6780\mu^3 + 0.1221\mu^2 - 0.009229\mu + 0.0002430 = 0 \qquad (6.68)$$

For each root of $\mu$, the polarization state $\vec{g}_T(\mu)$ and its intensity, which is obtained by solving Eq. 4.61., are displayed in Table 6.2 and illustrated in Fig. 6.4.

**Table 6.2. Roots, Intensities and Stokes Vectors for the Characteristic Polarization States of the Adjustable Intensity Optimization**

| Root Number | Root $\mu$ | Intensity $pg_{s0}$ | $g_{T0}$ | $g_{T1}$ | $g_{T2}$ | $g_{T3}$ |
|---|---|---|---|---|---|---|
| 1 | 0.8019 | 0.9677 | 1 | -0.5807 | 0.3378 | 0.7408 |
| 2 | 0.5739 | 0.6942 | 1 | 0.6134 | -0.4203 | -0.6686 |
| 3 | 0.3548 | 0.5923 | 0.997 | 0.3894 | -0.8702 | 0.2918 |
| 4 | 0.3170 | 0.5342 | 1 | 0.0221 | 0.6619 | -0.7492 |
| 5 | 0.0734 | 0.2776 | 1.006 | 0.8710 | 0.4486 | 0.2301 |
| 6 | 0.0633 | 0.2404 | 1.009 | -0.6812 | -0.4777 | -0.5707 |

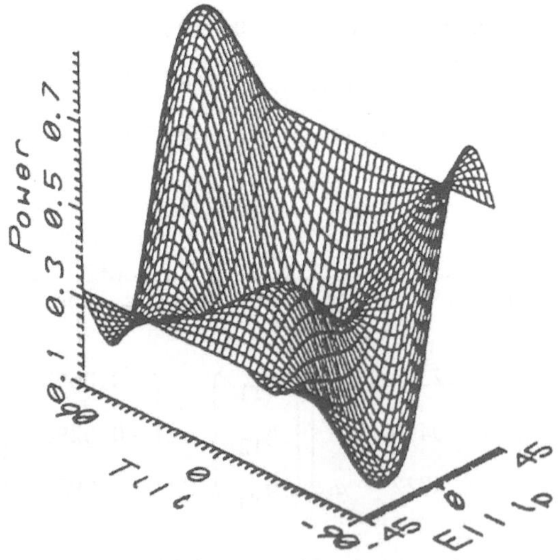


Fig. 6.4 Polarimetric Dependence of Ajustable intensity $pg_{S0}$: The general topology of the power density (surface) plote is determined by the six stationary points corresponding to six characteristic roots of Eq. 6.68.

From inspection of Fig. 6.4, we observe that we are obtaining six "characteristic polarization states", similar to the coherent case consisting of two power maxima ($X_1$,$X_2$), two power minima ($C_1$, $C_2$) and two "saddle-point" extrema ($S_1$, $S_2$) which, may be related to the XPOL-Nulls ($X_1$, $X_2$), the CO-POL-Nulls ($C_1$, $C_2$) and the XPOL- Saddle-points ($S_1$, $S_2$) of the coherent case, respectively. The last finding regarding the saddle-point extrema requires further analysis. For further details, we refer the reader to pertinent treatise of van Zyl [65] and those presented in [7].

The noise-like energy was minimized and the receivable intensity was maximized by numerical analysis as shown in Table 6.3.

**Table 6.3. The Intensities and the Stokes Vectors of the Optimal Polarization States of the Different Types of Energy Density**

| Method | Intensity | $g_{T0}$ | $g_{T1}$ | $g_{T2}$ | $g_{T3}$ |
|---|---|---|---|---|---|
| Maximum of $pg_{S0}$ | 0.9677 | 1 | -0.5807 | 0.3375 | 0.7408 |
| Minimum of $(1 - p)g_{S0}$ | -0.07 | 1 | -0.6108 | 0.3893 | 0.6894 |
| Maximum of $1/2(1 + p)g_{S0}$ | 0.9311 | 1 | -0.5602 | 0.3090 | 0.7686 |

## 7. COMPARISON OF RESULTS

We have demonstrated that there exist several different approaches for determining the characteristic polarization states in radar polarimetry. The general polarization ratio $\rho$ transformation method [1,62] of Section 6.1.3, although mathematically cumbersome, provides the most complete method for determining the five pairs of characteristic polarization states for the coherent 2x2 Sinclair matrix [S(A,B)]. On the other hand, the Langrangian multiplier method [64,63] is most suitable for determining the optimal polarization states of the 4x4 Stokes reflection matrices $[\bar{M}_c]$, $[\bar{M}_x]$ and $[\bar{M}_2]$ for the "degenerate coherent Mueller matrix case", as defined in Section 4.6 and derived in Section 6.2, and verified for one common example (Fig. 6.3). In addition, there exist the optimization procedures derived directly from the alternate presentation of the Mueller matrix expressed in terms of the Pauli spin matrices as derived in Section 4.5, or in terms of the associated covariance matrix (See II-9), which deserve extensive further analyses. It is, thus, shown and proven that Kennaugh's target characteristic operator theory [37] of the coherent case is correct and that Huynen's polarization fork concept is valid in the coherent case which was also established in Kanareykin etal [35,36] as summarized in [0-3] and [0-4].

For the general partially coherent case, no complete analytic optimization procedure for determining the optimum polarization states yet exists; however, in Section 6.2, it was shown that for the partially polarized case, for which the wave incident on the scatterer is completely polarized, a solution can be found [64] as was also demonstrated in the Soviet literature [**0-3,0-4**] and especially by Zhivotovskiy in [65] and in [**IV-3**]. In all of the cases investigated, it was demonstrated that for the partially polarized case there exsit five pairs of characteristic polarization states. However — in accordance with (6.59) — whereas for the coherent case (p = 1), the absolute (normalized) power maximum at co-pol max ($\rho_{cm1}$) and the absolute (normalized) power minimum at the co-pol null ($\rho_{cn1,2}$) locations, respectively become

$$P^c_{max} = (\rho_{cm1})/m^2 = 1 \tag{7.1}$$

$$P^c_{cn1,2} = (\rho_{cn1,2})/m^2 = 0 \quad , \tag{7.2}$$

we find that for the partially polarized case ($0 > P > 1$), the maximum normalized value will always be reduced by the value $(1 - p)/2$ and that the achievable minimal normalized power can never be less than $(1 - p)/2$; and that according to (6.59d) for the completely unpolarized case ($p = 0$), the minimally and maximally achievable normalized powers become equal and in the limit approach $g_{S0} \to .5$; i.e., the power density plot is flat in the extreme unpolarized case as is schematically illustrated in Fig. 7.1.

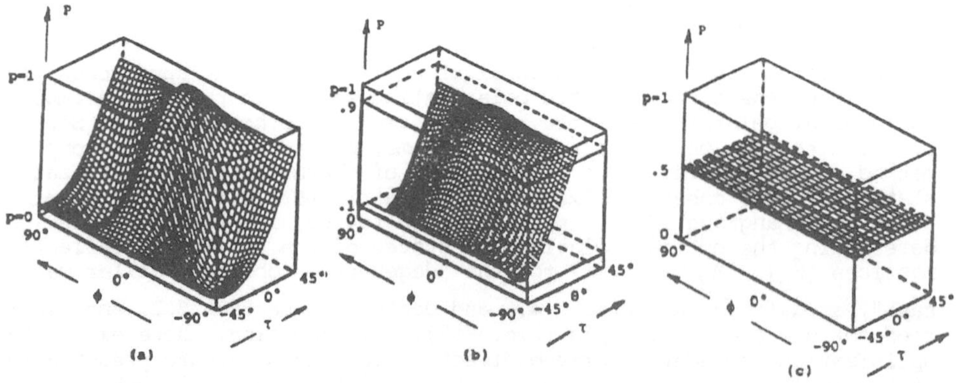


Fig. 7.1: Dependence of received power density plots on degree of polarization p: (a) p = 1, (b) p = .8, and (c) p = 0.

Thus, from the comparison of our results, we conclude that the optimal polarization state concept will — in principle — also play an important role for treating the partially (dual orthogonal) polarization radar reception problem. This holds true especially when the scattering center decorrelation times $\tau_s$, are small, as for example for hydrometeoric scatter (~ .1 ms to 30 ms at 3GHz) and the scattering matrix acquisition time period $\tau_m << \tau_s$ is much less, say less than 20 $\mu$s for above case, requiring ultra-fast electronic polarization state switching capabilities which are now technologically available. Thus, for each "instantaneous incremental measurement window $I_m$ the polarization fork may be established, and the time-sequential changes will lead to the clustering of the respective co/cross-pol nulls as is demonstrated in [VIII-9] and is also alluted to in [0-4].

## 8. CONCLUSION

A succinct overview was presented on the historical development of radar polarimetry, highlighting the important contributions including concise definitions and descriptions of the polarization state, the radar scattering matrices, and the polarimetric optimization problem. Based on such a critical assessment, the approximate formalism for handling basic polarimetric radar concepts was introduced. In a next step, we provided the scattering matrix formulations together with the transformation laws for measurements obtained with coherent dual polarization (scattering matrix) radars using different sets of antenna polarization bases such as linear horizontal/vertical (H,V) circular left/right (L,R) or any other orthogonal elliptic (A,B) transmit/recieve systems leading directly to polarimetric radar target and clutter phenomenology. Although the basic target versus clutter optimization procedures for determining the maximum and minimum transmit and receive polarization states have been given, no extensive applications will be shown here because of lack of space. Instead, we wish to refer to specific state-of-the-art review texts such as by Beckmann and Spizzichino on electromagnetic scattering from rough surfaces [3,4] and related Russian texts [11,25,36,43,50,52, 62], by Long [45], and by Ulaby, Moore and Fung [58]. Target down- and cross-range polarimetric mapping applications are discussed in [5,6,7, 14,18,19,30,37] and especially summarized in [24] and [48] using the coherent approach. For the partially polarized case [12,23] we refer to [41,29,32] which will also be covered in great detail in the forthcoming proceedings on radar polarimetry by Boerner et al [7], Root and Matkin [52]. It would be most illuminating to introduce the application of the Polarimetric Matched Filter Concept introduced in [42] to POL-SAR image analysis as will be discussed in great detail [6,7] and [51,52].

## 9. ACKNOWLEDGEMENTS

This paper is dedicated to the Seventieth birthday of Dr. Jean Richard Huynen, one to the distinguished pioneers of polarimetric radar target phenomenology. A sincere note of thenks is extended to Dr. Henry W. Mullaney, Dr. James W. Mink, Dr. Karl H. Steinbach, Dr. Walter A. Flood, Mr. Lloyd W. Root, Mr. Otto Kessler and Mr. James G. Smith for their continual interest in this research. A special thanks is extended to Mr. Richard W. Foster and Ms. Mirian R. Mailey for their skillful typing of this manuscript and the preparation of figures.

Financial support for this research was provided by the US Office of Naval Research, Contract Nos. N00014-80-C-0773 and N00014-90-J-1405, the US Army Research Office, Contract No. DAAL-03-89-K-0116, and US Senior Scientist Fellow Award of the Alexander von Humbolt Stiftung, Bonn, FRG.

## 10. REFERENCES

[1] A.P. Agrawal and W-M. Boerner, "Re-Development of Kennaugh's Target Characteristic Polarization State Theory Using the Polarization Transformation Ratio Formalism for the Coherent Case", IEEE Trans. GSRS, Vol. 27, No. 1, pp. 2-14, January (1989)

[2] R.M.A. Azzam and N.M. Bashara, Ellipsometry and Polarized Light, Amsterdam: North Holland, (1977)

[3] P. Beckmann, The Depolarization of Electromagnetic Waves, Boulder, CO: The Golem Press, (1968)

[4] P. Beckmann, and A. Spizzichino, The Scattering of Electromagnetic Waves from Rough Surfaces", New York: MacMillan, (1963)

[5] W-M. Boerner, "Polarization Utilization in Electromagnetic Inverse Scattering", Chpt. 7 in Inverse Scattering Problems in Optics, ed. by H.P. Baltes, Vol. 2; Topics in Current Physics, Vol. 20, Heidelberg: Springer Verlag, July (1980), pp. 237-305

[6] W-M. Boerner, et al. eds., Inverse Methods in Electromagnetic Imaging, Proc. NATO Advanced Res. Workshop on IMEI, Bad Windsheim, FR. Germany, Sept. 18-24, 1983, NATO ASI Series, Series C, Math. & Phys. Sci., Vol.143, Dordrecht, Holland: D. Reidel Publ. Co., (1985)

[7] W-M. Boerner, et al. (eds), Direct and Inverse Method in Radar Polarimetry, Proc. NATO-ARW-DIMRP (W-M. Boerner, Director), 1988 Sept. 18-24 Bad Windsheim FRG, NATO-ASI-Series C, (Math. & Phys. Sci.), Dordrecht/Boston: D. Reidel Publ. Co., (1989)

[8] W-M. Boerner and H.P.S. Ahluwalia, "On a Set of Continuous Wave Electromagnetic Inverse Boundary Conditions", Can.J. Phys., 50(23), pp. 3023-3061, Dec. 15, (1972) (also see: IEEE Trans. AP-21(5), pp. 663-672, May 1973; IEEE Trans. AP-22(5), pp. 673-682, May 1974; Can. J. Phys. 53, pp. 1404-1407, May 1975)

[9] W-M. Boerner, A.K. Jordan and I.W. Kay, "Introduction to the Special Issue on Inverse Methods in Electromagnetics", in IEEE Trans., AP-29, Guest Editors, W-M. Boerner, A.K. Jordan, I.W. Kay, March 1989, pp. 185-189

[10] W-M. Boerner, B-Y. Foo, H.J. Eom, "Interpretation of the Polarimetric Co-Polarization Phase Term ($\phi_{HH}-\phi_{VV}$) in High Resolution SAR Imaging Using the JPL CV-990 Polarimetric L-Band SAR Data", Special IGARSS '85 Issue of the IEEE Trans. GE-25, No. 1, pp. 77-82, January (1987)

[11] V.V. Bogorodsky, D.B. Kanareykin and A.E. Kozlov, Polarization of the Scattered Radio Radiation of the Earth Covers, Leningrad: Gidsometeorizdat, (in Russian) (1981)

[12] M. Born and E. Wolf, Principles of Optics, 3rd Ed., New York: Pergamon Press, (1964)

[13] C-Y. Chan, "Studies on the Power Scattering Matrix of Radar Targets", M.Sc. Thesis, Dept. of Electr. Engr. & Comp.Sci., University of Illinois at Chicago, Chicago, IL, (1981)

[14] S.K. Chaudhuri, W-M. Boerner, "A Polarimetric Model for the Recovery of High-Frequency Scattering Centers from Bistatic-Monostatic Scattering Matrix Data", IEEE Trans. AP-A, p-35, No. 1, January 1987.

[15] H.C. Chen, Theory of Electromagnetic Waves, New York: McGraw-Hill Book Company, (1983)

[16] D. Clarke and J.F. Grainger, Polarized Light and Optical Measurement, Oxford: Pergamon Press, (1971)

[17] R.E. Collin, Antennas and Radio Wave Propagation, New York: McGraw Hill, N.Y.,(1985)

[18] M. Davidovitz and W-M. Boerner, "Reduction of Bistatic Scattering Matrix Measurements for Inversely Symmetric Radar Targets", IEEE Trans. AP-31, No. 2, March (1983)

[19] M. Davidovitz and W-M. Boerner, "Extension of Kennaugh's Optimal Polarization Concept to the Asymmetric Matrix Case", IEEE Trans. AP-34(4), pp. 569-574, Apr. (1986)

[20] G.A. Deschamps, "Part 2: Geometrical Representation of the Polariza- tion of a Plane Electromagnetic Wave", Proc. IRE, Vol. 39, May (1951), pp. 540-544

[21] G.A. Deschamps and P.E. Mast, "Poincaré Sphere Representation of Partially Polarized Fields", IEEE Trans. AP-21(4), (1973), pp. 474-478

[22] B-Y. Foo, S.K. Chaudhuri and W-M. Boerner, "A High Frequency Inverse Scattering Model to Recover the Specular Point Curvatures from Polarimetric Scattering Data", IEEE Trans. AP-32, No. 11, pp. 1174- 1178, Nov. (1984)

[23] T. Gehrels, ed., "Planets, Stars and Nebulae Studied With Photopolari- metry", Tucson, Arizona: The University of Arizona Press, (1974) (Extensive Lists of Important References)

[24] D. Giuli, "Polarization Diversity in Radar", Proc. IEEE, Vol. 74(2), pp. 245-269, Feb. (1986)

[25] M.M. Gorshkov, Ellipsometry, Moscow: Sovetskoye Radio Press, (in Russian), (1974)

[26] C.D. Graves, "Radar Polarization Power Scattering Matrix", Proc. IRE, Vol. 44, Feb. (1956), pp. 248-252

[27] Gregory E. Heath, "Bistatic Scattering Reflection Asymmetry, Polariza- tion Reversal Asymmetry, and Polarization Reversal Reflection Symmetry", IEEE Trans. AP-29, No. 3, May (1987)

[28] W.A. Hiltner, Polarization Measurements, Actron. Technique, Chicago: Chicago U. Press, (1962)

[29] J.W. Hovenier, H.C. van de Hulst and C.V.M. van der Mee, "Conditions for the Elements of the Scattering Matrix", J. Astron. and Astrophysics, Vol. 157, pp. 301-310, (1986)

[30] J.R. Huynen, "Phenomenological Theory of Radar Targets", Ph.D. Dissertation, Technical University, Delft, The Netherlands, (1970)

[31] IEEE Standard Number 149-1979: Standard Test Procedures 1973, Revision of IEEE Stds. 145-1969, Definitions of Terms for Antennas, Published by the Institute of Electrical and Electronics Engineers, Inc., New York, (1979)

[32] G.A. Ioannidis and D.E. Hammers, "Optimum Antenna Polarization for Target Discrimination in Clutter", IEEE Trans. AP-27, May (1979), pp. 357-363

[33] A. Ishimaru, Wave Propagation and Scattering in Random Media, New York: Academic Press, Inc., (1978)

[34] R.C. Jones, "A New Calculus for the Treatment of Optical Systems", I. Description and Discussion, pp. 488-493, II. Proof of the Three General Equivalence Theorems, pp. 493-499, III. The Sohnke Theory of Optical Activity, pp. 500-503, J. Opt. Soc. Am., July 31 1941 (also see: W. Swindell, Polarized Light, Stroudsburg, PA: Halsted Press/John Wiley & Sons, 1975, pp. 186-240)

[35] D.B. Kanareykin, N.F. Pavlov and V.A. Potekhin, The Polarization of Radar Signals, Moscow: Sovyet Radio, Chap. 1-10 (in Russian), (1966); (English Translation of Chpts. 10-12: Radar Polarization Effects, New York: CCM Int. Corp., G.Collier and McMillan, (900 Third Ave, New York, N.Y. 10023)

[36] D.B. Kanareykin, V.A. Potekhin, and M.F. Shisikin, Maritime Polarimetry, Leningrad: Sudostroyenie, (1968)

[37] E.M. Kennaugh, "Polarization Properties of Radar Reflections", M.Sc. Thesis, Dept. of Elec. Engr., The Ohio State University, Columbus, OH, 43212, (1952); (also see: Proc. of the R&D Board Symposium on Radar Reflection Studies, Sept. 1950)

[38] G.P. Können, Polarized Light in Nature, English Translation, Cambridge, U.K.: Cambridge University Press, (1985)

[39] A.B. Kostinski and W-M. Boerner,"On Foundations of Radar Polarimetry", IEEE AP-34, No. 12, pp. 1395-1404, also see: comments by H. Mieras, pp. 1470-1471, and author's reply, pp. 1471-1473, Dec. (1986)

[40] A.B. Kostinski and W-M. Boerner, "On the Polarimetric Contrast Optimization", IEEE Trans. AP-35, No. 8, pp. 988-991, August (1987)

[41] A.B. Kostinski, B.D. James and W-M. Boerner, "On the Optimal Reception of Partially Polarized Waves", J. Optical Society of America, Part A, Optics & Image Sciences, Series 2, Vol. 5, No. 1, pp. 58-64, Jan. 1988

[42] A.B. Kostinski, B.D. James and W-M. Boerner, "Polarimetric Matched Filter for Coherent Imaging", Can J. Phys., Vol. 66, Issue 10, Special Issue on Coherent Imaging in Optics, pp. 871-877, Oct. (1988)

[43] A.L. Kozlov, "Radar Contrast of Two Objects", Izvestiya Vuz., Radioelektronika, Vol. 22, No. 7, July (1979), pp. 63-67

[44] J.D. Kraus, Electromagnetics, New York: McGraw-Hill Book Company, (1984)

[45] M.W. Long, Radar Reflectivity of Land and Sea, Lexington, MA: Lexington Books, D.C. Heath and Company, (1975)

[46] H. Mott, Polarization in Antennas & Radar, Englewood Cliffs, N.Y.: John Wiley & Sons, Inc., (1986)

[47] F. Pearson II, Map Projections: Theory and Applications, Boca Raton, Florida, CRC Press, Inc. (1990) (Also see C.H. Deetz and D.S. Adams, Elements of Map Projection, Washington: United States Government Printing Office, (1945)).

[48] A.J. Poelman and J.R.F. Guy, "Polarization Information Utilization in Primary Radar: An Introduction & Update to Activities at Shape Technical Center", Proc. NATO-ARW on IMEI, Bad Windsheim, FR. Germany, Sept. 18-24, (1983), Session RP.5, Section III: Paper No. III.2, pp. 521-572

[49] H. Poincaré, Théorie Mathématique de la Lumière, II-12, Paris: Georges Carré Publ. Co., (1892), pp. 282-285

[50] A.P. Rodimov and V.V Popovski, "Statistical Theory of Polarimetric Temporal Signal and Clutter Processing in Communication (propagation paths and lines)", Moscow, Vol. 21 in Series of Statistical Communications, Moscow: Radio & Comm., (1984) (in Russian)

[51] I.W. Root, Chairman, Workshop on "Polarimetric Radar Technology", June 25 -26, 1982, USA-MICOM-DRSMI-REG, Redstone Arsenal, Proc., Vol. 1, published by GACIAC IITRI, 10 W. 35th St., Chicago, IL 60615, GACIAC pr-81-02, Feb. (1981)

[52] L.W. Root and Matkin, Chairman/Editors, Proceedings, (Third) Polarimetric (Technology) Workshop, Redstone Arsenal, AL., 1988 August 16-18, GACIAC IIT-RI, 10 W. 35th St., Chicago, IL 60612, (1989)

[53] G.E. Schilov, Linear Algebra, New York: Dover, (1977)

[54] W.A. Shurcliff, Polarized Light, Cambridge, MA: Harvard Press, 1962)

[55] G. Sinclair, "The Transmission and Reception of Elliptically Polarized Waves", Proc. IRE, Vol. 38, Feb. 1950, pp. 148-151

[56] M.E. Skolnik, Radar Handbook, New York: McGraw-Hill, (1970)

[57] G. Strang, Linear Algebra and Its Applications, New York: Academic, (1976)

[58] J.A. Stratton, Electromagnetic Theory, New York: McGraw-Hill, (1941)

[59] W. Swindell, Polarized Light, Stroudsburg, PA: Halsted Press, (1975)

[60] F.T. Ulaby, R.K. Moore and A.K. Fung, Microwave Remote Sensing: Vols. 1-3, Reading, MA: Addison-Wesley, (1981)

[61] G. Wanielik, "Signal Description/Simulation/Processing with Polarimetric Radar System" (in German: Signal Beschreibung-Simulation und-Verarbeitung bei einem polarimetrischen Radar) Dr. Ing. Dissertation, Universität Karlsruhe, Karlsruhe, FRG, (Fortschritts- Berichte VDI, Reihe 10, Nr. 97, Düsseldorf: VDI-Verlag, 1988)

[62] A-Q. Xi and W-M. Boerner, "The Characteristic Radar Target Polarization State Theory for the Coherent Monostatic and Reciprocal Case Using the Generalized Polarization Transformation Ratio Formulation", to appear in: AEÜ, Vol. 44, No. 4, pp. 273-281, July/Aug. 1990

[63] Y. Yamaguchi, K. Sasagawa, M. Sengoku, T. Abe, W-M. Boerner, W-L. Yan and A-Q. Xi, "Characteristic Polarization States of Coherently Reflected Waves Based on the Stokes Vector Formulation", Japan Journal for Electronics and Communications Engineering (JECE), Vol._____, Sept. 1990, in print.

[64] Yan and Boerner

[65] L.A. Zhivotovskiy, "Optimum Polarization of Radar Signals", Radio Eng. and Electronic Phys., (1814), 1973, pp. 630-632

[66] J.J. van Zyl, "On the Importance of Polarization in Radar Scattering Problems", Ph.D. Dissertation, California Institute of Technology, Pasadena, CA, January (1986)

# AN ALTERNATIVE APPROACH TO FOUNDATIONS OF RADAR POLARIMETRY

Zbigniew H. Czyż

Telecommunications Research Institute, Antenna Department
Poligonowa 30
00-991 Warsaw
Poland

ABSTRACT. Transformation properties of asymmetric radar coherent scattering matrices are thoroughly examined in two-dimensional complex space of polarization and phase vectors and in four-dimensional real space of Stokes polarization four-vectors, simultaneously. Polarization and phase vectors, having Cayley-Klein rotation parameters as their components, are represented by Argand phasors tangent to the Poincare sphere in their polarization points. Two such unit vectors form an orthonormal basis which can be rotated by three Euler angles together with the rectangular coordinate system of three Stokes parameters of the Poincare sphere. Some "characteristic" coordinate system (CCS) has been found in which polarization transformation properties of an asymmetric radar scattering matrix can easily be described as inversion and rotation of the Poincare sphere. It is strongly believed that such geometrical models of scattering matrices can best describe their transformation properties also in the case of noncoherent scattering. As an example the decomposition of depolarizing symmetrical matrix, with two equal minimum radar crossections, into four nondepolarizing constituents has been presented.

## 1. INTRODUCTION

Papers on optimal polarizations for asymmetric scattering matrix were published in 1986 by Davidovitz and Boerner [1] and this author [2], independently. The main difference in approach to the problem was using in [2] the unitary transformation matrix dependent on three angular parameters: $\gamma$, $\delta$, and $\varepsilon$, and not two only: $\gamma$ and $\delta$ ($\varepsilon = -\delta$) as in [1].

As a consequence of applying three rotation parameters, the characteristic coordinate system, CCS, has been uniquely determined in which two null polarizations are placed in one of its coordinate planes, QV, symmetrically against one of coordinate axis, OQ, in its negative part. Very simple expressions have been also obtained for coordinates of all other special polarization points.

It has been shown in [2] and [3] that the whole 16-element power scattering matrix can be reconstructed, sometimes in two solutions,



*W.-M. Boerner et al. (eds.), Direct and Inverse Methods in Radar Polarimetry, Part 1, 247–266.*

when the coordinates of the inversion point inside the Poincare sphere of unit radius are known in CCS. Owing to the proposed definition of CCS it was easier to find the allowed regions in which the inversion point can be situated and, considering the boundaries of these regions, to classify the asymmetric scattering matrices according to their inherent polarization properties, independent of the CCS three Euler angles.

Next consequence of introducing the three parameter rotation matrix was joining in one complex vector both polarization and phase of an antenna or a wave and representing the polarization and phase (PP) vector as Argand phasor tangent to the Poincare sphere in the polarization point.

But still the problem remained unsolved, how to express the voltage received by an antenna from a wave or how to present the transmission equation between two antennas. Of course, Hermitian product would be desirable. That is why the new concept had to be proposed, to distinguish between complex amplitudes of waves, or antenna heights, and the PP vectors and to introduce the reversal transformation of PP vectors for antennas and waves in reversed spatial coordinate system, i.e. rotated by 180° about some, say, vertical axis.

Of course, the equation for transmission between two antennas will still have the form of ordinary product of two complex antenna heights. But it will be also Hermitian product of two PP vectors with conjugate factor reversed against the coordinate system.

What should be emphasized as additional advantage of such procedure is that both the change-of-basis and reversal transformations of PP vectors do not depend on the direction in which an antenna or a wave is oriented against the spatial coordinate system.

The important feature of methods applied by this author in investigation of polarization problems is the use of amplitude and power equations simultaneously with those of power type preferred. Most of problems can be better solved by using power formulation though some require an amplitude approach.

The radar power scattering matrix introduced is different from Mueller matrix which is usually applied in the literature but more suitable for one-way propagation investigations.

The most important feature of all transmission equations used here is their transposability for opposite direction of propagation. It can be accomplished when the form of an antenna polarization vector does not depend on whether the antenna has been used for transmission or for reception.

## 2. COHERENT SCATTERING

### 2.1. Polarization and Phase Vectors

Define the PP unit column vector $[u]$ by its components a and b, the Cayley-Klein rotation parameters, in the rectangular right-handed Oxyz coordinate system as follows:

$$[u] = \begin{bmatrix} u_y \\ u_x \end{bmatrix} = \begin{bmatrix} a \\ b \end{bmatrix}; \qquad \begin{aligned} a &= \cos\gamma \; e^{-j(\delta+\varepsilon)} \\ b &= \sin\gamma \; e^{j(\delta-\varepsilon)} \end{aligned} \tag{1}$$

Omitting square brackets for single symbol matrices, the two complex electric field column vectors of two plane TEM waves of unit magnitude and the same polarization and spatial phase for the time $t = 0$, propagating in opposite directions, can be presented as follows:

$$E^{+}(t,z) = u \; e^{j(\omega t-kz)}, \qquad E^{-}(t,z) = u^{*} \; e^{j(\omega t+kz)} \tag{2}$$

where the asterisk denotes complex conjugation.

These waves can be interpreted geometrically as two elliptic helices, $\mathrm{Re}\, E^{+}(t,z)$ and $\mathrm{Re}\, E^{-}(t,z)$, moving in opposite directions, but coinciding in time $t = 0$:

$$\mathrm{Re}\, E^{+}(0,z) = \mathrm{Re}\,(u\, e^{-jkz}) = \mathrm{Re}\,(u^{*}\, e^{jkz}) = \mathrm{Re}\, E^{-}(0,z) \tag{3}$$

It is clearly seen from the above that $\gamma$ and $\delta$ parameters describe polarization of both waves, while $\varepsilon$ parameter denotes their common spatial phase for $t = 0$.

Similarly, the unit complex heights of antennas which radiate the waves (2) will take the form of column vectors:

$$h^{+} = u, \qquad \text{and} \qquad h^{-} = u^{*}. \tag{4}$$

As seen from expressions (2), one should distinguish between PP vectors and complex amplitudes of waves or complex antenna heights. For instance, introducing the radar amplitude scattering matrix $A$ in the coordinate systems as in Figure 1, the complex amplitude of a scattered wave can be presented, for $t = z_1 = z_2 = 0$, as

$$E^{S-} = \sqrt{\sigma^{T}} \; u^{S*} = A\, u^{T}; \qquad \sigma^{T} \text{ - real, positive} \tag{5}$$

where $u^{T}$ and $u^{S*}$ are unit amplitudes of transmitted and scattered waves, respectively, whereas the same $u^{T}$, and $u^{S}$ without asterisk, denote the PP unit vectors of those waves.

Thus, the physical meaning of the eigenvalue problem, which can be formulated as

$$\sqrt{\sigma} \; u^{*} = A\, u \; ; \qquad \sigma \text{ - real, positive} \tag{6}$$

becomes entirely clear.

In turn, a one to one correspondence is postulated between the unit PP vector, $u^{P}$, and the unit Argand phasor, of the form $\exp(-j2\varepsilon^{P})$, in the complex plane tangent to the Poincaré polarization sphere in point P with angular coordinates $2\delta$ and $2\gamma$ as shown in Figure 2.

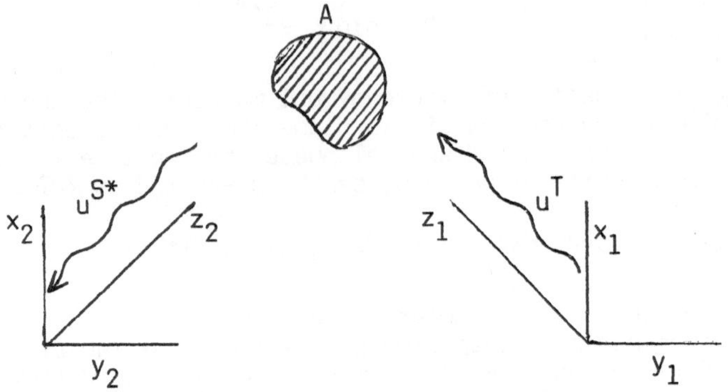


Figure 1. Unit complex amplitudes of illuminating and scattered waves in their coordinate systems

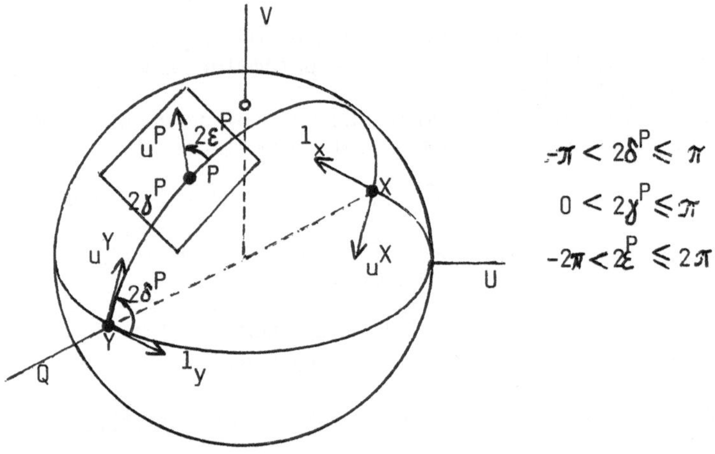


Figure 2. Poincaré sphere representation of the $(1_y, 1_x)$ basis and PP unit vector, $u^P$

No separate symbol for "phasor in point P" but $u^P$ for PP vector will be introduced.

Observe that two PP Argand phasors which differ in $2\varepsilon$ by $2\pi$ denote PP vectors of opposite phases. Thus, these Argand phasors, though coincide, should also be regarded of opposite sign. So, special rules for addition and multiplication of phasors on the Poincaré sphere, which can be derived from corresponding rules for PP vectors, should be presented.

At first, the Hermitian product of two PP vectors will be considered. The definition (1) yields

$$\widetilde{u}^A u^{B*} = e^{-j(\mathcal{E}^A - \mathcal{E}^B)}\left[\cos(\gamma^A - \gamma^B)\cos(\delta^A - \delta^B) - j\cos(\gamma^A + \gamma^B)\sin(\delta^A - \delta^B)\right]$$

$$= \cos\frac{AB}{2} e^{-j2\Delta^{AB}} \tag{7}$$

where

$$\cos^2\frac{AB}{2} = \frac{1}{2}\left[1 + \cos 2\gamma^A \cos 2\gamma^B + \sin 2\gamma^A \sin 2\gamma^B \cos(2\delta^A - 2\delta^B)\right] \tag{8}$$

and

$$2\Delta^{AB} = \mathcal{E}^A - \mathcal{E}^B + \tan^{-1}\frac{\cos(\gamma^A + \gamma^B)}{\cos(\gamma^A - \gamma^B)}\tan(2\delta^A - 2\delta^B)$$

$$= \mathcal{E}^A - \mathcal{E}^B + \delta^A - \delta^B - \frac{1}{2}E$$

$$= \mathcal{E}^A - \mathcal{E}^B + \frac{1}{4}(E' - E) \tag{9}$$

The phase of the product, $2\Delta^{AB}$, can be represented on the Poincaré sphere as one half of an angle $4\Delta^{AB}$ between the two phasors, $u^A$ and $u^B$, shifted parallel to one point along the great circle arc joining polarization points A and B as shown in Figure 3.

$$E = \hat{A} + \hat{B} + 2\delta^A - 2\delta^B - \pi$$

$$E' = 2\delta^A - 2\delta^B - E$$

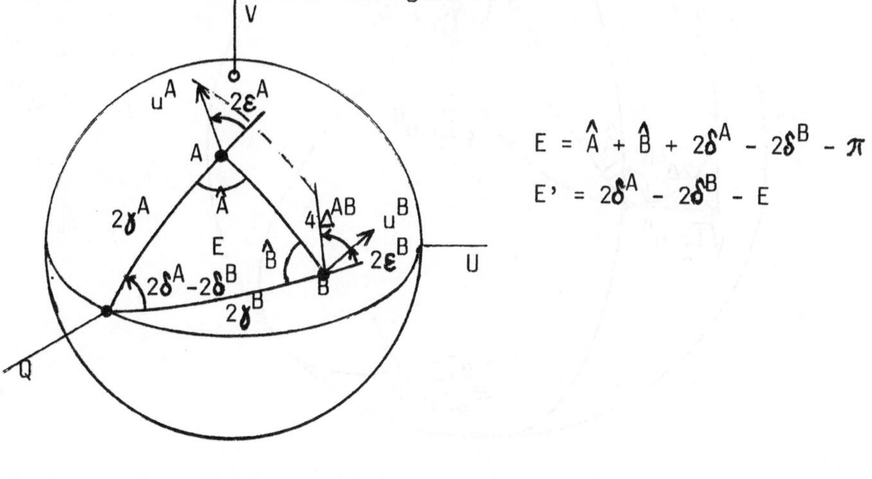

Figure 3. To the Poincaré sphere interpretation of the two PP vectors Hermitian product

The sum of the two PP vectors can be presented as follows

$$\sqrt{I}\, u^C = \sqrt{I_1}\, u^A + \sqrt{I_2}\, u^B = (\sqrt{I_1}\, \widetilde{u^A}\, u^{C*} + \sqrt{I_2}\, \widetilde{u^B}\, u^{C*})\, u^C$$

$$= \sqrt{I_1} \cos \frac{AC}{2}\, u^C\, e^{-j\alpha} + \sqrt{I_2} \cos \frac{BC}{2}\, u^C\, e^{j\beta} \tag{10}$$

where:

$$\alpha = 2\,\Delta^{AC}, \qquad \beta = -2\,\Delta^{BC}$$

and (see Figure 4)

$$\alpha + \beta = 2\,\Delta^{AC} - 2\,\Delta^{BC} = 2\,\Delta^{AB} - \frac{1}{2} E_{ABC} = \pi - \hat{C} \tag{11}$$

and where $I$, $I_1$, and $I_2$ can be interpreted as intensities of the sum wave and its constituents.

Note that the sum of the $u^{C'}$ components of $\sqrt{I_1}\, u^A$ and $\sqrt{I_2}\, u^B$ vectors in point C', opposite to C, shall vanish. This yields

$$\sqrt{I_1} \sin \frac{AC}{2} = \sqrt{I_2} \sin \frac{BC}{2} \tag{12}$$

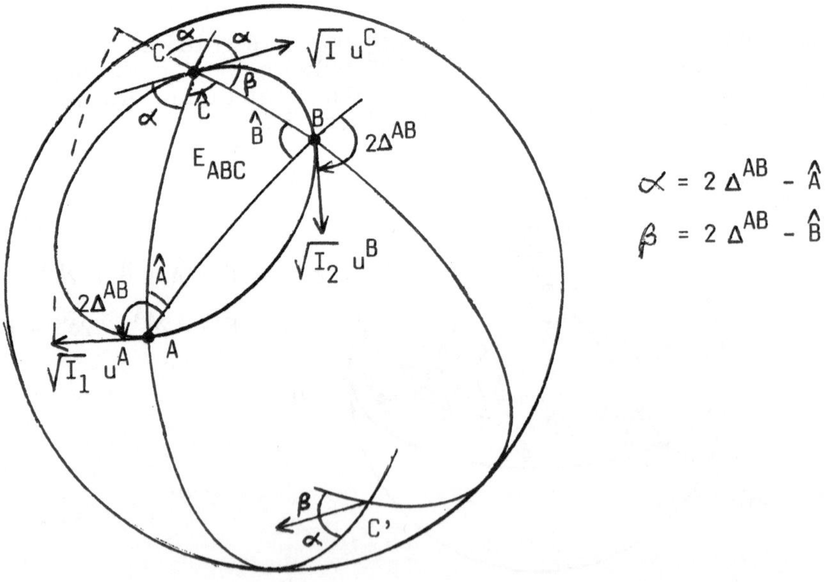


Figure 4. The sum of two PP vectors on the Poincaré sphere

The PP unit vector $u^C$ of the sum (10) can be represented on the Poincaré sphere as an Argand phasor tangent to the small circle ABC which in turn is tangent to such $u^A$ and $u^B$ phasors which form equal in magnitude but opposite in sign $2\Delta^{AB}$ angles with the AB great circle arc of the sphere. If $2\Delta^{AB}$ is positive then $u^A$ is retarded in phase with respect to $u^B$ and point C is on the left side of AB arc, as is the case in Figure 4.

Furthermore, it follows from (10), (11), and (12) that

$$I = I_1 \cos^2 \frac{AC}{2} + I_2 \cos^2 \frac{BC}{2} + 2\sqrt{I_1 I_2} \cos \frac{AC}{2} \cos \frac{BC}{2} \cos (\alpha + \beta)$$

$$0 = I_1 \sin^2 \frac{AC}{2} + I_2 \sin^2 \frac{BC}{2} - 2\sqrt{I_1 I_2} \sin \frac{AC}{2} \sin \frac{BC}{2}$$

and, after addition and using equations of spherical geometry,

$$I = I_1 + I_2 + 2\sqrt{I_1 I_2} \cos \frac{AB}{2} \cos 2\Delta^{AB} \tag{13}$$

Obtaining this well known result from equation (10) additionally confirmes correctness of Argand phasors representation of PP vectors.

Now, the definition (1) of PP vector can be considered as a special case of the above summation rule. Instead of $u^A$ and $u^B$ phasors of Figure 4, you have to take (see Figure 2):

$$u^Y = \begin{bmatrix} 1 \\ 0 \end{bmatrix} e^{-j\delta^P}, \qquad \text{and} \qquad u^X = \begin{bmatrix} 0 \\ 1 \end{bmatrix} e^{j\delta^P}$$

with:

$$2\Delta^{AB} = 0, \qquad \alpha = 0, \qquad \beta = 0$$

$$\sqrt{I_1} = \cos \frac{AC}{2} = \cos \gamma^P, \qquad \sqrt{I_2} = \cos \frac{BC}{2} = \sin \gamma^P$$

And finally, the $2\varepsilon^P$ angle can be taken arbitrarily. That is why any PP vector can be presented as a sum of two orthonormal versors, $1_y$ and $1_x$, with complex coefficients in the form of Cayley-Klein parameters.

Also, any other orthonormal basis of complex versors represented by Argand phasors parallel to any great circle of the polarization sphere can be obtained by rotation the pair $1_y$, $1_x$ by three Euler angles $2\delta$, $2\gamma$, and $2\varepsilon$.

## 2.2. Amplitude and Power Transmission Equations

Using Kronecker multiplication rules for matrices:

$$u \otimes u^* = \begin{bmatrix} a \\ b \end{bmatrix} \otimes \begin{bmatrix} a^* \\ b^* \end{bmatrix} = \begin{bmatrix} aa^* \\ ab^* \\ ba^* \\ bb^* \end{bmatrix} \tag{14}$$

$$A \otimes A^* = \begin{bmatrix} A_2 & A_3 \\ A_4 & A_1 \end{bmatrix} \otimes \begin{bmatrix} A_2^* & A_3^* \\ A_4^* & A_1^* \end{bmatrix} = \begin{bmatrix} A_2A_2^* & A_2A_3^* & A_3A_2^* & A_3A_3^* \\ A_2A_4^* & A_2A_1^* & A_3A_4^* & A_3A_1^* \\ A_4A_2^* & A_4A_3^* & A_1A_2^* & A_1A_3^* \\ A_4A_3^* & A_4A_1^* & A_1A_4^* & A_1A_1^* \end{bmatrix} \tag{15}$$

$$(A\ B) \otimes (C\ D) = (A \otimes C)\ (B \otimes D) \tag{16}$$

$$\widetilde{(A \otimes B)\ C} = \tilde{C}\ (\tilde{A} \otimes \tilde{B}) \tag{17}$$

and unitary 4 x 4 matrix

$$U = \frac{1}{\sqrt{2}} \begin{bmatrix} 1 & 1 & 0 & 0 \\ 0 & 0 & 1 & -j \\ 0 & 0 & 1 & j \\ 1 & -1 & 0 & 0 \end{bmatrix} \quad ; \qquad U\ \tilde{U}^* = \begin{bmatrix} 1 & 0 & 0 & 0 \\ 0 & 1 & 0 & 0 \\ 0 & 0 & 1 & 0 \\ 0 & 0 & 0 & 1 \end{bmatrix} \tag{18}$$

one can define the Stokes polarization unit column four-vector

$$P = \tilde{U}^*\ (u \otimes u^*) = \frac{1}{2} \begin{bmatrix} aa^* + bb^* \\ aa^* - bb^* \\ ab^* + ba^* \\ j(ab^* - ba^*) \end{bmatrix} = \frac{1}{\sqrt{2}} \begin{bmatrix} 1 \\ \cos 2\gamma \\ \sin 2\gamma \cos 2\delta \\ \sin 2\gamma \sin 2\delta \end{bmatrix} = \frac{1}{\sqrt{2}} \begin{bmatrix} 1 \\ q \\ u \\ v \end{bmatrix} \tag{19}$$

with normalized Stokes parameters: q, u, and v as its components, and radar power scattering matrix

$$F = \tilde{U}\ (A \otimes A^*)\ U \tag{20}$$

The following transformation rules will also be defined.
For u nad P vectors:
- reversal transformation,

$$u^{o*} = C^o\, u \; ; \quad C^o = \begin{bmatrix} -1 & 0 \\ 0 & 1 \end{bmatrix}, \quad C^o\, C^o = \begin{bmatrix} 1 & 0 \\ 0 & 1 \end{bmatrix} \tag{21}$$

$$P^o = D^o\, P \; ; \quad D^o = \tilde{U}\,(C^o \otimes C^o)\, U$$

$$= \begin{bmatrix} 1 & & & \\ & 1 & & \\ & & -1 & \\ & & & 1 \end{bmatrix}, \quad D^o\, D^o = \begin{bmatrix} 1 & & & \\ & 1 & & \\ & & 1 & \\ & & & 1 \end{bmatrix} \tag{22}$$

- change-of-basis transformation,

$$u^{P'} = \tilde{C}^*\, u^P \; ; \quad C = \begin{bmatrix} a & -b^* \\ b & a^* \end{bmatrix}, \quad C\, \widetilde{C^*} = \begin{bmatrix} 1 & 0 \\ 0 & 1 \end{bmatrix} \tag{23}$$

$$P^{P'} = \tilde{D}\;\; P^P \; ; \quad D = \tilde{U}^*\,(C \otimes C^*)\, U, \quad D^{-1} = \tilde{D} \tag{24}$$

- both transformations,

$$u^{Po,*} = \tilde{C}\, u^{Po} = \tilde{C}\, C^o\, C\, u^{P'} = u^{P',o*} \tag{25}$$

$$P^{Po'} = \tilde{D}\, P^{Po} = \tilde{D}\, D^o\, D\, P^{P'} = P^{P',o} \tag{26}$$

For A and F matrices:
- change from radar to propagation matrix,

$$A^o = C^o\, A \tag{27}$$

$$F^o = D^o\, F = \tilde{U}^*\,(A^o \otimes A^{o*})\, U \tag{28}$$

- change-of-basis transformation,

$$A' = \tilde{C}\, A\, C \tag{29}$$

$$F' = \tilde{D}\, F\, D \tag{30}$$

- both transformations,

$$A^{o'} = \tilde{C}^*\, A^o\, C = \tilde{C}^*\, C^o\;\; C^*\, A' = A'^o \tag{31}$$

$$F^{o'} = \tilde{D}\;\; F^o\, D = \tilde{D}\;\; D^o\;\; D\;\; F' = F'^o \tag{32}$$

Rotation matrices, C and D, can be separated into factors depending on single angular parameters:

$$C = \begin{bmatrix} e^{-j\delta} & 0 \\ 0 & e^{j\delta} \end{bmatrix} \begin{bmatrix} \cos\gamma & -\sin\gamma \\ \sin\gamma & \cos\gamma \end{bmatrix} \begin{bmatrix} e^{-j\varepsilon} & 0 \\ 0 & e^{j\varepsilon} \end{bmatrix} \tag{33}$$

$$D = \begin{bmatrix} 1 & 0 & & \\ 0 & 1 & & \\ 0 & 0 & \cos 2\delta & -\sin 2\delta \\ 0 & 0 & \sin 2\delta & \cos 2\delta \end{bmatrix} \begin{bmatrix} 1 & 0 & 0 & 0 \\ 0 & \cos 2\gamma & -\sin 2\gamma & 0 \\ 0 & \sin 2\gamma & \cos 2\gamma & 0 \\ 0 & 0 & 0 & 1 \end{bmatrix} \begin{bmatrix} 1 & 0 & 0 & 0 \\ 0 & 1 & 0 & 0 \\ 0 & 0 & \cos 2\varepsilon & -\sin 2\varepsilon \\ 0 & 0 & \sin 2\varepsilon & \cos 2\varepsilon \end{bmatrix} \tag{34}$$

Derivation of transformation rules listed can be performed very easily when starting from amplitude transmission equations and then applying Kronecker multiplication. As an example, the transmission between two antennas will be considered (see Figure 5).

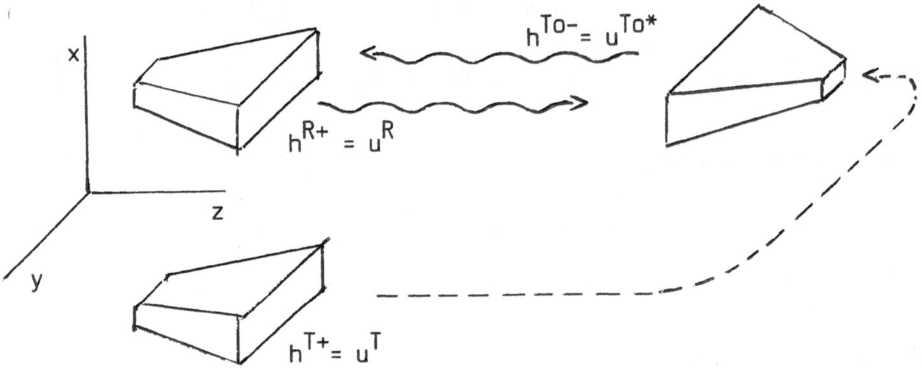


Figure 5. Unit complex hights of two antennas in their transmission coordinate system

The received voltage, $V_r$, and received power, $P_r$, can be given as:

$$V_r = \tilde{h}^{R+} h^{To-} = \tilde{u}^{R} u^{To} = \tilde{u}^{R} C^{o} u^{T}$$

$$= \tilde{u}^{Ro} u^{T} = \tilde{h}^{Ro-} h^{T+} \tag{35}$$

and

$$P_r = V_r V_r^* = V_r \otimes V_r^* = (\tilde{u}^{R} \otimes \tilde{u}^{R*})\, U^* \tilde{U}\, (u^{To*} \otimes u^{To}) = P^{R} P^{To}$$

$$= (\tilde{u}^{R} \otimes \tilde{u}^{R*})\, U^* \tilde{U}\, (C^{o} \otimes C^{o})\, U \tilde{U^*}\, (u^{T} \otimes u^{T*}) = \tilde{P}^{R} D^{o} P^{T} \tag{36}$$

The Hermitian product in $V_r$ preserves its form not only under reversal of coordinate system, as in (35), but also under change-of-basis transformation:

$$V_r = \widetilde{u^R}\, u^{To*} = \widetilde{u^R}\, C^* \tilde{C}\, u^{To*} = \widetilde{u^{R,}}\, u^{To,*} \tag{37}$$

The power equivalent of the middle of above expressions is

$$P_r = (\widetilde{u^R \otimes u^{R*}})\, U^* \tilde{U}\, (C^* \otimes C)\, U^* \tilde{U}\, (\tilde{C} \otimes \widetilde{C^*})\, U^* \tilde{U}\, (u^{To*} \otimes u^{To})$$

$$= \widetilde{P^R}\, D\, \tilde{D}\, P^{To} = \widetilde{P^{R,}}\, P^{To,} \tag{38}$$

Similarly, for two-way transmission the following equations hold:

$$V_r = \widetilde{u^R}\, A\, u^T = \widetilde{u^{Ro*}}\, A^o\, u^T = \widetilde{u^{R,}}\, A^,\, u^{T,} = \widetilde{u^{Ro,*}}\, A^{o,}\, u^{T,} \tag{39}$$

$$P_r = \widetilde{P^R}\, F\, P^T = \widetilde{P^{Ro}}\, F^o\, P^T = \widetilde{P^{R,}}\, F^,\, P^{T,} = \widetilde{P^{Ro,}}\, F^{o,}\, P^{T,} \tag{40}$$

Special attention should be given to the case of scattering through the intervenient mediums. The transmission equation proposed for bistatic radar system is

$$V_r = \widetilde{A_2^o\, u^R}\; A\; A_1^o\; u^T = \widetilde{u^R}\, \widetilde{A_2^o}\, A\; A_1^o\, u^T$$

$$= \widetilde{u^R}\, \widetilde{A_2}\, C^o\, A\, C^o\, A_1\, u^T \tag{41}$$

with its power equivalent

$$P_r = \widetilde{P^R}\, \widetilde{F_2^o}\, F\;\; F_1^o\, P^T \tag{42}$$

Observe that in monostatic radar system case in which $\tilde{A} = A$ and $A_2^o = A_1^o$, and for $u^R = u^T$, the $A_1^o$ matrix should not necessarily be symmetric.

## 2.3. Radar Coherent Scattering Matrices

Introducing for brevity:

$$M_k = A_k\, A_k^*, \qquad S_{ki} = S_{ik} = \frac{1}{2}\,(A_i\, A_k^* + A_k\, A_i^*)$$

and

$$-D_{ki} = D_{ik} = \frac{j}{2}\,(A_i\, A_k^* - A_k\, A_i^*) \tag{43}$$

mutual dependences will be given between elements of amplitude and power radar scattering matrices in their general form

$$A = \begin{bmatrix} A_2 & A_3 \\ A_4 & A_1 \end{bmatrix}, \qquad F = \begin{bmatrix} a_1 & b_1 & b_3 & b_5 \\ c_1 & a_2 & b_4 & b_6 \\ c_3 & c_4 & a_3 & b_2 \\ c_5 & c_6 & c_2 & a_4 \end{bmatrix} \tag{44}$$

From (20) one gets:

$$a_1 = \frac{1}{2}(M_2 + M_3 + M_4 + M_1)$$

$$a_2 = \frac{1}{2}(M_2 - M_3 - M_4 + M_1)$$

$$a_3 = S_{34} + S_{12}$$

$$a_4 = S_{34} - S_{12}$$

$$b_1 = \frac{1}{2}(M_2 - M_3 + M_4 - M_1) \qquad c_1 = \frac{1}{2}(M_2 + M_3 - M_4 - M_1)$$

$$b_2 = D_{12} + D_{34} \qquad c_2 = D_{12} - D_{34}$$

$$b_3 = S_{32} + S_{14} \qquad c_3 = S_{42} + S_{13}$$

$$b_4 = S_{32} - S_{14} \qquad c_4 = S_{42} - S_{13}$$

$$b_5 = D_{32} + D_{14} \qquad c_5 = D_{42} + D_{13}$$

$$b_6 = D_{32} - D_{14} \qquad c_6 = D_{42} - D_{13}. \tag{45}$$

and vice wersa:

$$A_2 = \sqrt{\frac{1}{2}(a_1 + a_2 + b_1 + c_1)}, \qquad A_3 = \left[b_3 + b_4 - j(b_5 + b_6)\right]/(2A_2)$$

$$A_4 = \left[ c_3 + c_4 - j\,(c_5 + c_6)\right]/(2A_2), \quad A_1 = \left[ a_3 - a_4 - j\,(b_2 + c_2)\right]/(2A_2) \qquad (46)$$

Only 7 of 16 elements (45) are independent for nondepolarizing matrix F. Possibly complete set of mutual dependences between F matrix elements, called conditions for preserving the complete polarization (PCP), important for practical applications is presented below:

$$a_1a_2 + a_3a_4 = b_1c_1 + b_2c_2$$

$$a_1a_3 + a_2a_4 = b_3c_3 + b_6c_6$$

$$a_1a_4 + a_2a_3 = b_5c_5 + b_4c_4$$

$$a_1b_1 - a_2c_1 = c_3c_4 + c_5c_6, \qquad a_1c_1 - a_2b_1 = b_3b_4 + b_5b_6$$

$$a_1b_2 - a_2c_2 = c_3b_5 - b_4c_6, \qquad a_1c_2 - a_2b_2 = b_3c_5 - c_4b_6$$

$$a_1b_3 - a_3c_3 = c_1b_4 + c_2c_5, \qquad a_1c_3 - a_3b_3 = b_1c_4 + b_2b_5$$

$$a_1b_4 - a_4c_4 = c_1b_3 - b_2c_6, \qquad a_1c_4 - a_4b_4 = b_1c_3 - c_2b_6$$

$$a_1b_5 - a_4c_5 = c_1b_6 + b_2c_3, \qquad a_1c_5 - a_4b_5 = b_1c_6 + c_2b_3$$

$$a_1b_6 - a_3c_6 = c_1b_5 - c_2c_4, \qquad a_1c_6 - a_3b_6 = b_1c_5 - b_2b_4$$

$$a_2b_3 + a_4c_3 = b_1b_4 + b_2c_5, \qquad a_2c_3 + a_4b_3 = c_1c_4 + c_2b_5$$

$$a_2b_4 + a_3c_4 = b_1b_3 - c_2c_6, \qquad a_2c_4 + a_3b_4 = c_1c_3 - b_2b_6$$

$$a_2b_5 + a_3c_5 = b_1b_6 + c_2c_3, \qquad a_2c_5 + a_3b_5 = c_1c_6 + b_2b_3$$

$$a_2b_6 + a_4c_6 = b_1b_5 - b_2c_4, \qquad a_2c_6 + a_4b_6 = c_1c_5 - c_2b_4$$

$$a_3b_1 + a_4c_1 = b_3c_4 + c_5b_6, \qquad a_3c_1 + a_4b_1 = c_3b_4 + b_5c_6$$

$$a_3b_2 + a_4c_2 = b_3b_5 - b_4b_6, \qquad a_3c_2 + a_4b_2 = c_3c_5 - c_4c_6$$

$$a_1^2 - b_1^2 - b_3^2 - b_5^2 = a_1^2 - c_1^2 - c_3^2 - c_5^2$$

$$= -c_1^2 + a_2^2 + b_4^2 + b_6^2 = -b_1^2 + a_2^2 + c_4^2 + c_6^2$$

$$= -c_3^2 + c_4^2 + a_3^2 + b_2^2 = -b_3^2 + b_4^2 + a_3^2 + c_2^2$$

$$= -c_5^2 + c_6^2 + c_2^2 + a_4^2 = -b_5^2 + b_6^2 + b_2^2 + a_4^2$$

$$b_1^2 + b_3^2 + b_5^2 = c_1^2 + c_3^2 + c_5^2$$

$$b_1^2 + b_4^2 + b_6^2 = c_1^2 + c_4^2 + c_6^2$$

$$b_2^2 + b_3^2 - b_4^2 = c_2^2 + c_3^2 - c_4^2$$

$$b_2^2 - b_5^2 + b_6^2 = c_2^2 - c_5^2 + c_6^2. \tag{47}$$

The matrices A and F have their most simple form in the characteristic coordinate system corresponding to the $1_y$, $1_x$ basis rotated by three Euler angles:

$$2\delta^K = \arg \rho^K,$$

$$2\gamma^K = 2 \tan^{-1}|\rho^K|$$

$$2\varepsilon^K = \frac{1}{2} (\arg A_2' - \arg A_1') - 2\delta^K$$

with

$$\rho^K = (R_1 - \sqrt{R_1^2 + |R_2|^2})/R_2, \qquad R_1 = A_2A_2^* - A_1A_1^*, \tag{48}$$

$$R_2 = - A_1(A_3^* + A_4^*) - A_2^*(A_3 + A_4)$$

According to transformation (29) one gets

$$\begin{bmatrix} A_2 & A_3 \\ A_4 & A_1 \end{bmatrix} \longrightarrow \begin{bmatrix} A_2^K & B_1^K + j\, B_2^K \\ -B_1^K - j\, B_2^K & A_1^K \end{bmatrix} e^{+j\mu^K} \tag{49}$$

with

$$A_2^K = |A_2'|, \qquad A_1^K = |A_1'|; \qquad A_2^K \geqslant A_1^K$$

$$B_1^K + j\, B_2^K = A_3'\, e^{j\mu^K}$$

where

$$A_2' = A_2 + (A_3 + A_4)\, \rho^K + A_1\, \rho^{K2}/N$$

$$A_1' = A_2\rho^{K*2} - (A_3 + A_4)\rho^{K*} + A_1 /N$$

$$A_3' = -A_2\rho^{K*} + A_3 - A_4\rho^{K}\rho^{K*} + A_1\rho^{K} /N$$

$$N = 1 + \rho^{K}\rho^{K*}$$

and where the general phase of A matrix is

$$\mu^{K} = \frac{1}{2}(\arg A_2' + \arg A_1')$$

In their CCS,A and F matrices have the form:

$$A = \begin{bmatrix} A_2 & B_1+jB_2 \\ -B_1-jB_2 & \end{bmatrix} ; \quad \begin{matrix} A_2 \geqslant A_1 & \text{- real,} \\ & \text{positive} \\ B_1, B_2 & \text{- real,} \end{matrix} \quad F = \begin{bmatrix} a_1 & b_1 & b_3 & b_5 \\ b_1 & a_2 & b_4 & b_6 \\ -b_3 & -b_4 & a_3 & 0 \\ -b_5 & -b_6 & 0 & a_4 \end{bmatrix} \qquad (50)$$

with:

$$\begin{aligned} a_1 &= \frac{1}{2}(A_2^2 + A_1^2) + B_1^2 + B_2^2, & b_3 &= B_1 (A_2 - A_1) \\ a_2 &= \frac{1}{2}(A_2^2 + A_1^2) - B_1^2 - B_2^2, & b_4 &= B_1 (A_2 + A_2) \\ a_3 &= A_1A_2 - B_1^2 - B_2^2, & b_5 &= -B_2 (A_2 + A_1) \\ a_4 &= -A_1A_2 - B_1^2 - B_2^2 \leqslant 0, & b_6 &= -B_2 (A_2 - A_1) \\ b_1 &= \frac{1}{2}(A_2^2 - A_1^2) \geqslant 0, \end{aligned} \qquad (51)$$

and with

$$a_2 = a_1 + a_3 + a_4 \qquad (52)$$

Not only the form (50) of the matrix is simple but, as it has been shown in [2] and [3], it is also possible in CCS to reconstruct the whole F matrix, though sometimes in two solutions, from the elements of its first row only.

## 2.4. Geometrical Model of Coherent Scattering Matrix

As the geometrical model of the coherent scattering matrix A or F the Poincaré sphere of unit radius will be considered in CCS with the so called "inversion" point, I, inside the sphere and some special polarization points on it. T point will represent the transmit polarization.

It follows from the scattering power equation

$$\sigma^T P^S = F P^T \tag{53}$$

that the scattered power is proportional to the square of chord (I T):

$$\sigma^T = (I\ T)^2\ \sigma_0/4 \tag{54}$$

Here $\sigma_0$, the so called "effective crossection", is given in any coordinate system by

$$\sigma_0 = 2\,(a_1 + \sqrt{a_1^2 - b_0^2}); \qquad b_0^2 = b_1^2 + b_3^2 + b_5^2 \tag{55}$$

while the coordinates of I point are: $Q^I = -2b_1/\sigma_0$, $U^I = -2b_3/\sigma_0$, and $V^I = -2b_5/\sigma_0$. The received power, $P_r$, for the same polarization of transmit and receive antenna can also be interpreted on the Poincaré sphere as

$$P_T^T = \widetilde{P^T}\ F\ P^T = (O_1T)^2\ (O_2T)^2\ (A_1 + A_2)^2/16 \tag{56}$$

where $O_1$ and $O_2$ are Kennaugh's "null" polarizations, and $A_1$ and $A_2$ are elements of A matrix in the characteristic polarizations basis.

The transformation of transmit polarization, T, to that of scattered wave, T", can be shown geometrically in two steps. Firstly, as inversion $T \rightarrow T'$ through the point I, and secondly, as rotation of the Poincaré sphere by some angle $2\gamma^P$ about some AB diameter which is situated in the UV plane of the OQUV CCS (see Figure 6).

The angle and axis of rotation after inversion can be given by coordinates in the CCS of some point P which, after the rotation, takes place of the characteristic polarization K situated on the OQ axis. These coordinates are:

$$\begin{bmatrix} q^P \\ u^P \\ v^P \end{bmatrix} = \begin{bmatrix} \cos 2\ ^P \\ \sin 2\ ^P \cos 2\ ^P \\ \sin 2\ ^P \sin 2\ ^P \end{bmatrix} = \frac{1}{a_0(a_1 + a_0)} \begin{bmatrix} a_3a_4 - a_0a_2 \\ b_4\ (a_4 - a_0) \\ b_6\ (a_3 - a_0) \end{bmatrix} \tag{57}$$

with

$$a_0^2 = a_1^2 - b_0^2$$

The coordinates in the CCS of some special polarization points such as: polarizations of maximum and minimum scattered power, M and N, polarization of scattered waves when illuminating ones are M and N, M" and N", Kennaugh's null polarizations, $O_{1,2}$, the eigenpolarizations, $E_{1,2}$, and the characteristic polarizations, K and L, are given in Table 1.

TABLE I

| | M | N | M" | N" | $O_{1,2}$ | $E_{1,2}$ | K,L |
|---|---|---|---|---|---|---|---|
| q | $\frac{b_1}{b_o}$ | $\frac{-b_1}{b_o}$ | $\frac{b_1}{b_o}$ | $\frac{-b_1}{b_o}$ | $\frac{A_1 - A_2}{A_1 + A_2}$ | $\pm\frac{\sqrt{b_1^2-b_4^2-b_6^2}}{b_1}$ | $\pm 1$ |
| u | $\frac{b_3}{b_o}$ | $\frac{-b_3}{b_o}$ | $\frac{-b_3}{b_o}$ | $\frac{b_3}{b_o}$ | 0 | $\frac{-b_4}{b_1}$ | 0 |
| v | $\frac{b_5}{b_o}$ | $\frac{-b_5}{b_o}$ | $\frac{-b_5}{b_o}$ | $\frac{b_5}{b_o}$ | $\frac{\mp 2\sqrt{A_1 A_2}}{A_1 + A_2}$ | $\frac{-b_5}{b_1}$ | 0 |

The scattered powers for some of these polarizations are:

$$\sigma^{M,N} = a_1 \pm b_o, \quad \sigma^{O_{1,2}} = B_1^2 + (B_2 \pm \sqrt{A_1 A_2})^2, \quad \sigma^{E_{1,2}} = a_2 \mp \sqrt{b_1^2 - b_4^2 - b_6^2} \tag{58}$$

The inversion point I is always situated on the MN diameter. But not all places inside the Poincaré sphere are permitted for the I point. It cannot be located outside of the "small sphere" of unit diameter

$$Q^2 + U^2 + V^2 = -Q \tag{59}$$

except in regions bounded by hyperbolae in the planes Q = const,

$$V = \mp \left[\sqrt{(Q^2 + U^2)(1 - Q^2)} - |U|\right]/Q, \quad \text{for } Q < 0 \text{ and } |V| > |U| \tag{60}$$

For the I point on the surface of the small sphere there is only one null polarization point $O_1 = O_2$. For I on the OQ axis the scattering matrix is symmetric and $\sigma_o = (A_1 + A_2)^2$. For I point on the surface of the Poincaré sphere, in QV plane, there is only one scattered polarization. For the scattering matrix which transforms every polarization into its orthogonal the I point is in the centre of the Poincaré sphere. For I point in the pole of the sphere the M and $O_1$ points coincide and, as a result, the polarization corresponding to the maximum scattered power is orthogonal to that of illuminating wave.

The coordinates of I point depend on first row elements of F matrix only and for the inversion point in CCS given the whole matrix can be restored though not uniquely for some of these points. Geometrical constructions have been also elaborated (see e.g.[2] and [3]) which enable one to find all special polarizations for I point given in the CCS.

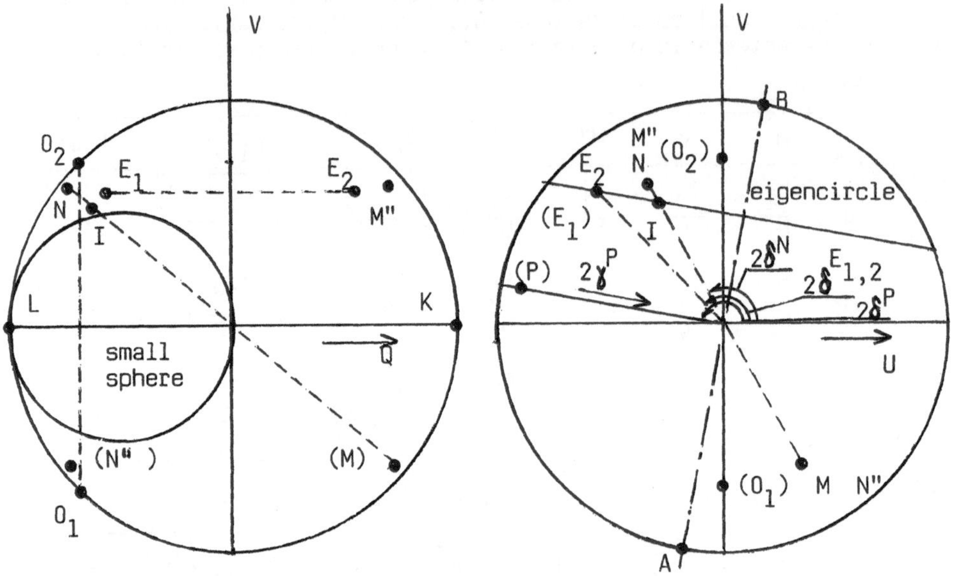


Figure 6. Geometrical model of the coherent scattering matrix - an example. The points in parentheses are on the invisible side of the sphere

Observe that transposing the matrix changes signs of its $B_1$ and $B_2$ elements thus causing rotation of the model by 180° about the characteristic KL axis. If polarizations of transmit and receive antennas are M and M", respectively, maximum received power will be obtained. From the model it is clearly seen that nothing will change if the direction of propagation will be reversed. Similarly, practical conclusion follows that no cross-talk will result in two-channel transmission system operating with scattering when using $E_1$ and $E_2$ as transmit polarizations and orthogonal as receive ones because for opposite direction of propagation these orthogonal polarizations become eigenpolarizations again.

## 3. NONCOHERENT SCATTERING

The partial polarization results when incoherent signals of different polarizations are being added. That is why any depolarizing matrix can be considered as sum of a number of nondepolarizing power radar scattering matrices.

For example, let the target in monostatic radar system be depolarizing and, with the same transmit-receive antenna, let the minimum received powers observed for two different polarizations are of equal value.

Symmetric power radar scattering matrix for such target

$$F = \begin{bmatrix} a_1 & b_1 & b_3 & b_5 \\ b_1 & a_2 & b_4 & b_6 \\ b_3 & b_4 & a_3 & b_2 \\ b_5 & b_6 & b_2 & a_4 \end{bmatrix} \tag{61}$$

is physically realizable if

$$a_1 = a_2 + a_3 + a_4, \qquad a_1^2 \geqslant b_1^2 + b_3^2 + b_5^3 \tag{62}$$

and, as can be shown,

$$a_2 + \frac{b_3 b_4 + b_5 b_6}{b_1} = a_3 + \frac{b_1 b_4 + b_5 b_2}{b_3} = a_4 + \frac{b_1 b_6 + b_3 b_2}{b_5} \tag{65}$$

For such depolarizing matrix a CCS also exists and the F matrix takes the form

$$F'' = \begin{bmatrix} a_1'' & b_1'' & & \\ b_1'' & a_2'' & & \\ & & a_3'' & \\ & & & a_4'' \end{bmatrix}, \quad \text{with} \quad a_1'' = a_2'' + a_3'' + a_4'' \tag{64}$$

after rotation of the OQUV coordinate system by three Euler angles:

$$2\delta = \tan^{-1}(b_5/b_3) \pm \frac{\pi}{2}(1 - \operatorname{sgn} b_3), \qquad 2\gamma = \cos^{-1}(b_1/b_o)$$

$$2\varepsilon = \frac{1}{2}\left[\tan^{-1}\frac{2\,b_2'}{a_3' - a_4'}\right] \pm \frac{\pi}{2}(1 - \operatorname{sgn}(a_3' - a_4')) \tag{65}$$

where

$$b_2' = b_o\,(b_3 b_6 - b_4 b_5)/(b^2 + b^2)$$

and

$$a_3' - a_4' = a_2 - a_2'' + \left[(a_3 - a_4)(b_3^2 - b_5^2) + 4 b_2 b_3 b_5\right]/(b_3^2 + b_5^2).$$

The primed elements, $a'_3$, $a'_4$, and $b'_2$, belonge to the F' matrix obtained from F after two rotations only, by the angels $2\delta$ and $2\gamma$ .

The elements of the F" matrix are as foolows:

$$a''_1 = a_1, \qquad b''_1 = b_o, \qquad a''_2 = a_2 + (b_3b_4 + b_5b_6)/b_1$$

$$a''_3 = \frac{1}{2}\left[a_1 - a''_2 + (a'_3 - a'_4)/\text{cis } 4\varepsilon\right]$$

$$a''_4 = \frac{1}{2}\left[a_1 - a''_2 - (a'_3 - a'_4)/\cos 4\varepsilon\right] \tag{66}$$

One of the possible decompositions of the matrix (64) into nondepolarizing components is

$$F'' = \begin{bmatrix} a_{1o} & b_{1o} & & \\ b_{1o} & a_{1o} & & \\ & & a_{3o} & \\ & & & -a_{3o} \end{bmatrix} + a_{1A}\begin{bmatrix} 1 & & & \\ & 1 & & \\ & & -1 & \\ & & & 1 \end{bmatrix} + \frac{a_{1B}}{2}\left(\begin{bmatrix} 1 & 1 & & \\ & 0 & & \\ 1 & 1 & & \\ & & & 0 \end{bmatrix} + \begin{bmatrix} 1 & -1 & & \\ & 0 & & \\ -1 & 1 & & \\ & & & 0 \end{bmatrix}\right)$$

$$= F_o + F_A + F_{B1} + F_{B2}. \tag{67}$$

The elements of the above matrices are:

$$a_{1o} = \left[(a''_2 - a''_4)^2 + b''^2_1\right] / \left[2(a''_2 - a''_4)\right], \qquad b_{1o} = b''_1$$

$$a_{3o} = \left[(a''_2 - a''_4)^2 - b''^2_1\right] / \left[2(a''_2 - a''_4)\right]$$

$$a_{1A} = (a''^2_2 - a''^2_4 - b''^2_1) / \left[2(a''_2 - a''_4)\right], \qquad a_{1B} = a''_1 - a''_2$$

Observe that the polarizations of minimum received power, $P^T_T$, are null polarizations of the $F_o$ matrix.

REFERENCES


1 M. Davidovitz and W-M. Boerner, "Extension of Kennaugh's optimal polarization concept to the asymmetric Scattering matrix case", IEEE Trans. Antennas and Propagation. vol. AP-33. no 4, pp. 569-574, Apr. 1986

2 Z.H. Czyż, "Polarization of radar scatterings", in Polish, Prace Przemysłowego Instytutu Telekomunikacji, Suplement Nr 5, Warsaw 1986

3 Z.H. Czyż, "Bistatic radar target classification by polarisation properties", ICAP 87, IEE Conf. Publ. No 274, Part I, pp. 545-548

# UNIQUENESS OF TARGET DECOMPOSITION THEOREMS IN RADAR POLARIMETRY

S.R. Cloude
Department of Mathematics and Computer Science
University of Dundee
Dundee DD1 4HN
Scotland

ABSTRACT. A key question in radar polarimetry is the existence of a target dichotomy ie. the existence of more than one target decomposition theory. In this paper we show, by using a coherency matrix approach, that there is no target dichotomy and subsequently only one unique target decomposition. We illustrate its usefulness by giving two examples; the scattering of light by small anisotropic particles and rough surface scattering under physical optics.

## 1. INTRODUCTION

Target decomposition (TD) theorems in radar polarimetry have an interpretation akin to that of generalised spectrum analysis. Both take a general object (for the latter it is a signal while for the former, a target matrix descriptor) and decompose it into the weighted sum of several composite functions (or target matrices). The component or 'basis' functions (targets) are selected because of some simplifying feature such as ease of analysis or physical significance and the aim of both spectrum and TD analysis is to formalise the various methods for achieving this decomposition and to investigate such things as uniqueness and complexity.

There are three matrix methods for modelling radar targets (coherent scattering matrix, Stokes reflection or Mueller matrix and target covariance or coherency matrix) and three corresponding formulations of TD analysis. All use matrix algebra to take a general matrix of the appropriate class and decompose it into a weighted sum of 'simpler' matrices.

The mathematics of group theory aids our choice of basis matrices: many groups of mathematical physics (especially the rotation groups) have important matrix representations which not only provide an immediate choice of 'basis' matrices, but also a geometrical interpretation of the basis elements themselves (see Murnaghan 1963). One such important set are the 2x2 Pauli matrices (as used in the quantum mechanical description of particle spin) which provide a formal link between unitary transformations in $C_2$, a two dimensional



*W.-M. Boerner et al. (eds.), Direct and Inverse Methods in Radar Polarimetry, Part 1*, 267–296.

complex space, and orthogonal transformations in $R_3$, a three dimensional real space (see Arfken 1970). The Pauli matrices underly the geometry of the Poincare sphere and the concept of a Stokes 4-vector, providing a geometrical interpretation of the target Mueller matrix as a Lorentz transformation of the Stokes vector (see Schmeider 1969, Cloude 1986, 1987 and Appendix 1). The Pauli matrices also have important interpretation as elementary scattering processes in coherent decomposition theorems and form an important link between the Mueller and covariance matrices in partially coherent theorems (see Appendix 1).

The concept of target decomposition was introduced by Huynen in his PhD thesis of 1970. It was, however, Chandrasekhar in his book on radiative transfer, who first gave a physical example of such a decomposition. He considered the lateral scattering of light by small anisotropic particles and derived an expression for the phase matrix for such a problem. He decomposed this matrix into two components: one which had an equivalent single target interpretation (namely dipole scattering) and the second which represented noise or a random target ie. one which scatters every incident state as a randomly polarised wave. However, he did not attempt a generalisation of this result and it was Huynen in 1970 who first set about a systematic study of TD theorems.

A key feature of TD theory is the concept of a target dichotomy; it is well known that a statistical description of wave states leads to two different but equally valid statistical models of wave propagation: the wave dichotomy (see van de Hulst 1981). One such model decomposes the wave into a noncoherent sum of two pure states while the other describes the wave as a single pure state corrupted by random noise. The latter is more commonly described in the literature but, if anything, the former is more important, as it is related to an eigenvector analysis of the wave coherency matrix and as such is easily extended to multidimensional problems.

TD theory seeks to provide an extension of these ideas to target scattering and in particular to investigate the existence and uniqueness of target models based on the 'average plus noise' and 'noncoherent sum' concepts.

Huynen concentrated on finding a target theorem analagous to the pure state plus noise model, arguing for the existence of such a model using the Mueller matrix formulation of target scattering. He found such a decomposition and put forward arguments for its uniqueness and physical applicability based on properties of roll invariance. He subsequently extended his result to model the time variation of [S], the coherent scattering matrix, and to calculate detection probabilities and other secondary parameters of interest to radar system designers (Giuli 1987).

Unfortunately, as was mentioned earlier, any useful TD theorem must be unique (ie. given a target matrix we must be able to perform the appropriate decomposition eg. single target plus noise, in only one way). Further, the decomposition must be invariant under physically reasonable transformations such as change of polarisation reference base. It is here that we find a problem. Huynen's theorem is not unique in the sense described above. Given a Mueller matrix [M], there are an infinite number of ways of performing this single state plus noise analysis and Huynen chose just one (recently, Barnes (1988) has shown that, even with the restriction of roll invariance there are two other decompositions which yield a single target plus residue). Even the concept of roll invariance itself applies only to a restrictive class of targets, so there is no general physical justification for adopting such a strategy when devising a TD theory.

The key to a useful decomposition is how one defines a noise target; Chandrasekhar's noise term was truly random in that it showed no polarisation dependent scattering. This is not true of Huynen's N-target (his term for his noise component) which is polarisation dependent and but one of an infinite set of such 'residue' matrices. A key concept here is the behaviour of a TD theorem under small perturbations (by, for example white noise). Any useful TD theorem should be able to isolate a single target from a noise background (for small enough perturbations). Unfortunately, Huynen's theorem does not show this property, mixing noise and target to yield a decomposition that does not allow separation of the two.

We have seen that single target plus noise theorems are not unique and further, are unable to separate targets from noise perturbations. There is however a second statistical model for targets based on the eigenvectors of a target covariance matrix (the analogue of the two pure state wave decomposition). As we shall see, this decomposition is unique and provides useful insight into the physical processes underlying the observed scattering matrix fluctuations. In particular, it is stable under small perturbations and permits isolation of the target component from noise. With this the only unique TD theorem we then conclude that there is no target dichotomy.

In this paper we seek to quantify and formalise these results. We begin by outlining the formulation of wave and target decomposition theorems and then proceed to two important examples of their application: Chandrasekhars' scattering by small particles and a physical optics model for scattering from rough surfaces (a problem used by Huynen in his thesis to illustrate his decomposition theorem). Finally, we conclude with some discussion on the applicability of TD theorems to radar target discrimination and identification and attempt to highlight key areas for future research.

## 2. DECOMPOSITION THEOREMS

The analysis of statistical effects in polarimetry is based on the concept of a coherency or covariance matrix (Cloude 1986, Borgeaud 1986, Swartz 1988). Such a matrix is fundamental in the development of multivariate statistics and represents, through correlation and variance, the second order statistical properties of two or more random variables. The covariance matrix is hermitian and positive semi-definite and, as such, may be diagonalised to yield a set of orthogonal eigenvectors and nonnegative real eigenvalues. This 'decomposition' has important implications for the physical interpretation of statistical wave and target phenomena and in particular leads to a so called wave dichotomy: two different but unique statistical models for wave states. A key feature of modern target studies is the question of the existence of a target dichotomy.

In this section we examine the nature of wave and target decomposition theorems and investigate the connection between the covariance and Mueller matrix formulations of radar scattering.

The statistical analysis of partially polarised waves is based on a 2x2 hermitian coherency matrix, formed from the wave spinor $\underset{\sim}{E}$ as

$$[J] = \langle \underset{\sim}{E}.\underset{\sim}{E}^{\dagger} \rangle \tag{2.1}$$

Under a change of polarisation base, 2.1 transforms as

$$[J]' = [U_2].[J].[U_2]^{\dagger} \tag{2.2}$$

where $[U_2]$ is a 2x2 unitary matrix. From 2.2 it results that an orthogonal base can always be found for which the coherency matrix is diagonal ie.

$$[J] = \begin{bmatrix} \lambda_1 & 0 \\ 0 & \lambda_2 \end{bmatrix} \tag{2.3}$$

where $\lambda_1 >= \lambda_2$. Diagonalisation of [J] leads to the following two statistical models for a partially polarised wave (the wave dichotomy):

1) The pure state decomposition

According to this model we write [J] in the form

$$[J] = \lambda_1 \underset{\sim}{E}_1.\underset{\sim}{E}_1^{\dagger} + \lambda_2 \underset{\sim}{E}_2.\underset{\sim}{E}_2^{\dagger} \tag{2.4}$$

and interpret the partially polarised wave in terms of a noncoherent mixture of two pure orthogonal polarisation states (the eigenvectors $\underset{\sim}{E}_1$ and $\underset{\sim}{E}_2$) occuring with statistical weight $\lambda_1$ and $\lambda_2$. Such a model has been used to analyse the interaction of partially polarised waves with targets by treating the scattering as two independent coherent problems and adding the results noncoherently (see Cloude 1987).

The degree of polarisation $D_p$ of the wave may be defined in terms of the eigenvalues as

$$D_p = \frac{\lambda_1 - \lambda_2}{\lambda_1 + \lambda_2} \tag{2.5}$$

An alternative measure of 'disorder' of the wave is the wave entropy

$$H_w = \sum_{i=1}^{2} - P_i \log_2 P_i \tag{2.6}$$

where $P_i$ can be interpreted as a probability and is defined as $P_i = \lambda_i/(\lambda_1 + \lambda_2)$. According to this definition of entropy, a pure state ($D_p = 1$) has zero uncertainty or $H_w = 0$ while a randomly polarised wave ($D_p = 0$) has maximum entropy or $H_w = 1$.

2) The pure state plus noise decomposition

An alternative interpretation of 2.3 is to express [J] as the sum of two components; one corresponding to a pure state and the other a randomly polarised ($D_p = 0$) or noise term. The latter has only 1 degree of freedom (the wave intensity) while pure states require 3 degrees of freedom (amplitude, inclination and ellipticity angles). Since [J] has 4 degrees of freedom, we can always expect to obtain such a decomposition and, from the diagonal form of [J], write it as

$$[J] = \begin{bmatrix} \lambda_1 - \lambda_2 & 0 \\ 0 & 0 \end{bmatrix} + \begin{bmatrix} \lambda_2 & 0 \\ 0 & \lambda_2 \end{bmatrix} \tag{2.7}$$

While this model is more commonly used (especially when expressed in terms of the Stokes vectors), both models are equally valid and unique, in the sense that no other decomposition of the appropriate type (two pure states or pure state plus noise) is possible given a particular [J].

Turning now to target analysis, we begin with the coherent scattering matrix [S], and express it in terms of a four element target vector $\underset{\sim}{k}$ defined by

$$\underset{\sim}{k} = \tfrac{1}{2}\, \mathrm{Tr}([S].\underset{\sim}{\sigma}) \tag{2.8}$$

where $\underset{\sim}{\sigma}$ are the four Pauli matrices

$$\sigma_o = \begin{bmatrix} 1 & 0 \\ 0 & 1 \end{bmatrix} \quad \sigma_1 = \begin{bmatrix} 1 & 0 \\ 0 & -1 \end{bmatrix} \quad \sigma_2 = \begin{bmatrix} 0 & 1 \\ 1 & 0 \end{bmatrix} \quad \sigma_3 = \begin{bmatrix} 0 & -i \\ i & 0 \end{bmatrix}$$

The Pauli matrices have special geometrical significance (see Appendix 1) and provide a fundamental link between the Mueller and covariance matrices, but we are not restricted to these 4 matrices for an expansion of the matrix [S]. A second useful set of basis matrices are those derived from a straightforward lexicographic expansion of [S] (as used by Borgeaud 1986) ie.

$$\gamma_o = \begin{bmatrix} 1 & 0 \\ 0 & 0 \end{bmatrix} \quad \gamma_1 = \begin{bmatrix} 0 & 1 \\ 0 & 0 \end{bmatrix} \quad \gamma_2 = \begin{bmatrix} 0 & 0 \\ 1 & 0 \end{bmatrix} \quad \gamma_3 = \begin{bmatrix} 0 & 0 \\ 0 & 1 \end{bmatrix}$$

In general we can use any set of 4 complex matrices satisfying constraints of completeness and normalisation. Even with these constraints there are 15 degrees of freedom involved in 'changing the target base' of $\underset{\sim}{k}$. We may write this change of base in the form

$$\underset{\sim}{k}' = [U_4]\ \underset{\sim}{k} \tag{2.9}$$

where $[U_4]$ is a 4x4 unitary matrix preserving $\underset{\sim}{k}^{\dagger}.\underset{\sim}{k}$ or the span of [S]. For example, we may relate the $\sigma$ and $\gamma$ expansions as

$$\underset{\sim}{k}_{\sigma} = \begin{bmatrix} 1 & 0 & 0 & 1 \\ 1 & 0 & 0 & -1 \\ 0 & 1 & 1 & 0 \\ 0 & -i & i & 0 \end{bmatrix} \underset{\sim}{k}_{\gamma} \tag{2.10}$$

The expansion of 2.8 may be termed a coherent decomposition theorem since we are decomposing [S] into the (complex) weighted sum of 4 component targets. This decomposition is particularly useful when the Pauli matrices are interpreted as elementary reflection properties of the target. If we use the (h,v) wave coordinates, they have the following interpretation:

a) $\sigma_0$ has the form of the 2x2 identity. This does not imply that all polarisation states are eigenstates, as antenna coordinate transformations require that left and right circular are copolar nulls. This target is characteristic of reflection from 'odd' bounce ray paths (as supplied for example by a trihedral corner reflector).

b) $\sigma_1$ is also diagonal but has a 180° phase difference between diagonal terms. This phase difference turns left and right circular polarisation into eigenstates and ± 45° linear into copolar nulls. This target component is characteristic of even bounce ray paths (or a dihedral corner reflector).

c) $\sigma_2$ may be obtained from $\sigma_1$ by rotation of the (h,v) reference axes through 45°. This means that it has a similar physical interpretation in terms of even bounce ray paths and may be modelled by a dihedral corner with its axis at 45° to the horizontal. This component is required in the decomposition to account for rotation dependence of even bounce reflection.

d) $\sigma_3$ is an unusual target component in that all incident polarisations are transformed into their orthogonal state (ie. the whole Poincare sphere represents copolar nulls). It signifies departure from symmetry in [S] and as such is not required for backscatter problems.

We could clearly use another set of 4 basis matrices for our decomposition and consider their elementary properties on the Poincare sphere, but the Pauli matrices are particularly useful as many targets (especially at high frequencies) are composed of elementary scatterers of the sort described above. The Pauli target base then becomes the natural one to use. Further, boundary problems involving the Fresnel reflection and transmission coefficients (such as physical optics) give rise to the Pauli matrices as a natural choice for target base. We will consider some specific examples in detail in the following sections.

The completeness of the Pauli matrices has important consequences for target identification studies based on measurement of [S]. Since reflection from radar targets is often dominated by specular backscatter from odd and even bounce corner reflectors, it follows that the resultant scattering matrix can take any form, depending critically on the phasing of the various target components, so we can expect even simple composite targets to appear anywhere in target space. This means that identification based on single frequency, single aspect polarisation measurements (even full scattering matrix) is impossible. A sensible conclusion is that polarisation must then be combined with some other feature such as azimuthal or range resolution for effective target discrimination and identification.

A key advantage of using the target vector $\underset{\sim}{k}$ instead of [S] is that we may easily define a target covariance matrix for the case when [S] is time varying. As for wave states we define this target coherency matrix as

$$[T] = \langle \underset{\sim}{k}.\underset{\sim}{k}^{\dagger} \rangle \qquad (2.11)$$

This 4x4 hermitian matrix contains information on variance and correlation for all complex elements of [S]. It is well known that the target Mueller matrix [M] (or Stokes reflection matrix) also contains information on second order statistics but, on first inspection, its relation to [T] is not clear. By using group theory (see Cloude 1986), it has been shown that the elements of [T] may be interpreted as invariants under rotations in a 6 dimensional target space and that, when [T] is expressed in the Pauli target base, the elements of [M] are obtained as linear combinations of the elements of [T], such that (see Appendix 1)

$$m_{ij} = \tfrac{1}{2}\, \mathrm{Tr}([T].\beta_{4i+j}) \qquad (2.12)$$

where $\beta$ are a set of 16 4x4 matrices (the generators of the group SU(4) plus the unit matrix) shown in figure 1. These matrices represent a generalisation of the Pauli matrices for use in $C_4$, the 4 dimensional complex space.

Under a change of target base [T] transforms as

$$[T]' = [U_4]\,[T]\,[U_4]^{\dagger} \qquad (2.13)$$

and, as for wave states, we have the important result that we can always find a set of normalised, orthogonal reference matrices for the expansion of [S], such that the resultant random fluctuations are uncorrelated ie.

$$[T] = \begin{bmatrix} \lambda_1 & 0 & 0 & 0 \\ 0 & \lambda_2 & 0 & 0 \\ 0 & 0 & \lambda_3 & 0 \\ 0 & 0 & 0 & \lambda_4 \end{bmatrix} \qquad (2.14)$$

where the choice of matrix base states is dictated by the eigenvectors of [T]. Note that while diagonal [T] has important physical interpretation, diagonalisation of [M] has no such clear physical significance.

As an important example of 2.14 consider the case of a target described by a time invariant scattering matrix, for which $\lambda_2 = \lambda_3 = \lambda_4 = 0$ and the eigenvector corresponding to $\lambda_1$ is then the Pauli matrix ordering of [S]. At the other extreme, consider a truly random target for which there exists no correlation between elements of [S] and $\lambda_1 = \lambda_2 = \lambda_3 = \lambda_4$ ie. [T] has degenerate eigenvalues.

As a formal measure of target disorder we define target entropy as

$$H_T = \sum_{i=1}^{4} -P_i \log_4 P_i \qquad (2.15)$$

with $P_i = \lambda_i/(\lambda_1+\lambda_2+\lambda_3+\lambda_4)$. Note that $H_T = 0$ for random targets and $H_T = 1$ for simple targets.

Equation 2.14 provides the starting point for an investigation of a target dichotomy. We begin by considering the most obvious interpretation of 2.14 as the noncoherent sum of four simple targets (the eigenvectors of [T]) weighted by the eigenvalues.

1) The 4-target theorem

We may write 2.14 as

$$[T] = \lambda_1 \underset{\sim}{k}_1 \cdot \underset{\sim}{k}_1^{\dagger} + \lambda_2 \underset{\sim}{k}_2 \cdot \underset{\sim}{k}_2^{\dagger} + \lambda_3 \underset{\sim}{k}_3 \cdot \underset{\sim}{k}_3^{\dagger} + \lambda_4 \underset{\sim}{k}_4 \cdot \underset{\sim}{k}_4^{\dagger} \qquad (2.16)$$

and model the stochastic target as the noncoherent sum of four simple targets (each of $\underset{\sim}{k}_1$, $\underset{\sim}{k}_2$, $\underset{\sim}{k}_3$ and $\underset{\sim}{k}_4$ has a corresponding single scattering matrix representation and hence is termed simple). Note that the condition for physical realisability, so complicated in Mueller matrix algebra, is simply a statement of the non negative nature of the eigenvalues of [T]. This provides a scheme for identifying a physically realisable Mueller matrix by calculating the corresponding covariance matrix and then testing to see whether all eigenvalues are non negative.

This decomposition is analagous to the pure state wave decomposition and, like the latter, is uniquely defined given [T]. We will consider examples of the usefulness of this decomposition in later sections but note for the moment that it provides a systematic method for determining the 'best' estimate of [S] in a noise background and for identifying dominant scattering mechanisms in scattering from random media.

2) Simple target plus noise decomposition

Returning to 2.14, we see that a random target $(H_T = 0)$ must have degenerate eigenvalues, (one degree of freedom). A simple target on the other hand has 7 degrees of freedom (ignoring an absolute phase term). We can then see that, with [T] having 16 degrees of freedom, it is impossible to obtain a single target plus noise decomposition, such an idea simply does not account for all the degrees of freedom in the covariance matrix.

Despite this, it is possible to extract a simple target (7 degrees of freedom) from [T] and ascribe the remainder to fluctuation components (but note that these components will be polarisation dependent). Huynen has considered one such decomposition in which he chose to identify the following two component decomposition:

$$[M] = [M_S] + [M_N] \tag{2.17}$$

where $[M_S]$ is the Mueller matrix for a simple target and $[M_N]$ the remainder or N-target (as Huynen chose to call it). By virtue of the 1-1 correspondence between [M] and [T], we may now express 2.17 in terms of covariance matrices

$$[T] = [T_S] + [T_N] \tag{2.18}$$

where $[T_S]$ is a simple target ie. has only one nonzero eigenvalue and $[T_N]$ is the remainder term, Huynen's definition of which requires it to have a zero first row and column (corresponding to a lack of what Huynen called target symmetry but which is now identified as a zero coefficient for the Pauli matrix $\sigma_o$). With this definition, and noting that we are considering backscatter only, we can write 2.18 as

$$\begin{bmatrix} x & x & x & 0 \\ x & x-\alpha_1 & x-\alpha_2 & 0 \\ x & x-\alpha_2^* & x-\alpha_3 & 0 \\ 0 & 0 & 0 & 0 \end{bmatrix} + \begin{bmatrix} 0 & 0 & 0 & 0 \\ 0 & \alpha_1 & \alpha_2 & 0 \\ 0 & \alpha_2^* & \alpha_3 & 0 \\ 0 & 0 & 0 & 0 \end{bmatrix} \tag{2.19}$$

where x represents the appropriate element of [T], $\alpha_1$ and $\alpha_3$ are real and $\alpha_2$ complex (ie $[T_N]$ has 4 degrees of freedom). Since [T] has 9 degrees of freedom, this leaves 5 degrees of freedom for $[T_S]$, the correct number for a symmetric matrix [S]. This means that, assuming the general form of 2.19, we can uniquely determine the $\alpha_i$ by constraining $[T_S]$ to be simple. This is how Huynen proved the 'uniqueness' of his decomposition. With the benefit of 2.19, we can however see the problem with decompositions of this type.

While 2.18 is uniquely defined <u>given</u> 2.19, the choice of $[T_N]$ in 2.19 is arbitrary. We could for example choose $[T_N]$ to be of the form

$$\begin{bmatrix} \alpha_1 & 0 & \alpha_2 & 0 \\ 0 & 0 & 0 & 0 \\ \alpha_2^* & 0 & \alpha_3 & 0 \\ 0 & 0 & 0 & 0 \end{bmatrix}$$

and still satisfy the simple target requirements of $[T_S]$, although of course the simple target will be very different from that found using Huynen's theory.

We can generalise this result by noting that decompositions of the form 2.18 are not unique because $[T_S]$ has, at most, 7 degrees of freedom, leaving $[T_N]$ with 9, and there are an infinite number of ways of defining a 4x4 hermitian matrix with 9 degrees of freedom.

Huynen attempted to give physical meaning to his decomposition by pointing out that his N-target remained invariant under roll of the observed target (note that he considered only backscatter). There are two problems with such an argument: firstly, Barnes has recently shown that (Barnes 1988) even with the constraint of roll invariance, there are still three choices of remainder which satisfy the TD theorem. These three options arise as eigenvectors of the rotation operator governing roll invariance and while one such eigenvector corresponds to Huynen's theorem, the other two yield very different single targets in the decomposition. A second problem is the assumption that rotation of the antenna coordinates is equivalent to fixing the coordinates and rolling the target about the line of sight. This is only true for simple symmetric objects while for more complex targets, rotation about the line of sight brings into view new scattering centres, previously hidden, and which cannot be predicted by adjustments in the receiver coordinate base.

The roll invariant constraint amounts to a requirement that any TD theorem be robust under a change of wave base ie. that it yields the same dominant single target whatever wave base states are chosen for measurement. In this respect we see that the 4 target theorem, being based on eigenvectors of [T], satisfies not only roll invariance but the much more stringent constraint of invariance under arbitrary change of base. When combined with the previously stated robustness of the 4 target theorem to noise and the nonuniqueness of Huynen's simple target, we must conclude that there is no target dichotomy.

## 3. SMALL PARTICLE SCATTERING

To illustrate the coherency matrix formalism, consider the case of scattering by small anisotropic particles. The scattering matrix for one such particle may be written in terms of a symmetric polarisability tensor and scattering angle $\theta$ (where $\theta = 180^\circ$ is backscatter) as (see Chandrasekhar 1950, Van de Hulst 1981)

$$[S] = \begin{bmatrix} P_{11}\cos\theta - P_{21}\sin\theta & P_{13}\cos\theta - P_{23}\sin\theta \\ -P_{13} & -P_{33} \end{bmatrix} \qquad (3.1)$$

where $P_{ij}$ are given by

$$P_{11} = \alpha_1 l_1^{\;2} + \alpha_2 l_2^{\;2} + \alpha_3 l_3^{\;2}$$

$$P_{13} = \alpha_1 l_1 n_1 + \alpha_2 l_2 n_2 + \alpha_3 l_3 n_3$$

$$P_{23} = \alpha_1 m_1 n_1 + \alpha_2 m_2 n_2 + \alpha_3 m_3 n_3$$

$$P_{33} = \alpha_1 n_1^{\;2} + \alpha_2 n_2^{\;2} + \alpha_3 n_3^{\;2}$$

and $\alpha_1$, $\alpha_2$, $\alpha_3$ are principal values of polarisability (defined with respect to the preferred dielectric coordinates of the particle) and l,m,n are the direction cosines relating source and dielectric coordinates. Note that the incident field is given by $\underset{\sim}{E} = (E_1, 0, E_3)$. From [S] we can calculate $\underset{\sim}{k}$ as

$$k_o = P_{11}\cos\theta - P_{21}\sin\theta - P_{33} \qquad (3.2)$$

$$k_1 = P_{11}\cos\theta - P_{21}\sin\theta + P_{33} \qquad (3.3)$$

$$k_2 = P_{13}\cos\theta - P_{23}\sin\theta - P_{13} \qquad (3.4)$$

$$k_3 = i(P_{13}\cos\theta - P_{23}\sin\theta + P_{13}) \qquad (3.5)$$

With knowledge of [S] for a single particle we then introduce statistical information on the number and distribution of particles. The simplest assumption to take is that of a large number of identical particles with random orientations, an assumption which allows a noncoherent summation to be made and leads to many terms in $\langle[T]\rangle$ averaging to zero, leaving only the following

$$\langle[T]\rangle = \begin{bmatrix} \langle k_o^{\;2}\rangle & \langle k_o k_1^{\;*}\rangle & 0 & 0 \\ \langle k_o^{\;*} k_1\rangle & \langle k_1^{\;2}\rangle & 0 & 0 \\ 0 & 0 & \langle k_2^{\;2}\rangle & 0 \\ 0 & 0 & 0 & \langle k_3^{\;2}\rangle \end{bmatrix} \qquad (3.6)$$

where

$$\langle k_o^{\;2}\rangle = L(1+\cos^2\theta) + M\sin^2\theta - 2N\cos\theta \qquad (3.7)$$

$$\langle k_1^{\;2}\rangle = L(1+\cos^2\theta) + M\sin^2\theta + 2N\cos\theta \qquad (3.8)$$

$$\langle k_2^{\;2}\rangle = 2M(1-\cos\theta) \qquad (3.9)$$

$$\langle k_3^{\;2}\rangle = 2M(1+\cos\theta) \qquad (3.10)$$

$$\langle k_o k_1^{\;*}\rangle = (M-L)\sin^2\theta \qquad (3.11)$$

The three parameters, L, M and N are defined as

$$L = 3\beta + 2\gamma = \overline{P_{ii}^{\ 2}} \tag{3.12}$$

$$M = \beta - \gamma = \overline{P_{ij}^{\ 2}} \tag{3.13}$$

$$N = \beta + 4\gamma = \overline{P_{ii}P_{jj}} \tag{3.14}$$

and $\beta$ and $\gamma$ are related to $\alpha_1$, $\alpha_2$ and $\alpha_3$ by

$$15\beta = \alpha_1^{\ 2} + \alpha_2^{\ 2} + \alpha_3^{\ 2} \tag{3.15}$$

$$15\gamma = \mathrm{Re}(\alpha_1\alpha_2^{\ *} + \alpha_2\alpha_3^{\ *} + \alpha_3\alpha_1^{\ *}) \tag{3.16}$$

To decompose the problem into its constituent targets, we first diagonalise <[T]>. By inspection, we find expressions for the eigenvalues as

$$\lambda_{1,2} = L(1+\cos^2\theta) + M\sin^2\theta \pm \sqrt{4N^2\cos^2\theta + (M-L)^2\sin^4\theta} \tag{3.17}$$

$$\lambda_3 = 2M(1 - \cos\theta) \tag{3.18}$$

$$\lambda_4 = 2M(1 + \cos\theta) \tag{3.19}$$

The eigenvectors are given by $(\cos\phi,\sin\phi,0,0)$, $(-\sin\phi,\cos\phi,0,0)$, $(0,0,1,0)$ and $(0,0,0,1)$, where the angle $\phi$ is determined from

$$\tan\phi = \frac{(M-L)\sin^2\theta}{-2N\cos\theta + (4N^2\cos^2\theta + (M-L)2\sin^4\theta)^{\frac{1}{2}}} \tag{3.20}$$

To illustrate the significance of these expressions, consider two special cases:

a) Lateral scattering ($\theta = 90^\circ$)

In this case <[T]> has the form

$$\langle[T]\rangle = \begin{bmatrix} L+M & M-L & 0 & 0 \\ M-L & L+M & 0 & 0 \\ 0 & 0 & 2M & 0 \\ 0 & 0 & 0 & 2M \end{bmatrix} \tag{3.21}$$

The upper quadrant of this matrix may be easily diagonalised to yield an eigenvalue spectrum (L,M,M,M). The eigenvector corresponding to L is (1,1,0,0) and corresponds to the simple matrix for a vertical dipole. The degeneracy of the remaining three eigenvalues indicates the presence of a random or noise target (ie. one which generates randomly polarised waves for arbitrary incident polarisation). We thus model the lateral scattering problem as the noncoherent sum of two targets: a vertical dipole of magnitude L together with a random target of magnitude M. The ratio of target components is given by

$$\text{dipole/noise} = (3\beta+2\gamma)/(\beta-\gamma) \tag{3.22}$$

If the particles are isotropic ($\beta=\gamma$) then the random target component is zero and we obtain the well known Rayleigh scattering law. This target decomposition was first developed by Chandrasekhar (1950) using a phase matrix formulation. The coherency matrix provides a clearer interpretation of results and allows generalisation to other scattering problems.

b) Backscatter ($\theta = 180^o$)

In this case $\langle[T]\rangle$ becomes

$$\langle[T]\rangle = \begin{bmatrix} L+N & 0 & 0 & 0 \\ 0 & L-N & 0 & 0 \\ 0 & 0 & 2M & 0 \\ 0 & 0 & 0 & 0 \end{bmatrix} \tag{3.23}$$

The matrix is diagonal in the Pauli base with a dominant eigenvector (1,0,0,0) corresponding to an eigenvalue L+N. Note that L-N = 2M, so the remaining two eigenvalues are degenerate, signifying a random target component of strength 2M. Again, for isotropic particles we have the result that M=0 and backscatter modelled by a single target component, namely $\sigma_o$. In general however, the scattering is a noncoherent mixture of $\sigma_o$ and noise, the ratio of the two being given by

$$\sigma_o/\text{noise} = (2\beta+3\gamma)/(\beta-\gamma) \tag{3.24}$$

Note that the fourth eigenvalue is zero because the scattering matrix is symmetric for backscatter (this is a general result based on reciprocity and means that for backscatter problems [T] is 3x3 hermitian).

We now consider the case of scattering from ellipsoidal particles with semiaxes a, b and c. In this case the anisotropy is due to particle shape rather than material properties. The particle polarisability is given by (see Van de Hulst 1981)

$$4\pi\alpha_i = \frac{V}{L_i + 1/(\epsilon-1)} \tag{3.25}$$

where V is the volume of the particle ($V=4\pi abc/3$), $\epsilon$ the relative permittivity (assumed constant but may be complex) and the three coefficients $L_i$ are given by cyclic permutations of

$$L_1 = \int_0^{\infty} \frac{abc\ ds}{2(s+a^2)^{3/2}(s+b^2)^{1/2}(s+c^2)^{1/2}} \tag{3.26}$$

Note that $L_1 + L_2 + L_3 = 1$.

Shown below are typical values for spheroidal particles (b=c) of $\epsilon = 2$ and unit volume. Three types of particle are considered; prolate spheroids (rods), oblate spheroids (discs) and spheres with L, M and N shown normalised to the sphere.

| | b/a | $L_1$ | $L_2$ | $L_3$ | $\alpha_1$ | $\alpha_2$ | $\alpha_3$ | L | M | N |
|---|---|---|---|---|---|---|---|---|---|---|
| rods | 0 | 0 | 1/2 | 1/2 | 1 | 2/3 | 2/3 | 1.093 | 0.013 | 1.067 |
| spheres | 1 | 1/3 | 1/3 | 1/3 | 3/4 | 3/4 | 3/4 | 1.000 | 0.000 | 1.000 |
| discs | $\infty$ | 1 | 0 | 0 | 1/2 | 1 | 1 | 1.274 | 0.030 | 1.215 |

Figure 2 shows the eigenvalue level as a function of scattering angle for rod scatterers (note that the eigenvalues are expressed as a percentage of the maximum eigenvalue). Note the behaviour at $\theta=90^o$ (showing degeneracy as expected) and at $\theta=180^o$ where the fourth eigenvalue becomes zero. Figure 3 shows the eigenvalue levels for discs (the levels for a sphere are all zero percent) and we see the same general type of behaviour as Figure 2 but at a higher level ie. the disc scattered light is less polarised than that from the rods.

Finally, figure 4 shows that the eigenvector associated with the maximum eigenvalue does not remain constant with scattering angle but rotates from $90^o$ for forward scatter through $45^o$ for lateral scattering to $0^o$ for backscatter. The eigenvector rotation for discs and rods are shown and, although there is some small difference, they display the same general behaviour with scattering angle.

## 4. ROUGH SURFACE SCATTERING

When considering polarisation effects in the scattering of EM waves from rough surfaces, the simplest useful theory is that of physical optics or the tangent plane approximation. In this theory we begin with the Stratton-Chu vector diffraction integral for the scattered field and approximate the tangential E and H fields by making two assumptions;

1) The current induced by the incident field is locally the same as that on an infinite tangent plane at the point.

2) The induced current is zero in the geometrical shadow region of the object.

With these two assumptions (valid only when the smallest radius of curvature of the surface is much greater than a wavelength ie. a high frequency approximation) the Stratton-Chu integral is considerably simplified and may be used to obtain the scattering matrix for arbitrary scattering angle and surface permittivity.

In the limit of infinite surface conductivity it can be shown that the integral factors into the product of two terms (see Beckmann 1968); a vector function dependent only on the incident and scattered wave vectors and a scalar surface integral determining the amplitude of the scattered wave. As an important special case, physical optics predicts no depolarisation of waves under backscatter from perfectly conducting objects.

When consideration is given to backscatter from dielectric surfaces however, the physical optics solution does predict depolarisation, quantified by evaluation of the following symmetric scattering matrix

$$[S] = \begin{bmatrix} E_p^+ & E_c \\ E_c & E_p^- \end{bmatrix} \tag{4.1}$$

where

$$E_p^+ = E\iint \cos\theta[D + S\cos 2\beta]\, e^{-2i\underset{\sim}{k}\cdot\underset{\sim}{r}} ds \tag{4.2}$$

$$E_p^- = E\iint \cos\theta\,[D - S\cos 2\beta]\, e^{-2i\underset{\sim}{k}\cdot\underset{\sim}{r}} ds \tag{4.3}$$

$$E_c = E\iint \cos\theta\; S\,\sin 2\beta\; e^{-2i\underset{\sim}{k}\cdot\underset{\sim}{r}} ds \tag{4.4}$$

There are two angles of importance in these expressions, both determined by surface geometry; $\theta$ the local aspect angle and $\beta$ the rotation of local normal $\underset{\sim}{n}$ about the incident wave vector $\underset{\sim}{k}$. The factors S and D are respectively the sum and difference of Fresnel reflection coefficients ($S = R^+ + R^-$ and $D = R^+ - R^-$) which themselves are functions of $\theta$ and surface permittivity $\epsilon$ (which may be complex)

$$R^+ = \frac{\epsilon\cos\theta - \sqrt{\epsilon - \sin^2\theta}}{\epsilon\cos\theta + \sqrt{\epsilon - \sin^2\theta}} \tag{4.5}$$

$$R^- = \frac{\cos\theta - \sqrt{\epsilon - \sin^2\theta}}{\cos\theta + \sqrt{\epsilon - \sin^2\theta}} \tag{4.6}$$

While evaluation of 4.1 requires integration over a surface, we may usefully define a local scattering matrix $[S_L]$ from 4.2, 4.3 and 4.4 as

$$[S_L] = A\cos\theta \begin{bmatrix} \cos\beta & -\sin\beta \\ \sin\beta & \cos\beta \end{bmatrix} \begin{bmatrix} R^+ & 0 \\ 0 & -R^- \end{bmatrix} \begin{bmatrix} \cos\beta & \sin\beta \\ -\sin\beta & \cos\beta \end{bmatrix} \tag{4.7}$$

where $A = Ee^{-2i\underset{\sim}{k}\cdot\underset{\sim}{r}}$. From this we define a local target vector $k_L$

$$k_L = A(D\cos\theta,\ S\cos\theta\cos2\beta,\ S\cos\theta\sin2\beta,\ 0) \tag{4.8}$$

We may then define the ijth element (where $1 \leq i,j \leq 3$) of a coherency matrix relating points 1 and 2 on the surface as

$$\langle k_1 k_2^* \rangle_{ij} = \iiiint \langle k_{1i} k_{2j}^* \exp(2i\underset{\sim}{k}.(\underset{\sim}{r}_2-\underset{\sim}{r}_1)\rangle dS_1 dS_2 \tag{4.9}$$

We can considerably simplify this expression by making two assumptions;

1) Assume $\langle k_1 k_2^* \rangle_{ij} = \langle k_{1i} k_{2j}^* \rangle \langle \exp(..)\rangle$

2) Assume $\langle k_1 k_2^* \rangle_{ij} = \langle k_1 k_1^* \rangle_{ij}$

Beckmann provides a detailed discussion of the validity of these two assumptions but they basically require the surface to be rough (rms height of surface must be greater that $\lambda$) and restrict the range of incident angles to $s\tan\theta_1 << 1$ where $s$ is the rms slope of the surface and $\theta_1$ the angle between the incident wave vector and mean surface normal. With these assumptions we may write the surface coherency matrix as

$$[T] = B \begin{bmatrix} \langle D^2\cos^2\theta\rangle & \langle DS^*\cos2\beta\cos^2\theta\rangle & \langle DS^*\sin2\beta\cos^2\theta\rangle \\ \langle SD^*\cos2\beta\cos^2\theta\rangle & \langle S^2\cos^2\theta\cos^22\beta\rangle & \langle S^2\sin2\beta\cos2\beta\cos^2\theta\rangle \\ \langle SD^*\sin2\beta\cos^2\theta\rangle & \langle S^2\sin2\beta\cos2\beta\cos^2\theta\rangle & \langle S^2\sin^22\beta\cos^2\theta\rangle \end{bmatrix} \tag{4.10}$$

where

$$B = \iiiint \langle \exp(2i\underset{\sim}{k}.(\underset{\sim}{r}_2-\underset{\sim}{r}_1)\rangle dS_1 dS_2$$

As we are interested primarily in polarisation properties of the scattered wave, we avoid evaluation of B and consider the form of [T] when the surface height statistics are stationary isotropic normal. Under these conditions the surface is characterised by two surface slope distributions, $Z_X$ and $Z_Y$, which are zero mean normal and statistically independent. Hence

$$P(Z_X,Z_Y) = P(Z_X)P(Z_Y)$$

where

$$P(Z_X) = \exp(-Z_X^2/2s^2)/\sqrt{2\pi}\,s$$

$$P(Z_Y) = \exp(-Z_Y^2/2s^2)/\sqrt{2\pi}\,s$$

and $s$ is the rms slope of the surface. Knowing $Z_X$ and $Z_Y$ we may calculate $\sin\beta$, $\cos\beta$ and $\cos\theta$ from

$$\cos\beta = \frac{\sin\theta_1 - Z_X\cos\theta_1}{[1 + Z_X^2 + Z_Y^2 - (Z_X\sin\theta_1 + \cos\theta_1)^2]^{\frac{1}{2}}} \tag{4.11}$$

$$\sin\beta = \frac{Z_Y}{[1 + Z_X^2 + Z_Y^2 - (Z_X\sin\theta_1 + \cos\theta_1)^2]^{\frac{1}{2}}} \tag{4.12}$$

$$\cos\theta = \frac{Z_X\sin\theta_1 + \cos\theta_1}{[1 + Z_X^2 + Z_Y^2]^{\frac{1}{2}}} \tag{4.13}$$

Using these expressions we may then evaluate $\langle G(\theta,\beta,\epsilon)\rangle$, the elements of [T] as

$$\langle G\rangle = \iint G(\theta,\beta,\epsilon)P(Z_X,Z_Y)dZ_X dZ_Y \tag{4.14}$$

Note that to allow for selfshading on the surface we must restrict $-90^o \leq \theta \leq 90^o$ and ignore any contributions from $Z_X$ and $Z_Y$ which yield values of $\theta$ outside this range.

For normal incidence ($\theta_1=0^o$), there is symmetry in the x and y directions and [T] reduces to diagonal form in the Pauli base. The surface is then modelled as a two component target; a $\sigma_o$ component of unit magnitude and a noise component (the remaining eigenvalues being equal if $Z_X$ and $Z_Y$ have the same distribution) of magnitude

$$\text{noise}/\sigma_o = \frac{\langle S^2\cos^2\theta\cos^2 2\beta\rangle}{\langle D^2\cos^2\theta\rangle} \tag{4.15}$$

We expect the noise level to increase as s increases and $\epsilon$ decreases (in the limit $\epsilon \to \infty$, $S \to 0$). Figure 5 shows the variation of noise level (as a percentage of the $\sigma_o$ component) for rms slope values from $1^o$ to $30^o$ and permittivities in the range $2 \to 16$ (refractive index $1.414 \to 4$). We can see that the noise level shows the qualitative behaviour expected with increasing noise level for increasing surface slope and decreasing permittivity.

A question of practical importance is the possibility of estimating the surface rms slope from measurements of backscattered polarisation. From our target model for the surface we see that the optimum choice of transmitter polarisation is circular (a copolar null for the $\sigma_o$ component) with a copolar backscattered intensity measurement yielding the noise level directly (when normalised to the received crosspolar component this may then be compared directly with figure 5.

We can see from figure 5 that, without knowledge of $\epsilon$, we cannot make an estimate of surface roughness from backscatter measurements.

We therefore consider the form of [T] for oblique incidence on the surface to investigate the possibility of estimating s and $\epsilon$ from measurement.

For an angle of incidence $\theta_1$ in the xz plane, the coherency matrix reduces to

$$[T] = \begin{bmatrix} \langle D^2\cos^2\theta\rangle & \langle DS^*\cos2\beta\cos^2\theta\rangle & 0 \\ \langle SD^*\cos2\beta\cos^2\theta\rangle & \langle S^2\cos^2\theta\cos^22\beta\rangle & 0 \\ 0 & 0 & \langle S^2\cos^2\theta\sin^22\beta\rangle \end{bmatrix} \tag{4.16}$$

We see that the matrix is not diagonal in the Pauli base and so the dominant target in the decomposition may not be $\sigma_0$. For real $\epsilon$, the eigenvectors undergo a rotation through $\gamma$ such that the dominant eigenvector becomes $(\cos\gamma, \sin\gamma, 0, 0)$. Figure 7 shows $\gamma$ as a function of incidence angle for an rms slope of $10^o$, and $\epsilon=2$. For normal incidence $\gamma=0^o$ and the dominant target is $\sigma_0$. As we move away from normal incidence $\gamma$ increases, passing through $45^o$ (corresponding to a dominant dipole component) and tending to $90^o$ (ie. $\sigma_1$ dominance) as we move towards grazing incidence. Note that assumption 2 may not be valid for angles well away from normal and this, combined with the increasing significance of volume scattering, means that results for large $\theta_1$ may be misleading.

The eigenvalue spectrum for oblique incidence also shows some interesting structure (see figure 6). At normal incidence the eigenvalues are degenerate as discussed earlier but, as the angle of incidence increases, the eigenvalues first increase together but then split, the third eigenvalue showing a peak at the Brewster angle for the surface. If this result is confirmed experimentally, it means that estimates may be made of both $\epsilon$ and s from measurements of the eigenvalue spectrum as a function of incidence angle.

In summary, we have seen that physical optics predicts depolarisation effects for backscatter from a rough surface. At normal incidence the results agree with intuition and provide a surface model of a specular component ($\sigma_0$) with diffuse background scattering showing no polarised structure (verified by the existence of degenerate eigenvalues for [T]). Further, the ratio of diffuse to specular component increases as surface roughness increases and $\epsilon$ decreases, in the limit as $\epsilon\to\infty$ there being no depolarisation.

As the angle of incidence increases, the surface model becomes more complicated, with the dominant component changing with angle and the eigenvalue spectrum becoming nondegenerate (signifying a polarisation dependent diffuse component). The eigenvalue levels also increase with angle, indicating that the diffuse background becomes larger for angles away from normal incidence. By examination of the eigenvalue spectrum for real $\epsilon$ we have seen a characteristic peak at the Brewster angle, a result which means that estimates may be made of both $\epsilon$ and rms slope from backscatter measurements obtained over a range of angles. However, it remains to be confirmed experimentally that the physical optics model used for the simulations is valid for large incidence angles.

5. CONCLUSIONS

The scattering of polarised waves from stochastic targets may be handled in two ways: the first by a 4x4 Mueller matrix relating incident and scattered wave Stokes vectors and the second by forming a covariance or target coherency matrix. The two methods are related by the $\beta$ matrices of figure 1 and are entirely equivalent. Both describe the second order statistics of the stochastic target and both have potential use in target detection, discrimination and identification.

The covariance matrix has significant advantages over the Mueller matrix in terms of identifying correlated scattering mechanisms in the target. It also provides a better starting point for the analysis of target decomposition theorems and can be used to show that there is no target dichotomy.

An important objective of future research is to obtain the polarimetric statistics for radar targets of interest such as military vehicles and ground clutter. While such an objective has long been identified, we now have the advantage of a fuller appreciation of the important relation between covariance formulations of target scattering and the Mueller matrix. Indeed, it is now possible to design radar systems using a Stokes vector receiver and to relate the output from these radars directly to second order target statistics.

It seems clear from the complexity of polarimetric measurement and theory that we should obtain target data using both theoretical and experimental investigations. The covariance formulation has the distinct advantage that it provides a useful common format for a description of polarimetric statistics, lending itself to both theoretical and experimental determination.

As an example of theoretical application, we saw in section 4 that, while physical optics predicts no depolarisation for backscatter from perfectly conducting scatterers, for backscatter from rough dielectric surfaces, [T] is nontrivial and has the general form

$$[T] = \begin{bmatrix} \alpha & \beta & 0 & 0 \\ \beta^* & \gamma & 0 & 0 \\ 0 & 0 & \delta & 0 \\ 0 & 0 & 0 & 0 \end{bmatrix} \qquad (5.1)$$

with 5 degrees of freedom. Only for normal incidence is $\beta=0$ and the Pauli matrices the eigentargets for this problem. For nonnormal incidence, shadowing causes $\beta$ to be finite and changes the dominant target component of the surface.

[T] has the same form as 5.1 for scattering from small anisotropic particles, giving, for lateral scattering, a new formulation of the classic result of Chandrasekhar. Both the above results are based on the simplest possible statistics, namely a normal surface with equal rms slopes and random particle orientation. Evidently, if either of these is invalid, we can expect a different form for [T].

For volume scattering, theoretical predictions are more difficult to make. Borgeaud (1986) has reviewed the application of Born techniques to volume scattering and shown that, under the first Born approximation (where multiple scattering effects are ignored) [T] has the form

$$[T] = \begin{bmatrix} \alpha & \beta & 0 & 0 \\ \beta & \delta & 0 & 0 \\ 0 & 0 & 0 & 0 \\ 0 & 0 & 0 & 0 \end{bmatrix} \tag{5.2}$$

with only 4 degrees of freedom. Borgeaud (1986) and Swartz (1988) have both reported that the first order Born approximation is a poor model, even for relatively homogeneous radar clutter, and have shown that the second order Born approximation is much more accurate when compared with experiment. The second Born approach differs from the first by allowing the possibility of cross polarisation due to second order multiple scattering. The second order Born method leads to a covariance matrix of the form shown in 5.1.

Novak (1987) has used 5.1 as the basis for an investigation of polarimetric detection algorithms, investigating the performance of optimum detectors, matched filters and a range of suboptimum algorithms. He makes the further assumption that the elements of [S] have a complex Gaussian PDF and concludes that optimum techniques offer no significant advantage over fixed polarisation detection. The techniques used in Novak's paper illustrate the usefulness of the covariance formulation for linking theoretical scattering models to statistical detection theory and matched filtering.

For random media where multiple scattering effects are important, the phenomenon of enhanced backscatter (EBS) has been observed experimentally (Albada 1985,1987). This effect is due to coherent addition of reciprocal path lengths in random media and is observed to be polarisation dependent. Albada (1987) has analysed EBS using a particulate model for the random media and ascribed the vector nature of the problem to anisotropy of particle scattering diagrams. Future work is aimed at deriving the form of the covariance matrix for this important phenomenon.

In conclusion we make the following observations:

1) While there exists a wave dichotomy for the statistical modelling of partially polarised waves, there does not exist a target dichotomy, there being only one unique target decomposition theorem, that based on an eigenvector expansion of [T].

2) Small particle scattering and rough surface scattering under physical optics both display interesting polarisation features which require experimental validation.

3) While theoretical techniques for radar clutter and target analysis predict a sparse coherency matrix, they usually represent the result of considerable averaging. On the other hand, fast polarisation switches and signal processing permit measurement of polarisation on much finer time scales than implied in these averages. Therefore, while such averaging is a useful starting point in polarimetric scattering theory, it is only the first stage towards a fuller understanding of the polarisation properties of stochastic targets.

## APPENDIX : POLARISATION AND GEOMETRY

A key feature of polarisation algebra is the link between mathematical descriptors of polarisation and the geometry of n-dimensional space. The construction of this geometry has more than just formal significance in that target identification studies seek always to identify invariants under various transformations. The geometrical model then provides the quickest and most elegant way of identifying these invariant features. In this section we review the geometrical description of waves and targets, with particular emphasis on identification of invariants.

We begin with wave descriptors, the geometrical character of which is well known and represented by the geometry of the Poincare sphere. This mapping of wave states onto the surface of a sphere is a result of the spinor nature of the electric field descriptor and the mapping into 3 dimensions follows from the quaternion algebra of Hamilton. According to this mapping we can represent a rotation in 3-space by either a real 3x3 orthogonal matrix or a 2x2 complex unitary matrix.

If we consider a 3-vector $\underset{\sim}{r}$ in a space with orthogonal axes 1,2 and 3, then an arbitrary rotation may be represented as the composite product of three rotations (the Euler angles) ie.

$$r' = [O_3]\,\underset{\sim}{r} \tag{A.1}$$

where $[O_3]$ can be written as the product of three matrices

$$[O_{12}]=\begin{bmatrix}\cos\theta & -\sin\theta & 0\\ \sin\theta & \cos\theta & 0\\ 0 & 0 & 1\end{bmatrix}\quad [O_{23}]\begin{bmatrix}1 & 0 & 0\\ 0 & \cos\theta & -\sin\theta\\ 0 & \sin\theta & \cos\theta\end{bmatrix}\quad [O_{13}]=\begin{bmatrix}\cos\theta & 0 & -\sin\theta\\ 0 & 1 & 0\\ \sin\theta & 0 & \cos\theta\end{bmatrix}$$

According to quaternion algebra, each of these component rotations has associated with it a 2x2 matrix (the Pauli matrices) usually written as $\underset{\sim}{\sigma} = (\sigma_1,\sigma_2,\sigma_3)$ where $\sigma_1$ is associated with $O_{23}$ etc. and

$$\sigma_1 = \begin{bmatrix} 1 & 0 \\ 0 & -1 \end{bmatrix} \quad \sigma_2 = \begin{bmatrix} 0 & 1 \\ 1 & 0 \end{bmatrix} \quad \sigma_3 = \begin{bmatrix} 0 & -i \\ i & 0 \end{bmatrix}$$

Equation 2.1 may then be written as

$$(\underset{\sim}{r}.\underset{\sim}{\sigma})' = \exp[-i\mu(\underset{\sim}{\sigma}.\underset{\sim}{n})]\ (\underset{\sim}{r}.\underset{\sim}{\sigma})\ \exp[i\mu(\underset{\sim}{\sigma}.\underset{\sim}{n})] \tag{A.2}$$

where $\mu = \theta/2$, $\underset{\sim}{n}$ is the direction vector for the axis of rotation and $(\underset{\sim}{r}.\underset{\sim}{\sigma})$ is called a spin matrix or quaternion and has explicit form

$$(\underset{\sim}{r}.\underset{\sim}{\sigma}) = \begin{bmatrix} r_1 & r_2 - ir_3 \\ r_2 + ir_3 & -r_1 \end{bmatrix}$$

The exponential matrix is 2x2 unitary ($\underset{\sim}{n}$ is real so $\underset{\sim}{\sigma}.\underset{\sim}{n}$ is hermitian and the exponential of an hermitian matrix is always unitary), defined by the following series expansion

$$\exp[X] = I + X + X^2/2! + .....X^n/n! \tag{A.3}$$

A spinor $\underset{\sim}{S}$ is defined by extracting half the transformation of equation 2 such that under a rotation of $\underset{\sim}{r}$, $\underset{\sim}{S}$ transforms as

$$\underset{\sim}{S}' = \exp[-i\mu(\underset{\sim}{\sigma}.\underset{\sim}{n})]\ \underset{\sim}{S} \tag{A.4}$$

We can now see how the Poincare sphere mapping arises by identifying $\underset{\sim}{S}$ with the electric field descriptor $\underset{\sim}{E}$ and the exponential matrix with a unitary change of base.

To identify invariants and verify the relationship between 1 and 2, we consider an infinitessimal transformation through $\delta\theta$, when only the first two terms of 3 need be used ie.

$$(\underset{\sim}{r}.\underset{\sim}{\sigma})' = [I - i\delta\mu(\underset{\sim}{\sigma}.\underset{\sim}{n})]\ (\underset{\sim}{r}.\underset{\sim}{\sigma})\ [I + i\delta\mu(\underset{\sim}{\sigma}.\underset{\sim}{n})] \tag{A.5}$$

Expanding and collecting terms to first order in $\delta\mu$ we obtain the following important form of 2

$$(\underset{\sim}{r}.\underset{\sim}{\sigma})' = (\underset{\sim}{r}.\underset{\sim}{\sigma}) + i\delta\mu\ [(\underset{\sim}{r}.\underset{\sim}{\sigma})(\underset{\sim}{\sigma}.\underset{\sim}{n}) - (\underset{\sim}{\sigma}.\underset{\sim}{n}).(\underset{\sim}{r}.\underset{\sim}{\sigma})] \tag{A.6}$$

From 6 we see that invariants can be identified from products of the Pauli matrices which commute ie. for which $\sigma_i\sigma_j - \sigma_j\sigma_i = 0$. The commutation properties of $\sigma$ are easily derived and conveniently written in the form

$$\sigma_i\sigma_j + \sigma_j\sigma_i = i\sigma_\kappa \qquad i,j,k = 1,2,3 \text{ and cyclic permutation}$$

$$\sigma_i^2 = I$$

We see that there is only one invariant from 6, namely the component of $\underset{\sim}{r}$ along the direction n, as expected from our understanding of rotations in 3 dimensions. This result may also be obtained by noting that the product of 2 spin matrices $\underset{\sim}{a}.\underset{\sim}{\sigma}$ and $\underset{\sim}{b}.\underset{\sim}{\sigma}$ may be written

$$(\underset{\sim}{a}.\underset{\sim}{\sigma})(\underset{\sim}{b}.\underset{\sim}{\sigma}) = \underset{\sim}{a}.\underset{\sim}{b} + i(\underset{\sim}{a}x\underset{\sim}{b})\underset{\sim}{\sigma} \tag{A.7}$$

and that the cross product term is antisymmetric under exchange of $\underset{\sim}{a}$ and $\underset{\sim}{b}$. This leaves only $\underset{\sim}{a}.\underset{\sim}{b}$ as a transformation invariant under 6.

In summary, we have seen that wave states may be associated with a 3 vector $\underset{\sim}{r}$ and that change of polarisation base is equivalent to a rotation of $\underset{\sim}{r}$. By studying infinitessimal rotations we showed that equation 2 guaranteed invariance of the component of $\underset{\sim}{r}$ along $\underset{\sim}{n}$ as expected from 1.

In polarisation algebra the spin matrix corresponds to the wave coherency matrix. Note that the general (or Lorentz) spin matrix is augmented by the wave intensity times the 2x2 identify (included in the Pauli ring as $\sigma_o$) and the intensity and 3-vector $\underset{\sim}{r}$ then form a Stokes 4-vector. Since $\sigma_o$ commutes with all Pauli matrices, its coefficient remains invariant under 6 and hence we are justified in identifying its coefficient as a scalar.

While wave states can be mapped into a 3 dimensional space, targets, with 8 degrees of freedom, require a higher dimensional representation. It has been shown [4.11] that they can be mapped into a 6 dimensional real space where rotations require 15 generalised Euler angles. Further, we may write rotation of a 6-vector $\underset{\sim}{t}$ either in terms of a 6x6 real orthogonal matrix or a 4x4 complex unitary matrix. It is this latter representation which provides a link with polarisation theory.

We begin by considering rotation of a 6-vector $\underset{\sim}{t}$ such that

$$\underset{\sim}{t} = [O_6]\ \underset{\sim}{t} \tag{A.8}$$

where $[O_6]$ may be written as the product of 15 component plane rotations of the form

$$[O_{12}] = \begin{bmatrix} \cos\theta & -\sin\theta & 0 & 0 & 0 & 0 \\ \sin\theta & \cos\theta & 0 & 0 & 0 & 0 \\ 0 & 0 & 1 & 0 & 0 & 0 \\ 0 & 0 & 0 & 1 & 0 & 0 \\ 0 & 0 & 0 & 0 & 1 & 0 \\ 0 & 0 & 0 & 0 & 0 & 1 \end{bmatrix}$$

Associated with each of the 15 plane rotations $O_{ij}$ is a 4x4 hermitian matrix $\beta_{ij}$ (see figure 1) eg.

$$[\beta_{12}] = \begin{bmatrix} 0 & 0 & i & 0 \\ 0 & 0 & 0 & 1 \\ -i & 0 & 0 & 0 \\ 0 & 1 & 0 & 0 \end{bmatrix}$$

The $\beta_{ij}$ form an important set of matrices (written as $\underset{\sim}{\beta}$), analagous to the Pauli matrices of quaternion algebra. The set $\underset{\sim}{\beta}$ are generators of the group SU(4), the group of 4x4 unitary matrices with unit determinant. It follows that we may write an arbitrary 4x4 unitary matrix $[U_4]$ in the form

$$[U_4] = \exp[-i(\underset{\sim}{n}.\underset{\sim}{\beta})] \tag{A.9}$$

If we define a spin matrix as $\underset{\sim}{m}.\underset{\sim}{\beta}$, $\underset{\sim}{m}$ is a 15 element vector and the spin matrix has the general form

$$S_{11} = M_5+M_{10}+M_{15} \qquad S_{12} = (M_1+M_4)+i(M_{14}-M_{11})=S_{21}^*$$

$$S_{13} = (M_2+M_8)+i(M_7-M_{13})=S_{31}^* \qquad S_{14} = (M_3+M_{12})+i(M_9-M_6)=S_{41}^*$$

$$S_{22} = M_5-M_{10}-M_{15} \qquad S_{23} = (M_6+M_9)+i(M_5-M_{12})=S_{32}^*$$

$$S_{24} = (M_7+M_{13})+i(M_8-M_2)=S_{42}^* \qquad S_{33} = M_{10}-M_{15}-M_5$$

$$S_{34} = (M_{11}+M_{14})+i(M_1-M_4)=S_{43}^* \qquad S_{44} = M_{15}-M_{10}-M_5$$

where $S_{ij}$ are the elements of a 4x4 Hermitian matrix. The most important consequence of this formulation is that when the spin matrix is identified with a target covariance matrix, then the vector $\underset{\sim}{m}$ is a lexicographic representation of the target Mueller matrix. This then provides a formal link between the covariance matrix and Stokes vector formulation of target scattering (note that the Mueller matrix has 16 elements with $m_{oo}$, the first element, a scalar, included in the $\beta$ formulation by analogy with the Pauli ring by defining $\beta_o$ as a 4x4 unit matrix).

To identify invariants of $\underset{\sim}{m}$, consider a transformation of the form

$$\underset{\sim}{m}.\underset{\sim}{\beta} = \exp[-i(\underset{\sim}{n}.\underset{\sim}{\beta})]\ (\underset{\sim}{m}.\underset{\sim}{\beta})\ \exp[i(\underset{\sim}{n}.\underset{\sim}{\beta})] \tag{A.10}$$

We may then define a 4-spinor $\underset{\sim}{k}$ by extracting half this transformation as

$$\underset{\sim}{k}' = \exp[-i(\underset{\sim}{n}.\underset{\sim}{\beta})]\ \underset{\sim}{k} \tag{A.11}$$

In section 2 we identified $\underset{\sim}{k}$ as a target descriptor obtained by expansion of the scattering matrix in terms of the Pauli matrices, while the exponential matrix is identified as a change of base matrix for this expansion. We now return to 10 and seek to identify invariants of the target spin matrix.

By considering infinitessimal rotations and using the series form for the unitary matrix we may again identify invariants from products of the $\underset{\sim}{\beta}$ matrices which commute. The commutation properties of $\underset{\sim}{\beta}$ are more difficult to represent than those of the Pauli matrices and are shown in tabular form in table I. The first column shows the plane under consideration and the top row the elements of $\underset{\sim}{m}$. The invariant elements are indicated by crosses. For example, under a 26 plane rotation, $m_1$, $m_4$, $m_5$, $m_8$, $m_9$, $m_{12}$ and $m_{13}$ are all invariant. Under composite rotations eg. in the 12 and 23 planes, the number of invariants reduces to 3 (in this case $m_3$, $m_{13}$ and $m_{14}$). Note that there are 8 invariants per plane rotation (as opposed to only 2 for waves states), the 7 indicated in table 1 plus the scalar $m_o$.

In summary, we have seen that both wave and target states may be given geometrical interpretation in terms of rotation invariants in n dimensional space. Wave states map into 3 dimensions (the Poincare sphere) where invariants under rotation generate 3 of the Stokes parameters, while targets are mapped into a 6 dimensional space where rotation invariants are identified as 15 elements of the target Mueller matrix. In both cases additional scalar invariants are also identified: the wave intensity for wave states and the first element of the Mueller matrix, trace of target spin matrix or span of scattering matrix for targets.

REFERENCES

M.P. van Albada, and A.Lagendijk 'Observation of Weak Localisation of Light in a Random Media', Phys. Rev. Letters, Vol.**55**, No.24, Dec.1985, pp 2692-2695.

M.P. van Albada and A. Lagendijk 'Vector character of light in weak localisation...', Phys. Rev. B., Vol.**36**, No.4, August 1987, pp 2353-2356.

G. Arfken 'Mathematical Methods for Physicists', Ch.4, Academic Press, 1970.

R.M. Barnes 'Roll Invariant Decompositions for the Polarisation Covariance Matrix', Internal report, Lincoln Laboratory, M.I.T., Lex. MA 02173-0073.

P. Beckmann 'The depolarisation of EM Waves', Golen Press, 1969.

M. Borgeaud, R.T. Shin and J.A. Kong 'Theoretical models for Polarimetric Radar Clutter', Journal of EM Waves and Applications, Vol. **1**, No.1, pp 73-91.

J. Byrne 'A Classification of Electron and Optical Polarisation transfer Matrices', J. Phys. B. (Atom.Molec.Phys.), Vol. **4**, 1971, pp 940-953.

S. Chandrasekhar 'Radiative Transfer' Dover 1960.

S. Cloude 'Radar Target Decomposition theorems', Electronics letters, Vol. **21**, No.1, Jan. 1985, pp22-24.

S. Cloude 'Group Theory and Polarisation Algebra', OPTIK, Vol. **75**, No.1, 1986, pp26-36.

S. Cloude 'Polarimetry: the characterisation of polarisation effects in EM scattering', PhD. thesis, University of Birmingham, Oct. 1986.

S. Cloude 'Optimisation of Signal to Clutter using polarisation Diversity', Electronics letters, Vol. **24**, No.4, Feb. 1988, pp194-195.

D. Giuli 'Polarisation Diversity in Radars', Proc. IEEE, Vol.**74**, No.2, Feb. 1986, pp245-269.

H. van de Hulst 'Light Scattering by small Particles', Dover press 1980.

J.R. Huynen 'Phenomenological Theory of radar targets' PhD. thesis, Drukkery Bronder-Offset N.V. Rotterdam 1970.

J.R. Huynen 'Component Fluctuations for fluctuating radar targets', Proc. IEEE trans AES-11 No.6, Nov. 1975, pp1316-1332.

F. Murnaghan 'The Unitary and Rotation Groups' Lectures on Applied Maths, Vol.**3**, Spartan Books, 1963.

L.M. Novak et al, 'Studies of Target Detection Algorithms which use Polarimetric radar data', Proc. 21st Conf. on Signals, Systems and Computers, Pacific Grove, CA, Nov. 2-4, 1987.

R. Schmeider, 'Stokes Algebra Formalism', J. Opt. Soc. Am., Vol. **59**, March 1969, pp297-302.

A. Swartz et al 'The Optimal Polarisations for Achieving Maximum Contrast in Radar Images' to appear in Journal of Geophysical Research, Solid Earth and Planets, 1988.

| m / $O_{ij}$ | 1 | 2 | 3 | 4 | 5 | 6 | 7 | 8 | 9 | 10 | 11 | 12 | 13 | 14 | 15 |
|---|---|---|---|---|---|---|---|---|---|---|---|---|---|---|---|
| 26 | x | | | x | x | | | x | x | | | x | x | | |
| 24 | | x | | x | | x | | x | | x | | x | | x | |
| 46 | | | x | x | | | x | x | | | x | x | | | x |
| 35 | x | x | x | x | x | x | x | | | | | | | | |
| 14 | x | | | x | x | | | | | x | x | | | x | x |
| 16 | | x | | x | | x | | | x | | x | | x | | x |
| 12 | | | x | x | | | x | | x | x | | | x | x | |
| 15 | x | x | x | | | | | x | x | x | x | | | | |
| 34 | x | | | | | x | x | x | x | | | | | x | x |
| 36 | | x | | | x | | x | x | | x | | | x | | x |
| 23 | | | x | | x | x | | x | | | x | | x | x | |
| 13 | x | x | x | | | | | | | | | x | x | x | x |
| 45 | x | | | | | x | x | | | x | x | x | x | | |
| 56 | | x | | | x | | x | | x | | x | x | | x | |
| 25 | | | x | | x | x | | | x | x | | x | | | x |

TABLE I : Invariants of Mueller Matrix under Plane Rotations in R6

$$\underset{0}{\begin{pmatrix}1&0&0&0\\0&1&0&0\\0&0&1&0\\0&0&0&1\end{pmatrix}}\quad\underset{1}{\overset{(26)}{\begin{pmatrix}0&1&0&0\\1&0&0&0\\0&0&0&i\\0&0&-i&0\end{pmatrix}}}\quad\underset{2}{\overset{(24)}{\begin{pmatrix}0&0&1&0\\0&0&0&-i\\1&0&0&0\\0&i&0&0\end{pmatrix}}}\quad\underset{3}{\overset{(46)}{\begin{pmatrix}0&0&0&1\\0&0&i&0\\0&-i&0&0\\1&0&0&0\end{pmatrix}}}$$

$$\underset{4}{\overset{(35)}{\begin{pmatrix}0&1&0&0\\1&0&0&0\\0&0&0&-i\\0&0&i&0\end{pmatrix}}}\quad\underset{5}{\overset{(14)}{\begin{pmatrix}1&0&0&0\\0&1&0&0\\0&0&-1&0\\0&0&0&-1\end{pmatrix}}}\quad\underset{6}{\overset{(16)}{\begin{pmatrix}0&0&0&-i\\0&0&1&0\\0&1&0&0\\i&0&0&0\end{pmatrix}}}\quad\underset{7}{\overset{(12)}{\begin{pmatrix}0&0&i&0\\0&0&0&1\\-i&0&0&0\\0&1&0&0\end{pmatrix}}}$$

$$\underset{8}{\overset{(15)}{\begin{pmatrix}0&0&1&0\\0&0&0&i\\1&0&0&0\\0&-i&0&0\end{pmatrix}}}\quad\underset{9}{\overset{(34)}{\begin{pmatrix}0&0&0&i\\0&0&1&0\\0&1&0&0\\-i&0&0&0\end{pmatrix}}}\quad\underset{10}{\overset{(36)}{\begin{pmatrix}1&0&0&0\\0&-1&0&0\\0&0&1&0\\0&0&0&-1\end{pmatrix}}}\quad\underset{11}{\overset{(23)}{\begin{pmatrix}0&-i&0&0\\i&0&0&0\\0&0&0&1\\0&0&1&0\end{pmatrix}}}$$

$$\underset{12}{\overset{(13)}{\begin{pmatrix}0&0&0&1\\0&0&-i&0\\0&i&0&0\\1&0&0&0\end{pmatrix}}}\quad\underset{13}{\overset{(45)}{\begin{pmatrix}0&0&-i&0\\0&0&0&1\\i&0&0&0\\0&1&0&0\end{pmatrix}}}\quad\underset{14}{\overset{(56)}{\begin{pmatrix}0&i&0&0\\-i&0&0&0\\0&0&0&1\\0&0&1&0\end{pmatrix}}}\quad\underset{15}{\overset{(25)}{\begin{pmatrix}1&0&0&0\\0&-1&0&0\\0&0&-1&0\\0&0&0&1\end{pmatrix}}}$$

FIGURE 1 : GENERATORS OF PLANE ROTATIONS IN R6

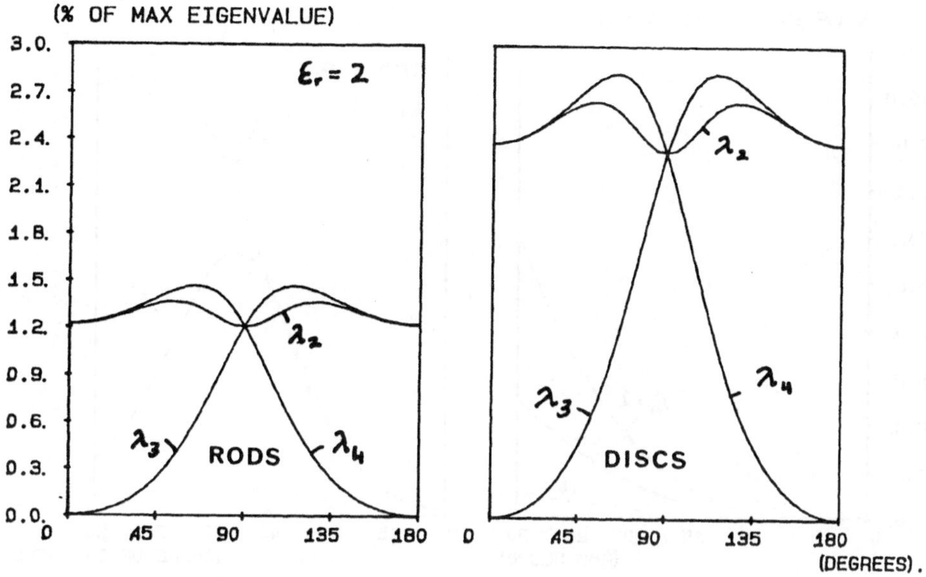


FIGURES 2/3: EIGENVALUES vs SCATTERING ANGLE.

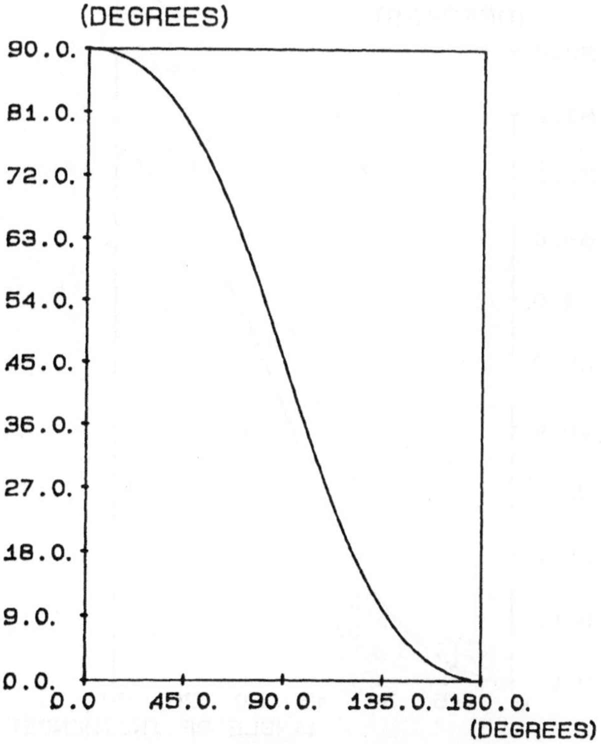


FIGURE 4: EIGENVECTOR ROTATION.

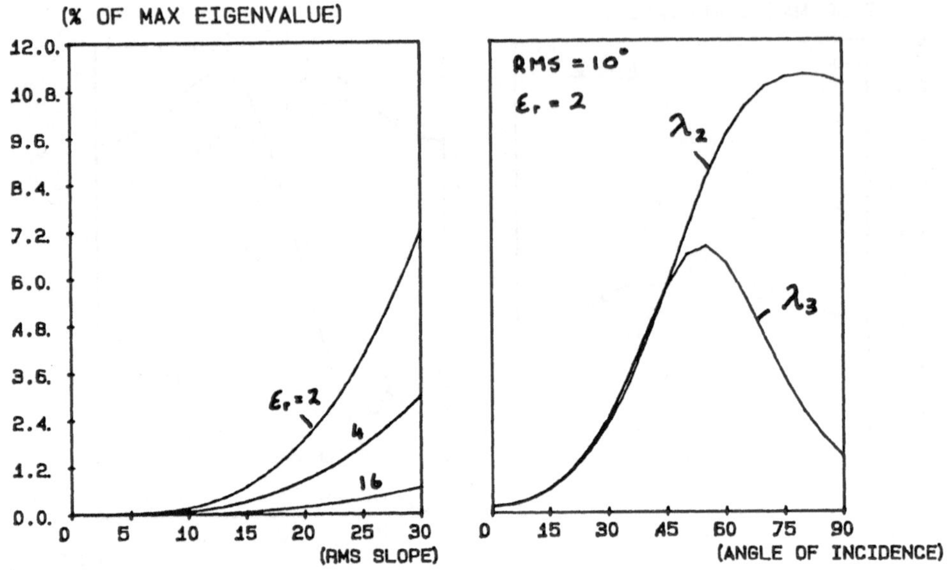


FIGURES 5/6 : EIGENVALUES FOR GAUSSIAN SURFACE.

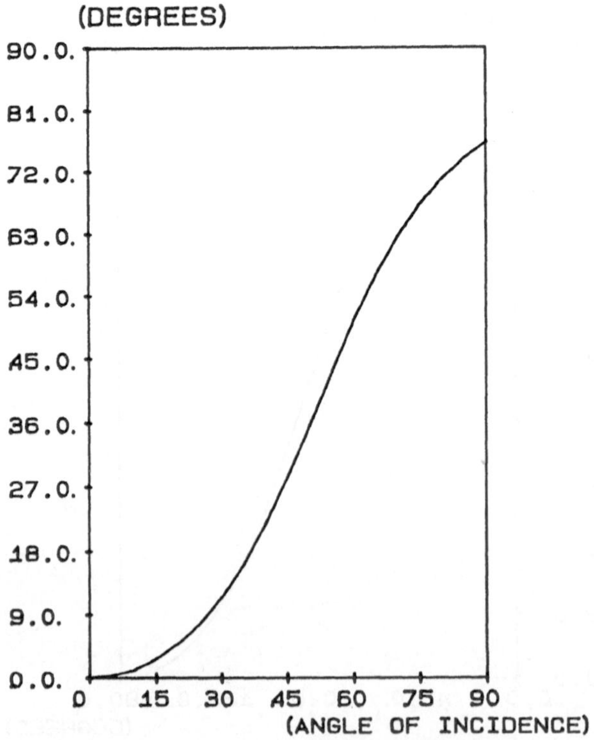


FIGURE 7 : EIGENVECTOR ROTATION.

# DETERMINATION OF THE CHARACTERISTIC POLARIZATION STATES OF THE RADAR TARGET SCATTERING MATRIX [S(AB)] FOR THE COHERENT, MONOSTATIC AND RECIPROCAL PROPAGATION SPACE USING THE POLARIZATION RATIO $\rho$ TRANSFORMATION FORMULATION

**An-Qing Xi and Wolfgang-Martin Boerner**
**Communications & Sensing Laboratory,**
**Department of Electrical Engineering & Computer Science,**
**University of Illinois at Chicago, Chicago, IL 60680**

**Abstract:** A problem originating in radar polarimetry is considered for which the radar target is to be characterized by its coherent polarization state properties, given complete coherent backscattering scattering matrix data sets at one frequency and for one target aspect angle. First, the Jones vector formalism for the coherent monostatic case, together with Sinclair's backscattering matrix [S(AB)] for the general polarization basis (AB) are introduced. Using the unitary change of polarization state transformation, the concept of the characteristic polarization states of a scatterer, first introduced by Kennaugh and Huynen, is presented. The generalized unitary transformation matrix formulation under the change of basis transformation, expressed in terms of the generalized polarization ratio $\rho(AB)$, is developed for emphasizing the unique properties of the interrelation among the existing characteristic polarization states. For the monostatic reciprocal case ($S_{AB} = S_{BA}$), treated here, it is shown that there exist in total five pairs of characteristic polarization states: The orthogonal cross-polarization null and co-polarization maximum state pairs, being identical and sharing one main circle with the co-polarization null and the orthogonal cross-polarization maximum state pairs, the latter being at right angles to the cross-polarized null pairs; and, a newly identified pair: the orthogonal cross-polarization saddle point extrema which are normal to the plane (main circle) spanned by the other four pairs. With this complete and unique mathematical description of Huynen's polarization fork concept, it is now possible to study the polarimetric radar target optimization problem more rigorously. Various examples are provided and interpreted by comparing the unique result with previous incomplete analyses. In conclusion, the relevance of these canonical results to optical polarimetry are highlighted and interpreted.

## 1. INTRODUCTION

The use of characteristic polarization states in radar target applications has been a subject of recurring interest in recent years. The concept of characteristic polarization states was first introduced by Kennaugh [1], who demonstrated that there exist radar polarization states for which the radar receives minimum/maximum power. This min/max polarization state theory was extended by Huynen [2] and by us [3]-[6].



*W.-M. Boerner et al. (eds.), Direct and Inverse Methods in Radar Polarimetry, Part 1*, 297–349.

In spite of the extensive studies of this theory, it has been found that in the available literature, the characteristic polarization state theory has not to this point been carried out rigorously and systematically and the question of uniqueness and validity was also occasionally raised [5]. Different approaches were applied for determining these characteristic polarization states by using the voltage equation [1], the eigenvalue problem of the power scattering matrix [2],[4], and the unitary transformation of the scattering matrix technique [3] to solve for four [3] or six [7] of the eight characteristic polarization states.

Recently, Kostinski and Boerner [5] have critically reviewed the fundamentals of radar polarimetry, and pointed out that several ambiguities exist in the available literature particularly in relation to the use of the unitary transformation matrix, and the interpretation of the characteristic polarization state theory. Because of the fact that this theory is being used in an increasingly wider set of current investigations, a rigorous redevelopment using exclusively the generalized ρ-formulation and including our new findings is desirable in order to apply it more effectively in radar target phenomenology, identification and classification applications. Especially, a generalized ρ-formulation for the reciprocal monostatic and the coherent scatterer case is considered here.

## 2. FUNDAMENTALS OF RADAR POLARIMETRY

### 2.1 Polarization States and the Poincaré Sphere

A plane electromagnetic wave $\vec{E}(HV)$, propagating in the +z - direction with $\exp\{j(\omega t-kz)\}$ can be expressed in terms of the two orthonormal components in the $\hat{h}_H$ and $\hat{h}_V$ polarization basis (HV) as

$$\vec{E}(HV) = E_H\hat{h}_H + E_V\hat{h}_V = \cos\alpha\, e^{j\delta_H}\,\hat{h}_H + \sin\alpha\, e^{j\delta}\,\hat{h}_V$$

$$= e^{j\delta_H}(\cos\alpha\hat{h}_H + \sin\alpha\, e^{j\delta}\,\hat{h}_V) \tag{1}$$

where $E_H$ and $E_V$ are the complex horizontally and vertically polarized components of the electric field, and the angles $\alpha$ and $\delta$ are defined in (2c). The polarization state $\vec{E}(HV)$ of the wave, also known as the Jones vector [8], is described in the form of an ellipse as shown in Fig. (1a). The parameters $\phi$ and $\tau$ are the tilt and the ellipticity angles of the ellipse given by

$$\tan 2\phi = 2|E_H||E_V|\cos\delta/(|E_H|^2-|E_V|^2) \quad , \tag{2a}$$

$$\sin 2\tau = 2|E_H||E_V|\sin\delta/(|E_H|^2+|E_V|^2) \quad , \tag{2b}$$

and $\delta(HV) = \delta_V - \delta_H$, the phase difference between the two orthogonal components $E_H$ and $E_V$, is the relative polarization phase. The polariza-

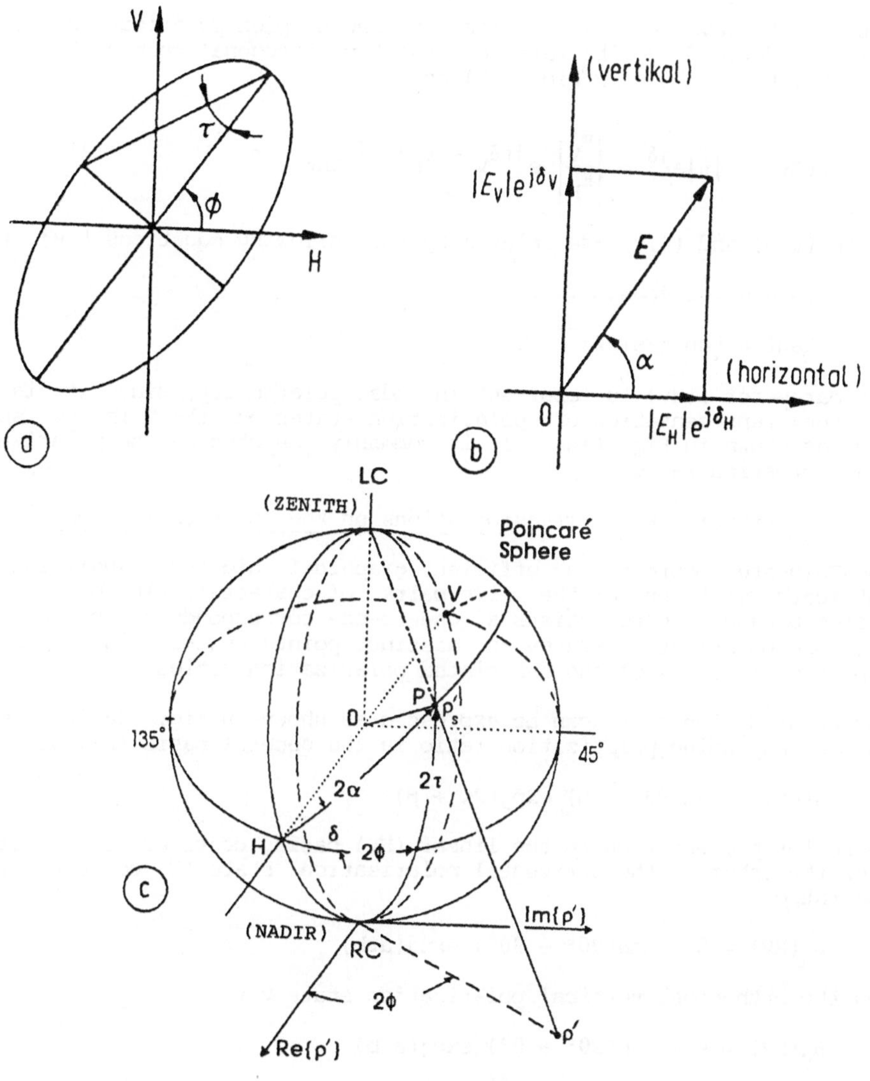


Fig. 1a Parametric presentation of the polarization ellipse
Fig. 1b Representation of the polarization ratio $\rho(HV)$ in the horizontal-vertical (H-V) basis: $\rho_H(HV) = 0$ and $\rho_V(HV) = \infty$
Fig. 1c Representation of a polarization state on the Poincaré sphere

tion of the wave can also described by the complex polarization ratio $\rho$, which is defined by the ratio of the two orthogonal components [9] in the (HV) basis shown in Fig. (1b) as

$$\rho(HV) = |\rho|e^{j\delta} = \left|\frac{E_V}{E_H}\right| e^{j(\delta_V - \delta_H)} = \tan\alpha\ e^{j\delta} \quad , \tag{2c}$$

where $(\alpha,\delta)$ and $(\phi,\tau)$ are related by the following equations [10], [11]:

$$\cos2\alpha = \cos2\phi\cos2\tau \quad , \tag{3a}$$

$$\tan\delta = \tan2\tau/\sin2\phi \quad . \tag{3b}$$

The parameter $\rho(HV)$ is important in radar polarimetry, since it is used for the representation of polarization states on the Poincaré sphere [10] as shown in Fig. (1c). It is commonly denoted as the polarization transformation ratio.

## Polarization Ratio Transformations on the Polarization Sphere

The Poincaré sphere is an efficient graphical aid for visualizing the intricate relations of the five pairs of characteristic polarization states because there exists a one-to-one correspondence between all possible polarization states as distinct points on the Poincaré sphere expressed in terms of the associated polarization ratios.

Any polarization state can be expressed as shown in Fig. 2a in terms of the corresponding polarization ratio in the general basis (AB) as

$$\rho(AB) = \tan(90^\circ - \eta)\ \mathrm{exp}j(2\nu + \beta) \tag{4}$$

where for presentation in the linear (HV) basis, centered at the origin 0 of the sphere, the horizontal polarization state (H) corresponds to the ratio

$$\rho_H(HV) = 0 = \tan(90^\circ - 90^\circ)\ \mathrm{exp}j(arb) \quad , \tag{5}$$

and the orthogonal vertical polarization state V to

$$\rho_V(HV) = \infty = \tan(90^\circ - 0^\circ)\ \mathrm{exp}j(arb) \quad , \tag{6}$$

with unspecified phase, i.e., $(2\nu + \beta) \hat{=}$ arbitrary, for these two special cases. The two points H and V, corresponding to orthogonal polarization ratios, $\rho_H\rho_V^* = -1$, are located at a pair of antipodal points on the Poincaré sphere. Referring to these points as the polarization basis (HV), with radius V0 as the "angle origin", then all other points corresponding to different polarization states can be located straightforwardly on the Poincaré sphere. According to the Poincaré sphere presentation rules [4], [10]:

(i) All points corresponding to polarization ratios $\rho(HV)$ with the

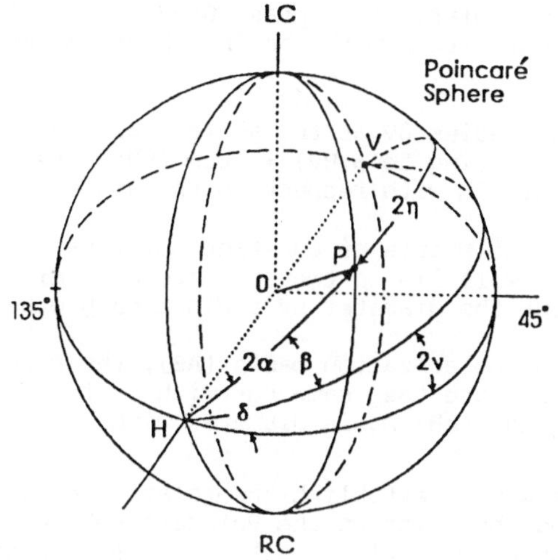


Fig. 2a: The polarization ratio ρ transformation on the polarization sphere

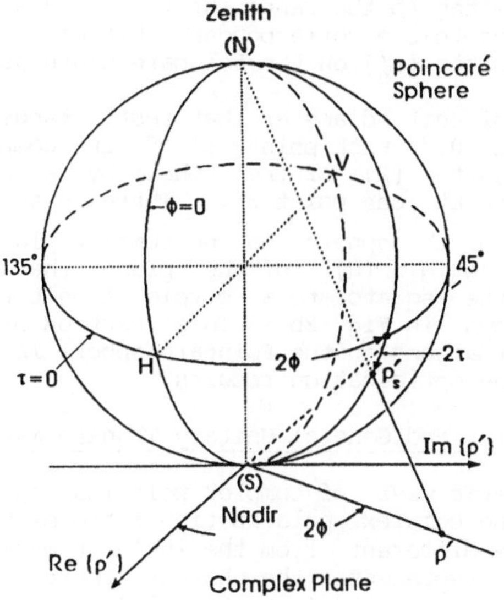


Fig. 2b: Correspondence of points on the complex polar plane ρ′ with the polarization sphere $\rho'_S$

same relative polarization phase $\delta(HV)$ = const. lie on one and the same great circle with the (HV) basis diameter joining H and V; (7)

(ii) Choosing the radius 0V as the angle orgin in the (HV) basis, any point corresponding to $\rho(HV) = \tan(90° - \eta)\, \exp j(2\nu + \beta)$ will be at an angle $2\eta$ with respect to the angle origin; (8)

(iii) Any two sets of points of constant polarization phases but with difference $\beta$ will lie on two separate main circles rotated by an angle $\beta$ about the diameter HV against each other; (9)

(iv) For any other polarization basis (AB), the Poincaré sphere presentation rules are the same but with reference points A and B consituting the (AB) basis diameter with $\rho_A(AB) = 0$ and $\rho_B(AB) = \infty$ (10)

Another more direct way to establish the relation between the polarization ratio $\rho(AB)$ and the point on the Poincaré sphere can be obtained by the mapping from the complex plane to the Poincaré sphere. The polarization ratio $\rho(AB)$ which is a complex number can also be expressed using the tilt angle $\phi$ and ellipticity angle $\tau$ as [26] for any basis (AB):

$$\rho(AB) = \frac{\tan\phi + j\tan\tau}{1 - j\tan\phi\tan\tau} \tag{11}$$

When $\phi$ and $\tau$ are limited to the range $-\frac{1}{2}\pi \leqslant \phi < \pi/2$ and $-\frac{1}{4}\pi \leqslant \tau \leqslant \frac{1}{4}\pi$, (11) provides an one-to-one correspondence between points ($\rho'$) on the complex plane and points ($\rho'_s$) on the Poincaré shere as shown in Fig. 2a.

Consider a sphere of unit diameter that rests tangent to the complex plane at its origin 0. Each point $\rho'$ of the complex plane can be connected to the Zenith (Z) of the sphere by a straight line that intersects the sphere at one point $\rho'_s$. There is a one-to-one mapping between every point of the sphere and the complex plane. The Nadir (N) corresponds to the "origin (0)" of the plane, the Zenith (Z) to the "infinity ($\infty$)" and the equator to a circle of unit radius centered at the origin 0 as shown in Fig. 2b. This section of the polarization ratio $\rho(AB)$ transformations on the Poincaré sphere is necessary for the interpretation of the optimization results.

## 2.2 Scattering Matrix and General Unitary Transformation Matrix

When an electromagnetic wave of complex polarization ratio $\rho(HV)$ illuminates a target, the complex polarization ratio of the scattered wave will, in general, be different from the incident wave. This change of polarization states represents the characteristic properties of the target expressed by the scattering matrix [S] for fixed frequency and aspect angle. In the monostatic reciprocal case, the normalized scattering matrix [S(HV)] with relative phase in the orthogonal linear polarization basis (HV) is given as [1]:

$$\vec{E}_s(HV) = [S(HV)]\, \vec{E}_i(HV) \tag{12}$$

where $\vec{E}_i(HV)$ and $\vec{E}_s(HV)$ denote the incident and the scattered fields, respectively, and

$$[S(HV)] = \begin{bmatrix} S_{HH} & S_{HV} \\ S_{VH} & S_{VV} \end{bmatrix} \quad (13)$$

with $S_{HV} = S_{VH}$ for the reciprocal monostatic case.

A 2x2 general unitary transformation matrix [T] for transferring a linear polarization basis (HV) to another orthonormal polarization basis (AB) can be written as [12]:

$$\underline{h}(AB) = [T]\underline{h}(HV) \quad (14a)$$

$$[T] = \begin{bmatrix} \cos\alpha\, e^{j\psi_1} & \sin\alpha\, e^{j\psi_2} \\ -\sin\beta\, e^{j\psi_3} & \cos\beta\, e^{j\psi_4} \end{bmatrix} \quad (14b)$$

Since [T] is an unitary matrix, it satisfies the following condition:

$$[T]^{*T} = [T]^{-1} \quad , \quad (15a)$$

and

$$[T]^{*T}[T] = [I] \quad ,$$

$$\det\{[T]^{*T}[T]\} = \det\{[T]^{*T}\}\det\{[T]\} = \{\det[T]\}^2 = 1 \quad ,$$

that is

$$|\det\{[T]\}| = 1 \quad . \quad (15b)$$

Applying (15b) to (14b) gives:

$$\psi_1 + \psi_4 = \psi_2 + \psi_3 \quad , \quad (16a)$$

and

$$\alpha = \beta \quad . \quad (16b)$$

and (14b) can be written in terms of $\rho(HV)$ as:

$$[T] = \frac{1}{\sqrt{1+\rho\rho^*}} \begin{bmatrix} e^{j\psi_1} & \rho e^{j\psi_1} \\ -\rho^* e^{j\psi_4} & e^{j\psi_4} \end{bmatrix} \quad , \quad (17a)$$

$$\cos\alpha = \frac{1}{\sqrt{1+\rho\rho^*}} \quad , \tag{17b}$$

where $\rho(HV) = \tan\alpha\, e^{j\delta}$, $\delta = \psi_2 - \psi_1 = \psi_4 - \psi_3$ and $\det\{[T]\} = e^{j(\psi_1+\psi_4)}$ .

Note that [T] is the transformation matrix for the polarization bases transformation from (HV) to (AB). The vector transformation matrix [U] and the basis transformation matrix [T] are related by $[U]=[T]^T$, where [U] is related to the field vectors $\vec{E}(HV)$ and $\vec{E}'(AB)$ [3], so that

$$\vec{E}(HV) = [U]\vec{E}'(AB) \quad , \tag{18a}$$

$$[U] = \frac{1}{\sqrt{1+\rho\rho^*}}\begin{bmatrix} e^{j\psi_1} & -\rho^* e^{j\psi_4} \\ \rho e^{j\psi_1} & e^{j\psi_4} \end{bmatrix} \quad , \tag{18b}$$

and with $\Psi_1 = \frac{1}{2}(\psi_1 - \psi_4)$ and $\Psi_2 = \frac{1}{2}(\psi_1 + \psi_4)$ we find that:

$$[U] = \frac{1}{\sqrt{1+\rho\rho^*}}\begin{bmatrix} \exp(j\Psi_1) & -\rho^*\exp(-j\Psi_1) \\ \rho\exp(+j\Psi_1) & \exp(-j\Psi_1) \end{bmatrix}\exp(j\Psi_2) \quad . \tag{18c}$$

2.3 Transformation of the Scattering Matrix

Substituting (18a) into the voltage equation $V = \vec{E}^T[S]\vec{E}$ [1], the voltage equation in a new basis can be written as:

$$V = V' = \vec{E}'^T[U]^T[S][U]\vec{E}' = \vec{E}'^T[S']\vec{E}' \quad . \tag{19}$$

This gives the scattering matrix in the new basis as

$$[S'(AB)] = [U]^T[S][U] \quad , \tag{20a}$$

$$[S'(AB)] = \begin{bmatrix} S'_{AA} & S'_{AB} \\ S'_{BA} & S'_{BB} \end{bmatrix} \qquad S'_{AB} = S'_{BA} \tag{20b}$$

Now substituting (18b) into (20), the elements of the transformed scattering matrix in the new basis (AB) can be written as:

$$S'_{AA} = (1+\rho\rho^*)^{-1}(\rho^2 S_{VV} + 2\rho S_{HV} + S_{HH})\, e^{2j\psi_1} \quad , \tag{21a}$$

$$S'_{AB} = S'_{BA} = (1+\rho\rho^*)^{-1}(\rho S_{VV} + (1-\rho\rho^*)S_{HV} - \rho^* S_{HH})e^{j(\psi_1+\psi_4)} \quad , \tag{21b}$$

$$S'_{BB} = (1+\rho\rho^*)^{-1}(\rho^{*2} S_{HH} - 2\rho^* S_{HV} + S_{VV})\, e^{2j\psi_4} \quad . \tag{21c}$$

Note that there is no constraint to the unitary transformation matrices, [T] and [U]; they constitute a four parameter family of unitary matrices and can be used in general cases of complex (elliptical) polarization states. Here and throughout the paper, we will always carry the two phase factors $\psi_1$ and $\psi_4$. The reasons are:

(i) $\psi_1$ and $\psi_4$ are the phases of the elements of the unitary transformation matrix [U], we cannot remove them without a proof of doing so;

(ii) During the basis transofrmation, the new form of the scattering matrix will depend on $\psi_1$ and $\psi_4$ as shown in (20b) and (21). One of the Huynen geometric parameters - the target skip angle $\upsilon$ will depend on $\psi_1$ and $\psi_4$ [2]. As a well known fact, $\upsilon$ should be unique for a given target for a fixed radar frequency and a given target aspect angle, whereas, $\psi_1$ and $\psi_4$ will depend on the chosen measurement basis (AB).

One special case [7] is the transformation from the linear (HV) polarization basis (l) to the circular (LR) one (c), and vice-versa, given as:

$$[T_{lc}(HV\to LR)] = \frac{1}{\sqrt{2}}\begin{bmatrix} 1 & j \\ 1 & -j \end{bmatrix} \tag{22a}$$

$$[T_{cl}(LR\to HV)] = \frac{1}{\sqrt{2}}\begin{bmatrix} 1 & 1 \\ -j & j \end{bmatrix} = [T_{lc}]^{-1} \tag{22b}$$

For the specific case of transformation from (HV) to (LR) bases, we find that $\rho = j$, $\psi_1 = 0$ and $\psi_4 = -\pi/2$, for which the unitary transformation matrix reduces to:

$$[T] = \frac{1}{\sqrt{1+\rho\rho^*}}\begin{bmatrix} 1 & \rho \\ j\rho^* & -j \end{bmatrix} \quad . \tag{23}$$

The elements of the scattering matrix in the new basis now have the following form by substituting $\psi_1 = 0$ and $\psi_4 = -\pi/2$ into (21):

$$S'_{AA} = (1+\rho\rho^*)^{-1}(S_{HH} + 2\rho S_{HV} + \rho^2 S_{VV}) \quad , \tag{24a}$$

$$S'_{AB} = S'_{BA} = -j(1 + \rho\rho^*)^{-1}(\rho S_{VV} + (1 - \rho\rho^*)S_{HV} - \rho^* S_{HH}) \quad , \qquad (24b)$$

$$S'_{BB} = -(1 + \rho\rho^*)^{-1}(\rho^{*2} S_{HH} - 2\rho^* S_{HV} + S_{VV}) \quad . \qquad (24c)$$

These equations agree with the formulae in [7], and (23) is the basis transformation matrix for (HV)→(LR). However, in general, $\psi_1$ and $\psi_4$ must remain undetermined until the specific transformation from the original (HV) to the new (AB) bases have been accomplished as carried out in Section 2.4. This formulation of not specifying $\psi_1$ and $\psi_4$ initially, counter to [7], is necessary for determining the correct presentation of the incident (transmitted) wave fields $\vec{E}_i(HV)$ or $\vec{E}_T(HV)$, which then in a last step can be specified using (B-8) as shwon in Appendix B.

## 2.4 The Power Expressions in the New Basis

The received power of the backscattered signal is proportional to the square of the amplitude of the voltage [1]. If the proportionality constant is taken to be unity, then with $\underline{h} = \underline{E}/||\underline{E}||$, the power expression can be written as:

$$P = |V|^2 = |\underline{h}_r^T[S]\underline{h}_t|^2 = |\underline{h}_r'^T[S']\underline{h}_t'|^2 \quad , \qquad (25)$$

where $\underline{h}_r$ and $\underline{h}_t$ are the effective vector antenna heights of the transmitting and receiving antennas [7] and (′) indicates the new basis.

The co-polarized power can be calculated for $\underline{h}_r = \underline{h}_t = \underline{h}$ and $\underline{h}_r' = \underline{h}_t' = \underline{h}'$, where

$$P^c = |V^c|^2 = |\underline{h}^T[S]\underline{h}|^2 = |\underline{h}'^T[S']\underline{h}'|^2 \quad . \qquad (26a)$$

Similarly, the cross-polarized power can be calculated for $\underline{h}_r = \underline{h}_t^{\perp} = \underline{h}_{\perp}$ and $\underline{h}_r' = \underline{h}_t'^{\perp} = \underline{h}_{\perp}'$, where

$$P^x = |V^x|^2 = |\underline{h}_{\perp}^T[S]\underline{h}|^2 = |\underline{h}_{\perp}'^T[S']\underline{h}'|^2 \quad . \qquad (26b)$$

The optimal polarization problem is to find the polarization states such that for the target of known scattering matrix [S], the power return to the receiver is maximized or minimized. Mathematically speaking, optimization of the bilinear form of (25) is easiest when the matrix [S] is diagonal. To diagonalize [S], we need to find the parameter $\rho$ in (21) such that the off-diagonal elements $S'_{AB} = S'_{BA}$ of (21b) vanish, i.e.,

$$S'_{AB} = S'_{BA} = (1+\rho\rho^*)^{-1}(\rho S_{VV} + (1-\rho\rho^*)S_{HV} - \rho^* S_{HH})e^{j(\psi_1+\psi_4)} = 0 \qquad (27a)$$

or

$$\rho S_{VV} + (1-\rho\rho^*)S_{HV} - \rho^* S_{HH} = 0 \quad . \tag{27b}$$

Eq. (27b) is the same as equation (14b) in [7], although we kept the two parameters $\psi_1$ and $\psi_4$ contrary to [7]. The result is the same as in [7]

$$\rho_{1,2} = \frac{-B \pm \sqrt{B^2 - 4AC}}{2A} \quad , \tag{28}$$

where $A = S_{HH}^* S_{HV} + S_{HV}^* S_{VV}$, $\quad B = |S_{HH}|^2 - |S_{VV}|^2$, $\quad C = -A^*$ .

We find that $\rho_1$, $\rho_2$ are orthogonal, i.e.,

$$\rho_1 \rho_2^* = -1 \quad . \tag{29}$$

In Appendix A, using the Graves power scattering matrix [G], defined in [5], its optimization and the relations to our approach are derived.

Choosing $\rho_1$ or $\rho_2$ in (21), the scattering matrix [S'(AB)] is diagonolized in the new basis and is rewritten in the following form

$$[S'(AB)] = \begin{bmatrix} S'_{AA} & 0 \\ 0 & S'_{BB} \end{bmatrix} = \begin{bmatrix} \lambda_1 & 0 \\ 0 & \lambda_2 \end{bmatrix} = [S_d] \quad , \tag{30}$$

where two pairs of possible λ's can be determined from:

$$\lambda_{11} = S'_{AA}(\rho_1) = (1+\rho_1\rho_1^*)^{-1}(S_{HH} + 2\rho_1 S_{HV} + \rho_1^2 S_{VV})e^{2j\psi_1} \tag{31a}$$

$$\lambda_{12} = S'_{BB}(\rho_1) = (1+\rho_1\rho_1^*)^{-1}(\rho_1^{*2} S_{HH} - 2\rho_1^* S_{HV} + S_{VV})e^{2j\psi_4} \tag{31b}$$

$$\lambda_{21} = S'_{BB}(\rho_2) = (1+\rho_2\rho_2^*)^{-1}(\rho_2^{*2} S_{HH} - 2\rho_2^* S_{HV} + S_{VV})e^{2j\psi_4} \tag{31c}$$

$$\lambda_{22} = S'_{AA}(\rho_2) = (1+\rho_2\rho_2^*)^{-1}(S_{HH} + 2\rho_2 S_{HV} + \rho_2^2 S_{VV})e^{2j\psi_1} \quad . \tag{31d}$$

Note that during the transformation, we used the unitary transformation matrix [U], so the column vectors of [U] are just the new basis vectors. There are two sets of basis vectors that we can choose depending on which $\rho_i(i=1,2)$ of $\rho_{1,2}(HV)$ had been chosen in diagonalizing the scattering matrix. There exist two possible systems (basis and scattering matrix):

(i) for choosing $\rho_1$, the new basis vectors are

$$\bar{V}_A = \frac{e^{j\psi_1}}{\sqrt{1+\rho_1\rho_1^*}}\begin{bmatrix}1\\ \rho_1\end{bmatrix} \tag{32a}$$

$$\bar{V}_B = \frac{e^{j\psi_4}}{\sqrt{1+\rho_1\rho_1^*}}\begin{bmatrix}-\rho_1^*\\ 1\end{bmatrix} \tag{32b}$$

$$[U(\rho_1)] = [\ \bar{V}_A\ \ \bar{V}_B] \tag{32c}$$

and the diagonal elements of the scattering matrix become

$$[S'(AB)] = \begin{bmatrix}\lambda_{11} & 0\\ 0 & \lambda_{12}\end{bmatrix},\quad \lambda_{11} = S'_{AA}(\rho_1)\ ,\quad \lambda_{12} = S'_{BB}(\rho_1)\ , \tag{32d}$$

(ii) for choosing $\rho_2$, the new basis vectors are

$$\bar{V}_{A'} = \frac{e^{j\psi_4}}{\sqrt{1+\rho_2\rho_2^*}}\begin{bmatrix}-\rho_2^*\\ 1\end{bmatrix} \tag{33a}$$

$$\bar{V}_{B'} = \frac{e^{j\psi_1}}{\sqrt{1+\rho_2\rho_2^*}}\begin{bmatrix}1\\ \rho_2\end{bmatrix} \tag{33b}$$

$$[U(\rho_2)] = [\ \bar{V}_{B'}\ \ \bar{V}_{A'}] \tag{33c}$$

and the diagonal elements of the scattering matrix become

$$[S'(A'B')] = \begin{bmatrix}\lambda_{22} & 0\\ 0 & \lambda_{21}\end{bmatrix},\quad \lambda_{21} = S'_{B'B'}(\rho_2)\ ,\quad \lambda_{22} = S'_{A'A'}(\rho_2). \tag{33d}$$

To find the optimal polarization ratios, we can either choose (32) or (33) as the new basis vectors. The final results are the same, because they do not depend on which polarization basis was chosen. So far we have established the power expression (25) in the new basis in which the scattering matrix is in daigonal form. We are now ready to discuss the optimal polarization problem.

# 3. THE OPTIMAL POLARIZATION STATES IN THE COHERENT CASE

## 3.1 The possible polarization ratios of the optimal polarization states

In mathematics, the extrema of a function can be found at the critical points of the function. The functions of the power return to the co-pol and cross-pol channels of the receiver are determined from the bilinear form (26) to become:

(i) For the function of the power returned to the cross-pol channel

$$P^x = |V^x|^2 = |\underline{h}_\perp'^T[S_d]\underline{h}'|^2$$

$$= (1+\rho'\rho'^*)^{-2}(|\lambda_1|^2\rho'\rho'^* - \lambda_1\lambda_2^*\rho'^{*2} - \lambda_1^*\lambda_2\rho'^2 + |\lambda_2|^2\rho'\rho'^*), \qquad (34)$$

where $\rho'$ is the polarization ratio of the transceiver in the new basis. The critical points are some $\rho'$s with which the first derivative of $P^x$ with respect to $\rho'$ and $\rho'^*$ vanishes. These critical points found in function $P^x$ are:

$$\rho'_{xn1} = 0 \qquad (35a)$$

$$\rho'_{xn2} = \infty \qquad (35b)$$

$$\rho'_{xm1,2} = \pm j\left(\frac{\lambda_1\lambda_2^*}{\lambda_1^*\lambda_2}\right)^{\frac{1}{4}} = \pm e^{j(2\nu+\pi/2)} \qquad (35c)$$

$$\rho'_{xs1,2} = \pm \left(\frac{\lambda_1\lambda_2^*}{\lambda_1^*\lambda_2}\right)^{\frac{1}{4}} = \pm e^{j2\nu} \quad . \qquad (35d)$$

(ii) For the function of the power returned to the co-pol channel:

$$P^c = |V^c|^2 = |\underline{h}'^T[S_d]\underline{h}'|^2$$

$$= (1+\rho'\rho'^*)^{-2}(|\lambda_1|^2 + \lambda_1\lambda_2^*\rho'^{*2} + \lambda_1^*\lambda_2\rho'^2 + |\lambda_2|^2\rho'^2\rho'^{*2}) \qquad (36)$$

The critical points are determined from

$$\rho'_{cm1} = \rho'_{xn1} = 0, \qquad \rho'_{cm2} = \rho'_{xn2} = \infty \quad , \qquad (37a,b)$$

$$\rho'_{cn1,2} = \pm\left(-\frac{\lambda_1}{\lambda_2}\right)^{\frac{1}{2}} = \pm \cot\gamma e^{j(2\nu+\pi/2)} = \pm \tan(90^\circ-\gamma)e^{j(2\nu+\pi/2)} \quad . \qquad (37c)$$

The right most expressions of (35)-(37) can be expressed in terms of the Huynen parameters [2] as:

$$\lambda_1 = |\lambda_1|e^{j\phi_1}, \quad \lambda_2 = |\lambda_2|e^{j\phi_2} \quad , \tag{37d}$$

$$\phi_1-\phi_2 = 4\nu \quad , \quad (-45° \leqslant \upsilon \leqslant 45°) \tag{37e}$$

$$\tan\gamma = \sqrt{\frac{|\lambda_2|}{|\lambda_1|}} \quad , \quad (0° \leqslant \upsilon \leqslant 45°) \tag{37f}$$

where $\nu$ – target skip angle ,
$\gamma$ – target characteristic angle .
$j = +\sqrt{-1} = \exp(+j\pi/2)$

Note that six of the characteristic polarization ratios ($\rho'_{xn1,2}$, $\rho'_{cn1,2}$ and $\rho'_{xm1,2}$) are the same as in [7]. They were named cross-pol nulls, co-pol nulls and cross-pol maxima, respectively. One additional pair of critical points ($\rho'_{xs1,2}$) can be denoted by cross-pol saddlepoint extrema because these are the saddlepoints of the function of power return in the cross-pol channel of the receiver which here was shown by the method of analysis for special cases.

For a rigorous proof, the second and higher order partial derivatives must be found which is considered elsewhere in more detail and not presented here because of the lengthy derivations.

To determine the optimal polarization states, we need to determine the polarization states for which the radar receives maximum or minimum power in the co-pol and cross-pol channels.

Without loss of generality, we assume $|\lambda_2| < |\lambda_1| = m$, where the following results are based on a comparison of the power return to the receiver.

## 3.2 The Cross-pol Nulls and Co-pol Maxima

The power return in the cross-pol channel will be zero if the transceiver is adjusted to the polarization ratio $\rho'_{xn1}$ or $\rho'_{xn2}$.

In case of $\rho'_{xn1} = 0$, the effective vector antenna height referred to the new basis becomes

$$\underline{h}'_{xn1} = \frac{e^{j\psi_1}}{\sqrt{1+\rho'_{xn1}\rho'^{*}_{xn1}}}\begin{bmatrix} 1 \\ 0 \end{bmatrix} = e^{j\psi_1}\begin{bmatrix} 1 \\ 0 \end{bmatrix} \quad , \tag{38a}$$

and similarly as $\rho'_{xn2} = \infty$

$$\underline{h}'_{xn2} = \frac{e^{j\psi_4}}{\sqrt{1+\rho'_{xn1}\rho'^{*}_{xn1}}}\begin{bmatrix} 0 \\ 1 \end{bmatrix} = e^{j\psi_4}\begin{bmatrix} 0 \\ 1 \end{bmatrix} \quad , \qquad (38b)$$

because $e^{j\psi_1}$ and $e^{j\psi_4}$ do not contribute to the power, we find

$$P^{x}_{xn1} = \left| [0 \quad 1] \begin{bmatrix} \lambda_1 & 0 \\ 0 & \lambda_2 \end{bmatrix} \begin{bmatrix} 1 \\ 0 \end{bmatrix} \right|^2 = 0 \qquad (39a)$$

$$P^{x}_{xn2} = \left| [1 \quad 0] \begin{bmatrix} \lambda_1 & 0 \\ 0 & \lambda_2 \end{bmatrix} \begin{bmatrix} 0 \\ 1 \end{bmatrix} \right|^2 = 0 \quad , \qquad (39b)$$

correspondingly, in the co-pol channel, the receiver will achieve maximum power returns as $\rho' = \rho'_{xn1} = 0$

$$P^{c}_{xn1} = \left| [1 \quad 0] \begin{bmatrix} \lambda_1 & 0 \\ 0 & \lambda_2 \end{bmatrix} \begin{bmatrix} 1 \\ 0 \end{bmatrix} \right|^2 = |\lambda_1|^2 = m^2 \quad ; \qquad (40a)$$

whereas the power return in the co-pol channel is not maximum for $\rho' = \rho'_{xn2} = \infty$, but yields a submaximum

$$P^{c}_{xn2} = \left| [0 \quad 1] \begin{bmatrix} \lambda_1 & 0 \\ 0 & \lambda_2 \end{bmatrix} \begin{bmatrix} 0 \\ 1 \end{bmatrix} \right|^2 = |\lambda_2|^2 = m^2 \tan^4\gamma \qquad (40b)$$

which is based on the assumption of $|\lambda_2| < |\lambda_1|$.

Note that a co-pol submax state for which $\rho'_{cm2} = \infty$, in fact, represents an unstable extremum of the power function $P^c$ of (36), if we discard the assumption of $|\lambda_1| \geqslant |\lambda_2|$. When the target characteristic angle $\gamma \geqslant 45°$ or $|\lambda_2| \geqslant |\lambda_1|$, $\rho'_{cm2} = \infty$ represents the co-pol max of $P^c$. With a decrease of $\gamma$, $\rho'_{cm2} = \infty$, would be a co-pol saddle point of $P^c$ at a certain angle of $\gamma$. When the target characteristic angle $\gamma = 0$, then the specific state $\rho'_{cm2} = \infty$ would become a co-pol null which is the case for $|\lambda_2| = 0$ for which the three characteristic polarization states $\rho'_{cn1,2}$ and $\rho'_{cm2}$ coincide producing one co-pol null state, e.g., for linear or helical targets [7].

### 3.3 The Co-pol Nulls

The power return in the co-pol channel will vanish if the polarization ratios of the transceiver in the new basis are $\rho'_{cn1,2} = \pm \left(-\frac{\lambda_1}{\lambda_2}\right)^{\frac{1}{2}}$ so that

$$P^c_{cn1,2} = |\underline{h}'^T[S_d]\underline{h}'|^2 = \frac{1}{\sqrt{1+\rho'_{cn1,2}\rho'^*_{cn1,2}}}|\rho'^2_{cn1,2}\lambda_2 + \lambda_1|^2 = 0 \quad . \quad (41)$$

But the power return to the cross-pol channel $\rho' = \rho'_{cn1,2}$ is not a maximum, since

$$P^x_{cn1,2} = |\underline{h}'^T_\perp[S_d]\underline{h}'|^2 = |\lambda_1||\lambda_2| = m^2\tan^2\gamma \quad (42)$$

is less than the power return of $P^x_{xm1,2}$ shown below.

### 3.4 The Cross-Pol Maxima

The power return to the cross-pol channel is maximum over all other cases if according to (35c)

$$\rho' = \rho'_{xm1,2} = \pm j\left(\frac{\lambda_1\lambda_2^*}{\lambda_1^*\lambda_2}\right)^{\frac{1}{4}} = \pm e^{j(2\nu+\pi/2)} \quad , \qquad \text{so that}$$

$$P^x_{xm1,2} = |\underline{h}'^T_\perp[S_d]\underline{h}'|^2 = 1/4(|\lambda_1| + |\lambda_2|)^2 = \frac{m^2}{4\cos^4\gamma} \quad . \quad (43)$$

But the corresponding power return in the co-pol channel is neither maximum nor a minimum

$$P^c_{xm1,2} = |\underline{h}'^T_\perp[S_d]\underline{h}'|^2 = 1/4(|\lambda_1| - |\lambda_2|)^2 = \frac{m^2\cos^2 2\gamma}{4\cos^4\gamma} \quad . \quad (44)$$

In summary, we have so far determined six optimal polarization states

$\rho'_{xn1}$ - cross-pol null and co-pol maximum

$\rho'_{xn2}$ - cross-pol null and co-pol extrema

$\rho'_{cn1,2}$ - co-pol nulls

$\rho'_{xm1,2}$ - cross-pol maxima .

### 3.5 Cross-Pol Saddlepoint Extrema

The remaining two polarization states $\rho'_{xs1,2}$ are called cross-pol saddlepoint extrema since the power return in the cross-pol channel has the following property:

At the vicinity of these two states $\rho'_{xs1,2}$, the power return to the cross-pol channel will on the complex polarization plane increase in some directions symmetric to the point, in some other orthogonal directions the received power will decrease depending on both the modulus and the phase of $\rho'$. The power return in the cross-pol and co-pol channel for these two states are:

$$P^{x}_{xs1,2} = \tfrac{1}{4}(|\lambda_1| - |\lambda_2|)^2 = \frac{m^2\cos^2 2\gamma}{4\cos^4\gamma} \quad , \tag{45}$$

$$P^{c}_{xs1,2} = \tfrac{1}{4}(|\lambda_1| + |\lambda_2|)^2 = \frac{m^2}{4\cos^4\gamma} \quad . \tag{46}$$

The importance of these two polarization states may be due to the changing phase of the voltage of the scattered wave and requires further analysis.

### 3.6 Presentation of the Characteristic Polarization States on the Poincaré Sphere

Based on the rules of Section 2.1 and due to the representation of the characteristic polarization states obtained in Section 3.5, we are now able to determine the interrelation of locations of $X_1$, $X_2$ (x-pol null or co-pol max); $C_1$, $C_2$ (co-pol null); $S_1$, $S_2$ (x-pol max); and $T_1$, $T_2$ (x-pol saddle) on the Poincaré sphere.

First, we establish a new basis (AB) such that the orthogonal cross-pol null pair $\rho_1$ and $\rho_2$ of (28), located at $X_1$ and $X_2$, represent the basis points $X_1$ with $\rho'_{xn1}(AB) = \rho_A = 0$ and $X_2$ with $\rho'_{xn2}(AB) = \rho_B = \infty$ according to the results of (35a,b) and (7)-(10) as shown in Fig. 3. The co-pol null pair in the new basis $\rho'_{cn1,2}(AB)$ are then located at $C_1$ and $C_2$ in Fig. 3. Because the corresponding polarization ratios can be expressed in the new basis (AB) as:

$$\rho'_{xn1}(AB) = 0 = \tan(90° - 90°)\ \exp\ \mathrm{j\,arb}\ (= \rho_A) \tag{47a}$$

$$\rho'_{xn2}(AB) = \infty = \tan(90° - 0°)\ \exp\ \mathrm{j\,arb}\ (= \rho_B) \tag{47b}$$

$$\rho'_{cn1}(AB) = \tan(90° - \gamma)\ \exp\ j(2\nu + \frac{\pi}{2}) \tag{47c}$$

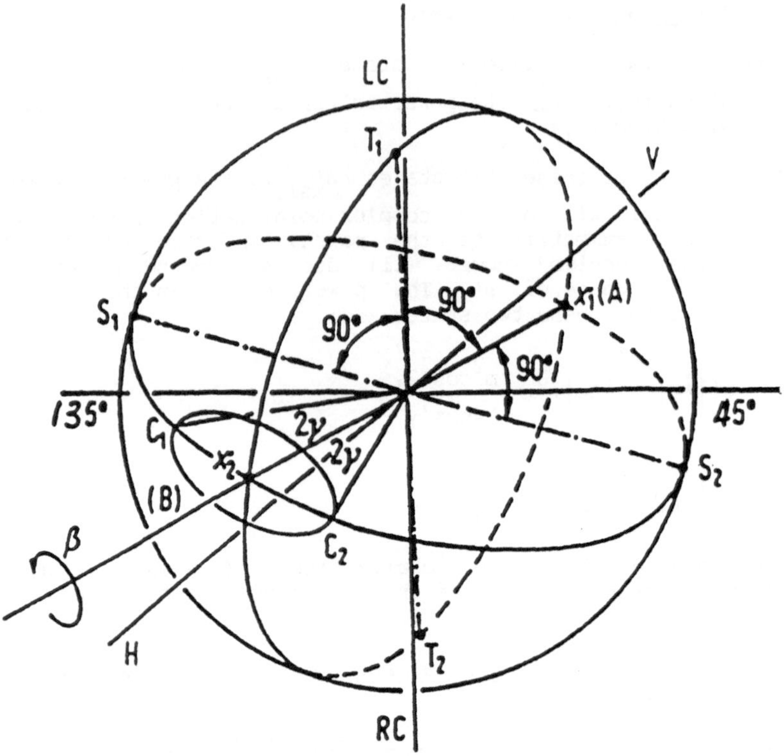


Fig. 3: Representation of the characteristic polarization states on the Poincaré sphere ($X_1$: cross-pol null and co-pol max; $X_2$: cross-pol null and co-pol extremum; $C_{1,2}$: co-pol nulls; $S_{1,2}$: cross-pol maxima; $T_{1,2}$: cross-pol saddlepoints; $\gamma$: target characteristic angle) presented in the new basis (AB)

$$\rho'_{cn2}(AB) = \tan(90° + \gamma) \exp j(2\nu + \tfrac{\pi}{2}) \quad , \tag{47d}$$

it follows that all four possess the same phase $(2\nu + \frac{\pi}{2})$ so that they lie on the same great circle about the new basis $(X_1X_2)$ diameter joining $X_1$ and $X_2$. Further, according to the magnitudes, $\tan(90° \pm \gamma)$, the angles $\sphericalangle C_1 0 X_2$ and $\sphericalangle X_2 0 C_2$ are equal to $2\gamma$, the line joining the cross-pol nulls $X_1$ and $X_2$ bisects the angle $\sphericalangle C_1 0 C_2$ spanned by the co-pol nulls at the origin of the sphere. These four points form the "polarization fork", with $X_1$ being the true maximum polarization representing the handle and the points $C_1$, $X_2$ and $C_2$ representing the fork with respect to the common origin 0, being consistent with the standardization used by [2] for "Huynen's Fork". In a next step, the pair of orthogonal cross-pol maxima (35c) can be expressed as

$$\rho'_{xm1} = \exp j(2\nu + \tfrac{\pi}{2}) = \tan(90° - 45°) \exp j(2\nu + \tfrac{\pi}{2}) \tag{47e}$$

$$\rho'_{xm2} = -\exp j(2\nu + \tfrac{\pi}{2}) = \tan(90° + 45°) \exp j(2\nu + \tfrac{\pi}{2}) \quad , \tag{47f}$$

which are represented by $S_1$ and $S_2$, respectively, on the Poincaré sphere of Fig. 3. Since the $\rho'_{xm1,2}$ are orthogonal and possess the same phase as $\rho'_{cn1,2}$ and $\rho'_{xn1,2}$, the points $S_1$ and $S_2$ lie on the same great circle with $X_1$, $X_2$ and $C_1$, $C_2$, where the diameter $S_1S_2$ is perpendicular to the basis diameter $X_1X_2$ because of the angle $2\eta = \pm\,\pi/2$, which is consistent with the "tree" concept portrayed in Agrawal and Boerner [7].

The additional pair of orthogonal polarization ratios $\rho'_{xs1,2}$ (35d), represented by $T_1$ and $T_2$ on the Poincaré sphere of Fig. 3, are given by

$$\rho'_{xs1} = \exp j2\nu = \tan(90° - 45°) \exp j2\nu \tag{47g}$$

$$\rho'_{xs2} = -\exp j2\nu = \tan(90° + 45°) \exp j2\nu \tag{47h}$$

where $\rho'^{*}_{xs1}\rho'_{xs2} = -1$.

The family of four points $T_1$, $T_2$, $X_1$ and $X_2$ lie on one great circle because they possess the same phase. Because the phase difference between $\rho'_{xs1,2}$ and $\rho'_{xm1,2}$ is $\pi/2$ and $2\eta = \pm\,\pi/2$ so that the location of $T_1$ and $T_2$ on the Poincaré sphere can be found by rotating $S_1$ and $S_2$ by an angle of $\pi/2$ about the base diameter $X_1X_2$, i.e., the three pairs $X_1X_2$, $S_1S_2$ and $T_1T_2$ are mutually perpendicular to each other as shown explicitly in Fig. 3.

Using the mapping from the complex plane to the Poincaré sphere, the polarization fork can be easily constructed as shown in Fig. 4 where $\rho' = 0$ as the common origin.

The cross-pol NULL(co-pol max) $\rho'_{xn1} = \rho'_{cm1} = 0$ is located at the origin of the complex plane, so the mapping point is X1 (south pole: N) on the Poincaré sphere. And $\rho'_{xn2} = \rho'_{cn2} = \infty$ is at the infinity of the complex plane, so the corresponding point is X2 (north pole: Z). $\overline{X_1X_2}$ forms a diameter of the sphere and is perpendiculer to the complex plane.

According to the expression of the cross-pol max and cross-pol saddles, they all lie on the unit circle and are the end points of two orthogonal diameters. So their corresponding points lie on the equator of the sphere as $S_1$, $S_2$, $T_1$ and $T_2$ with $\overline{S_1S_2}$ and $\overline{T_1T_2}$ perpendicular to each other. The co-pol nulls of $\rho'_{cn1,2}$ lie on the same straight line with $\rho'_{xm1,2}$ on the plane and symmetric about the origin 0, so their corresponding points on the sphere $C_1$ and $C_2$ lie on the same great circle with $X_1$, $X_2$, $S_1$ and $S_2$, symmetric about the diameter $\overline{X_1X_2}$. The complete Fork is shown in Fig. 4.

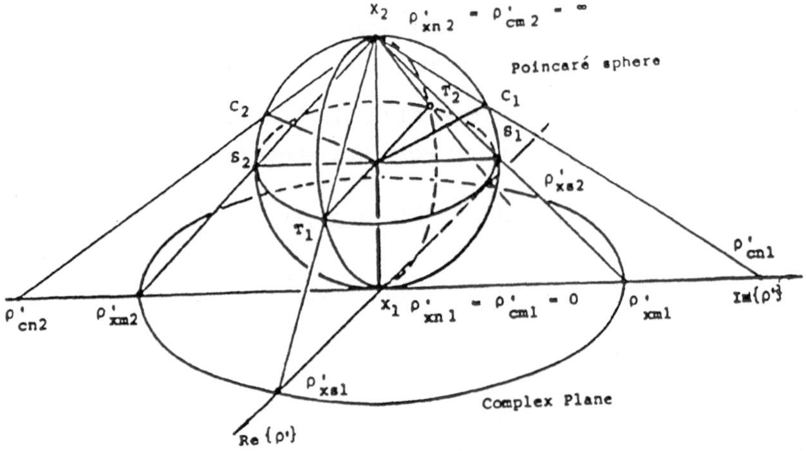


Fig. 4: Correspondence of $\rho'_{xn1,2} = \rho'_{cm1,2}$; $\rho'_{xm1,2}$; $\rho'_{xs1,2}$; and $\rho'_{cn1,2}$ on the complex plane with $X_{1,2}$; $T_{1,2}$; $S_{1,2}$ and $C_{1,2}$ on the Poincaré sphere

## 4. THE EQUIVALENCE OF OUR RESULTS WITH HUYNEN'S STANDARDIZED POLARIZATION FORK CONCEPT

The polarization fork presentation of Fig. 3, which was derived directly from the scattering matrix [S], in a next step is transformed into Huynen's standardized Polarization Fork concept as expressed by his target scattering matrix [T] in [2] and shown in Fig. 5.

### 4.1 The Standardized Polarization Fork of Huynen

Here, we emphasize the importance of the three parameters, $\rho_1$(or $\rho_2$) of (28), $\lambda_1$ and $\lambda_2$ of (31), with which we can determine the complete con

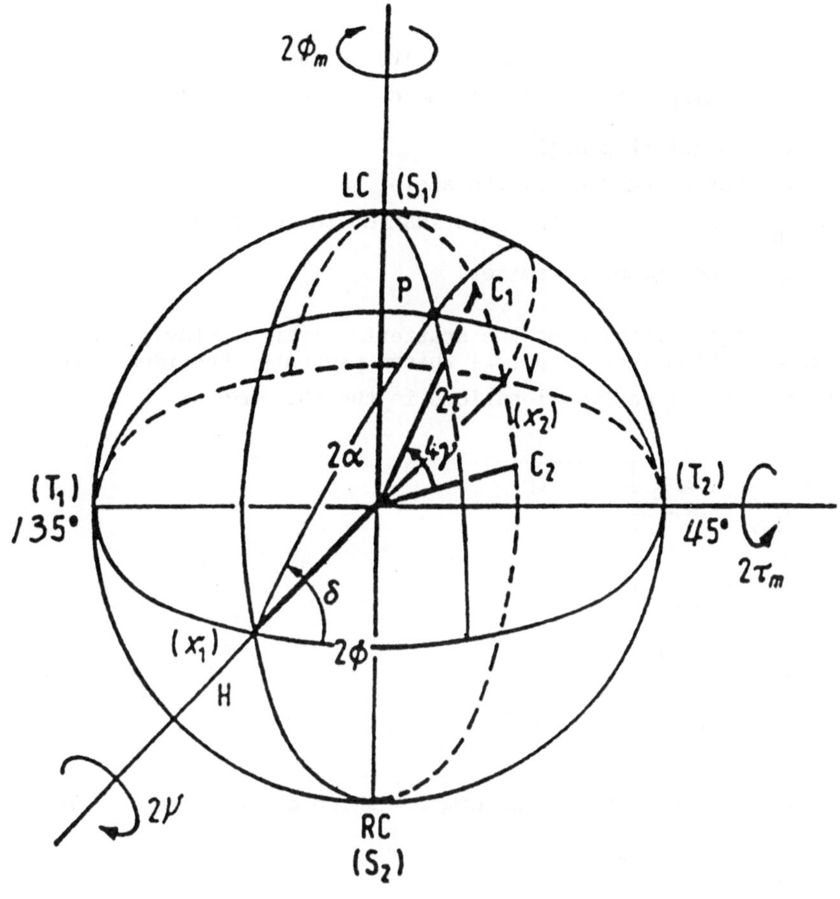


Fig. 5: The standardized "Polarization Fork of Huynen" with definition of Huynen's geometrical parameters presented in the "old basis"

$\phi$ - target orientation or tilt angle
$\upsilon$ - target skip angle
$\tau$ - target ellipticity angle
$\gamma$ - target characteristic angle
$\rho = \tan\alpha \exp j\delta$ - polarization ratio

figuration of the "polarization fork" and the power return density plots for a specified target at a specific aspect angle and fixed frequency with known scattering matrix [S] on the Poincaré sphere, since all the seven geometrical parameters (see Fig. 5) can be obtained directly from $\rho_1$, $\lambda_1$ and $\lambda_2$ as:

m – target size

$\phi$ – target orientation (or tilt) angle

$\tau_m$ – ellipticity angle of the co-pol maximum

$\nu$ – target skip angle

$\gamma$ – target characteristic angle

$\delta_m$ – the phase of the polarization ratio of the co-pol max

$\alpha_m$ – the spinor parameter

During diagonalization of the scattering matrix [S(HV)], we obtained the diagonalization factor $\rho_1(\rho_2)$ which presents the identical cross-pol null and co-pol maximum locations in the (HV) basis:

$$\rho_{1,2} = \frac{-B \pm \sqrt{B^2 - 4AC}}{2A} \quad , \tag{48}$$

where $A = S^*_{HH}S_{HV} + S^*_{HV}S_{VV}$,

$B = |S_{HH}|^2 - |S_{VV}|^2$,

$C = -A^*$ .

The polarization ratio $\rho_1$ is complex which can be expressed by

$$\rho_1 = \tan\alpha_m \, e^{j\delta_m} \quad . \tag{49}$$

The diagonal elements $\lambda_1, \lambda_2$ of the scattering matrix $[S_d]$, the geometric parameters $\nu$, $\gamma$ are given by (37d), (37e) and (37f), respectively.

From formula (49), we have:

$$\delta_m = \arg\{\rho_1\} \tag{50}$$

$$\alpha_m = \tan^{-1}(|\rho_1|) \tag{51}$$

From relations between $(\phi,\tau)$ and $(\alpha,\delta)$ [12], we have:

$$2\phi_m = \tan^{-1}(\tan 2\alpha_m \cos\delta_m) \qquad (-90° \leqslant \phi \leqslant 90°) \tag{52}$$

$$2\tau_m = \sin^{-1}(\sin 2\alpha_m \sin\delta_m) \qquad (-45° \leqslant \tau_m \leqslant +45°) \tag{53}$$

The last parameter is m which is given by: $m = |\lambda_1|$ (54)

## 4.2 Comparison of Our [S] and Huynen's [T] Matrix Decompositions

Having determined the parameters $2\phi_m$, $2\nu$ and $2\tau_m$, then the locations of the cross-pol nulls on the standardized Poincaré sphere of Fig. 5 can be identified in terms of the polarization ratios $\rho_H = \rho_1$ and $\rho_V = \rho_2$, as defined in the new basis (AB) of the previous section. The standardized Polarization Fork of Fig. 5 is then obtained from that of Fig. 3 by rotating the $X_1X_2$, $S_1S_2$ and $T_1T_2$ basis axes by $-2\nu$, $-2\phi_m$ and $-2\tau_m$, respectively, into the HV, LR and 45° 135° basis alignments; such that the remaining six characteristic polarization ratios are then defined by:

$$\rho'_{cn1,2} = \pm \tan(90° - \gamma)e^{j(2\nu + \pi/2)} \quad \text{– co-pol nulls} \tag{55a}$$

$$\rho'_{xm1,2} = \pm e^{j(2\nu + \pi/2)} \quad \text{– cross-pol maximum (L,R)} \tag{55b}$$

$$\rho'_{xs1,2} = \pm e^{j2\nu} \quad \text{– cross-pol saddles (45°,135°)} \tag{55c}$$

and can be located on the Poincaré sphere by these expressions. This estabishes Huynen's standardized "polarization fork" of Fig. 5.

Inversely, by determining the parameters $\phi_m$, $\tau_m$, $\delta_m$, $\nu$, $\gamma$ and $m = |\lambda_1|$ from radar measurements, we can determine the scattering matrix of the specific target: Simply find the polarization ratio $\rho_1$ of the cross-pol null (co-pol max) by

$$\rho_1 = \tfrac{1}{2}[(1-M)\tan 2\phi_m + j(1+M)\sin 2\tau_m] \tag{56}$$

where $$M = \frac{1 - \cos 2\phi_m \cos 2\tau_m}{1 + \cos 2\phi_m \cos 2\tau_m} = \tan^2\alpha_m$$

and the scattering matrix can be calculated with (56) according to (20a) as:

$$[S] = [U(\rho_1)]^*[S_d][U(\rho_1)]^{*T} =$$

$$= \frac{1}{(1+\rho_1\rho_1^*)}\begin{bmatrix} e^{-j\psi_1} & -\rho_1 e^{-j\psi_4} \\ \rho_1^* e^{-j\psi_1} & e^{-j\psi_4}\end{bmatrix}\begin{bmatrix} me^{j2\nu} & 0 \\ 0 & m\tan^2\gamma e^{-j2\nu}\end{bmatrix}\begin{bmatrix} e^{-j\psi_1} & \rho_1^* e^{-j\psi_1} \\ -\rho_1 e^{-j\psi_4} & e^{-j\psi_4}\end{bmatrix}\exp(j\zeta) \tag{57}$$

The quanity $\zeta$ is called the absolute phase of the target [2]; it disappears with power measurements, and it may be altered arbitrarily by moving the radar target along the line-of-sight direction, leaving the target's attitude otherwise unaltered. We note, that our result (57) of presenting the scattering matrix [S] in terms of the Huynen characteristic parameters ($m$, $\phi_m$, $\tau_m$, $\nu$, $\gamma$, $\delta_m$, $\alpha_m$) is an alternative presentation to the target scattering matrix [H] given by Huynen [2]* as follows:

$$[H] = [U^*(\psi, \tau_m, \nu)]\, m \begin{bmatrix} 1 & 0 \\ 0 & \tan^2\gamma \end{bmatrix} [U^{*\prime}(\psi, \tau_m, \nu)] \exp(j\zeta) \tag{58a}$$

and
$$[U(\psi, \tau_m, \nu)] = e^{\psi[J]}\, e^{+\tau_m[K]}\, e^{+\nu[L]} \quad , \tag{58b}$$

where $j[J]$, $j[K]$ and $j[L]$ are the Pauli spin matrices and $[I]$ is the identity matrix respectively defined by:

$$[I] = \begin{bmatrix} 1 & 0 \\ 0 & 1 \end{bmatrix}; \; [J] = \begin{bmatrix} 0 & -1 \\ 1 & 0 \end{bmatrix}; \; [K] = \begin{bmatrix} 0 & j \\ j & 0 \end{bmatrix}; \; [L] = \begin{bmatrix} -j & 0 \\ 0 & j \end{bmatrix} \tag{58c}$$

$$[K]^2 = [I], \qquad [L] = [J][K] = -[K][J], \qquad [L]^2 = [I] \tag{58d}$$

and

$$e^{\psi[J]} = \begin{bmatrix} \cos\psi & -\sin\psi \\ \sin\psi & \cos\psi \end{bmatrix}, \; e^{\tau_m[K]} = \begin{bmatrix} \cos\tau & j\sin\tau \\ j\sin\tau & \cos\tau \end{bmatrix}, \; e^{\nu[L]} = \begin{bmatrix} e^{-j\nu} & 0 \\ 0 & e^{j\nu} \end{bmatrix} \tag{58e}$$

Note that the same sets of parameters $\phi_m(\psi)$, $\tau_m$, $\nu$, $\gamma$, $m$ are contained in the expressions of both (57) and (58). The arrangement of the characteristic polarization states according to Huynen [2], as displayed in Fig. 5, and in comparison to Fig. 3, in the case of [H] in (58) is obtained by rotating with $2\psi$, $2\tau_m$, $2\nu$ about three orthogonal axes; whereas, in our case of [S] in (57), by a transformation along a great circle defined by the polarization ratio $\rho$ given in $[U^*(\rho_1)]$ with a subsequent rotation by $2\nu$ about the new base diameter $(X_1X_2)$. In order to further clarify the equivalence, the tranformation matrix relation of (57) is rewritten as:

---

* Huynen uses the symbol [T] for target scattering matrix which is here changed to [H], because [T] is here defined according to (17) to denote the transformation matrix.

$$[S] = [U^*(\rho_1)]\exp(\nu[L]^*)m\begin{bmatrix}1 & 0\\ 0 & \tan^2\gamma\end{bmatrix}\exp(\nu[L]^*)^T[U^*(\rho_1)]^T \exp(j\zeta) \quad , \text{(59a)}$$

$$[U^*(\rho_1)] = \frac{1}{\sqrt{1+\rho_1\rho_1^*}}\begin{bmatrix} e^{-j\psi_1} & -\rho_1(\phi_m,\tau_m)e^{-j\psi_4} \\ \rho_1^*(\phi_m,\tau_m)e^{-j\psi_1} & e^{-j\psi_4}\end{bmatrix} \quad \text{(59b)}$$

which is equivalent to (18c) for $\psi_1 = \Psi_1 + \Psi_2$ and $\psi_4 = \Psi_2 - \Psi_1$.

$$\exp\{\nu[L]\} = \begin{bmatrix}\exp(-j\nu) & 0 \\ 0 & \exp(j\nu)\end{bmatrix} \quad \text{(59c)}$$

In order to obtain Huynen's formulation, the expression of (59b) can be expressed as:

$$[U(\rho_1)] = \exp\{\psi[J]\}\ \exp\{\tau_m[K]\} \quad \text{(60)}$$

which implies that in our case we only need to transform along one main circle, whereas, Huynen requires two rotations to accomplish the same. Hence, we have established by rigorous mathematical derivations that our presentations of (57) and (59) are equivalent to Huynen's of (58) for which a mathematical proof is provided next.

The proof of equation (60) is straightforward, since $\rho_1$ can be expressed in the following form

$$\rho_1 = \tan\alpha_m e^{j\delta m} = \frac{1}{2}\,\frac{\sin2\phi_m\cos2\tau_m + j\sin2\tau_m}{\cos^2\phi_m\cos^2\tau_m + \sin^2\phi_m\sin^2\tau_m} \quad \text{(61)}$$

by using (3b) and (56).

The right-hand side of (60) can be rewritten as

$$\exp\{\psi[J]\}\exp\{\tau_m[K]\} = \begin{bmatrix} U_{11} & U_{12} \\ U21 & U22\end{bmatrix} \quad \text{(62)}$$

where
$$U_{11} = \cos\psi\cos\tau_m - j\sin\psi\sin\tau_m$$

$$U_{12} = -\sin\psi\cos\tau_m + j\cos\psi\sin\tau_m$$

$$U_{21} = \sin\psi\cos\tau_m + j\cos\psi\sin\tau_m$$

$$U_{22} = \cos\psi\cos\tau_m + j\sin\psi\sin\tau_m$$

with
$$U_{12} = -U_{21}^*, \qquad U_{22} = U_{11}^*,$$

$$|U_{11}|^2 + |U_{12}|^2 = |U_{12}|^2 + |U_{22}|^2 = 1. \tag{63}$$

The ratios of the elements in each column of (62) result in

$$\frac{U_{21}}{U_{11}} = \frac{\sin\psi\cos\tau_m + j\cos\psi\sin\tau_m}{\cos\psi\cos\tau_m - j\sin\psi\sin\tau_m}$$

$$= \frac{1}{2}\,\frac{\sin2\psi\cos2\tau_m + j\sin2\tau_m}{\cos^2\psi\cos^2\tau_m + \sin^2\psi\sin^2\tau_m} = \rho_1 \tag{64}$$

$$\frac{U_{12}}{U_{22}} = -\frac{U_{21}^*}{U_{11}^*} = -\rho_1^* \tag{65}$$

and by combining (63), (64) and (65), we have

$$|U_{11}| = |U_{22}| = \frac{1}{\sqrt{1 + \rho_1\rho_1^*}} \tag{66}$$

and from applying (64), (65) and (66) to (62), the required result follows:

$$\exp\{\psi[J]\}\exp\{\tau_m[K]\} = \frac{1}{\sqrt{1 + \rho_1\rho_1^*}}\begin{bmatrix} \exp\{j\psi_{11}\} & -\rho_1^*\exp\{-j\psi_{11}\} \\ \rho_1\exp\{j\psi_{11}\} & \exp\{-j\psi_{11}\} \end{bmatrix} \tag{67}$$

The relation between $\psi_{11}$ and $\psi$, $\tau_m$ is given by

$$\tan\psi_{11} = -\tan\psi\tan\tau_m \tag{68}$$

which was pre-assumed in Huynen's formulation during the construction of the unitary transformation matrix [U] in [2], whereas in our case (see (21a-c)), we left $\psi_1$ and $\psi_4$ arbitrary. The precise determination of $\rho_1$ or equivalently $\psi$ and $\tau_m$, as well as $\psi_1$ and $\psi_4$ are dependent on the specific scattering matrix [S] under consideration as shown in the derivations (B-5), (B-6) of App. B, where

$$\psi_1 = -\frac{\delta_2}{2} - \frac{\pi}{4} \tag{69a}$$

$$\psi_4 = +\frac{\delta_2}{2} - \frac{\pi}{4} \tag{69b}$$

and $\delta_2$ is the phase of $\rho_2$ associated with the second cross-pol null in (28), referenced to the old basis (HV).

## 5. EXAMPLE

Using the formulas for computer numerical evaluation presented in App. B, the following examples are evaluated:

### 5.1: Example (1)

Scattering matrix $[S_1] = \begin{bmatrix} 2j & 0.5 \\ 0.5 & -j \end{bmatrix}$

(a) The characteristic polarization ratios referenced to the old basis (HV):

| | Real | Imaginary | Phase | Modulus |
|---|---|---|---|---|
| $\rho_{xn1}$ | 0.0000 | 0.4142 | 90.0000 | 0.4142 |
| $\rho_{xn2}$ | 0.0000 | -2.4142 | -90.0000 | 2.4142 |
| $\rho_{xm1}$ | 1.0000 | 0.0000 | 0.0000 | 1.0000 |
| $\rho_{xm2}$ | -1.0000 | 0.0000 | 180.0000 | 1.0000 |
| $\rho_{xs1}$ | 0.0000 | -0.4142 | -90.0000 | 0.4142 |
| $\rho_{xs2}$ | 0.0000 | 2.4142 | 90.0000 | 2.4142 |
| $\rho_{cn1}$ | -1.3229 | -0.5000 | -159.2952 | 1.4142 |

(b) The characteristic polarization ratios referenced to the new basis (AB):

| | Real | Imaginary | Phases | Modulus |
|---|---|---|---|---|
| $\rho_{xn1}$ | 0 | | arb. | 0 |
| $\rho_{xn2}$ | $\infty$ | | arb. | $\infty$ |
| $\rho_{xm1}$ | 0.0000 | 1.0000 | 90.0000 | 1.0000 |
| $\rho_{xm2}$ | 0.0000 | -1.0000 | -90.0000 | 1.0000 |
| $\rho_{xs1}$ | 1.0000 | 0.0000 | 0.0000 | 1.0000 |
| $\rho_{xs2}$ | -1.0000 | 0.0000 | -180.0000 | 1.0000 |
| $\rho_{cn1}$ | 0.0000 | -1.6684 | -90.0000 | 1.6684 |
| $\rho_{cn2}$ | 0.0000 | 1.6684 | 90.0000 | 1.6684 |

(c) The Stokes vectors referenced to the old basis (HV):

| | $g_0$ | $g_1$ | $g_2$ | $g_3$ |
|---|---|---|---|---|
| $\rho_{xn1}$ | 1.0000 | 0.7071 | 0.0000 | 0.7071 |
| $\rho_{xn2}$ | 1.0000 | -0.7071 | 0.0000 | -0.7071 |
| $\rho_{cn1}$ | 1.0000 | -0.3333 | -0.8819 | -0.3333 |
| $\rho_{cn2}$ | 1.0000 | -0.3333 | 0.8819 | -0.3333 |
| $\rho_{xm1}$ | 1.0000 | 0.0000 | 1.0000 | 0.0000 |
| $\rho_{xm2}$ | 1.0000 | 0.0000 | -1.0000 | 0.0000 |
| $\rho_{xs1}$ | 1.0000 | 0.7071 | 0.0000 | -0.7071 |
| $\rho_{xs2}$ | 1.0000 | -0.7071 | 0.0000 | 0.7071 |

(d) Ellipticity ($\tau°$) and Tilt angle ($\phi$) of the characteristic polarization ratios:

| | $\phi°$ | $\tau°$ |
|---|---|---|
| $\rho_{xn10}$: | 0.0000 | 22.5000 |
| $\rho_{xn20}$: | 90.0000 | -22.5000 |
| $\rho_{cn10}$: | 34.6476 | -9.7356 |
| $\rho_{cn20}$: | -34.6476 | -9.7356 |
| $\rho_{xm10}$: | 45.0000 | 0.0000 |
| $\rho_{xm20}$: | -45.0000 | 0.0000 |
| $\rho_{xs10}$: | 0.0000 | -22.5000 |
| $\rho_{xs20}$: | 90.0000 | 22.5000 |

(e) The characteristic parameters of the CO-POL-MAX:

| $2\nu$ | $\gamma$ | $\delta_m$ | $\alpha$ | $\phi$ | $\tau_m$ |
|---|---|---|---|---|---|
| 0.000 | 30.937 | 90.000 | 22.500 | 0.000 | 22.500 |

(f) The phases of the unitary transformation matrix {(HV) → (AB)}

$$\text{for } [U] = \begin{bmatrix} \exp\{\psi_1\} & -\rho^* \exp\{\psi_4\} \\ \rho \exp\{\psi_1\} & \exp\{\psi_4\} \end{bmatrix}$$

$\psi_1 = 0.0000$  $\psi_4 = -90.0000$

(g) The power return to the Co-pol and cross-pol channel.

| $P^c_{xn1}$ | $P^c_{xn2}$ | $P^x_{cn1,2}$ | $P^x_{xm1,2}$ | $P^c_{xm1,2}$ | $P^x_{xs1,2}$ | $P^c_{xs1,2}$ |
|---|---|---|---|---|---|---|
| 4.871 | 0.629 | 1.750 | 2.250 | 0.500 | 0.500 | 2.250 |

The results for Example (1) are presented in Figs. 6a and 6b.

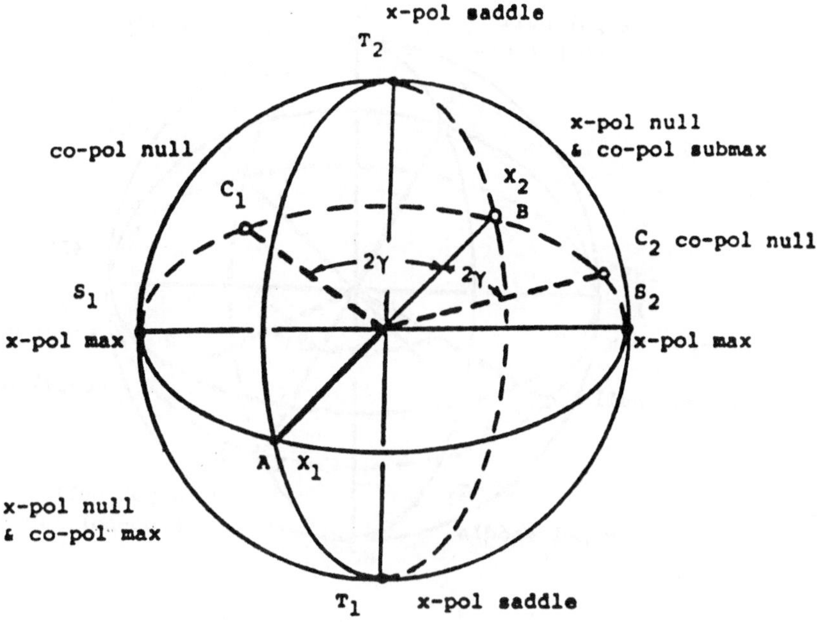


Fig. 6a: The characteristic polarization states on the Poincaré sphere of Example (1) referenced to the new basis (AB) for the scattering matrix $[S_1] = \begin{bmatrix} 2j & 0.5 \\ 0.5 & -j \end{bmatrix}$ with the characteristic polarization ratios:

$\rho'_{xn1} = \rho'_{cm1} = 0$, $(X_1)$, $\rho'_{xn2} = \rho'_{cm2} = \infty$ $(X_2)$, $\rho'_{xm1,2} = \pm j$ $(S_1, S_2)$, $\rho'_{xs1,2} = \pm 1$ $(T_1, T_2)$, $\rho'_{cn1,2} = \pm 1.668j$ $(C_1, C_2)$.

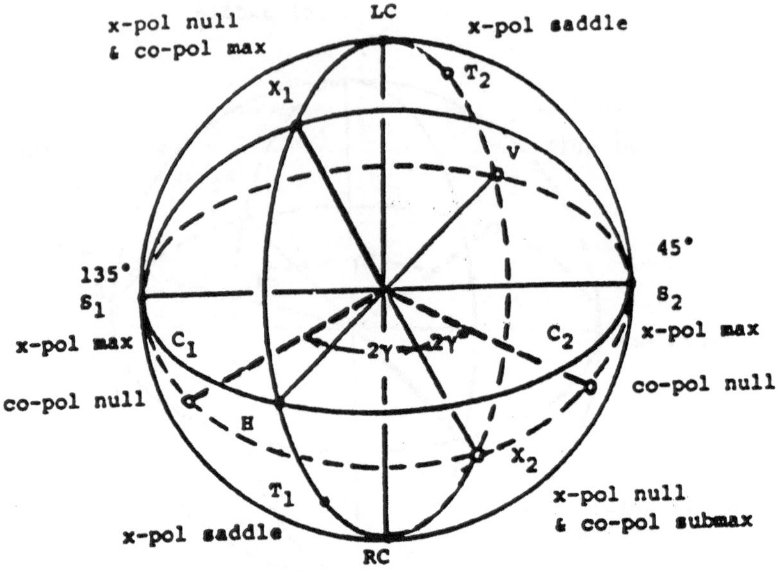


Fig. 6b: The characteristic polarization states on the Poincaré sphere referenced to the old basis (HV) for the scattering matrix $[S_1]$ of Example (1) with the characteristic polarization ratios:

$\rho_{xn1} = \rho_{cm1} = 0.414\ e^{j90°}(X_1)$, $\rho_{xn1} = \rho_{cm2} = 2.414\ e^{-j90°}(X_2)$,

$\rho_{xm1,2} = \pm 1(S_1, S_2)$, $\rho_{xs1} = 0.414\ e^{-j90°}(T_1)$, $\rho_{xs2} = 2.414\ e^{j90°}(T_2)$,

$\rho_{cn1} = 1.414\ e^{-j159°}(C_1)$, $\rho_{cn2} = 1.414\ e^{-j20.7°}(C_2)$.

and the geometric parameters:

$\nu = 0.0$, $\gamma = 30.9$, $\delta_m = 90.0$, $\alpha_m = 22.5$, $\phi = 0.0$, and $\tau_m = 22.5$

## 5.2 EXAMPLE (2):

$$[S_2] = \begin{bmatrix} -3 & j \\ j & 2j \end{bmatrix}$$

(a) The characteristic polarization ratios reference to the old basis (HV)

| | Real | Imaginary | Phase(°) | Modulus |
|---|---|---|---|---|
| $\rho_{xn1}$ | 0.2904 | 0.4356 | 56.3095 | 0.5235 |
| $\rho_{xn2}$ | -1.0596 | -1.5894 | -123.6901 | 1.9102 |
| $\rho_{xm1}$ | 0.5143 | -0.5316 | -45.9483 | 0.7396 |
| $\rho_{xm2}$ | -0.9401 | 0.9717 | 134.0517 | 1.3520 |
| $\rho_{xs1}$ | -0.3347 | -0.1388 | -157.4804 | 0.3624 |
| $\rho_{xs2}$ | 2.5491 | 1.0569 | 22.5196 | 2.7596 |
| $\rho_{cn1}$ | 0.4409 | -0.7971 | -61.0497 | 0.9109 |
| $\rho_{cn2}$ | -1.4409 | 0.7971 | 151.0497 | 1.6467 |

(b) The characteristic polarization ratios reference to the new basis:

| | Real | Imaginary | Phase(°) | Modulus |
|---|---|---|---|---|
| $\rho_{xn1}$ | 0 | | arb. | 0 |
| $\rho_{xn2}$ | $\infty$ | | arb. | $\infty$ |
| $\rho_{xm1}$ | 0.3563 | 0.9344 | 69.1276 | 1.0000 |
| $\rho_{xm2}$ | -0.3563 | -0.9344 | -110.8724 | 1.0000 |
| $\rho_{xs1}$ | 0.9344 | -0.3563 | -20.8724 | 1.0000 |
| $\rho_{xs2}$ | -0.9344 | 0.3563 | 159.1276 | 1.0000 |
| $\rho_{cn1}$ | 0.4981 | 1.3062 | 69.1276 | 1.3980 |
| $\rho_{cn2}$ | -0.4981 | -1.3062 | -110.8724 | 1.3980 |

(c) The Stokes vectors referenced to the old basis (HV):

| | $g_0$ | $g_1$ | $g_2$ | $g_3$ |
|---|---|---|---|---|
| $\rho_{xn1}$: | 1.0000 | 0.5698 | 0.4558 | 0.6838 |
| $\rho_{xn2}$: | 1.0000 | -0.5698 | -0.4558 | 0.6838 |
| $\rho_{cn1}$: | 1.0000 | 0.0930 | 0.4820 | -0.8712 |
| $\rho_{cn2}$: | 1.0000 | -0.4612 | -0.7764 | 0.4295 |
| $\rho_{xm1}$: | 1.0000 | 0.2928 | 0.6648 | -0.6872 |
| $\rho_{xm2}$: | 1.0000 | -0.2928 | -0.6648 | 0.6872 |
| $\rho_{xs1}$: | 1.0000 | 0.7679 | -0.5918 | -0.2454 |
| $\rho_{xs2}$: | 1.0000 | -0.7679 | 0.5918 | 0.2454 |

(d) Ellipticity ($\tau^o$) and Tilt angles ($\phi$):

| | $\phi^o$ | $\tau^o$ |
|---|---|---|
| $\rho_{xn10}$: | 19.3299 | 21.5692 |
| $\rho_{xn20}$: | 19.3299 | -21.5692 |
| $\rho_{cn10}$: | 39.5369 | -30.3018 |
| $\rho_{cn20}$: | 29.6463 | 12.7182 |
| $\rho_{xm10}$: | 33.1158 | -21.7051 |
| $\rho_{xm20}$: | 33.1158 | 21.7051 |
| $\rho_{xs10}$: | -18.8106 | -7.1015 |
| $\rho_{xs20}$: | -18.8106 | 7.1015 |

(e) The characteristic parameters of COL-POL-MAX:

| $2\nu$ | $\gamma$ | $\delta_m$ | $\alpha_m$ | $\phi$ | $\tau_m$ |
|---|---|---|---|---|---|
| 0.000 | 35.577 | 56.310 | 27.632 | 19.330 | 21.569 |

(f) The phases of the unitary transformation matrix from (HV) to (AB) basis for the unitary transformation matrix:

$$[U] = \begin{bmatrix} \exp\{\psi_1\} & -\rho^* \exp\{\psi_4\} \\ \rho \exp\{\psi_1] & \exp\{\psi_4\} \end{bmatrix}$$

$$\psi_1 = 16.8450 \qquad \psi_4 = -106.8450$$

(g) The power return to the co-pol and cross-pol channel:

| $P^c_{xn1}$ | $P^c_{xn2}$ | $P^x_{cn1,2}$ | $P^x_{xm1,2}$ | $P^c_{x1,2}$ | $P^x_{xs1,2}$ | $P^c_{xs1,2}$ |
|---|---|---|---|---|---|---|
| 11.887 | 3.113 | 6.083 | 6.791 | 0.709 | 0.709 | 5.791 |

The results of Example (2) are presented in Figs. 7a and 7b.

### 5.3 Interpretation of Results and Comparison with Yan and Boerner [14]

In our examples, we first transferred the scattering matrix from the old basis (HV) to a new basis (AB) in which the scattering matrix is in diagonal form; secondly, we used the "Critical point" method to find the characteristic polarization ratios for both the cross-pol channel and the co-pol channel, then the corresponding ratios were obtained by back transformation to the original basis (HV).

To compare the results of Examples 1 and 2 with the results obtained for the same matrices by Yan and Boerner [14], we changed the polarization states to their corrsponding Stokes vector by using formulae (B-10a) -

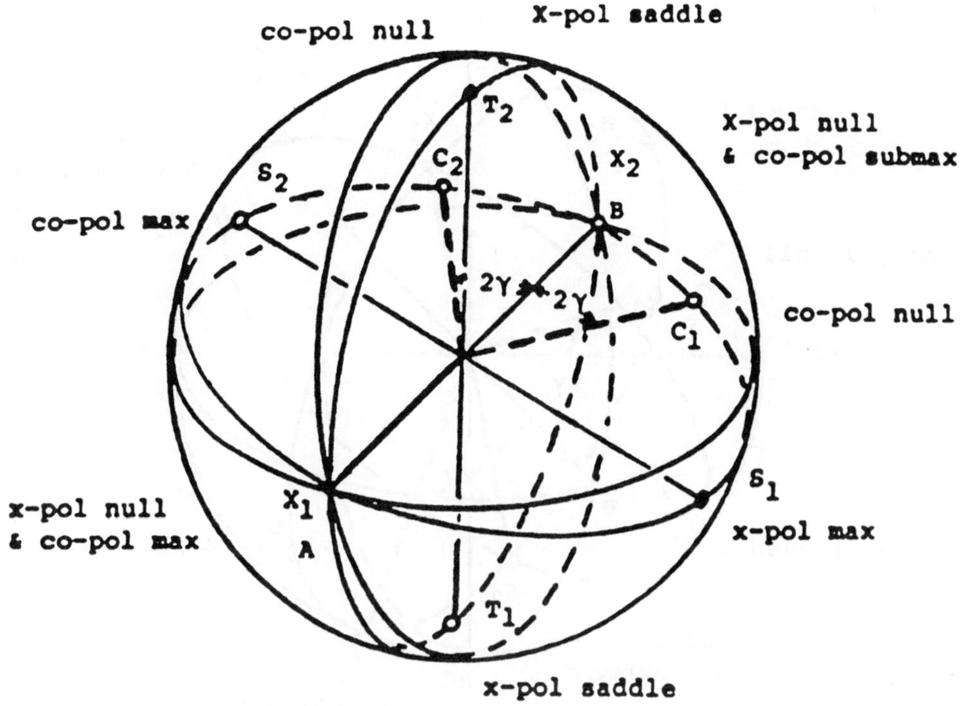


Fig. 7a: The characteristic polarization states on the Poincaré sphere of Example (2) referenced to the new basis (AB) for the scattering matrix $[S_2] = \begin{bmatrix} -3 & j \\ j & 2j \end{bmatrix}$ with the characteristic polarization ratios:

$\rho'_{xn1} = \rho'_{cm1} = 0\ (X_1)$, $\rho'_{xn2} = \rho'_{cm2} = \infty\ (X_2)$, $\rho'_{xm1} = e^{j69°}\ (S_1)$.

$\rho'_{xm2} = e^{-j110°}\ (S_2)$, $\rho'_{xs1} = e^{j20.9°}\ (T_1)$, $\rho'_{xs2} = e^{j159°}\ (T_2)$,

$\rho'_{cn1} = 1.39\ ej69°\ (C_1)$ and $\rho'_{cn2} = 1.398\ e{-}j111°\ (C_2)$.

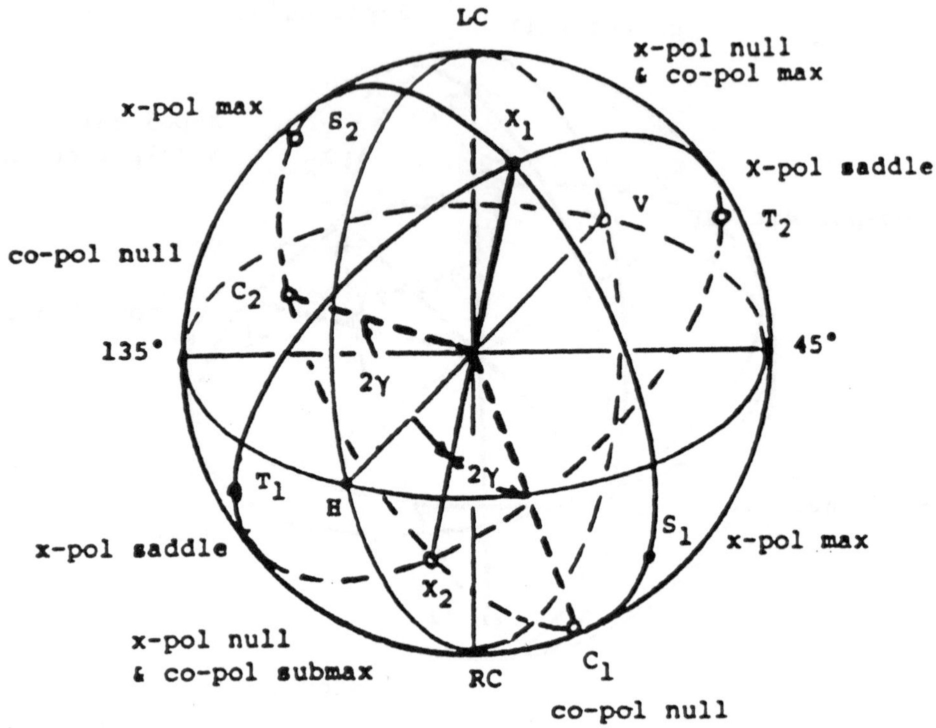


Fig. 7b: The characteristic polarization states on the Poincaré sphere referenced to the old basis (HV) for the scattering matrix $[S_2]$ of Example (2) with the characteristic polarization ratios:

$\rho_{xn1} = \rho_{cm1} = 0.523\ e^{j56°}(X_1)$, $\rho_{xn2} = \rho_{cm2} = 1.91\ e^{-j124°}(X_2)$,

$\rho_{xm1}=0.74e^{-j46°}(S_1)$, $\rho_{xm2}=1.35e^{j134°}(S_2)$, $\rho_{xs1}=0.36\ e^{-j157°}(T_1)$,

$\rho_{xs2}=2.76e^{j23°}(T_2)$, $\rho_{cn1}=0.9e^{-j61°}(C_1)$ and $\rho_{cn2}=1.64e^{j151°}(C_2)$.

and the geometric parameters:

$\nu = 0.0$, $\gamma = 35.6$, $\delta_m = 56.3$, $\alpha_m = 27.6$, $\phi = 19.3$ and $\tau_m = 21.6$.

(B-10e). We can see that the result is exactly the same for the first example of [14] in which the corresponding Mueller matrix [M] and the method of Lagrange multipliers were used. For the second example, we obtained the total of eight different characteristic polarization ratios or, equivalently, eight different Stokes vectors as in example (1), but in [14], only six different Stokes vectors were found which are the same as six out of eight of ours. The fact that two are missing may be caused by the numerical method used in [14]. Sometimes, solutions can be lost by using numerical methods although numerical computer evaluations are necessary for most of the complex scientific and engineering problems.

## 6. CONCLUSIONS

It was demonstrated with the use of the generalized polarization transformation ratio $\rho$ formulation that Huynen's polarization fork concept is correct, and it indeed presents a unique tool in radar polarimetry. In particular, our derivation clearly demonstrates the usefulness of the polarization transformation ratio $\rho$ formulation which in complete agreement with Huynen's formulation [2] shows, given a measured matrix [S], that the Huynen target characteristic parameters m, $\phi_m$, $\tau_m$, $\nu$, $\gamma$, $\delta_m$ and $\alpha_m$ can be uniquely determined; or inversely, given the parameters m, $\phi_m$, $\tau_m$, $\nu$ and $\gamma$, the corresponding scattering matrix [S] can be reconstructed. Hence, the resulting Huynen fork concept represents a unique example of a polarimetric radar inverse problem.

The optimal polarizations associated with a symmetric scattering matrix were first discovered by Kennaugh [1] during 1948-1952. He found that a copolarized transmitting and receiving radar antenna can be used to obtain the maximum echo return possible, and that there exists only one polarization for which this maximum is achieved. Kennaugh also showed that there are two polarizations, the co-pol nulls, for which the backscattered wave is orthogonally polarized to the radar antenna, thus rendering the radar "blind" to the incoming wave.

Based on Kennaugh's work, Huynen [2] introduced the "Polarization Fork" concept, and solved the same problem by using the so called "eigenvalue" method and gave a precise relation between the target scattering matrix [T] and the target geometrical parameters as presented in (58).

Under the consideration of the missing links of the existing optimal polarization theory, Agrawal and Boerner [7] used a different approach, the restricted unitary transformation matrix formulation in terms of the polarization ratio $\rho$, and found six distinct characteristic polarization states based on Kennaugh's theory, two co-pol maxima (identical to cross-pol nulls), two co-pol nulls and two cross-pol maxima from which a "polarization tree interpretation" on the Poincaré sphere was introduced. Their exposition was, however, still incomplete.

The current paper is essentially an extension of and based on Agrawal and Boerner [7] and Kennaugh's optimization theory [1] using a more general unitary transformation matrix approach also formulated in terms of the polarization ratio ρ. Eight distinct characteristic polarization states were found to exist by this method, the two additional states called cross-pol saddlepoint extrema are also orthogonal to each other, like the co-pol maxima (cross-pol nulls) and the cross-pol maxima. Especially for the Poincaré sphere representation, the three diameters, which join the three pairs of characteristic polarization states, are mutually at right angles to each other which together with the two co-pol nulls form the complete "Polarization Fork." By using the general unitary transformation matrix, the result obtained can be reduced to agree with Huynen's "Standardized Polarization Fork" concept and the scattering matrix expressed in terms of the target geometrical parameters.

Kennaugh and Huynen's work is so central to radar polarimetry because they discovered and interpreted the two co-pol nulls and showed that once these two characteristic states are located on the Poincaré sphere, the complete "Fork" can be determined for the monostatic reciprocal case, according to the positions of the two co-pol nulls and its interelation with the three distinct pairs of orthogonal target characteristic polarization states.

## 7. ACKNOWLEDGEMENTS

We wish to express our gratitude to Mrs. Wei-Ling Yan and Dr. Yoshio Yamaguchi for proof-reading and commenting on the manuscript, and to Dr. J. Richard Huynen for his continual encouragement in carrying out this research. Also, a sincere note of thanks is extended to Dr. Henry W. Mullaney, Dr. James W. Mink, Dr. Karl H. Steinbach, Dr. Walter A. Flood, Mr. Lloyd W. Root, Mr. Otto Kessler and Mr. James J. Smith for their continual interest in this research, and a special thanks to Ms. Mirian R. Mailey for her skillful typing of the final manuscript.

The financial support for this research was provided by the US Office of Naval Research, Contract Nos. N00014-80-C-0773 and N00014-90-J-1405, the US Army Research Office, Contract No. DAAL-03-89-K-0116, and a US Senior Scientist Fellow Award of the Alexander von Humboldt Stiftung, Bonn, FRG.

## 8. REFERENCES

[1] E.M. Kennaugh, "Effects of Type of Polarization on Echo Characteristics", The Ohio State University, Antennas Laboratory, Columbus, OH, Reports 381-1 to 394-24, 1949-1954 and Report 389-12 (M.Sc. Thesis) in particular.

[2] J.R. Huynen, "Phenomenological Theory of Radar Targets", Ph.D. dissertation, Rotterdam: Drukkerij Bronder-Offset N.V., 1970.

[3] W-M. Boerner and M.B. El-Arini et.al., "Polarization Dependence in Electroamgnetic Inverse Problems", IEEE Trans. A&P, Vol. AP-29, pp. 262-271, 1981.

[4] M. Davidovitz and W-M. Boerner, "Exentsion of Kennaugh's Optimal Polarization Concept to the Asymmetric Matrix Case", IEEE Trans. A&P, Vol. AP-34, No. 4, pp. 569-574, Apr. 1986,.

[5] A.B. Kostinski and W-M. Boerner, "On Foundations of Radar Polarimetry", IEEE A&P, Vol. AP-34, No. 12, pp. 1395-1404, Dec. 1986.

[6] H. Mieras, "Optimum Polarizations of Simple Compound Targets", IEEE Trans. A&P, Vol. AP-31, No.11, pp.996-999, Nov. 1983.

[7] A.P. Agrawal and W-M. Boerner, "Redevelopment of Kennaugh's Target Characteristic Polarization State Theory Using the Polarization Transformation Ratio Formalism for the Coherent Case", IEEE Trans. A&P, Vol. AP-27, No. 1, pp. 2-14, Jan 1989.

[8] R.M.A. Azzam & N.M. Bashara, "Elliposmetry and Polarized Light", Amsterdam: North-Holland Publishing Company, 1979.

[9] IEEE Standard Number 169-1979, "Test Procedures for Antennas", Published by the Institute of Electrical and Electronic Engineers, Inc., New York" IEEE Publ., 1979.

[10] G.A. Deschamps, "Part-2: Geometrical Representation of the Polarization of Plane Electromagnetic Wave", Proc. IRE, Vol. 39, pp. 540-544, 1951.

[11] H. Mott, "Polarization in Antennas and Radar", New York: John Wiley & Sons, 1986.

[12] A.L. Maffet, "Scattering Matrices", in "Methods of Radar Cross Section Analysis", J.W. Crispin and K.B. Siegel, Eds., New York: Academic, 1968.

[13] C.D. Graves, "Radar Polarization Power Scattering Matrix," Proc. IRE, Vol. 44, pp. 248-252, Feb. 1956.

[14] W-L. Yan and W-M. Boerner, "Optimal Polarization States Determina tion of the Stokes Reflection Matrix $[\bar{M}_p]$ for the Coherent Case, and of the Mueller [M] for the Partially Polarized Case", Journal of Electromagnetic Waves and Applications, Vol. 5, No. 7(?) in Print, pp. X01-X12, July 1991.

## 9. APPENDICES

### APPENDIX A: EQUIVALENCE PROOF

The polarization ratio $\rho_{1,2}$ of (28) are the cross-pol nulls and co-pol-maxima in the old basis (HV). The power corresponding to the cross-pol-nulls in the cross-pol channel are zeros and in the co-pol channel are $P_1^c = |\lambda_1|^2$ and $P_2^c = |\lambda_2|^2$ with $\lambda_1$ and $\lambda_2$ represented by (30) & (31a,b).

The result is the same as that given in Kostinski and Boerner [5] in which the Graves power matrix $[G] = [S]^+[S]$ as given in [13] together with the "three stages" method was used to solve the optimal problem in the bistatic case.

The total normalized power in the scattered wave according to [5] is given by

$$P_W = \vec{E}_R^{+} \vec{E}_R = \vec{E}_T^{+}[S]^{+}[S]\vec{E}_T \equiv \vec{E}_T^{+}[G]\vec{E}_T \qquad \text{(A-1)}$$

where $+ \equiv (*)^T$ denotes the hermitian conjugate and [G] is a hermitian matrix. The corresponding eigenvalue problem of (A-1) is

$$\{[G] - \mu[I]\}\vec{E}_{T.opt} = 0 \qquad \text{(A-2)}$$

and the corresponding characteristic equation is

$$\mu^2 - (g_{11} + g_{22})\mu + (g_{11}g_{22} - g_{12}g_{21}) = 0 \qquad \text{(A-3)}$$

where $g_{ij}$ (i,j = 1,2) are the elements of the power scattering matrix [G] with

$$g_{11} = |S_{HH}|^2 + |S_{VH}|^2$$

$$g_{12} = S_{HH}^{*} S_{HV} + S_{VV}S_{VH}^{*}$$

$$g_{21} = S_{HH}S_{HV}^{*} + S_{VV}^{*}S_{VH} = g_{12}^{*} \qquad \text{(A-4)}$$

$$g_{22} = |S_{VV}|^2 + |S_{HV}|^2$$

In order to verify our result, we assume for the monostatic case that $S_{HV} = S_{VH}$ in (A-4).

The eigenvalues can be obtained by solving (A-3) as:

$$\mu_{1,2} = \frac{g_{11} + g_{22} \pm \sqrt{(g_{11} + g_{22})^2 - 4(g_{11}g_{22} - g_{12}g_{21})}}{2}$$

$$= \frac{g_{11} + g_{22} \pm \sqrt{(g_{22} - g_{11})^2 + 4g_{12}g_{21}}}{2}$$

Substituting the eigenvalues $(\mu_1,\mu_2)$ of (A-5) into (A-2) and solving for the components of $\vec{E}_{T\cdot opt}$ for $\mu_1$

$$(g_{11} - \mu_1)E_{H1} + g_{12} E_{V1} = 0 \qquad \text{(A-6a)}$$

$$g_{21} E_{H1} + (g_{22} - \mu_1) E_{V1} = 0 \tag{A-6b}$$

Then assuming $\mu_1$ and $\mu_2$ are different eigenvalues, equations (A-6a,b) are equivalent. Setting $E_{H1} = 1$, then

$$E_{V1} = \frac{\mu_1 - g_{11}}{g_{12}} = \frac{g_{22} - g_{11} + \sqrt{(g_{22} - g_{11})^2 + 4g_{12}g_{21}}}{2g_{12}}$$

$$\equiv \frac{-B + \sqrt{B^2 - 4AC}}{2A} = \rho_1 \tag{A-7a}$$

which is exactly the same as (28).

For $\mu = \mu_2$ , the same routine yields:

$$E_{V2} = \frac{g_{22} - g_{11} - \sqrt{(g_{22} - g_{11})^2 + 4g_{12}g_{21}}}{2g_{12}} = \rho_2 \tag{A-7b}$$

So the normalized optimal (co-pol maximum and submaximum) polarization states of the wave are

$$\vec{E}_{T\cdot opt1} = \frac{1}{\sqrt{1 + E_{V1}E_{V1}^*}} \begin{bmatrix} 1 \\ E_{V1} \end{bmatrix} = \frac{1}{\sqrt{1 + \rho_1\rho_1^*}} \begin{bmatrix} 1 \\ \rho_1 \end{bmatrix} \tag{A-8a}$$

$$\vec{E}_{T\cdot opt2} = \frac{1}{\sqrt{1 + E_{V2}E_{V2}^*}} \begin{bmatrix} 1 \\ E_{V2} \end{bmatrix} = \frac{1}{\sqrt{1 + \rho_2\rho_2^*}} \begin{bmatrix} 1 \\ \rho_2 \end{bmatrix} \tag{A-8b}$$

yielding the same result as given in (32).

The "three stages" method using the Graves power matrix is a powerful tool for solving the optimal polarization problem in the bistatic case subject to the matching condition $\vec{h}_r = \vec{E}_R^*$, i.e., the vector antenna height of the receiver is the same as the conjugate of the scattered wave, which decreases the free degrees by one.

However, we may lose some information under this limitation; for example, the polarization states of zero power return to the receiver. But if one was to discard this matching condition ($\vec{h}_r = \vec{E}_R^*$ with $|\vec{E}_R| = 1$), then there exists no equivalent Graves power matrix [G], and we need to find other ways for treating the optimization problem for the general bistatic case.

**APPENDIX B: LIST OF FORMULAS FOR NUMERICAL EVALUATION:**

By using the "critical point" method, we can solve via computer-numerical calculations, the "optimization problem" for the monostatic reciprocal case. The formulas used in our program are:

(1) Definition of the diagonalizing factor $\rho_{1,2}$:

$$\rho_{1,2} = \frac{-B \pm \sqrt{B^2 - 4AC}}{2A} \tag{B-1}$$

where $A = S_{HH}^* S_{HV} + S_{HV}^* S_{VV}$, $B = |S_{HH}|^2 - |S_{VV}|^2$, $C = -A^*$

and the scattering matrix is given as:

$$[S] = \begin{bmatrix} S_{HH} & S_{HV} \\ S_{VH} & S_{VV} \end{bmatrix}, \quad S_{HV} = S_{VH} \tag{B-2}$$

(2) The diagonal elements of the scattering matrix in the new basis (AB) can be calculated as:

$$\lambda_1 = S'_{AB}(\rho_1) = (1 + \rho_1\rho_1^*)^{-1}(S_{HH} + 2\rho_1 S_{HV} + \rho_1^2 S_{VV})e^{j\psi_1} \tag{B-3}$$

$$\lambda_2 = S'BB(\rho_1) = (1 + \rho_1\rho_1^*)^{-1}(\rho_1^{*2} S_{HH} - 2\rho_1^* S_{HV} + \rho_1^2 S_{VV})e^{j\psi_4} \tag{B-4}$$

(3) $\psi_1$ and $\psi_4$ are the phases of the unitary transformation matrix [U] with

$$[U] = \frac{1}{\sqrt{1 + \rho_1\rho_1^*}} \begin{bmatrix} e^{j\psi_1} & -\rho_1^* e^{j\psi_4} \\ \rho_1^* \, e^{j\psi_1} & e^{j\psi_4} \end{bmatrix}$$

which can be determined uniquely using

$$\psi_1 = -\frac{\delta_2}{2} - \frac{\pi}{4} \tag{B-5}$$

$$\psi_4 = \frac{\delta_2}{2} - \frac{\pi}{4} \tag{B-6}$$

where $\delta_2$ is the phase of $\rho_2$ found in (B-1). The proof of formulae (B-5) and (B-6) are shown in appendix C.

(4) The characteristic polarization states in reference to the new basis (AB):

(i) cross-pol null $\rho'_{xn1} = 0 \quad = \rho'_{cm1}$ (B-7a)

and co-pol max: $\rho'_{xn2} = \infty \quad = \rho'_{cm2}$ (B-7b)

(ii) cross-pol maxima

$$\rho'_{xm1,2} = \pm j\left(\frac{\lambda_1\lambda_2^*}{\lambda_1^*\lambda_2}\right)^{1/4} \tag{B-7c}$$

(iii) cross-pol saddle

$$\rho'_{xs1,2} = \pm\left(\frac{\lambda_1\lambda_2^*}{\lambda_1^*\lambda_2}\right)^{1/4} \tag{B-7d}$$

(iv) co-pol nulls

$$\rho'_{cn1,2} = \pm\left(-\frac{\lambda_1}{\lambda_2}\right)^{1/2} \tag{B-7e}$$

(5) The characteristic polarization states reference to the old basis (AB) can be determined by transferring back using the unitary transformation matrix as:

$$\rho_{old} = \frac{(\cos\psi_1 + j\sin\psi_1) - \rho_1^*\rho'_{new}(\cos\psi_4 + j\sin\psi_4)}{\rho_1(\cos\psi_1 + j\sin\psi_1) + \rho'_{new}(\cos\psi_4 + j\sin\psi_4)} \tag{B-8}$$

(6) The formulae for the power returns to the cross-pol and co-pol channels at the characteristic polarization ratios:

$$P^x_{xn1,2} = 0 \tag{B-9a}$$

$$P^c_{xn1} = |\lambda_1|^2 \tag{B-9b}$$

$$P^c_{xn2} = |\lambda_2|^2 \tag{B-9c}$$

$$P^c_{cn1,2} = 0 \tag{B-9d}$$

$$P^{x}_{cn1,2} = |\lambda_1||\lambda_2| \tag{B-9e}$$

$$P^{x}_{xm1,2} = \tfrac{1}{4}(|\lambda_1| + |\lambda_1|)^2 \tag{B-9-f}$$

$$P^{c}_{xm1,2} = \tfrac{1}{4}(|\lambda_1| - |\lambda_1|)^2 \tag{B-9g}$$

$$P^{x}_{xs1,2} = \tfrac{1}{4}(|\lambda_1| - |\lambda_2|)^2 \tag{B-9h}$$

$$P^{c}_{xs1,2} = \tfrac{1}{4}(|\lambda_1| + |\lambda_2|)^2 \tag{B-9i}$$

(7) The Stokes vector $\vec{g}$ referenced to the old basis:

$$g_0 = \frac{1}{1 + \rho_i\rho_i^*} + \frac{\rho_i\rho_i^*}{1 + \rho_i\rho_i^*} = 1 \tag{B-10a}$$

$$g_1 = \frac{1}{1 + \rho_i\rho_i^*} - \frac{\rho_i\rho_i}{1 + \rho_i\rho_i^*} = \frac{1 - \rho_i\rho_i^*}{1 + \rho_i\rho_i^*} \tag{B-10b}$$

$$g_2 = \frac{1 + |\rho_i|}{1 + \rho_i\rho_i^*} \cos(\arg\{\rho_i\}) \tag{B-10c}$$

$$g_3 = \frac{1 + |\rho_i|}{1 + \rho_i\rho_i^*} \sin(\arg\{\rho_i\}) \tag{B-10d}$$

$$\vec{g} = \begin{bmatrix} g_0 \\ g_1 \\ g_2 \\ g_3 \end{bmatrix} \tag{B-10e}$$

where $\rho_i$ is the corresponding characteristic polarization ratio in the old basis.

(8) The tilt angle $\phi$ and the ellipticity angle $\tau$. There are two ways to find the parameters $\phi$ and $\tau$;

(i) using the components of the Stokes vector $\vec{g}$:

$$\phi = \tan^{-1}\left(\frac{g_2}{g_1}\right) \tag{B-11a}$$

$$\tau = \sin^{-1}(g_3) \tag{B-11b}$$

(ii) using polarization ratio $\rho$'s

$$\delta = \arg\{\rho_i\} \tag{B-12a}$$

$$\alpha = \tan^{-1}\{|\rho_i|\} \tag{B-12b}$$

$$\phi = \tfrac{1}{2}\tan^{-1}[\tan(2\alpha)\sin(\delta)] \tag{B-12c}$$

$$\tau = \tfrac{1}{2}\sin^{-1}[\sin(2\alpha)\sin(\delta)] \tag{B-12d}$$

Both ways reduce to the same result.

(9) Huynen's geometrical parameters:

(i) target size m

$$m = |\lambda_1| \tag{B-13a}$$

(ii) target skip angle $\nu$

$$\nu = \tfrac{1}{4}(\phi_1 - \phi_2) \tag{B-13b}$$

where $\lambda_i = |\lambda_i| e^{j\phi_i}$ , (i = 1, 2)

(iii) target characteristic angle $\gamma$

$$\gamma = \tfrac{1}{4}\tan^{-1}\left\{\left|\frac{\lambda_2}{\lambda_1}\right|^{\frac{1}{2}}\right\} \tag{B-13c}$$

(iv) the phase of co-pol MAX (cross-pol null) $\delta$

$$\delta_m = \arg\{\rho_1\} \tag{B-13d}$$

(v) the spinor parameter $\alpha_m$

$$\alpha_m = \tan^{-1}(|\rho_1|) \tag{B-13e}$$

(vi) target orientation or tilt angle $\phi$

$$\phi = \tfrac{1}{2}\tan^{-1}(\tan 2\alpha_m \cos\delta_m) \tag{B-13f}$$

(vii) ellipticity angle of the co-pol MAX (cross-pol null) $\tau_m$

$$\tau_m = \tfrac{1}{2}\sin^{-1}(\sin 2\alpha_m \sin\delta_m) \qquad \text{(B-13g)}$$

The results of two examples using the above formulae in fortran language are lised in Section 5-1. Be careful in treating special cases like the singular scattering matrix or the scattering matrix given in diagonal form. The two scattering matrices are:

$$\text{(a)} \quad [S] = \begin{bmatrix} 2j & 0.5 \\ 0.5 & -j \end{bmatrix}$$

$$\text{(b)} \quad [S] = \begin{bmatrix} -3 & j \\ j & 2j \end{bmatrix}$$

As proved in Appendix D, there are at most six characteristic polarization ratios for the cross-pol channel and four for the co-pol channel, of which two pairs are alike; thus, eight distinct characteristic polarization ratios in total .

**APPENDIX C: PROOF OF EQUATIONS (B-5) AND (B-6)**

In Section 2.4, we mentioned that to find the characteristic polarization states for a given transmitter-target-receiver geometry, the simplest way is to diagonalize the scattering matrix [S] by using the unitary transformation matrix [U] which will be determined uniquely for the case under consideration.

There are three parameters in [U], i.e., $\rho$, $\psi_1$ and $\psi_4$, of which $\rho$ can be obtained from (28), and $\psi_1$, $\psi_4$ can be determined uniquely from (B-5) and (B-6) which are:

$$\psi_1 = -\frac{\delta_2}{2} - \frac{\pi}{4} \qquad \text{(C-1)}$$

$$\psi_4 = \frac{\delta_2}{2} - \frac{\pi}{4} \qquad \text{(C-2)}$$

where $\delta_2$ was defined in Appendix B (B-12a).

In order to obtain the expressions (C-1) and (C-2), we first analyze conditions which should be satisfied for $\psi_1$ and $\psi_4$.

Since [U] is an unitary matrix, the following condition should be satisfied, i.e.,

$$\det\{[U]\} = e^{j(\psi_1 + \psi_4)} \qquad \text{(C-3)}$$

For the special case of transformation from the (HV) basis to the (LR) basis, the basis transformation matrix is given by (22a):

$$[T_{1c}(HV \rightarrow LR)] = \frac{1}{\sqrt{2}}\begin{bmatrix} 1 & j \\ 1 & -j \end{bmatrix}$$

where $[T] = [U]^{-1}$, and

$$\det\{[T_{1c}]\} = e^{-j\pi/2} \quad \text{or} \quad \det\{[U_{1c}]\} = e^{-j\pi/2} \tag{C-4}$$

Transformation HV → LR is a special case of the general basis transformation. It should be included in the general transformation. In other words, the relation of (C-4) could be derived from the general relation (C-3), i.e.,

$$\psi_1 + \psi_4 = -\frac{\pi}{2} \tag{C-5}$$

Note, that $\psi_1$ and $\psi_4$ cannot be chosen arbitrarily, because to do so will cause the target characteristic parameter $\nu$, the target skip angle, become uncertain and this is not allowed.

Equation (C-5) is one relation between $\psi_1$ and $\psi_4$, we need one more relation to determine $\psi_1$ and $\psi_4$ uniquely. Remember that we have two diagonalization factors $\rho_1$ and $\rho_2$ which are the co-pol max and co-pol submax reference to the old basis, and both can be used to diagonalize the scattering matrix [S].

For choosing $\rho_1$, the new basis vectors are:

$$\vec{V}_A = \frac{e^{j\psi_1}}{\sqrt{1 + \rho_1\rho_1^*}}\begin{bmatrix} 1 \\ \rho_1 \end{bmatrix} \tag{C-6a}$$

$$\vec{V}_B = \frac{e^{j\psi_4}}{\sqrt{1 + \rho_1\rho_1^*}}\begin{bmatrix} -\rho_1^* \\ 1 \end{bmatrix} \tag{C-6b}$$

For choosing $\rho_2$ , the new basis vectors are:

$$\vec{V}'_A = \frac{e^{j\psi_4}}{\sqrt{1 + \rho_2\rho_2^*}}\begin{bmatrix} -\rho_2^* \\ 1 \end{bmatrix} \tag{C-7a}$$

$$\vec{V}'_A = \frac{e^{j\psi_1}}{\sqrt{1 + \rho_2\rho_2^*}} \begin{bmatrix} 1 \\ \rho_2 \end{bmatrix} \tag{C-7b}$$

By using the orthogonal relation of $\rho_1$ and $\rho_2$ as:

$$\rho_1\rho_2{}^* = -1$$

which gives

$$|\rho_1||\rho_2| = 1 \tag{C-8a}$$

$$|\delta_1 - \delta_2| = \pi \tag{C-8b}$$

The two unitary transformations have the following relations:

$$[U(\rho_2)] = [U(\rho_1)][P] \tag{C-9}$$

where

$$[P] = \begin{bmatrix} 0 & -e^{-j\Psi} \\ e^{j\Psi} & 0 \end{bmatrix} \tag{C-10a}$$

$$\Psi = \psi_1 - \psi_4 + \delta_2 \tag{C-10b}$$

and $\delta_2$ is the phase of $\rho_2 = |\rho_2|e^{j\delta_2}$ .

Eq. (C-9) means that the unitary transformation $[U(\rho_2)]$ is equivalent to two consecutive unitary transformations $[U(\rho_1)]$ followed by $[P]$.

The disgonal form of the scattering matrix after transformation using $[U(\rho_1)]$ to the new basis of (C-6) becomes:

$$[U(\rho_1)]^T[S(HV)][U(\rho_1)] = \begin{bmatrix} S'_{AA}(\rho_1) & 0 \\ 0 & S'_{BB}(\rho_1) \end{bmatrix}$$

$$= \begin{bmatrix} \lambda_{11} & 0 \\ 0 & \lambda_{12} \end{bmatrix} = \begin{bmatrix} |\lambda_1|e^{j\phi_1} & 0 \\ 0 & |\lambda_1|e^{j\phi_2} \end{bmatrix} \tag{C-11}$$

and by transformation using $[U(\rho_2)]$ in the basis (C-7) becomes

$$[U(\rho_2)]^T[S(HV)][U(\rho_2)] = [P]^T \begin{bmatrix} \lambda_{11} & 0 \\ 0 & \lambda_{12} \end{bmatrix} [P]$$

$$= \begin{bmatrix} \lambda_{22} & 0 \\ 0 & \lambda_{21} \end{bmatrix} = \begin{bmatrix} |\lambda_2| e^{j(\phi_2+2\Psi)} & 0 \\ 0 & |\lambda_1| e^{j(\phi_1-2\Psi)} \end{bmatrix}$$

$$= \begin{bmatrix} |\lambda_2| e^{j\Phi_2} & 0 \\ 0 & |\lambda_1| e^{j\Phi_1} \end{bmatrix} \tag{C-12}$$

From (C-12), we can see that

$$\lambda_{21} = \lambda_{11} e^{-j2\Psi}, \tag{C-13a}$$

$$\lambda_{22} = \lambda_{12} e^{-j2\Psi}, \tag{C-13b}$$

and

$$\Phi_1 = \phi_1 - 2\Psi \quad , \quad \Phi_2 = \phi_2 + 2\Psi \tag{C-14}$$

Recall that the target skip angle $\nu = \frac{1}{4}(\phi_1 - \phi_2)$. Applying this definition to (C-11) and (C-12), yields

$$\nu_1 = \frac{1}{4}(\phi_1 - \phi_2) = \nu \tag{C-15}$$

$$\nu_2 = \frac{1}{4}(\phi_1 - \phi_2) - \Psi \tag{C-16}$$

Then the uniqueness of the characteristic parameter $\nu$

$$\Psi = \psi_1 - \psi_4 + \phi_2 = 0$$

or

$$\psi_1 - \psi_4 = -\delta_2 \tag{C-17}$$

Combining (C-17) and (C-5), the expressions for determining the phases $\psi_1$ and $\psi_4$ are obtained as

$$\psi_1 = -\frac{\delta_2}{2} - \frac{\pi}{4} \tag{C-18a}$$

$$\psi_4 = \frac{\delta_2}{2} - \frac{\pi}{4} \tag{C-18b}$$

With $\Psi = 0$, the two pairs of $\lambda_s$ $\lambda_{11}$, $\lambda_{12}$ and $\lambda_{21}$, $\lambda_{22}$ have the relation:

$$\lambda_{21} = \lambda_{11} = \lambda_1 \tag{C-19a}$$

$$\lambda_{22} = \lambda_{12} = \lambda_2 \tag{C-19b}$$

The diagonal form of the scattering matrix after unitary transformations by $[U(\rho_1)]$ and $[U(\rho_2)]$ are

$$[S_d(\rho_1)] = \begin{bmatrix} \lambda_1 & 0 \\ 0 & \lambda_2 \end{bmatrix} \tag{C-20a}$$

$$[S_d(\rho_2)] = \begin{bmatrix} \lambda_2 & 0 \\ 0 & \lambda_1 \end{bmatrix} \tag{C-20b}$$

respectively. this results also satisfies the unique solution for the optimal polarization state problem for a given target with fixed radar frequency and aspect angle.

## APPENDIX D: THE CHARACTERISTIC POLARIZATION STATES OF THE CO-POL AND X-POL CHANNELS

In Section 3, we used the mathematical extremum principle to prove that there exist at most four characteristic polarization ratios (one co-pol max, one co-pol submax and two co-pol nulls) for the co-pol channel and at most six characteristic polarization ratios (two cross-pol max, two cross-pol nulls and two cross-pol saddles) for the cross-pol channel for the reciprocal monostatic case in which the scattering matrix is symmetric. In this appendix, we will give the detailed mathematical proofs.

### D.1: The power expressions

The power expression for the co-pol channel for basis (AB), in which the scattering matrix is diagonal, is given by:

$$\begin{aligned} P^C &= |\vec{h}'^T [S_d]\vec{h}'|^2 \\ &= (1+\rho'\rho'^*)^{-2}(|\lambda_1|^2 + \lambda_1\lambda_2^*\rho'^{*2} + \lambda_1^*\lambda_2\rho'^2 + |\lambda_2|^2\rho'^2\rho'^{*2}) \\ &= (1+\rho'\rho')^{-2}(a + C\rho'^{*2} + C^*\rho'^2 + b\rho'^2\rho^{*\prime 2}) \end{aligned} \tag{D-1}$$

where, $a = |\lambda_1|^2$, $b = |\lambda_2|^2$, are real and

$C = \lambda_1\lambda_2^*$, $C^* = \lambda_1^*\lambda_2$ are complex or may be real

The power expression for the cross-pol channel in basis (AB) is:

$$P^x = |\vec{h}_\perp^T[S_d]\vec{h}|^2$$

$$= (1 + \rho'\rho'^*)^{-2}(|\lambda_1|^2\rho'\rho'^* - \lambda_1\lambda_2^*\rho'^2 + |\lambda_2|^2\rho'\rho'^*)$$

$$= (1 + \rho'\rho'^*)^{-2}(a\rho'\rho' - C\rho'^{*2} - C^*\rho'^2 + b\rho'\rho'^*) \quad \text{(D-2)}$$

D.2: <u>The Equivalence of the derivatives of $P^c$ and $P^x$ with respect to $\rho' = x + jy$ and $\rho'^* = x - jy$ and to $x = 1/2\ (\rho' + \rho^{*\prime})$ and $y = (\rho' - \rho^{*\prime})/(2j)$</u>

In the power expressions (D-1) and (D-2), there are two real variables x and y which come from the complex polarization ratio $\rho' = x + jy$, so the critical points (extrema and saddle points) of function (D-1) and (D-2) should be those points for which the first derivatives of $P^c$ or $P^x$ with respect to x and y vanish, i.e., the points satisfy the following conditions:

$$\frac{\partial P(x,y)}{\partial x} = 0 \quad \text{(D-3a)}$$

$$\frac{\partial P(x,y)}{\partial y} = 0 \quad \text{(D-3b)}$$

Only at those points, the maximum, minimum and saddle point states can occur.

To simplify the proof, we use the equivalent condition to (D-3) as:

$$\frac{\partial P(\rho',\rho'^*)}{\partial \rho'} = 0 \quad \text{(D-4a)}$$

$$\frac{\partial P(\rho',\rho'^*)}{\partial \rho'^*} = 0 \quad \text{(D-4b)}$$

The equivalency of (D-3) and (D-4) can be proved straight-forwardly:

$$\frac{\partial P}{\partial x} = \frac{\partial P}{\partial \rho'}\frac{\partial \rho'}{\partial x} + \frac{\partial P}{\partial \rho'^*}\frac{\partial \rho'^*}{\partial x} = \frac{\partial P}{\partial \rho'} + \frac{\partial P}{\partial \rho'^*} \quad \text{(D-5a)}$$

$$\frac{\partial P}{\partial y} = \frac{\partial P}{\partial \rho'}\frac{\partial \rho'}{\partial y} + \frac{\partial P}{\partial \rho'^*}\frac{\partial \rho'^*}{\partial y} = j\frac{\partial P}{\partial \rho'} - j\frac{\partial P}{\partial \rho'^*} \quad \text{(D-5b)}$$

To satisfy condition (D-3) and from (D-5), we have:

$$\frac{\partial P}{\partial \rho'} + \frac{\partial P}{\partial \rho'^{*}} = 0 \qquad \text{(D-5c)}$$

$$\frac{\partial P}{\partial \rho'} - \frac{\partial P}{\partial \rho'^{*}} = 0 \qquad \text{(D-5d)}$$

That leads to the required result as:

$$\frac{\partial P}{\partial \rho'} = 0 \qquad \text{(D-5e)}$$

$$\frac{\partial P}{\partial \rho'^{*}} = 0 \qquad \text{(D-5f)}$$

D.3: <u>The critical points of the power function ($P^C$) for the co-pol channel (D-1)</u>

The first partial derivatives of (D-1) with respect to $\rho'$ and $\rho'^*$ are

$$\frac{\partial P^C}{\partial \rho'} = \frac{2C^*\rho' + 2b\rho'\rho'^{*2} - 2a\rho'^* - 2C\rho^{*3}}{(1 + \rho'\rho'^*)^3} \qquad \text{(D-6a)}$$

$$\frac{\partial P^C}{\partial \rho'^{*}} = \frac{-2C^*\rho'^3 - 2a\rho' + 2b\rho'^*\rho'^2 + 2C\rho'^*}{(1 + \rho'\rho'^*)^3} \qquad \text{(D-6b)}$$

The first two critical points which satisfy the vanishing conditions (D-5e) and (D-5f) are

$$\rho' = 0 \qquad \text{(D-7)}$$

Since the numerators of (D-6) will vanish but not the denominator, another vanishing location exists at

$$\rho' = \infty \qquad \text{(D-8)}$$

because the numerator of (D-6) is of degree three in the combination of $\rho'$ and $\rho'^*$, but of degree six for the denominator. To find the remaining critical points of the power expression $P^C$ which satisfy the condition (D-5e,f) we solve the following system which is obtained from the first derivative of $P^C$ in (D-6)

$$-C\rho'^{*3} - a\rho'^* + b\rho'\rho'^{*2} + C^*\rho' = 0 \qquad \text{(D-9a)}$$

$$-C^*\rho'^3 - a\rho' + b\rho'^*\rho'^2 + C\rho'^* = 0 \qquad \text{(D-9b)}$$

Multiplying (D-9a) to $\rho'$ and minus (D-9b) by $\rho'^*$ with $\rho'$ non-zero, we have

$$-C\rho'^{*3}\rho' + C^*\rho'^{*2} + C^*\rho'^*\rho'^3 - C\rho'^{*2} = 0 \qquad \text{(D-10)}$$

Factorizing (D-10), we have

$$(1 + \rho'\rho'^*)\ (C^*\rho'^2 - C\rho'^{*2}) = 0 \qquad \text{(D-11)}$$

which gives

$$C^*\rho'^2 - C\rho'^{*2} = 0 \qquad \text{(D-12)}$$

The solution can be obtained from (D-12) which gives

$$\rho'^* = \pm \left(\frac{C^*}{C}\right)^{\frac{1}{2}} \rho' \qquad \text{(D-13)}$$

To obtain the exact critical points, substitute

$$\rho'^* = \left(\frac{C^*}{C}\right)^{\frac{1}{2}} \rho'$$

into (D-10) which gives

$$\rho'^2 = \frac{-C^* + a\sqrt{\frac{C^*}{C}}}{b\,\frac{C^*}{C} - C^*\sqrt{\frac{C^*}{C}}}$$

or $$\rho' = \pm \left(-\frac{\lambda_1}{\lambda_2}\right)^{\frac{1}{2}} \qquad \text{(D-14)}$$

Then substituting $\rho'^* = -\left(\frac{C^*}{C}\right)^{\frac{1}{2}} \rho'$ into (D-10), after solving and simplifying, we obtain the same solution as (D-14), i.e.

$$\rho' = \pm \left(-\frac{\lambda_1}{\lambda_2}\right)^{\frac{1}{2}} \qquad \text{(D-15)}$$

Since (D-15) and (D-14) are exactly the same, we can conclude that there exist at most four critical points for function $P^C$ which are

$$\rho'_{cm1} = 0 \qquad \text{(D-16)}$$

$$\rho'_{cm2} = \infty \tag{D-17}$$

$$\rho'_{cn1,2} = \pm \left(-\frac{\lambda_1}{\lambda_2}\right)^{\frac{1}{2}}$$

D.4: <u>The critical points for the cross-pol channel.</u>

The first partial derivatives of $P^x$ of (D-2) with respect to $\rho'$ is:

$$\frac{\partial P^x}{\partial \rho'} = \frac{-a\rho'\rho'^{*2} + 2C\rho'^{*3} - b\rho'\rho'^{*2} + a\rho'^{*} - 2C^{*}\rho' + b\rho'^{*}}{(1 + \rho'\rho'^{*})^3} \tag{D-18}$$

Two critical points can be easily found by the same reason as in the last section for the function of $P^C$ which are

$$\rho'_{xn1} = 0 \tag{D-19}$$

$$\rho'_{xn2} = \infty \tag{D-20}$$

To find the other critical points, we proceed with the same procedure as in Section D.3 and obtain the following equation:

$$C\rho'\rho'^{*3} - C^{*}\rho'^{3}\rho'^{*} - C^{*}\rho'^{2} + C\rho'^{*2} = 0 \tag{D-21}$$

By factorizing (D-21), we have

$$(1 + \rho'\rho'^{*})(C\rho'^{*2} - C^{*}\rho'^{2}) = 0 \tag{D-22}$$

which results into the following conditions

$$C\rho'^{*2} - C^{*}\rho'^{2} = 0 \tag{D-23}$$

From (D-23) we obtain

$$\rho'^{*} = \pm \left(\frac{C^{*}}{C}\right)^{\frac{1}{2}} \rho' \tag{D-24}$$

Substituting $\rho'^{*} = \left(\frac{C^{*}}{C}\right)^{\frac{1}{2}} \rho'$ into (D-21), we have

$$\rho'_{xs1,2} = \pm \left(\frac{\lambda_1 \lambda_2^{*}}{\lambda_1^{*} \lambda_2}\right)^{\frac{1}{4}} \tag{D-25}$$

and for $\rho'^* = -\left(\frac{C^*}{C}\right)^{\frac{1}{2}}\rho'$, we obtain

$$\rho'_{xm1,2} = \pm j \left(\frac{\lambda_1 \lambda_2^*}{\lambda_1^* \lambda_2}\right)^{\frac{1}{4}} \tag{D-26}$$

(D-19), (D-20), (D-25) and (D-26) are six different critical points of the function $P^x$ of (D-2), so there are a total of, at most six, characteristic polarization ratios for the cross-pol channel. They are the cross-pol nulls $\rho'_{xn1}$ and $\rho'_{xn2}$, the cross-pol saddle points $\rho'_{xs1,2}$ and the cross-pol maxima $\rho'_{xm1,2}$.

Note that the cross-pol null $\rho'_{xn1}$ is the same as the co-pol max $\rho'_{cm1}$ and the cross-pol null $\rho'_{xn2}$ is the same as the co-pol submax $\rho'_{cm2}$. So the total number of different characteristic polarization states for the monostatic backscattering reciprocal case are eight.

# OPTIMAL POLARIZATION STATES DETERMINATION OF THE STOKES REFLECTION MATRICES $[\bar{M}_p]$ FOR THE COHERENT CASE, AND OF THE MUELLER MATRIX [M] FOR THE PARTIALLY POLARIZED CASE

WEI-LING YAN AND WOLFGANG-MARTIN BOERNER
UNIVERSITY OF ILLINOIS AT CHICAGO,
DEPARTMENT OF ELECTRICAL ENGINEERING AND COMPUTER SCIENCE
COMMUNICATIONS & SENSING LABORATORY
840 W. TAYLOR STREET, SEL-4210, M/C 154
CHICAGO, ILLINOIS/USA 60607

**ABSTRACT:** A problem originating in radar polarimetry is considered for which the radar target is to be characterized by its 4x4 Mueller matrix [M] properties in terms of a partially polarized wave treatment for the monostatic reciprocal and non-reciprocal cases. In order to compare our results with previous coherent treatments of optimizing the corresponding 2x2 Sinclair matrix [S], first the vector formalism for the coherent case is introduced. The coherent formulation is then extended via the coherency matrix (vector) approach to the partially polarized case with the aid of the Kronecker expansion matrix and the Stokes vector formulism. Distinction of the Stokes Reflection Matrices is made between the one-antenna and the two-antenna cases. For the one-antenna case, the reception of the optimal energy densities and powers is accomplished separately for the co-polarized and the cross-polarized channels which is a natural approach in radar polarimetry. For the two-antenna case, the reception of the optimal power is accomplished by adjusting the polarization state of the receiving antenna to match or mismatch the polarization state of the scattered wave. The "degen- erate Mueller matrix case" for purely coherent wave reception is analyzed first to facilitate comparison with previous methods of optimizing the corresponding Sinclair matrix [S]. Here, the Lagrangian multiplier method is used for determining the characteristic optimal polarization states subject to the constraint that the incident wave is purely polarized.

It is then shown that for the more general partially polarized scattered wave case, there exist next to the total energy density, three specific types of energy densities: (i) the adjustable purely polarized intensity, (ii) the noise-like unpolarized intensity, and (iii) the receivable polarized intensity; for which separate optimization methods are developed. Whereas, for the first category, a closed solution may be found; for the second two criteria numerical solutions must be used. The paper is concluded with a comparison of the various results obtained here with those obtained by other methods.

## I. INTRODUCTION AND FORMULATION OF THE PROBLEM

### 1.1 Background:

Although a coherent treatment of the radar target scattering problem may



*W.-M. Boerner et al. (eds.), Direct and Inverse Methods in Radar Polarimetry, Part 1*, 351–385.

suffice at microwave frequencies, with increasing frequency and commencing with the sub-millimeter wave spectral region, a partially polarized wave treatment of the scattering process is preferrable, and may strictly be required.

In the optical and radar polarimetric literature until very recently, only methods for optimizing the Jones vector or Sinclair matrix were available as demonstrated in Boerner [1], which were developed from the original formulation of Kennaugh [2] and its extension by Huynen [3]. The optimization procedure for the coherent case was completely solved only recently in Boerner and Xi [4] using a unitary transformation matrix approach expressed in terms of the polarization transformation ratio $\rho$ presentation. There it was shown, in complete agreement with Huynen's polarization fork concept, that there exist for the monostatic, reciprocal ($S_{AB} = S_{BA}$) case, treated here, in total, five pairs of characteristic polarization states: The antipodal (orthogonal) cross-polarization null and co-polarization maximum state pairs, being identical and sharing one main circle with the co-polarization null and the antipodal (orthogonal) cross-polarization maximum state pairs, the latter being orthogonal to the cross-polarization max pairs; and, a newly identified pair: the antipodal (orthogonal) cross-polarization saddle point extrema, which are normal to the plane (main circle: the target characteristic circle) spanned by the other four pairs. With this complete and unique mathematical description of Kennaugh's loosely described target characteristic polarization state concept [2] and Huynen's polarization fork concept with its introduction of the target characteristic parameters [3], it is now possible to study the polarimetric radar target optimization problem more rigorously. Specifically, we require to obtain a closed mathematical optimization method for the Mueller matrix, which in the limit of the degenerate coherent case, results precisely in the same set of four distinct pairs of optimal polarization states determined in Boerner and Xi [4], via an optimization of the scattered energy density or the received power. For the coherent case, such an approach was given in Kostinski, and Boerner [5] via the "three-stage-procedure" which, however, can only determine one pair of characteristic polarization states. Similarly, in both Kostinski, James and Boerner [6] and in van Zyl [7], the optimization for the partially polarized wave case of the 4x4, Mueller matrix is treated but only in an incomplete manner. Both of these methods also only allow the determination of one pair of extrema, namely, the co-polarization maxima pair. Thus, a more general approach is required for determining the optimal polarization states directly from the optimization of the Mueller matrix. This is the main objective of this paper.

## 1.2 Formulation of the Problem

With increasing frequency and in spectral regions above and beyond the millimeterwave bands, a coherent treatment does not suffice and a more general partially polarized wave treatment must at least be attempted as the truly partially polarized wave approach still cannot be resolved [1,8]. Thus, in this paper we explore ways and means of solving the optimal reception problem for the partially polarized case using the coherency matrix (vector), or the Stokes vector formalism and the 4x4

Mueller matrix description.

In order to compare with the "degenerate coherent wave" cases, in Section 2 first the Jones vector $\vec{E}$, the coherent polarization transformation ratio ρ, and Sinclair's scattering matrix [S] are introduced from which the coherency matrix [J], the coherency vector $\vec{J}$, the Stokes vector $\vec{g}$ and the degenerate Mueller matrix [M] are derived via the Kronecker expansion matrix formalism. Great care is taken in developing a unique notation, and a normalized field vector $\vec{e}$, strictly required for the representation of the polarization state, is introduced.

In Section 3, the coherent case for both the Sinclair and degenerate Mueller matrix optimization are presented. First, the received power expression which not only depends on the scattered wave but also on the receiving antenna height $\vec{h}$ is introduced. So, it follows that two kinds of coherent wave reception cases need to be addressed, one being the one-antenna (mono- static transceiver) case and the other is the two - (separate) - antenna case. For the one-antenna case, one seeks a way of dealing separately with the received power in the co-polarized and the cross-polarized channels,respectively. For the two-antenna case, the optimum received power polarization states are found by adjusting the polarization state of the receiver antenna height to match or to mismatch the polarization state of the scattered wave. In order to obtain the complete five pairs of polarization state extrema of the received power the Lagrangian multiplier method is used because the variables of the received power can be subjected to a single polarimetric constraint. Several examples describing specific target scattering matrices are used to illustrate the method and to show the equivalence with previous optimization methods for the 2x2 Sinclair matrix.

In Section 4, the optimization procedure for the general partially polarized scattered wave treatment is considered. It is shown that next to the total received power, there exist three categories of energy densities which can be optimized separately: (i) the adjustable coherent intensity; (ii) the noise-like noncoherent energy density; and (iii) the receivable intensity, the latter being the first with the addition of one half of the second. Specific optimization criteria and constraints for these types of energy density optimizations are given; resulting in a parametric closed solution for optimizing the adjustable energy density, and numerical computer intensive solutions for the other two categories. The paper is concluded in Section 5 with a comparison of the results obtained here with those obtained by other methods providing specific examples for verification. Still unresolved, highly important tasks are identified and possible methods of solution are outlined.

## II. REPRESENTATIONS AND FORMULATIONS IN RADAR POLARIMETRY

### 2.1 The Representation of a Completely Polarized Wave

The electric field strength $\vec{\varepsilon}(z,t)$ of a single monochromatic plane wave traveling in the z-direction can be expressed in terms of a complex electric field vector $\vec{E}(z,t)$ [8],

$$\vec{\varepsilon}(z,t) = \mathrm{Re}\{\ \vec{E}\ (z,t)\} = \mathrm{Re}\{\vec{E}\ e^{j(\omega t-kz)}\} \quad . \tag{1}$$

By suppressing the harmonic time-space dependence term, $\exp[j(\omega t - kz)]$, we obtain the Jones vector $\vec{E}$ which can be decomposed into any two orthogonal components in the transverse plane as

$$\vec{E} = \begin{bmatrix} E_x \\ E_y \end{bmatrix} = \begin{bmatrix} |E_x|\ e^{j\phi_x} \\ |E_y|\ e^{j\phi_y} \end{bmatrix} = e^{j\phi_x} \begin{bmatrix} |E_x| \\ |E_y|\ e^{j\phi} \end{bmatrix} , \tag{2}$$

where $|E_x|$ and $|E_y|$ represent the magnitudes of the two orthogonal components, and $\phi_x$ and $\phi_y$ represent the initial absolute phases of the two oscillations, respectively. The Jones vector fully describes the characteristic properties of the polarization state of a monochromatic plane wave [9]. To facilitate the analysis, it is often useful to use the normalized Jones vector $\vec{e}$, which is a polarization state vector defined by

$$\vec{e} = \frac{\vec{E}}{|\vec{E}|} \quad . \tag{3}$$

Another standard representation of the polarization state is the polarization ratio $\rho$ defined by

$$\rho(XY) = \frac{E_y}{E_x} = \frac{|E_y|}{|E_x|}\ e^{j(\phi_y-\phi_x)} = |\rho|\ e^{j\phi} , \tag{4}$$

where $\rho(XY)$ is also a complex number and depends on the polarization basis (XY) chosen. A normalized, completely polarized electromagnetic plane wave $\vec{e}$ can then be specified by the complex polarization ratio $\rho$ as

$$\vec{e} = \begin{bmatrix} e_x \\ e_y \end{bmatrix} = e_x \begin{bmatrix} 1 \\ \rho \end{bmatrix} = \frac{e^{j\phi_x}}{\sqrt{1 + \rho\rho^*}} \begin{pmatrix} 1 \\ \rho \end{pmatrix} \tag{5}$$

## 2.2 Scattering Matrix [S]

The polarization state of an electromagnetic plane wave, scattered by a target, differs from that of the incident wave. This change in polarization states represents the characteristic properties of the target. If the target is deterministic and time-invariant, then this target can be characterized by a 2x2 coherent scattering matrix [S], also know as the Sinclair matrix [10] for the backscatter case and Jones matrix for the forward scatter case [9,11]. In this case, the scattered wave $\vec{E}_S$ is completely polarized for a completely polarized incident wave $\vec{E}_T$ so that by the scattering process

$$\vec{E}_S = [S]\,\vec{E}_T \quad . \tag{6}$$

## 2.3 Voltage Equation

In a given direction, both the transmitted wave and the received voltage of an antenna may depend on the effective antenna height $\vec{h}$. The value of the transmitted field at a large distance r from the antenna is given by [2, 12] and [13]

$$\vec{E} = \frac{Z_o I}{2\lambda r}\,\vec{h} \quad , \tag{7}$$

where $Z_o$ is the impedance of the antenna and I the terminal antenna current. So, the effective antenna height of a receiving antenna, according to (7), can be written as

$$\vec{h}_R = \frac{2\lambda r}{Z_o I_R}\,\vec{E}_R = \frac{2\lambda r}{Z_o I_R}\,|\vec{E}_R|\,\vec{e}_R = C_R\,\vec{e}_R \quad , \tag{8}$$

where $C_R$ is a constant depending on the receiving antenna and $\vec{e}_R$ represents the polarization state vector of the receiving antenna. Note that the $\vec{E}_R$ is not a reflected field from the target, but a wave which would be radiated by the receiving antenna if it were acting as a transmitter in the direction of the target.

The value of the terminal voltage induced by an arbitrary scattered wave $\vec{E}_S$, which in turn can be related to the incident field $\vec{E}_T$ through the scattering matrix [S], is then defined according to [2,14] by

$$V_R = \vec{h}_R^T\,\vec{E}_S = \vec{h}_R^T\,[S]\,\vec{E}_T = C_R\,\vec{e}_R^T\,[S]\,\vec{E}_T \quad , \tag{9}$$

where the superscript T denotes the transpose. Note that the effective antenna height of the receiver, $\vec{h}_R$, is a complex vector and so is the voltage, $V_R$. Because we are interested in finding the optimum polarization states, but not the expression of the accurate value of the range-dependent received power, the constant multiplier $C_R$ in (9) is suppressed, so that we define the complex reduced voltage as

$$V = \vec{e}_R^T \, [S] \, \vec{E}_T \quad , \tag{10}$$

which will be used throughout the remainder of this paper.

## 2.4. The Representation of a Partially Polarized Wave

The previous sub-sections deal with the completely polarized waves; i.e., waves for which $|E_x|$, $|E_y|$, and $\phi$ are constants. Consider a quasi-monochromatic wave at a fixed point in space whose temporal behavior is described by the Jones vector [15]

$$\vec{E}(t) = \begin{bmatrix} E_x(t) \\ E_y(t) \end{bmatrix} = \begin{bmatrix} |E_x(t)| e^{j\phi_x(t)} \\ |E_y(t)| e^{j\phi_y(t)} \end{bmatrix} e^{j\omega t} = e^{j(\phi_x(t)+\omega t)} \begin{bmatrix} |E_x(t)| \\ |E_y(t)| e^{j\phi(t)} \end{bmatrix} \tag{11}$$

where the time dependence of $|E_x(t)|$, $|E_y(t)|$, and $\phi(t)$ causes a spectral spread $\Delta f$ such that $\Delta f << \frac{\omega}{2\pi}$ . If the measurement duration $\tau$ satisfies $\tau << \frac{2\pi}{\Delta f}$ , or $|E_x(t)|$, $|E_y(t)|$, and $\phi(t)$ are very slowly varying functions of time, then the wave described by (11) can be considered completely polarized. On the other hand, for a long measurement time $\tau >> (2\pi)/\Delta f$, the wave is partially polarized. Then, the Stokes vector formalism, introduced by Sir George Stokes in 1852 [16,17], must be used. For a monochromatic wave, the Stokes vector is defined as follows:

$$\vec{g} = \begin{pmatrix} g_0 \\ g_1 \\ g_2 \\ g_3 \end{pmatrix} = \begin{pmatrix} |E_x|^2 + |E_y|^2 \\ |E_x|^2 - |E_y|^2 \\ 2\,|E_x||E_y|\cos\phi \\ 2\,|E_x||E_y|\sin\phi \end{pmatrix} \quad . \tag{12}$$

For a completely polarized wave, there are only three independent parameters, which are related in the following manner:

$$g_0^2 = g_1^2 + g_2^2 + g_3^2 \quad . \tag{13}$$

The Stokes vector is sufficient to characterize the magnitude, phase, and polarization of a wave. For a partially polarized wave, whose $|E_x(t)|$, $|E_y(t)|$ and $\phi(t)$ are functions of t, Stokes introduced a somewhat different four-parameter representation which is closely related to (12). The general Stokes vector is composed of the four quantities [15]:

$$\vec{g} = \begin{pmatrix} \langle E_x E_x^* \rangle + \langle E_y E_y^* \rangle \\ \langle E_x E_x^* \rangle - \langle E_y E_y^* \rangle \\ 2\langle |E_x||E_y|\cos\phi \rangle \\ 2\langle |E_x||E_y|\sin\phi \rangle \end{pmatrix} , \tag{14}$$

where

$$\langle \cdots \rangle \equiv \lim_{T\to\infty} \left[ \frac{1}{2T} \int_{-T}^{+T} (\cdots) dt \right] .$$

In order to manipulate the time average, we introduce the polarization coherency matrix [J] as [15, 17]

$$[J] = \langle \vec{E}\,\vec{E}^{+} \rangle = \begin{bmatrix} \langle E_x E_x^* \rangle & \langle E_x E_y^* \rangle \\ \langle E_y E_x^* \rangle & \langle E_y E_y^* \rangle \end{bmatrix} = \begin{bmatrix} J_{xx} & J_{xy} \\ J_{yx} & J_{yy} \end{bmatrix} \tag{15}$$

and the coherency vector $\vec{J}$ as

$$\vec{J} = \begin{pmatrix} J_{xx} \\ J_{xy} \\ J_{yx} \\ J_{yy} \end{pmatrix} = \begin{pmatrix} \langle E_x E_x^* \rangle \\ \langle E_x E_y^* \rangle \\ \langle E_x^* E_y \rangle \\ \langle E_y E_y^* \rangle \end{pmatrix} \tag{16}$$

where the superscript$(.^{+})$ denotes the conjugate $(.^{*})$and transpose $(.^{T})$, i.e., $^{+} \hat{=} (.^{*})^{T}$. Then the general Stokes vector can be expressed in terms of $\vec{J}$ as

$$\vec{g} = \begin{pmatrix} J_{xx} + J_{yy} \\ J_{xx} - J_{yy} \\ J_{xy} + J_{yx} \\ jJ_{xy} - jJ_{yx} \end{pmatrix} = [A]\,\vec{J} , \tag{17}$$

where [A] denotes a Kronecker expansion matrix [18], given by

$$[A] = \begin{pmatrix} 1 & 0 & 0 & 1 \\ 1 & 0 & 0 & -1 \\ 0 & 1 & 1 & 0 \\ 0 & j & -j & 0 \end{pmatrix} \quad . \tag{18}$$

A partially polarized wave may be regarded as the sum of a completely unpolarized wave and a completely polarized wave. Thus, we may write $\vec{g}$ for a partially polarized wave as follows [6]:

$$\begin{bmatrix} g_0 \\ g_1 \\ g_2 \\ g_3 \end{bmatrix} = \begin{bmatrix} (1-p)\, g_0 \\ 0 \\ 0 \\ 0 \end{bmatrix} + \begin{bmatrix} p\, g_0 \\ g_1 \\ g_2 \\ g_3 \end{bmatrix} \tag{19}$$

where p is the degree of polarization which is defined as the ratio of the completely polarized power to the total power and can be written as:

$$p = \frac{\sqrt{g_1^{\,2} + g_2^{\,2} + g_3^{\,2}}}{g_0} \tag{20}$$

Using (17) and (20), we can obtain an expression for the degree of polarization p in terms of the coherency matrix [J] as follows:

$$p = \sqrt{1 - \frac{4(\det[J])}{(J_{xx} + J_{yy})^2}} \tag{21}$$

We will deal with the coherent case in which the scattered wave is completely polarized ($p = 1$) in Section III, and with the incoherent case which scatters partially polarized waves ($p < 1$) in Section IV.

### 2.5 The Mueller Matrix [M]

If the scatterer properties change with time, or a number of independent incoherent scatterers are of interest, the scattering process is described [19] by the Mueller matrix [M]. The scattering process can be defined by a linear equation similar to (6).

$$\vec{g}_S = [M]\, \vec{g}_T \quad , \tag{22}$$

where $\vec{g}_T$ and $\vec{g}_S$ represent the Stokes vectors of the transmitted wave and the scattered wave, respectively. Through simple algebra (see Appendix

A), we obtain the relationship between the "degenerate" Mueller matrix [M] and the scattering matrix [S] for the coherent case [20]:

$$[M] = [A]\,([S] \otimes [S]^*)\,[A]^{-1} \quad , \tag{23}$$

where $\otimes$ is the direct (Kronecker) product [18]. Note that [A], given by (18), is not a unitary matrix, and

$$[A]^{-1} = \frac{1}{2}\begin{pmatrix} 1 & 1 & 0 & 0 \\ 0 & 0 & 1 & -j \\ 0 & 0 & 1 & j \\ 1 & -1 & 0 & 0 \end{pmatrix} = \frac{1}{2}\,[A^*]^T \quad . \tag{24}$$

## III. THE RECEIVED POWER AND ITS OPTIMIZATION FOR THE COHERENT CASE

For the coherent case, the target is independent of time and the scattered wave $\vec{E}_S$ is completely polarized if the transmitted wave $\vec{E}_T$, incident on the target, is completely polarized. There are two physical quantities that determine the received power in rader polarimetry: the first is the polarization state of the scattered wave and the second is the polarization state of the receiving antenna. The polarization state of the receiving antenna is fixed for the one-antenna case, but it can be changed for the two-antenna case. Therefore we will consider the one-antenna and the two-antenna case, separately [20].

### 3.1 The received power for the one-antenna monostatic case

In this case, one antenna is used simultaneously both as the transmitter and the receiver, as shown in Fig. 1, and care must be taken in treating separately the co-channel (co-polarized) and the cross-channel (cross-polarized) cases.

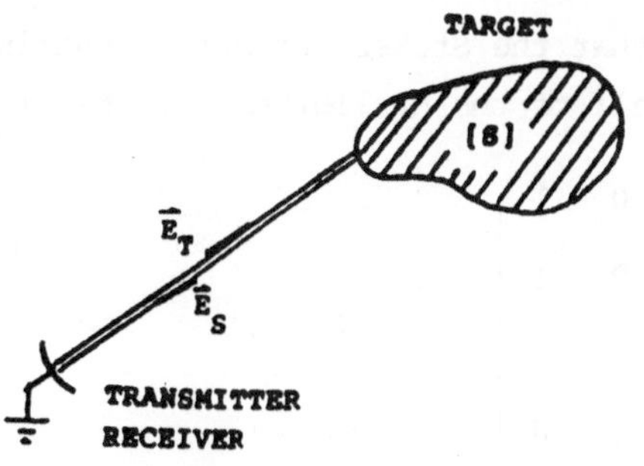


Fig. 1. Monostatic case for one-antenna configuration

a. **Co-polarized channel**

For the co-polarized channel, the polarization state of the effective antenna height $\vec{h}_R$ is the same as that of the transmitting antenna, i.e.,

$$\vec{e}_R = \vec{e}_T \quad . \tag{25}$$

Then the voltage equation given by (10) becomes

$$V_c = \vec{e}_T^{\,T} [S] \vec{E}_T \quad , \tag{26}$$

and the received power in the co-pol channel can be determined by

$$P_c = |V|^2 = |\vec{e}_T^{\,T} [S] \vec{E}_T|^2 \quad . \tag{27}$$

Expanding the power equation in the Stokes vector representation (Appendix B), yields

$$P_c = \vec{g}_T^{\,T} ([A]^{-1})^T ([S] \otimes [S]^*) [A]^{-1} \vec{g}_T \quad . \tag{28}$$

If we define a matrix $[\bar{M}_c]$ as

$$[\bar{M}_c] = ([A]^{-1})^T ([S] \otimes [S]^*) [A]^{-1} \quad , \tag{29}$$

then (28) becomes

$$P_c = \vec{g}_T^{\,T} [\bar{M}_c] \vec{g}_T \tag{30}$$

where $[\bar{M}_c]$ is denoted by Stokes reflection matrix in the co-pol channel. The above equation can also be used as a definition [20] of the Mueller matrix according to Huynen [3] and van Zyl [7].

An important property is that the Stokes reflection matrix $[\bar{M}_c]$ of (29) is not identical to one half of the Mueller matrix [M] of (23), because

$$[A^{-1}]^T = \frac{1}{2} \begin{pmatrix} 1 & 0 & 0 & 1 \\ 1 & 0 & 0 & -1 \\ 0 & 1 & 1 & 0 \\ 0 & -j & j & 0 \end{pmatrix} \neq \frac{1}{2} [A] \tag{31}$$

The two types of Mueller matrices can be related to one another by the following equation,

$$[\bar{M}_C] = \frac{1}{2} \begin{pmatrix} 1 & 0 & 0 & 0 \\ 0 & 1 & 0 & 0 \\ 0 & 0 & 1 & 0 \\ 0 & 0 & 0 & -1 \end{pmatrix} [M] \quad . \tag{32}$$

This fact was often overlooked in the literature.

**b. Cross-polarized channel**

For the cross-polarized channel, the polarization state of the effective antenna height is orthogonal to the polarization state of the transmitting antenna, i.e.,

$$\vec{e}_R = \vec{e}_{T\perp} \tag{33}$$

where the subscript $\perp$ denotes orthogonal state. The voltage equation given by (10) becomes

$$V_x = \vec{e}_{T\perp}^{\,T} \, [S] \, \vec{E}_T \tag{34}$$

A normalized, completely polarized transmitting wave $\vec{E}_T$ can be expressed according to (5) as

$$\vec{E}_T = \frac{e^{j\phi_x}}{\sqrt{1 + \rho\rho^*}} \begin{pmatrix} 1 \\ \rho \end{pmatrix} \quad , \tag{35}$$

and the related orthogonal polarization state of the transmitted wave can be written (see Appendix C) by

$$\vec{E}_{T\perp} = \frac{e^{j\phi_y'}}{\sqrt{1 + \rho\rho^*}} \begin{pmatrix} -\rho^* \\ 1 \end{pmatrix} \quad . \tag{36}$$

The received power $P_x$ in the cross-polarizated channel, according to (34), can be written (see Appendix C) as

$$P_x = |V|^2 = \left| \frac{e^{j\phi_y'}}{\sqrt{1 + \rho\rho^*}} (-\rho^* \quad 1) \, [S] \, \frac{e^{j\phi_x}}{\sqrt{1 + \rho\rho^*}} \begin{pmatrix} 1 \\ \rho \end{pmatrix} \right|^2$$

$$= \vec{g}_T^{\,T} \, [\bar{M}_x] \, \vec{g}_T \quad , \tag{37}$$

where

$$[\bar{M}_x] = ([A]^{-1})^T \left[ \begin{pmatrix} 0 & 0 & 0 & 1 \\ 0 & -1 & 0 & 0 \\ 0 & 0 & -1 & 0 \\ 1 & 0 & 0 & 0 \end{pmatrix} ([S] \otimes [S]^*) \right] [A]^{-1} = \frac{1}{2} \begin{bmatrix} 1 & 0 & 0 & 0 \\ 0 & -1 & 0 & 0 \\ 0 & 0 & -1 & 0 \\ 0 & 0 & 0 & 1 \end{bmatrix} [M] \quad (38)$$

which is the second Stokes reflection matrix $[\bar{M}_x]$ defined for the cross-polarized channel [20], and must be strictly distinguished from $[\bar{M}_c]$ and [M].

### 3.2 The received power for the matched two-antenna case

In the bistatic or monostatic separated two-antenna case, one antenna serves as the source of the transmitted wave, and the other as the receiver, as shown in Fig. 2.

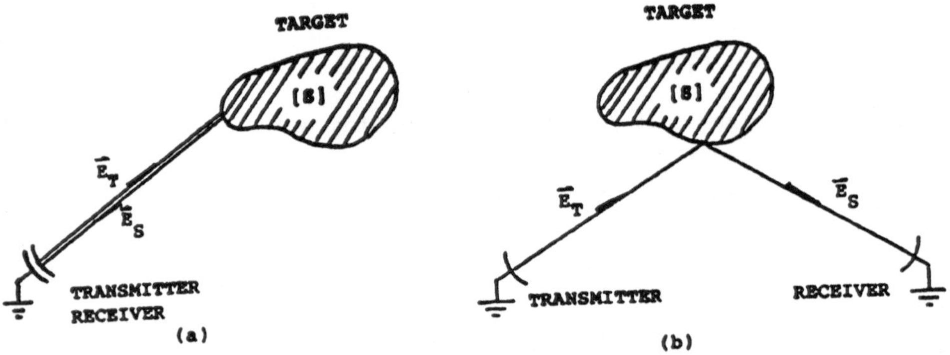


Fig. 2 Two-antenna configurations for (a) the bistatic case; and (b) the monostatic separated antennas case.

The optimal received voltage condition, in polarimetry, is given for the case in which the polarization state of the receiving antenna is matched to that of the scatteres field $\vec{E}_S$,

$$\vec{e}_R = \vec{e}_S^{\,*} \quad (39)$$

where $\vec{e}_R$ is the polarization state vector of the receiver, and $\vec{e}_S$ is the polarization state vector of the scattered wave. Then the optimum receiving voltage becomes

$$V_2 = \vec{e}_S^{*T}[S]\vec{E}_T = \left(\frac{[S]\vec{E}_T}{|[S]\vec{E}_T|}\right)^{*T}[S]\vec{E}_T = \frac{\vec{E}_T^{+}[S]^{+}[S]\vec{E}_T}{|[S]\vec{E}_T|} = \frac{\vec{E}_S \cdot \vec{E}_S}{|\vec{E}_S|} \tag{40}$$

So, the received power in the matched two-antenna case can be derived as follows,

$$P_2 = |V_2|^2 = \frac{|\vec{E}_S \cdot \vec{E}_S|^2}{|\vec{E}_S|^2} = |\vec{E}_S|^2 \quad . \tag{41}$$

Using the Stokes vector representation, yields

$$P_2 = g_{S0} = M_{00}g_{T0} + M_{01}g_{T1} + M_{02}g_{T2} + M_{03}g_{T3} \ , \ g_{T0} = 1 \tag{42}$$

This expression shows that the received power $P_2$ is the total energy density contained in the scatter wave $\vec{g}_S$. We can rewrite the above equation as

$$P_2 = \vec{g}_T^{T} \, [\bar{M}_2] \, \vec{g}_T \tag{43}$$

where

$$[\bar{M}_2] = \begin{pmatrix} 1 & 0 & 0 & 0 \\ 0 & 0 & 0 & 0 \\ 0 & 0 & 0 & 0 \\ 0 & 0 & 0 & 0 \end{pmatrix} [M] = [\bar{M}_c] + [\bar{M}_x] \tag{44}$$

which yields the third Stokes reflection matrix $[\bar{M}_2]$ in the matched two-antenna case as introduced above.

### 3.3 Power optimization and numerical illustrations

**a. The method of Lagrange multipliers [21].**

The problem of a search for the optimal received power is to find $\vec{g}_T$ such that the received power P is maximized for the given [S]. For simplicity, we always choose the transmitted wave $\vec{g}_T$ as a normalized, completely polarized wave; i.e., the components of $\vec{g}_T$ satisfy the following constraint,

$$g_{T0} = \sqrt{g_{T1}^2 + g_{T2}^2 + g_{T3}^2} = 1 \tag{45}$$

The received power P is a function of $g_{T1}$, $g_{T2}$, and $g_{T3}$ according to the previous sub-sections, wich is subject to the constraint of (45). This requirement dictates the use of the method of Lagrange multipliers to find the extremum of P [20].

Let our equation of constraint be given by

$$\Phi(g_{T1}, g_{T2}, g_{T3}) = \sqrt{g_{T1}^2 + g_{T2}^2 + g_{T3}^2} - 1 = 0 \,, \tag{46}$$

then the method of Lagrange multipliers for finding the extreme value of the power P $(g_{T1}, g_{T2}, g_{T3})$ results in

$$\left\{ \begin{aligned} \frac{\partial P}{\partial g_{T1}} - \mu\frac{\partial \Phi}{\partial g_{T1}} = 0 \\ \frac{\partial P}{\partial g_{T2}} - \mu\frac{\partial \Phi}{\partial g_{T2}} = 0 \\ \frac{\partial P}{\partial g_{T3}} - \mu\frac{\partial \Phi}{\partial g_{T3}} = 0 \end{aligned} \right. \quad . \tag{47}$$

When (45) and (47) are satisfied, $dP(g_{T1}, g_{T2}, g_{T3}) = 0$, and P is an extremum.

**b. First example**

Let the scattering matrix [S] be given and in correspondence with Agrawal and Boerner [22], Xi and Boerner [23] and with Yamaguchi et al [24] by

$$[S] = \begin{bmatrix} 2j & \frac{1}{2} \\ \frac{1}{2} & -j \end{bmatrix} \,, \tag{48}$$

for which the corresponding "degenerate" Mueller matrix [M] is obtained by evaluating (23) as

$$[M] = \begin{pmatrix} 2.75 & 1.5 & 0 & 1.5 \\ 1.5 & 2.25 & 0 & 0.5 \\ 0 & 0 & -1.75 & 0 \\ -1.5 & -0.5 & 0 & -2.25 \end{pmatrix} \tag{49a}$$

Similarly, the Stokes reflection matrices $[\bar{M}_c]$, $[\bar{M}_x]$, and $[\bar{M}_2]$, which must be clearly distinguished as shown in [20] and [24] from the Mueller matrix [M] given by (23), are obtained by solving (29), (38), and (44), respectively, as

$$[\bar{M}_c] = \begin{pmatrix} 1.375 & 0.75 & 0 & 0.75 \\ 0.75 & 1.125 & 0 & 0.25 \\ 0 & 0 & -0.875 & 0 \\ 0.75 & 0.25 & 0 & 1.125 \end{pmatrix} \quad (49b)$$

$$[\bar{M}_x] = \begin{pmatrix} 1.375 & 0.75 & 0 & 0.75 \\ -0.75 & -1.125 & 0 & -0.25 \\ 0 & 0 & 0.875 & 0 \\ -0.75 & -0.25 & 0 & -1.125 \end{pmatrix} \quad (49c)$$

$$[\bar{M}_2] = \begin{pmatrix} 2.75 & 1.5 & 0 & 1.5 \\ 0 & 0 & 0 & 0 \\ 0 & 0 & 0 & 0 \\ 0 & 0 & 0 & 0 \end{pmatrix} \quad (49d)$$

First, consider the received power in the co-pol channel. By substituting (49b) into (30), yields

$$P_c = 1.375 + 1.5g_{T1} + 1.5g_{T3} + 1.125g_{T1}^2 + 0.5g_{T1}g_{T3} - 0.875g_{T2}^2 + + 1.125g_{T3}^2.$$

Subject to the constraint equation (46)

$$\Phi = \sqrt{g_{T1}^2 + g_{T2}^2 + g_{T3}^2} - 1 = 0$$

for optimizing $P_c$, and applying the Lagrangian multiplier method, yields

$$
\begin{cases}
\dfrac{\partial P_c}{\partial g_{T1}} - \mu\dfrac{\partial \Phi}{\partial g_{T1}} = 1.5 + 2.25g_{T1} + 0.5g_{T3} - \mu g_{T1} = 0 \\[2ex]
\dfrac{\partial P_c}{\partial g_{T2}} - \mu\dfrac{\partial \Phi}{\partial g_{T2}} = -1.75g_{T2} - \mu g_{T2} = 0 \\[2ex]
\dfrac{\partial P_c}{\partial g_{T3}} - \mu\dfrac{\partial \Phi}{\partial g_{T3}} = 1.5 + 0.5g_{T1} + 2.25g_{T3} - \mu g_{T3} = 0
\end{cases}
\qquad (50)
$$

Solving (50) and (46) yields four solutions for the polarization state $\vec{g}_T$; where two solutions correspond to the two maxima and two correspond to the nulls, listed in Table 1, respectively.

**Table 1. Results for First Example**

| | | $g_{T0}$ | $g_{T1}$ | $g_{T2}$ | $g_{T3}$ | $P_c$ | $P_x$ |
|---|---|---|---|---|---|---|---|
| co-pol | max | 1 | 0.707 | 0 | 0.707 | 4.87 | 0 |
| | | 1 | -0.707 | 0 | -0.707 | 0.63 | 0 |
| | null | 1 | -0.333 | 0.882 | -0.333 | 0 | 1.75 |
| | | 1 | -0.333 | -0.882 | -0.333 | 0 | 1.75 |
| | max | 1 | 0 | 1 | 0 | 0.55 | 2.25 |
| | | 1 | 0 | -1 | 0 | 0.55 | 2.25 |
| x-pol | turning point | 1 | 0.707 | 0 | -0.707 | 2.25 | 0.5 |
| | | 1 | -0.707 | 0 | 0.707 | 2.25 | 0.5 |
| | null | 1 | 0.707 | 0 | 0.707 | 4.87 | 0 |
| | | 1 | -0.707 | 0 | -0.707 | 0.63 | 0 |

Second, for the received power in the x-pol channel, we obtain by substituting (49c) into (37),

$$P_x = 1.375 - 1.125g_{T1}^2 - 0.5g_{T1}g_{T3} + 0.875g_{T2}^2 - 1.125g_{T3}^2$$

and similarly by applying the Lagrangian multiplier method

$$\begin{cases} \dfrac{\partial P_x}{\partial g_{T1}} - \mu\dfrac{\partial \Phi}{\partial g_{T1}} = -2.25\ g_{T1} - 0.5g_{T3} - \mu g_{T1} = 0 \\[2ex] \dfrac{\partial P_x}{\partial g_{T2}} - \mu\dfrac{\partial \Phi}{\partial g_{T2}} = 1.75g_{T2} - \mu g_{T2} = 0 \\[2ex] \dfrac{\partial P_x}{\partial g_{T3}} - \mu\dfrac{\partial \Phi}{\partial g_{T3}} = -0.5g_{T1} - 2.25\ g_{T3} - \mu g_{T3} = 0 \end{cases} \tag{51}$$

Solving (51) and (46) yields the six polarization states $\vec{g}_T$ shown in Table 1. These solutions correspond to the two cross-pol maxima, two cross-pol turning points and two cross-pol nulls latter were identical to the co-pol max extrema.

Polarization dependence of both channel powers is plotted in Figs. 3a,c as a function of ellipticity and tilt angles. The characteristic polar ization states are displayed on the corresponding Poincaré spheres of Figs. 3b,d. Fig. 3f shows the generalized polarization fork. There are five pairs of characteristic polarization states: the orthogonal co-pol max pair are identical to the orthogonal x-pol null pair, and they are located on the antipodal points of the main great circle; the orthogonal x-pol max pair and the co-pol null pair share the same main great circle with co-pol max and x-pol null pairs, and the former are orthogonal to the co-pol max and x-pol null pairs; the orthogonal x-pol saddle points are normal to the plane (main great circle), and orthogonal to the co-pol max and x-pol null pairs as is shown in greater detail in Xi and Boerner [23].

Optimizing the received power P for the matched two-antenna case, by using (43) and (49d) yields

$$P_2 = 2.75 + 1.5g_{T1} + 1.5g_{T3}$$

$$\begin{cases} \dfrac{\partial P_2}{\partial g_{T1}} - \mu\dfrac{\partial \Phi}{\partial g_{T1}} = 1.5 - \mu g_{T1} = 0 \\[2ex] \dfrac{\partial P_2}{\partial g_{T2}} - \mu\dfrac{\partial \Phi}{\partial g_{T2}} = -\mu g_{T2} = 0 \\[2ex] \dfrac{\partial P_2}{\partial g_{T3}} - \mu\dfrac{\partial \Phi}{\partial g_{T3}} = 1.5 - \mu g_{T3} = 0 \end{cases} \tag{52}$$

Solving (52) and (46) yields two solutions, $(1, 0.707,0,0.707)^T$ and $(1,-0.707,0,-0.707)^T$, which are the same as the co-pol max pair of Table 1, because we are dealing with a symmetric scattering matrix of the re-

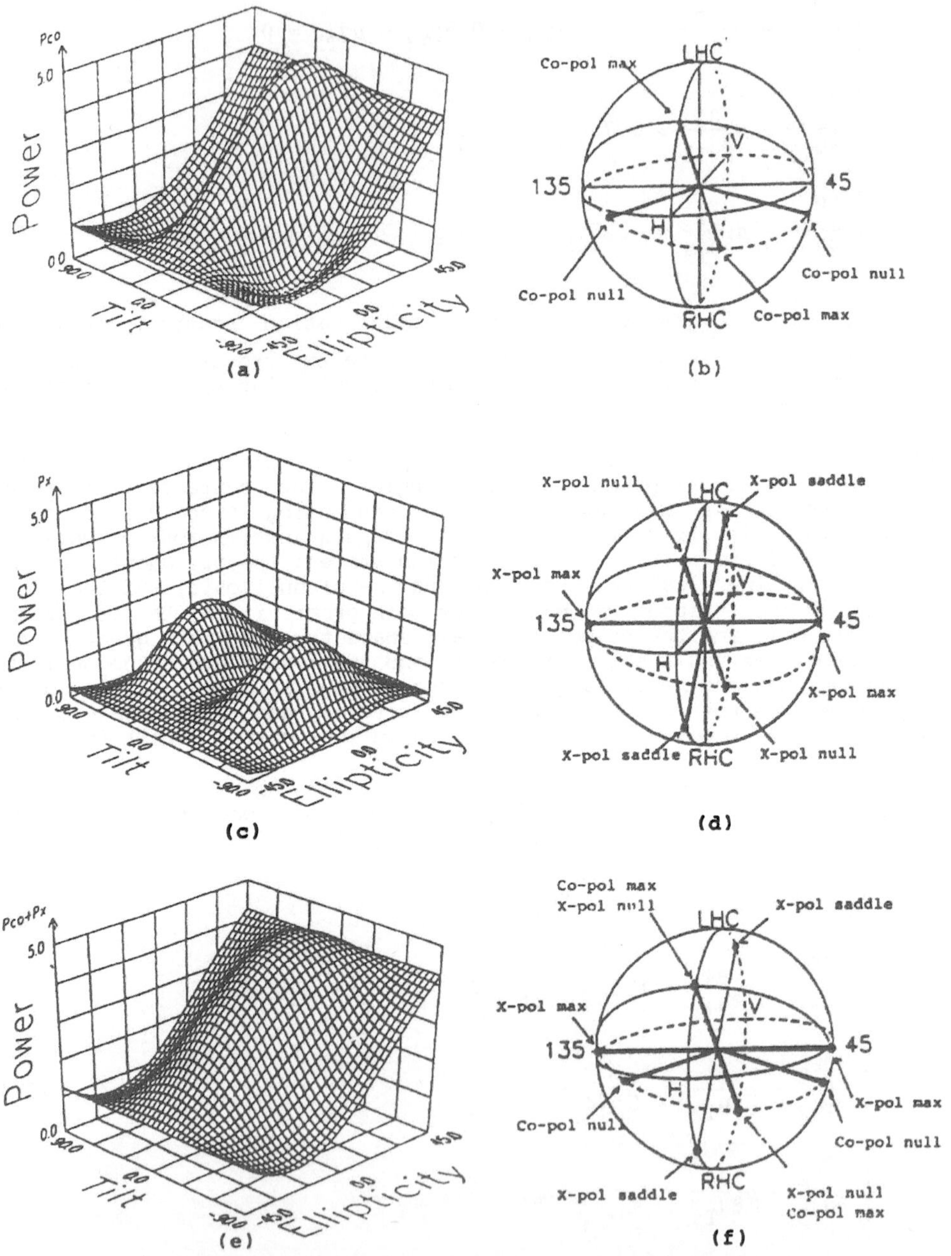


Fig. 3 Polarization characteristics for the scattering matrices [S] and [M] for the First Example:
(a) Co-polarized power specturm; (b) Co-polarization states;
(c) X-polarized power spectrum; (d) X-polarization states;
(e) Power spectrum for matched two-antenna case; (f) Generalized Polarization Fork

ciprocal monostatic case. The power spectrum of the matched two-antenna case is shown in Fig. 3e.

c. **Second example**

Let the scattering matrix [S] be given by

$$[S] = \begin{bmatrix} -3 & j \\ j & 2j \end{bmatrix} , \tag{53}$$

then the "degenerate" Mueller matrix [M] is obtained by solving (23)

$$[M] = \begin{pmatrix} 7.5 & 2.5 & 2 & 3 \\ 2.5 & 5.5 & -2 & 3 \\ 2 & -2 & 1 & 6 \\ -3 & -3 & -6 & -1 \end{pmatrix} . \tag{54}$$

Similarly, the Stokes reflection matrices $[\bar{M}_c]$, $[\bar{M}_x]$, and $[\bar{M}_2]$ are obtained using (29), (38), and (44), respectively, as

$$[\bar{M}_c] = \begin{pmatrix} 3.75 & 1.25 & 1 & 1.5 \\ 1.25 & 2.75 & -1 & 1.5 \\ 1 & -1 & 0.5 & 3 \\ 1.5 & 1.5 & 3 & 0.5 \end{pmatrix}$$

$$[\bar{M}_x] = \begin{pmatrix} 3.75 & 1.25 & 1 & 1.5 \\ -1.25 & -2.75 & 1 & -1.5 \\ -1 & 1 & -0.5 & -3 \\ -1.5 & -1.5 & -3 & -0.5 \end{pmatrix}$$

$$[\bar{M}_2] = \begin{pmatrix} 7.5 & 2.5 & 2 & 3 \\ 0 & 0 & 0 & 0 \\ 0 & 0 & 0 & 0 \\ 0 & 0 & 0 & 0 \end{pmatrix}$$

Using the Lagrangian multiplier method for optimizing the $P_c$ (equation

(30)), and $P_x$ (equation (37)), yields the result listed in Table 2.

**Table 2. Results for Second Example**

| | | $g_{T0}$ | $g_{T1}$ | $g_{T2}$ | $g_{T3}$ | $P_c$ | $P_x$ |
|---|---|---|---|---|---|---|---|
| co-pol | max | 1 | 0.5698 | 0.4558 | 0.6837 | 11.89 | 0 |
| | | 1 | -0.5698 | -0.4558 | -0.6837 | 3.11 | 0 |
| | null | 1 | 0.0930 | 0.4820 | -0.8712 | 0 | 6.08 |
| | | 1 | -0.4612 | -0.7764 | 0.4295 | 0 | 6.08 |
| x-pol | max | 1 | -0.2927 | -0.6648 | 0.6872 | 0.71 | 6.79 |
| | | 1 | 0.2927 | 0.6648 | -0.6872 | 0.71 | 6.79 |
| | turning point | 1 | -0.7677 | 0.5917 | 0.2453 | 6.79 | 0.71 |
| | | 1 | 0.7677 | -0.5917 | -0.2453 | 6.79 | 0.71 |
| | null | 1 | 0.5698 | 0.4558 | 0.6837 | 11.89 | 0 |
| | | 1 | -0.5698 | -0.4558 | -0.6837 | 3.11 | 0 |

Optimizing the received power $P_2$ in the matched two-antenna case (43) yields two solutions which are the same as the co-pol max pair, as illustrated in Fig.4. Note, that the results obtained by different methods, simultaneously developed in [22], [23] and [24], are identical.

## IV. OPTIMIZATION OF THE ENERGY DENSITY IN THE SCATTERED WAVE FOR THE INCOHERENT CASE

In the general case, a scatterer is a time-varying object of which shape and aspect angle vary with time, or it is a collection of randomly distributed scattering centers [6]. When a monochromatic (completely polarized) wave illuminates it, the scattered wave is no longer completely polarized, its amplitude and phase are functions of time. The energy received by the receiving antenna should be averaged. Therefore, the scattering properties, in general, cannot be characterized by a 2x2 complex scattering matrix [S], but by a 4x4 real Mueller matrix [M]. Correspondingly, the plane wave is represented by the Stokes vector $\vec{g}$. Let us use the Stokes vector $\vec{g}_T = (g_{T0}, g_{T1}, g_{T2}, g_{T3})^T$ and $\vec{g}_S = (g_{S0}, g_{S1}, g_{S2}, g_{S3})^T$ to represent the transmitted and scattered waves, respectively.

The scattered wave $\vec{g}_S$ is generally a partially polarized wave. It may

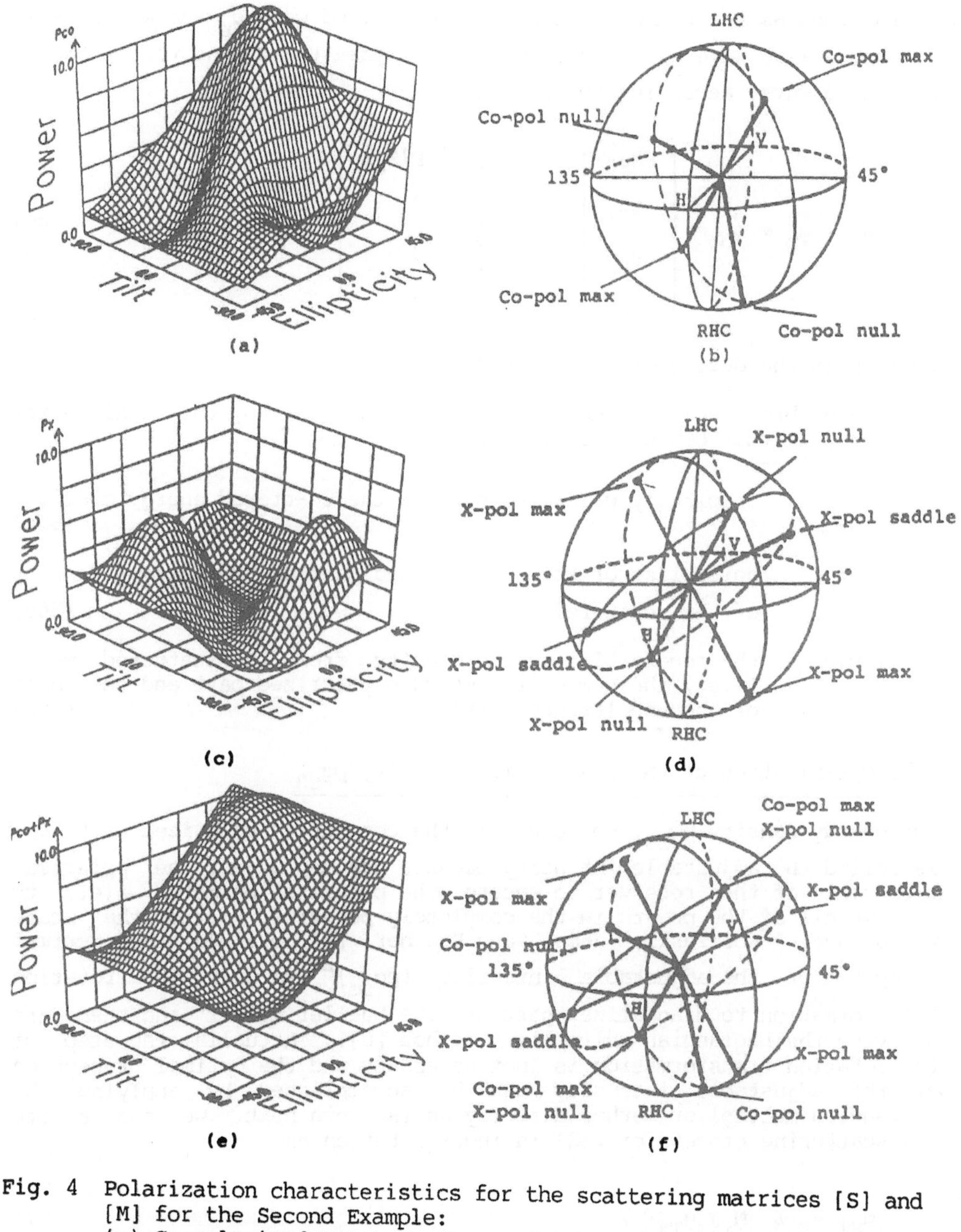


Fig. 4 Polarization characteristics for the scattering matrices [S] and [M] for the Second Example:
(a) Co-polarized power specturm; (b) Co-polarization states;
(c) X-polarized power spectrum; (d) X-polarization states;
(e) Power spectrum for matched two-antenna case; (f) Generalized Polarization Fork

be regarded as the sum of a completely polarized wave $\vec{g}_p$ and a completely unpolarized wave $\vec{g}_u$. Thus, we may decompose the Stokes vector of the scattered wave, according to (19) as:

$$\vec{g}_S = \vec{g}_p + \vec{g}_u = \begin{pmatrix} pg_{S0} \\ g_{S1} \\ g_{S2} \\ g_{S3} \end{pmatrix} + \begin{pmatrix} (1-p)g_{S0} \\ 0 \\ 0 \\ 0 \end{pmatrix} , \tag{55}$$

where p is the degree of polarization.

There are three types of energy density terms, next to the total energy density $g_{S0}$, that can be optimized [20]:

| | | |
|---|---|---|
| $pg_{S0}$ | completely polarized part of the scattered energy density; | (56a) |
| $(1 - p)g_{S0}$ | noise like or the unpolarized part of the energy density of the scattered wave; | (56b) |
| $\frac{1}{2}(1 + p)g_{S0}$ | the receivable energy density from the scattered wave. (i.e., the sum of the matched polarized part and one half of the unpolarized part). | (56c) |

### 4.1. Optimization of the adjustable intensity $pg_{S0}$

The energy density $pg_{S0}$, contained in the completely polarized part $g_p$, is called the adjustable intensity because one may adjust the polarization state of the receiver to ensure the polarization match, i.e., to receive all of the power in the completely polarized part of the scattered wave. Kostinski, James and Boerner [6] obtained the optimal reception of the adjustable intensity $(pg_{so})^2$ by using the rotation transformation to diagnalize part of the Mueller matrix and then are applying the Lagrangian multiplier method [6]. Actually, the step of the rotation transformation is not necessary and the optimal reception of the adjustable intensity can be accomplished by applying the Lagrangian multiplier method directly as is shown next. We can rewrite the scattering process of (22) in index notation as

$$g_{Si} = \sum_{j=0}^{3} M_{ij}\, g_{Tj} \quad , \tag{57}$$

where j = 0, 1, 2, 3. The first term in (55) is a completely polarized wave which has the following property:

$$pg_{S0} = (g_{S1}^2 + g_{S2}^2 + g_{S3}^2)^{1/2} \quad .$$

So, in the index notation, the adjustable intensity can be expressed as

$$pg_{S0} = \left( \sum_{i=1}^{3} g_{Si}^2 \right)^{1/2} = \left[ \sum_{i=1}^{3} \left( \sum_{j=0}^{3} M_{ij} g_{Tj} \right)^2 \right]^{1/2} \tag{58}$$

where the $g_{Ti}$'s are the elements of the Stokes vector of the transmitted wave. The adjustable intensity $pg_{S0}$ is a function of the $g_{Ti}$'s. Let the $g_{Ti}$'s be the variables, then the partial derivative of $(pg_{S0})^2$ with respect to $g_{Tk}$ can be derived as

$$\frac{\partial (pg_{S0})^2}{\partial g_{Tk}} = \sum_{i=1}^{3} \frac{\partial g_{Si}^2}{\partial g_{Tk}} = 2 \sum_{i=1}^{3} g_{Si} M_{ik} = 2 \sum_{i=1}^{3} \sum_{j=0}^{3} M_{ij} M_{ik} g_{Tj} \tag{59}$$

For optimizing the adjustable intensity, we apply the method of Lagrangian multipliers, which yields

$$\left\{ \begin{aligned} \frac{\partial (pg_{S0})^2}{\partial g_{T1}} - \mu \frac{\partial \Phi}{\partial g_{T1}} &= 2 \sum_{i=1}^{3} \sum_{j=0}^{3} M_{ij} M_{i1} g_{Tj} - \mu\, g_{T1} = 0 \\ \frac{\partial (pg_{S0})^2}{\partial g_{T2}} - \mu \frac{\partial \Phi}{\partial g_{T2}} &= 2 \sum_{i=1}^{3} \sum_{j=0}^{3} M_{ij} M_{i2} g_{Tj} - \mu\, g_{T2} = 0 \\ \frac{\partial (pg_{S0})^2}{\partial g_{T3}} - \mu \frac{\partial \Phi}{\partial g_{T3}} &= 2 \sum_{i=1}^{3} \sum_{j=0}^{3} M_{ij} M_{i3} g_{Tj} - \mu\, g_{T3} = 0 \end{aligned} \right. \tag{60}$$

where $\Phi$ is the constraint equation of (46). Equation (60) is a set of inhomogeneous linear equations in $g_{T1}(\mu)$, $g_{T2}(\mu)$, and $g_{T3}(\mu)$. Then the straightforward solutions for the $g_{Ti}(\mu)$ are three functions of $\mu$. Substituting $g_{Ti}(\mu)$, $(i = 1,2,3)$ into the constraint condition of (46) leads to a sixth-order polynomial equation of $\mu$. A solution of $\mu$ can be found by numerical calculation. For each $\mu$ value, we calculate $g_{T1}$, $g_{T2}$, $g_{T3}$, and $pg_{S0}$ according to the formula in (58). The largest (or smallest) intensity is used to choose the optimal intensity, the corre sponding $\vec{g}_T$ is the optimal polarization state of the transmitted wave.

## 4.2. Minimizing the noise-like energy density term: $(1-p)g_{SO}$

An unpolarized wave can always be represented by an incoherent sum of any two orthogonal completely polarized waves of equal intensity [25], which leads to a 50% efficiency for the reception of the unpolarized wave. We would like to minimize the noise-like energy so that as much energy as possible may be received.

The total energy density of the unpolarized part of the scattered wave is given by

$$(1 - p)g_{S0} = g_{S0} - pg_{S0} = \sum_{j=0}^{3} M_{0j}\, g_{Tj} - \sqrt{\sum_{i=1}^{3} \Big( \sum_{j=0}^{3} M_{ij}\, g_{Tj} \Big)^2} \qquad (61)$$

There is no simple way of giving the analytic closed form solution for the minimum solusion. Also, the Lagrangian multiplier method becomes far too involved, and instead computer numerical analysis must be used.

### 4.3. Maximizing the receivable intensity in the scattered wave: $1/2(1 + p)g_{so}$

The total receivable energy density consists of two parts: 100% reception efficiency for the completely polarized part of the scattered wave and 50% reception efficiency for the unpolarized part. We may write the following expression for the total receivable intensity:

$$1/2(1 + p)g_{S0} = pg_{S0} + 1/2(1 - p)g_{S0}$$

$$= \frac{1}{2} \sum_{j=0}^{3} M_{0j}\, g_{Tj} + \frac{1}{2} \sqrt{\sum_{i=1}^{3} \Big( \sum_{j=0}^{3} M_{ij}\, g_{Tj} \Big)^2} \qquad (62)$$

Also, this equation can only be solved using numerical analysis.

### 4.4. Numerical Examples

Consider the following Mueller matrix for the partially polarized case, which corresponds to experimental data taken for a combined collimator-radiometer system [24]. Note, for this partially polarized case, no equivalent scattering matrix exists.

$$M = \begin{bmatrix} 0.7599 & -0.0623 & 0.0295 & 0.1185 \\ -0.0573 & 0.4687 & -0.1811 & -0.1863 \\ 0.0384 & -0.1714 & 0.5394 & 0.0282 \\ 0.1240 & -0.2168 & -0.0120 & 0.6608 \end{bmatrix} \qquad (63)$$

By applying the method of Lagrangian multipliers for optimizing the adjustable intensity of (60), a set of inhomogeneous linear equations is obtained as follows,

$$\begin{pmatrix} 0.2961-\mu & -0.1747 & -0.2354 \\ -0.1747 & 0.3293-\mu & 0.0410 \\ -0.2354 & 0.0410 & 0.4722-\mu \end{pmatrix} \begin{pmatrix} g_{T1} \\ g_{T2} \\ g_{T3} \end{pmatrix} = \begin{pmatrix} 0.0603 \\ -0.0296 \\ -0.0937 \end{pmatrix}$$

Solving the linear equations results in three solutions $g_{T1}(\mu)$, $g_{T2}(\mu)$, and $g_{T3}(\mu)$ which are functions of $\mu$. Substituting the three solutions into the constraint equation of (45) leads to a sixth-order polynomial equation in $\mu$,

$$\mu^6 - 2.1842\mu^5 + 1.7813\mu^4 - 0.6780\mu^3 + 0.1221\mu^2 - 0.009229\mu + 0.0002430 = 0 \quad (64)$$

For each root of $\mu$, the polarization state $\vec{g}_T(\mu)$ and its intensity, which is obtained by solving (58), are displayed in Table 3.

**Table 3. Roots, Intensities and Stokes Vectors for the Characteristic Polarization States of the Adjustable Intensity Optimization**

| Root Number | Root $\mu$ | Intensity $pg_{s0}$ | $g_{T0}$ | $g_{T1}$ | $g_{T2}$ | $g_{T3}$ |
|---|---|---|---|---|---|---|
| 1 | 0.8019 | 0.9677 | 1 | -0.5807 | 0.3378 | 0.7408 |
| 2 | 0.5739 | 0.6942 | 1 | 0.6134 | -0.4203 | -0.6686 |
| 3 | 0.3548 | 0.5923 | 0.997 | 0.3894 | -0.8702 | 0.2918 |
| 4 | 0.3170 | 0.5342 | 1 | 0.0221 | 0.6619 | -0.7492 |
| 5 | 0.0734 | 0.2776 | 1.006 | 0.8710 | 0.4486 | 0.2301 |
| 6 | 0.0633 | 0.2404 | 1.009 | -0.6812 | -0.4777 | -0.5707 |

The polarization dependence of the adjustable intensity is illustrated in Fig. 5, in which the adjustable intensity is plotted as a function of ellipticity and tilt. The general topology of the surface is determined by the six stationary points corresponding to the six roots of (64), with the global maximum at tilt angle = 75.1°, and ellipticity angle = 24.1°, and with the global minimum at tilt angle = 72.4°, and ellipticity angle = -17.3°.

The noise-like energy was minimized and the receivable intensity was maximized by numerical analysis as shown in Table 4.

## V. COMPARISON OF RESULTS

There exist many different methods for calculating the optimal polarization states in radar polarimetry today. Boerner and Xi found five pairs

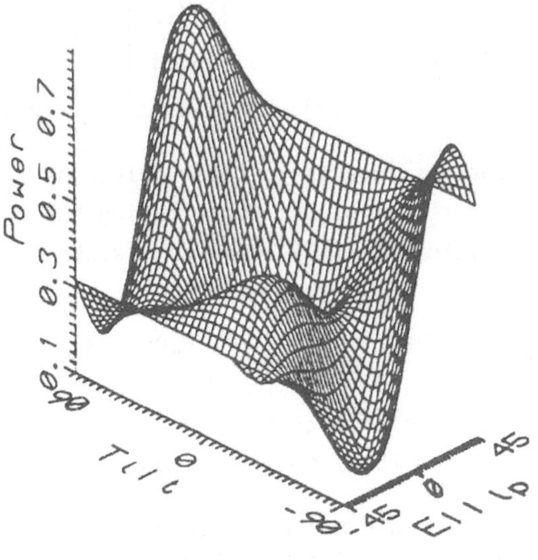


Fig. 5: Polarization dependence of the adjustable intensity.

**Table 4. The Intensities and the Stokes Vectors of the Optimal Polarization States of the Different Types of Energy Density**

| Method | Intensity | $g_{T0}$ | $g_{T1}$ | $g_{T2}$ | $g_{T3}$ |
|---|---|---|---|---|---|
| Maximum of $pg_{S0}$ | 0.9677 | 1 | -0.5807 | 0.3375 | 0.7408 |
| Minimum of $(1 - p)g_{S0}$ | -0.07 | 1 | -0.6108 | 0.3893 | 0.6894 |
| Maximum of $1/2(1 + p)g_{S0}$ | 0.9311 | 1 | -0.5602 | 0.3090 | 0.7686 |

of characteristic polarization states for the coherent monostatic and reciprocal case using the polarization transformation ratio formulation in 1990 [4,23]. Agrawal and Boerner found eight characteristic polarization states using the polarization transformation ratio formalism for the coherent case in 1989 [22]. Kostinski and Boerner introduced a three-step method in 1986 [5], which can solve the maximum reception for the matched two-antenna case very well, besides they also found a method for solving the optimal reception of partially polarized waves in 1988 [6], which also can be used for determining the polarization states of extrema for the coherent co-pol channel, but not for the x-pol channel. Now, we compare the characteristic polarizaiton states obtained in this paper with those using the different methods mentioned above.

For the coherent case, the comparison of the corresponding characteristic polarization states of the first example of Section 3 with the result obtained by other methods is made as follows, in Table 5.

**Table 5. Comparison of Results**

| Authors Name | | | Xi & Boerner | Agrawal& Boerner | Yamaguchi& Boerner | Kostinski& Boerner | Kostinski& James&Boerner |
|---|---|---|---|---|---|---|---|
| Reference | | | [23] | [22] | [24] | [5] | [6] |
| co-pol | max | $\vec{g}_T$ | same | same | same | same | same |
| | | $\vec{g}_T$ | same | same | same | ---- | same |
| | null | $\vec{g}_T$ | same | same | same | ---- | ---- |
| | | $\vec{g}_T$ | same | same | same | ---- | ---- |
| x-pol | max | $\vec{g}_T$ | same | same | same | ---- | ---- |
| | | $\vec{g}_T$ | same | same | same | ---- | ---- |
| | Saddle point | $\vec{g}_T$ | same | ---- | same | ---- | ---- |
| | | $\vec{g}_T$ | same | ---- | same | ---- | ---- |
| | null | $\vec{g}_T$ | same | same | same | ---- | ---- |
| | | $\vec{g}_T$ | same | same | same | ---- | ---- |

For the incoherent case of Section 4, our results for optimizing the adjustable intensity are the same as those obtained by Kostinski, et al [6], with the error being less then one per mill.

## VI. CONCLUSIONS

We have developed optimization procedures for determining the characteristic polarization state vectors using the method of Lagrangian multipliers with the assumption that the target backscattering information is given in terms of the 4x4 Mueller matrix for either the coherent or partially coherent radar cases. One of the objectives was to show that the optimization of the "degenerate" Mueller matrix provides identical results to those obtained for optimizing the corresponding Sinclair matrix [S] in the limit of purely coherent backscatter. Therefore, for the coherent case, we first assumed that the Sinclair matrix [S] is given from which the Stokes reflection matrices $[\bar{M}_c]$ and $[\bar{M}_x]$ for the co-polarized and cross-polarized channels, respectively, are derived for the one-antenna and $[\bar{M}_2]$ for the two-antenna matched cases. Then applying the Lagrangian multiplier method directly on the resulting formulation $[\bar{M}_c]$ and $[\bar{M}_x]$, it is shown that the five pairs of characteristic polarization states can also be distinctly determined from the Mueller matrix providing a result identical to that obtained by Xi and Boerner [23] and in Yamaguchi and Boerner [24]. However, only the monostatic reciprocal and non-reciprocal cases are treated; and for the general bistatic case, much work yet is to be done. We may expect meaningful results only if the bistatic target/antenna coordinate system is chosen properly together with the corresponding polarization bases formulation as discussed in Boerner et al [25].

For the partially polarized case, a factorized solution for determining the optimal adjustable intensity $pg_{So}$ was found to exist; whereas, for the determination of the noise-like noncoherent energy density and the receivable polarized energy density, computer-numerical methods must be used and a general analytical factorized solution, in case it does exist, must still be formulated for the latter two cases.

Finally, it is shown from a comparison of results, obtained for the selected set of target scattering matrices, with the results obtained by using other optimization methods such as described in [4], [22], [24] that those are in very good agreement. However, for the general partially coherent, as distinguished from the partially polarized case [26], much work yet is to be done. For example, recently Cloude [27,28] proposed a different approach of decomposing the scattering matrix into its basic measurement polarization states which yet needs to be checked in greater detail. A more systematic study is in progress and results will be reported elsewhere.

## VII. ACKNOWLEDGEMENT

The financial support for this research was provided by the US Office of Naval Research, Contract Nos. N00014-80-C0773 and N00014-90-J-1405, the US Army Research Office, Contract No. DAAL-03-89-K-0116. The skillful typing of the final manuscript by Ms. Mirian R. Mailey and Mr. Richard W. Foster is acknowledged, and we are grateful to Dr. Yoshio Yamaguchi for providing comments and contributing to the illustrations and to Dr. Ernst Lüneburg for his careful review of the paper.

## VIII. REFERENCES

[1] W-M. Boerner, "Polarization Utilization in Electromagnetic Inverse Scattering", Chpt. 7 in Inverse Scattering Problems in Optics, ed. by H.P. Baltes, Vol. 2; Topics in Current Physics, Vol. 20, Heidelberg: Springer Verlag, July (1980), pp. 237-305.

[2] E.M. Kennaugh, "Polarization Properties of Radar Reflections", M.Sc. Thesis, Report 389-12, Dept. of Elec. Engr., The Ohio State University, Columbus, OH, 43212, (1952 March 01); (also see: Proc. of the R&D Board Symposium on Radar Reflection Studies, Sept., 1950).

[3] J.R. Huynen, "Phenomenological Theory of Radar Targets", Ph.D. Dissertation, Technical University, Delft, The Netherlands,

(1970). Revised edition (1978) available from the author.

[4] W-M. Boerner and A-Q. Xi, "The Characteristic Radar Target Polarization State Theory for the Coherent Monostatic and Reciprocal Case Using the Generalized Polarization Transformation Ratio Formulation" Archiv der Elektrischen Übertragung: AEÜ Vol. 44, No. 4, pp 273-281, July 1990.

[5] A.B. Kostinski and W-M. Boerner, "On Foundations of Radar Polarimetry", IEEE AP-34, No. 12, pp. 1395-1404, also see: comments by H. Mieras, pp. 1470-1471, and author's reply, pp. 1471-1473, Dec. (1986).

[6] A.B. Kostinski, B.D. James and W-M. Boerner, "On the Optimal Reception of Partially Polarized Waves", J. Optical Society of America, Part A, Optics & Image Sciences, Series 2, Vol. 5, No. 1, pp. 58-64, Jan. 1988.

[7] J.J. van Zyl, "On the Importance of Polarization Radar Scattering Problems", Ph.D. Dissertation, California Institute of Technology, Pasadena, CA, January (1986).

[8] W-M. Boerner, Introduction to Radar Polarimetry--with assessments of the historical development and of the current State-of-the-Art, Section [0-1] in: W-M. Boerner et al, eds., Inverse Methods in Electromagnetic Imaging, Proc. NATO-ARW-DIMRP, 1988 Sept. 18- 24, Bad Windsheim, FRG. NATO-ASI-Series C, (Math. & Phys. Sci.), Dordrecht/Boston: D. Reidel Publ. Co., 1991.

[9] P. Beckmann, The Depolarization of Electromagnetic Waves, Boulder, CO: The Golem Press, (1968).

[10] G. Sinclair, "The Transmission and Reception of Elliptically Polarized Waves", Proc. IRE, Vol. 38, Feb. 1950, pp. 148-151.

[11] R.C. Jones, A New Calculus for the Treatment of Optical Systems, Journal of the Optical Society of America: Part I(Descriptions and Discussions of the Calculus) JOSA31, pp 488-493, 1941; Part III (The Söhnken Theory of Optical Activity), JOSA31, pp 500-503, 1941; Part IV (More General Formulations and Descriptions of Another Calculus)JOSA37, pp. 107-110, 1947.

[12] R.E. Collin, Antennas and Radio Wave Propogation, New York, McGraw Hill, 1985.

[13] C.H. Papas, Theory of Electromagnetic Wave Propagation, New York: McGraw-Hill, Inc., 1965.

[14] C-Y. Chan, "Studies on the Power Scattering Matrix of Radar Targets", M.Sc. Thesis, Dept. of Electr. Engr. & Comp. Sci., University of Illinois at Chicago, Chicago, IL (1981).

[15] M. Born and E. Wolf, Principles of Optics, 3rd Ed., New York: Pergamon Press, (1964).

[16] J.D. Kraus, Electromagnetics, New York: McGraw-Hill Book Company, (1984).

[17] A. Ishimaru, Wave Propagation and Scattering in Random Media, Vol. 1, Single Scattering and Transport Theory, New York: Academic Press, 1978.

[18] A. Graham, Kronecker Products and Matrix Calculus: with Applications, New York: Ellis Horwood, Ltd., 1981.

[19] P. Beckmann, and A. Spizzichino, "The Scattering of Electromagnetic Waves from Rough Surfaces", New York, MacMillan, (1963).

[20] W-M. Boerner, W-Y. Yan and A-Q. Xi, Basic Principles of Radar Polarimetry, in J.W. Battles, Polarimetric Technology Handbook, Chicago: GACIAC, IIT-RI, 1991 (Chapter 2).

[21] H.C. Chen, Theory of Electromagnetic Waves -- A coordinate-free approach, New York: McGraw-Hill Book Co., Series in Electrical Engineering, 1983.

[22] A.P. Agrawal and W-M. Boerner, "Re-Development of Kennaugh's Target Characteristic Polarization State Theory Using the Polarization Transformation Ratio Formalism for the Coherent Case", IEEE Trans. GSRS, Vol. 27, No. 1, pp. 2-14, January, (1989). Also see: A.P. Agrawal, A polarimetric rain backscatter model devel- oped for coherent polarization diversity radar applications, Ph.D. dissertation, Communications, EECS Dept., University of Illinois at Chicago, Chicago, IL, 1986.

[23] A-Q. Xi and W-M. Boerner, "Determination of the Characteristic Polarization STates of the Target Scattering Matrix [S(AB)] for the Coherent, Monostatic and Reciprocal Propagation Space Using the Polarization Transformation Ratio Formulation", J. Opt. Sci. Am., Part A, Optics & Image Sciences, Series 2, Vol.8, 1991 (in print)

[24] Y. Yamaguchi, K. Sasagawa, M. Sengoku, T. Abe, W-L. Yan and W-M. Boerner, On characteristic polarization states of coherent reflected waves based on the Stokes vector formulation in radar polarimetry, Proc. Institute of Electronics Information Communication Engineering of Japan, Antennas & Propagation Section, AP90-35, pp. 23-30, 1990 July 19.

[25] W-M. Boerner, et. al., eds., Inverse Methods in Electromagnetic Imaging, Proc. NATO Advanced Res. Workshop on IMEI, Bad Windsheim, FR. Germany, Sept. 18-24, 1983, NATO-ASI-Series C, Math. & Phys. Sci., Vol. 143, Holland/Dordrecht: D. Reidel Publ. Co., (1985).

[26] H.C. Ko, On the reception of quasi-monochromatic, partially polarized radio waves, Proc. IRE Vol. 50, pp. 1950-1957, 1962.

[27] S.R. Cloude, A comparison of optimization methods in radar polarimetry, IEEE Trans. Ant. & Propag., Subm. January 1990 (also

see: J. Saillard, editor, Proceedings of the International Workshop on Radar Polarimetry: JIPR, IRESTE, University of Nantes, France, 1990 March 20-22).

[28] S.R. Cloude, Uniqueness of target decomposition theorems in radar polarimetry, in: W-M. Boerner, et al. (eds), Direct and Inverse Method in Radar Polarimetry, Proc. NATO-ARW-DIMRP (W-M. Boerner, Director), 1988 Sept. 18-24, Bad Windsheim FRG, NATO-ASI-Series C, (Math. & Phys. Sci.), Dordrecht/Boston: D. Reidel Publ. Co., (1990).

## IX. APPENDICES

### APPENDIX A: Deriving the Relationship between [M] and [S] of (23)

The scattering matrix [S] is given by (6) as follows

$$\begin{pmatrix} E_{Sx} \\ E_{Sy} \end{pmatrix} = \begin{pmatrix} S_{xx} & S_{xy} \\ S_{yx} & S_{yy} \end{pmatrix} \begin{pmatrix} E_{Tx} \\ E_{Ty} \end{pmatrix} = \begin{pmatrix} S_{xx}E_{Tx} + S_{xy}E_{Ty} \\ S_{yx}E_{Tx} + S_{yy}E_{Ty} \end{pmatrix} \quad , \quad (A.1)$$

And the Mueller matrix [M] is defined in (22) as

$$\vec{g}_S = [M]\vec{g}_T \quad .$$

The above equation can be written in terms of the coherency vector $\vec{J}$ according to (17) ($\vec{g} = [A]\vec{J}$)

$$[A]\vec{J}_S = [M][A]\vec{J}_T$$

$$\vec{J}_S = [A]^{-1}[M][A]\vec{J}_T \quad . \quad (A.2)$$

Then the $\vec{J}_S$ vector may be calculated from the scattering matrix [S] using the definition of the vector $\vec{J}$ of (16)

$$\vec{J}_S = \begin{pmatrix} E_{Sx}E_{Sx}^* \\ E_{Sx}E_{Sy}^* \\ E_{Sx}^*E_{Sy} \\ E_{Sy}E_{Sy}^* \end{pmatrix} = \begin{pmatrix} (S_{xx}E_{Tx} + S_{xy}E_{Ty})(S_{xx}^*E_{Tx}^* + S_{xy}^*E_{Ty}^*) \\ (S_{xx}E_{Tx} + S_{xy}E_{Ty})(S_{yx}^*E_{Tx}^* + S_{yy}^*E_{Ty}^*) \\ (S_{xx}^*E_{Tx}^* + S_{xy}^*E_{Ty}^*)(S_{yx}E_{Tx} + S_{yy}E_{Ty}) \\ (S_{yx}E_{Tx} + S_{yy}E_{Ty})(S_{yx}^*E_{Tx}^* + S_{yy}^*E_{Ty}^*) \end{pmatrix}$$

$$= ([S] \otimes [S]^*)\,\vec{J}_T \quad . \quad (A.3)$$

Comparing (equation (A.2) and equation (A.3)) yields the relationship between [M] and [S]

$$[M] = [A]([S] \boxtimes [S]^*)[A]^{-1} \quad . \tag{A.4}$$

**Appendix B: Expanding the Power Equation in the Co-pol Channel of (28)**

Expanding (27) yields

$$P_c = \left| (e_{Tx} \quad e_{Ty}) \begin{pmatrix} S_{xx} & S_{xy} \\ S_{yx} & S_{yy} \end{pmatrix} \begin{pmatrix} E_{Tx} \\ E_{Ty} \end{pmatrix} \right|^2 \tag{B.1}$$

From (3), we know the polarization state vector of the transmitter, $\vec{e}$, is the normalized transmitted wave $\vec{E}_T$ , and by assuming in the early sections that the transmitted wave is always chosen as a normalized wave, i.e., $\vec{E}_T=\vec{e}_T$. It means that we should use $\vec{e}_T$ instead of $\vec{E}_T$ in all expressions. But customarily, we use $\vec{E}_T$ instead of $\vec{e}_T$ in the following equations. Therefore, (equation (B.1)) may be written as

$$P_c = \left| (E_{Tx} \quad E_{Ty}) \begin{pmatrix} S_{xx} & S_{xy} \\ S_{yx} & S_{yy} \end{pmatrix} \begin{pmatrix} E_{Tx} \\ E_{Ty} \end{pmatrix} \right|^2$$

$$= (E_{Tx}S_{xx}E_{Tx} + E_{Tx}S_{xy}E_{Ty} + E_{Ty}S_{yx}E_{Tx} + E_{Ty}S_{yy}E_{Ty})(E_{Tx}^*S_{xx}^*E_{Tx}^*$$

$$+ E_{Tx}^*S_{xy}^*E_{Ty}^* + E_{Ty}^*S_{yx}^*E_{Tx}^* + E_{Ty}^*S_{yy}^*E_{Ty}^*)$$

$$= \begin{pmatrix} E_{Tx}E_{Tx}^* \\ E_{Tx}E_{Ty}^* \\ E_{Tx}^*E_{Ty} \\ E_{Ty}E_{Ty}^* \end{pmatrix}^T ([S] \boxtimes [S]^*) \begin{pmatrix} E_{Tx}E_{Tx}^* \\ E_{Tx}E_{Ty}^* \\ E_{Tx}^*E_{Ty} \\ E_{Ty}E_{Ty}^* \end{pmatrix} = \vec{J}_T^{\,T} ([S] \boxtimes [S]^*) \vec{J}_T \tag{B.2}$$

Substituting $\vec{J} = [A]^{-1}\vec{g}$, given by (17), into (B.2) yields

$$P_c = \vec{g}_T^{\,T}([A]^{-1})^T([S] \boxtimes [S]^*)[A]^{-1}\vec{g}_T \tag{B.3}$$

**Appendix C: Deriving the Power Equation in the Cross-pol Channel of (37) and (38)**

Assume a transmitted wave to be a normalized, completely polarized wave represented by

$$\vec{E}_T = \begin{bmatrix} E_x \\ E_y \end{bmatrix} = E_x \begin{bmatrix} 1 \\ \rho \end{bmatrix} = |E_x|e^{j\phi_x} \begin{pmatrix} 1 \\ \rho \end{pmatrix} , \tag{C.1}$$

then, from the normalization condition

$$|\vec{E}_T|^2 = E_x^*(1 \quad \rho^*)\, E_x \begin{pmatrix} 1 \\ \rho \end{pmatrix} = |E_x|^2(1 + \rho\rho^*) = 1 \quad ,$$

we obtain

$$|E_x| = \frac{1}{\sqrt{1 + \rho\rho^*}} \quad .$$

Substituting $|E_x|$ into (equation (C.1)) yields

$$\vec{E}_T = \frac{e^{j\phi_x}}{\sqrt{1 + \rho\rho^*}} \begin{pmatrix} 1 \\ \rho \end{pmatrix} \quad . \tag{C.2}$$

Let the orthonormal polarization state to $\vec{E}_T$ be $\vec{E}_{T\perp}$, written as

$$\vec{E}_{T\perp} = \begin{pmatrix} E'_x \\ E'_y \end{pmatrix} \quad , \tag{C.3}$$

then, from the orthogonal condition

$$\vec{E}_{T\perp} \cdot \vec{E}_T = (\,E_x'^* \quad E_y'^*\,) \begin{pmatrix} E_x \\ E_y \end{pmatrix} = E_x'^* E_x + E_y'^* E_y = 0 \quad ,$$

we can obtain

$$E_x'^* = - \frac{E_y}{E_x} E_y'^* = - \rho\, E_y'^* \quad , \tag{C.4}$$

and

$$E'_x = - \rho^* E'_y \quad .$$

Substituting (equation (C.4)) into the normalization condition yields

$$\vec{E}_{T\perp} \cdot \vec{E}_{T\perp} = (E_x'^* \quad E_y'^*) \begin{pmatrix} E'_x \\ E'_y \end{pmatrix} = (- \rho E_y'^* \quad E_y'^*) \begin{pmatrix} -\rho^* E'_y \\ E'_y \end{pmatrix}$$

$$= \rho\rho^* |E'_y|^2 + |E'_y|^2 = 1 \quad ,$$

from which, we obtain

$$|E_y'| = \frac{1}{\sqrt{1 + \rho\rho^*}} \quad ,$$

then the orthonormal polarization state of $\vec{E}_{T\perp}$ can be represented as

$$\vec{E}_{T\perp} = \begin{pmatrix} -\rho^* E_y' \\ E_y' \end{pmatrix} = \frac{e^{j\phi_y'}}{\sqrt{1 + \rho\rho^*}} \begin{pmatrix} -\rho^* \\ 1 \end{pmatrix} \quad . \tag{C.5}$$

The received voltage in the cross-pol channel of (34) may be rewritten as

$$V_x = \frac{e^{j\phi_y'}}{\sqrt{1 + \rho\rho^*}} \begin{pmatrix} -\rho^* & 1 \end{pmatrix} [S] \frac{e^{j\phi_x}}{\sqrt{1 + \rho\rho^*}} \begin{pmatrix} 1 \\ \rho \end{pmatrix} \quad . \tag{C.6}$$

then the received power in the cross-pol channel can be expanded by

$$P_x = \left| \frac{e^{j\phi_y'}}{\sqrt{1 + \rho\rho^*}} \begin{pmatrix} -\rho^* & 1 \end{pmatrix} \begin{pmatrix} S_{xx} & S_{xy} \\ S_{yx} & S_{yy} \end{pmatrix} \frac{e^{j\phi_x}}{\sqrt{1 + \rho\rho^*}} \begin{pmatrix} 1 \\ \rho \end{pmatrix} \right|^2$$

$$= \frac{1}{1 + \rho\rho^*} \left| (-\rho^* S_{xx} - \rho^* S_{xy} \rho + S_{yx} + S_{yy}\, \rho) \right|^2$$

$$= |(-E_y^* E_x S_{xx} - E_y^* S_{xy} E_y + S_{yx} E_x E_x^* + S_{yy} E_y E_x^*)|^2$$

$$= \begin{pmatrix} E_x E_x^* \\ E_x E_y^* \\ E_x^* E_y \\ E_y E_y^* \end{pmatrix}^T \left[ \begin{pmatrix} 0 & 0 & 0 & 1 \\ 0 & -1 & 0 & 0 \\ 0 & 0 & -1 & 0 \\ 1 & 0 & 0 & 0 \end{pmatrix} ([S] \otimes [S]^*) \right] \begin{pmatrix} E_x E_x^* \\ E_x E_y^* \\ E_x^* E_y \\ E_y E_y^* \end{pmatrix}$$

$$= \vec{J}_T^{\,T} \left[ \begin{pmatrix} 0 & 0 & 0 & 1 \\ 0 & -1 & 0 & 0 \\ 0 & 0 & -1 & 0 \\ 1 & 0 & 0 & 0 \end{pmatrix} ([S] \otimes [S]^*) \right] \vec{J}_T$$

$$= \vec{g}_T^T \, ([A]^{-1})^T \left[ \begin{pmatrix} 0 & 0 & 0 & 1 \\ 0 & -1 & 0 & 0 \\ 0 & 0 & -1 & 0 \\ 1 & 0 & 0 & 0 \end{pmatrix} ([S] \otimes [S]^*) \right] [A]^{-1} \vec{g}_T$$

$$= \vec{g}_T^{\,T} \, [\bar{M}_x] \, \vec{g}_T \qquad \text{(C.7)}$$

Then the second Stokes reflection matrix $[\bar{M}_x]$ is

$$[\bar{M}_x] = ([A]^{-1})^T \left[ \begin{pmatrix} 0 & 0 & 0 & 1 \\ 0 & -1 & 0 & 0 \\ 0 & 0 & -1 & 0 \\ 1 & 0 & 0 & 0 \end{pmatrix} ([S] \times [S]^*) \right] [A]^{-1} \qquad \text{(C.8)}$$

# COMMENTS ON TARGET DECOMPOSITION THEOREMS

## Part I: Based on Eigenvalues

J. Richard Huynen

P. Q. Research
10531 Blandor Way
Los Altos Hills, CA 94022 U. S. A.

Some recent publications (3) have prompted the following analysis on the subject of radar target decompositions. There still seems to be considerable misunderstanding and confusion which are related to the physical significance of individual target parameters, the question of uniqueness, and the physical relevance of certain mathematical procedures. What is emphasized here are problems of target discrimination, target classification and identification. Other problems such as those related to target detection are not considered here and may well require a different approach.

In what follows I will use "Holm", "Barnes" and "Huynen" types of decompositions to distinguish three cases. The "Holm" and "Barnes" methods are based on the eigenvalues generated by the hermetian coherency matrix (or density matrix as Holm prefers to call it, which is often used in quantum mechanics [QM]).

Now the reference to QM in this case is rather unfortunate, because <u>there</u> the eigenvalues of the observable hermetian operator do have a very strong connection with reality; it is in fact the <u>measured value</u> for the observable! But with <u>radar targets</u>, unfortunately, there seems to be very little direct connection with physically related observables! At least that is my own impression.

Of course it is always possible to cook up special cases where "eigenvalue" and physical object reality are obviously related. Such is the



*W.-M. Boerner et al. (eds.), Direct and Inverse Methods in Radar Polarimetry, Part 1*, 387–399.

case for a sphere or diplane. But there seems to be no such connection in general.

In contrast, the "Huynen" decomposition is not based on eigenvalues but claims to give a direct insight into physical target structure based upon symmetry and non-symmetry arguments. In this theory, a distinction is made between parameters which refer to "target symmetry" ($A_O$, $B_O$ C, D, $\Psi$), "target non-symmetry" (N-target defined by $B_O$, B, E, F), and "coupling" ($H_\psi$ and G). Each group is further subcategorized (i.e., C = shape factor, F is helicity, etc.). A more complete discussion of these parameters is planned for the future. What I want to do next is to show how the three types of decomposition differ from or are similar to each other.

The Holm-method has the further complication that the relative values of the eigenvalues determine the type of target decomposition. Consider the Holm-decomposition for the following six cases:

$$\text{I} \quad \begin{bmatrix} a & & \\ & b & \\ & & c \end{bmatrix} = \begin{bmatrix} 0 & 0 & 1 \\ 0 & 1 & 0 \\ 1 & 0 & 0 \end{bmatrix} \begin{bmatrix} c & & \\ & b & \\ & & a \end{bmatrix} \begin{bmatrix} 0 & 0 & 1 \\ 0 & 1 & 0 \\ 1 & 0 & 0 \end{bmatrix}$$

$$\text{II} \quad \begin{bmatrix} a & & \\ & b & \\ & & c \end{bmatrix} = \begin{bmatrix} 0 & 0 & 1 \\ 1 & 0 & 0 \\ 0 & 1 & 0 \end{bmatrix} \begin{bmatrix} b & & \\ & c & \\ & & a \end{bmatrix} \begin{bmatrix} 0 & 1 & 0 \\ 0 & 0 & 1 \\ 1 & 0 & 0 \end{bmatrix}$$

$$\text{III} \quad \begin{bmatrix} a & & \\ & b & \\ & & c \end{bmatrix} = \begin{bmatrix} 0 & 1 & 0 \\ 0 & 0 & 1 \\ 1 & 0 & 0 \end{bmatrix} \begin{bmatrix} c & & \\ & a & \\ & & b \end{bmatrix} \begin{bmatrix} 0 & 0 & 1 \\ 1 & 0 & 0 \\ 0 & 1 & 0 \end{bmatrix}$$

$$\text{IV} \quad \begin{bmatrix} a & & \\ & b & \\ & & c \end{bmatrix} = \begin{bmatrix} 1 & 0 & 0 \\ 0 & 0 & 1 \\ 0 & 1 & 0 \end{bmatrix} \begin{bmatrix} a & & \\ & c & \\ & & b \end{bmatrix} \begin{bmatrix} 1 & 0 & 0 \\ 0 & 0 & 1 \\ 0 & 1 & 0 \end{bmatrix}$$

$$\text{V}\quad \begin{bmatrix} a & & \\ & b & \\ & & c \end{bmatrix} = \begin{bmatrix} 0 & 1 & 0 \\ 1 & 0 & 0 \\ 0 & 0 & 1 \end{bmatrix} \begin{bmatrix} b & & \\ & a & \\ & & c \end{bmatrix} \begin{bmatrix} 0 & 1 & 0 \\ 1 & 0 & 0 \\ 0 & 0 & 1 \end{bmatrix}$$

$$\text{VI}\quad \begin{bmatrix} a & & \\ & b & \\ & & c \end{bmatrix} = \begin{bmatrix} 1 & 0 & 0 \\ 0 & 1 & 0 \\ 0 & 0 & 1 \end{bmatrix} \begin{bmatrix} a & & \\ & b & \\ & & c \end{bmatrix} \begin{bmatrix} 1 & 0 & 0 \\ 0 & 1 & 0 \\ 0 & 0 & 1 \end{bmatrix}$$

On the left are what appears to be six cases for the _same_ type of target. From the theory of hermitian matrices we know that a, b, c must be non-negative real numbers. The only difference in the six cases is in the _value_ of these numbers. According to the Holm recipe (3, page 307) the coherency matrix is written in the form:

$$\rho = U \begin{bmatrix} \varepsilon_1 & & \\ & \varepsilon_2 & \\ & & \varepsilon_3 \end{bmatrix} U^{-1}$$

where $\varepsilon_1 < \varepsilon_2 < \varepsilon_3$ and U is a unitary transformation. This form is displayed on the right-hand side in the six cases above, i.e. for case I, this implies: $c < b < a$ ; for case II, $b < c < a$ , etc. Now Holm writes $\rho = \rho_0 + \rho_1 + \rho_2$ where

$$\rho_0 = U \begin{bmatrix} 0 & & \\ & 0 & \\ & & \varepsilon_3 - \varepsilon_2 \end{bmatrix} U^{-1}$$ is a single target,

$$\rho_1 = U \begin{bmatrix} 0 & & \\ & \varepsilon_2 - \varepsilon_1 & \\ & & \varepsilon_2 - \varepsilon_1 \end{bmatrix} U^{-1}$$ is a mixed target called "variance", and

$$\rho_2 = U \begin{bmatrix} \varepsilon_1 & & \\ & \varepsilon_1 & \\ & & \varepsilon_1 \end{bmatrix} U^{-1} = \varepsilon_1 I \text{ is a "noise" target.}$$

Using this recipe, we arrive at the following six target-decompositions:

$$\text{I} \quad \begin{bmatrix} a & & \\ & b & \\ & & c \end{bmatrix} = \begin{bmatrix} a-b & & \\ & 0 & \\ & & 0 \end{bmatrix} + \begin{bmatrix} b-c & & \\ & b-c & \\ & & 0 \end{bmatrix} + \begin{bmatrix} c & & \\ & c & \\ & & c \end{bmatrix}$$

$$\text{II} \quad \begin{bmatrix} a & & \\ & b & \\ & & c \end{bmatrix} = \begin{bmatrix} a-c & & \\ & 0 & \\ & & 0 \end{bmatrix} + \begin{bmatrix} c-b & & \\ & 0 & \\ & & c-b \end{bmatrix} + \begin{bmatrix} b & & \\ & b & \\ & & b \end{bmatrix}$$

$$\text{III} \quad \begin{bmatrix} a & & \\ & b & \\ & & c \end{bmatrix} = \begin{bmatrix} 0 & & \\ & b-a & \\ & & 0 \end{bmatrix} + \begin{bmatrix} a-c & & \\ & a-c & \\ & & 0 \end{bmatrix} + \begin{bmatrix} c & & \\ & c & \\ & & c \end{bmatrix}$$

$$\text{IV} \quad \begin{bmatrix} a & & \\ & b & \\ & & c \end{bmatrix} = \begin{bmatrix} 0 & & \\ & b-c & \\ & & 0 \end{bmatrix} + \begin{bmatrix} 0 & & \\ & c-a & \\ & & c-a \end{bmatrix} + \begin{bmatrix} a & & \\ & a & \\ & & a \end{bmatrix}$$

$$\text{V} \quad \begin{bmatrix} a & & \\ & b & \\ & & c \end{bmatrix} = \begin{bmatrix} 0 & & \\ & 0 & \\ & & c-a \end{bmatrix} + \begin{bmatrix} a-b & & \\ & 0 & \\ & & a-b \end{bmatrix} + \begin{bmatrix} b & & \\ & b & \\ & & b \end{bmatrix}$$

$$\text{VI} \quad \begin{bmatrix} a & & \\ & b & \\ & & c \end{bmatrix} = \begin{bmatrix} 0 & & \\ & 0 & \\ & & c-b \end{bmatrix} + \begin{bmatrix} 0 & & \\ & b-a & \\ & & b-a \end{bmatrix} + \begin{bmatrix} a & & \\ & a & \\ & & a \end{bmatrix}$$

| ↓ | ↓ | ↓ | ↓ |
|---|---|---|---|
| Mixed target | Single target | Mixed "Variance" | "Noise" target |

In the first two cases, the single target is a sphere type; in cases III and IV, a straight diplane type; and in cases V and VI, a diplane slanted at 45°. Hence the left-hand target shows a split personality, one time appearing as a sphere, the next time as a diplane, depending on the values of a, b and c! Further discussion on the Holm's interpretation follows below.

Next we discuss the "Barnes"-decomposition.

$$\underset{\substack{\downarrow\\ \text{mixed}\\ \text{target}}}{\begin{bmatrix} a & & \\ & b & \\ & & c \end{bmatrix}} = \underset{\substack{\downarrow\\ \text{single}\\ \text{sphere}}}{\begin{bmatrix} a & & \\ & 0 & \\ & & 0 \end{bmatrix}} + \underset{\substack{\downarrow\\ \text{single}\\ \text{diplane}}}{\begin{bmatrix} 0 & & \\ & b & \\ & & 0 \end{bmatrix}} + \underset{\substack{\downarrow\\ \text{single}\\ \text{tilted}\\ \text{diplane}}}{\begin{bmatrix} 0 & & \\ & 0 & \\ & & c \end{bmatrix}} \qquad \text{(B)}$$

On the left-hand side are shown three single targets, i.e. a sphere (a), a diplane (b), and a slanted diplane (c). Clearly this decomposition is unique and in <u>this</u> case has a well-defined connection with physical reality.

However, as soon as we look at the <u>general</u> result in (B), a less clear picture emerges.

$$\rho = U\begin{bmatrix} a & & \\ & b & \\ & & c \end{bmatrix}U^{-1} = U\begin{bmatrix} a & & \\ & 0 & \\ & & 0 \end{bmatrix}U^{-1} + U\begin{bmatrix} 0 & & \\ & b & \\ & & 0 \end{bmatrix}U^{-1} + U\begin{bmatrix} 0 & & \\ & 0 & \\ & & c \end{bmatrix}U^{-1} \qquad \text{(B')}$$

On the left-hand side is a general mixed coherency matrix, while on the right-hand side are shown three <u>single</u> targets. Although the Barnes-decomposition is unique, the connection of parameters to physical reality has been lost.

It is not clear what each single target on the right in B' represents. The eigenvalues a, b, and c have in general no clear physical significance.

However, there is another objection that can be made to this kind of approach. We know that on the left-hand side of B' the $\rho$ is given by 9 independent parameters. Three are shown (a, b and c) hence the U matrix must have 6 independent parameters which are also shown on the right-hand sice of B'. Because one single target on the right has 5 independent parameters, there must be 4 dependency relationships for each single target. Hence not only do we find in total 12 dependency relations for the three single targets taken separately, but there is an additional inter-target redundancy as well. From 9 parameters on the left we generate 3 x 9 = 27 parameters on the right! All this does not improve clarity of insight into physical target reality!

We now discuss the "Huynen" decomposition.

$$\begin{bmatrix} a & & \\ & b & \\ & & c \end{bmatrix} = \begin{bmatrix} a & & \\ & 0 & \\ & & 0 \end{bmatrix} + \begin{bmatrix} 0 & & \\ & b & \\ & & c \end{bmatrix} \tag{H}$$

| ↓ | ↓ | ↓ |
|---|---|---|
| mixed target | single sphere | mixed N-target |

This shows a unique decomposition into a single target plus "N-target" residue. The term N-target is derived from non-symmetrical targets. It exhibits physically the purest counterpart to a symmetrical target. The parameter 2Ao or (a) in the coherency matrix above in the generator of target symmetry.

Hence if $A_o = 0$, the target is a pure non-symmetrical target, or N-target. Hence $2A_o$ is not the generator of "the most sphere-like" target as mentioned in (3) page 306. The symmetric target is characterized by having a plane of reflection symmetry, and hence it has a strong connection to physical reality! It includes all roll-symmetric objects at any position in space, such as a cone, a cylinder and numerous other objects which are not at all "spherical". In terms of parameters, the symmetrical target is defined by $\tau_m = 0$ and, after elimination of $\Psi$ dependence, also by $B_o = B$, i.e. by parameters $A_o$, $B_o$, C, D, and $\Psi$; hence $E = F = G = H = 0$ for the symmetric target. On the other hand, a pure non-symmetrical target or N-target is defined by $\tau_m = \pm 45°$, i.e. by $B_o$, B, E, F, and $A_o = C = D = B = H = 0$.

The N-target has an additional interesting and important property. It is defined by a group-structure ( SU(2) ) which defines the Stokes-vector for a wave! Hence our knowledge of the Stokes-vector is directly applicable to the N-target. (This is discussed in great detail in [6]. This reference is not mentioned in [3], which may account for some of the misconceptions.)

First we generate $(H^1)$ as was done for the Barnes case:

$$\rho = U \begin{bmatrix} a & & \\ & b & \\ & & c \end{bmatrix} U^{-1} = \begin{bmatrix} \text{single} \\ \text{general} \\ \text{target} \end{bmatrix} + \begin{bmatrix} \text{mixed} \\ \text{N-} \\ \text{target} \end{bmatrix} \qquad (H')$$

The result (H') is called the Huynen-target decomposition theorem. Its physical realizability was proven in his thesis (4). In addition, the decomposition is unique, contrary to claims of the opposite by Cloude (5) and others. What has to be understood is that the N-target's SU(2) structure (i.e. $B_o$, B, E, F) is invariant to tilt $\Psi$. (This is simply because if $A_o = 0$ for one angle of tilt, it must be true for all angles!) It is easy to show that no other decomposition into two physically realizable object structures can have this property.

Another source for confusion is that the result (H') is often interpreted in a narrow sense, such that the "single target" is considered the main component of the decomposition where the N-target is considered only as "noise". In the original work, the term N-target noise was often used by this author, which perhaps contributed to this type of confusion and misunderstanding. Obviously one may have cases (such as mentioned in Holm [3]) where $A_o = 0$ and the main interest is focused on the N-target structure itself. We can use the same equation (H') with a broad interpretation of the result.

If the main focus is on the N-target, we may employ the property mentioned above that the N-target $(B_o, B, E, F) = (B_o, \underline{B})$ has the character of a Stokes-vector (see thesis p. 157 equ. [33.2] and [33.3] which already uses this notion; see also [b] where the SU[2] imbedding property of the N-target is discussed in detail). Hence we find for the N-target the familiar mixed Stokes-vector decomposition into a coherent part and an unpolarized part:

$$( B_o ; \underline{B} ) = ( |\underline{B}| ; \underline{B} ) + ( B_o - |\underline{B}| ; 0 )$$

Notice that the "unpolarized part" is defined by a single parameter $B_o - |\underline{B}|$ , while $B = E = F = 0$ for the "unpolarizing" N-target representation. We thus arrive at a Stokes (Mueller) matrix representation for the "unpolarizing target", which was already formulated much earlier (see thesis [4] pp. 198-199). We repeat here its unpolarizing scattering behavior:

$$\begin{bmatrix} B_o - |\underline{B}| & & & \\ & B_o - |B| & & \\ & & 0 & \\ & & & 0 \end{bmatrix} \begin{bmatrix} \underline{p_o} \\ 0 \\ p_2 \\ p_3 \end{bmatrix} = (B_o - |B|)\, p_o \begin{bmatrix} \underline{1} \\ 0 \\ 0 \\ 0 \end{bmatrix}$$

$\downarrow$ Huynen-noise target $\qquad \downarrow$ Lin-pol wave $\qquad \downarrow$ Unpolarized wave

We notice that this "unpolarizing target" behavior bears little resemblance to Holm's "noise" target.

Substituting the N-target decomposition into (H') we arrive at

$$\rho = \begin{bmatrix} \text{Single} \\ \text{general} \\ \text{target} \end{bmatrix} + \begin{bmatrix} \text{Single} \\ \text{N-} \\ \text{target} \end{bmatrix} + \begin{bmatrix} \text{Unpol.} \\ \text{residue} \\ \text{component} \end{bmatrix} \qquad (H'')$$

$\downarrow$ $A_o \neq 0$ $\qquad \downarrow$ $A_o = 0$ $\qquad \downarrow$ $B_o \neq 0$, all other $= 0$

We now claim that (H") applies to all cases whether $A_o = 0$ or not. In fact, the Holm example of a mixture of diplanes (3) p. 307 also fits this scheme. However, the great advantage of decompositions (H') or (H") is the non-proliferation of parameters!

For (H') we have the formula (9 par.) = (5 par.) + (4 par.); and for (H"), (9 par.) = (5 par.) + (3 par.) + (1 par.). The (H') decomposition is based solely on physical properties of target symmetry and non-symmetry and the SU(2) structure of the N-target (6). No such physical basis can be found with the other type of decompositions.

There are many other details that can be discussed and which are important, i.e. the basic "linearity" of the Huynen decomposition as compared to others. However, for the sake of brevity the present comments may suffice.

# COMMENTS ON TARGET DECOMPOSITION THEOREMS

## Part II: Real and Exotic Worlds

## 1. INTRODUCTION

In 1970 this author introduced the first account of polarimetric radar decomposition theorems. Recently others have contributed new proposals for extensions and/or modifications, notably by S. R. Cloude, W. A. Holm and R. M. Barnes. It is the purpose of this report to review these proposals and to show their merit or lack thereof for real world target applications. The target decomposition arises when one considers a target mixture from an ensemble of single targets. The mixture can no longer be represented by a single radar target which has an associated target scattering matrix, instead it belongs to a wider class of targets, i.e. the class of target mixtures.

Notice that in ordinary observation a group of objects is not commonly perceived as a single new object. Certain features are collected from the target mixture, to form an average target which does have an associated SM. The rest forms a kind of target residue. The original target decomposition theorem showed that a target mixture $\langle M \rangle$ with nine independent parameters could be uniquely split into an average single target $M_o$, with 5 independent parameters and an N-target residue with 4 remainder parameters. The N-target has the most non-symmetric characteristic. Hence what is favoured in the general case is target symmetry which components are preserved in the average single target $M_o$.

Broadening our scope, we can view a target decomposition as representing a VIEW OF A WORLD. This viewpoint may be related to the world we normally encounter, or it may be applicable to an unconventional or "exotic" world. For example, a world dominated by dinosaurs would be an exotic world in our present view, although at a certain time it was conventional. Hence the so-called Huynen-decomposition described above favours objects with symmetry although in this world most objects also have some non-symmetry. All this is expressed by the rule for non-symmetry: $(RL) = A_o = o$ which determines the non-symmetric N-target residue.

Recently R. M. Barnes has advocated two other methods of target decomposition [7] where the residue was characterized by $(RR) = o$ or $(LL) = o$. These "Barnes" decompositions have the roll-invariant property as required by the Huynen-decomposition (The method of target decomposition is independent of roll angle $\psi$) [7]. But let us now examine the significance of the Barnes decomposition in terms of the world view it represents.

## 2. R. M. BARNES DECOMPOSITIONS I and II

The two models Barnes presents fix a target decomposition by the property that the target residue must satisfy $(RR) = o$ or $(LL) = o$. Now according to well-known equations [4] $(RR) = 2(B_o + F)$ and $(LL) = 2(B_o - F)$ where $B_o \geq o$ is a measure of 'target structure', (in some literature (RR) and (LL) are both called 'even bounce' behavior, whereas (RL) is 'odd-bounce' behavior, this author considers such nomenclature 'flawed' at best) whereas $F$ is a measure of HELICITY. For most 'ordinary' real world targets helicity is small (only a right- or left-wound helix would produce large helicity) and if present is equally distributed between positive and negative values.

Hence the Barnes decomposition which gives a residue according to $(LL) = 2\ (B_o - F) = o$ produces a bias toward right-wound helices where $B_o^N = F^N$ (where N stands for residue target). In other words, no matter what F is for the mixture-target the $F^N$ is always positive (or zero)! Hence there is in this world a bias AGAINST positive helicity and hence a PREFERENCE for negative helicity! In this world positive and negative wound helicities would NOT be found in equal distributions, hence this must refer to not our world but an EXOTIC world

The other Barnes case with residue $(RR) = o$ leads to a world which has a preference for positive helicity and hence is the exotic world which is a mirror image of the previous one.

We conclude that the Barnes I & II target decompositions lead to two EXOTIC worlds where there is a preference either for right-wound helices or for left-wound helices.

## 3. THE S. R. CLOUDE PROPOSAL

The Cloude proposal for target decomposition is based upon the mathematical generalization of the Huynen-case. An immediate criticism is the 'proliferation of parameters' which is the result of this approach. Out of nine independent parameters of the mixed Mueller matrix are generated 18 parameters on the decomposition side. The single average target has only 5 independent parameters, with 4 dependencies intermixed. The residue target is left with 4 independent and 5 dependencies. This does not make much physical sense unless the extra dependent parameters can be made zero!

In addition, the process has to be faithful for rotations (or tilt) about the line of sight [see 7]. Hence if null-parameters are generated for one rotation angle $\Psi$, it must do so for all $\Psi$'s. This reduces the infinite amount of possible Cloude-decompositions to the two Barnes-decompositions and the Huynen case. It was shown above that two Barnes-types lead to 'exotic' worlds and hence the only decomposition which corresponds to the real world of basic symmetry is the Huynen-decomposition!

REFERENCES


1. Holm, W.A. and Barnes, R.M., On Radar Polarization Mixed Target State Decomposition Techniques", IEEE 1988 National Radar Conference.
2. Holm, W.A., "Polarimetric Fundamentals and Techniques", in Eaves, J.L. and Reedy, E.K., Principles of Modern Radar, Van Nostrand Reinhold Co., 1987, pp. 621-645.
3. Holm, W.A., "MMW Radar Signal Processing Techniques", in Currie, N.C. and Brown, C.E., Principles and Applications of Millimeter-Wave Radar, Artech House, 1987, pp. 241-312.
4. Huynen, J.R., Phenomenological Theory of Radar Targets, Ph.D. Thesis, Technical University, Delft, The Netherlands, 1970.
5. Cloude, S.R., "Target Decomposition Theorems in Radar Scattering", Electronic Letters, 1985, 21, pp. 22-24.
6. Huynen, J.R., "Toward a Theory of Perception for Radar Targets", in Boerner, W.M. et Al., Inverse Methods in Electromagnetic Imaging, Reidel, 1985, pp. 797-822.
7. Barnes, R. M., "Roll-Invariant Decompositions for the Polarization Covariance Matrix," Polarimetric Technology Workshop, Redstone Arsenal, Alabama, August 1988 (to be published).

# T-PULSE FOR EXCITING SINGLE MODES OF RADAR TARGETS

Yingbo Hua, Ercument Arvas
Tapan K. Sarkar
Department of Electrical and Computer Engineering
Syracuse University
Syracuse, New York 13244-1240 USA

ABSTRACT. We present an approach called T-pulse approach to excite a single mode of a target. The T-pulse is a time-limited signal that has maximum energy in a given frequency band. While the T-pulse method can be applied to any highly resonant target for the purpose of target identification and discrimination, an application of the T-pulse method to a synthesized thin-wire target is presented to illustrate its ability of exciting single modes.

## I. INTRODUCTION

In this paper, we present the T-pulse approach, which deals with an optimal time-limited and band-pass EM pulse to excite a target. The resulting response is a preselected single mode scattered response from the target which may be used for the purpose of target identification and discrimination (TID). From the target discrimination point of view, a preselected target may have a mode which is singularly different from all other modes of different targets under consideration. If the T-pulse is tuned to a preselected mode and is applied to a target, the desired single mode response is obtained. In the T-pulse approach, we synthesize the exciting T-pulse basically by tuning the center frequency of the T-pulse according to the resonant frequencies of a desired target.

Some approaches [1-4] for radar target identification requires using an EM pulse to excite a target and then retrieving many aspect-independent poles of the target from scattered field. There are two extreme cases for this approach. The first extreme case is that an impulse, or wide-band, signal is used to excite the target, and all the poles (or modes) in the impulse response are estimated. Ideally, this approach is very fast in the sense that "all" poles of the target are obtained together in a short period of time. However, the accuracy of

---

This work has been supported in part by Aeritalia Corporation and by the Office of Naval Research under contract N00014-79-C-0598.



*W.-M. Boerner et al. (eds.), Direct and Inverse Methods in Radar Polarimetry, Part 1*, 401–425.

the estimated poles are sometimes limited by the fact that there are too many poles in the scattered field so that any estimation method would be very sensitive to noise in measured field. This observation suggests the other extreme case in which only one mode is excited by some time-limited pulse. If the one-mode scattered field can be obtained successfully, then the corresponding pole can be estimated very accurately by SVD or eigensturcture based Prony's, Pisarenko's and Pencil-of-Functions methods [7].

We should mention that the exciting pulse synthesis has been a problem under study for the purpose of TID for many years. Two different examples of synthesized exciting pulses are the K-pulse [8] and the E-pulse [6].

For the K-pulse method a time limited pulse is generated by synthesizing a waveform which cancels all the poles of the target. The resulting transfer function then is an all-zero function and hence time limited. The difficulty in this approach is that the time limited pulse has very large bandwidth.

Similarly for E-pulse technique a wide band pulse is necessary to perform the various discriminations. In the following, we present the principles of the T-pulse method and illustrate its potential usefulness by applying T-pulses to a synthesized thin-wire target.

## 2. THE T-PULSE METHOD

We assume that the target under consideration is highly resonant so that all the modes are approximately isolated infrequency domain (or along the imaginary axis of the S-plane). The T-pulse method is to apply a time-limited signal, which has maximal energy in a given narrow (frequency) band, to excite a target. If the narrow band is in the effective band of a mode of the target, then the corresponding mode will be dominant in the scattered field after the exciting pulse dies out. It is known that, in general, each mode of a target has frequency components over an infinite bandwidth. By effective band, we mean that the frequency component in that band is dominant. If the narrow band is not an effective band of any mode, then the scattered field will be very weak after the exciting pulse dies out.

It is well known that time-limited signal can not be band-limited in strict sense. But there are time-limited signals with maximal energy in a given band. Such signals can be very good approximations of the (imagined) time-limited and band-limited signals. For brevity, we call such signals T-pulses to distinguish them from the K-pulses [8] and the E-pulses [6]. Since any band-pass signal can be obtained by modulating (multiplying by a sinusoidal function) a baseband signal [9], the problem is reduced to finding a time-limited signal with maximal energy in a given baseband. In other words, we need to find a time-limited signal E(t) for $0 \leq t \leq D$ such that the in-band ($-w_b \leq w \leq w_b$) ener

$$J = 1/2\pi \int_{-w_b \leq w \leq w_b} \left| \int_{0 \leq t \leq D} E(t) \exp(-jwt)\, dt \right|^2 dw$$

is maximized subject to the constraint that the total energy of E(t) is fixed. In this paper, we use the discrete version of the optimal signals which can be easily generated both in theory, through a computer program, and in real experimental conditions. The mathematical details of the design technique for discrete optimal signals are described in [10].

## 3. EXPERIMENT ON A SYNTHESIZED THIN-WIRE TARGET

Now we describe the following experiment. Since we will work on discrete data, the sampling time, denoted by T, is chosen to be 0.5 ns = 0.15 light-meters. First, we let a narrow Gaussian pulse excite, from broadside, a 2-meters thin-wire dipole (radius=0.01m). The current induced at the center of the dipole is computed to a prespecified degree of accuracy [11]. Then we observe the spectrum of the excited current after the exciting pulse dies out. From the spectrum we detect several resonant frequencies for references. Secondly, we apply, at different times, several optimal T-pulses as the derivatives of the electric field pulses impinging on the dipole, and then observe the induced currents at the center of the dipole, which are obtained from [11].

### 1) Impulse response:

We assume that a narrow Gaussian pulse whose derivative is shown in Fig. 1 impinges on a 2-m dipole from broadside. The exciting pulse becomes zero after t=6 light-meters. The synthesized (or computed) current induced at the center of the dipole is shown in Fig.2. In Fig. 3 the DFT (Discrete Fourier Transform) amplitude spectrum of the first 128 points (samples) of the exciting pulse of Fig. 1. It is seen that the spectrum has a wide pass band. In Fig. 4, the amplitude spectrum of the tail of the induced current, i.e., for $t \geq 6$ light-meters, is shown. From Fig. 4, we can observe four resonant frequencies which are at k=4, 13, 24 and 26 or f=0.0625, 0.203, 0.375 and 0.406 (in GHz). Note that the frequency f is related to the frequency index k by f=k/128T for 128-points DFT [12]. We should mention that, for ideal 2-m dipole, the resonant frequencies should be f=0.075, 0.225, 0.0375,....,which are computed from f = ic/2L where c is the light velocity; L is the dipole length; and i is an odd integer. The first three resonant frequencies detected from Fig. 4 appear to approximately match the first three ideal frequencies, although the fourth resonant frequency k=26 should not appear. The exact reason for the fourth resonant frequency in Fig. 4 is still not clear. However, this phenomenon does not affect the central theme presented in this paper, which is that pure single modes can be excited by optimal time-limited band-pass signals.

2) Optimal T-Pulses:

As we presented in Section 2, the optimal signals can be designed by an algorithm available in [10]. In Fig. 5 is a baseband optimal signal of duration D = 127T = 63.5 ns = 19.05 light-meters, and with in-band energy percentage 99.995% in the bandwidth 2/64 GHz. In Figs. 6-8 are three modulated optimal signals which are obtained by multiplying the base-band optimal signal as in Fig.5 by sine functions of frequencies k=9, 13 and 25, respectively. The DFT amplitude spectrums of the base-band optimal signal and the three optimal band-pass signals are shown in Figs. 9-12, respectively. As one sees, the side lobes in the spectrums of these optimal signals are almost 50 dB down from the main lobe peak, and half width of the main lobe is about two times the reciprocal of 128T.

3) Responses to Optimal T-Pulses:

We let the optimal signals presented above be the derivatives of the impinging electric field pulses. Note that the electric field pulses, which are the integrals of those optimal signals, are also narrow-band-pass signals.

Fig. 13 shows the current response excited by the pulse (Fig. 6) with center frequency (or called preselected resonant frequency) k=9 which is not equal to any one of the resonant frequencies of the dipole. As one observes, after the exciting pulse dies out, i.e., for $t \geq 19.05$ light-meters, the induced current has a very weak tail reponse, which indicates that the preselected resonant frequency is close to no actual resonant frequency of the target. Fig. 14 shows a (normalized) DFT amplitude spectrum of the weak tail response of the current. From this spectrum, we observe two peaks at k=4 and 26 and no peak at k=9. The strange peak at k=26 may be an indication that there are perhaps numerical errors in the synthesized current response which are in a periodic form of frequency k=26.

Fig. 15 shows the current response excited by the pulse (Fig.7) with center frequency k=13 which is equal to the second resonant frequency of the dipole (see Fig. 4). As one observes, the current response has a strong tail response for $t \geq 19.05$ light-meters, which indicates that the preselected frequency is fairly close to a resonant frequency of the target. Fig. 16 shows a DFT amplitude spectrum of the current tail response. The single strong peak occurs just at the preselected frequency k=13, i.e., a preselected single mode is excited.

Fig. 17 shows the current response excited by the pulse (Fig.8) with center frequency k=25 around which the (synthesized) target has two resonances (see Fig. 4). This current also has a strong tail for $t \geq 19.05$ light-meters. The interesting modulation phenomenon in the current tail is clearly explained by Fig. 18, which shows the DFT amplitude spectrum of that tail. The two dominant frequencies K=24 and 26 in the current tail produce the beat phenomenon as they are within the pass band of the exciting pulse (see Fig. 12).

## 4. CONCLUSION

An optimal time-limited and band-pass pulse (T-pulse) approach for exciting single mode (or several modes if needed) of an object has been presented. While this approach performs well, in principle, for all highly resonant targets, examples for a synthesized thin-wire target have been shown.

## REFERENCES

[1] A. J. Berni, 'Target identification by natural resonance estimation,' IEEE Trans. on Aero. and Electro. Syst. Vol. AES **11**, No.2, March 1975, pp. 147-154.

[2] M. Van Blaricum and R. Mittra, 'A Technique for extgracting the poles and residues of a system directly from its transient response,' IEEE Trans. Ant. Prop. Vol. **AP-23**, Nov. 1975. pp.777-781.

[3] L. C. Chan, D. L. Moffatt and L. Peters, 'A characterization of subsurface radar target,' Proc. IEEE, Vol. **67**, pp. 991-1000, July 1979.

[4] C. W. Chuang and D. L. Moffatt, 'Natural Resonances of radar target via Prony's method and target discrimination,' IEEE Trans. on Aerospace and Electronics, Vol. **AES-12**, pp. 583-589, Sept. 1976.

[5] K-M. Chen and D. Westmorland, 'Radar waveform synthesis for exciting single-mode backscatters from a sphere and application for target discrimination,' Radio Science; Vol. **17**, No. 3, pp. 574-588, May - June 1982.

[6] E. J. Rothwell, K-M Chen and D. P. Nyquist, 'Extraction of the natural frequencies of a radar target from a measured response using E-pulse techniques,' IEEE Trans. Ant., Prop, Vol. **AP-35**, No. 6, June 1987, pp.715-720.

[7] Y. Hua, 'On techniques for estimating parameters of exponentially damped/undamped sinusoids in noise,' Ph.D. dissertation, Syracuse University, Jan. 1988.

[8] E. M. Kennaugh, 'The K-pulse concept,' IEEE Trans. Antennas and Prop., Vol **AP-29**, No. 2, pp 327-331, Mar. 1981.

[9] A. Papoulis, Signal Analysis, New York, McGraw Hill, 1979.

[10] Y. Hua and T. K. Sarkar, 'Design of optimum discrete finite duration orthogonal Nyquist signals,' Tech. Report, Department of ECE, Syracuse University, Jan. 1986. Also in IEEE Trans. on Acoustics, Speech, and Signal Processing, Vol. **36**, No. 4, pp. 606-608, April 1988.

[11] S. M. Rao and T. K. Sarkar and S. A. Dianet, 'The application of the conjugate gradient method to the solution of transient elctromagnetic scattering from the wires,' Radio Science, Vol. **19**, No. 5, pp. 1319-1326, Sept-Oct 1984.

[12] L. R. Rabiner and B. Gold, Theory and Application of Digital Signal Processing, Prentice-Hall, Inc., 1975.

FIGURES

1. Derivative function of a narrow Gaussian electric field pulse impinging on a thin wire. It dies out completely for $t \geq 6$ light-meters or 40T. T is sampling interval and equal to 0.15 light-meters.

2. Synthesized current response at the center of a thin-wire activated by a narrow Gaussian electric field pulse (Fig. 1) from the broadside.

3. DFT amplitude spectrum of the first 128 samples of the pulse of Fig. 1.

4. DFT amplitude spectrum of 128 samples of the current response for $40T \leq t \leq 167T$.

5. Optimal baseband signal of duration 127T and with 99.995% energy in the band [-1/64T, 1/64T]. (The computing frequency resolution was N =30 [10]).

6. Modulated signal of the baseband signal of Fig. 5. The center frequency is k=9.

7. Modulated signal of the baseband signal of Fig. 5. The center frequency is k=13.

8. Modulated signal of the bseband signal of Fig. 5. The center frequency is k=25.

9. DFT amplitude spectrum of the baseband signal of Fig. 5.

10. DFT amplitude spectrum of the modulated signal of Fig. 6.

11. DFT amplitude spectrum of the modulated signal of Fig. 7.

12. DFT amplitude spectrum of the modulated signal of Fig. 8.

13. Synthesized current response at the center of the dipole activated by the pulse of Fig. 6.

14. DFT amplitude spectrum of the 128 samples of the current responses of Fig. 13 for $129T \leq t \leq 256T$.

15. Synthesized current response at the cneter of the dipole activated by the pulse of Fig. 7.

16. DFT amplitude spectrum of 128 samples of the current response of Fig. 15 for $129T \leq t \leq 256T$.

17. Synthesized current response at the cneter of the dipole activated by the pulse of Fig. 8.

18. DFT amplitude spectrum of 128 samples of the current response of Fig. 17 for $129T \leq t \leq 256T$.

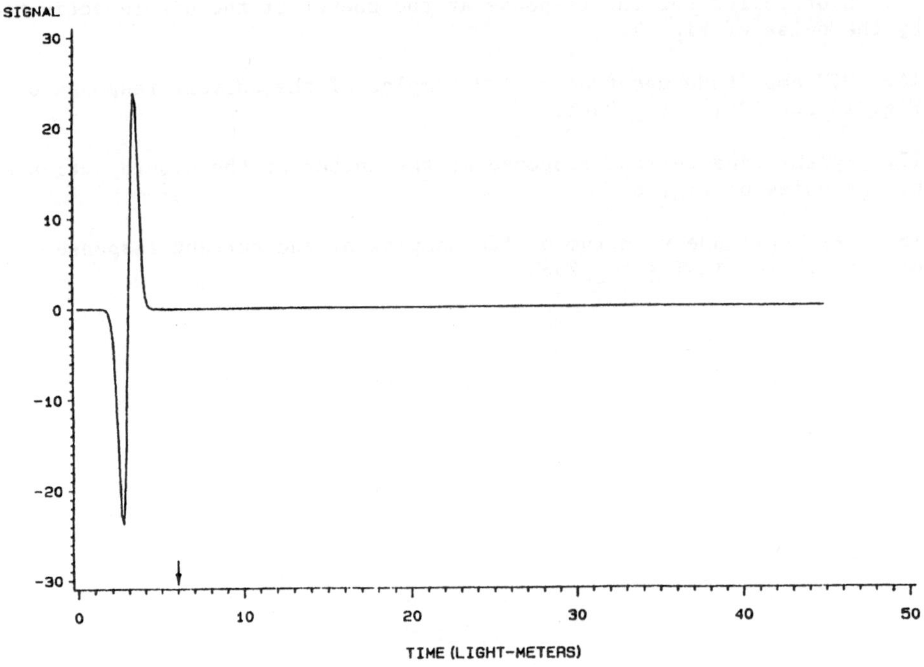


Figure 1.

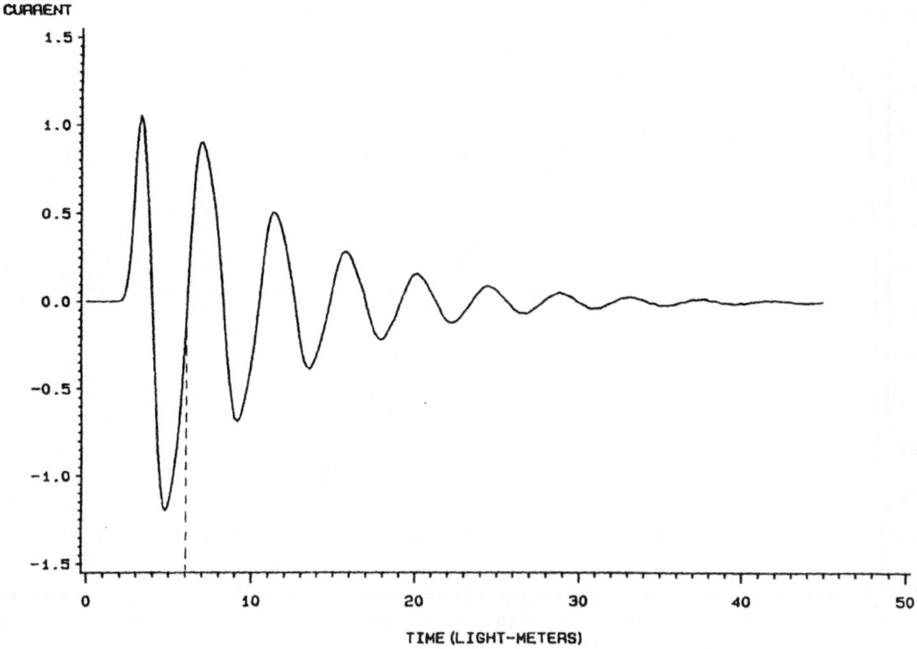


Figure 2.

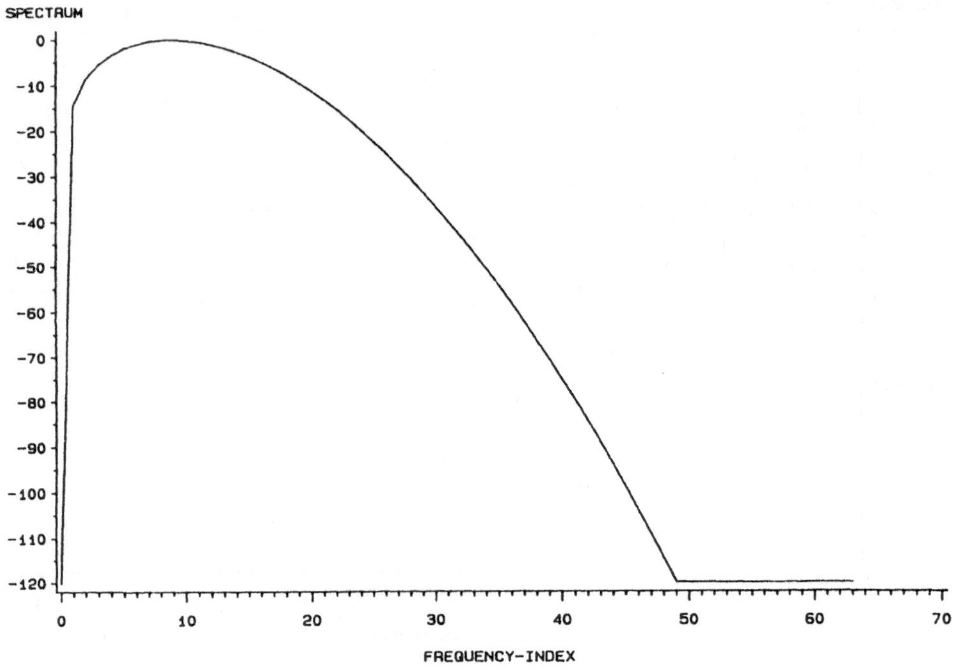
SPECTRUM
0
-10
-20
-30
-40
-50
-60
-70
-80
-90
-100
-110
-120
0
10
20
30
40
50
60
70
FREQUENCY-INDEX

Figure 3.

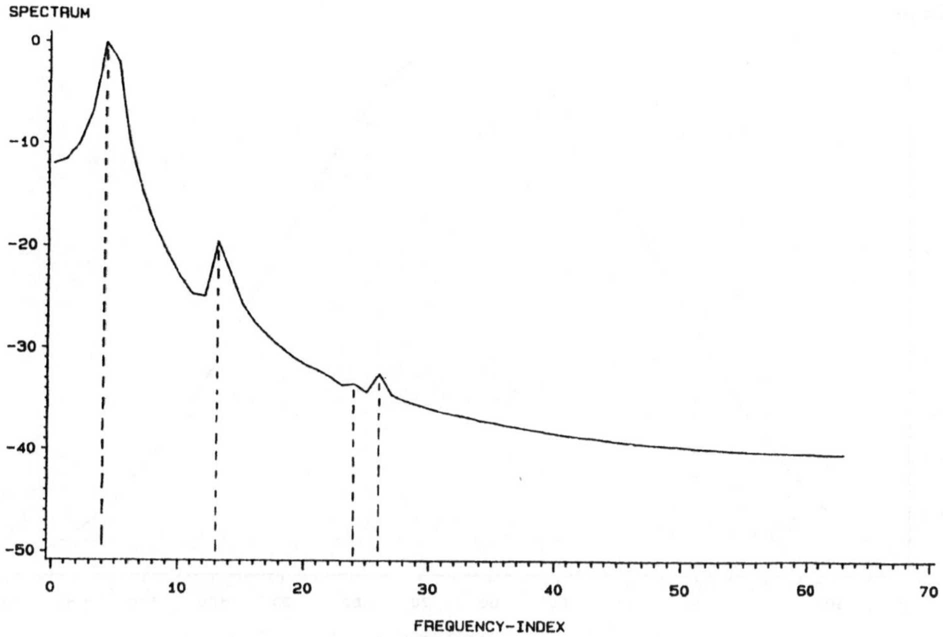


Figure 4.

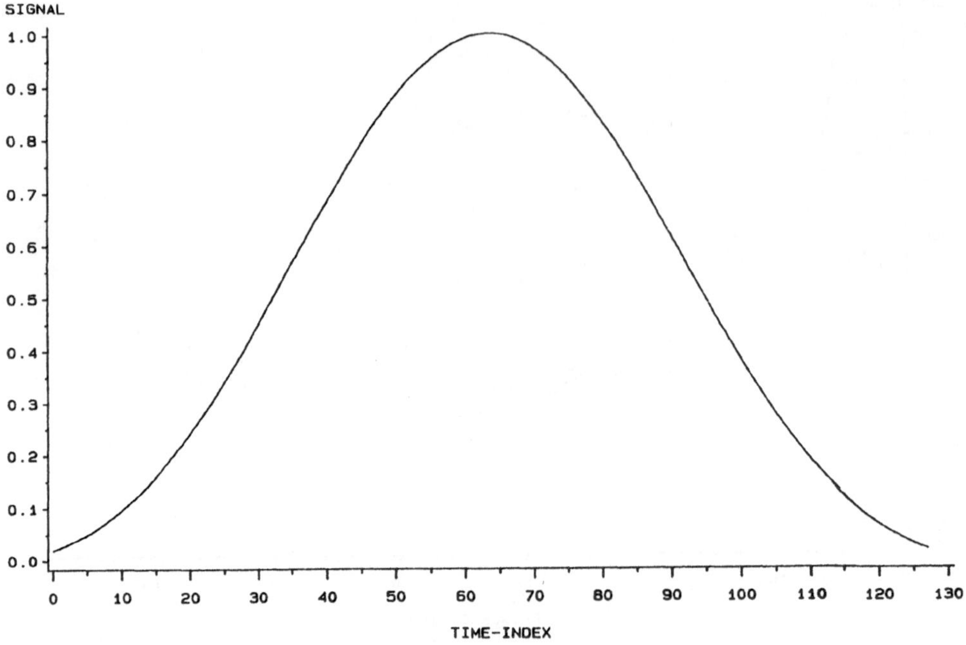
SIGNAL
1.0
0.9
0.8
0.7
0.6
0.5
0.4
0.3
0.2
0.1
0.0
0
10
20
30
40
50
60
70
80
90
100
110
120
130
TIME-INDEX

Figure 5.

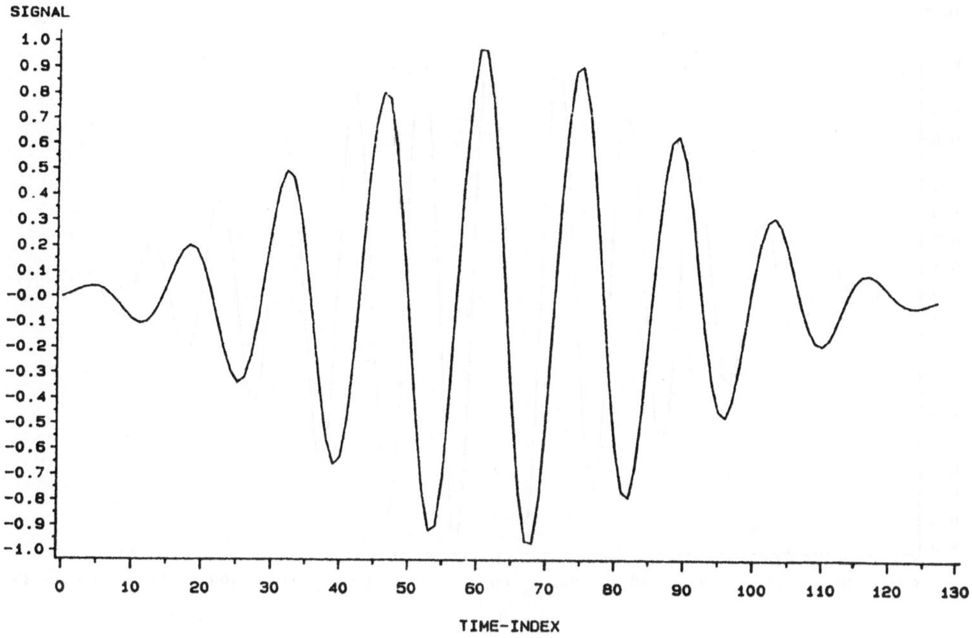
SIGNAL
1.0
0.9
0.8
0.7
0.6
0.5
0.4
0.3
0.2
0.1
-0.0
-0.1
-0.2
-0.3
-0.4
-0.5
-0.6
-0.7
-0.8
-0.9
-1.0
0
10
20
30
40
50
60
70
80
90
100
110
120
130
TIME-INDEX

Figure 6.

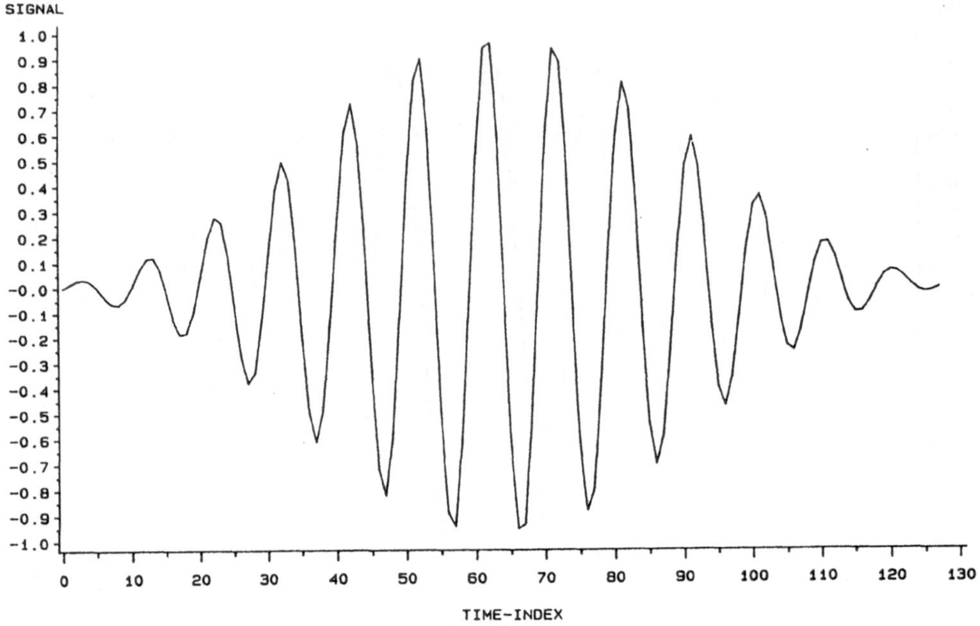
SIGNAL
1.0
0.9
0.8
0.7
0.6
0.5
0.4
0.3
0.2
0.1
-0.0
-0.1
-0.2
-0.3
-0.4
-0.5
-0.6
-0.7
-0.8
-0.9
-1.0
0
10
20
30
40
50
60
70
80
90
100
110
120
130
TIME-INDEX

Figure 7.

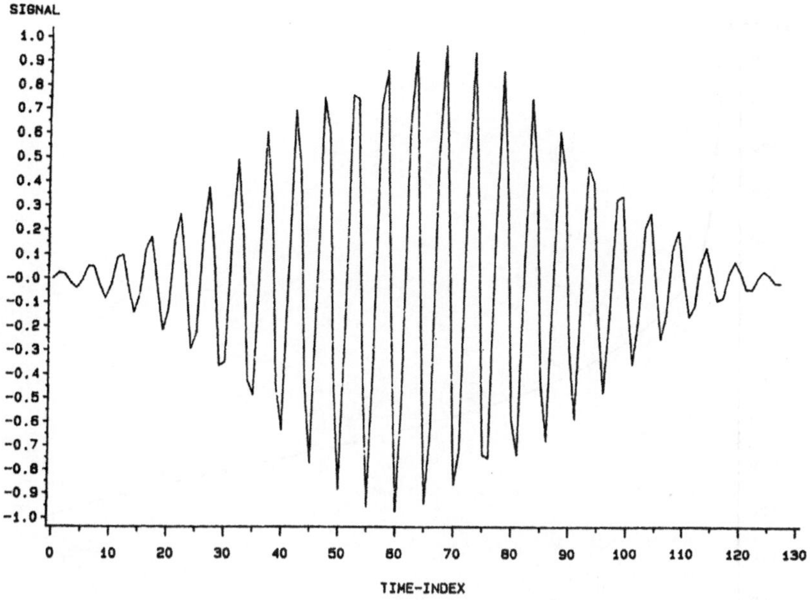
SIGNAL
1.0
0.9
0.8
0.7
0.6
0.5
0.4
0.3
0.2
0.1
-0.0
-0.1
-0.2
-0.3
-0.4
-0.5
-0.6
-0.7
-0.8
-0.9
-1.0
0
10
20
30
40
50
60
70
80
90
100
110
120
130
TIME-INDEX

Figure 8.

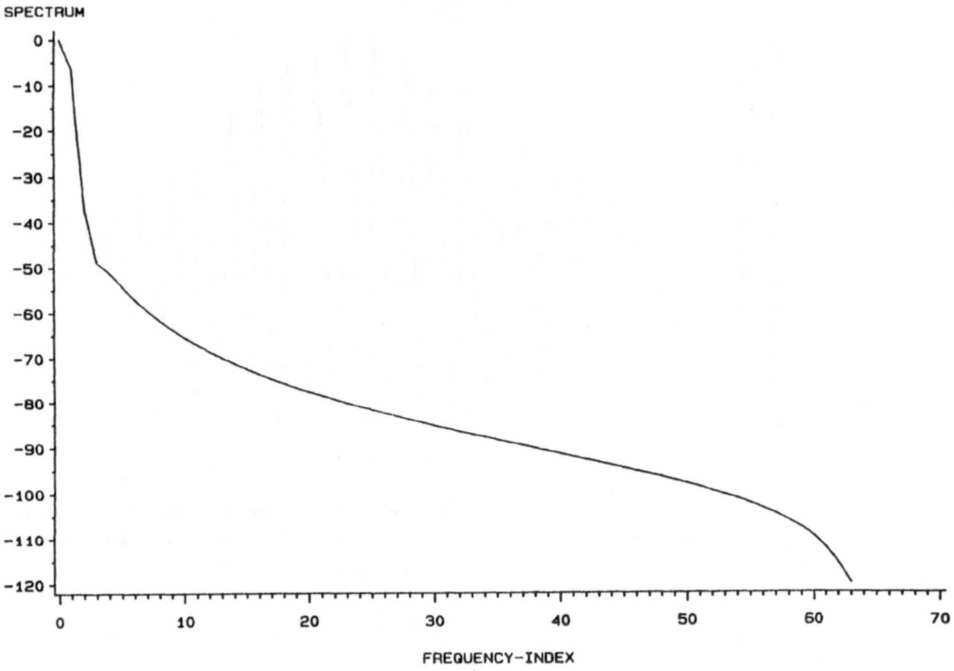
SPECTRUM
0
-10
-20
-30
-40
-50
-60
-70
-80
-90
-100
-110
-120
0
10
20
30
40
50
60
70
FREQUENCY-INDEX

Figure 9.

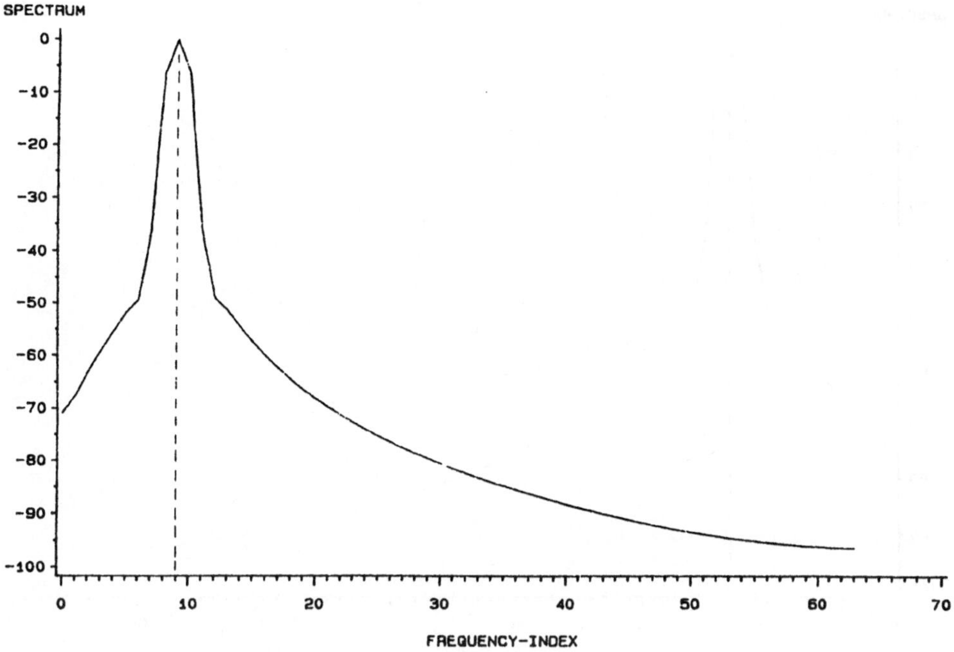
SPECTRUM
0
-10
-20
-30
-40
-50
-60
-70
-80
-90
-100
0
10
20
30
40
50
60
70
FREQUENCY-INDEX

Figure 10.

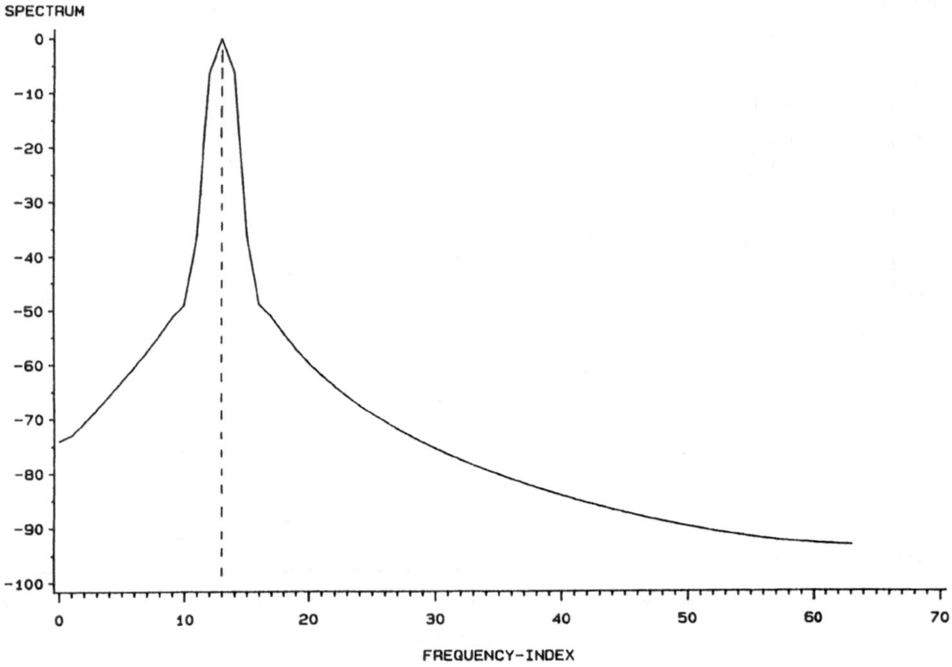
SPECTRUM
0
-10
-20
-30
-40
-50
-60
-70
-80
-90
-100
0
10
20
30
40
50
60
70
FREQUENCY-INDEX

Figure 11.

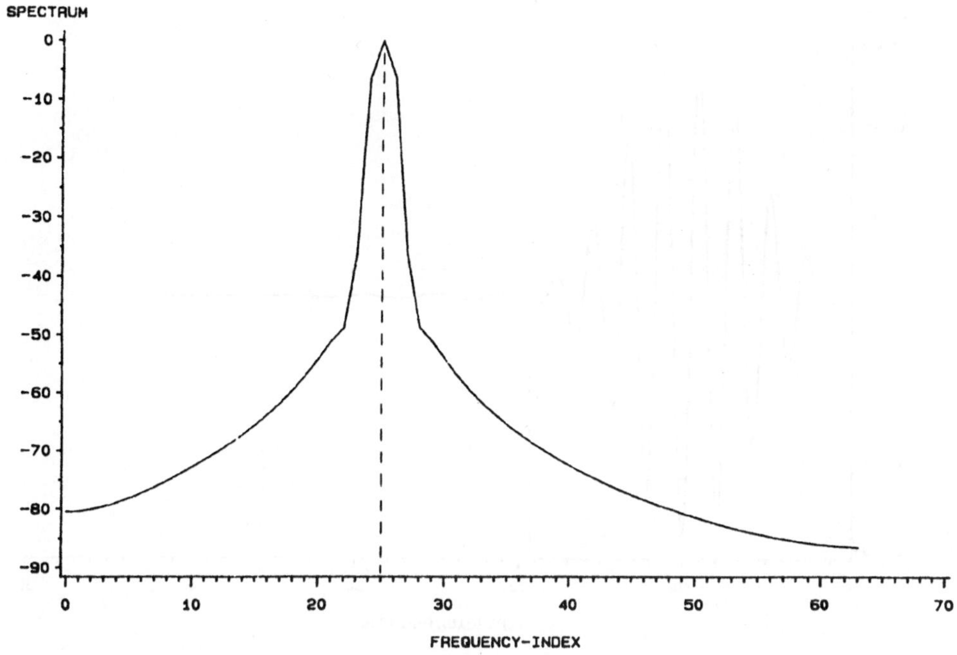
SPECTRUM
0
-10
-20
-30
-40
-50
-60
-70
-80
-90
0
10
20
30
40
50
60
70
FREQUENCY-INDEX

Figure 12.

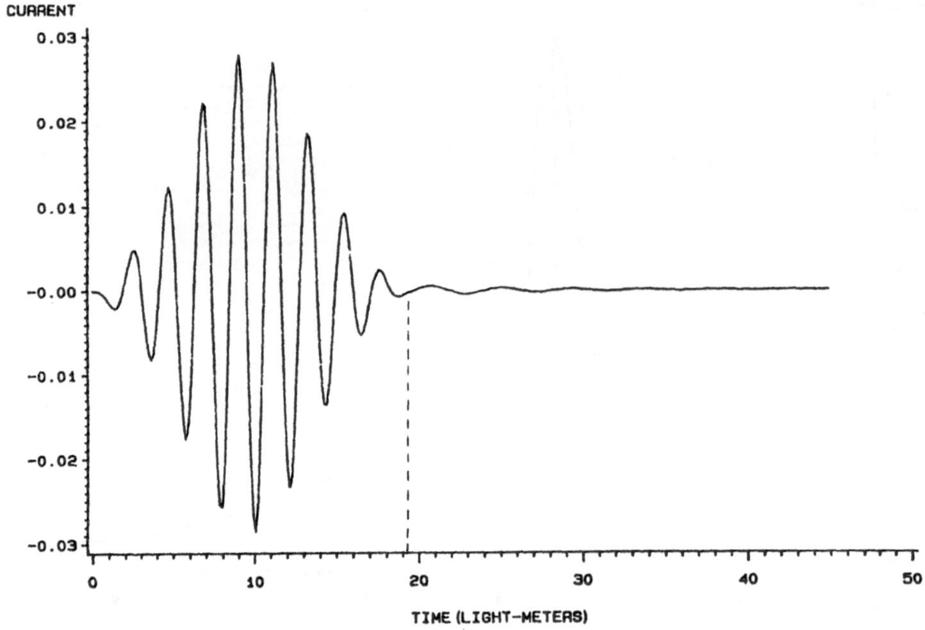
CURRENT
0.03
0.02
0.01
-0.00
-0.01
-0.02
-0.03
0
10
20
30
40
50
TIME (LIGHT-METERS)

Figure 13.

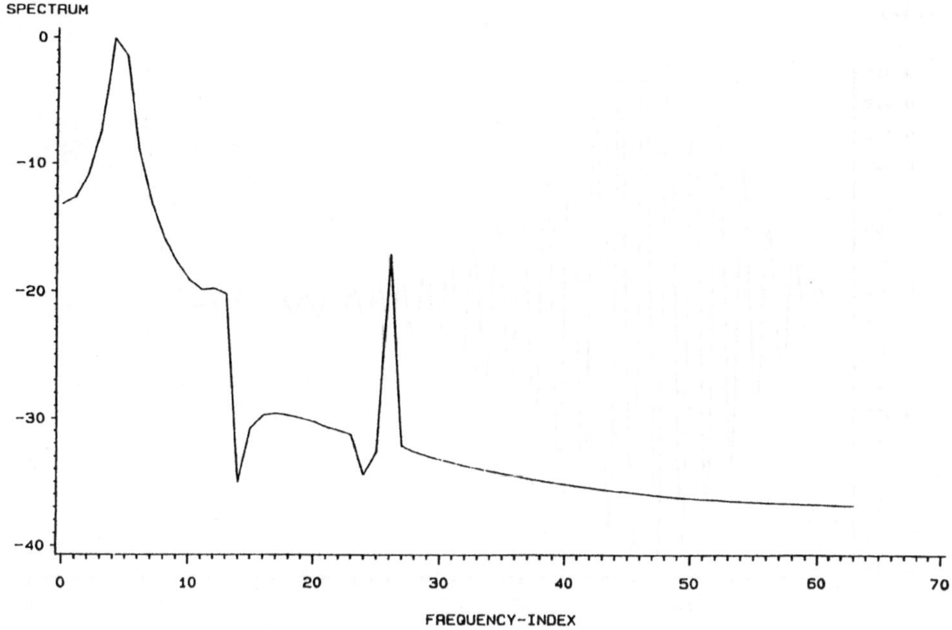


Figure 14.

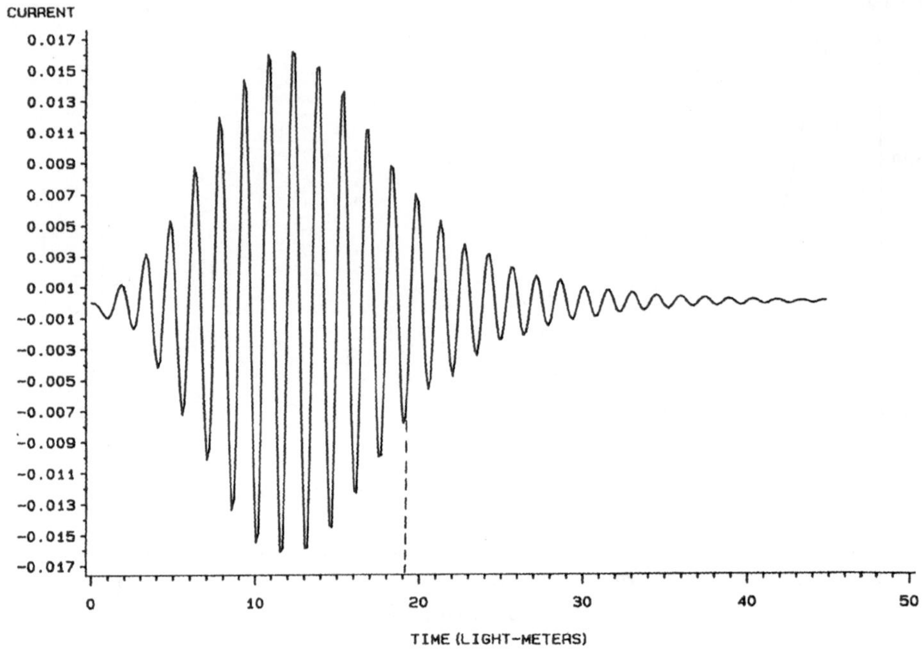
CURRENT
0.017
0.015
0.013
0.011
0.009
0.007
0.005
0.003
0.001
-0.001
-0.003
-0.005
-0.007
-0.009
-0.011
-0.013
-0.015
-0.017
0
10
20
30
40
50
TIME (LIGHT-METERS)

Figure 15.

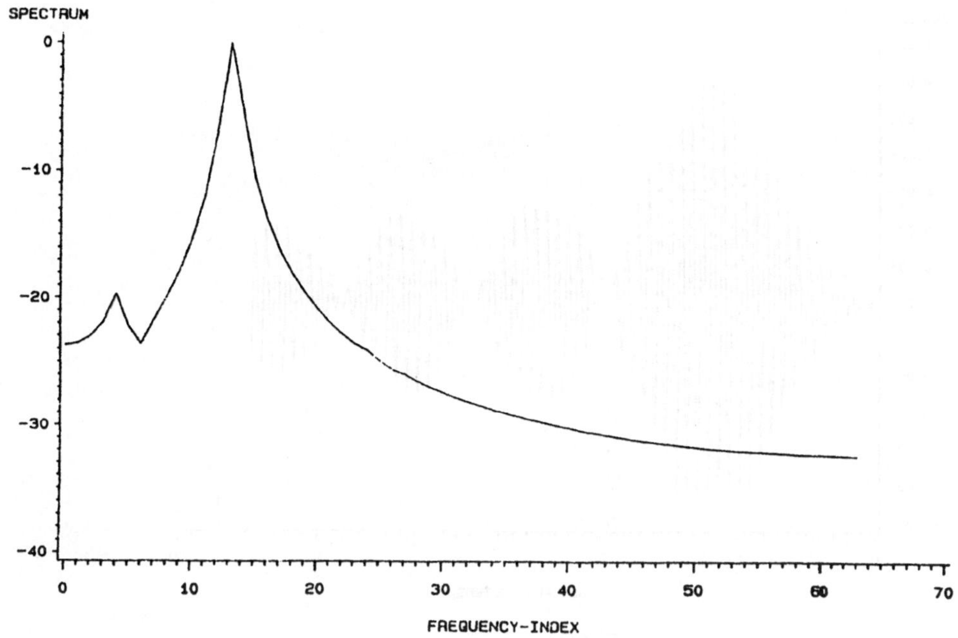


Figure 16.

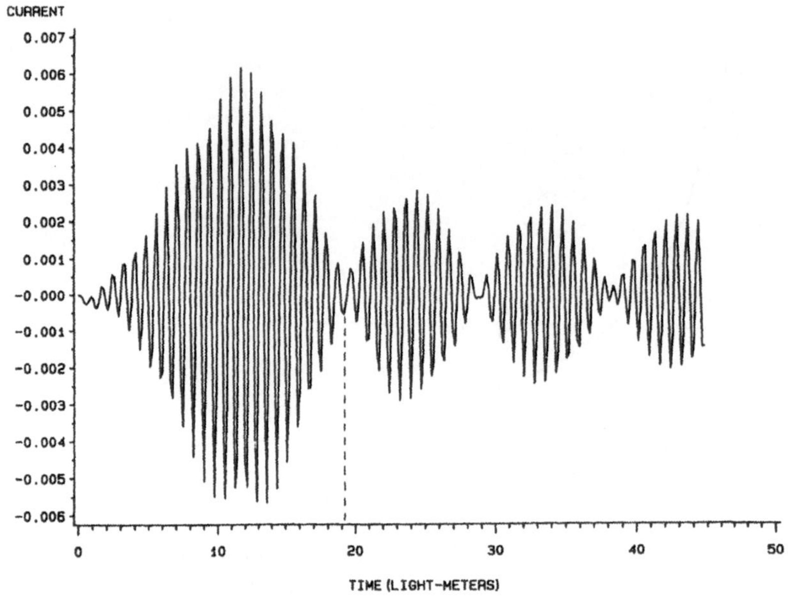
CURRENT
0.007
0.006
0.005
0.004
0.003
0.002
0.001
-0.000
-0.001
-0.002
-0.003
-0.004
-0.005
-0.006
0
10
20
30
40
50
TIME (LIGHT-METERS)

Figure 17.

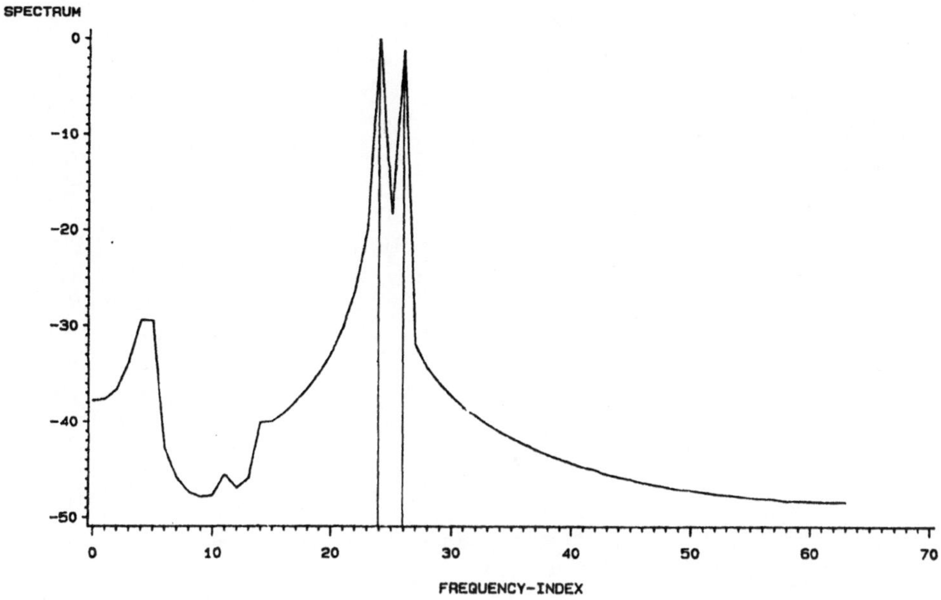


Figure 18.

# SEM AND EEM SCATTERING MATRICES, AND TIME-DOMAIN SCATTERER POLARIZATION IN THE SCATTERING RESIDUE MATRIX

Carl E. Baum
Weapons Laboratory/NTAAB
Kirtland AFB, NM 87117
USA

ABSTRACT. This paper extends the singularity expansion method (SEM) and eigenmode expansion method (EEM) to the representation of the scattering matrix (far field) for a finite size scatterer in free space. Part 1 takes the reciprocity relation between incident and scattered fields and applies it to the EEM and SEM terms for the scattered fields. Besides the symmetry relations due to reciprocity there is found a simple relation relating the backscattering coupling coefficients. At the SEM poles (natural frequencies) the case of non-degenerate modes gives a scattering residue matrix which converts an incoming wave into a scattered wave with complex polarization determined only by the scatterer orientation with respect to the observer of the scattered field. Including a symmetry plane in the scatterer and having directions of incidence and scattering near the symmetry plane, and having suitable polarizations allows one to separate the scatterer natural frequencies into two sets. Part 2 considers the properties of the backscattering residue matrix corresponding to the SEM poles or natural frequencies of a scatterer. For the case of non degenerate natural modes this matrix becomes a single dyad which is characterized by a complex two-component vector. Introducing various symmetries in the scatterer, combined with choice of observer location, gives other special properties to the residue matrix.

## PART 1. SCATTERING, RECIPROCITY, SYMMETRY, EEM, AND SEM

### 1.1 Introduction

In the theory of electromagnetic scattering, reciprocity plays an important role [1-11, 1-12]. As long as the scatterer is perfectly conducting or is comprised of reciprocal media (media characterized by symmetrical constitutive parameter matrices) there is scattering reciprocity. In this context, reciprocity means the equivalence between incident and scattered waves, i.e., a wave incident at some direction and polarization on a scatterer produces a far scattered field at some scattering direction and polarization. The result is the same upon interchange of the directions of incidence and scattering with the polarizations remaining unchanged [1-17 (chap. 2)]. This basic scattering reciprocity is considered via integral equations in sections 1.2 and 1.3.

The scattered fields are expanded in terms of eigenmodes of the integral equation (eigenmode expansion method or EEM) in section 1.4. Here the reciprocity is evident in terms of the various symmetric products involving the eigenmodes which decompose the solution into a sum over the eigenmode index $\beta$. In section 1.5 this is extended to the pole terms in the singularity expansion method (SEM) which further decompose the solution according to the natural-frequency index $\alpha$. Considered on a pole-by-pole basis the residues factor according to terms dependent separately on the various directions (incidence, scattering, and two polarizations). An important result is that the normalized backscattering coupling coefficient is the square of the normalized coupling coefficient for the natural current modes.



*W.-M. Boerner et al. (eds.), Direct and Inverse Methods in Radar Polarimetry, Part 1*, 427–486.

Section 1.6 discusses the application of scattering reciprocity to the forward scattering theorem. Specifically the total scattering (or extinction) of the incident wave is invariant to reversal of the direction of incidence.

If the scatterer has a symmetry plane P, then the geometrical symmetry combines with the reciprocity symmetry to give additional symmetries in the results. Section 1.7 discusses this and introduces the case of mirror scattering (a special case of bistatic scattering). Further specializing the direction of incidence as parallel to P, then defining polarizations parallel and perpendicular to P, the backscattered fields have no cross-polarized components. Furthermore, parallel (or "vertical") polarization only excites one set of natural modes (the symmetric modes), and the scattered fields only contain the corresponding natural frequencies $s_{sy,\alpha'}$. Likewise perpendicular (or horizontal) polarization only excites the remaining set of natural modes (the antisymmetric modes) and the scattered fields only contain the $s_{as,\alpha'}$.

1.2 Surface Current Density on Perfectly Conducting Scatterer

As indicated in Fig. 1.2.1 let us assume there is some perfectly conducting scatterer with volume V and boundary surface S. The surface current density $\vec{J}_S$ on the scatterer is related to the incident electric field $\vec{E}^{(inc)}$ through the impedance or E-field integral equation [1-4, 1-6] as

$$\tilde{\vec{E}}^{(inc)}(\vec{r}_S,s) = \left\langle \tilde{\vec{\vec{Z}}}_t(\vec{r}_S,\vec{r}_S\,';s);\tilde{\vec{J}}_S(\vec{r}_S\,';s)\right\rangle$$

$$\vec{r}_S,\vec{r}_S\,' \in S$$

$$\tilde{\vec{\vec{Z}}}_t(\vec{r}_S,\vec{r}_S\,';s) = \vec{\vec{1}}_t(\vec{r}_S)\cdot\tilde{\vec{\vec{Z}}}(\vec{r}_S,\vec{r}_S\,';s)\cdot\vec{\vec{1}}_t(\vec{r}_S\,')$$

$$= -s\mu_o\vec{\vec{1}}_t(\vec{r}_S)\cdot\tilde{\vec{\vec{G}}}_o(\vec{r}_S,\vec{r}_S\,';s)\cdot\vec{\vec{1}}_t(\vec{r}_S\,')$$

$$= -\frac{Z_o\gamma^2}{4\pi}\vec{\vec{1}}_t(\vec{r}_S)\cdot\left\{\left[-2\zeta^{-3}-2\zeta^{-2}\right]e^{-\zeta}\vec{1}_R\vec{1}_R\right.$$

$$\left.+\left[\zeta^{-3}+\zeta^{-2}+\zeta^{-1}\right]e^{-\zeta}\left[\vec{\vec{1}}-\vec{1}_R\vec{1}_R\right]\right\}\cdot\vec{\vec{1}}_t(\vec{r}_S\,')$$

$$R \equiv \left|\vec{r}_S-\vec{r}_S\,'\right|$$

$$\vec{1}_R = \frac{\vec{r}_S-\vec{r}_S\,'}{\left|\vec{r}_S-\vec{r}_S\,'\right|} \quad (for\ \vec{r}_S \neq \vec{r}_S\,')$$

$$\zeta \equiv \gamma R \qquad (1.2.1)$$

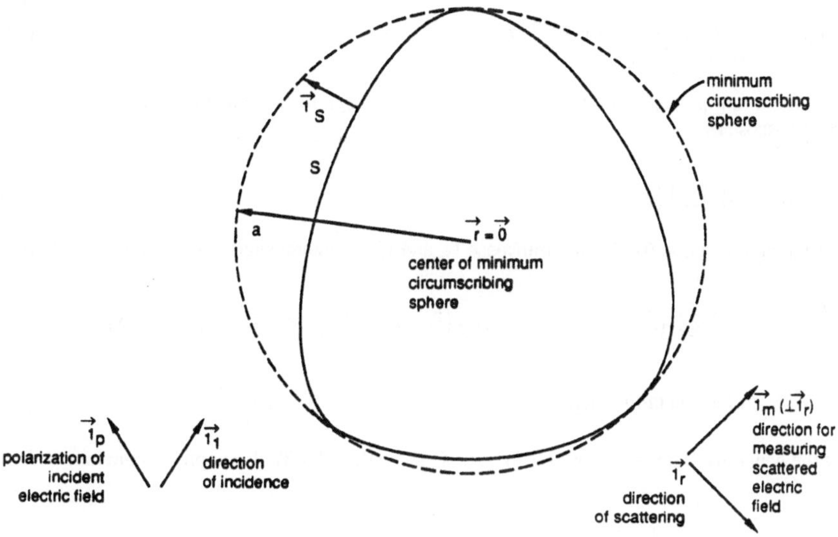


Figure 1.2.1. Incident and Scattered Waves.

$$\vec{\vec{1}} \equiv \vec{1}_x \vec{1}_x + \vec{1}_y \vec{1}_y + \vec{1}_z \vec{1}_z \quad \text{(identity dyad)}\,.$$

$$\vec{\vec{1}}_t(\vec{r}_S) \equiv \vec{\vec{1}} - \vec{1}_S(\vec{r}_S)\vec{1}_S(\vec{r}_S) \quad \text{(transverse dyad to S at } \vec{r}_S)$$

$$\vec{1}_S(\vec{r}_S) \equiv \text{ outward unit normal to S at } \vec{r}_S$$

$$\gamma = \frac{s}{c} \quad \text{(propagation constant)}$$

$$s = \Omega + j\omega \quad \text{(complex frequency)}$$

$$c = (\mu_0 \epsilon_0)^{-\frac{1}{2}} \quad \text{(speed of light)}$$

$$Z_0 = \left(\frac{\mu_0}{\epsilon_0}\right)^{\frac{1}{2}}$$

$\sim \equiv$ Laplace transform (two-sided)

Note that the delta function at $\vec{r}_S = \vec{r}\,'_S$ has been removed by the transverse dyads [1-13]. The formal solution to the integral equation is, of course,

$$\vec{\vec{J}}_s(\vec{r}_s,s) = \left\langle \tilde{\vec{\vec{Z}}}_t^{-1}(\vec{r}_s,\vec{r}_s\,';s);\ \vec{\vec{E}}^{(inc)}\,\vec{r}_s\,';s \right\rangle \tag{1.2.2}$$

An important property of this impedance kernel is its symmetry which takes the form of a generalized transpose as

$$\tilde{\vec{\vec{Z}}}_t(\vec{r}_s,\vec{r}_s\,';s) = \tilde{\vec{\vec{Z}}}_t^{T}(\vec{r}_s\,',\vec{r}_s;s) \tag{1.2.3}$$

i.e, besides taking the matrix (or dyadic) transpose $\vec{r}_s$ and $\vec{r}_s\,'$ are interchanged. The inverse is defined via

$$\left\langle \tilde{\vec{\vec{Z}}}_t(\vec{r}_s,\vec{r}_s\,'';s);\ \tilde{\vec{\vec{Z}}}_t^{-1}(\vec{r}_s\,'',\vec{r}_s\,';s)\right\rangle = \vec{\vec{1}}_t(\vec{r}_s)\,\delta_s(\vec{r}_s-\vec{r}_s\,') = \left\langle \tilde{\vec{\vec{Z}}}_t^{-1}(\vec{r}_s,\vec{r}_s\,'';s);\ \tilde{\vec{\vec{Z}}}_t(\vec{r}_s\,'',\vec{r}_s\,';s)\right\rangle$$

$$\delta_s \equiv \text{surface delta function} \tag{1.2.4}$$

By taking the generalized transpose of (1.2.4) one can show that (1.2.3) also applies to the inverse kernel, i.e.,

$$\tilde{\vec{\vec{Z}}}_t^{-1}(\vec{r}_s,\vec{r}_s\,';s) = \tilde{\vec{\vec{Z}}}_t^{-1^T}(\vec{r}_s\,',\vec{r}_s;s) \tag{1.2.5}$$

This strictly applies only where the inverse exists, i.e., for $(\vec{r}_s \neq \vec{r}_s\,')$ and away from natural frequencies $s_\alpha$.

This concept of symmetry in the kernel is very important, leading to the idea of reciprocity [1-10, 1-11, 1-12] which will be used later. While the formulation here is in terms of a perfectly conducting body it also applies to any body comprised of reciprocal media, even anisotropic media. One can have sheet impedances on s or volume distributions of $\vec{\vec{\epsilon}}$, $\vec{\vec{\mu}}$ and $\vec{\vec{\sigma}}$ (as long as these dyadic functions of space and frequency are symmetric, i.e., reciprocal). In the volume case the surface integrals are replaced by volume integrals.

Now diagonalize $\tilde{\vec{\vec{Z}}}_t$ (i.e., expand in eigenmodes) via

$$\left\langle \tilde{\vec{\vec{Z}}}_t(\vec{r}_s,\vec{r}_s\,';s);\vec{\vec{j}}_{s\beta}(\vec{r}_s\,',s)\right\rangle = \tilde{Z}_\beta(s)\,\vec{\vec{j}}_{s\beta}(\vec{r}_s;s)$$

$$= \left\langle \vec{\vec{j}}_{s\beta}(\vec{r}_s\,',s);\ \tilde{\vec{\vec{Z}}}_t(\vec{r}_s\,',\vec{r}_s;s)\right\rangle$$

$$\tilde{Z}_\beta(s) \equiv \text{eigenimpedances}$$

$$\vec{\vec{j}}_{s\beta}(\vec{r}_s,s) \equiv \text{eigenmodes}$$

$$\left\langle \vec{\vec{j}}_{s\beta_1}(\vec{r}_s\,',s);\vec{\vec{j}}_{s\beta_2}(\vec{r}_s\,',s)\right\rangle = 1_{\beta_1,\beta_2} \ \text{(biorthonormal)} \tag{1.2.6}$$

This gives

$$\tilde{\tilde{Z}}_t(\vec{r}_s,\vec{r}_s\,';s) = \sum_{\beta} \tilde{Z}_\beta(s)\tilde{\vec{j}}_{s\beta}(\vec{r}_s,s)\tilde{\vec{j}}_{s\beta}(\vec{r}_s\,',s)$$

$$\tilde{\tilde{Z}}_t^{-1}(\vec{r}_s,\vec{r}_s\,';s) = \sum_{\beta} \tilde{Z}_\beta^{-1}(s)\tilde{\vec{j}}_{s\beta}(\vec{r}_s,s)\tilde{\vec{j}}_{s\beta}(\vec{r}_s\,',s)$$

$$\vec{\vec{1}}_t(\vec{r}_s)\delta_s(\vec{r}_s-\vec{r}_s\,') = \sum_{\beta} \tilde{\vec{j}}_{s\beta}(\vec{r}_s,s)\tilde{\vec{j}}_{s\beta}(\vec{r}_s\,',s) \tag{1.2.7}$$

The symmetry (generalized transpose) here is apparent. The surface current density is now

$$\tilde{\vec{J}}_s(\vec{r}_s,s) = \sum_{\beta} \tilde{Z}_\beta^{-1}(s)\left\langle \tilde{\vec{E}}^{(inc)}(\vec{r}_s\,',s);\tilde{\vec{j}}_{s\beta}(\vec{r}_s\,',s)\right\rangle \tilde{\vec{j}}_{s\beta}(\vec{r}_s,s) \tag{1.2.8}$$

The SEM form of the solution is [1-4, 1-16]

$$\begin{aligned}
\tilde{\vec{J}}_s(\vec{r}_s,s) &= E_o \sum_{\alpha} \tilde{f}(s_\alpha)\eta_\alpha \vec{j}_{s_\alpha}(\vec{r}_s)[s-s_\alpha]^{-1} \\
&\quad + \text{singularities of } \tilde{f}(s) \\
&\quad + \text{possible entire function} \\
\tilde{f}(s) &= \text{incident waveform (Laplace transformed)} \\
\left\langle \tilde{\tilde{Z}}(\vec{r}_s,\vec{r}_s\,';s_\alpha);\vec{j}_{s_\alpha}(\vec{r}_s\,')\right\rangle &= \vec{0} \\
&= \left\langle \vec{j}_{s_\alpha}(\vec{r}_s\,');\tilde{\tilde{Z}}(\vec{r}_s\,',\vec{r}_s;s_\alpha)\right\rangle \\
s_\alpha &\equiv \text{natural frequency} \\
\vec{j}_{s_\alpha}(\vec{r}_s) &\equiv \text{natural mode} \\
\eta_\alpha &= \frac{E_o^{-1}\tilde{f}^{-1}(s_\alpha)\left\langle \tilde{\vec{E}}^{(inc)}(\vec{r}_s\,',s_\alpha);\vec{j}_{s_\alpha}(\vec{r}_s\,')\right\rangle}{\left\langle \vec{j}_{s_\alpha}(\vec{r}_s);\frac{\partial}{\partial s}\tilde{\tilde{Z}}_t(\vec{r}_s,\vec{r}_s\,';s)\Big|_{s=s_\alpha};\vec{j}_{s_\alpha}(\vec{r}_s\,')\right\rangle} \\
&\equiv \text{coupling coefficient} \\
E_o &\equiv \text{scaling constant (volts / meter) for incident field}
\end{aligned} \tag{1.2.9}$$

Note that coupling coefficients here are taken with no s dependence, i.e., class 1.

Now let the incident field be a plane wave with electric field as

$$\tilde{\vec{E}}^{(inc)}(\vec{r},s) = E_o \tilde{f}(s)\, \vec{1}_p\, e^{-\gamma \vec{1}_1 \cdot \vec{r}}$$

$$\vec{1}_1 \equiv \text{direction of incidence}$$

$$\vec{1}_p \equiv \text{polarization}$$

$$\vec{1}_1 \cdot \vec{1}_p = 0 \tag{1.2.10}$$

As in fig. 1.2.1, phase or arrival time is chosen by the location of coordinate center $\vec{r} = \vec{0}$, here taken for general finite sized scatterers as the center of the minimum circumscribing sphere of radius a [1-6, 1-14]. The surface current density is then

$$\tilde{\vec{J}}_s(\vec{r}_s, s) = E_o \tilde{f}(s) \left\langle \tilde{\vec{\vec{Z}}}_t^{-1}(\vec{r}_s, \vec{r}_s'; s);\, \vec{1}_p\, e^{-\gamma \vec{1}_1 \cdot \vec{r}_s'} \right\rangle$$

$$= E_o \tilde{f}(s) \left\langle \tilde{\vec{\vec{Z}}}_t^{-1}(\vec{r}_s, \vec{r}_s'; s), e^{-\gamma \vec{1}_1 \cdot \vec{r}_s'} \right\rangle \cdot \vec{1}_p$$

$$= E_o \tilde{f}(s) \vec{1}_p \cdot \left\langle e^{-\gamma \vec{1}_1 \cdot \vec{r}_s'}, \tilde{\vec{\vec{Z}}}_t^{-1}(\vec{r}_s', \vec{r}_s; s) \right\rangle \tag{1.2.11}$$

noting the symmetry of the kernel. In the EEM form this is

$$\tilde{\vec{J}}_s(\vec{r}_s, s) = E_o \tilde{f}(s) \sum_\beta \tilde{Z}_\beta^{-1}(s) \left[ \vec{1}_p \cdot \left\langle e^{-\gamma \vec{1}_1 \cdot \vec{r}_s'}, \tilde{\vec{j}}_{s_\beta}(\vec{r}_s', s) \right\rangle \right] \tilde{\vec{j}}_{s_\beta}(\vec{r}_s, s) \tag{1.2.12}$$

The SEM form is

$$\tilde{\vec{J}}_s(\vec{r}_s, s) = E_o \sum_\alpha \tilde{f}(s_\alpha)\, \eta_\alpha(\vec{1}_1, \vec{1}_p) \vec{j}_{s_\alpha}(\vec{r}_s) [s - s_\alpha]^{-1}$$

+ singularities of $\tilde{f}(s)$

+ possible entire function

$$\eta_\alpha(\vec{1}_1, \vec{1}_p) = \frac{\vec{1}_p \cdot \left\langle e^{-\gamma_\alpha \vec{1}_1 \cdot \vec{r}_s'}, \vec{j}_{s_\alpha}(\vec{r}_s') \right\rangle}{\left\langle \vec{j}_{s_\alpha}(\vec{r}_s); \frac{\partial}{\partial s} \tilde{\tilde{Z}}_t(\vec{r}_s, \vec{r}_s'; s)\Big|_{s=s_\alpha}; \vec{j}_{s_\alpha}(\vec{r}_s') \right\rangle}$$

$$\gamma_\alpha \equiv \frac{s_\alpha}{c} \tag{1.2.13}$$

Noting that $\vec{1}_p$ is always orthogonal to $\vec{1}_1$ we have a transverse dyad $\vec{1}_1$ with the property

$$\vec{\vec{1}}_1 \equiv \vec{\vec{1}} - \vec{1}_1 \vec{1}_1 \equiv \text{incident transverse dyad}$$

$$\vec{\vec{1}}_1 \cdot \vec{1}_p = \vec{1}_p, \vec{1}_1 \cdot \vec{\vec{1}}_1 = \vec{0} \tag{1.2.14}$$

This transverse dyad can be included wherever $\vec{1}_p$ is used. Since it is only a function of $\vec{1}_1$ then including it in integrals where $\vec{1}_1$ is already present it does not complicate matters. It can be thought to simplify things by removing the components "parallel" to $\vec{1}_1$ which don't enter the answer anyway. Including this we have alternate forms for (1.2.11) in regular form

$$\tilde{\vec{J}}_s(\vec{r}_s, s) = E_o \tilde{f}(s) \left\langle \tilde{\tilde{Z}}_t^{-1}(\vec{r}_s, \vec{r}_s'; s); \vec{\vec{1}}_1 e^{-\gamma \vec{1}_1 \cdot \vec{r}_s'} \right\rangle \cdot \vec{1}_p$$

$$= E_o \tilde{f}(s) \vec{1}_p \cdot \left\langle \vec{\vec{1}}_1 e^{-\gamma \vec{1}_1 \cdot \vec{r}_s'}; \tilde{\tilde{Z}}_t^{-1}(\vec{r}_s', \vec{r}_s; s) \right\rangle \tag{1.2.15}$$

for (1.2.12) in EEM form

$$\tilde{\vec{J}}_s(\vec{r}_s, s) = E_o \tilde{f}(s) \sum_\beta \tilde{Z}_\beta^{-1}(s) \left[ \vec{1}_p \cdot \left\langle \vec{\vec{1}}_1 e^{-\gamma \vec{1}_1 \cdot \vec{r}_s'}; \vec{j}_{s_\beta}(\vec{r}_s', s) \right\rangle \right] \vec{j}_{s_\beta}(\vec{r}_s, s) \tag{1.2.16}$$

for (1.2.13) in SEM form

$$\eta_\alpha(\vec{1}_1,\vec{1}_p) = \frac{\vec{1}_p\cdot\left\langle \vec{1}_1 e^{-\gamma_\alpha \vec{1}_1\cdot\vec{r}_s'};\vec{j}_{s_\alpha}(\vec{r}_s')\right\rangle}{\left\langle \vec{j}_{s_\alpha}(\vec{r}_s);\frac{\partial}{\partial s}\tilde{\tilde{Z}}_t(\vec{r}_s,\vec{r}_s';s)\Big|_{s=s_\alpha};\vec{j}_{s_\alpha}(\vec{r}_s')\right\rangle} \tag{1.2.17}$$

Note that $\vec{1}_1$ just weights the integral over the surface-current-density modes to remove any $\vec{1}_1$ component.

## 1.3 Scattered Fields

The far scattered electric field is calculated from the surface current density as [1-1]

$$\tilde{\vec{E}}_f(\vec{r},s) = \frac{-s\mu_0 e^{-\gamma r}}{4\pi r}\left\langle \vec{1}_r e^{-\gamma\vec{1}_r\cdot\vec{r}_s};\tilde{\vec{J}}_s(\vec{r}_s,s)\right\rangle$$

$$\vec{1}_r \equiv \vec{1} - \vec{1}_r\vec{1}_r \equiv \text{scattered transverse dyad}$$

$$\vec{1}_r\cdot\vec{1}_m = \vec{1}_m,\ \vec{1}_r\cdot\vec{1}_r = \vec{0} \tag{1.3.1}$$

Note again as in fig. 1.2.1 that the coordinate center has to be specified, here as the center of the minimum circumscribing sphere. Note that (1.3.1) is for the far field, i.e., the leading $r^{-1}$ term as $r\to\infty$ with the property that

$$\vec{1}_r\cdot\tilde{\vec{E}}_f(\vec{r},s) = 0 \tag{1.3.2}$$

As indicated in fig. 1.2.1 we have

$\vec{1}_r \equiv$ scattering direction

$\vec{1}_m \equiv$ direction for measuring (sampling) scattered electric field

$$\vec{1}_r\cdot\vec{1}_m = 0 \qquad \text{(constraint on } \vec{1}_m\text{)} \tag{1.3.3}$$

Note that we also then have

$$\vec{1}_m \cdot \vec{E}_f(\vec{r}, s) = \frac{-s\mu_o e^{-\gamma r}}{4\pi r} \vec{1}_m \cdot \left\langle \vec{1}_r e^{\gamma \vec{1}_r \cdot \vec{r}_s} ; \vec{J}_s(\vec{r}_s, s) \right\rangle$$

$$= \frac{-s\mu_o e^{-\gamma r}}{4\pi r} \vec{1}_m \cdot \left\langle e^{\gamma \vec{1}_r \cdot \vec{r}_s}, \vec{J}_s(\vec{r}_s, s) \right\rangle \tag{1.3.4}$$

Combining the far field results with (1.2.15) for the surface current density gives

$$\vec{1}_m \cdot \vec{E}_f(\vec{r}, s) = \frac{-E_o \tilde{f}(s) s\mu_o e^{-\gamma r}}{4\pi r} \vec{1}_m \cdot \left\langle \vec{1}_r e^{\gamma \vec{1}_r \cdot \vec{r}_s} ; \tilde{\tilde{Z}}_t^{-1}(\vec{r}_s, \vec{r}_s\,'; s); \vec{1}_1 e^{-\gamma \vec{1}_1 \cdot \vec{r}_s\,'} \right\rangle \cdot \vec{1}_p$$

$$= \frac{-E_o \tilde{f}(s) s\mu_o e^{-\gamma r}}{4\pi r} \vec{1}_p \cdot \left\langle \vec{1}_r e^{-\gamma \vec{1}_r \cdot \vec{r}_s} ; \tilde{\tilde{Z}}_t^{-1}(\vec{r}_s, \vec{r}_s\,'; s); \vec{1}_r e^{\gamma \vec{1}_r \cdot \vec{r}_s\,'} \right\rangle \cdot \vec{1}_m \tag{1.3.5}$$

Note the symmetry (reciprocity) in this formula where we have used the generalized transpose result in (1.2.5) for the inverse kernel. Here we have the result that the interchange

$$\left(\vec{1}_1, \vec{1}_p\right) \leftrightarrow \left(-\vec{1}_r, \vec{1}_m\right) \tag{1.3.6}$$

leaves the result unchanged. Referring to fig. 1.2.1, the scattering reciprocity result is that an incident plane wave characterized by $\left(\vec{1}_1, \vec{1}_p\right)$ giving a scattered by $\left(\vec{1}_r, \vec{1}_m\right)$ at a distance r implies that a second incident wave characterized by $\left(-\vec{1}_r, \vec{1}_m\right)$ gives the same scattered field with directions $\left(-\vec{1}_1, \vec{1}_p\right)$ at the same distance r.

Note that this symmetry is merely a result of reciprocity, the generalized transpose symmetry of the operator, including any reciprocal loading of the scatterer. While the foregoing consider scattering of a plane wave, the result is even more general. One could have any two (reciprocal) antennas at distances $r_1$ and $r_2$ from the scattering coordinate center. The ratio of received (open-circuit) voltage at one antenna to source current into the other antenna is independent of interchange of roles of transmitting and receiving antennas. This is one form of the reciprocity theorem and is often termed reaction [1-10].

A special case of (1.3.5) is that of backscattering with backscattered electric field measured in the same direction as the incident field so that

$$\vec{1}_r \equiv -\vec{1}_1, \quad \vec{1}_m \equiv \vec{1}_p \tag{1.3.7}$$

In this case (1.3.5) becomes

$$\vec{1}_p \cdot \tilde{\vec{E}}_f(\vec{r},s) = \frac{-E_o \tilde{f}(s) s \mu_o}{4\pi r} \vec{1}_p \cdot \left\langle \vec{1}_1 e^{-\gamma \vec{1}_1 \cdot \vec{r}_s} ; \tilde{\vec{Z}}_t^{-1}(\vec{r}_s, \vec{r}_s'; s); \vec{1}_1 e^{-\gamma \vec{1}_1 \cdot \vec{r}_s'} \right\rangle \cdot \vec{1}_p \tag{1.3.8}$$

which might be termed self-reciprocity, i.e., the incident-to-scattered-field transfer function is not only the same on interchange of incident and scattered (reciprocity), but also in the case of (1.3.7) in effect these are in some sense the same thing, (i.e., which is called incident and which is called scattered). Note that since this reciprocity applies to all frequencies, then by inverse Laplace (2-sided) or Fourier transformation it applies to arbitrary transient waveforms as well, the scattering taking the form of a convolution operator.

Another special case of (1.3.5) is cross-polarized backscattering so that

$$\vec{1}_r \equiv -\vec{1}_1, \quad \vec{1}_m \cdot \vec{1}_p = 0 \tag{1.3.9}$$

In this case we have from (1.3.5) that incident with polarization $\vec{1}_p$ into backscattering with polarization $\vec{1}_m$ is the same as with $\vec{1}_p$ and $\vec{1}_m$ interchanged. This can be interpreted as a symmetric backscattering matrix.

While the reciprocity here is stated in terms of electric fields and currents, one could consider source currents at one antenna producing magnetic fields at the other. With a sign interchange there is a similar reciprocity relation (or "magnetic reaction") as discussed in [1.8]. Furthermore, while these results have been stated in terms of (complex) frequency domain, since they apply for all frequencies they apply for time-domain waveforms as well (i.e., consider the inverse Laplace transform).

## 1.4 EEM Scattered Fields

Applying (1.3.5) in the context of EEM as in (1.2.8) and (1.2.16) gives

$$\vec{1}_m \cdot \tilde{\vec{E}}_p(\vec{r},s) = \frac{-E_o \tilde{f}(s) s \mu_o e^{-\gamma r}}{4\pi r} \sum_\beta \tilde{Z}_\beta^{-1}(s) \vec{1}_m \cdot \tilde{\vec{C}}_{f_\beta}(\vec{1}_r, \vec{1}_1; s) \cdot \vec{1}_p$$

$$\tilde{\vec{C}}_{f_\beta}(\vec{1}_r, \vec{1}_1; s) = \tilde{\vec{C}}_{r_\beta}(\vec{1}_r, s)\, \tilde{\vec{C}}_\beta(\vec{1}_1, s)$$

$$\tilde{\vec{C}}_\beta(\vec{1}_1, s) = \left\langle \vec{1}_1 e^{-\gamma \vec{1}_1 \cdot \vec{r}_s'} ; \tilde{\vec{j}}_{s_\beta}(\vec{r}_s', s) \right\rangle$$

$$\tilde{\vec{C}}_{r_\beta}(\vec{1}_r, s) = \left\langle \vec{1}_r e^{\gamma \vec{1}_r \cdot \vec{r}_s'} ; \tilde{\vec{j}}_{s_\beta}(\vec{r}_s, s) \right\rangle$$

$$\vec{1}_m \cdot \tilde{\vec{C}}_{f\beta}\left(\vec{1}_r, \vec{1}_1; s\right) \cdot \vec{1}_p = \left[\vec{1}_m \cdot \vec{C}_{r_\beta}\left(\vec{1}_r, s\right)\right]\left[\vec{1}_p \cdot \vec{C}_\beta\left(\vec{1}_1, s\right)\right]$$

$$= \left[\vec{1}_m \cdot \left\langle \vec{1}_r e^{\gamma \vec{1}_r \cdot \vec{r}_s}; \tilde{\vec{j}}_{s_\beta}\left(\vec{r}_s, s\right)\right\rangle\right]\left[\vec{1}_p \cdot \left\langle \vec{1}_1 e^{-\gamma \vec{1}_1 \cdot \vec{r}\,'_s}; \tilde{\vec{j}}_{s_\beta}\left(\vec{r}\,'_s, s\right)\right\rangle\right]$$

$$= \left\langle \vec{1}_m e^{\gamma \vec{1}_r \cdot \vec{r}_s}; \tilde{\vec{j}}_{s_\beta}\left(\vec{r}_s, s\right)\right\rangle \left\langle \vec{1}_p e^{-\gamma \vec{1}_1 \cdot \vec{r}\,'_s}; \tilde{\vec{j}}_{s_\beta}\left(\vec{r}\,'_s, s\right)\right\rangle \tag{1.4.1}$$

Note in this form how $\vec{1}_m$ and $\vec{1}_p$ factor out (in dot-product sense) with factors depending separately on $\vec{1}_r$ and $\vec{1}_1$, so that the four unit vectors all appear in separate factors. In this form reciprocity is apparent as

$$\vec{1}_m \cdot \tilde{\vec{C}}_{f_\beta}\left(\vec{1}_r, \vec{1}_1; s\right) \cdot \vec{1}_p = \vec{1}_p \cdot \tilde{\vec{C}}_{f_\beta}\left(-\vec{1}_1, -\vec{1}_r; s\right) \cdot \vec{1}_m \tag{1.4.2}$$

Note that $\tilde{\vec{C}}_{r_\beta}$ and $\tilde{\vec{C}}_\beta$ are simply related via

$$\tilde{\vec{C}}_{r_\beta}\left(\vec{1}_r, s\right) = \tilde{\vec{C}}_\beta\left(-\vec{1}_r, s\right) \tag{1.4.3}$$

so there is really only one vector function to consider. In the dyadic then we have

$$\tilde{\vec{C}}_{f_\beta}\left(\vec{1}_r, \vec{1}_1; s\right) = \tilde{\vec{C}}_{r_\beta}\left(\vec{1}_r, s\right)\tilde{\vec{C}}_\beta\left(\vec{1}_1, s\right) = \tilde{\vec{C}}_\beta\left(-\vec{1}_r, s\right)\tilde{\vec{C}}_\beta\left(\vec{1}_1, s\right)$$

$$= \tilde{\vec{C}}_\beta\left(-\vec{1}_r, s\right)\tilde{\vec{C}}_{r_\beta}\left(-\vec{1}_1, s\right) = \tilde{\vec{C}}_{r_\beta}\left(\vec{1}_r, s\right)\tilde{\vec{C}}_{r_\beta}\left(-\vec{1}_1, s\right)$$

$$= \tilde{\vec{C}}^{T}_{f_\beta}\left(-\vec{1}_1, -\vec{1}_r; s\right) \tag{1.4.4}$$

which is scattering reciprocity in terms of only the directions of incidence and scattering.

For backscattering we have in dyadic form

$$\vec{1}_r = -\vec{1}_1$$

$$\tilde{\vec{C}}_{r_\beta}(\vec{1}_r, s) = \tilde{\vec{C}}_\beta(\vec{1}_1, s)$$

$$\begin{aligned}\tilde{\vec{\vec{C}}}_{f_\beta}(-\vec{1}_1, \vec{1}_1; s) &= \tilde{\vec{C}}_\beta(\vec{1}_1, s)\tilde{\vec{C}}_\beta(\vec{1}_1, s) \\ &= \tilde{\vec{\vec{C}}}^{\,T}_{f_\beta}(-\vec{1}_1, \vec{1}_1; s)\end{aligned} \tag{1.4.5}$$

In scalar form we have for in-line or parallel polarization

$$\vec{1}_m = \vec{1}_p, \quad \vec{1}_r = -\vec{1}_1$$

$$\begin{aligned}\vec{1}_m \cdot \tilde{\vec{\vec{C}}}_{f_\beta}(\vec{1}_r, \vec{1}_1; s) \cdot \vec{1}_p &= \vec{1}_p \cdot \tilde{\vec{\vec{C}}}_{f_\beta}(-\vec{1}_1, \vec{1}_1; s) \cdot \vec{1}_p \\ &= \left[\vec{1}_p \cdot \tilde{\vec{C}}_\beta(\vec{1}_1, s)\right]^2\end{aligned} \tag{1.4.6}$$

which is a way of saying that for each eigenmode the coupling to the target is the same as the coupling of the target currents to the backscattered fields, at least in terms of angular dependence, i.e., $\vec{1}_1$ and $\vec{1}_p$. Note that in (1.4.1) there is a sum over $\beta$ for arbitrary frequencies, so this factorization by $\vec{1}_p$ and $\vec{1}_1$ does not apply to arbitrary frequencies for the scattered or backscattered fields.

## 1.5 SEM Pole Scattered Fields

Applying (1.3.5) in the context of SEM poles as in (1.2.9) and (1.2.17) gives

$$\begin{aligned}\vec{1}_m \cdot \tilde{\vec{E}}_f(\vec{r}, s) = \frac{E_o}{4\pi r} e^{-\gamma r} \sum_\alpha \tilde{f}(s_\alpha)\, \eta_{f_\alpha}(\vec{1}_1, \vec{1}_p; \vec{1}_r, \vec{1}_m)[s - s_\alpha]^{-1} \\ + \text{ singularities of } \tilde{f}(s) \\ + \text{ possible entire function}\end{aligned}$$

$$\eta_{f\alpha}\left(\vec{1}_1,\vec{1}_p;\vec{1}_r,\vec{1}_m\right)=\eta_\alpha\left(\vec{1}_1,\vec{1}_p\right)\eta_{r\alpha}\left(\vec{1}_r,\vec{1}_m\right)$$

$\equiv$ far coupling coefficient (class 1)

$$\eta_\alpha\left(\vec{1}_1,\vec{1}_p\right)=\frac{\vec{1}_p\cdot\left\langle \vec{1}_1 e^{-\gamma\vec{1}_1\cdot\vec{r}_s'};\vec{j}_{s\alpha}\left(\vec{r}_s'\right)\right\rangle}{\left\langle \vec{j}_{s\alpha}\left(\vec{r}_s\right);\frac{\partial}{\partial s}\tilde{\tilde{Z}}_t\left(\vec{r}_s,\vec{r}_s';s\right)\Big|_{s=s_\alpha};\vec{j}_{s\alpha}\left(\vec{r}_s'\right)\right\rangle}$$

$\equiv$ coupling coefficient (class 1)

$$\eta_{r\alpha}\left(\vec{1}_r,\vec{1}_m\right)=-s_\alpha\mu_o\vec{1}_m\cdot\left\langle \vec{1}_r e^{\gamma_\alpha\vec{1}_r\cdot\vec{r}_s'};\vec{j}_{s\alpha}\left(\vec{r}_s'\right)\right\rangle$$

$\equiv$ recoupling coefficient (class 1) (1.5.1)

The coupling coefficient gives the coupling of the incident field to the natural mode. The recoupling coefficient couples the natural mode to the far scattered field.

It is convenient to normalize these coupling coefficients so that they have peak magnitude of 1 over all angles concerned and assume the value +1 (real) at the angles concerned (or one set of these angle combinations). For $\eta_\alpha$ let this angle combination be $\vec{1}_{1_0}$ and $\vec{1}_{p_0}$ giving

$$\sup_{\vec{1}_1,\vec{1}_p}\left|\eta_\alpha\left(\vec{1}_1,\vec{1}_p\right)\right|=\left|\eta_\alpha\left(\vec{1}_{1_0},\vec{1}_{p_0}\right)\right|$$

$$\eta_\alpha^{(n)}\left(\vec{1}_1,\vec{1}_p\right)\equiv\frac{\eta_\alpha\left(\vec{1}_1,\vec{1}_p\right)}{\eta_\alpha\left(\vec{1}_{1_0},\vec{1}_{p_0}\right)}$$

$$\sup_{\vec{1}_1,\vec{1}_p}\left|\eta_\alpha^{(n)}\left(\vec{1}_1,\vec{1}_p\right)\right|=1=\eta_\alpha^{(n)}\left(\vec{1}_{1_0},\vec{1}_{p_0}\right) \tag{1.5.2}$$

In another form, noting that the 3-term symmetric product that is the denominator of $\eta_\alpha$ is not a function of $\vec{1}_1$ or $\vec{1}_p$, we have

$$\eta_\alpha^{(n)}\left(\vec{1}_1,\vec{1}_p\right)=\frac{\vec{1}_p\cdot\left\langle\vec{1}_1 e^{-\gamma\alpha\vec{1}_1\cdot\vec{r}_s'};\vec{j}_{s\alpha}\left(\vec{r}_s'\right)\right\rangle}{\vec{1}_{p_o}\cdot\left\langle\vec{1}_{1_o} e^{-\gamma\alpha\vec{1}_{1_o}\cdot\vec{r}_s'};\vec{j}_{s\alpha}\left(\vec{r}_s'\right)\right\rangle}$$

$$=\frac{\left\langle\vec{1}_p e^{-\gamma\alpha\vec{1}_1\cdot\vec{r}_s'};\vec{j}_{s\alpha}\left(\vec{r}_s'\right)\right\rangle}{\left\langle\vec{1}_{p_o} e^{-\gamma\alpha\vec{1}_{1_o}\cdot\vec{r}_s'};\vec{j}_{s\alpha}\left(\vec{r}_s'\right)\right\rangle} \tag{1.5.3}$$

For $\eta_{r\alpha}$ we have a maximum magnitude at angles $\vec{1}_{r_o}$ and $\vec{1}_{m_o}$ giving

$$\sup_{\vec{1}_r,\vec{1}_m}\left|\eta_{r\alpha}\left(\vec{1}_r,\vec{1}_m\right)\right|=\left|\eta_{r\alpha}\left(\vec{1}_{r_o},\vec{1}_{m_o}\right)\right|$$

$$\eta_{r\alpha}^{(n)}\left(\vec{1}_r,\vec{1}_m\right)\equiv\frac{\eta_{r\alpha}\left(\vec{1}_r,\vec{1}_m\right)}{\eta_{r\alpha}\left(\vec{1}_{r_o},\vec{1}_{m_o}\right)}$$

$$\sup_{\vec{1}_r,\vec{1}_m}\left|\eta_{r\alpha}^{(n)}\left(\vec{1}_1,\vec{1}_p\right)\right|=1=\eta_{r\alpha}^{(n)}\left(\vec{1}_{r_o},\vec{1}_{m_o}\right)$$

$$\eta_{r\alpha}^{(n)}\left(\vec{1}_r,\vec{1}_m\right)=\frac{\vec{1}_m\cdot\left\langle\vec{1}_r e^{\gamma\alpha\vec{1}_r\cdot\vec{r}_s'};\vec{j}_{s\alpha}\left(\vec{r}_s'\right)\right\rangle}{\vec{1}_{m_o}\cdot\left\langle\vec{1}_{r_o} e^{\gamma\alpha\vec{1}_{r_o}\cdot\vec{r}_s'};\vec{j}_{s\alpha}\left(\vec{r}_s'\right)\right\rangle}$$

$$=\frac{\left\langle\vec{1}_m e^{\gamma\alpha\vec{1}_r\cdot\vec{r}_s'};\vec{j}_{s\alpha}\left(\vec{r}_s'\right)\right\rangle}{\left\langle\vec{1}_{m_o} e^{\gamma\alpha\vec{1}_{r_o}\cdot\vec{r}_s'};\vec{j}_{s\alpha}\left(\vec{r}_s'\right)\right\rangle} \tag{1.5.4}$$

Comparing (1.5.4) to (1.5.3) we have the simple result (for at least one choice of angles)

$$\vec{1}_{m_o} = \vec{1}_{p_o}, \vec{1}_{r_o} = -\vec{1}_{1_o}$$

$$\eta_{\alpha}^{(n)}\left(\vec{1}_1, \vec{1}_p\right) = \eta_{r\alpha}^{(n)}\left(-\vec{1}_1, \vec{1}_p\right) \tag{1.5.5}$$

which is found by simply substituting $\vec{1}_r \rightarrow -\vec{1}_1$ and $\vec{1}_m \rightarrow \vec{1}_p$ in (1.5.4) giving a formula identical to (1.5.3).

Except then for an appropriate interchange of direction vectors the normalized recoupling and coupling coefficients are the same. So, knowing one we know the other. In a general form we can normalize the far coupling coefficient in (1.5.1) as

$$\eta_{f\alpha}^{(n)}\left(\vec{1}_1, \vec{1}_p; \vec{1}_r, \vec{1}_m\right) = \eta_{\alpha}^{(n)}\left(\vec{1}_1, \vec{1}_p\right)\eta_{r\alpha}^{(n)}\left(\vec{1}_r, \vec{1}_p\right) \tag{1.5.6}$$

One symmetry to observe is the basic reciprocity relation

$$\eta_{f\alpha}^{(n)}\left(-\vec{1}_r, \vec{1}_m; -\vec{1}_1, \vec{1}_p\right) = \eta_{\alpha}^{(n)}\left(-\vec{1}_r, \vec{1}_m\right)\eta_{r\alpha}^{(n)}\left(-\vec{1}_1, \vec{1}_p\right)$$

$$= \eta_{r\alpha}^{(n)}\left(\vec{1}_r, \vec{1}_m\right)\eta_{\alpha}^{(n)}\left(\vec{1}_1, \vec{1}_p\right)$$

$$= \eta_{f\alpha}^{(n)}\left(\vec{1}_1, \vec{1}_p; \vec{1}_r, \vec{1}_m\right)$$

$$\eta_{f\alpha}\left(-\vec{1}_r, \vec{1}_m; -\vec{1}_1, \vec{1}_p\right) = \eta_{f\alpha}\left(\vec{1}_1, 1_p; \vec{1}_r, \vec{1}_m\right) \tag{1.5.7}$$

Noting also

$$\eta_{\alpha}\left(\vec{1}_1, -\vec{1}_p\right) = -\eta_{\alpha}\left(\vec{1}_1, \vec{1}_p\right), \quad \eta_{\alpha}^{(n)}\left(\vec{1}_1, -\vec{1}_p\right) = -\eta_{\alpha}^{(n)}\left(\vec{1}_1, \vec{1}_p\right)$$

$$\eta_{r\alpha}\left(\vec{1}_r, -\vec{1}_m\right) = -\eta_{r\alpha}\left(\vec{1}_r, \vec{1}_m\right), \quad \eta_{r\alpha}^{(n)}\left(\vec{1}_r, -\vec{1}_m\right) = -\eta_{r\alpha}^{(n)}\left(\vec{1}_r, \vec{1}_m\right) \tag{1.5.8}$$

we can note that there are at least two choices each for $\vec{1}_{p_o}$ and $\vec{1}_{m_o}$ which results from

$$\eta_{\alpha}^{(n)}\left(\vec{1}_{1_o}, -\vec{1}_{p_o}\right) = -\eta_{\alpha}^{(n)}\left(\vec{1}_{1_o}, \vec{1}_{p_o}\right) = -1$$

$$\eta_{r\alpha}^{(n)}\left(\vec{1}_{r_o}, -\vec{1}_{m_o}\right) = -\eta_{r\alpha}^{(n)}\left(\vec{1}_{r_o}, \vec{1}_{m_o}\right) = -1 \tag{1.5.9}$$

which are of course now both of magnitude 1 on inversion of $\vec{1}_{p_o}$ and $\vec{1}_{m_o}$. Applying (1.5.8) in (1.5.7) we also have the symmetry

$$\eta_{f_\alpha}^{(n)}\left(-\vec{1}_r, -\vec{1}_m; -\vec{1}_1, -\vec{1}_p\right) = \eta_{f_\alpha}^{(n)}\left(\vec{1}_1, \vec{1}_p; \vec{1}_r, \vec{1}_m\right) \tag{1.5.10}$$

For the special case of backscattering with measurement parallel to the incident field we have

$$\vec{1}_r \equiv -\vec{1}_1, \vec{1}_m \equiv \vec{1}_p$$

$$\eta_{b_\alpha}^{(n)}\left(\vec{1}_1, \vec{1}_p\right) \equiv \eta_{f_\alpha}^{(n)}\left(\vec{1}_1, \vec{1}_p; -\vec{1}_1, \vec{1}_p\right)$$

$$= \eta_{\alpha}^{(n)}\left(\vec{1}_1, \vec{1}_p\right) \eta_{r_\alpha}^{(n)}\left(-\vec{1}_1, \vec{1}_p\right)$$

$$\left[\eta_{\alpha}^{(n)}\left(\vec{1}_1, \vec{1}_p\right)\right]^2 \tag{1.5.11}$$

Got that? The normalized backscattering coupling coefficient is exactly the square of the normalized coupling coefficient (a complex function of $\vec{1}_1$ and $\vec{1}_p$). Furthermore we have

$$\sup_{\vec{1}_1, \vec{1}_p} \left|\eta_{b_\alpha}^{(n)}\left(\vec{1}_1, \vec{1}_p\right)\right| = \sup_{\vec{1}_1, \vec{1}_p} \left|\eta_{\alpha}^{(n)}\left(\vec{1}_1, \vec{1}_p\right)\right|^2$$

$$= 1$$

$$= \eta_{b_\alpha}^{(n)}\left(\vec{1}_{1_o}, \vec{1}_{p_o}\right) \tag{1.5.12}$$

From the symmetry concerning inversion of $\vec{1}_p$ in (1.5.8) we also have

$$\eta_{b_\alpha}^{(n)}(\vec{1}_1, -\vec{1}_p) = \left[\eta_\alpha^{(n)}(\vec{1}_1, -\vec{1}_p)\right]^2$$

$$= \left[\eta_\alpha^{(n)}(\vec{1}_1, \vec{1}_p)\right]^2$$

$$= \eta_{b_\alpha}^{(n)}(\vec{1}_1, \vec{1}_p)$$

$$\eta_{b_\alpha}^{(n)}\left(\vec{1}_{1_o}, -\vec{1}_{p_o}\right) = \eta_{b_\alpha}^{(n)}\left(\vec{1}_{1_o}, \vec{1}_{p_o}\right) = 1 \tag{1.5.13}$$

So what this says is that measurements on the scatterer to obtain the normalized coupling coefficient for the natural-mode surface current density as a function of $\vec{1}_1$ and $\vec{1}_p$ are simply related to backscatter coupling coefficients (normalized) by a simple square. Note that $\vec{r} = \vec{0}$ needs to be maintained while rotating $\vec{1}_1$ and $\vec{1}_p$ (or equivalently the scatterer orientation) for these results to apply. The reference time for the backscatter signal is not taken from the first backscatter signal to arrive from a backscatter pulse, but the time for the backscatter signal to arrive from some reference position on the scatterer, such as the center of the minimum circumscribing sphere in fig. 1.2.1. Of course one could use the time of the first backscatter signal provided an angular-dependent correction is made.

The backscattered electric field can be considered as having two components, one parallel to $\vec{1}_p$, and one parallel to $\vec{1}_1 \times \vec{1}_p$. Defining

$$\vec{1}_c \equiv \vec{1}_1 \times \vec{1}_p \tag{1.5.14}$$

then we can have the cross polarized backscattering coupling coefficient as

$$\eta_{c\alpha}^{(n)}(\vec{1}_1, \vec{1}_p, \vec{1}_c) \equiv \tilde{\eta}_{f\alpha}^{(n)}(\vec{1}_1, \vec{1}_p; -\vec{1}_1, \vec{1}_c)$$

$$= \eta_{\alpha}^{(n)}(\vec{1}_1, \vec{1}_p)\, \eta_{r\alpha}^{(n)}(-\vec{1}_1, \vec{1}_c)$$

$$= \eta_{\alpha}^{(n)}(\vec{1}_1, \vec{1}_p)\, \eta_{\alpha}^{(n)}(\vec{1}_1, \vec{1}_c)$$

$$= -\eta_{\alpha}^{(n)}(\vec{1}_1, -\vec{1}_p)\, \eta_{\alpha}^{(n)}(\vec{1}_1, \vec{1}_c)$$

$$= -\eta_{\alpha}^{(n)}(\vec{1}_1, \vec{1}_p)\, \eta_{\alpha}^{(n)}(\vec{1}_1, -\vec{1}_c)$$

$$= \eta_{\alpha}^{(n)}(\vec{1}_1, -\vec{1}_p)\, \eta_{\alpha}^{(n)}(\vec{1}_1, -\vec{1}_c) \tag{1.5.15}$$

This gives the symmetries

$$\eta_{c\alpha}^{(n)}(\vec{1}_1, \vec{1}_p, \vec{1}_c) = \eta_{c\alpha}^{(n)}(\vec{1}_1, \vec{1}_c, \vec{1}_p) =$$

$$\eta_{c\alpha}^{(n)}(\vec{1}_1, -\vec{1}_p, -\vec{1}_c) = \eta_{c\alpha}^{(n)}(\vec{1}_1, -\vec{1}_c, -\vec{1}_p) =$$

$$-\eta_{c\alpha}^{(n)}(\vec{1}_1, -\vec{1}_p, \vec{1}_c) = -\eta_{c\alpha}^{(n)}(\vec{1}_1, \vec{1}_p, -\vec{1}_c) =$$

$$-\eta_{c\alpha}^{(n)}(\vec{1}_1, \vec{1}_c, -\vec{1}_p) = -\eta_{c\alpha}^{(n)}(\vec{1}_1, -\vec{1}_c, \vec{1}_p) \tag{1.5.16}$$

Note that in general we have the inequality

$$\left|\eta^{(n)}_{c\,\alpha}\left(\vec{1}_1,\vec{1}_p,\vec{1}_c\right)\right| = \left|\eta^{(n)}_{\alpha}\left(\vec{1}_1,\vec{1}_p\right)\right| \left|\eta^{(n)}_{\alpha}\left(\vec{1}_1,\vec{1}_c\right)\right|$$

$$\leq \left\{\sup_{\vec{1}_p} \left|\eta^{(n)}_{\alpha}\left(\vec{1}_1,\vec{1}_p\right)\right|\right\} \left\{\sup_{\vec{1}_c} \left|\eta^{(n)}_{\alpha}\left(\vec{1}_1,\vec{1}_c\right)\right|\right\}$$

$$= \left\{\sup_{\vec{1}_p} \left|\eta^{(n)}_{\alpha}\left(\vec{1}_1,\vec{1}_p\right)\right|\right\}^2$$

$$= \sup_{\vec{1}_p} \left|\eta^{(n)}_{b\,\alpha}\left(\vec{1}_1,\vec{1}_p\right)\right| \qquad (1.5.17)$$

which holds for all $\vec{1}_c$ for every choice of $\vec{1}_1$. So the cross polarized backscattering coupling coefficient is bounded in magnitude by the maximum of the backscattering (parallel polarized) coupling coefficient for every $\vec{1}_1$.

One can also express the far coupling coefficient in dyadic terms, anaolgous to the EEM form in section 1.4, as

$$\vec{1}_m \cdot \tilde{\vec{E}}_f\left(\vec{r}_s,s\right) = \frac{-E_o s\mu_o}{4\pi r}\, e^{-\gamma r} \sum_{\alpha} \tilde{f}(s_\alpha)\left\langle \vec{j}_{s\,\alpha}\left(\vec{r}_s\right); \frac{\partial}{\partial s}\, \tilde{\vec{\vec{Z}}}_t\left(\vec{r}_s,\vec{r}_s^{\,\prime};s\right)\Big|_{s=s_\alpha} ; \vec{j}_{s\,\alpha}\left(\vec{r}_s^{\,\prime}\right)\right\rangle^{-1}$$

$$\vec{1}_m \cdot \vec{\vec{C}}_{f\alpha}\left(\vec{1}_r,\vec{1}_1\right) \cdot \vec{1}_p\left[s-s_\alpha\right]^{-1}$$

$$+ \text{ singularities of } \tilde{f}(s)$$

$$+ \text{ possible entire function}$$

$$\vec{\vec{C}}_{f\,\alpha}\left(\vec{1}_r,\vec{1}_1\right) = \vec{C}_{r\alpha}\left(\vec{1}_r\right)\vec{C}_{\alpha}\left(\vec{1}_1\right)$$

$$\vec{C}_{\alpha}\left(\vec{1}_1\right) = \left\langle \vec{1}_1 e^{-\gamma_\alpha \vec{1}_1\cdot\vec{r}_s^{\,\prime}} ; \vec{j}_{s\,\alpha}\left(\vec{r}_s^{\,\prime}\right)\right\rangle$$

$$\vec{C}_{r\,\alpha}\left(\vec{1}_r\right) = \left\langle \vec{1}_r e^{\gamma_\alpha \vec{1}_r\cdot\vec{r}_s} ; \vec{j}_{s\,\alpha}\left(\vec{r}_s\right)\right\rangle$$

$$\eta_{f_\alpha}(\vec{1}_1,\vec{1}_p;\vec{1}_r,\vec{1}_m) = -s_\alpha\mu_o\left\langle \vec{j}_{s_\alpha}(\vec{r}_s);\frac{\partial}{\partial s}\tilde{\vec{Z}}_t(\vec{r}_s,\vec{r}_s';s)\Big|_{s=s_\alpha};\vec{j}_{s_\alpha}(\vec{r}_s')\right\rangle^{-1}\vec{1}_m\cdot\vec{C}_{f_\alpha}(\vec{1}_r,\vec{1}_1)\cdot\vec{1}_p$$

$$\vec{1}_m\cdot C_{f_\alpha}(\vec{1}_r,\vec{1}_1)\cdot\vec{1}_p = \left[\vec{1}_m\cdot\left\langle\vec{1}_r e^{\gamma_\alpha\vec{1}_r\cdot\vec{r}_s};\vec{j}_{s_\alpha}(\vec{r}_s)\right\rangle\right]\left[\vec{1}_p\cdot\left\langle\vec{1}_1 e^{-\gamma_\alpha\vec{1}_1\cdot\vec{r}_s'};\vec{j}_{s_\alpha}(\vec{r}_s')\right\rangle\right]$$

$$= \left\langle\vec{1}_m e^{\gamma_\alpha\vec{1}_r\cdot\vec{r}_s};\vec{j}_{s_\alpha}(\vec{r}_s)\right\rangle\left\langle\vec{1}_p e^{-\gamma_\alpha\vec{1}_1\cdot\vec{r}_s'};\vec{j}_{s_\alpha}(\vec{r}_s')\right\rangle \tag{1.5.18}$$

Note that the natural modes are not necessarily normalized in the same manner as the orthonormalized eigenmodes in section 1.4. In current form, however, the various pole terms (class 1) are not functions of the complex frequency.

Note that $\vec{C}_{r_\alpha}$ and $\vec{C}_\alpha$ are simply related via

$$\vec{C}_{r_\alpha}(\vec{1}_r) = \vec{C}_\alpha(-\vec{1}_r) \tag{1.5.19}$$

In dyadic form then we have

$$\vec{C}_{f_\alpha}(\vec{1}_r,\vec{1}_1) = \vec{C}_{r_\alpha}(\vec{1}_r)\vec{C}_\alpha(\vec{1}_1)$$

$$= \vec{C}_\alpha(-\vec{1}_r)\vec{C}_{r_\alpha}(-\vec{1}_1)$$

$$= \vec{C}^T_{f_\alpha}(-\vec{1}_1,-\vec{1}_r) \tag{1.5.20}$$

which is again reciprocity in terms of only the directions of incidence and scattering.

The present analysis is related to the earlier concept of far natural modes [1-1]. Except for a constant scaling factor these are the same as the $\vec{C}_{r_\alpha}(\vec{1}_r)$. In the present analysis the relation of these to the recoupling coefficient is interesting due to the exhibition of various symmetries.

For backscattering we have in dyadic form

$$\vec{1}_r = -\vec{1}_1$$

$$\vec{C}_{r_\alpha}\left(\vec{1}_r\right) = \vec{C}_\alpha\left(\vec{1}_1\right)$$

$$\vec{C}_{f_\alpha}\left(-\vec{1}_1,\vec{1}_1\right) = \vec{C}_\alpha\left(\vec{1}_1\right)\vec{C}_\alpha\left(\vec{1}_1\right)$$

$$\vec{C}^{T}_{f_\alpha}\left(-\vec{1}_1,\vec{1}_1\right) \tag{1.5.21}$$

In scalar form we have for in-line or parallel polarization

$$\vec{1}_m = \vec{1}_p, \vec{1}_r = -\vec{1}_1$$

$$\vec{1}_m \cdot \vec{C}_{f_\alpha}\left(\vec{1}_r,\vec{1}_1\right)\cdot\vec{1}_p = \vec{1}_p \cdot \vec{C}_{f_\alpha}\left(-\vec{1}_1,\vec{1}_1\right)\cdot\vec{1}_p$$

$$= \left[\vec{1}_p \cdot \vec{C}_\alpha\left(\vec{1}_1\right)\right]^2 \tag{1.5.22}$$

In contradistinction to the EEM results in section 4, the SEM results here when considered pole by pole at each natural frequency $s_\alpha$ do allow the factorization by $\vec{1}_1$ and $\vec{1}_p$.

Note that $\vec{C}_\alpha\left(\vec{1}_1\right)$ is dot multiplied by the incident field. This gives a coefficient to $\vec{C}_{r_\alpha}\left(\vec{1}_r\right)$ which represents the polarization of the scattered field at the observer. This is a residue vector which can be thought of as the amplitude and polarization of the $\alpha^{\text{th}}$ pole contribution at the observer. In time domain the complex exponential $e^{s_\alpha t}$ gives the $\alpha^{\text{th}}$ pole contribution the form of a decaying elliptical-spirally polarized wave.

## 1.6 Application to Forward Scattering

For forward scattering with measurement parallel to the incident field we have

$$\vec{1}_r = \vec{1}_1, \vec{1}_m = \vec{1}_p \tag{1.6.1}$$

In this case (1.3.5) becomes

$$\vec{1}_p \cdot \tilde{\vec{E}}_f(\vec{r}, s) = \frac{-E_o \tilde{f}(s) s\mu_o}{4\pi r} e^{-\gamma r} \vec{1}_p \cdot \left\langle \vec{1}_1 e^{\gamma \vec{1}_1 \cdot \vec{r}_s} ; \tilde{\tilde{Z}}_t^{-1}(\vec{r}_s, \vec{r}_s'; s) ; \vec{1}_1 e^{-\gamma \vec{1}_1 \cdot \vec{r}_s'} \right\rangle \cdot \vec{1}_p$$

$$= \frac{-E_o \tilde{f}(s) s\mu_o}{4\pi r} e^{-\gamma r} \vec{1}_p \cdot \left\langle \vec{1}_1 e^{-\gamma \vec{1}_1 \cdot \vec{r}_s} ; \tilde{\tilde{Z}}_t^{-1}(\vec{r}_s, \vec{r}_s'; s) ; \vec{1}_1 e^{\gamma \vec{1}_1 \cdot \vec{r}_s'} \right\rangle \cdot \vec{1}_p \tag{1.6.2}$$

which is invariant to inversion of $\vec{1}_1$ (i.e., to interchanging coming and going). In EEM form the terms in (1.4.1) become

$$\tilde{\tilde{C}}_\beta(\vec{1}_1, s) = \tilde{\tilde{C}}_{r_\beta}(-\vec{1}_1, s)$$

$$\tilde{\tilde{C}}_{f_\beta}(\vec{1}_1, \vec{1}_1; s) = \tilde{\tilde{C}}_{r_\beta}(\vec{1}_1, s)\, \tilde{\tilde{C}}_\beta(\vec{1}_1, s) = \tilde{\tilde{C}}_\beta(-\vec{1}_1, s)\, \tilde{\tilde{C}}_\beta(\vec{1}_1, s)$$

$$\tilde{\tilde{C}}^T_{f_\beta}(\vec{1}_1, -\vec{1}_1; s) \tag{1.6.3}$$

In SEM form we have $\tilde{\vec{C}}_\alpha$ results as in (1.6.3) and for the coupling coefficients

$$\eta_{f_\alpha}(\vec{1}_1, \vec{1}_p; \vec{1}_1, \vec{1}_p) = \eta_\alpha(\vec{1}_1, \vec{1}_p)\, \eta_{r_\alpha}(\vec{1}_1, \vec{1}_p)$$

$$= \eta_{r_\alpha}(-\vec{1}_1, \vec{1}_p)\, \eta_\alpha(-\vec{1}_1, \vec{1}_p)$$

$$= \eta^{(n)}_{f_\alpha}(-\vec{1}_1, \vec{1}_p; -\vec{1}_1, \vec{1}_p) \tag{1.6.4}$$

For the normalized coupling coefficients this is

$$\eta^{(n)}_{f_\alpha}(\vec{1}_1, \vec{1}_p; \vec{1}_1, \vec{1}_p) = \eta^{(n)}_\alpha(\vec{1}_1, \vec{1}_p)\, \eta^{(n)}_{r_\alpha}(\vec{1}_1, \vec{1}_p)$$

$$= \eta^{(n)}_\alpha(\vec{1}_1, \vec{1}_p)\, \eta^{(n)}_\alpha(-\vec{1}_1, \vec{1}_p)$$

$$= \eta^{(n)}_{r_\alpha}(-\vec{1}_1, \vec{1}_p)\, \eta^{(n)}_{r_\alpha}(\vec{1}_1, \vec{1}_p)$$

$$= \eta^{(n)}_{f_\alpha}(-\vec{1}_1, \vec{1}_p; -\vec{1}_1, \vec{1}_p) \tag{1.6.5}$$

Now in what is known as the forward-scattering theorem [1-7, 1-9, 1-15], the forward-scattered fields are related to the absorption and scattering cross sections, i.e. the total power removed from the incident wave (extinction cross section). Here we have found that forward scattering is invariant to inversion of $\vec{1}_1$. Therefore, the extinction cross section (sum of absorption and scattering cross sections) is also invariant to inversion of $\vec{1}_1$.

In [1-7] the forward-scattering theorem is extended into time domain for arbitrary incident transient waveforms with some limitations concerning boundedness and late-time behavior. It was observed that for a step-function incident wave one had to be careful concerning the order of taking r to infinity and t to infinity. The result was (in present notation)

$$W_s + W_\infty^{(sc)} \equiv \text{absorbed plus scattered energy}$$

$$= \frac{1}{2}\epsilon_o E_o^2\left\{\vec{1}_p \cdot \overleftrightarrow{P}_o \cdot \vec{1}_p \pm \left[\vec{1}_1 \times \vec{1}_p\right] \cdot \overleftrightarrow{M}_o \cdot \left[\vec{1}_1 \times \vec{1}_p\right]\right\}$$

$$E_o \equiv \text{magnitude of incident step} - \text{function electric field}$$

$$\overleftrightarrow{P}_o \equiv \text{low} - \text{frequency electric polarizability of scatterer}$$

$$\overleftrightarrow{M}_o \equiv \text{low} - \text{frequency magnetic polarizability of scatterer}$$

$$+ \Leftrightarrow r \to \infty \text{ before } t \to \infty$$

$$- \Leftrightarrow t \to \infty \text{ before } r \to \infty \tag{1.6.6}$$

The thing to note here is that this formula is also invariant to inversion of $\vec{1}_1$. In any event due to the fact that forward scattering is invariant to inversion of $\vec{1}_1$ and applies to all frequencies (and hence transients) then time-domain absorbed plus scattered energy is invariant to inversion of $\vec{1}_1$. Note that if the scatterer is lossless then scattered energy is also invariant to inversion of $\vec{1}_1$.

## 1.7 Inclusion of Reflection Symmetry in Scatterer

Besides the symmetries inherent in reciprocal scattering one can have geometric symmetries in the scatterer. A common such symmetry is reflection symmetry with respect to a symmetry plane P as in fig. 1.7.1. Here a typical aircraft is (at least approximately) such a symmetrical scatterer. As discussed in [1-3, 1-5] all electromagnetic parameters (including incident and scattered fields, currents on the scatterer, etc.) can be divided into two separate parts which do not couple to each other; these two parts are designated symmetric (sy) and antisymmetric (as).

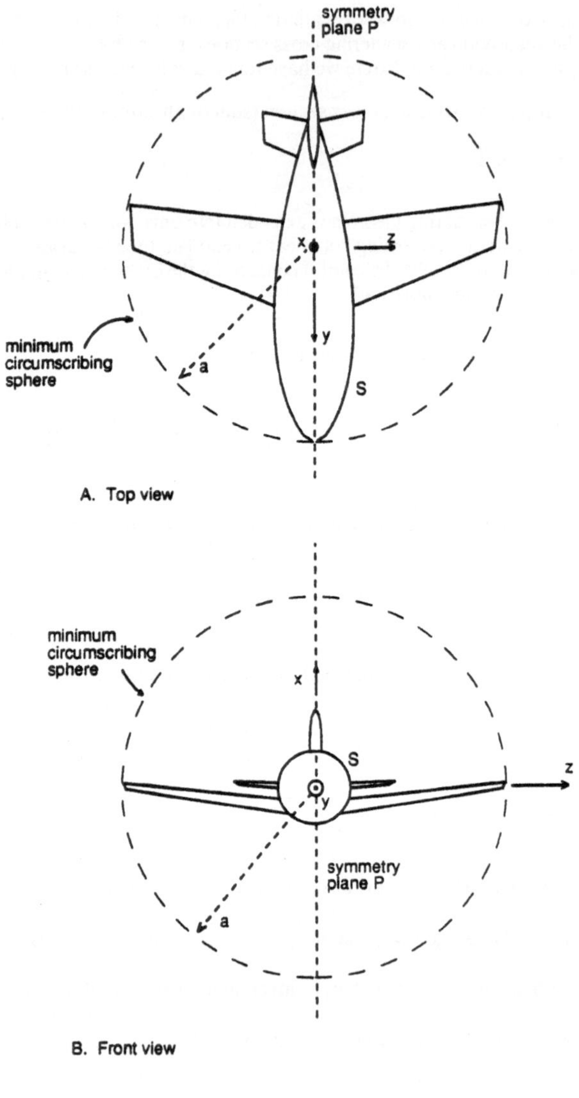


Figure 1.7.1 Typical Aircraft with Symmetry Plane.

Defining the coordinate center ($\vec{r} = \vec{0}$) as the center of the minimum circumscribing sphere (radius-a) let there be a cartesian coordinate system as indicated. With

$$\vec{1}_z = \vec{1}_p \equiv \text{unit normal vector to P} \\ = \vec{1}_x \times \vec{1}_y \tag{1.7.1}$$

we have a reflection dyad

$$\vec{\vec{R}} = \begin{pmatrix} 1 & 0 & 0 \\ 0 & 1 & 0 \\ 0 & 0 & -1 \end{pmatrix} = \vec{1}_x\vec{1}_x + \vec{1}_y\vec{1}_y - \vec{1}_z\vec{1}_z \tag{1.7.2}$$

which associates with every position

$$\vec{r} = x\vec{1}_x + y\vec{1}_y + z\vec{1}_z \tag{1.7.3}$$

a mirror position

$$\vec{r}_m \equiv \vec{\vec{R}} \cdot \vec{r} = x\vec{1}_x + y\vec{1}_y - z\vec{1}_z \tag{1.7.4}$$

For a perfectly conducting object with surface S this means that for every position $\vec{r} \in S$ then also $\vec{r}_m \in S$. This applies to non-perfectly-conducting objects as well by applying the symmetry requirement to the constitutive parameters $\varepsilon, \mu$, and $\sigma$, and even if they take the form of $3 \times 3$ matrices (tensors) [1-3, 1-5].

The fields, currents, etc., are decomposed into symmetric and antisymmetric parts, some examples of which are [1-3, 1-5]

$$\begin{aligned} \vec{E}_{\substack{sy\\as}}(\vec{r},t) &= \tfrac{1}{2}\left\{\vec{E}(\vec{r},t) \pm \vec{\vec{R}} \cdot \vec{E}(\vec{r}_m,t)\right\} \\ \vec{H}_{\substack{sy\\as}}(\vec{r},t) &= \tfrac{1}{2}\left\{\vec{H}(\vec{r},t) \mp \vec{\vec{R}} \cdot \vec{H}(\vec{r}_m,t)\right\} \\ \vec{J}_{s_{\substack{sy\\as}}}(\vec{r},t) &= \tfrac{1}{2}\left\{\vec{J}_s(\vec{r},t) \pm \vec{\vec{R}} \cdot \vec{J}_s(\vec{r}_m,t)\right\} \\ \rho_{s_{\substack{sy\\as}}}(\vec{r},t) &= \tfrac{1}{2}\left\{\vec{\rho}_s(\vec{r},t) \pm \vec{\rho}_s(\vec{r}_m,t)\right\} \end{aligned} \tag{1.7.5}$$

$$k_{\substack{sy\\as}}(\vec{r},t) = \frac{1}{2}\left\{k(\vec{r},t) \mp k(\vec{r}_m,t)\right\}$$

where k is the equivalent magnetic charge density on S [1-2]. Note "sy" takes the upper sign and "as" takes the lower sign.

Let us consider the various terms that appear in the modal expansions. The eigenmodes have the symmetries as in (1.7.5) so that the modes in (1.2.6) can be divided as

$$\vec{j}_{s_{\substack{sy\\as}},\beta'}(\vec{r}_s,s) = \pm\overleftrightarrow{R}\cdot\vec{j}_{s_{\substack{sy\\as}},\beta'}(\vec{r}_{s_m},s)$$

$$\beta = \left\{\begin{matrix}sy\\as\end{matrix},\beta'\right\} \equiv \text{partitioned index set for eigenmodes} \tag{1.7.6}$$

and similarly for the natural modes in (1.2.9) as

$$\vec{j}_{s_{\substack{sy\\as}},\alpha'}(\vec{r}_s) = \pm\overleftrightarrow{R}\cdot\vec{j}_{s_{\substack{sy\\as}},\alpha'}(\vec{r}_{s_m})$$

$$\alpha = \left\{\begin{matrix}sy\\as\end{matrix},\alpha'\right\} \equiv \text{partitioned index set for natural modes} \tag{1.7.7}$$

Referring to fig. 1.7.2 let the incident wave be characterized by the incidence direction $\vec{1}_1$ with two orthogonal polarization vectors $\vec{1}_v$ and $\vec{1}_h$ with

$$\begin{aligned}
&\vec{1}_h // (y,z) \text{ plane}\\
&\vec{1}_h\cdot\vec{1}_x = 0\ ,\quad \vec{1}_h\cdot\vec{1}_1 = 0\\
&\vec{1}_v\cdot\vec{1}_h = 0\ ,\quad \vec{1}_v\cdot\vec{1}_1 = 0\\
&\vec{1}_1\times\vec{1}_v = \vec{1}_h\ ,\quad \vec{1}_v\times\vec{1}_h = \vec{1}_1\ ,\quad \vec{1}_h\times\vec{1}_1 = \vec{1}_v\\
&\vec{1}_h \equiv \text{horizontal polarization}\\
&\vec{1}_v \equiv \text{"vertical" polarization - not exactly vertical, but perpendicular to } \vec{1}_h
\end{aligned} \tag{1.7.8}$$

Here the term vertical is used loosely, and is defined as above. Note that there is a set of mirror unit vectors (using the "electric" reflection rule in (1.7.5)) as

$$\vec{1}_{1_m} \equiv \overleftrightarrow{R}\cdot\vec{1}_1\ ,\quad \vec{1}_{1_v} \equiv \overleftrightarrow{R}\cdot\vec{1}_v\ ,\quad \vec{1}_{h_m} \equiv \overleftrightarrow{R}\cdot\vec{1}_h \tag{1.7.9}$$

Note that this is a left-handed system due to the reflection as

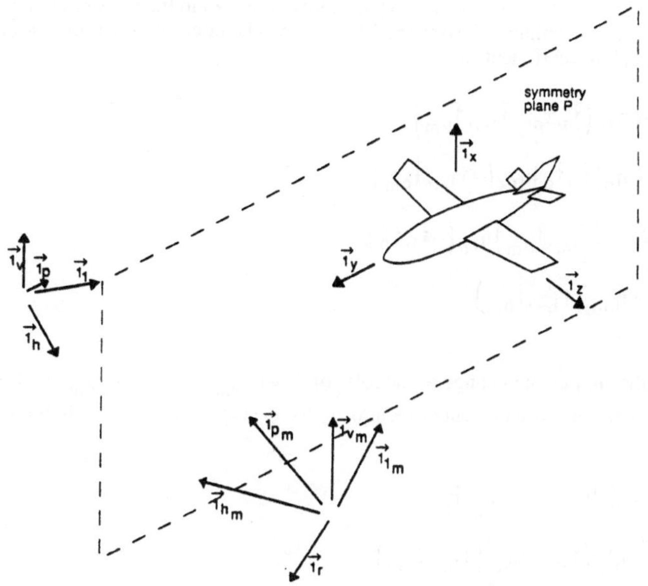


Figure 1.7.2. Incident and Scattered Fields at Symmetric Positions and Orientations with Respect to Symmetry Plane.

$$\vec{1}_{1_m} \times \vec{1}_{v_m} = -\vec{1}_{h_m} \ , \ \vec{1}_{v_m} \times \vec{1}_{h_m} = -\vec{1}_{1_m} \ , \ \vec{1}_{h_m} \times \vec{1}_{1_m} = -\vec{1}_{v_m} \tag{1.7.10}$$

The transverse dyads reflect as

$$\begin{aligned} \vec{\vec{1}}_{1_m} &= \vec{\vec{1}} - \vec{1}_{1_m}\vec{1}_{1_m} = \vec{\vec{1}} - \left(\vec{\vec{R}}\cdot\vec{1}_1\right)\left(\vec{\vec{R}}\cdot\vec{1}_1\right) \\ &= \vec{\vec{1}} - \vec{\vec{R}}\cdot\vec{1}_1\vec{1}_1\cdot\vec{\vec{R}} \\ &= \vec{\vec{R}}\cdot\vec{\vec{1}}_1\cdot\vec{\vec{R}} \end{aligned} \tag{1.7.11}$$

The other previously used general polarization vectors $\vec{1}_p$ and $\vec{1}_m$ also reflect as

$$\vec{1}_{p_m} = \vec{\vec{R}}\cdot\vec{1}_p \ , \ \vec{1}_{m_m} = \vec{\vec{R}}\cdot\vec{1}_m \tag{1.7.12}$$

Now let us apply the bistatic scattering formulae for the SEM coupling coefficients in section 5 to the case that

$$\vec{1}_r = -\vec{1}_{1_m} = -\vec{\vec{R}}\cdot\vec{1}_1 \ , \ \vec{1}_m = -\vec{1}_{p_m} = \vec{\vec{R}}\cdot\vec{1}_p \tag{1.7.13}$$

i.e., with scattering direction such that the receiver is in a mirror position to the transmitter and the receiver polarization is mirror to the transmitter polarization. This case can be referred to as mirror scattering. Then let us define a mirror coupling coefficient as

$$\begin{aligned}\eta_{m_\alpha}(\vec{1}_1,\vec{1}_p) &\equiv \eta_{f_\alpha}(\vec{1}_1,\vec{1}_p;-\vec{1}_{1_m},\vec{1}_{p_m}) \\ &= \eta_\alpha(\vec{1}_1,\vec{1}_p)\eta_{r_\alpha}(-\vec{1}_{1_m},\vec{1}_{p_m}) \\ &= \eta_\alpha(\vec{1}_{1_m},\vec{1}_{p_m})\eta_{r_\alpha}(-\vec{1}_1,\vec{1}_p) \\ &= \eta_{m_\alpha}(\vec{1}_{1_m},\vec{1}_{p_m}) \end{aligned} \tag{1.7.14}$$

This result follows on the simple interchange of the roles of $\vec{1}_1 \leftrightarrow \vec{1}_{1_m}$ and $\vec{1}_p \leftrightarrow \vec{1}_{p_m}$ in (1.5.1) with the choices in (1.7.13). It is also just a statement of reciprocity for the chosen directions. In normalized form this is

$$\begin{aligned}\eta_{m_\alpha}^{(n)}(\vec{1}_1,\vec{1}_p) &= \eta_{f_\alpha}^{(n)}(\vec{1}_1,\vec{1}_p;-\vec{1}_{1_m},\vec{1}_{p_m}) \\ &= \eta_\alpha^{(n)}(\vec{1}_1,\vec{1}_p)\eta_\alpha^{(n)}(\vec{1}_{1_m},\vec{1}_{p_m}) \\ &= \eta_{m_\alpha}^{(n)}(\vec{1}_{1_m},\vec{1}_{p_m}) \end{aligned} \tag{1.7.15}$$

This better exhibits the symmetry between coupling and recoupling coefficients.

Now consider the symmetry in the coupling coefficients because of the symmetry of the natural modes in (1.7.7). Consider the recoupling coefficient in (1.5.1) applied to the present case giving

$$\begin{aligned}\eta_{r_\alpha}(-\vec{1}_{1_m},\vec{1}_{p_m}) &= -s_\alpha\mu_0\vec{1}_{p_m}\cdot\left\langle \vec{1}_{1_m}e^{-\gamma_\alpha\vec{1}_{1_m}\cdot\vec{r}_s'};\vec{j}_{s_\alpha}(\vec{r}_s')\right\rangle \\ &= -s_\alpha\mu_0\left\langle \vec{1}_{p_m}e^{-\gamma_\alpha\vec{1}_{1_m}\cdot\vec{r}_s'};\vec{j}_{s_\alpha}(\vec{r}_s')\right\rangle \\ &= -s_\alpha\mu_0\left\langle \vec{1}_p\cdot\overleftrightarrow{R}e^{-\gamma_\alpha\vec{1}_1\cdot\overleftrightarrow{R}\cdot\vec{r}_s'};\pm\overleftrightarrow{R}\cdot\vec{j}_{s_{\substack{sy\\as}},\alpha}(\vec{r}_{s_m}')\right\rangle \\ &= \mp s_\alpha\mu_0\left\langle \vec{1}_p\,e^{-\gamma_\alpha\vec{1}_1\cdot\vec{r}_{s_m}'};\vec{j}_{s_{\substack{sy\\as}},\alpha}(\vec{r}_{s_m}')\right\rangle \\ &= \mp s_\alpha\mu_0\vec{1}_p\cdot\left\langle \vec{1}_1e^{-\gamma_\alpha\vec{1}_1\cdot\vec{r}_{s_m}'};\vec{j}_{s_{\substack{sy\\as}},\alpha}(\vec{r}_{s_m}')\right\rangle \end{aligned} \tag{1.7.16}$$

Regarding the variable of integration over S as $\vec{r}_{s_m}'$ then this is proportional to ± the numerator for the coupling coefficient $\eta_\alpha(\vec{1}_1, \vec{1}_p)$. Noting that the three-term symmetric-product denominator is just a complex number we now have

$$\begin{aligned}\eta_{\substack{sy\\as},\alpha'}(\vec{1}_1,\vec{1}_p) &= \pm\eta_{r_{\substack{sy\\as},\alpha'}}(-\vec{1}_{1_m},\vec{1}_{p_m})\frac{1}{-s_\alpha\mu_o}\left\langle \vec{j}_{s_\alpha}(\vec{r}_s);\frac{\partial}{\partial s}\tilde{\vec{Z}}_t(\vec{r},\vec{r}_s';s)\Big|_{s=s_\alpha};\vec{j}_{s_\alpha}(\vec{r}_s')\right\rangle^{-1}\\ &= \pm\eta_{\substack{sy\\as},\alpha'}(\vec{1}_{1_m},\vec{1}_{p_m})\end{aligned} \tag{1.7.17}$$

In normalized form this is just

$$\begin{aligned}\eta^{(n)}_{\substack{sy\\as},\alpha'}(\vec{1}_1,\vec{1}_p) &= \pm\,\eta^{(n)}_{r_{\substack{sy\\as},\alpha'}}(-\vec{1}_{1_m},\vec{1}_{p_m})\\ &= \pm\eta^{(n)}_{\substack{sy\\as},\alpha'}(\vec{1}_{1_m},\vec{1}_{p_m})\end{aligned} \tag{1.7.18}$$

These results are just the obvious extension of those for the natural modes in (1.7.7). Noting from (1.5.8) a symmetry on inversion of the polarization we also have

$$\eta^{(n)}_{\substack{sy\\as},\alpha'}(\vec{1}_1,\vec{1}_p) = \eta^{(n)}_{\substack{sy\\as},\alpha'}(\vec{1}_{1_m},\pm\vec{1}_{p_m}) \tag{1.7.19}$$

Thus an alternate way to look at coupling to antisymmetric modes is to reverse $\vec{1}_{p_m}$.

Applying these results to (1.7.15) gives

$$\begin{aligned}\eta^{(n)}_{m_{\substack{sy\\as},\alpha'}}(\vec{1}_1,\vec{1}_p) &= \eta^{(n)}_{m_{\substack{sy\\as},\alpha'}}(\vec{1}_{1_m},\vec{1}_{p_m})\\ &= \pm\left[\eta^{(n)}_{\substack{sy\\as},\alpha'}(\vec{1}_1,\vec{1}_p)\right]^2\\ &= \pm\left[\eta^{(n)}_{\substack{sy\\as},\alpha'}(\vec{1}_{1_m},\vec{1}_{p_m})\right]^2\end{aligned} \tag{1.7.20}$$

So the mirror coupling coefficient is also expressible as the square of a coupling coefficient, reminiscent of the result in (1.5.11) for the backscattering coupling coefficient. There is a difference in the signs, however, depending on whether symmetric or antisymmetric modes are being considered. In terms of the backscattering coupling coefficient in (1.5.11) the mirror coupling coefficient is

$$\eta^{(n)}_{m\,{sy \atop as},\alpha'}\left(\vec{1}_1,\vec{1}_p\right)=\pm\eta^{(n)}_{b\,{sy \atop as},\alpha'}\left(\vec{1}_1,\vec{1}_p\right)=\pm\eta^{(n)}_{b\,{sy \atop as},\alpha'}\left(\vec{1}_{1m},\vec{1}_{pm}\right) \tag{1.7.21}$$

This shows that mirror scattering is simply related to backscattering with a sign dependent on mode type.

To understand this a little better consider the orientation of the polarization $\vec{1}_p$ in fig. 1.7.2. This can be taken successively as one of two orthogonal polarizations $\vec{1}_v$ and $\vec{1}_h$. Consider the case that $\vec{1}_1$ is parallel to P as in fig. 1.7.3, i.e.,

$$\begin{aligned}&\vec{1}_1\cdot\vec{1}_z=0\ ,\ \ \vec{1}_{1m}=\vec{1}_1\\&\vec{1}_v=\vec{1}_{vm}\ ,\ \ \vec{1}_h=-\vec{1}_{hm}\end{aligned} \tag{1.7.22}$$

noting the opposite ways that $\vec{1}_v$ and $\vec{1}_h$ reflect. From (1.7.18), (1.7.19) and (1.5.8) we have

$$\begin{aligned}&\eta^{(n)}_{as,\alpha'}\left(\vec{1}_1,\vec{1}_v\right)=-\eta^{(n)}_{as,\alpha'}\left(\vec{1}_1,\vec{1}_v\right)=0\\&\eta^{(n)}_{sy,\alpha'}\left(\vec{1}_1,\vec{1}_h\right)=-\eta^{(n)}_{sy,\alpha'}\left(\vec{1}_1,\vec{1}_h\right)=0\end{aligned} \tag{1.7.23}$$

So vertical only couples to symmetric (and only scatters symmetric) and horizontal only couples to antisymmetric (and only scatters antisymmetric) provided $\vec{1}_1//P$. This is summarized in the backscattering formulae (as in (1.5.11))

$$\begin{aligned}\eta^{(n)}_{b\,{sy \atop as},\alpha'}\left(\vec{1}_1,\vec{1}_{v \atop h}\right)&=\left[\eta^{(n)}_{{sy \atop as},\alpha'}\left(\vec{1}_1,\vec{1}_{v \atop h}\right)\right]^2\\&=\eta^{(n)}_{f\,{sy \atop as},\alpha'}\left(\vec{1}_1,\vec{1}_{v \atop h};-\vec{1}_1,\vec{1}_{v \atop h}\right)\\&=\pm\eta^{(n)}_{m\,{sy \atop as},\alpha'}\left(\vec{1}_1,\vec{1}_{v \atop h}\right)\end{aligned} \tag{1.7.24}$$

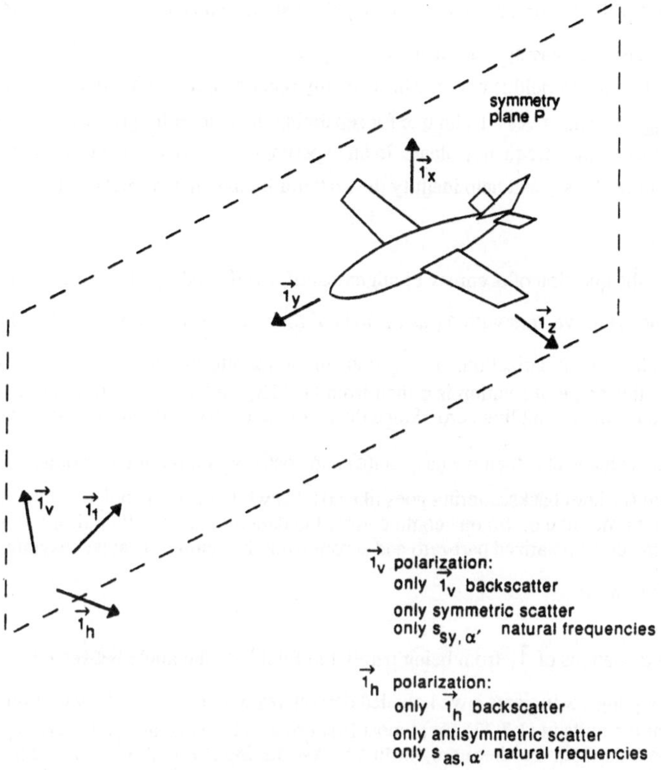


Figure 1.7.3. Incident Fields Propagating Parallel to Symmetry Plane.

$$
\begin{aligned}
\eta^{(n)}_{f_{\substack{sy\\as}},\alpha'}\left(\vec{1}_1,\vec{1}_{\substack{v\\h}};-\vec{1}_1,\vec{1}_{\substack{h\\v}}\right) &= \eta^{(n)}_{f_{\substack{sy\\as}},\alpha'}\left(\vec{1}_1,\vec{1}_{\substack{h\\v}};-\vec{1}_1,\vec{1}_{\substack{v\\h}}\right)\\
&= \eta^{(n)}_{f_{\substack{sy\\as}},\alpha'}\left(\vec{1}_1,\vec{1}_{\substack{v\\h}};-\vec{1}_1,\vec{1}_{\substack{h\\v}}\right)\\
&= 0
\end{aligned}
$$

Note that in the various pairs of symbols the upper symbols go together as do the lower symbols.

Consider for $\vec{1}_1//P$ that first one uses vertical polarization. The backscattered field is not only vertically polarized, it also has only symmetric natural frequencies $s_{sy,\alpha'}$. Now second use horizontal polarization. The backscattered field is not only horizontally polarized, it also has only antisymmetric natural frequencies $s_{as,\alpha'}$. This gives a technique for separating the natural frequencies of a scatterer into two separate sets in the complex frequency plane. In an experimental configuration this can be used to more accurately determine the $s_\alpha$ and help identify the scatterer by having two separate pole patterns at which to look.

Now there is the question of accuracy of alignment of $\vec{1}_v//P$ and $\vec{1}_h \perp P$. To the extent that there is some rotation of these unit vectors with $\vec{1}_1$ as the axis of rotation, this can be detected in the backscattered signals by cross polarization, i.e., $\vec{1}_v$ transmission scattering into $\vec{1}_h$ reception, and conversely. Suppose this angle of rotation is $\phi$ then from (1.7.23) we have the definition of $\phi = 0$ by the angle require to make the cross coupling zero. Since the coupling coefficients are proportional to the dot product of $\vec{1}_v$ and $\vec{1}_h$ with a vector then we have $\sin(\phi)$ and $\cos(\phi)$ variation of the various $\eta_\alpha^{(n)}$. The vertical and horizontal (in-line) backscattering goes like $\cos^2(\phi)$ while the cross polarized part goes like $\cos(\phi)\sin(\phi)$, or to first order like $\phi$. So one could correct the data by numerically adjust $\phi$ to rotate the coordinates to make the cross-polarized part zero and separate out the symmetric and antisymmetric parts and thereby the $s_{sy,\alpha'}$ and $s_{as,\alpha'}$.

Concerning deviations of $\vec{1}_1$ from being parallel to P let $\xi$ be the angle between $\vec{1}_1$ and P. Assuming that the coupling coefficients have bounded derivatives with respect to $\vec{1}_1$ variation, then the deviations away from the nulls in (1.7.23) are at most first order in $\xi$. This means that cross polarized backscattering is first order in $\xi$. Furthermore, in-line backscattering (for both vertical and horizontal polarizations) deviates from the $\xi = 0$ result by an error which is of order $\xi^2$ (due to the square for the backscattering coupling coefficient $\eta_b^{(n)}$ in (7.24)).

So for deviations of both direction of incidence and polarization that are small with respect to their ideal orientations with respect to P as in fig. 1.7.3, the errors are also small. This experimental configuration may then prove of practical significance.

## 1.8. Concluding Remarks

Symmetry then is a powerful concept in simplifying otherwise (more) complex problems. Reciprocity is a fundamental symmetry in the Maxwell equations for the case of suitably simple media (symmetric constitutive matrices). This leads to symmetry between incident and scattered far fields with special results for backscattering and forward scattering. Geometrical symmetry in a scatterer gives additional simplification to scattering, such as for the case that incidence and scattering directions are parallel to a symmetry plane.

These results can be used to guide scattering experiments. For SEM poles the coupling coefficients for surface currents are simply related to backscattering coupling coefficients. Furthermore a symmetry plane in a scatterer can be used to separate the natural frequencies into two sets in the backscattering.

PART 2. SEM BACKSCATTERING

2.1 Introduction

The previous part has considered the general electromagnetic scattering problem with some implications of reciprocity and symmetry, and applied these to the forms that appear in the eigenmode expansion method (EEM) and singularity expansion method (SEM). Looking at the far-field scattering of an incident plane wave one can define a scattering matrix (in complex frequency domain) which exhibits a symmetry (reciprocity) between the incident and scattered fields. In the case of backscattering this matrix is symmetric in the usual case of defined orthogonal horizontal and vertical polarizations. This part further develops the backscattering results for the SEM representation.

Our starting point is (1.5.18) from part 1 which reads

$$\vec{1}_m \cdot \tilde{\vec{E}}_f(\vec{r},s) = -\frac{E_o}{4\pi r} e^{-\gamma r} \mu_o \sum_\alpha s_\alpha \tilde{f}(s_\alpha) \left\langle \vec{j}_{s_\alpha}(\vec{r}_s); \frac{\partial}{\partial s} \tilde{\vec{Z}}_t(\vec{r}_s, \vec{r}_s\,'; s)\bigg|_{s=s_\alpha}; \vec{j}_{s_\alpha}(\vec{r}_s\,')\right\rangle^{-1}$$

$$\vec{1}_m \cdot \vec{C}_{f_\alpha}(\vec{1}_r, \vec{1}_1) \cdot \vec{1}_p [s - s_\alpha]^{-1}$$

$$+ \text{ singularities of } \tilde{f}(s)$$

$$+ \text{ possible entire function}$$

$$\vec{C}_{f_\alpha}(\vec{1}_r, \vec{1}_1) = \vec{C}_{r_\alpha}(\vec{1}_r)\vec{C}_\alpha(\vec{1}_1)$$

$$\vec{C}_\alpha(\vec{1}_1) = \left\langle \vec{1}_1 e^{-\gamma_\alpha \vec{1}_r \cdot \vec{r}_s\,'}; \vec{j}_{s_\alpha}(\vec{r}_s\,')\right\rangle$$

$$\vec{C}_{r_\alpha}(\vec{1}_r) = \left\langle \vec{1}_r e^{\gamma_\alpha \vec{1}_r \cdot \vec{r}_s}; \vec{j}_{s_\alpha}(\vec{r}_s\,')\right\rangle$$

$$\eta_{f_\alpha}(\vec{1}_1, \vec{1}_p; \vec{1}_r, \vec{1}_m) = -s_\alpha \mu_o \left\langle \vec{j}_{s_\alpha}(\vec{r}_s); \frac{\partial}{\partial s} \tilde{\vec{Z}}_t(\vec{r}_s, \vec{r}_s\,'; s)\bigg|_{s=s_\alpha}; \vec{j}_{s_\alpha}(\vec{r}_s\,')\right\rangle^{-1} \vec{1}_m \cdot \vec{C}_{f_\alpha}(\vec{1}_r, \vec{1}_1) \cdot \vec{1}_p$$

$$\vec{1}_m \cdot \vec{C}_{f_\alpha}(\vec{1}_r, \vec{1}_1) \cdot \vec{1}_p = \left[\vec{1}_m \cdot \left\langle \vec{1}_r e^{\gamma_\alpha \vec{1}_r \cdot \vec{r}_s}; \vec{j}_{s_\alpha}(\vec{r}_s)\right\rangle\right]\left[\vec{1}_p \cdot \left\langle \vec{1}_1 e^{-\gamma_\alpha \vec{1}_1 \cdot \vec{r}_s\,'}; \vec{j}_{s_\alpha}(\vec{r}_s\,')\right\rangle\right]$$

$$= \left\langle \vec{1}_m e^{\gamma_\alpha \vec{1}_r \cdot \vec{r}_s}; \vec{j}_{s_\alpha}(\vec{r}_s)\right\rangle \left\langle \vec{1}_p e^{-\gamma_\alpha \vec{1}_1 \cdot \vec{r}_s\,'}; \vec{j}_{s_\alpha}(\vec{r}_s\,')\right\rangle \qquad (2.1.1)$$

where

$\vec{1}_1 \equiv$ direction of incidence

$\vec{1}_p \equiv$ polarization of incident electtric field

$\vec{1}_1 \cdot \vec{1}_p = 0$

$\overleftrightarrow{1}_1 \equiv \overleftrightarrow{1} - \vec{1}_1 \vec{1}_1 \equiv$ incident transverse dyad

$\vec{1}_r \equiv$ scattering direction

$\vec{1}_m \equiv$ direction for measuring (sampling) scattered electric field

$\vec{1}_r \cdot \vec{1}_m = 0$ (constraint on $\vec{1}_m$)

$\overleftrightarrow{1}_r \equiv \overleftrightarrow{1} - \vec{1}_r \vec{1}_r \equiv$ scattered transverse dyad

$r \equiv$ distance from observer of scattered field to scatterer(center of minimum circumscribing sphere)

$E_o \equiv$ scaling constant (V / m) for incident field at scatterer

$f(t) \equiv$ waveform of incident field

$s \equiv \Omega + j\omega \equiv$ Laplace – transform (two sided) variable

$\equiv$ complex frequency

$\overleftrightarrow{Z}_t \equiv$ Kernel of E – field (or impedance) integral equation

$\vec{j}_{s_\alpha} \equiv$ natural mode of scatterer

$s_\alpha \equiv$ natural frequency of scatterer

$\tilde{\eta}_{f_\alpha} \equiv$ far coupling coefficient

(2.1.2)

Noting the important symmetry (reciprocity) relationship

$$\vec{C}_{r_\alpha}\left(\vec{1}_r\right) = \vec{C}_\alpha\left(-\vec{1}_r\right)$$

$$\overleftrightarrow{C}_{f_\alpha}\left(\vec{1}_r, \vec{1}_1\right) = \vec{C}_\alpha\left(-\vec{1}_r\right)\vec{C}_\alpha\left(\vec{1}_1\right) \tag{2.1.3}$$

then the single vector function $\vec{C}_\alpha$ can be used to characterize the symmetric dyadic residue.

As noted in part 1 the integrals defining $\vec{C}_\alpha$, $\vec{C}_{r_\alpha}$, and the three-term symmetric-product denominator in (2.1.1) are taken over the surface S of the scatterer and involve surface-current-density natural modes. This, however, is not a restriction in that volume integrals are also allowed with volume current-density natural modes. It is only necessary that the media comprising the scatterer be reciprocal, i.e., have symmetric constitutive-parameter matrices.

For backscattering we have

$$\vec{1}_r \equiv -\vec{1}_1$$

$$\overleftrightarrow{C}_{b_\alpha}\left(\vec{1}_1\right) \equiv \overleftrightarrow{C}_{f_\alpha}\left(-\vec{1}_1, \vec{1}_1\right) = \vec{C}_\alpha\left(\vec{1}_1\right)\vec{C}_\alpha\left(\vec{1}_1\right) \tag{2.1.4}$$

which is evidently a symmetric dyad (required by reciprocity). Furthermore introducing the standard unit vectors as in fig. 2.1.1

$$\vec{1}_h = \text{horizontal polarization}$$

$$\vec{1}_v = \vec{1}_h \times \vec{1}_1 = \vec{1}_r \times \vec{1}_h = \text{"vertical" polarization}$$

$$\vec{1}_v \times \vec{1}_h = \vec{1}_1$$

$$\vec{1}_1 \times \vec{1}_v = \vec{1}_h \tag{2.1.5}$$

Here horizontal is usually taken as parallel to the horizon and vertical is interpreted loosely. Taken in the sequence $\vec{1}_v, \vec{1}_h, \vec{1}_1$ this is a right handed system with $\vec{1}_1$ as the direction from the observer to the scatterer (direction of incidence). Later considering the h, v plane with positive sense of rotation from $\vec{1}_h$ toward $\vec{1}_v$ then $\left(\vec{1}_h, \vec{1}_v, -\vec{1}_1\right)$ *or* $\left(\vec{1}_h, \vec{1}_v, \vec{1}_r\right)$ form a right handed system.

Defining

$$W_\alpha \equiv w_\alpha^2 \equiv -s_\alpha \mu_o \left\langle \vec{j}_{s_\alpha}(\vec{r}_s); \frac{\partial}{\partial s} \tilde{\tilde{Z}}_t(\vec{r}_s, \vec{r}_s{}', s)\Big|_{s=s_\alpha}; \vec{j}_{s_\alpha}(\vec{r}_s{}') \right\rangle^{-1} \tag{2.1.6}$$

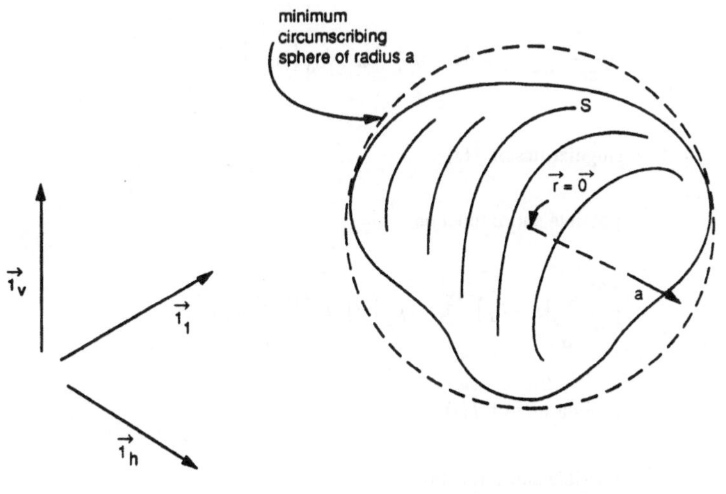


Figure 2.1.1. Backscattering from General Scatterer

we can in turn define

$$\vec{c}_\alpha\left(\vec{1}_1\right) \equiv w_\alpha \vec{C}_\alpha\left(\vec{1}_1\right)$$

$$\vec{c}_{r_\alpha}\left(\vec{1}_r\right) \equiv w_\alpha \vec{C}_{r_\alpha}\left(\vec{1}_r\right) = \vec{c}_\alpha\left(-\vec{1}_r\right)$$

$$\vec{\vec{c}}_{f_\alpha}\left(\vec{1}_r,\vec{1}_1\right) \equiv W_\alpha \vec{\vec{C}}_{f_\alpha}\left(\vec{1}_r,\vec{1}_1\right)$$

$$= \vec{c}_{r_\alpha}\left(\vec{1}_r\right)\vec{c}_\alpha\left(\vec{1}_1\right)$$

$$\vec{\vec{c}}_{b_\alpha}\left(\vec{1}_1\right) \equiv W_\alpha \vec{\vec{C}}_{b_\alpha}\left(\vec{1}_1\right)$$

$$= \vec{c}_\alpha\left(\vec{1}_1\right)\vec{c}_\alpha\left(\vec{1}_1\right) \tag{2.1.7}$$

This for is what is experimentally observable in (2.1.1) using far-field measurements. For symmetry one might as well normalize the two vectors in the dyad (particularly for backscattering) the same way. The relative normalization of the two is arbitrary since it is actually the dyadic product which is observable in the far field.

This form for $\vec{\vec{c}}_{b_\alpha}$ is convenient in that it is effectively the backscattering matrix for SEM measurement purposes. Rewriting (2.1.1) we have for backscattering

$$\vec{1}_m \cdot \tilde{\vec{E}}_f(\vec{r},s) = \frac{E_o e^{-\gamma r}}{4\pi r} \sum_\alpha \tilde{f}(s_\alpha)[s-s_\alpha]^{-1} \vec{1}_m \cdot \vec{\vec{c}}_{b_\alpha}\left(\vec{1}_1\right) \cdot \vec{1}_p$$

$$+ \text{ singularities of } \tilde{f}(s)$$

$$+ \text{ possible entire function}$$

$$= \frac{e^{-\gamma r}}{4\pi r} \sum_\alpha [s-s_\alpha]^{-1} \vec{1}_m \cdot \vec{\vec{c}}_{b_\alpha}\left(\vec{1}_1\right) \cdot \tilde{\vec{E}}^{(inc)}\left(\vec{O},s_\alpha\right)$$

$$+ \text{ singularities of } \tilde{f}(s)$$

$$+ \text{ possible entire function}$$

$$\tilde{\vec{E}}^{(inc)}(\vec{r},s) = E_o \tilde{f}(s) \vec{1}_p e^{-\gamma \vec{1}_1 \cdot \vec{r}} \tag{2.1.8}$$

So allowing for the delay $e^{-\gamma r}$ and $1/(4\pi r)$ expansion to the observer then the pole terms of the backscattering matrix are just $[s-s_\alpha]^{-1}\vec{c}_{b_\alpha}(\vec{1}_1)$. Allowing for the $1/(4\pi r)$ each such term has dimension meters. The reside matrix $\vec{c}_{b_\alpha}$ then has dimensions m/s giving $\vec{c}_\alpha$ dimensions (m/s)$^{1/2}$.

In terms of coupling coefficients we have

$$\eta_{f_\alpha}\left(\vec{1}_1,\vec{1}_p,\vec{1}_r,\vec{1}_m\right)=\eta_\alpha\left(\vec{1}_1,\vec{1}_p\right)\eta_{r_\alpha}\left(\vec{1}_r,\vec{1}_m\right) \qquad \text{(far coupling coefficient)}$$

$$\eta_\alpha\left(\vec{1}_1,\vec{1}_p\right)=-\frac{W_\alpha}{s_\alpha\mu_o}\vec{1}_p\cdot\vec{C}_\alpha\left(\vec{1}_1\right) \qquad \text{(coupling coefficient)}$$

$$=-\frac{w_\alpha}{s_\alpha\mu_o}\vec{1}_p\cdot\vec{c}_\alpha\left(\vec{1}_1\right)$$

$$\eta_{r_\alpha}\left(\vec{1}_r,\vec{1}_m\right)=-s_\alpha\mu_o\vec{1}_m\cdot\vec{C}_{r_\alpha}\left(\vec{1}_r\right) \qquad \text{(recoupling coefficient)}$$

$$=-s_\alpha\mu_o\vec{1}_m\cdot\vec{C}_\alpha\left(-\vec{1}_r\right)$$

$$=-\frac{s_\alpha\mu_o}{w_\alpha}\vec{1}_m\cdot\vec{c}_\alpha\left(-\vec{1}_r\right)$$

$$\eta_{f_\alpha}\left(\vec{1}_1,\vec{1}_p;\vec{1}_r,\vec{1}_m\right)=\vec{1}_m\cdot\vec{c}_\alpha\left(-\vec{1}_r\right)\vec{c}_\alpha\left(\vec{1}_1\right)\cdot\vec{1}_p$$

$$=W_\alpha\vec{1}_m\cdot\vec{C}_{f_\alpha}\left(\vec{1}_r,\vec{1}_1\right)\cdot\vec{1}_p \qquad (2.1.9)$$

For backscattering this becomes

$$\eta_{f_\alpha}\left(\vec{1}_1,\vec{1}_p;-\vec{1}_1,\vec{1}_m\right)=\vec{1}_m\cdot\vec{c}_{b_\alpha}\left(\vec{1}_1\right)\cdot\vec{1}_p \qquad (2.1.10)$$

where $\vec{1}_m$ *and* $\vec{1}_p$ assume values of $\vec{1}_h$ *and* $\vec{1}_v$ in all combinations.

Now interpret our backscattering residue dyad as a 2 x 2 matrix noting that $\vec{c}_\alpha(\vec{1}_1)$ has no component in the $\vec{1}_1$ direction. Labelling the two components of vectors (transverse) to $\vec{1}_1$ by subscripts h and v we have

$$\vec{c}_{b_\alpha}\left(\vec{1}_1\right) = \vec{c}_\alpha\left(\vec{1}_1\right)\vec{c}_\alpha\left(\vec{1}_1\right) = \begin{pmatrix} c_{h_\alpha}\left(\vec{1}_1\right) \\ c_{v_\alpha}\left(\vec{1}_1\right) \end{pmatrix} \begin{pmatrix} c_{h_\alpha}\left(\vec{1}_1\right) \\ c_{v_\alpha}\left(\vec{1}_1\right) \end{pmatrix}$$

$$= \begin{pmatrix} c_{h_\alpha}^2\left(\vec{1}_1\right) & c_{h_\alpha}\left(\vec{1}_1\right)\, c_{v_\alpha}\left(\vec{1}_1\right) \\ c_{v_\alpha}\left(\vec{1}_1\right)\, c_{h_\alpha}\left(\vec{1}_1\right) & c_{v_\alpha}^2\left(\vec{1}_1\right) \end{pmatrix}$$

$$= \begin{pmatrix} \vec{1}_h \cdot \vec{c}_{b_\alpha}\left(\vec{1}_1\right) \cdot \vec{1}_h & \vec{1}_h \cdot \vec{c}_{b_\alpha}\left(\vec{1}_1\right) \cdot \vec{1}_v \\ \vec{1}_v \cdot \vec{c}_{b_\alpha}\left(\vec{1}_1\right) \cdot \vec{1}_h & \vec{1}_v \cdot \vec{c}_{b_\alpha}\left(\vec{1}_1\right) \cdot \vec{1}_v \end{pmatrix}$$

$$= \left(c_{b_{n,m}}\left(\vec{1}_1\right)\right)_\alpha$$

$$= \begin{pmatrix} \eta_{f_\alpha}\left(\vec{1}_1, \vec{1}_h; -\vec{1}_1, \vec{1}_h\right) & \eta_{f_\alpha}\left(\vec{1}_1, \vec{1}_v; -\vec{1}_1, \vec{1}_h\right) \\ \eta_{f_\alpha}\left(\vec{1}_1, \vec{1}_h; -\vec{1}_1, \vec{1}_v\right) & \eta_{f_\alpha}\left(\vec{1}_1, \vec{1}_v; -\vec{1}_1, \vec{1}_v\right) \end{pmatrix} \qquad (2.1.11)$$

The various forms of this matrix are of course symmetric by reciprocity.

In this two-dimensional h, v space we have the identity (transverse identity)

$$\vec{1}_1 = \vec{1}_r = \vec{1} - \vec{1}_1\vec{1}_1 = \vec{1} - \vec{1}_r\vec{1}_r = \begin{pmatrix} 1 & 0 \\ 0 & 1 \end{pmatrix} \qquad (2.1.12)$$

with rotation by $\pi / 2$ in the positive sense (from $\vec{1}_h$ toward $\vec{1}_v$) as

$$\vec{1}_r \times \;= -\vec{1}_1 \times \;= \begin{pmatrix} 0 & -1 \\ 1 & 0 \end{pmatrix} \bullet \qquad (2.1.13)$$

Rotation by an angle $\psi$ in the positive sense is

$$\begin{pmatrix} \cos(\psi) & -\sin(\psi) \\ \sin(\psi) & \cos(\psi) \end{pmatrix} \bullet \qquad (2.1.14)$$

## 2.2 Representation of $\vec{c}_\alpha\left(\vec{1}_1\right)$

Assuming that at the natural frequency $s_\alpha$ of interest there is only one natural mode $\vec{j}_{s_\alpha}$ (no model degeneracy), then the observer has a backscattering residue dyad $\vec{c}_{b_\alpha}$ characterized by a single vector $\vec{c}_\alpha$. This is in general a complex valued vector. Restricting our attention to the two-dimensional h, v plane there are two vector components, each complex valued. This can also be thought of as four real numbers. The magnitude is straight forward as

$$\left|\vec{c}_\alpha(\vec{1}_1)\right|^2 = \vec{c}_\alpha(\vec{1}_1)\cdot\vec{c}_\alpha^{\,*}(\vec{1}_1) = \left|c_{h_\alpha}(\vec{1}_1)\right|^2 + \left|c_{v_\alpha}(\vec{1}_1)\right|^2$$

$$= \mathrm{Re}^2\left[c_{h_\alpha}(\vec{1}_1)\right] + \mathrm{Im}^2\left[c_{h_\alpha}(\vec{1}_1)\right] + \mathrm{Re}^2\left[c_{v_\alpha}(\vec{1}_1)\right] + \mathrm{Im}^2\left[c_{v_\alpha}(\vec{1}_1)\right] \tag{2.2.1}$$

This vector is an observable characteristic of the scatterer so its orientation or direction is also of interest. This is traditionally characterized by a polarization ellipse [2-13]. In such a context an incoming electromagnetic wave (monochomatic, i.e. single $\omega$) is viewed as an electric vector rotating in the h, v plane with the tip of the electric vector forming an ellipse. By establishing the lengths of the major and minor axes of the ellipse and orientation of the major axis in the h, v plane this incoming wave is characterized except for an arbitrary phase (in effect 3 of 4 real numbers established). Viewed as a complex vector this incoming wave has the same form as $\vec{c}_\alpha$.

Then analogous to the traditional use $\vec{c}_\alpha$ can also be so characterized by a polarization ellipse. As in fig. 2.2.1, let us introduce an angle $\psi$ in the positive sense in the h, v plane. This polarization ellipse is contained in many articles. A convenient form in [2-6] is applied in our case as

$$\vec{c}_\alpha = c_{h_\alpha}\vec{1}_h + c_{h_v}\vec{1}_v = \begin{pmatrix} c_{h_\alpha} \\ c_{v_\alpha} \end{pmatrix} = \begin{pmatrix} c_h \\ c_v \end{pmatrix}_\alpha$$

$$= a_\alpha \begin{pmatrix} \cos(\psi_\alpha) & -\sin(\psi_\alpha) \\ \sin(\psi_\alpha) & \cos(\psi_\alpha) \end{pmatrix} \cdot \begin{pmatrix} \cos(\tau_\alpha) \\ j\sin(\tau_\alpha) \end{pmatrix} e^{j\Delta_\alpha}$$

$$a_\alpha, \psi_\alpha, \tau_\alpha, \Delta_\alpha \quad \text{all real scalars} \tag{2.2.2}$$

Here the result is cast in terms of one magnitude and three angles. Note in the polarization ellipse the "overall" phase angle $\Delta_\alpha$ is not included, but can be specified by a number. Also all these parameters (including $\vec{1}_h$ and $1_v$) are functions of $\vec{1}_1$ which is now suppressed for convenience.

To visualize this better, define a real direction $\vec{1}_\alpha^{(r)}$ in the h, v plane such that $\left|\vec{1}_\alpha^{(r)}\cdot\vec{c}_\alpha\right|$ is maximized. This will determine the orientation of the major axis of the ellipse. Consider the radial unit vector in the h, v plane

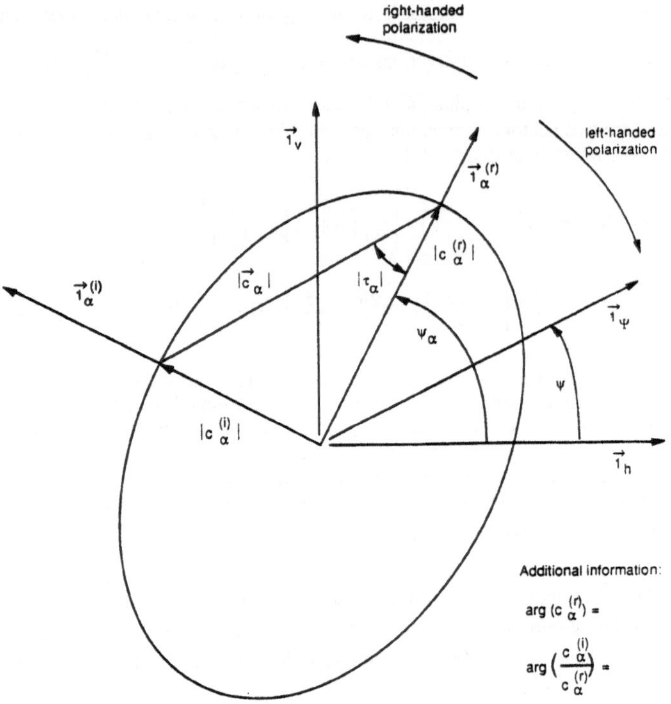


Figure 2.2.1. Polarization Ellipse for $\vec{c}_\alpha$.

$$\vec{1}_\psi = \cos(\psi)\vec{1}_h + \sin(\psi)\vec{1}_v = \begin{pmatrix} \cos(\psi) \\ \sin(\psi) \end{pmatrix} \tag{2.2.3}$$

and the corresponding angular unit vector

$$\vec{1}_\psi = -\sin(\psi)\vec{1}_h + \cos(\psi)\vec{1}_v = \begin{pmatrix} -\sin(\psi) \\ \cos(\psi) \end{pmatrix}$$

$$= \begin{pmatrix} 0 & -1 \\ 1 & 0 \end{pmatrix} \cdot \vec{1}_\psi = -\vec{1}_1 \times \vec{1}_\psi \tag{2.2.4}$$

So now maximize

$$\left|\vec{1}_\Psi \cdot \vec{c}_\alpha\right|^2 = \vec{1}_\Psi \cdot \vec{c}_\alpha^{\,*} \vec{c}_\alpha \cdot \vec{1}_\Psi$$

$$= \left|c_{h_\alpha}\right|^2 \cos^2(\psi) + \left[c_{h_\alpha} c_{h_\alpha}^* + c_{h_\alpha}^* c_{v_\alpha}\right]\cos(\psi)\sin(\psi)$$

$$+ \left|c_{v_\alpha}\right|^2 \sin^2(\psi) \tag{2.2.5}$$

and use this to define $\vec{1}_\alpha$ as $\vec{1}_\psi$ and $\psi_\alpha$ as $\psi$ corresponding to the maximum. Note the Hermetian dyad $\vec{c}_\alpha^{\,*}\vec{c}_\alpha$ for which we are finding the maximum projection along a real unit vector. Differentiating with respect to $\psi$ and setting to zero gives (after a little algebra) the well-known result

$$\tan(2\psi_\alpha) = \frac{c_{h_\alpha} c_{v_\alpha}^* + c_{h_\alpha}^* c_{v_\alpha}}{\left|c_{h_\alpha}\right|^2 - \left|c_{v_\alpha}\right|^2} = \frac{2\mathrm{Re}\left[c_{h_\alpha} c_{v_\alpha}^*\right]}{\left|c_{h_\alpha}\right|^2 - \left|c_{v_\alpha}\right|^2} \tag{2.2.6}$$

Over the range of $-\pi < \psi_\alpha < \pi$ there are 4 solutions, in general, of (2.2.6) corresponding to 2 opposite directions for for the major axis and 2 opposite directions for the minor axis. To resolve this form

$$c_\alpha^{(r)} \equiv \vec{1}_\alpha^{(r)} \cdot \vec{c}_\alpha$$

$$c_\alpha^{(i)} \equiv \vec{1}_\alpha^{(i)} \cdot \vec{c}_\alpha$$

$$\vec{1}_\alpha^{(r)} \equiv \begin{pmatrix} \cos(\psi_\alpha) \\ \sin(\psi_\alpha) \end{pmatrix}$$

$$\vec{1}_\alpha^{(i)} \equiv \begin{pmatrix} \cos\left(\psi_\alpha + \frac{\pi}{2}\right) \\ \sin\left(\psi_\alpha + \frac{\pi}{2}\right) \end{pmatrix} = \begin{pmatrix} -\sin(\psi_\alpha) \\ \cos(\psi_\alpha) \end{pmatrix} = \begin{pmatrix} 0 & -1 \\ 1 & 0 \end{pmatrix} \cdot \vec{1}_\alpha^{(r)} \tag{2.2.7}$$

and constrain

$$\left|c_\alpha^{(r)}\right| \geq \left|c_\alpha^{(i)}\right| \tag{2.2.8}$$

to select a major axis for $\vec{1}_\alpha^{(r)}$ and a minor axis for $\vec{1}_\alpha^{(i)}$. It remains but to choose which of two values of $\psi_\alpha$ separated by $\pi$ to specify. Note that reversing $\vec{1}_\alpha^{(r)}$ (rotation by $\pi$) changes the complex number $c_\alpha^{(r)}$ to $-c_\alpha^{(r)}$, so one could choose that $\psi_\alpha$ which made, say, $\mathrm{Re}\left[c_\alpha^{(r)}\right] \geq 0$. The usual convention is to choose $0 \leq \Psi_\alpha < \pi$, as in the example in fig. 2.2.1. Note that in a measurement situation absolute phase may not be available so that one may only have, say, $\left|c_\alpha^{(r)}\right|$.

Also, note that the sign of $\vec{c}_\alpha$ is ambiguous since the dyad $\vec{c}_\alpha \vec{c}_\alpha$ is what we obtain from the scattering measurement, this being the same as $(-\vec{c}_\alpha)(-\vec{c}_\alpha)$. The usual convention is then not a limitation. Even if we have absolute phase by knowing r sufficiently accurately, or have relative phases, referencing $\vec{c}_\alpha$ to the phase of some other selected $\vec{c}_{\alpha_o}$, there is still this sign ambiguity inherent in the dyad.

Referring back to the form in (2.2.2), this can be now put in the form

$$\vec{c}_\alpha = \begin{pmatrix} \cos(\psi) & -\sin(\psi_\alpha) \\ \sin(\psi_\alpha) & \cos(\psi_\alpha) \end{pmatrix} \cdot \begin{pmatrix} c_\alpha^{(r)} \\ c_\alpha^{(i)} \end{pmatrix} = c_\alpha^{(r)} \vec{1}_\alpha^{(r)} + c_\alpha^{(i)} \vec{1}_\alpha^{(i)}$$

$$|\vec{c}_\alpha|^2 = \left|c_\alpha^{(r)}\right|^2 + \left|c_\alpha^{(i)}\right|^2$$

$$\frac{c_\alpha^{(i)}}{c_\alpha^{(r)}} = j\tan(\tau_\alpha), \quad -\frac{\pi}{4} \leq \tau_\alpha \leq \frac{\pi}{4}$$

$$\frac{\text{minor axis}}{\text{major axis}} = \left|\frac{c_\alpha^{(i)}}{c_\alpha^{(r)}}\right|$$

$$c_\alpha^{(r)} = a_\alpha \cos(\tau_\alpha) e^{j\Delta_\alpha}$$

$$c_\alpha^{(i)} = j a_\alpha \sin(\tau_\alpha) e^{j\Delta_\alpha}$$

$$\arg\left(\frac{c_\alpha^{(i)}}{c_\alpha^{(r)}}\right) = \pm\frac{\pi}{2} \quad \left(\text{for } c_\alpha^{(i)} \neq 0\right) \tag{2.2.9}$$

This form of $\vec{c}_\alpha$ has it represented as two complex numbers times orthogonal real unit vectors. The two complex numbers are orthogonal also in the complex-plane sense (have imaginary ratio). So this decomposition is like a double complex number (with components designated by "r" and "i" superscripts).

Note that there is an IEEE standard for the sense of rotation of a polarization based on radiation and reception by a helical antenna, right or left handedness being defined in the usual screw-thread sense [2-13]. Right handed elliptical polarization has for an incoming plane wave the electric vector in the h, v plane rotating in the sense of increasing $\psi$. Left handed elliptical polarization has the sense of decreasing $\psi$. Noting time dependence $e^{st}$ as $e^{j\omega t}$ with $\omega$ positive (or $s_\alpha = \Omega_\alpha + j\omega_\alpha$ with $\omega_\alpha$ positive) then we have

$$\arg\left(\frac{c_\alpha^{(i)}}{c_\alpha^{(r)}}\right) = \left\{\begin{array}{l} +\frac{\pi}{2} \Rightarrow \text{left handed elliptical polarization} \\ -\frac{\pi}{2} \Rightarrow \text{right handed elliptical polarization} \end{array}\right\} \tag{2.2.10}$$

For special cases we have

$$\left|c_\alpha^{(i)}\right| = \left|c_\alpha^{(r)}\right| \Rightarrow \text{circular polarization}$$

$$c_\alpha^{(i)} = 0 \Rightarrow \text{linear polarization} \tag{2.2.11}$$

One can also use this information to represent the polarization on what is called the Poincare sphere [2-6, 2-13].

## 2.3 Properties of $\tilde{c}_{b_\alpha}$ for Non-Degenerate Modes

Now look at some of the properties of $\tilde{c}_{b_\alpha}$ assuming that at the $s_\alpha$ of interest the scatterer has only one natural mode, this being the typically encountered case. Since $\tilde{c}_{b_\alpha}$ can be represented as a single dyad then as discussed previously we need only know $\vec{c}_\alpha$, a vector in the h, v plane which is characterized by two complex numbers $c_{h_\alpha}$ and $c_{v_\alpha}$, or equivalently four real numbers. This can be compared to the usual backscattering matrix which is a 2 x 2 matrix in the h, v plane (as in (1.3.8) of Part 1). Applying reciprocity makes the matrix symmetric which means it requires in general three complex numbers (or equivalently six real numbers) to characterize it.

This single vector $\vec{c}_\alpha$ which characterizes the residue scattering matrix leads to an interesting result. From (2.1.11) such a backscattering matrix is singular as

$$\det\left(\tilde{c}_{b_\alpha}\left(\vec{1}_1\right)\right) = \det\left(\left(c_{b_{n,m}}\left(\vec{1}_1\right)\right)\right) = \det\left(\vec{c}_\alpha\left(\vec{1}_1\right)\vec{c}_\alpha\left(\vec{1}_1\right)\right) = 0 \text{ for all } \vec{1}_1 \tag{2.3.1}$$

This is a necessary condition for a single natural mode characterizing $\vec{c}_{b_\alpha}$ and can be used as a test on data (allowing for noise). Having found a zero-determinant residue matrix then (2.1.11) can be used to construct $\vec{c}_\alpha$.

Note that $\vec{c}_{b_\alpha}$ is already in diagonal form. One normalized eigenvector (right and left) is $\vec{c}_\alpha / [\vec{c}_\alpha \cdot \vec{c}_\alpha]^{1/2}$ with eigenvalue $\vec{c}_\alpha \cdot \vec{c}_\alpha$. The second eigenvalue is zero and corresponding eigenvector is orthogonal to $\vec{c}_\alpha$.

Now $\vec{c}_{b_\alpha}$ and $\vec{c}_\alpha$ are associated with $s_\alpha$ and $\vec{j}_{s_\alpha}$. Having found these we also have $s_\alpha^*, \vec{j}_{s_\alpha}^*, \vec{c}_{b_\alpha}^*$, and $\vec{c}_\alpha^*$ since we are dealing with the Laplace transform of real-valued time functions and operators. For each $s_\alpha$ not on the real axis of the s plane there is another (a separate value of $\alpha$) that we find automatically together with the corresponding $\vec{c}_\alpha^*$.

For the case that $s_\alpha$ is on the negative real axis we have real $\vec{j}_{s_\alpha}$, real $\vec{c}_\alpha$, real $\vec{c}_{b_\alpha}$ and real $W_\alpha$. However $w_\alpha$ can be real or imaginary depending on the sign of $W_\alpha$. From section 2 we then have

$$\vec{c}_\alpha(\vec{1}_1) = c_\alpha^{(r)}(\vec{1}_1)\,\vec{1}_\alpha(\vec{1}_1), \text{ real or imaginary for all } \vec{1}_1 \tag{2.3.2}$$

Since this comes from the residue matrix we have

$$\vec{c}_{b_\alpha}(\vec{1}_1) = \vec{c}_\alpha(\vec{1}_1)\vec{c}_\alpha(\vec{1}_1) = c_\alpha^{(r)^2}(\vec{1}_1)\vec{1}_\alpha^{(r)}(\vec{1}_1)\vec{1}_\alpha^{(r)}(\vec{1}_1)$$

$$c_\alpha^{(r)^2}(\vec{1}_1) = \vec{c}_\alpha(\vec{1}_1)\cdot\vec{c}_\alpha(\vec{1}_1) \text{ (real eigenvalue)}$$

$$\vec{1}_\alpha^{(r)}(\vec{1}_1) = \left[\vec{c}_\alpha(\vec{1}_1)\cdot\vec{c}_\alpha(\vec{1}_1)\right]^{-1/2}\vec{c}_\alpha(\vec{1}_1) \text{ (real eigenvector)} \tag{2.3.3}$$

Note also in this case of real $s_\alpha$ that $\vec{c}_\alpha$ is characterized by two real numbers (or one real and one imaginary) as contrasted to two complex numbers.

Since $\vec{c}_\alpha$ is an integral over the natural mode $\vec{j}_{s_\alpha}$, we can use the orientation of $\vec{c}_\alpha$ to tell something about the orientation of $\vec{j}_{s_\alpha}$ and hence of the scatterer. From (2.1.1) and (2.1.7) we have

$$\vec{c}_\alpha\left(\vec{1}_1\right) = w_\alpha\left\langle \vec{1}_1 e^{-\gamma_\alpha \vec{1}_1\cdot\vec{r}_s'} ; \vec{j}_{s_\alpha}\left(\vec{r}_s'\right)\right\rangle$$

$$= w_\alpha \vec{1}_1\left\langle e^{-\gamma_\alpha \vec{1}_1\cdot\vec{r}_s'} ; \vec{j}_{s_\alpha}\left(\vec{r}_s'\right)\right\rangle$$

$$= w_\alpha\left\langle e^{-\gamma_\alpha \vec{1}_1\cdot\vec{r}_s'} , \vec{1}_1 \cdot \vec{j}_{s_\alpha}\left(\vec{r}_s'\right)\right\rangle \tag{2.3.4}$$

Noting especially the last form, only components of $\vec{j}_{s_\alpha}$ orthogonal to $\vec{1}_1$ contribute to $\vec{c}_\alpha$.

Consulting fig. 2.3.1, look at some thin slice at constant $\vec{1}_1 \cdot \vec{r}_s'$ through the scatterer. While for a perfectly conducting scatterer $\vec{1}_1 \cdot \vec{j}_{s_\alpha}$ is locally tangential (parallel) to the scatterer, it can vary in phase over this slice. In some cases of interest there is a predominant orientation of $\vec{1}_1 \cdot \vec{j}_{s_\alpha}$ over the slice in some particular direction. Integrating (summing) over all such slices there is the phase and amplitude factor $e^{-\gamma_\alpha \vec{1}_1\cdot\vec{r}_s'}$ which weights the contributions from the different slices. Provided there is a predominant orientation to the $\vec{c}_\alpha$ contributions from each slice, then this predominant orientation will appear in $\vec{c}_\alpha$. One general kind of scatterer to which such considerations are simply applicable is a long slender conductor. Then the low-order natural modes can be thought of as currents running along the body. Then if the body is straight so the currents can be thought of as passing in the same direction, the projection of direction is given by $\vec{c}_\alpha$ at the observer. The observer then has the scatterer orientation as projected on the h, v plane.

Looking at fig. 2.3.2 one can see how $\vec{c}_\alpha$ can be used in some cases to tell shapes and orientations of scatterers. Suppose one is backscattering from a typical aircraft with side-on illumination. Consider, for example, one of the lowest order symmetric modes [2-2, 2-3, 2-10]. From the side $\vec{c}_\alpha$ will be dominated by fuselage currents and so will point (in the sense of $\vec{1}_\alpha^{(r)}$) approximately parallel to the fuselage. So in this case $\vec{c}_\alpha$ is related to the pitch angle (nose up or down) of the aircraft. For comparison consider some higher order symmetric natural frequency which might be associated with the vertical stabilizer (at least approximately, depending on details of the scattering shape). In such a case one could have $\vec{c}_\alpha$ more vertical.

Note that with side-on illumination the dominant antisymmetric resonances (associated with wings and horizontal stabilizers) will be suppressed, being small for $\vec{1}_1$ perpendicular to the symmetry plane of the aircraft provided that the antisymmetric currents are predominantly parallel to $\vec{1}_1$.

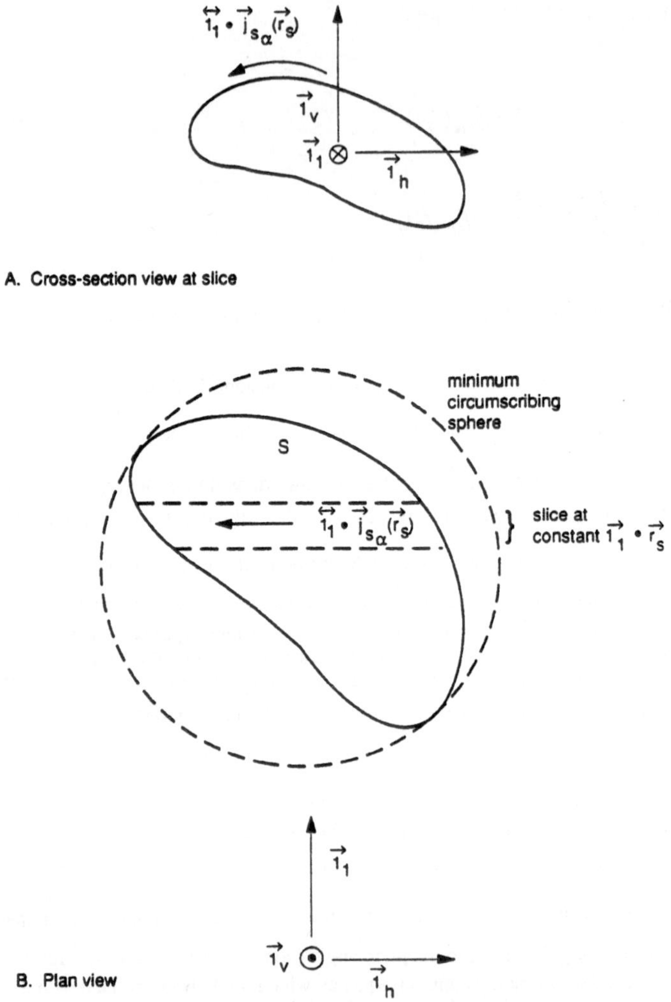


Figure 2.3.1. Relation of $\vec{c}_\alpha(\vec{1}_1)$ to the Natural Mode.

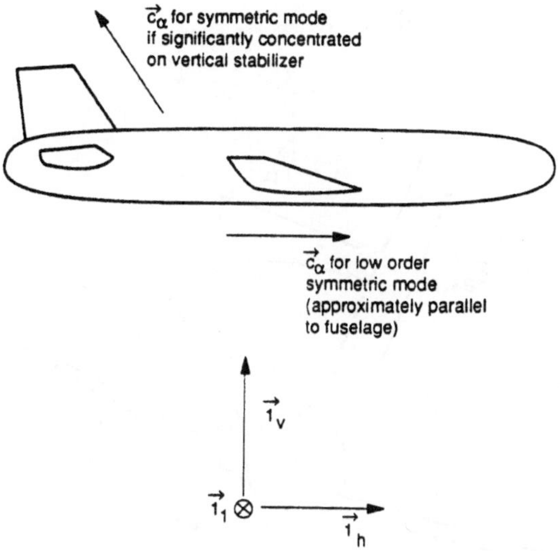


Figure 2.3.2. Typical Aircraft Viewed from Side.

Excessive presence of antisymmetric $\vec{c}_\alpha$ in the backscattering then tells something about the yaw angle of the aircraft (nose side to side) and/or roll angle of the aircraft, depending on the orientation of the $\vec{c}_\alpha$ for the low-order antisymmetric modes.

## 2.4. Observer on Symmetry Plane of Scatterer

Now let the scatterer have a symmetry plane [2-1, 2-3]. This separates the natural modes into two kinds, labelled symmetric (subscript sy) and antisymmetric (subscript as). As in fig. 2.4.1, let there be a symmetry plane P through the scatterer with unit normal $\vec{1}_P$. Then we have a reflection dyad

$$\overleftrightarrow{R}_P = \overleftrightarrow{1} - 2\,\vec{1}_P\vec{1}_P \tag{2.4.1}$$

which reflects coordinates through the symmetry plane as

$$\vec{r}_m = \overleftrightarrow{R}_P \cdot \vec{r} \tag{2.4.2}$$

The point symmetry group is

$$R_P = \{(R_P),(1)\} \quad , \quad (R_P)^2 = (1) \quad (identity) \tag{2.4.3}$$

Symmetric and antisymmetric natural modes are defined by their reflection properties as

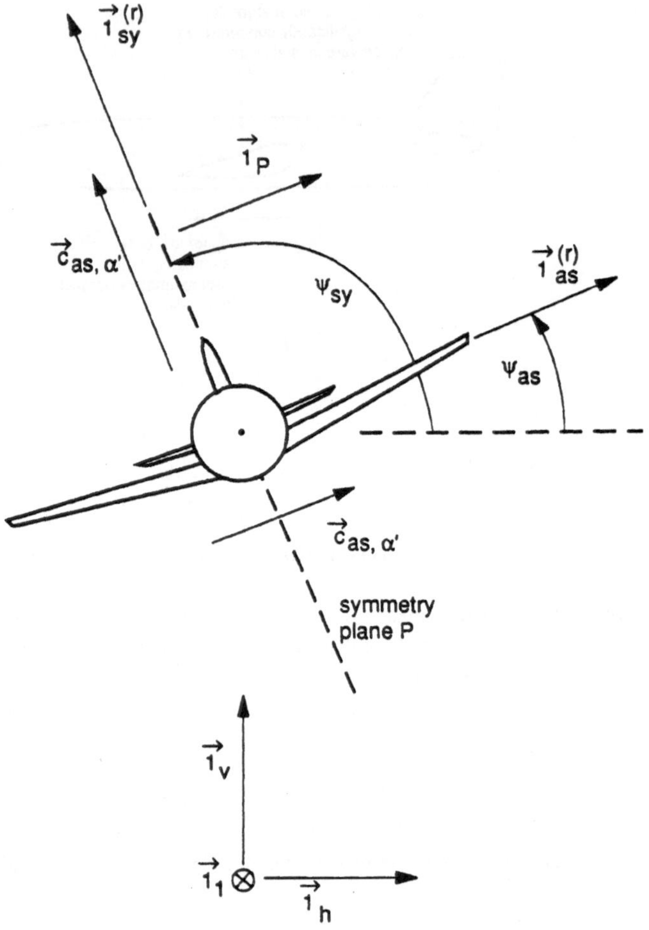


Figure 2.4.1. Symmetry Plane Through Scatterer and Observer.

$$\vec{j}_{s_{\substack{sy\\as}},\alpha'}\left(\vec{r}_{s_m}\right) = \pm\vec{\vec{R}}_P \cdot \vec{j}_{s_{\substack{sy\\as}},\alpha'}\left(\vec{r}_s\right) \tag{2.4.4}$$

where the natural-mode index is now partitioned as

$$\alpha = \left(\begin{matrix} sy \\ \\ as \end{matrix}, \alpha'\right) \tag{2.4.5}$$

For a general observer the backscattering residue matrices will exhibit this symmetry as $\vec{1}_1$ is reflected through P. (See some discussion concerning this in Part 1.) For present purposes, however, confine the observer position to lie on P so that

$$\vec{1}_1 \cdot \vec{1}_P = 0 \tag{2.4.6}$$

Then we have the result that the $\vec{c}_\alpha$ are all either perpendicular to or parallel to P (as well as being perpendicular to $\vec{1}_P$). This follows from the fact that on P a symmetric electric field is parallel to P while an antisymmetric electric field is perpendicular to P [2-1, 2-3]. Thus we have

$$\vec{c}_{sy,\alpha'} = c_{sy,\alpha'}^{(r)} \vec{1}_{sy}^{(r)} \quad , \quad \vec{1}_{sy}^{(r)} = -\vec{1}_1 \times \vec{1}_P$$

$$\vec{c}_{as,\alpha'} = c_{as,\alpha'}^{(r)} \vec{1}_{as}^{(r)} \quad , \quad \vec{1}_{as}^{(r)} = \vec{1}_P \tag{2.4.7}$$

Thus the $\vec{c}_\alpha$ are all now linearly polarized and are characterized in general by one complex and one real number (or equivalently three real numbers). Furthermore there are only two polarization vectors to cover all (assuming no modal degeneracy). Thus the observer can determine from measurements of the $\vec{c}_\alpha$ the presence of such a scatterer symmetry plane passing through itself (at least in a necessary-condition sense).

Referred to our basis in the h, v plane we have angles $\psi_{sy}$ and $\psi_{as}$ for $\psi_\alpha$ describing the linear polarizations via

$$\vec{1}_{sy}^{(r)} = \cos(\psi_{sy})\vec{1}_h + \sin(\psi_{sy})\vec{1}_v$$

$$\vec{1}_{as}^{(r)} = \cos(\psi_{as})\vec{1}_h + \sin(\psi_{as})\vec{1}_v$$

$$\psi_{sy} = \psi_{as} \pm \frac{\pi}{2} \tag{2.4.8}$$

Considering a typical aircraft with nose-on or tail-on illumination, then of course $\psi_{as}$ is the roll angle.

## 2.5 Body of Revolution

Suppose now our scatterer has an axis of revolution as illustrated in fig. 2.5.1. This means that the scatterer can be rotated about this axis by some angle, say $\phi$, with no observable change. This is $C_\infty$ symmetry with group elements.

$$(C_\infty)_\phi \equiv \text{rotation by } \phi$$

$$(C_\infty)_\phi^L = (C_\infty)_{L\phi} \equiv \text{rotation by } L\phi$$

$$(C_\infty)_o = (C_\infty)_{2\pi} = (C_\infty)_{2\pi L'} = (1) \equiv \text{identity}$$

$$L' \equiv \text{an integer } (+,\ -,\ or\ 0) \tag{2.5.1}$$

This symmetry assures that the backscattering to the observer will be independent of $\phi$. Note the $2\pi$ periodicity so that increasing $\phi$ by $2\pi$ does not give a new group element.

Now consider a plane P containing this axis and the observer location. Let us further assume that this is a symmetry plane. This result is not implied by the symmetry axis if one allows the scatterer to contain anisotropic (yet still reciprocal) materials. If the scatterer is perfectly conducting then $C_\infty$ symmetry implies reflection symmetry as in (2.4.3), but which we now term $R_a$ symmetry ("a" denoting the fact that this plane contains the symmetry axis) with

$$R_a = \{(R_a), (1)\}\ ,\ (R_a)^2 = (1) \tag{2.5.2}$$

Note that this differs from other notations [2-11] which term this $R_v$ symmetry, but this conflicts with the standard horizontal and vertical (h and v) designation for polarization of the transmitted and received electric fields. In this context one might also use $R_t$ (instead of $R_h$) to denote a symmetry plane perpendicular to (or "transverse" to) the symmetry axis.

If anisotropic materials are allowed let us require that the associated constitutive-parameter matrices also reflect with the appropriate symmetry with respect to P [2-1]. With this restriction the scatterer still has $R_a$ symmetry, which when adjoined to the $C_\infty$ symmetry, gives $C_{\infty a}$ symmetry and all planes passing through the axis are symmetry planes. This high order of symmetry has elements

$$C_{\infty a} = \left\{(C_\infty)_\phi, (C_\infty)_\phi (R_a) \mid \phi \text{ real}\right\} \tag{2.5.3}$$

For every $\phi$ there are then two kinds of group elements: rotation by $\phi$ and a combination of rotation and reflection. Note that $C_{\infty a}$ and $(R_a)$ do not in general commute. However, we have

$$(C_\infty)_\phi (R_a) = (R_a)(C_\infty)_{-\phi} = (R_a)(C_\infty)_{2\pi-\phi} \tag{2.5.4}$$

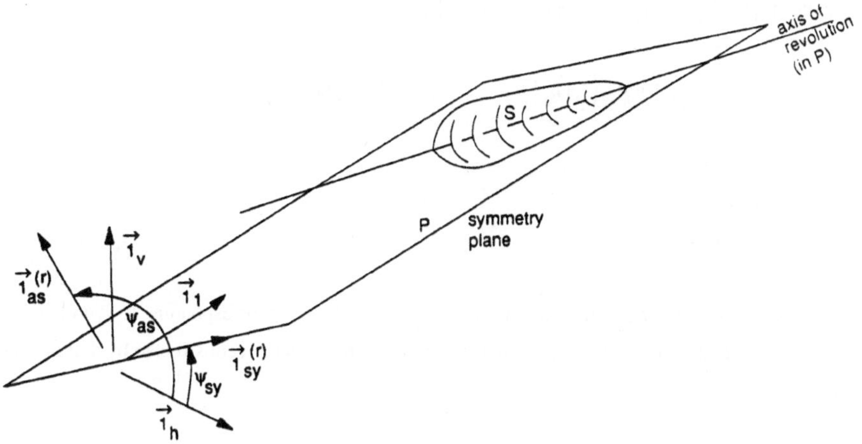


Figure 2.5.1. Body of Revolution with Symmetry Plane Through Observer.

where the defining symmetry plane is taken on $\phi = 0$.

Referring back to section 4 then the effect of the symmetry plane is to make the $\vec{c}_\alpha$ be some symmetric and some antisymmetric. So the observer will see these two sets as in (2.4.7) except for perhaps some shift to relative to $\psi_{as}$ as

$$\psi_{sy} = \psi_{as} - \frac{\pi}{2} \tag{2.5.5}$$

for the orientation of P in the h, v plane in fig. 2.4.1. This relates to the choice of convention for polarization angles in section 2.2.

Now as the scatterer is rotated through some angle $\phi$, the observer see no change, i.e. $\vec{c}_{b_\alpha}$ is unchanged. As one rotates the scatterer and the associated natural modes $\vec{j}_{s_\alpha}$ of the scatterer, and hence the associated $\vec{c}_\alpha$, this may raise some concern. However, the resolution is simple by noting that there is a modal degeneracy for such a body of revolution. The $\phi$ dependence of the currents on the body is expandable as a Fourier series giving cos $(m\phi)$ and sin $(m\phi)$ terms. The index m belongs to certain natural frequencies $s_\alpha$. There are then two natural modes for each $s_\alpha$ (except for m = 0 when there is but one). These two modes have separate $\vec{c}_\alpha(\vec{1}_1)$. There are then two different $\alpha$ index sets for the same $s_\alpha$.

For a particular $s_\alpha$ let us consider symmetric (sy) and antisymmetric (as) excitation. Given the fact that a symmetric incident field must give only a symmetric backscattered field, and similarly in the antisymmetric case, then we have

$$\tilde{c}_{b_\alpha}\left(\vec{1}_1\right) = \tilde{c}_{sy,\alpha'}\left(\vec{1}_1\right)\tilde{c}_{sy,\alpha'}\left(\vec{1}_1\right) + \tilde{c}_{as,\alpha'}\left(\vec{1}_1\right)\tilde{c}_{as,\alpha'}\left(\vec{1}_1\right)$$

$$= c_{b_{sy,\alpha'}}\left(\vec{1}_1\right)\vec{1}_{sy}\left(\vec{1}_1\right)\vec{1}_{sy}\left(\vec{1}_1\right) + c_{b_{as,\alpha'}}\left(\vec{1}_1\right)\vec{1}_{as}\left(\vec{1}_1\right)\vec{1}_{as}\left(\vec{1}_1\right)$$

$$\alpha = \begin{pmatrix} sy \\ & ,\alpha' \\ as \end{pmatrix} \tag{2.5.6}$$

This is a diagonal form for $\tilde{c}_{b_\alpha}$ with eigenvalues $c_{b_{sy,\alpha'}}$ and $c_{b_{as,\alpha'}}$, corresponding to symmetric and antisymmetric excitaiton, respectively. From measurements this can be reconstructed by diagonalizing the measured $\tilde{c}_{b_\alpha}$.

Note that

$$\det\left(\tilde{c}_{b_\alpha}\left(\vec{1}_1\right)\right) = \det\left(\left(c_{b_{n,m}}\left(\vec{1}_1\right)\right)\right) = c_{b_{sy,\alpha'}}\left(\vec{1}_1\right)c_{b_{as,\alpha'}}\left(\vec{1}_1\right)$$

$$= c^{(\alpha)}_{b_{h,h}}\left(\vec{1}_1\right)c^{(\alpha)}_{b_{v,v}}\left(\vec{1}_1\right) - c^{(\alpha)^2}_{b_{h,v}}\left(\vec{1}_1\right) \tag{2.5.7}$$

which is in general non zero, in contradistinction to the case of a single $\tilde{c}_\alpha$ as in (2.3.1).

Noting that

$$tr\left(\tilde{c}_{b_\alpha}\left(\vec{1}_1\right)\right) = tr\left(\left(c_{b_{n,m}}\left(\vec{1}_1\right)\right)\right) = c_{b_{sy,\alpha'}} + c_{b_{as,\alpha'}}$$

$$= c_{b_{h,h}}\left(\vec{1}_1\right) + c_{b_{v,v}}\left(\vec{1}_1\right) \tag{2.5.8}$$

then with (2.5.7) both eigenvalues are readily determined. The eigenvectors are normalized and real as $\vec{1}_{sy,\alpha'}$ and $\vec{1}_{as,\alpha'}$. The scatering residue matrix is now characterized by two complex numbers (the eigenvalues) and one real angle $\psi_{sy}$ (since by (2.5.5) $\psi_{as}$ is constrained). This is equivalently five real numbers. This is still one less than the three complex (or six real) numbers needed to generally characterize a scattering matrix in the h, v plane.

For the results of (2.5.6) to apply it is merely necessary that the two natural modes belonging to $s_\alpha$ produce two linearly independent $\vec{c}_\alpha$ (i.e. non parallel). These two vectors then span the h, v plane (i.e. any vector can be expressed as a linear combination of the two). The $\vec{1}_{sy}$ and $\vec{1}_{as}$ are merely a convenient diagonalizing basis. As can be seen in special cases (e.g. a symmetry plane perpendicular to the symmetry axis ($R_t$ symmetry) with this plane also through the observer), the two $\vec{c}_\alpha$ can become parallel giving a representation involving a single dyad.

For the case of m=0 (no $\phi$ variation in cylindrical coördinates based on the axis of rotation) there is only one natural mode for each $s_\alpha$. The results of section 2.4 then apply and $\vec{c}_{b_\alpha}$ is characterized by a single vector $\vec{c}_\alpha$ which is either parallel to P or perpendicular to it.

Note that the symmety axis of the scatterer is assumed to not pass through the observer for these results. Otherwise P is not uniquely specified.

## 2.6 Observer on Symmetry Axis of Scatterer

Continuing our consideration of symmetry implications in backscattering let us now orient the scatterer such that its symmetry axis points at the observer. As indicated in fig. 2.6.1 A our body of revolution is described in a cylindrical coordinate system $(\psi, \ \phi, \ z)$ where the z axis is the symmetry axis passing through the observer so that

$$\vec{1}_z = \vec{1}_1 \tag{2.6.1}$$

Now as a body of revolution the group elements in (2.5.1) have matrix (dyadic) representations [2-11]

$$(C_\infty)_\phi \to (C_{n,m}(\phi)) \equiv \begin{pmatrix} \cos(\phi) & -\sin(\phi) \\ \sin(\phi) & \cos(\phi) \end{pmatrix} \quad \equiv \text{rotation by } \phi$$

$$(C_\infty)_\phi^L \to (C_{n,m}(\phi))^L \equiv \begin{pmatrix} \cos(\phi) & -\sin(\phi) \\ \sin(\phi) & \cos(\phi) \end{pmatrix}^L = \begin{pmatrix} \cos(L\phi) & -\sin(L\phi) \\ \sin(L\phi) & \cos(L\phi) \end{pmatrix}$$

$$= (C_{n,m}(L\phi)) \quad \equiv \text{rotation by } L\phi$$

$$(C_\infty)_0 \to (C_{n,m}(0)) = (C_{n,m}(2\pi)) = \begin{pmatrix} 1 & 0 \\ 0 & 1 \end{pmatrix} = \vec{1}_z \equiv \vec{1}_x, \vec{1}_x + \vec{1}_y \vec{1}_y$$

$$= \vec{1}_h \vec{1}_h + \vec{1}_v \vec{1}_v \equiv \text{identity} \tag{2.6.2}$$

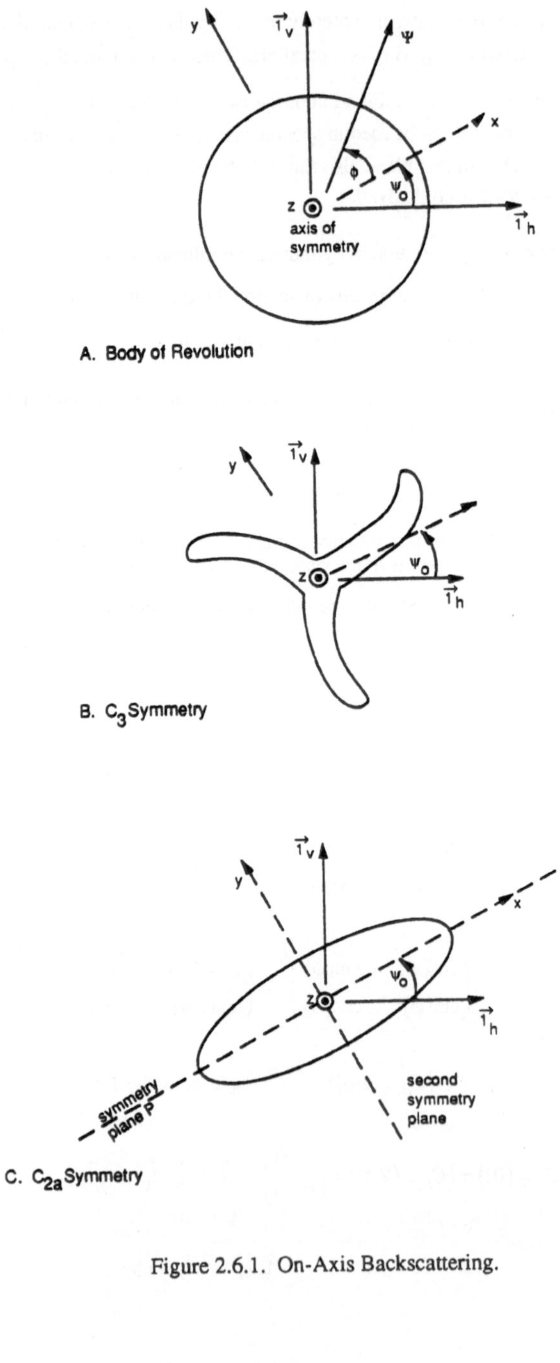


Figure 2.6.1. On-Axis Backscattering.

Note in fig. 2.6.1 A that $\phi = 0$ can be taken at any angle, say $\psi_o$, in the h, v plane. This is a two dimensional representation since we are not considering any symmetries (such as $R_t$) involving the z coordinate.

Note that the rotation matrix is real and unitary since

$$\left(C_{n,m}(\phi)\right)^T = \begin{pmatrix} \cos(\phi) & \sin(\phi) \\ -\sin(\phi) & \cos(\phi) \end{pmatrix} = \left(C_{n,m}(\phi)\right)^{-1}$$

$$= \left(C_{n,m}(-\phi)\right) \tag{2.6.3}$$

Let us not yet assume that there is $R_a$ symmetry (axial symmetry planes). What can be inferred from the $C_\infty$ symmetry? Now if we consider some incident polarization (real) characterized by some unit vector, say $\vec{1}_o$, in the h, v plane and scatter axially the field we obtain an amplitude and polarization characterized by $\vec{\vec{c}}_{b_\alpha} \cdot \vec{1}_o$. Rotating the scattering by $\phi$ gives backscatter $\left(C_{n,m}(\phi)\right) \cdot \vec{\vec{c}}_{b_\alpha} \cdot \vec{1}_o$. Now reverse the order. Rotate and then scatter by $\vec{\vec{c}}_{b_\alpha}$ giving $\vec{\vec{c}}_{b_\alpha} \cdot \left(C_{n,m}(\phi)\right) \cdot \vec{1}_o$. Since $\vec{1}_o$ is arbitrary we have

$$\left(C_{n,m}(\phi)\right) \cdot \vec{\vec{c}}_{b_a} = \vec{\vec{c}}_{b_\alpha} \cdot \left(C_{n,m}(\phi)\right) \tag{2.6.4}$$

i.e., the rotation and scattering <u>commute</u>. Viewed another way this is a coordinate rotation, in one case before and another after the scattering.

To see the constraints this imposes on $\vec{\vec{c}}_{b_\alpha}$ write (2.6.4) out in components as

$$\begin{pmatrix} \cos(\phi) & -\sin(\phi) \\ \sin(\phi) & \cos(\phi) \end{pmatrix} \cdot \begin{pmatrix} c_{b_{1,1}} & c_{b_{1,2}} \\ c_{b_{2,1}} & c_{b_{2,2}} \end{pmatrix}_\alpha = \begin{pmatrix} c_{b_{1,1}}\cos(\phi) - c_{b_{2,1}}\sin(\phi) & c_{b_{1,2}}\cos(\phi) - c_{b_{2,2}}\sin(\phi) \\ c_{b_{1,1}}\sin(\phi) + c_{b_{2,1}}\cos(\phi) & c_{b_{1,2}}\sin(\phi) + c_{b_{2,2}}\cos(\phi) \end{pmatrix}_\alpha$$

$$\begin{pmatrix} c_{b_{1,1}} & c_{b_{1,2}} \\ c_{b_{2,1}} & c_{b_{2,2}} \end{pmatrix}_\alpha \cdot \begin{pmatrix} \cos(\phi) & -\sin(\phi) \\ \sin(\phi) & \cos(\phi) \end{pmatrix} = \begin{pmatrix} c_{b_{1,1}}\cos(\phi) + c_{b_{1,2}}\sin(\phi) & -c_{b_{1,1}}\sin(\phi) + c_{b_{1,2}}\cos(\phi) \\ c_{b_{2,1}}\cos(\phi) + c_{b_{2,2}}\sin(\phi) & -c_{b_{2,1}}\sin(\phi) + c_{b_{2,2}}\cos(\phi) \end{pmatrix}_\alpha \tag{2.6.5}$$

Equating each of the matrix elements in the two forms of the matrix products we have

$$\left[c_{b_{1,1}}^{(\alpha)} \quad - c_{b_{2,2}}^{(\alpha)}\right] \quad \sin(\phi) = 0$$

$$\left[c_{b_{1,2}}^{(\alpha)} \quad + c_{b_{2,1}}^{(\alpha)}\right] \quad \sin(\phi) = 0 \tag{2.6.6}$$

For $\sin(\phi) \neq 0$ we then have

$$c_{b_{1,1}}^{(\alpha)} = c_{b_{2,2}}^{(\alpha)}$$

$$c_{b_{2,1}}^{(\alpha)} = - c_{b_{1,2}}^{(\alpha)}$$

$$\overleftrightarrow{c}_{b_\alpha} = c_{b_{1,1}}^{(\alpha)}\begin{pmatrix}1 & 0\\ 0 & 1\end{pmatrix} + c_{b_{1,2}}^{(\alpha)}\begin{pmatrix}0 & 1\\ -1 & 0\end{pmatrix} \tag{2.6.7}$$

so that the rotation invariance allows $\overleftrightarrow{c}_{b_\alpha}$ to be the linear combination of an identity and a rotation.

Now constrain the scatterer to be reciprocal, i.e. comprised of reciprocal media, in addition to symmetry. Then we have

$$c_{b_{1,2}}^{(\alpha)} = c_{b_{2,1}}^{(\alpha)} = 0 \quad , \quad \overleftrightarrow{c}_{b_\alpha} = \overleftrightarrow{c}_{b_\alpha}^{T}$$

$$\overleftrightarrow{c}_{b_\alpha} = c_{b_\alpha} \overleftrightarrow{1}_z \tag{2.6.8}$$

i.e., just a complex constant times the identify. This requires just one complex number (or two real numbers) to specify, no angles being needed. Note that while the on-axis $C_\infty$ discussion has been centered around scattering residue matrices it applies to scattering matrices at arbitrary complex frequencies as well.

Note that we have not assumed a symmetry plane containing the axis, i.e. $R_a$ symmetry for this result to hold, as has been done previously [2-5]. It only relies on $C_\infty$ symmetry and reciprocity. One might ask if this necessarily implies $R_a$ symmetry, but it does not. This can be seen through examples. Let the scatterer contain anisotropic materials (say conducting in some preferred direction). Let the currents spiral around the symmetry axis. Approximations involving N arms on, say, a conical spiral scatterer meet this condition.

Consider now some other symmetries with respect to the z axis for on-axis backscattering. Begin as in fig. 2.6.1 B with $C_N$ symmetry, the illustration being for $C_3$ symmetry. As in [2-4, 2-11] we have

$$C_N = \{(C_N)_1, (C_N)_2, \ldots, (C_N)_{N-1}, (1)\}$$

$$(C_N)_n \equiv \text{rotation by } \frac{2\pi n}{N} \text{ (positive } \phi \text{ direction)}$$

$$(C_N)_n = (C_N)_1^n$$

$$(C_N)_1^N = (C_N)_N = (1) \tag{2.6.9}$$

This group is a subgroup of $C_\infty$. It has a two-dimensional matrix representation like (2.6.2) as

$$(C_N)_1 \to \left(C_{n,m}\left(\frac{2\pi}{N}\right)\right) = \begin{pmatrix} \cos\left(\frac{2\pi}{N}\right) & -\sin\left(\frac{2\pi}{N}\right) \\ \sin\left(\frac{2\pi}{N}\right) & \cos\left(\frac{2\pi}{N}\right) \end{pmatrix} \equiv \text{rotation by } \frac{2\pi}{N}$$

$$(C_N)_n \to \left(C_{n,m}\left(\frac{2\pi n}{N}\right)\right) = \begin{pmatrix} \cos\left(\frac{2\pi n}{N}\right) & -\sin\left(\frac{2\pi n}{N}\right) \\ \sin\left(\frac{2\pi n}{N}\right) & \cos\left(\frac{2\pi n}{N}\right) \end{pmatrix} = \left(C_{n,m}\left(\frac{2\pi}{N}\right)\right)^n \equiv \text{rotation by } \frac{2\pi n}{N}$$

$$(C_N)_0 \to (C_{n,m}(0)) = (C_{n,m}(2\pi)) = \begin{pmatrix} 1 & 0 \\ 0 & 1 \end{pmatrix} = \vec{1}_z = \vec{1}_x\vec{1}_x + \vec{1}_y\vec{1}_y = \vec{1}_h\vec{1}_h + \vec{1}_v\vec{1}_v \tag{2.6.10}$$

As indicated in fig. 2.6.1 B one might associate some angle $\psi_o$ (corresponding to $\phi = 0$ and some plane which is in general not a symmetry plane). Let this scatterer have $C_N$ symmetry. Then (2.6.4) is replaced by

$$\left(C_{n,m}\left(\frac{2\pi n}{N}\right)\right) \cdot \tilde{c}_{b_\alpha} = \tilde{c}_{b_\alpha} \cdot \left(C_{n,m}\left(\frac{2\pi n}{N}\right)\right) \tag{2.6.11}$$

i.e., $\phi$ for rotation symmetry is now discrete. Considering (2.6.5) and the result (2.6.6) let us require

$$\sin\left(\frac{2\pi n}{N}\right) \neq 0 \text{ for at least one of } n = 1, 2, \ldots, N \tag{2.6.12}$$

This requirement is satisfied for N≥3. As can be seen for N=1 (n=1) and N=2 (n=1 and n=2) the sine function is always zero. From this it follows as in (2.6.8) that

$$\tilde{c}_{b_\alpha} = c_{b_\alpha} \overleftrightarrow{1}_z \text{ for } C_N \text{ symmetry with } N \geq 3 \tag{2.6.13}$$

Note the example in fig. 2.6.1 B does not have $R_a$ symmetry; it is not required for this result.

Note that $C_2$ is not adequate for this result. As a special case in fig. 2.6.1 C let $C_2$ symmetry be adjoined by $R_a$ symmetry, giving $C_{2a}$ symmetry. The existence of one symmetry plane (say defining $\psi_o$) implies a second (at $\psi_o + \pi / 2$). However, as illustrated, the extension of the scatterer in the direction of these two symmetry planes need not be the same. Then the results of section 2.4 apply, giving two different eigenvalues to $\tilde{c}_{b_\alpha}$ with unit eigenvectors oriented by $\psi_o$ and $\psi_o + \pi / 2$.

## 2.7. Concluding Remarks

So it appears that the backscattering matrix can be cast in the form of a residue matrix which is closely related to the scatterer properties. For non-degenerate natural modes this is a dyad which is characterized by a single two-component complex vector. For cases of symmetry in the scatterer and observer location various additional properties appear in the $\tilde{c}_{b_\alpha}'$.

Some of these results are reminiscent of low-frequency scattering [2-12]. The polarizability tensors are real valued for perfect conductors. The scattering matrix, being real valued, is characterized by three real (not complex) numbers. This is a reduction comparable to that for the case of a symmetry plane on the $\tilde{c}_\alpha$ as in section 2.4.

The previous part has considered the bistatic form of the SEM residue matrices, noting various symmetries in coupling to the incident field and recoupling to the scattered field based on reciprocity, as well as some effects of a symmetry plane. Various other symmetries can be explored in the scattering residue matrices, including effects of reciprocity and scatterer symmetry such as is done for the usual scattering matrices [2-7, 2-8, 2-11]. Hopefully further simplifications will result.

References (Part 1)


1-1. C. E. Baum, Singularity Expansion of Electromagnetic Fields and Potentials Radiated from Antennas or Scattered From Objects in Free Space, Sensor and Simulation Note 179, May 1973.

1-2. C. E. Baum, Measurement of the Surface Curl of the Surface Current Density, Sensor and Simulation Note 286, November 1984, and Electromagnetics, 1986, pp. 145-160.

1-3. C. E. Baum, Interaction of Electromagnetic Fields With an Object Which Has an Electromagnetic Symmetry Plane, Interaction Note 63, March 1971.

1-4. C. E. Baum, Emerging Technology for Transient and Broad-Band Analysis and Synthesis of Antennas and Scatterers, Interaction Note 300, November 1976, and Proc. IEEE, 1976, pp. 1598-1616.

1-5. C. E. Baum, A Priori Application of Results of Electromagnetic Theory to the Analysis of Electromagnetic Interaction Data, Interaction Note 444, February 1985, and Radio Science, 1987, pp. 1127-1136.

1-6. C. E. Baum, On the Eigenmode Expansion Method for Electromagnetic Scattering and Antenna Problems, Part II: Asymptotic Expansion of Eigenmode-Expansion Parameters in the Complex-Frequency Plane, Interaction Note 472, November 1988.

1-7. C. E. Baum, Scattering of Transient Plane Waves, Interaction Note 473, August 1988.

1-8. C. E. Baum, Electromagnetic Reciprocity and Energy Theorems for Free Space Including Sources Generalized to Numerous Theorems, to Combined Fields, and to Complex Frequency Domain, Mathematics Note 33, December 1973.

1-9. H. C. Van De Hulst, On the Attenuation of Plane Waves by Obstacles of Arbitrary Size and Form, Physica, Vol. 15, 1949, pp. 740-746.

1-10. V. H. Rumsey, Reaction Concept in Electromagnetic Theory, Physical Review, 1954, pp. 1483-1491.

1-11. M. H. Cohen, Application of the Reaction Concept to Scattering Problems, IRE Trans. Antennas and Propagation, 1955, pp. 193-199.

1-12. D. S. Saxon, Tensor Scattering Matrix for the Electromagnetic Field, Physical Review, 1955, pp. 1771-1775.

1-13. A. D. Yaghjian, Electric Dyadic Green's Functions in the Source Region, Proc. IEEE, 1980, pp. 248-263.

1-14. C. E. Baum and L. W. Pearson, On the Convergence and Numerical Sensitivity of the SEM Pole-Series in Early-Time Scattering Response, Electromagnetics, 1981, pp. 209-228.

1-15. K. S. H. Lee, Relations Between Electric and Magnetic Polarizabilities and Other Related Quantities, Radio Science, 1987, pp. 1235-1238.

References (Part 1) (Concluded)

1-16. C. E. Baum, The Singularity Expansion Method, in L. B. Felson (ed.), Transient Electromagnetic Fields, Springer-Verlag, 1976.

1-17. Y. T. Lo and S. W. Lee (eds.), Antenna Handbook, Van Nostrand Reinhold, 1988.

References (Part 2)

2-1. C. E. Baum, Interaction of Electromagnetic Fields With an Object Which Has an Electromagnetic Symmetry Plane, Interaction Note 63, March 1971.

2-2. C. A. Lin and J. T. Cordaro, Singularity Expansion Method Parameter Measurement, Interaction Note 409, May 1981, and Determination of the SEM Parameters for an Aircraft Model from the Transient Surface Current, Electromagnetics, pp. 65-75, 1983.

2-3. C. E. Baum, A Priori, Application of Results of Electromagnetic Theory to the Analysis of Electromagnetic Interaction Data, Interaction Note 444, February 1985, and Radio Science, 1987, pp. 1127-1136.

2-4. C. E. Baum, Combining RF Sources Using $C_N$ Symmetry, Circuit and Electromagnetic System Design Note 37, June 1989.

2-5. S. L. Borison, Diagonal Representation of the Radar Scattering Matrix for an Axially Symmetric Body, IEEE Trans. Antennas and Propagation, pp. 176-177, 1965.

2-6. W. -M. Boerner, M. B. El-Arini, C. -Y. Chan, and P. M. Mastoris, Polarization Dependence in Electromagnetic Inverse Problems, IEEE Trans. Antennas and Propagation, pp. 262-271, 1981.

2-7. G. E. Heath, Bistatic Scattering Reflection Asymmetry, Polarization Reversal Asymmetry, Polarization Reversal Reflection Symmetry, IEE Trans. Antennas and Propagation, pp. 429-434, 1981.

2-8. G. E. Heath, Properties of the Linear Polarization Bistatic Scattering Matrix, IEEE Trans. Antennas and Propagation, pp. 523-525, 1981.

2-9. M. Davidovitz and W. -M. Boerner, Reduction of Bistatic Scattering Matrix Measurements for Inversely Symmetric Radar Targets, IEEE Trans. Antennas and Propagation, pp. 237-242, 1983.

2-10. C. D. Taylor, Calculation of Natural Resonances for Perpendicular Crossed Wires Parallel to an Imperfect Ground, Electromagnetics, pp. 41-64, 1983.

2-11. M. Hamermesh, Group Theory and Its Application to Physical Problems, Addison-Wesley, 1962.

2-12. R. E. Kleinman, Low Frequency Electromagnetic Scattering, in P. L. E. Uslenghi (ed.), Electromagnetic Scattering, Academic Press, 1978.

2-13. J. D. Kraus, Antennas, 2nd. ed., McGraw Hill, 1988.

# TRANSIENT POLARIZATION

NEIL F. CHAMBERLAIN
*The Ohio State University*
*ElectroScience Laboratory*
*1320 Kinnear Road*
*Columbus, Ohio 43212 USA*

ABSTRACT. The theory of *Transient Polarization* is introduced. This new theory provides a basis for the polarimetric description of deterministic, plane electromagnetic waves of arbitrary bandwidth. The formalism of transient polarization is developed as a generalization of monochromatic polarization. Transient polarization responses (TPR), which are formed from bandlimited impulse responses, describe the full vector nature of wideband backscatter. Scattering centers may be identified (subject to adequate resolution) from the TPR, along with a transient polarization state. When the scattering matrix is configured in the appropriate polarization bases, the transient polarization state parameters relate directly (subject to adequate resolution) to the geometric structure of the scattering object. TPRs derived from synthesized data are presented. TPR parametrization is shown to provide a concise and geometrically meaningful representation of the full-polarization wideband scattering matrix.

## 1. Introduction

The purpose of this paper is to establish a theory of wideband polarization. The resulting formalism, which is called *transient polarization*, provides a basis for the polarimetric description of deterministic, plane electromagnetic waves of arbitrary bandwidth. As such, the theory of transient polarization is developed as a generalization of monochromatic polarization, and uses a notation which is consistent with the latter.

The discussion presented here is limited to the case of deterministic electromagnetic (EM) waves. EM waves with *random* polarization may be characterized in polarimetric terms by means of the Stoke's calculus [1]. Such waves are not considered further in the present discussion. The motivation for developing a theory of wideband polarization stems from a need to describe the vector nature of scattered electromagnetic far-fields when the incident EM pulse width is on the order of, or less than, the overall dimensions of the scattering object. In such cases, the tenets of monochromatic (or quasi-monochromatic) polarization are no longer adequate for a complete polarimetric description of object scattering, and it is the purpose of this paper to address this problem.

The term *monochromatic* implies an EM wave with a spectral content limited to one frequency. The conventional definition of polarization strictly applies only to these continuous waves. In practice, it is impossible to generate continuous



*W.-M. Boerner et al. (eds.), Direct and Inverse Methods in Radar Polarimetry, Part 1*, 487–501.

waves because the source of radiation must be switched on and off at some time. Thus we consider the wave as a pulse, which has a non-zero bandwidth. If the bandwidth is sufficiently small compared to the center frequency, then there will be many oscillations of the "carrier frequency" between the edges of the pulse. As long as we do not ask for a description of the electric field vector of the wave near the pulse edges, we may use the theory of monochromatic polarization to describe the vector behavior of this quasi-monochromatic EM wave.

As an example, consider the pulse produced by an airport surveillance radar operating at 2.5 GHz, which has a width on the order of $1\mu s$. There are approximately 2500 cycles of the electric field vector between the pulse edges, and therefore a quasi-monochromatic polarization description is appropriate in this case.

We take the view that polarization is a *global* property of the quasi-monochromatic EM wave, because the same polarization state is found nearly everywhere in the wave (or pulse), and the effects of the pulse edges are small compared to the middle portion of the pulse. Various substructures of the object contribute to the overall scattered signature, but these individual contributions are not discernable (resolvable) in the single frequency case. On the other hand, a very short (ultra-wideband) pulse may have only one or two cycles of the electric field vector between pulse edges. In this case (and depending on the incident pulse duration), the individual scattering responses of substructures may be discernable. Accordingly, there is a vector representation of these substructure signatures, which we call transient polarization in the current discussion. As such, the theory of transient polarization permits a local description of electric field behavior within a (wideband) pure or scattered EM wave.

## 2. Monochromatic Polarization

To develop the theory of transient polarization, we first restate (from [1]) the well--known formulation of monochromatic polarization. Consider a monochromatic wave with frequency $f_0$, travelling in free space in the positive $r$ direction. The electric field vector with general elliptical polarization may be written as

$$\mathbf{e}(t) = \hat{h}E_h \cos(2\pi f_0 t - kr + \phi_h) + \hat{v}E_v \cos(2\pi f_0 t - kr + \phi_v), \tag{1}$$

where $\hat{h}$ and $\hat{v}$ are unit vectors in the horizontal and vertical directions respectively. The direction of propagation $\hat{r}$, is perpendicular to the $hv$ plane, so that $\hat{r} = \hat{v} \times \hat{h}$, and $R$ is the distance of the wavefront from the source. The propagation constant is denoted by $k = 2\pi/\lambda$, where $\lambda$ is the wavelength.

Using phasor notation, the electric field may be represented by a complex vector as follows:

$$\begin{aligned} \widetilde{\mathbf{E}} &= \hat{h}\widetilde{E}_h + \hat{v}\widetilde{E}_v, & (2)\\ \widetilde{E}_h &= E_h e^{j\phi_h}, & (3)\\ \widetilde{E}_v &= E_v e^{j\phi_v}, & (4) \end{aligned}$$

and the real CW electric field may be recovered from the phasor electric field by the following relation:

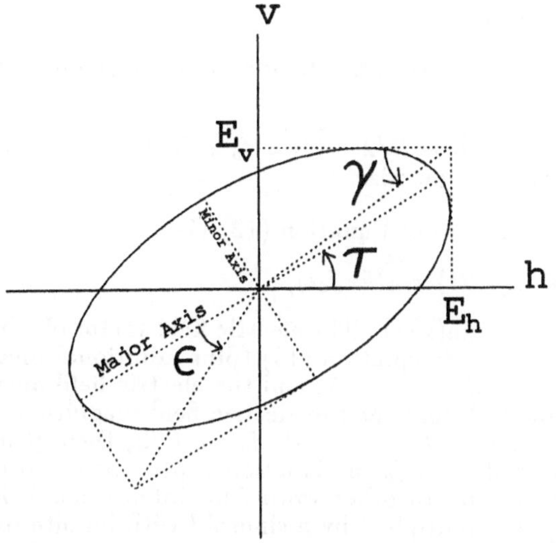


Figure 1: Elliptically polarized electric field of a monochromatic plane-wave.

$$\mathbf{e}(t) = \mathrm{Re}\{\widetilde{\mathbf{E}}e^{j(2\pi f_0 t - kr)}\}. \tag{5}$$

The polarization state of the wave expressed in Equation (1) may be stated in terms of two parameters; $\epsilon$ (ellipticity angle) and $\tau$ (tilt angle),

$$\epsilon = \frac{1}{2}\sin^{-1}\{\sin 2\gamma \sin\delta\}, \tag{6}$$

$$\tau = \frac{1}{2}\tan^{-1}\{\tan 2\gamma \cos\delta\}, \quad \text{where} \tag{7}$$

$$\gamma = \tan^{-1}\left\{\frac{E_v}{E_h}\right\}, \tag{8}$$

$$\delta = \phi_v - \phi_h. \tag{9}$$

The tilt and ellipticity parameters describe the familiar polarization ellipse, which is the locus of the tip of the CW electric field vector in the plane of the wavefront. We now designate a polarimetric descriptor $\wp$, which expresses both the amplitude and polarization properties of the CW wave as follows:

$$\wp = \{A, \epsilon, \tau\}, \quad \text{where} \tag{10}$$

$$A = \sqrt{E_h^2 + E_v^2} \tag{11}$$

A polarization ellipse with relevant parameters is shown in Figure 1. The polarization state of this wave may be mapped as a single point onto a spherical surface by considering $\epsilon$ and $\tau$ as latitude and longitude coordinates. This mapping is called the polarization (or Poincaré) sphere. The radius of the sphere may be thought of as being equal to the amplitude $A$.

## 3. Wideband EM plane-waves

Now consider a plane-wave with a flat bandlimited spectrum described in Equation (12)

$$\tilde{E}(f) = \begin{cases} 1/2 & f_0 - B/2 < |f| < f_0 + B/2 \\ 0 & \text{otherwise.} \end{cases} \quad (12)$$

The inverse Fourier transform of Equation (12) yields

$$e(t) = B\text{sinc}\{\pi Bt\} \cos(2\pi f_0 t), \quad (13)$$

where $\text{sinc}(x)$ abbreviates $\sin(x)/x$. The electric field vector of a bandlimited wave would thus have the form of Equation (13) (plus additional phase terms). Note that when $B \to 0$, then $\text{sinc}(\pi Bt) \to 1$, and the electric field magnitude in Equation (13) has the same CW form as the electric field magnitude in Equation (1). Furthermore, notice that if $B \to \infty$ with $f_0 \to B/2$, then $B\text{sinc}(\pi Bt) \to \delta(t)$, and the "carrier term" due to $f_0$ in Equation (13) appears to oscillate with an infinitesimally small period. In other words, for infinite bandwidth, the electric field becomes an impulse multiplied by a sinusoid with infinite frequency.

To further illustrate this point, consider an electric field with left circular (LC) polarization. As a purely monochromatic wave, the field may be written (IEEE convention [6]) in phasor form as

$$\tilde{\mathbf{E}} = \hat{h} + j\hat{v}. \quad (14)$$

As a wave with arbitrary bandwidth, the field would be written as

$$\mathbf{e}(t) = B\text{sinc}\{\pi Bt\} \{\hat{h}\cos(2\pi f_0 t - kr) + \hat{v}\cos(2\pi f_0 t + \pi/2 - kr)\}. \quad (15)$$

For infinite bandwidth, with the constraint that $f_0 = B/2$, the electric field vector is seen to rotate with zero period within the duration of the impulse. When $f_0 \to B/2$, it is easy to show as a limiting argument that the vector has time for one complete revolution between the first nulls ($t_n = \pm 1/B$) of the sinc function. In other words, for infinite bandwidth, the vector is instantaneously at all possible orientations. The "circularly polarized" impulse is, of course, a hypothetical construction; no such wave could be propagated. It is interesting, however, to consider the effect of launching such a wave at an object. Assuming that the object is located in free space, the impulse would traverse the object at the speed of light, subjecting each point on the object to all possible orientations of the electric field vector. For highly conducting objects, the back-scattered fields depend only on object geometry. Therefore the scattered electric field locus due to an incident "circularly polarized" impulse would be indicative of local object geometry. This concept is developed in the next section.

It is clear, then, that even very wideband EM waves may be written in a form which makes them amenable to a monochromatic-like polarization description. A wave characterized by Equation (15), for example, would be described as left circular pulse. For near-zero bandwidth, we say the wave has left circular polarization. For very wide bandwidths (or more correctly, for $B/f_0 \to 2$), the sinc function multiplier simply constrains the description "left circular" to a smaller interval of time, and we can still meaningfully assign a "polarization state" of $\epsilon = 45°$ and $\tau$ as indeterminable, to this wave.

## 4. TPR Formulation

In the previous section, a wave of arbitrary bandwidth was represented in terms of a "carrier function" multiplied by some bandwidth-dependent envelope function. In the current section, we extend this approach and develop a general polarimetric representation of scattered plane-waves [2]. We assume that the bandwidth of the incident wave is sufficient to resolve significant scattering substructures, although the analysis can be applied to EM waves of any bandwidth.

The incident (wideband) pulse which elicits the wideband scattering response may be formulated in any desired polarization basis. However, we chose the incident pulse to be left circular (or right circular) in the manner of Equation (15), so that each substructure is subjected to all orientations of the electric field vector in the plane of the wave. Similarly, we chose to receive the back-scattered wave with a pair of (very wideband) linear antennas oriented in the $\hat{h}$ and $\hat{v}$ directions respectively. This choice of receive antennas (polarizations) enables the scattered electric field locus to be displayed in the same manner as monochromatic polarization.

In the far-region of a scattering body, the received phasor electric fields may be written in terms of the incident phasor electric fields as follows:

$$\begin{bmatrix} \tilde{E}_h^r(f) \\ \tilde{E}_v^r(f) \end{bmatrix} = \begin{bmatrix} \tilde{S}_{hh}(f) & \tilde{S}_{hv}(f) \\ \tilde{S}_{vh}(f) & \tilde{S}_{vv}(f) \end{bmatrix} \begin{bmatrix} \tilde{E}_h^i(f) \\ \tilde{E}_v^i(f) \end{bmatrix} \frac{1}{\sqrt{4\pi R^2}}, \tag{16}$$

where $r$ denotes *received* and $i$ denotes incident, and where R denotes the distance from the antennas to the object. In this formulation, the incident fields are assumed to be true plane-waves, while the received electric fields are assumed to be approximately plane in the vicinity of the receive antenna. Object scattering is thus represented as a linear transformation of incident electric fields, called the *scattering matrix*. The complex scattering coefficients $\tilde{S}_{ab}$, where $ab \in \{hh, hv, vh, vv\}$, relate the received to incident electric field components. The index $a$ indicates the receive polarization, and the index $b$ indicates the transmit polarization. This indexing scheme is opposite to the usual convention, but obeys the rules of matrix multiplication. For the case of back-scatter, we have $\tilde{S}_{hv} = \tilde{S}_{vh}$ (by reciprocity), which is also called monostatic scattering.

The scattering coefficients are functions of both frequency and object orientation. They have dimensions of length, and are independent of range. Radar cross-section is calculated from the scattering coefficient by

$$\sigma_{ab} = |\tilde{S}_{ab}|^2 \quad m^2. \tag{17}$$

The scattering matrix equation in (16) is presented in terms of linear $h$ and $v$ polarizations. We refer to this as an $hv$:$hv$ polarization transformation basis because incident fields in the $hv$ basis are transformed into received (scattered) fields in the $hv$ basis. The scattering matrix may be expressed in terms of any polarization transformation basis. We choose an $hv$:$\ell r$ basis, for the reasons outlined above. In this case the transmit/incident fields are left circular ($\ell$) or right circular ($r$) and the scattered/received fields are $h$ or $v$.

The received electric fields in the $hv$:$\ell r$ basis are therefore written as

$$\begin{bmatrix} \tilde{E}_h^r(f) \\ \tilde{E}_v^r(f) \end{bmatrix} = \begin{bmatrix} \tilde{S}_{h\ell}(f) & \tilde{S}_{hr}(f) \\ \tilde{S}_{v\ell}(f) & \tilde{S}_{vr}(f) \end{bmatrix} \begin{bmatrix} \tilde{E}_\ell^i(f) \\ \tilde{E}_r^i(f) \end{bmatrix} \frac{1}{\sqrt{4\pi R^2}}, \tag{18}$$

and the new scattering matrix is obtained by means of a simple unitary matrix transformation of the original $hv{:}hv$ scattering matrix:

$$\begin{bmatrix} \tilde{S}_{h\ell}(f) & \tilde{S}_{hr}(f) \\ \tilde{S}_{v\ell}(f) & \tilde{S}_{vr}(f) \end{bmatrix} = \begin{bmatrix} \tilde{S}_{hh} + j\tilde{S}_{hv} & \tilde{S}_{hh} - j\tilde{S}_{hv} \\ \tilde{S}_{hv} + j\tilde{S}_{vv} & \tilde{S}_{hv} - j\tilde{S}_{vv} \end{bmatrix} \frac{1}{\sqrt{2}}. \tag{19}$$

In order to avoid the inconvenience of carrying the $1/\sqrt{4\pi R^2}$ factor, we effect a normalization of the frequency responses $\tilde{S}_{ab}(f)$, with the result that these responses are dimensionless, and the factor is removed. We may therefore think of the input and output components of Equation (18) as abstract signals, and change notation to emphasize this point,

$$\begin{bmatrix} \tilde{R}_h(f) \\ \tilde{R}_v(f) \end{bmatrix} = \begin{bmatrix} \tilde{S}_{h\ell}(f) & \tilde{S}_{hr}(f) \\ \tilde{S}_{v\ell}(f) & \tilde{S}_{vr}(f) \end{bmatrix} \begin{bmatrix} \tilde{I}_\ell(f) \\ \tilde{I}_r(f) \end{bmatrix}. \tag{20}$$

We refer to $\tilde{R}(f)$ as the response waveforms and $\tilde{I}(f)$ as the input waveforms. Taking the inverse Fourier transform of (20) yields

$$\begin{bmatrix} r_h(t) \\ r_v(t) \end{bmatrix} = \begin{bmatrix} s_{h\ell}(t) & s_{hr}(t) \\ s_{v\ell}(t) & s_{vr}(t) \end{bmatrix} \otimes \begin{bmatrix} i_\ell(t) \\ i_r(t) \end{bmatrix}, \tag{21}$$

where the symbol $\otimes$ denotes convolution. The responses $r(t)$ are called *transient responses*. The responses $s_{ab}(t)$ are called the *impulse responses* [4]. Finally, the responses $i(t)$ are called the input pulse functions. They express the bandlimited nature of the transmitted pulse. In the limit that the input pulses are delta functions, then the transient responses are impulse responses.

The transient responses can be written explicitly in terms of the input pulse polarization,

$$\begin{bmatrix} r_{h(\ell)}(t) & r_{h(r)}(t) \\ r_{v(\ell)}(t) & r_{v(r)}(t) \end{bmatrix} = \begin{bmatrix} s_{h\ell}(t) & s_{hr}(t) \\ s_{v\ell}(t) & s_{vr}(t) \end{bmatrix} \otimes \begin{bmatrix} i_\ell(t) & 0 \\ 0 & i_r(t) \end{bmatrix}. \tag{22}$$

The transient polarization responses (TPR) are defined in terms of the transient responses $r_{a(b)}$ as follows:

$$\vec{r}_\ell(t) = \hat{h} r_{h(\ell)}(t) + \hat{v} r_{v(\ell)}(t), \tag{23}$$

$$\vec{r}_r(t) = \hat{h} r_{h(r)}(t) + \hat{v} r_{v(r)}(t), \tag{24}$$

where $\vec{r}_\ell(t)$ is the TPR due to an LC pulse $\vec{i}_\ell(t)$, and $\vec{r}_r(t)$ is the TPR due to an RC pulse $\vec{i}_r(t)$. Note that each of the the vector functions, $\vec{r}_\ell$ and $\vec{r}_r$, describes 3-dimensional waveform. One of the dimensions is time, which corresponds to the down-range dimension, the direction of which is denoted $\hat{r}$. The other two dimensions correspond to the plane formed by the unit vectors $\hat{h}$ and $\hat{v}$, which is normal to $\hat{r}$. An example of a TPR is given in Section 6.

## 5. TPR Parameterization

The parameterization of the transient polarization response allows us to define transient polarization states. The form of the time response of Equation (23) is quite similar to the electric field expression of Equation (1); both equations define a vector time response in terms of horizontal and vertical components. It is simply a matter of arranging the expressions for the transient responses into a form suitable for polarimetric description.

The transient response components $r_{a(b)}(t)$ are the inverse Fourier transforms of bandlimited frequency responses $\tilde{R}_{a(b)}(f)$,

$$r_{a(b)}(t) = \mathcal{F}^{-1}\{\tilde{R}_{a(b)}(f)\}. \tag{25}$$

A real transient response has a frequency response which consists of a positive frequency response $\tilde{R}^{+}_{a(b)}(f)$, and a negative frequency response $\tilde{R}^{-}_{a(b)}(f)$, which is the complex conjugate of the former. The transient response may be written as

$$r_{a(b)}(t) = 2 \cdot \mathrm{Re}\{\mathcal{F}^{-1}\{\tilde{R}^{+}_{a(b)}(f)\}\}. \tag{26}$$

The transient responses may also be reconstructed from demodulated versions of the bandpass positive frequency responses. These lowpass frequency responses $\tilde{R}^{+}_{a(b)}(f - f_0)$, have complex time responses $\tilde{\alpha}_{a(b)}(t)$, assuming that the frequency responses are not symmetric about $f_0$;

$$\tilde{\alpha}_{a(b)}(t) = \mathcal{F}^{-1}\{\tilde{R}^{+}_{a(b)}(f - f_0)\}. \tag{27}$$

Therefore, from the frequency shift theorem [5]

$$\tilde{R}^{+}_{a(b)}(f) = \mathcal{F}\{e^{j2\pi f_0 t}\tilde{\alpha}_{a(b)}(t)\}. \tag{28}$$

It follows then, from (26) through (28) that

$$r_{a(b)}(t) \;=\; 2 \cdot \mathrm{Re}\{e^{j2\pi f_0 t}\, \tilde{\alpha}_{a(b)}(t)\}, \tag{29}$$

which can be rewritten as

$$r_{a(b)}(t) \;=\; 2|\tilde{\alpha}_{a(b)}(t)| \left\{\cos\left(2\pi f_0 t + \beta_{a(b)}(t)\right)\right\}, \qquad \text{where} \tag{30}$$

$$\beta_{a(b)}(t) \;=\; \tan^{-1}\left\{\frac{\tilde{\alpha}_{a(b)} - \tilde{\alpha}^{*}_{a(b)}}{j(\tilde{\alpha}_{a(b)} + \tilde{\alpha}^{*}_{a(b)})}\right\} \tag{31}$$

defines a time varying "phase" function. (The symbol $*$ denotes complex conjugation.) We now have, in Equation (30), a form of the transient response which may be interpreted as an amplitude and phase modulated "carrier". However, for very wideband signals, it is possible that variations in the "amplitude" and "phase" terms might be equal to, or faster than, variations in the "carrier" term. Hence it is not necessarily appropriate to think of these signals in the same way as amplitude or phase modulated signals commonly encountered in communication systems.

Substituting the expression for the transient responses of Equation (30) (with the appropriate indices) in Equations (23) and (24), it is then possible to calculate a set of time-varying polarimetric parameters, in the same way that was done for monochromatic polarization in Equations (2) through (10). The following parameters result:

$$A_{\substack{\ell \\ r}}(t) = 2\sqrt{|\tilde{\alpha}_{\substack{h(\ell) \\ h(r)}}(t)|^2 + |\tilde{\alpha}_{\substack{v(\ell) \\ v(r)}}(t)|^2}, \tag{32}$$

$$\epsilon_{\substack{\ell \\ r}}(t) = \frac{1}{2}\sin^{-1}\{\sin 2\gamma_{\substack{\ell \\ r}}(t)\sin\delta_{\substack{\ell \\ r}}(t)\}, \tag{33}$$

$$\tau_{\substack{\ell \\ r}}(t) = \frac{1}{2}\tan^{-1}\{\tan 2\gamma_{\substack{\ell \\ r}}(t)\cos\delta_{\substack{\ell \\ r}}(t)\}, \qquad \text{where} \tag{34}$$

$$\gamma_{\substack{\ell \\ r}}(t) = \tan^{-1}\left\{\frac{|\tilde{\alpha}_{\substack{v(\ell) \\ v(r)}}(t)|}{|\tilde{\alpha}_{\substack{h(\ell) \\ h(r)}}(t)|}\right\}, \tag{35}$$

$$\delta_{\substack{\ell \\ r}}(t) = \beta_{\substack{v(\ell) \\ v(r)}}(t) \pm \pi/2 - \beta_{\substack{h(\ell) \\ h(r)}}(t). \tag{36}$$

The polarimetric descriptors $\wp_\ell(t)$ and $\wp_r(t)$ can be formulated as in Equation (10), except that for the wideband case, the constituent parameters are functions of time

$$\wp(t) = \{A(t), \epsilon(t), \tau(t)\}. \tag{37}$$

TPR PARAMETERS OF CANONICAL OBJECTS

Using the above parameterization, TPR parameters of simple, idealized scattering objects are shown in Table 1. Note that because these objects have constant amplitude frequency responses (and, implicitly, linear phase frequency responses), their time responses are characterized by a single scattering center. This is somewhat unrealistic for practical scattering objects, but illustrates simply the relation between scattering matrix and TPR parameters. The polarimetric parameters $\wp_\ell(t)$ and $\wp_r(t)$ correspond to the location of the object in the transient response, which may also be seen as the peak amplitude in the envelope response $A(t)$.

Table 1, demonstrates that the TPR parameters correspond to 3 key attributes of a scattering structure. The size of the structure is reflected in the $A$ parameter, the symmetry of the object is reflected in the $\epsilon$ parameter, and the orientation of the object about the direction of propagation is reflected in the $\tau$ parameter. For more symmetrical objects ($|\epsilon| \to 45°$), tilt angle ($\tau$) is of diminishing importance. Thin, linear-type structures are characterized by $\epsilon \to 0°$. Furthermore, notice from Table 1 that, if we are prepared to ignore the left and right helices, then it makes little difference to the TPR parameters where the transmitted pulse is LC or RC.

## 6. TPR Example

Consider the fictitious aircraft model illustrated in Figure 2. This model is composed of four point-type scatterers, (as in Table 1), located at positions $p1$,

Table 1: Elementary scattering matrices and their TPR parameterizations.

| | Scattering Matrix | | Parameterization | |
|---|---|---|---|---|
| Object | $\mathbf{S}_{hv:hv}(f)$ | $\mathbf{S}_{hv:\ell r}(f)$ | $\wp_\ell$ Tx. LC pulse $A, \epsilon, \tau$ | $\wp_r$ Tx. RC pulse $A, \epsilon, \tau$ |
| Horizontal Dipole | $\begin{bmatrix} 1 & 0 \\ 0 & 0 \end{bmatrix}$ | $\begin{bmatrix} 1 & 1 \\ 0 & 0 \end{bmatrix} \frac{1}{\sqrt{2}}$ | $\frac{1}{\sqrt{2}}$, 0, 0 | $\frac{1}{\sqrt{2}}$, 0, 0 |
| Vertical Dipole | $\begin{bmatrix} 0 & 0 \\ 0 & 1 \end{bmatrix}$ | $\begin{bmatrix} 0 & 0 \\ j & -j \end{bmatrix} \frac{1}{\sqrt{2}}$ | $\frac{1}{\sqrt{2}}$, 0, 90 | $\frac{1}{\sqrt{2}}$, 0, 90 |
| 45° Dipole | $\begin{bmatrix} 1 & 1 \\ 1 & 1 \end{bmatrix} \frac{1}{2}$ | $\begin{bmatrix} e^{j\frac{\pi}{4}} & e^{-j\frac{\pi}{4}} \\ e^{j\frac{\pi}{4}} & e^{-j\frac{\pi}{4}} \end{bmatrix} \frac{1}{2}$ | $\frac{1}{\sqrt{2}}$, 0, 45 | $\frac{1}{\sqrt{2}}$, 0, 45 |
| Flat Plate | $\begin{bmatrix} 1 & 0 \\ 0 & 1 \end{bmatrix}$ | $\begin{bmatrix} 1 & 1 \\ j & -j \end{bmatrix} \frac{1}{\sqrt{2}}$ | 1, −45, — | 1, +45, — |
| Corner Reflector | $\begin{bmatrix} 1 & 0 \\ 0 & -1 \end{bmatrix}$ | $\begin{bmatrix} 1 & 1 \\ -j & j \end{bmatrix} \frac{1}{\sqrt{2}}$ | 1, +45, — | 1, −45, — |
| 45° Corner Reflector | $\begin{bmatrix} 0 & 1 \\ 1 & 0 \end{bmatrix}$ | $\begin{bmatrix} 1 & 1 \\ -j & j \end{bmatrix} \frac{1}{\sqrt{2}}$ | 1, 45, — | 1, −45, — |
| Right Helix | $\begin{bmatrix} 1 & -j \\ -j & -1 \end{bmatrix} \frac{1}{2}$ | $\begin{bmatrix} 1 & 0 \\ -j & 0 \end{bmatrix} \frac{1}{\sqrt{2}}$ | 1, −45, — | 0, —, — |
| Left Helix | $\begin{bmatrix} 1 & j \\ j & -1 \end{bmatrix} \frac{1}{2}$ | $\begin{bmatrix} 0 & 1 \\ 0 & j \end{bmatrix} \frac{1}{\sqrt{2}}$ | 0, —, — | 1, 45, — |

Table 2: Scattering matrices characterizing the substructures of a fictitious aircraft object; 0° az., 0° el., 10° roll.

| Location | $\mathbf{S}_{hv:hv}(f)$ | Description |
|---|---|---|
| $p_1$ | $\mathbf{S}_1 = \begin{bmatrix} 1 & 0 \\ 0 & 1 \end{bmatrix}$ | Nose |
| $p_2$ | $\mathbf{S}_2 = \begin{bmatrix} \cos^2 10° & \frac{1}{2}\sin 20° \\ \frac{1}{2}\sin 20° & \sin^2 10° \end{bmatrix} 10e^{j4\pi 30f/c}$ | Wing |
| $p_3$ | $\mathbf{S}_3 = \begin{bmatrix} \cos^2 10° & \frac{1}{2}\sin 20° \\ \frac{1}{2}\sin 20° & \sin^2 10° \end{bmatrix} 2e^{j4\pi 50f/c}$ | Horizontal Stabilizers |
| $p_4$ | $\mathbf{S}_4 = \begin{bmatrix} \sin^2 10° & -\frac{\sin 20°}{2} \\ -\frac{\sin 20°}{2} & \cos^2 10° \end{bmatrix} 3e^{j4\pi 51f/c}$ | Vertical Stabilizer |

$p2$, $p3$, and $p4$. These points are located at 0, 30, 50, and 51 meters respectively. The bandwidth of the scattering coefficients is (artificially) constrained to provide an effective resolution of 6 meters. Therefore, all points are resolvable except the last two.

The object is rolled 10° from the horizontal plane about the direction of propagation; which is directed from the "nose" of the aircraft to the "tail" of the aircraft. The scattering matrices for these individual scattering structures are shown in Table 2.

The total scattering matrix in the *hv:hv* basis, $\mathbf{S}_{hv:hv}$, is given simply by the sum of the individual matrices of Table 2: i.e.,

$$\mathbf{S}_{hv:hv}(f) = \sum_{i=1}^{4} \mathbf{S}_i(f). \tag{38}$$

The corresponding scattering matrix in the *hv:ℓr* basis, $\mathbf{S}_{hv:\ell r}$, is calculated from the unitary matrix transformation of Equation (19), and the total matrix is transformed to the time-domain by means of a fast Fourier transform algorithm. The TPR due to an incident LC pulse $\vec{r}_\ell$ is constructed from the appropriate components of the time-domain scattering matrix. This response is plotted in Figure 2. The parameter responses $A(t)$, $\epsilon(t)$ and $\tau(t)$ are displayed in Figure 3. The parameter values corresponding to the scattering center maxima of the envelope response are tabulated in Table 3.

Notice that the tilt angle of the response associated with the *wing* structure is 3.5° and not the expected value of 10°. This discrepancy is most noticeable when the tilt angle is close to one of the axes (i.e., approximately a multiple of 90°). The same fictitious object with a roll angle of 40° yields a response with tilt = 40° at the location of the *wing* structure. The ellipticity of the wing structure (−1.6°) is indicative of a thin structure.

Second, note that the peak amplitudes of $A_\ell(t)$ are in proportion to the size of the various substructures as represented in the magnitudes of the terms of the

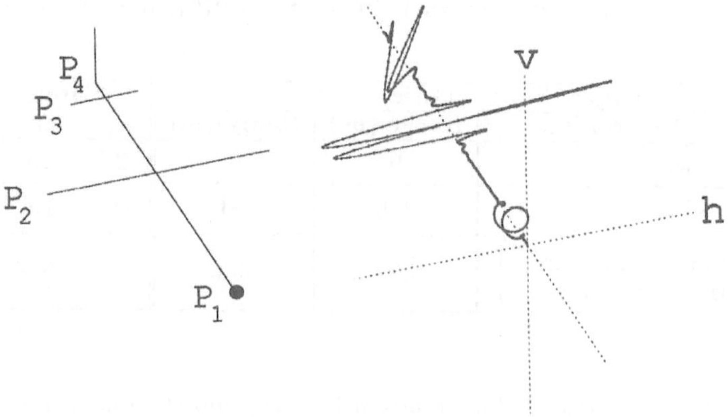


**Figure 2: Fictitious canonical aircraft object and TPR $\vec{r}_\ell(t)$.**

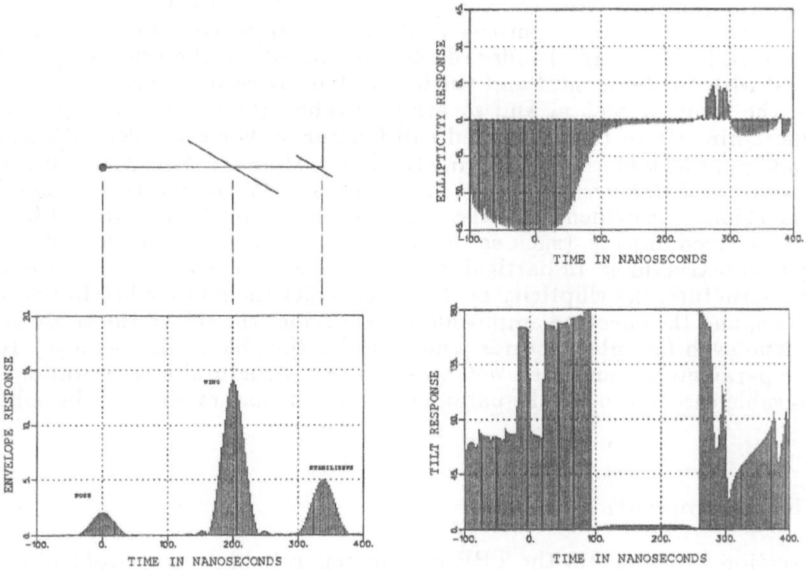


Figure 3: Parameter responses $A(t)$, $\epsilon(t)$ and $\tau(t)$ of a fictitious canonical aircraft object.

Table 3: Polarimetric parameters extracted from the TPR $\vec{r}_\ell$ of a fictitious aircraft object.

| Object Substructure | | | | Parameters | | |
|---|---|---|---|---|---|---|
| Location | Description | Range (ns) | Range (m) | $A_\ell$ | $\epsilon_\ell$ | $\tau_\ell$ |
| $p_1$ | Nose | 0 | 0 | 2 | $-45°$ | $124°$ |
| $p_2$ | Wing | 200 | 30 | 14 | $-1.6°$ | $3.5°$ |
| $p_3, p_4$ | Vert. Stabilizer<br>Horiz. Stabilizer } | 337 | 50.6 | 5 | $-6.5°$ | $61°$ |

individual scattering matrices. This relation follows from the linear nature of the Fourier transform.

Third, note that the *vertical stabilizer* and *horizontal stabilizer* structures combine so that the corresponding TPR tilt angle (60°) is representative of neither a vertical nor a horizontal structure. The polarimetric parameters for these interfering substructures are actually dependent on their down-range separation. For example, when the structures at $p_3$ and $p_4$ have the same down-range location of 50 meters, the polarimetric parameters are $A_\ell = 5, \epsilon_\ell = 34°, \tau_\ell = 100°$. Allowing for the roll angle of 10°, these parameters suggest a structure with vertical and horizontal components, with the vertical component moderately larger than the horizontal one. In fact, the ellipticity angle of 34° can be computed as arctangent of 2/3, the ratio of the amplitudes of the components of the scattering matrices associated with the horizontal and vertical stabilizers respectively.

When the structures at $p_3$ and $p_4$ are resolvable, the polarimetric parameters relate more directly to the associated substructures. For example, with $p_3$ at 50 meters and $p_4$ at 56 meters, the polarimetric parameters are $A_\ell = 3, \epsilon_\ell = 0°, \tau_\ell = 3°$ for the *horizontal stabilizer* and $A_\ell = 4, \epsilon_\ell = 3°, \tau_\ell = 95°$ for the *vertical stabilizer*.

In conclusion, it is evident that the polarimetric parameters extracted from the TPRs of idealized objects, (such as the one discussed above), are characteristic of the object substructures. In particular, the tilt angle $\tau(t)$ specifies the orientation of the substructure, the ellipticity angle $\epsilon(t)$ specifies the relative height and width dimensions, and the envelope amplitude $A(t)$ specifies the size of the substructure. This is true even for substructures which are irresolvable in down-range. In this case, the parameters $\epsilon$ and $\tau$ do not have a direct structural interpretation, as in the resolvable case, although the parameters remain characteristic of the substructures.

## 7 TPR Decomposition

In this section we show that the TPR may be separated into *elliptical* and *circular* components. The respective components for a LC transmitted pulse are symmetrical to the components of a RC transmitted pulse. Based on the frequency-domain relation of Equation (19), the following time-domain relation may be derived,

$$\begin{bmatrix} \tilde{\alpha}_{h(\ell)}(t) & \tilde{\alpha}_{h(r)}(t) \\ \tilde{\alpha}_{v(\ell)}(t) & \tilde{\alpha}_{v(r)}(t) \end{bmatrix} = \begin{bmatrix} \tilde{\alpha}_{h(h)} + j\tilde{\alpha}_{h(v)} & \tilde{\alpha}_{h(h)} - j\tilde{\alpha}_{h(v)} \\ \tilde{\alpha}_{h(v)} + j\tilde{\alpha}_{v(v)} & \tilde{\alpha}_{h(v)} - j\tilde{\alpha}_{v(v)} \end{bmatrix} \frac{1}{\sqrt{2}}, \tag{39}$$

where it is emphasized that the $\alpha$ coefficients in the $hv$ basis are time functions. It is now possible to express the TPRs due to left or right circular transmitted pulses in terms of the $hv$:$hv$ basis $\tilde{\alpha}$ and $\beta$ parameters,

$$\begin{aligned} \vec{r}_{\substack{\ell \\ r}}(t) &= \hat{h}\left\{\sqrt{2}|\tilde{\alpha}_{h(h)}(t)|\cos(2\pi f_0 t + \beta_{h(h)}(t))\right\} \mp \\ &\quad \hat{v}\left\{\sqrt{2}|\tilde{\alpha}_{v(v)}(t)|\sin(2\pi f_0 t + \beta_{v(v)}(t))\right\} + \\ &\quad \hat{h}\left\{\sqrt{2}|\tilde{\alpha}_{h(v)}(t)|\cos(2\pi f_0 t + \beta_{h(v)}(t))\right\} \mp \\ &\quad \hat{v}\left\{\sqrt{2}|\tilde{\alpha}_{h(v)}(t)|\sin(2\pi f_0 t + \beta_{h(v)}(t))\right\}. \end{aligned} \tag{40}$$

The transient responses may be written in terms of their "elliptical" and "circular" components thus,

$$\vec{r}_\ell(t) = \vec{r}_{\ell_e}(t) + \vec{r}_{\ell_c}(t), \tag{41}$$

$$\vec{r}_r(t) = \vec{r}_{r_e}(t) + \vec{r}_{r_c}(t), \tag{42}$$

where $\vec{r}_{\ell_e}(t)$ is the elliptical component of $\vec{r}_\ell(t)$, and $\vec{r}_{\ell_c}(t)$ is the circular component of $\vec{r}_\ell(t)$. The components are given by the following relations:

$$\begin{aligned} \vec{r}_{\substack{\ell_e \\ r_e}}(t) &= \hat{h}\left\{\sqrt{2}|\tilde{\alpha}_{h(h)}(t)|\cos(2\pi f_0 t + \beta_{h(h)}(t))\right\} \mp \\ &\quad \hat{v}\left\{\sqrt{2}|\tilde{\alpha}_{v(v)}(t)|\sin(2\pi f_0 t + \beta_{v(v)}(t))\right\}, \end{aligned} \tag{43}$$

$$\begin{aligned} \vec{r}_{\substack{\ell_c \\ r_c}}(t) &= \hat{h}\left\{\sqrt{2}|\tilde{\alpha}_{h(v)}(t)|\cos(2\pi f_0 t + \beta_{h(v)}(t))\right\} \mp \\ &\quad \hat{v}\left\{\sqrt{2}|\tilde{\alpha}_{h(v)}(t)|\sin(2\pi f_0 t + \beta_{h(v)}(t))\right\}. \end{aligned} \tag{44}$$

Referring to Equations (43) and (44), the polarimetric descriptors $\wp_{\ell_e}(t)$, $\wp_{r_e}(t)$, $\wp_{\ell_c}(t)$ and $\wp_{r_c}(t)$ of the TPR components $\vec{r}_{\ell_e}$, $\vec{r}_{r_e}$, $\vec{r}_{\ell_c}$, and $\vec{r}_{r_c}$, are defined by the following parameters:

$$A_{\substack{\ell_e \\ r_e}}(t) = \sqrt{2|\tilde{\alpha}_{h(h)}(t)|^2 + 2|\tilde{\alpha}_{v(v)}(t)|^2}, \tag{45}$$

$$A_{\substack{\ell_c \\ r_c}}(t) = 2|\tilde{\alpha}_{h(v)}(t)|, \tag{46}$$

$$\epsilon_{\substack{\ell_e \\ r_e}}(t) = \frac{1}{2}\sin^{-1}\{\sin(2\gamma_{\substack{\ell_e \\ r_e}}(t))\sin(\delta_{\substack{\ell_e \\ r_e}}(t))\}, \tag{47}$$

$$\epsilon_{\substack{\ell_c \\ r_c}}(t) = \mp\pi/4, \tag{48}$$

$$\tau_{\substack{\ell_e \\ r_e}}(t) = \frac{1}{2}\tan^{-1}\{\tan(2\gamma_{\substack{\ell_e \\ r_e}}(t))\cos(\delta_{\substack{\ell_e \\ r_e}}(t))\}, \quad \text{where} \tag{49}$$

$$\gamma_{\substack{\ell_e \\ r_e}}(t) = \tan^{-1}\left\{\frac{|\tilde{\alpha}_{v(v)}(t)|}{|\tilde{\alpha}_{h(h)}(t)|}\right\}, \tag{50}$$

$$\gamma_{\ell_c \atop r_c}(t) = \pi/4, \tag{51}$$
$$\delta_{\ell_e \atop r_e}(t) = \beta_{v(v)}(t) - \beta_{h(h)}(t) \mp \pi/2, \tag{52}$$
$$\delta_{\ell_c \atop r_c}(t) = \mp\pi/2. \tag{53}$$

It is now possible to summarize the polarimetric properties of the component responses of the TPRs:

| | |
|---|---|
| **1** $A_{\ell_e}(t) = A_{r_e}(t)$, | **4** $A_{\ell_c}(t) = A_{r_c}(t)$, |
| **2** $\epsilon_{\ell_e}(t) = -\epsilon_{r_e}(t)$, | **5** $\epsilon_{\ell_c}(t) = -\epsilon_{r_c}(t) = -\pi/4$, |
| **3** $\tau_{\ell_e}(t) = -\tau_{r_e}(t)$, | **6** $\tau_{\ell_c}(t)$, and $\tau_{r_c}(t)$ are indefinite. |

From the above derivations, it is clear that the transient polarization states $\wp_{\ell_e}(t)$ and $\wp_{r_e}(t)$ are symmetric ($\wp_{\ell_c}(t)$, and $\wp_{r_c}(t)$ are also symmetric).

## 8. Concluding Points

Transient polarization was developed as a means of describing the polarimetric character of wideband EM plane-waves. The formulation is a generalization of monochromatic polarization. As such, it is possible to describe the vector nature of EM plane-waves of arbitrary bandwidth, using basically the same algebra employed in the description of monochromatic EM waves. A further advantage of this approach is that familiar polarimetric mappings, such as the Poincaré sphere, may be used to display the polarization-related information of wideband EM waves.

For waves with sufficient bandwidth to resolve major scattering centers, the TPR is seen a succession of transitive polarization states, which are characteristic of local object geometry. As bandwidth decreases, the transient polarization states become mixed (due to unresolved structures), until finally only a single (monochromatic) polarization exists.

Analysis of synthesized data corresponding to a simple canonical aircraft-like object illustrates how the polarimetric parameters extracted from the TPR relate to object geometry. This analysis has been extended to measured data of real objects, and has been found to be consistent with the above findings [2].

Decomposition of the TPR shows that for a transmitted LC or RC pulse, the received waveforms contain elliptical and circular components which are symmetrical with respect to the different transmit types. As a result, the TPR due to an LC pulse contains essentially the same polarimetric information as a TPR due to an RC pulse (ignoring helical-type objects, which are rarely encountered in many practical situations). An important application of this finding is in radar target identification [3], because (virtually) all of the full polarization scattering matrix information can be acquired in a single measurement, by transmitting a wideband LC or RC pulse.

When used as classification features, TPR parameters were found to facilitate reliable recognition of aircraft radar targets [2], [3]. A significant feature of the TPR parameters is that they do not depend on absolute backscatter amplitude or absolute backscatter phase (being formulated in terms of amplitude ratios and phase differences). This is an advantage when absolute amplitude/phase information is unavailable. A further advantage of the TPR parameter representation, in connection with radar target recognition, is that the structural interpretation of the parameters gives an indication of object geometry independent of reference signatures.

## 6. Acknowledgements

This research was supported by the Naval Research Laboratories under contract N00014-87-K-2011, which is gratefully acknowledged. I thank Dr. Art Ksienski, Dr. Eric Walton, and Dr. Fred Garber for their invaluable guidance during the course of this work.

## 7. References

[1] Kraus, J.D. (1984) *Electromagnetics*, McGraw-Hill Book Company, New York.

[2] Chamberlain, N.F. (1989) "Recognition and Analysis of Aircraft Targets by Radar, Using Structural Pattern Representations Derived from Polarimetric Signatures," Ph.D. Dissertation, The Ohio State University ElectroScience Lab., Columbus, Ohio, June 1989.

[3] Chamberlain, N.F., Walton, E.K., and Garber F.D. (1990) "Radar Target Identification of Aircraft Using Polarization-Diverse Features," accepted for publication in *IEEE Transactions on Aerospace and Electronic Systems.*

[4] Kennaugh E.M., and Cosgriff R.L. (1958), "The Use of Impulse Response in Electromagnetic Scattering Problems," 1958 IRE National Convention Record, Part 1, pp. 72–77.

[5] Papoulis, A. (1962) *The Fourier Integral and its Applications*, McGraw-Hill Book Company, New York.

[6] Jay, F. (Editor) (1988) *IEEE Standard Dictionary of Electric and Electrical Terms*, 4th Edition, Wiley Interscience, New York.

# Physical Optics Polarization Correction of Backscattering From Conducting Smooth Surfaces

ALON SCHATZBERG

A.J. DEVANEY ASSOCIATES
26 EDMUNDS ROAD
WELLESLEY, MA 02181

ABSTRACT

It is well known that the Luneburg-Kline series expansion of the scattered field can in principle provide a full high frequency solution for scattering from curved surfaces. There are however situations when this approach becomes impractical, and a polarization corrected physical optics approach is more appropriate. In this paper we present the physical optics approach to calculation of the cross (ortho) polarized backscattered field from a smooth conducting surface. The formalism is then applied to derivation of an improved time domain polarization correction to the impulse response of a conducting target.

## 1. INTRODUCTION

The issue of obtaining polarization information in high frequency scattering is of interest, for scattering and inverse scattering applications. In monostatic scattering it is well known through both, classical geometrical and physical optics, that the scattered field is co-polarized with the incident field, an approximation that warrants a scalar analysis. Cross (or ortho) polarization information is absent in these approximations. Clearly a more refined analysis is necessary in order to reveal this information.

Polarization correction to high frequency single scattered field from smoothly curved surfaces has been obtained in the past through evaluation of the next-to-the-leading term in the Luneburg-Kline (L-K) expansion of the scattered fields in conducting scatterers [1] and in dielecteric scatterers [2]. This can be classified as a generalized geometrical optics approach. The scattered field correction term to plane wave incidence has been found to depend on the angle of incidence, polarization relative to the incidence plane, and the local curvature of the scatterer at the specular reflection point.

In particular for monostatic scattering the cross polarized term has been found to depend on the difference between the principal curvatures at the specular reflection point.

Consequently if the method were further applied, higher order scattering terms could be derived, and are expected to depend on higher order local derivatives of the scattering



W.-M. Boerner et al. (eds.), Direct and Inverse Methods in Radar Polarimetry, Part 1, 503–516.

surface at the specular reflection point. Inclusion of multiple scattering terms can in principle result finally in a complete description of the high frequency scattered field from curved surfaces.

The natural question arises therefore: if the above method solves the high frequency scattering problem and is already developed, what is the justification of a physical optics approach? Well, if needed, the task of evaluating even higher order scattering terms than those already evaluated in [1],[2] is formidable, and one would rather avoid this if an alternative is available. Why need to evaluate higher order deffraction terms anyway? We will answer this by imagining a particular class of scatterers where the next-to-the-leading order L-K term does not provide the cross polarization information: Consider a perfectly conducting scatterer which is cylindrically symmetric at its front tip but otherwise deviates from that symmetry as depicted in Fig.1 by a topographic map. We assume a head-on plane wave incidence, this is, incidence at a direction parallel to the axis. We are interested in the backscattered field. In this case, as mentioned above, the cross polarized correction term has been found to depend on the difference of the principal curvatures at the specular reflection point. In our case the scatterer is locally axially symmetric at its front part and will therefore not exhibit cross polarizing behavior to the next-to-the-leading order of the L-K expansion. We will see that through refined physical optics analysis one circumvents the need to dig deeper into the L-K series.

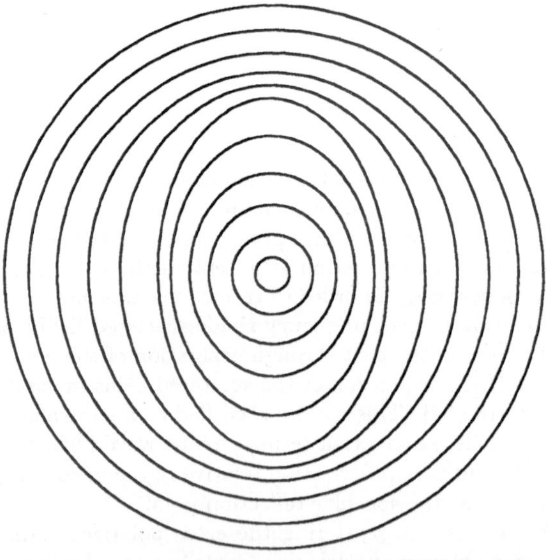

Figure 1: Topographic description of a scattering object for which the polarization corrected geometrical optics approach to scattering becomes impractical.

In the physical optics approach the induced current on the entire scatterer surface is evaluated first. Then the surface current is used in a propagation integral to evaluate the scattered field. Depending on the approximation to which the induced current is evaluated, we get an approximation to the scattered field. So it all comes down to evaluation of the induced current. The classical physical optics approximation uses the so called physical optics induced current. This is the induced current on an infinite plane surface coincident with the tangential plane at the point under consideration [3]. The polarization corrected physical optics approximation involves corrections to the physical optics induced current. One possible approach to correct the induced current is to derive it as a by-product of the generalized geometrical optics expansion for the scattered field, as was indeed done in [1]. Another approach (simpler for the purpose) is to iterate the magnetic field integral equation (MFIE) about the physical optics solution with evaluation the various correction terms in the high frequency limit. This is in principle the underlying idea employed in [4,5] where the work was carried out in time domain. The physical optics correction term in [4,5] is appropriate for normal incidence angle and coincides with later more general results [1,6].

Rigorous high frequency expansion of the induced current through iterating the MFIE has been carried out in [6]. A different MFIE approach has been suggested in [7]: this is through expanding the MFIE into a series of integral equations with known solutions.

In this paper we work along the lines of [6] to derive the high frequency power series expansion of the induced current on the scatterer. The iteration of the MFIE results in a power series that appears to be in integer powers of $K/k$ where $K$ is a characteristic curvature of the scatterer and $k = 2\pi/\lambda$ is the wavenumber of the incident plane wave. The zero'th (leading) order term in this expansion is the physical optics current. The first (next-to-the-leading) order term is the source of the physical optics correction field. We believe that best use of the correction term is for predicting the cross (ortho) polarized scattered field which, with the absence of this term would be zero. Regretfully in our formulation we have to limit the evaluation of the current correction term to points on the scatterer in the illuminated region, away from light-shadow transition regions. This is even though the method can in principle be applied also to those regions resulting in exotic expressions involving incomplete Airy functions [8]. This is not supposed however to be a real obstacle for monostatic scattering applications. The transition region is known to occupy a width of order of wavelengths which is of negligible area relative to the participating area in the scattering process in large bodies. Exclusion of this region from the radiation integral will therefore result in only a small error. In certain cases we may be even luckier: consider for example again the object of Fig.1. This object happens to be cylindrically symmetric also at the light-shadow transition region. By symmetry argument the corrected induced current in this region is not expected to radiate a cross-polarized field so that its evaluation can be discarded to begin with.

Our treatment is not restricted to linear polarizations nor are the incident and scattered polarizations described in terms of canonic linear polarizations. We rather avoid using scattering matrices and employ in turn a "coordinate free" approach by working in terms of co- and ortho-polarizations. We prefer the term "ortho-polarization" over "cross-polarization", to indicate that the treatment addresses general elliptic polarizations rather than just linear polarizations.

In chapter 2 we derive the first order corrected current. In chapter 3 this current is employed in a radiation integral, where the result is cast into the co-polarized and ortho-polarized constituents. In chapter 4 we check how the physical optics result of ch. 3 reduces to the geometrical optics result. Chapter 5 is concerned with application of the physical optics result to time domain application, and a corrected impulse response expression is derived. this result is compared against a known polarization corrected expression and is found to reduce to it under certain simplifying approximations.

## 2. First order correction to the physical optics induced current

In reference to Fig.2, an incident (locally plane) wave $\mathbf{H}^{inc}$, is propagating in the $z$ direction, impinging upon the perfectly conducting scatterer which is assumed to be of smooth shape. The source of the incident field is at $z = -z_0$. The phase reference is at the source.

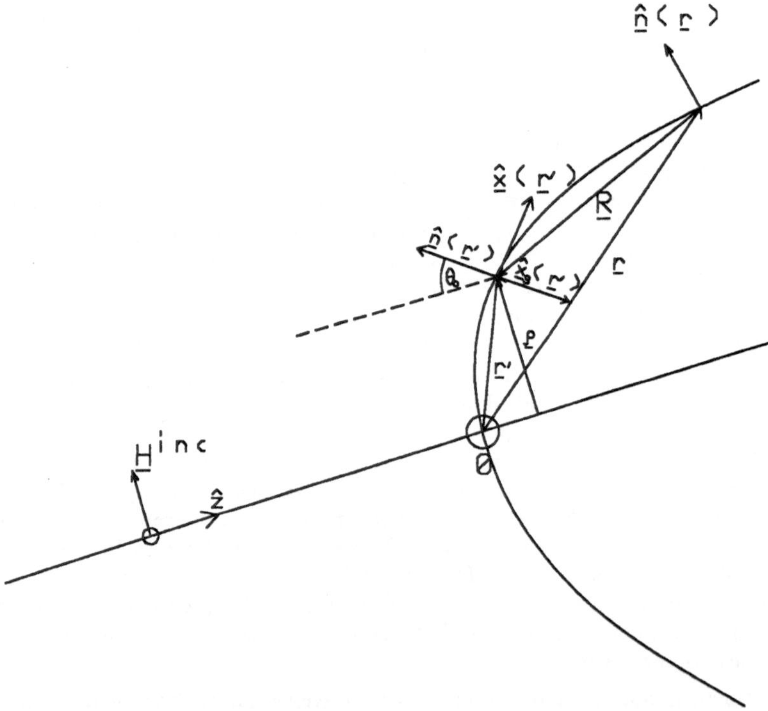


Figure 2: The incident field and the scattering object. The coordinate origin, $O$, is at the specular reflection point. $\mathbf{r}'$ is an arbitrary point on the scatterer surface in the illuminated region, where the induced surface current is beeing evaluated. $\mathbf{x}$ is a tangential polar radial vector at $\mathbf{r}'$.

The induced surface current is known (see Appendix A.) to satisfy the magnetic field integral equation,

$$\mathbf{J}^s(\mathbf{r}') = 2\hat{\mathbf{n}}(\mathbf{r}') \times \mathbf{H}^{inc} + \frac{1}{2\pi}\hat{\mathbf{n}}(\mathbf{r}') \times P \oint d^2r\, L\mathbf{J}^s(\mathbf{r}) \times \hat{\mathbf{R}}(\mathbf{r}',\mathbf{r}) \tag{1}$$

where,

$$L \equiv \frac{1-ikR}{R^2}e^{ikR}, \qquad \mathbf{R} \equiv \mathbf{r}' - \mathbf{r}, \qquad R \equiv |\mathbf{R}| \qquad \hat{\mathbf{R}} \equiv \frac{\mathbf{R}}{R} \tag{2}$$

The integration is carried out over the entire scatterer surface, and $P$ stands for the principal value sense.

The physical optics approximation, $\mathbf{J}_0$, to $\mathbf{J}^s$ is known to be equal to the first term in the RHS of Eq.(1), on the illuminated portion, $\mathbf{L}$, of the scatterer surface, and equal to zero in the shadow region $\mathbf{L}'$. The first order correction term, $\mathbf{J}_1$, to $\mathbf{J}_0$, is obtained by,

$$\tilde{\mathbf{J}}_1(\mathbf{r}') = \frac{1}{2\pi}\hat{\mathbf{n}}(\mathbf{r}') \times P \oint d^2r\, L\mathbf{J}_0(\mathbf{r}) \times \hat{\mathbf{R}}(\mathbf{r}',\mathbf{r}), \quad \mathbf{J}_1(\mathbf{r}') = k^{-1}\lim_{k\to\infty}\{k\tilde{\mathbf{J}}_1(\mathbf{r}')\} \tag{3}$$

Higher order correction terms, $J_i$, to the physical optics field, can be obtained recursively through,

$$\tilde{\mathbf{J}}_i(\mathbf{r}') = \frac{1}{2\pi}\hat{\mathbf{n}}(\mathbf{r}') \times P \oint d^2r\, L\mathbf{J}_{i-1}(\mathbf{r}) \times \hat{\mathbf{R}}(\mathbf{r}',\mathbf{r}), \quad \mathbf{J}_i(\mathbf{r}') = k^{-1}\lim_{k\to\infty}\{k\tilde{\mathbf{J}}_i(\mathbf{r}')\}, \quad i = 1,2\ldots \tag{4}$$

We will however focus on $\mathbf{J}_1(\mathbf{r}')$. The $\mathbf{J}_i$ terms in the illuminated region are of order of $(K/k)^i$.

The procedure outlined above is identified as the Neumann series approach [9] to solve the integral equation Eq.(1).

We now use Eq.(3) to derive an expression for $\tilde{\mathbf{J}}_1$. Substitute the value of $\mathbf{J}_0$ in Eq.(3),

$$\begin{aligned}\tilde{\mathbf{J}}_1(\mathbf{r}') =& \frac{1}{\pi}e^{ikz_0}\hat{\mathbf{n}}(\mathbf{r}') \times P \int_{\mathbf{L}} d^2r\, Le^{ikz(\mathbf{r})}[\hat{\mathbf{n}}(\mathbf{r}) \times \mathbf{H}_0] \times \hat{\mathbf{R}}(\mathbf{r}',\mathbf{r}) = \\ & \frac{1}{\pi}e^{ikz_0}P \int_{\mathbf{L}} d^2r\, Le^{ikz(\mathbf{r})}\{[\hat{\mathbf{n}}(\mathbf{r}') \times \mathbf{H}_0][\hat{\mathbf{n}}(\mathbf{r}) \cdot \hat{\mathbf{R}}(\mathbf{r}',\mathbf{r})] - [\hat{\mathbf{n}}(\mathbf{r}') \times \hat{\mathbf{n}}(\mathbf{r})][\mathbf{H}_0 \cdot \hat{\mathbf{R}}(\mathbf{r}',\mathbf{r})]\}\end{aligned} \tag{5}$$

where we have used,

$$\mathbf{H}^{inc}(\mathbf{r}) = \mathbf{H}_0 e^{ik[z_0+z(\mathbf{r})]} \tag{6}$$

The integral in Eq.(5) is carried out only over the illuminated portion $\mathbf{L}$, and yields a value for $\tilde{\mathbf{J}}_1$ everywhere on the scatterer. Eq.(5) is now evaluated asymptotically in the zero wavelength limit. This result is composed from two terms: the stationary point contribution, and the $\mathbf{L} : \mathbf{L}'$ boundary line discontinuity contribution. We will disregard the boundary line contribution (which is dominant for $\mathbf{r}'$ in the transition and shadow regions), and consider only the stationary point which happens to be at $\mathbf{r}'$, for $\mathbf{r}'$ in the $\mathbf{L}$

region. The boundary line contribution is interpreted as a geometrical optics diffraction term. We will therefore obtain a valid approximation for $\mathbf{J}_1$ in the $\mathbf{L}$ region away from its boundary with $\mathbf{L}'$. For this purpose we need the local values of the integrand amplitude term as $\mathbf{r}$ tends to $\mathbf{r}'$. The surface function of the scatterer is expanded about $\mathbf{r}'$, with the aid of a local right handed cartezian coordinate system $x_{123}$, with $x_3$ along the inward pointing normal, and $x_1, x_2$ along the local principal directions.

$$x_3(\mathbf{x}) = \frac{1}{2}(K_1 x_1^2 + K_2 x_2^2) + O(x^3), \quad x \equiv \sqrt{x_1^2 + x_2^2}, \quad x \to 0 \tag{7}$$

where $K_1, K_2$, are the local principal curvatures of the surface. We obtain,

$$\hat{\mathbf{n}}(\mathbf{r}) \cdot \hat{\mathbf{R}}(\mathbf{r}', \mathbf{r}) = -\frac{1}{2x}(K_1 x_1^2 + K_2 x_2^2) + O(x^2), \quad \mathbf{x} \equiv x_1 \hat{\mathbf{x}}_1 + x_2 \hat{\mathbf{x}}_2, \quad x \equiv |\mathbf{x}| \tag{8}$$

$$\hat{\mathbf{n}}(\mathbf{r}') \times \hat{\mathbf{n}}(\mathbf{r}) = K_2 x_2 \hat{\mathbf{x}}_1 - K_1 x_1 \hat{\mathbf{x}}_2 + O(x^2) \tag{9}$$

$$z(\mathbf{r}) = z(\mathbf{r}') - \hat{\mathbf{z}} \cdot \mathbf{R}(\mathbf{r}', \mathbf{r}) = z(\mathbf{r}') + \hat{\mathbf{z}} \cdot \mathbf{x} + O(x^2), \quad R = x + O(x^3) \tag{10}$$

$$d^2 r = d^2 x[1 + O(x^2)], \quad \hat{\mathbf{n}}(\mathbf{r}') \times \mathbf{H}_0 = -\hat{\mathbf{x}}_3 \times \mathbf{H}_0 + O(x), \quad \mathbf{H}_0 \cdot \hat{\mathbf{R}} = -\mathbf{H}_0 \cdot \hat{\mathbf{x}} + O(x^2) \tag{11}$$

Substitute these into Eq. (5),

$$\tilde{\mathbf{J}}_1(\mathbf{r}') = \frac{1}{2\pi} e^{ik[z_0 + z(\mathbf{r}')]} P \int d^2 x \, \frac{1 - ikx}{x} e^{ikx(1 + \hat{\mathbf{z}} \cdot \hat{\mathbf{x}}) + O(x^2)} [1 + O(x)] \cdot \left[ \frac{1}{x^2}(K_1 x_1^2 + K_2 x_2^2)(\hat{\mathbf{x}}_3 \times \mathbf{H}_0) + \frac{2}{x}(\mathbf{H}_0 \cdot \hat{\mathbf{x}})(K_2 x_2 \hat{\mathbf{x}}_1 - K_1 x_1 \hat{\mathbf{x}}_2) \right] \tag{12}$$

Introduce spherical-polar variables,

$$x_1 = x \cos\phi, \quad x_2 = x \sin\phi, \quad \hat{\mathbf{z}} = \hat{\mathbf{x}}_3 \cos\theta_0 + (\hat{\mathbf{x}}_1 \cos\phi_0 + \hat{\mathbf{x}}_2 \sin\phi_0) \sin\theta_0 \tag{13}$$

and use them in the integral,

$$\tilde{\mathbf{J}}_1(\mathbf{r}') = e^{ik[z_0 + z(\mathbf{r}')]} \mathbf{H}_0 \cdot \frac{1}{2\pi} \int_0^{2\pi} d\phi \, [(K_1 \cos^2\phi + K_2 \sin^2\phi)(\hat{\mathbf{x}}_1 \hat{\mathbf{x}}_2 - \hat{\mathbf{x}}_2 \hat{\mathbf{x}}_1) + 2(\hat{\mathbf{x}}_1 \cos\phi + \hat{\mathbf{x}}_2 \sin\phi)(\hat{\mathbf{x}}_1 K_2 \sin\phi - \hat{\mathbf{x}}_2 K_1 \cos\phi) \int_0^\infty dx \, x \frac{1 - ikx}{x} e^{ikx[1 + \sin\theta_0 \cos(\phi - \phi_0)] + O(x^2)} [1 + O(x)] \tag{14}$$

The interpretations of $\theta_0, \phi_0$, are as follows: $\theta_0$ is the angle between the inward normal of the surface element, and the positively oriented $z$ axis. $\phi_0$ is the angle of one principal direction at $\mathbf{r}'$ to the meridian line at $\mathbf{r}'$. $\theta_0$, $\phi_0$ are both functions of $\mathbf{r}'$ and $\hat{\mathbf{z}}$.

Integration with the $x, \phi$ variables takes care inherently of the principal value sense of the integral in Eq.(12). In Eq.(14) we also introduced dyadic notation for compactness of the representation.

The $x$ integral is readily performable, yielding,

$$\tilde{\mathbf{J}}_1(\mathbf{r}') = -\frac{1}{ik}[1+O(k^{-1})]e^{ik[z_0+z(\mathbf{r}')]}\mathbf{H}_0 \cdot \frac{1}{2\pi}\int_0^{2\pi} d\phi \frac{2+\sin\theta_0\cos(\phi-\phi_0)}{[1+\sin\theta_0\cos(\phi-\phi_0)]^2} \cdot$$
$$[\hat{\mathbf{x}}_1\hat{\mathbf{x}}_2+\hat{\mathbf{x}}_2\hat{\mathbf{x}}_1)(K_2\sin^2\phi - K_1\cos^2\phi) + \hat{\mathbf{x}}_1\hat{\mathbf{x}}_1 K_2\sin 2\phi - \hat{\mathbf{x}}_2\hat{\mathbf{x}}_2 K_1 \sin 2\phi \tag{15}$$

We face three different types of angular integrations, which are also performable, resulting in,

$$\mathbf{J}_1(\mathbf{r}') = -\frac{1}{ik}e^{ik[z_0+z(\mathbf{r}')]}\mathbf{H}_0 \cdot \mathcal{D} \tag{16}$$

where,

$$\mathcal{D} \equiv (K_2 I_2 - K_1 I_1)(\hat{\mathbf{x}}_1\hat{\mathbf{x}}_2+\hat{\mathbf{x}}_2\hat{\mathbf{x}}_1) + 2K_2 I_3 \hat{\mathbf{x}}_1\hat{\mathbf{x}}_1 - 2K_1 I_3 \hat{\mathbf{x}}_2\hat{\mathbf{x}}_2 \tag{17}$$

with $I_1$, $I_2$, $I_3$,

$$\begin{aligned}
I_1 &\equiv \frac{1}{2\pi}\int_0^{2\pi} d\phi \frac{2+\sin\theta_0\cos(\phi-\phi_0)}{[1+\sin\theta_0\cos(\phi-\phi_0)]^2}\cos^2\phi \\
&= \frac{1}{2\cos^3\theta_0}(1+\cos^2\theta_0+\sin^2\theta_0\cos 2\phi_0) \\
I_2 &\equiv \frac{1}{2\pi}\int_0^{2\pi} d\phi \frac{2+\sin\theta_0\cos(\phi-\phi_0)}{[1+\sin\theta_0\cos(\phi-\phi_0)]^2}\sin^2\phi \\
&= \frac{1}{2\cos^3\theta_0}(1+\cos^2\theta_0-\sin^2\theta_0\cos 2\phi_0) \\
I_3 &\equiv \frac{1}{2\pi}\int_0^{2\pi} d\phi \frac{2+\sin\theta_0\cos(\phi-\phi_0)}{[1+\sin\theta_0\cos(\phi-\phi_0)]^2}\sin\phi\cos\phi \\
&= \frac{1}{2\cos^3\theta_0}\sin^2\theta_0\sin 2\phi_0
\end{aligned} \tag{18}$$

Note that from Eq.(16) it is found that indeed, $\mathbf{J}_1$ in the illuminated region is of order of $K/k$, where $K$ is the characteristic curvature of the scatterer. It is straightforward to show by induction that $\mathbf{J}_i$ of Eq.(4) in the illuminated region is of order of $(K/k)^i$. (The diffraction wave terms originating from the $\mathbf{L} : \mathbf{L}'$ boundary line will be interweaved as noninteger powers in the $K/k$ power series expansion, mentioned above.)

Another important notion is the following: from Eq.(18) it is evident that something uncalled for happens as $\theta_0$ approaches $\pi/2$. This is not because of the approaching to the the vicinity of the transition region, which for itself suggests the introduction of diffraction fields, but rather because the geometrical approximation to $R$ in Eq.(10) assumes that the scatterer surface is locally plane. This approximation introduces a problem near grazing angle incidence and has to be corrected for these regions. We exclude this region in this paper.

Eq.(16) manifests the dependence of the induced current correction term $\mathbf{J}_1$, on the incidence angle $\theta_0$. Eq.(16) is in agreement with the result obtained in [6] aside of a global sign reversal (we believe that our result is the correct one) and coincides with the results in [4,5] only at incidence angle $\theta_0 = 0$.

## 3. The first order correction to the physical optics backscattered far field

The backscattered far field of $\mathbf{J}_1$ is given through,

$$\mathbf{H}_1^s = -ik\frac{e^{ikz_0}}{4\pi z_0}\oint d^2r'\, e^{ik\hat{\mathbf{z}}\cdot\mathbf{r}'}\hat{\mathbf{z}}\times\mathbf{J}_1(\mathbf{r}') \tag{19}$$

Substituting Eq.(16) in Eq.(19) yields,

$$\mathbf{H}_1^s = -ik\frac{e^{ikz_0}}{2\pi z_0}\int_{\mathbf{L}} d^2r'\, e^{ik\hat{\mathbf{z}}\cdot\mathbf{r}'}\frac{\mathbf{H}_0\cdot\mathcal{D}\times\hat{\mathbf{z}}}{2ik} \tag{20}$$

Eq.(20) presents the first order correction to the physical optics backscattered field.

Eq.(20) is not explicit as far as ortho-polarization goes. To obtain this information, the backscattered field has to be decomposed into a co-polarized ($cp$) and an ortho-polarized ($op$) component.

We define the $cp$ and $op$ polarizations as follows,

$$\mathbf{H}_{cp}\equiv\mathbf{H}_0,\quad \mathbf{H}_{op}\equiv-\hat{\mathbf{z}}\times\mathbf{H}_0^* \tag{21}$$

where the asterisk denotes complex conjugation. The above defined polarizations are orthogonal in the sense that an antenna which is polarization matched to one of them, will be completely mismatched to the other [10]. The $cp$ component is co-polarized in the sense that it has the polarization of the zero order physical optics field. The minus sign in the definition of $\mathbf{H}_{op}$ is immaterial, and has been introduced for improved appearance of following equations.

The definition of orthogonality of the vectors in Eq.(21) over the field of complex polarization vectors, implies the following inner product,

$$<\mathbf{V},\mathbf{W}>\equiv\mathbf{V}\cdot\mathbf{W}^* \tag{22}$$

Having defined an inner product, we can now represent an arbitrary polarization vector V as a superposition of a $cp$ and a $op$ component.

$$\mathbf{V} = C\mathbf{H}_{cp} + X\mathbf{H}_{op} = C\mathbf{H}_0 - X(\hat{\mathbf{z}}\times\mathbf{H}_0^*) \tag{23}$$

where,

$$C = \frac{<\mathbf{V},\mathbf{H}_{cp}>}{<\mathbf{H}_{cp},\mathbf{H}_{cp}>} = \frac{\mathbf{V}\cdot\mathbf{H}_0^*}{\mathbf{H}_0\cdot\mathbf{H}_0^*},\quad X = \frac{<\mathbf{V},\mathbf{H}_{op}>}{<\mathbf{H}_{op},\mathbf{H}_{op}>} = \frac{-\mathbf{V}\cdot\hat{\mathbf{z}}\times\mathbf{H}_0}{\mathbf{H}_0\cdot\mathbf{H}_0^*} \tag{24}$$

Applying this to the four vector components in Eq.(20) results in,

$$\begin{aligned}\mathbf{H}_{1cp}^s &= ik\frac{e^{2ikz_0}}{2\pi z_0}\mathbf{H}_{cp}\int_{\mathbf{L}} d^2r'\, e^{2ik\hat{\mathbf{z}}\cdot\mathbf{r}'}\cdot\\ &\qquad\frac{(K_1I_1-K_2I_2)(C_1+C_2)+2K_1I_3C_3-2K_2I_3C_4}{2ik}\\ \mathbf{H}_{1op}^s &= ik\frac{e^{2ikz_0}}{2\pi z_0}\mathbf{H}_{op}\int_{\mathbf{L}} d^2r'\, e^{2ik\hat{\mathbf{z}}\cdot\mathbf{r}'}\cdot\\ &\qquad\frac{(K_1I_1-K_2I_2)(X_1+X_2)+2K_1I_3X_3-2K_2I_3X_4}{2ik}\end{aligned} \tag{25}$$

where the $C_i$ and $X_i$ constants are readily evaluated and given below, (we denote $\mathbf{H}_0 \cdot \hat{\mathbf{x}}_i = H_i$)

$$\begin{aligned} C_1 &= \frac{|H_1|^2 \cos\theta_0 - H_1 H_3^* \sin\theta_0 \cos\phi_0}{|\mathbf{H}_0|^2} & X_1 &= \frac{H_1 H_2}{|\mathbf{H}_0|^2} \\ C_2 &= \frac{-|H_2|^2 \cos\theta_0 + H_2 H_3^* \sin\theta_0 \sin\phi_0}{|\mathbf{H}_0|^2} & X_2 &= \frac{H_1 H_2}{|\mathbf{H}_0|^2} = X_1 \\ C_3 &= \frac{H_1^* H_2 \cos\theta_0 - H_2 H_3^* \sin\theta_0 \cos\phi_0}{|\mathbf{H}_0|^2} & X_3 &= \frac{H_2^2}{|\mathbf{H}_0|^2} \\ C_4 &= \frac{-H_1 H_2^* \cos\theta_0 + H_1 H_3^* \sin\theta_0 \sin\phi_0}{|\mathbf{H}_0|^2} & X_4 &= \frac{H_1^2}{|\mathbf{H}_0|^2} \end{aligned} \tag{26}$$

It can be readily verified that for an axially symmetric scatterer with axial incidence the monostatic scatterer is completely co-polarized as expected from the configuration.

For reference recall that the leading order physical optics field is,

$$\mathbf{H}_0^s = ik \frac{e^{2ikz_0}}{2\pi z_0} \mathbf{H}_{cp} \int_{\mathbf{L}} d^2r' \, e^{2ik\hat{\mathbf{z}}\cdot\mathbf{r}'} \cos\theta_0 \tag{27}$$

It can be readily verified that for axially symmetric scatterers the ortho-polarized terrm is zero as expected. Eq.(25) is the main result of this paper.

## 4. The geometrical optics backscattered field.

It is interesting to verify that the above result does reduce to the geometrical optics result. The above physical optics formulation, the *op* part of Eq.(25) can be evaluated asymptotically in the zero wavelength limit. Assuming that the coordinate origin is affixed at a specular point of the scatterer, the contribution of that point is readily evaluated to yield, ($\theta_0 = 0$)

$$\mathbf{H}_{opgo}^s = \frac{-e^{2ikz_0}}{2\sqrt{K_1 K_2} z_0} \mathbf{H}_{op} \frac{2(K_1 - K_2) H_1 H_2}{2ik|\mathbf{H}_0|^2} \tag{28}$$

The same process applied to the *cp* part of Eq.(25) is expected to provide only a part of the $K/k$ order geometrical optics correction term. This is since an additional contribution of the same order results from taking the second term in the asymptotic expansion of the $\mathbf{J}_0$ radiation integral.

In Eq.(28) we have not accounted the first order diffraction wave contribution, which arises from the surface current discontinuity at the $\mathbf{L} : \mathbf{L}'$ boundary.

In view of Eq.(28) it is interesting to have for reference the leading order *g.o.* field,

$$\mathbf{H}_{0go}^s = \frac{-e^{2ikz_0}}{2\sqrt{K_1 K_2} z_0} \mathbf{H}_{cp} \tag{29}$$

## 5. Application: Time Domain Physical Optics Polarization Correction to Monostatic Scattering from Convex Scatterers.

Our work is particularly suitable for time domain applications since in time domain the light-shadow boundary diffraction fields which we were unwilling to consider, are separated on a temporal scale from the radiation from the currents we corrected.

To derive the impulsive plane wave response, the $\omega$ variable in the time harmonic results has to be Fourier transformed to $t$, typically,

$$\mathbf{H}(t) = \frac{1}{2\pi}\int_{-\infty}^{\infty} d\omega\, e^{-i\omega t}\mathbf{H}(\omega) \tag{30}$$

In the following we denote in Eqs.(25) for brevity,

$$\begin{aligned} S_{cp} = S_{cp}(\phi, z) &\equiv (K_1 I_1 - K_2 I_2)(C_1 + C_2) + 2K_1 I_3 C_3 - 2K_2 I_3 C_4 \\ S_{op} = S_{op}(\phi, z) &\equiv (K_1 I_1 - K_2 I_2)(X_1 + X_2) + 2K_1 I_3 X_3 - 2K_2 I_3 X_4 \end{aligned} \tag{31}$$

We now Fourier transform Eqs.(25) into time domain,

$$\begin{aligned} \mathbf{H}^{s}_{1\,cp}(t) &= \frac{1}{4\pi z_0}\mathbf{H}_{cp}\int_{\mathbf{L}} d^2r'\, S_{cp}(\phi, z)\delta\left[t - \frac{2(z_0 + z)}{c}\right] \\ \mathbf{H}^{s}_{1\,op}(t) &= \frac{1}{4\pi z_0}\mathbf{H}_{op}\int_{\mathbf{L}} d^2r'\, S_{op}(\phi, z)\delta\left[t - \frac{2(z_0 + z)}{c}\right] \end{aligned} \tag{32}$$

What we need now is a convenient coordinate system for specifying the $d^2r'$ integration. The coordinate system of our choice is cylindrical $\rho\phi z$. The surface function is described through,

$$\rho = \rho(\phi, z), \qquad z > z_0 \tag{33}$$

The unit normal vector to the surface is,

$$\hat{\mathbf{n}}(\phi, z) = \frac{\hat{\rho}(\phi) - \frac{1}{\rho}\frac{\partial\rho}{\partial\phi}\hat{\phi}(\phi) - \frac{\partial\rho}{\partial z}\hat{\mathbf{z}}}{\sqrt{1 + \frac{1}{\rho^2}\left(\frac{\partial\rho}{\partial\phi}\right)^2 + \left(\frac{\partial\rho}{\partial z}\right)^2}} \tag{34}$$

$\theta_0 = \theta_0(\phi, z)$, the incidence angle, is given through, $\cos\theta_0 = -\hat{\mathbf{n}}\cdot\hat{\mathbf{z}}$. The surface element $d^2r'$ can be expressed in terms of $z, \phi$ differentials,

$$d^2r' = \frac{\rho\, dz\, d\phi}{\hat{\mathbf{n}}\cdot\hat{\rho}} = \frac{\rho\, dz\, d\phi\, \partial\rho/\partial z}{\cos\theta_0} \tag{35}$$

We use this in Eq.(32),

$$\begin{aligned} \mathbf{H}^{s}_{1\,cp}(t) &= \frac{1}{4\pi z_0}\mathbf{H}_{cp}\int_0^{2\pi} d\phi \int dz\, \rho\, \frac{\partial\rho}{\partial z}\frac{S_{cp}(\phi, z)}{\cos\theta_0(\phi, z)}\delta\left[t - \frac{z_0 + z}{c/2}\right] \\ \mathbf{H}^{s}_{1\,op}(t) &= \frac{1}{4\pi z_0}\mathbf{H}_{op}\int_0^{2\pi} d\phi \int dz\, \rho\, \frac{\partial\rho}{\partial z}\frac{S_{op}(\phi, z)}{\cos\theta_0(\phi, z)}\delta\left[t - \frac{z_0 + z}{c/2}\right] \end{aligned} \tag{36}$$

The $z$ integral is trivial,

$$\mathbf{H}^{s}_{1\,cp}(t) = \frac{c}{8\pi z_0}\mathbf{H}_{cp}\frac{d}{dz}\int_0^{2\pi} d\phi \frac{\rho^2(\phi,z)}{2}\frac{S_{cp}(\phi,z)}{\cos\theta_0(\phi,z)}\bigg|_{z=\frac{ct}{2}-z_0}$$

$$\mathbf{H}^{s}_{1\,op}(t) = \frac{c}{8\pi z_0}\mathbf{H}_{op}\frac{d}{dz}\int_0^{2\pi} d\phi \frac{\rho^2(\phi,z)}{2}\frac{S_{op}(\phi,z)}{\cos\theta_0(\phi,z)}\bigg|_{z=\frac{ct}{2}-z_0} \tag{37}$$

Upon transforming to equivalent time according to, $z = \frac{ct}{2} - z_0$, we finally obtain,

$$\mathbf{H}^{s}_{1\,cp}(t) = \frac{1}{4\pi z_0}\mathbf{H}_{cp}\frac{d}{dt}\int_0^{2\pi} d\phi \frac{\rho^2(\phi,z)}{2}\frac{S_{cp}(\phi,z)}{\cos\theta_0(\phi,z)}\bigg|_{t=\frac{z_0+z}{c/2}}$$

$$\mathbf{H}^{s}_{1\,op}(t) = \frac{1}{4\pi z_0}\mathbf{H}_{op}\frac{d}{dz}\int_0^{2\pi} d\phi \frac{\rho^2(\phi,z)}{2}\frac{S_{op}(\phi,z)}{\cos\theta_0(\phi,z)}\bigg|_{t=\frac{z_0+z}{c/2}} \tag{38}$$

This relation is valid as long as z is entirely in the illuminated region. The integral is a function of $z$ and contains geometric information from the vicinity of the scattering surface slice as sliced with a plane perpendicular to the z axis. This is the polarization correction of the classical physical optics result [11,12]. As we now show, this result can be viewed as a refined polarization correction to a known result [13,14,15]. We evaluate Eq.(38) at times corresponding to $z$ in the vicinity of the specular point $z_0$. For this vicinity we approximate $\theta_0$, the incidence angle to be zero, $\theta_0 \approx 0$. We also consider the principal curvatures to be approximately constant and equal to those of the specular reflection point.

From Eq.(17) we find, $I_1 \approx I_2 \approx 1$, $\quad I_3 \approx 0$. From Eq.(26) we find,

$$C_1 \approx \frac{|H_1|^2}{|\mathbf{H}_0|^2}, \quad C_2 \approx \frac{-|H_2|^2}{|\mathbf{H}_0|^2}, \quad X_1 = X_2 \approx \frac{H_1 H_2}{|\mathbf{H}_0|^2} \tag{39}$$

Substitute these into Eq.(31),

$$S_{cp} \approx (K_1 - K_2)\frac{|H_1|^2 - |H_2|^2}{|\mathbf{H}_0|^2}, \quad S_{op} \approx (K_1 - K_2)\frac{2H_1 H_2}{|\mathbf{H}_0|^2} \tag{40}$$

$S_{cp}$ and $S_{op}$ are approximately constant as far as the $\phi$ integration is concerned in Eq.(38). The remaining $d\phi$ integration is identified as $A(t)$, the silhouette area of the scatterer as delineated by the incident wavefront travelling at half the free space velocity of light.

$$\mathbf{H}^{s}_{1\,cp}(t) \approx \frac{1}{4\pi z_0}\mathbf{H}_{cp}S_{cp}\frac{dA(t)}{dt}\bigg|_{t=\frac{z_0+z}{c/2}}$$

$$\mathbf{H}^{s}_{1\,op}(t) \approx \frac{1}{4\pi z_0}\mathbf{H}_{op}S_{op}\frac{dA(t)}{dt}\bigg|_{t=\frac{z_0+z}{c/2}} \tag{41}$$

We now combine the *cp* and *op* components,

$$\mathbf{H}_1^s(t) \approx \frac{1}{2\pi z_0}\frac{dA(t)}{dt}\frac{K_1 - K_2}{2}\left[\frac{|H_1|^2 - |H_2|^2}{|\mathbf{H}_0|^2}\mathbf{H}_0 - \frac{2H_1H_2}{|\mathbf{H}_0|^2}\hat{\mathbf{z}} \times \mathbf{H}_0^*\right] \tag{42}$$

In terms of $\hat{\mathbf{x}}_1, \hat{\mathbf{x}}_2$ polarizations we finally obtain (provided that $\mathbf{H}_0$ is real, this is we are talking about linear polarization),

$$\mathbf{H}_1^s(t) \approx \frac{1}{2\pi z_0}\frac{dA(t)}{dt}\frac{K_1 - K_2}{2}(H_1\hat{\mathbf{x}}_1 - H_2\hat{\mathbf{x}}_2) \tag{43}$$

This last result identifies with existing formulation [15]. We conclude that the existing formulation Eq.(43) is valid only for $t$ corresponding to $z$ in the vicinity if the leading edge of the scatterer, whereas for later $t$ Eq.(38) is more appropriate.

## 6. Discussion

In this paper we have obtained a first order correction to the physical optics field through a local correction to the physical optics induced current.

The geometrical optics specular point correction field has been obtained through the zero wavelength asymptotic evaluation of the physical optics integral.

We have applied the physical optics result to derive a Kennaugh-Kosgriff type polarization correction to the impulse response of a scatterer.

No higher order derivatives of the area function of the scatterer appear to be involved in the results. The curvature information on the entire scattering surface is equivalent to higher order derivative information on the surface at one point, consistent with statements in [16].

## Acknowledgement

This work has been performed under an SBIR Phase I grant to A.J.Devancy Associates, from the Office of Naval Research.

I would like to acknowledge the stimulating discussions with Dr. Anthony J. Devaney, related to the material reported herein.

## References

1. S.W. Lee, "Electromagnetic reflection from a conducting surface: geometrical optics solution," IEEE Trans. Ant. Propag., AP-23, 2, 1975, 184-191.

2. H. Ansorge, "Electromagnetic reflection from a curved dielectric interface," IEEE Trans. on Antennas and Propagation, AP-34, 6, 842-845, 1986.

3. R. F. Harrington, "Time-Harmonic Electromagnetic Fields," McGraw-Hill, 1961.

4. C. L. Bennett, A. M. Auckenthaler, R. S. Smith, J. D. DeLorenzo, "Space-time integral equation approach to the large body scattering problem," RADC-CR-73-70, AD763794. Sperry Rand Research Center, Sudbury, MA, May 1973.

5. C. L. Bennett, "Time domain inverse scattering," IEEE Trans. Ant. Propag., AP-29, 1981, 213-219.

6. H. Chaloupka, H.J. Meckelburg, "Improved high frequency current approximation for curved conducting surfaces," AEÜ, 39, 1985, 245-250.

7. G. S. Brown, "An integral equation based Luneburg-Kline development for the current induced on a conducting surface," paper presented before the 1987 National Radio Science Meeting, Boulder, CO, January 1987.

8. L. Levey, L.B. Felsen "On incomplete Airy functions and their application to diffraction problems," Radio Science, 4, 10, 1969, 959-969.

9. J. Mathews, R. L. Walker, "Mathematical Methods of Physics," Benjamin/Cummings Publishing, 1970.

10. G. Sinclair, "The transmission and reception of elliptically polarized waves," IRE Proc., 148-151, February 1950.

11. E. M. Kennaugh, R. L. Cosgriff, "The use of impulse response in electromagnetic scattering problems," IRE National convention Record, Part I., 1958, 72-77.

12. C. L. Bennett, A. M. Auckenthaler, R. S. Smith, J. D. DeLorenzo, "Space-time integral equation approach to the large body scattering problem," RADC-CR-73-70, AD763794. Sperry Rand Research Center, Sudbury, MA, May 1973.

13. C. L. Bennett, "Time domain inverse scattering," IEEE Trans. Ant. Propag., AP-29, 1981, 213-219.

14. W.M Boerner and C.M. Ho, "Analysis of physical optics far field inverse scattering for the limited data case using Radon theory and polarization information," Wave Motion, 3, 1981, 311-333.

15. S.K. Chaudhuri and W.M Boerner, "Polarization utilization in profile inversion of a perfectly conducting prolate spheroid," IEEE Trans. Ant. Propag., AP-25, 1977, 505-511.

16. A. K. Dominek, L. Peters Jr., W. D. Burnside, "An additional physical interpretation in the Luneburg-Kline expansion," IEEE Trans. Ant. Propag., AP-35, 1987, 406-411.

## Appendix A: Derivation of the Integral Equation for the Induced Surface Current

The expression for the surface current is derived from an expression for the total field at an arbitrary point in space. This arbitrary point is specialized to a point on the scatterer surface, and the appropriate boundary conditions are applied. First step is to replace the perfect conductor by the currents induced on its surface. This should not alter the fields throughout.

The scattered field has to nullify the total field within the conducting surface.

$$\begin{aligned}
\mathbf{H}^s(\mathbf{r}') &= \frac{1}{\mu_0}\nabla\times\mathbf{A}(\mathbf{r}') = \frac{1}{4\pi}\nabla\times\oint d^2r\,\mathbf{J}^s(\mathbf{r})\frac{e^{ikR}}{R} \\
&= -\frac{1}{4\pi}\oint d^2r\,\mathbf{J}^s(\mathbf{r})\times\nabla\frac{e^{ikR}}{R} = -\frac{1}{4\pi}\oint d^2r\,\mathbf{J}^s(\mathbf{r})\times\hat{\mathbf{R}}(ik-\frac{1}{R})\frac{e^{ikR}}{R} \\
&= \frac{1}{4\pi}\oint d^2r\left(\frac{1}{R^2}-\frac{ik}{R}\right)\mathbf{J}^s(\mathbf{r})e^{ikR}\times\hat{\mathbf{R}} \qquad (A.1)
\end{aligned}$$

with $R \equiv |\mathbf{r}'-\mathbf{r}|$. Now we need to evaluate the field just on the exterior to the scattering surface. We consider an observation point close to the surface relative to the local radii of curvature there, and to wavelength. We can break the surface integral into two parts: first is the surface integral over the entire surface excluding a small patch in the vicinity of the observation point, and second is the integration over the patch which is assumed to be of linear dimensions large relative to the distance to the observation point, yet small relative to the local radii of curvature, and wavelength. The patch contribution is readily known to be given by the field of an infinite plane sheet of constant current. When the distance of the observation point to the surface is shrinked to zero, we also allow the patch dimensions to shrink, rendering the integral over the remainder of the surface a principal value integral.

The plane sheet contribution is,

$$\mathbf{H}_1^s = \pm\frac{1}{2}\mathbf{J}^s\times\hat{\mathbf{n}} \qquad (A.2)$$

where the + sign relates to the side into which $\hat{\mathbf{n}}$ points. We can therefore write Eq. (A.1) in the form,

$$\mathbf{H}^s(\mathbf{r}') = \frac{1}{4\pi}P\oint d^2r\left(\frac{1}{R^2}-\frac{ik}{R}\right)\mathbf{J}^s(\mathbf{r})e^{ikR}\times\hat{\mathbf{R}}\pm\frac{1}{2}\mathbf{J}^s\times\hat{\mathbf{n}} \qquad (A.3)$$

where the + sign relates to the exterior to the scatterer. In the interior of the scatterer we get,

$$\mathbf{H}^s(\mathbf{r}') = \frac{1}{4\pi}P\oint d^2r\left(\frac{1}{R^2}-\frac{ik}{R}\right)\mathbf{J}^s(\mathbf{r})e^{ikR}\times\hat{\mathbf{R}}-\frac{1}{2}\mathbf{J}^s\times\hat{\mathbf{n}} = -\mathbf{H}^{inc}(\mathbf{r}') \qquad (A.4)$$

We now eliminate the plane sheet contribution in the exterior, with the aid of Eq. (A.4),

$$\mathbf{H}(\mathbf{r}') = \mathbf{H}^s(\mathbf{r}')+\mathbf{H}^{inc}(\mathbf{r}') = 2\mathbf{H}^{inc}(\mathbf{r}')+\frac{1}{2\pi}P\oint d^2r\left(\frac{1}{R^2}-\frac{ik}{R}\right)\mathbf{J}^s(\mathbf{r})e^{ikR}\times\hat{\mathbf{R}} \qquad (A.5)$$

We are now in the position to obtain an expression for the surface current: the field discontinuity across the scatterer surface should be supplemented by the surface current,

$$\mathbf{J}^s(\mathbf{r}') = \hat{\mathbf{n}}(\mathbf{r}')\times\mathbf{H}(\mathbf{r}') = 2\hat{\mathbf{n}}(\mathbf{r}')\times\mathbf{H}^{inc}(\mathbf{r}')+\frac{1}{2\pi}\hat{\mathbf{n}}(\mathbf{r}')\times P\oint d^2r\,L\mathbf{J}^s(\mathbf{r})\times\hat{\mathbf{R}} \qquad (A.6)$$

where,

$$L \equiv \left(\frac{1}{R^2}-\frac{ik}{R}\right)e^{ikR} \qquad (A.7)$$

# POLARIZATION CORRECTION AND EXTENSION OF THE KENNAUGH-COSGRIFF TARGET-RAMP RESPONSE EQUATION TO THE BI-STATIC CASE AND APPLICATIONS TO ELECTROMAGNETIC INVERSE SCATTERING

Bing-Yuen Foo, Sujeet K. Chaudhuri & Wolfgang-M. Boerner
University of Illinois at Chicago Chicago, IL 60680-4348

**ABSTRACT:** An analytical time-domain expression is derived for the early time impulse response for smooth, convex, perfectly conducting scatterers under the physical optics approximation for the bistatic case. The physical optics bistatic early time impulse responses can still be interpreted as cross-sectional areas of the scatterer similar to the work of Kennaugh for the monostatic case. A crude polarization correction to the leading edge of the physical optics impulse response is obtained for the bistatic case, leading to a simple asymptotic relation between the specular principal curvature difference and certain co-polarized phase terms in the bistatic scattering matrix. Applications to direct scattering are discussed. Profile reconstruction from bistatic data with a priori knowledge of the validity range of physical optics in the time domain is proposed and tested with the sphere.

## 1. INTRODUCTION

The concept of the impulse response as applied to electromagnetic scattering problems was first introduced by Kennaugh and Cosgriff [1,2], who interpreted the far-field impulse response, $E^S(\bar{r},t)$, for the monostatic backscattered case and under the physical optics approximation, as the second derivative of the projected monostatic area function A(t) of the scatterer,

$$E^S(\bar{r},t) = \frac{1}{2\pi rc} \frac{\partial^2 A(t)}{\partial t^2} \tag{1}$$

where $E^S(\bar{r},t)$ is the far-field impulse response. The silhouette area A(t) is formed by delineating the scatterer with a fictitious plane moving at one half of the propagation velocity. By integrating twice, (1) implies that the ramp response is thus proportional to the cross-sectional area of the scatterer, assumed to be perfectly conducting, convex shaped, smooth scattering body. The monostatic ramp response has been utilized in inverse scattering for profile reconstruction [3]-[6]. The main objective of this paper is to investigate extensions to the bistatic case and polarization correction.



*W.-M. Boerner et al. (eds.), Direct and Inverse Methods in Radar Polarimetry, Part 1*, 517–535.

In this paper, it is shown that as in the monostatic case, the ramp response is also proportional to the cross-sectional area of the scatterer in two principal polarizations, but the area is formed by a fictitious plane delineating the scatterer along the direction which bisects the angle between the transmitter, the scatterer and the receiver. Moreover, the fictitious plane is moving at a speed which is greater than that of the monostatic case and depends on the angular separation of the transmitter and the receiver in the far-field. The two principal polarizations are those for which the incident magnetic and electric fields are perpendicular to the plane formed by the transmitter, the receiver and the scatterer (denoted by TM and TE, respectively). Any other case can be considered as the superposition of the two. In any case, the extension to Kennaugh's formula does not give the proper depolarization effects in general. A crude polarization correction to the physical optics approximation can be derived by extending Bennett's leading edge correction [10] for the monostatic case. The correction in the bistatic case is found to take the form of the first time derivative of the cross-sectional area along the bisector direction, and to be proportional to the specular principal curvature difference. By considering the total field due to physical optics and its correction and transforming into the frequency domain, analytical expressions of the bistatic scattering matrix elements can be obtained, leading to an asymptotic relation between the specular principal curvature difference and the phase difference of the copolarized elements of the bistatic scattering matrix.

The polarization correction and the bistatic phase-curvature relation are discussed in Section 4, following the formulation and derivation of the exten- sion of the Kennaugh formula [2] in Sections 2 and 3. Numerical verification is carried out only for the test sphere in Section 5, because neither bistatic theoretical nor measurement data for general ellipsoidal conducting scatterers are available to us. Applications to direct scattering for bistatic radar cross section estimation are discussed in Section 6. Since the inverse problem of profile reconstruction was formulated [3] as the classical Radon problem [7] of reconstruction from projections, in Section 7 the bistatic scattering data of the sphere are processed to produce the image of the unit sphere with the help of certain a priori information. Conclusions and recommendations for future work are discussed in Section 8.

We wish to acknowledge that one of the reviewers was so kind to refer the authors to an early report on a similar treatment of E.M. Kennaugh [2] which was not published in the open literature and unknown to the authors. However, from a close examination of [2], it can be shown that our approach is different because we also include the depolarization effects although leading essentially to same results for the co-polarized bistatic physical optics impulse response interpretation, (see eqs. (13) and (14)).

## 2. FORMULATION

It is intended to find the bistatic impulse response and to interpret it in terms of cross-sectional area functions, under the physical optics approximation. The H-field impulse response is formulated as the curl of the vector potential, which can be expressed in terms of the induced surface currents on the surface of the scatterer. In this way, an integral expression for the far-field bistatic impulse response was obtained by Bennett in [8],

$$\bar{H}^s(\bar{r},t) = \frac{1}{4\pi rc} \iint_S \left. \frac{\partial \bar{J}(\bar{r}',\tau)}{\partial \tau} \right|_{\tau=t-R/c} \times \hat{a}_r \, dS' \tag{2}$$

where $\bar{J}$ is the induced surface current due to an incident impulsive plane wave; $\bar{r}$ and $\bar{r}'$ are the position vectors of the observation and integration points, respectively; $R = |\bar{r} - \bar{r}'|$; S is the surface of the scatterer; and $\hat{a}_r$ is the unit vector along the receiver direction. Once the physical optics currents are substituted in (2), the scattered field in any direction $\hat{a}_r$ can then be determined. The physical optics assumption approximates the currents for the illuminated side as

$$\bar{J} = 2\, \hat{n} \times \bar{H}^i \tag{3}$$

where $\hat{n}$ is the outward unit normal vector; $\bar{H}^i$ is the incident magnetic field; and the currents on the shadow side as zero. Now it is to interpret the integral in (2) with (3).

## 3. EXTENSION TO THE BISTATIC CASE

Fig. 1 depicts an impulse traveling towards the scatterer for the TM case. The origin is arbitrarily chosen as the point where the incident impulse first touches the scatterer, at which instant, time is referenced as zero; the incident H-field is then represented by the Dirac delta function,

$$\bar{H}^i = \hat{y}\, \delta(t+z) \tag{4}$$

where the unit of time is chosen as the light-meter as in [8] to normalize the propagation speed c to unity.

By substituting (3) and (4) into (2), the physical optics bistatic impulse response can be expressed as

$$r\bar{H}^s(\bar{r},t) = \frac{1}{2\pi} \frac{\partial}{\partial t} \iint_S [\hat{n} \times \hat{y}\, \delta(\tau+z)] \Big|_{\tau=t-R} \times \hat{a}_r \, dS' \tag{5}$$

The integral in (5) mathematically sums up contributions from integration points on S (which becomes the illuminated side) for which the argument of the delta function vanishes, i.e.

$$\tau + z = 0 \qquad \text{or} \qquad t - R + z = 0 \tag{6}$$

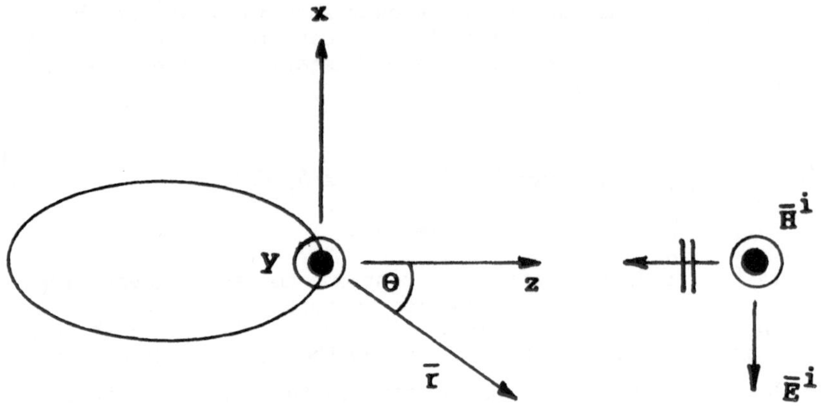


Fig. 1: Scattering coordinate system

Equation (6) requires that the receiver at time t measures the scattered returns from those integration points on S such that the sum of the time taken for the incident impulse traveling along the negative $\hat{z}$-direction to excite those integration points and the time taken for the returns to reach the receiver after excitation, must equal t. This is physically plausible, and helps in finding out those integration points which contribute to the receiver at a given time, under the physical optics assumption and in response to an incident impulse. Fig. 2 shows two points P and Q on S. Although P is first excited by the incident impulse along the negative $\hat{z}$-direction, this can be exactly compensated by the positional advantage of Q along the receiver direction so that contributions from P and Q arrive at the far-field receiver simultaneously. This can only happen if P and Q lie on a plane whose normal bisects the positive z axis and $\hat{a}_r$.

To obtain a more explicit constraint on such an integration point (x,y, z), (6) can be written with some trigonometry and in terms of a dot product as

$$(x,y,z) \cdot (\sin\frac{\theta}{2},\ 0,\ -\cos\frac{\theta}{2}) = \frac{(t-r)}{2\cos\frac{\theta}{2}} \tag{7}$$

where θ is the bistatic angle. According to (7), all integration points contributing at time t, must lie on a plane whose displacement p from the origin along the bisector of the transmitting and receiving directions is

$$p = \frac{(t-r)}{2\cos\frac{\theta}{2}} \tag{8}$$

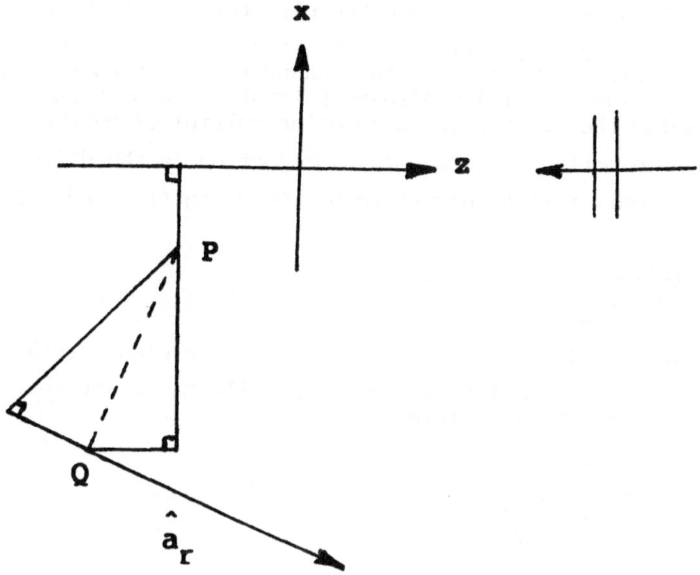


Fig. 2: Illustration of the bisector direction

implying that the plane moves towards the scatterer along the bisector direction with a velocity of

$$\frac{dp}{dt} = \frac{1}{2\cos\frac{\theta}{2}} \tag{9}$$

By expanding the vector triple product in (5) and using the scaling property of the delta function, (5) can be written as

$$r\bar{H}^S(\bar{r},t) = \hat{y}\,\frac{1}{2\pi}\,\frac{\partial}{\partial t}\iint_S \frac{\delta\left(p - \frac{t-r}{2\cos\frac{\theta}{2}}\right)}{\left|2\cos\frac{\theta}{2}\right|}\,\hat{n}\cdot\hat{a}_r\,dS' \tag{10}$$

in which all the projective surface elements $\hat{n}\cdot\hat{a}_r\,dS'$, at a fixed p, can be grouped together. For the portion of S, bounded by a differential dp, the integral $\iint_{\Delta S} \hat{n}\cdot\hat{a}_r\,dS'$ over the bounded area of surface slice $\Delta S$, between the two bounding contours at distances p and p + dp from the origin, can be broken down into two terms,

$$\iint_{\Delta S} \hat{n}\cdot\hat{a}_r\,dS' = \cos\frac{\theta}{2}\iint_{\Delta S} \hat{n}\cdot\hat{b}\,dS' - \sin\frac{\theta}{2}\iint_{\Delta S} \hat{n}\cdot\hat{b}_\perp\,dS' \tag{11}$$

where $\hat{b}$ is a unit outward vector along the bisector direction, and $\hat{b}_\perp$ is orthogonal to $\hat{b}$ and $\hat{y}$ (Fig. 2). If the bounded portion is neither entirely nor partially submerged in the shadow region, then the second term of the right side of (11) vanishes at least for smooth and convex scatterers, leaving the first term as a cosine multiplied by the projection of the bounded portion onto $\hat{b}$. This projection is the differential of the cross-sectional area $A_p$ normal to $\hat{b}$. On using (11) and defining

$$A(t) = A_p \left(\frac{t - r}{2 \cos \frac{\theta}{2}}\right) \tag{12}$$

where $A_p(p)$ is the cross-sectional area formed by the plane (transverse to the bisector direction) at displacement p with the scatterer, (10) can be re-cast into the desired form

$$r\bar{H}^s(\bar{r},t) = \hat{y} \frac{\cos \frac{\theta}{2}}{2\pi} \frac{\partial 2A(t)}{\partial t^2} \tag{13}$$

In the TE case, the incident electric field assumes the form of the delta function in the $\hat{y}$ direction, and a result similar to (13) can be obtained after some more algebraic manipulation, where

$$r\bar{E}^s(\bar{r},t) = -\hat{y} \frac{\cos \frac{\theta}{2}}{2\pi} \frac{\partial 2A(t)}{\partial t^2} \tag{14}$$

Both (13) and (14) lead to a geometrical interpretation for the physical optics bistatic impulse response: for both the TM and TE cases, the impulse response takes the form of the second derivative of the cross-sectional area formed by the smooth, convex perfectly conducting scatterer and a fictitious plane delineating the scatterer along the bisector direction at a speed given by (9), for times so early that the fictitious plane has not touched the shadow region. By setting $\theta$ to zero, (13) and (14) reduce to the Kennaugh-Cosgriff formula for the monostatic case. We note here that Kennaugh obtained the same result for (13) and (14) in [2] for his bistatic extension of the physical optics ramp response formula using a slightly different approach. However, he did not pursue the depolarization effects for the bistatic case in any further detail.

Since the physical optics approximation ignores the interaction between induced currents (i.e. only source effects are accounted for), (13) and (14) are expected to be more accurate for earlier times. For times so late that the fictitious plane has already touched the shadow region, the impulse response may not be interpreted as a cross-sectional area function because the integrals of (11) are over an incomplete bounded portion and the second term of the right side of (11) may not vanish. The effect of increasing the bistatic angle can be viewed as advancing the critical time at which the fictitious plane just touches the shadow region and beyond which (13) and (14) do not hold. Thus, the larger the bistatic angle, the smaller the time range for which (13) and (14) are valid. As the bistatic angle reaches such a value that the fictitious plane moving along the bisector direction can no longer delineate the smooth, convex scatterer strictly in the illuminated region, (13) and (14) will become invalid. The limit of the bistatic angle depends on the location of the shadow boundary, which in turn depends on the geometry and orientation of the scatterer, and no definite limit can be specified mathematically.

The physical optics approximation is known to suffer depolarization deficiency for the monostatic case; for the bistatic case, (13) and (14) implies that the directions of the incident H-field and the E-field are not altered on scattering in the TM case and the TE case, respectively. This is certainly true for the sphere, but not in general; i.e., the physical optics assumption needs to be corrected to give the proper depolarization contributions [9].

## 4. LEADING EDGE POLARIZATION CORRECTION TO PHYSICAL OPTICS AND A BISTATIC PHASE-CURVATURE RELATIONSHIP

An approximate polarization correction to the physical optics approximation was first derived by Bennett [10]. By considering the leading edge of the monostatic far field impulse response of the scatterer, only the very front portion of the illuminated region, encompassing the specular point, needs to be taken into account. Under this space-time integral equation approach [8], Bennett integrated the contributions due to the specular portion, which can be described in terms of spatial derivatives of position vectors and hence associated with geometrical quantities such as curvatures; and he concluded that for the leading edge of the impulse response, the polarization correction to physical optics approximately takes the form of the first time derivative of the silhouette area function, and is proportional to the specular principal curvature difference. This result [10] is extended to the bistatic case in [9], by considering the very front portion along the bisector direction instead. For the bistatic case, a derivation parallel to Bennett's leads to the correction to the impulse response $\bar{H}^s_{corr}$,

$$r\bar{H}^s_{corr} = -\frac{K_{ub} - K_{vb}}{4\pi}\frac{\partial A}{\partial t}\,[\hat{a}_r\,(\cos\psi\cos 2\psi_b\cos\frac{\theta}{2} + \sin\psi\sin 2\psi_b)$$

$$+\,\hat{y}\,(\cos\psi\sin 2\psi_b\cos\frac{\theta}{2} - \sin\psi\cos 2\psi_b)\cos\frac{\theta}{2}\,] \qquad (15)$$

in which the area function A(t) along the bisector direction defined by (12) is used in lieu of the monostatic silhouette area function, and $K_{ub}$ and $K_{vb}$ are principal curvatures at the specular point touched by the fictitious plane moving along the bisector direction. In addition, this correction also depends on the bistatic angle $\theta$, the incident polarization angle $\psi$, and the orientation angle $\psi_b$ of the principal curves at the specular point with respect to the transmitter-scatterer-receiver plane, as defined in Fig. 3. The vectorial depolarization effects manifested by (15) are described in terms of two unit basis vectors $\hat{a}_r$ and $\hat{y}$, which are both perpendicular to the receiving direction $\hat{a}_r$ and so are $(\hat{b}, \hat{y})$ and $(\hat{a}_t, \hat{y})$ to the bisector (outward) and transmitting directions, respectively, with $\hat{y}$ perpendicular to the transmitter-scatterer-receiver plane.

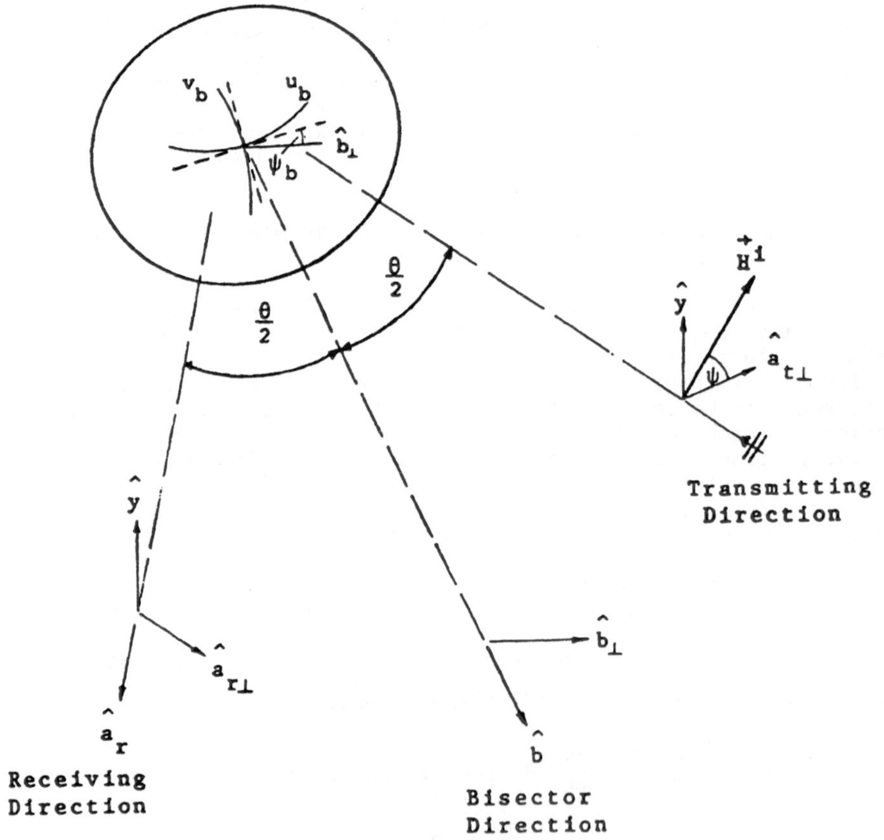


Fig. 3: Coordinate system for polarization correction

By adding the fields due to physical optics and the derived polarization correction, the total impulse response can be investigated in four special π cases, in which incident magnetic field is either along $\hat{a}_t$ ($\psi$=0) π or $\hat{y}$($\psi$=-0), and the receiver polarization is either along $\hat{a}_r$ or $\hat{y}$. The resulting 2 responses are designated by $S_{HH}$, $S_{VV}$, $S_{VH}$ and $S_{HV}$, with the r dependence removed. The first and second subscripts respectively denote receiving and transmitting polarizations, with 'H' associated with $\hat{a}_t$ or $\hat{a}_r$ , and 'V' with $\hat{y}$.

Then,

$$S_{HH} = \frac{\cos\frac{\theta}{2}}{2\pi}\,\frac{\partial^2 A}{\partial t^2} - \frac{K_{ub} - K_{vb}}{4\pi}\,\frac{\partial A}{\partial t}\,\cos 2\psi_b \cos\frac{\theta}{2} \tag{16}$$

$$S_{VV} = \frac{\cos\frac{\theta}{2}}{2\pi}\,\frac{\partial^2 A}{\partial t^2} + \frac{K_{ub} - K_{vb}}{4\pi}\,\frac{\partial A}{\partial t}\,\cos 2\psi_b \cos\frac{\theta}{2} \tag{17}$$

$$S_{VH} = -\frac{K_{ub} - K_{vb}}{4\pi}\,\frac{\partial A}{\partial t}\,\sin 2\psi_b \cos^2\frac{\theta}{2} \tag{18}$$

$$S_{HV} = -\frac{K_{ub} - K_{vb}}{4\pi}\,\frac{\partial A}{\partial t}\,\sin 2\psi_b \tag{19}$$

Equations (16) to (19) imply that in general $S_{HV}$ is not equal to $S_{VH}$ except for the following cases: (i) monostatic case, (ii) locally spherical scatterer, (iii) one of the principal directions at the specular point is parallel to the transmitter-scatterer-receiver plane. In the last two cases both $S_{HV}$ and $S_{VH}$ will vanish. In the first case, (16) to (19) reduce to Bennett's results. It is also noted that by interchanging the roles of the transmitter and the receiver, the reciprocity condition can be satisfied. Moreover, the discussion here is restricted to smooth, convex-shaped, perfectly conducting scatterers. Furthemore, as the bistatic angle θ is increased from zero, the virtually flat, specular portion of the scatterer (determined by the bisector direction) to which early time scattering is attributed will approach the boundary of the shadow region. Beyond this limit, (15) to (19) are not valid anymore.

On working in the frequency domain and repeating the algebra of our previous studies in [16], (16)-(19) can lead to a high frequency asymptotic relation between the specular principal curvature difference $K_{ub} - K_{vb}$ and the phase difference $\phi_d$ of the copolarized elements $S_{HH}$ and $S_{VV}$, denoted as relative polarimetric copolarization phase here,

$$\frac{K_{ub} - K_{vb}}{2k} = \frac{1}{\cos 2\psi_b}\,\tan\frac{\phi_d}{2} \tag{20}$$

where k is the wave number. This bistatic relation can be reduced to the monostatic relation in [15]. Since neither theoretical nor measurement data for bistatic scattering of shapes other than the sphere are available, (20) is not numerically tested yet.

## 5. NUMERICAL VERIFICATION

### A. Prediction of the Leading Edge in the Impulse Response

Equations (13) and (14) reveal that once the area function is known, the impulse response can be found by double differentiation, which implies that the initial part of the leading edge is an impulse. To see how well physical optics approximates the leading edge of the bistatic impulse response, the sphere of unit radius is taken as a test case. The physical optics impulse response for the sphere is convolved with a short incident Gaussian pulse, and the response is compared with that obtained by taking the inverse Fourier transform of the product of the Gaussian pulse spectrum and the Mie series. Figs. 4, 5 and 6 depict the responses computed from both physical optics and from the Mie series, for the monostatic case and the 45 degree TM and TE cases. For the sphere, the origin of Fig. 1 is shifted to the sphere center for convenience; moreover, the far field dependence $e^{-jkr}$ is suppressed, so that the initial peak in the response would have been impulsive at exactly t = -2 light-meters for the monostatic case had the incident Gaussian pulse been an impulse (i.e. the response would have become the impulse response). For the bistatic case, (8) predicts a shift $\Delta t$ of the initial impulse in the impulse response towards the positive time axis for the unit sphere,

$$\Delta t = 2\ (1-\cos \frac{\theta}{2}) \qquad (21)$$

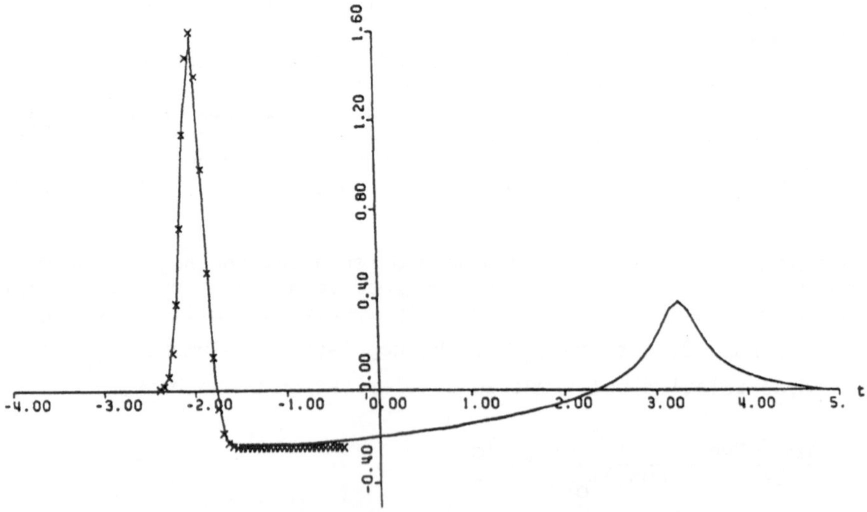


Fig. 4: Responses due to Short Gaussian Pulse for the Monostatic Case
——— from the Mie Series
xxxxx from Physical Optics

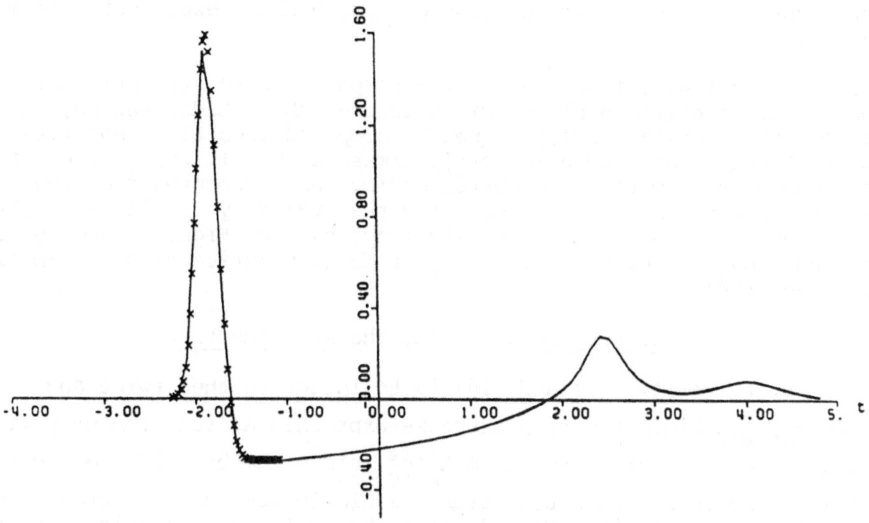


Fig. 5: Responses due to Short Gaussian Pulse for the 45 Degree TM Case
——— from the Mie Series
xxxxx from Physical Optics

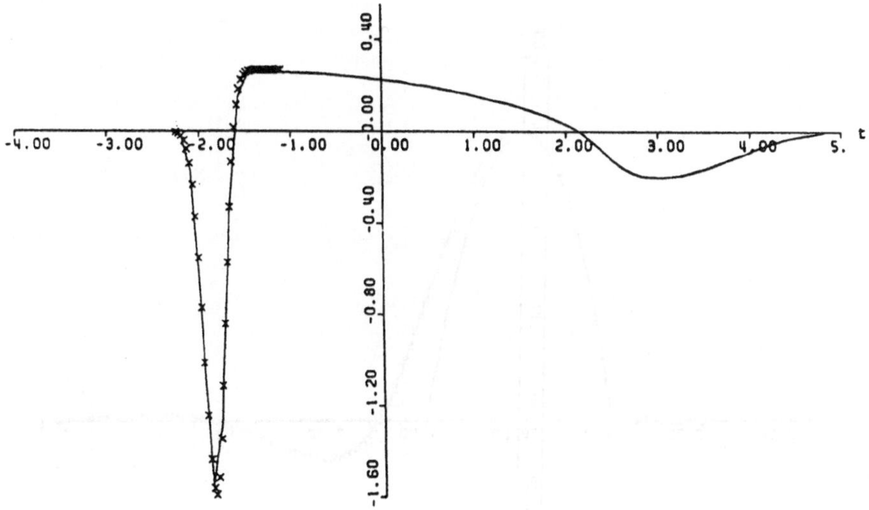


Fig. 6: Responses due to Short Gaussian Pulse for the 45 Degree TE Case
——— from the Mie Series
xxxxx from Physical Optics

Measurement of the shift of the peak of Fig. 5 shows excellent agreement with (21).

Since (13) and (14) define the impulse response up to the critical time at which the fictitious plane first touches the shadow region, those plots of the Gaussian pulse response computed from (13) and (14) in Figs. 4, 5 and 6 are shown for early times only. It can be seen that those responses do match the corresponding ones computed from the Mie series for early times, thus verifying the validity of (13) and (14). The responses computed from physical optics in Figs. 5 and 6 are indistinguishable from each other except for a negative sign, as implied by (13) and (14).

B. Comparison of the Ramp Response and the Area Function

Another way to verify (13) and (14) is to integrate them twice and compare $\frac{2\pi}{\cos \Theta/2}$ times the ramp response expressed as function of p with the cross-sectional area function $A_p(p)$ along the bisector direction. The ramp response is generated from the Mie Series through transformation to the time domain. Fig. 7 shows how well the comparison is for the 30 degree TM case. The vertical line in Fig. 7 indicates the critical time beyond which (13) and (14) are not supposed to hold, despite that the comparison seems good even beyond.

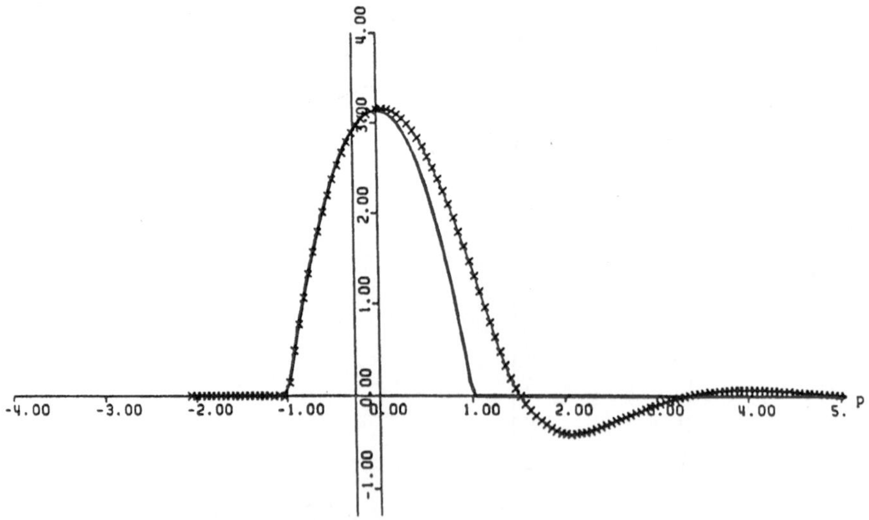


Fig. 7: Ramp Response (TM: 30°)
—— Exact Area
xxxxx Normalized Ramp Response

## 6. APPLICATIONS TO DIRECT SCATTERING

In this section and Section 7, (13) and (14) are applied to direct and inverse scattering. The restrictions of t and θ as discussed in Section 3 are noted here.

### A. Estimation of Bistatic Cross-Section at High Frequencies

The bistatic cross section of the unit sphere can be quickly estimated by extending the time range of validity of (13) and (14) beyond the critical time according to the moment condition [1] and then transforming the extended physical optics impulse response to the frequency domain. This was done by Kennaugh in [1] for the monostatic case. The impulse response according to (13) is an impulse followed by a step function, up to the critical time, which is -sin θ light-meters for the unit sphere. If the step function is allowed to continue beyond the critical time to the zero time, then the area under the impulse response will be zero and the zero moment condition will be satisfied. The extended impulse response in this case is

$$r\bar{H}^S(\bar{r},t) = \hat{y}\,\frac{1}{4\cos\frac{\theta}{2}}\,\{2\cos\tfrac{\theta}{2}\,\delta(t+2\cos\tfrac{\theta}{2}) - [u(t+2\cos\tfrac{\theta}{2})-u(t)]\} \qquad (22)$$

The resulting frequency response is plotted for the 45 degree TM case in Fig. 8 and matches the corresponding exact frequency response at large values of ka, where a is the unit radius. For low frequencies, physical optics is not expected to provide good matching as it is a high frequency technique. For smooth, convex shapes other than the sphere,

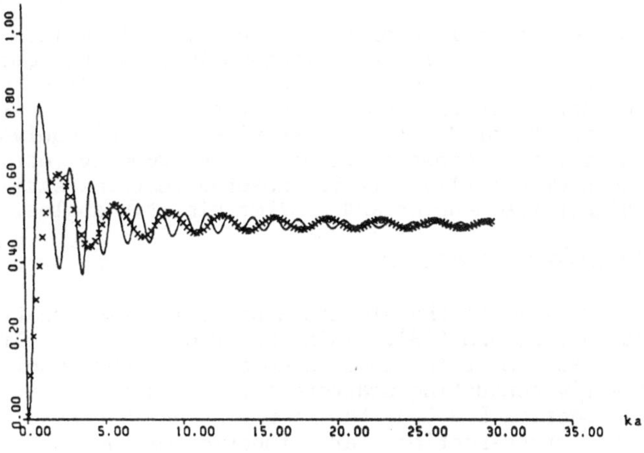


Fig. 8: Amplitude of Frequency Response (45 Degree TM)
—— from the Mie Series
xxxxx from extended Physical Optics

extrapolation and higher order moment conditions may assist in the artificial extension of (13) and (14) into the shadow region.

B. Impulse Response Augmentation Technique [10]

The impulse response augmentation technique is one for an accurate and complete determination of the monostatic impulse response in the entire time domain, and was first suggested by Bennett [10]. The technique requires the time domain solution to scattering due to a short pulse, which can be obtained by marching on in time [10]; it also requires the leading edge solution as input, which is exactly what (13) and (14) can provide for the bistatic case. In this way, it is anticipated that both the bistatic impulse response and the bistatic radar cross section can be accurately computed for all times and for all frequencies, respectively.

C. High Frequency Bistatic-Monostatic Equivalence

Apparently there is a difficulty in taking the Fourier transform of (13) and (14) to obtain the frequency response, since (13) and (14) are incomplete in the sense that the impulse response shape is not defined after the critical time. However, for smooth, convex shaped, perfectly conducting scatterers, the frequency response at high frequencies depends mainly on the early time portion of the impulse response. If the Fourier transform of the early time portion is taken, then the application of the scaling theorem of Fourier transforms to (13) or (14) yields a high frequency estimation of the frequency response $rH_k(\bar{r},k)$,

$$rH_k(\bar{r},k) = \pm \frac{1}{4\pi} (j2k \cos\frac{\theta}{2})^2 \alpha(2k \cos\frac{\theta}{2}) \tag{23}$$

where $\alpha(k)$ is the Fourier transform of $A(t)$. Hence, (23) implies that the bistatic radar cross section, for high frequencies, can be obtained from the monostatic radar cross section measured on the bisector by replacing the wave number k in the monostatic cross section with k cos $\theta/2$. This is in support of Kell's monostatic-bistatic equivalence statement, by which bistatic cross sections can be roughly estimated from monostatic cross sections [11]. It is expected that the estimation will be better for higher frequencies and smaller bistatic angles.

## 7. APPLICATIONS TO INVERSE SCATTERING

A detailed literature review of the electromagnetic inverse scattering problem can be found in [12] and [13]. This paper concerns itself with the utilization of Kennaugh's ramp response concept in the imaging of smooth, convex, perfectly conducting scatterers with bistatic scattering data. It is shown in Section 5 B that the ramp response gives an excellent estimation of the cross-sectional area function of the scatterer, at least up to the critical time. By rotating the scatterer by 180 degrees, two complementary ramp responses can be combined to estimate the complete area function of the scatterer except for the region between the two critical times. Fig. 9 displays the ramp response

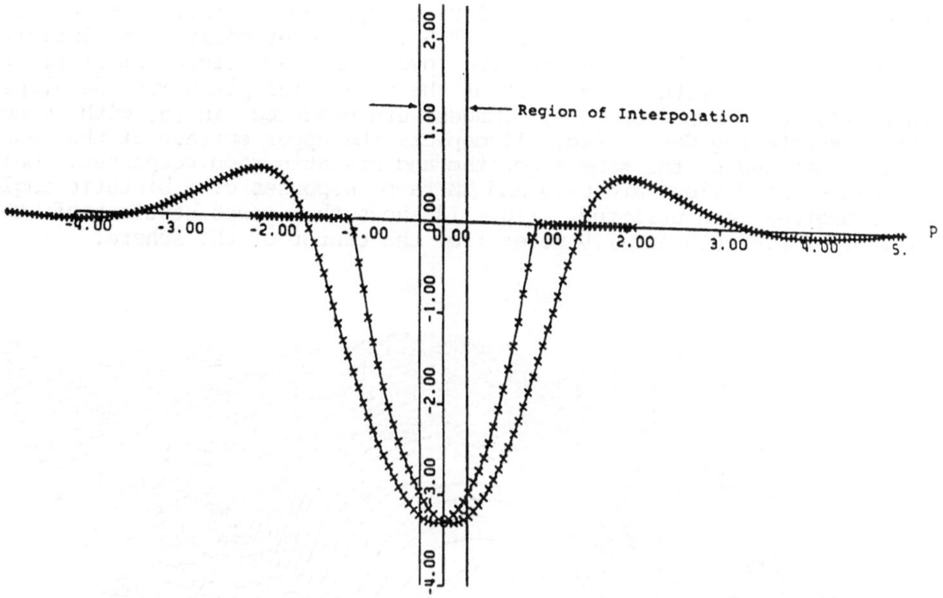


Fig. 9: Two Complementary Ramp Responses (30 Degree TE)

(multiplied by $\frac{2\pi}{\cos\,\theta/2}$) in opposite views for the unit sphere, for the 30 degree TE case, with two vertical lines to indicate the region between the critical times. It is suggested here that interpolation by polynomial fitting be done in this region to obtain a complete estimation of the area function as shown in Fig. 10. The imaging problem can

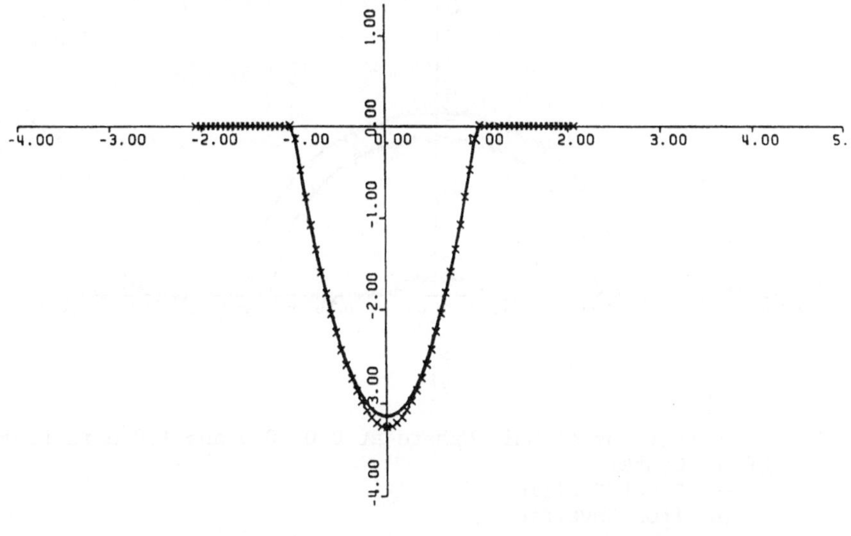


Fig. 10: Comparison of Interpolated Ramp Responses and Area Function (TE 30 Degree)

then be formulated as the classical Radon problem of profile reconstruction from cross-sectional areas [3,5]. Theoretically, an infinite number of views are required for the inversion. An algorithm of finite views for bodies with symmetry about the equatorial plane was developed in [3,14], and based on this, images were produced in [5] with monostatic scattering data. Fig. 11 depicts the upper surface of the unit sphere obtained by the same algorithm and bistatic ramp responses. Only 12 complemented views are used and TM ramp responses of a bistatic angle of 30 degrees are utilized. Fig. 12 shows the cross sections of the Fig. 11 at 0.0, 0.5 and 1.0 meter from the center of the sphere.

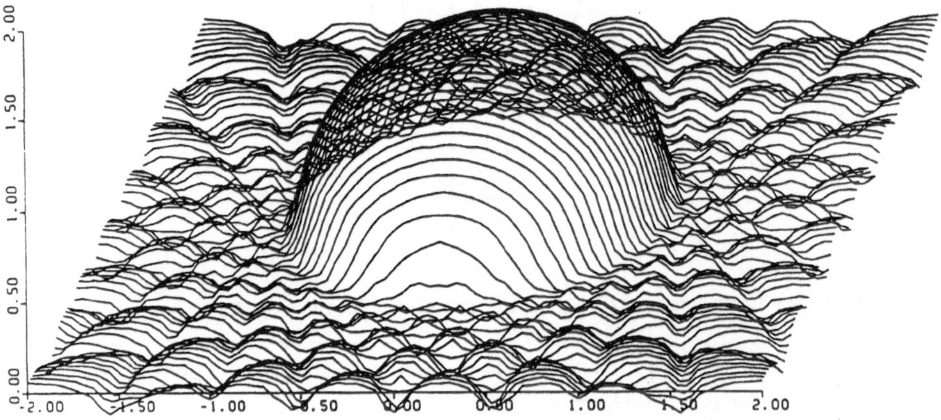


Fig. 11: Upper Surface of the Unit Sphere (TM 30 degree)

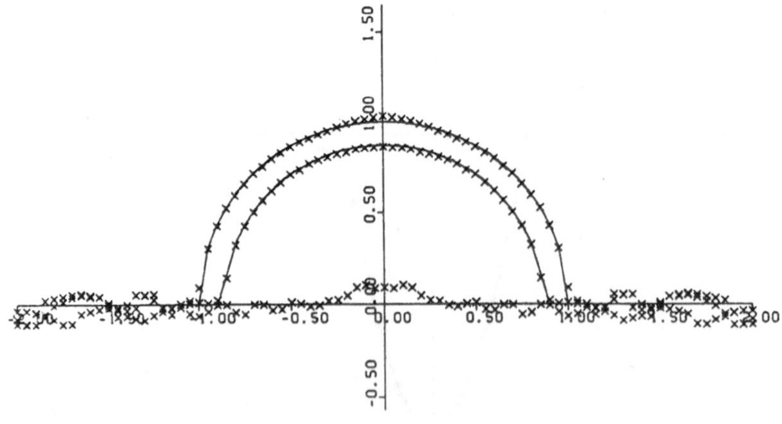


Fig. 12: Cross Sections of Unit Sphere at 0.0, 0.5 and 1.0 unit from Sphere Center
——— Exact Profile
xxxxx from Inversion

The image of Fig. 11 is produced from a bistatic ramp response synthesized from the Mie series. In practice, the ramp response can be synthesized from frequency domain scattering data at several harmonic frequencies [15]. One problem for the above inversion procedure is that the critical times are not known for unknown shapes, and must be given as a priori information, and the larger the bistatic angle is, the wider the range of interpolation.

## 8. CONCLUSIONS AND RECOMMENDATIONS

The Kennaugh-Cosgriff formula is extended to the bistatic case under the physical optics approximation. Applications to both direct and inverse scattering are discussed. A crude leading edge polarization correction to physical optics is derived, resulting in a bistatic phase-curvature relationship. As inherited from Bennett's approach, this polarization correction is valid for small curvature difference, i.e., for surfaces which produce negligible depolarization. For cases such as the depolarization by cylinder, it is suggested that Bennett's approach be generalized to cover these cases of significant depolarization as well. For the practical limited aperture situation of radar imaging with multistatic configuration, it is speculated that (13) and (14) may partially alleviate the limited aperture problem by providing information of area functions along aspects bisecting any pair of transmitter and receiver, in addition to area functions along monostatic aspects. Limited aperture multistatic radar imaging should be investigated in more detail, as has been attempted in [17].

## 9. ACKNOWLEDGMENTS

This work was supported, in parts, by the National Science and Engineering Research Council of Canada under Operating Grants A7240 and A0013; the U.S. Office of Naval Research under Grant N00014-80-C-0773 and N00019-82- C-0306, and Naval Air Systems Command Research Program under Grant N00019-80- C-0620, and the U.S. Army Research office under Grant D-AAG-28-80K-0027.

The authors would like to thank one of the reviewers to draw their attention to an early report of Prof. Edward M. Kennaugh which was not published in the open literature, unavailable to the authors before and during the preparation of this manuscript, and is now included as reference [2].

We wish to thank Drs. Henry W. Mullaney, Richard J. Brandt, Jim Smith, Michael A. Morgan and Arthur K. Jordan of the Office of Naval Research as well as Drs. Guenter Winkler, Robert J. Dinger and Brett Borden, NWC-Physics Division, for their continued interest in our research. In addition, the discussions with Drs. C. Leonard Bennett, Harry L. Mieras and Prof. Shi-Ming Lin are gratefully acknowledged. We wish to thank Mr. Richard W. Foster for skillfully typing the manuscript and preparing the figures.

## 10. REFERENCES

[1] E.M. Kennaugh and R.L. Cosgriff, 'The Use of Impulse Response in Electromagnetic Scattering Problems', in 1958 IRE National Conv. Rec., part I, pp. 72-77.

[2] E.M. Kennaugh, 'Interpretations of the Physical Optics Approximation in the Time Domain', Proceedings of the Second Symposium on Ground Identification of Satellites, the MITRE Corporation, Bedford, MA, 2-4 Oct. 1967, pp. 81-88.

[3] Y. Das and W-M. Boerner, 'On Radar Target Shape Estimation Using Algorithms for Reconstruction from Projections', IEEE Trans. A&P, Vol. 26(2), pp. 274-279, March 1978.

[4] S.K. Chaudhuri and W-M. Boerner, 'Polarization Utilization in Profile Inversion of a Perfectly Conducting Prolate Spheroid', IEEE Trans. Antennas Prop., Vol. AP-25, pp. 505-511, July 1977.

[5] W-M. Boerner, C-M. Ho and B-Y. Foo, 'Use of Radon's Projection Theory in Electromagnetic Inverse Scattering', IEEE Trans. A&P, Vol. 29, Special Issue on Inverse Methods in Electromagnetics, pp. 336-341, March 1981.

[6] C.L. Bennett, 'Time Domain Inverse Scattering', IEEE Trans. A&P, Vol. 29, Special Issue on Inverse Methods in Electromagnetics, pp. 213-219, March 1981.

[7] J. Radon, 'Uber die Bestimmung von Funktionen durch ihre Integralwerte längs gewisser Mannigfaltigkeiten', Ber. Verh. Sächs. Akad., Vol. 69, pp. 262-277, 1917.

[8] C.L. Bennett, 'A Technique for Computing Approximate Electromagnetic Impulse Response of Conducting Bodies', Ph.D. Dissertation, Purdue University, Lafayette, Indiana, Aug. 1968.

[9] B-Y. Foo, 'Application of Kennaugh's Ramp Response to Vector Electromagnetic Inverse Scattering in Monostatic and Bistatic Cases', Ph.D. thesis, University of Illinois at Chicago, Chicago, Illinois, December 1986.

[10] C.L. Bennett et al, 'Space Time Integral Equation Approach to the Large Body Scatteing Problem', final report on contract F30602-71-C-0162, RADC-CR-73-70, AD763794, Sperry Rand Research Center, Sudbury, MA, May 1973.

[11] R.E. Kell, 'On the Derivation of Bistatic RCS from Monostatic Measurements', IEEE Proc., pp. 983-988, Aug. 1965.

[12] W-M. Boerner, 'Polarization Utilization in Electromagnetic Inverse Scattering', in Ch. 7 of "Inverse Scattering Problems in Optics", Vol. 2, Springer Verlag, by H.P. Baltes, 1980. (ibid, Detailed State of the Art Review on 'Polarization Utilization in Electromagnetic Inverse Scattering: Microwave Imaging, Radar Target Mapping, Radon Shape Reconstruction and Remote Sensing', Communications Laboratory, University of Illinois at Chicago, Oct. 1978).

[13] W-M. Boerner et al, 'Inverse Methods in Electromagnetic Imaging', Proceedings of Sept. 83, 18-21 NATO-ARW(Bad Windsheim, FR Germany), NATO-ASI series C: Mathematical & Physical Sciences, Vol. 143, parts 1 & 2, Dordrecht/Boston/Lancaster, D. Reidel Publishing Company, Jan. 1985.

[14] Y. Das, 'Application of Concepts of Image Reconstruction from Projections and Radon Transform Theory to Radar Target Identification', Ph.D. thesis, University of Manitoba, Winnipeg, Manitoba, 1977.

[15] E.M. Kennaugh and D.L. Moffatt, 'The Use of Transient and Impulse Response Approximations in Electromagnetic Scattering Problems', Vol. I, Radar Target Identification, Dept. of Electrical Engineering, Ohio State University, Columbus, Ohio, Sept. 1977.

[16] B-Y. Foo, S.K. Chaudhuri and W-M. Boerner, 'A High Frequency Inverse Scattering Model to Recover the Specular Point Curvatures from Polarimetric Scattering Data', IEEE Trans. Antenna Prop., Vol. 32, #11, pp. 1174-1178, Nov. 1984.

[17] S.K. Chaudhuri and W-M. Boerner, 'A Polarimetric Model for the Recovery of the High-Frequency Scattering Centers from Bistatic-Monostatic Matrix Data, IEEE Trans. Ant. & Propag., Vol. AP-35(1), pp. 87-93, January 1987.

In Commemoration

# INTERPRETATION OF THE PHYSICAL OPTICS APPROXIMATION IN THE TIME DOMAIN FOR THE BISTATIC CASE*

Edward Morton Kennaugh (1922-1983)
ElectroScience Laboratory
(formerly Antenna Laboratory)
The Ohio State University
Columbus, Ohio 43212

**ABSTRACT:** In addition to providing a simple estimate for the radar cross-section of an object, the physical optics approximation also suggests a close relation between the geometric shape and the scattered waveform, when an object is illuminated by a polychromatic radar. The paper develops the relation between target geometry and scattering characteristics (predicted by physical optics for conducting bodies) for bistatic as well as monostatic radars. A simple interpretation of the bistatic case has not been previously available. This relation is developed in terms of the ramp response waveform of the object.

## I. INTRODUCTION AND FORMULATION OF THE PROBLEM

The physical optics or Kirchhoff approximation to the solution of scattering problems simplifies in the time domain. In particular, closed form expressions for the scattered field in any direction can be obtained when a plane incident wave strikes a perfectly conducting object. In this approximation, the surface current density $\underline{K}$, induced on a scattering object, is given by

$$\underline{K} = 2\hat{n} \times \underline{H}^i \quad \text{over illuminated portion of scatterer,}$$

$$\underline{K} \equiv 0 \quad \text{over shadowed portion of scatterer surface S.}$$

The unit outward-directed normal to S is $\hat{n}$ ; and, we shall find it convenient to denote the illuminated portion of S by $S_1$.

To apply this concept to transient scattering phenomena, let us assume a step function incident wave, traveling in the negative z-direction as shown in Fig. 1, with the scatterer at the origin of an XYZ coordinate frame. The most general case of incident wave polarization and direction of scattering can be treated by superimposing solutions for several special cases, as follows:

*This paper was presented in the Proc. Second Symposium on "Ground Identification of Satellites", MITRE Corp., Bedford, MA, 1967 Oct. 2-4, pp. 81-88; and released for publication as Commemorative Contribution for these Proceedings also with the kind permission of Mrs. Mary Kennaugh.



*W.-M. Boerner et al. (eds.), Direct and Inverse Methods in Radar Polarimetry, Part 1*, 537–544.

Case I: $\underline{E}^i$ in $\hat{y}$ direction, observation point in the XZ plane

Case II: $\underline{E}^i$ in $\hat{x}$ direction, observation point in the XZ plane.

In each case, two transverse components of the scattered field at great distances must be determined: the electric field intensity parallel and perpendicular to the planes of scattering. In any instance, the scatterer orientation with respect to the XYZ coordinate frame will be adjusted so as to place the observation point in the XZ plane.

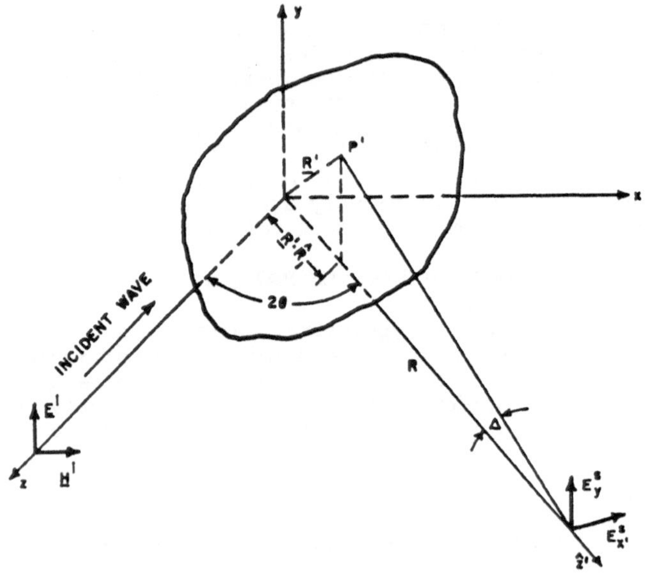


Fig. 1 Bi-static target scattering configuration for Case I

Case I

In Fig. 1, the situation for Case I is illustrated. The electric field at a distant point (R,2θ) is separated into a component $E^s_y$ perpendicular to the XZ plane and a component $E^s_{x'}$ in the XZ plane. These are determined from the vector potential $\underline{A}$ associated with the current distribution over $S_1$ and the coordinates R,2θ of the observation point. At large distances R, the transverse components of $\underline{E}^s$ are related to $\underline{A}$ by

$$E^s_y = - \frac{\partial A}{\partial t} \cdot \hat{y}$$

$$E_x^{s1} = -\frac{\partial A}{\partial t} \cdot \hat{x} \quad .$$

For large R, the vector potential is related to the surface current density $\underline{K}$ by

$$\underline{A}(\underline{R},\ t-R/c) = \frac{\mu_o}{4\pi R}\int_{S_1} \underline{K}\left(\underline{R}',\ t + \frac{\underline{R}' \cdot \hat{R}_1}{c}\right) ds$$

where $\hat{R}_1$ is a unit vector pointing in the direction of the observation point, and $\underline{R}'$ is a vector from the origin to a variable point on S. The incident plane wave is of the form

$$\underline{E}^i = U(t + z/c)\hat{y}$$

$$\underline{H}^i = \sqrt{\frac{\varepsilon_o}{\mu_o}}\ U(t + z/c)\hat{x} \quad .$$

On $S_1$, the surface current density is given by $2n \times \underline{H}^i$, and

$$\underline{K}(\underline{R}',t) \cdot \hat{y} = 2\sqrt{\frac{\varepsilon_o}{\mu_o}}\ U(t + z/c)\hat{n} \times \hat{x} \cdot \hat{y} = 2\sqrt{\frac{\varepsilon_o}{\mu_o}}\ U(t + z/c)\hat{n} \cdot \hat{z}$$

$$\underline{K}(\underline{R}',t) \cdot \hat{x}' = 2\sqrt{\frac{\varepsilon_o}{\mu_o}}\ U(t + z/c)\hat{n} \times \hat{x} \cdot \hat{x}' =$$

$$= 2\sqrt{\frac{\varepsilon_o}{\mu_o}}\ \sin 2\theta\ U(t + z/c)\hat{n} \cdot \hat{y} \quad .$$

Therefore,

$$\underline{A} \cdot \hat{y} = \frac{1}{2\pi Rc}\int_{S_1} U\left(t + \frac{z(1 + \cos 2\theta) + x \sin 2\theta}{c}\right) dS_z$$

$$\underline{A} \cdot \hat{x}' = \frac{\sin 2\theta}{2\pi Rc}\int_{S_1} U\left(t + \frac{z(1 + \cos 2\theta) + x \sin 2\theta}{c}\right) dS_y$$

where $dS_z$ and $dS_y$ denote the signed projections of the elements of $S_1$ onto planes perpendicular to the z and y axes, respectively. These projections are positive when the outward normal $\hat{n}$ makes an acute angle with the positive z or y axis; otherwise, they are taken as negative.

## II. THE BISTATIC TIME DOMAIN SOLUTION

The integrals for each of the two components of $\underline{A}$ reduce to integrals over portions of $S_1$, in each case determined by those points over $S_1$ for which

$$t + \frac{z(1 + \cos 2\theta) + x \sin 2\theta}{c} \geq 0 \quad .$$

As shown in Fig. 2, such portions of $S_1$ are determined by a cutting

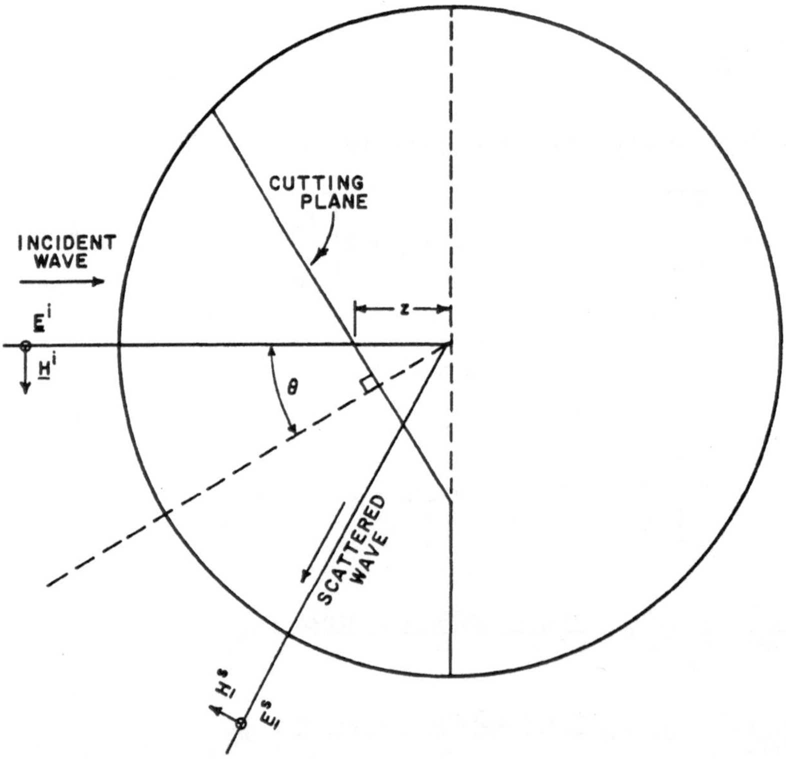


Fig. 2 Bistatic scattering by a sphere

plane whose normal coincides with the bisector of the angle $2\Theta$, and which moves along this bisector so that at time t

$$-ct = z(1 + \cos 2\Theta) + x \sin 2\Theta = 2 \cos\Theta(z\cos\Theta + x\sin\Theta)$$

Thus, the cutting plane moves along the bisector with an apparent velocity $c/2\cos\Theta$. The projections of that portion of $S_1$ defined by the cutting plane, on the XY or XZ plane can thus be considered as functions of time $S_z(t)$ and $S_y(t)$, respectively. These are related to the vector potential by

$$\hat{y} \cdot \underline{A}(t - R/c) = \frac{1}{2\pi Rc} S_z(t)$$

$$\hat{x}' \cdot \underline{A}(t - R/c) = \frac{1}{2\pi Rc} S_y(t)\sin 2\Theta$$

Since the incident plane wave has been assumed of step function form, the time-dependent far field electric field intensity, after suitable normalization, becomes the step response waveform for the scattering object [1]. Thus,

$$_{11}F_U(t - R/c) = \frac{2R}{c} \underline{E}^s \cdot \hat{y} = -\frac{2R}{c}\frac{\partial}{\partial t}\underline{A} \cdot \hat{y}$$

$$_{12}F_U(t - R/c) = \frac{2R}{c} \underline{E}^s \cdot \hat{x}' = -\frac{2R}{c}\frac{\partial}{\partial t}\underline{A} \cdot \hat{x}' \quad .$$

Since the step response waveform can also be obtained by time-differentiation of of the ramp response waveform,

$$_{11}F_R(t - R/c) = -\frac{1}{\pi c^2} S_z(t)$$

$$_{12}F_R(t - R/c) = -\frac{1}{\pi c^2} S_y(t)\sin 2\Theta ,$$

where $_{11}F_R(t-R/c)$ denotes the component of the ramp response waveform parallel to the incident electric field intensity, and $_{12}F_R(t-R/c)$ denotes the component lying in the plane of scattering, orthogonal to the incident electric field intensity. Note that $S_z(t)$ is a monotonic function, since the projections of the elements of $S_1$ on the XY plane are always of postive sign, $S_1$ being the illuminated portion of S. On the other hand, the function $S_y(t)$ is so constrained, and may exhibit maxima or even sign changes, since there will generally be portions of $S_1$ which make obtuse as well as acute angles between the normal $\hat{n}$ and the positve y axis. If the scattering plane is a plane of symmetry for the scatterer, $S_y(t)$ vanishes identically.

Solutions for Case II

Let us now consider Case II, for which the incident E vector lies in the XZ plane. In this case, the incident field is given by:

$$\underline{E}^i = U(t + z/c)\hat{x}$$

$$\underline{H}^i = -\sqrt{\frac{\varepsilon_o}{\mu_o}}\, U(t + z/c)\hat{y}$$

In a manner similar to that for Case I, the important components of the surface current density $\underline{K}$ on $S_1$ are found to be:

$$\underline{K}(\underline{R}',t) \cdot \hat{y} = -2\sqrt{\frac{\varepsilon_o}{\mu_o}}\, U(t + z/c)\, \hat{n} \times \hat{y} \cdot \hat{y} \equiv 0$$

$$\underline{K}(\underline{R}',t) \cdot \hat{x}' = -2\sqrt{\frac{\varepsilon_o}{\mu_o}}\, U(t + z/c)\, \hat{n} \times \hat{y} \cdot \hat{x}'$$

$$= 2\sqrt{\frac{\varepsilon_o}{\mu_o}}\, U(t + z/c)\, \hat{n} \cdot \hat{z}' \quad .$$

The associated ramp response waveforms are given by:

$${}_{21}F_R(t - R/c) \equiv 0$$

$${}_{22}F_R(t - R/c) = -\frac{1}{\pi c^2} S_z'(t) \quad .$$

Here, $S_z(t)$ is the signed projection of the elements of $S_1$ defined by the cutting plane at time t, projected upon a plane perpendicular to the scattering direction. Thus, in computing $S_z'(t)$, a projection on a plane perpendicular to the line of sight to the receiver (observer) is used, while in computing $S_z(t)$ (in Case I) a projection on a plane perpendicular to the line of sight to the source is required. If a portion of the surface $S_1$ is "visible" to both source and receiver, it makes identical contributions to $S_z(t)$ and $S_z'(t)$. If, however, the cutting plane interacts the shadow boundary of the surface S or parts of its intersection with S are not visible form the receiving point, the values of $S_z'(t)$ and $S_z(t)$ will be different.

## III. INTERPRETATION OF RESULTS

The scattering matrix formed by the ramp response waveforms can be written in terms of the area functions:

$$[S] = -\frac{1}{\pi c^2}\begin{bmatrix} S_z(t) & S_y(t)\sin 2\theta \\ 0 & S_z'(t) \end{bmatrix}$$

Note that this is not a symmetric matrix, except when $S_y(t)$ vanishes, or for the case of backscattering (in which case it also becomes a scalar matrix). The reciprocity theorem requires that an interchange of transmitter and receiver positions in Fig. 1 should lead to a scattering matrix which is the transpose of that given above. However, this is not obtained in all cases for the physical optics approximation, since $S_z(t)$ and $S_z'(t)$ do not remain unchanged. The off-diagonal elements, if non-zero, do not transpose as required, also.

While reciprocity is violated in general, by this approximation, there are important classes of targets for which the response waveforms satisfy reciprocity initially, or for short times after these begin. Consider, for example, a convex object such that the cutting plane, moving along the bisector of the two lines of sight, does not intersect either shadow boundary of $S_1$ for some time increment $\Delta$.* It then follows that for this time interval, $S_y(t) = 0$, while $S_z(t)$ and $S_z'(t)$ are identical, and do not vary with source-receiver interchange. For such a time interval, the ramp response of the target is proportional to the instantaneous cross-sectional area excited from the scatterer by the cutting plane, multiplied by $\cos\theta$. In such cases, the bistatic ramp response waveform is similar to the monostatic ramp response waveform when viewed along the bisector. The monostatic ramp response waveform is compressed along time and amplitude scales by the factor $\cos\theta$ to obtain the bistatic waveform.

## IV. CONCLUSIONS

A number of interesting observations may be drawn from these transient response waveforms and their dependence upon suitable cross-sections of the scattering object. One must be aware, however, that this approximation is most accurate for high frequencies, or short times, but may be greatly in error elsewhere. The derivation of more accurate ramp response waveforms by experimental or theoretical techniques is urgently needed to validate conclusions tentatively derived on the basis of physical optics, between the geometry of a scattering object and its transient response waveforms.

---

*"Either" shadow boundary of $S_1$ implies that $S_1$ for both location of source and receiver is considered.

## V. ACKNOWLEDGEMENT

*The work reported in this paper was supported in part by Contract F19628-67-C0308, between Electronic Systems Division, Air Force Systems Command, L.G. Hanscom Field, Bedford, Massachusetts and the Ohio State University, Research Foundation.

## BIBLIOGRAPHY

1. Kennaugh, E.M. and D.L. Moffatt, "Transient and Impulse Response Approximations," Proc. IEEE, Vol. 53, No. 8, pp. 893-901, August 1965.

# INVERSE GTD VIA POLARIMETRIC LINEAR PREDICTION

H. Chaloupka
University of Wuppertal, Department of Electrical Engineering
PO Box 100 127
D 5600 Wuppertal 1
FR Germany

ABSTRACT: For targets which allow the radar backscattering to be modeled by the superposition of GTD-contributions, the solution of the inverse problem consists in a decomposition of the scattering data into the individual contributions (inverse GTD). Based on a known frequency and aspect angle dependency of the data, estimates for the positions of the "scattering centers" as well as their partial scattering matrices have to be deduced. With respect to instationary ("moving")scattering centers the aspect angle width of the synthetic aperture has to be restricted to a relatively "small" value, resulting in insufficient resolving power if Fourier transform techniques is used. To overcome this problem a high-resolution method based on the linear-prediction (LP) approach is proposed. This method is based on a new multivariate version of the Kumaresan-Prony LP algorithm which allows to take advantage of the full polarization information (polarimetric LP). Simulations indicate that utilization of polarization information increases the resolving power considerably in comparison to the scalar version.

## 1. INTRODUCTION

The scattering by targets which are large in terms of the wavelength is essentially a localized phenomenon and may be well modeled by the interference of isolated contributions. For a wide class of targets the geometrical theory of diffraction (GTD) [1] turns out to be an adequate model. In accordance to GTD the (total) far field scattering matrix is approximated by the sum of partial scattering matrices belonging to the isolated contributions. Whereas, due to interference, the total scattering matrix exhibits a rapid scintillation with aspect angle and frequency the partial scattering matrices are only slowly varying functions of these variables. Solving the inverse GTD problem means to decompose the total scattering matrix for each aspect angle and frequency into a sum of partial scattering matrices.
If a "small" number of partial scattering matrices is sufficient to describe the back-scattering in a given aspect angle range the method may be used as a mean for "data reduction" which can form the basis for applications like target identification and estimation of target orientation. Other applications may take advantage of the fact, that some data contained in the partial scattering matrices are simply related to some geometric properties of the target.



*W.-M. Boerner et al. (eds.), Direct and Inverse Methods in Radar Polarimetry, Part 1*, 545–554.

## 2. PROBLEM STATEMENT

### 2.1 Representation of the scattering and the 2-D inverse problem

Since a generalization of the proposed concept to other configurations (e.g. bistatic configuration , 3-D imaging system) is straightforward the present paper is restricted to the 2-D monostatic imaging configuration shown in Fig.1. A complete representation of the backscattering is given by the monostatic scattering matrix

$$(\tilde{\rho}) = \begin{bmatrix} \rho_{11} & \rho_{12} \\ \rho_{12} & \rho_{22} \end{bmatrix} \quad , \tag{1}$$

which relates the scattered to the incident field components (directions $\vec{e}_1$ and $\vec{e}_2$). Considering the positive real eigenvalues $\lambda_{1,2}$ of the matrix $(\rho)(\rho)^*$, where the asterisk denotes complex conjugation, and taking the scattering amplitude $|\rho|$ to be the positive square root of the eigenvalue with the largest magnitude

$$|\rho| = \sqrt{\max(\lambda_{1,2})} \ , \tag{2}$$

one can write in vector notation

$$(\rho) = \begin{bmatrix} \rho_{11} \\ \rho_{22} \\ \rho_{12} \end{bmatrix} = |\rho| \exp(-j\,\Phi) \begin{bmatrix} p_{11} \\ p_{22} \\ p_{12} \end{bmatrix} = \rho \ (p) \ . \tag{3}$$

The vector (p) is characteristic for the polarization properties ("polarization vector") of the target.
$(\rho)$ is a function of frequency $\omega$ and aspect direction $\vec{u}_i(\Theta)$. Both "observation parameters" are combined to the "observation vector"

$$\vec{K} = 2\,\frac{\omega}{c}\,\vec{u}_i = 2\,\frac{\omega}{c}\,(\sin\Theta\,\vec{u}_x + \cos\Theta\,\vec{u}_z) = (K_x\,\vec{u}_x + K_z\,\vec{u}_z) \tag{4}$$

and assumed to vary within the "aperture" $|\Theta| \leq \Delta\Theta/2$ and $\omega_o - \Delta\omega \leq \omega \leq \omega_o + \Delta\omega/2$ or equivalently $-\Delta K_x \leq K_x \leq \Delta K_x$ and $K_o - \Delta K_z \leq K_z \leq K_o + \Delta K_z$.

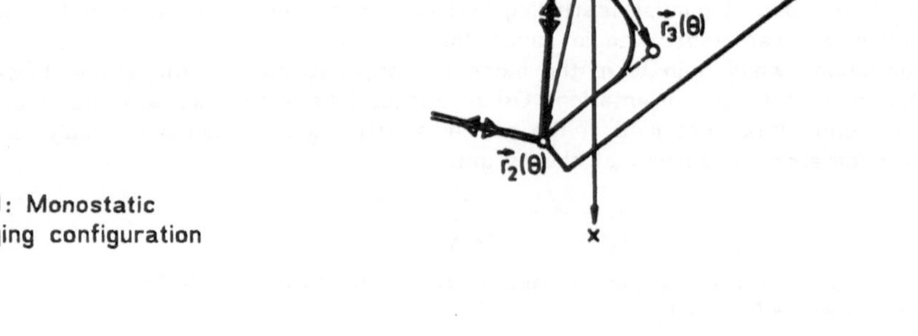

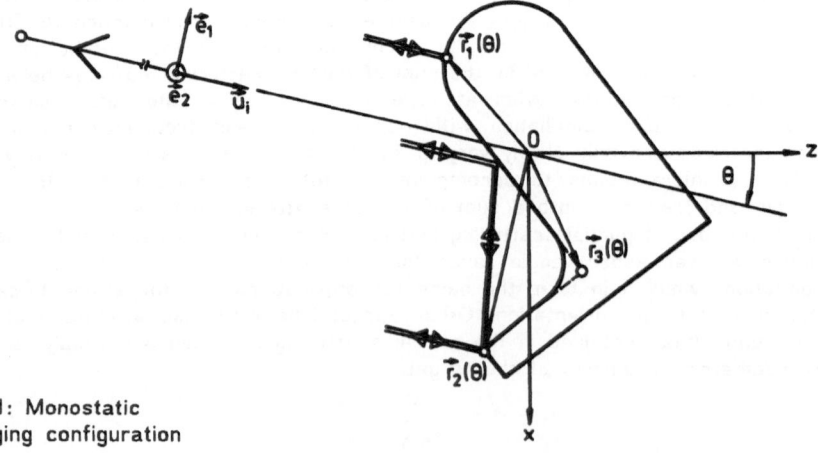

Fig. 1: Monostatic imaging configuration

If high-frequency scattering is represented asymptotically by GTD, it is modeled as being due to the interference of isolated contributions (single and multiple specular reflections, edge diffractions, creeping waves etc.). To each of these individual contributions a partial scattering amplitude $|\rho|_\mu$ and partial polarization vector $(p)_\mu$ can be assigned to, leading to

$$(\rho) = |\rho(\vec{K})| \exp(-j\Phi(\vec{K}))(p(\vec{K})) = \sum_{\mu=1}^{M} |\rho(\vec{K})|_\mu \exp(-j\Phi_\mu(\vec{K}))\ (p(\vec{K}))_\mu \quad . \tag{5}$$

In contrast to the functions $|\rho(\vec{K})|$, $\Phi(\vec{K})$ and $(p(\vec{K}))$ which represent the total scattering, $|\rho(\vec{K})|_\mu$, $\Phi_\mu(\vec{K})$ and $(p(\vec{K}))_\mu$ are only slowly varying functions of $\vec{K}$. If the synthetic aperture is "sufficiently small" $|\rho(\vec{K})|_\mu$ and $(p(\vec{K}))_\mu$ can be taken to be constant and $\Phi_\mu(\vec{K})$ can be replaced by the Taylor expansion

$$\Phi_\mu(\vec{K}) \approx \Phi_\mu(\vec{K}_o) + \vec{r}_\mu \cdot (\vec{K} - \vec{K}_o) \quad \text{with } \vec{r}_\mu = \text{grad } \Phi_\mu(\vec{K}_o) = x_\mu \vec{u}_x + z_\mu \vec{u}_z \quad . \tag{6}$$

$\vec{r}_\mu$ represents the position of a "real" scattering point if it belongs to a single scattering contribution like specular reflection (e. g. point $\vec{r}_1$ in Fig. 1) and a "fictitious" scattering point if it belongs to multiple reflections (e. g. point $\vec{r}_4$ in Fig. 1), creeping waves etc.. We shall use the term "scattering center for the position $\vec{r}_\mu$" and assign a partial scattering amplitude and polarization vector to each scattering center:

$$(\rho(K_x, K_z)) \approx \sum_{\mu=1}^{M} |\rho|_\mu \exp(-j(K_x x_\mu + K_z z_\mu))(p)_\mu \quad . \tag{7}$$

The problem to be solved is the determination of the $\vec{r}_\mu$'s, $(p)_\mu$'s and $|\rho|_\mu$'s from the knowledge of $(\rho(K_x, K_z)) = (\rho(\omega,\Theta))$ in a suitably chosen aperture. With respect to the angular width of the aperture $\Delta\Theta$ one has to take care of the following contradiction:

(a) For scattering centers (s.c.) which have a fixed spatial position within a wide region of the aspect angle (e.g. scattering by an edge) the resolution power can be increased by increasing $\Delta\Theta$.

(b) Other scattering centers like specular reflection points "move" with the aspect direction. Hence, $\Delta\Theta$ has to be chosen small enough to ensure the validity of the approximation according to eq. (6).

Because of this incompatibility the use of high-resolution technique instead of Fourier transform technique offers significant advantages.

## 2.2 Transformation to 1-D multivariate spectral estimation problems

The problem defined in section 2.1 represents a 2-D spectral estimation problem which is multivariate due to the vector properties of $(\rho)$ (see eq.(7)). It may be separated into two 1-D estimation steps which have to be performed successively [2]. We assume sample values of $(\rho(K_x, K_z))$ to be given on a rectangular $K_x$-$K_z$-grid with $\delta K$ as increment. If the measurements are performed in equidistant $\omega$- and $\Theta$-steps interpolation to the $K_x$- $K_z$-grid is necessary except for the case of "very small" $\Delta\Theta$ where $K_x \approx 2\omega\Theta/c$ and $K_z \approx 2\omega/c$ holds (see eq.(4), not "focused" aperture [3]). Typically as a consequence of restrictions due to scattering center movements as well as practical measurement range limits $\Delta K_x/\Delta K_z = \Delta\Theta\ \omega_o/\Delta\omega$ is often much smaller than one. In this case it is sufficient to employ the high-resolution method in the $K_x$-direction only. This can be realized by first performing a 1-D Fourier transform with respect to $K_z$ resulting in

$$(\rho(K_x = n\delta K,\ z)) = \sum_{\mu=1}^{M} |\rho|_\mu\ \chi(z-z_\mu)\exp(-j(n\,\delta K\cdot x_\mu)\,(p)_\mu \quad , \tag{8}$$

where $\chi(z)$ is the 1-D point-spread function.
Alternatively high-resolution techniques can be used in both the $K_x$- and $K_z$-direction. One of a number of various methods [2] is
($\alpha$) first to determine the parameter of the $K_z$-dependence by the linear-prediction method described below, and
($\beta$) to use these parameters to "extend" the bandlimited data into the domain $|K_z-K_O|>\Delta K_z/2$.
($\gamma$)The extended data may serve as input to a 1-D Fourier transform in the $K_z$-direction leading again to a result as given in eq. (8).

For this reason the remainder of this paper will be devoted to the 1-D estimation problem with special emphasis to the utilization of polarimetric information. For this purpose the notation of eq. (8) is simplified by writing $(\rho(n\delta K, z))=(\rho(n))$ and $|\rho|_\mu\chi(z-z_\mu) = \rho_\mu$. Taking the presence of additive multivariate noise $(w(n))$ into account the input data to the estimator are given by the samples $(\rho(n))$ with

$$(\rho(n)) = \sum_{\mu=1}^{M} \rho_\mu\ \exp(-jn\cdot\delta K\cdot x_\mu)\ (p)_\mu\ +\ (w(n)) \tag{9}$$

Here $(\rho(n))$ are known vectors with 3 complex-valued components related to the two copolar and one crosspolar element of the scattering matrix. The number M of the different scattering contributions ("centers") cannot assumed to be known "a priori" and therefore has to be estimated. The problem consists in (a) the estimation of the cross-range positions $x_\mu$ of the scattering centers and (b) the determination of the coresponding scattering amplitudes and $\rho_\mu$ and polarization vectors $(p)_\mu$. Here the main attention has to be directed to step (a), because once the $x_\mu$'s are accurately known, $\rho_\mu(p)_\mu$ can be determined by a least square fit to the data. Due to space-limitations step (b) is not explicitely explained in this paper.

## 3. POSITION ESTIMATION BY MEANS OF A POLARIMETRIC LP METHOD

### 3.1 Introduction of a multivariate LP processing scheme

Fig. 2 illustrates a special processing scheme for the polarimetric scattering data obtained with the synthetic aperture (Fig. 1). L consecutive samples $(\rho(\nu))$ are linearly transformed by means of 3x3 weighting matrices $(G)_\nu$ onto 2 new vectors $(r^f)$ and $(r^b)$ according to (see upper part of Fig. 2)

$$(r^f) = (r^f_{11}, r^f_{22}, r^f_{12})^t = -\sum_{\nu=1}^{L} (G)_\nu\ (\rho(n-\nu)) \tag{10 a}$$

and (see lower part of Fig. 2)

$$(r^b) = (r^b_{11}, r^b_{22}, r^b_{12})^t = -\sum_{\nu=1}^{L} (G)^*_\nu\ (\rho(n-L+\nu)) \tag{10 b}$$

$(()^t$= transposed, $()^*$= conjugated, $()^+$= transjugated).

Furthermore $(r^f)$ and $(r^b)$ are substracted from $(\rho(n))$ and $(\rho(n-L))$ to give

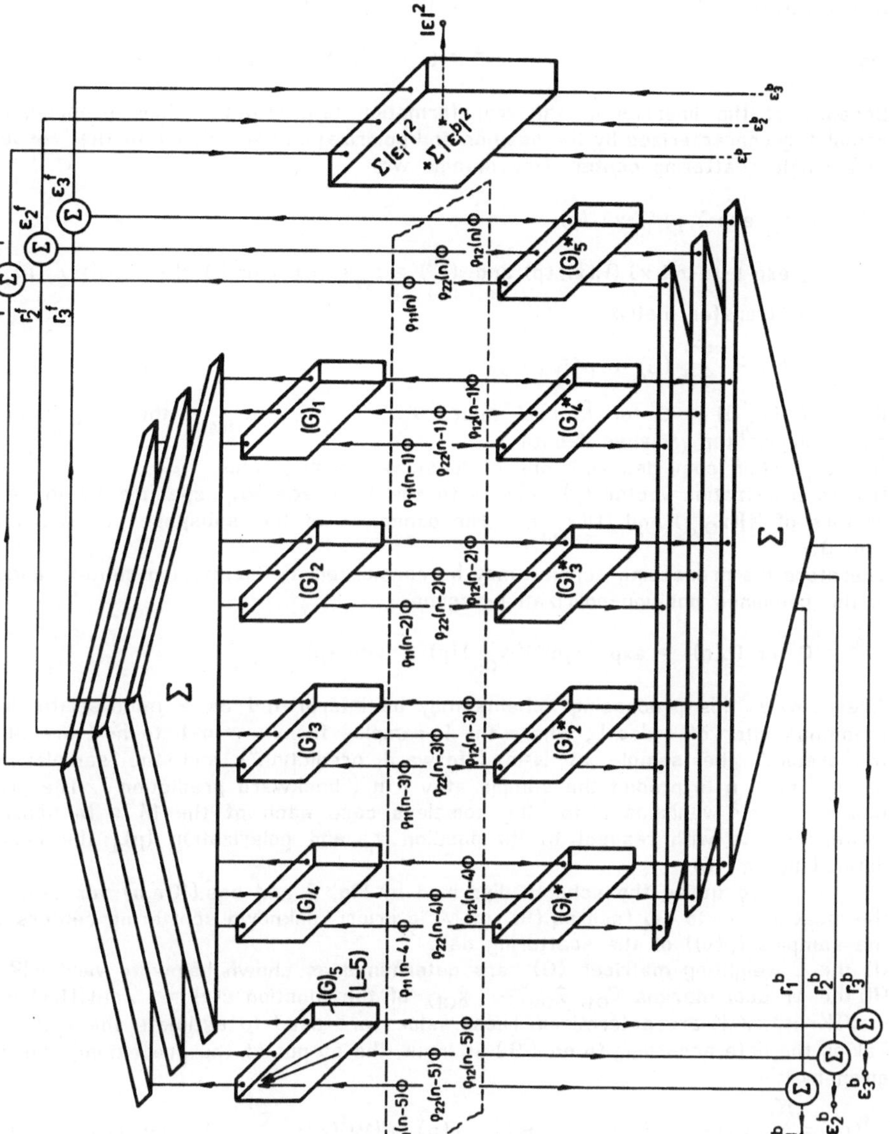


Fig.2 : Multivariate linear-prediction filter ("scattering center canceller")

$$(\varepsilon^f) = (\rho(n)) - (r^f) \qquad \text{and} \qquad (\varepsilon^b) = (\rho(n-L)) - (r^b) \tag{11}$$

and

$$|\varepsilon|^2 = (\varepsilon^f)^+(\varepsilon^f) + (\varepsilon^b)^+(\varepsilon^b). \tag{12}$$

Because of the linearity of the transformation (eq. (10a,b)) their properties are completely characterized by the position and polarization dependence of the "response" to a single scattering center. One obtains with

$$(\rho(\nu)) = \rho_o \exp[-j\nu\cdot\delta K\cdot x] (p) \tag{13}$$

$$(\varepsilon^f) = \rho_o \exp[-jn\cdot\delta K\cdot x] (H(x))(p) \text{ and } (\varepsilon^b) = \rho_o \exp[-j(n-L)\cdot\delta K\cdot x] (H(x))^*(p) \tag{14}$$

with the "transfer matrix"

$$(H(x)) = (1) + \sum_{\nu=1}^{L} (G)_\nu \exp[j\nu\cdot\delta K\cdot x] . \tag{15}$$

If $\det(H(x_{O\mu})) = 0$ for $\tilde{M}$ real values $x_{01}, x_{02}, \dots, x_{O\tilde{M}}$, the response to a scattering center vanishes if both

(I) its position coincides with one of these values $x_{O\mu}$, and

(II) its polarization vector (p) belongs to the subspace $V_{O\mu}$ spanned by the eigenvectors of $(H(x_{O\mu}))$ and $(H(x_{O\mu}))^*$. The dimension of this subspace can reach from 1 to 3.

Therefore the processing scheme may be considered as "scattering center canceller" in the combined position-polirization-regime:

$$|\varepsilon|^2 = 0 \text{ for } (\rho(n)) = \exp(-jn\cdot\delta K\cdot x_{O\mu})(p) \quad \text{with } (p) \in V_{O\mu} . \tag{16}$$

Alternatively, this processing scheme may be interpreted as a multivariate linear-prediction filter of order L, where the L samples from $\nu$ = n-L to n−1 are utilized to "predict" the sample at N=n ("forward prediction") and the samples from $\nu$ = n-L+1 to n to predict the sample at $\nu$ = n ("backward prediction"). The prediction error $|\varepsilon|^2$ vanishes if for the noiseless case each of the $\tilde{M} \leq M$ scattering centers meets with respect to its position $x_\mu$ and polarization $(p)_\mu$ the condition defined by eq.(16).

To utilize the scheme discribed by Fig. 2 and eqs.(10a,b) for estimating the locacions $x_1$ to $x_M$ (see eq.(9)) of M "a priori" unknown scattering centers from the samples $(\rho(\nu))$ of the scattering data

(I) the L weighting matrices $(G)_\nu$ are determined as shown below to yield $|\varepsilon|^2 = 0$.

(II) the $\tilde{M}$ local maxima $\hat{x}_{01}, \hat{x}_{02}, \cdots, \hat{x}_{O\tilde{M}}$ of the function $s(x) = 1/\det(H(x))|^2$ with $-\pi/\delta K \leq x \leq \pi/\delta K$ are determined. These values are used to estimate the $x_\mu$.

Taking the data according to eq.(9) as input, the output of the precessing scheme is given by

$$(\varepsilon^f(n)) = \sum_{\mu=1}^{M} \exp(-jn\cdot\delta K\cdot x_\mu)\,\rho_\mu (H(x_\mu))(p)_\mu + (W^f(n)) = \sum_{\mu=1}^{M} (\varepsilon^f)_\mu + (W^f(n)) \tag{17a}$$

$$(\varepsilon^b(n)) = \sum_{\mu=1}^{M} \exp(-j(n-L)\cdot\delta K\cdot x_\mu)\,\rho_\mu (H(x_\mu))^*(p)_\mu + (W^b(n)) = \sum_{\mu=1}^{M} (\varepsilon^b)_\mu + (W^b(n)) \tag{17b}$$

### 3.2 Scattering data with infinite signal-to-noise ratio

In this case $(W^f(n)) = (W^b(n)) = (0)$ holds in eqs. (17 a,b). To allow that independently of the polarization vectors $(p)_\mu$ the equation $\det(H(x))$ has at least M different solutions $L \geq M$ must hold. To ensure, furthermore, that from $(\varepsilon^f) = (\varepsilon^b) = (0)$ each individual term in the sum of eqs. (17 a,b) can be deduced to be zero $((\varepsilon^f)_\mu = (\varepsilon^b)_\mu = 0$ for all $\mu)$ at least M independent equations are required. Therefore the processing scheme has to be applied to at least M/2 displaced data sets of length L+1. We shall choose the smallest possible value of the complete data length N given by

$$N = L + M/2 \quad . \tag{18}$$

We are therefore dealing with a multivariate generalization of the Kumerasan - Prony case [4,5] which offers computational advantages.
In this (theoretically) noiseless case the obtained function $s_1(x)$ will have poles at the searched position coordinates $x_1$ to $x_M$.

The L weighting matrices $(G)_\nu$ are obtained by applying the equation $(\varepsilon^f) = (\varepsilon^b) = (0)$ to the M/2 data sets. To find a compact formulation for the resulting system of equations the matrices $(R)_\mu$ (dimension 3xM/2) are introduced which are composed of the data $(\rho(\nu))$ via

$$(R)_\mu = [\,(\rho(\mu)), (\rho(\mu+1)), \cdots, (\rho(\mu-1+M/2))\,] \tag{19}$$

With this notation one obtains

$$\begin{bmatrix} (R)^t_L \,, (R)^t_{L-1} \,, \cdots \,, (R)^t_1 \\ (R)_2 \,, (R)^+_3 \,, \cdots \,, (R)^+_{L+1} \end{bmatrix} \begin{bmatrix} (G)_1 \\ (G)_2 \\ \cdot \\ \cdot \\ \cdot \\ (G)_{L-1} \\ (G)_L \end{bmatrix} = - \begin{bmatrix} (R)_{L+1} \\ (R)^*_1 \end{bmatrix} \tag{20}$$

Eq. (20) represents 3xM linear equations for 3x3L unknown coefficients $G^\mu_{nm}$. Since $L \geq M$ this system of equations is underdetermined. The minimum Euclidean norm solution is found by the pseudoinverse of the equation matrix which is via

$$(W) = \begin{bmatrix} (W)_{11} & (W)_{12} \\ (W)^+_{12} & (W)_{22} \end{bmatrix} = \begin{bmatrix} \sum_1^L (R)_{L+1-\nu}(R)_{L+1-\nu} & \sum_1^L (R)_{L+1-\nu}(R)_{1+\nu} \\ \sum_1^L (R)_{1+\nu}(R)_{L+1-\nu} & \sum_1^L (R)_{1+\nu}(R)_{1+\nu} \end{bmatrix} \tag{21}$$

given by

$$-(G)_\nu = \{(R)^*_{L+1-\nu}(W_{11}) + (R)_{\nu+1}(W_{12})^t\}(R)^t_{L+1} + \{(R)^*_{L+1-\nu}(W_{12}) + (R)_{N+1}(W_{22})\}(R)^+_1 \tag{22}$$

Eq. (22) is the searched solution for the weighting matrices.

### 3.3 Scattering data with relatively low SNR

If the measured data are corrupted by additive noise from $(\varepsilon^f(n)) = (\varepsilon^b(n)) = (0)$ it can no longer be deduced that each single contribution $(\varepsilon^f)_\mu$ and $(\varepsilon^b)_\mu$ in eq. ( 17 a,b ) is zero. These contributions have now to differ from 0 to compensate the noise contributions $(W^f(n))$ and $(W^b(n))$. So the values where $\det(H) = 0$ will with decreasing SNR move away from the real values $x_{0\mu} = x_\mu$ to the complex values $\zeta_{0\mu} = x_{0\mu} + j\,\xi_{0\mu}$ .

In the case of a single scattering center (M=1) $\hat{x}_{01} = x_{01}$ will hold and this value is in the mean equal to the searched value $x_1$ [ 4-6 ]. In contrast to the noiseless case the maximum of the function $s_1(x)$ becomes finite for $\hat{x} = x_{01}$ .

If a second scattering center is present the mean value of $\hat{x}_{01}$ will differ from $x_1$ . The bias $x_1 - \hat{x}_{01}$ origins from 2 different effects:

($\alpha$) Even if $x_{01} = x_1$ would hold in the mean, the position $\hat{x}_{01}$ of the maximum would deviate from $x_1$ since the finite - valued maximum is displaced by the superposition of the contribution by the other scattering center.

($\beta$) Due to noise the poles $\zeta_{0\mu}$ corresponding to the different scattering centers are mutually " coupled " [ 4-6 ], so that the presence of the second scattering center leads to a noise dependent displacement of $\zeta_{01}$ .

For decreasing SNR below a certain threshold level the two distinct maxima corresponding to 2 different scattering centers will vanish and only one maximum remains. The smallest distance of two close scattering centers which for a given SNR leads to two maxima defines the resolution power. It is very important to note that the " coupling " between the poles is the dominant effect for a finite resolving power.

With the proposed multivariate LP method this coupling between 2 close poles is lowered if the polarization properties of the corresponding scattering centers differ from each other. Therefore the polarimetric LP method can offer a considerably higher resolution power in comparison to scalar LP methods. This statement will be illustrated in the next section by means of a numerical example.

### 3.4 Numerical Illustration

Improvements due to the utilization of polarization information in the LP method can be demonstrated by means of a numerical example. Two scattering centers with partial scattering matrices (arguments in deg)

$$(\rho)_1 = \rho_0 \begin{bmatrix} \cos 10 & 0 \\ 0 & \sin 10 \end{bmatrix} \qquad \text{and} \qquad (\rho)_2 = \rho_0 \begin{bmatrix} \cos(10+\alpha) & 0 \\ 0 & \sin(10+\alpha) \end{bmatrix}$$

are situated at a cross - range distance of $\Delta x/\lambda_0 = 3.65$. Note that the difference in the polarization properties are in this simple example represented by $\alpha$.

$N = 25$ measurement data are assumed to be taken in the aspect angle range of $\Delta\Theta = 2.4$ deg with an increment of $\delta\Theta = 0.1$ deg and a SNR of about 5 dB. With conventional Fourier transform technique the distance should be much larger, namely $\Delta x/\lambda_0 \approx 12$ to allow the scattering centers to be resolved under these circumstances [3].

In Fig. 3 the function $s_1(x) = 1/|\det(H(x))|^2$ is used as a basis for cross-range estimation. Whereas scattering centers with equal polarization properties ($\alpha = 0$) cannot be resolved (Fig. 3a), they are resolved nearly bias-free for $\alpha = 30$ deg (Fig. 3b). With a scalar instead of the proposed polarimetric LP method the result would in both cases ($\alpha = 0$ and $\alpha = 30$ deg) look like Fig. 3a.
The function $s_2(x) = 1/|\varepsilon(x,\alpha)|^2$ which allows the cross-range and the polarization to be resolved simultaneously is given in Fig. 4.

a.)

b.)

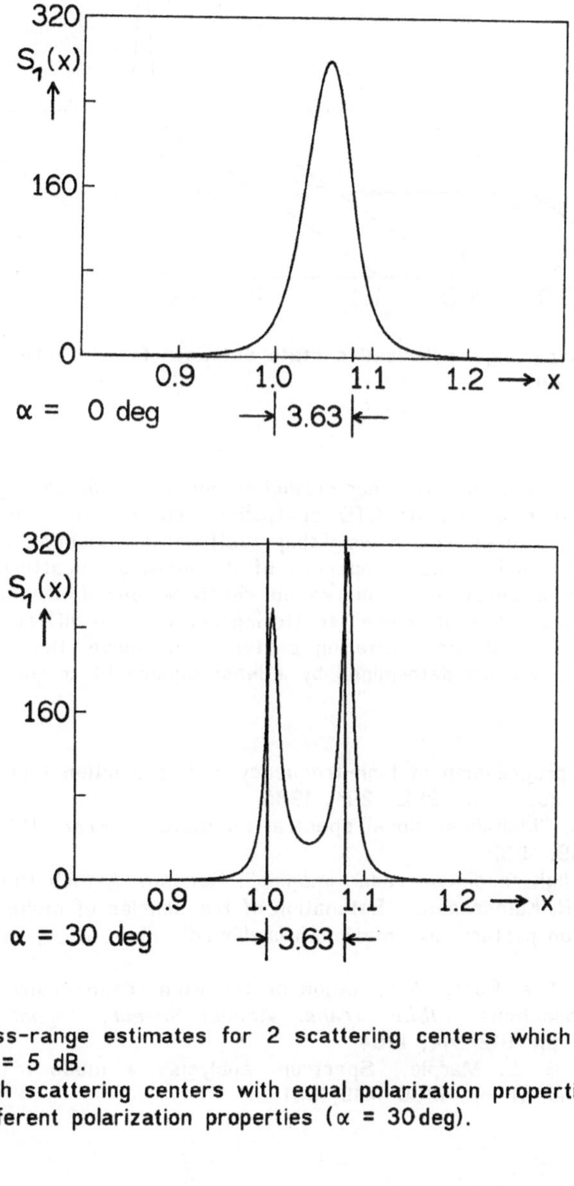


Fig. 3: Cross-range estimates for 2 scattering centers which are $\Delta x = 3.65\ \lambda_0$ apart. SNR = 5 dB.
Fig. 3a: Both scattering centers with equal polarization properties ($\alpha$=0).
Fig. 3b: Different polarization properties ($\alpha$ = 30deg).

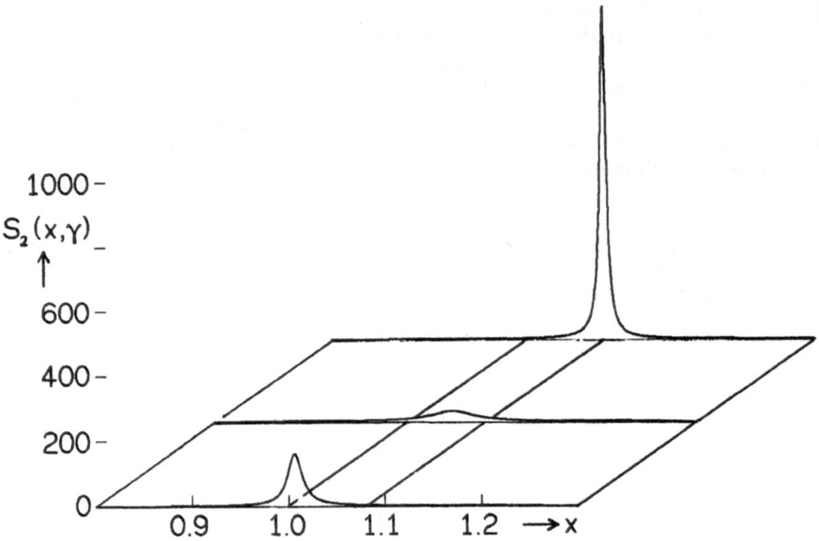


Fig. 4: Combined cross-range/polarization-state estimate for 2 scattering centers. $\alpha = 30$ deg, SNR = 5 dB.

## 4. CONCLUSIONS

A method of employing a multivariate linear prediction concept to derive high-resolution estimates of the position of GO or GTD scattering centers has been presented. Instead of the commenly used scalar weighting coefficients weighting matrices are used to account for the polarization properties of the individual scattering contributions. This results in a considerable increased resolving power if the "a priori" unknown polarization properties of "close" scattering centers are different from each other. Once the positions of the scattering centers are known, the corresponding partial scattering matrices are determined by a least square fit to the data.

## REFERENCES

[1] E. F. Knott, " A progression of high-frequency RCS prediction techniques," *Proc. IEEE* , vol. 73, no. 2, pp. 252 -264, 1985

[2] J. H. Mc Clellan, "Multidimensional spectral estimation", *Proc. IEEE* , vol. 70, no. 9, pp. 1029 -1039, 1982

[3] D. L. Mensa , "High resolution radar imaging", Dedham: Artech House, 1981

[4] D. W. Tufts and R. Kumaresan, " Estimation of frequencies of multiple sinusoids: Making linear prediction perform like maximum likelihood", *Proc. IEEE*, vol. 70, no. 9, pp. 975-989, 1982

[5] L. D. Bacon and R. e. Post, "A limitation of the Kumaresan-Prony algorithm in direction finding applications", *IEEE Trans. Acoust.,Speech, Signal Processing*, vol. ASSP-32, no. 4, pp. 912-914, 1984

[6] S. M. Kay and S. L. Marple, "Spectrum analysis- a modern perspective", *Proc. IEEE*, vol. 69, no. 11, pp. 1380-1419,1981

# A VALIDATION ANALYSIS OF HUYNEN'S TARGET-DESCRIPTOR INTERPRETATIONS OF THE MUELLER MATRIX ELEMENTS IN POLARIMETRIC RADAR RETURNS USING KENNAUGH'S PHYSICAL OPTICS IMPULSE RESPONSE FORMULATION*

**Sujeet K. Chaudhuri, Bing-Yuen Foo and W-M. Boerner**
University of Illinois at Chicago
UIC-EECS/CL, 840 W. Taylor St.
SEL-4210, M/C 154
Chicago, IL 60680-4348

**ABSTRACT:** By applying corrections to Kennaugh's impulse response formulation for smooth, convex targets, Huynen's descriptors have been shown to relate to specular geometry at high frequencies. A time-gating technique has been developed to separate the specular signature from the creeping wave contribution in the total backscattered return. This enables approximately equivalent high-frequency data to be extracted from bandlimited scattering measurements. It is shown that the validation of the target-descriptor interpretations of the Mueller matrix elements with the isolated specular contribution data is acceptable, whereas the validation of these interpretations with the total backscattered data seems doubtful.

## 1. INTRODUCTION

It is known that high frequency radar interrogation may disclose fine geometrical structures of targets, whereas low frequency interrogation may recover such coarse information as target sizes and volumes. Any target descriptor which is intended to describe the geometry has to be defined on a high frequency basis and thus its validity becomes questionable at low frequencies. In the phenomenological target-descriptor approach of Huynen [1,3], the frequency range of validity has not been established, nor have the descriptors been rigorously related to target geometry by using electromagnetic theory. In view of these unresolved problems, the main objective of this paper is to show, based on electromagnetic theory, that Huynen's phenomenological target-descriptors, at high frequencies, can be closely related to specular geometry (in particular, specular curvature) and target orientation. Numerical analyses with measured polarimetric backscattering data are shown, and application of the descriptors in radar target discrimination or identification is suggested.

*S.K. Chaudhuri, B.-Y. Foo and W-M. Boerner, "A Validation Analysis of Huynen's Target-Descriptor Interpretations of the Mueller Matrix Elements in Polarimetric Radar Returns Using Kennaugh's Physical Optics Impulse Response Formulation", IEEE Trans. on Ant. & Propag., Vol. AP-34, No. 1, pp. 11-20, January 1986



*W.-M. Boerner et al. (eds.), Direct and Inverse Methods in Radar Polarimetry, Part 1*, 555–579.

In the following, first the relationship between the elements of the scattering matrix ([S]) and the Mueller matrix ([M]) is established. Next, by using these expressions along with the first order corrected physical optics impulse response formulation, certain approximate relations between the Mueller matrix elements and scatterer physical characteristics are established. Now it is possible to compare the phenomenological target-descriptor interpretations of these Mueller matrix elements with the above electromagnetically established geometry related expressions.

During the numerical validation studies the experimental backscattering data on ellipsoidal structures are used. These data primarily consist of two contributions: one specular contribution and the other creeping wave contribution. A time-domain scheme is developed to separate these two individual components in the measured frequency-domain broadband backscattered data. It is shown that the validation of the target-descriptor interpretations of the Mueller matrix elements with the isolated specular contribution data is acceptable, whereas the validation of these interpretations with the total backscattered data is doubtful.

## 2. THE SCATTERING MATRIX AND THE MUELLER MATRIX

The scattering matrix [S] can be defined by the incident $[E^i]$ and the scattered $[E^s]$ fields:

$$[E^s] = [S][E^i] \quad ,$$

where

$$[S] = \begin{bmatrix} S_{VV} & S_{VH} \\ S_{HV} & S_{HH} \end{bmatrix} \tag{1}$$

The matrix [S] contains two co-polarized elements and two cross-polarized elements. In the monostatic backscattering case, $S_{VH} = S_{HV}$ by reciprocity of the propagation in an isotropic medium. The subscripts V and H denote a pair of orthogonal polarization bases (e.g., H $\hat{=}$ horizontal, V $\hat{=}$ vertical). Properties of the relative phase scattering matrix [S] were investigated in the pioneering studies of Kennaugh (1949-1952) [10], used and further interpreted by Huynen [3], as summarized in Boerner [11].

The voltage induced at the receiving antenna is given by:

$$V^r = [S][E^i]\cdot[E^r] \tag{2}$$

in which $[E^r]$ is the polarization vector of the receiving antenna. The power received is given by:

$$P^r = [M]g[E^i]\cdot h[E^r] \tag{3}$$

in which $g[E^i]$ and $h[E^r]$ are the Stokes vectors of $[E^i]$ and $[E^r]$ respectively, and [M] denotes the 4x4 Mueller matrix. The elements of [M] and [S] are related through Huynen's algebraic approach [12], in which [S] is represented by algebraic variables $\underline{a}$, $\underline{b}$ and $\underline{c}$, as follows:

$$[S] = \begin{bmatrix} \underline{a}+\underline{b} & \underline{c} \\ \underline{c} & \underline{a}-\underline{b} \end{bmatrix} \tag{4}$$

In [12], [M] is given as:

$$[M] = \begin{bmatrix} A_o+B_o & F & C_\psi & H_\psi \\ F & -A_o+B_o & G_\psi & D_\psi \\ C_\psi & G_\psi & A_o+B_\psi & E_\psi \\ H_\psi & D_\psi & E_\psi & A_o-B_\psi \end{bmatrix},$$

where

$$\left.\begin{aligned} A_o &= \frac{1}{2}\,|\underline{a}|^2 \\ B_o &= \frac{1}{2}\,(|\underline{b}|^2 + |\underline{c}|^2) \\ B_\psi &= \frac{1}{2}\,(|\underline{b}|^2 - |\underline{c}|^2) \\ C_\psi + jD_\psi &= \underline{a}^*\underline{b} \\ E_\psi + jF &= \underline{b}\,\underline{c}^* \\ H_\psi + jG_\psi &= \underline{a}\,\underline{c}^* \end{aligned}\right\} \tag{5}$$

with * denoting complex conjugation.

Using (1), (4) and (5) for the purely coherent case, the Huynen descriptors are expressed in terms of the elements of [S] as follows [12]:

$$A_o = \frac{1}{8}(|S_{VV}|^2 + |S_{HH}|^2 + 2|S_{VV}||S_{HH}|\cos\phi_B) \tag{6a}$$

$$B_o = \frac{1}{8}(|S_{VV}|^2 + |S_{HH}|^2 - 2|S_{VV}||S_{HH}|\cos\phi_B) + \frac{1}{2}|S_{HV}|^2 \tag{6b}$$

$$B_\psi = \frac{1}{8}(|S_{VV}|2 + |S_{HH}|^2 -2|S_{VV}||S_{HH}|\cos\phi_B) - \frac{1}{2}|S_{HV}|^2 \quad (6c)$$

$$C_\psi = \frac{1}{4}(|S_{VV}|^2 - |S_{HH}|^2) \quad (6d)$$

$$D_\psi = \frac{1}{2}(|S_{VV}||S_{HH}|\sin\phi_B) \quad (6e)$$

$$E_\psi + jF = \frac{1}{2}(|S_{VV}|-|S_{HH}|e^{-j\phi_B})|S_{HV}|e^{j\phi_A} \quad (6f)$$

$$H_\psi + jG_\psi = \frac{1}{2}(|S_{VV}| + |S_{HH}|e^{-j\phi_B})|S_{HV}|e^{j\phi_A} \quad (6g)$$

where $\phi_A = \phi_{VV} - \phi_{VH}$ and $\phi_B = \phi_{VV} - \phi_{HH}$, and the $\phi_i$'s being the corresponding relative phases of the elements of [S].

In [3], the sum $A_o + B_o$ is considered as a rough measure of half the total power in the wave returned from the target. $A_o$ is associated with regular, smooth, spherical types of surface scattering, which contributes to specular returns such as from a sphere. $B_o$ may be considered as a measure of all the target's non-symmetric, irregular, rough-edged, non-spherical depolarizing components of scattering. $C_\psi$ is related to oblongity and F to helicity [3].

If the specular region is smooth and convex, and if the incident field is along one of the principal directions, then the cross-component $|S_{HV}| = 0$ [4,8], and

$$B_o = B_\psi \quad (7)$$

$$E_\psi = F = G_\psi = H_\psi = 0$$

In particular, in the simple case of a sphere, $B_o$ and $B_\psi$ vanish because of symmetry. In contrast, for an ellipsoid-like target, a non-zero value for $B_o$ is expected. Thus $B_o$ may serve to discriminate between spherical and ellipsoidal targets.

## 3. INTERPRETATION OF HUYNEN'S DESCRIPTORS IN TERMS OF SPECULAR CURVATURES

### (i) First Order Polarization Correction to Physical Optics [5]

For a perfectly conducting scatterer, the time-domain surface current $\vec{J}$ induced on the scatterer by the incident field can be given by the space-time integral equation [5]:

$$\vec{J}(\vec{r},t) = 2\hat{a}_n \times \vec{H}_i(\vec{r},t) + \frac{1}{2\pi} \iint \hat{a}_n \times \{[\frac{1}{R^2} + \frac{1}{RC}\frac{\partial}{\partial \tau}] \vec{J}(\vec{r}',\tau) \times \hat{a}_R\}ds' \quad (8)$$

where $\vec{r}$ is the position vector to the observation point, $\vec{r}'$ is that to an integration point, $\vec{R} = R\,\hat{a}_R = \vec{r} - \vec{r}'$, $\tau$ is the retarded time $t - R/C$ (C - free space propagation velocity), and $\hat{a}_n$ is the unit normal to the surface (s) at the observation point. The first term of (8) is the source term, and is also the Physical Optics approximation for the illuminated side. The integral term represents the contribution of retarded currents at all integration points. In [5] Bennett et al., simplified the integral term as the first order correction to Physical Optics, under the leading edge assumptions. The surface integration was carried out only over a small patch around the specular point, and the far-field impulse response can thus be derived from the surface current approximation [4,6]

$$r_o \vec{H}_s(\vec{r},t) = \frac{1}{2\pi}\frac{\partial^2}{\partial t^2} A(t)\,\hat{a}_{H_i} + \frac{K_u - K_v}{4\pi}\frac{\partial}{\partial t} A(t)\{(\hat{a}_{H_i}\cdot\hat{a}_u)\hat{a}_u - (\hat{a}_{H_i}\cdot\hat{a}_v)\hat{a}_v\} \quad (9)$$

where A(t) is the silhouette area of the scatterer as delineated by the incident impulsive plane wave moving at half the speed of light, $r_o$ is the radar range, $\hat{a}_{H_i}$ is the unit vector in the direction of the incident magnetic field $\hat{H}_i$; $\hat{a}_u$ and $\hat{a}_v$ are the unit vectors along directions of the principal curvatures $K_u$ and $K_v$ at the specular point (see Fig. 1).

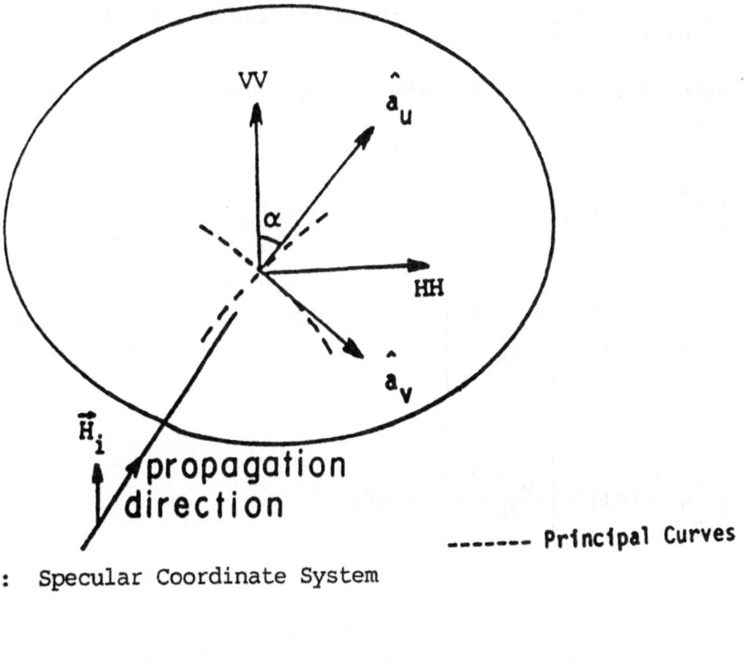


Fig. 1: Specular Coordinate System

The expression (9) was derived for a smooth, convex scatterer.

The first term in (9) is the Kennaugh-Cosgriff formula [9], which is polarization-independent. The second term gives the first order polarization correction and takes the functional form of the first derivative of the area function A(t) and is proportional to the principal curvature difference ($K_u - K_v$). The polarization correction to the Physical Optics approximation is essential, if the high frequency scattering matrix elements, containing polarization information, are to be related to target geometry.

(ii) Scattering Matrix and Mueller Matrix Elements in Terms of Curvature Difference

By taking the Fourier transform of the impulse response $\vec{H}_s$ in (9), the elements of the monostatic scattering matrix [S] can be rewritten as [4,8]:

$$S_{VV} = \frac{1}{2\pi}(jk)^2 A(k) - (jk) A(k) \frac{K_u - K_v}{4\pi k} \cos 2\alpha \tag{10}$$

$$S_{HH} = \frac{1}{2\pi}(jk)^2 A(k) + (jk) A(k) \frac{K_u - K_v}{4\pi k} \cos 2\alpha \tag{11}$$

$$S_{HV} = (jk) A(k) \frac{K_u - K_v}{4\pi k} \sin 2\alpha = S_{VH} \tag{12}$$

where A(k) is the Fourier transform of the silhouette area, and $\alpha$ is the polarization angle between the incident field and the principal direction of $K_u$ (Fig. 1)

By substituting (10-12) in (6), the Mueller matrix descriptors can be written as [2]:

$$A_o = \frac{1}{8} \frac{k^4}{\pi^2} |A(k)|^2 \tag{13}$$

$$B_o = \frac{1}{8} \frac{k^4}{\pi^2} |A(k)|^2 \left( \frac{K_u - K_v}{2k} \right)^2 \tag{14}$$

$$B_\psi = \frac{1}{8} \frac{k^4}{\pi^2} |A(k)|^2 \left( \frac{K_u - K_v}{2k} \right)^2 \cos 4\alpha \tag{15}$$

$$D_\psi = \frac{-k^4}{4\pi^2} |A(k)|^2 \left(\frac{K_u - K_v}{2k}\right) \cos 2\alpha \tag{16}$$

$$E_\psi = \frac{-k^4}{8\pi^2} |A(k)|^2 \left(\frac{K_u - K_v}{2k}\right)^2 \sin 4\alpha \tag{17}$$

$$G_\psi = \frac{k^4}{4\pi^2} |A(k)|^2 \left(\frac{K_u - K_v}{2k}\right) \sin 2\alpha \tag{18}$$

$$C_\psi,\ F,\ H_\psi \to 0 \text{ with increasing frequency} \tag{19}$$

It is now obvious here that $B_o$ is a measure of non-spherical symmetry, and hence serves to differentiate between spherical and non-spherical target shapes. The orientation invariance of $B_o$ as shown by equation (14) enhances its practicality.

The high frequency assumption (equivalent to the specular region contributions only) is already implied in this interpretation of the Huynen descriptors, as (10-12) are derived on the assumptions of leading edge (high frequency) conditions. It can be observed that equations (6a to 6g) do not explicitly show the frequency dependence of the descriptors, whereas equations (13-18) do. Morever in [6], it has been shown that $A(k)$ takes the functional form of $\frac{1}{k^2}$ in the case of ellipsoids for large values of k. Consequently, all the descriptors except $A_o$ eventually tend to zero with increasing frequency. This is not surprising as the target then virtually looks like a flat plate, (specular "patch"), and there is no polarization dependence in the optical region. It is conjectured that these arguments can be extended to any target with smooth, convex specular sections.

The target orientation explicitly appears in the form of the polarization angle $\alpha$. $A_o$ and $B_o$, which are related to total power, are orientation invariant. Other orientation invariances, namely, $B_\psi^2 + E_\psi^2$ and $D_\psi^2 + G_\psi^2$, are also separately satisfied as required in [1,3]. It is important to note that target orientation with respect to polarization basis directions can be recovered practically and accurately by taking the ratios $B_\psi / B_o$ and $E_\psi / B_o$:

$$\cos 4\alpha = B_\psi / B_o \text{ , or} \tag{20}$$

$$\sin 4\alpha = E_\psi / B_o \qquad (21)$$

A smooth convex scatterer can be roughly modeled as a combination of two scattering centers, one at the specular region, which gives rise to the specular return, and the other being the creeping waves which circumnavigate and return energy characteristic of their paths. At low resonance frequencies, the contributions of both the specular return and the creeping wave return are comparable and interfere with each other. This interference masks correlation between fine geometry and Huynen's descriptors. The correlation between the specular geometry or orientation and the Huynen's descriptors at very high frequencies is expected to be good. This is because at high frequencies, it is known that the creeping waves decay and the specular return dominates the backscattered signal. In order to extract the specular geometry information, it is thus suggested that the specular contribution and the creeping wave contribution in the total backscattered signal be separated. A method for such separation will be discussed next, and will be applied to the band-limited measured data. Based on the separated specular data, which are approximately equivalent to high frequency data, Huynen's descriptors will be computed, and the results will be compared with the theoretical predictions of equations (13-19).

## 4. SEPARATION OF THE SPECULAR RETURN FROM THE CREEPING WAVE RETURN BY SIGNAL PROCESSING

The purpose is to decompose the measured scattering matrix, which is in the frequency domain, by time-gating, into the specular component and the creeping wave components. For the simple canonical shapes of the sphere and the ellipsoid, complete scattering matrix data are available from the ElectroScience Laboratory of the Ohio State University (ESL-OSU) [13]. In general, the data consist of amplitudes and phases, from 1 to 12 GHz, in steps of 10 MHz, with different aspect angles. The range of the spectrum of a "first derivative Gaussian" pulse (functional dependence: $gte^{-gt^2}$) is made to coincide with that of the measured data, by adjusting the narrow pulse width. By multiplying the pulse spectrum with the measured complex spectrum and inverse Fourier transforming the product, the specular return and the creeping wave return are distinctly separated in the time domain, provided that the first derivative Gaussian (FDG) pulse is sufficiently narrow relative to the target size. Either returns can thus be time-gated, Fourier-transformed, and finally divided by the spectrum of the FDG pulse to form new elements of the separated matrix [S]. Obviously for very broad-band measurements, the FDG pulse is very narrow relative to the target size, and hence the creeping wave has a large delay with respect to the specular return, so that the time-gating process becomes trivial. In practice, measurements can be band-limited, and the target can be small, and partial overlapping may occur for the creeping wave and specular returns in the time domain. Yet by locating the transition point in the overlapping region through the approximate knowledge of the "space width" of the interrogating pulse (~ pulse width × free space propagation velocity) and the creeping wave path length, reasonably good results have been obtained by this technique. Figure 2(i) corresponds to the former case (very broadband data), in which ka ranges from 0.8 to 9.6 (1 to 12 GHz), where a is the smallest semi-axis of the ellipsoidal target; the semi-axes are 6": 3": 1.5". Figure 2(ii) corresponds to the latter case in which ka ranges from 0.2 to 1.2 (2 to 12 GHz), the semi-axes of the ellipsoid are $\frac{1.5''}{2} : \frac{1.5''}{4} : \frac{1.5''}{8}$. Nose-on aspect and vertical polarization data were taken for both cases.

The separated co-polarized specular contributions are depicted in Figure 3. Due to the nature of the deconvolution involved in the signal processing, the end portions are not accurate. However, other than the end frequencies, the whole range shows that the amplitudes are steady in general. The reason is that with the creeping wave removed, the target behaves as if it were the specular portion, which resembles a small flat plate. Therefore, there is less frequency dependence at low frequencies than if the data were not processed. Thus the separation technique enables equivalent high frequency information to be approximated from low frequency band-limited measurements.

The other component of [S], due to creeping wave contributions only, is shown in Figure 4. Once again, the end portions are not accurate, but the whole range, in general, is in accordance with the fact that the creeping waves decay as the frequency increases. Figure 5 shows that the two components add up to the original measured scattering matrix [S] for the case of horizontal polarization. At high frequencies, the measured [S] approaches the specular [S]. Obviously, if the Huynen descriptors are computed with measured [S] at low frequencies, the fluctuations in contrast to the steady, flat-plate behavior at high frequencies will not lead to an observable correlation with the target specular geometry.

## 5. APPLICATION OF SPECULAR [S] IN THE VALIDATION ANALYSIS

### (i) Comparison between the experimentally processed and theoretically calculated target-descriptors

The target-descriptor Mueller matrix elements, normalized with respect to $A_o$, are computed from the specular [S] using the expressions given in (6). These values are compared with the corresponding theoretical values calculated using (13) to (19), and are shown in Fig. 6. The scattering geometry used for these data involves broad-side plane wave incidence on a conducting ellipsoid of semi-axes 6" : 3" : 1.5", with the polarization angle $\alpha$=30°. The agreement between the theoretical and experimental results in Fig. 6 is good and acceptable. It is emphasized here that the theoretical values are derived from high frequency assumptions, thus as the frequency increases the difference between the theoretical and experimental results decreases.

### (ii) $A_o$ as a measure of specular size

Since $A_o$ is approximately related to the vector, i.e., involving two

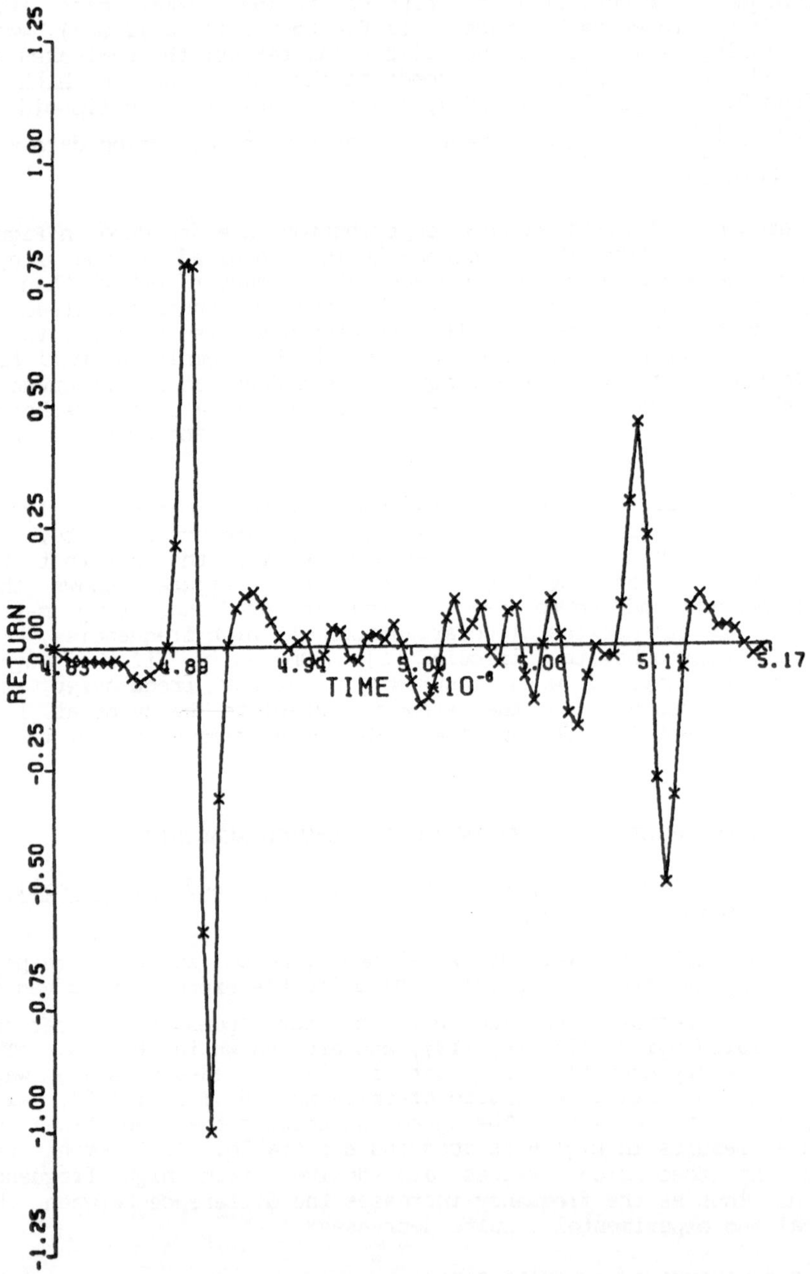


Fig. 2i: Return Due to a Relatively Narrow FDG Pulse

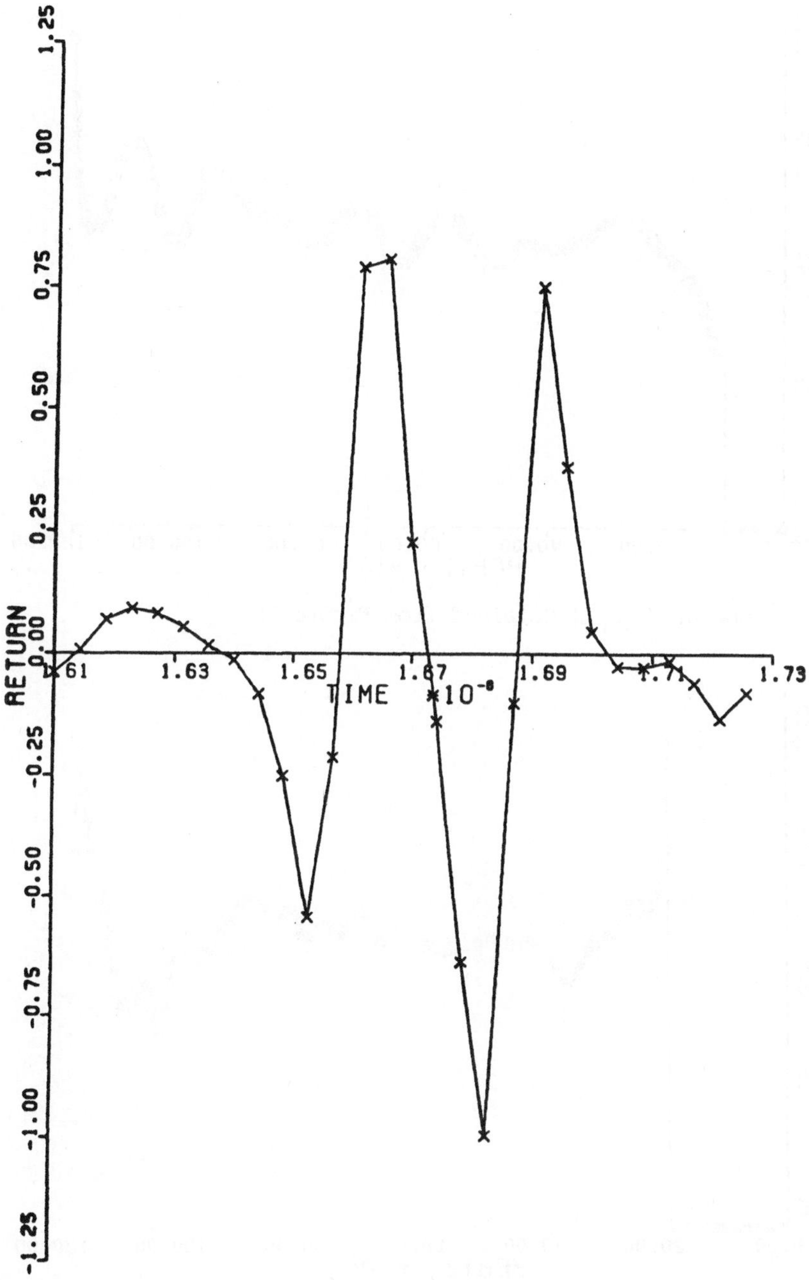


Fig. 2ii: Return Due to a Relatively Wide FDG Pulse

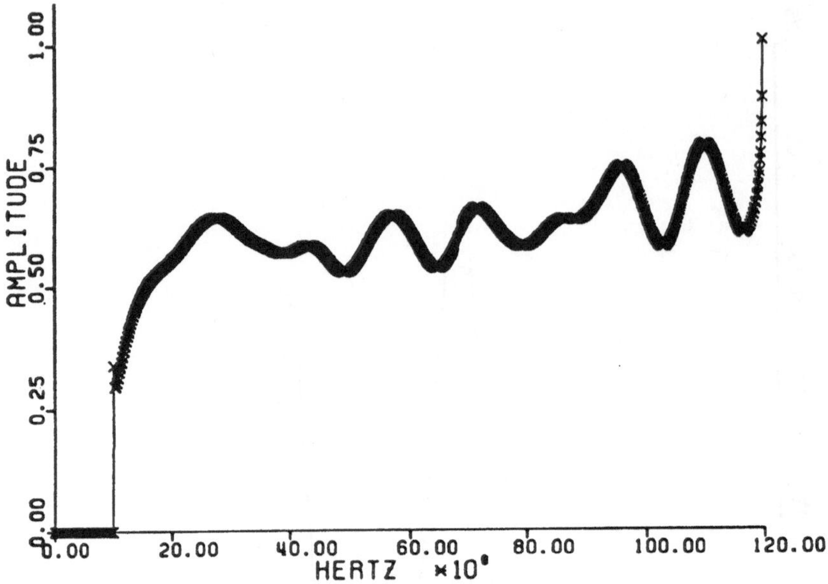


Fig. 3i: Specular $|S_{VV}|$ Obtained from Figure 2i

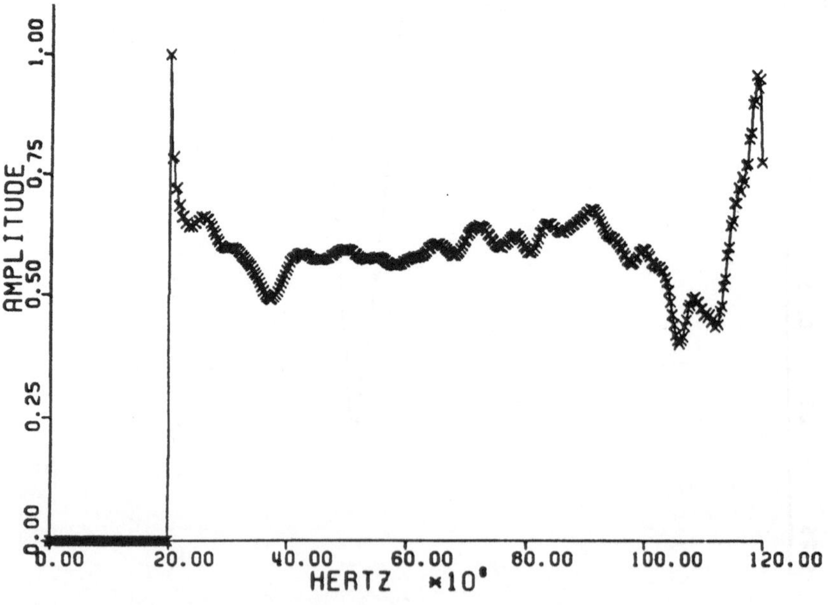


Fig. 3ii: Specular $|S_{VV}|$ Obtained from Figure 2ii

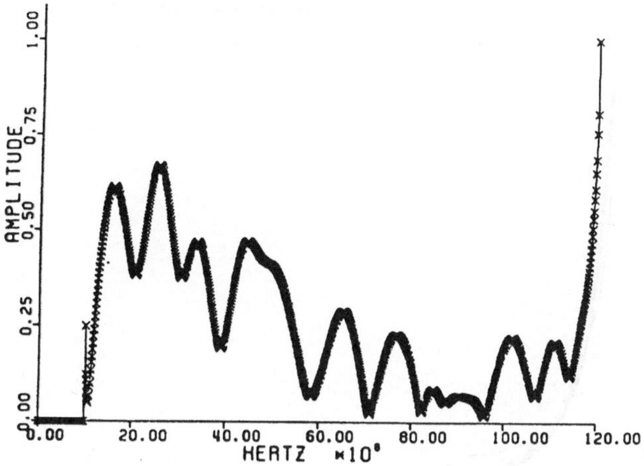


Fig. 4: Creeping Wave Component of $|S_{VV}|$ Obtained from Figure 2i

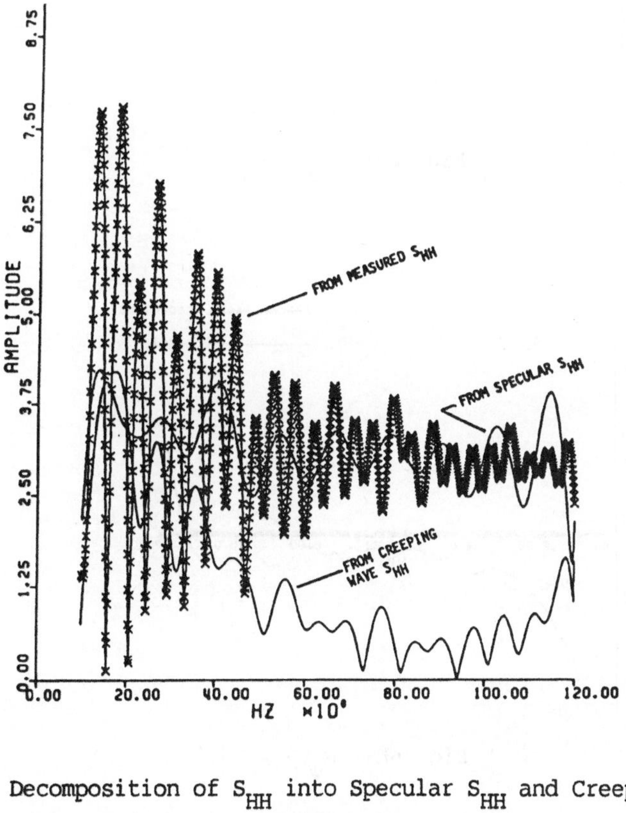


Fig. 5: Decomposition of $S_{HH}$ into Specular $S_{HH}$ and Creeping Wave Component (Nose-on Incidence)

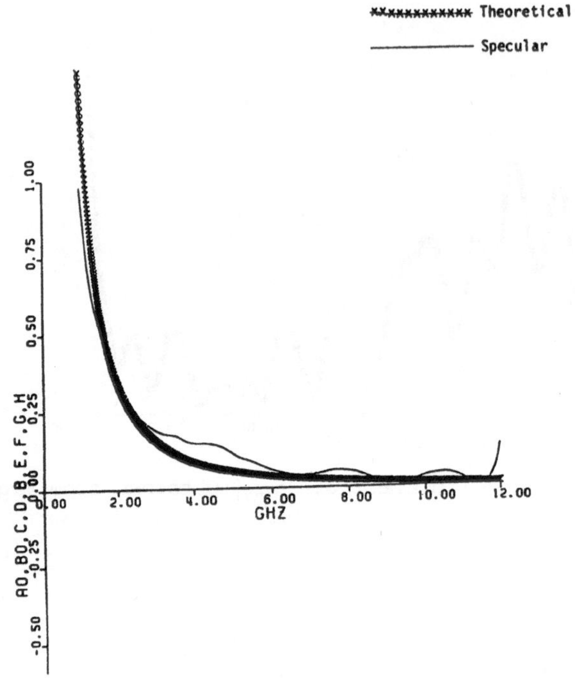


Fig. 6a: $B_o / A_o$

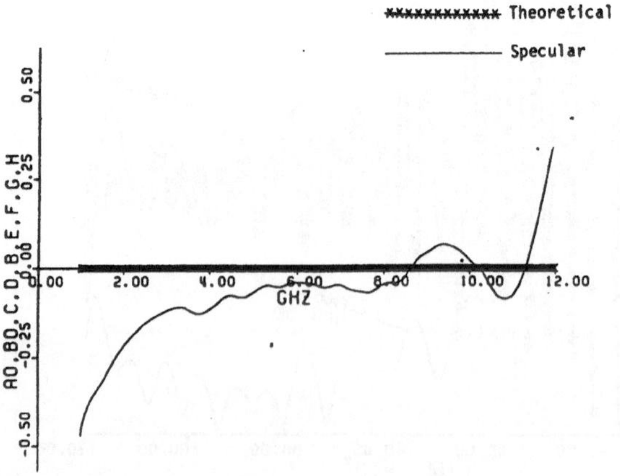


Fig. 6b: $C_\psi / A_o$

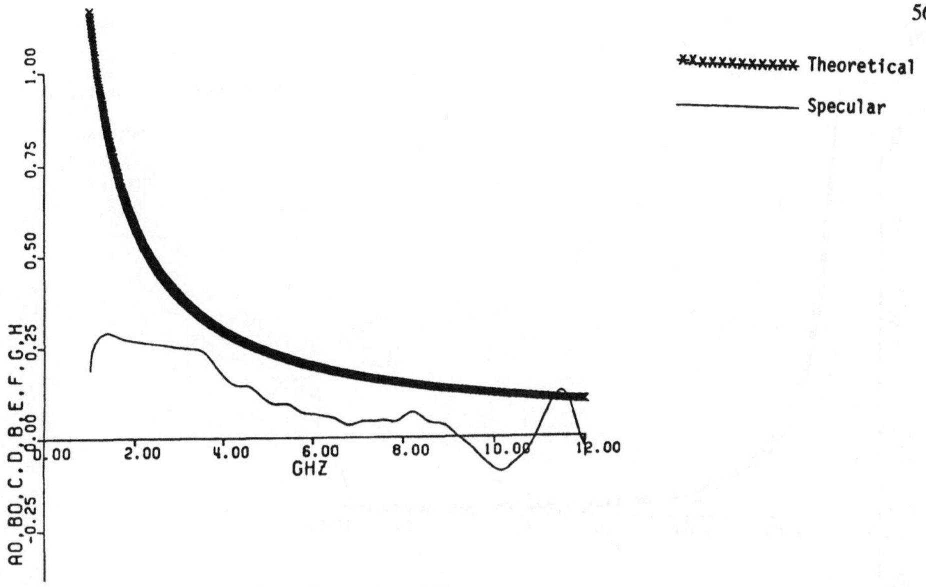


Fig. 6c: $D_{\psi} / A_{o}$

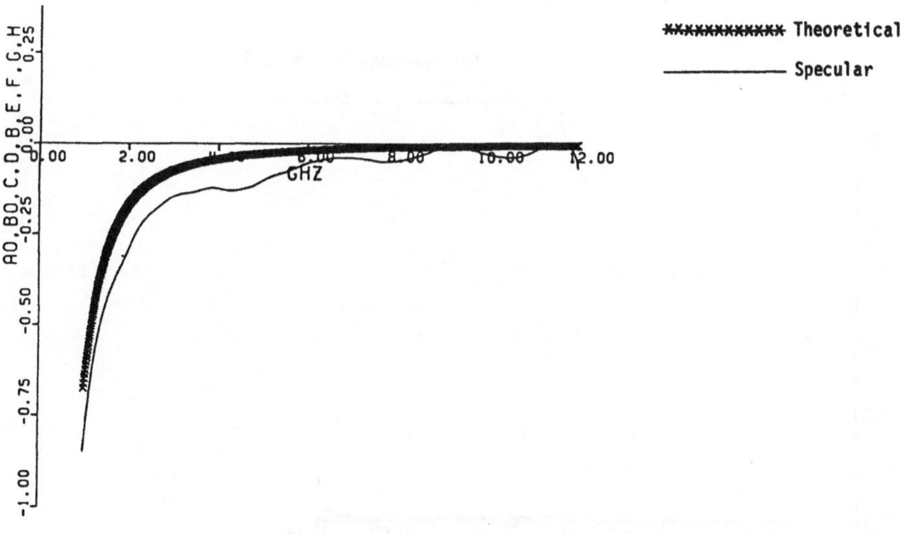


Fig. 6d: $B_{\psi} / A_{o}$

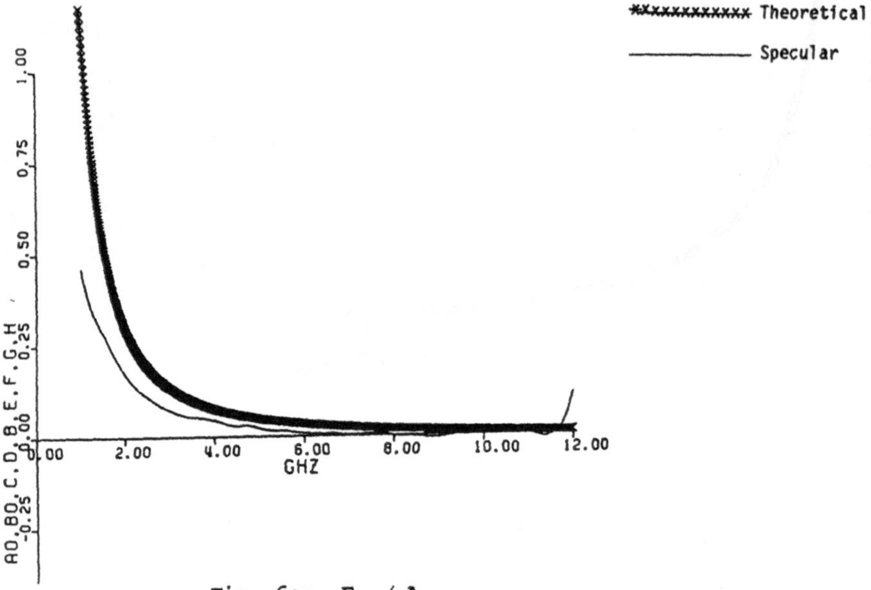


Fig. 6e: $E_\psi / A_o$

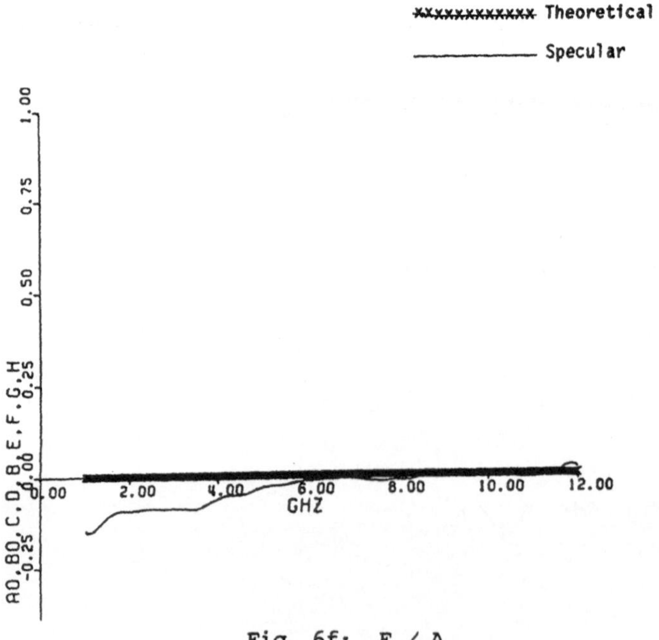


Fig. 6f: $F / A_o$

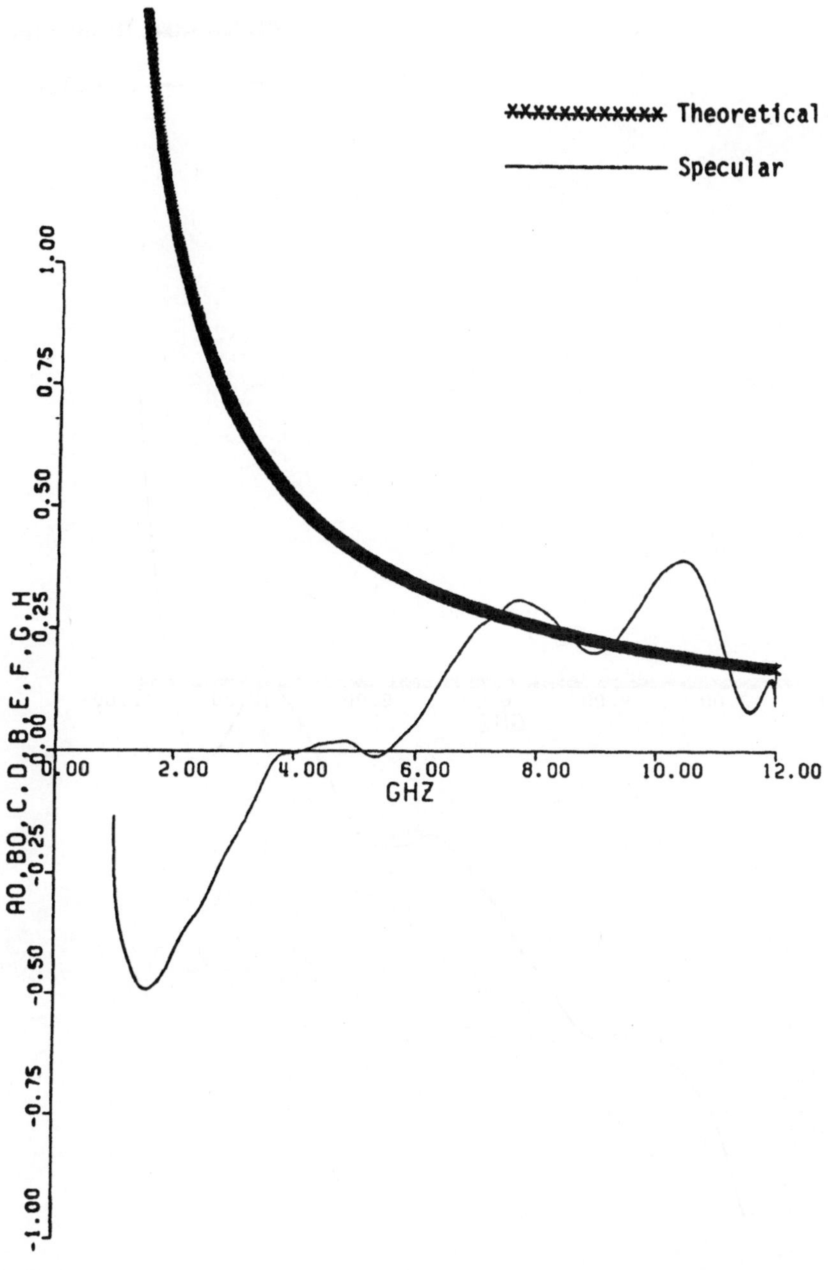


Fig. 6g: $G_{\psi} / A_{o}$

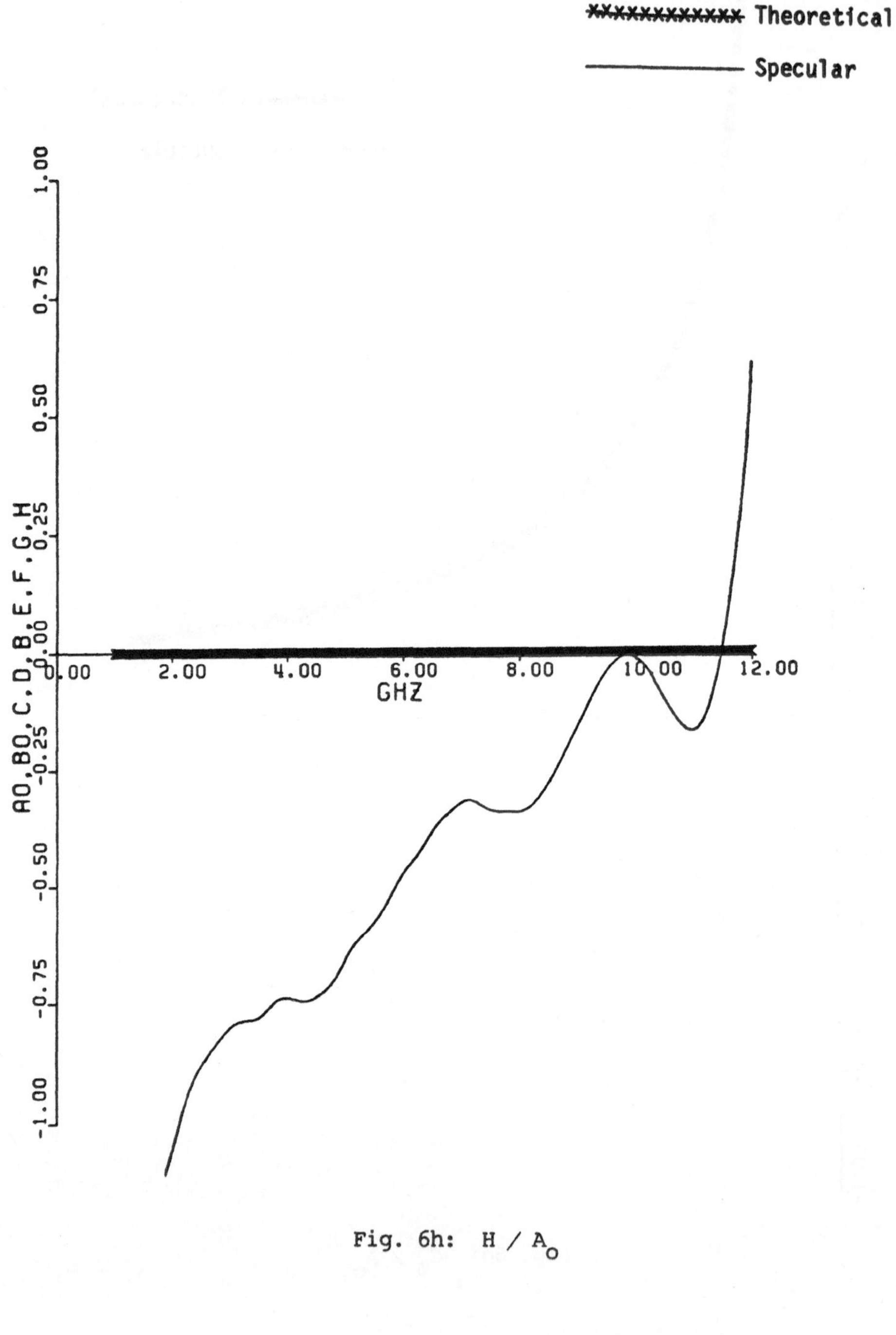
xxxxxxxxxxxx Theoretical
Specular
AO,BO,C,D,B,E,F,G,H
1.00
0.75
0.50
0.25
0.00
-0.25
-0.50
-0.75
-1.00
0.00
2.00
4.00
6.00
8.00
10.00
12.00
GHZ

Fig. 6h: H / $A_o$

orthogonal polarizations, larger values of $A_o$ are associated with larger sizes of the specular region. This is verified by comparing $A_o$ of a sphere of diameter 1.5" and $A_o$ of an ellipsoid of the semi-axes $\frac{1.5"}{2}$: $\frac{1.5"}{4}$: $\frac{1.5"}{8}$ for nose-on incidence (Figure 7).

(iii) $B_o/A_o$ as a measure of electric specular curvature difference

From equations (13, 14),

$$\frac{B_o}{A_o} = \left(\frac{K_u - K_v}{2k}\right)^2 \tag{22}$$

The normalization of $B_o$ with respect to $A_o$ is introduced to remove the factor A(k). At any rate, $B_o$ or $B_o/A_o$ should be zero for a sphere, and should deviate from zero for ellipsoidal targets. In Figure 8, the values of $B_o/A_o$ are plotted for the ellipsoid of 6" : 3" : 1.5", with nose-on incidence.

The specular curvature difference $K_u - K_v$, in this case, is 78.74 per meter as computed from Minkowski's support functions [6]. In Fig. 8, the theoretical values predicted by equation (22) is also compared with the measured values, showing good agreement at high frequencies. At low frequencies, equation (22), based on the high-frequency leading edge assumption, does not hold, and thus causes discrepancies. Again, at the high-frequency end, the deconvolution process introduces inaccuracy of $B_o/A_o$ computed from the specular [S].

The $B_o/A_o$ validation curve for another case of incidence (broad-side incidence, $\alpha = 30°$) on the above ellipsoid is given in Fig. 6(i). There the agreement between the theoretical predictions and the processed measurement data is excellent. This clearly indicates that the correction term added to the physical optics in (9) holds more accurately for smaller curvature difference at the specular point (for Fig. 6(i) $K_u - K_v = 49.21$ per meter).

(iv) Orientation recovery from $B_\psi/B_o$ or $E_\psi/B_o$

In some of the measurement data described so far, the incident field is along one of the principal directions of the ellipsoid. Under such circumstances the cross-polarized element of [S] is zero provided the target is sufficiently smooth so that finite principal curves exist at the specular point. On the other hand, if the incident field is not along one of the principal directions, then the cross-polarized elements are non-zero, and the orientation can be recovered using (20) or (21). In Fig. 9, both values of $\alpha$ computed from (20) and (21) are shown separately, and match the value of the polarization angle set up in the

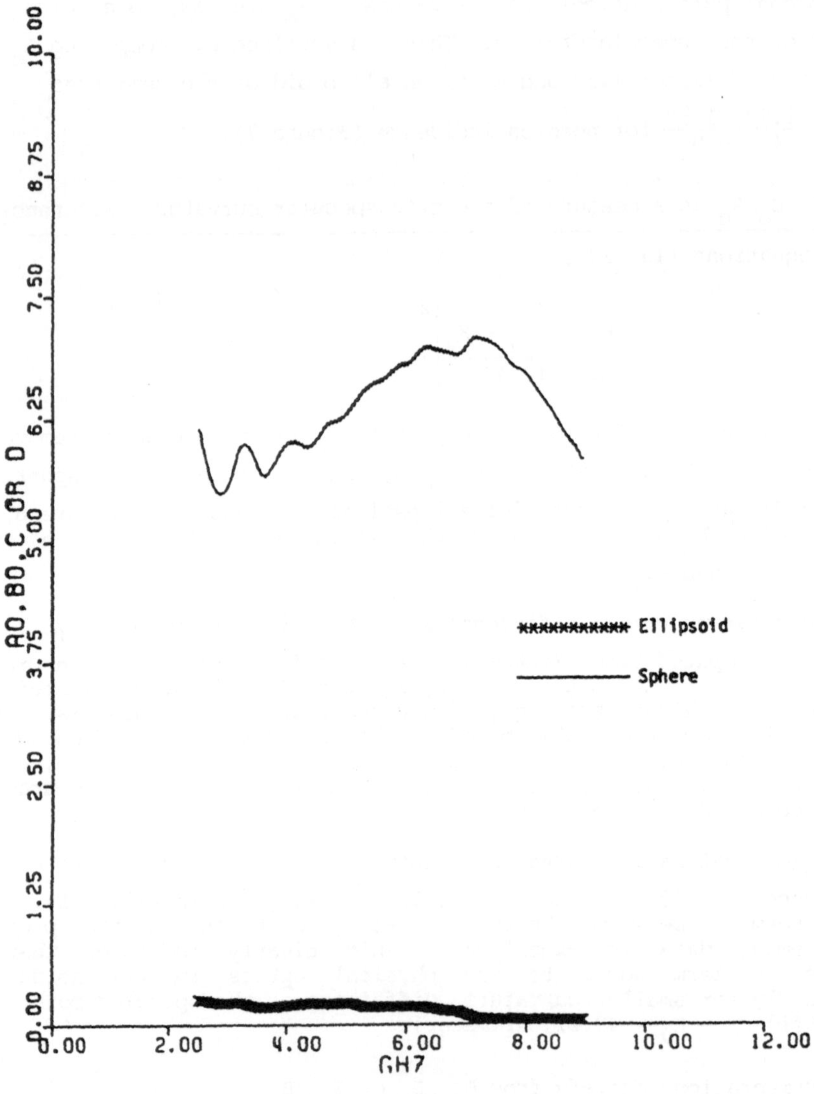


Fig. 7: Comparison of $A_O$

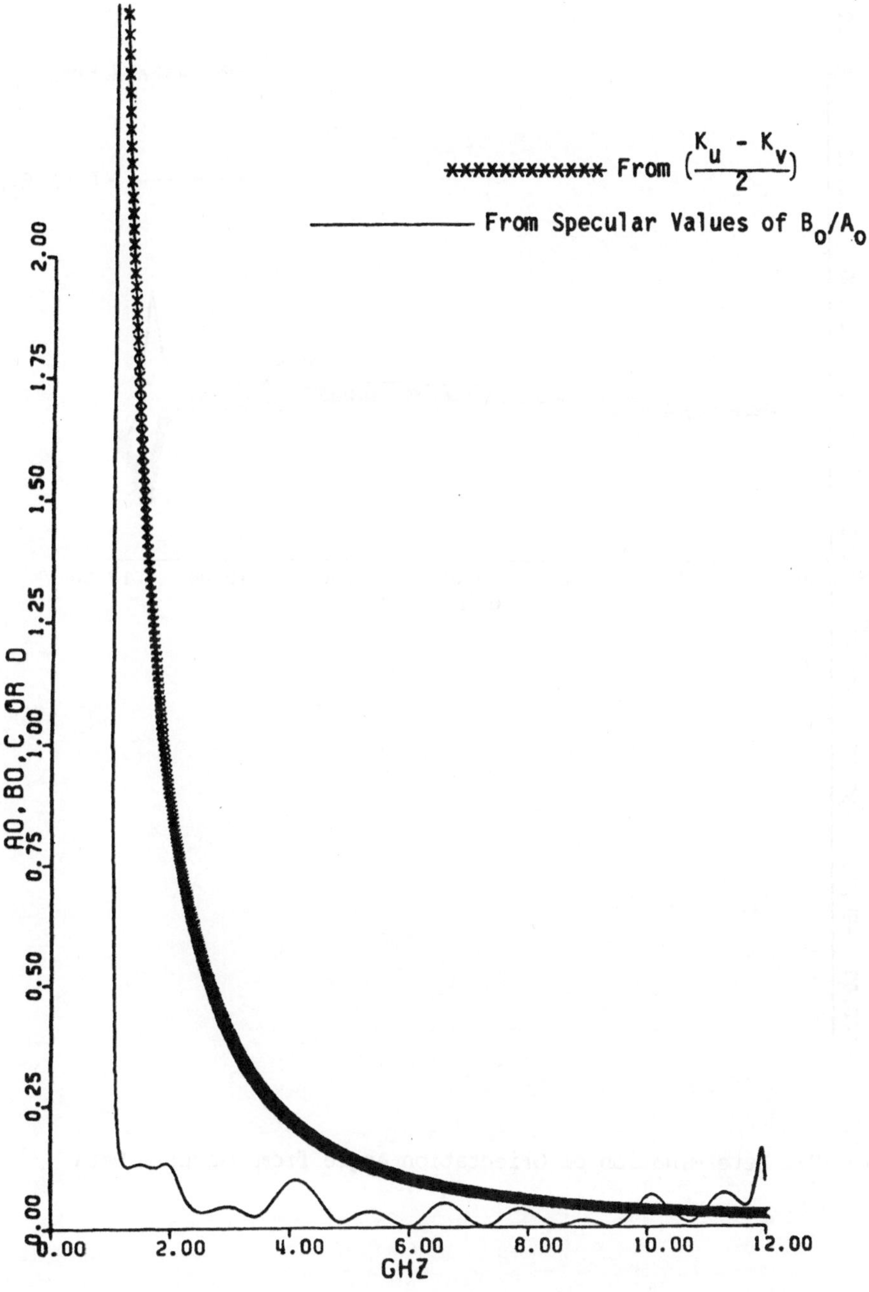


Fig. 8: $B_o$ / $A_o$ (Broadside Incidence, Tilt Angle is 30°)

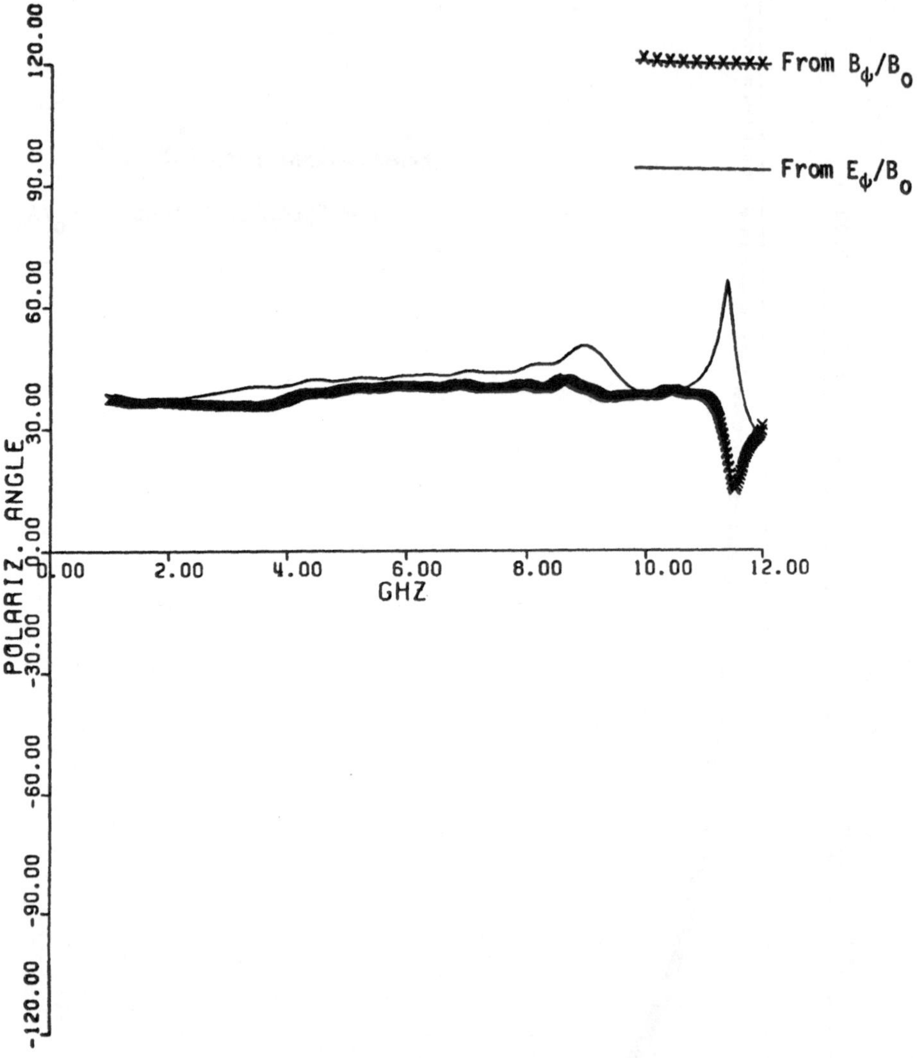


Fig. 9: Determination of Orientation Angle from Measured Data

ESL-OSU measurement arrangement. The conditions in which measurements were made are the same as those of Fig. 6.

## 6. CONCLUSIONS

The target descriptors are related to specular geometrical information such as specular size, curvature difference and orientation. Although the discussion in this paper is restricted to smooth, convex objects, the proposed time-gating technique can be applied to a certain class of objects with one or more specular regions which may include edges, or wedges, etc. Each specular region serves as a scattering center to interfere with others, but the response of each center, time-gated out of the total time-domain return, should provide Mueller matrix target descriptors characteristic of its geometry. Thus objects of complex shapes may be studied through modeling and decomposition into simpler scattering centers; and separately for each scattering center, the target characteristic polarimetric operators must be analyzed for the object classification and identification algorithms.

The creeping wave component, which can be isolated by the separation technique described, should deserve more investigation in the future as it certainly contains information about the geometry within the shadow region. This multi-frequency inverse scattering approach requires a broad-band measurement. Not only broad-band measurements are required, but also the phases in the scattering matrix must be coherent and as accurate as possible to enable curvature differences to be determined correctly. In contrast, the orientation angle has been quite accurately determined with the existing ESL-OSU data. By knowing the curvature difference sufficiently accurately, the target profile can be reconstructed through the application of differential geometry [6,7].

## ACKNOWLEDGMENTS

The research reported in this study would not have been possible without the pioneering contributions of the late Professor Edward M. Kennaugh, Professor Emeritus, The Ohio State University, and of Dr. J. Richard Huynen, whose many constructive comments are gratefully acknowledged. The interest shown by our contract sponsors, Dr. Richard G. Brandt, ONR, Pasadena Branch; Dr. Günter Winkler, NWC, China Lake, CA.; and by Dr. James W. Mink, ARO, Research Triangle Park, N.C., is recalled with pleasure. Thanks to their support we were able to obtain, with the collaboration of Dr. Jonathan Young and Prof. David L. Moffatt, the superb measurement data collected on the OSU-ESL, Columbus, OH. near-field/far-field measurement range under the capable supervision of Dr. Eric Walton, whose many comments contributed to our data interpretation.

**REFERENCES**

[1] J.R. Huynen, Towards a Theory of Perception for Radar Targets with Application to the Analysis of Their Data Base Structures & Presentations, Proc. NATO-Advanced Research Workshop on "Inverse Methods in Electromagnetic Imaging" (Bad Windsheim, FR Germany, Sept. 18-24, 1983), D. Reidel Publ. Co., Dordrecht, Holland, pp. 797-822.

[2] B.Y. Foo, S.K. Chaudhuri, W-M. Boerner, Polarization Correction to Kennaugh's Target Impulse Response Formulation and its Application to the Interpretation of Huynen's Mueller Matrix Parameters in Radar Polarimetry", Proc. IEEE/AP-S URSI International Symposium, 25-29 June 1984, Vol. 1, pp. 31-34.

[3] J.R. Huynen, Phenomenological Theory of Radar Targets, Ph.D. Dissertation, Technical University, Delft, The Netherlands, 1970.

[4] B.Y. Foo, S.K. Chaudhuri, W-M. Boerner, A High Frequency Inverse Scattering Model to Recover the Specular Point Curvature from Polarimetric Scattering Data, IEEE Trans. AP-32(11), November 1984, pp. 1174-1178.

[5] C.L. Bennett, et al., Space Time Integral Equation Approach to the Large Body Scattering Problems, Sperry Research Center, Sudbury, MA., Final Report on Contract No. F30602-71-C-0162, AD763794, May 1973.

[6] S.K. Chaudhuri, Utilization of Polarization-Depolarization Characteristics in Electromagnetic Inverse Scattering, Ph.D. Thesis, Univ. of Manitoba, Canada, 1977.

[7] B. Borden, Application of the Christoffel-Hurwitz Inversion Identity to Electromagnetic Imaging, Proc. NATO Advanced Research Workshop, 18-24 Sept. 1983, pp. 251-259 (also see IEEE Trans on Ant. & Prop., Vol. AP-32, June 1984, pp. 651-654.

[8] B.Y. Foo, A High Frequency Inverse Scattering Model to Recover the Specular Point Curvature from Polarimetric Scattering Data, M.Sc. Thesis, Univ. of Ill. at Chgo., May 1982.

[9] E.M. Kennaugh and R.L. Cosgriff, The Use of Impulse Response in Electromagnetic Scattering Problems, "IRE National Convention Record, Part I, 1958.

[10] E.M. Kennaugh, Research Studies on the Polarization Properties of Radar Targets, Collection of Selected Research Reports published by Prof. E.M. Kennaugh, ElectroScience Laboratory, Ohio State Univ., Columbus, OH., 43212, Commemorative Vols. I and II, July 1984.

[11] W-M. Boerner, et al., Editors, Proceedings, NATO Advanced Research Workshop on "Inverse Methods in Electromagnetic Imaging", (Bad Windsheim, F.R. Germany, Sept. 18-24, 1983), Vols. I & II, D. Reidel Publ. Co., Dordrecht, Holland, 1984 (see pp. 1-32, 33-42 and 609-628).

[12] J.R. Huynen, Polarization Discrimination with Applications to Target Classification and Identification, Final Report, Boeing Aerospace Corporation, Kent Space Research Center, Contract No. CH 6891, 1983.

[13] E.K. Walton and J.D. Young, The Ohio State University Compact Radar Cross-Section Measurement Range, Submitted to IEEE Trans. Antennas & Propagation, May 1984. (Also see Digest of the IEEE International Conference on Instrumentation and Measurements, Long Beach, CA., pp. 127-131, Jan. 1984.)

# THEORY AND MEASUREMENT OF SURFACE-TORSION

J. RICHARD HUYNEN

P. Q. Research
10531 Blandor Way
Los Altos Hills, CA 94022
U. S. A.

## ABSTRACT

A geometric theory of surface torsion is presented. A high frequency smooth surface patch is characterized by zeroth order scattering and a polarization dependent first-order correction term. For symmetrical surfaces, a well-known result due to Bennett produces a term proportional to ( $K_x - K_y$ ), where $K_x$ and $K_y$ are the principal curvatures at the specular point on any convex surface patch. Such general surface patch may be locally represented at the specular point by an elliptical parabolololoid. The surface is cut into elliptical slices along the normal direction. Each slice is given a rotation $\alpha(u) = - \alpha_0 \ln u$, where $z = u^2 \leq 1$ measures a small distance along the negative normal from the point of specular return. It is shown that only this rotation produces surface torsion. The torsion parameter $\alpha_0$ can be measured by the full polarimetric radar backscattered return. A sample measurement is numerically evaluated. The technique may be useful for target classification and identification of single man-made objects with intersecting curved surfaces (see example in this paper), as well as for terrain and clutter targets. The type of phenomenon discussed here is well-known to natural biologists who have coined the term "phyllotaxis" to the processes by which leaves and flower buds in plants unfold. The study of phyllotaxis has a rich history which dates back two thousand years. The interesting fact is that the logarithmic spiral obtained by geometry above, also appears prominently in phyllotaxis, based on empirically derived biological studies [ 38 - 44 ]. The author hopes that the present study may be useful not only to man-made target identification, by also may be useful to natural scientists, who hereby obtain a new method by which to study phyllotaxis-related phenomena.



*W.-M. Boerner et al. (eds.), Direct and Inverse Methods in Radar Polarimetry, Part 1*, 581–623.

# I. INTRODUCTION

The present work is inspired by the EM theory for time-domain backscatter from a specular smooth metallic and convex surface patch, as was pioneered by Kennaugh [1,2,3,5] followed by the work of Bennett [8,9,11,12,15], Mieras [23,26,29] and others: Chaudhuri [13,14,16,20], Boerner [21,22], and Foo [24,33]. First we present the geometric theory of surface torsion. It is shown how at each specular point the geometry is given by the two principal curvatures $K_x$ and $K_y$, and also by a surface torsion parameter, $T_{os}$. In order to produce surface torsion, a very special type of rotation, $\alpha = -\alpha_o \ln u$, is required (where $z = u^2$ is distance along the negative normal from the tip). This type of rotation relates in biology to the process of unfolding of leaves and flower buds in plants which is called "phyllotaxis" by natural scientists.

Next the time-domain EM theory is developed along the lines pioneered by Bennett, Mieras, and others. First for the case without torsion, any locally convex surface patch is represented by an elliptical paraboloidal surface geometry. The high frequency radar backscattering is characterized by a zeroth order specular term and a polarization dependent first order correction term, proportional to $(K_x - K_y)$.

Next the surface is cut into elliptical slices along the negative normal direction $z$ at the specular point. Each slice is given a rotation $\alpha(u)$ along the normal and this produces surface torsion. It will be shown that the torsion parameter may be measured by full polarimetric backscatter return from the specular point. This links the polarimetric work of Kennaugh [1,2], Huynen [6,10], and others with the time-domain scattering results of Bennett [9] and Mieras [29].

Most current theories of target scattering are designed to deal with purely symmetrical targets. Surface torsion is a primary example of a non-symmetrically shaped object. Another example is a helix. However, this study shows that helicity is **not** related to surface torsion. Helicity is produced by a line element **twist**, whereas $T_{os}$ results from a local **surface deformation** phenomenon. As we approach the tip, $z \to 0$ and $u \to 0$, we notice that the rotation angle $\alpha(u)$ approaches infinity. All that means physically is that the surface transforms into a circular cap at the tip.

A general non-symmetrical target is called an N-target. Most targets have a combination of symmetrical as well as non-symmetrical scattering behavior. A host of target parameters derived from the Stokes matrix and easily obtained from the full measured target scattering-matrix [32] are given carefully chosen designations: $A_o$, $B_o$, B, C, D, E, F, G, and H. These have direct physical relevance in terms of real target symmetry and non-symmetry properties and the coupling between these [17]. Other useful radar target scattering presentations are by the 3×3 **target**-coherency matrix [32], which is not to be confused with the 2×2 **wave**-coherency matrix. The novel techniques thus presented provides a powerful tool for real-world target classification and identification of manmade targets, for terrain, sea state, and other clutter environments, as well as for biological studies in phyllotaxis.

## II. THE GEOMETRY OF SURFACE TORSION

It is a rather curious fact that surface torsion is not discussed in any great detail, in current texts on classical differential geometry [4] . One finds torsion of curves and of geometry in higher dimensions, and in the theory of materials, but practically none of surfaces. The consensus seems to be that the mixed term $\hat{n} \cdot \bar{r}_{uv}$, which produces torsion, can be made to disappear by proper choice of coordinate system $(u,v)$ on the surface. This however is not always the case, as we will show below. There is a **special case**, i.e., the logarithmic law of surface torsion which is of geometric interest as well as in EM theory.

We begin by describing a general smooth surface patch near the specular point $P$ by an elliptical paraboidal surface coordinate net $(u,v)$. The system $(\hat{n}, \hat{x}, \hat{y})$ forms a righthanded orthogonal set of coordinates at point P. Notice that the $\hat{z}$-axis is chosen along the negative $\hat{n}$ direction. Generally we are interested only in the specular region of the surface patch, where $u$ is small. Occasionally a normalized condition with $u = 1$ also is of interest. For $u =$ constant an elliptical slice on the surface is given by:

$$\frac{x^2}{a(u)^2} + \frac{y^2}{b(u)^2} = 1 \tag{2.1}$$

For the elliptical paraboloid $a(u) = a_o u$ and $b(u) = b_o u$ and the equation for the surface is given by:

$$\frac{x^2}{a_o^2} + \frac{y^2}{b_o^2} = z = u^2 \tag{2.2}$$

It is clear that in the $(x,z)$ and $(y,z)$ planes we have parabolas with focus $F = 2R$, where $K = \frac{1}{R}$ is the surface curvature at $P$ in that plane and $R$ is the radius of curvature. Hence

$$K_x = \frac{2}{a_o^2} \quad \text{and} \quad K_y = \frac{2}{b_o^2} \tag{2.3}$$

are the principal curvatures at the point P.

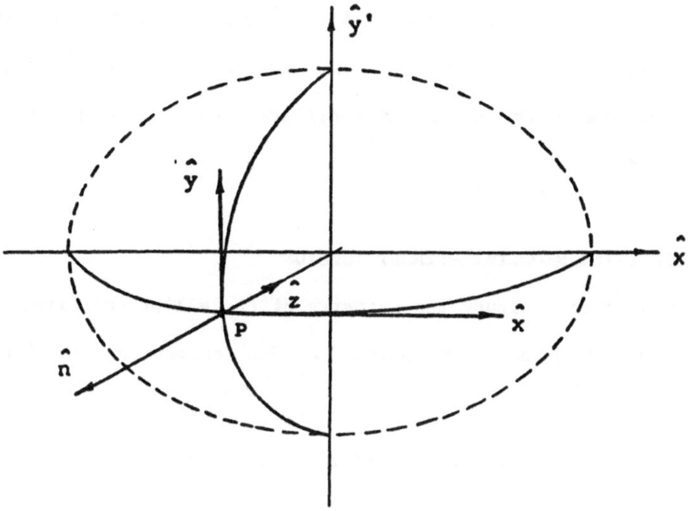


Fig. 1. Surface patch around specular point P.

The elliptical slices, given by (2.1) collapse into the single point of observation P, as $u \to 0$. This causes a singularity in the coordinate system $(u,v)$ which describes the net on the surface, but no singularity appears on the surface itself.

Another interesting feature is the introduction of a **gauge-function**, to describe the points on the elliptical slice. A gauge-function is a phase function $\beta(u,v)$, which is often used in quantum mechanics. For any point $\bar{r}_o$ on the patch, we have for the surface **without torsion**:

$$\bar{r}_o = \begin{bmatrix} x(u,v) \\ y(u,v) \\ z(u,v) \end{bmatrix} = \begin{bmatrix} a(u) \cos \beta (u,v) \\ b(u) \sin \beta (u,v) \\ z(u) \end{bmatrix} \tag{2.4}$$

We notice that $\beta(u,v)$ enters in (2.4) as a phase-function, but it does not appear in (2.1) or (2.2). Hence $\beta$ does not effect the **shape** of the surface but it does influence the choice of coordinate net $(u,v)$, by which the surface is covered.

## III. SURFACE PATCH GEOMETRY WITHOUT TORSION

In this section we study the geometry of the elliptical slices which make up the surface patch near the point P. For convenience we introduce the vector $\bar{a}$ :

$$\bar{r} = (x,y,z) = (\bar{a},z) \; ; \; \bar{a} = (x,y)$$

Without torsion the vector $\bar{a}$ is called $\bar{a}_0$ :

$$\bar{a}_0 = \begin{bmatrix} a(u) \cos \beta\,(u,v) \\ b(u) \sin \beta\,(u,v) \end{bmatrix} = \begin{bmatrix} a_0 \cos \beta \\ b_0 \sin \beta \end{bmatrix} u =$$

$$= \rho_0 \begin{bmatrix} \cos \phi \\ \sin \phi \end{bmatrix} u = \rho_0\, u\, e^{J\phi} \begin{bmatrix} 1 \\ 0 \end{bmatrix} \qquad (3.1)$$

Here $\rho_0\, u$ is simply the distance of the point on the ellipse from its center. Notice that $u$ acts as a **scale** for the figure. We adopted the matrix rotation operator:

$$e^{J\phi} = \begin{bmatrix} \cos \phi & -\sin \phi \\ \sin \phi & \cos \phi \end{bmatrix} = \cos \phi\, I + \sin \phi\, J \qquad (3.2)$$

where

$$I = \begin{bmatrix} 1 & 0 \\ 0 & 1 \end{bmatrix} \quad \text{and} \quad J = \begin{bmatrix} 0 & -1 \\ 1 & 0 \end{bmatrix} \qquad (3.3)$$

Similarly it is convenient to define the "orthogonal point" $\bar{a}_{o\perp}$ on the ellipse, obtained by increasing $\beta$ to $\beta + \pi/2$ ; as indicated in Figure 2:

$$\bar{a}_{o\perp} = \begin{bmatrix} -a_o \sin\beta \\ b_o \cos\beta \end{bmatrix} u = \rho_{o\perp} u\, e^{J\,\phi_\perp} \begin{bmatrix} 0 \\ 1 \end{bmatrix} \tag{3.4}$$

From (3.1) and (3.4) we obtain the identities:

$$\left.\begin{aligned} a_o \cos\beta &= \rho_o \cos\phi \\ b_o \sin\beta &= \rho_o \sin\phi \end{aligned}\right\} \qquad \left.\begin{aligned} a_o \sin\beta &= \rho_{o\perp} \sin\phi_\perp \\ b_o \cos\beta &= \rho_{o\perp} \cos\phi_\perp \end{aligned}\right\} \tag{3.5}$$

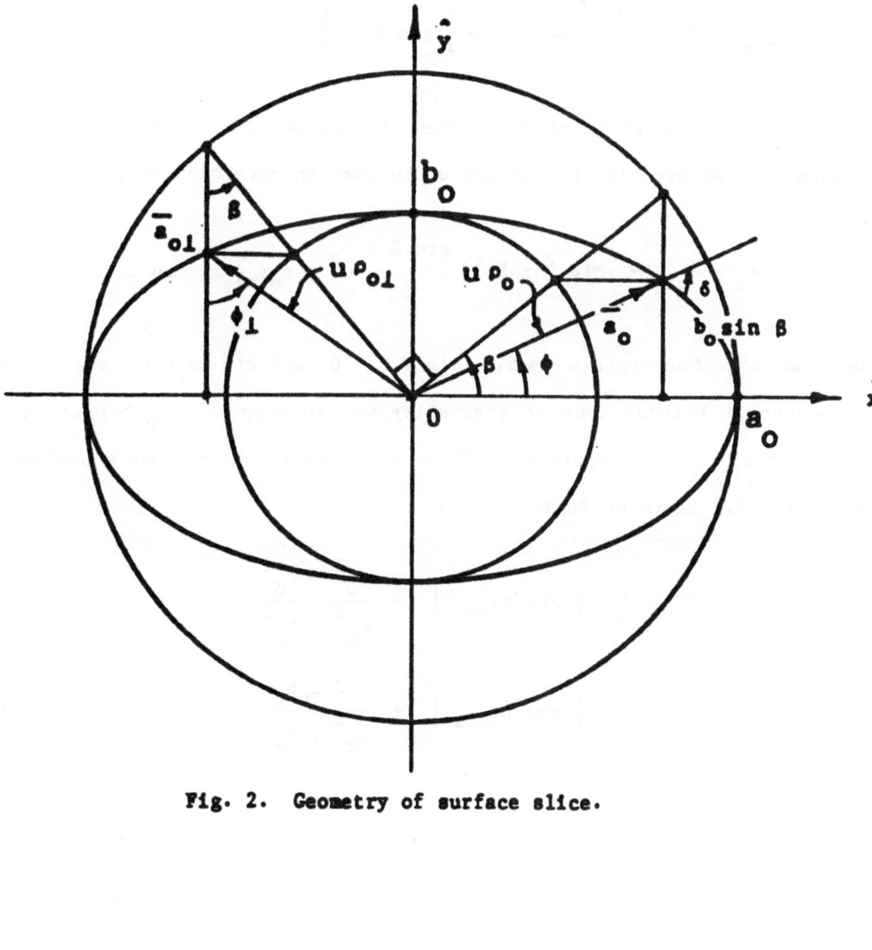


Fig. 2. Geometry of surface slice.

These relationships define a host of other useful identities:

$$\left.\begin{aligned} \rho_o^2 &= a_o^2 \cos^2 \beta + b_o^2 \sin^2 \beta \\ \rho_{o\perp}^2 &= a_o^2 \sin^2 \beta + b_o^2 \cos^2 \beta \end{aligned}\right\} \tag{3.6}$$

$$a_o b_o \sin 2\beta = \rho_o^2 \sin 2\phi = \rho_{o\perp}^2 \sin 2\phi_\perp \tag{3.7}$$

and

$$\left.\begin{aligned} a_o b_o &= \rho_o \rho_{o\perp} \cos \Delta\phi \\ -(a_o^2 - b_o^2) \frac{\sin 2\beta}{2} &= \rho_o \rho_{o\perp} \sin \Delta\phi \end{aligned}\right\} , \quad \Delta\phi = \phi - \phi_\perp \tag{3.8}$$

We can look upon (3.6) as normalized distances, for $u = 1$. Notice that "**orthogonal points**" in $\beta$ are not orthogonal in direction on the ellipse:

$$\bar{a}_o \cdot \bar{a}_{o\perp} = -(a_o^2 - b_o^2) \frac{\sin 2\beta}{2} u^2 = \rho_o \rho_{o\perp} u^2 \sin \Delta\phi \tag{3.9}$$

Only at the four points where $\sin 2\beta = 0$ and $\sin \Delta\phi = 0$ are directions orthogonal. Without loss of generality we can suppose: $a_o \geq b_o$. The angle $\Delta\phi = \phi - \phi_\perp$ has an important role in the theory. It reaches a maximum value at $\beta = 45°$. There we find:

$$\left| \sin \Delta\phi_{max} \right| = \frac{a_o^2 - b_o^2}{a_o^2 + b_o^2} \tag{3.10}$$

$$\left| \cos \Delta\phi_{max} \right| = \frac{2 a_o b_o}{a_o^2 + b_o^2} \tag{3.11}$$

For the computation of curvature the principal directions $\sin 2\beta = 0$ are of most interest and hence we find the important rule from (3.8): $a_o b_o = \rho_o \rho_{o\perp}$. The angle $\Delta\phi$ has the further significance as being complementary to the angle $\delta$ between radius and tangent of the ellipse:

$$\tan \delta = \cot \Delta\phi \tag{3.12}$$

Next we compute derivatives. From (3.5) we find:

$$\tan \phi = \frac{b_o}{a_o} \tan \beta \tag{3.13}$$

and hence:

$$\frac{d\beta}{d\phi} = \frac{a_o \cos^2 \beta}{b_o \cos^2 \phi} = \frac{a_o b_o}{\rho_{o\perp}^2 \cos^2 \Delta\phi} \tag{3.14}$$

The last step followed from (3.8). We also compute $d\rho_o/d\phi$ from (3.6), (3.8), and (3.14):

$$\frac{d\rho_o}{d\phi} = -\frac{\left(a_o^2 - b_o^2\right)}{2\,\rho_o} \sin 2\beta \frac{d\beta}{d\phi} = \frac{a_o b_o \sin \Delta\phi}{\rho_{o\perp} \cos^2 \Delta\phi} \tag{3.15}$$

Next we compute the normalized tangent vectors on the surface to the coordinate lines $u = \text{const}$, and $v = \text{const}$.

$$\hat{a}_{ou} = \begin{bmatrix} a_o \cos \beta - a_o u\, \beta_u \sin \beta \\ b_o \sin \beta + b_o u\, \beta_u \cos \beta \end{bmatrix} \frac{1}{\rho_o} \tag{3.16}$$

$$\hat{a}_{ov} = \begin{bmatrix} -a_o \sin \beta \\ b_o \cos \beta \end{bmatrix} \frac{1}{\rho_{o\perp}} \tag{3.17}$$

Equation (3.16) is normalized strictly only as $u \to 0$; which is the condition we are most interested in.

From (3.16) and (3.17) we find:

$$\hat{a}_{ou} \cdot \hat{a}_{ov} = \left[ -(a_o^2 - b_o^2) \frac{\sin 2\beta}{2} + \rho_{o\perp}^2 u \beta_u \right] \frac{1}{\rho_o \rho_{o\perp}} =$$

$$= \sin \Delta\phi + \frac{\rho_{o\perp}}{\rho_o} u \beta_u \tag{3.18}$$

where we used (3.8). Hence, as $u \to 0$, the only points, where the $(u,v)$ net is orthogonal, are indeed the four regions where $\sin \Delta\phi = 0$.

## IV. COORDINATES AND GAUGE CONDITIONS WITH TORSION

Torsion implies a specific rotation about the normal direction of the elliptical slices where $u$ = constant, as function of $u$. Hence for a point on the surface we find in $(\hat{x}, \hat{y}, \hat{z})$ coordinates:

$$\bar{r} = \begin{bmatrix} e^{J\alpha} \begin{bmatrix} a(u) \cos \beta(u,v) \\ b(u) \sin \beta(u,v) \end{bmatrix} \\ z(u) \end{bmatrix} = \begin{bmatrix} x \\ y \\ z \end{bmatrix} \tag{4.1}$$

For the elliptical paraboloid $a(u) = a_o u$ and $b(u) = b_o u$, $z(u) = u^2$ as before. The new feature is the rotation $\alpha(u)$ about the z-axis of the elliptical slice. The operator $e^{J\alpha}$ was given in (3.2) and (3.3). We have to use frequently the rule of differentiation:

$$\frac{d}{du} e^{J\alpha(u)} \begin{bmatrix} x(u) \\ y(u) \end{bmatrix} = e^{J\alpha} \begin{bmatrix} \dot{x} - \dot{\alpha} y \\ \dot{y} + \dot{\alpha} x \end{bmatrix} \tag{4.2}$$

With scalar multiplication the rotation operators cancel:

$$e^{J\alpha}\begin{bmatrix} x_1 \\ y_1 \end{bmatrix} \cdot e^{J\alpha}\begin{bmatrix} x_2 \\ y_2 \end{bmatrix} = x_1x_2 + y_1y_2 \tag{4.3}$$

As before, we define the $\bar{a}$ vector by the x- and y- components of $\bar{r}$ :

$$\bar{r} = (\ x,y,z\ ) = (\ \bar{a},z\ )\ \ ;\ \ \bar{a} = (\ x,y\ ) \tag{4.4}$$

The introduction of $\bar{a}$ is strictly for convenience and brevity, as $u \to 0$ the z-component of $\bar{r}$, $\bar{r}_u$, and $\bar{r}_v$ vanishes. However in order to be precise we start off with $\bar{r}$ in 3-dimensions. As in (3.1) we introduce the angle $\phi$:

$$\bar{r} = \begin{bmatrix} e^{J\alpha}\ \rho_o u\ e^{J\phi}\begin{bmatrix} 1 \\ 0 \end{bmatrix} \\ z(u) \end{bmatrix} = \begin{bmatrix} \rho_o u\ \ e^{J(\alpha+\phi)}\begin{bmatrix} 1 \\ 0 \end{bmatrix} \\ z(u) \end{bmatrix} \tag{4.5}$$

The notational benefits thus become apparent. We will need both (4.1) and (4.5) to calculate tangent vectors as $u \to 0$. Next we compute derivatives using Eq. (4.2)

$$\bar{r}_u = \begin{bmatrix} e^{J\alpha}\begin{bmatrix} a_o \cos\beta - (\ a_o\gamma\ u + b_o\ \dot{\alpha}\ u\ ) \sin\beta \\ b_o \sin\beta + (\ b_o\gamma\ u + a_o\ \dot{\alpha}\ u\ ) \cos\beta \end{bmatrix} \\ 2\ u \end{bmatrix} \tag{4.6}$$

$$\bar{r}_v = \begin{bmatrix} e^{J\alpha}\begin{bmatrix} -\ a_o \sin\beta \\ b_o \cos\beta \end{bmatrix} c(u,v)\ u \\ 0 \end{bmatrix} \tag{4.7}$$

Scalar multiplication is simple with this notation, the rotation operators cancel. In (4.6) and (4.7) we defined $\gamma(u,v) = \beta_u$ and $c(u,v) = \beta_v$. We will find that $c(u,v)$ acts as a multiplicative constant. For instance if $\beta = v$, $c = \beta_v = 1$, but for other gauge functions $c(u,v)$ is not constant, however, it seems to act as one!

Next we will find under which conditions $\bar{r}_n \cdot \bar{r}_v = 0$. This defines the orthogonal points on the surface net which resembles the net before torsion was applied.

$$\bar{r}_u \cdot \bar{r}_v = \left[ -(a_o^2 - b_o^2) \frac{\sin 2\beta}{2} + \rho_{o\perp}^2 u \gamma + \dot{\alpha} u a_o b_o \right] c u \qquad (4.8)$$

Hence $\bar{r}_u \cdot \bar{r}_v = 0$ if $\sin 2\beta = 0$ and

$$\gamma = \beta_u = -\dot{\alpha} \frac{a_o b_o}{\rho_{o\perp}^2} \qquad (4.9)$$

The last condition expresses the amount of adjustment the surface net $(u,v)$ requires after torsion, to become normal again.

Comparison with (3.14) gives the condition on $\phi$::

$$\phi_u = -\dot{\alpha} \cos^2 \Delta\phi \qquad (4.10)$$

hence:

$$\phi + \alpha = v \to 0 \quad \text{as} \quad \begin{matrix} u \to 0 \\ \Delta\phi \to 0 \end{matrix} \qquad (4.11)$$

We now investigate the implications of (4.11). From (4.5) we find:

$$\bar{a}_u = e^{J(\alpha + \phi)} \begin{bmatrix} \rho_o + u\,\rho_{ou} \\ (\dot{\alpha} + \phi_u)\rho_o u \end{bmatrix} \qquad (4.12)$$

and

$$\bar{a}_v = e^{J(\alpha + \phi)} \begin{bmatrix} \rho_{ov} \\ \rho_o\,\phi_v \end{bmatrix} u \qquad (4.13)$$

By using condition (4.11) we find the important property at $P$ :

$$\hat{a}_u \rightarrow \hat{x} \quad \text{and} \quad \hat{a}_v \rightarrow \hat{y} \quad \text{as} \quad u \rightarrow 0 \quad \text{and} \quad \Delta\phi \rightarrow 0 \tag{4.14}$$

Hence the net effect of the gauge condition is to assure that the surface net near the specular point remains the same as before the torsion was applied.

## V. SURFACE CURVATURE AND TORSION

It was indicated in Section IV that the surface torsion is introduced as a rotation about the line of observation, along the $\hat{n}$ direction of point P.

We now proceed to compute the curvatures and torsion parameter at the point P as $u \rightarrow 0$. The normal is defined as $\bar{n} = \bar{r}_u \times \bar{r}_v$

$$\bar{n} = \begin{bmatrix} -r_{u3}\, r_{v2} \\ r_{u3}\, r_{v1} \\ r_{u1}\, r_{v2} - r_{u2}\, r_{v1} \end{bmatrix} = \begin{bmatrix} e^{J\alpha} \begin{bmatrix} -b_o \cos\beta \\ -a_o \sin\beta \end{bmatrix} 2\, c\, u^2 \\ n_z \end{bmatrix} \tag{5.1}$$

$$n_z = \left[ a_o\, b_o + \dot{\alpha}\, u\, (a_o^2 - b_o^2)\, \frac{\sin 2\beta}{2} \right] c\, u \tag{5.2}$$

We notice that the gauge $\gamma(u,v)$ does not enter $\bar{n}$, also the $n_z$ component dominates as $u \rightarrow 0$, hence $|\bar{n}| = n_z$.

It is easy to check that $\bar{n} \cdot \bar{r}_u = \bar{n} \cdot \bar{r}_v = 0$. Next:

$$\bar{r}_{uv} = \begin{bmatrix} e^{J\alpha} \begin{bmatrix} -a_o \sin\beta - a_o\, u\, \gamma \cos\beta - \dot{\alpha}\, u\, b_o \cos\beta - a_o\, u\, \gamma_{v/c} \sin\beta \\ b_o \cos\beta - b_o\, u\, \gamma \sin\beta - \dot{\alpha}\, u\, a_o \sin\beta + b_o\, u\, \gamma_{u/c} \cos\beta \end{bmatrix} c \\ 0 \end{bmatrix} \tag{5.3}$$

$$\bar{n} \cdot \bar{r}_{uv} = \left[ a_o\, b_o\, u\, \gamma + \dot{\alpha}\, u\, \rho_{o\perp}^2 \right] 2\, c^2 u^2 \tag{5.4}$$

Substitution of (4.9) into (5.4) gives:

$$\bar{n} \cdot \bar{r}_{uv} = - 2 \dot{\alpha} u \left(a_o^2 - b_o^2\right) \cos 2 \beta c^2 u^2 \tag{5.5}$$

In order to complete the work, we need the additional surface parameters $|\bar{r}_u|$, $|\bar{r}_v|$, $\bar{r}_{uu}$, $\bar{r}_{vv}$, and $\bar{n} \cdot \bar{r}_{uu}$, $\bar{n} \cdot \bar{r}_{vv}$. We find by straightforward scalar multiplication of (4.6),

$$|\bar{r}_u| = \rho_o + \dot{\alpha} u \frac{(a_o^2 - b_o^2)}{\rho_{o\perp}} \frac{\sin 2 \beta}{2} \tag{5.6}$$

and

$$|\bar{r}_v| = \rho_{o\perp} c u \tag{5.7}$$

$$\bar{r}_{vv} = \begin{bmatrix} e^{J\alpha} \begin{bmatrix} -a_o \cos \beta - a_o c_{v/u}^2 \sin \beta \\ -b_o \sin \beta + b_o c_{v/u}^2 \cos \beta \end{bmatrix} \\ 0 \end{bmatrix} c^2 u \tag{5.8}$$

Hence:

$$\bar{n} \cdot \bar{r}_{vv} = 2 a_o b_o c^3 u^3 \tag{5.9}$$

We notice $c_v = \beta_{vv}$ disappears.

By straightforward differentiation of (4.6) we find:

$$\bar{r}_{uu} = \begin{bmatrix} e^{J\alpha} \begin{bmatrix} -2a_o\gamma \sin \beta - (a_o\gamma_u u + b_o\ddot{\alpha} u) \sin \beta - (a_o\gamma u + b_o\dot{\alpha} u)\gamma \cos \beta \\ - \dot{\alpha}[2b_o \sin \beta + (b_o\gamma u + a_o\dot{\alpha} u) \cos \beta] \\ 2b_o\gamma \cos \beta + (b_o\gamma_u u + a_o\ddot{\alpha} u) \cos \beta - (b_o\gamma u - a_o\dot{\alpha} u)\gamma \sin \beta \\ + \dot{\alpha}[2a_o \cos \beta - (a_o\gamma u + b_o\dot{\alpha} u) \sin \beta] \end{bmatrix} \\ 2 \end{bmatrix} \tag{5.10}$$

$$\bar{n} \cdot \bar{r}_{uu} = \Bigg[ - (a_o^2 - b_o^2) \frac{\sin 2\beta}{2} \ddot{\alpha} u^2 + a_o b_o \gamma^2 u^2 + 2 \rho_{o\perp}^2 \dot{\alpha} u \gamma u +$$

$$- (a_o^2 - b_o^2) \frac{\sin 2\beta}{2} \dot{\alpha} u + a_o b_o \dot{\alpha}^2 u^2 + a_o b_o \Bigg] 2 c u =$$

$$= 2 c u \Bigg[ a_o b_o - (a_o^2 - b_o^2) \frac{\sin 2\beta}{2} (\dot{\alpha} u + \ddot{\alpha} u^2) +$$

$$+ a_o b_o \left( \dot{\alpha}^2 u^2 + \gamma^2 u^2 + \frac{2\rho_{o\perp}^2}{a_o b_o} \dot{\alpha} u \gamma u \right) \Bigg] \tag{5.11}$$

Further evaluation is done after $\alpha(u)$ is determined. For the Torsion coefficient is found:

$$T_s = \frac{\bar{n} \cdot \bar{r}_{uv}}{|\bar{n}| \, |\bar{r}_u| \, |\bar{r}_v|} = \frac{2 \dot{\alpha} u}{a_o b_o} \left( \frac{\rho_{o\perp}}{\rho_o} - \frac{\rho_o}{\rho_{o\perp}} \right) \tag{5.12}$$

Let $\dot{\alpha} u = -\alpha_o$, then

$$\boxed{\alpha(u) = -\alpha_o \ \ell n \ u} \tag{5.13}$$

Then

$$T_s = 2 \alpha_o \left( \frac{a_o^2 - b_o^2}{a_o^2 b_o^2} \right) \cos 2 \beta \tag{5.14}$$

Since $u \leq 1$ a positive torsion rotation rate $\alpha_o$ gives positive angles $\alpha(u)$. Notice the important fact that the logarithmic law (5.13) produces a **constant** torsion coefficient in the neighborhood of point P. A discussion of the angle $\beta$ dependence follows below. First the local curvatures are calculated:

$$K_v = \frac{\bar{n} \cdot \bar{r}_{vv}}{|\bar{n}|\,|\bar{r}_v|^2} = \frac{2}{\rho_{o\perp}^2} \tag{5.15}$$

and:

$$K_u = \frac{\bar{n} \cdot \bar{r}_{uu}}{|\bar{n}|\,|\bar{r}_u|^2} = \frac{2}{\rho_o^2}\left[1 - \alpha_o^2\left(1 - \frac{a_o^2 b_o^2}{\rho_{o\perp}^4}\right)\right]$$

After using $\rho_o\ \rho_{o\perp} = a_o b_o$ we obtain:

$$K_u = \frac{2}{\rho_o^2} + 2\,\alpha_o^2\,\frac{(a_o^2 - b_o^2)}{a_o^2\ b_o^2}\ \cos 2\,\beta \tag{5.16}$$

From this we find the interesting property:

$$\frac{K_u - K_v}{2} = -\,(1 - \alpha_o^2)\left(\frac{a_o^2 - b_o^2}{a_o^2\quad b_o^2}\right)\cos 2\,\beta \tag{5.17}$$

and

$$T_s = \frac{-2\,\alpha_o}{1 - \alpha_o^2}\left(\frac{K_u - K_v}{2}\right) \tag{5.18}$$

The last equation links the surface torsion coefficient with the local curvatures. There can be no doubt about the existence of this geometric property, although most standard textbooks have ignored it. The logarithmic spiral (5.13) is unique to the surface torsion and it plays an important role in phyllotaxis [38]. We notice from (5.17) and (5.18) that the torsion coefficient as well as local curvature depend on the angle $\beta$. For $\sin 2\beta = 0$ the local curvatures become the principal curvatures and we can adopt, via (5.18) a similar definition for surface torsion.

Hence define: $$K_y = \frac{2}{b_o^2} \tag{5.19}$$

$$\frac{K_x - K_y}{2} = -(1 - \alpha_o^2)\left(\frac{a_o^2 - b_o^2}{a_o^2\, b_o^2}\right) \tag{5.20}$$

and $$T_{os} = 2\,\alpha_o\left(\frac{a_o^2 - b_o^2}{a_o^2\, b_o^2}\right) \tag{5.21}$$

This will be our definition of **principal** surface torsion. The factor $\cos 2\beta$ in (5.14) merely reflects the rotation of coordinate system as we move with angle $\beta$ around the specular point P.

We notice that as we approach the tip, as $u \to 0$ the angular rotation $-\alpha(u) \to \infty$, which means that the rotation close to the specular point becomes very large. At the tip a circular cap is formed which contributes to the so-called zeroth order scattering. Away from point P, but still locally we will find contributions to first-order scattering, which show dependence on $K_x$, $K_y$, and $T_{os}$. These topics are discussed next.

## VI. ELECTROMAGNETICS OF SURFACE TORSION

We follow the EM theory for high-frequency backscatter from a specular smooth metallic surface patch as developed by Bennett [9], Mieras [29] and summarized by Foo [24]. The far-field magnetic field is expressed by:

$$r_o\bar{H}^s(t_f) = \frac{1}{4\pi}\int \frac{\partial}{c\,\partial t}\left(\bar{J}(\bar{r},t) \times \hat{r}_o\right) dS \tag{6.1}$$

$$\bar{J}(\bar{r},t) = 2\,\hat{n} \times \bar{H}^i(r,t) +$$

$$+ \frac{\hat{n}}{2\pi} \times \int_S \left( \frac{1}{R^2} + \frac{\partial}{cR\,\partial t} \right) (\bar{J}(\bar{r}',\tau) \times \hat{a}_R)\, dS' \qquad (6.2)$$

where $t = t_f - z/c$, and

$$\tau = t - R/c\ , \quad \bar{R} = \bar{r} - \bar{r}' = R\,\hat{a}_R \qquad (6.3)$$

The far-field time frame $t_f$ is chosen such that an impulse, arriving at $t = 0$ at P, will arrive at the far-field scattering point at $t_f$. The incident field is taken as a ramp:

$$\bar{H}^i\,(\bar{r},t) = (t - z/c)\,u(t - z/c)\,\hat{H}^i \qquad (6.4)$$

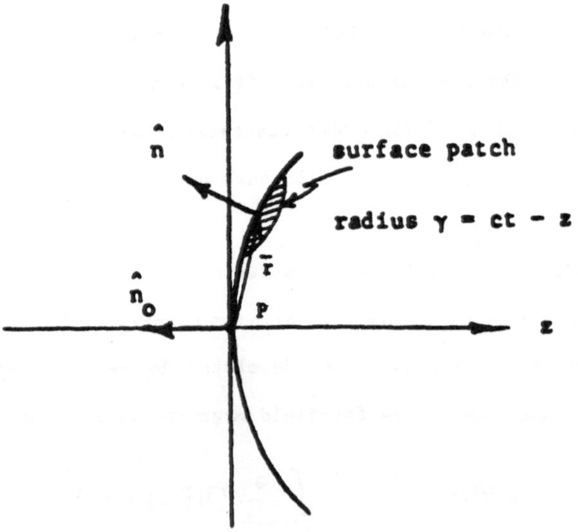


Fig. 3. Local geometry of specular point.

Early in time, the integration in (6.2) is limited to a patch approximating a circle with radius $\gamma = ct - z$. The currents on the surface are of two types:

$$\bar{J}(\bar{r},t) = \bar{J}_o(\bar{r},t) + \bar{J}_1(\bar{r},t) \tag{6.5}$$

and

$$\bar{J}_o(\bar{r},t) = 2\,\hat{n} \times \bar{H}^{\,i}(\bar{r},t) \tag{6.6}$$

$$\bar{J}_1(\bar{r},t) = \frac{1}{2\pi} \int \hat{n} \times \left( \frac{\bar{J}}{R^3} \times \bar{R} \right) dS' \tag{6.7}$$

It was assumed that the currents are locally constant.

Let:
$$\bar{J} = \bar{J}_u\,\hat{a}_u + \bar{J}_v\,\hat{a}_v \tag{6.8}$$

in surface coordinates.

Now we expand the distance between two points on the surface in a Taylor series, where $u$ and $v$ are small:

$$\bar{R} = \bar{r}' - \bar{r} = \bar{r}_{uo}u + \bar{r}_{vo}v + 1/2\,\bar{r}_{uuo}u^2 + 1/2\,\bar{r}_{vvo}v^2 + \bar{r}_{uvo}\,u\,v \tag{6.9}$$

also

$$\begin{aligned} \bar{r}_u &= \bar{r}_{uo} + \bar{r}_{uuo}\,u + \bar{r}_{uvo}\,v \\ \bar{r}_v &= \bar{r}_{vo} + \bar{r}_{uvo}\,u + \bar{r}_{vvo}\,v \end{aligned} \tag{6.10}$$

Hence

$$\begin{aligned} \hat{n} \cdot \bar{R} &= \tfrac{1}{2} L\,u^2 + \tfrac{1}{2} N\,v^2 + M\,u\,v \\ \hat{n} \cdot \hat{a}_u &= \frac{1}{|\bar{r}_u|}(L\,u + M\,v) \\ \hat{n} \cdot \hat{a}_v &= \frac{1}{|\bar{r}_v|}(M\,u + N\,v) \end{aligned} \tag{6.11}$$

where $L = \hat{n} \cdot \bar{r}_{uuo}$ ; $N = \hat{n} \cdot \bar{r}_{vvo}$; $M = \hat{n} \cdot \bar{r}_{uvo}$ (6.12)

Next we evaluate:

$$\hat{n} \times ( \hat{a}_u \times \bar{R} ) = ( \hat{n} \cdot \bar{R} ) \hat{a}_u - ( \hat{n} \cdot \hat{a}_u ) \bar{R} =$$

$$= ( \frac{1}{2} L u^2 + \frac{1}{2} N v^2 + M u v ) \hat{a}_u +$$

$$- ( L u + M v ) \frac{1}{| \bar{r}_u |} ( \hat{a}_u | \bar{r}_u | u + \hat{a}_v | \bar{r}_v | v ) =$$

$$= - \frac{1}{2} [ L u^2 - N v^2 ] \hat{a}_u - [ L u + M v ] \frac{| \bar{r}_v |}{| \bar{r}_u |} v \hat{a}_v \qquad (6.13)$$

Similarly we find:

$$\hat{n} \times ( \hat{a}_v \times \bar{R} ) = ( \hat{n} \cdot \bar{R} ) \hat{a}_v - ( \hat{n} \cdot \hat{a}_v ) \bar{R} =$$

$$= \frac{1}{2} [ L u^2 - N v^2 ] \hat{a}_v - [ M u + N v ] \frac{| \bar{r}_u |}{| \bar{r}_v |} u \hat{a}_u \qquad (6.14)$$

In order to evaluate (6.7), we have to select a patch, away from the specular point P, where locally the coordinate system $\hat{a}_u$ , $\hat{a}_v$ is **orthogonal**. Hence we select, following (4.8) the point $\bar{r}$, where $\beta = 0$ (or $\beta = 90°$). These are also the principal directions for the curvature (for point P ) on the surface. Having selected a representative surface patch (away from P ) we now insert (6.13) and (6.14) with (6.8) into (6.7)

If we put locally:

$$\left| \bar{r}_u \right| u = R \cos \theta$$
$$\left| \bar{r}_v \right| v = R \sin \theta \qquad (6.15)$$
$$ds = R \, dR \, d\theta$$

and

$$L = K_u \left| \bar{r}_u \right|^2 , \quad N = K_v \left| \bar{r}_v \right|^2 , \quad M = T_s \left| \bar{r}_u \right| \left| \bar{r}_v \right| \qquad (6.16)$$

Next perform the integration, the terms with $u^2$ and $v^2$ lead to a factor $\frac{\gamma}{2}$, where $\gamma$ is the radius of the patch defined above. The terms with u v integrate to zero. We find for the surface current:

$$\bar{J}_1(\bar{r},t) = J_{1u} \hat{a}_u + J_{1v} \hat{a}_v =$$
$$= -\frac{\gamma}{2} \left[ \left( \frac{K_u - K_v}{2} \right) \hat{a}_u + T_s \hat{a}_v \right] J_u +$$
$$+ \frac{\gamma}{2} \left[ \left( \frac{K_u - K_v}{2} \right) \hat{a}_v - T_s \hat{a}_u \right] J_v \qquad (6.17)$$

Next, we approximate the induced surface current as due to the incident field:

$$\bar{J} = 2 \hat{n} \times \bar{H}^i = 2 \hat{a}_v H_u^i - 2 \hat{a}_u H_v^i \qquad (6.18)$$

or

$$J_u = -2 H_v^i , \quad J_v = 2 H_u^i \qquad (6.19)$$

Hence:

$$\bar{J}_1 = \gamma \left[ \left( \frac{K_u - K_v}{2} \right) H_v^i - T_s H_u^i \right] \hat{a}_u +$$
$$+ \gamma \left[ \left( \frac{K_u - K_v}{2} \right) H_u^i + T_s H_v^i \right] \hat{a}_v \qquad (6.20)$$

This is the solution for surface current.

## VII. FIRST-ORDER SCATTERING FIELD

The first order scattering field is obtained by insertion of the surface current (6.20) into the far field equation (6.1).

This requires the formation of

$$J_1 \times \bar{n}_o = J_{1v}\,\hat{a}_u - J_{1u}\,\hat{a}_v =$$

$$= \gamma\left[\left(\frac{K_u - K_v}{2}\right) H_u^i + T_s\, H_v^i\right]\hat{a}_u +$$

$$+ \gamma\left[T_s\, H_u^i - \left(\frac{K_u - K_v}{2}\right) H_v^i\right]\hat{a}_v \qquad (7.1)$$

Notice that the effect due to surface torsion is represented in (7.1) by:

$$T_s = \frac{-2\,\alpha_o}{1 - \alpha_o^2}\left(\frac{K_u - K_v}{2}\right) \qquad (7.2)$$

If $K_x = 2/a_o^2$, $K_y = 2/b_o^2$, $a_o \geq b_o$, then $T_{os} \geq 0$ for the principal curvatures and torsion. But torsion also affects the direction of $\hat{a}_u$ and $\hat{a}_v$. From (4.12) and (4.13) we find for small $u$ :

$$\hat{a}_u = e^{J(\alpha + \phi)}\begin{bmatrix} 1 + u\,\rho_{ou}/\rho_o \\ (\dot{\alpha} + \phi_u)u \end{bmatrix} \qquad (7.3)$$

$$\hat{a}_v = e^{J(\alpha + \phi)}\begin{bmatrix} \rho_{ov}/\rho_o\phi_v \\ 1 \end{bmatrix} \qquad (7.4)$$

### A. CASE WITHOUT TORSION: $\alpha_o = 0$

First we investigate the case without torsion. The first order scattering field for monostatic backscatter in the direction $\hat{n}_o$ is given by:

$$r_o\,\overline{H}_1^s = \frac{1}{4\pi}\int \frac{\partial}{\partial t}\left(t-\frac{z}{c}\right)^2 u\left(t-\frac{z}{c}\right)\left(\frac{K_u-K_v}{2}\right)\left(H_{uo}^{\,i}\,\hat{a}_{uo} - H_{vo}^{\,i}\,\hat{a}_{vo}\right)(\hat{n}\cdot\hat{n}_o)\,dS$$

$$= \frac{1}{2\pi}\int \left(t-\frac{z}{c}\right) u\left(t-\frac{z}{c}\right)\left(\frac{K_u-K_v}{2}\right)\begin{bmatrix}1 & 0\\ 0 & -1\end{bmatrix}\begin{bmatrix}H_{uo}^{i}\\ H_{vo}^{i}\end{bmatrix}\cdot\begin{bmatrix}\hat{a}_{uo}\\ \hat{a}_{vo}\end{bmatrix}(\hat{n}\cdot\hat{n}_o)\,dS \tag{7.5}$$

This was the case for a ramp response as was discussed in Section VI. For $\alpha_o = 0$ the coordinate vectors reduce to; as $u \to 0$, Eq. (7.3) and (7.4):

$$\begin{bmatrix}\hat{a}_{uo}\\ \hat{a}_{vo}\end{bmatrix} = e^{-J\,\phi(u,v)}\begin{bmatrix}\hat{x}\\ \hat{y}\end{bmatrix} \tag{7.6}$$

We notice from (7.5) that due to matrix $\begin{bmatrix}1 & 0\\ 0 & -1\end{bmatrix}$ the effect due to $\phi$ does not cancel! A change to the $(\hat{x}, \hat{y})$ system results in:

$$r_o\,\overline{H}_1^s = \frac{1}{2\pi}\int \left(t-\frac{z}{c}\right) u\left(t-\frac{z}{c}\right)\frac{(K_x-K_y)}{2}\cos 2\beta\begin{bmatrix}1 & 0\\ 0 & -1\end{bmatrix}e^{-2\,J\,\phi}\begin{bmatrix}H_x^{i}\\ H_y^{i}\end{bmatrix}\cdot\begin{bmatrix}\hat{x}\\ \hat{y}\end{bmatrix}dS' \tag{7.7}$$

The impulse response is found by twice differentiating (7.7) which gives:

$$r_o\,\overline{H}_1^s = \frac{1}{2\pi}\int \delta\left(t-\frac{z}{c}\right)\frac{(K_x-K_y)}{2}\cos 2\beta\begin{bmatrix}1 & 0\\ 0 & -1\end{bmatrix}e^{-2\,J\,\phi}\begin{bmatrix}H_x^{i}\\ H_y^{i}\end{bmatrix}\cdot\begin{bmatrix}\hat{x}\\ \hat{y}\end{bmatrix}dS' \tag{7.8}$$

(If understood, the dot product reference to the coordinate system on the right in (7.8) may be omitted).

We now proceed to evaluate the integral in (7.8). This is done in more detail in Appendix A. We notice a dependency on $\cos 2\beta$ mutiplied with either $\cos 2\phi$ or $\sin 2\phi$ in the rotation matrix. We are interested in average values which contribute to the impulse response. In (7.7) and (7.8) $dS'$ is the projected surface element in $\bar{n}_0$ direction. Hence the average value over $\cos 2\beta \sin 2\phi$ is easily shown to vanish. This reduces the rotation matrix to a unit matrix and we have to evaluate:

$$\frac{1}{2\pi}\int_0^{2\pi} \cos 2\beta \cos 2\phi \, d\beta$$

(see Appendix A). Since there exists a relationship (3.13) between $\phi$ and $\beta$ for an ellipse, the integral can be evaluated exactly. To a good approximation the angles $\phi$ and $\beta$ can be considered equal for this case, which leads to a factor $\frac{1}{2}$ for the integral. Hence close to point P the fields are constant and the integral becomes:

$$r_0 \, \bar{H}_1^s = \frac{1}{4\pi}\left(\frac{d\,A(t)}{d\,t_f}\right)\begin{bmatrix} \frac{K_x - K_y}{2} & 0 \\ 0 & -\frac{K_x - K_y}{2} \end{bmatrix}\begin{bmatrix} H_x^i \\ H_y^i \end{bmatrix} \tag{7.9}$$

This is the form for the first-order correction due to Bennett [9] and Mieras [29] for the case without torsion. It shows the well-known dependence on differential projected area or silhouette area of the surface.

## B. CASE WITH TORSION $\alpha_0 \neq 0$

The previous case without torsion, is now generalized to the case with torsion. It will be found that the main contribution comes from the specular region around P. We now have for the surface current Eq. (6.20) and instead of (7.5) we obtain for the impulse response:

$$r_0 \, \bar{H}_1^s = \frac{1}{2\pi}\int_S \delta\left(t - \frac{z}{c}\right)\begin{bmatrix} \frac{K_u - K_v}{2} & T_s \\ T_s & \frac{-K_u + K_v}{2} \end{bmatrix}\begin{bmatrix} H_v^i \\ H_u^i \end{bmatrix} \cdot \begin{bmatrix} \hat{a}_u \\ \hat{a}_v \end{bmatrix} (\hat{n}\cdot\hat{n}_0)\, dS \tag{7.10}$$

where $\hat{a}_u$ and $\hat{a}_v$ are given by

$$\begin{bmatrix} \hat{a}_u \\ \hat{a}_v \end{bmatrix} = e^{-J\,\alpha(u)}\; e^{-J\,\phi} \begin{bmatrix} \hat{x} \\ \hat{y} \end{bmatrix} \tag{7.11}$$

Substitution into (7.10) gives:

$$r_o\,\bar{H}_1^s = \frac{1}{2\pi}\int_S \delta\left(t - \frac{z}{c}\right)\cos 2\beta \begin{bmatrix} \frac{K_x - K_y}{2} & T_{os} \\ T_{os} & \frac{-K_x + K_y}{2} \end{bmatrix} e^{-2\,J\,(\,\alpha + \phi\,)} \begin{bmatrix} H_x^i \\ H_y^i \end{bmatrix} \cdot \begin{bmatrix} \hat{x} \\ \hat{y} \end{bmatrix} dS' \tag{7.12}$$

The integration in (7.12) proceeds in two steps. The first is similar to the previous case, the integral:

$$\frac{1}{2\pi}\int_0^{2\pi} \cos 2\beta \cos 2\phi \; d\beta$$

gives a factor $\frac{1}{2}$. However the integral over $z$ has to be treated differently.

First we define an **aperture matrix**:

$$A\,(\,z\,) = \pi\, a_o b_o z\, e^{-J\,\alpha_o\,\ln z}\; \frac{e^{J\,\psi_o}}{\sqrt{1 + \alpha_o^2}} \tag{7.13}$$

for which we can find the Fourier transform (see Appendix B):

$$A\,(\,p\,) = A_o(\alpha_o)\, e^{J\,(\,\alpha_o\,\ln p + \pi/2\,)}\; e^{2\,J\,\psi_o} \tag{7.14}$$

where $\psi_o$ is defined later. $A_o(\alpha_o)$ is a scalar. The derivative of $A\,(\,z\,)$ is easily calculated:

$$\frac{dA(z)}{dz} = \pi a_o b_o e^{-J\alpha_o \ln z} \tag{7.15}$$

Hence the integral (7.12) can be evaluated:

$$r_o \bar{H}_1^s = \frac{1}{4\pi}\left(\frac{d A(z)}{d z}\right)_{z=ct} \begin{bmatrix} \frac{K_x - K_y}{2} & T_{os} \\ T_{os} & \frac{-K_x + K_y}{2} \end{bmatrix} \begin{bmatrix} H_x^i \\ H_y^i \end{bmatrix} \tag{7.16}$$

We have to be aware that the derivative (7.15) is a matrix operator in this case!

The effect due to torsion is shown by the off diagonal term:

$$T_{os} = \frac{-2\alpha_o}{1-\alpha_o^2}\left(\frac{K_x - K_y}{2}\right) \tag{7.17}$$

The bracketed matrix in (7.16) is real and it appears that the surface geometry with torsion would be independent of rotation. However, this is not the case as can be shown by a simple argument. We substitute (7.17) into (7.16) which gives:

$$r_o \bar{H}_1^s = \frac{1}{4\pi}\left(\frac{d A(t)}{d t}\right)\left(\frac{K_x - K_y}{2}\right)\left(\frac{1+\alpha_o^2}{1-\alpha_o^2}\right) e^{-2J\psi_o} \begin{bmatrix} 1 & 0 \\ 0 & -1 \end{bmatrix} \begin{bmatrix} H_x^i \\ H_y^i \end{bmatrix} \tag{7.18}$$

where

$$e^{-2J\psi_o} = \frac{1}{1+\alpha_o^2} \begin{bmatrix} 1-\alpha_o^2 & 2\alpha_o \\ -2\alpha_o & 1-\alpha_o^2 \end{bmatrix} \tag{7.19}$$

The notation used in (7.19) is useful, it is easily checked that $\det(e^{-2J\psi_o}) = 1$ and the righthand side has the form of a rotation operator. Further expression of (7.18) gives the desired result:

$$r_o \bar{H}_1^s = \frac{1}{4\pi}\left(\frac{d\,A(t)}{d\,t}\right)\left(\frac{K_x - K_y}{2}\right)\left(\frac{1+\alpha_o^2}{1-\alpha_o^2}\right) e^{-J\,\psi_o}\begin{bmatrix}1 & 0\\ 0 & -1\end{bmatrix} e^{J\,\psi_o}\begin{bmatrix}H_x^i\\ H_y^i\end{bmatrix} \tag{7.20}$$

where

$$e^{J\,\psi_o} = \frac{1}{\sqrt{1+\alpha_o^2}}\begin{bmatrix}1 & -\alpha_o\\ \alpha_o & 1\end{bmatrix} \tag{7.21}$$

The physical interpretation of (7.20) is that the surface with torsion, viewed at normal incidence in the direction $\hat{n}_o$ may be considered as an average surface **without** torsion with extra factor $\frac{1+\alpha_o^2}{1-\alpha_o^2}$, which has an average 'tilt' relative to the $(\hat{x},\hat{y})$ coordinate system of angle $\psi_o$ defined by (7.21) in terms of torsion-rate factor $\alpha_o$. The two points of view are completely equivalent.

Now the total target **scattering matrix** is defined, in terms of the local $(\hat{x}_o,\hat{y}_o)$ coordinates:

$$S(u,v) = \begin{bmatrix}a' + b' & c'\\ c' & a' - b'\end{bmatrix} \tag{7.22}$$

where, by using (7.13):

$$a' = \frac{1}{2\pi}\frac{d^2}{dt^2}A_T'(t)$$

$$b' = \frac{1}{4\pi}\frac{d}{dt}A_T(t)\left[\cos(\alpha(t)+\psi_o)\frac{K_x - K_y}{2} - \sin(\alpha(t)+\psi_o)\,T_{os}\right] \tag{7.23}$$

$$c' = \frac{1}{4\pi}\frac{d}{dt}A_T(t)\left[\sin(\alpha(t)+\psi_o)\frac{K_x - K_y}{2} + \cos(\alpha(t)+\psi_o)\,T_{os}\right]$$

where $T_{os}$ is principal **surface torsion**. Note that $A_T'(t)$ and $A_T(t)$ are different scalars.

For time-harmonic (single frequency) case the Fourier-transform leads to:

$$S_o(k) = \begin{bmatrix} a_o(k) + b_o(k) & c_o(k) \\ c_o(k) & a_o(k) - b_o(k) \end{bmatrix} \tag{7.24}$$

where

$$a_o(k) = \frac{1}{2\pi}(jk)^2 A_F'(k)$$

$$b_o(k) = \frac{1}{4\pi}(jk) A_F(k)\left[\sin\left(\alpha(k) - 2\psi_o\right)\frac{K_x - K_y}{2} - \cos\left(\alpha(k) - 2\psi_o\right) T_{os}\right]$$

$$c_o(k) = \frac{1}{4\pi}(jk) A_F(k)\left[\cos\left(\alpha(k) - 2\psi_o\right)\frac{K_x - K_y}{2} + \sin\left(\alpha(k) - 2\psi_o\right) T_{os}\right]$$

(7.25)

where $K_x$ and $K_y$ are the principal curvatures and $T_{os}$ the principal surface torsion. (Note that $a_o$, $b_o$ in (3.1) are **constant**, not to be confused with (7.24) above.)

We notice that $S_o(k)$ above expresses the target scattering matrix along the **surface** principal coordinate directions (defined previously by $\beta = 0°$ and $\beta = 90°$). The function $A(t)$ in (7.20) is usually called the projected area or silhouette function for the specular point under consideration. This would be for the case without torsion as in (7.9). However, for the new case with torsion, the scalar function has to be replaced by a term with a rotation matrix. Hence this new term could be called the **silhouette matrix** for the specular point in question.

# VIII. SCATTERING MATRIX AND MUELLER MATRIX COEFFICIENTS AND MEASUREMENTS

It is now a simple matter to transform the scattering matrix for local surface coordinates to the field $(\hat{x}_R, \hat{y}_R)$ **radar** coordinates. We assume that between the two coordinate systems there is an angle $\psi$ such that:

$$S_\psi = e^{J\,\psi}\, S_o\, e^{-J\,\psi} \tag{8.1}$$

Now assuming that

$$S_o = \begin{bmatrix} a_o + b_o & c_o \\ c_o & a_o - b_o \end{bmatrix} \tag{8.2}$$

where $a_o = a_o(k)$, etc., then

$$S_\psi = \begin{bmatrix} a_o + b_\psi & c_\psi \\ c_\psi & a_o - b_\psi \end{bmatrix} \tag{8.3}$$

we find:

$$\begin{aligned} b_\psi &= b_o \cos 2\psi - c_o \sin 2\psi \\ c_\psi &= b_o \sin 2\psi + c_o \cos 2\psi \end{aligned} \tag{8.4}$$

In the fixed radar coordinate system we thus have

$$S_{HH} = \frac{1}{2\pi}(jk)^2 A_F'(k) + \frac{1}{4\pi}(jk)\, A_F(k) \left[ \frac{K_x - K_y}{2} \cos 2\psi' - T_{os} \sin 2\psi' \right] \tag{8.5}$$

$$S_{VV} = \frac{1}{2\pi}(jk)^2 A_F'(k) - \frac{1}{4\pi}(jk)\, A_F(k) \left[ \frac{K_x - K_y}{2} \cos 2\psi' - T_{os} \sin 2\psi' \right] \tag{8.6}$$

$$S_{HV} = \frac{1}{4\pi}(jk)\, A_F(k) \left[ \frac{K_x - K_y}{2} \sin 2\psi' + T_{os} \cos 2\psi' \right] \tag{8.7}$$

where:

$$2\psi' = 2\psi + 2\psi_o - \alpha(k) + \frac{\pi}{2} \tag{8.8}$$

Hence we notice that the factor $2\psi_o$ in the alternate interpretation (7.18) cancels if put together with (8.8) as a total target orientation rotation angle about the line of sight.

The Stokes (or Mueller) matrix parameters are found from the relationships (Huynen [10]):

$$A_o = \frac{1}{2}|a_o|^2 = \frac{1}{8\pi^2}k^4 |A_F'(k)|^2 \tag{8.9}$$

$$B_o = \frac{1}{2}(|b_o|^2 + |c_o|^2) = \frac{k^2}{32\pi^2}|A_F(k)|^2\left[\frac{(K_x - K_y)^2}{4} + T_{os}^2\right] \tag{8.10}$$

$$B = \frac{1}{2}(|b_o|^2 - |c_o|^2) = \frac{k^2}{32\pi^2}|A_F(k)|^2\left[\frac{(K_x - K_y)^2}{4} - T_{os}^2\right] \tag{8.11}$$

$$a_o^* b_o = C + jD = \frac{-jk^3}{8\pi^2}|A_F'A_F|\frac{K_x - K_y}{2} \tag{8.12}$$

$$b_o c_o^* = E + jF = \frac{+k^3}{16\pi^2}|A_F|^2\frac{K_x - K_y}{2}T_{os} \tag{8.13}$$

$$a_o c_o^* = H + jG = \frac{+jk^3}{8\pi^2}|A_F'A_F|\,T_{os} \tag{8.14}$$

From which follows for the Stokes matrix parameters:

$$D = \frac{-k^4}{8\pi^2}|A_F'A_F|\frac{K_x - K_y}{2k} \tag{8.15}$$

$$E = \frac{k^4}{16\pi^2}|A_F|^2\frac{K_x - K_y}{2k}\,T_{os}/k \tag{8.16}$$

$$G = \frac{k^4}{8\pi^2}|A_F'A_F|\,T_{os}/k \tag{8.17}$$

We find that $D$ measures curvature difference and $G$ measures the torsion parameter, while $E$ depends on both. In fact,

$$\frac{G}{D} = \frac{+2\alpha_o}{1 - \alpha_o^2} \tag{8.18}$$

because of Eq. (7.1). The complete Stokes reflection matrix is [10]:

$$M = \begin{bmatrix} A_o + B_o & F & C & H \\ F & -A_o + B_o & G & D \\ C & G & A_o + B & E \\ H & D & E & A_o - B \end{bmatrix} \tag{8.19}$$

For the local surface patch geometry return, although quite general in our model (an elliptical paraboloid), still leaves $C = F = H = 0$. Here $C$ measures **'shape'** and for the extended surface patch $C = 0$. Only for a wire-like structure would $C \neq 0$. Also $F$ is a measure for **'helicity'**. Hence surface torsion is **different** from helicity. A helical coil would produce $F \neq 0$. Finally, $H = 0$, because $M$ in (8.19) refers to the local coordinates with principal axis. The following relationships between Stokes parameters exist (Huynen [10], p. 136):

$$\left.\begin{aligned} 2A_o(B_o + B) &= C^2 + D^2 \\ 2A_o(B_o - B) &= G^2 + H^2 \\ B_o^2 &= B^2 + E^2 + F^2 \\ 2A_o E &= -D\,G + C\,H \\ D(B_o - B) &= -E\,G + F\,H \\ G(B_o + B) &= -D\,E + C\,F \end{aligned}\right\} \text{ where } C = F = H = 0 \tag{8.20}$$

Based on the above formula (8.9 through 8.14), the Stokes parameters are easily obtained from full polarimetric scattering matrix measurements of $a_o(k)$, $b_o(k)$ and $c_o(k)$. The crucial first step in obtaining these data is to first eliminate the target orientation parameter $\psi'$. This can be done by transforming $H_\psi$ to $H = 0$ from the equations

$$\left.\begin{aligned} H_\psi &= C \sin 2\psi' \\ C_\psi &= C \cos 2\psi' \end{aligned}\right\} \text{for } (C \neq 0) \tag{8.21}$$

where $H_\psi = \mathrm{Re}(a_o c_\psi^*)$ and $C_\psi = \mathrm{Re}(a_o^* b_\psi)$. If $C = 0$, Eq. (8.21) does not work. For a set of conditions when also $G = 0$ we can use instead of (8.21):

$$\left.\begin{aligned} G_\psi &= -D \sin 2\psi' \\ D_\psi &= D \cos 2\psi \end{aligned}\right\} \tag{8.22}$$

to de-$\psi$ the Stokes matrix.

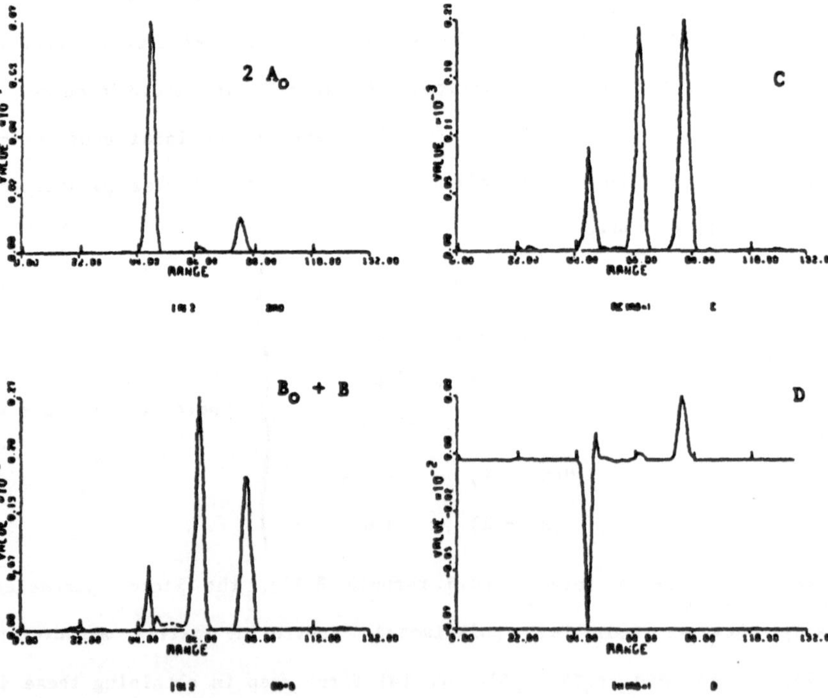


Fig. 4. Measured target parameters.

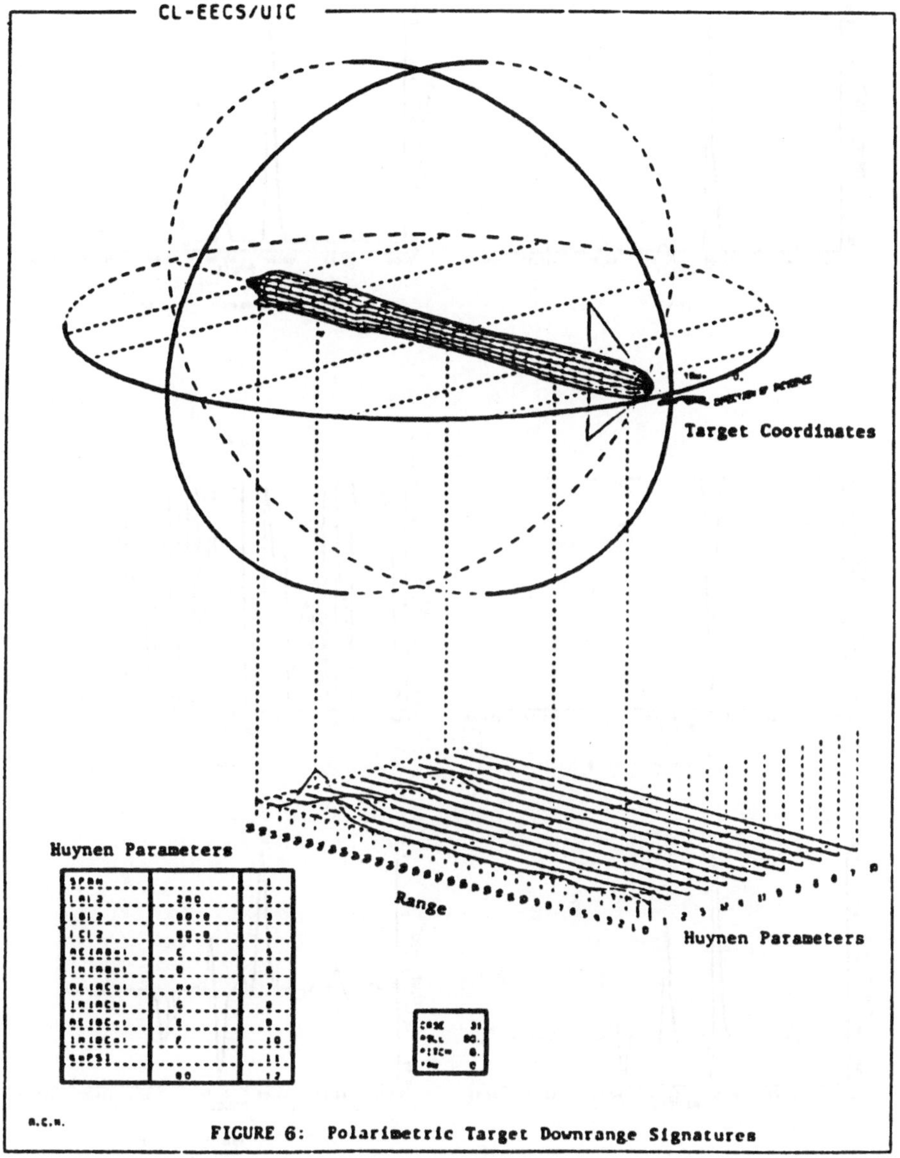


FIGURE 6: Polarimetric Target Downrange Signatures

A. C. MANSON AND W.-M. BOERNER Ref. 30

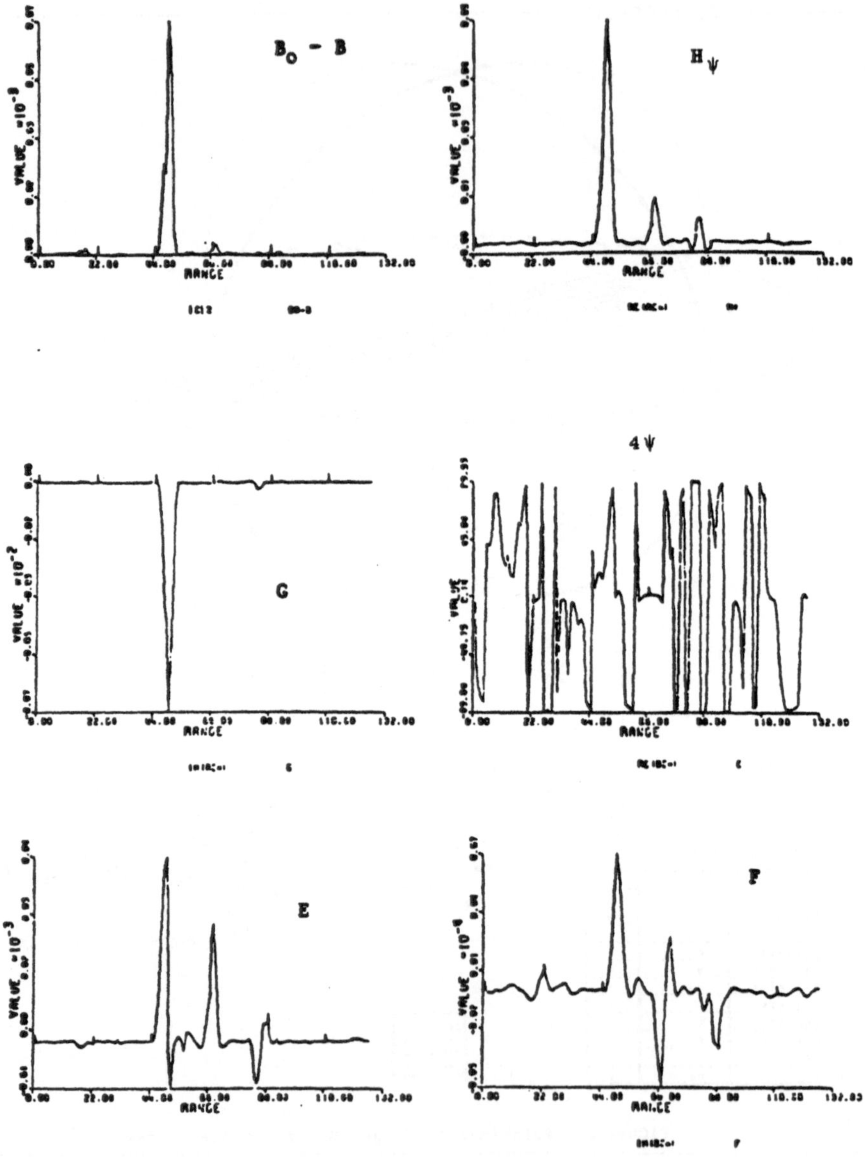


Fig. 5. Measured target parameters.

## IX. RESULT OF MEASUREMENTS

A set of measurements from down range imaging from a model missile body was taken as our data base. The Stokes matrix parameters are displayed in the graphs. We will use the data for $2A_o$, D, G, and E for the range point '50' on the horizontal axis. We read off the following values

$$2A_o = +.07 \times 10^{-1}$$
$$D = -.045 \times 10^{-2}$$
$$G = -.07 \times 10^{-2}$$
$$E = -.040 \times 10^{-3} \quad \text{(sign in graph is wrong)}$$
$$C = +.010 \times 10^{-3}$$
$$F = +.010 \times 10^{-4}$$

We notice that indeed C and F are small in this case. The surface patch consists of the smooth metallic fairing of the missile front wing with the body, creating conditions for surface torsion. The sign of E in the graph was in error.

The following results are obtained from these measurements

$$\frac{G}{D} = \frac{+2\alpha_o}{1-\alpha_o^2} = \frac{.07 \times 10^{-2}}{.045 \times 10^{-2}} = 1.56$$

Hence:

$$\boxed{\alpha_o = .55}$$

Another useful relationship is:

$$\frac{D}{2A_o} = \frac{+E}{-G} = -\frac{K_x - K_y}{4k} = -.06 \text{ (assume } A'(k) = A(k)\text{)}$$

The fact that E has the wrong sign in the diagram follows from the fact that $2A_o > 0$ ; $G < 0$ and hence D and E must have the same sign from the general equality: $2A_oE = -DG$. The wrong sign in E occurs quite frequently in the literature because in the original work (Huynen [10]), the E was defined with opposite sign compared to later publications.

The alternate interpretation of parameters leads to the values:

$$\frac{1 + \alpha_o^2}{1 - \alpha_o^2} = 1.87$$

$$\psi_o = 28.81°$$

Because $C = 0$ for this case: $\gamma = 45°$ (see Huynen [10], p. 54; notice $\gamma$ here is not $\gamma$ in (4.9) or in (6.17)), which leads to an ambiguity in sets of parameters. The fact that the torsion parameter is hidden in the total target orientation angle $\psi'$ by (8.8) makes it difficult to use a straightforward approach. In most cases C is not quite zero and the total target orientation can then be determined from (8.21). Perhaps multifrequency approaches are helpful to recover the parameters. (Notice that $\alpha(k)$ in (8.8) depends on frequency.) A third alternative is to have a 'priori knowledge of target position and orientation with controlled indoor or outdoor range measurements.

## IX. ACKNOWLEDGEMENTS

The author wishes to thank in particular Dr. Harry Mieras for his patient and thorough review of this work as it matured through various stages of development. Any errors or omissions remaining in the text are entirely the author's responsibility. Also he wishes to thank Mr. Andrew Jansons for compiling an updated list of references and assisting with the numerous details of bringing the report to completion.

The task of transcribing the handwritten text by word processor into a finished document fell into the able hands of Mrs. Ditter Peschcke-Koedt for which I express my heartfelt gratitude.

# APPENDIX A

## CALCULATION OF SURFACE INTEGRALS

CASE I: **WITHOUT TORSION** $\alpha = 0$

We evaluate the integral of (7.8):

$$I_o = \frac{1}{2\pi} \int \delta\left(t - \frac{z}{c}\right) \cos 2\beta \cos 2\phi \, dS' \tag{A.1}$$

$$dS' = dx\,dy = \begin{vmatrix} X_u & X_v \\ Y_u & Y_v \end{vmatrix} du\,dv =$$

$$= (X_uY_v - X_vY_u)\,du\,dv = (\bar{r}_u \times \bar{r}_v)_z\,du\,dv =$$

$$= |\bar{n}|\,du\,dv = a_o\,b_o\,\beta_v\,u\,du\,dv =$$

$$= \begin{vmatrix} 1 & 0 \\ \beta_u & \beta_v \end{vmatrix} a_o\,b_o\,u\,du\,dv = a_o\,b_o\,u\,du\,d\beta = \frac{a_o\,b_o}{2}\,d\beta\,dz \tag{A.2}$$

Hence, $$I_o = \frac{1}{2\pi} \int \delta\left(t - \frac{z}{c}\right) \frac{a_o\,b_o}{2}\,dz \int_0^{2\pi} \cos 2\phi \cos 2\beta \, d\beta \tag{A.3}$$

below is shown: $$\int_0^{2\pi} \cos 2\phi \cos 2\beta \, d\beta = \Pi \tag{A.4}$$

Since for an elliptical slice: $A = \pi\,a_o\,b_o\,u^2 = \pi\,a_o\,b_o\,z$, we have:

$$\frac{d\,A\,(\,z\,)}{d\,z} = \pi\,a_ob_o \tag{A.5}$$

Hence $$I_o = \frac{1}{4\pi}\left(\frac{d\,A\,(\,z\,)}{d\,z}\right)_{z=ct} \tag{A.6}$$

CASE II: **WITH TORSION** $\alpha_o \neq 0$

The operator integral to be evaluated, with torsion, (7.12) is:

$$I_1 = \frac{1}{2\pi} \int \delta\left(t - \frac{z}{c}\right) \cos 2\beta \, e^{2 J ( \alpha + \phi )} \, dS' \tag{A.7}$$

where as before $dS' = \frac{a_o \, b_o}{2} \, d\beta \, dz$ .

The integration over $\beta$ leads to result (A.4).

Hence

$$I_1 = \frac{1}{4\pi} \int \delta\left(t - \frac{z}{c}\right) ( \pi \, a_o \, b_o) \, e^{2 J \alpha(u)} \, dz$$

$$= \frac{1}{4\pi} \left[ \pi \, a_o \, b_o \, e^{- J \alpha_o \ln z} \right]_{z=ct} = \frac{1}{4\pi} \left( \frac{d A ( z )}{d z} \right)_{z=ct} \tag{A.8}$$

where the **matrix** A was defined in (7.13).

EVALUATION of :

$$\frac{1}{2\pi} \int_0^{2\pi} \cos 2\beta \cos 2\phi \, d\beta \tag{A.9}$$

Since $\tan \beta = \frac{a_o}{b_o} \tan \phi$, the integral above may be evaluated exactly:

$$\frac{1}{2\pi} \int_0^{2\pi} \cos 2\beta \cos 2\phi \, d\beta = \frac{2 \, a_o \, b_o}{(a_o + b_o)^2} \tag{A.10}$$

This shows dependence on ellipse axial ratio $p = b_o/a_o$ and hence on curvature ratio. In order to obtain an approximate value independent of axial ratio we can assume $a_o \approx b_o$ which gives the value 1/2 for (A.10).

## APPENDIX B

### FOURIER TRANSFORM PAIRS: $p = jk$ , $t = z$

$$z^{i\alpha_o} \longleftrightarrow \Gamma(1 + i\alpha_o)\, p^{-(1+i\alpha_o)} = A_o(p) + i\, B_o(p) \qquad \text{[Ref. 45]}$$

$$-z\, f(z) \longleftrightarrow \frac{dF(p)}{dp}$$

$$\frac{df(z)}{dz} \longleftrightarrow p\, F(p)$$

Let

$$\cos(\alpha_o\, \ell n\, z) \longleftrightarrow A_o(p)$$

$$\sin(\alpha_o\, \ell n\, z) \longleftrightarrow B_o(p)$$

$$-\frac{\alpha_o}{z} \sin(\alpha_o\, \ell n\, z) \longleftrightarrow p\, A_o(p)$$

$$\frac{\alpha_o}{z} \cos(\alpha_o\, \ell n\, z) \longleftrightarrow p\, B_o(p)$$

$$\alpha_o \sin(\alpha_o\, \ell n\, z) \longleftrightarrow A_o(p) + p\, A_o'(p) = \alpha_o B_o(p)$$

$$-\alpha_o \cos(\alpha_o\, \ell n\, z) \longleftrightarrow B_o(p) + p\, B_o'(p) = -\alpha_o A_o(p)$$

Solutions:

$$A_o(p) = \frac{C(\alpha_o)}{p} \sin(\alpha_o\, \ell n\, p)$$

$$B_o(p) = \frac{C(\alpha_o)}{p} \cos(\alpha_o\, \ell n\, p)$$

$$C(\alpha_o) = \frac{\Gamma(1 + i\alpha_o)}{i}$$

$$e^{-J \alpha_o \ell n\, z} \longleftrightarrow \frac{-C(\alpha_o)}{p}\, J\, e^{J \alpha_o \ell n\, p}$$

$$z\, e^{-J \alpha_o \ell n\, z} \longleftrightarrow \frac{-C(\alpha_o)}{p^2}\, J\, e^{J \alpha_o \ell n\, p}\, e^{J \psi_o} \sqrt{1 + \alpha_o^2}$$

**REFERENCES**

1. E.M. Kennaugh, "Effects of type of polarization on echo characteristics," O.S.U. Antenna Lab., Columbus, OH, Reports 389-1 to 389-24, 1949-1954.

2. E.M. Kennaugh, "Polarization properties of radar reflections," M.S. Thesis, Dept. Electrical Engineering, O.S.U., Columbus, OH, 1952 (O.S.U. Antenna Lab., Report 389-12, March 1, 1952).

3. E.M. Kennaugh and R.L. Cosgriff, "The use of impulse response in electromagnetic scattering problems," IRE National Convention Record, Part I, pp. 72-77.

4. D.J. Struik, "Lectures on classical differential geometry," Addison Wesley Publishing COmpany, Inc., Reading, MA and London, England, 1961.

5. E.M. Kennaugh and D.L. Moffat, "Transient and impulse response approximation," IEEE Proc., Vol. 53, No. 8, pp. 893-901, 1965.

6. J.R. Huynen, "Measurement of the target scattering matrix," IEEE Proc., Vol. 53, pp. 936-946, August 1965.

7. D.B. Kanareykin, N.F. Pavlov, and U.A. Potekhin, **The Polarization of Radar Signals**, Sovyetskoye Radio, Moscow, 1966. English translation of Chapters 10-12: **Radar Polarization Effects**, MacMillan, New York.

8. C.L. Bennett, "Time domain inverse scattering," IEEE Trans. Antennas Propagations, Special Issue on Inverse Methods in Electromagnetics, Vol. AP-29, pp. 213-219, March 1981.

9. C.L. Bennett, "A technique for computing approximate electromagnetic impulse responses of conducting bodies," Ph.D. Dissertation, Purdue University, Lafayette, Indiana, August.

10. J.R. Huynen, "Phenomenological theory of radar targets," Ph.D. Dissertation, Drukkerij Bronder-Offset, N.V. Rotterdam, 1970.

11. C.L. Bennett and E.K. Miller, "Some computational aspects of transient electromagnetics," 1972 Spring URSI Meeting, Washington, D.C., 13-15 April 1972.

12. C.L. Bennett, A.M. Auckenthaler, R.S. Smith, J.D. DeLorenzo, "Space-time integral equation approach to the large body scattering problems," Sperry Research Center, Sudbury, MA, Final Report on Contract No. F30602-71-C-0162, AD 763794, May 1973.

13. S.K. Chaudhuri and W.-M. Boerner, "A monostatic inverse scattering model based on polarization utilization," Applied Physics, Vol. 11, No. 4, pp. 337-350, December 1976.

14. S.K. Chaudhuri and W.-M. Boerner, "Polarization utilization in profile inversion of a perfectly conducting prolate spheroid," IEEE Trans. Antennas Propagation, Vol. AP-25, pp. 505-511, July 1977.

15. C.L. Bennett, R. Hieronymus and H. Mieras, "Impulse response target sutyd," Sperry Research Center, Final Report, 1977.

16. S.K. Chaudhuri, "Utilization of polarization-depolarization characteristics in electromagnetic inverse scattering," Ph.D. Dissertation, University of Manitoba, Winnipeg, Canada, 1977.

17. J.R. Huynen, "Phenomenological theory of radar target," in **Electromagnetic Scattering**, P.L.E. Uslenghi, Ed., Academic Press, New York, 1978, Chapter 11.

18. R.E. Barnhill, J.H. Brown, and I.M. Klucewicz, "A new twist in computer aided geometric design," Computer Graphics and Image Processing, Vol. 8, 1978, pp. 78-91.

19. I.D. Faux and M.J. Pratt, **Computational Geometry for Design and Manufacture**, John Wiley & Sons, New York, 1979.

20. S.K. Chaudhuri, "A time domain synthesis of electromagnetic backscattering by conducting ellipsoids," IEEE Trans. Antennas Propagations, Vol. AP-28, pp. 523-530, July 1980.

21. W.-M. Boerner, M.B. El-Arini, C.Y. Chan and P.M. Mastoris, "Polarization dependence on electromagnetic inverse problem," IEEE Trans. Antennas Propagation (Special Issue on Inverse Methods in Electromagnetics), Vol. AP-29, pp. 262-269, March 1981.

22. W.-M. Boerner, C.M. Ho and B.Y. Foo, "Use of Radon's projection theory in electromagnetic inverse scattering," IEEE Trans. Antennas Propagation (Special Issue on Inverse Methods in Electromagnetics), Vol. AP-29, pp. 262-269, March 1981.

23. H. Mieras and C.L. Bennett, "Space-time integral equation approach to dielectric targets," IEEE Trans. Antennas and Propgation, Vol. AP-30, No. 1, January 1982.

24. B.Y. Foo, "A high frequency inverse scattering model to recover the specular point curvature from polarimetric scattering data," M.Sc. Thesis, Electr. Engr. & Comp. Sci. Dept., University of Illinois at Chicago, IL, Communications Lab Report No. 82-05-21, May 21, 1982.

25. J.R. Huynen, "Polarization discrimination with applications to target classification," Proceedings of the Second Workshop on Polarimetric Radar Technology, Vol. !, U.S. Army Missle Command, Redstone Arsenal, AL, May 3-5, 1983, pp. 197-216.

26. H. Mieras, R.M. Barnes, G.M. Yachula, J.N. Nucknam, C.L. Bennett and W.-M. Boerner, "Polarization null characteristics of simple targets," Rome Air Development Center, Air Force Systems Command, Griffis Air Force Base, N.Y., Report No. RADC-TR-82-335, Final Technical Report, 1983.

27. H.J. Lugt, Vortex Flow in Nature and Technology, John Wiley & Sons, New York, 1983.

28. P.C. Gasson, **Geometry of Spatial Forms**, John Wiley & Sons, New York, 1983.

29. M. Mieras, "Local influence technique in time domain scattering," Ph.D. Dissertation, Northeastern University, Boston, MA, February 1984.

30. W.-M. Boerner, A.C. Manson and J.R. Huynen, "Radar target classification using polarimetric target slant range signatures," Naval Air Systems Command Final Rept. No. UIC-EECS/CL-EMID-83-06-15, Contract No. N00019-82-C-0306, 1983.

31. A.H. Barr, "Global and local deformations of solid primitives," Computer Graphics, Vol. 18, No. 3, July 1984, pp. 21-30.

32. J.R. Huynen, "Towards a theory of perception for radar targets," in Inverse Methods in Electromagnetic Imaging, W.M. Boerner, Ed., Reidel Publishing Co., Dordrecht, Holland, 1984.

33. B.-Y. Foo, S.K. Chaudhuri and W.-M. Boerner, "A high frequency inverse scattering model to recover the specular point curvature for polarimetric scattering data," IEEE Trans. A&P, 1984.

34. E.A. Lord and C.B. Wilson, **The Mathematical Description of Shape and Form**, John Wiley & Sons, New York, 1986.

35. S.K. Chaudhuri, B.-Y. Foo and W.-M. Boerner, "A validation analysis of Huynen's target-descriptor interpretations of the Mueller matrix elements in polarimetric radar returns using Kennaugh's physical optics impulse response formulation," IEEE Trans. on Ant. and Prop., Jan. 1986, pp. 1-20.

36. A.P. Agrawal, H.J. Eom and W.M. Boerner, "An analysis of polarimetric (VV, HH, and VH polarized) scattering matrix data of near-grazing sea clutter backscatter at X-Band," Proceedings 1986 Working Symposium on Oceanographic Data Systems, D. Steiger, Ed., IEEE Comp. Soc. Press, 1986, pp. 39-49.

37. W.A. Holm, "Polarimetric fundamentals and techniques," in **Principles of Modern Radar**, J.E. Eaves and E.K. Reedy, Eds., Von Nostrand Reinhold Co., 1987, Chapter 2, pp. 621-645.

38. R.L. Thomas, "The generative spiral in phyllotaxis theory," Ann. Bot., Vol. 45, 1980, pp. 237-249.

39. R. L. Thomas, "Orthostichy, parastichy and plastochrone ratio in a central theory of phyllotaxis," Ann. Bot., Vol. 39, 1975, pp. 455-489.

40. R. L. Thomas *et al.*, "Phyllotaxis in the oil palm: Arrangements of fronts on the trunk of mature palms," Ann. bot., vol. 33, 1969, pp. 1001-1008.

41. L. A. Bursill *et al.*, "Spiral lattice concepts," Modern Physics Letters B, Vol. 1, Nos. 5 & 6, pp. 1001-1008.

42. J. H. M. Thornley, "Phyllotaxis I. A mechanical model," Ann. Bot., Vol. 39, 1975, pp. 451-507.

43. J. H. M. Thornley, "Phyllotaxis II. A description in terms of intersecting logarithmic spirals," Ann. bot., Vol. 49, 1982, pp. 747-760.

44. R. Dixon, "The mathematical daisy," New Scientist, 17 Dec. 1981, pp. 752-795.

45. I. S. Gradshteyn and I. M. Ryznik, **Table of Integrals, Series, and Products**, Academic Press, Inc., New York, 1980.

46. J. R. Huynen, "The calculation and measurement of surface-torsion by radar," Report PQR No. 102, June 1988, available from the author P. Q. Research, 10531 Blandor Way, Los Altos Hills, CA 94022 (415)-941-2374.

# THE SOLUTION OF SCATTERING AND RADIATION PROBLEMS FOR COMPLEX CONFIGURATIONS BASED ON 3-D GRID AND PATCH MODELS

V. Stein

Deutsche Forschungsanstalt für Luft- und Raumfahrt e.V.
Institut für Hochfrequenztechnik, D-8031 Oberpfaffenhofen,
Federal Republic of Germany

ABSTRACT

The increasing interest in predicting the scattering and radiation characteristics of objects with a complicated structure has stimulated the development of several theories. A rigorous treatment of the electrodynamic problem requires the solution of a boundary value problem based on Maxwell's differential equations or on the equivalent integral equations. The application of rigorous methods for objects whose dimensions are large compared to the wavelength is limited by the required computer memory and execution time. Therefore, methods which solve the boundary value problem approximately come into consideration. Each type of solution method involves a typical model either of the surface or the volume of the structure and it's surrounding. So, geometric models consisting of canonical shapes, wire-grids, surface patches, and volume cells are described and the requirements of the specific solution methods are discussed. In some cases estimations for the necessary modeling accuracy are given. Methods which are based on geometrical optics principles require models where the surface parts which are illuminated by the incident wave and the surface parts which are hidden can be separated for each aspect angle. Such a procedure is discussed as well as the procedure to treat double reflections. Some computational examples for radiation and scattering processes are given and comparisons with measurements are made.



*W.-M. Boerner et al. (eds.), Direct and Inverse Methods in Radar Polarimetry, Part 1*, 625–674.

1. INTRODUCTION

In electrodynamics there is an increasing demand in predicting the radiated field of antennas installed on complicated structures or the scattered field of radar objects. Therefore, the extension of known methods and the development of new methods is stimulated to describe the interaction process between the electromagnetic wave and the structure under test. With the following figures some typical problems of electrodynamics are illustrated.

Fig. 1.1 shows a periscope within a sea surface. One is interested in the polarization dependent radar cross-section of the object in this specific surrounding. The radar cross-section is a far-field quantity which means that it is to be determined at a distance which exceeds $2D^2/\lambda$. The object with dimension D is large against the wavelength $\lambda$ of the incident electromagnetic wave.

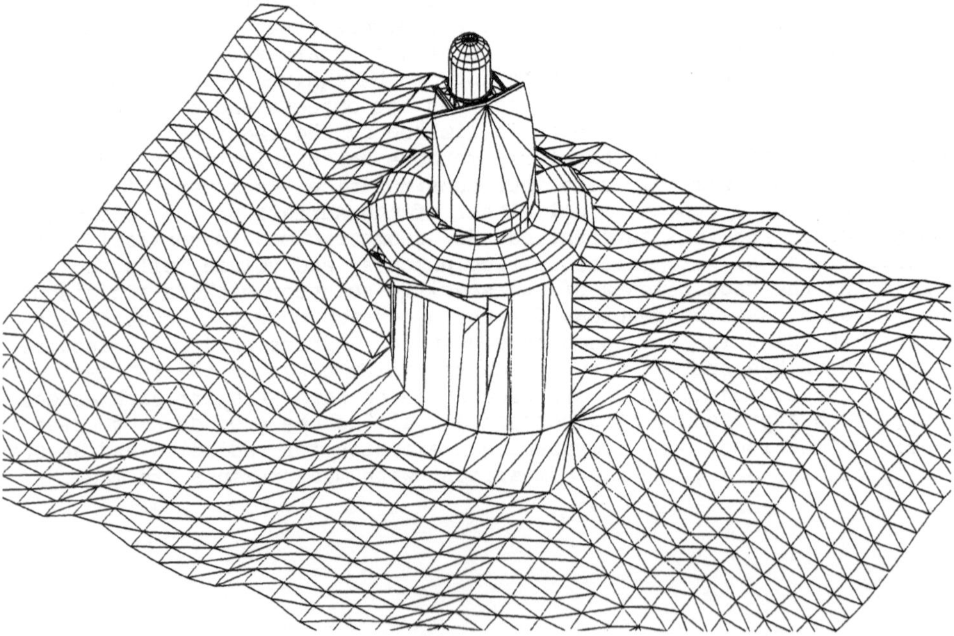

Fig. 1.1 Periscope within a moving sea surface.

Fig. 1.2 shows an airplane which is equipped for position finding purposes. Since the phase front of the incoming plane wave is distorted by the airplane structure, the question arises, where an appropriate place on the airplane for the installation of the direction finding antenna can be found. The magnitude of the distortion of the wave front in dependence from frequency, polarization and angle of incidence is of interest. The solution of this problem requires the computation of the scattered near-field, since the dimensions of the object are in the order of the wavelength and the position finding antenna is installed at a distance above the surface which amounts to a fraction of a wavelength.

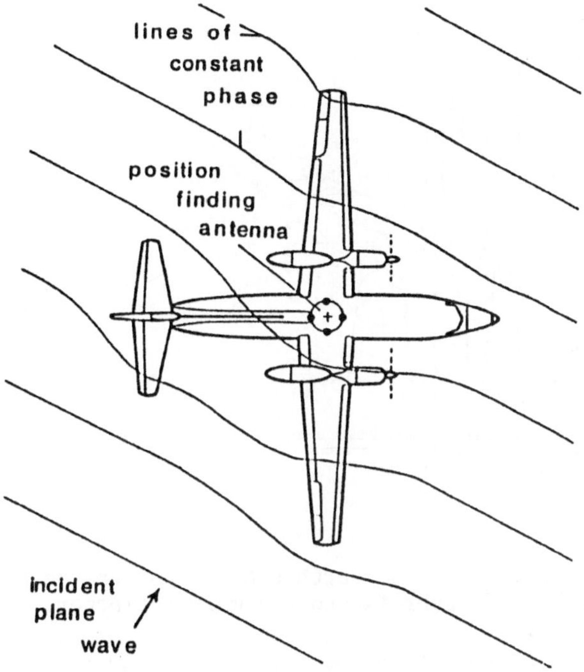


Fig. 1.2 Distortion of the plane wave due to the interaction between the incident electromagnetic wave and the structure.

Fig. 1.3 shows a reflector antenna for a satellite earth station. The radiation forming parts are the hornparabola, the subreflector and the main reflector. The spherical wave which propagates in the horn is reflected by the parabolic reflector of the horn towards the subreflector and transformed into a plane wave. Since the subreflector also is parabolic the wave is again transformed into a spherical wave. A third transformtion by the parabolic main reflector finally generates the radiated plane wave. It takes place if the focus of the main reflector coincides with the phase center of the spherical wave going out from the subreflector. The computation of the antenna characteristics is a typical far-field problem since the dimensions of such an antenna are much greater than the wavelength.

Fig. 1.4 finally shows a torus antenna mounted above a circular plate and excited by a small flat plate. Such a device is a very compact radiator since the circumference of the torus antenna is in the order of a wavelength. It finds use as a primary radiator in a central fed reflector. Usually one is not only interested in the far-field characteristics of such an antenna, but also in the input admittance which is a near-field quantity.

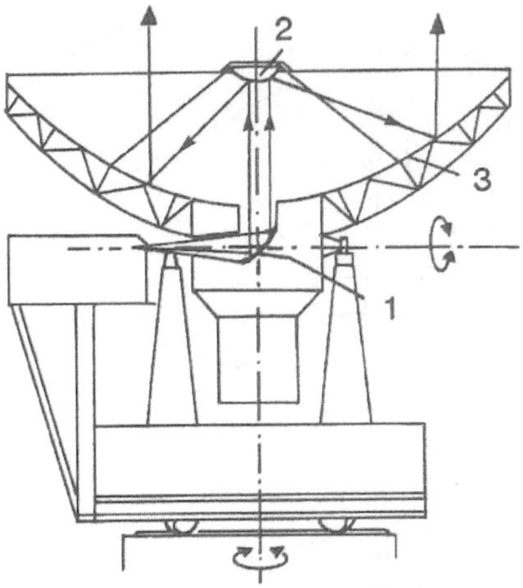


Fig. 1.3 Principle of a satellite earth station antenna (cassegrain type), 1 hornparabola, 2 subreflector, 3 main reflector.

Fig. 1.4 Torus antenna, capacitive coupled to the inner conductor of a coaxial line.

In the past only approximate theoretical or experimental methods were available for prediction purposes. While approximate methods can fail in giving sufficient accurate predictions of the interaction process, experimental methods require a high effort especially if changes of the structure must be carried out for optimization purposes. In these cases it is desirable to reduce the number of experimental studies to a few final measurements, which can be defined by the preceding theoretical analysis. This requires from the theory to develop and validate methods which are able to describe the interaction of electromagnetic waves with a complicated structure with an accuracy, which is sufficient for practical applications.

The progress in computer techniques permits in an increasing manner the use of theories, which because of their high numerical effort could not be considered in the past. There is a great variety of methods which in principle come into question. Each method has its specific advantages and drawbacks depending from the type of the problem. Among these methods there are as well as heuristic methods like the physical optics method, and the geometrical theory of diffraction as rigorous methods, which solve Maxwell's differential equations directly or undertake the solution of the equivalent integral equations. The choice of geometric models is influenced by the chosen solution method. Within the context of this paper it is not possible to describe the electrodynamic methods in detail. More informations may be obtained by the cited references. A rough classification of the theories can be given as follows.

In the radar case the distance between the source of the incident wave and the object is much greater than the dimensions of the object. The antenna case is characterized by the fact that the source of the incident wave is either within the radiating structure itself (e.g. a feeding gap of a dipole antenna) or in the immediate neighbourhood of the scattering object (e.g. an antenna on an airplane). From the standpoint of a rigorous solution of Maxwell's equations there is no difference between a radiation and a scattering problem. In the following the more general term scattering problem will be used for both types of problems.

If one is interested in the description of near-field characteristics, e.g. the scattered field in the immediate neighbourhood of the structure or the field at the surface of the structure itself, only rigorous methods can be used. If the dimension of the structure, however, exceeds the order of several wavelength (high frequency case), rigorous methods will fail in practice because of the required high computer effort. Fortunately the main contributions to the scattered field in this case originate from parts of the structure which are illuminated from the incident wave or which give rise to double or multiple reflections. These phenomena may be treated by geometrical optics. Wedges in the structure give rise to diffraction processes which can be described by an asymptotic evaluation of the rigorous theories for large distances and large dimensions of the structure. This concept leads to the development of heuristic methods for the high frequency case. So a very important criterium for the choice of the solution methods is the dimension of the object referred to the wavelength, that is the parameter $D/\lambda$.

The solution of Maxwell's equations for the electric field $\vec{E}$ and the magnetic field $\vec{H}$ can be carried out for the volume which surrounds the scatterer. For this purpose it is necessary not only to model the surface of the scatterer but also the surrounding volume. This leads to one type of geometric models. Maxwell's differential equations can be transduced in equivalent integral equations over the tangential fields at the surface of the scatterer. This leads to a class of geometric models which have to represent only the surface of the scatterer. Following the publications in electrodynamics the use of surface models is more common than the use of volume models. In the subsequent sections several geometric models are discussed which are typical for the individual electrodynamic solution procedures. Models which are suited for solutions with the integral equation or the physical optics method are emphasized.

## 2. MODELING WITH CANONICAL SHAPES

The structure is considered to be an ensemble of components, each of which can be geometrically approximated by a simple shape. A coarse model of an airplane for example may be established by a cylinder, flat plates, section of a sphere, see Fig. 2.1. For a certain class of canonical shapes, in particular wedges and smooth surfaces, theories are available to determine the scattered field for arbitrary aspect angles, from which the radar cross-section can be evaluated. Several problems will arise. The first problem consists in combining the contributions of the individual shapes by proper phasing to the total field or cross-section. A further problem arises by the fact that in dependence from the aspect angle one canonical substructure may hide completely or partially the other one. A third problem concerns the electromagnetic interaction of one canonical shape with the other.

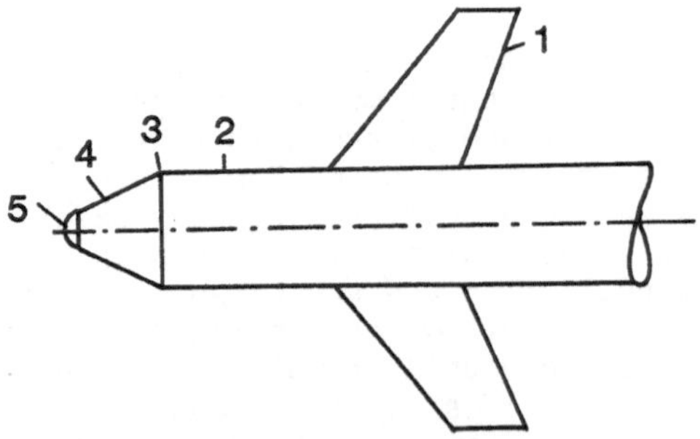


Fig. 2.1 Model of an airplane with simple shapes.
1 flat plate, 2 cylinder, 3 wedge, 4 conus, 5 sphere.

A procedure which overcomes the first problem consists in combining the contribution of the several canonical shapes by random phase [1]. This is based on the assumption that the different phases of the canonical structures are randomly distributed; then upon averaging over the phases one obtains for the radar cross-section $\sigma_t$ of the total object, which is composed of N canonical shapes with cross-sections $\sigma_j$, the expression

$$(2.1) \qquad \sigma_t = \sum_{j=1}^{N} \sigma_j \ .$$

In connection with this method of approach one can estimate the amount of probable deviation from the average cross-sectin $\sigma_t$ by employing the root mean square spread. This measure of the probable variation in cross-section due to relative-phase effects leads to the following bounds: $\sigma_t \pm s$, where

$$(2.2) \qquad s^2 = \left( \sum_{j=1}^{N} \sigma_j \right)^2 - \sum_{j=1}^{N} \sigma_j^2 \ .$$

The random-phase method is designed to give estimates of the amount by which the cross-section might deviate from the average value because of phase effects. If one is interested in finding an order-of-magnitude estimate of the cross-section as a function of aspect angle the random--phase procedure would be adequate. Since only cross-sections play a role one also can use experimental data for such shapes for which no theoretical values are available. A further advantage of this method concerns the computer effort since the evaluation of the radar cross-section for canonical shapes can be done analytically. The values of the individual shapes can also be calculated in advance for discrete elevation and azimuth angles and together with the experimental data for the non-canonical shapes properly arranged in a data file. For arbitrary aspect angles the cross-sections can be evaluated by interpolation schemes.

If one is interested not only to estimate the average value and the amount by which the cross-section deviates from the average value but also to predict the peaks and nulls of the radar cross-section in dependence from frequency, polarization and position of the observer point one has to combine the scattered field of the canonical shapes with their relative phases referred to a common reference point. This makes the use of experimental data more difficult.

At this time the most capable method in electrodynamics which is based on canonical shapes and solves the relative-phase problem is the geometrical theory of diffraction (GTD) [12]. The GTD is a ray optical technique and is, therefore, bound to such applications where the object's dimensions are much greater than a wavelength. The GTD makes use from the rigorous solutions for canonical shapes in such a way that a diffraction coefficient D is evaluated which connects the diffracted field with the incident field. This diffraction coefficient plays a similar role as the well-known reflection coefficient R in optics. The most important canonical shapes of the GTD are shown together with typical ray paths in Fig. 2.2. For the wedge and the smooth surface the diffraction coefficients are known in case of perfectly conducting bodies while the reflection coefficient can be determined also for nonperfectly conducting multilayered panels, see Sec. 7.1.

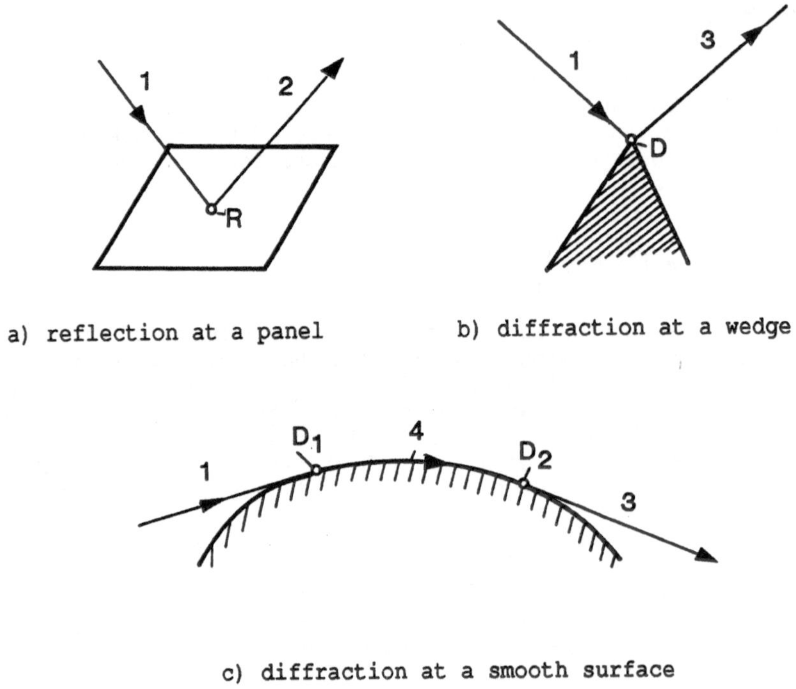
1
2
R
a) reflection at a panel
1
3
D
b) diffraction at a wedge
$D_1$
4
$D_2$
1
3
c) diffraction at a smooth surface

Fig. 2.2 Canonical shapes and interaction processes.
1 incident ray, 2 reflected ray, 3 diffracted ray, 4 creeping wave.

The rays which undergo reflection and diffraction at the structure and reach the observer point are to be determined from geometrical and differential geometrical considerations. This ray tracing procedure is illustrated at hand of Fig. 2.3. A source (slot antenna) is installed at the top of the fuselage of an airplane. For a given observer direction the following rays are depicted:

- direct ray,
- reflected ray, the reflection occurs at the surface of the left wing,
- wedge diffracted ray, the diffraction occurs at the trailing edge of

the left wing,

- surface diffracted ray (creeping wave) which encircles the fuselage and is then radiated toward the observer point.

The contribution of the several rays are summed up with correct amplitude and phase. If multiply reflected and diffracted rays are taken into account the complexity increases considerably. The ray tracing for arbitrary positions of source and observer is one of the main problems of this technique. Nevertheless, the GTD besides the physical optics method is the most promising approach for the solution of high frequency problems.

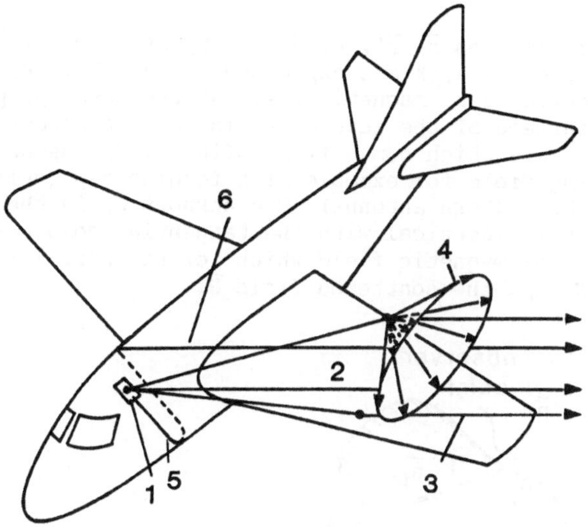


Fig. 2.3 Ray paths: 1 slot antenna, 2 direct ray, 3 reflected ray, 4 cone of diffracted rays, 5 geodesic path, 6 surface diffracted ray.

## 3. MODELS USED IN THE INTEGRAL EQUATION APPROACH

### 3.1 REMARKS TO THE ELECTRODYNAMIC THEORY

The electrodynamic processes which are expressed by the differential equations of Maxwell under satisfying the boundary conditions can be formulated by equivalent integral equations (IE) for the fields tangential to the surface of the scatterer. For a perfectly conducting scatterer one obtains the following integral equtions which each independently from the other can be applied to solve for the surface tangential field. One of the integral equations is denoted as the electric field integral equation (EFIE) and given by [12, 14]

$$(3.1)\qquad \vec{n}(\vec{r}) \times \vec{E}_e(\vec{r}) = -\frac{1}{4\pi j\omega\varepsilon}\,\vec{n}(\vec{r}) \times \int_F \Big((-1-jkR+k^2R^2)\,\vec{J}_F(\vec{r}\,') + (3+3jkR-k^2R^2)\,(\vec{J}_F(\vec{r}\,')\cdot\vec{e}_R)\cdot\vec{e}_R\Big)\,\frac{e^{-jkR}}{R^3}\,df' ,$$

the other is denoted as the magnetic field integral equation (MFIE) and expressed by

$$(3.2)\qquad \vec{J}_F(\vec{r}) = 2\vec{n}(\vec{r}) \times \vec{H}_e(\vec{r}) + \frac{1}{2\pi}\,\vec{n}(\vec{r})\times \int_F (1+jkR)\,(\vec{J}_F(\vec{r}\,') \times \vec{e}_R)\,\frac{e^{-jkR}}{R^2}\,df',$$

The geometric magnitudes $\vec{r}$, $\vec{r}\,'$, $\vec{R}$, $\vec{e}_R$, $\vec{n}$, F, df' are illustrated in Fig. 3.1. The magnitudes $\omega$, $\varepsilon$, k are explained in Sec. 7.1. $\vec{E}_e$ resp. $\vec{H}_e$ is the incident electric resp. magnetic field at the observer point which is situated on the surface of the scatterer. In radar problems $\vec{E}_e$, $\vec{H}_e$ is supported by a plane wave which comes from infinity. In the antenna case $\vec{E}_e$, $\vec{H}_e$ is the outgoing field for example of a feeding gap (cylinder antenna) or of an aperture (horn antenna). The unknown $\vec{J}_F$ is the electric surface current which is identical with the tangential component $\vec{n}(\vec{r}\,')\times\vec{H}(\vec{r}\,')$ of the total magnetic field which for his part is the sum of the incident field $\vec{H}_e$ and the scattered field $\vec{H}_s$.

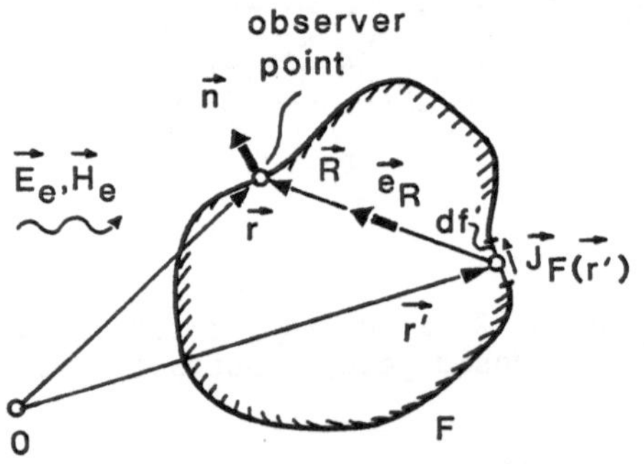


Fig. 3.1 Geometry for the evaluation of the EFIE, MFIE.

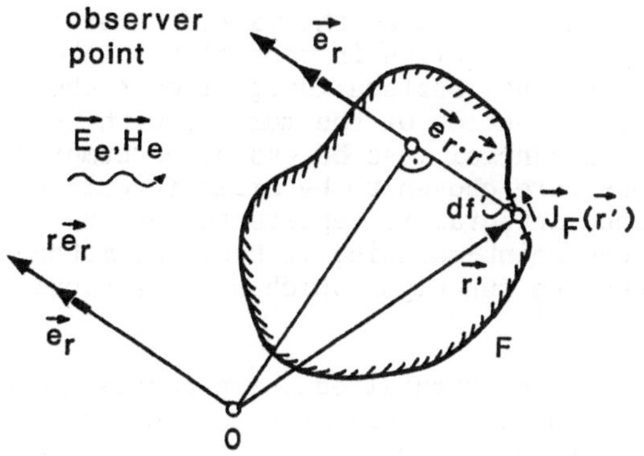


Fig. 3.2 Far-field geometry.

The EFIE is an integral equation of the first kind, the MFIE an integral equation of the second kind. In principle both integral equations may be used independently from another to determine the surface current. However, in practice it turns out that there are significant differences depending from the geometry of the scatterer. The MFIE has advantages if the structure consists of smooth surfaces (e.g. an airplane), whose radius of curvature are large compared to the wavelength. The first term on the right-hand side represents the physical optics current distribution, see Sec. 4. The contribution of the integral can be considered as a correction term to the physical optics theory. The MFIE cannot be used for structures which are very thin compared to the wavelength like thin cylinders or thin shells. In these cases the EFIE must be used. The use of the EFIE is also recommended if the scatterer shows a lot of structural details (e.g. a tank or a helicopter).

The integral equations can be transduced to linear systems of equations by the method of moments. For this purpose a set of known basis functions (expansion functions) with unknown complex coefficients are introduced according which the surface current is expanded. With a second set of known testing functions (weighting functions) both sides of the equations are multiplied and then integrated over the region of the structure. These steps result in a linear system of equations for the unknown coefficients of the basis functions. The right-hand side contains the known incident field depending from aspect angle and polarization. The matrix of the system of equations, frequently denoted as impedance matrix, depends only from the frequency and from the geometry of the structure.

The computer effort of the integral equation method is due to the calculation of the matrix elements and to the matrix inversion. The calculation of the matrix elements requires in general two integrations, one for the evaluation of the original surface integral over the basis functions, the other for the construction of the moments with the testing functions. Both integrals in general must be evaluated numerically. Very often the testing functions were chosen to be Dirac functions. In this case the value of the second integral is represented by the integrand itself and one speaks from the point-matching method. The method of Galerkin is characterized by testing functions which are identical with the basis functions.

An important feature of the integral equation approach consists in the fact, that besides the modeling of the structure no more geometric problems must be solved if the direction of the incident wave varies. This is due to the magnitude R in the integral equations which represents the geometric distance between the integration point and the observer point on the surface, which is independent from the direction of the incident wave. This is also true if there exists some part of the structure between these points.

If the matrix of the system of equations is inverted directly the surface currents can be calculated for each aspect angle and polarization of the incident wave by merely multiplying the right-hand column vector with the inverse. This solution method should, therefore, preferred if for only a few frequencies a lot of changes of the incident field are foreseen.

If the surface current is known a further integration has to be carried out in order to compute the scattered far-field, from which the scattering matrix T and the polarization dependent radar cross-sections may be derived. The expressions for the far-field are given by:

$$\vec{E}_s(\vec{r}) = \frac{j\omega\mu}{4\pi}\,\frac{e^{-jkr}}{r}\int_F (\vec{e}_r \times (\vec{e}_r \times \vec{J}_F(\vec{r}\,'))\; e^{jk\vec{e}_r\cdot\vec{r}\,'}\, df' \;. \tag{3.3}$$

The far-field geometry is represented in Fig. 3.2.

The components of the backscattered field can be related to the components of the incident field by a scattering matrix [T] in the following manner:

$$\begin{bmatrix} E_{sx} \\ E_{sy} \end{bmatrix} = [T] \begin{bmatrix} E_{ex} \\ E_{ey} \end{bmatrix} , \tag{3.4}$$

with

$$[T] = \frac{1}{\sqrt{4\pi\, r^2}} \begin{bmatrix} t_{xx} & t_{xy} \\ t_{yx} & t_{yy} \end{bmatrix} . \tag{3.5}$$

For this representation it has been assumed that the z-axis of a cartesian coordinate system with origin in the neighbourhood of the object is directed towards the radar observer. The elements $t_{ij}$ of the scattering matrix are given by

$$(3.6) \qquad t_{ij} = \sqrt{4\pi r^2}\ \frac{E_{si}}{E_{ej}}, \qquad r \to \infty, \quad i = x,y, \quad j = x,y .$$

The polarization dependent cross-sections then are given by

$$(3.7) \qquad \sigma_{ij} = t_{ij}\, t_{ij}^{*} .$$

An analytical evaluation of the scattering matrix is not possible in general. Since the integral equation method determines the current distribution at the surface of the structure a 3-D model of the surface only is required. There are two principal ways to establish surface models. One way consists in modeling a solid surface body with a grid of wires, the so-called wire-grid model. The other common approach breaks the surface up into patches or cells each having a continuous metallic surface. Both models are discussed in the following two sections in more detail.

## 3.2 WIRE-GRID MODEL

In the field of antennas there are a variety of structures which consist of wires like a corner reflector antenna, see Fig. 3.3 or a Yagi-Uda-antenna, see Fig. 3.4. For both antennas the wire technique is used to generate specific antenna characteristics.

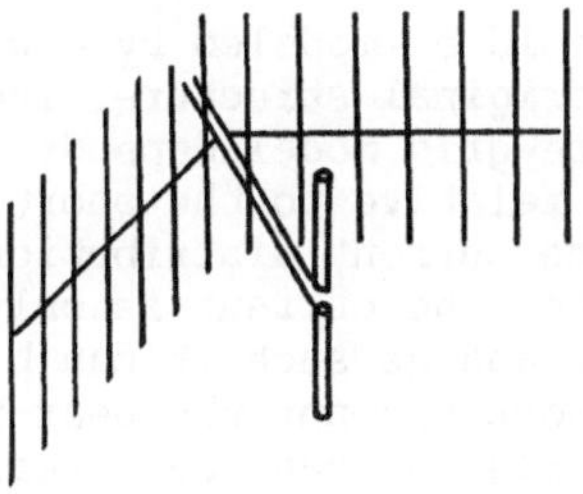

Fig. 3.3 Corner reflector antenna.

Wire-grid meshes also find many uses in applications where the effect of a solid conducting surface is required but the weight and/or wind resistance of the latter must be avoided. They may be used, for example, to fabricate radar antenna reflectors, see Fig. 3.5, and as shields to screen sensitive equipment from stray fields [8].

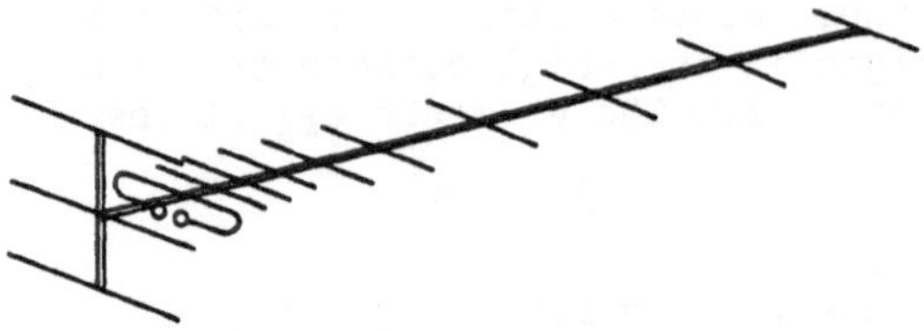

Fig. 3.4 Yagi-Uda array.

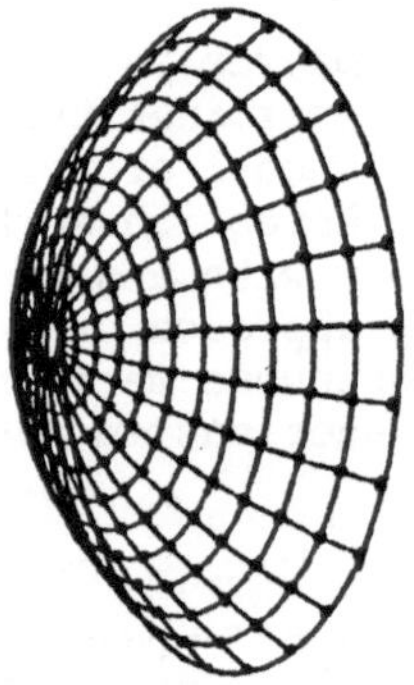

Fig. 3.5 Radar antenna reflector.

If any of the three structures would be modeled by a wire-grid, the model would perfectly agree with the original structure. The substitution of an arbitrary solid surface by a wire-grid model depends upon the fact, that as the mesh size becomes smaller relative to the shortest wavelength of concern, the mesh supports a surface current distribution which approaches that on the continuous surface. The current is only an approximation to the actual current, however, and as such it can be expected to reasonably predict the far-field but possibly not the near-fields. This is due to the fact that the grid supports an evanescent reactive field on both sides of its surface. An actual continuous surface is not capable of supporting such a field [18].

If thin (compared to the wavelength) wires are chosen to construct a wire-grid the current esentially has only an axial component. In this case the EFIE simplifies considerably. The MFIE, however, will fail in the thin-wire approximation. The thin-wire EFIE is given by

$$(3.8)\qquad \vec{s}\cdot\vec{E}_e(s) = -\frac{1}{4\pi j\omega\varepsilon}\,\vec{s}\cdot\int_L \Big((-1-jkR+k^2R^2)\,\vec{s}\,' + (3+3jkR-k^2R^2)(\vec{s}\,'\cdot\vec{e}_R)\vec{e}_R\Big)\, I(s')\,\frac{e^{-jkR}}{R^3}\,ds' ,$$

where $\vec{s}$ is the unit tangent vector of the wire at the observer point and $\vec{s}\,'$ is the unit tangent vector of the wire at the integration point. Fig. 3.6 shows two wires i and j from a wire-grid for which the interaction after discretization of Eq. (3.8) is computed. j is the wire with the integration point and i the wire with the observer point. For more details,

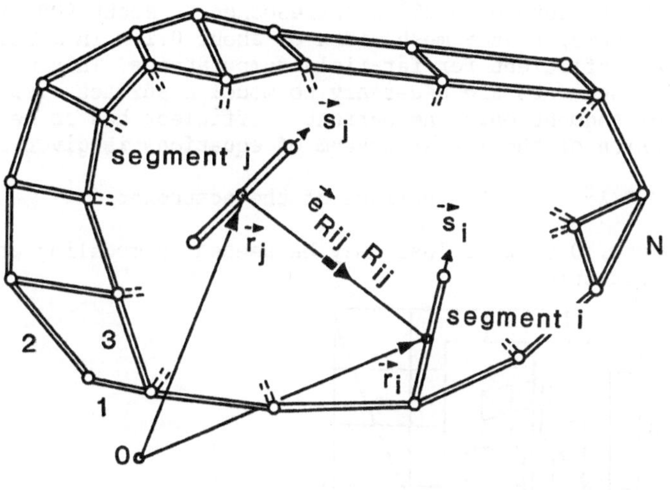


Fig. 3.6 Geometric situation of two wires i and j of the ensemble of N wires.

see [14].

Reference [10] was apparently the first report on the application of the thin-wire EFIE to the analyses of wire-grid models for circular disks and spheres. Satisfactory agreement was demonstrated between the wire-grid results and independent analytical or experimental back-scatter cross-section data presented as a function of frequency.

Since the number of the unknowns of the linear system of equations is identical with the number of wire segments an estimation of the minimum mesh width is of great interest. This is primarily dependent from the choice of basis and test functions.

A good estimation of the mesh width is obtained, if a source is positioned within a structure, which is modeled by a wire-grid with variable mesh width. The field in the exterior of the structure, which should be zero, is then computed in dependence from the mesh width.

For this theoretical experiment a cube with an edge length of 0.4 $\lambda$ has been chosen. In the center of the cube a point source was positioned. The diameter of the wire was 0.003 $\lambda$, so that the thin-wire approximation of the EFIE holds. The basis functions are chosen to be pulses, the boundary conditions were satisfied in the center of the wire (Dirac functions as testing functions).

Fig. 3.7 shows the electric field in the exterior in dependence from the mesh width. A mesh width of 0.057 $\lambda$ corresponds to about 600 wire segments. One can assume, that a mesh width of about 0.1 $\lambda$ in modeling a structure should be sufficient for far-field computations. This means, that about 200 wire segments are necessary to model a surface with a size of $\lambda \times \lambda$. Since per segment only one current coefficient has to be determined the dimension of the linear system of equations is given by

(3.9a) $\qquad N \approx 200\ F/\lambda^2 , \qquad F$ = surface of the scatterer.

Shorter wire segments, 0.05 $\lambda$ or less, may be needed in modeling critical regions of a structure.

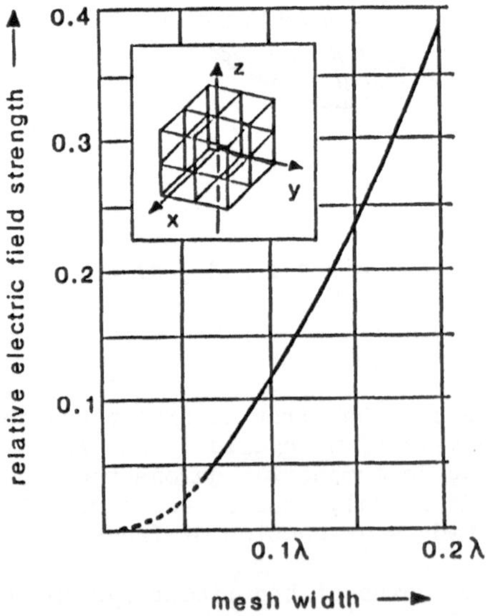


Fig. 3.7 Influence of the mesh width.

More sophisticated studies show, that the accuracy with which a wire-grid model simulates an actual surface depends on the computer code (i.e., expansion and weighting functions)used, the radius of the wire segments used, as well as the grid size. For example, with pulse basis functions it has been found that a grid spacing of about 0.1 λ to 0.2 λ yields good results. With the piecewise sinusoidal Galerkin method, it has been found that the grid size should not exceed λ/4 and that a suitable wire radius is a = w/25 where w denotes the width or length (whichever is greater) of the apertures [18]. A grid size of λ/4 would lead to the following expression for the dimension of the linear system of equations:

$$N \approx 32\, F/\lambda^2 \,. \qquad (3.9b)$$

From the above remarks concerning the substitution of a solid surface by a wire-grid system one concludes that the far-field properties of the structures such as the radar cross-section or the antenna radiation pattern can be predicted with sufficient accuracy. This is demonstrated at hand of the computed far-field pattern of an antenna system installed on a helicopter of the type BO 105.

Fig. 3.8 presents the details of the actual structure with the position of the two λ/4-monopoles operating in the VHF-band. Fig. 3.9 shows the wire-grid model. The antenna on the right-hand side of the flight direction was driven, the other was terminated. The in flight-measurement

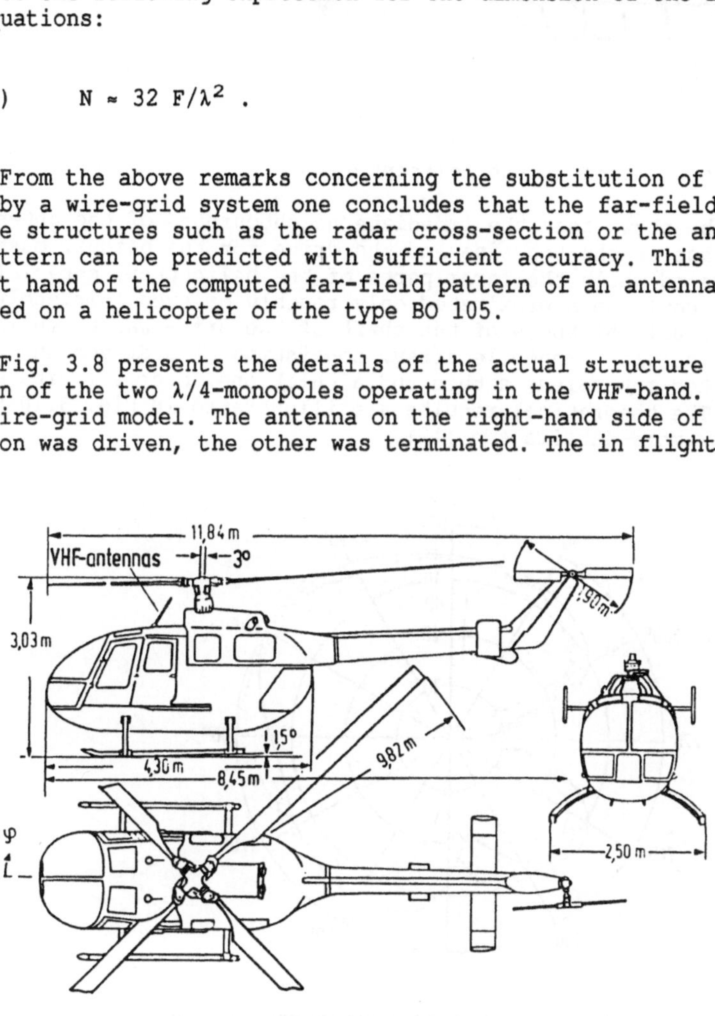


Fig. 3.8 Position of the VHF-antennas on the helicopter.

of the radiation pattern in the horizontal plane for a frequency of 117.6 MHz is presented in Fig. 3.10 by the dashed line. The flight direction is defined by $\varphi = 0°$ .

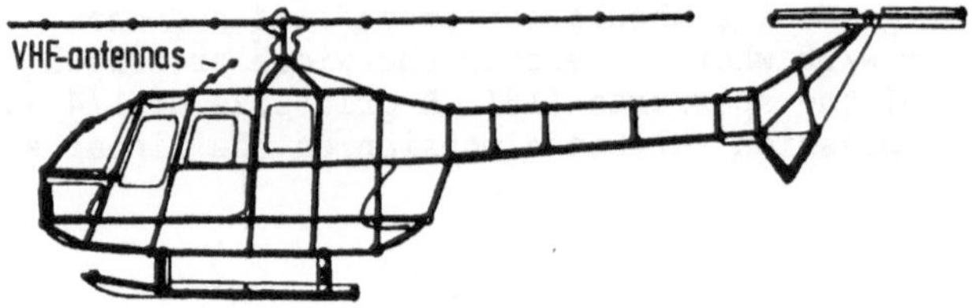


Fig. 3.9 Wire-grid model of the helicopter.

It could be shown that the immediate neighbourhood of the antennas, that is the shape of the top side and the drive for the blades, must carefully be modeled while the farer parts of the helicopter, especially the lower part could be approximated only roughly or even completely neglected. The actual thickness of the shaft of the drive was taken into account. In principle a monopole array, consisting of a driven and two parasitic excited monopoles, a thin and a thick one, over a finite plane, was analyzed. The theoretical results are illustrated in Fig. 3.10 by the solid line. For more details see [12].

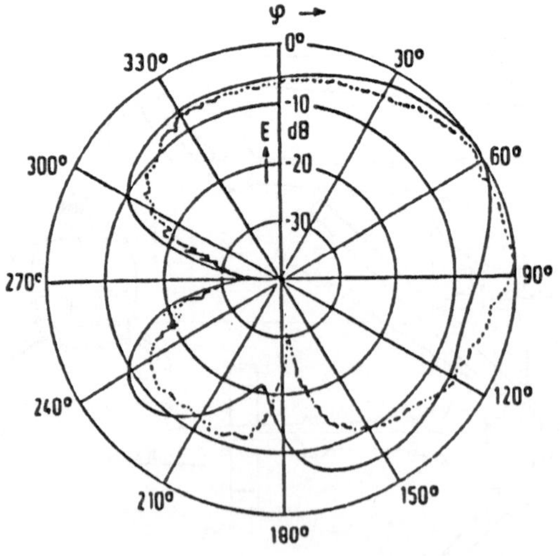


Fig. 3.10 Comparison between in flight-measurements (dashed line) and results of the IEM (solid line), frequency 117.6 MHz.

Such studies, while illustrating the applicability of wire-grid meshes as models for solid surfaces in terms of their far-field electromagnetic behaviour are not entirely convincing as to the use of wire-grid models to determine near-field quantities such as current distributions [8]. Preliminary studies in this regard to compare the results obtained with independent theoretical results or with experimental results are not yet conclusive. Such comparisons should do much to more clearly define areas of applications and limitations of wire-grid models. A special problem seems to be the stability of numerical results, that is the independence of the solution from the number of wires.

This is demonstrated at hand of a simple dipole-antenna of length 0.45 λ and diameter of 0.014 λ. Fig. 3.11 shows the input admittance (proportional to the current at the source point) which is a near-field quantity in dependence from the number of wires which are used to model this dipole [24]. The upper diagram represents the real part (conductance), the lower diagram the imaginary part (susceptance) for two feeding models. The dashed line is based on a voltage-source model, the solid line on a frill-current model. One realizes that the conductance for both source models tends to a finite value in dependence from the number of wires. The susceptance, however, does not reach a stable value in both cases. This behaviour may be due to either the source models or the thin-wire approximation of the dipole or to the specific combination of basis and testing functions or the neglected end cap contribution. Further studies are in progress.

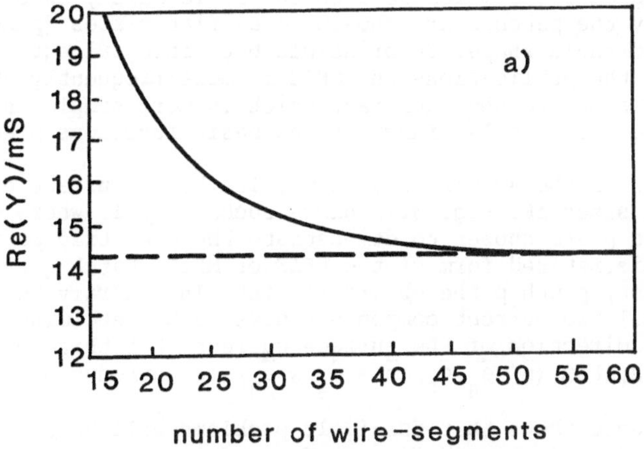


Fig. 3.11a Input conductance of a dipole antenna in dependence from the number of wire-segments; ___ magnetic frill current-source, --- voltage source.

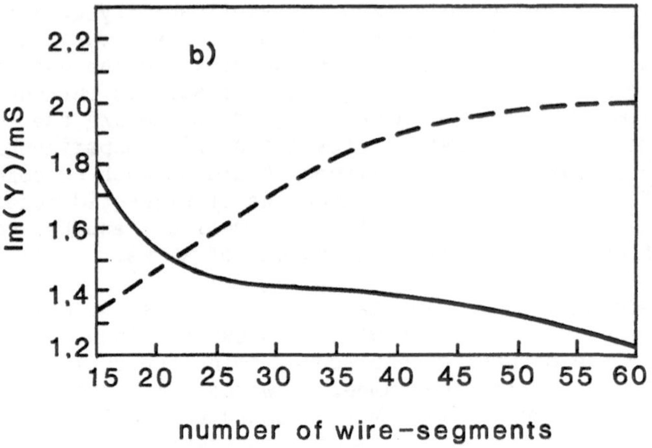


Fig. 3.11b Input suspectance of a dipole antenna in dependence from the number of wire-segments; ___ magnetic frill current-source, --- voltage-source.

### 3.3 SURFACE PATCH MODEL

The continuous surface model by patches is an alternative approach for the modeling of 3-D complex structures. It is mainly used for smooth surfaces. Mostly the patches are chosen to be flat plates (panels) of a triangle or quadrangle shape. In principle both integral equations may be used. Following the publications the MFIE is more frequently used than the EFIE. This is due to the integrand which is more simple in the case of the MFIE. The use of pulse functions as basis functions predominates.

The surface of the smooth body is modeled by $N^*$ surface patches with individual sizes $\Delta F$. Fig. 3.12 shows such a model, where two surface patches q and p are chosen to demonstrate the geometric parameters required in the discretized form of the EFIE or MFIE. Patch q contains the integration point, patch p the observer point. In contrary to the wires of the grid model two current components have to be determined in order to describe the direction of the surface current. For this purpose a local coordinate system ($\vec{n}$, $\vec{e}_\eta$,$\vec{e}_\xi$, $\vec{n} = \vec{e}_\xi \times \vec{e}_\eta$) is introduced for each panel.

The experience shows that far-field problems will be solved with sufficient accuracy if one models a surface of one square wavelengths by at least 20 to 30 patches. This means that the edge length of a surface patch should have an amount of less than about $\lambda/5$ (maximum size of about 0.04 square wavelengths). This estimation is based on pulse functions as basis functions and Dirac functions as testing functions. The dimension N (number of complex unknowns) of the linear system of equations to be solved is determined as follows: a patch size of about $\lambda/5$ results in 25 patches for one square wavelength of the scatterer's surface. For each patch 2 complex current coefficients have to be determined. Therefore, the dimension of the complex linear system of equations can be estimated by

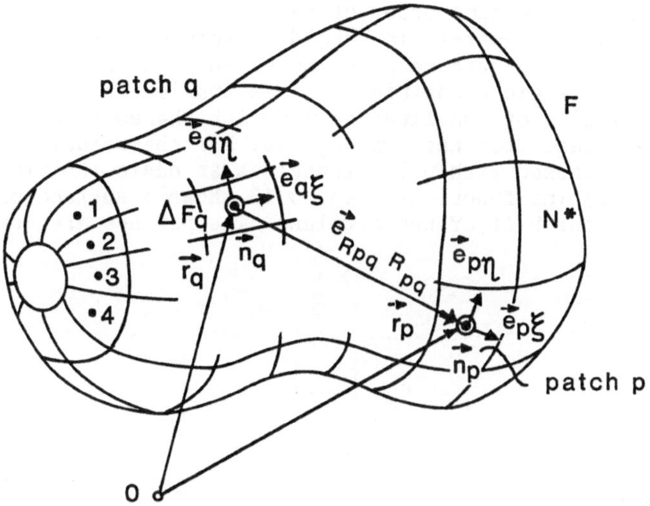


Fig. 3.12 Geometric situation of two patches p and q of the ensemble of $N^*$ patches used for a continuous surface model of a structure.

$$N = 2N^* \approx 50\ F/\lambda^2 \ . \qquad (3.10)$$

An estimation of the computer time $\tau$ for matrix inversion in dependence from the surface area F of a general 3-D structure is presented in Table 3.1. In the second column of the table the edge length $a/\lambda$ of an equivalent cube as a representative for a general 3-D structure is given and in the third column the number N of the complex unknowns which have to be determined are listed. It is assumed that an economic programming under introducing of block structures has been achieved. From this table one can realize that the limit of application of the integral equation method with respect to the structure's dimensions is drawn by the required computer effort.

Table 3.1 Matrix inversion time (computer Cray-1) in dependence from the surface area F of a general 3-D structure.

| $F/\lambda^2$ | $a/\lambda$ | N | $\tau$/s |
|---|---|---|---|
| 10 | 1.29 | 500 | < 50 |
| 20 | 1.82 | 1000 | 50 |
| 30 | 2.23 | 1500 | 180 |
| 40 | 2.58 | 2000 | 350 |
| 50 | 2.89 | 2500 | 700 |
| 60 | 3.16 | 3000 | 1200 |

The figure N is in the same order of magnitude as it is in the wire-grid model. In the case of near-field considerations much finer modeling has to be chosen. Since the surface patch model in principle should be suited to compute near-field quantities one has applied the surface patch model for the calculation of the lines of constant phases in the extreme near-field of an airplane from the type Do 228. For the solution of this problem the MFIE was chosen with pulse functions as basis functions and Dirac-functions as testing functions. Fig. 3.13 shows a surface patch model of 280 patches for half of the airplane. The patches are quadrangles and triangles.

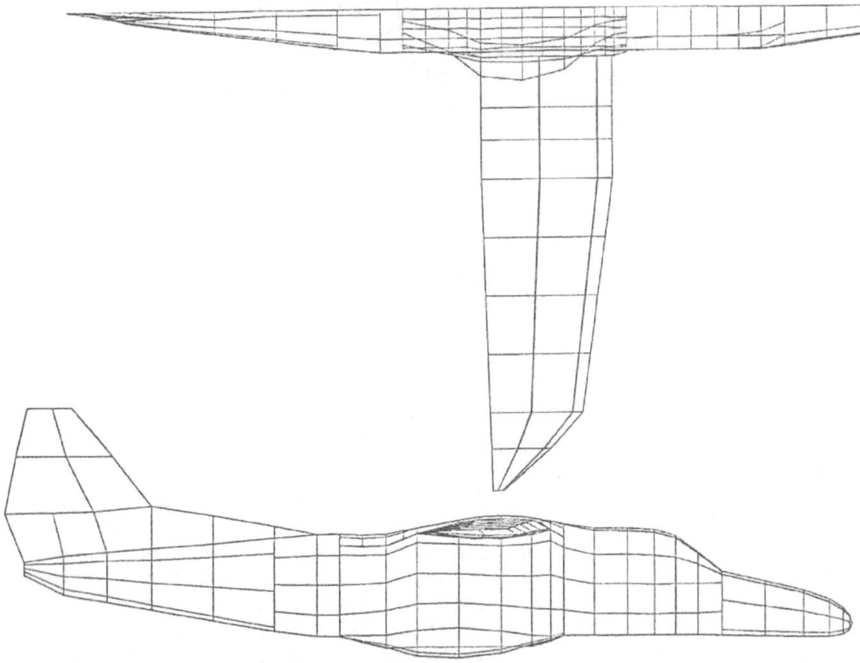

Fig. 3.13 Surface patch model of an airplane from the type Do 228.

The lines of constant phases, represented in Fig. 3.14, are calculated for a wavelength of $\lambda = 6.0$ m and a distance of $\lambda/30$ from the upper part of the fuselage. The plane wave is incident towards the nose of the airplane under an elevation angle of 30° over the horizontal plane. The polarization vector is vertical to this plane. The increment between the lines of constant phases has an amount of 10°.

Since direct experimental validations require a high effort the method was verified at hand of a cube with edge length $a = \lambda$. One side of the cube was subdivided in 5 × 5 quadratic panels for measurement purposes, see Fig. 3.15. In the middle of each panel a hole was bored by which probes for the surface fields could be positioned from the interior of the cube. For computational purposes the number of the panels was varied until a stable result was attained.

The comparison between measurement (circle) and theory (solid line) is drawn in Fig. 3.16. Fig. 3.16a represents the amplitude and Fig. 3.16b the phase of the surface current on the front of the cube.

The component of the current distribution is chosen to be parallel to the vector of the field incident vertically on the front. Amplitudes and phases are relative values. Parameter of the curves is the row of holes defined by the normalized z-coordinate.

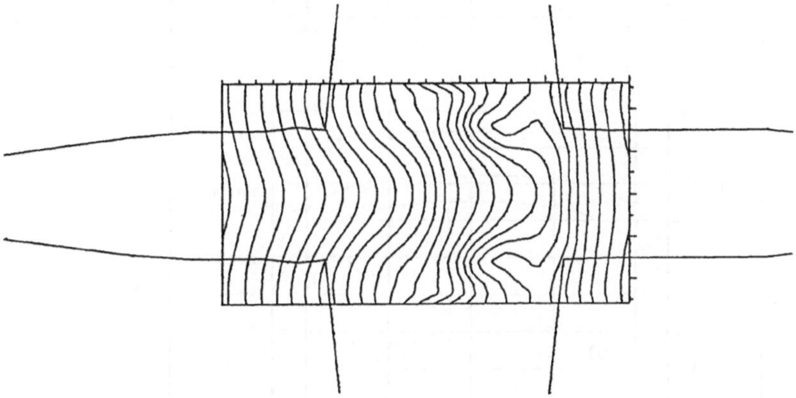

Fig. 3.14 Lines of constant phases in the near-field of an airplane.

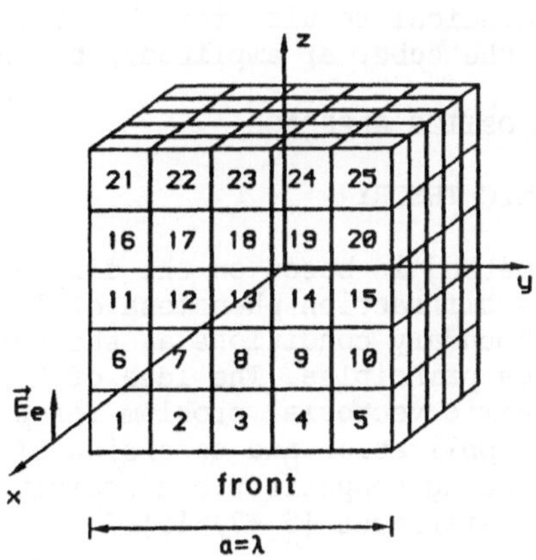


Fig. 3.15 Panel model of a cube.

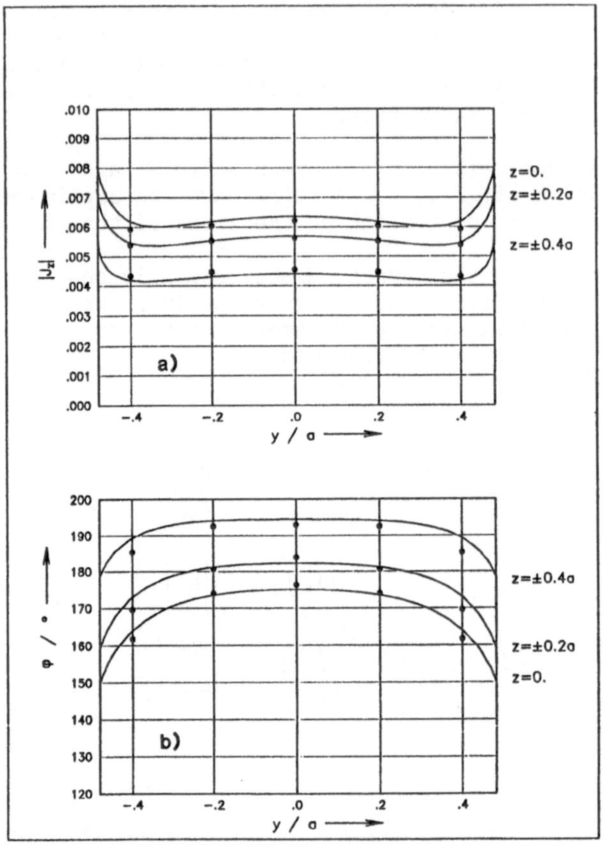


Fig. 3.16 Experimental and theoretical results for the current distribution on the front of the cube. a) amplitude, b) phase.

## 4. MODELS USED IN THE PHYSICAL OPTICS APPROACH

### 4.1 REMARKS TO THE ELECTRODYNAMIC THEORY

The physical optics (PO) method is based on the diffraction theory of Kirchhoff, who described the diffraction phenomena of light (scalar problem) by approximating the boundary conditions at the surface of the scatterer with the aid of optics principles. The idea of Kirchhoff has been extended to the electrodynamic vectorial problem for perfectly conducting bodies [3, 4, 11]. The application and extension of the PO-method for complicated structures including nonperfect conductivity and double reflections is reported in the references [5 -7, 15].

If the scatterer is perfectly conducting the following boundary conditions according the idea of Kirchhoff are valid for the shadow region of the structure:

$$\vec{n} \times \vec{E} = 0 \, , \tag{4.1}$$

$$\vec{n} \times \vec{H} = 0 \, , \tag{4.2}$$

while for the illuminated region the boundary conditions are given by

$$\vec{n} \times \vec{E} = 0, \tag{4.3}$$

$$\vec{n} \times \vec{H} = \vec{n} \times \vec{H}_e + \vec{n} \times \vec{H}_r = 2\vec{n} \times \vec{H}_e \, , \tag{4.4}$$

where $\vec{H}_r$ is the vector of the reflected magnetic field.

The boundary conditions for the electric field are exact, while the boundary conditions for the magnetic field would only be exact if the scatterer would consist of an infinitely extended plane, where $\vec{n} \times \vec{H}_r = \vec{n} \times \vec{H}_e$ in the illuminated region and $\vec{n} \times \vec{H} = 0$ in the shadow region holds. However, it can be assumed that the above given boundary conditions would approximate very well the actual ones, if the scatterer to be modeled is large compared to the wavelength. So a modeling of the structure by flat surface patches (panels) is a natural consequence of the formulation of the boundary conditions.

From Eq. (4.4) follows that the surface current is given by

$$\vec{J}_F = 2\vec{n} \times \vec{H}_e \tag{4.5}$$

which, therefore, is identical to the first term on the right-hand side of the rigorous integral equation, see Eq. (3.2). The evaluation of the far-field integral, Eq. (3.3), can be done analytically and leads to the following expression for the backscattering matrix of a single panel

$$[T_1] = - \frac{jke^{-jkr}}{2\pi r} \int_{F_p} e^{2jkz'} \, dx' \, dy' \begin{bmatrix} 1 & 0 \\ 0 & 1 \end{bmatrix} . \tag{4.6}$$

The scattering matrix of the total object then is the sum of the scattering matrices of the individual panels. If a panel has straight edges the phase integral can be solved analytically, so that the calculation of the scattered field requires only a minor computer effort.

If the scatterer is nonperfectly conducting one cannot formulate any exact boundary condition in the PO-sense. In the shadow region the boundary conditions again are approximated by Eqs. (4.1) and (4.12) while in the illuminated region

$$(4.7) \qquad \vec{n} \times \vec{E} = \vec{n} \times \vec{E}_e + \vec{n} \times \vec{E}_r ,$$

$$(4.8) \qquad \vec{n} \times \vec{H} = \vec{n} \times \vec{H}_e + \vec{n} \times \vec{H}_r$$

are assumed. $\vec{E}_r$ is the vector of the reflected electric field. Again these boundary conditions would be exact, if the scatterer could be represented by an infinitely extended nonperfectly conducting plane. Since the reflected field can be calculated from the incident field by multiplication with the reflection coefficients of Fresnel the scattered field again can be evaluated analytically from the integral representation for the far-field

$$(4.9) \quad \vec{E}_s(\vec{r}) = \frac{j\omega\mu}{4\pi}\frac{e^{-jkr}}{r}\int_F \left(\vec{e}_r\times(\vec{e}_r\times\vec{J}_F(\vec{r}\,'))+\sqrt{\frac{\varepsilon}{\mu}}\,(\vec{e}_r\times\vec{K}_F(\vec{r}\,'))\right) e^{jk\vec{e}_r\cdot\vec{r}\,'}\,df' ,$$

where the magnetic surface current is given by

$$(4.10) \qquad \vec{K}_F = -\vec{n} \times \vec{E} .$$

One receives for the scattering matrix of a single panel

$$(4.11) \quad [T_2] = -\frac{jke^{-jkr}}{2\pi r(1-n_z^2)}\int_{F_P} e^{2jkz'}\,dx'\,dy' \begin{bmatrix} R_H\, n_x^2 - R_E\, n_y^2 & (R_H+R_E)\, n_x\, n_y \\ (R_H+R_E)\, n_x\, n_y & R_H\, n_y^2 - R_E\, n_x^2 \end{bmatrix} .$$

$n_x$, $n_y$, $n_z$ are the components of the normal unit vector of the panel. $R_H$ resp. $R_E$ are the reflection coefficients at the surface of a multilayered panel for the case that the incident magnetic field resp. electric field is directed parallel to the surface of the panel, see Sec. 7.1 and [9].

The ansatz of PO implies that the current distribution will be constant in the amplitude over the surface of a panel and varies proportional to the phase of the incident field. Since this result differs from the result of more rigorous solutions it is necessary to estimate the deviations. For this purpose a strip with a width of $a = \lambda$ is considered, the edges of which are directed along the z-axis, which is vertical to the plane of the Fig. 4.1. The vector of the incident electric field, which hits the strip plane vertically is also directed along the z-axis, that is parallel to the edges of the plane. The current distribution over the strip computed with the IE-method [23] is represented by the solid line, whereas the constant amplitude of the PO-current is indicated by the dashed line.

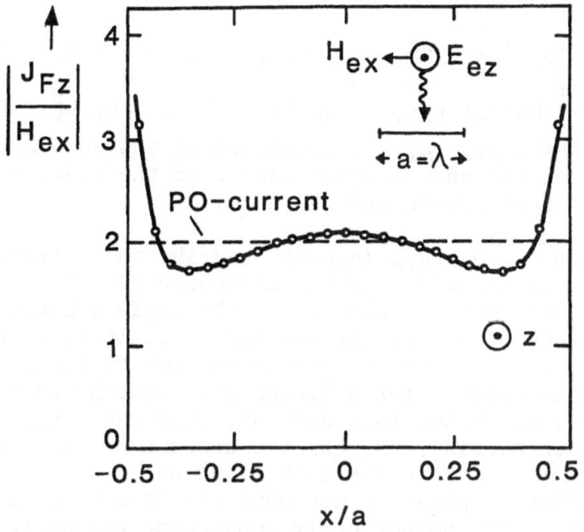


Fig. 4.1 Comparison of the current distributions calculated by the IE- and the PO-method.

One finds that the magnitude of the differences between the IE- and PO- result depends on the distance from the edge. In the middle of the panel the deviations are small while in approaching the edges the differences increase considerably. With increasing panel size referred to the wave length, the panel area in which edge current effects play a role are much smaller than the remaining panel area where the PO-current dominates. Therefore, the PO-field will represent the actual field with increasing accuracy if the panel size increases. That is the range of frequencies for which the PO-method may be applied joins on the frequency range of the IE-method.

The need to improve the PO-field for small panels gives rise to the development of an edge correction term. The evaluation of an edge correction term for panels with perfectly conducting faces is based on the rigorous solution of an infinitely extended wedge. This solution, however, is very complex so that an asymptotic representation ($kr \gg 1$) of the result is preferred, which leads to analytical expressions for the total scattered field consisting of the PO-field and an edge diffracted field. Since a correction term is required the PO-field in its asymptotic form must be removed from the total scattered field. Following this idea which is the basis of the physical theory of diffraction (PTD), one receives the backscattering marix for the edge of a perfectly conducting wedge as follows [16, 17]:

$$(4.12)\qquad [T_3] = -\frac{e^{-jkr}}{r}\, L\, \frac{\sin(kL\cos\beta)}{kL\cos\beta}\, \frac{e^{j2kr_{Mz}}}{1-t_z^2} \times$$

$$\times \begin{bmatrix} D_e t_x^2 + D_m t_y^2 - D_{em}\, t_x t_y & (D_e - D_m)\, t_x t_y + D_{em} t_x^2 \\ (D_e - D_m)\, t_x t_y - D_{em}\, t_y^2 & D_m t_x^2 + D_e\, t_y^2 + D_{em} t_x t_y \end{bmatrix},$$

with L = length of the edge, $r_{Mz}$ = z-coordinte of the edge mid-point, $t_x$, $t_y$, $t_z$ = components of the unit tangent vector of the edge, $\beta$ =angle between the direction of incidence and the edge.

The coefficients $D_e$, $D_m$, $D_{em}$ $(\beta,\psi,n)$ describe the difference between the asymptotic rigorous solution and the asmptotic PO-solution and depend besides $\beta$ from the outer wedge angle $n\pi$ and the angle $\psi$ between the plane of incidence and a face of the wedge, see Sec. 7.2. It is emphasized that the rigorous asymptotic solution is carried out for an infinite long wedge, while actually the results for a finite edge length are needed. That is corner effects are not taken into account. This can cause the unsymmetry of the above given matrix. However, the effects of an edge with finite length is comparable to the basis idea of PO which uses a reflection coefficient of an infinite plane to estimate the effects of a finite panel. A similar solution for nonperfectly conducting wedges is missing at this time.

The matrixes $[T_1]$, $[T_2]$ and $[T_3]$ represent the most important analytical tool of the PO. The illuminated elements of the surface patch model can be classified as follows: perfectly conducting panels, nonperfectly conducting panels, perfectly conducting doubly reflecting panels, nonperfectly conducting doubly reflecting panels, perfectly conducting edges. The doubly reflecting panels can be treated like the directly reflecting panels using the method discussed in Sec. 4.4. An independent summation of all matrixes $[T_1]$, $[T_2]$ and $[T_3]$ can be achieved. The sum of the individual summations then represents the scattering matrix of the total object.

The evaluation of the matrix elements and the summation of the matrizes can be done in a very economic way. Thus the computer effort for the electrodynamic calculations (about 0.05 s per panel and aspect angle) is much less than that of the IE-method. Another important difference concerns the size of the panels. The IE-method requires panel dimensions in the order of $\lambda/5$ even if a cube is modeled for example. The panels of the PO-method may be as large as the actual surface can be modeled with sufficient accuracy, see the following section. The cube, therefore, can be modeled accurately with only 6 panels. This fact is also suited to reduce the computer time considerably. However, in contrary to the IE-method the geometric problems are not at all solved with the creation of a surface model. According the geometrical optics idea a decision must be made whether a panel is illuminated, partially illuminated or hidden. The same is true for all edges. This decision must be made whenever the observer point changes. A similar geometric problem is connected with double and multiple reflections. It can be concluded that the ray-tracing problem inherent to the GTD, becomes also relevant with a PO-method of increasing complexity.

While the principle of PO is known since a long time only a limited experience exists in applying the method for complex structures. This requires an extensive comparison of the theoretical results with experimental results which has be done with good success for a variety of metallic objects [5 - 7, 15]. An estimation of the edge diffraction theory will be available in short. Tests to compare theory and experiment for nonperfectly conducting bodies are in progress [16, 17].

4.2 SIZE OF THE PANELS

In modeling a structure by panels the question arises according to which criterium the size of the panels has to be determined. On the one hand one would like to choose the panel size as large as possible in order to save computer time. On the other hand the admissible deviation between the true surface and the model surface is subject to the required accuracy of the electromagnetic magnitudes. A series of tests has shown that the deviation between the true surface and the model surface should not exceed a value of about

$$\Delta \approx \lambda/16 , \quad \text{see Fig. 4.2.} \tag{4.13}$$

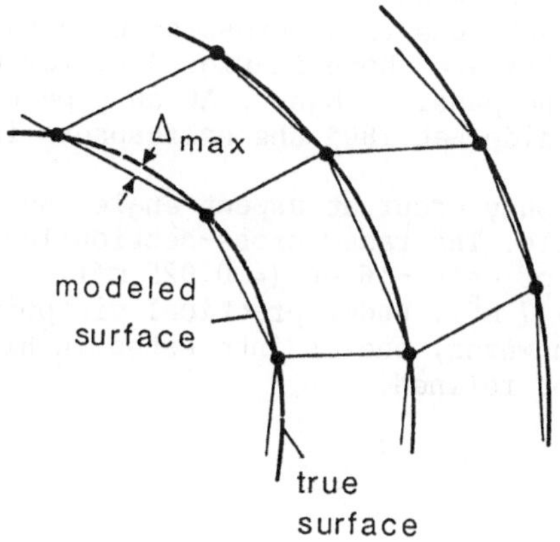


Fig. 4.2 Admissible deviation between the true and the modeled surface: $\Delta_{max} \approx \lambda/16$.

This criterium is well known from antenna measuring technique. If the admissible phase error over the aperture with diameter D of the antenna under test is assumed to be 22.5° ($\lambda/16$), then the far-field distance R must be chosen in such a way that $R > 2D^2/\lambda$. The true far-field pattern for the distance $R \to \infty$ then will differ only in a negligible amount from the measured one.

In order to estimate the errors which are generated by the differences between an actual surface and a panel model of it the following test was arranged. The test object consists of a cone, a cylinder and a half-sphere and is manufactured twice: one configuration with smooth surfaces and the other by modeling the smooth surfaces by panels, see Fig. 4.3.

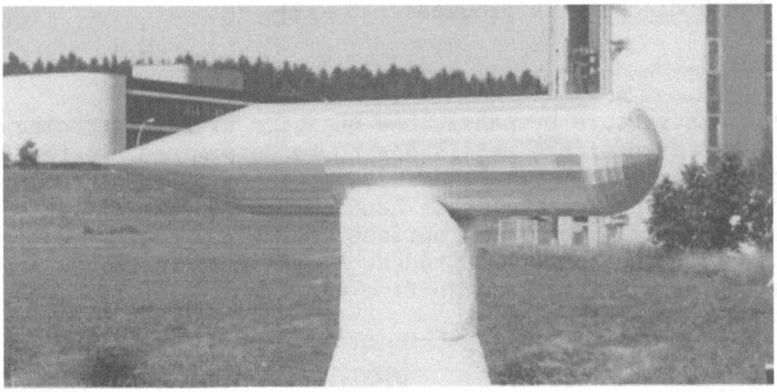

Fig. 4.3 Test object modeled by panels.

The geometric differences between these two objects does not exceed a value of $\lambda/16$. For both objects the radar cross-section has been determined by experiment. The results are shown in Fig. 4.4a for the smooth object and in Fig. 4.4b for the paneled object. At an aspect angle of 90° the object is seen from broadside, at 180° the half-sphere is seen.

Significant differences only occur at aspect angles near 180° where the half-sphere becomes visible. The radar cross-section level in this region, however, has a level of only -16 dB ($\approx 0.025\ m^2$) compared to the peak level of about 8.5 dB ($\approx 7\ m^2$). Under practical viewpoints this difference can be ignored. If, however, one is interested in higher accuracies the panel model has to be refined.

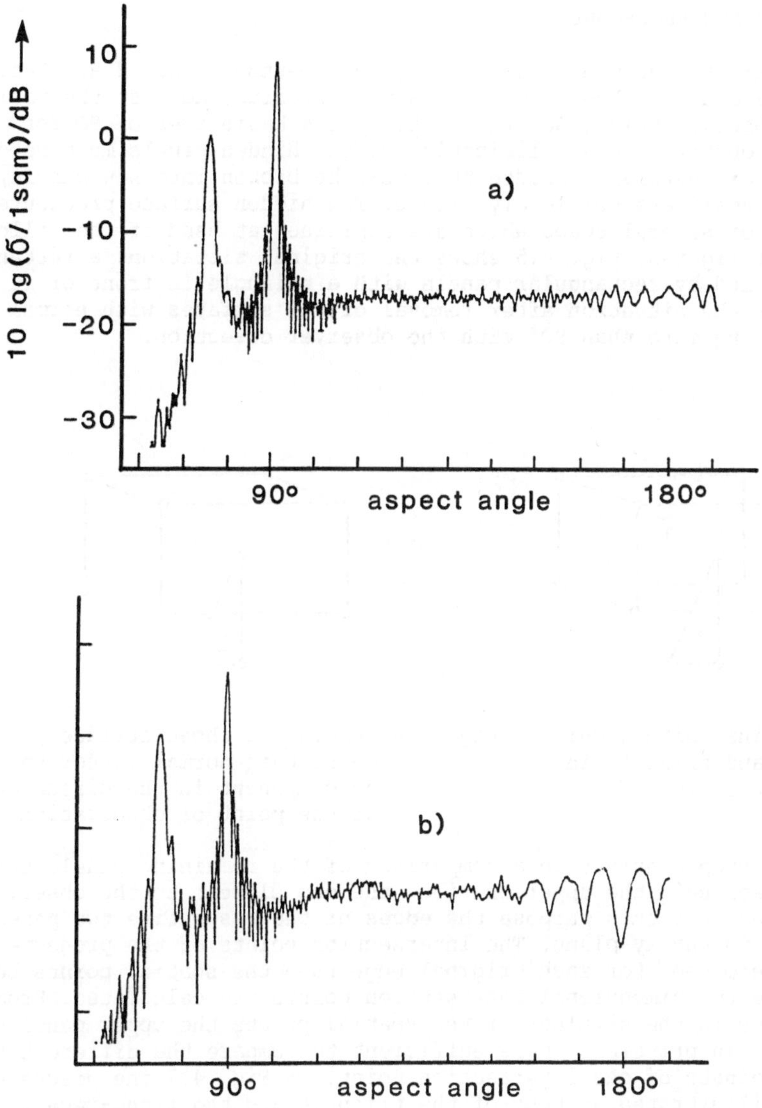


Fig. 4.4 Comparison of the measured radar cross-section for a smooth and a paneled test object. a) smooth object, b) paneled object. Length: 2450 mm, diameter: 440 mm, frequency: 15.5 GHz, polarization: hh.

## 4.3 HIDDEN SURFACE PROCEDURE

In contrary to the models of the rigorous methods, Sec. 3 and Sec. 5, the surface model of the PO-method is not invariant against the position of the observer point. According the optics basis idea of PO for each point of observation an elimination of the hidden panels must be done. If panels are partially hidden they must be broken into new panels, some of which again are completely hidden. The hidden surface procedure [13] consists of several steps which are explained at hand of the following series of figures. Fig. 4.5 shows the original situation: a rectangular box modeled by rectangular panels with a triangle in front of it. Fig. 4.6 shows the situation after removal of all surfaces with normal vectors including more than 90° with the observer direction.

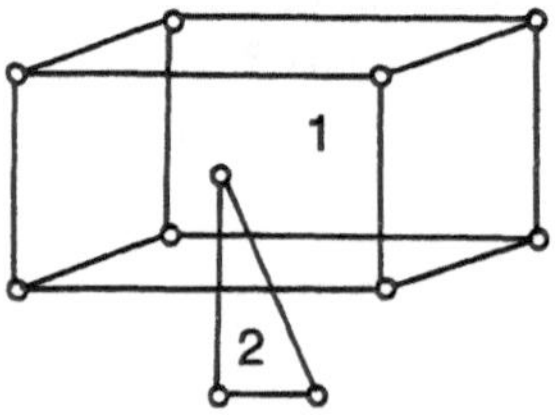


Fig. 4.5 Original situation: box and triangle in front of it.

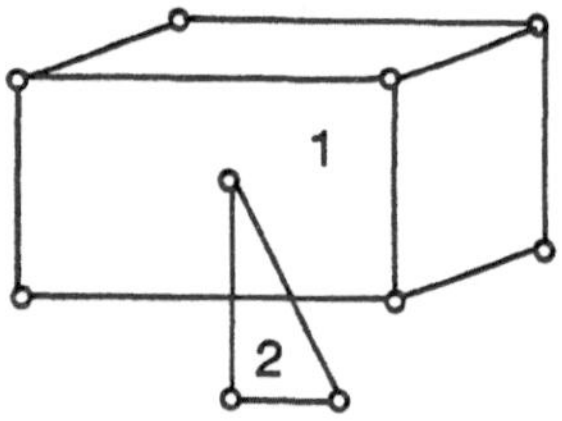


Fig. 4.6 Removal of those surfaces, whose unit normal vector has no component in the direction of the point of observation.

The next step consists in a comparison of the remaining panels two by two. For each pair the upper one i.e. the one closest to the observer has to be found. For this purpose the edges of the respective two panels are projected in the xy-plane. The intersection points of the projected edges are determined. For each original edge then the spatial points belonging to the two-dimensional intersection points are calculated. From the differences in the z-values of the spatial points the upper panel can be determined. In principle it is sufficient to compare the differences in the z-coordinate of one intersection point. In Fig. 4.7 the discussed procedure is illustrated at hand of the triangle and the front-face of the box.

If no intersection points can be found the following two situations may occur: one projected panel lies in the interior of the other, see Fig. 4.8, or both panels are wholly apart, see Fig. 4.9.

For the decision which situation is in question each corner point P of the second panel is inserted in the equations ($\vec{x} = \vec{x}_{PO} + \lambda\vec{x}_{P2} + \mu\vec{x}_{P4}$)

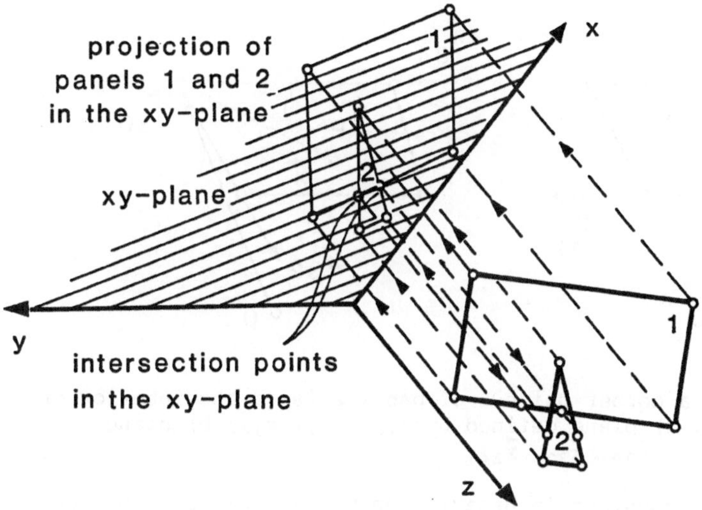


Fig. 4.7 Determination of the upper one of two panels.

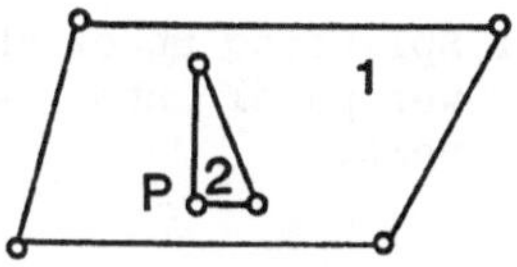


Fig. 4.8 No intersection point: panel 2 lies in the interior of panel 1.

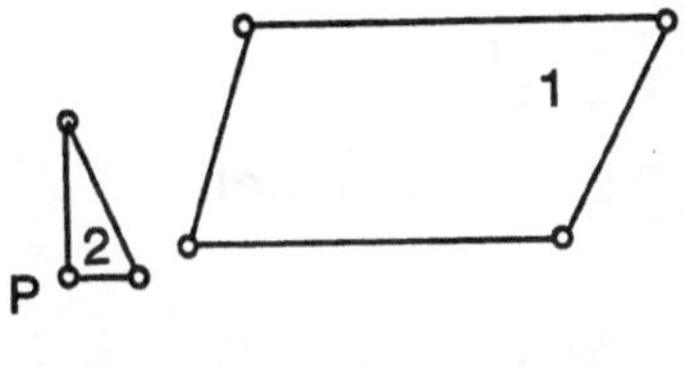


Fig. 4.9 No intersection point: the panels don't hide each other.

for two identical planes which are generated by the vectors of the adjacent edges of panel 1. This is illustrated by Fig. 4.10. If all of the parameters of the plane equations have positive values the edge point P and therewith the total projected panel 2 lies in the interior of the projected panel 1. Again from the difference of the z-coordinates of the projected corner points one can decide which of the panels is the upper one.

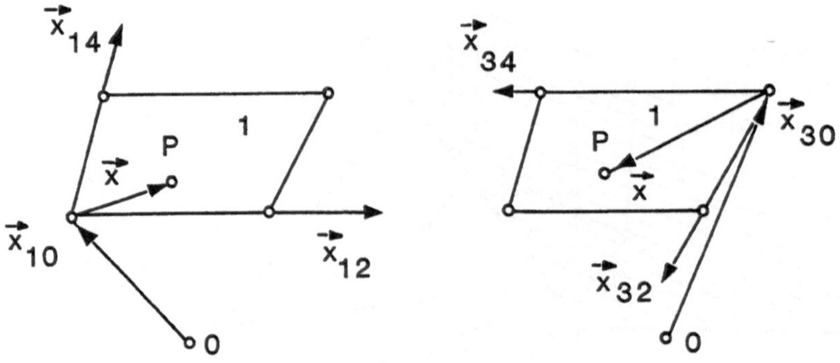


Fig. 4.10 Test if a corner point P of panel 2 is in the interior of panel 1. a) plane defined by $\vec{x}_{10}$, $\vec{x}_{12}$, $\vec{x}_{14}$; b) plane defined by $\vec{x}_{30}$, $\vec{x}_{32}$, $\vec{x}_{34}$.

The last step consists in breaking up the visible part of the lower panel in new panels which are visible and those which are hidden by the upper panel. This is illustrated by Fig. 4.11.

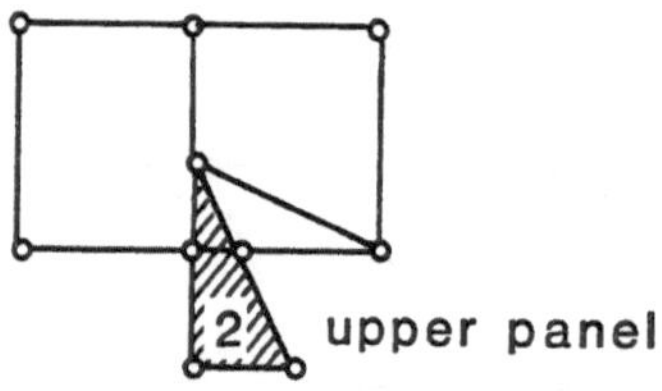


Fig. 4.11 Splitting up of the lower panel into new panels.

Fig. 4.12 shows the original box with the triangle in front of it after the discussed method has been applied to all panels. Fig. 4.13 just shows the remaining visible panels of the box.

Since all the panels of a structure must be compared two by two for each angle of observation one tries to minimize the number of comparisons by a pre-processing of the panel data. One efficient method is based on the construction of a minimum rectangle in the projection plane. This is illustrated at hand of Fig. 4.14.

A grid with variable mesh width which is parallel to the axis of the xy-plane is generated. For each panel the minimum rectangle which also is parallel to the axis is determined and all the elements of the grid which completely are hidden by the minimum rectangles are computed and associated to the individual panels. Only those panels of the structure must be compared which are associated to the same grid element.

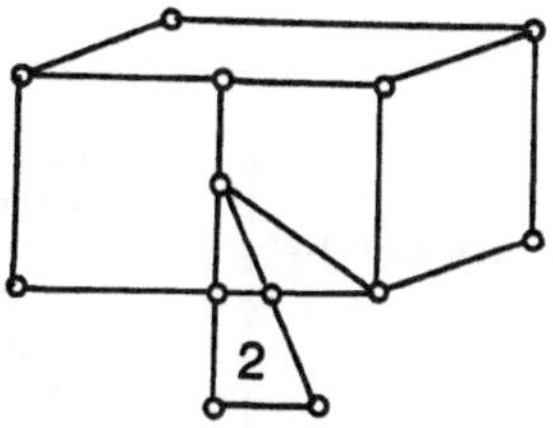

Fig. 4.12 Box and triangle after application of the hidden surface method.

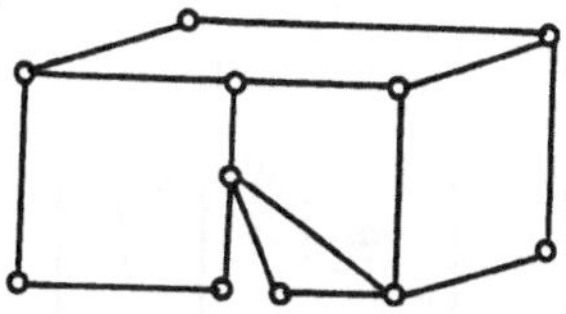

Fig. 4.13 Remaining panels of the box.

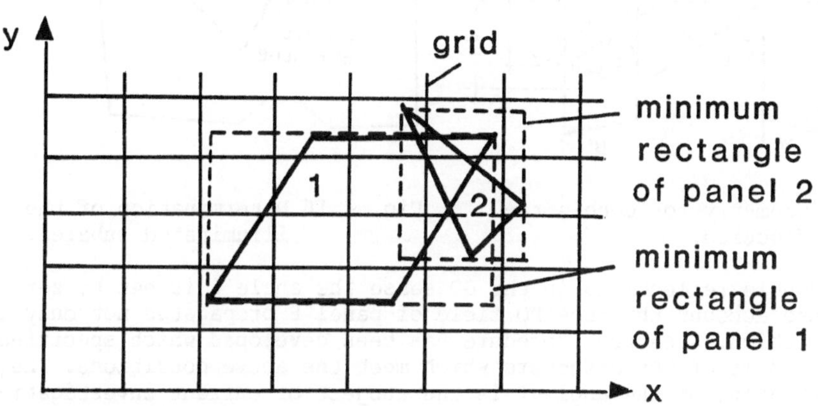


Fig. 4.14 Construction of minimum rectangles.

## 4.4 DOUBLY REFLECTING PANELS

Double reflection between two panels A and B occurs, see Fig. 4.15, if

1. both panels are visible from the radar transmitter/receiver,
2. the unit normal vectors $\vec{n}_a$ and $\vec{n}_b$ of the panels are vertical to each other, which is expressed by the formula

$$\arccos(\vec{n}_a \cdot \vec{n}_b) = \frac{\pi}{2} \pm \delta \, , \tag{4.14}$$

and

3. the line of intersection of the panel planes is vertical to the z-axis:

$$(4.15)\qquad \arccos((\vec{n}_a \times \vec{n}_b) \bullet \vec{e}_z) = \frac{\pi}{2} \pm \delta \ .$$

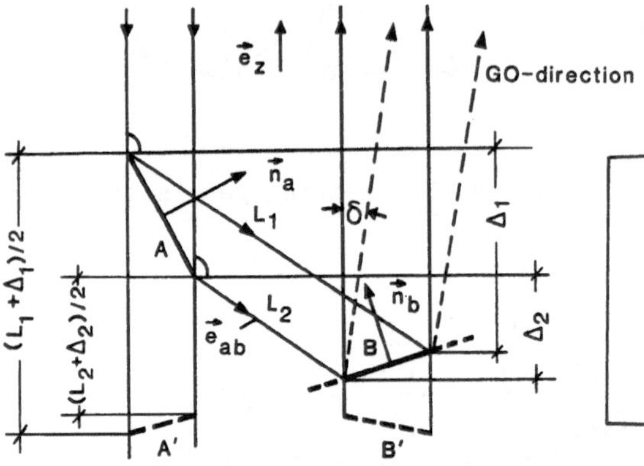


Fig. 4.15 Geometry for double reflection.

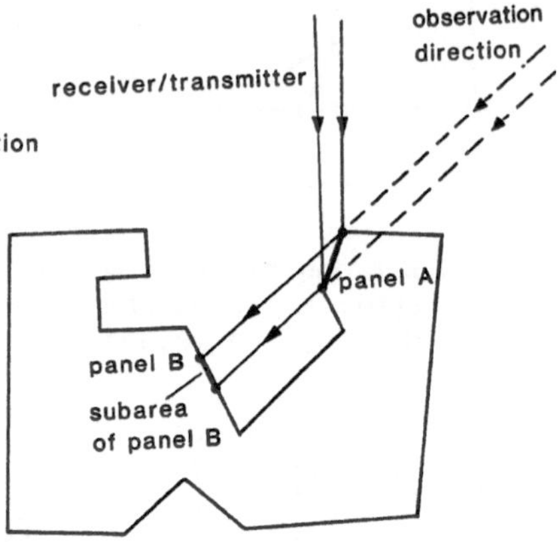


Fig. 4.16 Determination of the illuminated subarea.

For double reflections in the GO sense the angle $\delta$ is set to zero. $\delta$ takes into account that the PO field of panel B propagates not only in the GO direction. A search procedure has been developed which specifies all those panels of the structure which meet the above conditions. The appropriate value of the angle $\delta$ is the subject of current investigations.

Fig. 4.16 shows two doubly reflecting panels A and B where only a subarea of panel B is illuminated by the reflected rays emanating from panel A. The subarea of panel B is constructed by a further application of the previous explained hidden surface procedure, where the observation direction now is coincident with the direction of the reflections from panel A. The part of panel B which is shadowed by panel A is the desired subarea.

The backscattering matrix of each panel is described directly by Eq. (4.11). Additionally, each panel generates a reflected field at the surface of the other panel, which is scattered toward the receiver. This means that two doubly reflecting panels produce two backscattered fields and two field contributions due to double reflections. In Fig. 4.15 only the path from the transmitter via panel A and B back to the receiver is illustrated. The construction of the backscattered field is outlined as follows.

The field incident from the transmitter at panel A has to be decomposed into components parallel and vertical to the plane of incidence and multiplied with the appropriate reflection coefficients. From these components the reflected field of panel A and therewith the incident field at panel B is constructed, taking into account the appropriate path length. The same procedure is repeated at panel B to receive the reflected GO-field. From the incident and the reflected field the total field at the surface of panel B can be constructed and Eq. (4.9) can be applied to compute the scattered field and therewith the scattering matrix for the direction of the receiver.

For double reflection the phase integral in Eq. (4.9) can be treated in the same manner as in the case of a single panel by introducing the so-called virtual panels. A virtual panel is constructed in such a way that the path length transmitter → doubly reflecting panels → receiver is the same as the path length transmitter → virtual panel → receiver. In Fig. 4.15 the construction of the virtual panel A' is illustrated.

The theory has been tested at hand of a cube with additional surfaces which give rise to shadowing and double reflection effects, see Fig. 4.17. The comparison between theoretical and experimental results for the radar cross-section at a frequency of 16.66 GHz ($\lambda$ = 18 mm) is presented in Fig. 4.13 for vertical polarization and for aspect angles ranging from -45° to 135°. Within this range the interference of the two doubly reflecting parts of the structure takes place.

Comparing the measured with the computed results, one observes a rather good agreement down to levels of about -30 dB. The peaks at 0° and 90° are due to the GO reflections, the peaks between arise from the interference of the PO fields from the two doubly reflecting areas. The theoretical and experimental determined number of these peaks agree well.

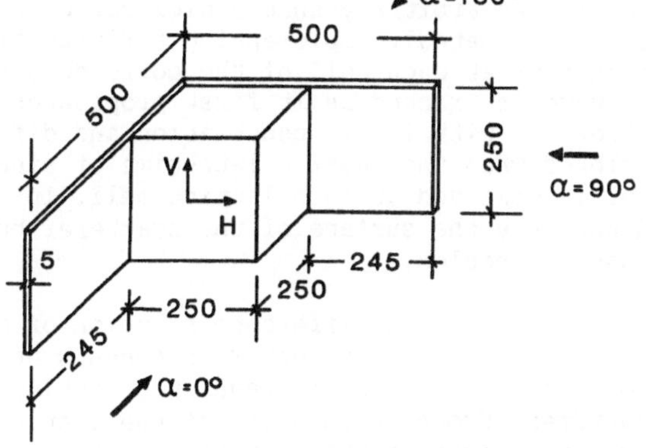


Fig. 4.17 Cube with additional shadowing surfaces, dimensions in mm.

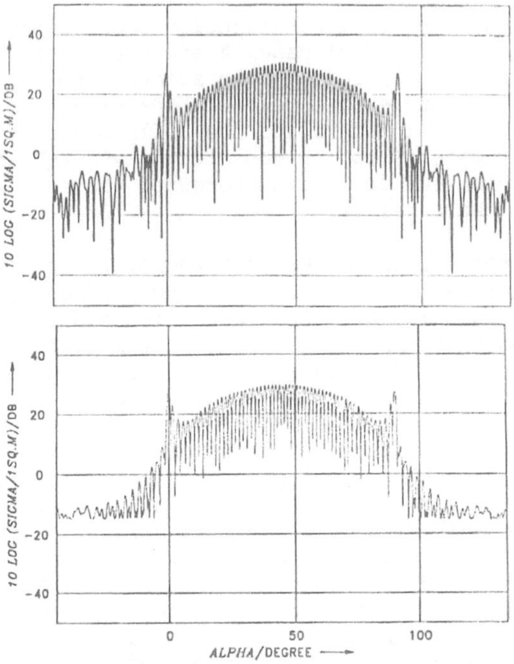


Fig. 4.18 Radar cross-section of a cube with additional shadowing surfaces. a) theoretical results, b) experimental results.

## 5. MODELS USED FOR THE TIME-DEPENDENT SOLUTION OF MAXWELL'S EQUATIONS

In the following two rather new methods are discussed which solve Maxwell's time-dependent curl equations numerically. These methods seem to be useful for studying propagation of an electromgnetic wave into a volume of space containing an arbitrary-shaped dielectric or conducting body. By time-stepping or repeatedly implementing a finite-difference analog of the curl equations at each cell of the corresponding space lattice, the incident wave is tracked as it first propagates to the structure and then interacts with it via penetration and diffraction. Wave tracking is completed when the desired late-time or sinusoidal steady-state behaviour is observed at each lattice cell. In contrary to the IE- or PO-method not only the surface of the scatterer but also the surrounding volume must be modeled.

A basic problem with any finite difference solution of Maxwell's equations is the treatment of the field vector components at the lattice truncation. Because of limited computer storage, the lattice must terminate close to the scatterer. Proper truncation of the lattice requires that any outgoing wave disappears at the lattice boundary without reflections during the continuous time stepping of the algorithm.

In the first reported method [19 - 21] Maxwell's equations

$$(5.1) \qquad \frac{\partial \vec{E}}{\partial t} = \frac{1}{\varepsilon} \nabla \times \vec{H} ,$$

$$(5.2) \qquad \frac{\partial \vec{H}}{\partial t} = - \frac{1}{\mu} \nabla \times \vec{E}$$

are solved directly.

The scatterer is enclosed in a rectangular volume, see Fig. 5.1. The various details of the structure are modeled with a resolution of one unit cell. Time-stepping is accomplished by an explicit finite-difference procedure. For a Cartesian cubic-cell space lattice, this procedure involves positioning the components of $\vec{E}$ and $\vec{H}$ about a unit cell of the lattice as shown in Fig. 5.1 and evaluating $\vec{E}$ and $\vec{H}$ at alternate half-time steps [22]. In this manner, centered difference expressions can be used for both the space and time derivatives to attain second-order accuracy in the space and time increments. This leads to a system of 6 finite-difference equations.

For example [19], the x-component of Eq. (5.2), written as

$$(5.3) \qquad \frac{\partial H_x}{\partial t} = \frac{1}{\mu} \left(\frac{\partial E_y}{\partial z} - \frac{\partial E_z}{\partial y}\right)$$

is implemented as the following time-stepping relation for $H_x$:

$$(5.4) \qquad H_x^{n+1/2} (i,j + 1/2, k + 1/2) = H_x^{n-1/2} (i,j + 1/2, k + 1/2) +$$

$$+ \frac{\delta t}{\mu(i,j + 1/2, k + 1/2)\ \delta} (E_y^n(i,j+1/2,k+1) - E_y^n(i,j+1/2,k) +$$

$$+ E_z^n (i,j,k+1/2) - E_z^n(i,j+1,k+1/2)) .$$

The space-time functional notation $F^n(i,j,k) = F(i\delta,j\delta,k\delta,n\delta t)$ is used, where $\delta = \delta x = \delta y = \delta z$ is the space increment, $\delta t$ is the time increment, and i,j,k and n are integers.

With the system of finite-difference equations the new value of a field vector component at any lattice point depends only on its previous value and on the previous values of the components of the other field vector at adjacent points. Therefore, at any given time step the computation of a field vector may proceed one point at a time.

The second time-domain approach [2] to be discussed is based on an integral form of Maxwell's equations. Integration of Eq. (5.1) resp. Eq. (5.2) over a volume V fixed in space with surface F yields

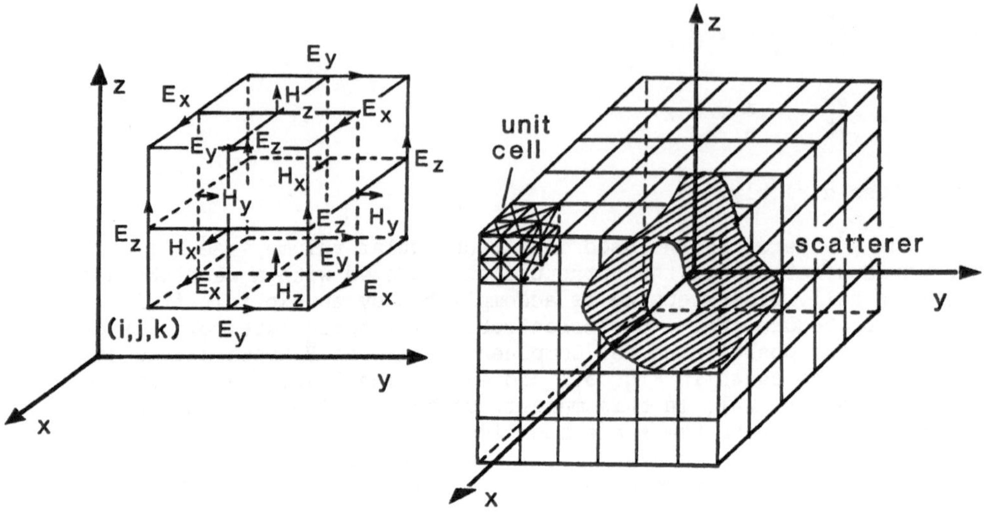


Fig. 5.1 Geometry of a scatterer and lattice arrangement (b), positions of the field components about a unit cell (a) [19].

(5.5)
$$\frac{d}{dt}\int_V \vec{E}\ dv = -\frac{1}{\varepsilon}\int_F \vec{H}\times d\vec{f}\ ,$$

(5.6)
$$\frac{d}{dt}\int_V \vec{H}\ dv = \frac{1}{\mu}\int_F \vec{E}\times d\vec{f}\ .$$

The computational domain around the structure is finite and is discretized by a grid aligned to the structure's surface. The grid consists of curved coordinate surfaces i = const., j = const. and k = const., the volume elements V(i,j,k) are general hexahedra (Fig. 5.2a) with surface vectors $\vec{f}_i(i,j,k)$, $\vec{f}_i(i+1,j,k)$, $\vec{f}_j(i,j,k)$, ... $\vec{f}_k(i,j,k+1)$. According to Eqs. (5.5) and (5.6) the vectors $\vec{H}(i,j,k)$, $\vec{E}(i,j,k)$ of the left-hand side are centered within V(i,j,k) and the vectors $\vec{H}_i(i,j,k)$, $\vec{E}_i(i,j,k)$, ... are centered within related faces (Fig. 5.2b).

The vectors are volume-averaged resp. face-averaged field quantities. Eq. (5.6) for example has the following centered difference expression for the space derivatives:

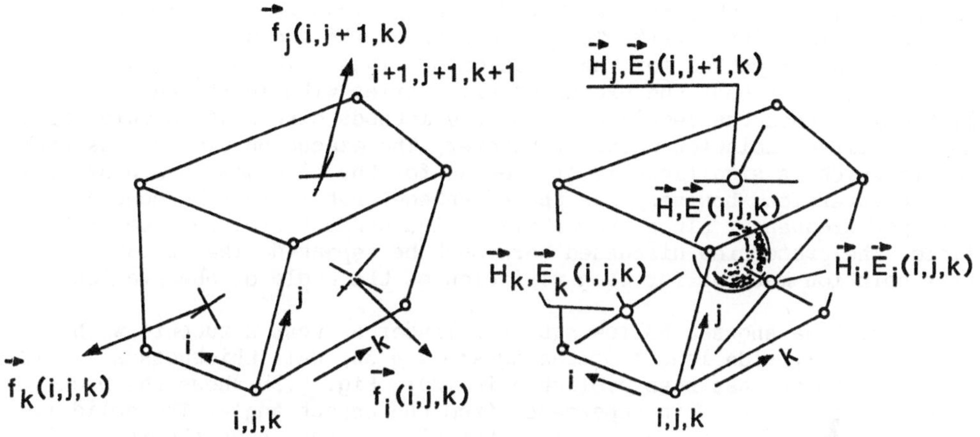


Fig. 5.2 Volume element V(i,j,k) and location of vectors [2].

$$(5.7) \qquad \frac{d}{dt}\vec{H}(i,j,k) = \frac{1}{V(i,j,k)}\,((\vec{E}_i \times \vec{f}_i)_{i+1,j,k} + (\vec{E}_i \times \vec{f}_i)_{i,j,k} + (\vec{E}_j \times \vec{f}_j)_{i,j+1,k} + (\vec{E}_j \times \vec{f}_j)_{i,j,k} + (\vec{E}_k \times \vec{f}_k)_{i,j,k+1} + (\vec{E}_k \times \vec{f}_k)_{i,j,k}) \ .$$

No difference expression for the time derivative is used. The surface vectors $\vec{H}_i$, $\vec{E}_j$, ... are computed from the volume vectors $\vec{H}$, $\vec{E}$ which are known from the preceding time step by linear interpolation in the index space (i,j,k). A similar expression holds for $d\vec{E}(i,j,k)/dt$. Thus Eqs. (5.5)and (5.6) are replaced by a set of ordinary differential equations with respect to time t. These equations are integrated by a Runge/Kutta-type procedure with appropriate step-size control.

Since in both approaches no matrix equations have to be solved, it is possible to split the grid into subdomains. In addition both computer codes are suited for adaption to parallel-processing and vector-array-processing computers. The required memory increases for both methods only linearly with N, the total number of field components to be determined. The value of N is a function of the normalized electric volume of a scatterer, e.g. $N \approx (N/\lambda)^3$ where D is a characteristic dimension of the scatterer. With L, the number of required time steps, the execution time is estimated to be proportional to $L(D/\lambda)^3$.

For the IE-method applied to perfectly conducting scatterers the number of unknowns varies with the surface of the scatterer, see Eqs.

(3.9) and (3.10), that is $N \approx (D/\lambda)^2$. The required memory for matrix inversion varies with $(D/\lambda)^4$. The estimation of the execution time depends from the chosen method for the solution of linear system of equations. For iterative methods the execution time varies with $(D/\lambda)^4$ which is little more than the requirements of the methods discussed in this Section. For direct solution methods, however, the execution time varies with $(D/\lambda)^6$ which is significantly greater as for the time-domain approach. A direct inversion, however, has the advantage that is must be done only once per frequency, since the matrix is independent from the incident field. The procedures discussed here must be repeated like iterative matrix solution methods for each variation of the angle of observation.

Fig. 5.3a shows a hollow square cylinder as test structure with length $L = 9\lambda$, side length of the square $a = 3\lambda$, wall thickness $s = \lambda/10$. This test object has been published in [20]. Fig. 5.3b shows the backscatter cross-section in dependence from the aspect angle. The solid lines represent experimental results, the circles represent theoretical results of the first method discussed in this section and the stars represent results of the second method [2].

A lattice cell size of approximtely $1/11\lambda$ was selected for the first time-domain approach. Each cylinder wall was formed by 96 × 32 × 1 cells, and the overall lattice size was 112 × 48 × 48 cells. 661 time steps were used, equivalent to 31 cycles of the incident field. For the application of the second approach a grid with 85 × 17 × 73 cells was used, see Fig. 5.4.

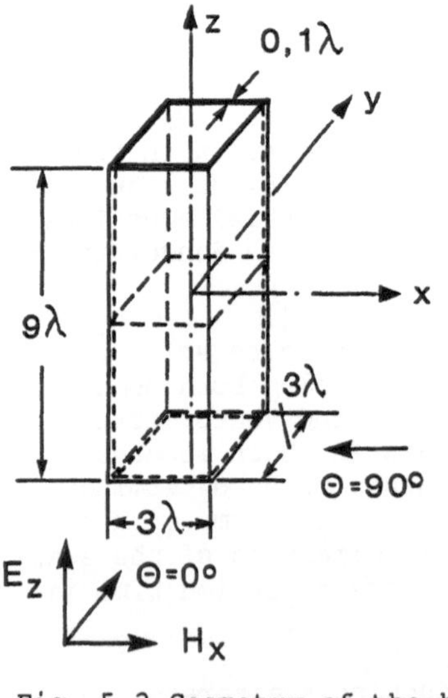


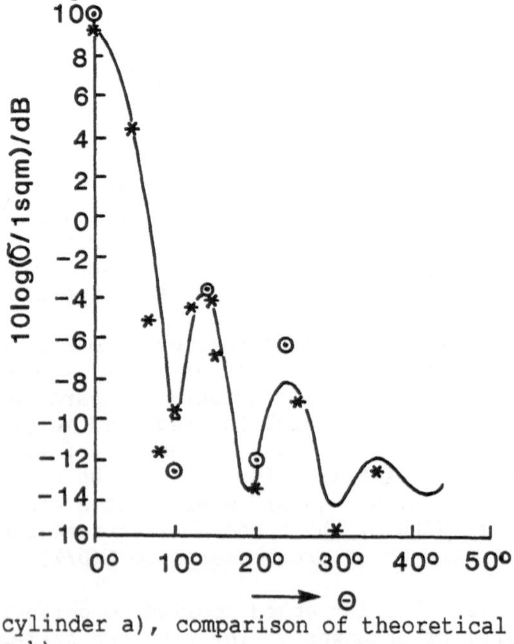


Fig. 5.3 Geometry of the hollow cylinder a), comparison of theoretical and experimental results b).

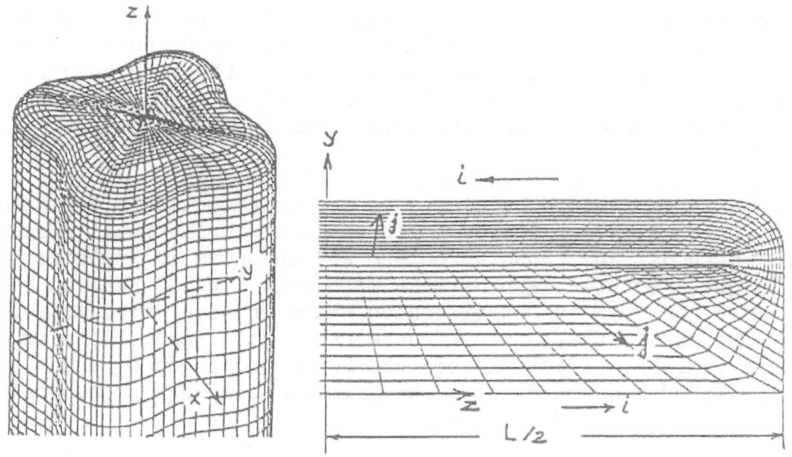


Fig. 5.4 Grid for the hollow cylinder.
a) outer grid surface, b) grid in a plane section k = const.

## 6. CONCLUSION AND REMARKS FOR FURTHER WORK

In this paper the basic ideas of a variety of methods in electromagnetics are outlined in short. The appropriate geometric model of each method is discussed in more detail. The requirements with respect to the type of the model and with respect to the accuracy are established. For some models and methods an estimation of the computer effort is given. For the following problems comparison between theory and experiment has been drawn: radiation characteristic of an antenna installed on a helicopter, current distribution of a cube, radar cross-section of a structure with doubly reflecting surfaces, radar cross-section of a hollow cylinder. Much more test objects are presented in the references cited.

Despite the fact that the number of validation tets increases in the publications there are some essential gaps in estimating the overall effectiveness of the methods for practical applications. In applying any of the methods for example the following problems are not satisfyingly clarified: incorporation of dielectric media into the model, examination of 3-D structures with empty and loaded cavities and of structures in the resonance region, investigation of near-field properties. Clear statements concerning the stability of results against the type of model (e.g. wire-grid or surface patch model) and the refinement of the model are missing. In the IE-method only a few results are available which are obtained by the independent use of either the EFIE or the MFIE for the same structure. The PO-method has to be extended to treat multiple reflections, creeping waves, bistatic and quasi near-field problems, and must be validated for these cases. For the two time-domain approaches discussed more results concerning cavity problems, internal resonances and nonperfectly conducting bodies should be available since these techniques seem to be advantageous in these special cases. Further studies should also deal with demonstration of the attained accuracy for coarse lattice sizes.

A further gap which should be closed is due to the classification of the methods in those which are suited to treat electrically small structures (rigorous methods) and those which are advantageous in treating electrically large structures (approximate methods). Studies seem to be useful to estimate the structure's dimension where one type of solution method can be substituted by the other one without significant loss in accuracy for the results. If for example the PO-method could be used instead of the IE-method one could solve the problem much more economically.

Finally the study of the efficiency of the different approaches at the same test objects (an example has been given in the paper) could lead to recommendations of high practical interest. The test objects could be selected with increasing complexity to study effects of edge and corner diffraction, creeping waves, cavity and resonance phenomena, perfectly and nonperfectly conducting bodies.

## 7. APPENDIX

### 7.1 REFLECTION COEFFICIENTS

The reflection coefficients at the surface of a panel consisting of N layers, see Fig. 7.1, are given by

$$R_H = \frac{Z^{(1)}\cos\alpha^{(1)} - Z_{iH}^{(2)}\cos\alpha^{(2)}}{Z^{(1)}\cos\alpha^{(1)} + Z_{iH}^{(2)}\cos\alpha^{(2)}}, \tag{7.1}$$

$$R_E = \frac{Z_{iE}^{(2)}\cos\alpha^{(1)} - Z^{(1)}\cos\alpha^{(2)}}{Z_{iE}^{(2)}\cos\alpha^{(1)} + Z^{(1)}\cos\alpha^{(2)}}, \tag{7.2}$$

where $Z_{iH}^{(2)}$, $Z_{iE}^{(2)}$ are computed by the following recurrence formulas:

$$Z_{iH}^{(n)} = \frac{Z_{iH}^{(n+1)}\cos\alpha^{(n+1)} + jZ^{(n)}\cos\alpha^{(n)}\tan(c^{(n)}d^{(n)})}{Z^{(n)}\cos\alpha^{(n)} + jZ_{iH}^{(n+1)}\cos\alpha^{(n+1)}\tan(c^{(n)}d^{(n)})}\, Z^{(n)}, \tag{7.3}$$

$$Z_{iE}^{(n)} = \frac{Z_{iE}^{(n+1)} \cos\alpha^{(n)} + jZ^{(n)} \cos\alpha^{(n+1)} \tan(c^{(n)} d^{(n)})}{Z^{(n)} \cos\alpha^{(n+1)} + jZ_{iE}^{(n+1)} \cos\alpha^{(n)} \tan(c^{(n)} d^{(n)})} Z^{(n)} , \tag{7.4}$$

$n = 2,3, \ldots, N-1$, where $Z_{iH}^{(n)}$, $Z_{iE}^{(n)}$ are input impedances of the nth layer for H- or E-polarization, respectively. For $n = N$, Eqs. (7.3) and (7.4) must be replaced by

$$Z_{iH}^{(N)} = Z_{iE}^{(N)} = Z^{(N)} . \tag{7.5}$$

The impedance of the nth layer is given by

$$Z^{(n)} = \sqrt{\frac{\mu_o}{\varepsilon_o}} \sqrt{\frac{\mu_r^{(n)}}{\varepsilon_r^{(n)}}} , \qquad \text{where } \varepsilon_r^{(n)} = \varepsilon_r'^{(n)} - j\varepsilon_r''^{(n)}$$

is the complex relative permittivity (dielectric constant) of the nth layer.

$$\mu_r^{(n)} = \mu_r'^{(n)} - j\mu_r''^{(n)}$$

is the complex relative permeability, and

$$c^{(n)} = k^{(n)} \cos\alpha^{(n)} ,$$

with $k^{(n)} = \omega \sqrt{\varepsilon_o \varepsilon_r^{(n)} \mu_o \mu_r^{(n)}}$ being the wavenumber and $d^{(n)}$ the thickness of the nth layer. $\omega = 2\pi f$ = angular frequency.

The value $\cos\alpha^{(n)}$ can be derived from the equation

$$\cos\alpha^{(n)} = \sqrt{1 - \frac{\mu_r^{(1)} \varepsilon_r^{(1)}}{\mu_r^{(n)} \varepsilon_r^{(n)}} \sin^2 \alpha^{(1)}} .$$

The last and the first layer are assumed to be of semi-infinite thickness. For more details see [7, 9].

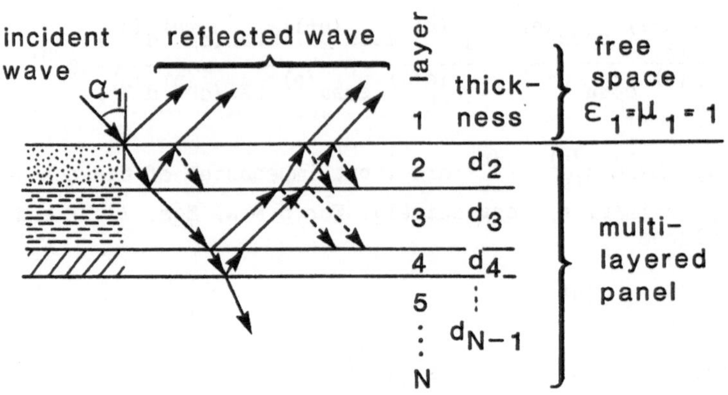


Fig. 7.1 Geometry of a multilayered panel.

## 7.2 PTD EDGE DIFFRACTION COEFFICIENTS

The coefficients of the backscattering matrix, see Eq. (4.12), are given by

$$D_v = D_{v1} - D_{v1}^{po} + D_{v2} - D_{v2}^{po}, \quad v = e,m,em, \tag{7.6}$$

$$D_{e1} = \frac{\sin\frac{\psi_e}{n}}{n} \frac{1}{\cos\frac{\pi-\alpha_1}{n} - \cos\frac{\psi_e}{n}}, \tag{7.7}$$

$$D_{e1}^{po} = -U(\pi-\psi_e)\frac{\sin\psi_e}{\mu_1 + \cos\psi_e}, \tag{7.8}$$

$$D_{m1} = \frac{\sin\frac{\psi_e}{n}}{n} \frac{\sin\frac{\pi-\alpha_1}{n}}{\sin\alpha_1} \frac{1}{\cos\frac{\pi-\alpha_1}{n} - \cos\frac{\psi_e}{n}}, \tag{7.9}$$

$$D_{m1}^{po} = -U(\pi-\psi_e)\frac{\sin\psi_e}{\mu_1 + \cos\psi_e}, \tag{7.10}$$

$$D_{em1} = \frac{1}{n}\cos\beta_e(\mu_1 + \cos\psi_e)\frac{\sin\frac{\pi-\alpha_1}{n}}{\sin\alpha_1} \frac{1}{\cos\frac{\pi-\alpha_1}{n} - \cos\frac{\psi_e}{n}}, \tag{7.11}$$

(7.12) $D^{po}_{em1} = 0$ ,

(7.13) $\mu_1 = \cos\psi_e - \frac{2}{\tan^2\beta_e}$ ,

(7.14) $\alpha_1 = \arccos\mu$ ,

(7.15) $U(x) = \left\{ \begin{matrix} 1 \\ 0 \end{matrix} \right\}$ for $\left\{ \begin{matrix} x > 0 , \\ x < 0 . \end{matrix} \right.$

The coefficients $D_{\nu 2}$ are obtained from the coefficients $D_{\nu 1}$ by the following substitutions:

$$\psi_e \to n\pi - \psi_e , \quad \beta_e \to \pi - \beta_e , \quad \mu_1 \to \mu_2 , \quad \alpha_1 \to \alpha_2 .$$

For more details, see [16, 17].

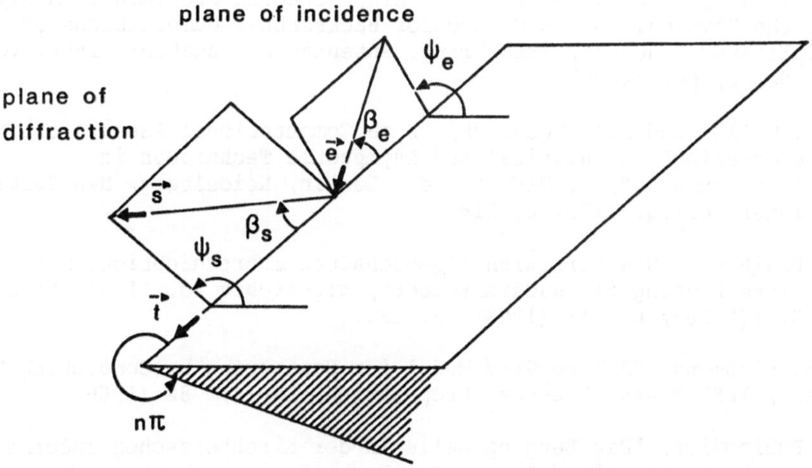


Fig. 7.2 Geometry of the bistastic wedge diffraction. In the backscattering case $\psi_s = \psi_e$, $\beta_s = \pi - \beta_e$.

8. REFERENCES


[1] J.W. Crispin and K.M. Siegel, "Methods of Radar Cross-Section Analysis", New York: Academic Press, 1968, Chapter 9.

[2] J. Grashof, "Finite-Volume Method for the Time-Dependent Maxwell Equations and Prediction of the Scattering from Perfectly Conduc-

ting Bodies", 3rd International IGTE Symposium, "Numerical Field Calculation in Electrical Engineering", Graz, Austria, 1988.

[3] H. Hönl, A.W. Maue and K. Westpfahl, "Theorie der Beugung" in "Handbuch der Physik", S. Flügge, Hrsg., Berlin, Göttingen, Heidelberg: Springer Verlag,1961, S. 218-219, 238-245, 268-289.

[4] G.F. Koch, "Die verschiedenen Ansätze des Kirchhoffschen Prinzips und ihre Anwendung auf die Beugungsdiagramme bei elektromagnetischen Wellen", A.E.Ü., Band 14, 1960, Heft 2, S. 77-98 und Heft 3, S. 132-153.

[5] D. Klement, J. Preißner und V. Stein, "Computation of the Scattering Matrix of Radar Targets: Concept of the Method and First Results", AGARD Conference Print No.364, 1984, pp. 20-1 to 20-23.

[6] D. Klement, "Scattering by Complicated Structures: Calculation of the Radar Properties of Metallic Targets by Means of Physical Optics", ESA-TT 946, Translation of DFVLR-FB 85-22, 1985.

[7] D. Klement, J. Preißner and V. Stein, "Special Problems in Applying the Physical Optics Method for Backscatter Computations of Complicated Objects", IEEE Trans. Antennas Propagation, 1988, Vol. 36, No. 2, pp. 228-237.

[8] E.K. Miller and F.J. Deadrick, "Some Computational Aspects of Thin-Wire Modeling" in"Numerical and Asymptotic Techniques in Electromagnetics", R. Mittra, ed., Berlin, Heidelberg, New York: Springer Verlag, 1975, p. 118.

[9] J. Preißner, "Die Reflexionseigenschaften mehrschichtiger Medien und ihre Eignung als Radarabsorber", NTZ-Archiv Bd. 11 (1989) H. 4, S. 175-182, Bd. 11 (1989), S. 298.

[10] J.H. Richmond, "A Wire-Grid Model for Scattering by Conducting Bodies", IEEE Trans. Antennas Propagation, AP-14, 782 (1966).

[11] A. Rubinowicz, "Die Beugungswelle in der Kirchhoffschen Theorie der Beugung", Berlin, Heidelberg, New York: Springer Verlag, Warszawa: Polnischer Verlag der Wissenschaften, 1966, S. 1-58.

[12] A. Schroth and V. Stein, "Moderne numerische Verfahren zur Lösung von Antennen- und Streuproblemen", München, Wien: R. Oldenbourg Verlag, 1985.

[13] R. Schwemmer, "Ein Algorithmus zur Ermittlung der sichtbaren Teile von Strukturen, die durch Dreiecke und Vierecke modelliert sind", DFVLR-Mitt. 88-03.

[14] V. Stein, "Numerical Modeling: Integral Equation Method", AGARD Lecture Series No. 152 on "Theoretical Aspects of Target Classification", 1987, Rome, Neubiberg, Noresund, pp. 10-1 to 10-20.

[15] V. Stein, "Physical Optics Method: Prediction of Radar Signatures", AGARD Lecture Series No. 152 on "Theoretical Aspects of Target Classification", Rome, Neubiberg, Noresund 1987, pp. 5-1 to 5-17.

[16] V. Stein, "Beziehungen zwischen bekannten Theorien zur Behandlung der Beugung am Keil im Hochfrequenzfall", DLR-FB 89-49, 1989.

[17] V. Stein, "Die Rückstreumatrix einer endlich langen Kante", report in preparation.

[18] W.L. Stutzmann and G.A. Thiele, "Antenna Theory and Design", New York, Chichester, Brisbane, Toronto: John Wiley and Sons, 1981, pp. 356-370.

[19] A. Taflove, "Radar Cross Section of General Three-Dimensional Scatterers", IEEE Trans. Electromagnetic Compatibility, Vol. EMS-25, No. 4, Nov. 1983, pp. 433-440.

[20] A. Taflove, K. Umashankar and T.G. Jurgens, "Validation of FD-TD Modeling of the Radar Cross Section of Three-Dimensional Structures Spanning Up to Nine Wavelength", IEE Trans. Antennas Propagation, Vol. AP-33, No.6, June 1985, pp. 662-666.

[21] K.Umashankar and A. Taflove, "A Novel Method to Analyze Electromagnetic Scattering of Complex Objects", IEEE Trans. Electromagnetic Compatibility, Vol. EMC-24, No. 4, Nov. 1982, pp. 397-405.

[22] K.S. Yee, "Numerical Solution of Initial Boundary Value Problems Involving Maxwell's Equations in Isotropic Media", IEEE Trans. Antennas and Propagation, Vol. AP-14, pp. 302-307, May 1966.

[23] D.R. Wilton and S. Govind, "Incorporation of Edge Conditions in Moment Method Solutions", IEEE Trans. Antennas Propagation, Vol. AP-25, No. 5, Nov. 1977, pp.345-850.

[24] M. Zink, "Bestimmung charakteristischer Strahlungseigenschaften einer Torusantenne", DFVLR-FB 87-46.

## 9. ACKNOWLEDGEMENT

The author wishes to thank his colleagues D. Klement, J. Preißner, R. Schwemmer, and Dr. E. Kemptner for the cooperation in applying and extending of the PO- and the IE-theory and the development of computer programs. The author wishes further to thank his colleagues Dr. B. Röde, K.H. Bethke, M. Kleintz and K.-H. Dreher for the precise measurements and data processing required for the validation of the theories. Thanks are also due to Mrs. M. Malchow for carefully typing the manuscript and G. Jacob for accurately drawing the illustrations.

The first computer program used for the solution of the hidden surface problem in context with the PO-method was developed by Professor Dr. B.G. Böge, Universität der Bundeswehr, Neubiberg. The computer program which now is in use was developed by R. Schwemmer. The computer program for the treatment of double reflections is based on an idea of D. Klement and was developed by Dr. Aßfalg and Dr. Mosebach, IABG, Ottobrunn. The work on PO has been sponsored in part by the Bundesamt für Wehrtechnik und Beschaffung, Koblenz.

The original version of this material was first published by the Advisory Group for Aerospace Research and Development, North Atlantic Treaty Organisation (AGARD/NATO) in Conference Proceedings CP 464 - Applications of Mesh Generation to Complex 3-D Configurations - published in 1990.

**PROPERTIES OF SCATTERING MATRIX STATISTICS: IMPROVED POLARIZATION DETECTION AND DISCRIMINATION ALGORITHMS FOR POL-RAD (TARGET RADIO-LOCATION) APPLICATIONS**

**Anatolij I. Kozlov, and Aleksander I. Logvin**
**Moscow Civil Aviation Engineering Institute**
**MIIGA, Kronshtadsky Boulevard 20**
**USSR, 125-493 MOSCOW**

**ABSTRACT:** For a class of fluctuating scatterers the probability distribution law of the changing multi-dimensional density for the complex elements of an associated scattering matrix is considered. Its invariance with respect to a change of polarization basis is shown. Similar results have been obtained for the correlation and covariance matrices. The transformations of the mean values and the dispersions of the scattering matrix elements are considered separately. The invariance of the associated dispersion sums of these elements is proven, and their extreme values are calculated. The concept of Graves generalized power matrix is introduced, and its transformation properties are established. It is shown that the elements of the related correlation matrix are not independent and that there exist corresponding limiting bounds for them.

## 1. INTRODUCTION

The most general descriptor of a radar (radio-location) target is its statistical instantaneous scattering matrix $[S_s(t)]$, of which the elements $S_{ij}$ $(i,j = 1,2)$ are randomly time-dependent. Because of their non-coherent random statistical nature, a four dimensional coherency-type vector $\vec{S}_s = (S_{11}(t),\ S_{12}(t),\ S_{21}(t),\ S_{22}(t))^T$ is introduced. The statistics of these elements of the scattering matrix $[S_s]$ are considered based on an analysis of the associated Poincaré sphere parameters under subjection to random time changes. This formulation provides an opportunity for simulating distribution laws for real and imaginary element components of the associated scattering matrices, their moduli and arguments, together with the degree of anisotropy, complete cross-section, generalized phase, and associated matrix determinants, which were derived in greater detail in [1] and [2]. The solutions to this modeling approach made it possible to determine numerically the characteristics of these random values (the mean value, the variances, the dispersion, the excess, etc.) and to calculate correlation moments and functions.

We were able to demonstrate in simulation exercises that the distribution laws obtained differ greatly from the universal statistical laws



*W.-M. Boerner et al. (eds.), Direct and Inverse Methods in Radar Polarimetry, Part 1*, 675–691.

obtained by classical non-polarimetric ways of simulation, but that they conform exceedingly better with experimental results obtained from rough, slightly hilly terrain covered with surface vegetation (bushes, groves, trees, etc.) and subjected to ambient meteorologic conditions [1]. The nature of the probability distribution density as a function of anisotropy, moduli and phases of the scattering matrix elements is modeled and calculated within the framework of object presentation as a system of independent scattering parameters, systematizing the obtained results on the basis of Pearson's distributions and checking their correspondence to commonly used probability densities. Calculations are carried out for specific types of distributed anisotropic environmental backscatter scenarios. The results are compared to the experimental polarimetric data which makes it possible to speak of rather stable laws for a wide class of distributed scatter center ensembles. This investigation, in turn, opens up some new possibilities - hitherto not investigated - for creating polarization-adaptive algorithms that can be applied to the effective suppression of interfering reflections.

## 2. MODELING OF PROBABILITY DENSITY FUNCTION

The most general characteristics of a radar (radiolocation) target is its statistical scattering matrix $[S_s(t)]$, for which the elements $S_{ij}(t),(i,j = 1,2)$ are randomly time-dependent. Let's associate a four-dimensional complex vector $\vec{S} = (S_{11}(t),\ S_{12}(t),\ S_{21}(t),\ S_{22}(t))^T$ with matrix [S] in which index "T" means matrix transposition operation. Study [1] shows that in transforming to the new polarization basis (PB), vector $\vec{S}_C$ is subjected to a linear transform $\vec{S}_H = [L]\vec{S}_C$ by means of an unitary matrix [L], depending on three independent parameters $\gamma$, $\eta$, $\xi$ (their geometrical properties are given in [1]). The real form of the matrix L has the following structure

$$[L] = \frac{1}{2}\begin{bmatrix} 2e^{2i\eta}\cos^2\gamma & e^{i(\eta+\xi)}\sin2\gamma & e^{i(\eta+\xi)}\sin2\gamma & 2e^{2i\xi}\sin^2\gamma \\ -e^{i(\eta-\xi)}\sin2\gamma & 2\cos^2\gamma & -2\sin^2\gamma & e^{i(\xi-\eta)}\sin2\gamma \\ -e^{i(\eta-\xi)}\sin2\gamma & -2\sin^2\gamma & 2\cos^2\gamma & e^{i(\xi-\eta)}\sin2\gamma \\ 2e^{-2i\xi}\sin^2\gamma & -e^{-i(\eta+\xi)}\sin2\gamma & -e^{-i(\eta+\xi)}\sin2\gamma & 2e^{-2i\eta}\cos^2\gamma \end{bmatrix} \quad (1)$$

Let the eight-dimensional joint probability distribution density function be defined by $W(S_{11}^C,\ S_{12}^C,\ S_{21}^C,\ S_{22}^C)$ which, when transformed to the new polarization basis, can be found as:

$$W(S_{11}^H,\ S_{12}^H,\ S_{21}^H,\ S_{22}^H) = W(S_{11}^C(\vec{S}_{11}),\ S_{12}^C(\vec{S}_{11}),\ S_{21}^C(\vec{S}_{11}),\ S_{22}^C(\vec{S}_{11}))\,|J| \quad (2)$$

with the $S_{ij}^C$ corresponding to the vector $\vec{S}_C$ components.

In this equation $|J| = |\det[L]^{-1}|$ is the Jakobian of transformation from the old basis to the new one, where in the following the determinant of the matrix [L] unitarily equals 1. This means that the first main property of the statistical scattering matrix elements is the invariance of the distribution densities of the elements of this matrix with respect to the polarization basis change. From this property follows a very essential practical consequence. If the distribution density $W(\vec{S}_C)$ is Gaussian, then due to the unitarity of matrix [L] on the one hand, and the condition (2) requirement on the other hand, the distribution density $W(\vec{S}_H)$ will also be Gaussian. Only such parameters as the expectation and the dispersion are subject to changes according to this law. In order to determine the one-dimensional moduli distribution density (their squared numbers) of the scattering matrix elements, it is necessary to accomplish a series of transformations and integrations which leads to the uni-parametric laws (Rayleigh and exponential) — provided the expectation of the initial Gaussian function equals zero. It means that in this case the one-dimensional law will be invariant to the polarization basis change as well.

The important characteristics of the given random process $\vec{S}(t)$ are its correlation matrix

$$[R] = [\overline{\vec{S}\vec{S}^{+}}] - [(\overline{\vec{S}})\cdot(\overline{\vec{S}^{+}})] \tag{3a}$$

and its co-variance matrix

$$[\tilde{R}] = \overline{[\vec{S}\vec{S}^{+}]} \tag{3b}$$

where the sign "+" means double operation: complex conjugation and transposition; and the line above means statistical averaging. Let's consider in what way matrices [R] and $[\tilde{R}]$ change if the polarization basis changes. Using equation

$$\vec{S}_H = [L]\vec{S}_C \, , \tag{4a}$$

we have

$$[R_H] = [L]\ [\overline{\vec{S}_C\ \vec{S}_C{}^{+}}]\ [L]^{+} - [L]\ [\overline{\vec{S}_C} \cdot \overline{\vec{S}_C{}^{+}}]\ [L]^{+} = [L][R_C][L]^{+} \tag{4b}$$

Thus, the correlation matrix in transformation to the new polarization basis is subject to the transform of similarity by means of the unitary matrix [L], defined by (1). Exp. (4b) shows that a covariance matrix $[\tilde{R}]$ has the same properties, with the determinants of the R and $[\tilde{R}]$ matrices being constant. After considering the above properties, we can arrive at the conclusion, that if the multi-dimensional law W(S) is Gaussian then all one-dimensional laws $W(S_{ij})$ are Gaussian too. In this case, if we change the polarization basis, then only mean values and second order moments will change. Speaking about mean values, their transform coincides with the transformation of the polarization basis of (1) for the elements $S_{ij}$ of the scattering matrix [S]. Second moments and dispersions of the elements $S_{ij}$ do not cause any difficulties. They correspond to the diagonal elements of matrix $[R_H]$ but those are not being presented here because of their awkwardness. However, a number of properties associated with the dispersion of the elements $S_{ij}$ should be emphasized. Direct calculations prove invariance of dispersion sums to the change of the polarization basis, i.e.,

$$(D_{11}^2)_H + (D_{22}^2)_H + 2(D_{12}^2)_H = (D_{11}^2)_\subset + (D_{22}^2)_\subset + 2(D_{12}^2)_\subset = D_\Sigma^2 \rightarrow \text{inv.} \qquad (5a)$$

The sum, in which the symbol $D_{ij}^2 = \overline{|S_{ij}|^2} - (\overline{|S_{ij}|})^2$ is introduced, is named "full dispersion" - $D_\Sigma^2$. It can now be shown that the knowledge of the dispersions $d_{ij}^2$ of any target scattering matrix element in a certain polarization basis makes it possible to estimate their maximum values using the inequalities:

$$D_\Sigma^2 + 2d_{12}^2 \leq 4D_{11}^2 \leq 2D_\Sigma^2 + 2(d_{11}+d_{22})\sqrt{(d_{11}-d_{22})^2 + 4d_{12}^2}$$

$$\left.\begin{matrix} d_{11}^2 + d_{22}^2 \\ 4d_{12}^2 \end{matrix}\right\} \leq 4\, D_{12\,max}^2 \leq 2D_\Sigma^2 - (d_{11} - d_{22})^2 \qquad (5b)$$

In order to determine more detailed limits on the estimation of dispersion changes of scattering matrix elements, it is necessary to introduce the dispersion increase coefficient $K = D_{11}^2/d_{11}^2$ and to normalize the dispersions $y = d_{12}^2/d_{11}^2$, $x = d_{22}^2/d_{11}^2$. If we use real expressions for the dispersions, then we can obtain the following equations for the limiting bounds

$$\left.\begin{aligned} 4\,K_{min} &= 1+x+4y \\ y &= (\sqrt{K_{max}} - 1)(\sqrt{K_{max}} - \sqrt{x}) \end{aligned}\right\} \qquad (6)$$

These dependencies are illustrated in Fig.1.

Thus, knowing the dispersions of the elements $S_{ij}$ in the initial or in the Descartes basis, it is possible to estimate the limits of changes in $K_{min}$ and $K_{max}$ using Fig.1. For example, if for a certain target $d_{22}^2/d_{11}^2 = 3$, $d_{12}^2/d_{22}^2 = 2$, then the change in polarization type will cause the change in $D_{11}^2/d_{11}^2$ within the interval from 2 to 8. The ratio for the increase of dispersion $D_{12}^2/d_{12}^2$ is obtained in the same way as

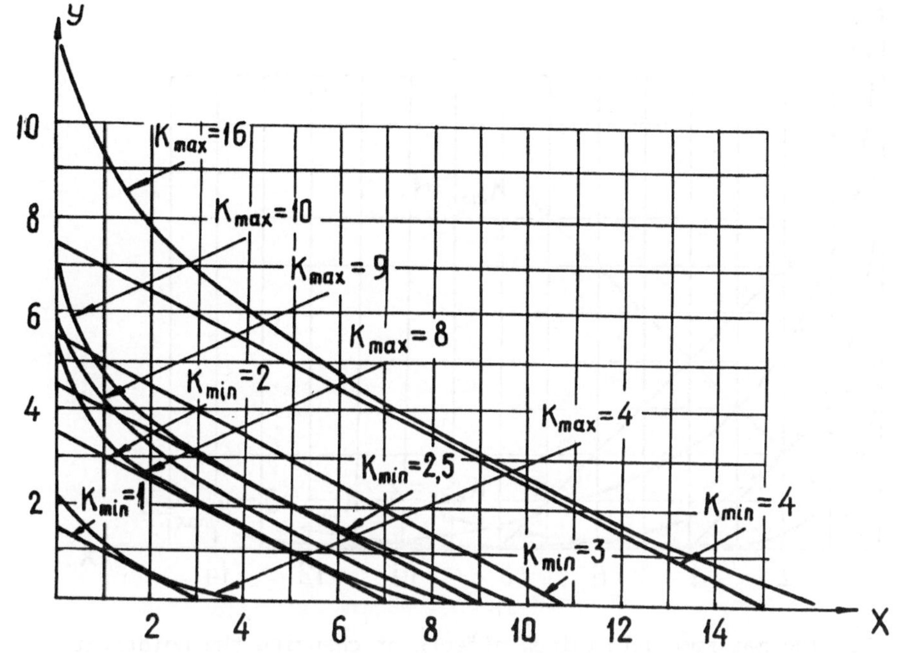


Fig. 1: The limiting bounds of radar cross-sections $\sigma_{11}^2$ and $\sigma_{22}^2$

$$\left.\begin{array}{l} y = 4\, K_{min} - x \\ \sqrt{x} + \sqrt{y} = 2\sqrt{K_{max} - 1} \end{array}\right\} , \tag{7}$$

where $y = d_{11}^2/d_{12}^2$ , $x = d_{22}^2/d_{12}^2$ and the appropriate curves are presented in Fig. 2. It might be logical to generalize the Graves matrix $[G_0]$ to the following law:

$$[G_0] = \overline{[\vec{S}^*\vec{S}]} - [\overline{\vec{S}}^* \cdot \overline{\vec{S}}] \tag{8a}$$

and to show that in the course of changing the polarization basis, the matrix $[G_0]$ is subject to the transform of similarity using the unitary matrix [Q] which has the size 2x2 (its true form is given in [1]). It means that there always exists such a polarization basis for which $[G_0]$ is diagonal. Let's call this polarization basis "zero PB" with diagonal elements $g_{11} = \delta_{11}^2 + \delta_{12}^2$, $g_{22} = \delta_{12}^2 + \delta_{22}^2$, where the $\delta_{ij}^2$ are the elements $S_{ij}$ in the "zero PB". The trace, $Tr\{[G_0]\}$, of the matrix $[G_0]$ is invariant to a polarization basis change (4). The determinant of matrix

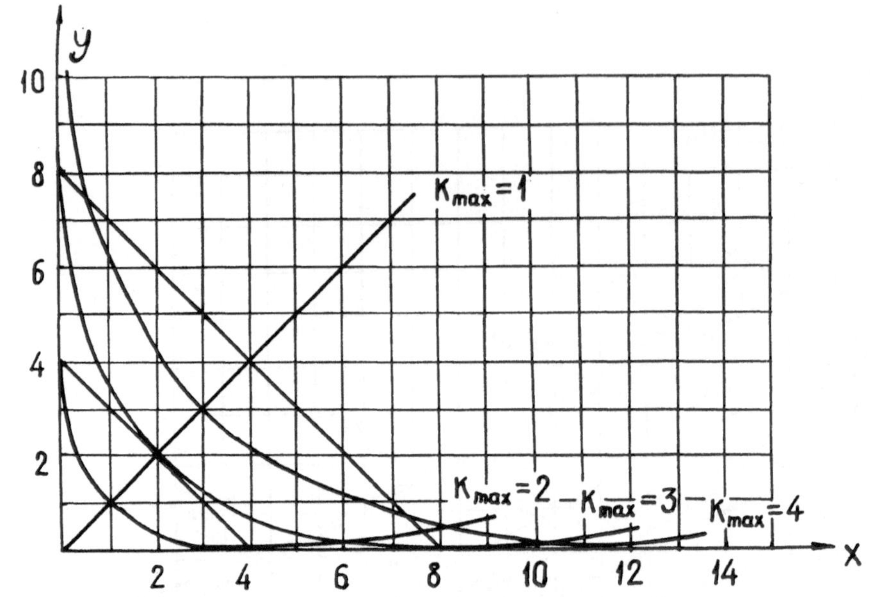


Fig. 2: The maximum and minimum effects of changing the polarization basis

$[G_0]$ has the same invariant properties.

$$\text{Trace}\{[G_0]\} = \text{invariant} \tag{8b}$$

$$\text{Det}\{[G_0]\} = \text{invariant} \tag{8c}$$

Having fulfilled the above-mentioned transform of similarity on matrix $[G_0]$ in the zero polarization basis, it is possible to obtain the true form of matrix $[G_0]$ in the new polarization basis, determined by parameters $(\gamma, \psi)$ in relation to the "zero PB" as:

$$[G_0(\gamma,\psi)] = \frac{D_\Sigma^2}{2}\begin{bmatrix}1 + q\cos 2\gamma & qe^{-2i\psi}\sin 2\gamma \\ qe^{2i\psi}\sin 2\gamma & 1 - q\cos 2\gamma\end{bmatrix}, \tag{9}$$

where $q = (\delta_{11}^2 - \delta_{22}^2)/D_\Sigma^2$ is the analogue of the degree of anisotropy of the radar (radiolocation) target determined in [1] and being invariant to a polarization basis change.

Using initial representations for the matrix $[G_0]$ with expressions for the dispersions $d_{ij}^2$, it is easy to obtain the very important dispersion relation:

$$d_{11}^2 - d_{22}^2 = D_\Sigma^2\, q \cos 2\gamma , \tag{10}$$

demonstrating that the maximally possible difference of the dispersions of the elements $S_{11}$ and $S_{22}$ is equal to $qD_\Sigma^2$, and that there exists such a polarization basis, for which $d_{11}^2 = d_{22}^2$.

In addition, special importance must be attached to the possibility for the estimation of the covariance coefficient $\tau_{ijkl}$ and its phase $\rho_{ijkl}$ determined by equation

$$\overline{S_{ijkl} S_{ijkl}^*} = \tau_{ijkl}\, \sigma_{ij}\, \sigma_{kl} \exp\{j\rho_{ijkl}\} \tag{11a}$$

When the polarization basis changes, the values $\tau_{ijkl}$ and $\rho_{ijkl}$ change as well. Their limits can be found from the non-negative requirements to values $\overline{|S_{ij}|^2}$ in an arbitrary polarization basis, which after some manipulations are reduced to the inequality [1,2],

$$R_{12}^2 + 2xR_{12} + y \leq 0 , \tag{11b}$$

where

$$2x = R_1^2 \cos 2\phi_1 + R_2^2 \cos 2\phi_2 - 2R_1R_2\cos(\phi_1+\phi_2) ;$$

$$y = R_1^2 + R_2^2 - 2R_1R_2\cos(\phi_1-\phi_2) - 1 ;$$

$$R_1\sqrt{\sigma_{11}^2 + \sigma_{22}^2} = \sigma_{11}\tau_{1112} ;$$

$$R_2\sqrt{\sigma_{11}^2 + \sigma_{22}^2} = \sigma_{22}\tau_{1222};$$

$$R_{12}(\sigma_{11}^2 + \sigma_{22}^2) = 2\sigma_{11}\sigma_{22}\tau_{1122};$$

$$2\phi_1 = \rho_{1122} - 2\rho_{1112};$$

$$2\phi_2 = \rho_{1122} - 2\rho_{1222} ,$$

Here, it should be emphasized that all of the $R_1$, $R_2$, $R_{12}$ are non-negative and less than one.

The given inequality analysis makes it possible to reduce its requirements to the limiting bounds, presented in Fig. 3. The same limiting bounds up to Fig. 3 belong to dispersions, correlation factors and their phases.

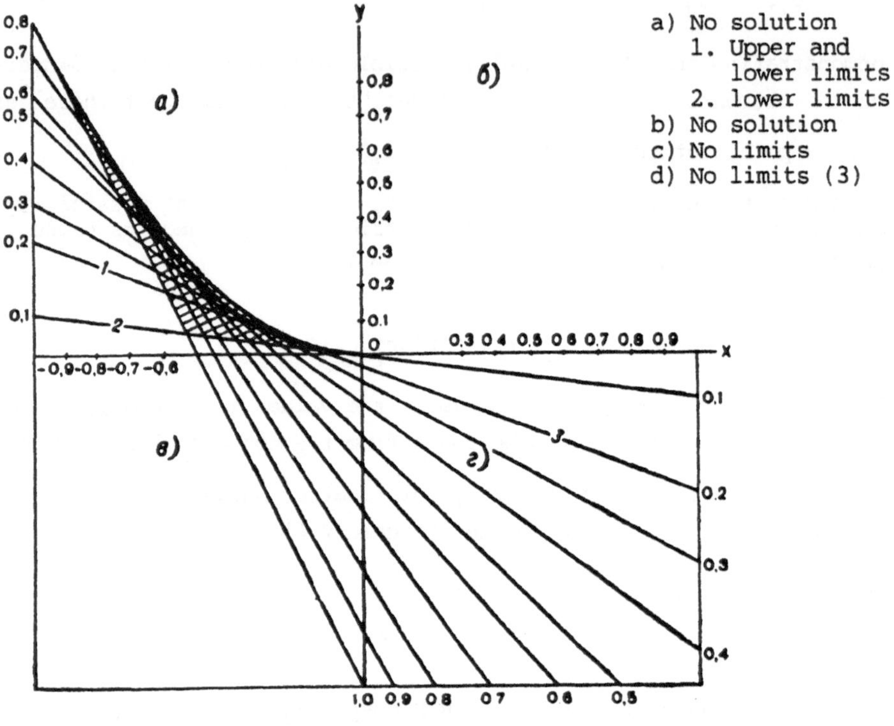


Fig. 3: The areas of permissible values for the correlation coefficient between elements of the dispersion matrix: a–d

## 3. SIMULATION OF SCATTERING MATRIX DATA

The problem of determining the probability density distribution of different elements of their scattering matrices or of their combinations is central for the radio-location of targets, for detection, and for discrimination. However, this problem is closely related to finding the satisfactory solutions for the associated diffraction and scattering problems. Because of the inherent complexities of the scattering scenarios, a search for the desired results to analytic integro-differential formulations of the complex scattering process becomes non-perspective and, in general, is to be ruled out. On the other hand the lack of experimental data reduces the opportunity of formulating convincing conclusions. The way we see it, the most promising way to solve the problem is by means of statistical simulation of the scattering matrix itself. To achieve this, first of all it is necessary to find an approach of formulating the desired scattering matrix elements. It could be done by constructing these elements through the scattering matrix eigenvalues $\lambda_1, \lambda_2$ and through parameters which characterize the location of the polarization basis being considered in relation to the eigen-polarization states: $\alpha, \eta, \psi, \gamma$. Let's use these formulas, which were determined in [1], where it was shown that

$$\left.\begin{aligned} S_{11}e^{-2i\psi} &= e^{2i(\alpha+\eta)}(\lambda_1\ e^{-2i\alpha}\cos^2\gamma + \lambda_2 e^{2i\alpha}\sin^2\gamma) \\ S_{12}e^{-2i\psi} &= 0.5(-\lambda_1\ e^{-2i\alpha} + \lambda_2 e^{2i\alpha})\sin 2\gamma \\ S_{22}e^{-2i\psi} &= e^{-2i(\alpha+\eta)}(\lambda_1 e^{-2i\alpha}\sin^2\gamma + \lambda_2 e^{2i\alpha}\cos^2\gamma) \end{aligned}\right\}. \qquad (12)$$

The main idea of this method, given in [1], consists in the generation of the definitions and approximations given in (11) according to the required statistical laws of random variables $\lambda_1$, $\lambda_2$, $\alpha$, $\eta$, $\psi$, $\gamma$ and by further calculation of the desired values $S_{ij}$ by use of (3).

Approximate distribution functions were chosen from Johnson's curves or Pearson's distributions [1]. This choice of distributions was accomplished according to the first four moments followed by the check on whether the corresponding hypothesis was satisfied.

The two-dimensional Gaussian law has been tested as the basic variance for the eigen-values $\lambda_1$ and $\lambda_2$, and, of the mean values of the scattering matrix; and the dispersions were allowed to vary within the range from 0 to 20, and the correlation factor varied between 0 to 1 distribution. For the other distribution parameters, a uniform distribution law and a Gaussian law were chosen. The parameters obtained by this method were used to form an extensive set of random numbers including 5,000 points, which provided the possibility for scattering matrix elements to have valid, efficient and non-shifted estimations.

The data obtained in the course of the statistical simulation are incorporated in Table I, in which I - is for the $\beta$-Pearson distribution of the I type; III is for the $\gamma$-distribution; IV is the for the fourth type of Pearson's distribution; VI is for a $\beta$-distribution of the second type; VII is for a Gaussian distribution; XII is a particular case of the first type $\beta$-distribution.

The table shows that the most general distribution is the $\beta$-distribution of the first type having the form depending on certain parameter values. If we apply the Gaussian law for $\alpha$, $\eta$, $\gamma$, then the results look more stable in accordance with a certain distribution type. The picture varies, when the uniform law is being used.

Real and imaginary parts of the distributions of scattering matrix elements most often correspond to Pearson's distribution of the fourth type, which in some particular cases degenerate into the t-distribution

**Table I.**
**Statistical Distribution Approximation Data**
**Types of Pearson's Distribution**

| Parameters | Uniform distribution $\alpha, \eta, \psi, \gamma$ | Gauss distribution $\alpha, \eta, \psi, \gamma$ | Experiment |
|---|---|---|---|
| $ReS_{11}$ | IV, t-distr., VII | I, IV | |
| $ImS_{11}$ | IV, VII | I, III | |
| $ReS_{12}$ | IV, t-distr. | I, IV | |
| $ImS_{12}$ | IV, t-distr., VII | I, IV, t-distr. | |
| $\|S_{11}\|$ | I, III, IV, VII | I | I, III, VII |
| $\|S_{12}\|$ | I, III, IV, VI | I | IV |
| $\|S_{11}\|^2$ | I, III, VI | I, XII | |
| $\|S_{12}\|^2$ | I, VI | I | |
| $\psi_{11}$ | I | I | III |
| $\psi_{12}$ | I | I | I |
| $\psi_{11}+\psi_{22}-2\psi_{12}$ | I | I | III |
| $\psi_{22} - \psi_{11}$ | I | I | III |
| det{[S]} | I, III | I, III | |
| q | I, III, XII | I, XII | III,IV,VIII |
| $\lambda_1^2 + \lambda_2^2$ | I, III, IV | I, III | IV, VI |

or the Gaussian distribution. In varying the two-dimensional Gaussian law parameters and in the course of transition from the angle parameters' uniform distribution, the fourth type of distribution is enriched by $\gamma$-and $\beta$-distributions of the first type.

Speaking about the distribution of scattering matrix elements arguments of the phases, $\psi_{11}$, $\psi_{22}$, $2\psi_{12}$ and their combinations $\psi_{22} - \psi_{11}$ and $\psi_{11} + \psi_{22} - 2\psi_{12}$, they can, possibly have at any distribution density of initial parameters' probabilities only a $\beta$-distribution of the first type and for its particular cases- uniform and triangular distributions.

Depending on the choice of two-dimensional Gaussian initial law parameters for the other values; the distributions turned out not to be concrete but certain parts of appropriate curve types may be associated with the non-uniform location of possible distribution areas along the whole Pearson's diagram. For $|S_{22}|$ and $|S_{11}|$ these areas coincide and stretch from a $\gamma$-distribution up to U-shape and $\beta$-distribution (Fig.4). For the degree of anisotropy the area of possible distribution types stretch from critical up to $\gamma$-distribution. The main requirement for the scattering matrix is relative uniformity of distributions along the whole Pearson's diagram from a $\beta$-distribution of the first type to a $\gamma$-distribution. For the full RCS the most typical is the F-shaped $\beta$-distribution. To illustrate this result, Figures 5-7 give several types of distributions. Column 4 gives the experimentally obtained results according to the types of distribution for a relatively even earth surface, covered with bushes and groves and with a limited number of hills.

This simulation method made it possible to study the correlation properties of random processes. The analysis showed that correlation coefficients for pairs of $\mathrm{Re}S_{11}$ and $\mathrm{Re}S_{22}$, and of $\mathrm{Im}S_{11}$ and $\mathrm{Im}S_{22}$ are practically unlimited. Concerning the other pairs made of real and imaginary parts of the correlation coefficients of the scattering matrix elements, their values may be accepted to be equal to zero. The determination of the distribution density for the full RCS, $\sigma_\Sigma^2$, and for the anisotropy degree is very important for practical radiolocation purposes. Using the Gaussian law as multidimensional distribution law for real and imaginary parts of scattering matrix elements, it is quite possible to determine the desired distribution densities by direct calculations:

$$W(\sigma_\Sigma^2) = \frac{sL\left(\sigma_\Sigma^2 \sqrt{\alpha^2 + \beta^2}\right)}{\sqrt{\alpha^2+\beta^2}} \cdot \frac{\exp\{-2(1-R^2)^{-1}\sigma_\Sigma^2\}}{d_1^2 d_2^2 (1-R^2)}$$

$$W(q) = 2\Omega(1-R^2)\left\{\frac{(1+\Omega) - q(1-\Omega)}{\{[(1+\Omega)-q(1-\Omega)]^2-4\Omega R^2(1-q^2)\}^{1.5}} + \right.$$

$$\left. + \frac{(1+\Omega)+q(1-\Omega)}{\{[(1+\Omega)+q(1-\Omega)]^2-4\Omega R^2(1-q^2)\}^{1.5}}\right\} , \qquad (13)$$

where $d_1^2$ , $d_2^2$ are random values of dispersions $\lambda_1$ , $\lambda_2$; R is the correlation coefficient between them; $\alpha = (1-R^2)^{-1}(d_2^{-2} - d_1^{-2})$; $\beta = R[(1-R^2) d_1 d_2]^{-1}$; $\Omega = d_1^2/d_2^2$ is the target asymmetry degree. Fig. 8 contains corresponding curves for distribution density $\sigma_\Sigma^2$ and for q.

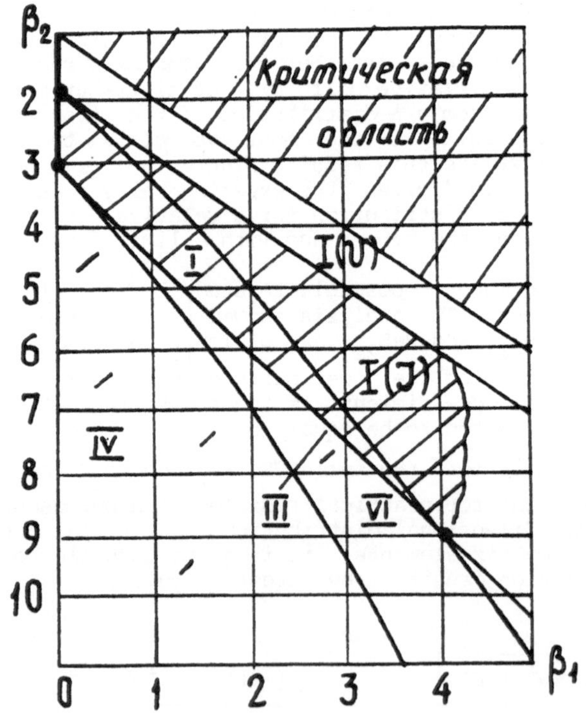


Fig. 4: The areas of supposed propagation for $|S_{11}|$ and $|S_{12}|$.

## CONCLUSIONS

The modelling results for simulating the statistical random behavior of the scattering matrix element their moduli, phase arguments, degree of

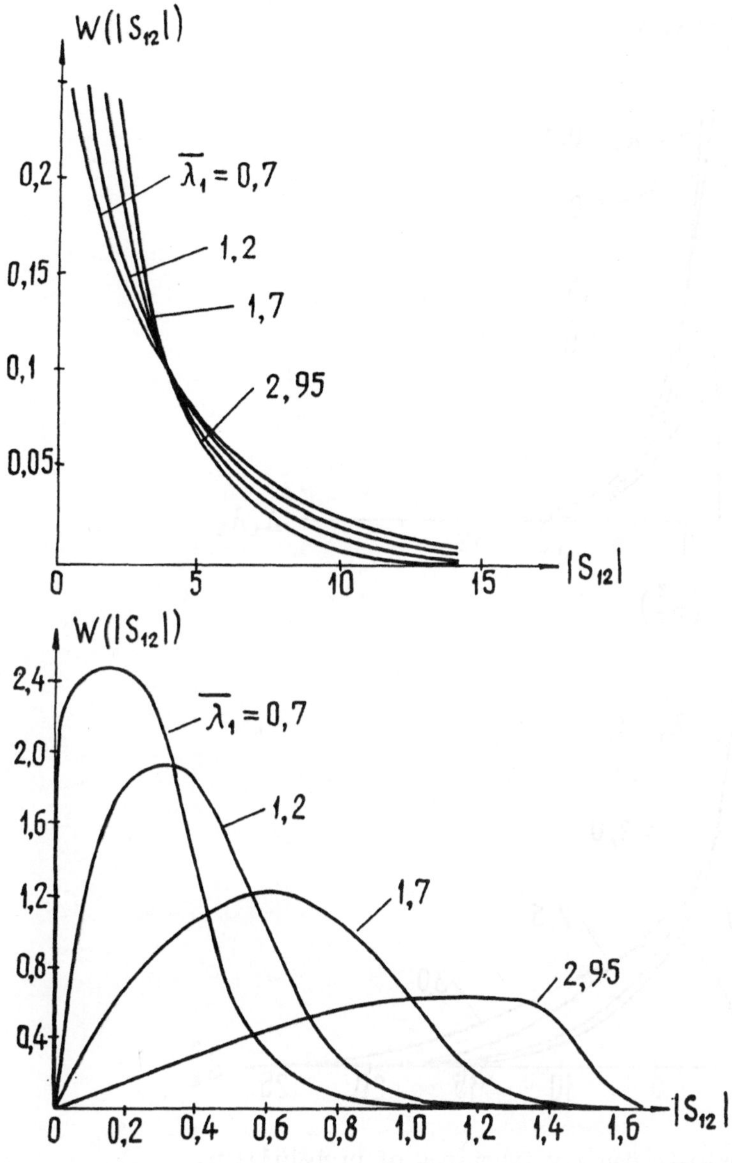


Fig. 5: Approximate density functions of probability $|S_{12}|$:

a) $\bar{\lambda}_2 = 0.5$; $d_{\lambda 1} = 20$;

b) $\bar{\lambda}_2 = 0.5$; $d_{\lambda 1} = 20$; $d_{\lambda 2} = 20$;

c) $\bar{\lambda} = 0.5$; $d_{\lambda 1} = 0.15$; $d_{\lambda 2} = 0.1$.

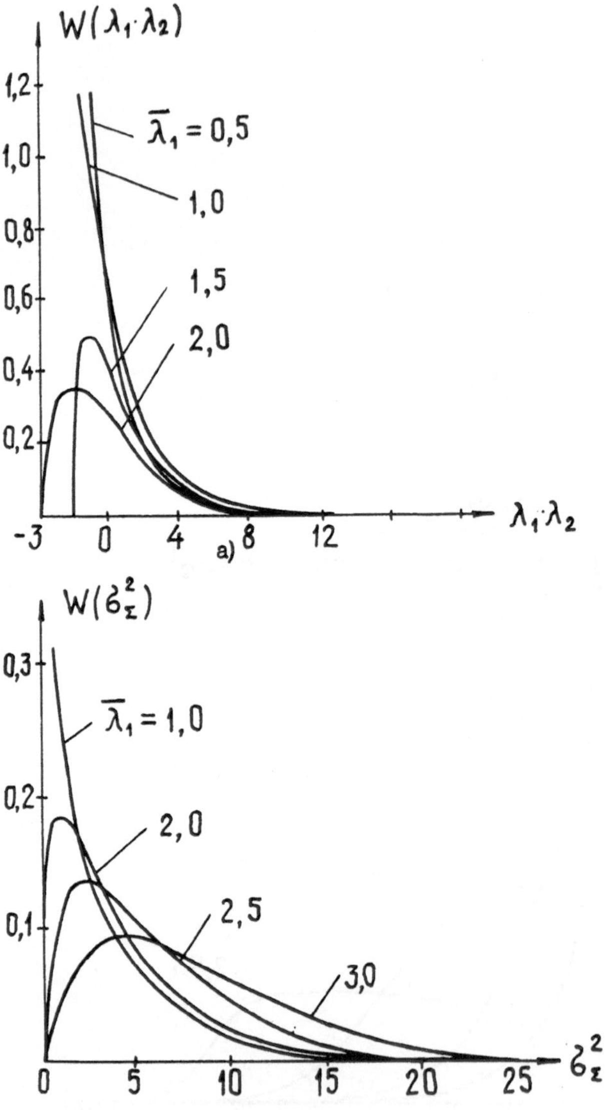


Fig. 6: Approximate density functions of probability:

a) $|\det S|$, $\bar{\lambda}_2 = 0.7$, $d_{\lambda 1} = 0.7$; $d_{\lambda 2} = 1.2$;

b) $\sigma_\Sigma^2$; $\bar{\lambda}_2 = 1$, $d_{\lambda 1} = 1$, $d_{\lambda 2} = 1$.

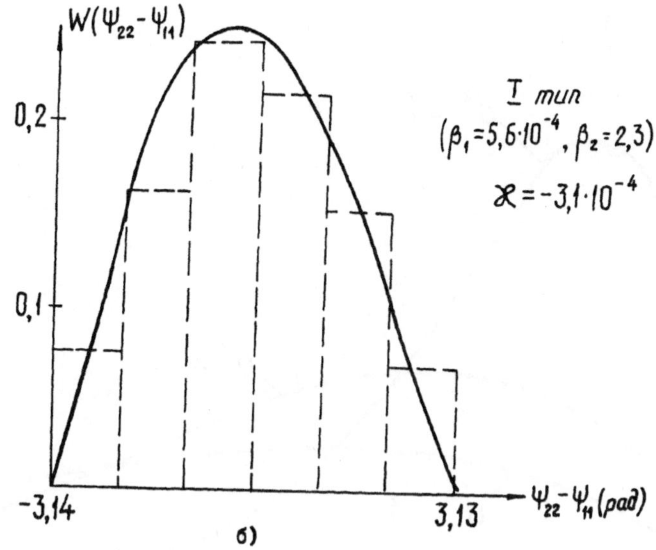


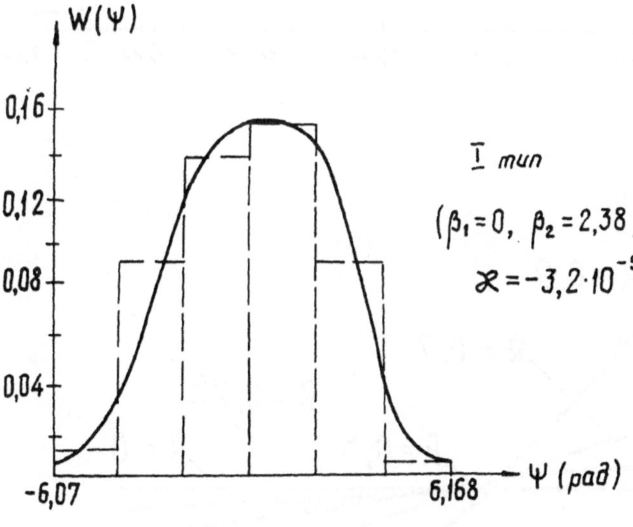


Fig. 7: Histograms and approximate density functions:

a) $\psi_{22} - \psi_{11}$; $\bar{\lambda}_1 = 1.0$; $\bar{\lambda}_2 = 0.1$; $d_{\lambda 1} = 0.7$; $d_{\lambda 2} = 1.2$

b) $\phi = \psi_{22} + \psi_{11} - 2\psi_{12}$; $\bar{\lambda}_1 = 1.5$; $\bar{\lambda}_2 = 0$; $d_{\lambda 1} = d_{\lambda 2} = 1$

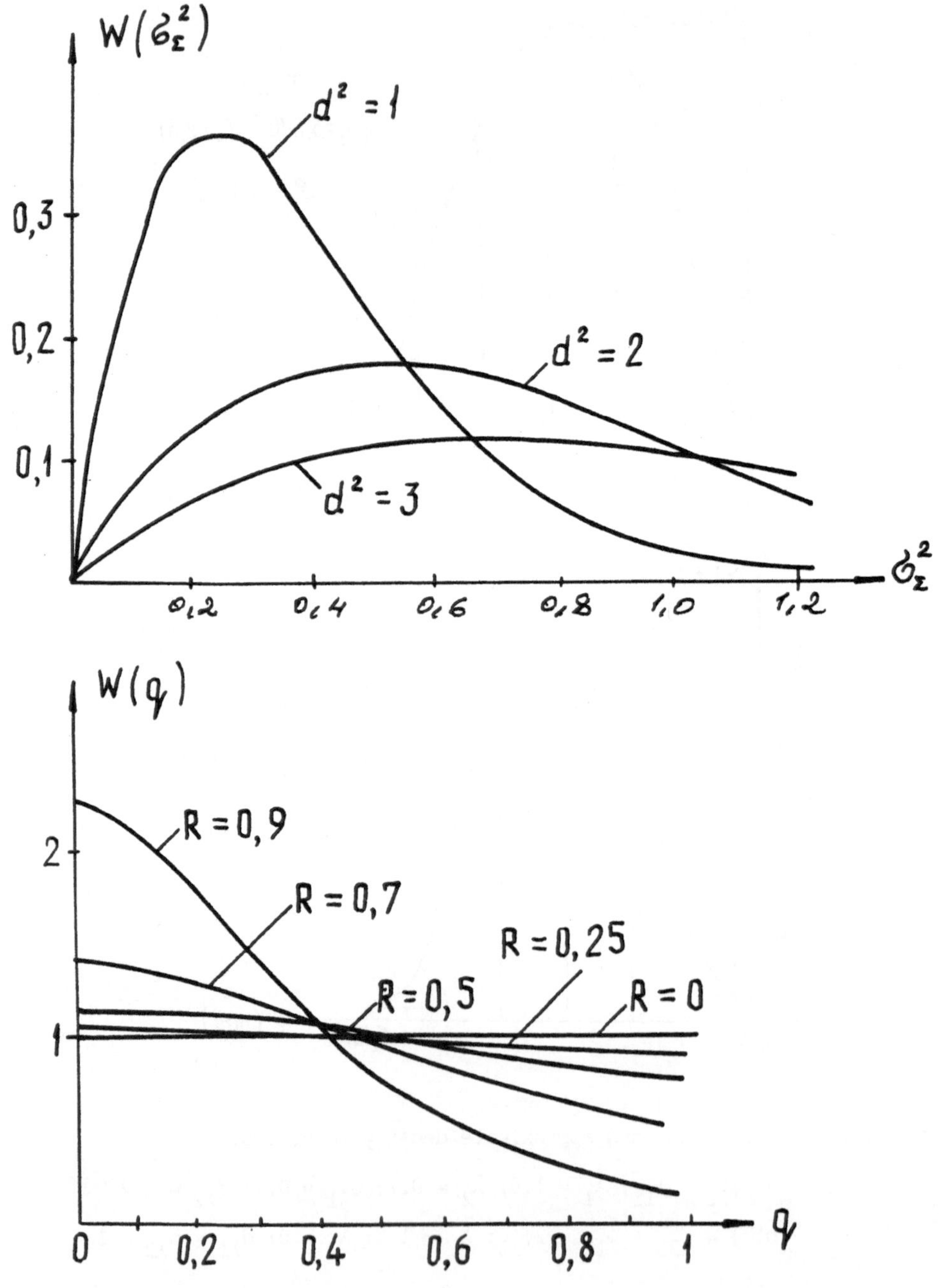


Fig. 8: The probability density functions: a) $\sigma_{\Sigma}^{2}$ ; b) q .

anisotropy, equivalent radar cross-section and generalized phase were presented. It is concluded that such multi-dimensional modeling approaches for obtaining more useful and efficient distribution functions for the interpretation of polarimetric measurement data should be advanced, hand in hand, with accumulating highly improved complete polarimetric measurement data sets.

**ACKNOWLEDGEMENTS**

The authors gratefully acknowledge the stimulating interaction with the workshop director and the editor of the proceedings, and thank him for incorporating one of many of the authors' papers on this subject into the proceedings.

**REFERENCES:**

1. Bogordsky, V.V., Kanareikin, D.B., Kozlov, A.I., 'Scattering radio radiation polarization of earth surface', L. Gidrometeoizdat, 1981, p.280.

2. Kozlov, A.I., 'Several properties of scattering matrix element parameters of radar (radiolocation) targets', Radioelectronika, 1979, Vol. XXII.

# OPTIMAL POLARIZATIONS FOR RECIPROCAL RANDOM TARGETS

*Karl Tragl*
*Institute for Radio Frequency Technology*
*German Aerospace Research Establishment DLR[1])*
*8031 Oberpfaffenhofen*
*Federal Republic of Germany*

ABSTRACT. A theory of optimal polarizations for radar backscattering from reciprocal random targets is developed, employing a revised covariance matrix approach. The polarization dependence of the covariance matrix is formulated in terms of unitary similarity transformations. Utilizing this transformation property, inequalities for the polarization dependent mean copolar and crosspolar power return are derived. Functional relations can be proved, this way, which relate the derivatives of these power functions with respect to the polarization parameters to the cross-correlation observables contained in the covariance matrix. Optimal polarizations for the power return in two orthogonally polarized radar channels can be determined. Polarizations which extremize mean copolar power are found to reduce the correlation of the backscattered wave components exactly to zero. In case of crosspolar optimal polarizations a particular correlation difference rather than individual interchannel correlations tends to zero. The presented polarimetric concept, finally, is demonstrated to generalize the well-established theory of characteristic polarizations for deterministic targets.

## 1. Introduction

The determination of characteristic polarization states in radar target applications is one of the main concerns in radar polarimetry. For deterministic scatterers, the theory of characteristic polarizations has been studied extensively in the past, see e.g. Kennaugh (1952), Huynen (1970), Boerner et al (1981), Agrawal et al (1989), based on a scattering matrix concept. Extending polarimetric radar methods and the theory of optimal polarizations to random target observations is currently a research topic of general interest, cf. recent publications by McCormick et al (1985), Kostinski et al (1988), van Zyl et al (1987) and Tragl et al (1989).

The present paper is concerned with backscattering from reciprocal random targets within an orthogonal polarization configuration. The dependence of mean co- and crosspolarized backscattered power upon the transmitter polarization is investigated. A revised covariance matrix approach is employed to study random target features, this matrix containing the relevant physical observables: scattered powers in the orthogonally polarized radar channels and cross-correlations between both

---

[1]) The German Aerospace Research Establishment, former DFVLR, has been renamed DLR since January 1989.



W.-M. Boerner et al. (eds.), *Direct and Inverse Methods in Radar Polarimetry, Part 1*, 693–706.

channels. The polarization dependence of the covariance matrix can be described in terms of unitary similarity transformations. This way, inequalities for the polarization dependent mean copolar and crosspolar power return are found. Furthermore, functional relations between the derivatives of these power functions with respect to the polarization parameters and correlation functions of the backscattered signals can be determined. The general problem of optimal polarizations, thus, can be solved quite naturally within this new approach.

Optimal polarizations for the copolar mean power return has been derived previously by van Zyl (1986), based on the Mueller matrix approach. McCormick et al (1985), on the other hand, indicated that copolar and crosspolar power minima should be located in the vicinity of those polarization states, for which the correlation between the co- and crosspolarized backscattered signals tends to zero. As a new result, we derive exact necessary conditions for the power maxima and minima in the co- and crosspolar radar channel, involving the correlation observables. The above-mentioned decorrelation property, thus, can be proved to hold exactly for copolar optimal power return, but has to be modified for the crosspolar one.

The well-established characteristic polarization theory for deterministic targets is demonstrated to represent just the limiting case of time-independent scattering within the general optimal polarization theory, as developed in the paper.

## 2. Some Fundamentals of Radar Polarimetry

### 2.1. REPRESENTATION OF ELECTROMAGNETIC PLANE WAVES

The electric field of a monochromatic plane wave, propagating in the positive z-direction with angular frequency $\omega_o$ and propagation constant $k_o$, can be represented by a two-component field vector in any polarization basis. Separating the harmonic time dependence and the spatial propagation part $exp[j(\omega_o t - k_o z)]$, the electric field can be expressed in terms of a complex polarization vector

$$\vec{E} = \begin{bmatrix} E_H \\ E_V \end{bmatrix} = \begin{bmatrix} a_H \cdot e^{(j\,\delta_H)} \\ a_V \cdot e^{(j\,\delta_V)} \end{bmatrix} , \qquad (1)$$

with real amplitudes $a_H$, $a_V$ and phases $\delta_H$, $\delta_V$. The vector components in (1) are related to the orthogonal, linear horizontal and vertical (H,V)-polarization basis; j denotes the imaginary unit with $j^2 = -1$.

The electric field vector at each fixed point in space traces out a polarization ellipse, as sketched in Figure 1. The size of the ellipse is proportional to the wave amplitude, which is related to the total power of the wave. The shape of the polarization ellipse can be characterized by two geometrical parameters: the ellipticity $\tau$ varying from -45° to +45° and the orientation angle $\varphi$ varying from 0° to 180°, see Deschamps (1951). Negative and positive values of $\tau$ denote right and left handedness of the polarization, respectively, as defined by the IEEE standard 149 (1979). The polarization state of an electromagnetic wave can also be characterized by the complex polarization ratio $\rho$, which is defined as the ratio of the wave components

$$\rho = \frac{E_V}{E_H} = \frac{a_V}{a_H}\, e^{j(\delta_V - \delta_H)} , \qquad (2)$$

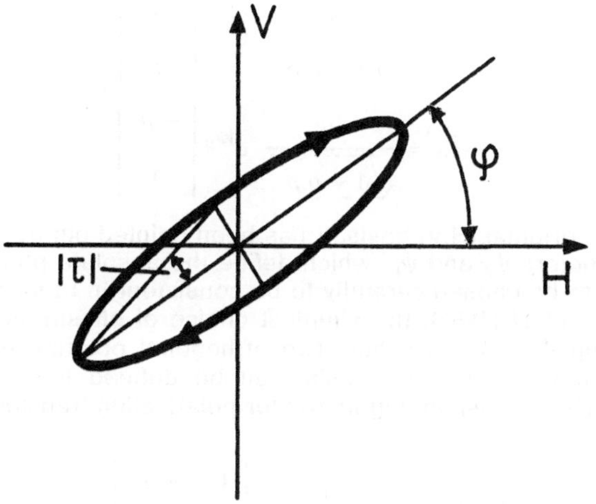


**Figure 1.** **Polarization ellipse** characterized by the geometrical polarization parameters ellipticity $\tau$ and orientation angle $\varphi$. The propagation z-direction is pointed at the observer.

related to the (H,V)-polarization basis, as detailed by Azzam et al (1987). The polarization ratio $\rho$ can be expressed in terms of the geometrical parameters by

$$\rho = \frac{\cos(2\tau)\ \sin(2\varphi) + j \cdot \sin(2\tau)}{1 + \cos(2\tau)\ \cos(2\varphi)} \quad . \tag{3}$$

The electric field vector $\vec{E}$, given by (1), refers to the (H,V)-polarization basis. Applying unitary transformations, the field vector can be referred to any other orthogonal elliptical (A,B)-polarization basis, see Azzam et al (1987), by

$$\vec{E}' = \mathbf{U}^{-1} \cdot \vec{E} \quad , \tag{4}$$

where **U** denotes a unitary transformation matrix[2]) and the prime indicates reference to the new basis. Any 2x2 unitary matrix, in general, can be written as

$$\mathbf{U} = \frac{1}{\sqrt{1 + \rho\,\rho^*}} \begin{bmatrix} e^{j\psi_A} & -\rho^*\, e^{j\psi_B} \\ \rho\, e^{j\psi_A} & e^{j\psi_B} \end{bmatrix} \quad , \tag{5}$$

with the asterisk denoting complex conjugation.

The columns of the matrix represent the orthogonal unit vectors of the new polarization basis:

---

[2]) Note that $\mathbf{U}^{-1} = \mathbf{U}^{+}$ for unitary matrices.

$$\vec{\varepsilon}_A = \frac{1}{\sqrt{1+\rho\rho^*}} e^{j\psi_A} \begin{bmatrix} 1 \\ \rho \end{bmatrix}$$
$$\vec{\varepsilon}_B = \frac{1}{\sqrt{1+\rho\rho^*}} e^{j\psi_B} \begin{bmatrix} -\rho^* \\ 1 \end{bmatrix} \quad , \tag{6}$$

in terms of the original (H,V)-basis. It has been pointed out by Kostinski et al (1986), that the parameters $\psi_A$ and $\psi_B$, which define the absolute phases of the new basis vectors, have to be chosen carefully to be consistent in radar applications. According to Boerner et al (1981), the simplest choice of phases is used in the present paper, implying $\psi_A = \psi_B = 0$. Thus, two orthogonal polarization vectors (6), which determine a new polarization basis, can be defined in terms of one complex parameter $\rho$. The corresponding matrix for polarization transformations, this way, is given by

$$\mathbf{U}(\rho) = \frac{1}{\sqrt{1+\rho\rho^*}} \begin{bmatrix} 1 & -\rho^* \\ \rho & 1 \end{bmatrix} \quad . \tag{7}$$

The particular choice of phases, used in the present paper, guarantees that $\rho = 0$ corresponds to the identity transformation $\mathbf{U}(0) = \mathbf{1}$.

## 2.2. COHERENT SCATTERING FROM DETERMINISTIC TARGETS

The polarization state of an electromagnetic plane wave scattered by a complex radar target, in general, differs from that of the incident wave. If the target's time scale of temporal fluctuations is much larger than the observation time of the measurement device, we call such a target, or target state, deterministic. In this case, the scattering characteristics can uniquely be expressed in terms of a polarimetric scattering (S-)matrix, which determines the information about the target for the particular scattering geometry and the employed radar frequency completely. The scattered field $\vec{E}_s$ is related to the incident wave $\vec{E}_i$ by the linear scattering equation

$$\vec{E}_S = \mathbf{S} \cdot \vec{E}_I \quad . \tag{8}$$

The amplitudes and phases of the scattered electromagnetic wave can be regarded constant during the measurement. Hence, the waves are completely polarized and fully coherent.

In the following, we are interested in the backscattering features of reciprocal radar targets. The scattering matrix in the common (H,V)-polarization basis is defined by

$$\mathbf{S} = \begin{bmatrix} S_{HH} & S_{HV} \\ S_{VH} & S_{VV} \end{bmatrix} \quad , \tag{9}$$

where $S_{HV} = S_{VH}$ for reciprocal targets, implying symmetry of the scattering matrix. Enforcing polarimetric invariance of the receiver voltage equation $V = \vec{E}^T \cdot \mathbf{S} \cdot \vec{E}$, the scattering matrix transforms under the change of polarization basis (4) as

$$\mathbf{S}' = \mathbf{U}^T(\rho) \cdot \mathbf{S} \cdot \mathbf{U}(\rho) \quad , \tag{10}$$

involving the unitary transformation matrix $\mathbf{U}(\rho)$, see Boerner et al (1981). The superscript T denotes the transposed of a matrix or a vector; reference to the new (A,B)-polarization basis is indicated by a prime. Equation (10) relates the scattering matrix representation in any orthogonal (A,B)-polarization basis, characterized by a complex polarization ratio $\rho$ according to (6), to the S-matrix (9) given in the particular (H,V)-basis. The expansion of equation (10) results in explicit transformation equations of the scattering matrix elements:

$$\begin{aligned} S_{AA}(\rho) &= \left[S_{HH} + 2\,\rho\, S_{HV} + \rho^2 S_{VV}\right] / (1 + \rho\,\rho^*) \\ S_{AB}(\rho) = S_{BA}(\rho) &= \left[-\rho^* S_{HH} + (1 - \rho\,\rho^*)\, S_{HV} + \rho\, S_{VV}\right] / (1 + \rho\,\rho^*) \\ S_{BB}(\rho) &= \left[\rho^{*2} S_{HH} - 2\,\rho^* S_{HV} + S_{VV}\right] / (1 + \rho\,\rho^*) \quad . \end{aligned} \tag{11}$$

Note that the symmetry of the scattering matrix (reciprocity of the target) is naturally preserved under a change of the polarization reference basis as defined by (4), (10).

## 2.3. BACKSCATTERING FROM RECIPROCAL RANDOM TARGETS

In most radar applications the observed targets are subject to temporal variations. In radar meteorology, for instance, the observed scattering volumes within precipitation clouds are filled with a large number of moving particles and, therefore, evidently are time dependent. If such a random target is illuminated by a monochromatic radar wave, the amplitudes and phases of the backscattered signals are time dependent as well. Hence, in extended time observations the scattered waves are only partially polarized with incoherent scattering contributions. Most of the considered random targets, however, generate a quasi-monochromatic field variation of the scattered waves, resulting in a finite bandwidth $\Delta\omega$ of the field components which is small compared to their center frequency ($\Delta\omega \ll \omega_0$). Thus, the physically well defined observables are given as time averages of functions which depend on the scattered fields, with an integration time much larger than typical time scales of the target's variations.

Several concepts may be used to describe the long-term statistics of quasi-monochromatic waves, e.g. the coherency matrix or the Stokes vector formalism, see e.g. Azzam et al (1987). Random targets, this way, can either be characterized by a polarization covariance matrix, which contains all second order moments of the scattering coefficients, or in terms of a Mueller matrix, which is related to the covariance matrix by linear transformations, see Azzam et al (1987). The present paper is concerned with the covariance matrix approach, because this matrix contains the relevant physical observables: the scattered powers in the orthogonally polarized radar channels and the cross-correlations between the channels.

The definition of a covariance matrix, first of all, requires a vector formulation of the target scattering features. Each instantaneous state of a time varying random target is completely characterized by a corresponding scattering matrix. In case of reciprocal targets, the crosspolar scattering components are equal and, therefore, only three independent scattering parameters are required to characterize the target state. Thus, we may introduce a target feature vector as

$$\vec{\Omega}^{T} = \left[ S_{HH} \quad \sqrt{2}\, S_{HV} \quad S_{VV} \right] \quad , \tag{12}$$

in the linear (H,V)-polarization basis. The factor $\sqrt{2}$ in front of the crosspolar component is implied by the required norm conservation under polarization transformations: the square of the norm $\vec{\Omega}^{+}\vec{\Omega}$ of a target feature vector[3]) equals the *span* of the corresponding scattering matrix, which is an important polarimetric invariant, emphasized by Boerner et al (1981). Target descriptions by means of scattering matrices or feature vectors are obviously equivalent.

The components of a target feature vector in any orthogonal elliptical basis, defined by the polarization ratio $\rho$ according to (7) can be related to the vector representation in the original (H,V)-basis by a linear transformation

$$\vec{\Omega}' = \mathbf{T}(\rho) \cdot \vec{\Omega} \quad , \tag{13}$$

with the prime denoting reference to the new basis system. The unitary polarization transformation matrix

$$\mathbf{T}(\rho) = \frac{1}{(1+\rho\rho^{*})} \begin{bmatrix} 1 & \sqrt{2}\,\rho & \rho^{2} \\ -\sqrt{2}\,\rho^{*} & (1-\rho\rho^{*}) & \sqrt{2}\,\rho \\ \rho^{*2} & -\sqrt{2}\,\rho^{*} & 1 \end{bmatrix} \tag{14}$$

of the target feature vector, referred to the particular (H,V)-basis, may directly be derived from the transformation equations (11) of the scattering components. A similar formulation has been used by Cloude (1986), studying interrelations of radar polarimetry with group theory.

Scattering problems, involving time varying random targets, require statistical methods of the theory of stochastic processes. The target feature vector in that case represents a multivariate random variable. The polarization dependence of averaged scattered power and signal correlations can be derived, introducing the polarization covariance matrix. This positive semi-definite Hermitian matrix contains the second order moments of the scattering coefficients and is defined by

$$\Sigma = \langle \vec{\Omega} \cdot \vec{\Omega}^{+} \rangle \quad , \tag{15}$$

where $\langle . \rangle$ denotes appropriate time averages of a stochastic variable.

In the linear (H,V)-polarization basis, the covariance matrix for reciprocal targets is explicitly given by

$$\Sigma = \begin{bmatrix} \langle |S_{HH}|^{2} \rangle & \sqrt{2}\,\langle S_{HH} S_{HV}^{*} \rangle & \langle S_{HH} S_{VV}^{*} \rangle \\ \sqrt{2}\,\langle S_{HV} S_{HH}^{*} \rangle & 2\langle |S_{HV}|^{2} \rangle & \sqrt{2}\,\langle S_{HV} S_{VV}^{*} \rangle \\ \langle S_{VV} S_{HH}^{*} \rangle & \sqrt{2}\,\langle S_{VV} S_{HV}^{*} \rangle & \langle |S_{VV}|^{2} \rangle \end{bmatrix} \quad . \tag{16}$$

[3]) The superscript + designates the transposed complex conjugated representation of a vector or a matrix.

The matrix contains nine independent parameters: three real quantities in the main diagonal, representing the mean copolar power values and twice the crosspolar one, and three complex interchannel correlation coefficients.

Under a change of polarization basis (4), (13), the covariance matrix transforms as

$$\Sigma' = \mathbf{T}(\rho) \cdot \Sigma \cdot \mathbf{T}(\rho)^{-1} \quad , \tag{17}$$

with the unitary polarization transformation matrix $\mathbf{T}(\rho)$, defined by (14). The unitary similarity transformation (17) is an important feature of the covariance matrix approach and guarantees quite interesting transformation, limitation and invariance properties of radar observables. The transposition congruence transformation (10) of the monostatic scattering matrix, on the other hand, gives rise to some peculiarities, e.g. resulting in pseudo-eigenvalue equations for the null polarizations, as discussed by Kostinski et al (1986) and Mieras (1986).

## 3. Polarimetric Power Signatures and Optimal Polarizations

### 3.1. POLARIZATION SIGNATURES OF MEAN BACKSCATTERED POWER

Suppose that the 3x3 covariance matrix $\Sigma$ of a reciprocal random target has been measured in the linear (H,V)-polarization basis. The mean power values of the co- and crosspolarized backscattered signals for linear horizontal and vertical polarizations are determined by the the main diagonal elements of the covariance matrix, according to (16). Applying polarization transformations (17) to this covariance matrix, the power values of the backscattered waves in any other orthogonal polarization configuration may be determined. This way, the well known polarization signatures, introduced by van Zyl (1986), of mean backscattered power in the co- and crosspolar radar channel can be synthesized.

The mean copolar power return for transmission and reception of polarization $\vec{\varepsilon}_A(\rho)$ is given by

$$P^A_{co}(\rho) = \langle |S_{AA}(\rho)|^2 \rangle \quad , \tag{18}$$

as a function of the complex ratio $\rho$, which defines the polarization state of the copolar radar channel according to (6) [4], and the measured covariance matrix elements (16). In the same way, the mean copolar power return for receiving and transmitting the polarization $\vec{\varepsilon}_B(\rho)$ is defined by

$$P^B_{co}(\rho) = \langle |S_{BB}(\rho)|^2 \rangle \quad . \tag{19}$$

The crosspolarized power return may also be determined by a complex polarization ratio $\rho$, defining the transmit polarization $\vec{\varepsilon}_A(\rho)$ or $\vec{\varepsilon}_B(\rho)$, and the receive polarization $\vec{\varepsilon}_B(\rho)$ or $\vec{\varepsilon}_A(\rho)$, respectively, yielding:

$$P_X(\rho) = \langle |S_{AB}(\rho)|^2 \rangle = \langle |S_{BA}(\rho)|^2 \rangle \quad , \tag{20}$$

as a function of the complex quantity $\rho$ and the measured covariance matrix elements (16). A common proportionality factor involving the radar parameters (transmitter power, antenna and receiver gains) has been set equal to one for sake of

[4]) Note that the particular choice of phases $\psi_A = \psi_B = 0$ is used in the present paper.

simplicity. Further important observables are the cross-correlations between the copolar and crosspolar radar channel:

$$R_x^A(\rho) = \langle S_{AA}(\rho)\, S_{AB}^*(\rho) \rangle \tag{21}$$

$$R_x^B(\rho) = \langle S_{BB}(\rho)\, S_{AB}^*(\rho) \rangle \quad . \tag{22}$$

These interchannel correlations represent the polarimetric correlation of the orthogonally polarized components of the backscattered waves. This way, $R_x^A(\rho)$ characterizes this correlation of the wave components if polarization $\vec{\varepsilon}_A(\rho)$ is transmitted by the radar. The same formulation holds in case of $R_x^B(\rho)$ and the corresponding transmit polarization $\vec{\varepsilon}_B(\rho)$. Detailed representations of the above-defined covariance observables can be derived by substituting the matrix (14) into transformation (17) and collecting terms. This results in the following formulas for mean power return in the copolar radar channel A, defined by (18):

$$\begin{aligned} P_{co}^A(\rho) = \Big\{ &\langle |S_{HH}|^2 \rangle + 2\rho^* \langle S_{HH} S_{HV}^* \rangle + \rho^{*2} \langle S_{HH} S_{VV}^* \rangle \\ &+ 2\rho \langle S_{HV} S_{HH}^* \rangle + 4\rho\rho^* \langle |S_{HV}|^2 \rangle + 2\rho\rho^{*2} \langle S_{HV} S_{VV}^* \rangle \\ &+ \rho^2 \langle S_{VV} S_{HH}^* \rangle + 2\rho^2\rho^* \langle S_{VV} S_{HV}^* \rangle + \rho^2\rho^{*2} \langle |S_{VV}|^2 \rangle \Big\} / (1+\rho\rho^*)^2 \quad , \end{aligned} \tag{23}$$

the copolar radar channel B, defined by (19):

$$\begin{aligned} P_{co}^B(\rho) = \Big\{ &\rho^2\rho^{*2} \langle |S_{HH}|^2 \rangle - 2\rho\rho^{*2} \langle S_{HH} S_{HV}^* \rangle + \rho^{*2} \langle S_{HH} S_{VV}^* \rangle \\ &- 2\rho^2\rho^* \langle S_{HV} S_{HH}^* \rangle + 4\rho\rho^* \langle |S_{HV}|^2 \rangle t - 2\rho^* \langle S_{HV} S_{VV}^* \rangle \\ &+ \rho^2 \langle S_{VV} S_{HH}^* \rangle - 2\rho \langle S_{VV} S_{HV}^* \rangle + |S_{VV}|^2 \rangle \Big\} / (1+\rho\rho^*)^2 \end{aligned} \tag{24}$$

and the crosspolar power function (20):

$$\begin{aligned} P_x(\rho) = \Big\{ &\rho\rho^* \langle |S_{HH}|^2 \rangle - \rho^*(1-\rho\rho^*) \langle S_{HH} S_{HV}^* \rangle - \rho^{*2} \langle S_{HH} S_{VV}^* \rangle \\ &- \rho(1-\rho\rho^*) \langle S_{HV} S_{HH}^* \rangle + (1-\rho\rho^*)^2 \langle |S_{HV}|^2 \rangle + \rho^*(1-\rho\rho^*) \langle S_{HV} S_{VV}^* \rangle \\ &- \rho^2 \langle S_{VV} S_{HH}^* \rangle + \rho(1-\rho\rho^*) \langle S_{VV} S_{HV}^* \rangle + \rho\rho^* \langle |S_{VV}|^2 \rangle \Big\} / (1+\rho\rho^*)^2 \quad . \end{aligned} \tag{25}$$

The interchannel correlation A, defined by (21), explicitly is given by

$$\begin{aligned} R_x^A(\rho) = \Big\{ &-\rho \langle |S_{HH}|^2 \rangle + (1-\rho\rho^*) \langle S_{HH} S_{HV}^* \rangle + \rho^* \langle S_{HH} S_{VV}^* \rangle \\ &- 2\rho^2 \langle S_{HV} S_{HH}^* \rangle + 2\rho(1-\rho\rho^*) \langle |S_{HV}|^2 \rangle + 2\rho\rho^* \langle S_{HV} S_{VV}^* \rangle \\ &- \rho^3 \langle S_{VV} S_{HH}^* \rangle + \rho^2(1-\rho\rho^*) \langle S_{VV} S_{HV}^* \rangle + \rho^2\rho^* \langle |S_{VV}|^2 \rangle \Big\} / (1+\rho\rho^*)^2 \end{aligned} \tag{26}$$

and the interchannel correlation B, defined by (22), reads

$$R_x^B(\rho) = \Big\{ -\rho\rho^{*2}\langle|S_{HH}|^2\rangle + \rho^{*2}(1-\rho\rho^*)\langle S_{HH}S_{HV}^*\rangle + \rho^{*3}\langle S_{HH}S_{VV}^*\rangle$$
$$+2\rho\rho^*\langle S_{HV}S_{HH}^*\rangle - 2\rho^*(1-\rho\rho^*)\langle|S_{HV}|^2\rangle - 2\rho^{*2}\langle S_{HV}S_{VV}^*\rangle \qquad (27)$$
$$-\rho\langle S_{VV}S_{HH}^*\rangle + (1-\rho\rho^*)\langle S_{VV}S_{HV}^*\rangle + \rho^*\langle|S_{VV}|^2\rangle\Big\}/(1+\rho\rho^*)^2 \quad .$$

The polarization dependence of the transmitted wave is parametrized in terms of the complex polarization ratio $\rho$[5]. For a given target covariance matrix, polarization signatures of mean power return in the orthogonally polarized radar channels and the interchannel correlations may be derived, this way, and graphically displayed.

In the considered case of monostatic reciprocal scattering, a very interesting polarization symmetry shows up. The crosspolar power function $P_x(\rho)$, see (20) and (25), is easily seen to be invariant with respect to substituting the polarization $\rho$ by its orthogonal one $\rho_{orth} = -1/\rho^*$, as defined by Azzam et al (1987). This symmetry relation

$$P_x(-1/\rho^*) = P_x(\rho) \qquad (28)$$

states a general formulation of the reciprocity theorem and proves that orthogonal polarizations attain the same crosspolar power values.

The polarization signatures of mean power, backscattered by a reciprocal random target, attain minima rather than zeros. The reason for this limitation becomes quite obvious in the presented covariance matrix approach. As mentioned above, the polarization dependent mean power values in the co- and crosspolar radar channels are determined by the main diagonal elements of the covariance matrix, represented in the respective polarization basis. The transformation of the covariance matrix from a given polarization basis to any arbitrary orthogonal one may be accomplished by the unitary similarity transformation (17). These transformations preserve the eigenvalues and, furthermore, the main diagonal elements of any transformed covariance matrix are bounded by the largest and the smallest eigenvalue, see e.g. Horn et al (1985). Since, the covariance matrix is Hermitian and positive semi-definite, it has three nonnegative real eigenvalues which can be ordered as

$$\lambda_{max} \geq \lambda_o \geq \lambda_{min} \geq 0 \quad . \qquad (29)$$

The polarization dependent mean power return, thus, is confined by the following inequalities:

Mean power return in the copolar radar channel, as function of the transmitter polarization $\rho$, is bounded by

$$\lambda_{min} \leq P_{co}^{(A,B)}(\rho) \leq \lambda_{max} \quad . \qquad (30)$$

Limits for the crosspolar power function are given by

$$1/2 \cdot \lambda_{min} \leq P_x(\rho) \leq 1/2 \cdot \lambda_{max} \quad . \qquad (31)$$

The minimum eigenvalue of the covariance matrix, in general, is nonvanishing and therefore, optimal polarizations can only produce minima rather than zeros.

---

5) Note that $P_{co}^A(0) = \langle|S_{HH}|^2\rangle$, $P_x(0) = \langle|S_{HV}|^2\rangle$ and so forth, indicating that the phase convention in (7) is reasonable and consistent.

But even the above mentioned limiting values for the power functions, given by the eigenvalues of the covariance matrix, may not be achieved by any of the possible polarization transformations. Since, in order to abtain an eigenvalue as one of the main diagonal elements in the transformed covariance matrix (17), one of the columns of the transformation matrix (14) must coincide with the corresponding normalized eigenvector. Each column of the transformation matrix, however, depends on two real parameters only and, therefore, cannot be matched to any of the eigenvectors of the covariance matrix, which in general depend on five real parameters[6]). For that reason, the limiting values for the power functions may not be achieved by any one of the polarization transformations.

### 3.2. OPTIMAL POLARIZATION THEORY

Mean backscattered power (18), (20) in the orthogonal radar channels is a continuous function of the transmitter polarization, characterized by the complex polarization ratio $\rho$. Optimal polarizations are defined to represent those states, which maximize or minimize the mean power scattered into a particular radar channel. The existence of such extrema in the power signatures is guaranteed by the Weierstrass theorem, quoted by Horn et al (1985), due to the compactness of the set of all possible polarizations (e.g. the Poincare sphere).

In principle, mean power return is a function of two real variables, real part $\alpha$ and imaginary part $\beta$ of the complex polarization ratio ($\rho = \alpha + j\,\beta$). The extremal values of the respective power function P, in the following designating either the copolar or crosspolar power, can be determined, see e.g. Leitmann (1962), by equating the first derivatives simultaneously to zero:

$$\frac{\partial P}{\partial \alpha} = 0$$
$$\frac{\partial P}{\partial \beta} = 0 \quad . \tag{32}$$

Introducing the partial complex derivative

$$\frac{\partial}{\partial \rho^*} := \frac{1}{2}\left(\frac{\partial}{\partial \alpha} + j \cdot \frac{\partial}{\partial \beta}\right) \quad , \tag{33}$$

as defined by Hille (1973), for instance, the set of real equations (32) can be rewritten in terms of one complex equation

$$\frac{\partial P}{\partial \rho^*} = 0 \quad . \tag{34}$$

The partial complex derivatives can be calculated, regarding the power function P as depending upon two independent variables $\rho$ and $\rho^*$[7]) , and differentiating with respect to $\rho^*$ as usual[8]).

---

[6]) This parametric restriction of the transformation matrix cannot be overcome by taking the absolute phases $\psi_A$ and $\psi_B$ of (5) into account, which has been set equal to zero.

[7]) Note that $\frac{\partial \rho}{\partial \rho^*} = 0$ and $\frac{\partial P}{\partial \rho^*} = \left(\frac{\partial P}{\partial \rho}\right)^*$ for real functions $P$.

[8]) It is found that: $\frac{\partial}{\partial \rho^*}(1 + \rho\rho^*)^{-2} = -2\rho(1 + \rho\rho^*)^{-3}$, and so forth.

The partial complex derivatives of the copolar and crosspolar power functions, as defined above, can easily be determined by straightforward differentiation of the power representations (23), (24) and (25). This way, it can be proved that:

$$\frac{\partial P_{co}^{A}(\rho)}{\partial \rho^{*}} = \frac{2}{(1+\rho\,\rho^{*})} \cdot R_{x}^{A}(\rho)$$

$$\frac{\partial P_{x}(\rho)}{\partial \rho^{*}} = -\frac{1}{(1+\rho\,\rho^{*})} \cdot \{R_{x}^{A}(\rho) - R_{x}^{B}(\rho)^{*}\} \tag{35}$$

$$\frac{\partial P_{co}^{B}(\rho)}{\partial \rho^{*}} = -\frac{2}{(1+\rho\,\rho^{*})} \cdot R_{x}^{B}(\rho)^{*} \quad ,$$

involving the interchannel correlations A and B, defined by (21), (22) and explicitly stated in (26) and (27).

The necessary condition (34), which determines the power extrema in the respective radar channels can now be formulated as follows.

- Copolar optimal polarizations $\rho_o^{co}$ are given as zero values of the interchannel correlation A:

$$R_{x}^{A}(\rho_{o}^{co}) = 0 \quad . \tag{36}$$

- Crosspolar optimal polarizations $\rho_o^{x}$, on the other hand, are solutions of

$$R_{x}^{A}(\rho_{o}^{x}) - R_{x}^{B^{*}}(\rho_{o}^{x}) = 0 \quad , \tag{37}$$

generating zero values in a particular correlation difference.

McCormick et al (1985) and Agrawal (1986) indicated that optimal polarizations for random target scattering should be located in the vicinity of nulls in the correlation between the co- and the crosspolar radar channel. In the present paper, the exact conditions have now been derived:

***For optimal polarizations which extremize the copolar mean power return, the correlation of the co- and the crosspolarized wave components tends exactly to zero. Crosspolar optimal states, on the other hand, are related to zero values of the particular correlation difference (37), rather than generating nulls in an individual interchannel correlation. Furthermore, crosspolar extrema always appear in pairs of orthogonal polarizations, due to the orthogonality symmetry (28) of the crosspolar power function.***

In the presented covariance matrix approach, functional relations have been derived, disclosing co- and crosspolar optimal polarizations as zero values in particular polarimetric correlation observables of the backscattered wave components. These correlation observables can be measured directly by radar and, furthermore, determine the second order statistics of the scattering components.

Utilizing the derived optimal polarization equations (36) and (37), a simple numerical iteration scheme (Newton algorithm) can be established to obtain the optimal polarization states for a given target covariance matrix, see Tragl (1990a).

### 3.3. OPTIMAL POLARIZATIONS FOR DETERMINISTIC TARGETS

Necessary conditions for the derivation of optimal polarizations have now been found for random target applications. Deterministic scatterers, as limiting cases of time varying random targets, must be included in the presented concept as well.

In case of time independent scattering coefficients, condition (36) for the copolar power extrema reduces to

$$S_{AA}(\rho)\, S^*_{AB}(\rho) = 0 \quad . \tag{38}$$

The solutions of this equation are that polarization states $\rho$, obviously, which either equate the copolar or the crosspolar scattering coefficient to zero. Comparing with the well known characteristic polarization theory, published by Agrawal et al (1989), these solutions correspond to the co- and crosspolar null polarizations, which generate zero and maximum power return in the copolar radar channel, respectively.

The crosspolar optimal polarization equation (37) yields for deterministic targets

$$S_{AA}(\rho)\, S^*_{AB}(\rho) = S_{AB}(\rho)\, S^*_{BB}(\rho) \quad . \tag{39}$$

The crosspolar null polarizations, with $S_{AB}(\rho) = 0$ and vanishing crosspolar power, evidently solve that equation. Furthermore, it can be proved by explicit calculations that the crosspolar maximum polarizations, as derived by Agrawal et al (1989), are the other solutions of equation (39). The well known orthogonality of the crosspolar characteristic polarizations in the deterministic case, forming pairs of nulls and maxima, is just a consequence of the orthogonality symmetry (28) of the crosspolar power function in the general case.

The widely used characteristic polarization theory for deterministic targets is demonstrated to represent the limiting case for time-independent scattering of the general optimal polarization theory for random targets, as expected.

## 4. Conclusion

In the present paper, radar backscattering from reciprocal random targets is studied, employing a revised covariance matrix approach. Unitary similarity transformations of the covariance matrix describe consistently the polarization dependence of second order radar observables. Limits for the polarimetric power functions can be related to the eigenvalues of the covariance matrix, thus, providing a valuable tool for the estimation of optimization limits. Functional relations can be determined, which relate the derivatives of the power functions, differentiated with respect to the polarization parameters, to the cross-correlation observables of the the covariance matrix. Necessary conditions for the power extrema in the co- and crosspolarized radar channel are stated. Hence, it can be proved that polarizations, which extremize the mean copolar power return, decorrelate the backscattered wave components completely. Crosspolar optimal polarizations, on the other hand, are related to zero values in a particular correlation difference. The characteristic polarization theory for deterministic targets, finally, is demonstrated to be consistently included in the developed covariance matrix approach.

The presented polarimetric concept has been applied extensively to meteorological as well as chaff radar data, aquired by the DLR weather radar[9]), yielding a variety of very interesting results, as published by Tragl (1990b).

[9]) A detailed description of the polarimetric DLR weather radar facility is published in the same issue.

ACKNOWLEDGMENT

This work was performed at the Institute for Radio Frequency Technology in the German Aerospace Research Establishment *DLR*. I should like to thank all my colleagues for many invaluable discussions. Especially, I should like to express my gratitude to Dr. A. Schroth for the continuous support of my work and Dr. E. Lueneburg for his constructive encouragement.

## Bibliography

Agrawal,A.P. (1986) *'A Polarimetric Rain Backscattering Model Developed for Coherent Polarization Diversity Radar Applications'*, Ph.D. dissertation, Communications Lab., EECS Dept., Univ. of Illinois at Chicago, U.S.A..

Agrawal,A.P., Boerner,W.-M. (1989) *'Redevelopment of Kennaugh's Target Characteristic Polarization State Theory Using the Polarization Transformation Ratio Formalism for the Coherent Case'*, IEEE Trans. Geosci. Remote Sensing, vol. GE-27, no.1, pp.2-14.

Azzam,R.M.A., Bashara,N.M. (1987) *'Ellipsometry and Polarized Light'*, North-Holland, Amsterdam.

Boerner,W.-M., Al-Arini,M.B., Chan,C.-Y., Mastoris,P.M. (1981) *'Polarization Dependence in Electromagnetic Inverse Problems'*, IEEE Trans. Antennas Propagat. vol. AP-29, no.2, pp.262-271.

Cloude,S.R. (1986) *'Polarimetry: The Characterisation of Polarisation Effects in Electromagnetic Scattering'*, Ph.D. dissertation, University of Birmingham, Great Britain.

Deschamps,G.A. (1951) *'Geometrical Representation of the Polarization of a Plane Electromagnetic Wave'*, Proc. IRE vol.39, pp.540-544.

Hille, E. (1973) *'Analytic Function Theory, Vol. I.'*, Chelsea Publishing Company, New York, pp. 81 ff..

Horn,R.A.; Johnson,C.R. (1985) *'Matrix Analysis'*, Cambridge University Press, Cambridge.

Huynen,J.R. (1970) *'Phenomenological Theory of Radar Targets'*, Ph.D. dissertation, TH Delft, The Netherlands.

IEEE (1979), *'Test Procedures for Antennas'*, IEEE-Standard no.149-1979.

Kennaugh,E.M. (1952), *'Polarization Properties of Radar Reflections'*, M.Sc. thesis, Dept. of Electr. Engr., The Ohio State University, Columbus, U.S.A..

Kostinski,A.B., Boerner,W.-M. (1986) *'On Foundations of Radar Polarimetry'*, IEEE Trans. Antennas Propagat. vol. AP-34, no.12, pp.1395-1404.

Kostinski,A.B.; James,B.D.; Boerner,W.-M. (1988), *'Optimal Reception of Partially Polarized Waves'*, J. Opt. Soc. Am. A 5, no.1, pp.58-64.

Leitmann,G. (1962) *'Optimization Techniques'*, Academic Press, London.

McCormick,G.C.; Hendry,A. (1985) *'Optimal Polarizations for Partially Polarized Backscatter'*, IEEE Trans. Antennas Propagat. vol. AP-33, no.1, pp.33-40.

Mieras,H. (1986) *'Comments on "On Foundations of Radar Polarimetry"'*, IEEE Trans. Antennas Propagat. vol. AP-34, no.12, pp.1470-1471.

Tragl,K.; Schroth,A.; Lueneburg,E. (1989), *'Optimal Polarizations for Distributed Random Targets - Statistical Analysis of Polarimetric DFVLR Radar Data'*, IEEE AP-S Internat. Symp. & URSI Radio Sci. Meeting 1989, San Jose, CA/USA (1989), IEEE Cat. No. CH-2654-2/89, pp.792-795.

Tragl,K. (1990a) *'Polarimetric Radar Backscattering from Reciprocal Random Targets'*, IEEE Trans. Geosci. Remote Sensing, will be published.

Tragl,K. (1990b) *'Polarimetrische Radarbeobachtung zeitveraenderlicher Zufallsziele'*, Ph.D. dissertation, Universitaet Kaiserslautern, West Germany, will be published.

van Zyl,J.J. (1986) *'On the Importance of Polarization in Radar Scattering Problems'*, Ph.D. Dissertation, California Institute of Technology, U.S.A..

van Zyl,J.J.; Papas,C.H.; Elachi,C. (1987), *'On the Optimum Polarizations of Incoherently Reflected Waves'*, IEEE Trans. Antennas Propagat. vol. AP-34, no.7, pp.818-825.

# A POLARIMETRIC MODEL FOR MULTI-PATH IMAGING/IDENTIFICATION ANALYSIS USING POLY-STATIC (BI-STATIC PLUS MONO-STATIC) SCATTERING MATRIX DATA

Sujeet K. Chaudhuri
University of Waterloo
Waterloo, Ont., Canada

Wolfgang-M. Boerner
University of Illinois at Chicago
Chicago, IL., U.S.A.

**ABSTRACT:** The objective of this investigation is to develop a model for target discrimination algorithms in a multi-path, broadband, multi-static environment using the complete broadband multi-static scattering matrix target phenomenology.

Electromagnetic scattering from a complex object, at high frequencies, is dominated by certain specular components. The location of these specular points on the complex object are known as the scattering centers. Generation of these scattering centers is dependent on the geometry of the object and its surroundings with respect to the aspect directions of the transmitters and the receivers. It is shown that the knowledge of the locations, and the local geometrices of these scattering centers are useful in developing multi-path target identification/discrimination algorithms.

In the current work, the multi-path returns are modeled as "image scattering centers". Since these are not actual scattering centers, the bistatic scattering matrix will show behaviour that is different from those of the physical scattering centers. Our approach is to analyze in detail the matrix properties for isolating the primary target scattering center signatures from multipath generated image signatures.

## 1. INTRODUCTION

Electromagnetic scattering from a complex object, at high frequencies (physical optics, geometric optics), is dominated by certain specular components. The locations of these specular points on the complex object are known as the scattering centers. Generation of these scattering centers is dependent on the geometry of the object with respect to the aspect directions of the transmitter and the receiver (general bistatic case). The knowledge of the locations and the local geometries of these scattering centers can be useful in developing air-target discrimination/identification algorithms. Therefore, a basic inverse scattering model to recover these scattering centers from poly-static polarimetric signatures of the target needs to be developed. The objective of this investigation is to develop multi-static narrow-band target discrimination algorithms in a multi-path broadband multi-static environment using the complete broadband multi-static scattering matrix



W.-M. Boerner et al. (eds.), Direct and Inverse Methods in Radar Polarimetry, Part 1, 707–738.

target phenomenology. Our approach is to analyze in detail the matrix properties for isolating the primary target scattering center signature from multi-path generated image signatures. Methods of multi-static scattering matrix metrology and the associated target matrix phenomenology by application of scattering matrix theory are considered.

## 2. PARAMETERS OF A SCATTERING CENTER

For the electromagnetic scattering analysis, the isolated scattering center is modeled as a conducting rectangular flat plate. This tangent plane approximation to the specular region has been used successfully in sea surface scattering models [Barrick and Peake, 1968, Kwoh and Lake, 1983].

The geometry of a scattering center with respect to a given coordinate system is shown in Fig. 1. In this figure $\hat{n}$ denotes the unit normal to the scattering center. The vector location of the scattering center with respect to the fixed reference coordinate system is given by $\underline{d}$.

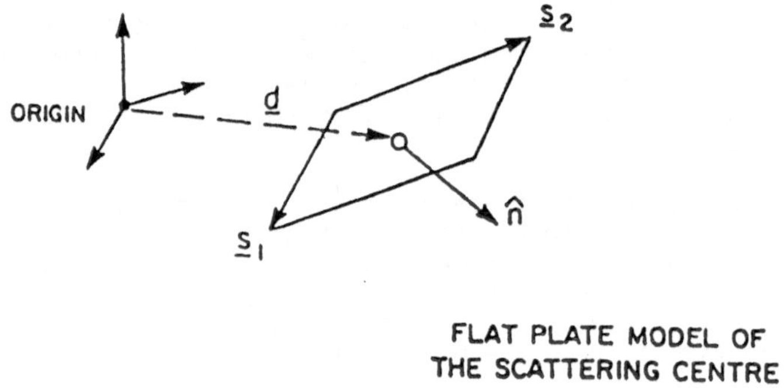


Fig. 1 Scattering Center Geometry

The vector sides $\underline{s}_1$( $= s_1\hat{a}_{s1}$) and $\underline{s}_2$( $= s_2\hat{a}_{s2}$) determine the local geometry of the scattering center, and they point in the directions of the principal curves of the scattering target at that specular point. The magnitudes of $\underline{s}_1$ and $\underline{s}_2$ are indicative of the radii of the curvature of these principal curves. Determination on $\hat{n}$, $\underline{s}_1$, $\underline{s}_2$, $\underline{d}$ from the multistatic polarimetric scattering matrix (with plane wave incidence) data in the context of the present work is known as the "scattering center recovery" problem.

## 3. COORDINATE SYSTEM FOR BISTATIC POLARIMETRIC SCATTERING DATA

There is a certain ambiguity associated with the measurement of the scattering matrix by a bistatic radar system. In the monostatic case the transmit-receive antenna polarizations are defined in the same system of coordinates. The bistatic configuration implies physical separation of the transmit and receive antennas; therefore, requiring two separate coordinate systems to define the antenna polarizations. In such a case, the idea of common transmit-receive antenna polarization has no physical meaning and, must, therefore, be given an artificial definition, based on the scattering matrix [S]. By choosing a fixed global spherical coordinate scheme, the transmitter and the receiver polarization basis vectors can be defined consistently. At each location the orthogonal polarization basis is given by the spherical unit vectors $\hat{a}_\theta$, and $\hat{a}_\phi$ at that point. Thus, the transmitter polarization is

$$\underline{h}_i = h_{\theta_i}\,\hat{a}_{\theta_i} + h_{\phi_i}\,\hat{a}_{\phi_i} ,$$

and the receiver polarization is

$$\underline{h}_r = h_{\theta_r}\,\hat{a}_{\theta_r} + h_{\phi_r}\,\hat{a}_{\phi_r} .$$

Now the bistatic scattering matrix [S], is defined as:

$$[\underline{h}_r] = [S][\underline{h}_i] \Rightarrow \begin{bmatrix} h_{\theta_r} \\ h_{\phi_r} \end{bmatrix} \begin{bmatrix} S_{\theta\theta} & S_{\phi\theta} \\ S_{\theta\phi} & S_{\phi\phi} \end{bmatrix} \begin{bmatrix} h_{\theta_i} \\ h_{\phi_i} \end{bmatrix}$$

The elements of the bistatic scattering matrix (BSM) above are complex, and in general, the matrix is not symmetric,i.e., $S_{\phi\theta} \neq S_{\theta\phi}$ [Davidovitz, 1983]. Thus, a total of eight parameters are required to give the complete bistatic scattering matrix (seven, if only the relative phases are of interest).

Two other conventional choices of the coordinate systems for BSM measurements have been used by Peake and Oliver [1971], and Heath [1983], respectively. The system used by Peake and Oliver is shown in Fig. 2.

There, the two orthogonal polarization states are specified by (i) the horizontal components $p_i^h$, $p_s^h$ in the x-y plane perpendicular to $\underline{k}_i$ and $\underline{k}_s$, respectively; and (ii) the vertical components $\underline{p}_i^v = \underline{p}_i^h \times \underline{k}_i$ and $\underline{p}_s^v = \underline{p}_s^h \times \underline{k}_s$. From this description it is clear that $(\underline{p}_i^h, \underline{p}_i^v; \underline{p}_s^h, \underline{p}_s^v)$ are the same vectors as $(\hat{a}_{\phi_i}, \hat{a}_{\theta_i}; \hat{a}_{\phi_r}, \hat{a}_{\theta_r})$ introduced in the previous paragraph. Therefore, the coordinate system proposed here is essentially the same as the one used by Peake and Oliver [1971].

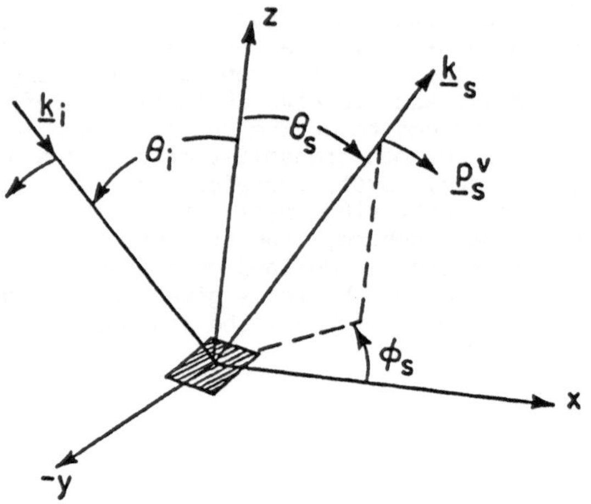


Fig. 2  Bistatic Scattering Matrix Coordinate System

The coordinate system used by Heath [1983] defines the vertical polarizations as the direction normal to the plane formed by $\underline{k}_i$ and $\underline{k}_s$ (i.e., $\underline{p}_i^v$ and $\underline{p}_s^v$ are same directions), and the horizontal polarization is in the plane of $\underline{k}_i$, $\underline{k}_s$ and normal to $\underline{p}_i^v$ and $\underline{k}_i$, $\underline{p}_s^v$ and $\underline{k}_s$, respectively. This system is useful only for the bistatic cases; for polystatic cases, since in general the planes defined by $\underline{k}_i$ and $\underline{k}_s$ corresponding to different receiver directions will be different, one cannot define a consistent polarization basis for the whole problem. Thus for the present application this coordinate system of Heath (1983) is <u>not</u> recommended.

## 4. DERIVATION OF THE BISTATIC SCATTERING MATRIX (BSM)

To obtain the BSM of the geometry in Fig. 1, first the physical optics approximation is used to obtain the induced current on the flat plate. This is done for a given direction of incidence with the transmitter polarization in $\hat{a}_{\theta_i}$ (or $\hat{a}_{\phi_i}$) direction. Next, these induced surface currents are used in the scattered far-field integral to obtain the vector scattered field at the receiver location. Finally, this vector scattered field is decomposed to construct the elements of the BSM. These analytical expressions are in terms of the scattering center geometry parameters $\hat{n}$, $\underline{s}_1$, $\underline{s}_2$, and $\underline{d}$. Therefore, these relations can be useful in some inverse scattering applications.

From electromagnetic theory, the induced surface current, $\underline{J}_i$, on the conducting flat plate model of the scattering center is given by

$$\underline{J}_s = 2\hat{n} \times \underline{H}_i - (\hat{n} \times \iint_{s'} \underline{J}'_s \times \nabla G ds') \qquad (1)$$

where $\underline{H}_i$ is the incident magnetic field, S' is the surface of the scatterer excluding the observation point, $\underline{r}$ and $\underline{r}'$ are the position vectors to the observation point and the integration points, respectively, G is the appropriate Green's function, and $\underline{J}_s'$ represent the surface current at the integration patch ds'. The first term on the right hand side of (1) is the physical optics term and the second term represents the interaction between the currents at the various points of the scatterer. In the physical optics (high frequency) formulation, the interaction term is assumed to be small, and only the first term is retained for obtaining $\underline{J}_s$.

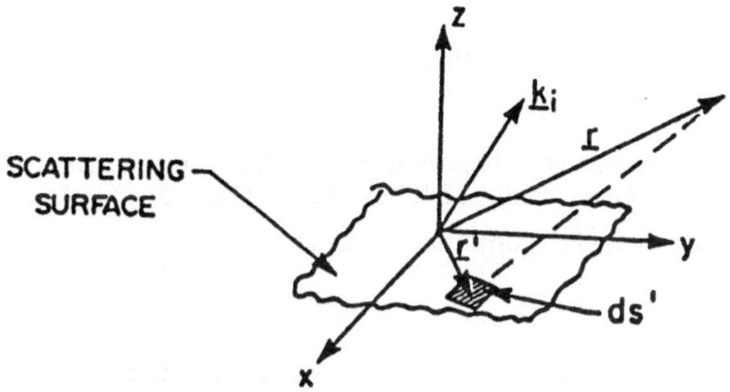


Fig. 3 Geometry for Calculating Vector Potential

The contribution to the radiation fields from the electric surface current, $\underline{J}_s$, is found by evaluating the magnetic vector potential

$$\underline{A} = \frac{e^{-jkr}}{4\pi r} \int_S\int \underline{J}_s (\underline{r}')e^{jkr\cdot\hat{r}'} ds', \qquad (2)$$

where, as shown in Fig. 3, S denotes the surface of the scatterer, and k = $2\pi/\lambda$, $\lambda$-wavelength of the operating frequency; this means $\underline{k}_i = k\hat{k}_i$, and $\underline{k}_s = k\hat{k}_s$. The far-zone scattered electric field is found from the components of

$$\underline{E}_s = - j\omega\mu(\underline{r}/r) \times (\hat{n} \times \underline{A}) \qquad (3)$$

For a general formulation of the scattering problem one should also consider the electric vector potential $\underline{F}$, which is associated with the magnetic current density. In the present case, however, $\underline{F}=0$, since the magnetic current density on a perfect electric conductor is zero (tangential electric field boundary condition).

Using the physical optics formulation ($\underline{J}_s = 2\hat{n} \times \underline{H}_i$), one can now write

$$\underline{A} = \frac{e^{-jkr}}{2\pi r} \hat{n} \times \int_s \underline{H}_i e^{jk\hat{r}\cdot r'} ds' = \frac{e^{-jkr}}{2\pi r} \hat{n} \times \underline{Q} , \tag{4}$$

where,
$$\underline{Q} = \int_s \underline{H}_i e^{jk\hat{r}\cdot r'} ds' \tag{5}$$

If the components of $\underline{Q}$ in $\hat{a}_{s1}$ and $\hat{a}_{s2}$ directions are written as $Q_{s1}$ and $Q_{s2}$, respectively, then using (3) and (4), the far scattered field is written as

$$\underline{E}_s = jk\eta \frac{e^{-jkr}}{2\pi r} [Q_{s2}\hat{a}_{s1} - Q_{s1}\hat{a}_{s2}], \tag{6}$$

where $\eta$ is the characteristic impedance of the propagation medium. In deriving (6) use has been made of the fact that $\hat{a}_{s1}$, $\hat{a}_{s2}$, $\hat{n}$ form an orthogonal unit vector basis (unitary matrix).

**ELEMENTS OF THE BSM**

In the global fixed spherical coordinate system of this work, a general polarization of the electric field associated with an incident plane wave is represented by

$$\underline{E}_i = \eta(E_{\theta_i}\hat{a}_{\theta_i} + E_{\phi_i}\hat{a}_{\phi_i}) \tag{7}$$

In order to calculate the elements of the BSM one considers two special cases of (7). In one case, $E_{\theta_i} = 0$, which yields the $S_{\phi\theta}$ and $S_{\phi\phi}$ elements of the BSM. To obtain the remaining two elements ($S_{\theta\phi}$ and $S_{\theta\theta}$) of the BSM, one considers $E_{\phi_i} = 0$. The detailed calculations are as follows:

(a) $E_{\theta_i} = 0, \quad E_{\phi_i} = \eta \Rightarrow \underline{E}_{\phi_i} = \eta\hat{a}_{\phi_i} [e^{-jk|\underline{r}_i - \underline{d}| + jk\hat{k}_i, \underline{r}'}]$

Using the plane wave equations, the corresponding magnetic field is given by

$$\underline{H}_\theta = \hat{a}_\theta [e^{-jk|\underline{r}_i - \underline{d}| + k\hat{k}_i, \underline{r}'}] \tag{8}$$

With (8) describing the incident magnetic field on the scattering geometry of Fig. 1, the integral in (5) is used to obtain the corresponding components of the $\underline{Q}$ vector in the directions $\hat{a}_{s1}$ and $\hat{a}_{s2}$.

These are

$$Q^{\phi}_{s1} = H^{\theta}_{s1} s_1 s_2 \text{ sinc} \left[\left(\frac{\underline{k}_s + \underline{k}_i}{2}\right) \cdot \underline{s}_1\right] \text{ sinc} \left[\left(\frac{\underline{k}_s + \underline{k}_i}{2}\right) \cdot \underline{s}_2\right], \quad (9a)$$

$$Q^{\phi}_{s2} = H^{\theta}_{s2} s_1 s_2 \text{ sinc} \left[\left(\frac{\underline{k}_s + \underline{k}_i}{2}\right) \cdot \underline{s}_1\right] \text{ sinc} \left[\left(\frac{\underline{k}_s + \underline{k}_i}{2}\right) \cdot \underline{s}_2\right], \quad (9b)$$

where $H^{\theta}_{s1} = \underline{H}_{\theta_i} \cdot \hat{a}_{s1}$, and $H^{\theta}_{s2} = \underline{H}_{\theta_i} \cdot \hat{a}_{s2}$ .

Now, by using (6) and (9) one obtains the co- and cross-polarized signals at the bistatic receiver site as follows:

$$E^{s}_{\phi\theta} = S_{\phi\theta} = \underline{E}_{s\phi} \cdot \hat{a}_{\theta r} = Q^{\phi}_{s2}(\hat{a}_{s1} \cdot \hat{a}_{\theta r}) - Q^{\phi}_{s1}(\hat{a}_{s2} \cdot \hat{a}_{\theta r}) \quad (10a)$$

$$E^{s}_{\phi\phi} = S_{\phi\phi} = \underline{E}_{s\phi} \cdot \hat{a}_{\phi r} = Q^{\phi}_{s2}(\hat{a}_{s1} \cdot \hat{a}_{\phi r}) - Q^{\phi}_{s1}(\hat{a}_{s2} \cdot \hat{a}_{\phi r}) \quad (10b)$$

(b) For the remaining elements of the BSM, consider $E_{\phi i}=0$, $E_{\theta i} = \eta$. This will yield, for the plane wave incidence:

$$\underline{H}_{\phi_i} = \hat{a}_{\phi_i} \left[e^{-jk|\underline{r}_i - \underline{d}| + jk\hat{k}_i \cdot \underline{r}'}\right]$$

Using the same method as used for $\phi$-polarization incidence, one gets:

$$Q^{\theta}_{s1} = H^{\phi}_{s1} s_1 s_2 \text{ sinc} \left[\left(\frac{\underline{k}_s + \underline{k}_i}{2}\right) \cdot \underline{s}_1\right] \text{ sinc} \left[\left(\frac{\underline{k}_s + \underline{k}_i}{2}\right) \cdot \underline{s}_2\right], \quad (11a)$$

$$Q^{\theta}_{s2} = H^{\phi}_{s2} s_1 s_2 \text{ sinc} \left[\left(\frac{\underline{k}_s + \underline{k}_i}{2}\right) \cdot \underline{s}_1\right] \text{ sinc} \left[\left(\frac{\underline{k}_s + \underline{k}_i}{2}\right) \cdot \underline{s}_2\right], \quad (11b)$$

where $H^{\phi}_{s1} = \underline{H}_{\phi_i} \cdot \hat{a}_{s1}$ , and $H^{\phi}_{s2} = H_{\phi_i} \cdot \hat{a}_{s2}$.

Now, the corresponding co- and cross-polarized components at the receiver are

$$E^{s}_{\theta\phi} = Q^{\theta}_{s2}(\hat{a}_{s1} \cdot \hat{a}_{\phi_r}) - Q^{\theta}_{s1}(\hat{a}_{s2} \cdot \hat{a}_{\phi_r}) \quad (12a)$$

$$E^s_{\theta\theta} = Q^\theta_{s2}(\hat{a}_{s1} \cdot \hat{a}_{\theta_r}) - Q^\theta_{s1}(\hat{a}_{s2} \cdot \hat{a}_{\theta_r}) \tag{12b}$$

Using (10) and (12) the complete BSM is given as

$$[S] = \begin{bmatrix} S_{\theta\theta} & S_{\phi\theta} \\ S_{\theta\phi} & S_{\phi\phi} \end{bmatrix} = \begin{bmatrix} E^s_{\theta\theta} & E^s_{\phi\theta} \\ E^s_{\theta\phi} & E^s_{\phi\phi} \end{bmatrix} \tag{13}$$

$$= \begin{bmatrix} \{Q^\theta_{s2}(\hat{a}_{s1} \cdot \hat{a}_{\theta_r}) - Q^\theta_{s1}(\hat{a}_{s2} \cdot \hat{a}_{\theta_r})\} & \{Q^\phi_{s2}(\hat{a}_{s1} \cdot \hat{a}_{\theta_r}) - Q^\phi_{s1}(\hat{a}_{s2} \cdot \hat{a}_{\theta_r})\} \\ \{Q^\theta_{s2}(\hat{a}_{s1} \cdot \hat{a}_{\phi_r}) - Q^\theta_{s1}(\hat{a}_{s2} \cdot \hat{a}_{\phi_r})\} & \{Q^\phi_{s2}(\hat{a}_{s1} \cdot \hat{a}_{\phi_r}) - Q^\phi_{s1}(\hat{a}_{s2} \cdot \hat{a}_{\phi_r})\} \end{bmatrix}$$

Substituting for $Q_{s1}$ and $Q_{s2}$ from (9) and (11), and making use of the vector relationships between $\hat{a}_{s1}$, $\hat{a}_{s2}$, and $\hat{n}$, the final form of the BSM of a flat rectangular conducting plate is given by

$$[S] = E_{0s} \begin{bmatrix} (\underline{H}_{\phi_i} \times \hat{a}_{\theta_r}) \cdot \hat{n} & (\underline{H}_{\theta_i} \times \hat{a}_{\theta_r}) \cdot \hat{n} \\ (\underline{H}_{\phi_i} \times \hat{a}_{\phi_r}) \cdot \hat{n} & (\underline{H}_{\theta_i} \times \hat{a}_{\phi_r}) \cdot \hat{n} \end{bmatrix} \tag{14}$$

with $$E_{0s} = s_1 s_2 \operatorname{sinc}\left[\left(\frac{\underline{k}_s + \underline{k}_i}{2}\right) \cdot \underline{s}_1\right] \operatorname{sinc}\left[\left(\frac{\underline{k}_s + \underline{k}_i}{2}\right) \cdot \underline{s}_2\right] \tag{15}$$

The form of the BSM in (14) is an interesting one from the imaging or recovery point of view, since all the elements of the matrix are independent of frequency and (15) involves only one of the unknowns ($\hat{n}$). Thus this form of the BSM will be exploited in the next section to recover the surface normal to the scattering center. It must be noted, however, that the radar measurables are the complex numbers $S_{\theta\theta}$, $S_{\theta\phi}$, $S_{\phi\theta}$, and $S_{\phi\phi}$, hence $E_{0s}$ in (14) is not known from the measurement and it has to be determined by calculations before $\hat{n}$ can be obtained.

## 5. RECOVERY OF THE SURFACE NORMAL $\hat{n}(n_1, n_2, n_3)$

Vector algebraic expansion of the elements of the BSM in (14) results into the following four equations:

$$S_{\theta\theta} = E_{0s}[n_1 \sin\theta_s \cos\phi_i + n_2 \sin\theta_s \sin\phi_i + n_3 \cos\theta_s \cos(\phi_s - \phi_i)] \tag{16}$$

$$S_{\theta\phi} = E_{0s}[n_3\sin(\phi_i - \phi_s)] \tag{17}$$

$$S_{\phi\theta} = E_{0s}[n_1(\sin\theta_s\cos\theta_i\sin\phi_i - \cos\theta_s\sin\theta_i\sin\phi_s) + n_2(\cos\theta_s\cos\phi_s\sin\theta_i - \sin\theta_s\cos\theta_i\cos\phi_i) + n_3\cos\theta_s\cos\theta_i\sin(\phi_i - \phi_s)] \tag{18}$$

$$S_{\phi\phi} = -E_{0s}[n_1\cos\phi_s\sin\theta_i + n_2\sin\phi_s\sin\theta_i + n_3\cos\theta_i\cos(\phi_s - \phi_i)] \tag{19}$$

here $(\theta_i, \phi_i)$, $(\theta_s, \phi_s)$ are the spherical coordinate representations of the $\hat{k}_i$ and $k_s$, respectively. From (17) one gets

$$n_3 = \left(\frac{1}{E_{0s}}\right)\left[\frac{S_{\theta\phi}}{\sin(\phi_i - \phi_s)}\right] \tag{20}$$

Using (16) and (17), and (19) and (17), the following two equations are obtained

$$n_1\cos\phi_i + n_2\sin\phi_i = R_5/E_{0s} \tag{21}$$

$$n_1\cos\phi_s + n_2\sin\phi_s = -R_6/E_{0s} \tag{22}$$

with $R_5 = \dfrac{S_{\theta\theta} - S_{\theta\phi}\cos\theta_s\cot(\phi_i - \phi_s)}{\sin\theta_s}$,

and $R_6 = \dfrac{S_{\phi\phi} + S_{\theta\phi}\cos\theta_i\cot(\phi_i - \phi_s)}{\sin\theta_i}$

Solving (20), (21), and (22) for $\hat{n}$, one gets

$$n_1 = \frac{1}{E_{0s}} \cdot \frac{R_5\sin\phi_s + R6\sin\phi_i}{\sin(\phi_s - \phi_i)} \tag{23a}$$

$$n_2 = -\frac{1}{E_{0s}} \cdot \frac{R_6\cos\phi_i + R_5\cos\phi_s}{\sin(\phi_s - \phi_i)} \tag{23b}$$

$$n_3 = \frac{1}{E_{0s}} \cdot \frac{-S_{\theta\phi}}{\sin(\phi_s - \phi_i)} \tag{23c}$$

By inspecting the equations (20) to (23) it is clear that the solution given in (23) is valid with the conditions,

$\phi_i \neq \phi_s$, and

$\theta_i$, $\theta_s \neq 0$ or $\pi$.

The first of the two conditions rules out the use of the monostatic scattering matrix (MSM) as a special case of the BSM in obtaining the solution in (23). The second condition is expected because, $\theta_i$, $\theta_s = 0$, or $\pi$ represent a radar site at the poles (north or south, respectively) of the fixed spherical coordinate system. At these two points $a_\theta$ and $\hat{a}_\phi$ directions are not defined uniquely, and consequently the mathematical representation of the system breaks down.

In the solution given in (23) the factor $E_{0s}$, as noted earlier, is still unknown from the radar measurables, $S_{\theta\theta}$, $S_{\theta\phi}$, $S_{\phi\theta}$, and $S_{\phi\phi}$. To determine $E_{0s}$, the expressions in (23) are substituted into the condition

$$|\hat{n}| = 1, \text{ i.e., } n_1^2 + n_2^2 + n_3^2 = 1.$$

This gives

$$E_{0s} = \pm \left[\left\{\frac{R_5^2 + R_6^2 + 2R_5R_6\cos(\phi_i - \phi_s)}{\sin^2(\phi_i - \phi_s)}\right\} + \frac{S_{\theta\phi}^2}{\sin^2(\phi_i - \phi_s)}\right]^{1/2} \qquad (24)$$

The $\pm$ sign of $E_{0s}$ from (24), substituted into (23) will simply result in two anti-parallel solutions $\pm\,\hat{n}$, which is inconsequential. If it is essential, then the proper sign of $E_{0s}$ in (24) can be decided by using the condition

$$\hat{n} \cdot \hat{k}_i < 0.$$

It is interesting to note that if one attempts to use the MSM for the recovery of $\hat{n}$ by substituting $\hat{k}_i = \hat{k}_s$ (i.e., $\theta_i = \theta_s$, $\phi_i = \phi_s$) into equations (16) to (19) one obtains:

$$S_{\theta\theta} = -\,S_{\phi\phi}\ ;\ S_{\theta\phi} = S_{\phi\theta} = 0 \text{ (when } \theta_i = \theta_s,\ \phi_i = \phi_s)$$

Obviously, now these elements of the scattering matrix cannot be used for the recovery of $\hat{n}$. This, in the opinion of the authors, is a clear demonstration of the usefulness of the BSM in the target recovery problems.

Another interesting situation is the special case of the BSM where

$\phi_i=\phi_s=\phi$, but $\theta_i \neq \theta_s$. This means that the transmitter and the receiver is placed along a polar great circle of the spherical coordinate system (therefore it can be referred to as the vertical stacking). For this configuration, from (17) and (18), we find that $S_{\theta\phi} = 0$, but $S_{\phi\theta} \neq 0$. This is expected, since now $\hat{a}_{\phi_i} = \hat{a}_{\phi_s}$, and $\hat{a}_{\theta_i} \neq \hat{a}_{\theta_s}$, which means half the polarization basis is like that for MSM and the other half follows the general definition given for BSM in Fig. 2.

For the vertically stacked configuration, the nonzero elements of BSM in equations (16), (18), and (19) reduce to the following ($S_{\theta\phi} = 0$):

$$S_{\theta\theta} = E_{0s}[n_1 \sin\theta_s \cos\phi + n_2 \sin\theta_s \sin\phi + n_3 \cos\theta_s] \quad (25)$$

$$S_{\phi\theta} = E_{0s}[n_1 \sin(\theta_s - \theta_i)\sin\phi - n_2 \sin(\theta_s - \theta_i)\cos\phi] \quad (26)$$

$$S_{\phi\phi} = - E_{0s}[n_1 \sin\theta_i \cos\phi + n_2 \sin\phi \sin\theta_i + n_3 \cos\theta_i] \quad (27)$$

Solving (25), (26), (27) for $\hat{n} = (n_1, n_2, n_3)$ one gets:

$$n_1 = \frac{1}{E_{0s}} \cdot \frac{S_{\theta\theta}\cos\theta_i \cos\phi + S_{\phi\theta}\sin\phi + S_{\phi\phi}\cos\theta_s \cos\phi}{\cos(\theta_s - \theta_i)} \quad (28a)$$

$$n_2 = \frac{1}{E_{0s}} \cdot \frac{S_{\theta\theta}\cos\theta_i \sin\phi - S_{\phi\theta}\cos\phi + S_{\phi\phi}\cos\theta_s \sin\phi}{\cos(\theta_s - \theta_i)} \quad (28b)$$

$$n_3 = \frac{1}{E_{0s}} \cdot \frac{S_{\theta\theta}\sin\theta_i + S_{\phi\phi}\sin\theta_s}{\cos(\theta_s - \theta_i)} \quad (28c)$$

Once again, before the solution in (28) can be useful, $E_{0s}$ has to be known. Applying $|\hat{n}| = 1$ condition on (28), the expression for $E_{0s}$ is obtained as

$$E_{0s} = \pm \frac{[(S_{\theta\theta}^2 + S_{\phi\theta}^2 + S_{\phi\phi}^2) + 2(S_{\theta\theta}S_{\phi\phi}\cos(\theta_s - \theta_i))]^{1/2}}{\cos(\theta_s - \theta_i)} \quad (29)$$

## 6. RECOVERY OF THE SCATTERING CENTER SURFACE GEOMETRY ($\underline{s}_1$ and $\underline{s}_2$)

In the BSM representation of (14) it is seen that the information regarding $\underline{s}_1$ and $\underline{s}_2$ is contained only in the common factor $E_{0s}$. The dependence of $E_{0s}$ on $\underline{s}_1$ and $\underline{s}_2$ is complicated due to the multivalued nature of the sinc function. Furthermore, only a partial information regarding $\underline{s}_1$ and $\underline{s}_2$ (namely their components in the $\frac{\underline{k}_s + \underline{k}_i}{2}$ direction, and their scalar magnitude) is contained in the factor $E_{0s}$. Therefore, a straight-forward recovery of the surface geometry from a single BSM (i.e. single frequency, single receiver) is not expected. Hence, in the following, this problem is approached from two viewpoints; one with the multi-frequency BSM input data, and the other with the multi-static (multi-receiver) single frequency input data. It is pointed out here that the partial information about $\underline{s}_1$ and $\underline{s}_2$ in $E_{0s}$, in conjunction with the knowledge of $\hat{n}$ (from Section 5) and the orthogonal relationship between $\hat{a}_{s1}$, $\hat{a}_{s2}$, $\hat{n}$ was found to be sufficient to construct $\underline{s}_1$ and $\underline{s}_2$ uniquely.

**(a) Multi-frequency BSM Approach**

Once $\hat{n}$ is known, $E_{0s}$ for the bistatic matrix at three harmonically related frequencies can be manipulated to extract information about $\underline{s}_1$ and $\underline{s}_2$.

Consider three frequencies corresponding to which the wave number k is given by $0.5k_o$, $k_o$, and $2k_o$, respectively. By using the method of Section 5, $E_{0s}$ at each of these frequencies can be calculated from the measured BSM. These calculated values can in turn be related to $\underline{s}_1$ and $\underline{s}_2$ by using (15), i.e.,

$$s_1 s_2 \mathrm{sinc}(\psi_1/2)\mathrm{sinc}(\psi_2/2) = E_{0s}|_{0.5ko} \tag{30a}$$

$$s_1 s_2 \mathrm{sinc}(\psi_1)\mathrm{sinc}(\psi_2) = E_{0s}|_{ko} \tag{30b}$$

$$s_1 s_2 \mathrm{sinc}(2\psi_1)\mathrm{sinc}(2\psi_2) = E_{0s}|_{2ko} \tag{30c}$$

$$\text{where } \psi_1 = \left(\frac{\underline{k}_{so} + \underline{k}_{io}}{2}\right) \cdot \underline{s}_1 , \quad \psi_2 = \left(\frac{\underline{k}_{so} + \underline{k}_{io}}{2}\right) \cdot \underline{s}_2 . \tag{30d}$$

Using (30a) and (30b), one gets

$$\cos(\psi_1/2)\cos(\psi_2/2) = \frac{E_{0s}|_{ko}}{E_{0s}|_{0.5ko}} = C_1 \tag{31a}$$

similarly, combination of (30b) and (30c) yields

$$\cos(\psi_1)\cos(\psi_2) = \frac{E_{0s}|_{2ko}}{E_{0s}|_{ko}} = C_2 \tag{31b}$$

By manipulating the non-linear system of equations in (31) one can solve for $\psi_1$ and $\psi_2$. These solutions are

$$\cos\frac{\psi_2}{2} = \pm\left[\left(\frac{1-C_1}{2}\right)^{\frac{1}{2}}\left\{\left(\frac{1}{2}+R\right) \pm \left(\frac{1}{4}+R\right)^{\frac{1}{2}}\right\}\right], \tag{32a}$$

$$\cos\frac{\psi_1}{2} = \frac{C_1}{\cos(\psi_2/2)} \tag{32b}$$

where $R = 2C_2^2/(1-C_1)$, and $C_1 \neq 1$, $C_1$, $C_2 < 1$.

By analyzing (31) one notes that when $C_1 = 1$, $C_2$ must be equal to 1, and

$$\psi_1,\ \psi_2 \rightarrow 0, \quad \text{or } 2n\pi, \qquad n = 1, 2, \ldots \quad .$$

Since all four solutions of (32a) are real, we must select the appropriate one before $\underline{s}_1$ and $\underline{s}_2$ can be constructed from the knowledge on $\hat{n}$, $\psi_1$, and $\psi_2$. To develop an appropriate criterion for this selection one needs to work with numerical data so that a quantitative behavior of these solutions can be observed.

**(b) Multi-angle BSM Approach**

From the practical point of view, the requirement of three harmonically related frequencies may not be desirable. Thus it is thought that the recovery of $\underline{s}_1$ and $\underline{s}_2$ using a single frequency, multi-angle BSM should also be attempted.

The main problem in this approach is that it may not be possible to generate a system of equations similar to (30) from the original expression in (15).

Consider three receiver sites identified by the reception direction, $\underline{k}_{s1}$, $\underline{k}_{s2}$, and $\underline{k}_{s3}$, respectively. Now, the arguments of the sinc functions in (15) become,

$$\left(\frac{\underline{k}_i + \underline{k}_{sj}}{2}\right)\cdot \underline{s}_1, \quad \text{and} \quad \left(\frac{\underline{k}_i + \underline{k}_{sj}}{2}\right)\cdot \underline{s}_2\ ; \qquad j = 1, 2, 3.$$

Unless these arguments for different values of j are harmonically related, one will generate six unknowns instead of two as is the case in equation (30). Consequently, the solution procedure presented earlier in this section cannot be applied.

One possible way to solve the multi-angle problem is to control the transmit directions (say $\underline{k}_{im}$) and the receive directions ($\underline{k}_{sj}$) such that for various combinations of m and j, the vector $(\underline{k}_{im} + k_{sj})/2$ yields

$$(\underline{k}_{io} + \underline{k}_{so})/2, \qquad (\underline{k}_{io} + \underline{k}_{so}), \qquad 2(\underline{k}_{io} + \underline{k}_{so}), \qquad \ldots, \tag{33}$$

where $\underline{k}_{io}$, and $\underline{k}_{so}$, as used in the earlier multifrequency case, represent some reference directions. To elaborate on this method, consider

$$\underline{k}_i = k_{i_x}\hat{a}_x + k_{i_y}\hat{a}_y + k_{i_z}\hat{a}_z \ ,$$

$$\underline{k}_s = k_{s_x}\hat{a}_x + k_{s_y}\hat{a}_y + k_{s_z}\hat{a}_z \ ,$$

$$\underline{s}_1 = s_{1_x}\hat{a}_x + s_{1_y}\hat{a}_y + s_{1_z}\hat{a}_z \ ,$$

where $\hat{a}_x$, $\hat{a}_y$, $\hat{a}_z$ represent unit vectors in x,y,z directions in the global spherical coordinate system. The argument of the first sinc function in (15) can now be written as

$$\left(\frac{\underline{k}_i + \underline{k}_s}{2}\right) \cdot \underline{s}_1 = s_{1x}\left(\frac{k_{ix} + k_{sx}}{2}\right) + s_{1y}\left(\frac{k_{iy} + k_{sy}}{2}\right) + s_{1z}\left(\frac{k_{iz} + k_{sz}}{2}\right). \tag{34}$$

In (34) it should be possible to manipulate the values of ($k_{sx}$, $k_{sy}$, $k_{sz}$) to meet the requirements specified in (33). Further development of this multi-angle approach should be undertaken.

Another adhoc, but straight-forward approach to the multi-angle problem is to generate $E_{0s}$ for a number of arbitrarily located receivers (the minimum number is expected to be 3) and then apply these to a suitable computer generated data fit to match the theoretical behavior of $E_{0s}$ given in (15). This can be accomplished by using a non-linear optimization routine, with $\underline{s}_1$ and $\underline{s}_2$ as the optimization variables. This computer aided approach can be initiated only when some suitable, theoretically and experimentally generated numerical multi-static scattering matrix data is made available.

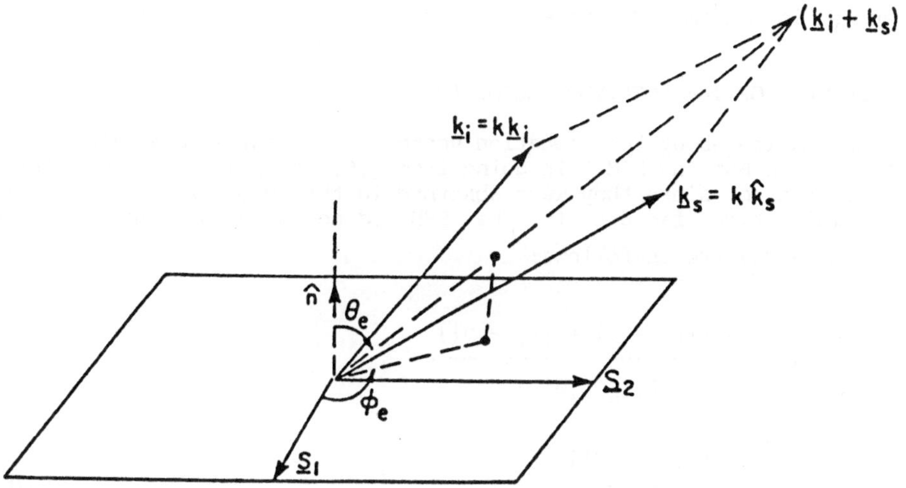


Fig. 4 Local Geometry of the Scattering Center

To construct $\underline{s}_1$ and $\underline{s}_2$ with the knowledge of $\psi_1$, $\psi_2$, and $\hat{n}$, one needs to consider the local geometry of the scattering center. This is shown in Fig. 4.

Using this scattering geometry, one can write

$$\cos\theta_e = (\underline{k}_{io} + \underline{k}_{so}) \cdot \hat{n} \tag{35}$$

Next, from the geometrical interpretation of $\psi_1$, and $\psi_2$ (i.e., projection of $\underline{s}_1$ and $\underline{s}_2$ onto the vector $\frac{\underline{k}_{io} + \underline{k}_{so}}{2}$, respectively) one can obtain

$$\psi_1\psi_2 = \frac{s_1 s_2}{4} \cos\phi_e \sin\phi_e (\cos^2\theta_e)$$

$$\text{or } \sin 2\phi_e = \frac{8\psi_1\psi_2}{(s_1 s_2)\cos^2\theta_e} \quad . \tag{36}$$

Using the expression for $E_{0s}\,|_{k_o}$ in (30b), with the knowledge of $\psi_1$ and $\psi_2$, one can calculate the value of $s_1 s_2$ (area of the rectangular scattering center). Once $s_1 s_2$ is known $\theta_e$ and $\phi_e$ can be calculated from (35) and (36), respectively. With $(\theta_e, \phi_e)$ and $\psi_1$, $\psi_2$, from the Fig. 4,

$\underline{s}_1$ and $\underline{s}_2$ is determined completely.

## 7. RECOVERY OF THE LOCATION VECTOR ($\underline{d}$)

The information about the location vector $\underline{d}$ is contained in the phase factor of the BSM and MSM. In going from (6) to (14) these phase terms were suppressed, since they were absorbed in the normalization factors. The normalization factors for the BSM (denoted by $N_B$) and the MSM (denoted by $N_M$) are as follows:

$$N_B = jk\eta \frac{e^{-jk[|\underline{r}_s - \underline{d}| + |\underline{r}_i - \underline{d}|]}}{2\pi|\underline{r}_s - \underline{d}|} \tag{37a}$$

$$N_M = jk\eta \frac{e^{-jk[2|\underline{r}_i - \underline{d}|]}}{2\pi|\underline{r}_i - \underline{d}|} \tag{37b}$$

The relationship between $\underline{r}_s$, $\underline{r}_i$, and $\underline{d}$ is shown in Fig. 5. It is to be noted that since the location of the radar sites with respect to the fixed global spherical coordinate system is known, the vectors $\underline{r}_i$ and $\underline{r}_s$ are known, and the only unknown in the normalization factors of (37) is $\underline{d}$.

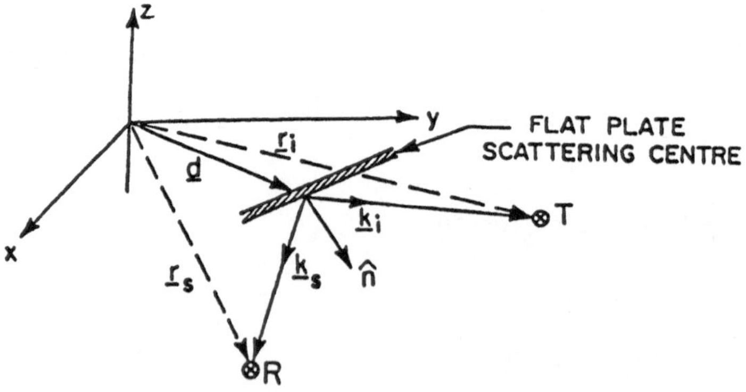


Fig. 5 Geometry for the Recovery of the Location Vector ($\underline{d}$)

Because of the multiple valued nature of the angular function measurements (i.e., $0 < \theta < 2\pi$, and $\theta + 2n\pi$ both will be measured as $\theta$ on a CW systems), the phase of $N_M$ in (37b) cannot be used directly to obtain $\underline{d}$.

To illustrate this point, consider $\phi_{M1}$ to be the measured phase of the MSM elements with the fixed origin as the phase reference. Then using (37b) one can write

$$2k|\underline{r}_i - \underline{d}| = \phi_{M1} + 2\pi l,$$

where l is an integer number which is unknown. If the operating frequency is $f_1$, with corresponding wavelength $\lambda_1$, then

$$|\underline{r}_i - \underline{d}| = \left(\frac{\phi_{M1}}{2\pi} + l\right)\frac{\lambda_1}{2} \tag{38}$$

In order to determine the unknown integer l, consider a second frequency $f_2$ which is very close to the frequency $f_1$ (say a few megahertz in a GHz range). Corresponding to this frequency the wavelength is $\lambda_2$, and the measured phase is $\phi_{M2}$; now from (37b)

$$2|\underline{r}_i - \underline{d}| \cdot \frac{2\pi}{\lambda_2} = \phi_{M2} + 2\pi m; \quad \text{m-integer.}$$

Substituting for $|\underline{r}_i - \underline{d}|$ from (38), one gets

$$(\phi_{M1} + 2\pi l)\frac{\lambda_1}{\lambda_2} = 2\pi m + \phi_{M2}$$

or

$$m - l\frac{\lambda_1}{\lambda_2} = \frac{\lambda_1/\lambda_2\ \phi_{M1} - \phi_{M2}}{2\pi} \tag{39}$$

Solution of (39), subjected to the restriction that the solutions m and l must be integers, provide some interesting conclusions. First, there will be an infinite set of solutions $[(m_i, l_i), i = 1,2.....]$. Second, the separation between the solution sets $(m_i, l_i)$ will depend on the "rational fraction" corresponding to $(\lambda_1/\lambda_2)$. It can be shown that if $\lambda_1$ and $\lambda_2$ are two real numbers very close to each other, then the separation between solution sets $(m_i, l_i)$ will be large. Consequently, the multiple values of $|\underline{r}_i - \underline{d}|$ obtained from (38) corresponding to the multiple values of $l(l_i;\ i=1,2....)$ will be spaced far apart. The knowledge of the approximate distance to the target space from the MSM radar site in conjunction with the above large solution separation will be useful in rejecting the unwanted solutions of (39). Once again, it is emphasized that a definite answer to the above suggestions can only be obtained by quantitative analysis with the theoretical and experimental input data.

Once $|\underline{r}_i - \underline{d}|$ is known, using the geometry of Fig. 5 one can determine the location of the tip of the vector $\underline{d}$, since the transmission direction $\hat{k}_i$ is known.

## 8. MULTIPLE SCATTERING CENTER RECOVERY (Two-Scattering-Centers Model)

The general scattering geometry for this case is shown in Fig. 6. The parameters describing the two centers with respect to the global spherical coordinate system are $(\hat{n}_1, \underline{s}_1, \underline{s}_2, \underline{d}_1)$, and $(\hat{n}_2, \underline{s}_3, \underline{s}_4, \underline{d}_2)$, respectively. Since the interaction between the centers is neglected, the contribution from each center to the net scattered field will be the same as if the scatterer was present alone in isolation. The BSM in this case will be given by

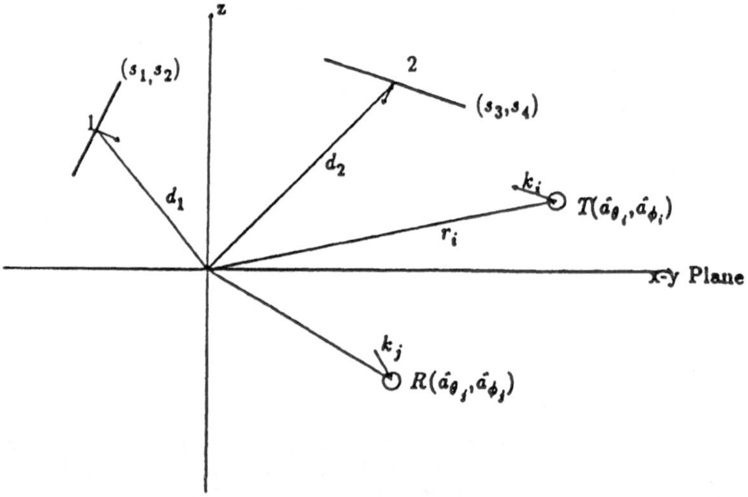


Fig. 6 Geometry for the Two Scattering Center Model

$$[S] = [S]_1 + [S]_2 = \begin{bmatrix} (S_{\theta\theta})_1 + (S_{\theta\theta})_2 & (S_{\phi\theta})_1 + (S_{\phi\theta})_2 \\ (S_{\theta\phi})_1 + (S_{\theta\phi})_2 & (S_{\phi\phi})_1 + (S_{\phi\phi})_2 \end{bmatrix}, \tag{40}$$

where subscript 1 and 2 refer to the center 1 and 2, respectively. Using (14), derived for an isolated center, one can now write the BSM for the two scattering center models as follows:

$$[S] = \begin{bmatrix} (\underline{H}_i^{\phi} \times \hat{a}_{\theta r}) & (\underline{H}_i^{\theta} \times \hat{a}_{\theta r}) \\ (\underline{H}_i^{\phi} \times \hat{a}_{\phi r}) & (\underline{H}_i^{\theta} \times \hat{a}_{\phi r}) \end{bmatrix} \cdot [N_1 E_{01} \hat{n}_1 + N_2 E_{02} \hat{n}_2] \,, \tag{41}$$

where $\cdot$ indicates the vector inner product; $N_1$, $N_2$ are the normalization factors for the scattering centers 1 and 2, respectively, and are given by (37a). Also,

$$E_{01} = s_1 s_2 \operatorname{sinc}\left[\left(\frac{\underline{k}_s + \underline{k}_i}{2}\right) \cdot \underline{s}_1\right] \operatorname{sinc}\left[\left(\frac{\underline{k}_s + \underline{k}_i}{2}\right) \cdot \underline{s}_2\right] \,, \tag{42a}$$

$$E_{02} = s_3 s_4 \operatorname{sinc}\left[\left(\frac{\underline{k}_s + \underline{k}_i}{2}\right) \cdot \underline{s}_3\right] \operatorname{sinc}\left[\left(\frac{\underline{k}_s + \underline{k}_i}{2}\right) \cdot \underline{s}_4\right] \,. \tag{42b}$$

If the radar sites are in the far-field region ($|\underline{r}_s| \gg |\underline{d}_1 - \underline{d}_2|$), then in the denominators of $N_1$ and $N_2$, one can write $|\underline{r}_s - \underline{d}_1| \approx |\underline{r}_s - \underline{d}_2| = R_o$ (see Fig. 6).

Now, using a normalization factor $N_o = j\frac{k\eta}{2\pi R_o}$, and (37a), the factors $N_1$ and $N_2$ become pure phase terms, i.e.,

$$N_1 = e^{-jk[|\underline{r}_s - \underline{d}_1| + |\underline{r}_i - \underline{d}_i|]} \tag{43a}$$

$$N_2 = e^{-jk[|\underline{r}_s - \underline{d}_2| + |\underline{r}_i - \underline{d}_2|]} \tag{43b}$$

In order to use the single scattering center recovery model as a building block for the solution of the multiple scattering centers recovery problem, consider the composite vector

$$\begin{aligned} N_1 E_{01} \hat{n}_1 + N_2 E_{02} \hat{n}_2 = & ( N_1 E_{01} n_{x1} + N_2 E_{02} n_{x2} ) \; \hat{a}_x \\ & + ( N_1 E_{01} n_{y1} + N_2 E_{02} n_{y2} ) \; \hat{a}_y \\ & + ( N_1 E_{01} n_{z1} + N_2 E_{02} n_{z2} ) \; \hat{a}_z \end{aligned}$$

where, $\hat{n}_1 = n_{x1} \hat{a}_x + n_{y1} \hat{a}_y + n_{z1} \hat{a}_z$

and $\quad \hat{n}_2 = n_{x2}\hat{a}_x + n_{y2}\hat{a}_y + n_{z2}\hat{a}_z.$

In general, for the transmit (i) and receive (j) direction combinations of $\underline{k}_i$, $\underline{k}_j$, one can write

$$\underline{C}^{ij} = C_x^{ij}\hat{a}_x + C_y^{ij}\hat{a}_y + C_z^{ij}\hat{a}_z = N_1^{ij}E_{01}^{ij}\hat{n}_1 + N_2^{ij}E_{02}^{ij}\hat{n}_2$$

$$=( N_1^{ij}E_{01}^{ij}n_{x1} + N_2^{ij}E_{02}^{ij}n_{x2} )\hat{a}_x$$

$$+( N_1^{ij}E_{01}^{ij}n_{y1} + N_2^{ij}E_{02}^{ij}n_{y2} )\hat{a}_y$$

$$+( N_1^{ij}E_{01}^{ij}n_{z1} + N_2^{ij}E_{02}^{ij}n_{z2} )\hat{a}_z . \tag{45}$$

Here $E_{01}^{ij}$, $E_{02}^{ij}$, $N_1^{ij}$, $N_2^{ij}$ are defined as in (42a), (42b), (43a), (43b), respectively, after replacing $\underline{k}_s$ by $\underline{k}_j$ in (42) and $\underline{r}_s$ by $\underline{r}_j$ in (43). Treating $\underline{C}^{ij}$ as the parameters of a fictitious scattering center, one can determine $C_x^{ij}$, $C_y^{ij}$, and $C_z^{ij}$ using (23a), (23b), and (23c), respectively [note that $\underline{C}^{ij}$ is equivalent to $E_{0s}\,\hat{n}$ in (23)]. Once $\underline{C}^{ij}$ is known from the BSM data of $(\underline{k}_i, k_j)$ combination, then using (45) one gets:

$$N_1^{ij}E_{01}^{ij}n_{x1} + N_2^{ij}E_{02}^{ij}n_{x2} = C_x^{ij} \tag{46a}$$

$$N_1^{ij}E_{01}^{ij}n_{y1} + N_2^{ij}E_{02}^{ij}n_{y2} = C_y^{ij} \tag{46b}$$

$$N_1^{ij}E_{01}^{ij}n_{z1} + N_2^{ij}E_{02}^{ij}n_{z2} = C_z^{ij} \tag{46c}$$

The ratios of (46a) to (46c), and (46b) to (46c) yield the following two equations.

$$\frac{n_{x1} + \left(\dfrac{N_2^{ij}E_{02}^{ij}}{N_1^{ij}E_{01}^{ij}}\right) n_{x2}}{n_{z1} + \left(\dfrac{N_2^{ij}E_{02}^{ij}}{N_1^{ij}E_{01}^{ij}}\right) n_{z2}} = \frac{C_x^{ij}}{C_z^{ij}} = C_1^{ij} , \tag{47a}$$

$$\frac{n_{y1} + \left(\frac{N_2^{ij}E_{02}^{ij}}{N_1^{ij}E_{01}^{ij}}\right) n_{y2}}{n_{z1} + \left(\frac{N_2^{ij}E_{02}^{ij}}{N_1^{ij}E_{01}^{ij}}\right) n_{z2}} = \frac{C_y^{ij}}{C_z^{ij}} = C_2^{ij} . \tag{47b}$$

The factor $\frac{N_2^{ij}E_{02}^{ij}}{N_1^{ij}E_{01}^{ij}}$ can be rewritten as a complex variable

$$R_{ij} = |R_{ij}|e^{j\theta_{ij}} = Re(R_{ij}) + jIm(R_{ij}) \text{ where, } |R_{ij}| = \frac{E_{02}^{ij}}{E_{01}^{ij}} , \ e^{j\theta_{ij}} = \frac{N_2^{ij}}{N_1^{ij}} .$$

This definition of $R_{ij}$ is possible due to the nature of the functions $E_{01}$, $E_{02}$ and $N_1$, $N_2$ described in (42) and (43), respectively. Rearranging (47a), one can write

$$[n_{x1} + Re(R_{ij})n_{x2}] + jIm(R_{ij})n_{x2}$$

$$= \{Re(C_1^{ij}[n_{z1} + Re(R_{ij})n_{z2}] - Im(C_1^{ij})Im(R_{ij})n_{z2}\}$$

$$+ j\{Im(C_1^{ij})[n_{z1} + Re(R_{ij})n_{z2}] + Re(C_1^{ij})Im(R_{ij})n_{z2}\} .$$

Equating the Real and Imaginary parts on both sides:

$$n_{x1} + Re(R_{ij})n_{x2} = Re(C_1^{ij})[n_{z1} + Re(R_{ij})n_{z2}] - Im(C_1^{ij})Im(R_{ij})n_{z2} \tag{48a}$$

$$Im(R_{ij})n_{x2} = Im(C_1^{ij})[n_{z1} + Re(R_{ij})n_{z2}] + Re(C_1^{ij})Im(R_{ij})n_{z2} . \tag{48b}$$

Solving (48) for $n_{x1}$, and $n_{x2}$ one gets,

$$n_{x1} = \frac{1}{Im(R_{ij})} [Re(C_1^{ij})Im(R_{ij}) - Re(R_{ij})Im(C_1^{ij})][n_{z1} + Re(R_{ij})n_{z2}]$$

$$- [Re(R_{ij})Re(C_1^{ij}) + Im(C_1^{ij})Im(R_{ij})]n_{z2} \text{ , and} \tag{49a}$$

$$n_{x2} = \frac{Im(C_1^{ij})}{Im(R_{ij})} [n_{z1} + Re(R_{ij})n_{z2}] + Re(C_1^{ij})n_{z2} \tag{49b}$$

Analyzing (47b) similar to (47a) above, one gets expressions for $n_{y1}$ and $n_{y2}$ as follows:

$$n_{y1} = \frac{1}{Im(R_{ij})}[Re(C_2^{ij})Im(R_{ij}) - Re(R_{ij})Im(C_2^{ij})][n_{z1} + Re(R_{ij})n_{z2}]$$

$$- [Re(R_{ij})Re(C_2^{ij}) + Im(C_2^{ij})Im(R_{ij})]n_{z2} \text{ , and} \tag{49c}$$

$$n_{y2} = \frac{Im(C_2^{ij})}{Im(R_{ij})} [n_{z1} + Re(R_{ij})n_{z2}] + Re(C_2^{ij})n_{z2} \tag{49d}$$

The system of equations in (49) reduces the number of unknowns by four, leaving $n_{z1}$, $n_{z2}$, $Re(R_{ij})$, $Im(R_{ij})$ as the unkwown parameters to be determined. In addition to (49), two other conditions to be satisfied are,

$$|\hat{n}_1| = n_{x1}^2 + n_{y1}^2 + n_{z1}^2 = 1, \tag{50a}$$

$$|\hat{n}_2| = n_{x2}^2 + n_{y2}^2 + n_{z2}^2 = 1. \tag{50b}$$

**Solution Scheme Using Non-linear Optimization:**

One effective numerical approach to solving the non-linear system of equations described by (49) and (50) is the use of an iterative non-linear optimization algorithm [Chaudhuri and Chow, 1981]. This algorithm essentially minimizes an error function $F(\underline{X})$ by iteratively adjusting the n-dimensional input variable vector $\underline{X}$. The optimization of the vector $\underline{X}$ is based on a gradient search method [Fletcher, 1970]; therefore, in addition to the error function (F), its gradient ($\nabla F$) needs to be defined. In the present problem the error function is defined as follows:

$$F = \sum_{i=1}^{M} \sum_{j=1}^{N} [(F_1^{ij})^2 + (F_2^{ij})^2] \quad , \tag{51}$$

where,

$$F_1^{ij} = [1 - (n_{z1}^2 + n_{y1}^2 + n_{x1}^2)] \quad ,$$

$$F_2^{ij} = [1 - (n_{z2}^2 + n_{y2}^2 + n_{x2}^2)] ,$$

and $(n_{x1}, n_{x2}, n_{y1}, n_{y2})$ are expressed by (49) for the $(\hat{k}_i, \hat{k}_j)$ combination of the transmit and receive directions. Thus the input variable vector for this optimization scheme is defined by $F(\underline{X}) = F[n_{z1}, n_{z2}, Re(R_{ij}), Im(R_{ij})]$.

A given combination of $\hat{k}_i$, $\hat{k}_j$ will produce eight unknowns $[(n_{x1},n_{y1}, n_{z1}), (n_{x2},n_{y2},n_{z2}), Re(R_{ij})Im(R_{ij})]$ in four equations (49a to 49d). An additional combination of $\hat{k}_i$, $\hat{k}_j$ will introduce only two new unknowns [corresponding values of $Re(R_{ij})$, and $Im(R_{ij})$], since $\hat{n}_1$ and $\hat{n}_2$ are not dependent on the transmit or receive directions. From this analysis one can see that for two combinations of $(\hat{k}_i, \hat{k}_j)$ ten unknown are generated, and minimization of (51) will satisfy ten independent equations [two sets of systems in (49) for two combinations of and $(\hat{k}_i, \hat{k}_j)$, and (50)]. Thus for the two scattering center problems, a minimum of two combinations of $(\hat{k}_i, \hat{k}_j)$ are expected to be required for an efficient and accurate optimization of (51). The effect of redundancy (BSM data for more than two combinations of $(\hat{k}_i, \hat{k}_j)$) on the optimized solution scheme can only be analyzed in conjunction with numerical experimentation.

The solution $\underline{X} = [n_{z1}, n_{z2}, Re(R_{ij}), Im(R_{ij})]$ obtained by minimizing the error function F in (51), will lead to the complete recovery of the two scattering centers through the following steps:

- Using the system of equations in (49) one can now obtain $(n_{x1}, n_{y1})$ and $(n_{x2}, n_{y2})$. Note that the values of $(n_{x1}, n_{y1})$ and $(n_{x2}, n_{y2})$ for various combinations of $\hat{k}_i$, $\hat{k}_j$, i.e., $R_{ij}$ will serve as a check of the optimization solution of (51).
- Once $R_{ij}$, $\hat{n}_1( = n_{x1}, n_{y1}, n_{z1})$, and $\hat{n}_2( = n_{x2}, n_{y2}, n_{z2})$ are known, the system of equations in (46) can be used to obtain $N_1^{ij}E_{01}^{ij}$ and

  $N_2^{ij}E_{02}^{ij}$, since $N_2^{ij}E_{02}^{ij}, = R_{ij}N_1^{ij}E_{01}^{ij}$.
- Because $E_{01}^{ij}$, and $E_{02}^{ij}$ are real numbers, and $N_1^{ij}$, $N_2^{ij}$ are complex numbers of unit magnitude [see (42) and (43), one can easily separate the complex valued $N_1^{ij}E_{01}$ and $N_2^{ij}E_{02}^{ij}$ into $[N_1^{ij}( =$ phase of

$N_1^{ij}E_{01}^{ij}$), $E_{01}^{ij}$( = magnitude of $N_1^{ij}E_{01}^{ij}$)] and [$N_2^{ij}$, $E_{02}^{ij}$], respectively.

- With the knowledge of $N_1^{ij}$, $E_{01}^{ij}$, $N_2^{ij}$, $E_{02}^{ij}$ for suitable combinations of ($\hat{k}_i$, $\hat{k}_j$)s, the techniques developed for the single scattering center can be used to obtain the parameters ($\underline{s}_1$, $\underline{s}_2$, $\underline{d}_1$) of each scattering center individually.

## 9. FORMULATION OF MULTIPATH SCATTERING FROM AN ISOLATED CENTER

In bistatic configurations when an isolated scattering center is located near a reflecting surface, the receivers will not only collect the direct scattered signal from the target but will also receive scattered energy "bounced" off the reflecting surface (see Fig. 7). This situation is referred to as the multi-path scattering problem. The energy reaching the receiver after bouncing off the interface (path-2 in Fig. 7) can be accounted for by introducing an image of the scattering center as shown in Fig. 7. In other words, the physical problem of an isolated center on the reflecting surface is equivalent to a model consisting of the pair of scattering centers, a (physical center), and a′ (image center) in the free space.

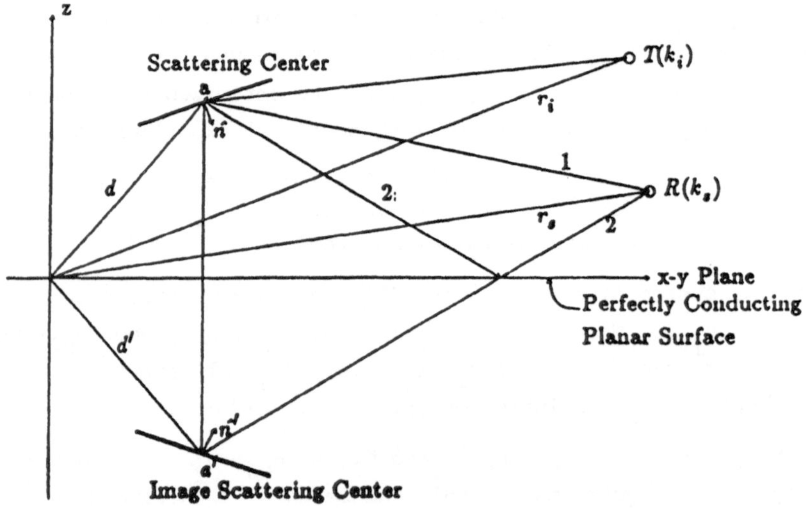


Fig. 7 Geometry for the Multipath Scattering Problem

Before the above model of the multi-path scattering problem is analyzed the following difference between the two-scattering-centers model and the present model must be observed:

- If a and a′ were two real scattering centers, the currents induced on the two centers will have different phases and different magnitudes. Thus the resulting BSM is identical to the two scattering center formulations given in Section 8, with the additional relation that (using the geometrical symmetry):

$$n_{x1} = n_{x2},\ n_{y1} = n_{y2},\ n_{z1} = -n_{z2},$$

$$s_{x1} = s_{x3},\ s_{y1} = s_{y3},\ s_{z1} = -s_{z3},$$

$$s_{x2} = s_{x4},\ s_{y2} = s_{y4},\ s_{z2} = -s_{z4}$$

$$d_{x1} = d_{x2},\ d_{y1} = d_{y2},\ d_{z1} = -d_{z2}.$$

- If on the other hand a′ was the image of a, then the currents induced on a′ in $\hat{a}_{s3}$ ($\underline{s}_3 = s_3\hat{a}_{s3}$) and $\hat{a}_{s4}$($\underline{s}_4 = s_4\hat{a}_{s4}$) directions will be negative (−) of that induced on a in $\hat{a}_{s1}$ ($\underline{s}_1 = s_1\hat{a}_{s1}$) and $\hat{a}_{s2}$($\underline{s}_2 = s_2\hat{a}_{s2}$) directions, respectively, i.e.,

$$\underline{J}_a = (J_{s1}\hat{a}_{s1} + J_{s2}\hat{a}_{s2})\, e^{j\psi a}\ , \tag{52a}$$

$$\underline{J}a' = (-J_{s1}\hat{a}_{s1} - j_{s2}\hat{a}_{s2})\, e^{j\psi a}\ . \tag{52b}$$

Thus the initial formulation of the problem in Section 8 has to be modified for the given multipath problem. In the following reformulation of the BSM, an isolated scattering unit will, in effect, represent the pair a and a′, i.e., the physical scattering center (a), and its mirror image (a′).

In Fig. 7, let ($\underline{s}_1$, $\underline{s}_2$, $\hat{n}$, $\underline{d}$), and ($\underline{s}'_1$, $\underline{s}'_2$, $\hat{n}'$, $\underline{d}'$) define the parameters of the physical center and the image center, respectively. From the geometrical symmetry one can write:

$$\underline{s}_1' = \underline{s}_1 - 2s_{1z}\hat{a}_z,\ \underline{s}_2' = \underline{s}_2 - 2S_{2z}\hat{a}z,$$

$$\hat{n}' = \hat{n} - 2n_{z1}\hat{a}_z,\ \underline{d}' = \underline{d} - 2d_{z1}\hat{a}_z\ . \tag{53}$$

As in the other formulations presented in this report, for the arbitrary plane wave incidence, the current induced on the physical center itself is given by,

$$\underline{J}_a = 2\hat{n}x\underline{H}_i = 2\hat{n}x\underline{H}_{i0}e^{-jk|\underline{r}_i - \underline{d}| + jk\hat{k}_i \cdot \underline{r}_a'}\ , \text{ or}$$

$$\underline{J}_a(\underline{r}'_a) = (J_{s1}\hat{a}_{s1} + J_{s2}\hat{a}_{s2})e^{-jk|\underline{r}_i - \underline{d}| + jk\hat{k}_i \cdot \underline{r}'_a} \tag{54a}$$

and the corresponding induced image current on a′ is,

$$\underline{J}_a'(\underline{r}'_{a'}) = (-J_{s1}\hat{a}'_{s1} - J_{s2}\hat{a}'_{s2})e^{-jk|\underline{r}_i - \underline{d}| + jk\hat{k}_i \cdot \underline{r}'_a} \tag{54b}$$

(Note, the exponential term is the same in $\underline{J}_a$ , $\underline{J}_{a'}$)

Here $\underline{r}'_{a'}$ and $\underline{r}'_{a'}$ are the position vectors to a general observation point on 'a' and its image on a′, with respect to the local coordinate systems (($\hat{a}_{s1}$, $\hat{a}_{s2}$, $\hat{n}$) on 'a' and ($\hat{a}_{s'2}$ , $\hat{a}_{s'1}$ , $\hat{n}$) on a′ as shown in Fig. 7, respectively. Using the image relationship, the correspondence between the position vectors $\underline{r}'_a$ and $\underline{r}'_{a'}$ can be expressed as follows:

If the local position vector to a point on 'a' ($\underline{r}'_a$) is given by

$$\underline{r}'_a = x_a\hat{a}_{s1} + y_a\hat{a}_{s2} , \tag{55a}$$

then the local position vector to the corresponding point on a′ will be given by

$$\underline{r}'_{a'} = x_a\hat{a}_{s'1} + y_a\hat{a}_{s'2} . \tag{55b}$$

Using the method described in Section 4, the vector potentials due to the currents given in (54a) and (54b) can be obtained. For a $\phi$-polarized incident wave ($\underline{E}_{\phi i}$, $\underline{H}_{\theta i}$ components only), the magnetic vector potential is given by (suppressing the normalized factor)

$$\underline{A}_a^{\phi} = (Q_{s1}^{\phi}\hat{a}_{s2} - Q_{s2}^{\phi}\hat{a}_{s1}), \text{ due to the physical center (a)}$$

$$\underline{A}_{a'}^{\phi} = -(Q_{s'1}^{\phi}\hat{a}'_{s2} - Q_{s'2}^{\phi}\hat{a}'_{s1}), \text{ due to the image center (a′).}$$

where,

$$Q_{s1}^{\phi} = H_{s1}^{\theta}s_1s_2\text{sinc}\left[\left(\frac{\underline{k}_s + \underline{k}_i}{2}\right) \cdot \underline{s}_1\right] \text{sinc}\left[\left(\frac{\underline{k}_s + \underline{k}_i}{2}\right) \cdot \underline{s}_2\right]$$

$$Q_{s2}^{\phi} = H_{s2}^{\theta}s_1s_2\text{sinc}\left[\left(\frac{\underline{k}_s + \underline{k}_i}{2}\right) \cdot \underline{s}_1\right] \text{sinc}\left[\left(\frac{\underline{k}_s + \underline{k}_i}{2}\right) \cdot \underline{s}_2\right]$$

$$Q^{\phi}_{s',1} = H^{\theta}_{s1} s_1 s_2 \mathrm{sinc}\left[\left(\frac{\underline{k}_s \cdot \underline{s}_1' + \underline{k}_i \cdot \underline{s}_1}{2}\right)\right] \mathrm{sinc}\left[\left(\frac{\underline{k}_s \cdot \underline{s}_2' + \underline{k}_i \cdot \underline{s}_2}{2}\right)\right]$$

$$Q^{\phi}_{s',2} = H^{\theta}_{s2} s_1 s_2 \mathrm{sinc}\left[\left(\frac{\underline{k}_s \cdot \underline{s}_1' + \underline{k}_i \cdot \underline{s}_1}{2}\right)\right] \mathrm{sinc}\left[\left(\frac{\underline{k}_s \cdot \underline{s}_2' + \underline{k}_i \cdot \underline{s}_2}{2}\right)\right]$$

with $H^{\theta}_{s1} = \underline{H}_{\theta i} \cdot \hat{a}_{s1}$ , $H^{\theta}_{s2} = \underline{H}_{\theta i} \cdot \hat{a}_{s2}$ .

Next, the scattered far-field can be written as

$$\underline{E}^{s}_{\phi} = -j\omega\mu\,(\underline{A}^{\phi}_{a} + \underline{A}^{\phi}_{a'})\ ,$$

and finally, using the BMS element definitions;

$$S_{\phi\theta} = \underline{E}^{s}_{\phi} \cdot \hat{a}_{\theta r}\ ,\quad S_{\phi\phi} = \underline{E}^{\phi}_{s} \cdot \hat{a}_{\phi\theta}\ .$$

Thus for the multipath BSM formulation,

$$S_{\phi\theta} = N_{0a}[Q^{\phi}_{s2}(\hat{a}_{s1}\cdot\hat{a}_{\theta r}) - Q^{\phi}_{s1}(\hat{a}_{s2}\cdot\hat{a}_{\theta r})] - N_{0a'}[Q^{\phi}_{s',2}(\hat{a}_{s',1}\cdot\hat{a}_{\theta r}) - Q^{\phi}_{s',1}(\hat{a}_{s',2}\cdot\hat{a}_{\theta r})] \tag{56a}$$

$$S_{\phi\phi} = N_{0a}[Q^{\phi}_{s2}(\hat{a}_{s1}\cdot\hat{a}_{\phi r}) - Q^{\phi}_{s1}(\hat{a}_{s2}\cdot\hat{a}_{\phi r})] - N_{0a'}[Q^{\phi}_{s',2}(\hat{a}_{s',1}\cdot\hat{a}_{\phi r}) - Q^{\phi}_{s',1}(\hat{a}_{s',2}\cdot\hat{a}_{\phi r})] \tag{56b}$$

where the normalization factors are

$$N_{0a} = \frac{e^{-jk|\underline{r}_s - \underline{d}|}}{|\underline{r}_s - \underline{d}|}\ , \text{ and}$$

$$N_{0a'} = \frac{e^{-jk|\underline{r}_s - \underline{d}'|}}{|\underline{r}_s - \underline{d}'|}\ ,$$

Also, a common normalization factor, $N_o = j\left(\frac{\eta}{\lambda}\right)e^{-jk|\underline{r}_i - \underline{d}|}$ has been suppressed throughout.

An identical analysis with the $\Theta$-polarized incident wave ($\underline{E}_{\Theta i}$, $H_{\phi i}$ components only) yields

$$S_{\Theta\phi} = N_{0a}[Q^{\Theta}_{s2}(\hat{a}_{s1}\cdot\hat{a}_{\phi r}) - Q^{\Theta}_{s1}(\hat{a}_{s2}\cdot\hat{a}_{\phi r})] - N_{0a'}[Q^{\Theta}_{s'2}(\hat{a}_{s'1}\cdot\hat{a}_{\phi r}) - Q^{\Theta}_{s'1}(\hat{a}_{s'2}\cdot\hat{a}_{\phi r})] \quad (56c)$$

$$S_{\Theta\Theta} = N_{0a}[Q^{\Theta}_{s2}(\hat{a}_{s1}\cdot\hat{a}_{\Theta r}) - Q^{\Theta}_{s1}(\hat{a}_{s2}\cdot\hat{a}_{\Theta r})] - N_{0a'}[Q^{\Theta}_{s'2}(\hat{a}_{s'1}\cdot\hat{a}_{\Theta r}) - Q^{\Theta}_{s'1}(\hat{a}_{s'2}\cdot\hat{a}_{\Theta r})] \quad (56d)$$

$$Q^{\Theta}_{s1} = H^{\phi}_{s1}s_1s_2\text{sinc}\left[\left(\frac{\underline{k}_s + \underline{k}_i}{2}\right)\cdot\underline{s}_1\right]\text{sinc}\left[\left(\frac{\underline{k}_s + \underline{k}_i}{2}\right)\cdot\underline{s}_2\right]$$

$$Q^{\Theta}_{s2} = H^{\phi}_{s2}s_1s_2\text{sinc}\left[\left(\frac{\underline{k}_s + \underline{k}_i}{2}\right)\cdot\underline{s}_1\right]\text{sinc}\left[\left(\frac{\underline{k}_s + \underline{k}_i}{2}\right)\cdot\underline{s}_2\right]$$

$$Q^{\Theta}_{s'1} = H^{\phi}_{s1}s_1s_2\text{sinc}\left[\left(\frac{\underline{k}_s\cdot\underline{s}_1 + \underline{k}_i\cdot\underline{s}'_1}{2}\right)\right]\text{sinc}\left[\left(\frac{\underline{k}_s\cdot\underline{s}_2 + \underline{k}_i\cdot\underline{s}'_2}{2}\right)\right]$$

$$Q^{\Theta}_{s'2} = H^{\phi}_{s2}s_1s_2\text{sinc}\left[\left(\frac{\underline{k}_s\cdot\underline{s}_1 + \underline{k}_i\cdot\underline{s}'_1}{2}\right)\right]\text{sinc}\left[\left(\frac{\underline{k}_s\cdot\underline{s}'_2 + \underline{k}_i\cdot\underline{s}'_2}{2}\right)\right]$$

Here $H^{\phi}_{s1} = \underline{H}_{\phi i}\cdot\hat{a}_{s1}$, and $H^{\phi}_{s2} = \underline{H}_{\phi i}\cdot\hat{a}_{s2}$

Substituting $H^{\Theta}_{s1} = H_{\Theta i}\cdot\hat{a}_{s1}$, and $H^{\Theta}_{s2} = \underline{H}^{\Theta}_{i}\cdot\hat{a}_{s2}$, and $\hat{a}_{s'1} = \hat{a}_{s1} - (2\hat{a}_{s1}\cdot\hat{a}_z)\hat{a}_z$, $\hat{a}_{s'2} = \hat{a}_{s2} - (2\hat{a}_{s2}\cdot\hat{a}_z)\hat{a}_z$ [see Fig. 7] in the expression for $S_{\phi\Theta}$, i.e., (56a), one gets

$$S_{\phi\Theta} = (N_{0a}E_{0s} - N_{0a'}E_{0s'})[\underline{H}_{\Theta i}\cdot\hat{a}_{s1})(\hat{a}_{s2}\cdot\hat{a}_{\Theta r}) - (\underline{H}_{\Theta i}\cdot\hat{a}_{s2})(\hat{a}_{s1}\cdot\hat{a}_{\Theta r})]$$

$$+ N_{0a'}E_{0s'}(\hat{a}_z\cdot\hat{a}_{\Theta r})[(\underline{H}_{\Theta i}\cdot\hat{a}_{s1})(\hat{a}_{s2}\cdot 2\hat{a}_z) - (\underline{H}_{\Theta i}\cdot\hat{a}_{s2})(\hat{a}_{s1}\cdot 2_{az})] ,$$

with $E_{0s} = s_1s_2\text{sinc}\left[\left(\frac{\underline{k}_s + \underline{k}_i}{2}\right)\cdot\underline{s}_1\right]\text{sinc}\left[\left(\frac{\underline{k}_s + \underline{k}_i}{2}\right)\cdot\underline{s}_2\right]$,

$$E_{0s'} = s_1 s_2 \text{sinc}\left[\left(\frac{\underline{k}_s \cdot \underline{s}_1' + \underline{k}_i \cdot \underline{s}_1}{2}\right)\right] \text{sinc}\left[\left(\frac{\underline{k}_s \cdot \underline{s}_2' + \underline{k}_i \cdot \underline{s}_2}{2}\right)\right] \quad .$$

Using the vector identity,

$$(\underline{A} \times \underline{B}) \cdot (\underline{C} \times \underline{D}) = (\underline{A} \cdot \underline{C})(\underline{B} \cdot \underline{D}) - (\underline{A} \cdot \underline{D})(\underline{B} \cdot \underline{C}) \ ,$$

and the fact that $\hat{a}_{s1} \times \hat{a}_{s2} = \hat{n}$ , one finds that

$$S_{\phi\theta} = (N_{0a}E_{0s} - N_{0a'}E_{0s'}) \ [(\underline{H}_{\theta_i} \times \hat{a}_{\theta_r}) \cdot \hat{n}]R$$

$$+ 2N_{0a'}E_{0s'}(\hat{a}_z \cdot \hat{a}_{\theta_r}) \ [(\underline{H}_{\theta_i} \times \hat{a}_z) \cdot \hat{n}]$$

and with similar vector manipulation one gets,

$$S_{\phi\phi} = (N_{0a}E_{0s} - N_{0a'}E_{0s'}) \ [(\underline{H}_{\theta_i} \times \hat{a}_{\phi_r}) \cdot \hat{n}]$$

$$+ 2N_{0a'}E_{0s'}(\hat{a}_z \cdot \hat{a}_{\phi_r}) \ [(\underline{H}_{\theta_i} \times \hat{a}_z) \cdot \hat{n}] \ ,$$

$$S_{\theta\phi} = (N_{0a}E_{0s} - N_{0a'}E_{0s'}) \ [(\underline{H}_{\phi_i} \times \hat{a}_{\phi_r}) \cdot \hat{n}]$$

$$+ 2N_{0a'}E_{0s'}(\hat{a}_z \cdot \hat{a}_{\phi_r}) \ [(\underline{H}_{\phi_i} \times \hat{a}_z) \cdot \hat{n}] \ ,$$

$$S_{\theta\theta} = (N_{0a}E_{0s} - N_{0a'}E_{0s'}) \ [\underline{H}_{\phi_i} \times \hat{a}_{\theta_r}) \cdot \hat{n}]$$

$$+ 2N_{0a'}E_{0s'}(\hat{a}_z \cdot \hat{a}_{\theta_r}) \ [(\underline{H}_{\phi_i} \times \hat{a}_z) \cdot \hat{n}] \ ,$$

From the definition of the coordinate system it is known that $a_z \cdot a_{\phi r} = 0$, always; hence the BSM for a "doublet center" (physical center plus image center) is given by

$$S_{\theta\theta} = (N_{0a}E_{0s} - N_{0a'}E_{0s'}) \ [(\underline{H}_{\phi i} \times \hat{a}_{\theta r}) \cdot \hat{n}]$$

$$+ 2N_{0a'}E_{0s'}(\hat{a}_z \cdot \hat{a}_{\theta r}) \ [(\underline{H}_{\phi i} \times \hat{a}_z) \cdot \hat{n}] \tag{57a}$$

$$S_{\phi\theta} = (N_{0a}E_{0s} - N_{0a'}E_{0s'}) \ [(\underline{H}_{\theta i} \times \hat{a}_{\theta r}) \cdot \hat{n}]$$

$$+ 2N_{0a'}E_{os'}(\hat{a}_z \cdot \hat{a}_{\theta r}) \ [(\underline{H}_{\theta i} \times \hat{a}_z) \cdot \hat{n}] \tag{57b}$$

$$S_{\theta\phi} = (N_{0a}E_{0s} - N_{0a'}E_{0s'}) [(\underline{H}_{\phi i} \times \hat{a}_{\phi r}) \cdot \hat{n}] \quad (57c)$$

$$S_{\phi\phi} = (N_{0a}E_{0s} - N_{0a'}E_{0s'}) [(\underline{H}_{\theta i} \times \hat{a}_{\phi r}) \cdot \hat{n}] \quad (57d)$$

From these expressions for the BSM elements of a doublet center it is clear that such a center cannot be recovered by the two scattering center recovery model given in Section 8. An alternate scheme to recover the parameters of a doublet center from its BSM model, given in (57), needs to be developed.

## 10. SUMMARY OF THE RESULTS

- For an isolated scattering center, the BSM at a single frequency is sufficient to obtain $\hat{n} = (n_1, n_2, n_3)$ in terms of $S_{\theta\theta}$, $S_{\phi\phi}$ and $S_{\theta\phi}$.
- An analytical solution for $\underline{s}_1$ and $\underline{s}_2$ has been obtained by recovering the $E_{0s}$ given in (15) at three related frequencies, namely $0.5k_o$, $k_o$, and $2k_o$.
- Analyzing the phase of the normalization factor of the MSM elements in (37) simultaneously at two arbitrary frequencies (which can be accommodated in a very narrow bandwidth) the vector location ($\underline{d}$) of the scattering center with respect to the origin of the fixed coordinate system can be recovered. From the target identification point of view this is deemed to be a very important parameter of recovery.
- The single scattering center recovery results have been extended to two scattering center models. This requires simultaneous processing of the multi-static scattering matrix data with a non-linear optimization algorithm. The separation between the scattering centers are asumed to be such that the secondary interaction between the various centers can be neglected.
- Finally the BSM for a multi-path scattering problem, involving an isolated scattering center over an infinite planar reflecting surface, has been derived. The difference between this multi-path situation (doublet center) and the "two-scattering-center" model has been clearly demonstrated. A scheme to recover the parameters of a doublet center from its BSM model, given in (57), is being studied currently.

## 11. CONCLUSIONS AND DIRECTIONS FOR FUTURE WORK

A high frequency bistatic/multi-static scattering matrix for the multi-path scattering center representation of a complex target has been developed. The secondary interactions between the scattering centers have been neglected on the basis of the assumption that the separations between these generated centers are large compared to the wavelength of the incident field. For the two-scattering-centers problem, an analytical extraction of information from the BSM data is not feasible, so a numerical method based on a non-linear optimization algorithm has been designed. The performance of this algorithm has to be studied with the theoretical and experimental test data.

For the doublet center model of the multipath scattering problem, analysis of the BSM in (57) should be carried out for the inversion applications. For this purpose, the distinct nature of the expressions for the elements $S_{\theta\phi}$ and $S_{\phi\phi}$ with respect to the expressions for $S_{\theta\theta}$ and $S_{\phi\theta}$ [see (57a, b, c, and d)], need to be exploited. Once the recovery model for the doublet center is established, it can be used as a building block towards the solution of the multiple doublet center problem in a manner similar to the development of the two-scattering-center model in Section 8 from the isolated scattering center results.

Experimental testing of the scattering center recovery models is important before any pattern recognition algorithms are designed. It is recommended that the measurement data on test targets be collected at a compact radar range facility. Some of the canonical test targets that should be used are:

- Flat rectangular plate
- Flate circular plate
- Ellipsoids

- Two flat plates, separated by a large distance compared to the wavelength, and held rigidly at certain angles (~0°, 15°, ... 45°, ... 90°).

- An ellipsoid and a flat plate (plate dimensions large compared to those of the ellipsoids), separated by a large distance compared to the wavelength, and held rigidly in various configurations simulating an actual multipath scattering problem.

Finally, in this paper the need for complete polarimetric scattering matrix information in a complex target identification scheme has been established clearly. A judicious combination of multi-frequency and multistatic scattering data should be the focus of future poly-static radar target imaging investigation.

**ACKNOWLEDGMENTS**

This research was supported in parts by the US Office of Naval Research under Contract #ONR N00014-80-C-0773, by NATO-IMOR Grant SA.5.2038-(189) 84; under General Dynamics Order 614972, under Westinghouse Order 86EJ-HA-3069, by NSERC, Canada Operating Grant A0013, and by NATO Grant SA.5-2-05 RG (85/0653).

We would like to acknowledge the continual interest shown in our research by Mr. James G. Smith, Office of Naval Research; Mr. Otto Kessler, Naval Air Development Center; Dr. Hans-Peter Schmid, General Dynamics, Pomona Division; Mr. Norman Powell, Westinghouse Electric Corporation; Mr. Lloyd W. Root, US Army Missile Command; and Dr. Craig Sinclair, NATO Scientific Affairs Division. We thank Ms. Chao-Hui Tuan and Richard W. Foster for skillfully typing the manuscript and preparing the figures.

**REFERENCES**

D.E. Barrick and W.H. Peake, "A Review of Scattering from Surfaces with Different Roughness Scales", Radio Science, Vol. 3, pp. 865-868, Aug. 1968.

S.K. Chaudhuri and W-M. Boerner, "A Polarimetric Model for the Recovery of High Frequency Scattering Centers from Bistatic - Monostatic Scattering Matrix Data", IEEE Trans. Ant. & Propag. Vol. AP-35, No. 1, pp. 87-93, Jan. 1987.

S.K. Chaudhuri and Y.L. Chow, "Nonlinear Optimization for Field, Scattering and SEM Problems", Proceedings, IEE International Conference on Antennas and Propagation, Univ. of York, England (full paper in the Refereed Conference Proceedings), pp. 378-382, April 1981.

M. Davidovitz, "Analysis of Certain Characteristic Properties of the Bistatic, Asymmetric Scattering Matrix", M.Sc. Thesis, University of Illinois at Chicago, Communications Lab., Report No.EMID-CL-1983-04-15-03, April 1983.

M. Davidovitz and W-M. Boerner, "Reduction of Bistatic Scattering Matrix Measurements for Inversely Symmetric Radar Targets", IEEE Trans. Vol. AP-31, (2), March 1983, pp. 237-242.

M. Davidovitz and W-M. Boerner, "Extension of Kennaugh's Optimal Polarization Concept to the Asymmetric Scattering Matrix Case", IEEE Trans. on A&P, Vol. 34(4), pp. 569-574, April 1986.

R. Fletcher, "A New Approach to Variable Metric Algorithms", Computer Journal, Vol. 13, pp. 317-322, 1970.

G.E. Heath, "Direct and Inverse Multistatic Physical Optics: The Significance of Reciprocity, Polarization, and Bistatic-Monostatic Equivalence, Linsoln Lab., M.I.T., Report, 1983.

D.S.W. Kwoh and B.M. Lake, "The Nature of Microwave Backscattering from Water Waves", AGARD Conference on Propagation Factors Affecting Remote Sensing by Radio Waves, AGARD Conf. Proceedings, No. 345, pp. 23-1 to 23-16, 1983.

W.H. Peake and T.L. Oliver, "The Response of Terrestrial Surfaces at Microwave Frequencies", ElectroScience Lab., The Ohio State University, Report No. AFAL-TR-70-301, AD884106, May 1971.

# Basic Polarimetric Measurements on Monostatic or Bistatic Radar Images

S.Riegger, W.Wiesbeck, D.Kähny
Institut für Höchstfrequenztechnik und Elektronik
University of Karlsruhe
Kaiserstr.12
7500 Karlsruhe 1
Federal Republic of Germany

## ABTRACT

This paper deals with the description of a high sophisticated Radar measurement system. In the first part a summary of the fundamental polarimetric definitions will be given. Then a vector error correction, which allows the precise measurement of the complex, polarimetric RCS matrix, is described. The system performance will be verified for simple targets. In the last part the application of such RCS-measurements will be demonstrated for complex targets as there are e.g. aircraft models.

## 1. INTRODUCTION

Radars in operational systems like in aircraft, ships or land based installations are until to day in most cases restricted to a single frequency and to one polarization. Only a few radars have frequency and polarization agility. These features are primarily used as ECCM measures, they are not part of advanced capabilities for polarimetric and multi-frequency target acquisition. This holds true in spite of the since many years available knowledge, that target acquisition, classification and identification can significantly be improved by coherent, polarimetric Radars [1,2]. On the other hand it must be recognized that appropriate hardware was not readily available until several years ago. Especially in the last four years laboratory type equipment suitable as subsystems for coherent, polarimetric Radars was increasingly offered. This equipment made it possible to configur wideband, coherent polarimetric Radar Cross Section (RCS) measurement systems, that allowed to prove theories developed up to now. Systems like this are primarily based on vector network analyzers. This paper describes an advanced laboratory type system with outstanding features based on a new calibration and verification philosophy. Results are presented for several targets as there are spheres, cylinders, chaff and model aircraft. It is shown, that by characteristic signatures like dispersive polarimetric signatures or polarization signatures target behavior can be interpreted.



*W.-M. Boerner et al. (eds.), Direct and Inverse Methods in Radar Polarimetry, Part 1*, 739–772.

## 2. DEFINITIONS

### 2.1 COMPLEX RCS

The scalar Radar cross section $\sigma$ is usually defined by the following equation, which relates the radiated field $\vec{\underline{E}}^r$ at a distance R to the incident field $\vec{\underline{E}}^i$ at the target:

$$\sigma = 4\pi \cdot \lim_{R\to\infty} R^2 \cdot \frac{\vec{\underline{E}}^r \cdot \vec{\underline{E}}^{r*}}{\vec{\underline{E}}^i \cdot \vec{\underline{E}}^{i*}} \tag{2.1}$$

("_" stands for complex quantities, "*" denotes conjugate complex). A given $\sigma$ therefore characterizes the target only for one aspect angle, one frequency, one polarization and contains no phase information.

For phase measurements and especially for absolute phase measurements a reference plane is required. Fig. 1 shows a schematic, arbitrary bistatic RCS-measurement configuration. By introducing a reference radius $R_0$ according to fig.1, a reference plane for the incident and scattered wave is defined. $R_0$ can be arbitrary but has to be fixed during calibration. The objects are centered at the origin of co-ordinates.

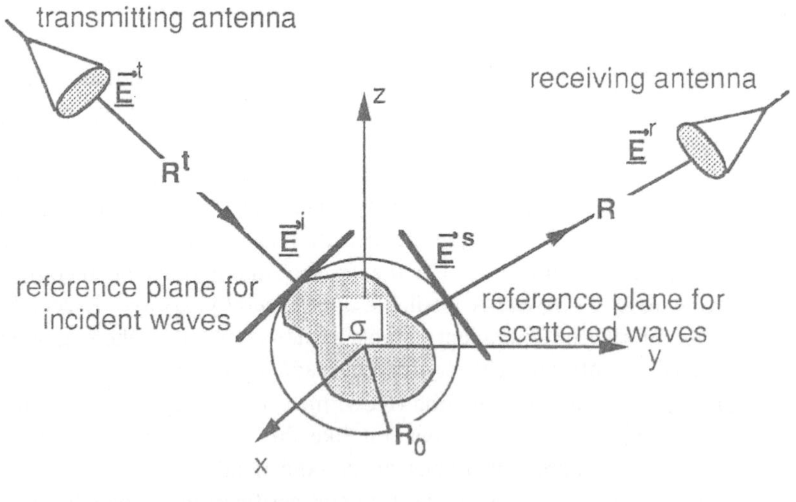


Fig.1 Bistatic RCS-measurement configuration for the definition of the reference planes

In addition to the radiated field $\vec{\underline{E}}^r$, used in (2.1) a scattered field $\vec{\underline{E}}^s$ is introduced. While the incident wave at $R_0$ is a plane wave (for $R_t \to \infty$), the scattered field $\vec{\underline{E}}^s$ is only a fictitious field given by the transformation:

$$\underline{\vec{E}}^{s} = \frac{\sqrt{4\pi R^2}}{\sqrt{4\pi R_0^2}} \cdot e^{j\beta\cdot(R-R_0)} \cdot \underline{\vec{E}}^{r}(R) \tag{2.2}$$

R has to be a far field range for object and antenna, β is the wave number. Theoretically, the field $\underline{\vec{E}}^s$ at $R_0$ can be seen to be generated by an isotropic radiation in the origin with the field strength $\underline{\vec{E}}^r$ at R. Using (2.2), a complex Radar Cross Section $\underline{\sigma}$ can be introduced.

$$\underline{\sigma} = \lim_{R\to\infty} 4\pi R^2 \; e^{j2\beta(R-R_0)} \left(\frac{\underline{E}^r}{\underline{E}^i}\right)^2$$
$$= 4\pi R_0^2 \cdot \left|\frac{\underline{E}^s}{\underline{E}^i}\right|^2 e^{j2\cdot\underbrace{(\varphi_s-\varphi_i)}_{\varphi}} \tag{2.3}$$

where $\varphi_i, \varphi_s$ denote the phase of the incident and scattered field at $R_0$. The magnitude of $\underline{\sigma}$ is identical with σ from (2.1).

The suitability of this phase definition can be shown for a mono-static measurement of a conducting sphere with radius $R_0$. According to the reflected waves physical phase at the sphere's front, φ becomes 180°. The definition used above is not only helpful for phase measurement of $\underline{\sigma}$ but also for the definition of a new kind of scattering matrix as shown below.

## 2.2 RCS-MATRIX AND OBJECT RELATED SCATTERING MATRIX

Measurement of the targets polarization information requires an adequate mathematical description. An arbitrary polarized plane wave can always be expressed by the linear combination of two orthogonal fields. The electrical field vector of a wave e.g. propagating along the z-axes (see fig. 1), originating from +∞, can be expressed by two linear (horizontal, vertical) polarized waves.

$$\underline{\vec{E}}^{i} = \underline{\vec{E}}^{i}_{h} + \underline{\vec{E}}^{i}_{v} = (\underline{E}^{i}_{x}\,\vec{e}_x + \underline{E}^{i}_{y}\,\vec{e}_y)\cdot e^{j(\omega t+\beta z)} \tag{2.4}$$

The change from the linear basis system to a general elliptical basis leads to unitary transformations [3,4], representing the basis of radar polarimetry. Unitary transformations conserve the total power of a wave, i.e. the norm of the electric field vector $\|\underline{\vec{E}}\|$ remains invariant under change-of-basis.

Unitary matrices satisfy the requirement

$$[\underline{U}]^{*T} = [\underline{U}]^{+} = [\underline{U}]^{-1} \tag{2.5}$$

with $|\det[\underline{U}]| = 1$

and are the generalization of orthogonal matrices ( $[U]^T=[U]^{-1}$ ) for complex numbers. Similar to [4], for transformation a unitary matrix $[\underline{Q}]$ will be used, which can be described as product of a complex ellipticity matrix $[\underline{H}(\alpha)]$ and real rotation matrix $[R(\psi)]$.
The rotation matrix $[R(\psi)]$ causes a rotation by the angle $\psi$ in a plane perpendicular to the direction of propagation.

$$\left[R(\psi)\right] = \begin{bmatrix} \cos\psi & -\sin\psi \\ \sin\psi & \cos\psi \end{bmatrix} \tag{2.6}$$

with det $[R]=+1$

The ellipticity matrix $[\underline{H}(\alpha)]$ varies the axial ratio of the polarization ellipse (see fig.2) from linear ($\alpha=0°$) over arbitrary elliptical up to circular polarization ($\alpha=\pm 45°$).

$$\left[\underline{H}(\alpha)\right] = \begin{bmatrix} \cos\alpha & j\sin\alpha \\ j\sin\alpha & \cos\alpha \end{bmatrix} \tag{2.7}$$

with det $[\underline{H}]=+1$

This results in a total transformation matrix $[\underline{Q}] = [R(\psi)] \cdot [\underline{H}(\alpha)]$

$$[\underline{Q}] = \begin{bmatrix} \cos\psi\cos\alpha - j\sin\psi\sin\alpha & -\sin\psi\cos\alpha + j\cos\psi\sin\alpha \\ \sin\psi\cos\alpha + j\cos\psi\sin\alpha & \cos\psi\cos\alpha + j\sin\psi\sin\alpha \end{bmatrix} \tag{2.8}$$

Thus, the operator $[\underline{Q}(\alpha,\psi)]$ performs a transformation from the linear (x,y)-basis-system to an arbitrary elliptical, but orthogonal basis-system " ' " as shown in (2.9). This transformation is fundamental for the theory of radar polarimetry applied in section 5.

$$\vec{\underline{E}} = \begin{bmatrix} \underline{E}_x \\ \underline{E}_y \end{bmatrix} = [\underline{Q}] \cdot \begin{bmatrix} \underline{E}_1' \\ \underline{E}_2' \end{bmatrix} = [\underline{Q}] \cdot \vec{\underline{E}}' \tag{2.9}$$

Fig. 2 shows the unit vectors of the new basis " ' " in the x,y-plane represented by their polarization ellipses, which are rotated by the angle $\psi$.
The target's polarization information can be described by two orthogonal incident and scattered waves respectively. In the following, this will be shown for linear polarizations. The normally defined polarization scattering matrix $[\underline{S}]$ is according to (2.4) and fig. 1:

$$\begin{bmatrix} \underline{E}_h^r \\ \underline{E}_v^r \end{bmatrix} = \begin{bmatrix} \underline{S}_{hh} & \underline{S}_{hv} \\ \underline{S}_{vh} & \underline{S}_{vv} \end{bmatrix} \cdot \begin{bmatrix} \underline{E}_h^i \\ \underline{E}_v^i \end{bmatrix} \tag{2.10}$$

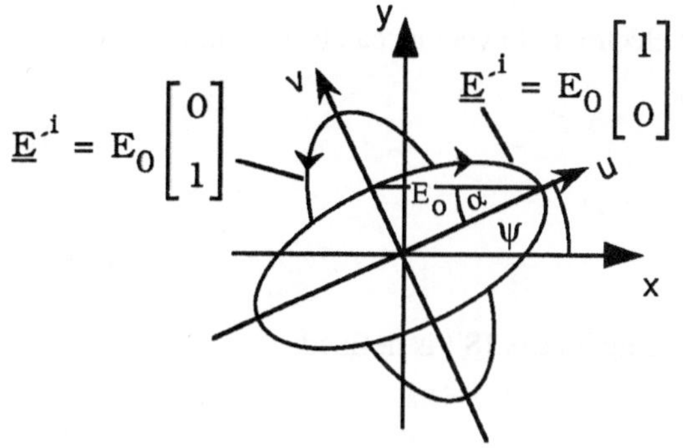


Fig.2 Polarization ellipse

The problem with this matrix is, that the parameters of the matrix are dependent in magnitude and phase on the distance R ($R \rightarrow \infty$) of the receiving antenna, i.e. this definition may be suitable for theoretical calculations of RCS but not for absolutely, calibrated measurements. Using the definition of the complex RCS $\underline{\sigma}$ (2.3) a complex RCS-matrix can be introduced, avoiding the drawback of (2.10).

$$[\underline{\sigma}] = \begin{bmatrix} \underline{\sigma}_{hh} & \underline{\sigma}_{hv} \\ \underline{\sigma}_{vh} & \underline{\sigma}_{vv} \end{bmatrix} \tag{2.11}$$

with $$\begin{bmatrix} \underline{E}_h^s \\ \underline{E}_v^s \end{bmatrix} = \frac{1}{\sqrt{4\pi R_0^2}} \cdot \left[\sqrt{\underline{\sigma}}\right] \cdot \begin{bmatrix} \underline{E}_h^i \\ \underline{E}_v^i \end{bmatrix}$$

Analogously an object related scattering matrix $[\underline{S}^o]$, independent on R is:

$$\left[\underline{S}^o\right] = \frac{1}{m} \cdot \left[\sqrt{\underline{\sigma}}\right] \tag{2.12}$$

## 2.3 POLARIMETRIC DEFINITIONS

In the previous section the object related polarization scattering matrix (2.12) was defined for linear polarization, adapted to the antenna system used. The knowledge of the complete scattering matrix allows one to investigate the scattering behavior for arbitrary transmitting and receiving polarization. Mathematically this will be done by the operator [Q], defined by (2.8).

Arbitrary polarized incident and scattered electric field vectors can be synthesized by

$$\underline{\vec{E}}^{i} = \left[\underline{Q}(\alpha_1, \psi_1)\right] \cdot \underline{\vec{E}}'^{i} \tag{2.13}$$

$$\underline{\vec{E}}^{s} = \left[\underline{Q}(\alpha_2, \psi_2)\right] \cdot \underline{\vec{E}}'^{s} \tag{2.14}$$

Starting from (2.10), a new, primed scattering matrix [S'] is defined

$$\underline{E}'^{s} = [\underline{S}'] \cdot \underline{\vec{E}}'^{i} \tag{2.15}$$

with $\left[\underline{S}'\right] = \left[\underline{Q}(\alpha_2 , \psi_2)\right]^{-1} \left[\underline{S}\right] \cdot \left[\underline{Q}(\alpha_1 , \psi_1)\right]$

In the following this matrix with its parameters $\alpha_1, \psi_1, \alpha_2, \psi_2$ and frequency is used as basis equation for Radar Polarimetry. Two special cases arise for $\alpha_1 = \pm\alpha_2$ and $\psi_1 = \psi_2$. For (+)-sign the same transformation is applied on the transmitting and receiving vector. The hereby defined matrix [S"] originates from a true change-of-basis-transformation.

$$[\underline{S}''] = [\underline{Q}]^{-1} \cdot [\underline{S}] \cdot [\underline{Q}] \tag{2.16}$$

For (-)-sign a matrix [S^] is defined:

$$\left[\underline{S}^{\wedge}\right] = \begin{bmatrix} \hat{\underline{S}}_{11} & \hat{\underline{S}}_{12} \\ \hat{\underline{S}}_{21} & \hat{\underline{S}}_{22} \end{bmatrix} = \left[\underline{Q}(-\alpha, \psi)\right]^{-1} \left[\underline{S}\right] \cdot \left[\underline{Q}(\alpha, \psi)\right] \tag{2.17}$$

using the unitarity of [Q] results in

$$\left[\underline{S}^{\wedge}\right] = \left[\underline{Q}\right]^{T} . \left[\underline{S}\right] . \left[\underline{Q}\right] \tag{2.18}$$

This definition is synonymous to the so called voltage equation [3,5,6], but is not a true change-of-basis transformation, because different transformations for transmitting and receiving vectors are used (see also [3,7]). Nevertheless, (2.17) will be used in the following sections in order to be compatible with the results from previous papers [2,5,6,8]. Only for linear polarizations, (i.e. if $[\underline{Q}]$ is a real matrix), are (2.16) and (2.18) identical.

An important tool for target characterization is the "**P**olarization **S**ignature" (PS), introduced by [6,9]. There, the scattering behavior is shown for arbitrary polarization state of the transmitting antenna, while the receiving antenna is co- or cross-polarized to the transmitting antenna. Using (2.17) the PS can be seen as 3D-plot of $|\hat{\underline{S}}_{22}|$ (co-polarized) and $|\hat{\underline{S}}_{12}|$ (cross-polarized) versus $\alpha$ and $\psi$ , i.e. a plot on a projection of the Poincaré Sphere.

With the "**D**ispersive **P**olarimetric **S**ignature" (DPS) a further dimension is introduced to the PS [10,11]. DPS's are shown as 3D-diagrams, with the scattering parameters $|\hat{\underline{S}}_{22}|$ or $|\hat{\underline{S}}_{12}|$ plotted versus frequency and rotation angle $\psi$. The ellipticity angle $\alpha$ remains fixed (e.g. $\alpha=0$). Instead of the scattering matrix, the complex RCS-matrix can be used, applying (2.12).

A further helpfull tool for interpretation of the scattering behavior are the characteristic properties as, e.g. Null- and Eigen- polarizations [4,5] and especially the invariant properties (regarding $\alpha$, $\psi$) of the scattering matrix.

Introducing the power scattering matrix [P] [12]

$$[P] = [\underline{S}^\wedge]^{*T} \cdot [\underline{S}^\wedge] \tag{2.19}$$

and keeping in mind, that $[\underline{Q}]$ is unitary, the trace of [P] is

$$P_1 = |\hat{\underline{S}}_{11}|^2 + |\hat{\underline{S}}_{22}|^2 + 2|\hat{\underline{S}}_{21}|^2 \tag{2.20}$$

In (2.20) reciprocity ($\underline{S}_{vh}=\underline{S}_{hv}$) is assumed. The trace represents the total received power and is independent of $\alpha$ and $\psi$. For the same reason, the determinant of $[\underline{S}^\wedge]$ is also independent of $\alpha$ and $\psi$, while the depolarization D

$$D = 1 - \frac{|\hat{\underline{S}}_{11} + \hat{\underline{S}}_{22}|^2}{2\,P_1} \tag{2.21}$$

is only independent of the rotation angle $\psi$. The depolarization is a measure for the extent and symmetry of the target scattering regions and the isolation between them. A body with a single symmetric scattering center such as a a sphere, does not rotate polarization and results in a depolarization D=0. D=0.5 is adequate for the scattering of a long and thin cylinder ( $\underline{S}_{hh} = 0$; $\underline{S}_{vh} = 0$; $\underline{S}_{vv} = 1$ ), i.e. cross-polarization varies with rotation angle $\psi$. The maximum possible cross-polarized content in the received signal of this cylinder is 50%. D=1 stands for complete cross-polarization (see dihedral corner reflector).At least two scattering centers are necessary.

## 3. MEASUREMENT SET-UP

The RCS measurement set up, used for coherent polarimetric measurements , is shown in fig. 3. The main block of the system is a Vector Network Analyzer (NWA HP 8510 B) with a 32 - bit Desktop Computer working as an external controller. The system consists of a synthesizer with a sweep and step capability of up to 26,5 GHz and a phase locked four-channel receiver (HP 8511 Opt. H41), which measures, unlike a FM-CW-Radar system, magnitude and phase of the RF. The system works in the frequency range from 45 MHz to 40 GHz. The bandwidth is limited by the antennas and external couplers.
The antenna system consists of two closely positioned, orthogonally polarized, wide band horn antennas. The reason for using two different antennas is to improve decoupling of the receive from the transmit signal. The measurement is performed quasi-monostatic.
The signal source feeds successively the vertical and the horizontal input of the transmitting antenna, while the receiving antenna is simultaneously connected with both orthogonal outputs to the receiver of the Vector Network Analyzer.
For all signals magnitude and phase are measured. The ratio of the corresponding complex quantities yields the uncorrected complex polarization scattering matrix at each measured frequency. The test environment is preferably an anechoic chamber with a remotely controlled rotary target mount (styrofoam).

## 4. SYSTEM ERROR CORRECTION AND CALIBRATION

Like with all measurement systems, systematic and statistical errors limit the accuracy of complex polarimetric RCS measurements. Statistical errors result from instabilities in time, frequency and noise. Systematic errors are introduced by frequency dependent components, like mismatches, coupling in the analyzer and the antennas as well as non perfect anechoic chambers and target mounts [13]. In the following a procedure for error determination and correction will be developed.

### 4.1. STATISTICAL AND SYSTEMATIC ERRORS

#### 4.1.1 STATISTICAL ERRORS

Statistical errors can be reduced by an internal averaging routine of the measurement system. The synthesizer locks at each frequency and recycles the measurement up to 4096 times. The NWA computes the arithmetico-geometrical mean of the measured data at each frequency. By this averaging the noise level is reduced up to 20 dB. This reduction of statistical errors is valid, as long as the whole configuration is not changed during a measurement cycle.

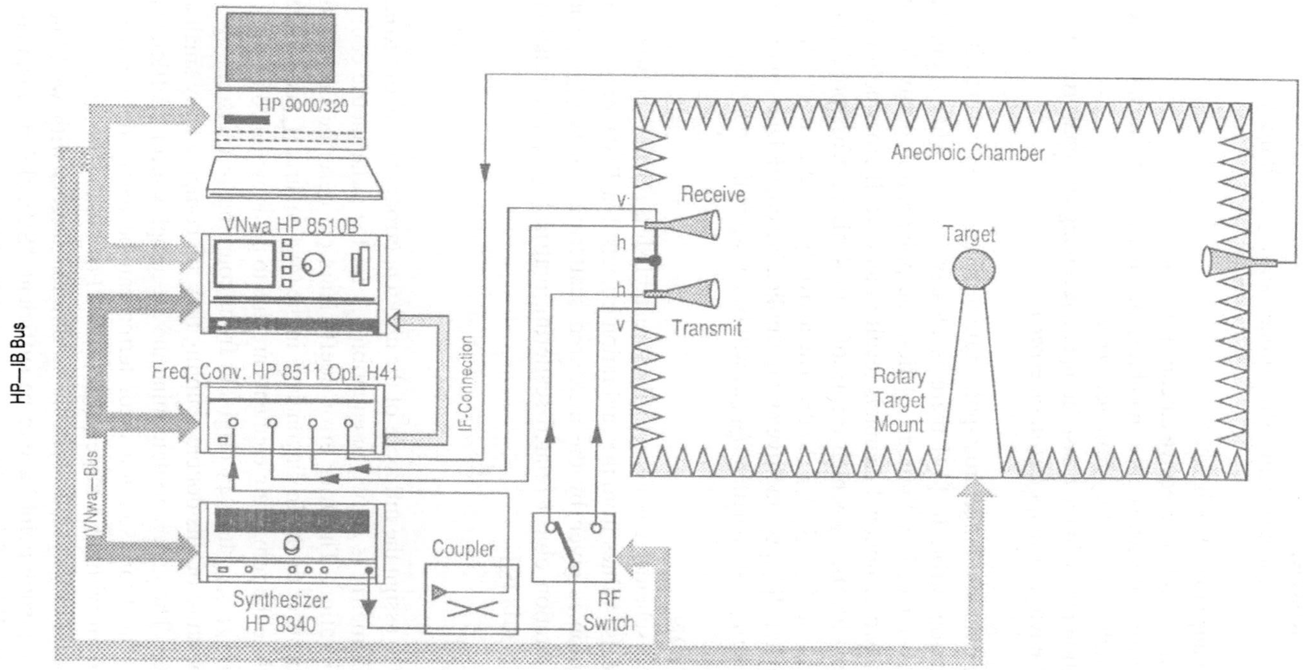


Fig. 3 Wideband, coherent, polarimetric RCS measurement system

### 4.1.2 SYSTEMATIC ERRORS

Systematic errors of the measurement system, as mentioned above, result from:

- reflections in the anechoic chamber,
- coupling between transmit and receive antenna ( *isolation error* ),
- frequency response and mismatches in cables, antennas, connectors, detectors, couplers and switches (*frequency response error* ),
- limited polarization purity in the transmitting and receiving antenna, cross-polarization of the test environment (*polarization error* ).

Up to now precise, high dynamic range cross-polarization measurements were difficult, because of the polarization error. In [14] there is clearly shown, what significant performance reduction is induced by limited isolation and polarization purity. Typical wideband antennas with two orthogonal channels result in an overall cross-polarization purity of approximately -20 dB, by which errors of 3 to 4 dB in cross-polarization measurement are induced for $(\underline{S}_{\xi\eta}/\ \underline{S}_{\xi\xi})$ = -10 dB. The above systematic errors influence the true scattering parameters $\underline{S}^0{}_{\xi\eta}$ by additive and multiplicative coefficients and lead to erroneous measured scattering parameters $\underline{S}^m{}_{\xi\eta}$ , as will be shown in the following.

### 4.2 ERROR EQUATION

For any error correction scheme it is necessarily to establish a relation between the measured values and precise values from the calibration objects. For Radar cross section measurements this relation is given by the measured scattering matrix [ $\underline{S}^m$ ] and a calculated, precise, calibration object related scattering matrix [ $\underline{S}^0$ ]. The relation is established via the error matrix $[\underline{C}]$.

$$[\underline{S}^m] = [\underline{C}_0] + [\underline{C}] \bullet [\underline{S}^c] \qquad (4.1)$$

The equation 4.1 allows to assign the influence of the different errors listed above to error coefficients $\underline{c}_{\xi\eta}$. The full matrix is shown in equation 4.2 for the polarimetric complex RCS measurement calibration. The four error coefficients $\underline{c}_{i0}$ describe the additive isolation error. The elements of the main diagonal $\underline{c}_{ii}$ in the (4x4)-matrix $[\underline{C}]$ consider the actual response error, while the remaining elements are due to polarization coupling. The encircled elements $\underline{c}_{12}$, $\underline{c}_{21}$, $\underline{c}_{34}$ and $\underline{c}_{43}$ originate from coupling in transmitting <u>and</u> receiving channel and because of this double coupling they are comparatively small and may usually be neglected. The task is to determine the 20 complex error coefficients at each frequency point. The four isolation error terms can be easily determined by measuring the empty room without a target, ($[\underline{S}^0]=[0]$). For the determination of the remaining 16 coefficients, mostly four suitable calibration targets with well-known matrices $[\underline{S}^{0i}]$ at each frequency point have to be sufficient. "Suitable" means, that the true, object related matrices $[\underline{S}^{0i}]$ of the calibration targets have to be linear independent at each frequency point. However it can be shown that det($\underline{C}$)≠0, although the coefficients $\underline{c}_{ij}$ are multiplicatively combined [15]. Using this non-linear relation and keeping in mind, that the double coupling coefficients are small, only three calibration targets lead to the determination of the complete matrix $[\underline{C}]$ with all 20 error coefficients. Inversion of (4.2) completes the calibration algorithm. The understanding of the error matrix can be improved by the following example for the cross-polarization which is explained in fig. 4. It shows the feeding system of the antennas, that is the basis for the above introduced coupling coefficients. The transmit antenna, in this case of the switch position, transmits a horizontally polarized signal, while the receiver simultaneously is connected with the vertical and horizontal channel of the receive antenna. For all possible polarization combinations an equation for the signal path can be determined.

Isolation error — Response error

$$\begin{bmatrix} \underline{S}^m_{vv} \\ \underline{S}^m_{hh} \\ \underline{S}^m_{vh} \\ \underline{S}^m_{hv} \end{bmatrix} = \begin{bmatrix} \underline{c}_{10} \\ \underline{c}_{20} \\ \underline{c}_{30} \\ \underline{c}_{40} \end{bmatrix} + \underbrace{\begin{bmatrix} \underline{c}_{11} & \underline{c}_{12} & \underline{c}_{13} & \underline{c}_{14} \\ \underline{c}_{21} & \underline{c}_{22} & \underline{c}_{23} & \underline{c}_{24} \\ \underline{c}_{31} & \underline{c}_{32} & \underline{c}_{33} & \underline{c}_{34} \\ \underline{c}_{41} & \underline{c}_{42} & \underline{c}_{43} & \underline{c}_{44} \end{bmatrix}}_{[\underline{C}]} * \begin{bmatrix} \underline{S}^0_{vv} \\ \underline{S}^0_{hh} \\ \underline{S}^0_{vh} \\ \underline{S}^0_{hv} \end{bmatrix} \qquad (4.2)$$

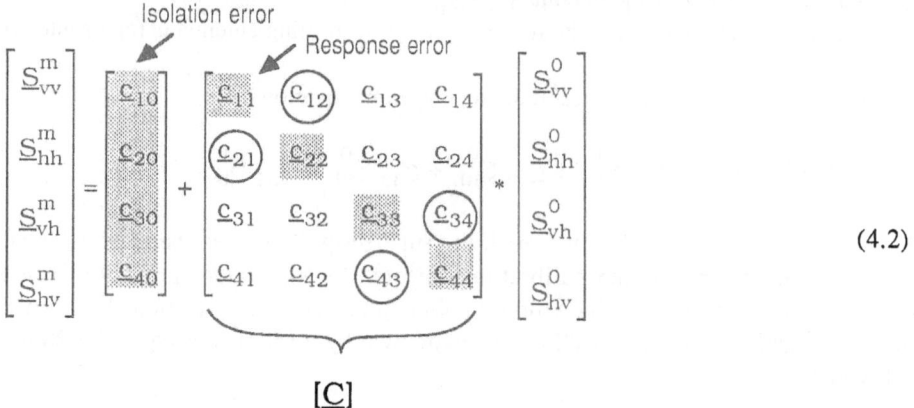

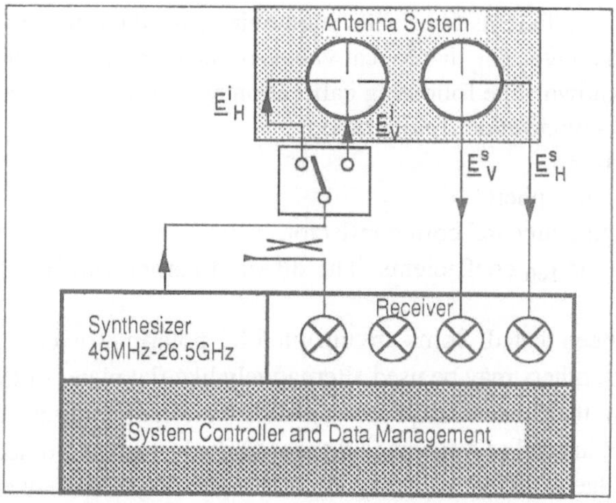


Fig. 4 Polarimetric antenna feeding System for 2 GHz–20 GHz

The error coefficients $\underline{c}_{\xi\eta}$ for the cross-polarized path have their origin as listed below according to fig. 4:

- the isolation error, resulting from the reflections of the anechoic chamber and the mutual coupling of the antennas is represented by an additive complex constant: $\underline{c}_{30}$
- the desired signal path, where the incident signal is depolarized from horizontal to vertical by the measured object, is falsified by the response error: $\underline{c}_{33} \cdot \underline{S}^0_{vh}$
- the transmit signal has a vertical polarized part, due to the polarization coupling in the transmit antenna from the horizontal to the vertical channel. This signal is scattered with $\underline{S}^0_{vv}$ of the object: $\underline{c}_{31} \cdot \underline{S}^0_{vv}$
- the transmitted signal with horizontal polarization is scattered with $\underline{S}^0_{hh}$ by the object. In the receiving antenna results a vertical received signal due to the polarization coupling of the receiving antenna: $\underline{c}_{32} \cdot \underline{S}^0_{hh}$
- the polarization coupling in the receiving and transmitting antenna is represented by: $\underline{c}_{34} \cdot \underline{S}^0_{hv}$

The sum of theses contributions result in one line from equation 4.2:

$$\underline{S}^m_{vh} = \underline{c}_{30} + \underline{c}_{31}\,\underline{S}^0_{vv} + \underline{c}_{32}\,\underline{S}^0_{hh} + \underline{c}_{33}\,\underline{S}^0_{vh} + \underline{c}_{34}\,\underline{S}^0_{hv} \qquad (4.3)$$

The superscript $^m$ denotes the measured, the superscript $^0$ denotes the true, calculated scattering parameters for the calibration objects. An equivalent procedure can be demonstrated for arbitrary polarization combinations. Thus, the relation between the measured $[\underline{S}^m]$ and the true $[\underline{S}^0]$ can be expressed by a set of four eqs. with 20 error coefficients.

## 4.3 CALIBRATION TARGETS

To determine the 20 error coefficients four calibration objects for the four measurements are necessary as explained above. The theoretical values of the complex scattering matrix of these objects must be known. The following calibration targets proved to be qualified for these measurements in a monostatic configuration:

- empty room,
- conducting sphere,
- conducting dihedral corner reflector.

The empty room renders the $\underline{c}_{i0}$ coefficients. The dihedral corner will be used in two positions.

Many other targets have been tested. Some are unsuited for polarimetric calibration like trihedral corner refelectors, others may be used alternatively like flat plates or cylinders. It has to be mentioned again, that linear independancy of the calibration targets is required. So a flat plate may be used instead of a sphere and the cylinder instead of a dihedral corner reflector. Flat plates have the advantage of a considerable higher RCS but they are difficult to position.

### 4.3.1 Conducting Sphere

The conducting sphere is used as an ideal calibration target. The first theoretical evaluation of the scattering of spherical objects was done by Mie in 1908 [16] . From this results the theoretical value of the RCS of a conducting sphere described in [17,18]. Equation (4.4) shows the exact solution

$$\sqrt{\underline{\sigma}} = -\frac{2\cdot\sqrt{\pi}}{\beta}\cdot\sum_{n=1}^{\infty}(-j)^{n-1}\ \frac{n\,(n+1)}{2}\ (\underline{A}_n - j\,\underline{B}_n) \tag{4.4}$$

with $A_n$, $B_n$ being the complex Mie coefficients, consisting of spherical Bessel and Hankel functions. The main advantage of a sphere is the non critical positioning because of its complete symmetry and the best agreement of theory and reality.

### 4.3.2 Metallic Dihedral Corner Reflector

The solution for the scattering of the metallic dihedral corner reflector is necessary for the last two calibration steps. In fig. 5 the geometrical configuration of the dihedral corner is shown.

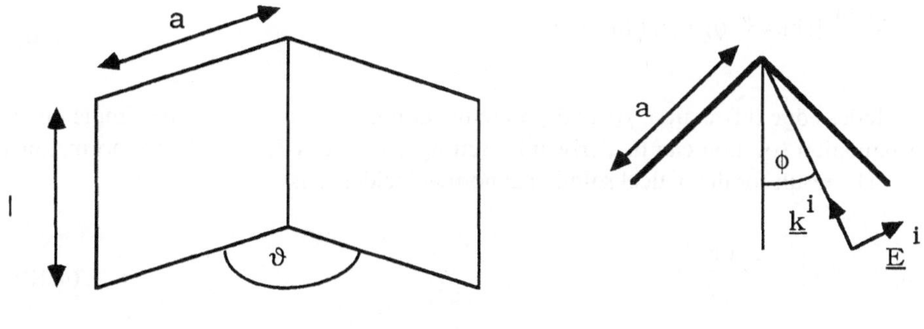


Fig. 5 Geometry and nomenclature for a dihedral corner reflector.

Simple optical approximations are given in most radar books. The disadvantage of these approximation techniques is, that edge diffractions are not included, although they contribute significantly, as will be shown. A solution, including edge diffraction, is derived from [19]. It is based on geometrical and physical optics. This solution is based on an ideal conducting reflector with infinitely thin plates. The reference plane for phase measurements is the symmetry axis of the corner reflector (see fig.5). For vertical polarization the electrical field vector is || to the symmetry axis, for horizontal polarization ⊥ to it. In equation 4.5 and 4.6 the term hh in brackets denotes the solution for horizontal transmission and horizontal reception.

$$\underline{\sigma}_{vv(hh)} = \frac{\lambda^2}{\pi} \left| \underline{A}_{vv(hh)} \right|^2 e^{j\,2\,\angle \underline{A}_{vv(hh)}} \tag{4.5}$$

with
$$\begin{aligned} \underline{A}_{vv} &= \underline{A}_1 + \underline{A}_2 \\ \underline{A}_{hh} &= \underline{A}_1 + \underline{A}_3 \end{aligned} \tag{4.6}$$

$$\begin{aligned} \underline{A}_1 = j\frac{k\,a\,l}{\lambda} \{ & \sin(45^\circ+\phi)\, e^{-j\,k\,a\cos(45^\circ+\phi)} \frac{\sin(k\,a\cos(45^\circ+\phi))}{k\,a\cos(45^\circ+\phi)} + \\ & + \sin(45^\circ-\phi)\, e^{-j\,k\,a\cos(45^\circ-\phi)} \frac{\sin(k\,a\cos(45^\circ-\phi))}{k\,a\cos(45^\circ-\phi)} \} \end{aligned} \tag{4.7}$$

$$\underline{A}_2 = -j\frac{k\,a\,l}{\lambda} \{ \sin(135^\circ+\phi) + \sin(135^\circ-\phi) \} \tag{4.8}$$

$$\underline{A}_3 = -j\frac{k\,a\,l}{\lambda} \{ \sin(45^\circ-\phi) + \sin(45^\circ+\phi) \} \tag{4.9}$$

$\underline{A}_1$ includes edge diffraction. $\underline{A}_2$ and $\underline{A}_3$ are the double bounced rays. The simple optical approximation solution can be derived by setting $\phi=0^\circ$ in (4.7) to (4.9), for normal incidence. The result for the optical solution at normal incidence is:

$$\left| \underline{\sigma}_{hh} \right| = \left| \underline{\sigma}_{vv} \right| = \frac{8\,\pi\,a^2\,l^2}{\lambda^2} \tag{4.10}$$

To determine the error coefficients in equation (4.2) it is necessary to use one calibration object, which depolarizes definitely the scattered field. If the dihedral corner reflector's symmetry axis is positioned parallel or perpendicular to the incident electrical field vector, it does not depolarize the field. On contrary, if the symmetry axis is turned 45° to the incident field vector, the reflected field is fully cross-polarized. Fig. 6 explains this behavior.

The incident field vector $\underline{E}^i$ is split into a vector parallel (v-polarization) and a vector perpendicular to the symmetry axis (h-polarization). By this, two simple scattering problems result:

- The vertical field vector is two times bounced and changes two times the direction, due to the boundary condition $\underline{E}_{tan}=0$ for ideal conducting planes. The scattered field vector has the same direction as the incident one.
- The horizontal scattered field vector changes the direction. The phase difference is $\pi$. Therefore the resulting received electrical field vector $\underline{E}^s$ is totally cross-polarized.

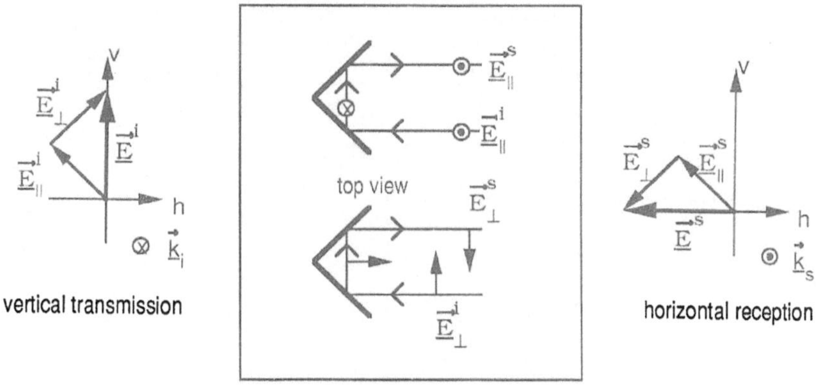


Fig. 6 Cross-polarization of the electrical field by a dihedral corner reflector

## 4.4 CALIBRATION PROCEDURE

The calibration procedure, using the calibration targets described in the previous section, runs in the steps demonstrated in the signal flow diagram in fig. 7.

In the first step the measurement parameters of the NWA are set, as there are for example: frequency range, the averaging factor, the number of frequency points. Also the type of calibration can be selected. In the next step the complex RCS matrix is computed for the intended calibration targets, as there are: metallic sphere and a dihedral corner reflector in two different positions. The four calibration measurements run as shown in the flow diagram with the targets shown below.

1. Empty Room

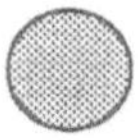

2. Conducting Sphere

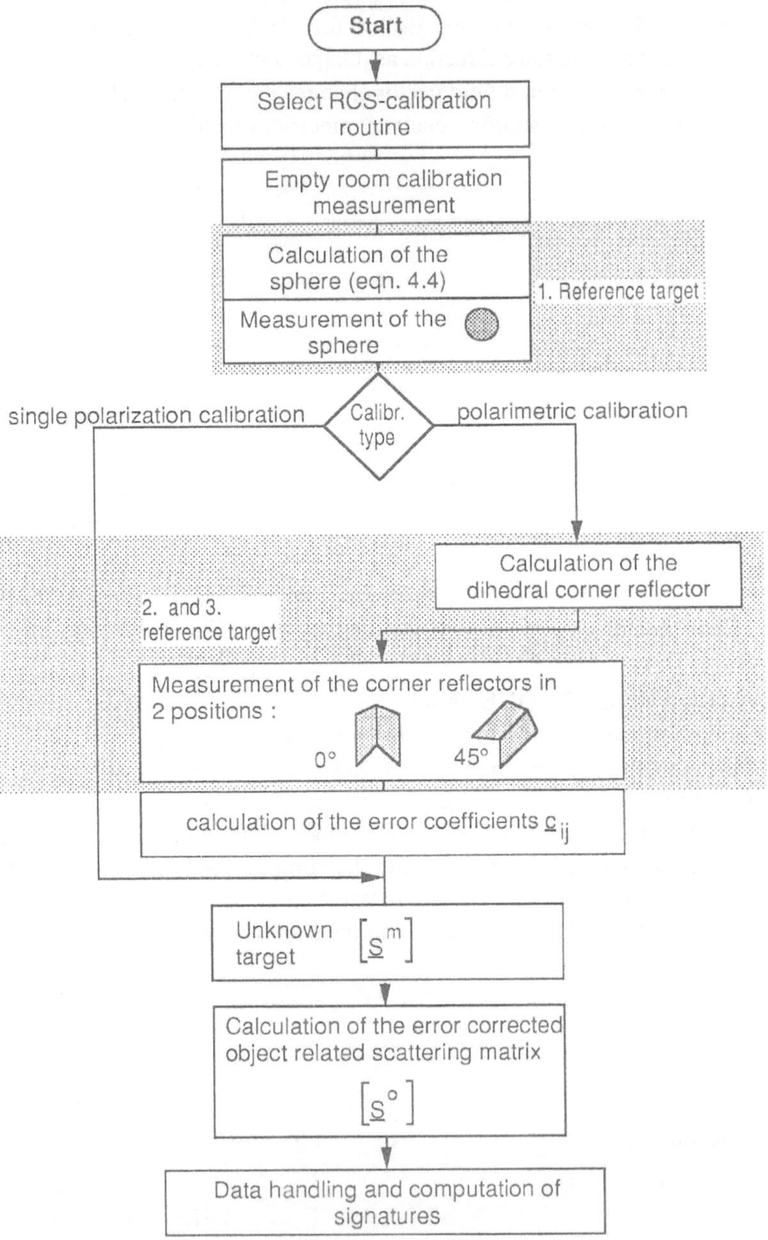


Fig. 7 Signal flow diagramm for the polarimetric RCS calibration

3. Dihedral Corner Reflector

(1. Aspect Angle) $E_i \uparrow$

(2. Aspect Angle) $E_i \uparrow$

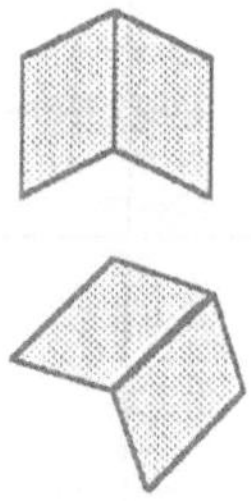

After the calibration procedure all elements of the error matrix [C] in (4.2) are known. The system is calibrated.

## 4.4 SYSTEM PERFORMANCE AND VERIFICATION

To show the system performance and especially the polarization decoupling measurement results for several objects are demonstrated. As a first test object a conducting sphere with a diameter different to the one of the calibration sphere (∅=36 mm) was selected. The theoretical RCS was computed in the frequency domain and compared to the measurement results (fig. 8). The difference between theory and measurement over the whole frequency range is less then 0.25 dB.

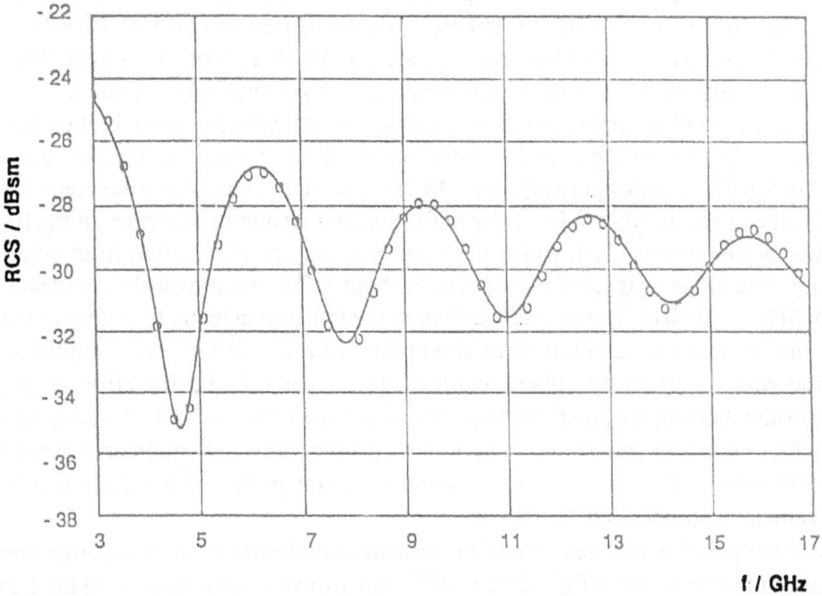


Fig. 8 RCS of a metallic sphere ( ∅=36 mm) in the frequency domain. (— measured; o calculated)

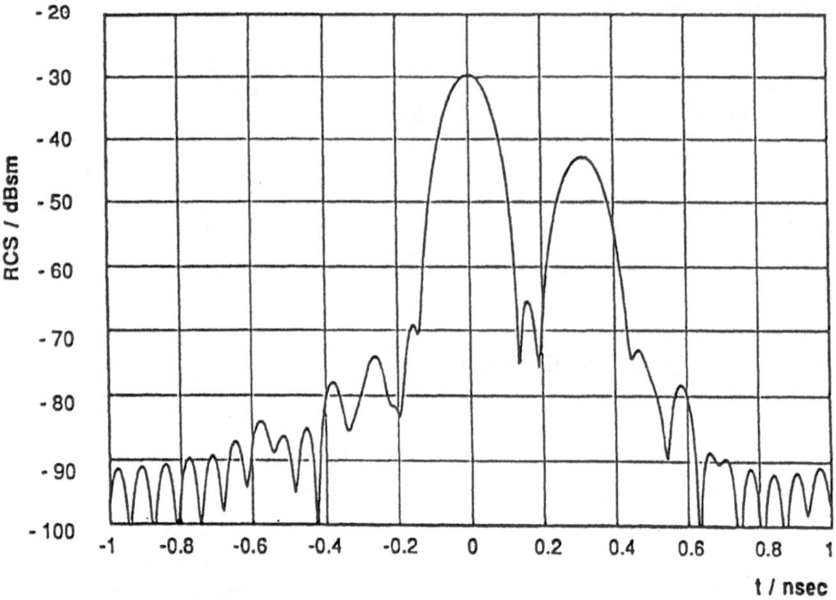


Fig. 9 RCS of a metallic sphere ( $\phi$=36 mm) in the time domain

The time domain representation, see fig.9, exhibits a major peak due to the optical reflections at the front surface of the sphere. The measured magnitude of -29.9 $dBm^2$ corresponds exactly to the cross-sectional area of 10.18 $cm^2$ of the sphere. The minor peak is due to the creeping waves and is located $[(2+\pi) \cdot r]$ behind the first one.

The improvement of the polarization decoupling due to the proposed calibration procedure is shown in fig.11 in comparison with the results from standard calibration procedures plotted in fig.10. In both figures $\underline{\sigma}_{vh}$ of a 15 cm diameter sphere was measured. The theoretical value of $\underline{\sigma}_{vh}$ for this object is by definition and also in reality zero. In fig.10 a simplified error model was used, which does not consider errors resulting from polarization coupling. The value of $|\underline{\sigma}_{vv}|$ of this sphere is about -17.5 dBsm over the frequency range from 3 GHz to 20 GHz. The signal coupling in the transmit antenna from the horizontal to the vertical channel is -25 db (polarization purity of the antenna). This coupled signal is scattered with $\underline{\sigma}_{vv}$ from the sphere and leads to a cross-polarization error of $|\underline{\sigma}_{vh}|$=-42 dBsm, from what a cross-polarization purity of about 25 dB results. If the error correction and calibration model, introduced in section 4 , is used, the polarization error is reduced to about -78 dBsm, thus a synthetic cross-polarization purity of 60 db is reached, the improvement is around 35 dB.

To verify the polarimetric precision of the system , a dielectric cylinder was measured. It is positioned in the x-y plane (fig. 12) at $\vartheta$=45° with a wave propagation along the z-axes.

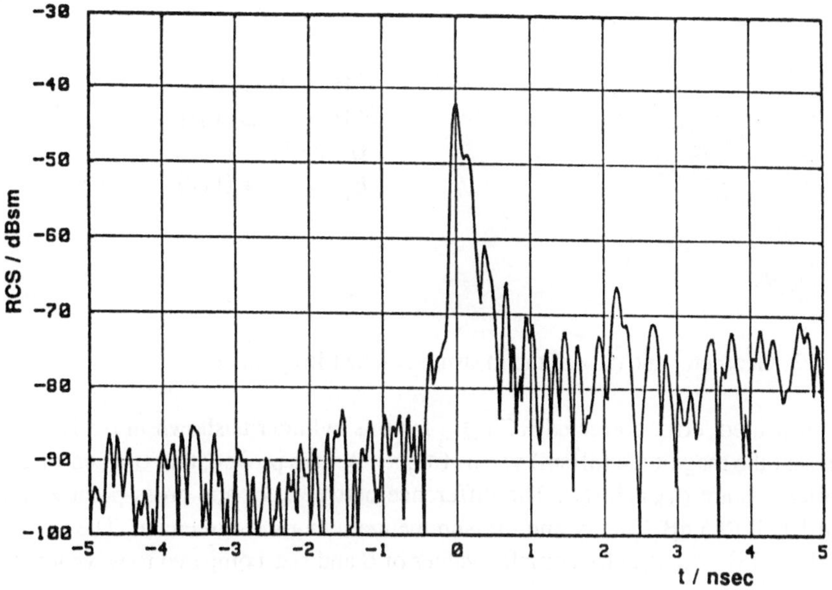


Fig.10 Time domain $|\underline{\sigma}_{vh}|$ of a conducting sphere with a diameter ∅=150 mm (simplified error model)

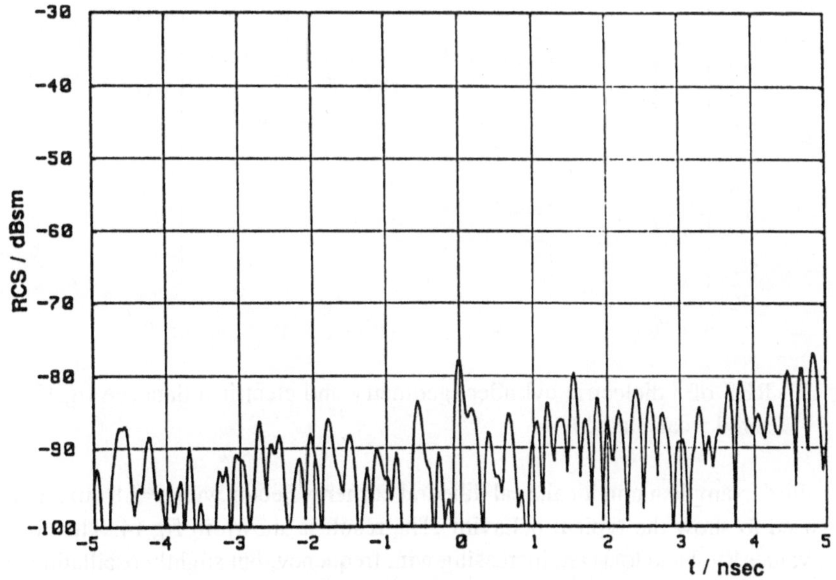


Fig.11 Time domain $|\underline{\sigma}_{vh}|$ of a conducting sphere with a diameter ∅=150 mm (20 term error model)

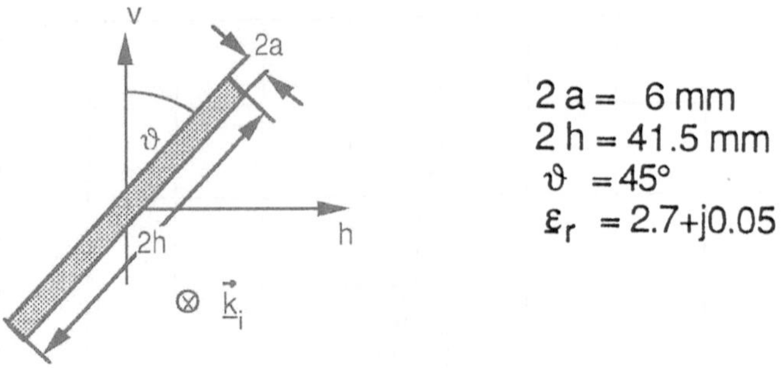


Fig. 12 Geometry and data of the measured dielectric cylinder

The measured, error corrected RCS $\underline{\sigma}_{\xi\eta}$ for this cylinder is shown in fig.13. The curves represent the magnitude (dBsm) versus frequency. It is proved, that $\underline{\sigma}_{vv}$ and $\underline{\sigma}_{hh}$ are really identical, as are $\underline{\sigma}_{vh}$ and $\underline{\sigma}_{hv}$. The difference between co- and cross-polarization varies from 10 dB to 5 dB. Theory and measurement are in good agreement. The used theory is based on [18], extended for complex values of σ and size compared to wave length.

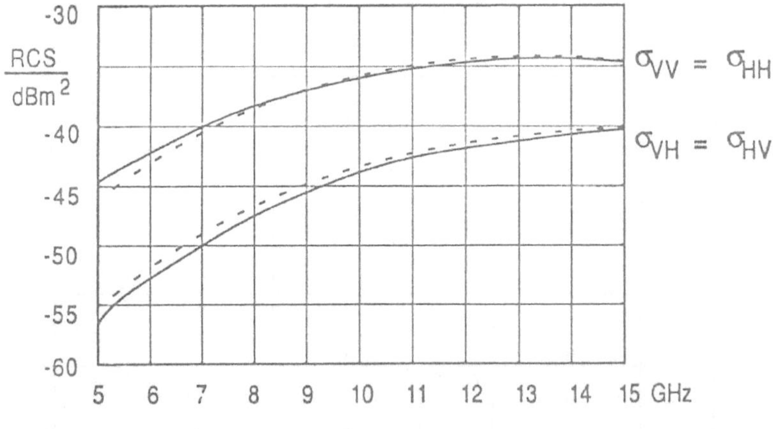


Fig.13 RCS of a dielectric cylinder (geometry and electrical data see fig.12), (-- measured, –calculated)

As a third example a non ideal, real dihedral corner reflector with a=9.8 cm and h=8.0 cm was used to show the typical behavior .The result is shown in fig.14. The co-polarized RCS values $|\underline{\sigma}_{vv}|$ and $|\underline{\sigma}_{hh}|$ are increasing with frequency, but slightly oscillating. This

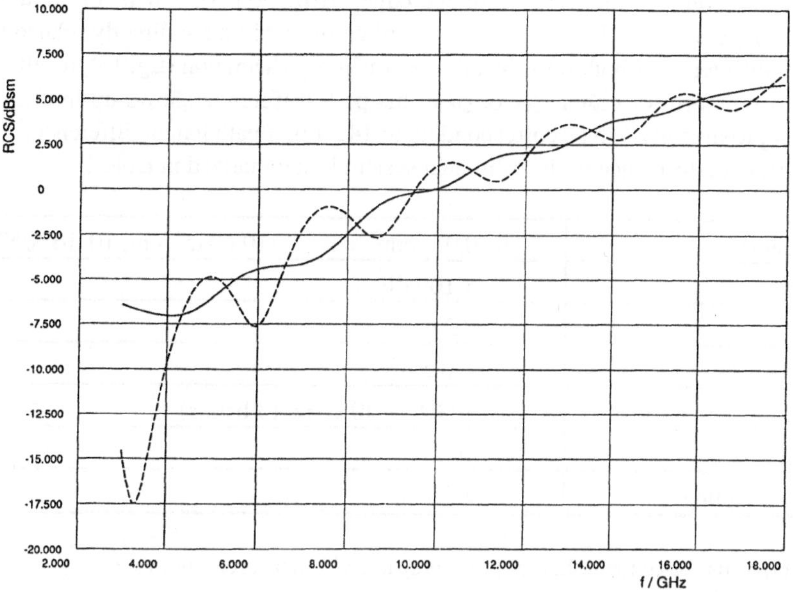


Fig.14 $|\underline{\sigma}_{vv}|$ (-) and $|\underline{\sigma}_{hh}|$ (--) of a dihedral corner reflector (a=9.8 cm; l=8.0 cm),edge thickness 0.2 mm

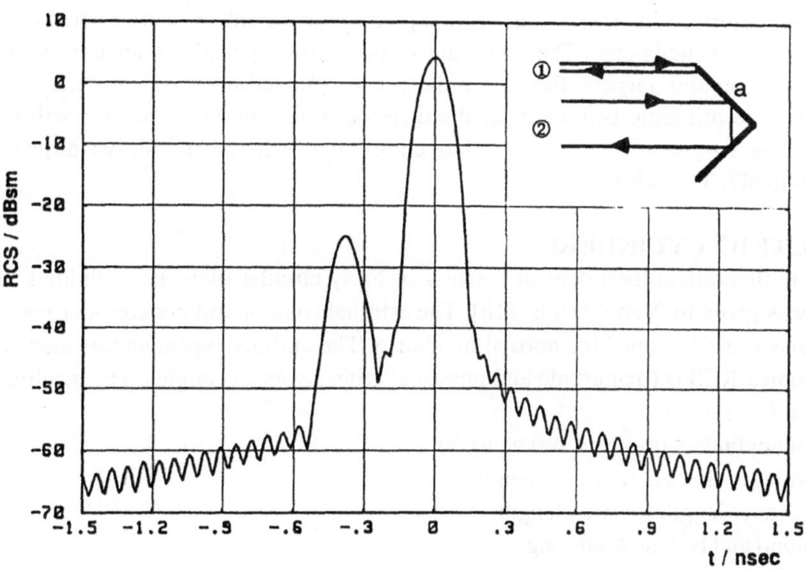


Fig. 15 Time domain representation of $\underline{\sigma}_{vv}$ (dihedral corner reflector)

oscillation results from the interference of the edge diffraction ((②) with the double bounced ray (①) (see fig.15). The repetition of minima or maxima are directly related to the path length difference of both rays. In a time domain representation (fig. 15), resulting from a fourier transform of either $\underline{\sigma}_{vv}$ or $\underline{\sigma}_{hh}$, this path difference shows up in a time difference of approximately 475 ps corresponding to 142.4 mm path length difference. The unique system performance of the calibrated system is summarized in table 1:

| | | |
|---|---|---|
| Minimum target size | 0.01 $mm^2$ | (10 GHz, 5 m, 10 dB S/N) |
| Dynamic range | > 100 dB | |
| Phase error | < 2 degree | |
| Amplitude error | < 0.5 dB | |
| Bandwidth | 2 GHz – 40 GHz, (3 bands) | |
| IF-Bandwidth | 10 kHz | |
| Polarization decoupling | > 55 dB | |

Table 1 Performance of the coherent polarimetric RCS measurement system

## 5. SYSTEM VERIFICATION AND APPLICATION

With the system described in the previous section precise measurements can be performed over a very wide frequency range. The following section shows typical applications of the system for some complex targets. First for a dielectric cylinder and a dihedral metallic corner reflector the principle behaviour of the different polarimetric signatures will be demonstrated. These signatures will then be derived for chaff material and for two airplane models (Boeing 747, Tornado).

### 5.1 DIELECTRIC CYLINDERS

An appropriate theoretical treatment of lossless or lossy circular dielectric cylinders of finite length was given by Van de Hulst [18]. The original solution did not consider phase information and was developed for normal incidence. The authors expanded this method to get the complex RCS $\underline{\sigma}$ (magnitude and phase) at arbitrary aspect angles. The resulting equation is valid for:

- thin cylinders (radius < wavelength)
- prolate cylinders (radius < length)
- cylinder length >≈ wavelength
- mono and bistatic scattering
- arbitrary polarization

The theoretical calculation is described in detail in [15].

It was experienced, that theoretical solutions are well in agreement with measurements. Measured results for a dielectric cylinder are presented in the following. They were selected because they make it easier to understand more complex targets.

In fig. 16 the trace, which is a measure for the total power received (see eq.2.20), is given for a dielectric cylinder with the length l=20 cm, diameter d=2 cm and a dielectric constant of $\underline{\varepsilon}$=3+j0.07, for normal incidence on the cylinder. The periodicity of the minima and maxima is related to the diameter of the cylinder, due to the interference of two waves. The first is reflected directly from the front surface of the cylinder and the second travels through the cylinder and is reflected from the back of the cylinder in the incident direction. The time delay is related to the periodicity.

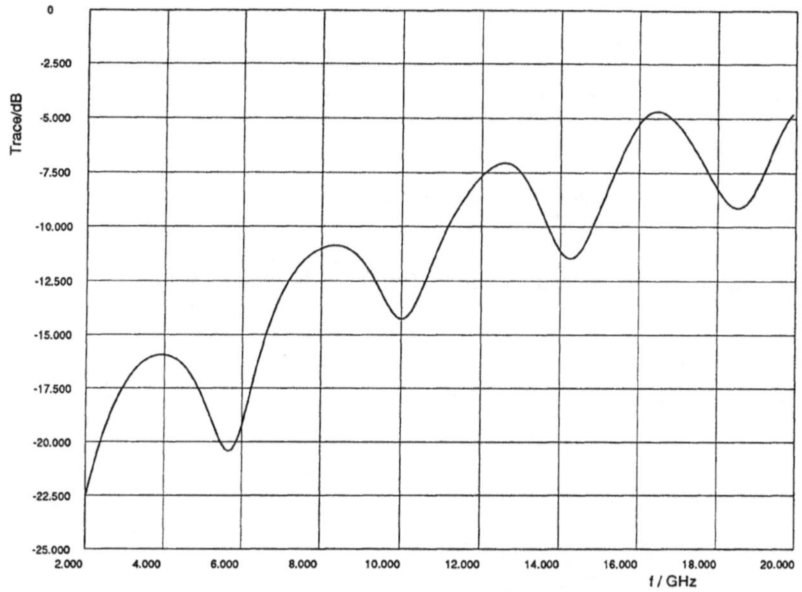


Fig.16 Trace $P_1$ of a dielectric cylinder (l=20cm; d=2cm; $\underline{\varepsilon}$=3+j0.07)

The depolarization D of the same object is shown in fig.17 (see eq.2.21). It reveals the same periodicity as the trace. At frequencies where the trace has its minimum the cylinder reflects its maximum cross-polarization.

The cross-polarized dispersive polarimetric signature (see section 2.3) is given in fig.18. Regarding the characteristic versus the rotation angle $\psi$, the maximum cross-polarization results for $\psi \pm 45°$ over the whole frequency range. This behaviour is in agreement with theory and typical for objects having a geometry with a preferred extension.

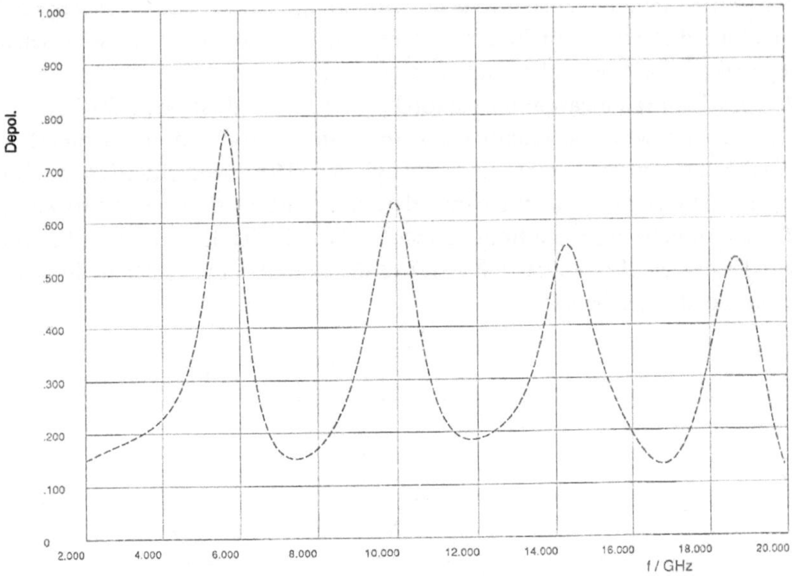


Fig.17 Depolarization D of a dielectric cylinder (l=20cm; d=2cm; $\underline{\varepsilon}$=3+j0.07)

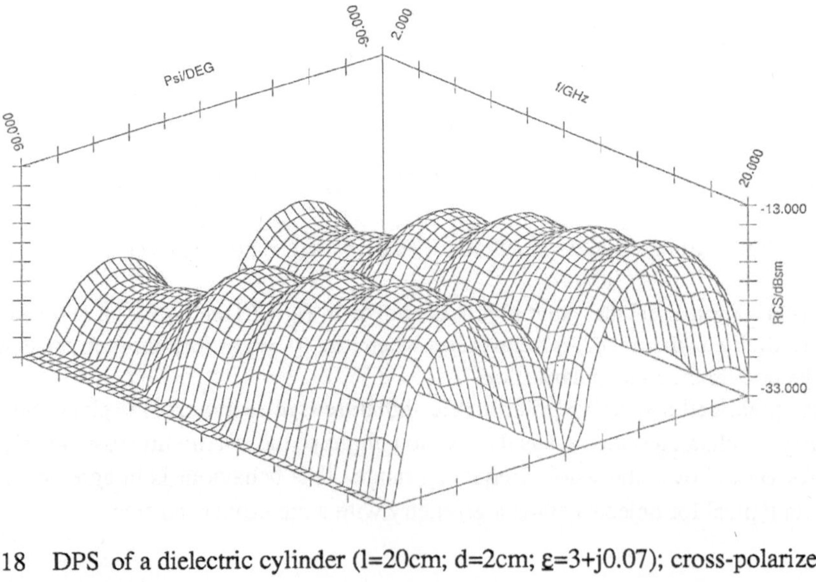


Fig.18 DPS of a dielectric cylinder (l=20cm; d=2cm; $\underline{\varepsilon}$=3+j0.07); cross-polarized, linear polarization

## 5.2 Dihedral metallic corner reflector

The scattering behaviour of this target is theoretically described in section 4.3.2. From the measurements the depolarization D is selected in fig.19. Versus the whole frequency range D is approximately one. This means and is in agreement with theory, that a corner reflector is able to change the total incident power into the orthogonal polarization. From the previous section it is known that this is for a rotation of 45° to the symmetry axis.

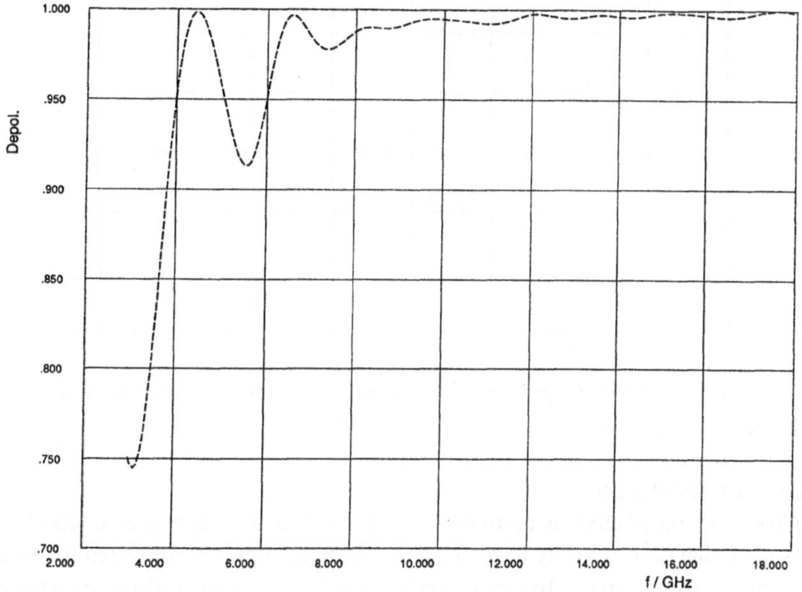


Fig.19 Depolarization D of a corner reflector ((a=9.8 cm; l=8.0 cm))

During the measurements of numerous targets the Null- and Eigen-polarizations showed to be very characteristic. As an example the frequency dependent movement of the co-polarization-minima is shown for a dihedral corner reflector. This plot is called Nyquist plot as it shows the dependency from the rotation angle $\psi$ and the ellipticity angle $\alpha$ as a function of frequency.

The Nyquist plot of the co-polarization Nulls versus frequency, gives the polarization state, where a minimum power is reflected in co-polarization (see section 2.3) for a given frequency. These co-polarization Nulls are in most cases equivalent to the maximum cross-polarization return.

Fig. 20 gives the plot for the dihedral corner reflector. It is obviously that the minimum co-polarization return, respectively the maximum cross-polarization return, results for linear polarization ($\alpha$=0°) and a rotation angle of ±45° ($\psi$=±45°), which corresponds to the theory evaluated in section 2.3. The small deviations from the ideal polarization state ($\alpha$=0°, $\psi$=±45°) result from the edge diffraction.

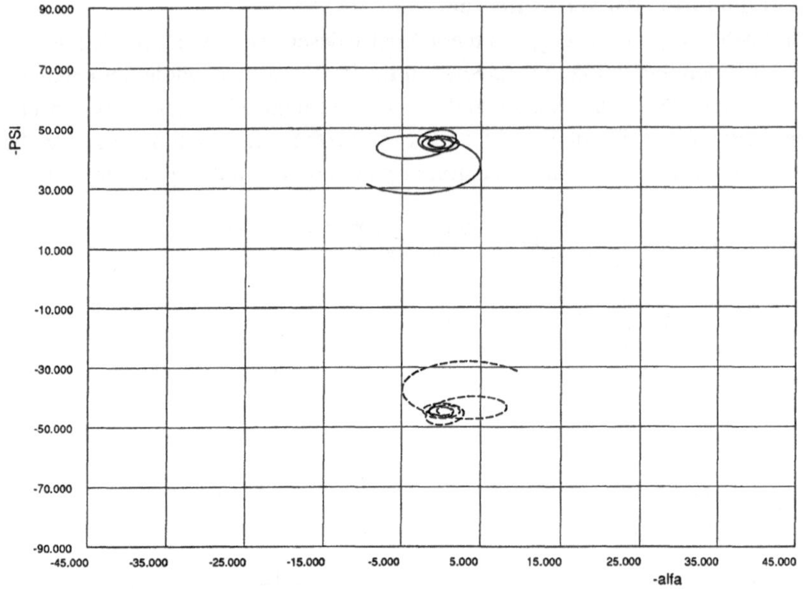


Fig.20 Nyquist plot of the co-polarization Nulls of a corner reflector (a=9.8 cm; l=8.0 cm)

### 5.3 CHAFF MATERIALS

The Radar Cross section of chaff is up to now solely examined in dispensed chaff clouds. This delivers data, that come closely to realistic applications, but it is not suited for examination of the chaff element itself. In chaff clouds there is always the influence of the dispenser and the environment, which may rapidly change. For chaff mainly metallic wire strips, or coated dielectric strips are used. Differences can be found in material, strip cross section, coating procedure and coating thickness. All these parameters influence the RCS. The only means to precise characterization of chaff material is to measure single elements. Numerous chaff strips, different materials, from several suppliers were measured. In fig. 21 and 22 two examples are shown, a silver coated polyamid chaff in fig. 21 and a glass cylinder evaporated with aluminium in fig. 22. For both the wave polarization is parallel to the element orientation. The maximum of the RCS for chaff, approximately 2 cm long and 25 μm thick is about -30 dBsm ± 3 dB at frequencies between 7 GHz and 10 GHz corresponding to the l/2 wave length resonance. The 10 dB bandwidth varies from 3 GHz to 4 GHz. In J-band the RCS is usually 15 dB below the maximum and there are also some materials having a further decrease of another 15 to 18 dB at 18 GHz. Nearly all chaff elements show a deep minimum in the center of the S-band. For straight elements the RCS for the perpendicular polarization is negligible. The same is true for cross-polarization, as long as the incident wave is polarized parallel to the strip. Because of the extreme thin elements, no creeping waves for the TE polarization can be distinguished, as it is possible for a thickness of several millimeters.

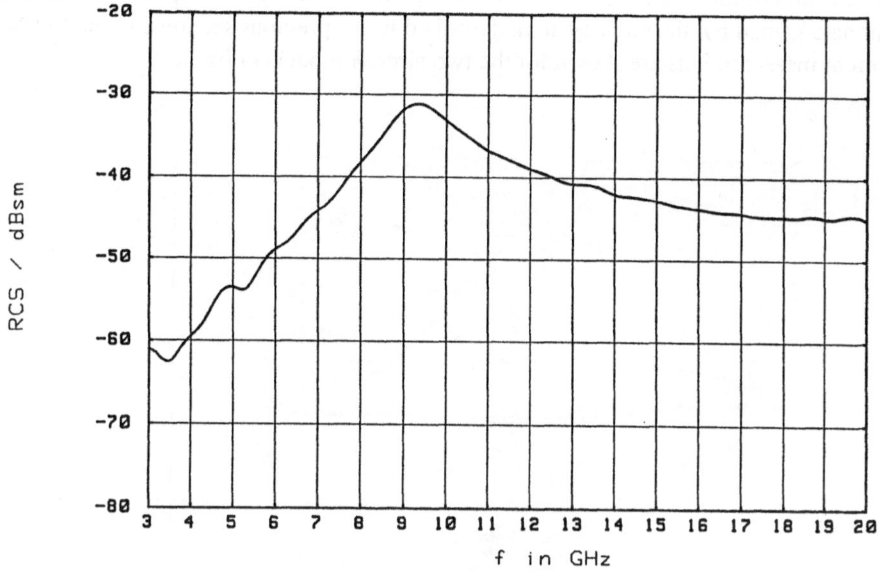


Fig.21 RCS of a chaff strip; material: silver coated polyamid; length 1.5 cm; Ø=0.1 mm

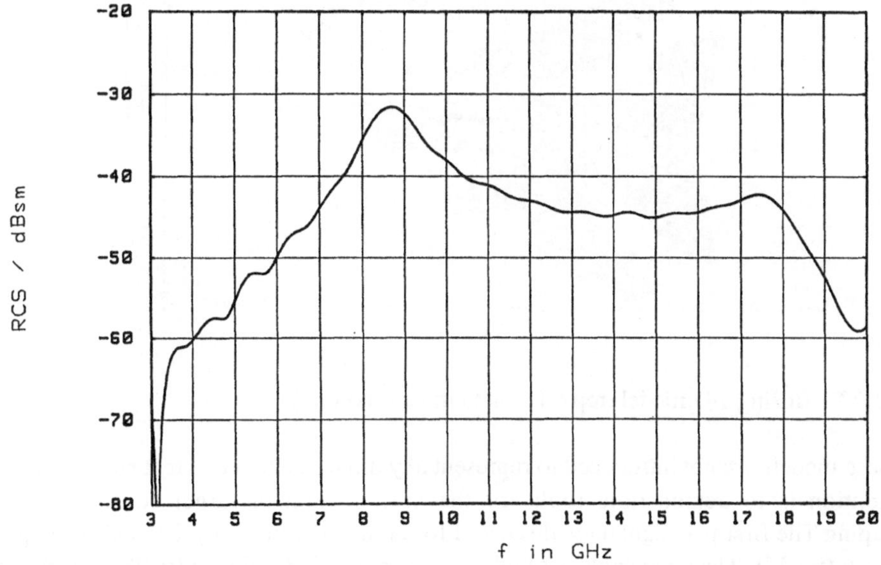


Fig.22 RCS of a chaff strip; material: Al evaporated glass; length 1.7 cm; Ø=0.024 mm

## 5.4 Airplane models

A task in radar laboratories is very often the optimization of targets, in the sense of minimization of the RCS. A powerful tool are polarimetric signatures, using the measurement data gained by the radar system, described in the previous sections. In the following different measurements are shown for the two aircraft models in fig. 22.

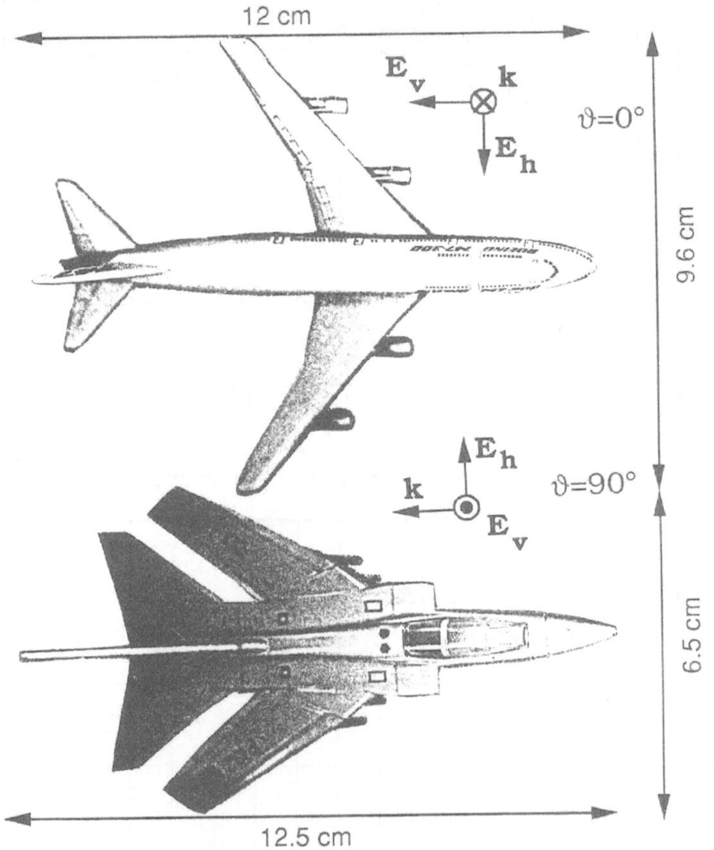


Fig.23 Boeing 747 model (top); Tornado model (bottom)

These models are not introduced to represent any direct relation to a real size aircraft. The intention is to demonstrate tools for the optimization of target configuration and shaping.The first two signatures discussed for both is the trace (fig.24) and the depolarization (fig.25). The view angle for both measurements is $\vartheta$=0° (see fig.23), which means the target is viewed from the top.

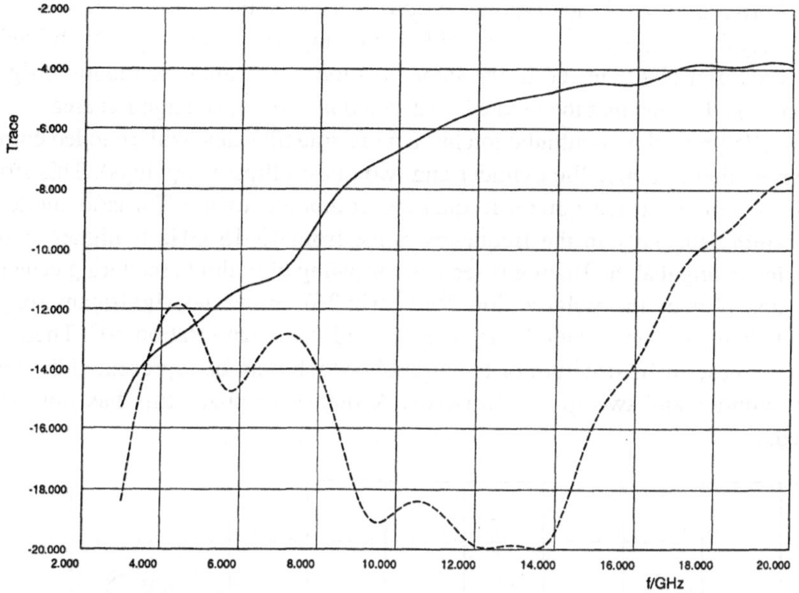


Fig.24 Trace $P_1$ of the Boeing747 model (dashed line) and Tornado model (plain line)

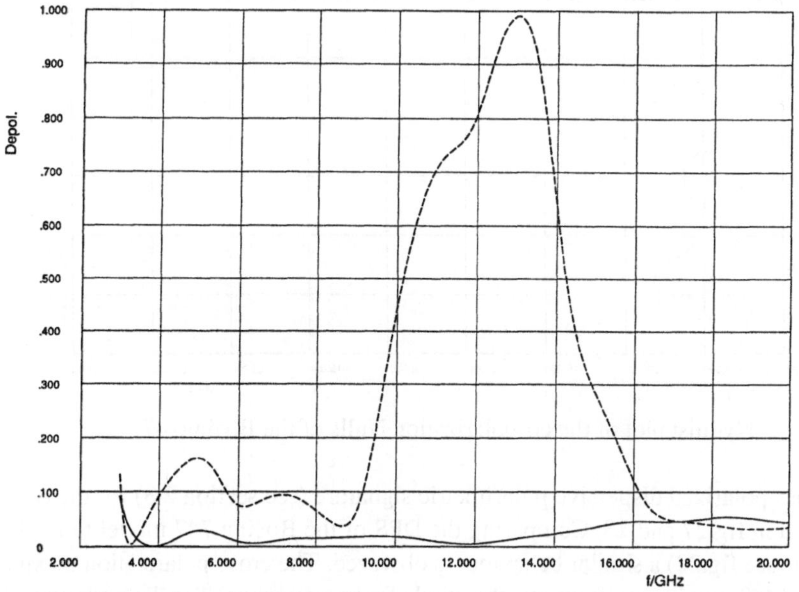


Fig.25 Depolarization D of the Boeing 747 model (dashed line) and Tornado model (plain line)

For the Tornado model the depolarization is versus the whole frequency range approximately 0, whereas the trace is steadily increasing with frequency. This behaviour is analogous to a flat plate viewed in the same position. Regarding the model in fig.23 by inspection it can be seen that the model in this position consists of large flat areas.
The Boeing 747 model, in contradiction has a more smooth shape with rounded corners. It could be regarded as a metallic cylinder and two quasi-ellipsoids (wings). This structure results in a reduced trace versus frequency, compared to the Tornado model. The depolarization especially in the frequency range from 13-15 GHz is higher as for the Tornado, indicating that the Boeing is seen as consisting of multiple scattering centers.
The Nyquist plot of the polarization Nulls (fig.26) shows no significant frequency dependant changes. The mean ellipticity is 36° and the mean rotation 50°. These values correspond to the main structure of the aircraft, which may be represented by a central metallic cylinder and two quasi-ellipsoids. A theory to prove this has not yet been developed.

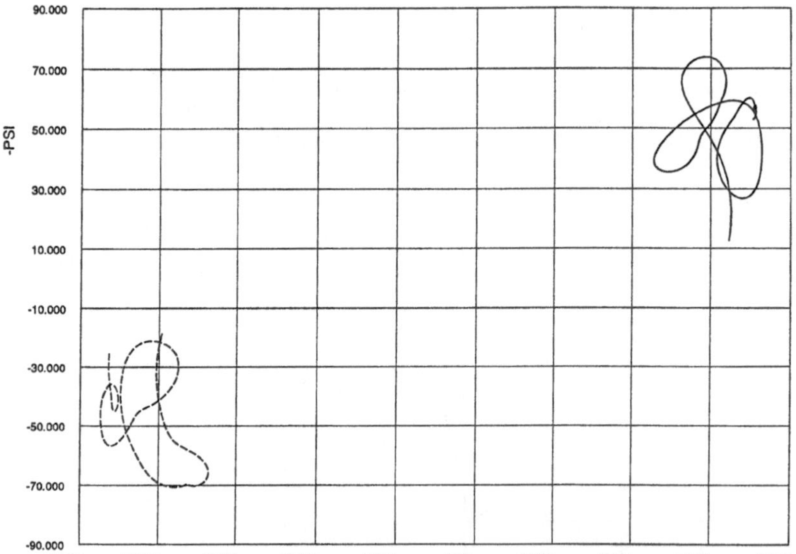


Fig.26 Nyquist plot of the co-polarization Nulls of the Boeing747

The cross-polarized dispersive polarimetric signature (see section 2.3) for the two models is plotted in fig.27 and 28. Comparing the DPS of the Boeing 747 model to the DPS of a cylinder (see fig.22) a similar behaviour is observed. The cross-polarization maxima result also for ±45° rotation angle versus the whole frequency range. The Tornado model shows especially in the frequency range above 10 GHz a significant different behaviour. The curve is more flat, indicating no preferred direction in geometry

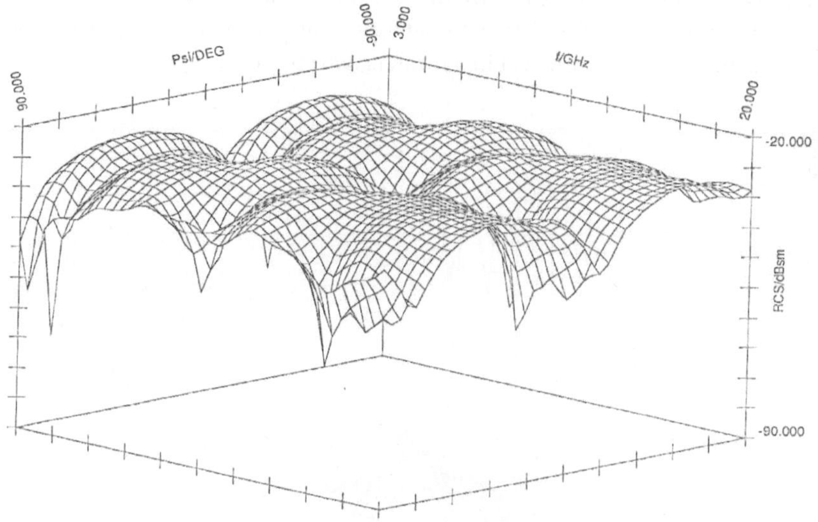


Fig.27 DPS of a Boeing 747 model ; cross-polarization; linear polarization

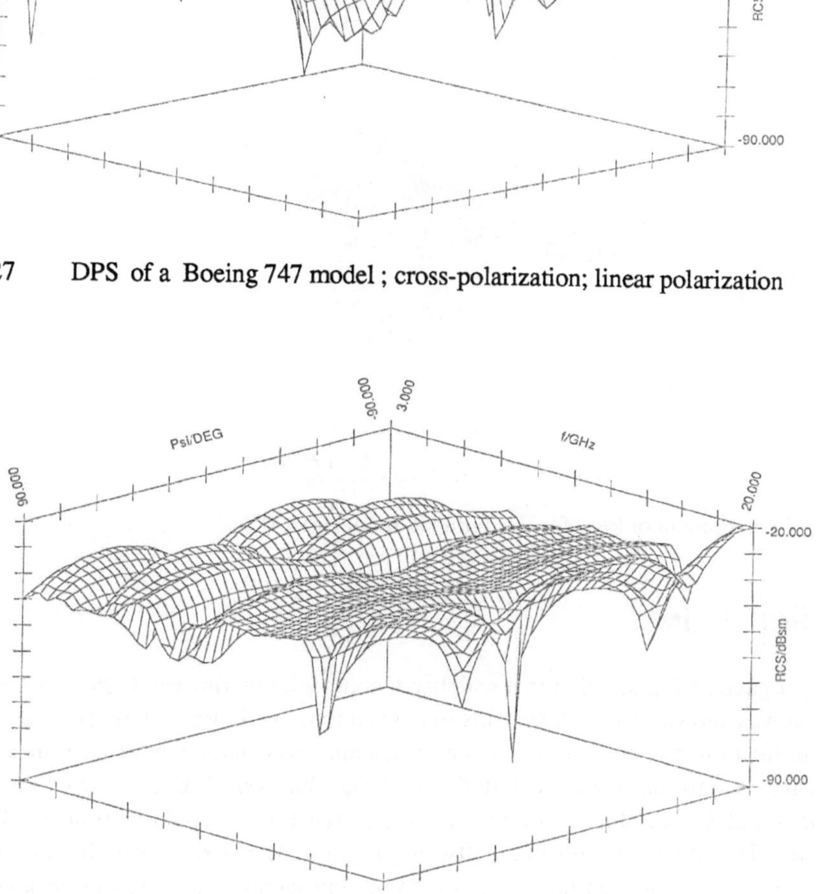


Fig.28 DPS of a Tornado model ; cross-polarization; linear polarization

For the Tornado model viewed nose on ($\vartheta$=90°) the time domain representation of $|\sigma_{hh}|$ is plotted in fig. 29. In this graph the different scattering centers can clearly be distinguished (dotted lines), as there are the nose, cabin, air inlets, wheels and wings.

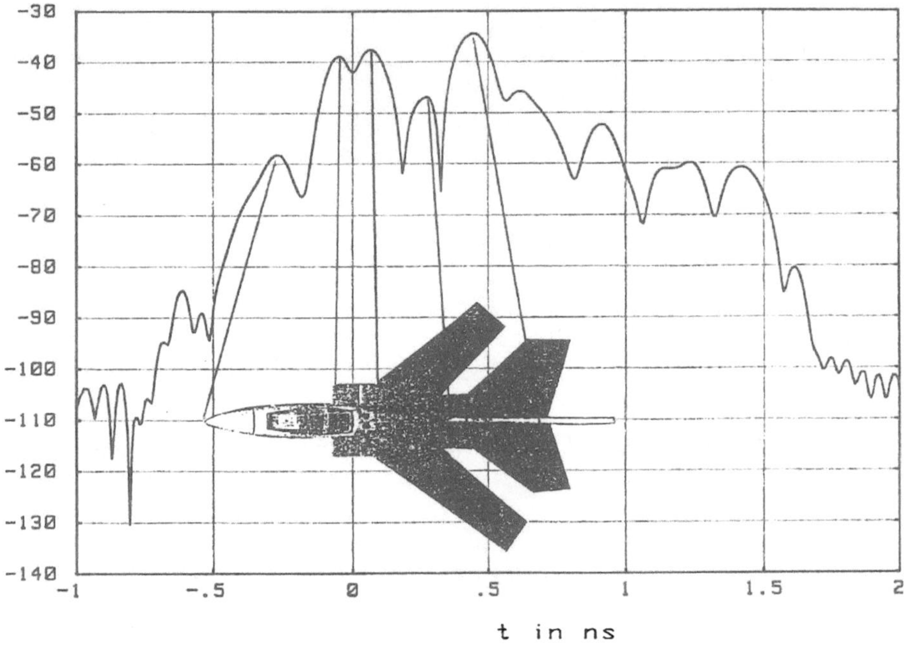


Fig.29 Time domain of $|\sigma_{hh}|$ for the Tornado model ($\vartheta$=90°)

## 6. CONCLUSION

This paper gave an introduction to a wideband, coherent, polarimetric RCS measurement system. It was shown, that with systems like this a new age in Radar target classification and identification is opened. In our laboratories numerous targets have been measured, from which a data base was installed. For target analysis 3-D and 2-D first order signatures and several higher order signatures like trace, depolarization, Null- and optimum polarization are built in a software program. The next steps will comprise an improvement for automized target classification and identification as well as a system modification for long range RCS measurements. It can be foreseen, that not for long civil and military Radars will increasingly make use of the knowledge gained in laboratories world wide.

## REFERENCES

[1] E. M. Kennaugh, "Research studies on the polarization properties of Radar targets," *Scientific reports*, Ohio State University, July 1984.

[2] J. R. Huynen, "Phenomenological theory of Radar targets," *Ph.D. thesis*, Tech. Univ. Delft,1970.

[3] A. B. Kostinski, W.-M. Boerner, "On foundations of Radar polarimetry," *IEEE Trans. Ant. Prop.: AP-34,* pp. 1395-1403, Dec. 1986.

[4] S. H. Bickel, "Some invariant properties of the polarization scattering matrix," *Proc. IEEE, Special Issue on Radar Reflectivity,* pp. 1070-1072, Aug. 1965.

[5] J. R. Huynen, "Measurement of the target scattering matrix," *Proc. IEEE, Special Issue on Radar Reflectivity,* pp. 936-946, Aug. 1965.

[6] H. A. Zebker, J. J. Van Zyl, D. N. Held, "Imaging Radar polarimetry from wave synthesis," in *Journal of Geophysical Research, Vol. 92, No. B1*, pp. 683-701, Jan. 1987.

[7] H. A. Mieras, A. B. Kostinski, W.-M. Boerner, "Comments on "On foundations of Radar polarimetry" ," *IEEE Trans. Ant. Prop.: AP-34,* pp. 1470-1473, Dec. 1986.

[8] W.-M. Boerner, M. B. El-Arini, C.-Y. Chan, P. M. Mastoris, "Polarization dependence in electromagnetic inverse problems," *IEEE Trans. Ant. Prop.: AP-29*, pp. 262-271, March 1981.

[9] J. J. Van Zyl, "On the importance of polarization in Radar scattering problems," *Ph.D. thesis,* California Institute of Technology, 1986.

[10] W. Wiesbeck, S. Riegger, "Measurement of the complex RCS-matrix for small metallic and dielectric objects," in *RADAR 87, International Conf*, London, Oct. 1987, pp. 360-364.

[11] D. Kähny, S. Riegger, G. Schöne, W. Wiesbeck, "Coherent polarimetric signatures of coniferous trees, a survey," *Fourth International Colloquium on Spectral Signatures of Objects in Remote Sensing, Conf. Proc.*, Aussois, Jan. 1988, pp.13-18

[12] C. D. Graves, "Radar polarization power scattering matrix," *Proceedings of the IRE,* vol. 44, pp. 248-252, Feb. 1956.

[13] S. Riegger , W. Wiesbeck, A. J. Sieber, "On the Origin of Cross-Polarization," *IGARSS'87 Conference Proceedings*, Ann Arbor, 1987, pp.577-580

[14] S. Riegger , W. Wiesbeck,"Single reference calibration for polarimetric Radar Cross Section measurements," to be published in 1989

[15] S. Riegger, "Analyse und Synthese polarimetrischer, dispersiver Radarstreuquerschnittssignaturen," *Ph. D. Thesis*, Institut für Höchstfrequenztechnik und Elektronik ; University of Karlsruhe (TH) 1988

[16] G. Mie, "Beiträge zur Optik trüber Medien, speziell kolloidaler Metallösungen", *Annalen der Physik*, 4. Folge, Band 25, 1908

[17] G.T.Ruck, D.E. Barrick, W.D. Stuart , C.K. Krichbaum ,"Radar cross section handbook," *Plenum Press* , New York-London , 1970

[18] H. C. van de Hulst , "Light Scattering by small particles," *Dover Publications*, Inc. New York, pp. 304-306, 1981

[19] E.F. Knott, "RCS reduction of dihedral corners," *IEEE Transactions on Antennas and Propagation*, vol. AP-25, pp. 406-409, 1977

# MEASUREMENT AND ANALYSIS IN RADAR POLARIMETRY

S.R. Cloude
Department of Mathematics and Computer Science
University of Dundee
Dundee DD1 4HN
Scotland.

ABSTRACT. In this paper we consider methods for the measurement and analysis of polarimetric target information. We concentrate on three main areas; the design of a single channel real time Stokes receiver, conditions for physical realisability of the Stokes reflection matrix and a generalisation of the optimum polarisation theory of Kennaugh and Graves to the case of stochastic targets.

## 1. INTRODUCTION

The extraction and analysis of polarisation information from radar systems poses severe technical and theoretical problems. Polarisation metrology, while a highly developed science in optics (Bickel 1985, Boyer 1979, Houston 1978, Jellison 1985, Talmage 1986), has only recently been developed for systems operating at centimetre and millimetre wavelengths (see Giuli 1986). Two particular research initiatives have been important in this development: first in antenna design, where considerable effort has been invested in reflector design for satellite communications and radio astronomy (see Clarricoats 1977), enabling the design of high gain low cross polarisation antennas with wide bandwidth for imaging and identification studies. The second area is in the development of fast polarisation rf switches, enabling full scattering matrix measurements on millisecond timescales. Nonetheless, there are still important areas to be resolved in polarisation metrology, particularly in the areas of calibration, correlation receiver design and synthetic aperture formation.

Theoretical justification for the considerable expense involved in vector radar systems stems from the general theory of null or characteristic polarisations developed by Kennaugh (1952), Graves (1956) and Gent (1954) in the early post war years. However, it was quickly realised that radar targets display highly dynamic polarisation characteristics, reducing somewhat the impact of this elegant coherent theory. This, combined with the difficulties of direct and indirect vector scattering theory, has left open the



W.-M. Boerner et al. (eds.), Direct and Inverse Methods in Radar Polarimetry, Part 1, 773–791.

question of widespread applicability of radar polarimetry.

Faced with these problems, there are two ways to proceed: either by the use of target averaging or adaptive systems, with the latter capable of exploiting the coherent theory, but on time scales less than target decorrelation time. In this respect, radar targets differ from those considered in optics and deserve special attention (Giuli 1986). The dynamics of radar scattering have vector decorrelation times on the order of milliseconds or even seconds (the advantage of polarisation in this respect is that one is concerned with the dynamics of relative rather than absolute time events, with the former generally yielding longer correlation times).

In this paper we consider techniques for the measurement and analysis of correlation phenomena in polarimetry, concentrating on two themes: fast measurement systems and generalisation of the null polarisation concepts to stochastic targets.

In view of the aforementioned statistical nature of radar targets, and the as yet undetermined decorrelation times, we consider the use of coherent receivers as inappropriate and consider instead the design of receivers for measurement of the full wave coherency matrix. The Stokes vector (see Bickel 1985) provides a full correlation measurement based on intensity rather than phase measurement and in section 2 we consider methods for the real time determination of all four Stokes parameters and hence, via multiple transmitter polarisations, measurement of the full Stokes reflection matrix (SRM).

We then consider the effects of experimental error on analysis of the SRM and develop conditions for physical realisability (see Huynen 1970, Kostinski 1988), by which we mean a set of quadratic constraints on the elements of the SRM such that it always yields a scattered wave with degree of polarisation less than or equal to unity. We develop these conditions using a target decomposition theorem outlined in an earlier paper of these proceedings (Cloude 1989), the main result of which is to relate the SRM to a target covariance or coherency matrix (CM). In this way we make an important link between measured parameters based on the Stokes vector and target correlation estimates.

Finally, we show how the eigenvectors of the covariance matrix may be used to devise a general analysis method for stochastic targets and used to decompose a measured target operator into physical and nonphysical components. We illustrate an application of this decomposition to published data on measurements for synthetic aperture radar (Van Zyl 1987) and optical system calibration (Howell 1979, Cloude 1986).

We then consider generalisation of the null polarisation concepts of the coherent theory to stochastic targets, and compare and contrast results obtained using the CM (Swartz 1988, Novak 1987, Cloude 1986) with others based on more direct analysis of the SRM (Van Zyl 1987, Kostinski 1988, Cloude 1988).

## 2. REAL TIME STOKES VECTOR MEASUREMENT

In this section we consider methods for the experimental measurement of the wave Stokes vector. Particular emphasis is to be placed on methods for real time determination of all four Stokes parameters, where 'real time' corresponds to measurement on time scales less than those expected for target decorrelation (typically milliseconds).

We show how frequency modulation techniques may be used to design a single channel receiver, and compare and contrast this design with others based on multiple channel reception. The receiver has been built and demonstrated using a He-Ne laser system operating at 0.632μm, and, while reference will be made to details of this system, it should be noted that many aspects of the design are generic, in the sense that they are applicable at both microwave and optical frequencies.

Real time measurement of the polarisation of light is an important objective in many optical systems. Examples include electro-optic modulation, polarisation preserving fibre sensors and coherent optical communications. However, despite the mature state of optical technology, most measurement systems tend to be of the canonical type; an arrangement of beamsplitters, polarising filters and detectors designed to obtain four independent intensity measurements from which g is subsequently calculated (see Talmage 1986, Thompson 1980, Howell 1979). These systems suffer from several drawbacks:

1) The need for multiple detectors makes differential drift, sensitivity and calibration a problem.
2) The Stokes parameters are calculated from linear combinations of intensity measurements, which often involve taking small differences between large measured values, and are prone to error.
3) The arrangement of beamsplitters and filters requires careful alignment, leading to a relatively complicated design.

While differences exist in hardware for microwave and optical realisations of the canonical design, the above drawbacks apply equally to both operating regimes and hence may be considered generic to the canonical approach.

In this section we describe a single detector system capable of real time measurement of all four elements of g. The system operates by encoding each Stokes parameter at a discrete frequency and using a combination of lock-in amplifiers to measure g. As it involves only a single receiver channel and requires little post processing and simpler calibration, it is inherently more accurate than the canonical approach (note that similar approaches have been used in optics for some time, see Berry 1977 and Boyer 1979).

We begin a brief review of the relevant theory by noting that we can relate the Stoke vector g to the wave coherency matrix [J] as

$$[J] = \begin{bmatrix} E_H E_H^* & E_H E_V^* \\ E_V E_H^* & E_V E_V^* \end{bmatrix} = \begin{bmatrix} g_0+g_1 & g_2-ig_3 \\ g_2+ig_3 & g_0-g_1 \end{bmatrix} \qquad (2.1)$$

where $g=(g_0,g_1,g_2,g_3)$ is the Stoke 4-vector, $E_H$ and $E_V$ the complex field amplitudes of the horizontally and vertically polarised components of the wave and [J] a 2x2 hermitian matrix characterising the correlation properties of the light. When light with vector $g$ passes through a filter, the transmitted light has a modified vector $g'$ given by

$$g' = [M]g \tag{2.2}$$

where [M] is a 4x4 real matrix called the Stokes Reflection matrix (a similar matrix, modified slightly for coordinate changes is called the Mueller matrix in optics, see Bickel 1985). This matrix may be obtained from the coherent scattering matrix (CSM) by the transformation (Cloude 1986)

$$[M] = \tfrac{1}{2}\,\mathrm{Tr}([S]^{+}\underset{\sim}{\sigma}[S]\underset{\sim}{\sigma}) \tag{2.3}$$

where Tr is the trace of the matrix product, [S] the 2x2 CSM and $\underset{\sim}{\sigma}$ the Pauli matrices. Note that this mapping is only defined one way ie. [S] to [M]. The mapping in the other direction requires introduction of a target covariance or coherency matrix [T], which is 4x4 hermitian with eigenvectors identified with a set of 4 CSM's such that

$$[M] = \lambda_1[M_1] + \lambda_2[M_2] + \lambda_3[M_3] + \lambda_4[M_4] \tag{2.4}$$

where the $\lambda$ are nonnegative eigenvalues of [T] and each $[M_i]$ corresponds to a phase unique CSM (ie. unique only down to an absolute phase angle).

As an example, consider a linear dichroic filter (polaroid at optical frequencies or a dipole antenna at RF). For an angle $\theta$ between the dichroic axis and reference horizontal we have

$$[S] = \begin{bmatrix} \cos^2\theta & \cos\theta\sin\theta \\ \cos\theta\sin\theta & \sin^2\theta \end{bmatrix} \tag{2.5}$$

and

$$[M] = \begin{bmatrix} 1 & \cos2\theta & \sin2\theta & 0 \\ \cos2\theta & \tfrac{1}{2}(1+\cos4\theta) & \tfrac{1}{2}\sin4\theta & 0 \\ \sin2\theta & \tfrac{1}{2}\sin4\theta & \tfrac{1}{2}(1-\cos4\theta) & 0 \\ 0 & 0 & 0 & 0 \end{bmatrix} \tag{2.6}$$

By setting $\theta = 0^\circ$, $45^\circ$ and $90^\circ$ we obtain intensity measurements ($g_0'$) of $g_0+g_1$, $g_0+g_2$ and $g_0-g_1$ respectively. This result forms the basis for the canonical approach outlined above. Note that $g_3$ cannot be determined using a linearly dichroic filter; to obtain $g_3$ we must use a circularly dichroic filter (circular polaroid in optics or helical antenna at RF), the CSM and SRM of which are

$$[S] = \begin{bmatrix} 1 & -i \\ i & 1 \end{bmatrix} \qquad (2.7)$$

and

$$[M] = \begin{bmatrix} 1 & 0 & 0 & 1 \\ 0 & 0 & 0 & 0 \\ 0 & 0 & 0 & 0 \\ 1 & 0 & 0 & 1 \end{bmatrix} \qquad (2.8)$$

To obtain a real time system, we first divide the incident wave into four (using a beamsplitter in optics, orthomode transducer or directed dipoles plus helix at rf) and use four matched square law detectors operating in parallel. There are in fact an infinite number of such groups of four polarising filters which may be used but all suffer from the drawbacks outlined earlier.

In order to avoid such designs, we use a single channel receiver, and resort to frequency or time division multiplexing for the creation of four information channels for the Stokes parameters (see Boyer 1979, Berry 1977). Here we choose a frequency modulation technique by employing a two component receiving filter: a fixed linear dichroic (the axis of which defines a vertical reference direction) and a rotating linear birefringent (the fundamental rotation frequency of which defines a basic modulation frequency for the system). In this way we obtain a periodic variation of output intensity which is uniquely determined by the Stokes vector of the incident wave. The emphasis here is placed on uniqueness, for we wish to design a system capable of unambiguous measurement of all four elements of $\underline{g}$.

To proceed, we note that the CSM and SRM of a linear birefringent filter with fast axis orientation $\theta$ and retardation $\delta$ are

$$[S] = \begin{bmatrix} \cos^2\theta+\sin^2\theta\exp(i\delta) & \sin\theta\cos\theta(\exp(i\delta)-1) \\ \sin\theta\cos\theta(\exp(i\delta)-1) & \sin^2\theta+\cos^2\theta\exp(i\delta) \end{bmatrix} \qquad (2.9)$$

and

$$[M] = \begin{bmatrix} 2 & 0 & 0 & 0 \\ 0 & \alpha+\beta\cos4\theta & \beta\sin4\theta & 2\sin2\theta\sin\delta \\ 0 & \beta\sin4\theta & \alpha-\beta\cos4\theta & -2\cos2\theta\sin\delta \\ 0 & -2\sin2\theta\sin\delta & 2\cos2\theta\sin\delta & 2\cos\delta \end{bmatrix} \qquad (2.10)$$

where $\alpha = 1+\cos\delta$ and $\beta = 1-\cos\delta$.

We now have four possible combinations to consider and use the notation XY to represent the option under consideration (X and Y permute B (birefringent) and D (dichroic) and Y is the first element encountered by the input light). The rotating element is signified by the over dot.

1) $B\dot{D}$

In this case the measured intensity is independent of $\delta$ and given by

$$g_o' = g_0 + g_1\cos\omega t + g_2\sin\omega t \tag{2.11}$$

where $\theta=\omega t$, $\omega=2\pi f$ and $f$ is the fundamental rotation frequency. We can recover three of the four Stokes parameters with this arrangement ($g_1$ and $g_2$ are in phase quadrature and hence require phase sensitive detection for discrimination) but cannot determine $g_3$.

2) $\dot{D}B$

In this case $g_0'$ is dependent on both $\theta$ and $\delta$ and is given by

$$g_0' = g_0 + A\cos\omega t + B\sin\omega t \tag{2.12}$$

where

$$A = g_1(\alpha+\beta\cos4\theta) + g_2\beta\sin4\theta + 2g_3\sin\delta\sin2\theta \tag{2.13}$$

$$B = g_1\sin4\theta + g_2(\alpha-\beta\cos4\theta) - 2g_3\cos2\theta\sin\delta \tag{2.14}$$

Again we have two signals in phase quadrature and, for fixed $\delta$, have only three independent measurements and hence cannot fully determine $g$.

3) $\dot{B}D$

This rather trivial arrangement yields a transmitted intensity independent of $\omega$, given by

$$g_0' = g_0 + g_1\cos\phi + g_2\sin\phi \tag{2.15}$$

where $\phi$ is a constant.

4) $D\dot{B}$

This case yields a measured intensity of the form

$$g_0' = 2g_0 + \alpha g_1 + g_1\beta\cos4\omega t + g_2\beta\sin4\omega t + 2g_3\sin2\omega t\sin\delta \tag{2.16}$$

and is the only combination which permits full determination of $g$ (as long as $\delta$ is not $0^o$ or $180^o$). In particular, if $\delta = 90^o$ (ie. B is a quarter wave plate) then the received intensity is

$$g_0' = 2g_0 + g_1(1+\cos4\omega t) + g_2\sin4\omega t + 2g_3\sin2\omega t \tag{2.17}$$

which enables determination of $g$ with a minimum of computation.

We have verified these results using a He-Ne gas laser operating at 0.632μm, with a mica quarter wave plate and HN22 linear polaroid (see figure 1). Precision quartz quarter and half wave plates are used to convert the linearly polarised laser output to arbitrary polarisation before collimating the light into the receiver. A single photodiode is used to measure the light intensity transmitted by the filter, with the mica plate rotating at 75Hz (4500 rpm). The output from the detector is sent to four lock-in amplifiers (note that the incident light is chopped at 1KHz to remove the dc term in the expression for $g_0'$). The outputs from these amplifiers are then digitised and stored in a small microcomputer for off line analysis and real time presentation on a polarisation chart. In this way the whole Poincare sphere may be mapped and errors determined for all incident wave states (restricted to waves with unity degree of polarisation, it being more difficult to obtain sources for calibration of partially polarised waves).

By measuring $g$ and switching the transmitter through 4 states, a set of 16 equations in the 16 unknowns of [M] may be obtained. With the SRM so determined, the target covariance matrix may be obtained and its eigenvectors used to obtain an estimate of the dominant scattering mechanism in the object. By measuring the SRM in real time, estimates may then be obtained for target decorrelation by examining the eigenvalues of [T] as a function of time.

## 3. CONDITIONS FOR PHYSICAL REALISABILITY

It is important to realise that from the group of all 4x4 real matrices (the group GL(4,R)), only a subgroup (to be called SRM(4,R)) are isomorphic to physical targets. This has important implications for polarimetric measurement, in that noise and experimental error often cause a measured matrix to lie in GL(4,R) but outside SRM(4,R). Clearly, it is important to identify whether a given matrix lies in or out of this subgroup and, if possible, to quantify how far removed it is from realisability. With such a distance measure, it is then possible to filter the matrix so as to obtain an estimate of the nearest member of SRM(4,R), which is then used for subsequent analysis in place of the measured matrix and, unlike the latter, guarantees physical realisability of secondary parameters such as eigenstates and copolar nulls. In this section we develop a definition of SRM(4,R) and devise such a filtering scheme based on eigenvectors of the covariance matrix.

The subgroup SRM(4,R) stems from the requirement that all Stokes vectors, $g$, of the scattered light have a degree of polarisation between zero and unity, independent of the incident wave vector $h$. Further, $g_0$, the first element of $g$ and intensity of the scattered wave must be nonnegative and (for passive media) less than or equal to $h_0$, the intensity of the incident wave. We can then write down the definition of SRM(4,R) as

$$g = [M]\, h \tag{3.1}$$

where [M] is an element of SRM(4,R), if and only if,

$$g_0^2 - g_1^2 - g_2^2 - g_3^2 \geq 0 \qquad (3.2)$$

$$0 \leq g_0 \leq h_0 \quad \text{(for active media the upper bound is not set)}$$

for all Stokes vectors h.

These conditions impose quadratic constraints on the elements of [M], the form of which have been studied by several authors (see Huynen 1970, Fry and Kattawar 1981, Kostinski 1988). However, it is not obvious from the form of these equations how to define a filtering scheme for the matrix. A better approach is to use the target covariance matrix (CM), mapped onto [M] by the sixteen generating matrices of SU(4) (Cloude 1986). The constraints on [M] are then equivalent to requiring a positive semi-definite CM (the lemma that all 2x2 minors of the CM be nonnegative then corresponds directly to the quadratic constraints developed for [M]).

We can summarise this result by saying that SRM(4,R) is defined as all 4x4 real matrices which have, under the SU(4) mapping from GL(4,R) to GL(4,C), positive semi-definite hermitian form.

We now have a method for determining the realisability of a measured matrix: form the corresponding covariance matrix; find its eigenvalues; if the eigenvalue spectrum is nonnegative then the matrix belongs to SRM(4,R); if one or more eigenvalues are negative then the matrix is nonphysical. Further, by isolating the eigenvectors corresponding to negative eigenvalues, we can decompose the matrix into physical and nonphysical components which allows us to define a measure of separability (the ratio of maximum negative to maximum positive eigenvalue) and to undertake a filtering scheme: the nearest member of SRM(4,R) may be taken as the matrix corresponding to that obtained from the sum of eigenvectors with positive eigenvalues.

To illustrate, we consider three examples taken from the literature: measurement of the transmission matrix for an optical quarter wave plate (Cloude 1986); matrix data published for calibration of a passive radiometer (Howell 1979) and finally, synthetic aperture radar data published for the urban San Francisco area (Van Zyl 1987).

By transmitting 4 different states and measuring the Stokes vector of the light transmitted through a chain of optical elements, [M] may be obtained by Gaussian elimination. Cloude (1986) has published examples of such measurements for quarter and half wave plates, a typical example for a quarter wave plate being

$$[M] = \begin{bmatrix} 0.997 & 0.003 & 0.017 & 0.035 \\ 0.002 & 1.000 & -0.030 & -0.001 \\ 0.076 & 0.022 & 0.091 & -1.013 \\ 0.010 & 0.034 & 0.976 & 0.232 \end{bmatrix} \qquad (3.3)$$

This matrix does not belong to SRM(4,R), as verified by converting to covariance form and diagonalising, the resulting eigenvalue spectrum being

$$\lambda = (2.07,\ -0.10,\ 0.08,\ 0.00) \tag{3.4}$$

Note the negative eigenvalue, caused by experimental error and implying that [M] is not physically realisable. In this case, however, the matrix is very close to SRM(4,R), with a distance measure of only 0.048, and it would seen justifiable to ignore all eigenvector contributions but those from the dominant eigenvalue. When this is done, the resulting estimate of the CSM for the quarter wave plate is well within the manufacturers specification (see Cloude 1986).

As an example of a cascaded system, consider the collimator/ radiometer described by Howell (1979). The collimator has a measured calibration matrix

$$[M] = \begin{bmatrix} 0.8488 & -0.0503 & 0.0294 & 0.0617 \\ -0.0503 & 0.8304 & 0.0913 & -0.0920 \\ 0.0294 & 0.0913 & 0.8277 & 0.0000 \\ 0.0617 & -0.0920 & 0.0000 & 0.7947 \end{bmatrix} \tag{3.5}$$

which is not physically realisable, as verified by passing elliptically polarised light with Stokes vector

$$g = (1,\ -0.5877,\ 0,\ 0.809) \tag{3.6}$$

The collimated light then has a degree of polarisation of 1.05. From this, we conclude that [M] does not belong to SRM(4,R), a result confirmed by converting to covariance form and diagonalising. The eigenvalue spectrum is

$$\lambda = (1.665,\ 0.148,\ 0.007,\ -0.113) \tag{3.7}$$

which again shows a negative eigenvalue as the source of the problem. We can filter this matrix and make it realisable by subtracting the eigenvector corresponding to this negative eigenvalue, a result which yields the new calibration matrix

$$[M]_{COLL} = \begin{bmatrix} 0.9054 & -0.0498 & 0.0290 & 0.0621 \\ -0.0498 & 0.8254 & 0.0572 & -0.0473 \\ 0.0290 & 0.0573 & 0.7938 & -0.0296 \\ 0.0621 & -0.0473 & -0.0295 & 0.7771 \end{bmatrix} \tag{3.8}$$

The light from the collimator is then passed through a radiometer system, the measured calibration matrix for which has the form

$$[M] = \begin{bmatrix} 0.8886 & -0.0131 & 0.0055 & 0.0786 \\ -0.0115 & 0.5762 & -0.2820 & -0.1668 \\ 0.0048 & -0.2809 & 0.6825 & 0.0026 \\ 0.0775 & -0.1672 & 0.0012 & 0.8061 \end{bmatrix} \tag{3.9}$$

This too lies outside SRM(4,R) with an eigenvalue spectrum of

$$\lambda = (1.482,\ 0.399,\ 0.176,\ -0.279) \tag{3.10}$$

Again we have a single negative eigenvalue implying that for some input states the output will have a degree of polarisation greater than unity (as for the collimator there are many input states satisfying this condition, an example being

$$g = (1,\ -0.509,\ 0.294,\ 0.809) \tag{3.11}$$

which yields a degree of polarisation of 1.07). Again we may filter this matrix and make it realisable by subtracting the eigenvector corresponding to the negative eigenvalue. The resulting matrix does belong to SRM(4,R) and is given by

$$[M]_{RAD} = \begin{bmatrix} 1.028 & -0.014 & 0.004 & 0.078 \\ -0.013 & 0.603 & -0.157 & -0.111 \\ 0.004 & -0.156 & 0.637 & 0.044 \\ 0.077 & -0.111 & 0.043 & 0.685 \end{bmatrix} \tag{3.12}$$

The overall system matrix is then found as the product of the two filtered components ie. $[M]_{SYS} = [M]_{RAD}, [M]_{COLL}$, given by

$$[M]_{SYS} = \begin{bmatrix} 0.936 & -0.066 & 0.029 & 0.125 \\ -0.054 & 0.495 & -0.087 & -0.111 \\ 0.032 & -0.094 & 0.496 & 0.023 \\ 0.119 & -0.126 & 0.010 & 0.541 \end{bmatrix} \tag{3.13}$$

which is guaranteed to belong to SRM(4,R) and hence be physically realisable, its eigenvalue spectrum is

$$\lambda = (1{,}256,\ 0.353,\ 0.197,\ 0.067) \tag{3.14}$$

Note that forming the product of the unfiltered matrices simply yields another matrix which lies outside SRM(4,R). This matrix has been published by Howell and has the form

$$[M] = \begin{bmatrix} 0.760 & -0.062 & 0.029 & 0.119 \\ -0.057 & 0.469 & -0.181 & -0.186 \\ 0.038 & -0.171 & 0.539 & 0.028 \\ 0.124 & -0.217 & -0.012 & 0.661 \end{bmatrix} \tag{3.15}$$

with a corresponding eigenvalue spectrum of

$$\lambda = (1.237,\ 0.353,\ 0.123,\ -0.194) \tag{3.16}$$

As an example of a matrix with multiple negative eigenvalues, consider the matrix published by Kostinski (1988) to illustrate their optimisation scheme.

$$[M] = \begin{bmatrix} 0.180 & 0.004 & -0.001 & -0.005 \\ 0.290 & 0.125 & -0.310 & -0.015 \\ -0.020 & -0.025 & -0.160 & -0.350 \\ 0.015 & 0.100 & -0.080 & -0.195 \end{bmatrix} \tag{3.17}$$

This matrix has an eigenvalue spectrum of

$$\lambda = (0.56,\ -0.29,\ 0.172,\ -0.082) \tag{3.18}$$

and shows a pair of negative eigenvalues, one of which has half the magnitude of the largest positive eigenvalue. We conclude that this matrix is well outside of SRM(4,R) and cannot be considered a 'reasonable' representation of its nearest neighbour in the set.

We make several observations from the above. Firstly, we see that physical realisability of [M] is dependent on the sign of the eigenvalues of the CM. Further, by using the SM, we can isolate the nonphysical components and hence filter them from further consideration (whether the remaining matrix elements form a faithful representation of the system is of course dependent on the magnitude of the negative eigenvalues, which ultimately depends on the quality of the experimental system). We may consider such a filtering process as a systematic method for choosing the nearest member of SRM(4,R) to our given matrix [M]. We can expect the fidelity of our filtered matrix to be inversely proportional to the distance between [M] and SRM(4,R). This we may measure by taking the ratio of maximum negative to maximum positive eigenvalue. For the collimator this is 0.068, for the radiometer, 0.189, and for the whole system 0.157. We may then say that the collimator matrix is better ie. nearer to SRM(4,R) than the radiometer matrix and that the overall system lies, as expected, midway between these two. Whether such values represent points close to or far away from SRM(4,R) depends of course on the application (we clearly want the ratio to be much less than unity) but this procedure does at least allow us to compare different systems, an important feature for radar calibration studies.

Finally we consider published radar data taken from a polarimetric SAR system operated by JPL in California, USA (Van Zyl 1987). The matrix is formed as an average of SRM's obtained from the CSM's for many pixels over the San Francisco area and has the form

$$[M] = \begin{bmatrix} 1.0000 & 0.0762 & 0.1399 & 0.0264 \\ 0.0762 & 0.7862 & 0.3832 & -0.0615 \\ 0.1399 & 0.3832 & -0.2302 & 0.0596 \\ 0.0264 & -0.0615 & 0.0596 & 0.4619 \end{bmatrix} \tag{3.19}$$

This has an eigenvalue spectrum of

$$\lambda = (1.384,\ 0.543,\ 0.072,\ 0) \tag{3.20}$$

and belongs to SRM(4,R) (as expected for a matrix calculated from averages over many CSM's). Note that, in calculating this spectrum, we have used transformations to account for the conjugate nature of radar backscatter (this involves forming the negative of the fourth row of [M] before converting to covariance form).

Although the SAR pixel size is relatively small, we can still expect several scattering centres per pixel and hence the pixel CSM obtained by the radar must be considered only a sample of a complex stochastic process, with averaging (of the CM or SRM not the CSM) needed before attempting image interpretation. The above matrix represents such an average and indicates that, while not equivalent to a single CSM, the urban area shows some degree of correlated polarisation response (the largest eigenvalue is over twice the magnitude of the second). The eigenvector of this dominant eigenvalue has a corresponding CSM of

$$[S] = \begin{bmatrix} 1.070 & 0.330 \\ 0.330 & 0.797(-168^{o}) \end{bmatrix} \tag{3.21}$$

which may be considered the 'dominant' scattering mechanism of this urban area. In the next section we consider the implications of this result for choosing optimum polarisation states.

## 4. OPTIMUM POLARISATION STATES

A key objective of radar polarimetry research is to use polarisation diversity to enhance or suppress reflections from scatterers, either in isolation or in contrast with some clutter background. The key idea that optimum polarisation states may be generic to radar targets was first developed by Kennaugh, Graves and Gent in the immediate post war years. A simple but practical example of optimum states is the use of circular polarisation for rain clutter suppression; the surprising conclusion of the work of these early researchers being that any target, whatever its shape and composition, has a set of optimum states, analagous to circular polarisation for spherical raindrops. There is however one important constraint on the targets considered, they must correspond to a single coherent scattering matrix [S]. Given this, the crosspolar nulls (all linear polarisations for spherical raindrops but only two pairs of orthogonal states in general) are obtained as eigenvectors of the Hermitian matrices $[S].[S]^+$ and $[S]^+.[S]$. In addition to these, copolar nulls (circular polarisation for spherical raindrops) were also generalised by derivation from the singular values of [S].

The problem with these ideas is of course the statistical nature of radar targets: far from being represented by a single coherent scattering matrix, most targets display a dynamic signature, with the elements of [S] remaining correlated for only milliseconds or less. If measurements are made on the order of these timescales then partially polarised waves result and the Stokes vector, rather than the coherent wave vector, is the more appropriate measure. A key question then remains as to the applicability of optimum polarisation concepts to stochastic targets.

We consider two main approaches to this problem: the first via direct analysis of the Stokes vector and Stokes reflection matrix. This path has been followed by several authors, most notably Cloude (1988), Kostinski (1988), Ioannidis and Hammers (1979) and Van Zyl (1987). These theories generally employ techniques for finding the extrema of a quadratic form and result in optimum states found as roots of a polynomial. The algebra can however become very heavy and there is no clear connection with the coherent theory of Graves and Kennaugh.

The second approach is to use analysis directly on the target covariance matrix, as developed for example by Swartz (1988) and Novak (1987). This latter technique results in an eigenvector solution to the problem and is easier to interface with standard multivariate statistical detection theory than the Stokes vector approach (see Novak 1987).

We leave details of all these techniques to the original references and concentrate instead on the connection between an eigenvector decomposition of the covariance matrix, the classical optimum state analysis of Graves and Kennaugh and these more recent optimisation schemes for stochastic targets.

We saw in the previous section that physical realisability is linked to the sign of eigenvalues of the covariance matrix. As mentioned earlier, it is important that negative eigenvalue contributions are removed before applying optimisation schemes. Failure to do so may result in optimum states that are not physically realisable.

The eigenvalue spectrum also yields important information on the applicability of optimisation techniques. At one extreme the CM will have only one nonzero eigenvalue ie. a spectrum of $(\lambda, 0, 0, 0)$ and the object has a single CSM representation with the optimisation of Graves and Kennaugh being appropriate. At the other extreme the matrix may have degenerate eigenvalues ie. a spectrum of $(\lambda, \lambda, \lambda, \lambda)$

and the object shows no correlated polarisation structure (ie. it returns a randomly polarised wave for all incident wave states). In this case there is no point in using optimisation theory, as all states are degenerate and polarimetry not applicable.

In general of course the scattering object will lie somewhere between these two extremes. As an example consider again the SAR data obtained for the urban San Francisco area. Van Zyl (1987) used this matrix to illustrate his optimisation routine based on the SRM. He determines the optimum copolar transmit state as

$$g = (1,\ 0.919,\ 0.3941,\ -0.012) \tag{5.3}$$

with a corresponding copolar signal strength of 2.141 (being 12% larger than that obtained using HH).

By examination of the eigenvalue spectrum (equation 3.20), we see a dominant target component corresponding to an eigenvalue of 1.384 with a second eigenvalue some 4dB smaller. As a useful, but suboptimum, estimate of the maximum copolar signal strength, we might try using the Kennaugh eigenstates of the CSM corresponding to the dominant eigenvector (see 3.21). The scattering matrix corresponding to the largest eigenvalue has orthogonal eigenstates

$$g = (1,\ \pm 0.922,\ \pm 0.345,\ \pm 0.171) \tag{5.4}$$

which yields a copolar signal strength of 2.118 (within 1% of the optimum). A similar procedure may be applied if the minimum copolar signal is required; in which case the copolar nulls, not the Kennaugh eigenstates, are chosen as optimum.

Although suboptimal, this eigenvector approach to optimisation of copolar signal strength has a clear physical interpretation (ie. to maximise or minimise the return from the dominant target component of [T]) and further, in the special case when [M] corresponds to a single matrix [S], the TD approach becomes identical to the classic solution outlined by Kennaugh and Graves.

A completely different approach to the optimisation problem is to consider not the return from a single target, but the signal to clutter ratio when the desired object is found against some clutter background. Ioannidis and Hammers (1979) considered this problem using a Mueller matrix formulation and, via a Lagrange multiplier technique, derived a set of optimum states for maximising the signal to clutter ratio. Cloude (1987) has developed a simpler optimisation scheme for maximising the total backscattered power signal to clutter ratio, again based on the Mueller matrix. Recently, Swartz, Yueh, Kong, Novak and Shin (1987) have produced an important solution of this problem based on a covariance matrix representation. The relative simplicity of their solution when compared with that of Ioannidis and Hammers illustrated again the usefulness of a covariance formulation of target scattering.

CONCLUSIONS

Figure 2 shows a flow chart summarising key steps involved in the measurement and analysis of polarisation information. The measurement of matrix descriptors proceeds either via the coherent approach of a dual channel receiver with fast switching of the transmitter between two orthogonal states to determine the coherent scattering matrix, or the partially coherent approach of real time measurement of the Stokes vector with fast switching of the transmitter through four states to determine the Stokes Reflection matrix. We have argued for the latter approach on the grounds that stochastic targets have important correlation properties which a coherent receiver is in danger of missing (unless measurements can be guaranteed on time scales much less than decorrelation times). With this in mind we described a single channel Stokes receiver system, capable of fast measurement and better accuracy than conventional parallel detector designs.

With raw target matrix information obtained, the next step is common to both schemes: determination of the target covariance matrix (if real time measurements are available then the CM may be formed for varying integration period to form a matrix of correlation functions rather than simple correlation values). The CSM is related to the CM via the 2x2 Pauli matrices $\sigma$, while the SRM is transformed into the CM via the 16 $\beta$ matrices of SU(4).

The eigenvalues and eigenvectors of the CM then provide the basis for decisions on analysis. The eigenvalue spectrum yields information on the effects of experimental error by revealing nonphysical components and isolating them, allowing a filtering scheme to be used to guarantee physical realisability. To illustrate, published data on optical and radar systems is used to show how such filtering is invariably required of real measurements. We also stress the importance of such a procedure for calibration of radar systems. A suitable scheme might be the measurement of matrices for a collection of calibration targets (a set of three corner reflectors will suffice for radar backscatter systems) and, by looking at the magnitude of the second and thid eigenvalues, systems may be compared for fidelity and noise immunity.

The ratio of positive eigenvalues gives information on the optimisation approach best suited to the target. The optimisation techniques considered vary in complexity and applicability and it is only by looking at the eigenvalues of the CM that the appropriate choice may be made. If the eigenvalue spectrum shows a dominant component then the classical analysis of Graves and Kennaugh may be the most suitable. If, on the other hand, the spectrum is degenerate then optimisation schemes are inappropriate.

The eigenvectors of the CM may be used to establish optimum polarisation states using either the theory of Kennaugh and Graves or the more recent techniques employing the SRM or CM.

REFERENCES

H.G. Berry "Measurement of the Stokes parameters of light". Applied Optics Vol.16, 1977, p3200.

W. Bickel and W.M.Bailey "Stokes vectors, Mueller matrices and polarised scattered light". American Journal of Physics, Vol.53 (5) May 1985, pp468-478.

G.R. Boyer et al "Automatic measurement of the Stokes vector of light". Applied Optics Vol. 18, 1979, p1217.

P.J.B. Clarricoats and G.T. Poulton "High Efficiency microwave reflector antennas - a review. Proc. IEEE Vol.65, No.10, Oct. 1977, pp1470-1504.

S. Cloude "Group Theory and polarisation Algebra". OPTIK Vol.75, No.1, 1986, pp26-36.

S. Cloude "Polarimetry: the characterisation of polarisation effects in EM scattering". PhD. thesis, University of Birmingham, Oct.1986.

S. Cloude "Optimisation of Signal to Clutter using polarisation Diversity". Electronics Letters, Vol.24, No.4, Feb. 1988, pp194-195.

S. Cloude "Uniqueness of Target Decomposition Theorems in Radar Polarimetry", in Direct and Inverse methods in Radar Polarimetry, Eds. W.M. Boerner et al, D. Reidel Publ. Comp. Dordrecht 1989.

E.S. Fry and G.W. Kattawar "Relationships between the elements of the Stokes matrix". Applied Optics, Vol. 20, pp2811-2814, 1981.

H. Gent "Elliptically Polarised Waves and their Reflection from Radar targets". TRE memo No. 584, Telecommunications Research Establishment, Gt. Malvern, Worcs. England, 1954.

D. Giuli "Polarisation Diversity in Radars". Proc. IEEE, Vol.74, No.2, Feb. 1986, pp245-269.

C.D. Graves "Radar Power polarisation Scattering matrix". Proc. IRE Feb. 1956, pp248-257.

J.D. Houston and A.I. Caswell "Four component polarisation measurements of Lidar atmospheric scattering". Applied Optics Vol.17, No.4, Feb. 1978, pp614-620.

B.J. Howell "Measurement of the polarisation effects of an instrument using partially polarised light". Applied Optics, Vol.18, 1979, pp809-812.

J.R. Huynen "Phenomenological theory of radar targets". Ph.D. thesis, Drukkery Bronder-Offset, N.V. Rotterdam 1970.

G. Ioannidis and D. Hammers "Optimum antenna polarisations for target discrimination in clutter". Proc. IEEE AP-27 No.3, May 1979, pp357-363.

G.E. Jellison and D.H. Lowndes "Time Resolved Ellipsometry". Applied Optics Vol.24, No.17, Sept. 1985, pp2948-2955.

J.C.Kemp "Piezo-optical birefringence modulators". Journal of Opt. Soc. Am. Vol.8, 1969, p1217.

E.M.Kennaugh "Polarisation Properties of Radar reflections". M.Sc. thesis, Ohio State University, 1952.

A.B. Kostinski and W.M. Boerner "On the Polarimetric contrast optimisation". Proc. IEEE AP-35 No.8, Aug. 1987.

A.B. Kostinski, B.D.James and W.M. Boerner "Optimal Reception of Partially Polarised Waves". Journal of Opt. Soc. Am. Vol.5 1988, p58.

L.M. Novak, M.B. Sechtin and M. Cardullo "Studies of target detection algorithms which use polarimetric radar data". Proc. 21st ASILOMAR Conf. on Signals, Systems and Computers, Pacific Grove CA, USA, Nov. 1987.

A. Swartz, H.A. Yueh, J.A. Kong, L.M. Novak and R.T. Shin "The Optimal Polarisations for Achieving Maximum Contrast in Radar Images". To appear in Journal of Geophysical Research, Solid Earth and Planets, 1988.

D. Talmage and P.J. Curran "Remote sensing using partially polarised light". Int. Journal of Remote Sensing, Vol.7, No.1, pp47-64.

R.C. Thompson, J.R.Bottiger and E.S. Fry "Measurement of Polarised Light interactions via the Mueller matrix". Applied Optics Vol.19, No.8, 1980, pp1323-1332.

J.J. van Zyl, C. Papas and C. Elachi "On the optimum Polarisations of incoherently reflected waves". Proc. IEEE AP-35 No.7, July 1987, pp818-825.

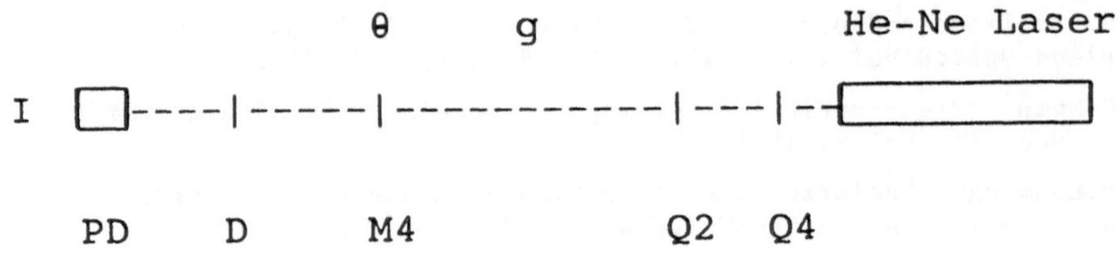


Q4 - Quartz quarter wave plate

Q2 - Quartz half wave plate

M4 - Spinning Mica quarter wave plate

D - HN22 Polaroid

PD - Photodiode detector

$$I = 2g_o + g_1(1+\cos 4\theta) + g_2 \sin 4\theta + 2g_3 \sin 2\theta$$

Figure 1 : Real Time Stokes Analyser

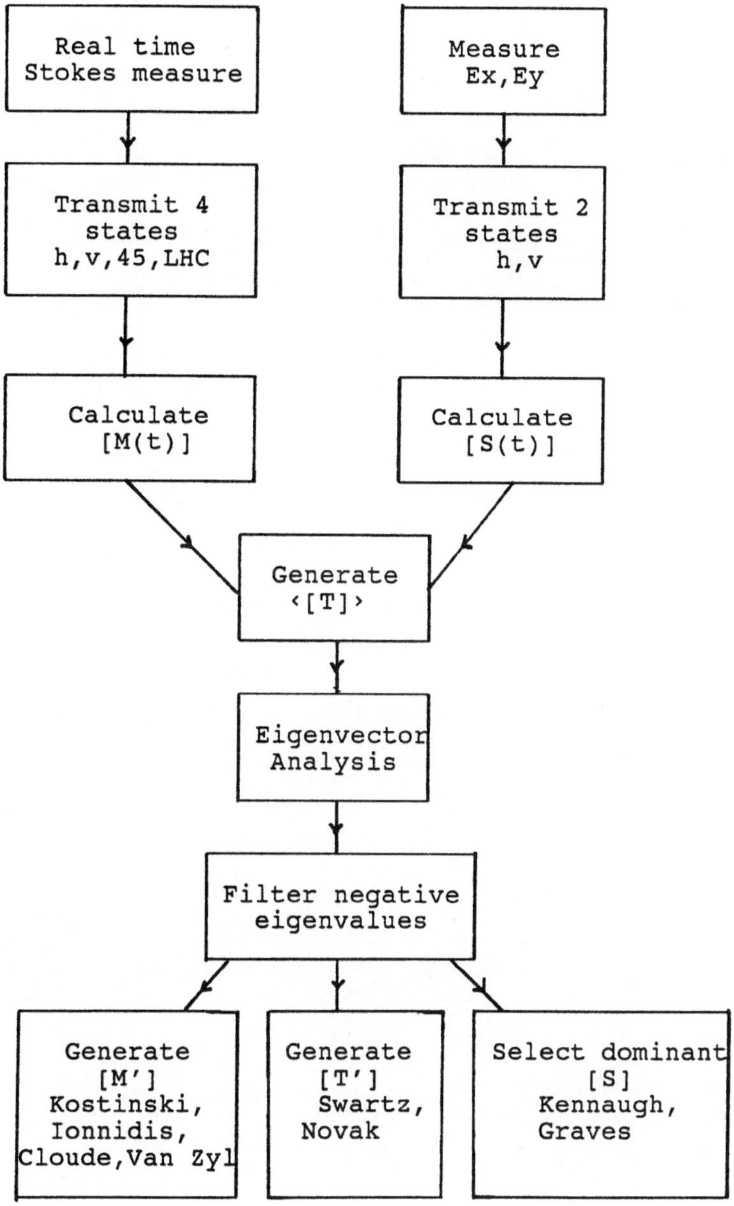


Figure 2: Flow diagram for Polarimetric Analysis

# A Complete Error Model for Polarimetric RCS- and Antenna Measurements

W. Wiesbeck
Institut für Höchstfrequenztechnik und Elektronik
University Karlsruhe
Kaiserstr. 12
7500 Karlsruhe
Federal Republic of Germany

**Abstract.** Polarimetric Radar and antenna measurements are influenced by numerous errors for example the frequency response, the coupling between the transmit channels, the coupling between the receive channels, the coupling from transmit to receive and finally the residual reflections of the environment. This paper presents an error model, based on signal flow graphs, that incorporates all possible errors. From the signal flow graph an error matrix is deduced. Based on the error model, a calibration procedure is developed, which determines the complex error coefficients of the error matrix. Approximate solutions and a complete solution are shown.

## 1. Introduction

In microwave measurements it is usual to describe the signal flow in systems with signal flow charts. These flow graphs can also be used to describe the influences of unwanted signals i.e. error signals in RF- and microwave systems. This technique has widely been used in the past for configurations with guided waves as a basis for error correction and system calibration. For systems implying free space propagation, signal flow graphs were used rather seldom because of the difficulty to determine the error paths. This was the reason why RCS- and antenna measurements suffer up to now proper calibration and error correction. With the introduction of polarimetric measurements the application of systematic error correction became inevitable. Starting with simple flow charts, derived from guided wave configurations, increasingly complex flow charts were developed. This



*W.-M. Boerner et al. (eds.), Direct and Inverse Methods in Radar Polarimetry, Part 1*, 793–807.

paper describes the first time a complete, but nevertheless rather schematic signal flow graph for fully polarimetric RF- and microwave systems. These flow graphs are well suited for the development of the proper algorithms for error correction and calibration [1]. High level calibration techniques containing all systematic errors, as introduced e.g. in [2], can advantageously be derived and also illustrated by means of signal flow charts. Moreover they clearly show the schematic of the system and the points of influence for errors like spurious signals, reflections or coupling.

## 2. Schematic Measurement System Configuration

Polarimetric RCS- [2] and RCS type antenna measurements [3] require arbitrary but orthogonal polarized antennas for transmit and receive, in total four channels. Fig. 1 shows a schematic diagram of a coherent multifrequency measurement system or CW-Radar.

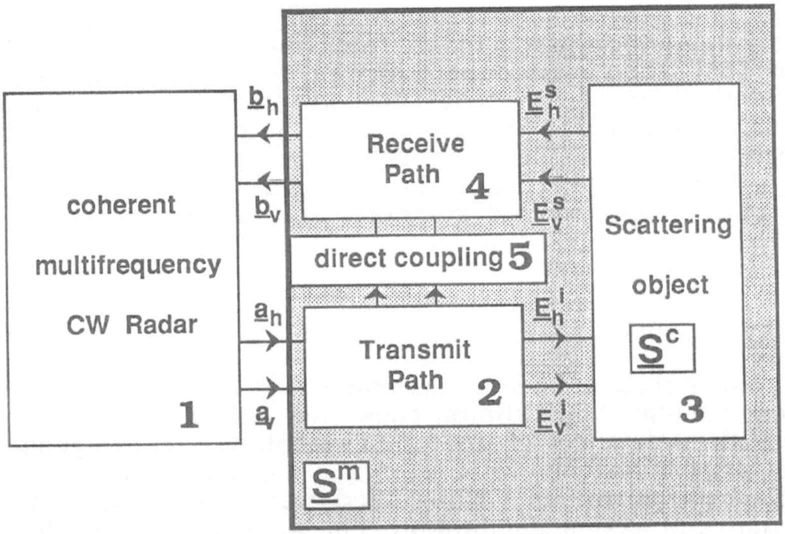


Fig. 1 Schematic diagram of the hardware RCS measurement system configuration

The measurement equipment is completely housed in box **1**. The object of interest is shown by **3**. The transmit and receive paths are represented by **2** and **4**, respectively. The system is polarimetric and the orthogonal waves are linear horizontal and vertical in this case, but any other arbitrary orthogonal polarizations may be used without affecting the principles of this paper. Since the regarded RCS measurement equipments or Radars are measuring voltage or power waves ( $\sqrt{\text{Watt}}$ ), it is suitable to describe the objects scattering behavior by a complex polarization scattering matrix $[\underline{S}^c]$, [4, 5]. The measurement equipment **1** has no direct access to the target. As can be seen in fig. 1 the radar system is only able to measure the transmitted power waves $a_h$, $a_v$ and the received waves $b_h$ and $b_v$, where h, v denote horizontal or vertical polarization respectively. The power waves are complex quantities. The measured scattering matrix $[\underline{S}^m]$, results from the incident and reflected power waves as follows:

$$\begin{bmatrix} \underline{b}_h \\ \underline{b}_v \end{bmatrix} = \begin{bmatrix} \underline{S}^m_{hh} & \underline{S}^m_{hv} \\ \underline{S}^m_{vh} & \underline{S}^m_{vv} \end{bmatrix} \begin{bmatrix} \underline{a}_h \\ \underline{a}_v \end{bmatrix} \tag{1}$$

## 3. Errors Introduced in Polarimetric Systems

Due to the influences on the transmit and receive paths the measured matrix (1) is subject to a number of error signals as there are in detail:

- errors induced by the frequency response and mismatches in hardware equipment like cables, switches, antennas, couplers as well as free space propagation, **2**, **4**,
- coupling in the receive channels, i. e. mainly in the antennas but also in the further microwave parts, **4**,
- coupling in the transmit channels, **2**,
- coupling from the transmit to the receive paths, **5**,
- residual reflections in the anechoic chamber, **5**.

These errors are in fig. 1 part of the blocks **2**, **4**,and **5**. The above transmit, receive and coupling errors may also be represented by their scattering matrices. This leads to a relation between the measured scattering matrix $[\underline{S}^m]$, and the correct target scattering matrix $[\underline{S}^c]$, see fig. 1:

$$[\underline{S}^m] = [\underline{I}] + [\underline{R}] * [\underline{S}^c] * [\underline{T}] \tag{2}$$

or showing all components

$$\begin{bmatrix} \underline{S}^m_{hh} & \underline{S}^m_{hv} \\ \underline{S}^m_{vh} & \underline{S}^m_{vv} \end{bmatrix} = \begin{bmatrix} \underline{I}_{hh} & \underline{I}_{hv} \\ \underline{I}_{vh} & \underline{I}_{vv} \end{bmatrix} + \begin{bmatrix} \underline{R}_{hh} & \underline{R}_{hv} \\ \underline{R}_{vh} & \underline{R}_{vv} \end{bmatrix} \begin{bmatrix} \underline{S}^c_{hh} & \underline{S}^c_{hv} \\ \underline{S}^c_{vh} & \underline{S}^c_{vv} \end{bmatrix} \begin{bmatrix} \underline{T}_{hh} & \underline{T}_{hv} \\ \underline{T}_{vh} & \underline{T}_{vv} \end{bmatrix} \tag{3}$$

Like fig. 1, the mathematical representation (3) shows that the true, correct target scattering matrix $[\underline{\mathbf{S}}^c]$ is erroneous by the additive matrix $[\underline{\mathbf{I}}]$ and the multiplicative matrices $[\underline{\mathbf{R}}]$ and $[\underline{\mathbf{T}}]$. The measured scattering matrix $[\underline{\mathbf{S}}^c]$ is subject to twelve error components, the isolation errors $\underline{I}_{\xi\eta}$, the transmit errors $\underline{T}_{\xi\eta}$ and the receive errors $\underline{R}_{\xi\eta}$, four each.

## 4. Graphical Representation of the Error Model

Equation (3) is the mathematical description for the polarimetric error model. With this equation it is still difficult to trace the physical error paths. A convenient way to show the relations between the measured matrix $[\underline{\mathbf{S}}^m]$ and the correct matrix $[\underline{\mathbf{S}}^c]$ is to describe the different signal and error paths in a flow chart according to Mason [1]. Problems arise with the graphical representation because of the many crossings of the flow graphs in a polarimetric system. Using a two dimensional representation would result in a badly arranged view graph that is not suitable for physical interpretations. A three dimensional arrangement of the signal flow lines yields in the surprisingly regular structure shown in fig. 2, where all signal paths lie on the edges or diagonals of a cube. This three dimensional description, defined as the *Error Cube*, is convincing from a physical point of view as will be demonstrated in the following. The polarimetric input and output waves to the *Error Cube*, representing $[\underline{\mathbf{S}}^m]$, $a_h$ , $a_v$ and $b_h$ , $b_v$ respectively, are located at the four front corners of the cube. The front plane of the error cube in fig. 2 describes the *Radar plane*. This front side contains the isolation errors $\underline{I}_{\xi\eta}$ (equal indices are for co-polarized, unequal indices for cross-polarized coupling), resulting from direct coupling between transmit and receive channels and residual reflections of the surroundings. The top plane is the *horizontal plane*, the bottom plane the *vertical plane* carrying only signal paths with horizontal and vertical polarization respectively. The transmit and receive paths (denoted $\underline{T}_{\xi\eta}$ and $\underline{R}_{\xi\eta}$ ) form the side walls of the cube, defined as the *transmit plane* and the *receive plane*. They include the wanted, co-polarized, direct signals as edges of the cube and the undesired, cross-polarized signal paths in the diagonals. These diagonals connect the vertical plane and the horizontal plane. The back wall of the cube represents the *target plane* with the correct target scattering matrix $[\underline{\mathbf{S}}^c]$. Incident to the target plane are the transmitted, co-polarized and the cross-polarized signals of the transmit plane. The reflected signals propagate in the receive plane. The co-polarized scattering coefficients $\underline{S}_{vv}$ and $\underline{S}_{hh}$ form the top and the bottom line respectively of the target plane , while the cross-polarized scattering coefficients

$\underline{S}_{vh}$ and $\underline{S}_{hv}$ form the diagonals. It has to be noted that all graphs are unidirectional in their planes.

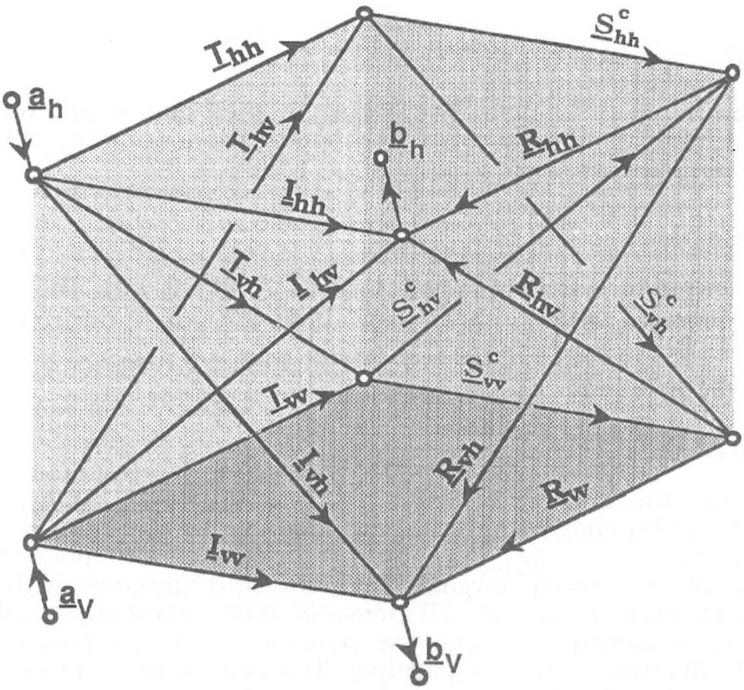


Fig 2 The signal flow graph for polarimetric RCS- and RCS type antenna measurements arranged as the *Error Cube*

## 5. Determination of the Target Scattering Matrix

As already mentioned, the Radar is not able to measure the complex target scattering matrix $[\underline{\mathbf{S}}^c]$, but only the matrix $[\underline{\mathbf{S}}^m]$. This is evident in fig. 1 and in equations (2) and (3). The determination of the unknowns in equation (3) as there are the 12 matrix coefficients $\underline{I}_{\xi\eta}$, $\underline{R}_{\xi\eta}$, $\underline{T}_{\xi\eta}$ needs a further investigation in the error cube as (3) is nonlinear (matrix product) and a simple inversion is not successful, as will be shown in the following. To solve for the correct target scattering matrix $[\underline{\mathbf{S}}^c]$ (3) has to be rearranged according to (4).

$$[\underline{\mathbf{S}}^c] = [\underline{\mathbf{R}}]^{-1} * \{[\underline{\mathbf{S}}^m] - [\underline{\mathbf{I}}]\} * [\underline{\mathbf{T}}]^{-1} \quad (4)$$

The determination of the unknowns is based on the measurement of reference targets for which the scattering matrices $[\underline{\mathbf{S}}^c]$ are well known. It is recommended to start with the simplest object, the empty room to determine the matrix elements of $[\underline{\mathbf{I}}]$ as $[\underline{\mathbf{S}}^c] = 0$ in this case.

$$[\underline{\mathbf{S}}^m] = [\underline{\mathbf{I}}] \tag{5}$$

The remaining 8 unknowns $\underline{R}_{\xi\eta}, \underline{T}_{\xi\eta}$ seem they can result from two additional measurements with two calibration targets. With the substitution

$$[\underline{\mathbf{M}}] = [\underline{\mathbf{S}}^m] - [\underline{\mathbf{I}}] \tag{6}$$

for the two additional calibration targets 1 and 2 with their known scattering matrices $[\underline{\mathbf{S}}^{c1}]$ and $[\underline{\mathbf{S}}^{c2}]$ equations (7) can be derived.

$$[\underline{\mathbf{M}}^1] = [\underline{\mathbf{R}}] * [\underline{\mathbf{S}}^{c1}] * [\underline{\mathbf{T}}] \tag{7a}$$

$$[\underline{\mathbf{M}}^2] = [\underline{\mathbf{R}}] * [\underline{\mathbf{S}}^{c2}] * [\underline{\mathbf{T}}] \tag{7b}$$

With the substitution (6) and the result from (5) $[\underline{\mathbf{M}}^1]$ and $[\underline{\mathbf{M}}^2]$ are known. It can be shown that it is not possible to solve the equations (7) for the unknowns $\underline{R}_{\xi\eta}, \underline{T}_{\xi\eta}$. A solution is even not possible with three or more calibration targets. This can be understood with a look at the error cube in fig. 2. All possible measurements result in a product of transmit and receive parameters $\underline{R}_{\xi\eta} * \underline{T}_{\xi\eta}$, it is not possible to determine them separately. Therefore it is advantageous to reorder equation (3). From multiplication of the matrix products in (3) and ordering the square matrix in vector form, ($[\underline{\mathbf{M}}]$, $[\underline{\mathbf{S}}^c]$, $[\underline{\mathbf{S}}^m]$, $[\underline{\mathbf{I}}]$ are vectors now), the following expression results:

$$\begin{bmatrix} \underline{S}^m_{vv} - \underline{I}_{vv} \\ \underline{S}^m_{hh} - \underline{I}_{hh} \\ \underline{S}^m_{vh} - \underline{I}_{vh} \\ \underline{S}^m_{hv} - \underline{I}_{hv} \end{bmatrix} = \begin{bmatrix} \underline{M}_{vv} \\ \underline{M}_{hh} \\ \underline{M}_{vh} \\ \underline{M}_{hv} \end{bmatrix} = \begin{bmatrix} \underline{R}_{vv}\underline{T}_{vv} & \underline{R}_{vh}\underline{T}_{hv} & \underline{R}_{vh}\underline{T}_{vv} & \underline{R}_{vv}\underline{T}_{hv} \\ \underline{R}_{hv}\underline{T}_{vh} & \underline{R}_{hh}\underline{T}_{hh} & \underline{R}_{hv}\underline{T}_{hh} & \underline{R}_{hh}\underline{T}_{vh} \\ \underline{R}_{vv}\underline{T}_{vh} & \underline{R}_{vh}\underline{T}_{hh} & \underline{R}_{vv}\underline{T}_{hh} & \underline{R}_{vh}\underline{T}_{vh} \\ \underline{R}_{hv}\underline{T}_{vv} & \underline{R}_{hh}\underline{T}_{hv} & \underline{R}_{hv}\underline{T}_{hv} & \underline{R}_{hh}\underline{T}_{vv} \end{bmatrix} \begin{bmatrix} \underline{S}^c_{vv} \\ \underline{S}^c_{hh} \\ \underline{S}^c_{vh} \\ \underline{S}^c_{hv} \end{bmatrix} \tag{8}$$

This equation is the basis for the determination of the propagation and coupling influences to the measurement system. Equation (8) can be written in the forms of equations (9) and (10) that have been used in [6]:

$$[\underline{M}] = [\underline{C}] * [\underline{S}^c] \tag{9}$$

The explicit formulation of equation (9) introduces the formal coefficients $\underline{c}_{ij}$:

$$\begin{bmatrix} \underline{S}^m_{vv} - \underline{c}_{10} \\ \underline{S}^m_{hh} - \underline{c}_{20} \\ \underline{S}^m_{vh} - \underline{c}_{30} \\ \underline{S}^m_{hv} - \underline{c}_{40} \end{bmatrix} = \begin{bmatrix} \underline{M}_{vv} \\ \underline{M}_{hh} \\ \underline{M}_{vh} \\ \underline{M}_{hv} \end{bmatrix} = \begin{bmatrix} \underline{c}_{11} & \underline{c}_{12} & \underline{c}_{13} & \underline{c}_{14} \\ \underline{c}_{21} & \underline{c}_{22} & \underline{c}_{23} & \underline{c}_{24} \\ \underline{c}_{31} & \underline{c}_{32} & \underline{c}_{33} & \underline{c}_{34} \\ \underline{c}_{41} & \underline{c}_{42} & \underline{c}_{43} & \underline{c}_{44} \end{bmatrix} \begin{bmatrix} \underline{S}^c_{vv} \\ \underline{S}^c_{hh} \\ \underline{S}^c_{vh} \\ \underline{S}^c_{hv} \end{bmatrix} \tag{10}$$

The coefficients $\underline{c}_{ij}$ are called error coefficients in the following. In equation (10) four error coefficients $\underline{c}_{i0}$ are additive, while 16 are multiplicative. The additive coefficients $\underline{c}_{i0}$ contain the isolation error and the residual reflections, they have been already determined with the empty room calibration in equation (5). The coefficients $\underline{c}_{ij}$ result from the transmit and receive scattering coefficients $\underline{T}_{\xi\eta}$ and $\underline{R}_{\xi\eta}$. Especially the elements in the diagonal of $[\underline{C}]$, $\underline{c}_{ii}$ represent the frequency response error of the system. The matrix elements $\underline{c}_{12}$, $\underline{c}_{21}$, $\underline{c}_{34}$ and $\underline{c}_{43}$ result from the cross-coupling in the two orthogonaly polarized channels. While in equation (3) the total number of unknowns is only 12, the number of unknown complex error coefficients is 16+4 = 20 in equation (10). The 16 error coefficients in the matrix $[\underline{C}]$ are the product of a receive coefficient $\underline{R}_{\xi\eta}$ and a transmit coefficient $\underline{T}_{\xi\eta}$. As already mentioned they are formed by only 8 transmission coefficients, but there is no simple relation for a separation of the $\underline{R}_{\xi\eta}$ and $\underline{T}_{\xi\eta}$ coefficients. On the other hand it is possible to represent the 16 unknown coefficients by seven out of them, because of their nonlinear relation. Starting from equation (10) step by step error coefficients $\underline{c}_{ij}$ can be eliminated by sequential substitution through other error coefficients $\underline{c}_{ij}$. Numerous solutions exist for this, but only a few are advantageous. The best solutions can be selected on the basis of experimental knowledge concerning the accuracy and magnitude of coefficients which can be determined by measurements. Having this knowledge one can set up a configuration of the error matrix $[\underline{C}]$ where the denominators of the dependent error coefficients are high in magnitude and with small errors in phase and amplitude, thus reducing possible sequential errors. Equation (11) gives one possible advantageous solution. Other solutions lead to larger sequential errors in the computation of the error coefficients $\underline{c}_{ij}$.

$$[\underline{C}] = \begin{bmatrix} \underline{c}_{11} & \dfrac{\underline{c}_{32}\underline{c}_{42}}{\underline{c}_{22}} & \dfrac{\underline{c}_{32}\underline{c}_{11}}{\underline{c}_{33}} & \dfrac{\underline{c}_{33}\underline{c}_{42}}{\underline{c}_{22}} \\ \dfrac{\underline{c}_{31}\underline{c}_{41}}{\underline{c}_{11}} & \underline{c}_{22} & \dfrac{\underline{c}_{33}\underline{c}_{41}}{\underline{c}_{11}} & \dfrac{\underline{c}_{31}\underline{c}_{22}}{\underline{c}_{33}} \\ \underline{c}_{31} & \underline{c}_{32} & \underline{c}_{33} & \dfrac{\underline{c}_{32}\underline{c}_{31}}{\underline{c}_{33}} \\ \underline{c}_{41} & \underline{c}_{42} & \dfrac{\underline{c}_{42}\underline{c}_{41}\underline{c}_{33}}{\underline{c}_{22}\underline{c}_{11}} & \dfrac{\underline{c}_{11}\underline{c}_{22}}{\underline{c}_{33}} \end{bmatrix} \quad (11)$$

This equations (11) leads to a homogeneous system of equations with 7 unknowns that is solvable. A solution with three linear independent calibration targets having the scattering matrices $[\underline{\mathbf{S}}^{c1}]$, $[\underline{\mathbf{S}}^{c2}]$ and $[\underline{\mathbf{S}}^{c3}]$ is shown in the following. The only condition imposed on the three calibration targets, not on the objects to measured later, is $\underline{S}_{vh} = \underline{S}_{hv}$.

For the three reference targets ( i= 1,2,3 ) the equations (12) result:

$$\begin{bmatrix} \underline{M}^{i}_{vv} \\ \underline{M}^{i}_{hh} \\ \underline{M}^{i}_{vh} \\ \underline{M}^{i}_{hv} \end{bmatrix} = \underbrace{\begin{bmatrix} \underline{c}_{11} & \underline{c}_{12} & (\underline{c}_{13} + \underline{c}_{14}) \\ \underline{c}_{21} & \underline{c}_{22} & (\underline{c}_{23} + \underline{c}_{24}) \\ \underline{c}_{31} & \underline{c}_{32} & (\underline{c}_{33} + \underline{c}_{34}) \\ \underline{c}_{41} & \underline{c}_{42} & (\underline{c}_{43} + \underline{c}_{44}) \end{bmatrix}}_{\text{(4x3)-matrix}} \begin{bmatrix} \underline{S}^{ci}_{vv} \\ \underline{S}^{ci}_{hh} \\ \underline{S}^{ci}_{vh} \end{bmatrix} \quad (12)$$

For the vv components results equation (13) from equations (12) :

$$\begin{bmatrix} \underline{M}^{1}_{vv} \\ \underline{M}^{2}_{vv} \\ \underline{M}^{3}_{vv} \end{bmatrix} = \underbrace{\begin{bmatrix} \underline{S}^{c1}_{vv} & \underline{S}^{c1}_{hh} & \underline{S}^{c1}_{vh} \\ \underline{S}^{c2}_{vv} & \underline{S}^{c2}_{hh} & \underline{S}^{c2}_{vh} \\ \underline{S}^{c3}_{vv} & \underline{S}^{c3}_{hh} & \underline{S}^{c3}_{vh} \end{bmatrix}}_{[\underline{V}]} \begin{bmatrix} \underline{c}_{11} \\ \underline{c}_{12} \\ \underline{c}_{13} + \underline{c}_{14} \end{bmatrix} \quad (13)$$

Accordingly for the hh components equation (14) can be derived from the equations (12):

$$\begin{bmatrix} \underline{M}^1_{hh} \\ \underline{M}^2_{hh} \\ \underline{M}^3_{hh} \end{bmatrix} = \begin{bmatrix} \underline{S}^{c1}_{vv} & \underline{S}^{c1}_{hh} & \underline{S}^{c1}_{vh} \\ \underline{S}^{c2}_{vv} & \underline{S}^{c2}_{hh} & \underline{S}^{c2}_{vh} \\ \underline{S}^{c3}_{vv} & \underline{S}^{c3}_{hh} & \underline{S}^{c3}_{vh} \end{bmatrix} \begin{bmatrix} \underline{c}_{21} \\ \underline{c}_{22} \\ \underline{c}_{23} + \underline{c}_{24} \end{bmatrix} \qquad (14)$$

The cross-polarized contributions from the measurements with the three reference targets result in equations (15) and (16). Up to now especially these contributions were difficult to determine.

$$\begin{bmatrix} \underline{M}^1_{vh} \\ \underline{M}^2_{vh} \\ \underline{M}^3_{vh} \end{bmatrix} = \begin{bmatrix} \underline{S}^{c1}_{vv} & \underline{S}^{c1}_{hh} & \underline{S}^{c1}_{vh} \\ \underline{S}^{c2}_{vv} & \underline{S}^{c2}_{hh} & \underline{S}^{c2}_{vh} \\ \underline{S}^{c3}_{vv} & \underline{S}^{c3}_{hh} & \underline{S}^{c3}_{vh} \end{bmatrix} \begin{bmatrix} \underline{c}_{31} \\ \underline{c}_{32} \\ \underline{c}_{33} + \underline{c}_{34} \end{bmatrix} \qquad (15)$$

$$\begin{bmatrix} \underline{M}^1_{hv} \\ \underline{M}^2_{hv} \\ \underline{M}^3_{hv} \end{bmatrix} = \begin{bmatrix} \underline{S}^{c1}_{vv} & \underline{S}^{c1}_{hh} & \underline{S}^{c1}_{vh} \\ \underline{S}^{c2}_{vv} & \underline{S}^{c2}_{hh} & \underline{S}^{c2}_{vh} \\ \underline{S}^{c3}_{vv} & \underline{S}^{c3}_{hh} & \underline{S}^{c3}_{vh} \end{bmatrix} \begin{bmatrix} \underline{c}_{41} \\ \underline{c}_{42} \\ \underline{c}_{43} + \underline{c}_{44} \end{bmatrix} \qquad (16)$$

The equations (13) to (16) can be solved for targets with $\det(\underline{V}) \neq 0$, i.e. if the calibration targets are linear independent and therefore also the line and column vectors of $[\underline{V}]$ are linear independent. The equations can be solved with Cramer's rule for the 7 primary error coefficients of equation (11). In the following the unknowns are determined step by step from the above equations. From equation (13) $\underline{c}_{11}$ can be derived:

$$\underline{c}_{11} = \frac{1}{\det(\underline{V})} \begin{vmatrix} \underline{M}^1_{vv} & \underline{S}^{c1}_{hh} & \underline{S}^{c1}_{vh} \\ \underline{M}^2_{vv} & \underline{S}^{c2}_{hh} & \underline{S}^{c2}_{vh} \\ \underline{M}^3_{vv} & \underline{S}^{c3}_{hh} & \underline{S}^{c3}_{vh} \end{vmatrix} \qquad (17)$$

The second error coefficient for co-polarization $\underline{c}_{22}$ results from equation (14):

$$\underline{c}_{22} = \frac{1}{\det(\underline{V})} \begin{vmatrix} \underline{S}^{c1}_{vv} & \underline{M}^{1}_{hh} & \underline{S}^{c1}_{vh} \\ \underline{S}^{c2}_{vv} & \underline{M}^{2}_{hh} & \underline{S}^{c2}_{vh} \\ \underline{S}^{c3}_{vv} & \underline{M}^{3}_{hh} & \underline{S}^{c3}_{vh} \end{vmatrix} \qquad (18)$$

The error coefficients determining the cross-polarization properties $c_{31}$ and $c_{32}$ can be extracted from equation (15):

$$\underline{c}_{31} = \frac{1}{\det(\underline{V})} \begin{vmatrix} \underline{M}^{1}_{vh} & \underline{S}^{c1}_{hh} & \underline{S}^{c1}_{vh} \\ \underline{M}^{2}_{vh} & \underline{S}^{c2}_{hh} & \underline{S}^{c2}_{vh} \\ \underline{M}^{3}_{vh} & \underline{S}^{c3}_{hh} & \underline{S}^{c3}_{vh} \end{vmatrix} \qquad (19)$$

$$\underline{c}_{32} = \frac{1}{\det(\underline{V})} \begin{vmatrix} \underline{S}^{c1}_{vv} & \underline{M}^{1}_{vh} & \underline{S}^{c1}_{vh} \\ \underline{S}^{c2}_{vv} & \underline{M}^{2}_{vh} & \underline{S}^{c2}_{vh} \\ \underline{S}^{c3}_{vv} & \underline{M}^{3}_{vh} & \underline{S}^{c3}_{vh} \end{vmatrix} \qquad (20)$$

$\underline{c}_{41}$ and $\underline{c}_{42}$ can be derived from equation (16):

$$\underline{c}_{41} = \frac{1}{\det(\underline{V})} \begin{vmatrix} \underline{M}^{1}_{hv} & \underline{S}^{c1}_{hh} & \underline{S}^{c1}_{vh} \\ \underline{M}^{2}_{hv} & \underline{S}^{c2}_{hh} & \underline{S}^{c2}_{vh} \\ \underline{M}^{3}_{hv} & \underline{S}^{c3}_{hh} & \underline{S}^{c3}_{vh} \end{vmatrix} \qquad (21)$$

$$\underline{c}_{42} = \frac{1}{\det(\underline{V})} \begin{vmatrix} \underline{S}^{c1}_{vv} & \underline{M}^{1}_{hv} & \underline{S}^{c1}_{vh} \\ \underline{S}^{c2}_{vv} & \underline{M}^{2}_{hv} & \underline{S}^{c2}_{vh} \\ \underline{S}^{c3}_{vv} & \underline{M}^{3}_{hv} & \underline{S}^{c3}_{vh} \end{vmatrix} \qquad (22)$$

The sum of $\underline{c}_{33} + \underline{c}_{34}$ follows from equation (15):

$$\underline{c}_{33} + \underline{c}_{34} = \frac{1}{\det(\underline{Y})} \begin{vmatrix} \underline{S}_{vv}^{c1} & \underline{S}_{hh}^{c1} & \underline{M}_{vh}^{1} \\ \underline{S}_{vv}^{c2} & \underline{S}_{hh}^{c2} & \underline{M}_{vh}^{2} \\ \underline{S}_{vv}^{c3} & \underline{S}_{hh}^{c3} & \underline{M}_{vh}^{3} \end{vmatrix} \tag{23}$$

In a comparison of $\underline{c}_{33}$ and $\underline{c}_{34}$ from equation (6) results:

$$\underline{c}_{33} = \underline{R}_{vv}\underline{T}_{hh} = \frac{\underline{c}_{31}\underline{c}_{32}}{\underline{c}_{34}} \tag{24}$$

With the substitution $\underline{b}_3 = \underline{c}_{33} + \underline{c}_{34}$ in equation (24) $\underline{c}_{33}$ results in :

$$\underline{c}_{33} = \frac{\underline{b}_3}{2} \pm \sqrt{\frac{\underline{b}_3^2}{4} - \underline{c}_{31}\underline{c}_{32}} \tag{25}$$

The sign in equation (25) can be determined by the condition $|\underline{c}_{33}| >> |\underline{c}_{34}|$, or from the phase angles $\angle\underline{b}_3 \approx \angle\underline{c}_{33}$. By this all independent error coefficients are determined, all others can be calculated from these 7 according to equation (11). With the known coefficients $\underline{c}_{ij}$ the scattering matrices $[\underline{\mathbf{S}}^m]$ of unknown targets can be measured and error corrected for the object related matrices $[\underline{\mathbf{S}}^c]$. Now restrictions concerning target characteristics are imposed. The matrix coefficients $\underline{S}_{\xi\eta}$ can be determined by equations (26) to (29):

$$\underline{S}_{vv}^{c} = \frac{1}{\det(\underline{C})} \begin{vmatrix} \underline{M}_{vv} & \underline{c}_{12} & \underline{c}_{13} & \underline{c}_{14} \\ \underline{M}_{hh} & \underline{c}_{22} & \underline{c}_{23} & \underline{c}_{24} \\ \underline{M}_{vh} & \underline{c}_{32} & \underline{c}_{33} & \underline{c}_{34} \\ \underline{M}_{hv} & \underline{c}_{42} & \underline{c}_{43} & \underline{c}_{44} \end{vmatrix} \tag{26}$$

$$\underline{S}_{hh}^{c} = \frac{1}{\det(\underline{C})} \begin{vmatrix} \underline{c}_{11} & \underline{M}_{vv} & \underline{c}_{13} & \underline{c}_{14} \\ \underline{c}_{21} & \underline{M}_{hh} & \underline{c}_{23} & \underline{c}_{24} \\ \underline{c}_{31} & \underline{M}_{vh} & \underline{c}_{33} & \underline{c}_{34} \\ \underline{c}_{41} & \underline{M}_{hv} & \underline{c}_{43} & \underline{c}_{44} \end{vmatrix} \tag{27}$$

$$\underline{S}^{c}_{vh} = \frac{1}{\det(\underline{C})} \begin{vmatrix} \underline{c}_{11} & \underline{c}_{12} & \underline{M}_{vv} & \underline{c}_{14} \\ \underline{c}_{21} & \underline{c}_{22} & \underline{M}_{hh} & \underline{c}_{24} \\ \underline{c}_{31} & \underline{c}_{32} & \underline{M}_{vh} & \underline{c}_{34} \\ \underline{c}_{41} & \underline{c}_{42} & \underline{M}_{hv} & \underline{c}_{44} \end{vmatrix} \tag{28}$$

$$\underline{S}^{c}_{hv} = \frac{1}{\det(\underline{C})} \begin{vmatrix} \underline{c}_{11} & \underline{c}_{12} & \underline{c}_{13} & \underline{M}_{vv} \\ \underline{c}_{21} & \underline{c}_{22} & \underline{c}_{23} & \underline{M}_{hh} \\ \underline{c}_{31} & \underline{c}_{32} & \underline{c}_{33} & \underline{M}_{vh} \\ \underline{c}_{41} & \underline{c}_{42} & \underline{c}_{43} & \underline{M}_{hv} \end{vmatrix} \tag{29}$$

From the equations (26) to (29) the errors can be calculated that are made if only unipolar scattering components are measured during calibration instead of performing a full polarimetric calibration. Especially the double coupling contributes direct to the co-polarization errors.

## 6. Graphical Representation of the Calibration Steps for the Determination of the Signal Paths in the Error Model

The above demonstrated theoretical solution for the determination of the error coefficients can also be shown with the signal flow graph, the *Error Cube*. In the following the calibration steps with three targets and the empty room (no target) are graphically represented. The characteristics of the selected targets are as follows:

- empty room (no target) $[\underline{S}^c] = [0]$,
- conducting sphere $\underline{S}_{vv} = \underline{S}_{hh}$ and $\underline{S}_{vh}, \underline{S}_{hv} = 0$,
- vertical dihedral corner reflector $\underline{S}_{vv} \neq \underline{S}_{hh}$ and $\underline{S}_{vh}, \underline{S}_{hv} = 0$,
- 45° inclined dihedral corner $\underline{S}_{vv}, \underline{S}_{hh} = 0$ and $\underline{S}_{vh}, \underline{S}_{hv} \neq 0$.

These above targets are useful for linear orthogonal polarizations, other arbitrary but orthogonal polarizations may require other calibration targets.

In fig. 3 the flow graph without a target is shown. As there is no target, there exist no transmit paths $\underline{T}_{\xi\eta}$ and no receive paths $\underline{R}_{\xi\eta}$. The four measured scattering parameters determine the $\underline{c}_{i0}$ coefficients, according to equation (10), in the Radar plane.

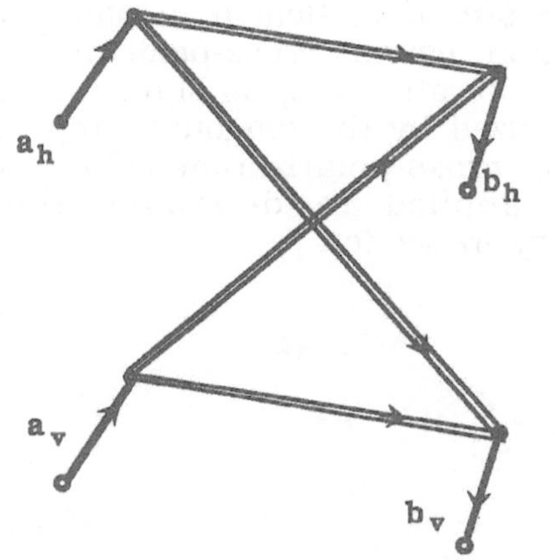


Fig. 3 Calibration of the Radar plane with the empty room

The next calibration step is performed with a sphere. As the sphere produces no cross-polarization the diagonals in the target plane do not show up in fig. 4. The identical flow graph describes the calibration with a vertical cylinder or dihedral corner reflector because of the identical conditions as listed above.

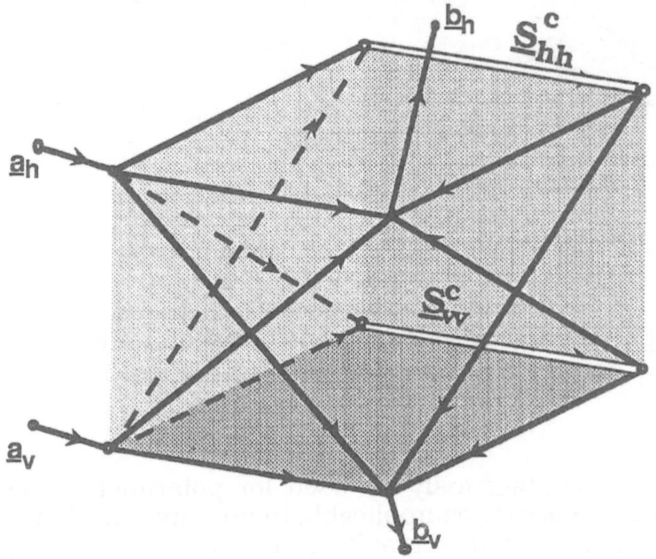


Fig. 4 Signal flow graph for a sphere as calibration target

Turning the dihedral corner 45° out of vertical in a plane perpendicular to path of wave propagation, causes cross-polarization, while the co-polarization is zero. This calibration step is shown in the flow graph in fig. 5 and it is characterized by the diagonals, representing the cross-polarization. With this cross-polarization calibration the necessary targets have been applied to determine the error coefficients according to the theory in section 5.

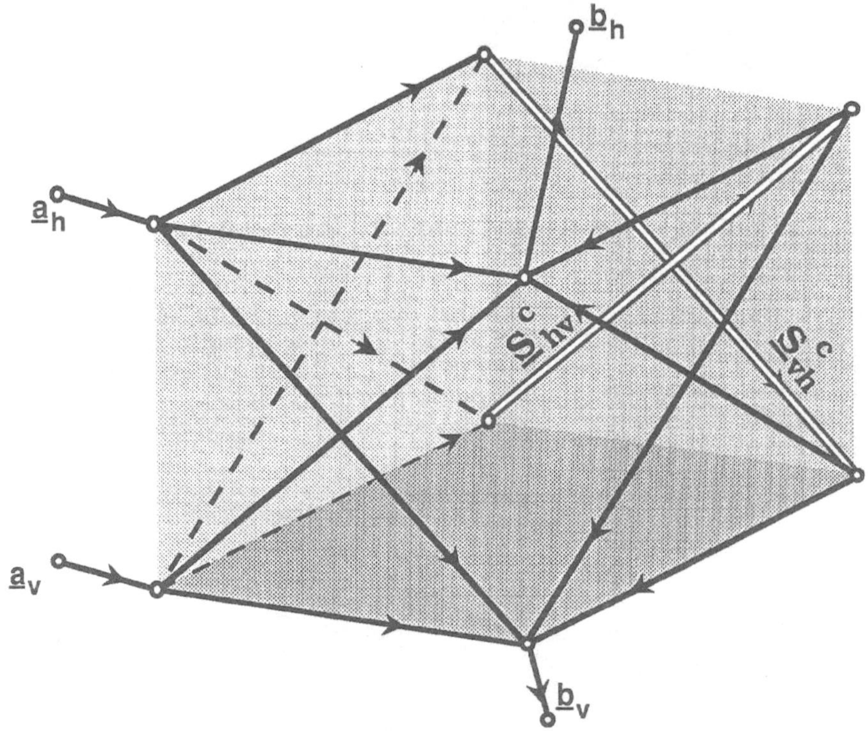


Fig. 5 Signal flow graph for targets with cross-polarization only

## 7. Conclusion

Signal flow graphs can advantageously be used for polarimetric free space transmission measurements, as applicable to antenna and Radar cross section measurements. The flow graphs can be represented in a three dimensional *Error Cube*. Although these graphs are rather complicated, they are a very useful tool for understanding the physical system behavior and for determining the places of error influence.

This enables one to improve the system characteristics and to select the proper calibration targets. In addition mathematical formulations for error correction and system calibration can be derived from the signal flow graphs. This has been shown for orthogonal linear polarizations, but the procedure can also be applied for any other orthogonal polarizations. In the laboratory polarimetric calibration has so significantly improved the accuracy of measurements that a reliable verification of theories in the area of scattering and Radar cross section computation in now no longer a problem.

## 7. References

[1] S. J. Mason, H. J. Zimmermann: "Electronic circuits, signals and systems", *J. Wiley*, New York, 1960

[2] G. Schöne, S. Riegger, E. Heidrich: "Wideband Polarimetric Radar Cross Section Measurement", *IEEE AP-S International Symposium & URSI Radio Science Meeting*, Syracuse, Conf. Proc. Vol. II, 537-540, June 1988

[3] E. Heidrich, W. Wiesbeck: "Features of advanced polarimetric RCS antenna measurements", *IEEE AP-S International Symposium & URSI Radio Science Meeting*, San Jose, Vol. II, 1026-1029, June 1989

[4] H. A. Zebker, J. J. Van Zyl, D. N. Held, "Imaging Radar polarimetry from wave synthesis," *Journal of Geophysical Research*, Vol. 92, No. B1, pp. 683-701, Jan. 1987

[6] S. Riegger, W. Wiesbeck: "Wide-Band Polarimetry and complex Radar Cross Section Signatures", *Proceedings of the IEEE*, Vol. 77, No. 5, pp. 649 - 658, May 1989

[6] S. Riegger: "Analyse und Synthese polarimetrischer, dispersiver Radarstreuquerschnitts-Signaturen", *Ph. D. thesis, University of Karlsruhe*, February 1988

## Acknowledgement

The author wishes to thank Dipl. Ing. Eberhardt Heidrich and Dr. Ing.. Sebastian Riegger for their extremely helpful contributions to this paper.

# AN EFFICIENT CALIBRATION SCHEME FOR THE FOUR-DETECTOR PHOTOPOLARIMETER USING A ROTATING QUARTER-WAVE RETARDER

R.M.A. Azzam
Department of Electrical Engineering
University of New Orleans
New Orleans, Lousiana 70148
USA

ABSTRACT: Rotating a quarter-wave retarder (QWR) in a linearly polarized light beam of fixed azimuth produces a wide range of elliptical polarization states that satisfy the condition of linear independence. A plurality of these states, obtained at equispaced angular positions of the QWR, are used to calibrate the four-detector photopolarimeter (FDP) leading to the determination of its instrument matrix $\underline{A}$. Experimental results demonstrate the application of this method to an FDP of 4 Si detectors at wavelength $\lambda$ = 632.8 nm. The use of a nonideal QWR with small but known imperfections is also discussed.

## 1. INTRODUCTION

Measurement of the state of polarization of light is essential to polarimetric remote sensing and its many applications in astrophysics, atmospheric sciences, physical and chemical analysis of surfaces, thin films, and of collections of scattering particles, etc. (1-12). Conventional optical polarimetry (13) is based on interposing polarizing optical elements (POEs) in the path of a light beam and measuring the emergent light flux using linear photodetection. For automated measurement of polarization, the POEs are typically subjected to some form of (photo-elastic, electro-optic, or magneto-optic) periodic modulation or mechanical rotation and the Stokes parameters of light are obtained by Fourier analysis of the detected signal.

Recently, we have introduced and implemented a novel scheme for measuring the general state of partial elliptical polarization of light, as represented by the four Stokes parameters $S_0$, $S_1$, $S_2$ and $S_3$, using an arrangement of four photodetectors (14-16). In this four-detector photopolarimeter (FDP), shown in Fig. 1, the light beam, whose state of polarization is to be measured, is relayed by partial specular reflection at oblique incidence from detector $D_0$ to $D_1$ to $D_2$ and the remaining radiation is captured by the last detector $D_3$. In Fig. 1, $p_i$ refers to the linear polarization direction parallel to the ith plane of incidence and $\alpha_1$, $\alpha_2$ are the angles with which plane of incidence rotates between successive reflections. It is essential that $\alpha_i$ (i = 1,2) $\neq$



W.-M. Boerner et al. (eds.), Direct and Inverse Methods in Radar Polarimetry, Part 1, 809–818.

0 or $90^{o}$, so that the light beam does not remain in one plane. The fraction of the radition absorbed by the kth detector, which is a function of the incident polarization, produces a proportional electrical signal $i_k$. Consequently, the output signal vector

$$\underline{I} = ( i_0 \quad i_1 \quad i_2 \quad i_3)^t \qquad (1)$$

of the FDP becomes linearly related,

$$\underline{I} = \underline{A}\ \underline{S}, \qquad (2)$$

to the input Stokes vector

$$\underline{S} = (S_0 \quad S_1 \quad S_2 \quad S_3)^t \qquad (3)$$

where t indicates the matrix transpose.

Explicit expressions have been derived (15) for the elements of the instrument matrix $\underline{A}$ in terms of the reflective detector surface parameters for a given light path, using the Mueller matrix calculus.

Practically, $\underline{A}$ is measured by calibration by recording the output signal vector $\underline{I}$ for a sufficient number of known input polarization states. When the minimum number of 4 states is used, the optimum choice of these 4 states corresponds to the vertices of a tetrahedron (maximum-volume pyramid) inscribed inside the Poincaré sphere (15, 16). However, to determine $\underline{A}$ accurately, the FDP is calibrated with a plurality of input states whose number is >>4. A procedure has already been described (17) for the determination of $\underline{A}$ using many linear states (represented by equispaced points along the equator of the Poincaré sphere) plus the right- and left-handed circular polarizations (the north and south poles of the sphere).

In this paper we offer an alternative calibration method that uses elliptical polarization states whose locus is a more complicated trajectory on the Poincaré sphere. However, the scheme is convenient in that these calibration states are generated by the angular adjustment of a single optical element, a quarter-wave retarder, which is rotated in a beam of linearly polarized light of zero reference azimuth. Experimental results are presented that demonstrate this method for one configuration of our prototype FDP (that consists of 4 windowless Si detectors) at the He-Ne laser of $\lambda$ = 632.8nm.

## 2. EXPERIMENTAL ARRANGEMENT

Figure 2 shows the basic calibration optics of the FDP. PSG is the polarization state generator that consists of a fixed (calcite) linear polarizer P whose transmission axis establishes the zero reference direction (the x axis of a transverse external laboratory xy coordinate system with respect to which $\underline{A}$ is actually measured) and a (crystalline quartz) quarter-wave retarder QWR which is rotated around the light beam as an axis. The output signal vector $\underline{I}$ of the FDP is normalized by taking the ratio $\underline{I}/i_r$, where $i_r$ the output signal of a reference detector $D_r$ that receives a polarization-independent fraction of incident radiation via the slightly tilted beam splitter BS. This normalization eliminates the effect of light-level fluctuations at the source, or small systematic changes of the light intensity that may accompany the rotation of the QWR.

## 3. TRAJECTORY OF THE CALIBRATION STATE ON THE POINCARE SPHERE

The normalized Stokes vector of light exiting the PSG and entering the FDP is given by

$$\underline{S} = \begin{bmatrix} 1 \\ s_1 \\ s_2 \\ s_3 \end{bmatrix} = \begin{bmatrix} 1 \\ \frac{1}{2} + \frac{1}{2}\cos 4C \\ \frac{1}{2}\sin 4C \\ \sin 2C \end{bmatrix} \qquad (4)$$

where C is the fast-axis azimuth of the QWR measured from the zero reference direction specified by the fixed polarizer transmission axis. It is easily verified that $s_1^2 + s_2^2 + s_3^2 = 1$, so that the endpoint of the reduced polarization vector $\underline{s} = (s_1, s_2, s_3)$ lies on the unit-radius Poincaré sphere ($|\underline{s}| = 1$). The trajectory of the endpoint of $\underline{s}$, which represents the calibration polarization state, as C is varied, is the double-lobed contour T sketched in Fig. 3. In this figure, LO represents the linear polarization state of zero azimuth, the equator E is the locus of all linear states, and the north and south poles $C_1$ and $C_2$ represent the right-handed and left-handed circular polarizations, respectively. The contour T is symmetrical with respect to the equator and its stereographic projection onto a plane tangent to the sphere at LO, with the center of projection being the point diametrically opposite to LO, gives the Lemniscate of Bernoulli (two-leaved rose) (18). Other interesting characteristics of this contour are considered elsewhere (19).

## 4. RESPONSE OF THE FDP AND DETERMINATION OF THE INSTRUMENT MATRIX $\underline{A}$

If we substitute Eq. (4) into Eq. (2) and expand the matrix product, we obtain the output response of the k th detector as

$$\begin{aligned} i_k(C) &= (a_{k0} + \tfrac{1}{2} a_{k1}) + a_{k3}\sin 2C \\ &+ \tfrac{1}{2} a_{k1}\cos 4C + \tfrac{1}{2} a_{k2}\sin 4C, \quad k = 0, 1, 2, 3. \end{aligned} \qquad (5)$$

Therefore, by recording $i_k(C)$ as a function of C at discrete angular positions of the QWR fast axis and subjecting the resulting data to the discrete Fourier transform, the elements $a_{kl}$ of the kth row (l = 0, 1, 2, 3) of $\underline{A}$ are all determined directly from the amplitudes of the cos 4C, sin 4C, sin 2C, and the dc term in Eq. (5).

Figures 4 & 5 ahow the results that we have obtained in applying this procedure to our prototype FDP in one spatial configuration at $\lambda = 632.8$ nm. The diamonds represent the normalized currents recorded as C is incremented from 0 to 180° in 10° steps. The continuous curves are obtained by a least-squares best-fit of the data to Fourier series of the form of Eq. (5). The residual rms errors in fitting the normalized currents $i_0$, $i_1$, $i_2$ and $i_3$ are 0.0011, 0.0051, 0.0064, and 0.0170, respectively. The resulting instrument matrix $\underline{A}$, obtained from the Fourier coefficients of Eq. (5), is given below

$$\underline{A} = \begin{bmatrix} 0.970 & 0.199 & -0.071 & 0.002 \\ 3.056 & -1.429 & 0.638 & 0.571 \\ 0.876 & -0.290 & 0.160 & -0.520 \\ 3.188 & -2.320 & -0.383 & -0.535 \end{bmatrix} \qquad (6)$$

This matrix satisfies all of the general theoretical constraints that we have recently established for the FDP's instrument matrix elements (20).

In the calibration procedure described here the QWR is assumed to be an exact quarter-wave retarder, so that the PSG produces the ideal Stokes parameters given by Eq. (4). This requires that the retardance of the QWR be $90^0$ to within $\pm$ $0.01^0$ or less, which is difficult to achieve.

The calibration method can be modified to apply when QWR has small but known (or separately measured) imperfections. In this case, the normalized Stokes parameters generated by the (imperfect) PSG are given by (17)

$$\begin{aligned} s_1 &= (\tfrac{1}{2} + \tfrac{1}{2}\cos 4C) + \Delta s_1 \\ s_2 &= (\tfrac{1}{2}\sin 4C) + \Delta s_2 \\ s_3 &= (\sin 2C) + \Delta s_3 \end{aligned} \qquad (7)$$

where the bracketed terms are the idealized parameters,

$$\begin{aligned} \Delta s_1 &= a_0 + a_2 \cos 2C + a_4 \cos 4C + b_4 \sin 4C + a_6 \cos 6C \\ \Delta s_2 &= d_2 \sin 2C + c_4 \cos 4C + d_4 \sin 4C + d_6 \sin 6C \\ \Delta s_3 &= e_2 \cos 2C + f_4 \sin 4C, \end{aligned} \qquad (8)$$

are the Stokes parameter perturbations generated by the QWR imperfections, and

$$\begin{aligned} a_0 &= - a_4 = - d_4 = - \delta/2 \\ a_2 &= - a_6 = d_2/3 = - d_6 = - f_4/2 = \tau/4 \\ b_4 &= - c_4 = e_2 = - 2\gamma \end{aligned} \qquad (9)$$

In Eqs. (9), $\delta$ is the deviation of the relative retardation of the QWR from $\pi/2$ (in radians), $\tau$ is the deviation of the relative amplitude transmittance along the fast and slow axes from 1, and $\gamma$ is the misalignment angle between the QWR fast axis and the polarizer transmission axis when their scales nominally read 0. All of the imperfection parameters $\delta$, $\tau$ and $\gamma$ can be measured by a separate test. Use of Eqs. (7)-(9) into Eq. (2) leads to the correct functional dependence of the four detector signals $i_k$ (k = 0, 1, 2, 3) on C. By fitting the resulting longer Fourier series to the experimental data, the elements of the instrument matrix $\underline{A}$ can be obtained free of the QWR imperfections.

## 5. BRIEF SUMMARY

The FDP is perhaps the simplest complete Stokes parameters photopolarimeter. Although we have constructed it to operate in the visible spectrum using Si detectors, its principle holds the same in other spectral ranges. All that is needed is to identify and use the appropriate type of detector (with specular front-surface reflection) which is suited to the spectral interval of interest (e.g. Ge or Hg Cd Te in the IR). In this paper we have described an efficient calibration scheme with which the all-important instrument matrix $\underline{A}$ can be determined by the angular rotation of a single optical element, a quarter-wave retarder (QWR). The procedure is simplest when the QWR is perfect, or nearly so, but the method can be extended to apply when the QWR is nonideal and its small imperfections are known.

## ACKNOWLEDGEMENTS

I wish to thank Ali G. Lopez for his assistance and to acknowledge the support of the National Science Foundation grant no. ECS 8520035.

## 6. REFERENCES

1. R.M.A. Azzam and D.L. Coffeen, Eds., Optical Polarimetry-Instrumentation and Applications, Proceedings of the Society of Photo-Optical Instrumentation Engineers, Vol. **112** (SPIE, Bellingham, WA, 1977).

2. R.M.A. Azzam and D.L. Coffeen, Guest Eds., Optical Polarimetry, Special Issue, Vol. **21**, No 1 of Optical Engineering (SPIE, Bellingham, WA 1981).

3. D. Moisil and G. Moisil, Theoria Si Pracatica Ellipsometriei (Editurna, Tehnica, Bucharest, 1973) in Romanian.

4. R.M.A. Azzam and N.M. Bashara, Ellipsometry and Polarized Light (North-Holland, Amsterdam, 1977).

5. A.V. Rzhanov, K.K. Svitashev, A.I. Semeneko, L. V. and V.K. Sokolov, Principles of Ellipsometry (NAUK, USSR, 1979) in Russian.

6. E. Passaglia, R.R. Stromberg and J. Kruger, Eds., Ellipsometry in the Measurement of Surfaces and Thin Films, NBS Misc. Publ. 256 (USGPO, Washington, D.C., 1964).

7. N.M. Bashara, A.B. Buckman and A.C. Hall, Eds., Recent Developments in Ellipsometry (North-Holland, Amsterdam, 1969).

8. N.M. Bashara and R.M.A. Azzam, Eds. Proceedings of the Third International Conference on Ellipsometry (North-Holland, Amsterdam, 1976).

9. R.H. Muller, R.M.A. Azzam and D.E. Aspnes, Eds., Proceedings of the Fourth International Conference on Ellipsometry (North-Holland, Amsterdam, 1980).

10. Proceedings of the International Conference on Ellipsometry and and Other Optical Methods for Surface and Thin Film Analysis, Journal de Physique, Vol. 44, Colloque C 10 (Les Editions de Physique, Paris, 1984).

11. T. Gehrels, Ed., Planets, Stars, and Nebulae Studied with Photopolarimetry (Univ. Arizona, Press, Tucson, 1977).

12. G.P. Kuiper, Planets and Satellites (Univ. Chicago Press, Chicago, 1961).

13. See, e.g., P.S. Hauge, 'Recent Developments in Instrumentation in Ellipsometry', Surface Sci., Vol. 96, 108-140 (1980).

14. R.M.A Azzam, "Arrangement of Four Photodetectors for Measuring the State of Polarization of Light', Opt. Lett., Vol. 10, 309-311 1985); U.S. Patent 4,681,450 (July 21, 1987).

15. R.M.A. Azzam, I.M. Elminyawi, and A.M. El-Saba, 'General Analysis and Optimization of the Four-Detector Photopolarimeter', J. Opt. Soc. Am. A, Vol. 5, 681-689 (1988).

16. R.M.A. Azzam, E.Masetti, I.M. Elminyawi, and F.G. Grosz, 'Construction, Calibration, and Testing of a Four-Detector Photopolarimeter', Rev. Sci. Instrum., Vol. 59, 84-88 (1988).

17. R.M.A. Azzam and A.G. Lopez, 'Accurate Calibration of the Four-Detector Photopolarimeter with Imperfect Polarizing Optical Elements', J. Opt. Soc. Am. A, Vol. 6, in print (1989).

18. See Ref. 4, p. 236.

19. R.M.A. Azzam, to be published.

20. R.M.A. Azzam, 'Instrument Matrix of the Four-Detector Photopolarimeter (FDP). Physical Meaning of its Rows and Columns and Constraints on Its Elements', J. Opt. Soc. Am. A, Vol. 7, in press (1990).

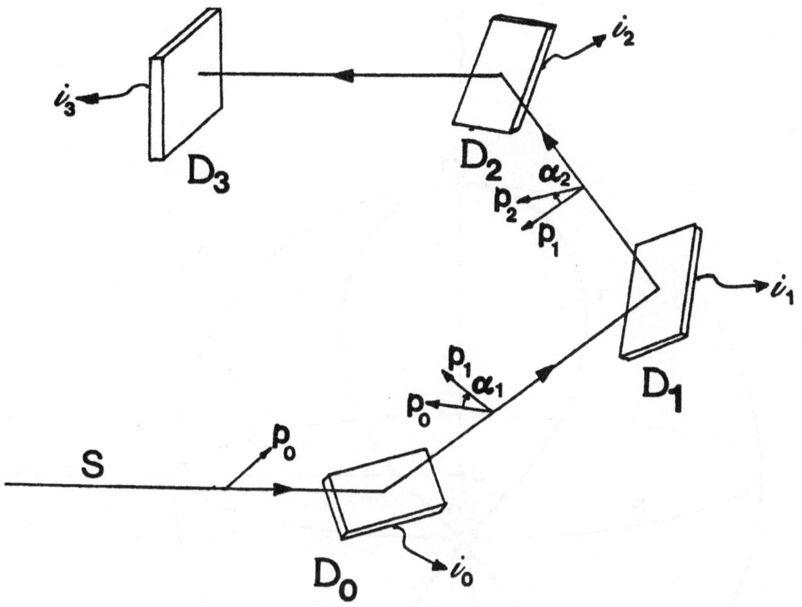


Figure 1 The four-detector photopolarimeter (FDP)

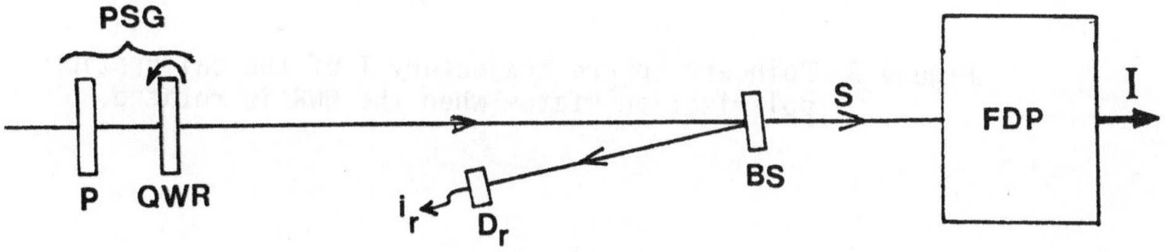


Figure 2 Calibration arrangement for the FDP

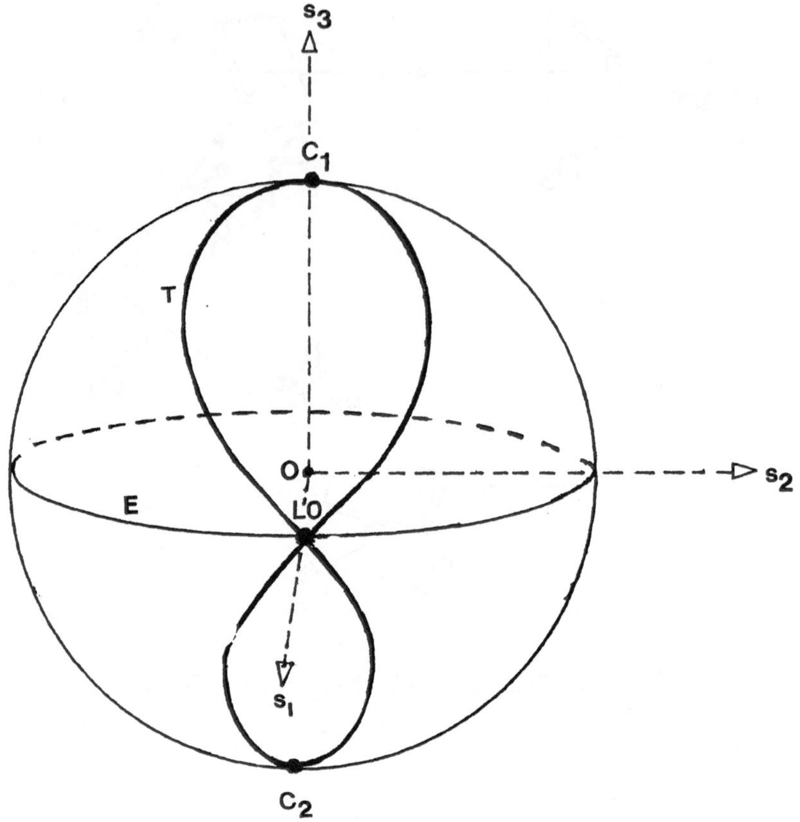


Figure 3 Poincaré sphere trajectory T of the calibration polarization states when the QWR is rotated.

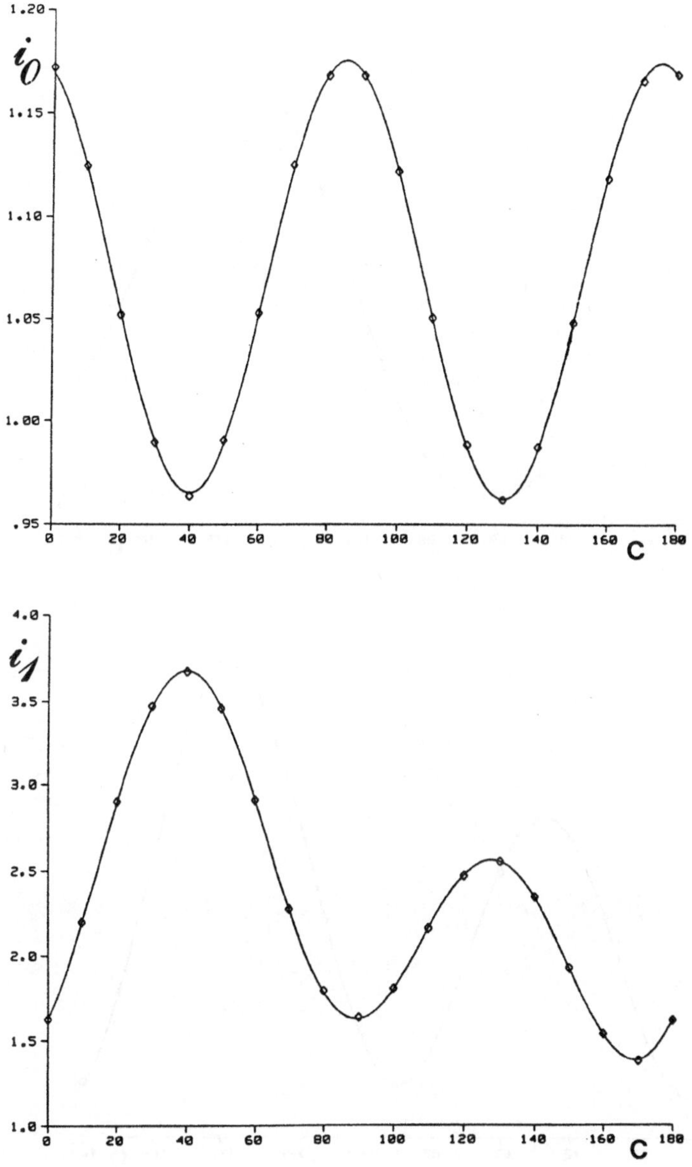


Figure 4 Normalized output signals of the first two detectors of the FDP plotted versus the QWR fast-axis azimuth C.

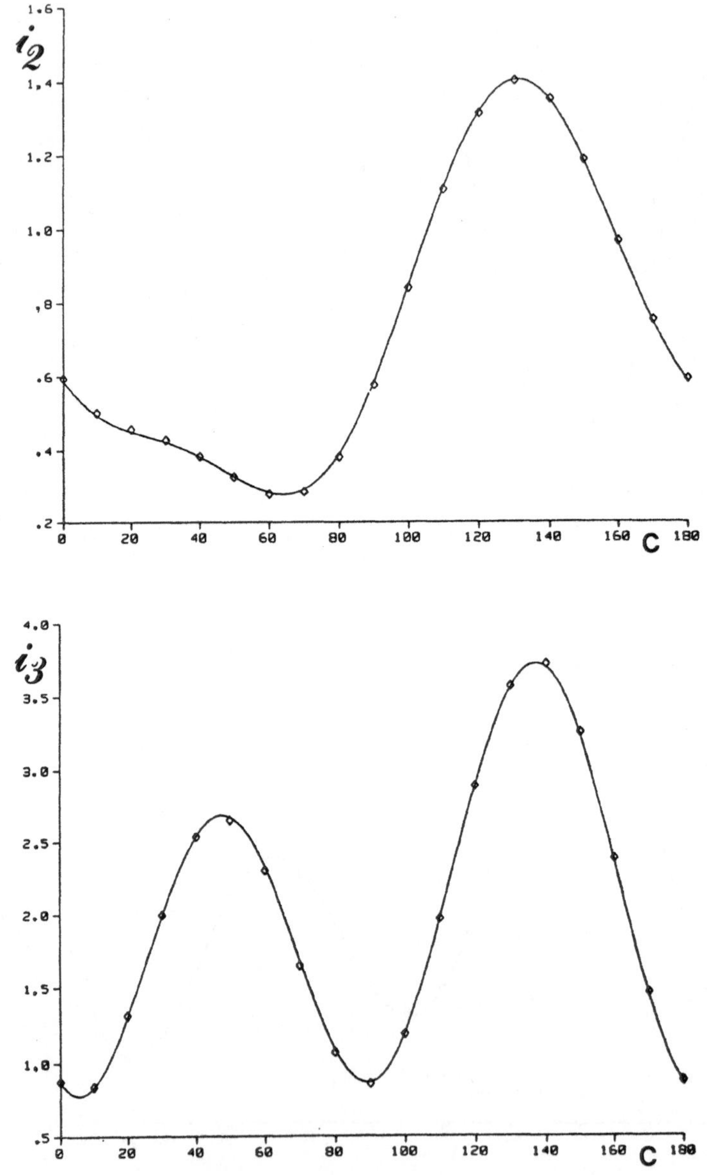


Figure 5 Normalized output signals of the last two detectors of the FDP plotted versus the QWR fast-axis azimuth C.

POLARIZATION STATE MEASUREMENTS DERIVED FROM A GENERAL SIX-PORT ANALYZER: A THEORETICAL AND EXPERIMENTAL APPROACH

Alireza Mahmoodshahi
Member of the Technical Staff
AT&T Bell Laboratories
Crawford Corner Road
Holmdel, NJ 07733 U.S.A.

and

Walter K. Kahn
EE & CS Department
The George Washington University
Washington, DC 20052 U.S.A.

ABSTRACT. A novel method for measuring the state of polarization of a plane wave utilizing only power measurements from a general six-port is presented. The six-port may be a waveguide junction, the interconnection of several junctions, or comprised of a distant source of polarized radiation and a sufficiently general receiving array of four-elements. It is shown that the system may be calibrated with five standard polarizations, independent of power level changes.

## 1. INTRODUCTION

This paper presents the theory of a novel method for measurement of the state of polarization of a plane wave. The plane wave may be considered to be incident on a dual mode input (two ports) of a physical six-port. One such device was constructed and tested [1]. Alternatively, the six-port may exist only conceptually. In the latter case a distant variable polarization source may be considered to form two ports of the conceptual six-port while an array of four (one-port) antennas provides the remaining four ports of the conceptual six-port.

Since Engen and Hoer [3] introduced the six-port network analyzer, numerous variations of such circuits, calibration and measurement algorithms have been developed [1-6, 12-17]. Originally, their algorithms were complicated and tedious. Oishi and Kahn [12,13] derived a Stokes vector representation for the six-port analyzer which simplified calibration and measurement. Six-port techniques have also previously been employed in connection with measurement of polarization [1,16]. Such six-port systems are especially attractive either for implementation for millimeter waves or in automated polarization measurement equipment because only power measurements (no distinct phase measurements or comparisons) are required to obtain the polarization information; furthermore, unlike some conventional techniques for measurement of polarization, six-port techniques do not involve moving parts or rotary joints. Finally, the (constituent parts or components utilized in the construction of) six-port systems are not required to meet any particular tight electrical specifications or tolerances. Once such a system is assembled or conceptually identified, it may be calibrated using five (or more) relatively arbitrarily chosen known input polarizations.

In the first general sort of implementation the wave to be analyzed is incident on a two element array or two-mode horn antenna constituting two input ports of the six-port; the required four power meters are connected at the remaining four ports of a specifically designed microwave network assembly forming the complete six-port analyzer. In this paper this configuration will be termed the "physical six-port configuration" to distinguish it from the "conceptual six-port configuration" to be described next. The experimental and theoretical details of polarization measurement with this type of six-port network analyzer are presented in reference [10].



*W.-M. Boerner et al. (eds.), Direct and Inverse Methods in Radar Polarimetry, Part 1*, 819–827.

In the second implementation the wave is incident on a general four-element array of antennas. The use of a receiver array composed of four general elementary antennas is entirely novel. This novel configuration suggests itself quite naturally in consequence of a simple but nevertheless fundamental conceptual shift. In the preceding configuration and other previous work the receiver sub- system itself constituted the six-port. In this alternative concept all of the transmitter, the propagation path and receiving antenna system taken as a whole constitutes the six-port. Thus the six-port exists only conceptually and is not identified with a specifically designed physical six-port junction. The theoretical analysis and calibration must take cognizance of this change in viewpoint but can substantially be carried over to fit this new situation. The conceptual six-port polarization analyzer was first presented at a polarimetric technology workshop [8].

## 2. THE PHYSICAL SIX-PORT POLARIZATION ANALYZER

The experimental and theoretical details of the polarization measurement using a six-port network analyzer are presented in [10]. This includes the theoretical basis for the calibration and measurement procedures. Various computer codes were written to automate calibration and measurements utilizing the experimental six-port analyzer assembled. Sample measurement sets were analyzed, and the overall performance compared with that of an idealized model. In addition, a theoretical model of the Six-Port Network Analyzer (SPNA) was devised. Based on this model, a simulation program was written to compute the error sensitivity of the SPNA in order to measure the polarization by introducing errors in one or any combination of two power meters.

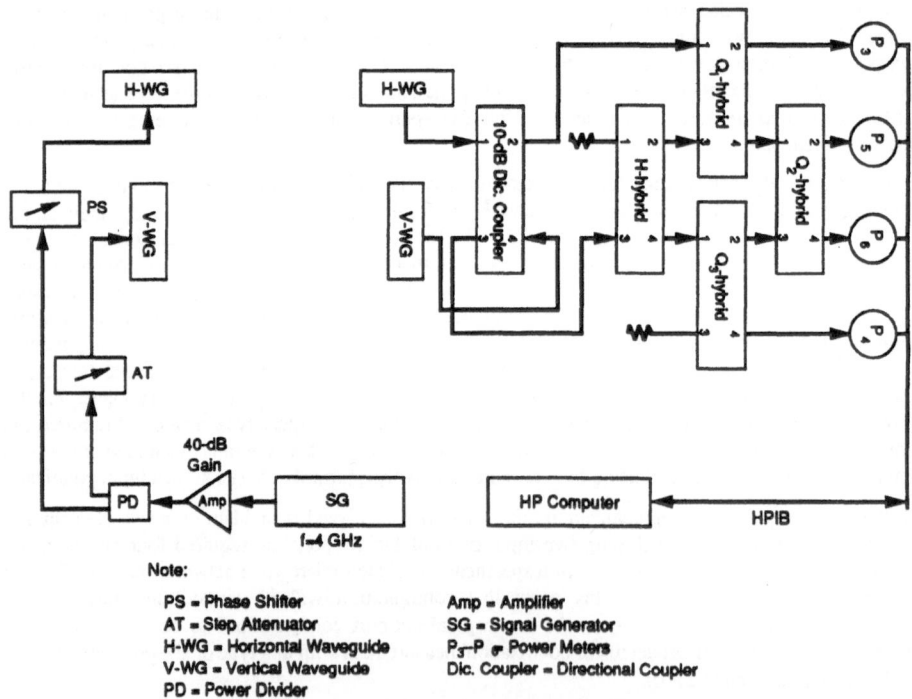

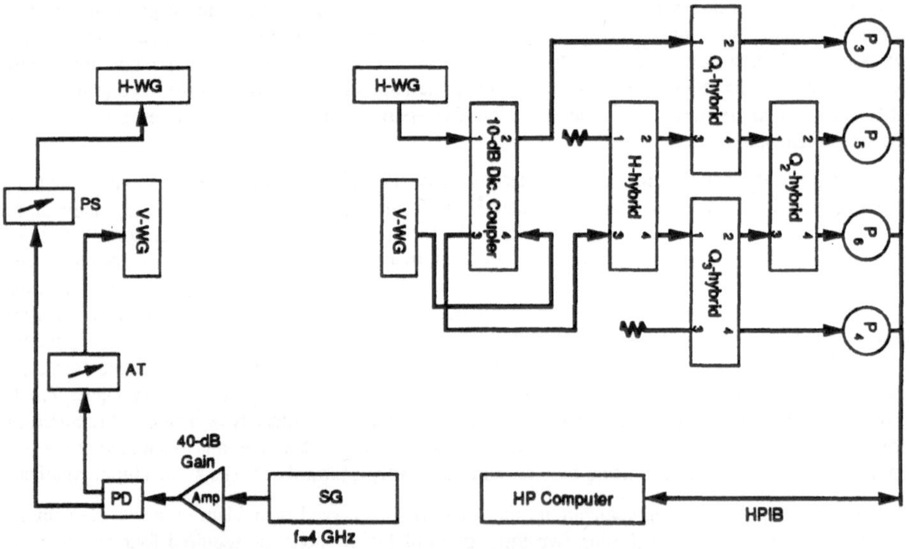


Note:

PS = Phase Shifter
AT = Step Attenuator
H-WG = Horizontal Waveguide
V-WG = Vertical Waveguide
PD = Power Divider

Amp = Amplifier
SG = Signal Generator
$P_3$-$P_6$ = Power Meters
Dic. Coupler = Directional Coupler

FIGURE 2.1 SPNA MEASUREMENT SYSTEM

A complete experimental system, including a transmitter with variable polarization, receiving equipment necessary for conventional measurement of the polarization ellipse, and the especially designed six-port circuit was set up in a microwave laboratory made available for this purpose by AT&T Bell Laboratories, Holmdel, NJ. The six-port design was based on a form of the device used by Engen [3]. The SPNA was calibrated; that is, the calibration matrix was measured employing five standard polarizations. As shown by Oishi and Kahn [12] the SPNA calibration matrix comprises four rows, each of which has the form of a Stokes vector. The outputs of four power meters are directly proportional to the scalar products of these four Stokes vectors and the Stokes vector representing the polarization of the incoming wave. These basic features of the measurement technique were verified. In addition, the Stokes vector form was made the basis of a self-consistent scheme for determining the relative sense of the calibration polarizations, which was not measured in advance by conventional means.

The SPNA consists of one 10-dB directional coupler, one H-hybrid and three quadrature hybrids. The 10-dB directional coupler has four ports. The port coupled to the input port is 10 dB down, and the direct port is 1.25 dB down from the input signal, while port four is isolated. The H-Hybrid is a four-port device. One of the ports is terminated while the other two ports provide equal power, which is 3 dB down from the input signal.

Figure 2.1 shows the SPNA test setup. This setup consists of a transmitter and a receiver, only about 10 feet apart. The transmitter antenna comprises two 4 - GHz rectangular waveguides placed next to each other at one end of the bench. One of them is oriented with narrow dimension horizontal, and the other is oriented with narrow dimension vertical. The POLARAD (model number 1206) provides the 4-GHz signal source. This signal is amplified using an Avantek amplifier with a gain of 40 dB and a noise figure of 1.5 dB. The output of the amplifier is split via a 10-dB HP splitter into two signals. One of the output ports is connected through a phase shifter to feed the horizontally polarized waveguide. The second output port is connected through a step attenuator to feed the vertically polarized waveguide. The receiver antenna also comprises two identical waveguides installed next to each other at the other end of the bench one of them is oriented with narrow dimension horizontal and the other one is oriented with narrow dimension vertical.

To calibrate the SPNA, five signals with different polarizations are used. It may appear that four sets of known polarizations would be sufficient for the calibration of the SPNA [1]. However, this is the case only if, in addition to polarization, the correct relative power of the four incident waves is known and the corresponding power level of the four sets of power meter readings noted. This is impracticable. Details of the calibration matrix are presented in [10] and in connection with the general viewpoint developed in Section 3.

Five polarizations were measured using conventional techniques. A dipole is used in a conventional technique to measure the polarization. It is connected to a calibrated power meter. To determine the polarization of the incoming wave, the dipole is rotated through minimum and maximum power levels. These levels occur at angles $\phi'$ and $\phi$ in relation to the x-axis respectively. The ratio of the minimum power to the maximum powr level is called the axial ratio. The inverse tangent of this axial ratio is called the ellipticity angle ($\tau$) and the angle at which the maximum occurs is called the orientation angle ($\phi$). The sense of polarization was not measured in our experiment. Five sets of power meter readings form the SPNA corresponding to each of these known polarizations were then obtained.

With this data two possible points of the Poincare sphere can be found for each polarization. The sense of polarization is still ambiguous. The special character of the calibration matrix in our procedure (each row has a form of stokes vector) allows the determination of a consistent set of signs for the ellipicity angle in the data. By trying all different possible senses for the five sets of data one easily selects the consistent set of signs. This does not, of course, avoid the final sense ambiguity of the measured polarization. This ambiguity can be resolved once and for all, if necessary, through a single independent conventional measurement of sense.

The calibration matrix derived from the first five sets of known polarizations must be the same or very close if we permute the roles of the five measurements within the calibration procedure. It has been shown that this is true for the SPNA measurement. The average calibration matrix is used. Having the calibration matrix, it is possible to measure any unknown polarization state. Table 2.1 describes the measured polarization using SPNA vs. the conventional technique. As can be seen from the table, these are in close agreement. Since the SPNA detects the correct relative sense of polarization, only a single independent measurement of sense will fix the absolute sense of polarization in all the data.

The fraction of the power which is rejected by the receiving antenna due to the polarization mismatch between the polarization of the incident wave and the polarization that the receiving antenna likes to receive is called the polarization deviation. The polarization deviation can be written as follows:

$$\textit{Polarization Deviation} = 1 - \cos^2 \frac{\Delta}{2}$$

Where, $\Delta$ = Angular separation between on the Poincare sphere between the polarization measured by the conventional technique and the one measured by the SPNA.

Table 2.1

MEASURED POLARIZATION VS. THE CONVENTIONAL TECHNIQUES

| | Case # | 6 | 7 | 8 | 9 | 10 |
|---|---|---|---|---|---|---|
| Conventional Technique | Ellipticity (Degrees)<br>Orientation (Degrees) | ±.428<br>165.00 | ±9.575<br>140.00 | ±.87<br>185.00 | ±14.86<br>175.00 | ±1.44<br>195.00 |
| SPNA With Average Calibration Matrix | Ellipticity (Degrees)<br>Orientation (Degrees) | 1.08<br>161.24 | 10.106<br>148.3 | -2.11<br>178.69 | 16.546<br>169.16 | -2.76<br>187.98 |
| Polarization Deviation [1] | | .004 | .02 | .01 | .008 | .02 |

## 3. THE CONCEPTUAL SIX-PORT POLARIZATION ANALYZER

The idea that underlies the conceptual six-port polarization analyzer is depicted in the upper portion of Fig. 3.1. When the characteristics of the array (including mutual interaction) are known or, alternatively after appropriate calibration, only the real average power received by power meters at each of the four antennas is required to infer the state of polarization of the incident wave uniquely.

### 3.1 THEORY

The six-port (transmitter, propagation path, four element array plus receivers) will be specified by the scattering matrix, $S = [Smn]$. Following our previous work [10-12], the incident and reflected wave vectors as well as the scattering matrix are partitioned to separate the transmitter port and receiver (power meter) port quantities:

$$\underline{b} = \begin{bmatrix} \underline{b}_\alpha \\ \underline{b}_\beta \end{bmatrix} \begin{bmatrix} S_{\alpha\alpha} & S_{\alpha\beta} \\ S_{\beta\alpha} & S_{\beta\beta} \end{bmatrix} = \begin{bmatrix} \underline{a}_\alpha \\ \underline{a}_\beta \end{bmatrix} = S\underline{a} \tag{3.1}$$

where

$$\underline{a}_\alpha = \begin{bmatrix} a_1 \\ a_2 \end{bmatrix} \quad \text{and} \quad \underline{a}_\beta = \begin{bmatrix} a_3 \\ a_4 \\ a_5 \\ a_6 \end{bmatrix}, etc.$$

The power detectors may be mismatched with reflection coefficients $Q_n$. Writing the diagonal matrix $Q = [Q_n \delta_{nm}]$, we have

$$\underline{a}_\beta = Q\underline{b}_\beta \tag{3.2}$$

This relation may be substituted in (3.1) to obtain

$$\underline{b}_\beta = (1 - S_{\beta\beta}Q)^{-1} S_{\beta\alpha}\underline{A}_\alpha \tag{3.3a}$$

$$= \underline{A}_\beta a_1 + \underline{B}_\beta a_2 \tag{3.3b}$$

The diagonal elements of the matrix $\overline{b_\beta b_\beta^\dagger}$ are the (average) power quantities $\hat{P}_n = b_n b_n^*$, incident on the detectors, $n = 3,4,5,6$. In matrix form

$$\underline{\hat{P}} = \begin{bmatrix} \hat{P}_1 \\ \hat{P}_2 \\ \hat{P}_3 \\ \hat{P}_4 \end{bmatrix} = \begin{bmatrix} |A_3|^2 & A_3B_3^* & B_3A_3^* & |B_3|^2 \\ |A_4|^2 & A_4B_4^* & B_4A_4^* & |B_4|^2 \\ |A_5|^2 & A_5B_5^* & B_5A_5^* & |B_5|^2 \\ |A_6|^2 & A_6B_6^* & B_6A_6^* & |B_6|^2 \end{bmatrix} \begin{bmatrix} |a_1|^2 \\ a_1a_2^* \\ a_2a_1^* \\ |a_2|^2 \end{bmatrix} \tag{3.4}$$

Notice that on the right the rows have the same structure as the column vector. We now introduce the Stokes vector $\underline{A}$ via the linear transformation $T$.

$$\begin{bmatrix} |a_1|^2 \\ a_1a_2^* \\ a_2a_1^* \\ |a_2|^2 \end{bmatrix} = \frac{1}{2} \begin{bmatrix} 1 & 0 & 0 & 1 \\ 0 & 1 & j & 0 \\ 0 & 1 & -j & 0 \\ 1 & 0 & 0 & -1 \end{bmatrix} \begin{bmatrix} |a_1|^2 + |a_2|^2 \\ 2Re(a_1a_2^*) \\ 2Im(a_1a_s^*) \\ |a_2|^2 - |a_2|^2 \end{bmatrix} = T\underline{A} \tag{3.5}$$

so that

$$\underline{\hat{P}} = \underline{\underline{M}}\,\underline{A} \tag{3.6}$$

The rows of $\underline{M}$ have the same form as the vector on the right of (3.5), i.e., the form of Stokes vectors. If the Stokes vector is divided by $|a_1|^2 + |a_2|^2$, the Stokes vector is termed normalized and written with prescript o as $^{\circ}\underline{A}$.

The 6-port may now be calibrated with five corresponding sets of known polarizations (normalized Stokes vectors) and relative power meter readings, i.e., $n = 1,2,\ldots,5$,

$$^{\circ}\underline{A}^{(n)} <-> \underline{P}^{(n)} . \tag{3.7}$$

### 3.2 SIMULATION

Assume that a particular configuration, scattering matrix S, leads via (3.3) to columns $\underline{A}_{\beta}$ and $\underline{B}_{\beta}$ proportional to:

$$\underline{A}_{\beta} = \begin{bmatrix} 0 \\ 1+j \\ 1 \\ 2 \end{bmatrix}, \quad \underline{B}_{\beta} = \begin{bmatrix} 2 \\ 1 \\ 1+j \\ 0 \end{bmatrix} .$$

From (3a) we see, when the power meters are well matched to the receiver (Q=0), the columns $\underline{A}_{\beta}$ and $\underline{B}_{\beta}$ are the columns of $S_{\beta\alpha}$. For a sufficiently general array of four antennas, neither the columns nor any two rows of $S_{\beta\alpha}$ may be linearly dependent [12,13].

Utilizing (3.4), we construct sets of power meter readings corresponding to any number of transmitter inputs $a_1$ and $a_2$. For calibration we require information that five sets of received power meter, say the first five sets, correspond to known polarizations. These are specified by normalized Stokes vectors. Seven complete sets are exhibited in Table 3.1. Note that the input power is not kept constant so that the normalized Stokes vectors stand in no fixed proportionality relative to the level of the power meter indications.

The matrix $\underline{M}$, the inverse of the calibration matrix, is then determined from the simultaneous relations

$$\underline{P} = \underline{M}\,^{\circ}\underline{A}\,\underline{D} \tag{3.8a}$$

$$\underline{P}^{(5)} = \underline{M}\,^{\circ}\underline{A}^{(5)} D_{55} . \tag{3.8b}$$

The matrices $\underline{P}$ and $^{\circ}\underline{a}$ have as columns four corresponding column vectors (3.7). $\underline{D}$ is a diagonal matrix. The fifth pair of columns is used in (3.8b). Eliminating $\underline{M}$, we obtain

$$\underline{P}^{(5)} = \left[\underline{P}\underline{D}^{-1}\,^{\circ}\underline{A}^{-1}\right]\,^{\circ}\underline{A}^{(5)} D_{55} . \tag{3.9}$$

When the scalar amplitude $D_{55}$ is chosen arbitrarily, again emphasizing the independence of amplitude level, this matrix equation is equivalent to four scalar equations in the four unknown diagonal elements of $\underline{D}$. The compact matrix form is written out below, with appropriate substitution from the first five entries in Table 3.1 and indicating the solution values obtained for the diagonal elements of $\underline{D}$.

$$\underset{\underline{P}}{\begin{bmatrix} 0 & 4 & 4 & 8 \\ 2 & 1 & 9 & 4 \\ 1 & 2 & 4 & 9 \\ 4 & 0 & 8 & 4 \end{bmatrix}} \underset{\underline{D}^{-1}}{\begin{bmatrix} 2 & 0 & 0 & 0 \\ 0 & 2 & 0 & 0 \\ 0 & 0 & \frac{2}{3} & 0 \\ 0 & 0 & 0 & \frac{2}{3} \end{bmatrix}} \underset{{}^{\circ}\underline{A}^{-1}}{\begin{bmatrix} \frac{1}{2} & -\frac{3}{4} & \frac{1}{4} & \frac{1}{2} \\ \frac{1}{2} & -\frac{3}{4} & -\frac{1}{4} & -\frac{1}{2} \\ 0 & \frac{3}{4} & -\frac{3}{4} & 0 \\ 0 & \frac{3}{4} & \frac{3}{4} & 0 \end{bmatrix}} = \underset{\underline{M}}{\begin{bmatrix} 4 & 0 & 0 & -4 \\ 3 & 2 & -2 & 1 \\ 3 & 2 & 2 & -1 \\ 4 & 0 & 0 & 4 \end{bmatrix}}$$

The calibration is now easily verified using the remaining two entries listed in Table 3.1 or with any number of additional data sets that the reader might construct for himself via Equation (3.3).

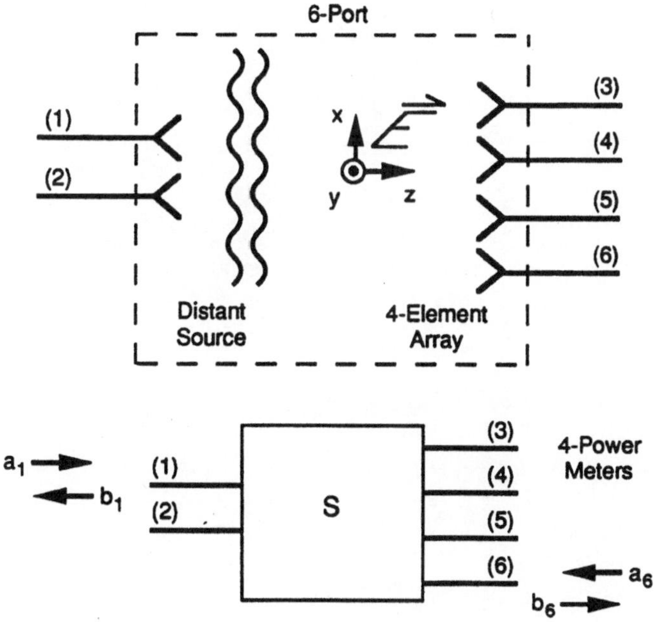


FIGURE 3.1 6-PORT AND EQUIVALENT CIRCUIT

**TABLE 3.1 SIMULATION**

| (n) | $a_1$ | $a_2$ | $°A_0$ | $°A_1$ | $°A_2$ | $°A_3$ | $P_3$ | $P_4$ | $P_5$ | $P_6$ |
|---|---|---|---|---|---|---|---|---|---|---|
| 1 | 1 | 0 | 1 | 0 | 0 | 1 | 0 | 2 | 1 | 4 |
| 2 | 0 | 1 | 1 | 0 | 0 | -1 | 4 | 1 | 2 | 0 |
| 3 | 1-j | 1 | 1 | 2/3 | -2/3 | 1/3 | 4 | 9 | 4 | 8 |
| 4 | 1 | 1-j | 1 | 2/3 | 2/3 | -1/3 | 8 | 4 | 9 | 4 |
| 5 | 1 | 1 | 1 | 1 | 0 | 0 | 4 | 5 | 5 | 4 |
| 6 | 1 | j | 1 | 0 | -1 | 0 | 4 | 5 | 1 | 4 |
| 7 | 1 | 3 | 1 | .6 | 0 | -.8 | 36 | 17 | 25 | 4 |

**4. REFERENCES**

[1] R. M. A. Azzam, I. M. Elminyawi, and A. M. El-Saba, "General Analysis and Optimization of the Four-Detector Photopolarimeter," J. Opt. Soc. Am., A, vol. 5, no. 5, pp. 681-689, May 1988.

[2] H. M. Cronson and L. Susman, "A Six-port Automatic Network Analyzer," IEEE Trans. Microwave Theory Tech., vol. MTT-25, pp. 1086-1091, Dec. 1979.

[3] G. F. Engen and C. A. Hoer, "Application of an Arbitrary 6-port Junction to Power Measurement Problems," IEEE Trans. Intstrim. Meas., vol. IM-21, No. 4, pp. 470-474, Nov. 1972.

[4] G. F. Engen, "Calibration of an Arbitrary Six-port Junction for Measurement of Active and Passive Circuit Parameters," IEEE Trans. Instrum. Meas., vol. IM-22, no. 4, pp. 295-299, Dec. 1973.

[5] G. F. Engen, "Calibrating the Six-port Reflectometer by Means of Sliding Terminations," IEEE Trans. Microwave Theory Tech., vol. MTT-26, pp. 951-957, Dec. 1978.

[6] C. A. Hoer, "The 6-port Coupler: A New Approach to Measuring Voltage, current, Power, Impedance, and Phase," IEEE Trans. Instrum. Meas., vol. IM-21, no. 4, pp. 466-470, Nov. 1972.

[7] J. R. Huynen, "Phenomenological Theory of Radar Targets," Dissertation, Technische Hogeschool Delft, Netherlands, December 1970.

[8] W. K. Kahn, "Theory of Polarization Derived from the Real Power Received by a General Four-element Antenna Array," Presented at The Polarimetric Workshop, Redstone Arsenal, Huntsville, AL, 16-18 August 1988; Proceedings of the Polarimetric Technology Workshop, Vol. I, pp. 361-369, May 1990. GACIAC, IIT Research Institute, Chicago, IL.

[9] A. B. Kostinski and W.-M. Boerner, "On Foundations of Radar Polarimetry," IEEE Transactions on Antennas and Propagation, vol. AP-34, 1986, pp. 1395-1404.

[10] A. Mahmoodshahi, "Polarization Measurement Using Six-Port Network Analyzer: A Theoretical and Experimental Approach," Dissertation, George Washington University, Washington, D.C., March 1988.

[11] A. Mahmoodshahi and W. K. Kahn, "Polarization Measurement Using Six-Port Network Analyzer: A Theoretical and Experimental Approach," Presented at the 1989 AP-S International Symposium, June 26-30, 1989, San Jose, California; Digest reference, Session 45-2, Vol. II, pp. 1018-1021.

[12] T. Oishi and W. K. Kahn, "Stokes Vector Representation of the Six-port Network Analyzer: Calibration and Measurement," IEEE MTT-S International Symposium Digest, pp. 503-506, 1985.

[13] T. Oishi, "The Investigation of Six-Port Systems Using Stokes Vector Representations," Dissertation, George Washington University, Washington, D.C., December 1987.

[14] P. I. Slomo and J. D. Hunter, "A Six-port Reflectometer and Its Complete Characterization by Convenient Calibration Procedures," IEEE Trans. Microwave Theory Tech., vol. MTT-30, pp. 186-192, Feb. 1982.

[15] L. Susman, "Calibration of a Six-Port Reflectometer Using Projective Geometry Concepts," Electronics Letters, vol. 20, no. 1, pp. 9-11, 1984.

[16] M. A. Wood, "A Theoretical Study of Calibration Procedures for Coherent and Non-coherent Polarimetric Radars," Royal Signals and Radar Establishment Report No. 86011, June 1986.

[17] D. Woods, "Analysis and Calibration Theory of the General 6-port Reflectometer Employing Four Amplitude Detectors," Proc. IEE, vol. 126, no. 2, pp. 221-228, Feb. 1979.

# UNDERSTANDING RECIPROCITY IN RADAR POLARIMETRY

M. R. Feinstein
Johns Hopkins University Applied Physics Laboratory
Johns Hopkins Road
Laurel, Maryland 20707-6099

ABSTRACT. Unconventional techniques are presented for analysis of the effects of two-way propagation on electromagnetic fields. Difficulties arising from the use of different coordinate systems for transmission and reception are demonstrated and discussed. The effects of linear processes and reciprocity are worked out in detail. The analysis uses the Jones vector representation for electromagnetic fields.

## 1. INTRODUCTION

Although polarization is a fundamental property of electromagnetic radiation, it is often treated as an afterthought in problems involving radar propagation and scattering. However, since variations in received polarization will have an effect on radar measurements and since polarization can be useful in locating and identifying scatterers, there is a real need for a natural extension of conventional analysis that includes polarization without engendering confusion.

Such extension turns out to require more care than one might expect; in fact, radar problems involving polarization are notorious for presenting unexpected subtleties and traps for the unwary. A particularly troublesome class of problems is the behavior of polarization in two-way propagation through some medium, depicted in Figure 1.

In the figure, a signal transmitted by an antenna at point A propagates to point B. The signal at B undergoes a linear scattering or, more generally, a linear process of some sort; it then returns to A, where it is received by the antenna from which it was originally transmitted. The basic question is how the process at B appears to an observer at A.

Difficulties arise in the analysis of two-way radar propagation because standard conventions can obstruct understanding, making relatively simple matters appear complex. In particular, the basic convention that the receive and transmit polarizations of a reciprocal antenna are equal and the associated convention that different



*W.-M. Boerner et al. (eds.), Direct and Inverse Methods in Radar Polarimetry, Part 1*, 829–844.

coordinate systems must be used for reception and transmission make the two-way propagation problem seem difficult, when, in fact, it is not. The matter of whether one should use different coordinate systems for reception and transmission has been the subject of some recent controversy in the literature[1].

Of course, the standard treatment, applied correctly, does work; but it does so at the cost of obscuring important aspects of the scattering and propagation problem. This, in turn, leads to unnecessary difficulty in analysis of multipath scattering and propagation problems and a miasma of doubt and uncertainty concerning the analysis of variations in polarization due to propagation and scattering.

These uncertainties can be resolved with methods introduced into optics 40 years ago by R. C. Jones[2]. Jones introduced the so-called Jones vector representation of electromagnetic fields in which

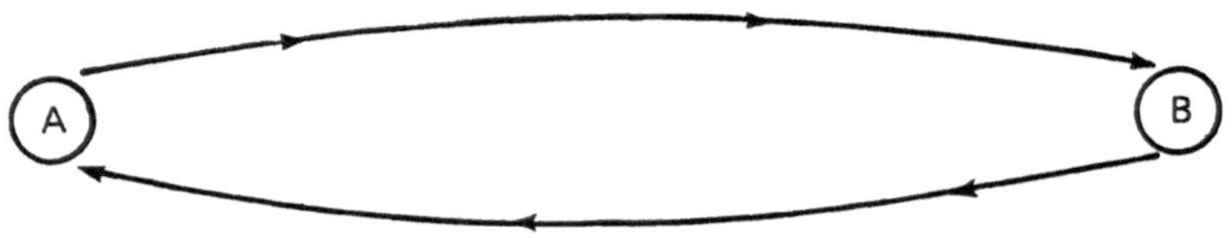


Figure 1: Two-Way Transmission

the electromagnetic field is represented by a two dimensional complex vector

$$\tilde{E} = \begin{bmatrix} E_H \\ E_V \end{bmatrix}$$

where $E_H$ and $E_V$ are complex numbers whose phase and magnitude are the phase and intensity of the field's horizontal and vertical components, respectively. It will be shown that this representation simplifies the analysis of multipath scattering and propagation. In particular, the "Jones vector version" of receive and transmit antenna properties will be spelled out in detail.

The basic physical principle involved in two-way propagation and in the relation between receive and transmit antenna properties is reciprocity. Indeed, in some sense, the fundamental issue here is how to cast the analysis of two-way propagation in a way that plainly exposes the consequences of reciprocity.

The next section of this paper defines the Jones vector representation of a field. This is followed by definitions of the basic quantities involved in the two-way propagation problem.

The discussion then turns to definition and proof of the reciprocity principle and the basic consequences of reciprocity; this is accompanied, in Appendix A, by noting conditions under which reciprocity does not hold.

The reciprocity principle is then applied to the receive and transmit properties of antennas; the difference between the Jones vector approach and the standard approach are discussed and the basic physical consequences of reciprocity in two-way propagation are demonstrated and discussed.

2. JONES VECTORS

If the amplitude and phase of the horizontal component of an electromagnetic signal at some point are $|E_H|$ and $\phi_H$ while the amplitude and phase of the vertical component at the same point are $|E_V|$ and $\phi_V$ then the Jones vector of the signal at that point is a two-dimensional complex column vector

$$\tilde{E} = \begin{bmatrix} |E_H| e^{i\phi_H} \\ |E_V| e^{i\phi_V} \end{bmatrix} \equiv \begin{bmatrix} E_H \\ E_V \end{bmatrix} .$$

It is important to note immediately that this definition makes no mention of the direction of propagation of the wave. The directions "horizontal" and "vertical" are parallel to two orthogonal axes of a three-dimensional Cartesian coordinate system, where the third axis points either parallel or anti-parallel to the direction of propagation. For example, the vector

$$\tilde{E}_{45} = \frac{E_S}{\sqrt{2}} \begin{bmatrix} 1 \\ 1 \end{bmatrix}$$

represents a signal, shown in Figure 2, whose electric field vector oscillates in the horizontal-vertical plane between extremes $E_S^+$ to $E_S^-$ along a line at 45 degrees to the horizontal. However, the signal may be propagating either into or out of the plane of the figure. In conventional terms, this Jones vector represents a signal whose tilt is either plus or minus 45 degrees, depending on the direction of propagation; the Jones vector representation does not distinguish between the two possibilities. Similarly, the vector

$$\tilde{E}_C = \frac{E}{\sqrt{2}} \begin{bmatrix} 1 \\ i \end{bmatrix}$$

represents a signal illustrated in Figure 3, whose Jones vector has a fixed magnitude and rotates clockwise about a circle in the

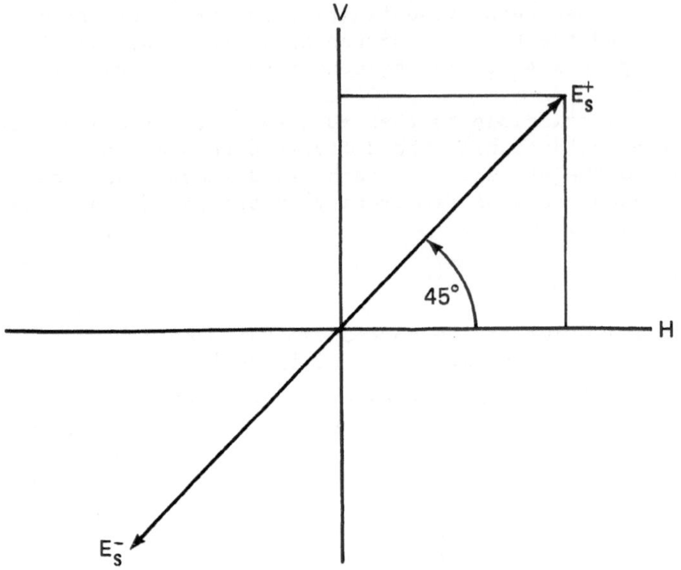


Figure 2: Linear Polarization, +45° from Horizontal

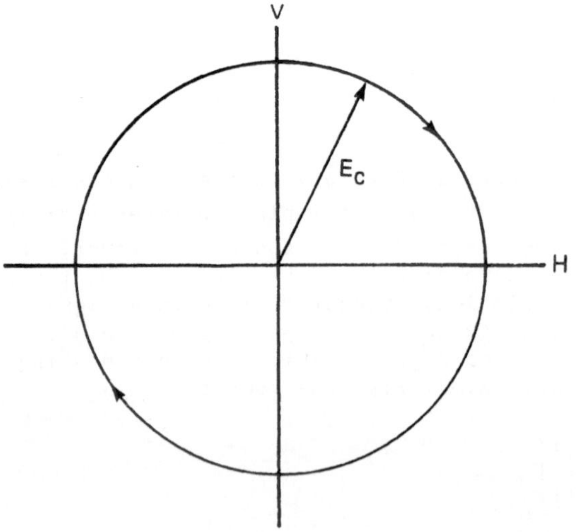


Figure 3: Circular Polarization

horizontal-vertical plane. Again, the signal may be propagating either into or out of the plane of the figure, so that according to conventional definitions, the wave may be either right- or left-hand circularly polarized, depending on the direction of propagation.

This indifference to the direction of propagation is a basic feature of the Jones vector approach, distinguishing its treatment of polarization from conventional treatments. As discussed below, standard analyses tend to create confusion because the usual definitions of polarization include implicit information about the direction of propagation of the wave. However, since, as discussed below, this information is, in fact, often irrelevant, one must carefully remove it at some point in the analysis. The Jones vector approach avoids this problem.

## 3. BASIC FEATURES OF TWO-WAY PROPAGATION

Two-way propagation problems are solved with two fundamental principles: linearity and reciprocity[3,4,5]. Linearity reduces the parameter space for describing propagation and scattering to a small set of complex numbers. For example, in the situation of Figure 1, a transmitted signal at A is represented by a Jones vector

$$\tilde{E}_{AT} = V_T \tilde{E}_A$$

where $V_T$ represents the intensity and phase of the input to the antenna and $\tilde{E}_A$ is the "transmit" Jones vector of the antenna at A. The Jones vector of the signal at B before the interaction at B takes place, $\tilde{E}_B$, is a linear function of the signal at A:

$$\tilde{E}_B = \approx{M}_{AB} \tilde{E}_{AT}$$

where $\approx{M}_{AB}$ is a "propagation matrix" of complex numbers that do not depend on $\tilde{E}_{AT}$. In the same way, the signal at B after the interaction, $\tilde{E}'_B$, is related to the signal subsequently received at A by the propagation matrix $\approx{M}_{BA}$. It is assumed that linearity also characterizes the process at B, relating the Jones vectors of the signal before and after the interaction:

$$\tilde{E}'_B = \approx{Q}_B \tilde{E}_B$$

where $\approx{Q}_B$ is a 2x2 complex matrix.

Thus, from linearity, the received signal at A, $E_{AR}$, is $\tilde{a}$

linear function of the signal transmitted from A, $\tilde{E}_{AT}$:

$$\tilde{E}_{AR} = V_T \, \tilde{\tilde{M}}_{BA} \, \tilde{\tilde{Q}}_B \, \tilde{\tilde{M}}_{AB} \, \tilde{E}_{AT} \quad .$$

One can define the monostatic scattering matrix, $\tilde{\tilde{S}}$, in terms of $\tilde{\tilde{M}}_{AB}$, $\tilde{\tilde{Q}}_B$, and $\tilde{\tilde{M}}_{BA}$ as

$$\tilde{\tilde{S}} \equiv \tilde{\tilde{M}}_{BA} \, \tilde{\tilde{Q}}_B \, \tilde{\tilde{M}}_{AB} \quad . \tag{1}$$

The assumption of reciprocity produces additional simplification by relating parameters for different signal paths. In particular, reciprocity relates reception and transmission characteristics of a given antenna, and relates matrices representing propagation back and forth along the same path.

## 4. A PROOF OF RECIPROCITY

Because the topic of reciprocity is sometimes the subject of disagreement, it is useful to give proofs of the reciprocity theorem and of some of the basic consequences of the reciprocity theorem. This will make the discussion unambiguous and will clarify the issue of the assumptions and basic physical laws on which reciprocity rests. The following standard proof can be found in Landau-Lifshitz[4] or Monteath[5].

Two different collections of external currents, $\vec{J}_{ex,1}$ and $\vec{J}_{ex,2}$, in a given environment will give rise to two different electromagnetic fields, $\vec{E}_1$, $\vec{B}_1$ and $\vec{E}_2$, $\vec{B}_2$. Each of these fields will be solutions of Maxwell's equations:

$$\vec{\nabla} \times \vec{E}_1 = -i\omega\vec{B}_1 \quad , \quad \vec{\nabla} \times (\frac{1}{\mu} \vec{B}_1) = i\omega\epsilon\vec{E}_1 + \vec{J}_{ex,1}$$

$$\vec{\nabla} \times \vec{E}_2 = -i\omega\vec{B}_2 \quad , \quad \vec{\nabla} \times (\frac{1}{\mu} \vec{B}_2) = i\omega\epsilon\vec{E}_2 + \vec{J}_{ex,2} \quad .$$

Taking the scalar product of the first two equations with $\frac{1}{\mu}\vec{B}_2$ and $\vec{E}_2$, respectively, and the scalar product of the second two equations with $-\frac{1}{\mu}\vec{B}_1$ and $-\vec{E}_1$, adding, then applying a vector identity gives

$$\vec{\nabla} \cdot [\vec{E}_1 \times (\frac{1}{\mu} \vec{B}_2) - \vec{E}_2 \times (\frac{1}{\mu} \vec{B}_1)] = -i\omega[\vec{B}_1 \cdot (\frac{1}{\mu} \vec{B}_2) - (\frac{1}{\mu} \vec{B}_1) \cdot \vec{B}_2]$$

$$-i\omega[\vec{E}_1 \cdot (\epsilon \vec{E}_2) - (\epsilon \vec{E}_1) \cdot \vec{E}_2]$$

$$+ [\vec{J}_{ex,1} \cdot \vec{E}_2 - \vec{J}_{ex,2} \cdot \vec{E}_1] .$$

Assuming the dielectric tensor, $\epsilon$, and the magnetic permeability tensor, $\mu$, are symmetric, integrating over all space, applying the divergence theorem, and discarding a surface integral at infinity gives

$$\int \vec{E}_2 \cdot \vec{J}_{ex,1} \, dV_1 = \int \vec{E}_1 \cdot \vec{J}_{ex,2} \, dV_2$$

where $V_1$ and $V_2$ are the volumes containing the source currents $\vec{J}_{ex,1}$ and $\vec{J}_{ex,2}$, respectively.

This is the reciprocity theorem. If the current sources are small and separated by a large distance, then the fields are nearly constant over the volumes of integration and can be moved outside the integrals, giving

$$\vec{E}_2 (1) \cdot \vec{P}_1 = \vec{E}_1 (2) \cdot \vec{P}_2 \qquad (2)$$

where $\vec{P}_1$ and $\vec{P}_2$ are the dipole moments of the source charge distributions, and $\vec{E}_2(1)$ and $\vec{E}_1(2)$ are the values of $\vec{E}_2$ and $\vec{E}_1$ at the positions of $\vec{P}_1$ and $\vec{P}_2$.

## 5. BASIC CONSEQUENCES OF RECIPROCITY

If the dipole moments in Equation (2) are produced by antennas that may receive as well as transmit, then the scalar products in the equation have physical meaning as the received amplitudes at the sites of the antennas. In this case, the reciprocity theorem is the statement that the value obtained in an electromagnetic measurement is unchanged if the source and observer are interchanged.

Specializing further, if one of the antennas is horizontal, while the other is vertical and the two antennas are co-located, then the scalar products in the reciprocity law are proportional to the off-diagonal elements of the monostatic scattering matrix defined in Equation (2), so that if reciprocity holds, then $\approx{S}$ is symmetric. Note, however, that this conclusion assumes that the linear process at the interaction site, denoted above as $\approx{Q}_B$, is entirely passive; i.e., does not involve any external currents.

As noted by Monteath, reciprocity is not a universal physical law. It rests on specific assumptions about the environment in which electromagnetic fields are propagating and may fail in cases where these assumptions do not hold. Appendix A has a discussion of the conditions under which reciprocity may fail.

## 6. THE RECIPROCITY PROBLEM

We claim that the problem with the conventional method for representing the effects of reciprocity in radar measurements is the requirement that the reception and transmission characteristics of a reciprocal antenna be identical and the consequent requirement that different coordinate systems must be used for signal reception and transmission. It is shown below that in the two-way propagation problem this constraint leads to an unpleasant situation: one must introduce a non-linear mathematical process (complex conjugation) to compare the polarizations of fields on the two different propagation paths in spite of the fact that no non-linear physical processes are taking place. This difficulty is aggravated by the fact that complex conjugation does play a significant role in two-way propagation.

The true role of complex conjugation in two-way propagation can be demonstrated by stating the reciprocity relation between receive and transmit characteristics of an antenna using a fixed coordinate system at the antenna. In this case,, the assumptions of linearity and reciprocity lead to the following two rules[3]:

RULE 1

Suppose the measured responses of an antenna to incident unit horizontal and vertical signals

$$\tilde{E}_H = \begin{bmatrix} 1 \\ 0 \end{bmatrix} \quad \tilde{E}_V = \begin{bmatrix} 0 \\ 1 \end{bmatrix}$$

are $R_H$ and $R_V$, respectively. Then the response of the antenna to a general incident signal

$$\tilde{E} = \begin{bmatrix} E_H \\ E_V \end{bmatrix}$$

is

$$R = R_H E_H + R_V E_V \quad .$$

This implies that the "matched" received signal; i.e., the received signal that gives maximum antenna response for fixed incident energy, has the Jones vector

$$\tilde{E}_{RM} \equiv V_R \tilde{E}_R = V_R \begin{bmatrix} R_H{}^* \\ R_V{}^* \end{bmatrix}$$

where * denotes complex congugation and $V_R$ is a complex constant determined by the phase and amplitude of the incident signal.

RULE II

If the same antenna is used to transmit a signal, in the same coordinate system the "matched" transmitted signal will have a Jones vector

$$\tilde{E}_{TM} \equiv V_T \tilde{E}_T = V_T \begin{bmatrix} R_H \\ R_V \end{bmatrix}$$

where $R_H$ and $R_V$ are the complex constants defined in RULE I, and $V_T$ is a complex constant determined by the phase and voltage of the input to the antenna. Note that the vectors $\tilde{E}_T$ and $\tilde{E}_R$ are complex conjugates. This is the core of the problem, because complex conjugation is not a linear relation. Thus, in the standard treatment, in which $\tilde{E}_T$ and $\tilde{E}_R$ have to be equal because the receive and transmit polarizations of an antenna are, by convention, equal, one must both use a different coordinate system for $\tilde{E}_T$ than was used for $\tilde{E}_R$ and introduce a spurious non-linearity to undo the apparent change in polarization produced by the change in coordinate system.

It can be argued that setting the receive and transmit polarizations of a reciprocal antenna equal to each other is intuitive and simplifies some computations. However, this apparent simplification in some cases is achieved at the cost of promoting confusion in other cases.

For example, since, as noted above, practically all effects of propagation and scattering are linear, a returned signal must be a linear function of any transmitted signal. But, since complex conjugation is not a linear process, the received signal matched to a given antenna is not a linear function of the signal transmitted by the same antenna, so it is clear immediately that there is no linear process that will always return a matched signal from any antenna. This basic fact is obscured by the conventions used in standard methods.

The view that the receive and transmit characteristics of a reciprocal antenna are different alters the explanation of some basic phenomena. For example, it is common knowledge that reflection of a circularly polarized wave from a flat conducting plate will return a signal cross-polarized to the original transmitter. The usual explanation of this fact is that reflection from the plate changes the original circular transmission to the opposite circular polarization; since the transmitting and receiving polarizations of the antenna are the same, the signal and the antenna will be cross-polarized.

In the Jones vector approach, the explanation of the same phenomenon is that the flat plate leaves the signal polarization unchanged; the returned signal is cross-polarized to the antenna because the receive and transmit polarizations of a circularly polarized antenna are orthogonal. The fundamental advantage of this view is that the reflection from the flat plate, the transmissions along the two signal paths and the processes of transmission and reception are now all represented unambiguously as linear processes. By contrast, in the conventional approach, a complex conjugation is required at some point.

## 7. RECIPROCITY IN TWO-WAY PROPAGATION

For the reasons given above, the analysis of two-way propagation is simplified with the use of Jones vector methods, which require that all quantities at a given point be referred to a fixed coordinate system without regard to the direction of propagation of a signal at that point. With this convention, reciprocity relates the matrices $\tilde{\tilde{M}}_{AB}$ and $\tilde{\tilde{M}}_{BA}$ for propagation back and forth along the same path. One has

$$\tilde{\tilde{M}}_{BA} = \tilde{\tilde{M}}_{AB}^{T}$$

that is, $\tilde{\tilde{M}}_{BA}$ is the matrix transpose of $\tilde{\tilde{M}}_{AB}$. It is shown in Appendix B that this formulation of reciprocity in terms of propagation matrices is equivalent to the formulation above in terms of receive and transmit properties of antennas. Putting together all the pieces, if an antenna at A transmits a Jones vector,

$$\tilde{E}_{TA} = V_T \tilde{E}_T$$

where $\tilde{E}_T$ is the "transmit" vector defined above, then in a two-way transmission the received amplitude is

$$V_R = V_T \, \tilde{E}_T^T \, \tilde{\tilde{M}}_{AB}^T \, \tilde{\tilde{Q}}_B \, \tilde{\tilde{M}}_{AB} \, \tilde{E}_T \quad .$$

Note that this equation is entirely linear, with no complex conjugation and with all quantities determined at a given point referred to a fixed coordinate system at that point. Setting

$$\tilde{E}_{EFF} = \approx{M}_{AB}\tilde{E}_T$$

gives

$$t_B \equiv \frac{V_R}{V_T} = \tilde{E}^T_{EFF}\, \approx{Q}_B\, \tilde{E}_{EFF} \quad .$$

The horizontal and vertical components of $\tilde{E}_{EFF}$ may be interpreted as the responses of the antenna A to unit horizontal and vertical signals from B.

As noted above, if the process $\approx{Q}_B$ is passive, not involving any external currents, then the monostatic scattering matrix (Equation 1) is symmetric, implying that the matrix $\approx{Q}_B$ is also symmetric.

A good illustration of the power and simplicity of Jones vector methods is the analysis of how interaction will vary if the scatterer is rotated. Two model processes of interest in this regard are the perfect reflector

$$\approx{Q}_R = Q_R^{(o)} \begin{bmatrix} 1 & 0 \\ 0 & 1 \end{bmatrix}$$

and the cross-reflector

$$\approx{Q}_X = Q_X^{(o)} \begin{bmatrix} 0 & 1 \\ -1 & 0 \end{bmatrix} \quad .$$

Note that the process $\approx{Q}_X$ cannot be produced by a passive scatterer.

For the perfect reflector, one has

$$t_R = (E^2_{EFF,H} + E^2_{EFF,V})Q_R^{(o)}$$

and for the cross-reflector, one has

$$t_X = 0 \quad .$$

The perfect reflector and the cross-reflector share an interesting property: their effects on a signal are independent of rotation in the horizontal-vertical plane[3]. That is, if the physical object that receives and returns the signal is rotated in the horizontal-vertical plane to some arbitrary angle from its original position, its effect on a signal is unchanged.

For the perfect reflector, this invariance can be proved by noting that its matrix is proportional to the unit matrix, which is invariant under rotation. For the cross-reflector, invariance follows from the fact that the received intensity $t_X$ is evidently a constant scalar, independent of orientation. From linearity, it then follows that interaction with a linear combination of a perfect reflector and a cross-reflector

$$\approx{Q}_{INV} = A\approx{Q}_R + B\approx{Q}_X$$

is also invariant under rotation. In fact, $\approx{Q}_{INV}$ is the most general form of rotationally invariant process, so a general interaction matrix

$$\approx{Q} = \begin{bmatrix} q_{11} & q_{12} \\ q_{21} & q_{22} \end{bmatrix}$$

can be broken up into rotationally invariant and rotationally non-invariant parts:

$$\approx{Q} = \approx{Q}_{INV} + \approx{Q}_{N-INV}$$

where

$$\approx{Q}_{INV} = \tfrac{1}{2} \begin{bmatrix} q_{11} + q_{22} & q_{12} - q_{21} \\ q_{21} - q_{12} & q_{11} + q_{22} \end{bmatrix}$$

$$\approx{Q}_{N-INV} = \tfrac{1}{2} \begin{bmatrix} q_{11} - q_{22} & q_{12} + q_{21} \\ q_{12} + q_{21} & q_{22} - q_{11} \end{bmatrix} .$$

# APPENDIX A

As noted above, reciprocity is not a fundamental physical law. Reciprocity rests on specific assumptions about the environment and the materials in which electromagnetic fields propagate, and may fail in cases where the assumptions do not hold.

One such assumption, explicit in the proof of reciprocity given above, is symmetry of the dielectric and magnetic permeability tensors $\epsilon$ and $\mu$. This symmetry is often taken to be the fundamental assumption on which reciprocity rests. However, Landau and Lifshitz[4] show that symmetry of these tensors is a consequence of deeper assumptions about the algebraic form of the free energy of a system. In particular, using the approach of Landau and Lifshitz, one can show that if the response of a system to external electric and magnetic fields is linear, then the dielectric and magnetic permeability tensors will be symmetric. It must be noted that in this context, "linear" describes large-signal linearity, rather than conventional small-signal linearity. This analysis explains why reciprocity fails for materials with large-signal non-linearity such as ferrites and plasmas, in which polarizability depends on an external magnetic field or on internal magnetization.

Monteath[5] states that reciprocity will only fail in cases of magnetic non-linearity, that is, in cases where $\epsilon$ and $\mu$ depend explicitly on a magnetic field. His demonstration relies on the fact that in free space, the carriers of positive and negative electric charge are distinguishable by their different masses, while North and South magnetic poles are not distinguishable in this way. However, such distinctions need not hold within materials. For example, within an intrinsic semiconductor, the carriers of positive and negative charge (electrons and holes) may or may not have the same mass. More generally, since elementary excitations within a material do not necessariy obey symmetries that hold in free space, there is no a priori reason why a material that demonstrates electric non-linearity could not be non-reciprocal.

Examples of "electric" non-linearity and consequent failure of reciprocity occur in optics. A good example is the $BaTiO_3$ phase-congugate mirror[6].

# APPENDIX B

Two different formulations of reciprocity are given above: one relating receive and transmit properties of an antenna, the other relating matrices for propagation back and forth along the same path. It is shown here that the two formulations are equivalent. To show that the rules relating receive and transmit characteristics of an antenna follow from the rule relating propagation matrices, consider the situation pictured in Figure 4a. At A, a signal, $\tilde{E}$, is received by a vertical antenna. The output of this antenna is connected directly into the input of a second, arbitrary antenna that emits an electromagnetic signal

$$\tilde{T} = E_V \begin{bmatrix} T_H \\ T_V \end{bmatrix} .$$

The matrix $\tilde{\tilde{M}}_{AB}$ relating $\tilde{E}$ to $\tilde{T}$ is found by noting that

$$\tilde{T} = E_V \begin{bmatrix} T_H \\ T_V \end{bmatrix} = \begin{bmatrix} 0 & T_H \\ 0 & T_V \end{bmatrix} \begin{bmatrix} E_H \\ E_V \end{bmatrix} .$$

For the reciprocal situation shown in Figure 4b, one sends an arbitrary signal

$$\tilde{R} = \begin{bmatrix} R_H \\ R_V \end{bmatrix}$$

into the antenna at B. Using reciprocity, the output of the vertical antenna at A is

$$\begin{bmatrix} E_H \\ E_V \end{bmatrix} = \tilde{\tilde{M}}_{BA} \begin{bmatrix} R_H \\ R_V \end{bmatrix} = \tilde{\tilde{M}}_{AB}^T \begin{bmatrix} R_H \\ R_V \end{bmatrix}$$

$$= \begin{bmatrix} 0 & 0 \\ T_H & T_V \end{bmatrix} \begin{bmatrix} R_H \\ R_V \end{bmatrix}$$

$$= \begin{bmatrix} 0 \\ R_H T_H + R_V T_V \end{bmatrix}$$

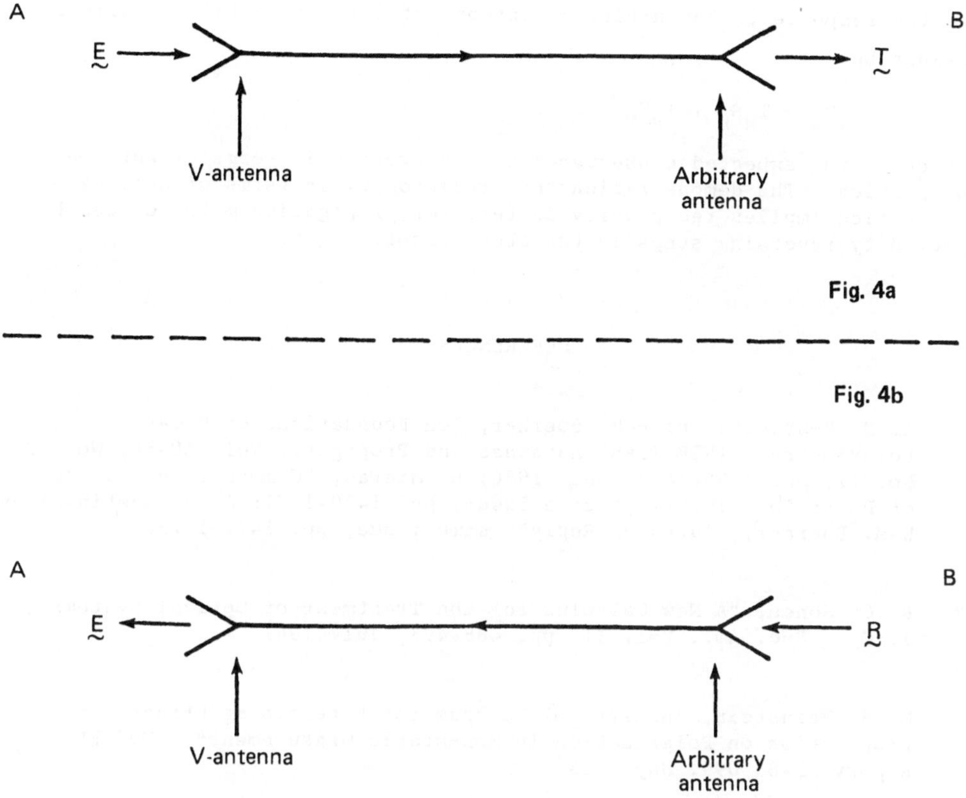


**Fig. 4 Application of reciprocity to transmit and receive characteristics of an antenna.**

so the response of the arbitrary antenna at B to the arbitrary signal $\tilde{R}$ must be

$$V_B = T_H R_H + T_V R_V$$

which is the expected consequence of reciprocity in terms of antenna properties. The demonstration that reciprocity in terms of antenna properties implies reciprocity in terms of propagation matrices can be proved by reversing steps in the above proof.

## REFERENCES


1. A. B. Kostinski and W-M. Boerner, "On Foundations of Radar Polarimetry", IEEE Tran. Antennas and Propagat., Vol. AP-34, No. 12, No. 12, pp. 1395-1404, Dec. 1986; H. Mieras, "Comments on 'Foundations of Radar Polarimetry'," same issue, pp. 1470-1471; A. B. Kostinski and W-M. Boerner, "Authors' Reply", same issue, pp. 1471-1473.

2. R. C. Jones, "A New Calculus for the Treatment of Optical Systems", J. Opt. Soc. Am., Vol. 31, pp. 488-493, July 1941.

3. M. R. Feinstein, "Analytical Methods for Determining Effects of Propagation on Polarization in Monostatic Measurement", JHU/APL Report FS-83-099, May 1983.

4. L. D. Landau and E. M. Lifshitz, "Electrodynamics of Continuous Media", pp. 288-9 (Pergamon Press 1960).

5. G. D. Monteath, "Applications of the Electromagnetic Reciprocity Principle", (Pergamon Press 1973).

6. J. Feinberg, "Photorefractive Nonlinear Optics", Physics Today, October 1988, pp. 46-52.

# ASSESSMENT OF CALIBRATION PROCEDURES FOR POLARIMETRIC RADARS

Lloyd W. Root
U. S. Army Missile Command
Research, Development and Engineering Laboratories
Redstone Arsenal, Alabama 35898
United States of America

ABSTRACT. Calibration standards, internal and external calibration techniques (both passive and active types) which are unique to polarimetric radars are discussed and followed by the author's experience in using them. A new even bounce conical corner reflector is introduced which has excellent potential for establishing an angle reference for passive reflector arrays. Active external calibration techniques are recommended where the active device used was calibrated in a controlled environment.

## 1. POLARIMETRIC RADAR CALIBRATION

### 1.1 Calibration Requirements

The optimum calibration technique is one that can be used during the measurement of polarimetric radar data so that relative measurements can be made between an unknown target and calibration standards. Each measured orthogonal or non-orthogonal degree of freedom, must have a calibration technique applied before, optimally during, or after the measurement. This assessment is limited to a calibration of the polarimetric properties of the radar and does not include the calibration of time space position data.

All calibration techniques have been divided into two general categories called 'Internal Calibration' and 'External Calibration'. Internal Calibration techniques involve the polarimetric radar's hardware which may inject known signals at various points (active types) so that the measurement and data recording system will record known signals. Or the Internal Calibration technique may use a passive type installed on the radar's transmission lines. One early example of such a passive device is the so called 'echo box'. The 'echo box' receives a sample of the transmitted signal (both polarimetric channels), and according to the 'Q' of this 'box', a known signal is returned with exponentially decreasing amplitude and range linear phase. By appropriate selection of hybrids, the differential phase



W.-M. Boerner et al. (eds.), Direct and Inverse Methods in Radar Polarimetry, Part 1, 845–860.

between each echo box may be made plus or minus phase quadrature.

External Calibration techniques involve free space propagation and also use 'active' or 'passive' polarimetric radar target simulating types. One example of an active type of external calibrating system is the coherent repeater. This active type receives the polarimetric radar's signal and delays (to displace the signal in 'range' or time to remove unwanted ground clutter), amplifies and re-transmits this signal with known characteristics. The active type may simply transmit a sample of the polarimetric radar's signal from a remote tower with known characteristics. External passive types of calibration consist of the well known corner reflectors or spheres being dropped from an aircraft, mounted in a drifting balloon, or on a calibration tower.

## 1.2 Calibration Standards

The term 'corner reflector' has been used in the past to decsribe a reflector made up of two or three mutually orthogonal planes. If any of these planes are not mutually orthogonal, then an additional descriptor such as 'acute' (less than 90 degrees) or 'obtuse' (greater than 90 degrees) must be used with the term 'corner'. Odd bounce radar reflectors such as a flat plate, sphere, Luneberg lens reflector, trihedral (triangle corner) or a polytrihedral are very common and have been used for years [1] with linearly polarized non-polarimetric monostatic radars. Circularly polarized non-polarimetric radars have had to find an even bounce reflector to calibrate the copolarized (and only) radar channel.

The even bounce dihedral corner reflector had only one serious spatial angular use problem. With the exception of the odd bounce flat plate, the odd bounce reflectors have comparatively large spatial angles over which they would provide a reasonable return (Sphere- all angles, Spherical Luneberg lens reflector- 150 degree cone {with good return over a hemisphere}, Triangular corner {trihedral}-76 degree cone {38 degree half-power angle}, Circular corner- 92 degree cone, and a square corner {corner cube}- 50 degree cone) [2]. The use of circularly polarized radars prompted efforts to obtain an even bounce reflector without the spatial positioning problems of the dihedral and without the spatial area efficiency problems of the biconical corner reflector. The spatial area efficiency is defined as the ratio of the effective aperture area to the physical area expressed in percent. The effective aperture area is that area 'A' used to find the radar cross section in the formula radar cross section is equal to four pi times the area 'A' squared divided by the wavelength squared, [3].

Efforts to produce a spatially broad coverage and spatial area efficient even bounce radar reflector, took three primary routes (modify existing corner reflectors, create new reflectors and use antennas as reflectors). First, one of the three 'bounces' that took place on a trihedral did not have to take place on a conductor with the associated 180 degree phase delay. Instead, a dielectric sheet was added to one of the three sides of the trihedral and was reported by Ed Kennaugh [4] on 1 October 1955. Later, John Hines [5], measured a circular corner reflector with one dielectric side at 9080 MHz with nearly identical results (even bounce) compared to the circular corner reflector without the dielectric sheet as an odd bounce radar reflector. The dielectric sheet produced a reduction in response for 45 degree linear polarization for the dielectric sheet circular corner placed in the horizontal plane. But no reduction occurred when the the incident linear polarization was parallel or perpendicular to the dielectric sheet. A trihedrial was filled with paraffin (dielectric constant=2.15) with one metal side removed . This was tested at 9090 MHz [6], and found to have a cross section about 1 to 1.5 dB below that of a triangular corner reflector with even bounce properties .

The second approach to the even bounce reflector came with what Ed Kennaugh called the 'Double-double-bounce' corner reflector [7]. This reflector was a 90 degree pyramidal horn or pyramidal corner reflector. Although theory predicted a complete null at the center of this reflector, unwanted reflections, possibly from the nearly flat edges of the four corners, filled in this null. The peak cross section of this reflector was 1.82 dB below (or about 2/3) that of a flat plate of the same aperture area or 1.7 dB above that obtained from a trihedral with the same aperture area.

Ed Kennaugh [8] introduced the metal plate 'Circularizer' with a flat plate reflector. The 'Circularizer' consisted of a series of parallel metal strips with a spacing and width such that the TE01 guide wavelength would apply a 90 degree phase lag to the wave parallel to the strips. The circularizer would be placed at a 45 degree angle with respect to a propagating linear polarized wave. The emerging wave would then be circularly polarized if no waveguide losses were encounted during the 90 degree phase lag. This 'circularizer' was the metal lens equavilant of the British Twist Reflector. In fact, there were small reflections along the edge of the parallel metal strips forming the 'guide' before the TE01 mode could be excited, so it was suggested to use a dielectric material in place of the metal strips with the reflector. The idea of enhancing the reflections from the edge of the metal strips by replacing the strips with wires embedded into a dielectric sheet (The British Twist Reflector method) was not considered.

The author accidentally discovered an elliptical cone reflector in 1978. During polarimetric radar calibrations at Redstone Arsenal, the author had a large (man size) 300 square meter bi-conical corner reflector mounted within the field of view of the MARFS (Multi-environment Active Radio Frequency Seeker) 17 GHz polarimetric radar when suddenly a gust of wind blew it over on its side, unintentionally forming a slightly dented (elliptical) conical corner reflector on each end. The resultant 20 or so dB increase in the even bounce radar cross section led the author (unpublished) to investigate conical horn antennas with different cone flare angles as even bounce reflectors. But this leads to the third approach for even bounce polarimetric radar calibration standards.

The third approach to an even bounce reflector probably began again with Ed Kennaugh [8] . Using an antenna with its transmission line terminals terminated in a moving short circuit or some other type load, an even bounce antenna such as the X-Band helical reflector reported by Ed Kennaugh, could provide the even bounce reflector function. The main problem with the helix was that when it was properly terminated with a ground plane, the ground plane itself caused a large odd bounce reflection which destroyed the circularity of the reflected wave. Another problem with the helix is that it does not respond to both right and left circular transmitted waves because of the physical turn direction of the helix. Conical ground planes were evidently not considered but they would have reduced the retro reflection problem.

Ed Kennaugh [9] introduced parasitic antennas as reflectors. He showed that a 6 wavelength polyrod with its end either open or short circuited, would act as an even bounce (right or left circular received would mean that right or left circular would be transmitted) reflector with a two way beamwidth of about 22 degrees (one way beamwidth was approximately 27 degrees). He re-introduced the familar effective receivng area times gain as the peak echo area or cross section. The gain squared times wavelength squared divided by four pi is therefore the radar cross section. His 6 wavelength shorted polyrod had a peak echo area of 225 square wavelengths at a 3 centimeter wavelength. That compared to a calculated peak echo area of 200 square wavelengths using the value of gain obtained from the antenna patterns (approximately 17 dB). The sidelobes that would normally be recorded in an antenna application were not present in the shorted antenna configuration for cross section. This same approach could be used on any antenna or antenna array. The addition of active devices such as varible phase shifters and attenuators would also enhance the dynamic calibration. The author has proposed to use a hollow magnetic dipole physically centered over a hollow electric dipole called a SID (Standard Integrated Dipoles), all embedded into a stryofoam airfoil (so it can be towed by an aircraft). Each dipole would contain rechargeable batteries, a laser fiber optics RF/EO (Radio Frequency to Electro Optical) transducers, delay lines followed by EO/RF transducers and RF amplifiers to re-transmit both the magnetic and electric dipole as a calibration tool (Provides both vertical and horizontal simulated external active calibration targets for both mono-static and bi-static radars).

The Van Atta Array [10] consists of an array of radiating elements which are interconnected in such a way that the phase distribution corresponding to a received wavefront is transformed into that required to form a transmitting beam in the same direction. The transformation is accomplished by connecting each element to its linearly opposed mate (with reference to the array center), all connecting lines having a length determined by their position in the linear array. The measured properties of such a device show that it provides angular coverage superior to that of a corner reflector. For example, a 4 by 4 dipole array with half wavelength spacing produced a half power return at an angle of 30 degrees (60 degree cone). Another advantage towards using antennas as calibrated reflectors over corner reflectors is that corner reflectors require a significant three dimensional space while antennas generally require a two dimensional space with little third dimensional depth (end fire arrays to the contrary).

Figure 1 schematically shows what happens when an electromagnetic wave hits a target. There are initial specular type reflections (which may or may not be directed back towards the radar) and a range delayed re-transmission because the discontinuity at the reflection point acts as a receiveing antenna and it re-radiates the resulting guided wave. These waves were first called 'creeping waves' in 1952 by Franz and Depperman [11]. The geometry of the discontinuity would make the horizontal and vertical portions travel different distances which would result in a change in the polarization of the return wave. These conditions are far more prevalent on targets than one might imagine but they are hidden due to their lower (Gain squared Wavelength squared divided by four pi) radar cross section.

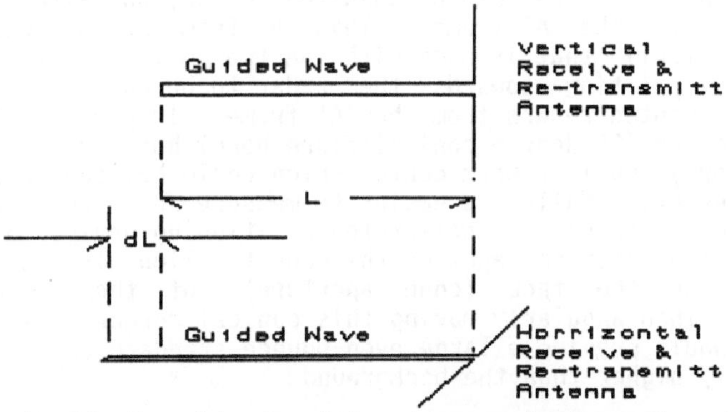


Figure 1. Vertical and horizontal antennas showing the differential phase 'dL' that could exist between the two guided paths.

Figure 2 shows the free space propagation equivalent to the conditions of figure 1 where the differential phase 'dL' is represented by the spacing between the two orthogonal wire grids. Since targets exhibit this characteristic, it follows that a calibration system must be able to simulate a single scattering center within a single polarimetric radar range cell.

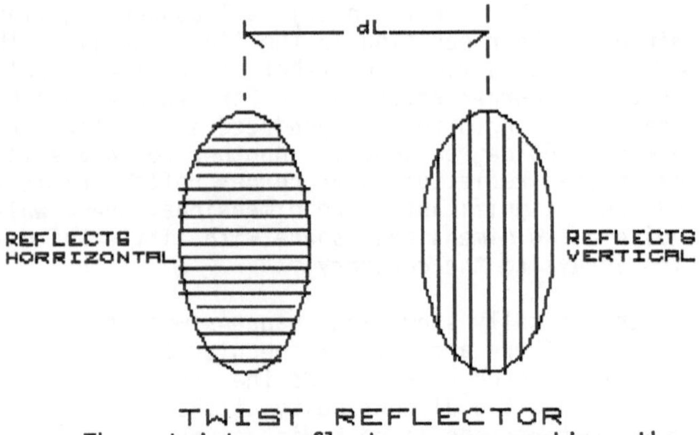


Figure 2. The twist reflector representing the propagation differential path length equivalance to the guided waves of figure 1.

This suggests an active calibration standard should be for field use. Then this active external calibration standard would be placed in an anechoic room and calibrated against a corner reflector standard such as a rotating dihedral and a trihedral. Depending upon the processing capability of the polarimetric radar, sets of odd and even bounce corner reflectors may have to be assembled on a metal 'A' frame that can be adjusted for any angle to diminish the reflections off the 'A' frame. This 'A' frame may be constructed of angle iron (steel) that is made with the two metal surfaces of the angle iron- pointing towards the radar to be calibrated to further reduce the unwanted return from the 'A' frame. In general, the conical corner reflector (90 degree conical flare horn) has a very deep (30 to 40 dB or more) null at beam center which could be used as a direction centering device. Nulls are easier to measure so after finding the null position, a transit or calibrated positioning device could be used with a small hole in the apex of the cone to align with two orthogonal cross hairs on the face (cone aperture) of the conical corner reflector. Then angularly moving this conical corner reflector to the peak value would provide a large even bounce standard that had a return significantly higher than the background.

Figures 3 and 4 show all of the passive reflectors that are currently used as external passive calibration standards. Table A shows the equations for the peak radar cross section for each of these reflectors. In general geometric ray optics has been used to obtain these equations similar to R. C. Spencer [12] . Many of the equations used to show how the radar cross section varies as a function of the attitude of the corner reflector are not quite correct and are being re-visited by Dr. Herb Holl in Huntsville Alabama.

In general, these equations are based on the assumption of an equivalant uniformly illuminated aperture where the ratio of its one dimensional size to the wavelength 'lambda' is very much greater than unity. The Luneberg reflector is simply a set of concentric dielectric spheres of varying dielectric constant with a reflecting surface on one hemisphere. K. M. Siegel [13] holds a patent on a Siegel-Luneberg reflector which consists of a Luneberg Lens with the entire surface coated with a thin partially conducting layer (instead of being hemispherically coated with a fully conducting layer). The final result being an isotropic scatterer. A good description of this and most of the equations found in Table A can be found in reference [13].

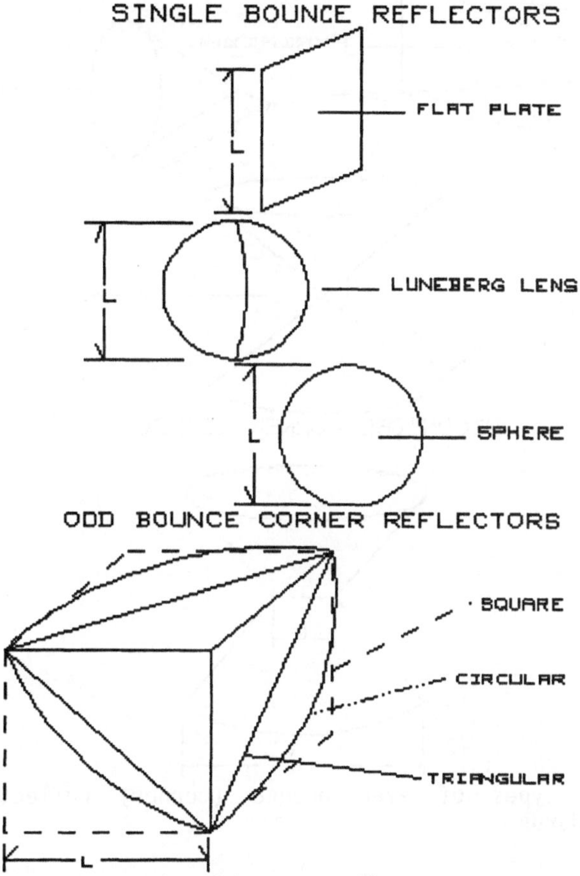


Figure 3. Six types of odd bounce reflectors used as a calibration standard.

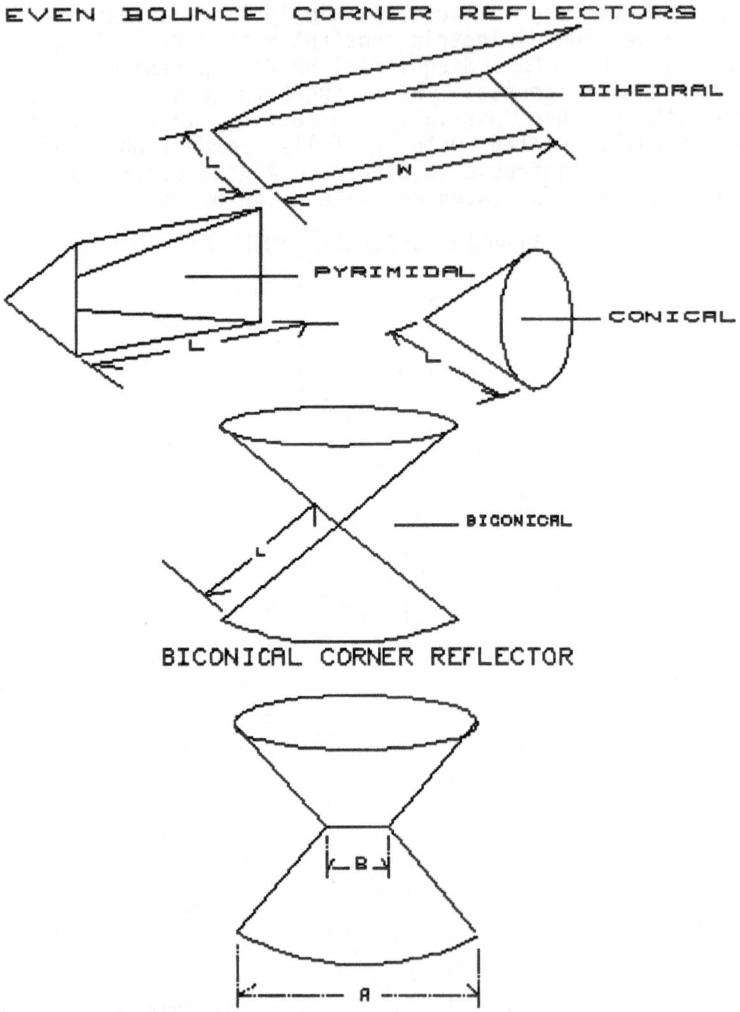


Figure 4. Four types of even bounce (corner) reflectors used as calibration standards.

Table A shows the radar cross section equations for these corner reflectors. The equation for the even bounce dielectric filled triangular corner reflector is the same as the odd bounce triangle corner with the 1 to 1.5 dB loss applied to the total cross section. The equation for the sphere also includes a near field approximation from V. H. Weston [14].

TABLE A

| REFLECTOR TYPE | RADAR CROSS SECTION | AREA EFFICIENCY % |
|---|---|---|
| General equation | $\sigma = 4\pi \left( \frac{Area}{\lambda} \right)^2$ | $\frac{\text{EFF AREA X 100}}{\text{AREA}}$ |
| Flat Plate | $\sigma = 4\pi \left( \frac{L^2}{\lambda} \right)^2$ | 100% |
| Luneberg Lens | $\sigma = 4\pi \left( \frac{\pi L^2}{4\lambda} \right)^2$ | 100% |
| Sphere (R=Range) | $\sigma = \frac{\pi L^2}{4} \left( 1 + \frac{L}{2R} \right)^{-2}$ | (R>>L) |
| Cylinder, r=radius | $\sigma = \frac{2\pi r L^2}{\lambda}$ | |
| Square Corner | $\sigma = 12\pi \left( \frac{L^2}{\lambda} \right)^2$ | 100% |
| Circular Corner | $\sigma = 4\pi \left( \frac{1.1138 L^2}{\lambda} \right)^2$ | 81.9% |
| Triangle Corner | $\sigma = \frac{4\pi}{3} \left( \frac{L^2}{\lambda} \right)^2$ | 66.7% |
| Dihedral Corner | $\sigma = 4\pi \left( \frac{2\sqrt{2}\, LW}{\lambda} \right)^2$ | 100% |
| Pyrimidal Corner | $\sigma = 4\pi \left( \frac{L^2}{1.233\lambda} \right)^2$ | 81.1% |
| Biconical Corner | $\sigma = \frac{4\pi}{9\lambda} \left( A\sqrt{2A-B} - B\sqrt{B} \right)^2$ | |
| Conical Corner | $\sigma = 4\pi \left( \frac{\pi L^2}{4.33\lambda} \right)^2$ : Not Verified | 92.4% |

The equation for the conical corner reflector has not been properly verified and is based on a peak cross section at 22.5 degrees. The spatial area efficiency is listed in the last column of Table A and is a function of wavelength for a single curved surface or wavelength squared for a doubly curved surface.

1.3 Calibration Problems

The primary problem with the calibration of a polarimetric radar in the operational environment is multipath or the pollution of the desired calibration signal with undesirable electromagnetic energy. This pollution can be represented as a multipath or interfering wave of relative amplitude 'A' and phase 'alpha', which adds to the wave of unity amplitude and zero relative phase according to equation 1.

$$1e^{j0} \pm Ae^{j\alpha} = B_{\pm}e^{j\beta_{\pm}} \qquad (1)$$

The resultant wave has an amplitude 'B' and is at a phase 'beta' such that B is the amplitude error and 'beta' is the phase error. If we call the parameter 'X' the amplitude error expressed in dB (X=20*Log[B]), and the parameter 'Y' as the interference ratio (Y=20*Log[A]), then equation 2 is the result.

$$Y=10*LOG\left[10^{(X/10)}+1-2*10^{(X/20)}*COS(\beta)\right] \qquad (2)$$

Now we plot in figures 5 and 6, the amplitude error vs the interference ratio 'Y' for fixed values of phase error 'beta' and vs the phase error 'beta' for fixed values of interference ratio 'Y'. Notice in figure 5, only a one degree phase error can still cause an amplitude error of 1 dB with an interference ratio of -18 dB. Figure 6 shows the regions in which a given interference ratio can provide serious phase errors. For example an interference ratio of -10 dB (one tenth the power of the wave to be measured) causes phase errors as much as 18 degrees with little or no amplitude error. Adjacent polarimetric range gate pollution causes serious errors depending upon the time filtering capability of the radar. Decisions as to odd or even bounce will be bounded by these errors. Even frequency agility with its FFT processing is bounded by these same errors.

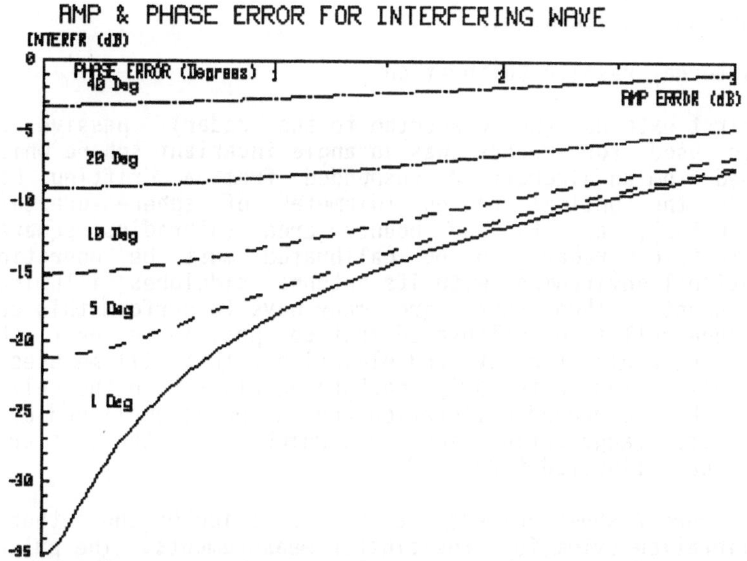


Figure 5. Interference ratio vs amplitude error for fixed values of phase error 'beta', where the interference ratio is the relative magnitude of the interfering wave cause by multipath or transmission line discontinuities.

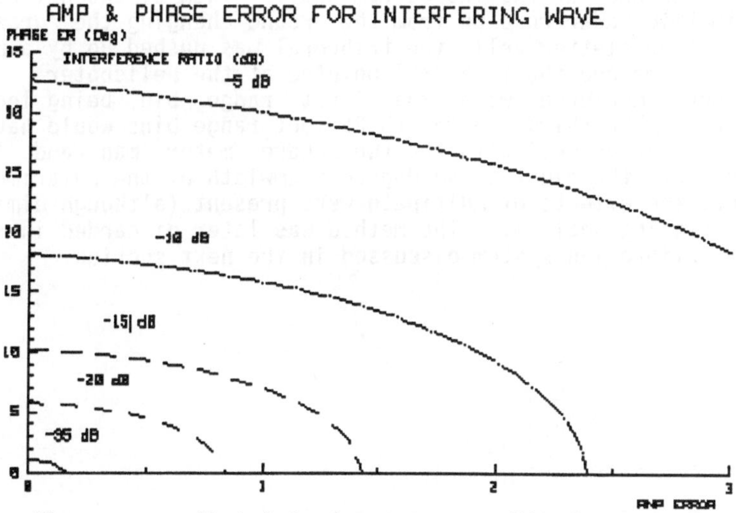


Figure 6. Phase error 'beta' in degrees vs amplitude error for fixed values of the interference ratio from equation 2.

## 2. CALIBRATION TECHNIQUES

### 2.1 External Passive Calibration

The first external (not connected to the radar) passive calibration device used for radars was an angle invariant sphere which could be dropped from an aircraft or suspended from a drifting balloon and provide the optical region (diameter of sphere greater than 10 wavelengths), as its odd bounce area (pi*radius squared). The polarimetric radar to be calibrated must be operating in its operational environment with its antenna sidelobes illuminating that environment. Then there are many ways to perform this calibration. The range cell to be calibrated must be positioned over the various ranges and angles (azimuth and elevation), that will be used to perform the calibration with only background clutter in the cell. This is called the background calibration and can be accomplished by early and late gate range gating techniques which sample the clutter before and after the calibrated target.

Figure 7 shows one example that was tried by the author to provide a calibration sytem for tree clutter measurements. The pulse width was 70 nanoseconds for a 35 foot range resoulution. The radar had 50 nanosecond range bins but because of adjacent range bin pollution, an 'A' frame was constructed out of wood (PVC Pipe was not strong enough without making a truss for each 'A' frame leg), that was 75 feet long. The bottom edge of the 'A' frame was hinged in a rotating joint staked to the ground at the two bottom ends. Then as the Helicoptor borne polarimetric radar rose up from the ground changing the grazing angle to the tree clutter cell, the trihedral was pushed up by a man keeping the 'A' frame and the trihedral pointed at the helicopter. Four range bins were measured with the first range bin being locked to the trihedral. The third and fourth 25 foot range bins would have measured the tree clutter cell with the two square meter man and 'A' frame. Because of the eight or so degree beamwidth of the polarimetric radar antenna, the effects of multipath were present (although diminished at higher grazing angles). The method was later discarded in favor of an active calibration system discussed in the next section.

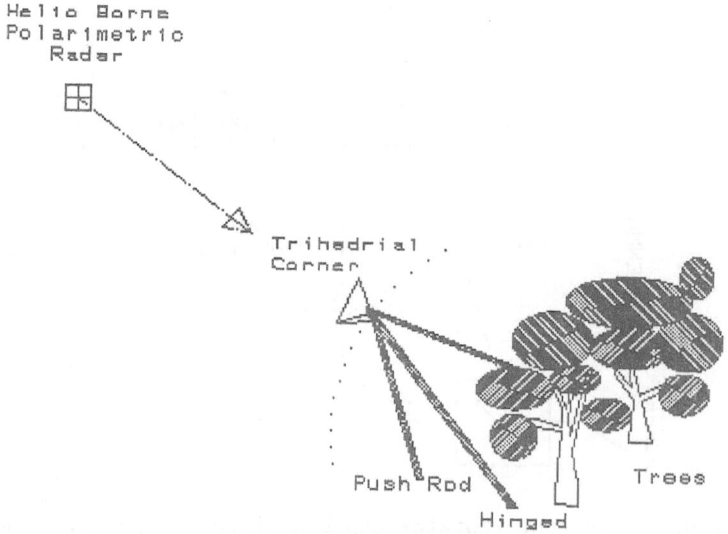


Figure 7. Illustration of a Trihedral corner reflector mounted on an 'A' frame hinged on the earth at each end with a man operated push rod to position (elevate) the reflector as a helicopter mounted polarimetric radar varies its monostatic viewing angle of the tree clutter.

## 2.2 External Active Calibration

Figure 8 shows a generalized schematic of a coherent repeater that was built by the author at the frequency of the MARFS polarimetric radar (17 GHz, alternately right and left circularly polarized transmitted waves). A 35 GHz version was also designed constructed and tested by the author and J. Cole [15]. Later a vertically polarized horn antenna was substituted for the circular receive horn and polarizer, orthomode transducer combination because multipath from the elevated MARFS polarimetric radar was less, especially near the Brewster angle. The receive antenna, first attenuator and amplifier combination was mounted on a gun stock so it could be held by a one square meter man and directed at the MARFS/Helicopter in the clutter cell of figure 7 replacing the passive calibration system. Anechoic material was used on the ground to cover the coherent repeater and as much of the man possible. The delay lines were set to move the tracked coherent repeater target four range bins 100 feet behind the clutter cell to be measured (and based on grazing angle, the coherent target was out of the clutter cells to be measured). Then an early gate measured the tree clutter referenced to the relative range stationary coherent repeater target.

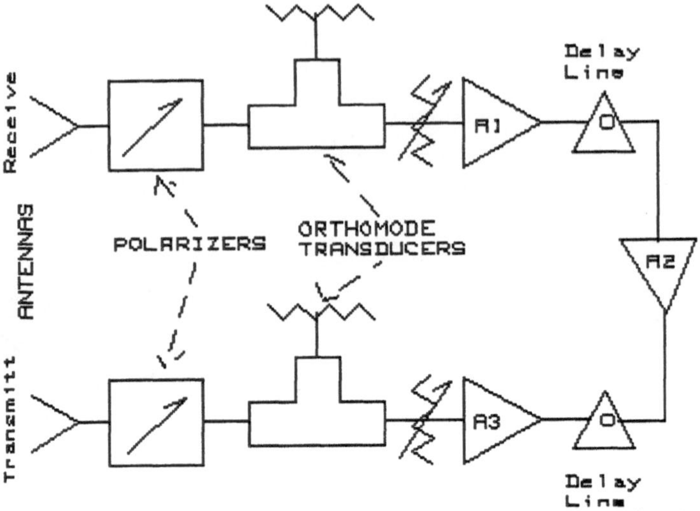


Figure 8. Coherent repeater capable of receiving any polarization by switching the polarizer and orthomode transducer. The input attenuator insures proper operation within the dynamic range of amplifier A1, the two delay lines and amplifier A2. The output attenuator relative to the setting of the input attenuator sets the amplitude of the delayed return and the transmitting antenna polarization is adjusted with the switchable polarizer and orthomode transducer combination.

## 2.3 Internal Passive Calibration

Besides the echo box approach, variable phase shifters and step attenuators have been used with internal active calibration systems. Radiometric techniques such as a 'Y' factor temperature comparison method between the polarimetric antenna and a passive or active noise source have been used to provide relative measurements. This accessment will not cover internal calibration methods which are hardware and system dependent.

## 2.4 Internal Active Calibration

The first internal (connected to the radar) active calibration device used for radars consisted of inserted signals to match incoming signals. Polarimetric radars have the additional problem of matching both phase and amplitude. There are a wide variety of internal calibration techniques where samples of the transmitted signal both in phase and amplitude are delayed and re-injected at various points within the polarimetric radar's measurement system. Again, this assessment will not cover hardware dependent internal calibration methods.

## 3. CONCLUSIONS

Wherever possible, an active external calibration technique should be used with a passive external calibration standard. The passive external calibration standard should be a triangle corner reflector for the odd bounce and the dielectric filled triangle corner reflector with one conducting side missing for the even bounce reflector. Care must be taken to insure the dielectric used has no dispersive characteristics. Additional care must be taken to insure that the effective beamwidth of the passive external calibration standard does not have a geometrical multipath condition which would pollute the calibration. Bistatic measurements should use the SID (Standard Integrated Dipoles) discussed in this paper.

For measurements which involve polarimetric seekers or angle measuring radars, the even bounce standard should be the conical corner reflector because of its deep null on boresight and its relatively broad coverage and large radar cross section (efficient use of the aperture). The conical corner can have a small sighting hole cut in the end with nylon cross hairs fastened on the front face with angle markings on the thread used for the cross hairs. The odd bounce field standard for angle measurements should be the sphere or the Luneberg lens reflector after it has been calibrated against a corner reflector in a controled enviornment.

## 4. REFERENCES

[1] S. D. Robertson, 'Targets for Microwave Radar Navigation', 'Bell System Technical Journel, Vol **26**, pp 852-869, 1947

[2] D. Levine and W. Welch, 'Spatial Coverage of Radar Reflectors'' IEEE Trans on Aerospace, Vol **AES2**, #2, April 1964, pp 160-165

[3] L. Ridenour, editor <u>Radar System Engineering</u> MIT Radiation Laboratory series Vol 1, 1946 , p 66

[4] E. Kennaugh Ohio 'Annual Engineering Report' State University Research Foundation-Antenna Laboratory Report 612-4,1 Oct 1955

[5] E. Kennaugh 'Interim Engineering Report' Ohio State University Research Foundation-Antenna Laboratory Report 612-7 dated 2 April 1956

[6] J. N. Hines, E. M. Kennaugh, and J. Upson 'Test of a New Type of Circularly Polarized Corner Reflector with Broad Coverage', Ohio State University Research Foundation-Antenna Laboratory Report # 612-8 dtd 2 July 1956,

[7] E. Kennaugh, 'Quarterly Progress Report' Ohio State University Research Foundation-Antenna Laboratory Report #612-2 dtd 16 March 1955

[8] E. Kennaugh 'Quarterly Progress Report' Ohio State University Research Foundation-Antenna Laboratory Report # 612-1, dtd 3 Jan 1955

[9] E. Kennaugh 'Interim Engineering Report' Ohio State University Research Foundation-Antenna Laboratory Report # 612-3 dtd 1 July 1955

[10] E. D. Sharp and M. A. Diab, 'Van Atta Reflector Array', IRE Trans on Antennas Propag Vol AP8, #4, pp 436-438 July 1960: L. C. Van Atta, U. S. Patent # 2908002, Serial # 514040; 'Electromagnetic Reflector' October 6, 1959

[11] Franz and Depperman, 'Theory of diffraction by a cylinder as affected by the surface wave', Ann. Physik Vol 10, p 361 (1952)

[12] R. C. Spencer 'Optical Theory of the Corner Reflector', Massachusetts Institute Of Technology, Radiation Laboratory Report # 433, March 1944

[13] J. W. Crispin Jr. and K. M. Siegel, Methods of Radar Cross-Section Analysis, Academic Press, New York, London 1968

[14] V. H. Weston, 'Exact Near-Field and Far Field Solution For The Backscattering of a Pulse from a Perfectly Conducting Sphere.'-University of Michigan Radiation Lab report # 2778-3-T (1959)

[15] L. Root and J. Cole, 'Test and Evaluation of a Coherent 35 GHz Radar Repeater', U. S. Army Missile Command, Advanced Sensors Directorate # RE-84-1, Jan 1984

# POLARIMETRIC TARGET AND CLUTTER SCATTERING MATRIX MEASUREMENTS AT CM TO SUB-MM WAVELENGTHS

G. N. Jepps
THORN EMI Electronics Limited,
Wells,
Somerset,
U.K.

ABSTRACT. The UK National Radar Scale Modelling Facility has been developed by THORN EMI Electronics Limited on behalf of the RSRE and other MOD establishments. Its function is to make radar scattering measurements on targets and shapes of interest, principally using sub-scale modelling, although full-scale measurements (for example of clutter) are also made.

Increasingly, intelligent use is being made of the target information contained in the detailed polarimetric structure of the radar signature. Consequently, in order to make reliable performance assessments of such new systems, reliable target radar scattering information is required. A wide range of polarimetric scattering matrix measurement equipments are available, operating across the frequency band 10 GHz to 1 THz. The higher frequency is useful for obtaining M-band equivalent data on 0.1 scale targets. All systems may be used to provide detailed data on the characteristics of target component scatterers.

The systems in use are described, and examples of measurements shown. Some informative displays derived from the measurement data are also presented.

## 1. INTRODUCTION

Radar signature data may be gathered either using the actual targets and radar frequencies of interest or, alternatively, by using scale (usually sub-scale) models together with appropriately scaled wavelengths. The two methods have their own respective merits and disadvantages, and should be seen as complementary rather than competitive. The use of model systems has the significant advantage of making possible economical measurements to a degree of precision and repeatability not practical with full-scale trials. However, whilst scaling of metallic targets and low-loss dielectrics is relatively straightforward, some materials pose difficulties. In addition, natural targets (ground clutter, for example) cannot be scaled convincingly.



*W.-M. Boerner et al. (eds.), Direct and Inverse Methods in Radar Polarimetry, Part 1*, 861–876.

During the past three decades considerable experience has been acquired by the UK National Radar Scale Modelling Facility of the techniques required for scale modelling (1). Increasingly, intelligent use is being made of the target information contained in the detailed polarimetric structure of the radar signature, and in the past 10 years the techniques of scale modelling have been extended to include a fully polarimetric measurement capability across the frequency band 10-1000 GHz. Some of the systems can operate in bistatic mode (up to 160 deg bistatic angle) as well as monostatic. Figure 1 summarises the polarization scattering matrix radars available and gives typical applications. At the lower end of the spectrum, the 10 GHz radar has been used for full-scale trials against ground clutter and other targets. At the higher end, scattering matrix measurements can be made on 0.01 scale ships and 0.1 scale land vehicles.

| MODEL FREQUENCY (GHZ) | TYPICAL APPLICATIONS | SCALING RATIO | FULL-SCALE FREQUENCY BAND |
|---|---|---|---|
| 10 | Ground Clutter | 1:1 | I |
| 15 | Shells | 1:1 | J |
| 35* | Missiles and aircraft | 4:1 | I |
| 35* | Missiles | 1:1 | K |
| 80* | Missiles and aircraft | 8:1 | I |
| 140* | Missiles and aircraft | 8:1 | J |
| 140* | Missiles, aircraft and land vehicles | 4:1 | K |
| 890 | Land vehicles | 10:1 | M |
| 890 | Ships | 100:1 | I |

* Denotes monostatic and bistatic capability

Figure 1 Available Scattering Matrix Radars and Typical Applications

In addition to the systems shown in Figure 1, many other polarization diverse systems are available. However, since these other systems do not allow quasi-simultaneous measurement of the scattering matrix components, they will not be discussed in this paper.

## 2. CM AND MM SCATTERING MATRIX RADARS

An introduction to the technique used is given in (2). Matrix measurement systems operating in the 10-140 GHz band use similar technology. Figure 2 shows the 80 GHz system, a block diagram of which is given in Figure 3. A pulsed microwave source is used as the

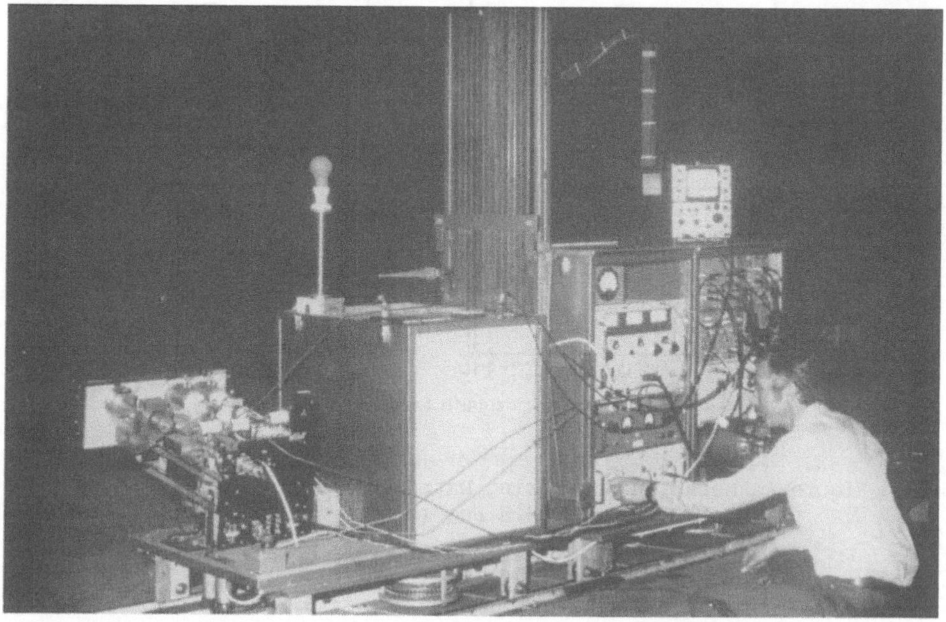

Figure 2  80 GHz Scattering Matrix System

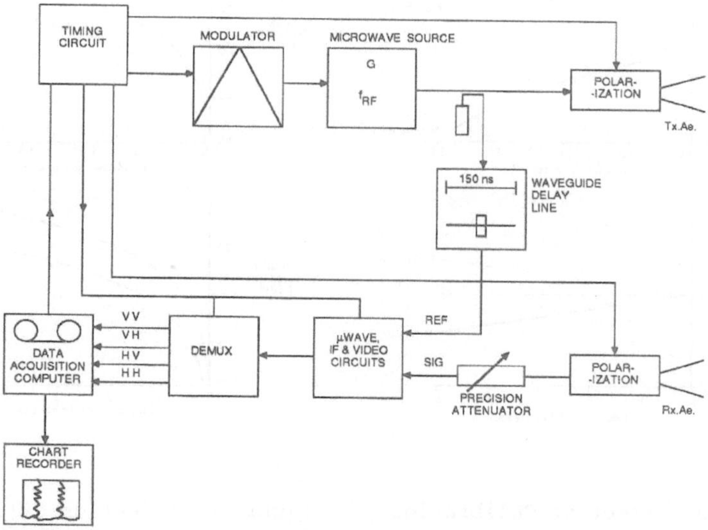


Figure 3  Block Diagram of 80 GHz Scattering Matrix Radar

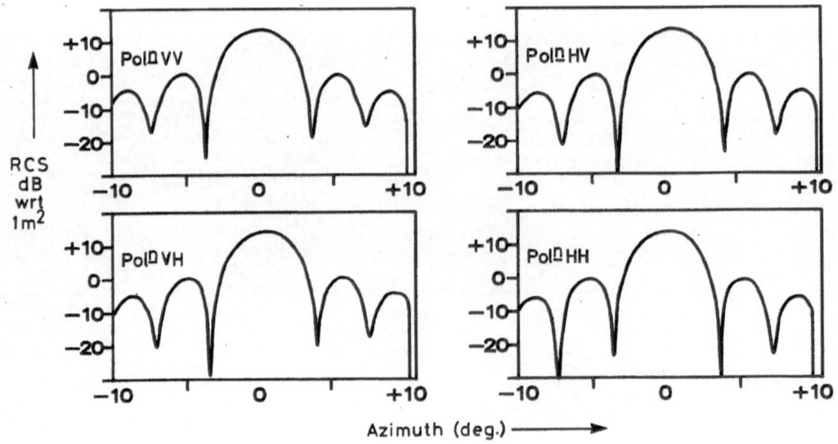


Figure 4 Measured Scattering Matrix Components for a Dihedral Rolled through 22.5 Deg (Linear Polarization)

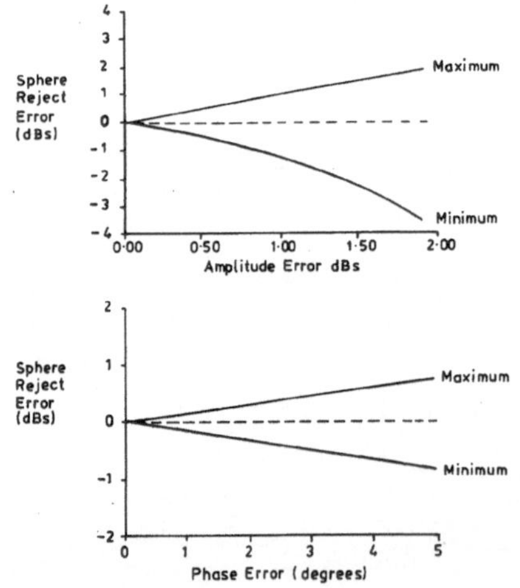


Figure 5 Effect of Calibration Errors on the Computed Sphere Reject Return

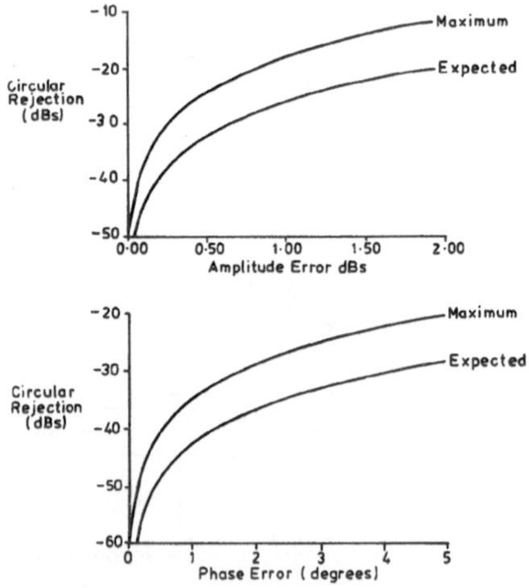


Figure 6 Effect on Calibration Errors on the Circular Rejection Ratio

transmitter, and a microwave delay line provides a phase reference. A (0, 90 deg) phase shifter in the reference path permits analogue signals representing each of the two phase quadrature components to be extracted. Each antenna assembly incorporates an electrically switched polarizer, controlled so as to produce the four-step polarization sequence VV, VH, HV, HH.

For monostatic systems, calibration is achieved by using a dihedral corner, with apex rolled through 22.5 deg from the vertical, viewed with a horizontal sightline. With this roll, the same amplitude return is produced on boresight in all four channels. Figure 4 shows a typical measurement made with the corner rotating about a vertical axis. The known 0 or 180 deg phase difference between the four channels permits phase calibration. Figure 5 shows the effect on the computed sphere reject return from the dihedral of small phase and amplitude errors in the calibration of the measured co-linear and crossed-linear channels. The curves show worst case errors. Figure 6 shows the corresponding circular rejection levels. Here the curves show worst case and mean (assuming uniformly distributed errors) rejection ratios.

Circular polarization data inferred from measurement of the linear basis scattering matrix have been compared with corresponding data obtained by direct measurement. Figure 7 shows derived and measured scattering data from a dihedral. The comparison is good, the only significant differences being in the rejection level some 30-40 dB

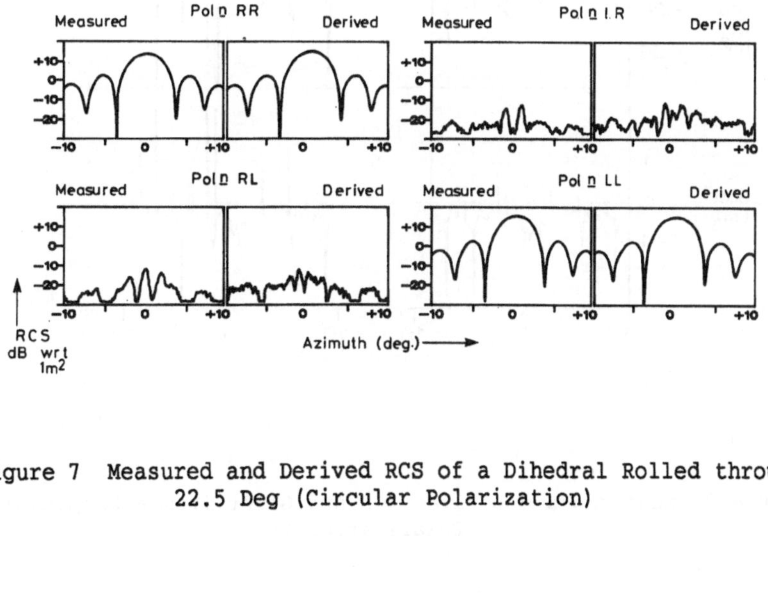


Figure 7 Measured and Derived RCS of a Dihedral Rolled through 22.5 Deg (Circular Polarization)

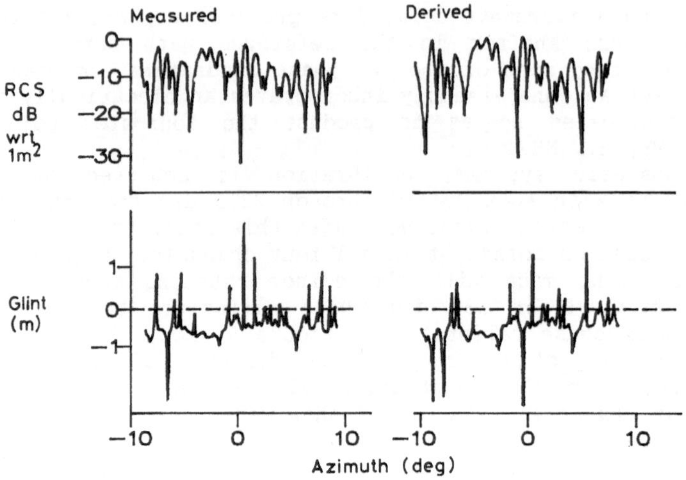


Figure 8 Measured and Derived RCS from a Complex Target, Polarization RR

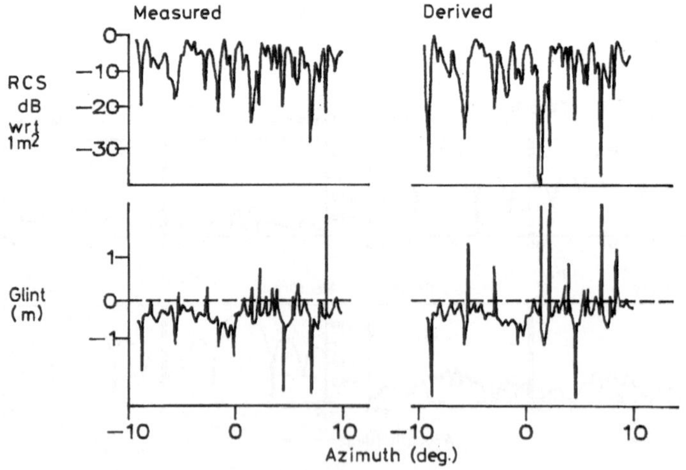


Figure 9 Measured and Derived RCS and Glint from a Complex Target, Polarization RL

below the peak return. Figures 8 and 9 show corresponding results for a more complex target. In this case target glint (a measure of the scattered wave phase front perturbation) is also shown. Again the comparison is excellent, with the principal differences being associated with the RCS null depths and the corresponding glint spikes.

Calibration of the more recently available bistatic scattering matrix radars is more difficult; the dihedral is an unsuitable calibration target. Frequently, a sphere is used and measured results calibrated against the exact Mie solution.

## 3. SUB-MM SCATTERING MATRIX RADAR

Very large targets such as ships require the use of high scaling ratios; 100:1 is typical. To gather data in the 3 to 30 GHz range on such targets, sub-mm wavelengths are needed. These wavelengths are also required for more moderately sized targets at higher full-scale equivalent frequencies. For example, 0.1 scaling of land vehicles at M-band implies wavelengths in the region of 0.3 mm.

THORN EMI Electronics Limited have developed a Hydrogen Cyanide laser system operating at 890 GHz (3). A photograph of the system, operating in non-coherent spot illumination mode, is shown in Figure 10. In this mode of operation, a narrow beam 1.5 cm wide at

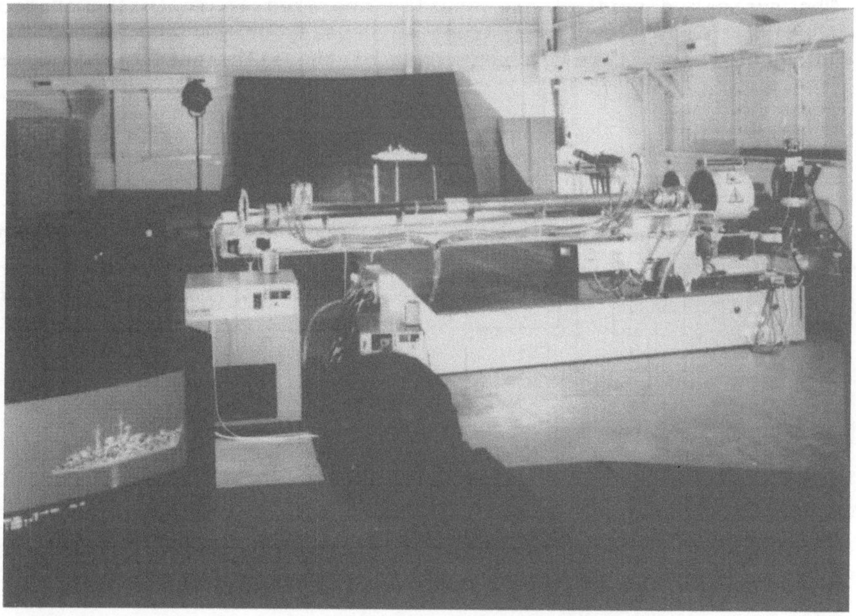

Figure 10 890 GHz Radar

the target is swept raster fashion across the target and an image of the target's radar echo centres obtained. The system also operates in full-illumination coherent mode, and in this mode of operation is currently being configured to gather scattering matrix data.

A block diagram of the system is shown in Figure 11. The vertically polarized signal produced by the laser is rotated through 0 or 90 deg depending on whether a vertically or horizontally polarized signal is required before being transmitted towards the target. The received signal is passed through a polarizer linked to the transmit polarizer by a switching wheel giving the desired VV, VH, HV, HH polarization sequence. (0, 90 deg) modulation is used to separate the two quadrature components of each channel. The scattering matrix data acquisition unit demodulates the received signal into eight baseband signals, corresponding to the two phase quadrature components of each of the four polarization combinations. At the heart of the system lie the two polarization transformers (Figure 12). Orthogonal linear components are separated and recombined using polarizing beamsplitters G1 and G2. Switching the relative phases between the two orthogonal components by 180 deg allows an initially linearly polarized wave to be rotated through 90 deg. Switching by 90 deg changes linear to circular polarization.

## 4. SYNTHETIC APERTURE TECHNIQUES

The systems described are usually operated in full-illumination mode, that is to say the model target is usually scaled and placed so that it lies entirely within the main beam of the illuminating radar. However,

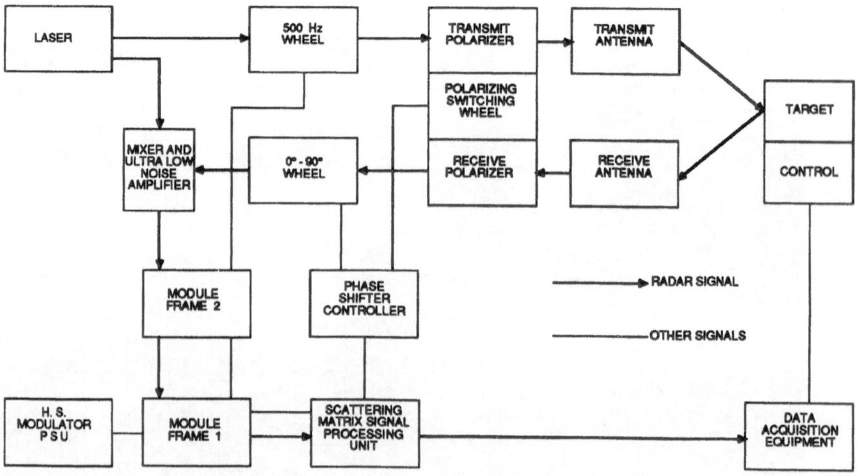


Figure 11 Block Diagram of 890 GHz Scattering Matrix Radar System

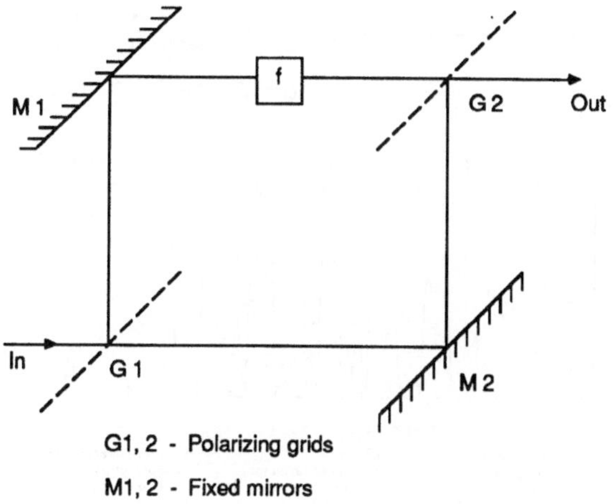


Figure 12 Polarization Transformer for 890 GHz Radar

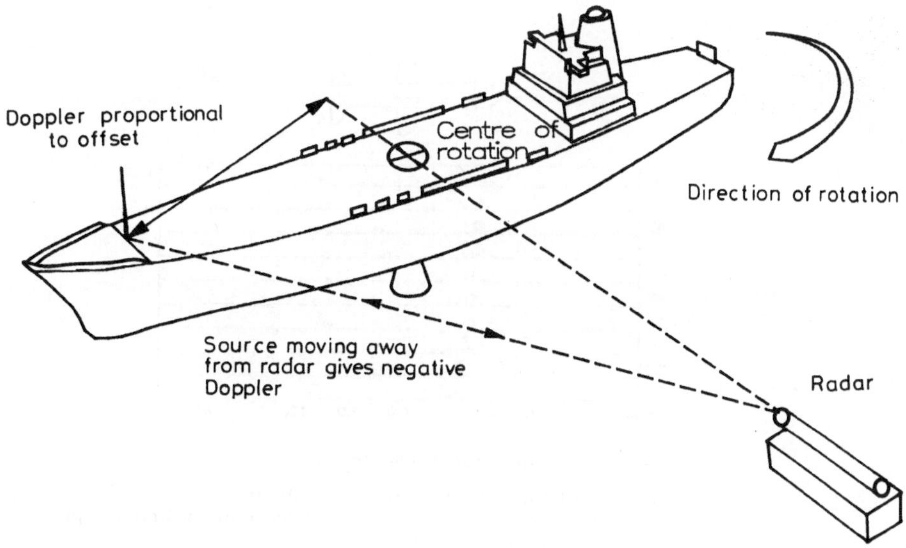


Figure 13 Principles of Angular Spectral Analysis

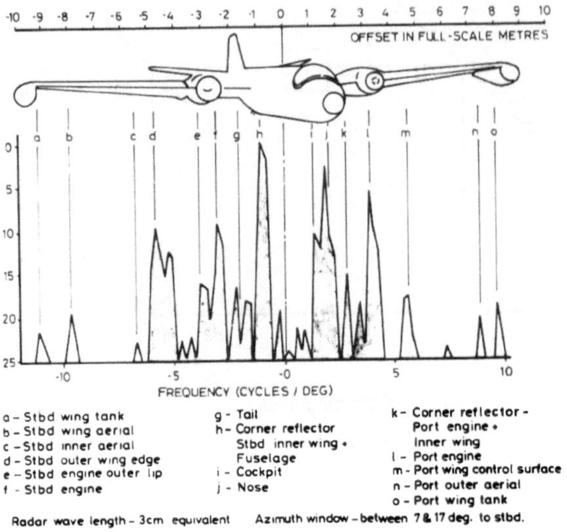


Figure 14 Principal Scatterers on a Model Aircraft

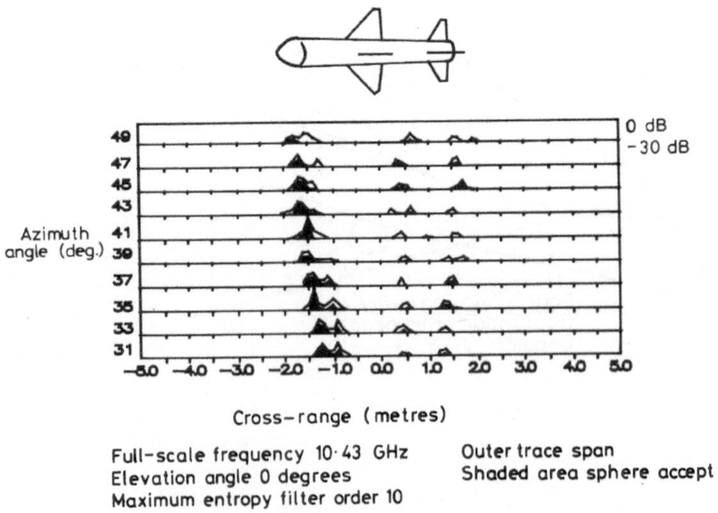


Figure 15 Polarimetric Cross-Range Spectra for a Missile

the use of synthetic aperture techniques (4,5) enables the radar characteristics of individual component scatterers to be measured.

The usual mode of data gathering entails the radar-target sightline being rotated about an axis fixed in the target. In this way an aperture much larger than the transmitting antenna may be synthesised and consequent signal processing may allow individual scatterers on a target to be resolved.

Different methods of target support are used depending on the radar frequency. For 10 GHz measurements the target will be suspended from an overhead gantry by thin dielectric strings and rotated. At 80 or 140 GHz, the target will be tethered and the radar moved along a high precision circular railway track (the track can be seen in Figure 2). At 890 GHz the target is mounted upon a tapered stalk and rotated. In all cases the radar-target centre distance is maintained to within a small fraction of a wavelength so that accurate phase measurements may be made.

The principle of the technique used to resolve individual component scatterers is shown in Figure 13. As the target rotates, reflectors towards the bow move away from the radar giving rise to negative doppler components in the received signal, whilst reflectors located towards the stern give rise to positive doppler components. The doppler frequency of the signal received from each reflector will be proportional to its distance from the centre of rotation resolved into a direction that is perpendicular both to the axis of rotation and to the sightline direction (i.e. to its cross-range).

Spectral analysis of the signal recorded as the sightline rotates will identify the different doppler frequencies present in the signal, and hence the cross-ranges of the respective scatterers. A reflector at d wavelengths cross-range will give rise to a signal with doppler frequency equivalent to $2d/\lambda$ cycles per radian of sightline rotation. Figure 14 shows an example of this form of processing applied to single polarization measurement data gathered using a 0.125 scale aircraft. A synthetic arc of 10 deg has been used to identify the 15 most significant scatterers on the target

With traditional FFT algorithms, an aperture size of $\Theta$ radians will discriminate reflectors separated in frequency by more than $1/\Theta$ cycles/radian (the line spacing), or $1/2\Theta$ wavelengths in cross-range. However, data adaptive spectral analysis methods such as the Maximum Entropy Method (6,7,8) can improve significantly on this resolution. The algorithms are capable of extension to vector or matrix data (9) and therefore may be used to process polarization scattering matrix data.

Figure 15 shows a set of 10 cross-range spectra obtained by analysing scattering matrix data gathered from a 0.1306 scale model of a generic missile. The model is fitted with a parabolic dish antenna inside a nose radome. Each spectrum has been obtained by analysing 2 deg of data and the total aspect window analysed covers 30 to 50 deg from head-on. The view shown of the missile corresponds to the mid-point of this angular range.

The outer trace in each spectrum shows the span of the scattering matrix whilst the inner shaded area shows the circular sphere accept component. The span is shown as it is invariant with change of polarization base and is a measure of the polarimetric 'size' of the scatterer. The set of spectra quite clearly show the relative amplitudes of the nose, main wing and tail assemblies. The first is clearly a predominantly sphere accept or odd-bounce return while the latter two have more complex polarimetric characteristics. Although this set of Maximum Entropy spectra are somewhat coarsely sampled at 0.1 m spacing, there is a clear improvement in resolution over the 0.4 m line spacing obtained with FFT methods.

## 5. DISPLAY OF POLARIMETRIC DATA

Measurements made using a polarimetric radar generate four channel complex data representing the target scattering matrix. Variation of this matrix from sample to sample may be due either to time variation of the target, to time variation of the aspect, or to both. Any direct display of the measured scattering matrix components tends to be too complex to reveal any useful time varying polarimetric properties of the target. However, each scattering matrix can be characterised by two null polarizations (10). These polarizations are those which produce zero signal in a common transmit/receive (i.e. co-polarized) radar system.

Any polarization can be characterised by a polarization ellipse as shown in Figure 16, which in turn can be described by two angles $\phi$ and $\tau$. $\phi$ is the inclination of the ellipse major axis to the horizontal

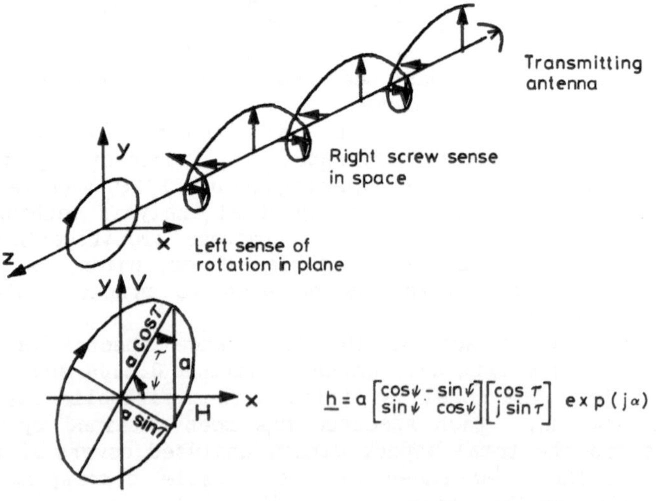


Figure 16 Radar Polarization

and tan $\tau$ is the ratio of minor/major axes. These angles can be used to represent any polarization on a polarization sphere. The latitude and longitude of any point on the sphere are $2\tau$ and $2\phi$ respectively. With this representation the poles of the sphere correspond to the circular polarizations, whereas the equator corresponds to linear polarizations.

A monostatic scattering matrix includes six parameters, i.e. three complex numbers. To reduce to the four parameter representation given by the two null polarizations, absolute amplitude and absolute phase are neglected.

To display null polarization data, plane projections of the polarization sphere are used. In the common projections the radial scale is proportional either to cos $2\tau$ or $1-|\tan \tau|$. The polar angle is $2\phi$. Different projections are possible and in particular equal area projections have been discussed in (11).

A slightly different projection which the author has found useful is used in Figure 17. The figure displays measured natural ground clutter scattering matrix data. On this projection the radial scale is linearly proportional to $90-|2\tau|$. The data shown correspond to 15 sec of observation of a 60 m x 30 m clutter patch containing grass, trees and cows. The frequency of the illuminating radar was in the I-band. One difficulty with this sort of presentation is that four charts are needed to display the data (left and right hemispheres, first and second null polarizations). A more serious difficulty is that there is no amplitude weighting.

An alternative 2-D projection is shown in Figure 18. On the plot the null polarizations are shown as solid lines emanating from the centre of the polarization sphere, weighted in length according to the span of the scattering matrix. Also, the maximum polarizations (which can of course be deduced from the null polarization pairs) are shown as dotted lines. This sort of display is very useful for obtaining a visual appreciation of the distribution of nulls and of their clustering. In the figure, the nulls and the maximum polarization are well separated. Colour coding the different polarizations enables such displays to be used even when the polarizations are more dispersed.

## 6. CONCLUSIONS

The UK National Scale Modelling facility has responded to the increasing demands for fully polarimetric radar scattering data by developing a range of scattering matrix radars operating between 10 GHz and just under 1000 GHz. The radars operating at the lower end of the scale (10-140 GHz) have been extensively used, whereas the 890 GHz system is a new development. The availability of these radars has led to an increased level of target understanding and it is expected that this increased understanding will be extensively exploited in future projects.

## 7. ACKNOWLEDGEMENTS

The author wishes to acknowledge the support of UK MOD(PE) and that of his colleagues at THORN EMI Electronics Limited, Wells, Somerset.

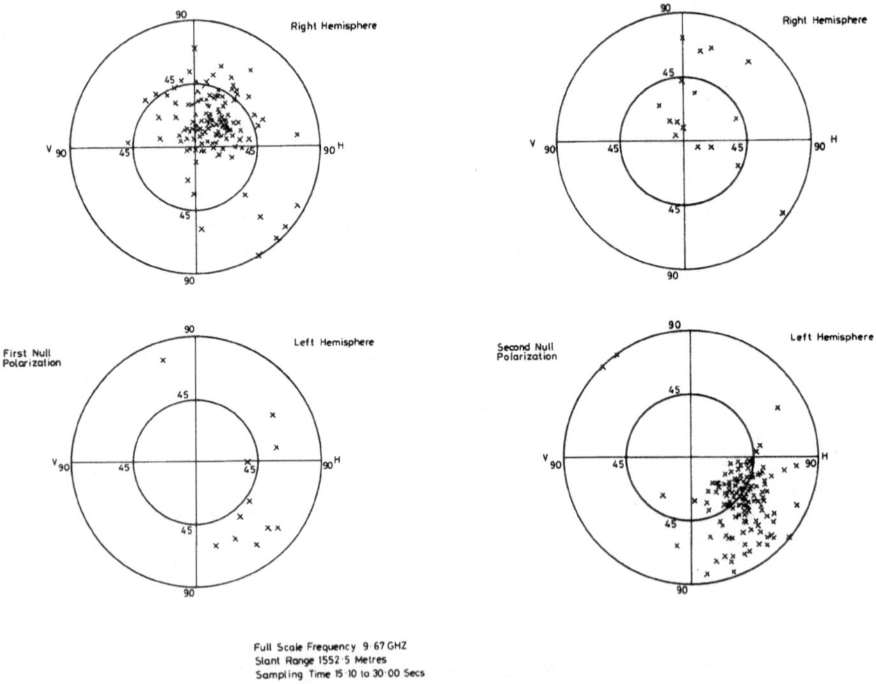


Figure 17 Measured Null Polarization Plots for Natural Ground Clutter

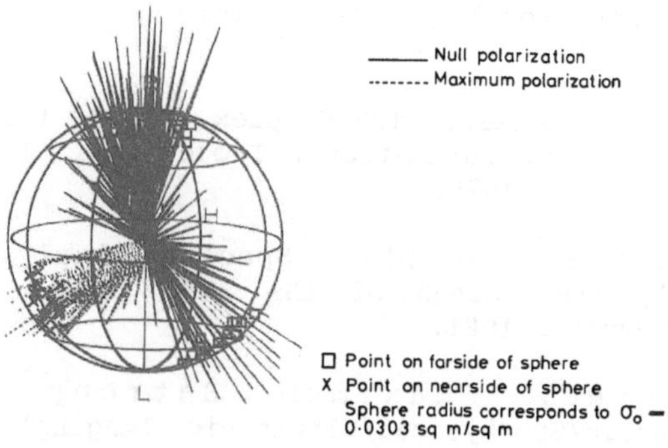


Figure 18 Polarization Forks on the Poincaré Sphere

## REFERENCES

(1) L.A. Cram and S.C. Woolcock, 'Review of two decades of experience between 30 GHz and 900 GHz in the development of model radar systems'; AGARD Conference Proceedings No. 245, pp 6-1 to 6-15, September 1978.

(2) D. Bird, M.W. Plaster, C.G.C. Wilson, 'Measurement of depolarization of radar scattering using scale modelling techniques'; Military Microwaves Conference Proceedings, Microwaves Exhibitors and Publishers, 1980 pp 60-65.

(3) S.E. Bathe, B.E. Prewer, G. Villis, 'Recent Advances in the use of Sub-Millimetric Wave Techniques to Obtain Backscatter Information at the UK National Scale Modelling Centre', Proceedings of the 17th European Microwave Conference, Rome, Italy, September 1987.

(4) L.A. Cram, 'Inverse Methods Applied to Microwave Target Imaging', Proceedings of the Nato Advanced Research Workshop on Inverse Methods in Electro-Magnetic Imaging, September 1983. D. Reidel Publishing Co. 1985, Volume 2, pp 841-870.

(5) D.L. Mensa, 'High Resolution Radar Imaging', Artech House, 1981.

(6) S. Haykin and S. Kesler, 'The Complex Form of the Maximum Entropy Method for Spectral Estimation', Proceedings of the IEEE Volume 64, pp 822-823, May 1976.

(7) S.M. Kay and S.L. Marple, 'Spectrum Analysis - a Modern Perspective', Proceedings of the IEEE Volume 69, no. 11, pp 1380-1419, November 1981.

(8) R. Bevensee, 'Maximum Entropy Methods in Electro-Magnetic/Geophysical/Ultrasonic Imaging', Proceedings of the NATO Advanced Research Workshop on Inverse Methods in Electro-Magnetic Imaging, September 1983. D. Reidel Publishing Co. 1985, Volume 1, pp 375-396.

(9) O.N. Strand, 'Multi-Channel Complex Maximum Entropy (Auto-regressive) Spectral Analysis', IEEE Transactions AC-22, pp 634-640, August 1977.

(10) E.M. Kennaugh, 'Polarization Properties of Radar Reflections', MSc thesis, Dept. Elec. Engineering, Ohio State University, Columbus, Ohio, 1952 (Antenna Lab. Report 389-12, March 1952).

(11) R.S. Raven, 'On the Null Representation of Polarization Parameters', Proceedings of the Nato Advanced Research Workshop on Inverse Methods in Electro-Magnetic Imaging, September 1983. D. Reidel Publishing Co. 1985, Volume 1, pp 629-642.

# POLARIMETRIC CALIBRATION AND REMOTE SENSING APPLICATIONS USING AN X-C-L-BAND SAR

D.R. Sheen, E.S. Kasischke, R.A. Shuchman, and R.G. Onstott
Radar Science Laboratory
Advanced Concepts Division
Environmental Research Institute of Michigan
P.O. Box 8618
Ann Arbor, MI 48107 USA

ABSTRACT. A polarimetric radar has been constructed by the Environmental Research Institute of Michigan (ERIM) for the Naval Air Development Center (NADC) and mounted on a P-3 airplane. This radar operates at X-band, C-band, and L-band and is able to produce fully polarimetric images at each of these frequencies. Over the past few months this system's performance has been verified and the system has been in the process of being calibrated. Polarimetric calibration involves determining the gain, phase shift, leakage, noise, and stability of each of the four channels used to compute the scattering matrix. Targets with known reflection characteristics have been deployed and used to calibrate the system. These targets include dihedral corner reflectors set at a variety of angles, trihedrals of many sizes, and Active Radar Calibrators (ARCs). Preliminary results indicate that the radar is yielding useful polarimetric data. The remote sensing applications of this radar will be discussed and preliminary backscatter data from a variety of media will be presented.

## 1. INTRODUCTION

A fully-polarimetric, X-C-L-band Synthetic Aperture Radar (SAR) has been developed by ERIM and mounted on a U.S. Navy P-3 airplane. This radar is capable in operating in a variety of modes including single frequency, fully-polarimetric modes. The radar system itself is described in References [1,2]. The fully-polarimetric mode gathers amplitude and phase data for each of the four traditional, linear polarization combinations (HH, HV, VH, and VV). The P-3/SAR has completed a testing and evaluation program and now a calibration effort is underway. Kasischke et. al., [3] have examined the power characteristics of the different channels of the radar including gains, linearity, stability, and noise. In this paper the polarimetric aspects of calibration will be examined. Polarimetric calibration involves determining the relative gains, phase shifts, and leakages between the channels.

Polarimetric calibration is accomplished by measuring targets with



*W.-M. Boerner et al. (eds.), Direct and Inverse Methods in Radar Polarimetry, Part 1*, 877–898.

known scattering matrices and computing the gains, phase shifts, and leakages between channels necessary for measured scattering matrices to be in agreement with actual scattering matrices. A few targets with different polarimetric characteristics are necessary to determine all of the various radar parameters [4,5]. Targets used so far include trihedrals, dihedrals at a variety of angles, and ARCs. In Section 2 a method will be examined which uses ARCs to calibrate the system [5]. In Section 3, phase characteristics of ground clutter will be discussed and methods to exploit these phase characteristics for calibration purposes will be described. In Section 4, some polarimetric images will be presented and the remote sensing capabilities of the radar discussed. In Section 5 conclusions will be made about present work and plans discussed for future work.

## 2. CALIBRATION USING ARCS

This section will detail a procedure to calibrate a polarimetric radar using three ARCs. The ARC has the advantage of being able to form different scattering matrices, all with good gain, and wide backscatter beamwidth. The only drawback to ARCs is their complexity and expense. The Radar Science Laboratory at ERIM now has 12 ARCs - four at X-band, four at C-band, and four at L-band. These ARCs are sufficient to carry out a complete polarimetric calibration for one area in the radar image using the procedure discussed in this paper.

### 2.1 The Measured Scattering Matrix

The P-3/SAR uses an orthogonal linear polarization base composed of horizontal and vertical transmit and receive polarizations. The polarization base will be denoted by the numerals 1 and 2 with 1 corresponding to horizontally polarized and 2 corresponding to vertically polarized. Because of small coupling and gain imbalances in any radar the actual transmitted field may vary slightly in polarization state, amplitude, and phase from what ideally would be transmitted. Likewise, the measured received field is distorted by channel couplings and gain differences. The goal of calibration is to precisely determine the gain imbalances and coupling in the radar system as a whole. Barnes [1,2] has formulated the problem assuming a linear system and solved it using in-scene passive targets. Shen [5] has solved the problem using Barnes' [4] formulation, but solved it using in-scene ARCs.

Following Shen's [5] and Barnes' [4] notation the measured scattering matrix E can be written in terms of the scattering matrix S.

$$[E] = Ae^{j\phi} [R]^T [S] [T] \tag{1}$$

Here, R and T are 2x2 complex matrices which describe the polarization distortion of the receive and transmit channels of the radar. $Ae^{j\phi}$ is function of round trip gain or loss and the phase shift due to path length and is necessary to achieve absolute calibration. Here only the relative calibration of the radar's polarimetric channels will be examined. In an ideal polarimetric system these R and T matrices should be close to identity matrices and not cause any difference between E and S.

## 2.2 Using ARC's to Calibrate

An ARC is a transponder or a receiving antenna attached to an amplifier which is attached to a transmitting antenna. The ARC is an especially useful tool for polarimetric calibration and has been discussed by Brunfeldt [6]. Depending on the polarizations of the transmit and receive antennas different scattering matrices can be formed. The ERIM ARCs have linearly polarized receive and transmit antennas which can be rotated individually and set at horizontal, vertical, or 45°.

An important assumption in using ARCs to determine the transmit and receive distortion of the radar is that the ARCs have distortion negligible compared to the radar. The ERIM ARC's have cross-channel coupling of $<$ -35 dB and the radar has cross-channel coupling of approximately -23 dB so that the ARCs should allow determination of the radar's distortion matrices.

The relative calibration of the four polarimetric channels of the radar can be completed using three ARCs in the configurations given in Table I. Table I shows the orientation of the transmit and receive antennas and the resulting scattering matrix for the ARC. Shen [5] has solved for the transmit and receive distortion matrices using three ARCs in these configurations. The measured scattering matrices are given by X,Y, and Z. The measured scattering matrices can be related to the elements of the distortion matrices R and T. Shen's [5] results are presented in Figure 1 and Figure 2. Either set of equations presented in Figure 1 or Figure 2 will allow relative calibration of the four channels.

Three ARCs configured in the way illustrated in Table I will allow a complete calibration including the channel coupling effects. ERIM now has four working ARCs at each frequency: X-, C-, and L-band. Of course to determine the spatial variation or time variation of the calibration more ARCs might be desireable.

## 2.3 A Simplified Calibration Procedure

Polarimetric data can be calibrated quickly if channel coupling is negligible. To carry out this calibration only two targets are needed. First, a large trihedral could be used to show that in fact the coupling is negligible. A large trihedral would be used so that coupling into the cross-polarized channel would be significant compared to ground clutter. The trihedral has the following theoretical scattering matrix (T):

$$T = G \begin{bmatrix} 1 & 0 \\ 0 & 1 \end{bmatrix} \qquad (2)$$

This results in a measured scattering matrix normalized to the HH channel of:

$$T_M = \begin{bmatrix} 1 & K_{12}\alpha \\ K_{21}\beta & K_{22} \end{bmatrix} \qquad (3)$$

If the K's are known, the relative magnitudes of the coupling factors ($\alpha$ and $\beta$) is given by the ratio of like to cross-polarized returns. If the cross polarized terms are small, say 30 dB down from the like polarized terms then the coupling coefficients could reasonably be neglected.

An ARC would be used as the second calibration target to determine the relative gains and phases of the channels relative to the HH channel. An ARC with the receive and transmit antennas set at 45° has the following scattering matrix:

$$R = \frac{G}{2} \begin{bmatrix} 1 & 1 \\ 1 & 1 \end{bmatrix} \qquad (4)$$

The measured scattering matrix relative to the HH channel neglecting coupling would be:

$$R_M = \begin{bmatrix} 1 & K_{12} \\ K_{21} & K_{22} \end{bmatrix} \qquad (5)$$

The relative phases and gains of each channel are given directly by $R_{M12}$, $R_{M21}$, and $R_{M22}$. Thus calibration can be accomplished with just this one ARC if the coupling can be neglected.

## 2.4 Correcting the Image

To calibrate the image the inverse of equation (1) must be carried out. This operation is as follows.

$$S = \frac{1}{A} e^{-j\phi} (R\ )^{T^{-1}} E\ (T)^{-1} \qquad (6)$$

If one is only interested in the relative values of the matrix S then the factor $1/A\ e^{-j\phi}$ can be neglected. The solutions for R and T presented previously could be used to correct an image for the radar's polarimetric distortion.

# 3. PHASE CALIBRATION AND CLUTTER

## 3.1 Minimum Calibration to Examine Clutter

The expected scattering matrices and correlation matrices for the clutter from distributed targets can be simplified somewhat from the general expressions for these matrices. First, the 12 and 21 elements of the scattering matrix should be the same in the monostatic case. Second, the like and cross-polarized channels are completely uncorrelated. This is because the cross-pol terms come entirely from higher-order scattering (2-bounce or greater) while the like-polarized terms usually come predominately from first-order scatter (single-bounce). For randomly positioned scatterers the higher-order scattering paths are independent of the first-order scattering paths and thus, the cross and like-polarized terms of the scattering matrix are uncorrelated. The final resultant average clutter correlation matrix for the monostatic case ($C_M$) as a function of the scattering matrix (f) has the following form:

$$C_M = \begin{bmatrix} \langle |f_{11}|^2 \rangle & 0 & \langle f_{11} f_{22} \rangle \\ 0 & \langle |f_{21}|^2 \rangle & 0 \\ \langle f_{11} f_{22} \rangle^* & 0 & \langle |f_{22}|^2 \rangle \end{bmatrix} \qquad (7)$$

The only phase difference which is critical is that between the 11 and 22 channel. This would indicate that a polarimetric radar would only have to be phase calibrated between the 22 and 11 channels to collect good data on different types of ground clutter. A trihedral would be sufficient to do this phase calibration. Of course, to examine other targets one would also need to phase calibrate the cross-polarized channels relative to the like-polarized channels.

## 3.2 Using Clutter Phase Statistics to Phase Calibrate the System

The fact that clutter has correlation between the two like-polarized channels and between the two cross-polarized channels can be used as an aid in phase calibration [7]. This promises to be very useful since calibration targets are often difficult to deploy and require some degree of ground support. Also, the phase calibration can change spatially in the image and many calibration targets would have to be deployed to examine the spatial variation of the phase. As outlined in the previous section it is not expected that the like- and cross-polarized channels would be correlated with each other. Thus, it will not be possible to get this relative phase shift. The method would only allow HH (horizontal transmit, horizontal receive) to be phase calibrated with respect to VV (vertical transmit, vertical receive) and likewise for VH (vertical transmit, horizontal receive) to be phase calibrated with HV (horizontal transmit, vertical receive).

In Figure 3, histograms of relative phase for some calibrated, X-band, P-3/SAR data from a forested region approximately 1 Km square are presented. These data were calibrated using in scene trihedrals and dihedrals and the image will be presented later. In Figure 3(a) the histogram of the phase of VV vs. HH is shown and one can see that it appears Gaussian with a width which is confined in the interval $-\pi$ to $\pi$ and a mean of 0. This is fortunate since a wider distribution of phases would exhibit aliasing and one could not use the clutter to phase calibrate. The phase distribution of HV vs. VH is given in Figure 3(b). In theory, for an ideal, calibrated, monostatic radar this should be a delta function at 0°. As illustrated in this Figure the HV vs. VH phase is not always 0 but is approximately a Gaussian centered at 0 with standard deviation much less than VV vs. HH. In Figure 3(c) a histogram of the phase between VV and VH is shown which indicates that these two polarizations have uniformly distributed phase difference and are uncorrelated. The other combinations of like and cross-polarizations were also examined and seen to be uncorrelated. Figure 3, indicates that the mean value of the phase difference between VV and HH as well as VH and HV is zero for clutter. In uncalibrated data, these two mean phase differences would not necessarily be zero but could be corrected by phase shifting one of the like-polarized channels and one of the cross-polarized channels.

One important assumption in using the clutter to calibrate the VV channel with respect to HH is that the VV and HH reflections come predominately from single bounce scattering and thus tend to be in phase. Figure 3(a) is a strong indication that for X-band scattering from trees this is the case. VV and HH may not always tend to be in phase. For example, at L-band the ground-trunk interface may act like a horizontal dihedral and be a significant source of scattering. In this case VV and HH would be 180° out of phase. For this type of clutter one could not assume an average phase difference of 0° between VV and HH.

Using the clutter phase statistics one could partially phase calibrate an image and could also examine how the relative phases shift spatially in the image. In Figure 4, the geometry of a SAR is illustrated. The antenna may have slightly different phase centers for Vertical (V) and Horizontal (H) polarization. This difference from H to V is given by the vector d and is exaggerated in the illustration. From Figure 4 it can be seen that one expects the pathlengths H to V and V to H to be the same. Thus, phase calibration of the HV channel relative to the VH channel should be constant spatially. However, this is not the case with the HH vs. VV channel because of the offset d. The phase variation across the image can be written as follows:

$$\theta_{VH} = \frac{4\pi}{\lambda}\left[\vec{d}\cdot\frac{\vec{R}}{|R|}\right] + \theta_C \tag{8}$$

where $\theta_{VH}$ is the phase of VV relative to HH, $\lambda$ is the radar wavelength, d is the offset from H to V, R is the vector from the radar to the pixel of interest, and $\theta_C$ is a constant phase offset resulting from different path lengths in the transmitter and receiver for HH and VV. R can be written in terms of the incidence angle $\theta_i$ as follows:

$$\frac{R}{|R|} = \sin \theta_i \hat{x} - \cos \theta_i \hat{y} \tag{9}$$

and

$$\theta_i = \cos^{-1} \left[ \frac{A}{R} \right] \tag{10}$$

where A is the altitude above ground. Using Equations 9 and 10, Equation 8 can be written in the following form:

$$\theta_{VH} = \frac{4\pi}{\lambda} \left[ d_x \left[ 1 - \left[ \frac{A}{R} \right]^2 \right]^{-1/2} - d_y \left[ \frac{A}{R} \right] \right] + \theta_c \tag{11}$$

The average phase difference as a function of range for some X-band tree data is presented in Figures 5 and 6. The data has a pixel size of 2.4 m in range, and 1.6 m in azimuth. The images are processed in 4096 x 4096 pixel images. A subset of the image which was 4096 pixels in range and 100 pixels in azimuth was extracted. The average phase difference for a given range was computed by averaging together the phases for 100 azimuth pixels. The HV-VH phase is plotted in Figure 5 and is seen to be zero across most of the image with exception of some near range points. The near range pints are in the proximity of the nadir reflection (at range 143) and probably noisy. The result is anticipated since the average phase of HV vs. VH is not expected to vary across the image. The HV vs. VH was calibrated using some 45° dihedrals in the image so that is why it is a constant 0 rather than some other constant. In Figure 6, the average phase of VV-HH is plotted as a function of range. This results in an interesting curve which was fitted to the theoretical curve given by equation (11) using a non-linear least-square curve fitting algorithm. The results are plotted with a dotted line in Figure 6. The parameters used in equation (11) are:

$$A = 2299 \text{ pixels}$$

$$R = 2156 + \text{range record number}$$

and fitted parameter values are:

$$d_x = .26\ \lambda$$
$$d_y = -.017\ \lambda$$
$$\theta_c = -63.3°$$

The fitted values are not unique because of phase wraparound, but do give the expected variation in phase across the image quite accurately. The spatial variation in phase of the image could easily be corrected by multiplying by a spatially varying, complex calibration factor.

This section has illustrated that radar clutter from distributed targets promises to be a useful to phase calibrate VH relative to HV and HH relative to VV. To calibrate the like-polarized channels relative to the cross-polarized channels a dihedral or ARC would have to be used.

## 4. EXPERIMENTAL RESULTS AND REMOTE SENSING APPLICATIONS

In this section some imagery will be presented. The P-3 SAR has not collected that much fully-polarimetric data. This is because many of the operating modes used to date have been multi-frequency modes not single-frequency, fully-polarimetric modes. The system cannot collect fully-polarimetric, multifrequency data simultaneously. At this time there are approximately 10 fully polarimetric data sets with the majority being X-band, a few L-band, and one C-band data set. The calibration effort has been carried out simultaneously with experiments. All of the data collected so far indicates that the system is functioning properly and can be calibrated both in amplitude and phase at all three frequencies.

The color images in Figure 7 and 8 are calibrated fully polarimetric images. In these images the intensity of the image corresponds to the intensity of the receive signal and the color corresponds to the phase. The color bar at the bottom of the images gives the phase from -180° on the left to +180° on the right. The polarimetric images are arranged with the HH channel in the upper left, VH in the upper right, HV in the lower left, and VV in the lower right. Figure 7, is a calibrated X-band image of some trihedrals and dihedrals in a wooded area in Alaska. The image is phase referenced to the HH channel. Figure 8 is a calibrated L-band image of 4 ARC's in a grassy area. The image is phase referenced to the VV image. One interesting thing to note is that red area in the right side of the VV image. This area is an area of dense small trees. The red color indicates a phase difference of 180° between the VV and HH elements. This would be consistent for scattering from horizontal dihedrals. Perhaps the ground-trunk interface is acting as a dihedral.

A power maximization operation can be performed on the images in Figure 7 and 8. This operation is described by DeGraaf [8]. The operation determines the backscattered power in each pixel of a polarimetric image assuming that the transmit and receive polarizations are chosen to maximize the power. This value is the equivalent to the largest eigenvalue of the conjugate transpose of the scattering matrix multiplied by the scattering matrix [8]. Figure 9(a) is the power maximized image corresponding to Figure 7. Figure 9(b) is the power maximized image corresponding to Figure 8.

A useful way of displaying the polarimetric characteristics of targets is their polarimetric signature [7,9]. In Figure 10 the co-polarized polarimetric signature for a horizontal dihedral from the X-band Alaska data is shown along with the theoretical signature. Visually the signatures are in good agreement indicating the radar is functioning properly. An average signature for Spruce Trees is given in Figure 11. This signature is computed for the average Stokes matrix for an area 25 x 25 pixels. The co-polarized and cross-polarized signatures are given respectively in Figures 11(a) and 11(b). The important thing to note about the tree signature is how it sits on a pedestal with no polarizations resulting in zero cross-section. This indicates a depolarizing type of scattering.

## 5. CONCLUSIONS

The P-3/SAR system is working well in the fully-polarimetric mode at X-,C-, and L-band. Preliminary work on the spatial variation of the phase calibration has been carried out for X-band tree clutter. The long term time stability of the phase calibration has not been investigated because of lack of data sets. Future work will involve studying the time stability of the phase calibration and also measurement of channel leakages using the ARCs. The system exhibits very high image quality and promises to be a powerful remote sensing tool both because of its multifrequency capabilities, its high resolution, and its fully-polarimetric capabilities. Future data sets should include polarimetric data from a variety of ground targets and allow verification of polarimetric scattering models and useful information for remote sensing applications.

## 6. ACKNOWLEDGEMENTS

A great many people contributed to the success of this initial polarization experiment. Our thanks go to the crew and operators of the P-3/SAR system (under the direction of C. Haney of NADC and A. Nichols of ERIM and to the data processors working behind the scenes at ERIM (under the direction of C. Wackerman and D. Gineris). Special thanks also to L. Johnston who developed and implemented software to do various types of polarimetric processing.

## 7. REFERENCES

[1] R.J. Sullivan, A.D. Nichols, R.F. Rawson, C.W. Haney, F.P. Darreff, J.J. Schanne, Jr., 'Polarimetric X/L/C-Band SAR,' Proc. 1988 IEEE Radar Conference, pp. 9-14.

[2] A. Kozma, A.D. Nichols, R.F. Rawson, S.J. Shackman, C.W. Haney, J.J. Schanne, Jr., 'Multifrequency-polarimetric SAR for remote sensing,' Proc. IGARSS '86 Symp., Vol. 1, September 1986, pp. 715-719.

[3] E.S. Kasischke, D.R. Sheen, and G.F. Fowler, 'Radiometric calibration of airborne SAR data,' Proc IGARSS '88 Symp., Edinburgh, Scotland, 13-16 September 1988.

[4] R.M. Barnes, 'Antenna polarization calibration using in-scene reflectors,' Project Report TT-65, Lincoln Laboratory, September 1986.

[5] Y. Shen, 'External polarimetric SAR calibration using polarimetric active radar calibrators,' JPL Memorandum #3343-88-010, January 1988.

[6] D.R. Brunfeldt and F.T. Ulaby, 'Active reflector for radar calibration,' IEEE Trans. on Geoscience and Remote Sensing, Vol. GE-22, No. 2, March 1984, pp. 165-169

[7] H.A. Zebker, J.J. van Zyl, and D.N. Held, 'Imaging radar polarimetry from wave synthesis,' J. of Geophysical Research, Vol. 92, No. B1, January 1987, pp. 683-701.

[8] S.R. DeGraaf, 'SAR image enhancement via adaptive polarization synthesis and detection performance,' Proceedings of 1988 Polarimetric Technology Workshop, Huntsville, Alabama, August 16-18 1988.

[9] J.J. van Zyl, H.A. Zebker, and C. Elachi, 'Imaging radar polarization signatures: Theory and observation,' Radio Science, Vol. 22, No. 4, July-August 1987, pp. 529-543.

Table I

| ARC | Receive Polarization | Transmit Polarization | S |
|---|---|---|---|
| $X_A$ | Horizontal | Vertical | $G_X \begin{bmatrix} 0 & 0 \\ 1 & 0 \end{bmatrix}$ |
| $Y_A$ | Vertical | Horizontal | $G_Y \begin{bmatrix} 0 & 1 \\ 0 & 0 \end{bmatrix}$ |
| $Z_A$ | 45° | 45° | $G_Z \begin{bmatrix} 1 & 1 \\ 1 & 1 \end{bmatrix}$ |

Table I. Configurations of three ARC's for polarimetric calibration.

Expressing R in terms of $R_{22}$ and T in terms of $T_{11}$:

$$R = R_{22} \begin{bmatrix} Q_1 & Q_1 \cdot Y_{22}/Y_{12} \\ X_{11}/X_{21} & 1 \end{bmatrix}$$

where

$$Q_1 = \frac{Z_{11}/Z_{21} - X_{11}/X_{21}}{1 - Z_{11}/Z_{21} \cdot Y_{22}/Y_{12}} \quad \text{or} \quad \frac{Z_{12}/Z_{22} - X_{11}/X_{21}}{1 - Z_{12}/Z_{22} \cdot Y_{22}/Y_{12}}$$

And

$$T = T_{11} \begin{bmatrix} 1 & X_{22}/X_{12} \\ Q_2 \cdot Y_{11}/Y_{12} & Q_2 \end{bmatrix}$$

where

$$Q_2 = \frac{Z_{12}/Z_{11} - X_{22}/X_{21}}{1 - Z_{12}/Z_{11} \cdot Y_{11}/Y_{12}} \quad \text{or} \quad \frac{Z_{22}/Z_{21} - X_{22}/X_{21}}{1 - Z_{22}/Z_{21} \cdot Y_{11}/Y_{12}}$$

Figure 1. Solution for transmitted and received distortion matrices R and T (Solutions by Y. Shen [5])
Expressing R in terms of $R_{22}$ and T in terms of $T_{11}$.

Expressing R in terms of $R_{11}$ and T in terms of $T_{22}$:

$$R = R_{11} \begin{bmatrix} 1 & Y_{22}/Y_{12} \\ Q_3 \cdot X_{11}/X_{21} & Q_3 \end{bmatrix}$$

where

$$Q_3 = \frac{Z_{21}/Z_{11} - Y_{22}/Y_{12}}{1 - Z_{21}/Z_{11} \cdot X_{11}/X_{21}} \quad \text{or} \quad \frac{Z_{22}/Z_{12} - Y_{22}/Y_{12}}{1 - Z_{22}/Z_{12} \cdot X_{11}/X_{21}}$$

And

$$T = T_{22} \begin{bmatrix} Q_4 & Q_4 \cdot X_{22}/X_{21} \\ Y_{11}/Y_{12} & 1 \end{bmatrix}$$

where

$$Q_4 = \frac{Z_{11}/Z_{12} - Y_{11}/Y_{12}}{1 - Z_{11}/Z_{12} \cdot X_{22}/X_{21}} \quad \text{or} \quad \frac{Z_{21}/Z_{22} - Y_{11}/Y_{12}}{1 - Z_{21}/Z_{22} \cdot X_{22}/X_{21}}$$

Figure 2. Solution for transmitted and received distortion matrices R and T (Solutions by Y. Shen [5])
Expressing R in terms of $R_{11}$ and T in terms of $T_{22}$.

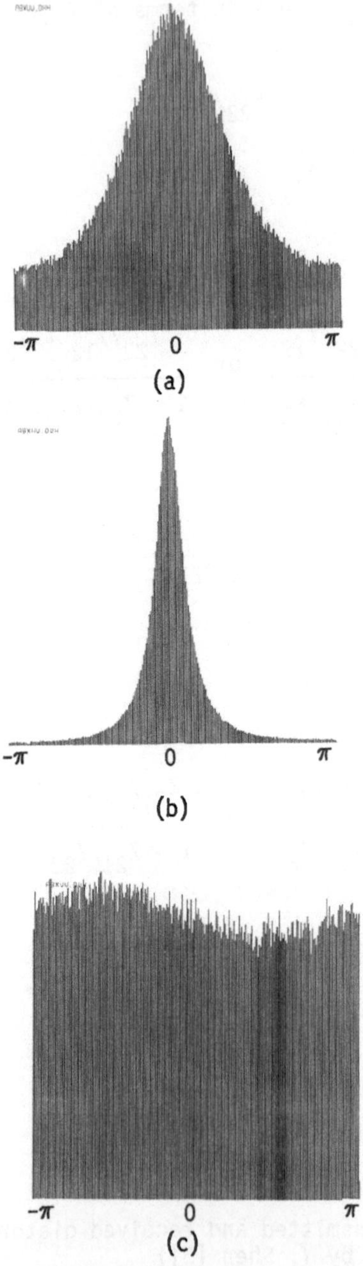


Figure 3. Histograms of phase differences for X-band tree clutter Data: (a) VV-HH (b) HV-VH, and (c) VV-VH.

RS-88-81

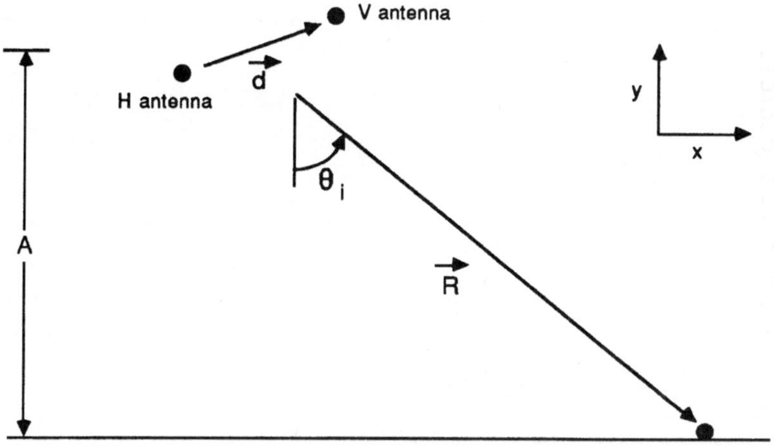


Geometry of SAR.

V and H antenna separation exaggerated and given by the vector d.

Figure 4. SAR geometry. Used to solve for VV-HH phase difference as a function of slant range R.

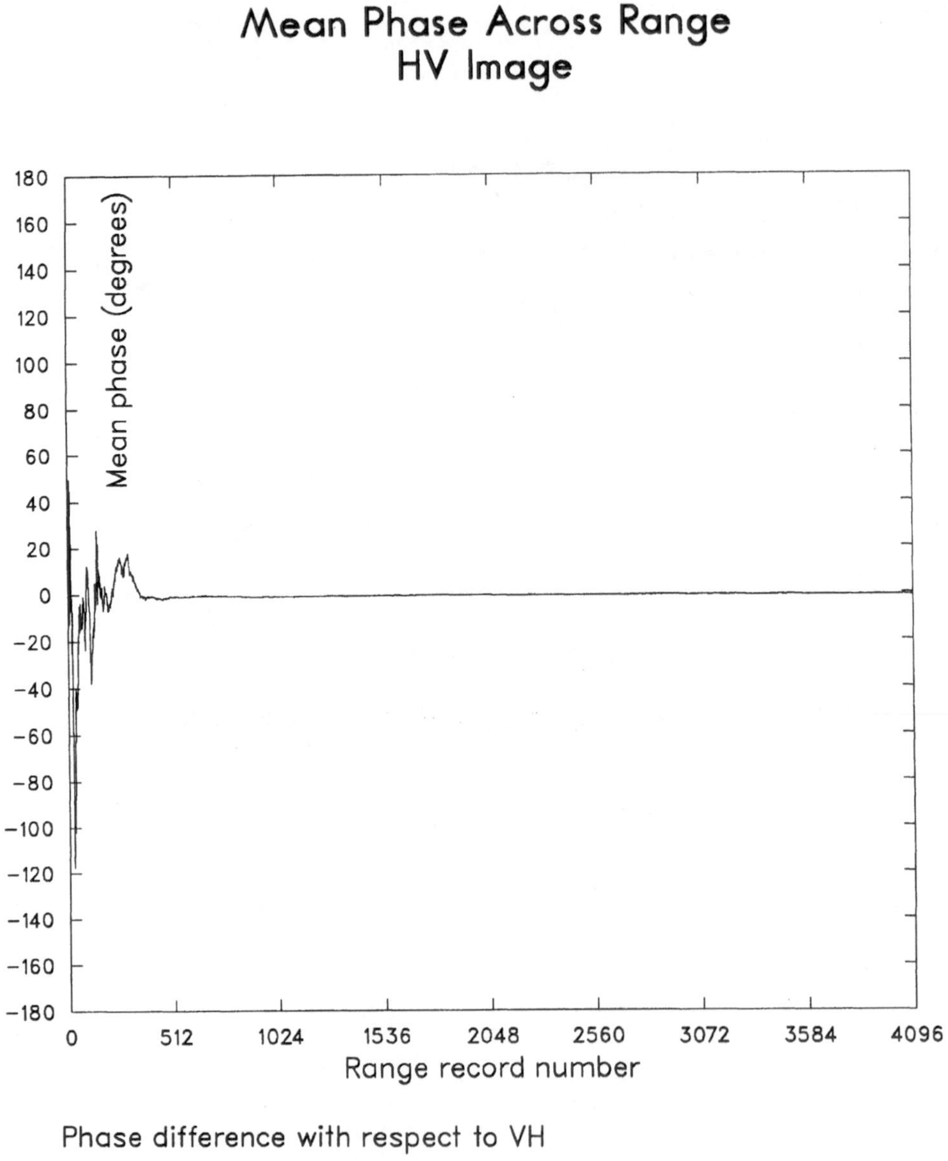


Figure 5. Average phase differences (HV-VH) as a function of slant range for X-band tree clutter data.

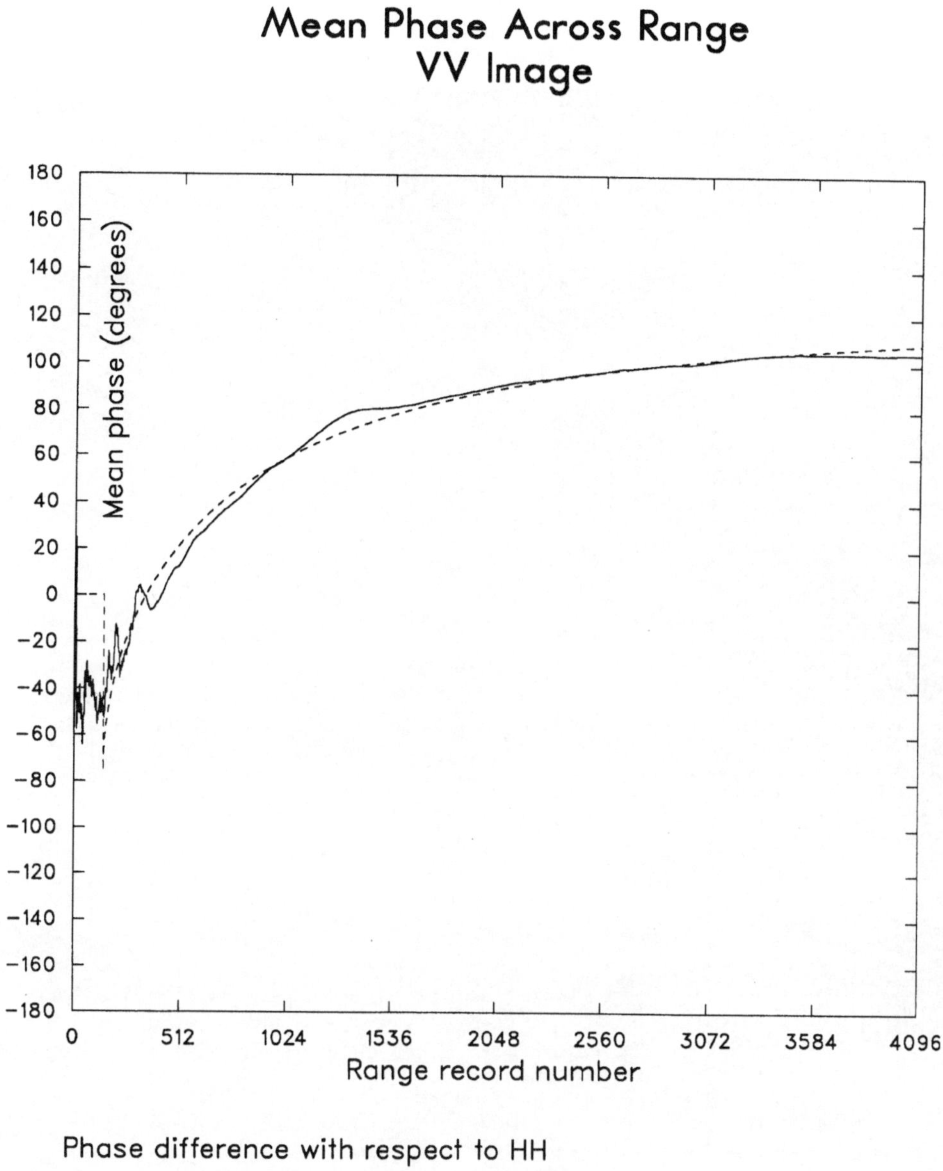


Figure 6. Average phase difference (VV-HH) as a function of slant range for X-band tree clutter data. The dotted line is a fitted curve assuming the H and V antennas have different phase centers.

Figure 7. X-band, polarimetric image phase referenced to the HH channel. Image is of a forested region in Alaska containing trihedrals and dihedrals.

Figure 8. L-Band Polarimetric Image Phase Referenced to the VV Channel. Image is of four ARC's.

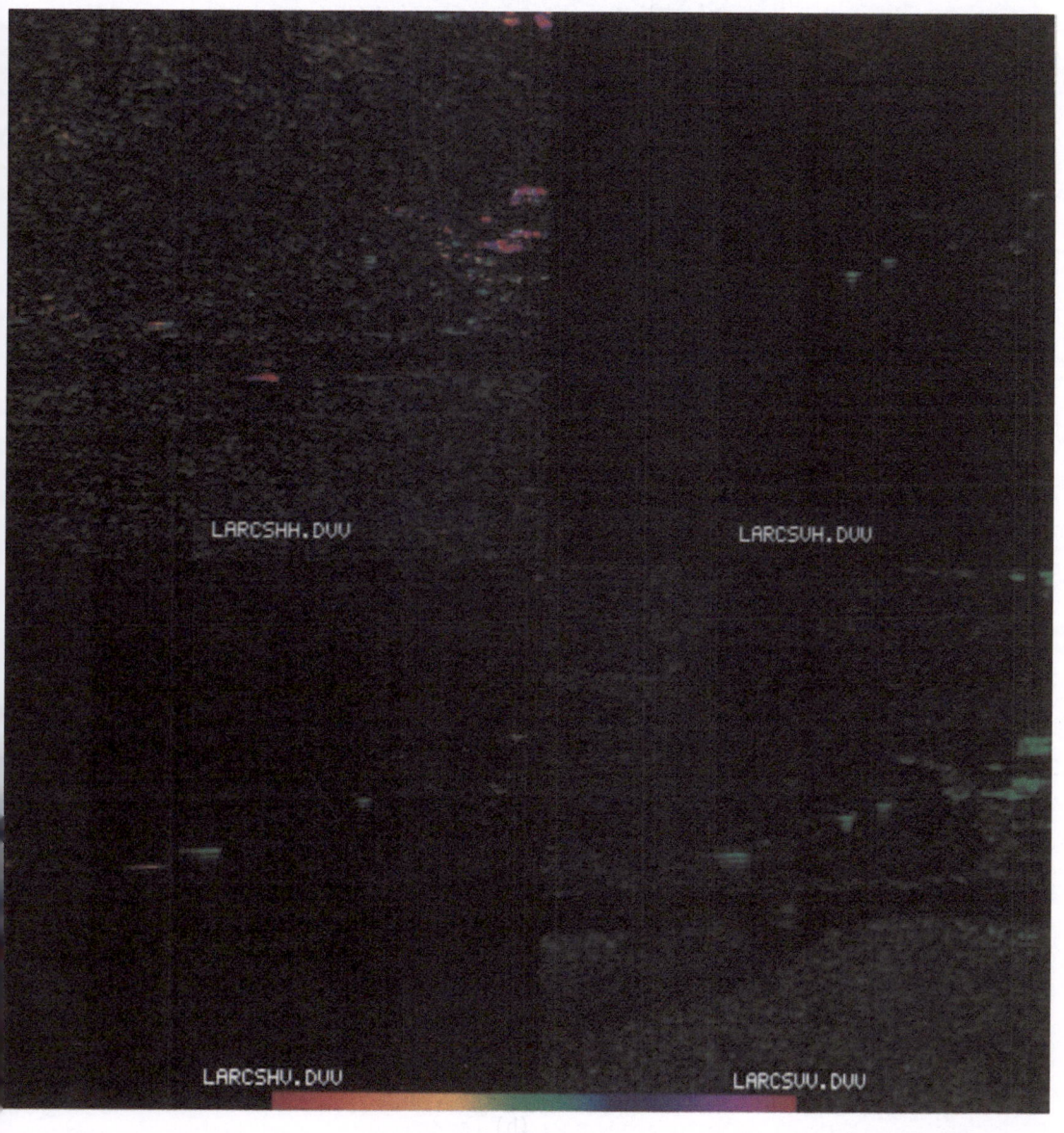

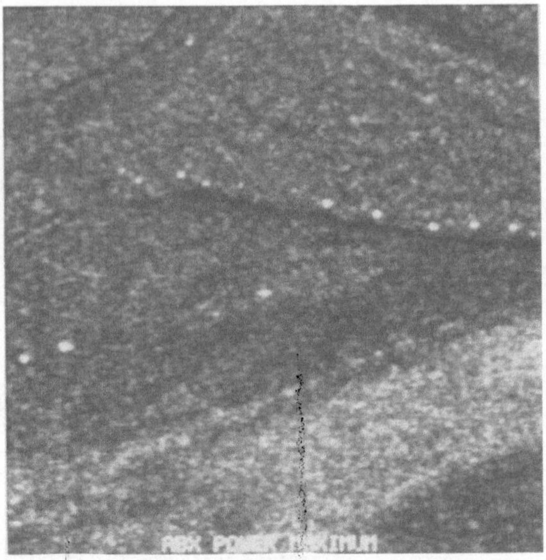


(a)

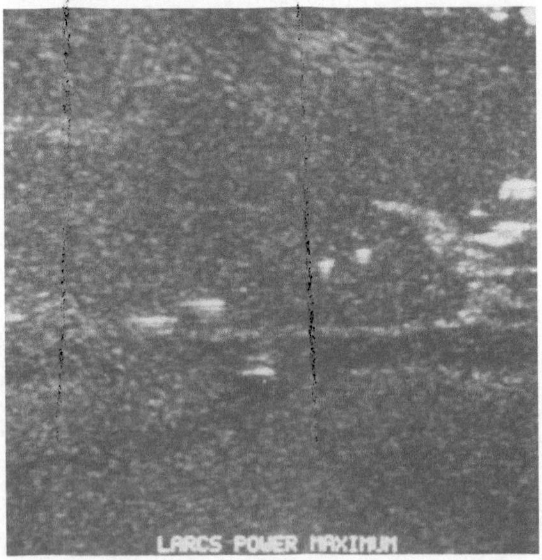


(b)

Figure 9. Power maximized images [8].
(a) Corresponds to Figure 7.
(b) Corresponds to Figure 8.

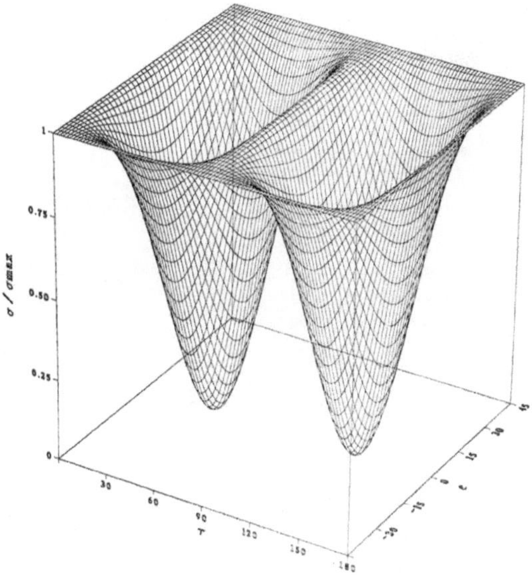


(a)

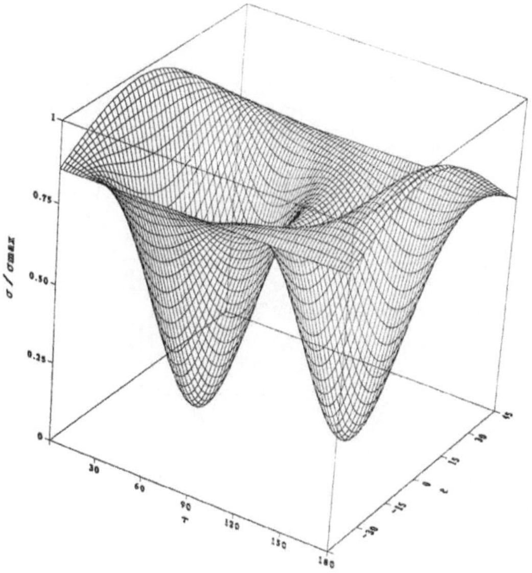


(b)

Figure 10. Co-polarized polarimetric signatures corresponding to a theoretical dihedral (a) and a measured dihedral (b) a calibrated X-band image.

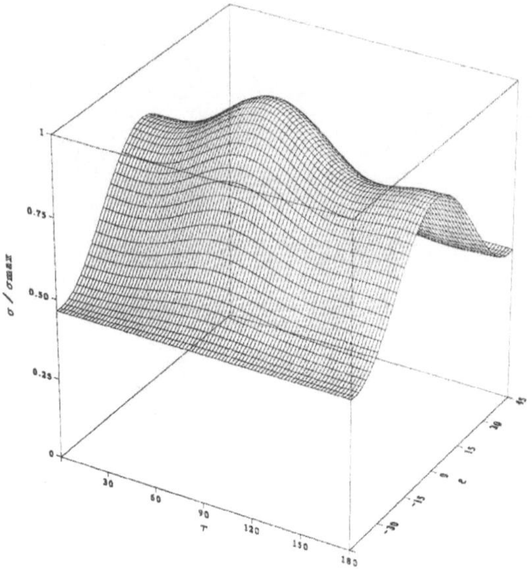


(a)

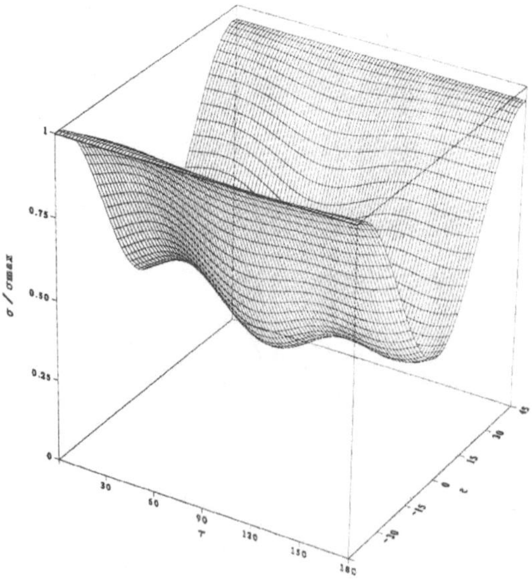


(b)

Figure 11. Co-polarized (a) and cross-polarized (b) signatures for X-band radar clutter from Spruce Trees.

# POLARIZATION CALIBRATION CONSIDERATIONS FOR ROUGH SURFACE INCOHERENT BACKSCATTERING OF SPHERICAL WAVES

Hyo Joon Eom
Dept. of Electr. Eng.
KAIST - Korea Advanced Institute of Science and Technology
400 Kusung-dong Yusung-gu
T305-701 Taejon, South Korea

Wolfgang-M. Boerner
Univ. of Illinois at Chicago
UIC-EECS/CSL, P.O. Box
P.O. Box 60680-4348 SEL-4210,
840 W. Taylor St. M/C 154
Chicago, IL 60607

**ABSTRACT:** The spherical electromagnetic wave backscattering from a random rough terrain is investigated. The incoherent backscattered power is computed to examine how the antenna beamwidth and wave sphericity influence the magnitude of copolarized terrain radar backscattering. The one-dimensional random rough surface in the Kirchhoff approximation is invoked to model a rough terrain. Incident and scattered wave phase fronts are assumed to be spherical, and the effects of antenna beamwidth and wave sphericity are accounted for in radar backscattering coefficient computation. It is found that incoherent backscattering power strongly depends upon the antenna beamwidth and wave sphericity in case the air/satellite borne SAR imaging radar operates at L band or lower frequencies.

## 1. INTRODUCTION

In the study of radar backscattering from naturally-occurring rough terrain, it is important to understand how the radio wave interacts with rough terrain. The scattered power, in general, consists of coherent and incoherent components [1]. The off-nadir backscattered power from rough terrain is mostly incoherent in nature due to the random structure of the terrain profile. The incoherent backscattered power strength is known to depend mostly upon rough terrain structures in terms of surface height, slope, and radar system parameters including antenna beam width, operating frequency, and radar altitude. The incoherent backscattered power is represented in terms of the backscattering coefficient, which is defined as a ratio of backscattered power to incident power. In the theoretical calculation of the incoherent backscattering coefficient, one often introduces the uniform-plane wave assumption for incident and scattered waves due mainly to its simplicity. The uniform-plane wave assumption allows one to approximate the actual spherical wave phase front as a planar one over the area illuminated by radar, thus, facilitating the scattering coefficient computation. One might suspect that the validity for the plane wave assumption seems to depend on the size of the illumination area relative to the surface height correlation length. The exact range of applicability for the plane wave assumption, however, remains quite ambiguous since many others such as antenna beamwidth, radar-altitude, frequency, and surface roughness may also be



*W.-M. Boerner et al. (eds.), Direct and Inverse Methods in Radar Polarimetry, Part 1, 899–907.*

important factors in the radar backscattering process. The purpose of this study lies in re-examining the plane wave assumption in radar scattering computation by evaluating the scattering coefficient in conjunction with spherical wave consideration. This study is, particularly, of great importance in SAR imaging data calibration where the swath width across the track associated with azimuth Doppler processing is often large enough for the wave sphericity effects to be significant.

Recently, Kojima [2,3] has considered random rough surface scattering problems with the Hermite-Gaussian beam assumption, where both coherent and incoherent scattered powers are evaluated. The problem of Gaussian beam interaction with rough surfaces is important in SAR (Synthetic Aperture Radar) imagery data calibration [4] where the swath width across the track is large enough for the wave sphericity to be significant.

In order to model terrain structures, and its interaction with electromagnetic waves, one often adopts random rough scattering theories. Amongst many existing theories, we will use the Kirchhoff approximation and furthermore, for simplicity, assume that the rough terrain may be represented with a one-dimensional random rough Kirchhoff surface. Hence, the problem at hand may be addressed as how to evaluate the incoherent backscattering power when the spherical wave impinges on the one-dimensional rough surface in the Kirchhoff approximation [5].

In the next section, a theory regarding radar backscatter from a one-dimensional Kirchhoff surface is presented and its theoretical angular behavior with a different choice of surface roughness is shown. The difference between backscatter with/and without consideration of antenna beamwidth is considered. A brief summary is given in a conclusion.

## II. INCOHERENTLY BACK-SCATTERED POWER

Under the Kirchhoff approximation, the scattered field $E_s$ from an irregular one-dimenisonal surface S due to a spherical incident wave $e^{ikR_1}/R_1$ is given according to [4] by (see Fig. 1)

$$E_s = \frac{-j}{4\pi} \int_S R(\vec{r})(\hat{n}\cdot\vec{q})e^{jk(R_1 + R_3)}/(R_1 R_3)d\vec{r} \tag{1}$$

where

| | |
|---|---|
| $R(\vec{r})$ | the Fresnel reflection coefficient at $\vec{r}$ |
| $\vec{r}$ | $= x\hat{x} + \zeta(x)\hat{z}$ |
| $\zeta(x)$ | the surface height |
| $\hat{n}$ | the unit vector normal to the surface S at $\vec{r}$ |
| $\vec{q}$ | $= -k\nabla(R_1 + R_3) = -k(\hat{R}_1 - \hat{R}_3)$ |

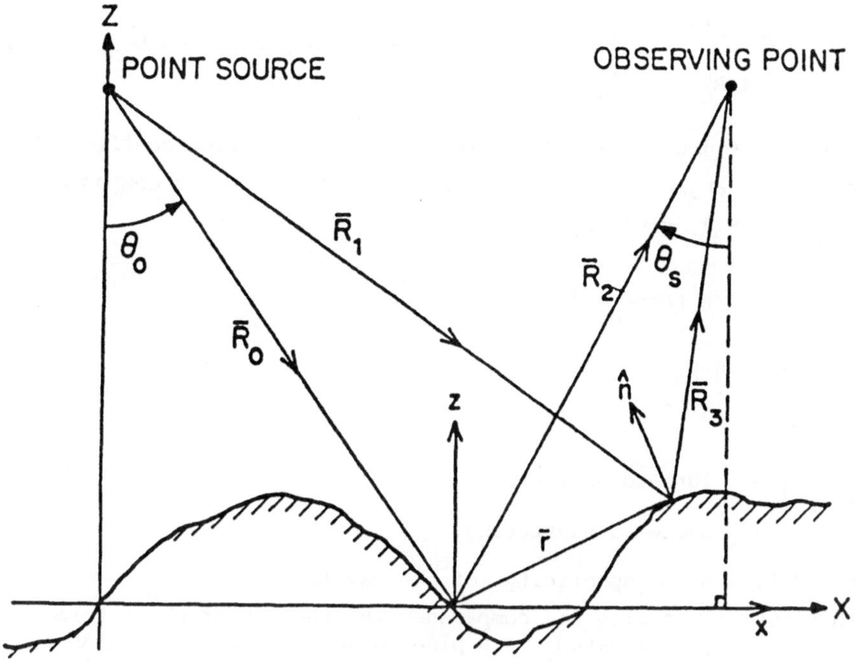


Fig. 1 Geometry of Scattering Problem

| | |
|---|---|
| $\hat{R}_1, \hat{R}_3$ | unit vectors in $\vec{R}_1, \vec{R}_3$ directions (Fig. 1) |
| $\lambda$, $k = 2\pi/\lambda$, | wavelength and wave number |
| $R_1$, $R_3$ | the path lengths from $\vec{r}$ to the point source and the observing point, respectively. |

By assuming Gaussian beams, $e^{-g_o{}^2x^2\cos^2\theta_o}$ and $e^{-g_s^2x^2\cos^2\theta_s}$ along $\vec{R}_o$ and $-\vec{R}_2$, respectively, $E_s$ in the plane of incidence becomes

$$E_s = \frac{-j}{4\pi}\int_{-\infty}^{\infty} R(\vec{r})(\hat{n}\cdot\vec{q})\ \frac{e^{jk(R_1 + R_3)}}{R_1R_3}\ e^{-g_o^2x^2\cos^2\theta_o - g_s^2x^2\cos^2\theta_s}dx \tag{2}$$

where $g_o = 1/R_o\beta_o, g_s = 1/R_2\beta_s : \beta_o, \beta_s$ are the one-sided beam-widths of the transmitter and receiver, respectively, when the surface slope is assumed small.

For simplicity let us further assume small surface heights and $R_o = R_2$, $g_o = g_s$, $\theta_o = \theta_s$; then, $E_s$ is given approximately by

$$E_s = \frac{-jka_o e^{2jkR_o}}{4\pi R_o^2} \int_{-\infty}^{\infty} e^{jkx^2/R_o - j[k_x x + k_z \zeta(x)]} e^{-2g_o^2 x^2 \cos^2\theta_o} dx \qquad (3)$$

where $a_o = 2R(\theta_o)\cos\theta_o$, $R(\theta_o)$ is the Fresnel reflection coefficient at the incidence angle $\theta_o$. $k_x$ and $k_z$ are the x and z components of $-k(\vec{R}_o/R_o - \vec{R}_2/R_2)$, given by

$$k_x = k \sin\theta_o(\cos\phi_s - 1)$$

$$k_z = 2k \cos\theta_o$$

where

$$\phi_s = \begin{cases} 0, & \text{forward scattering} \\ \pi, & \text{backward scattering.} \end{cases}$$

The term defining the sphericity of the wavefront is $x^2/R_o$. Note that when this term is negligible compared with the linear phase term, the incident wave may be treated as a plane wave. At the first glance, it may appear that a large $R_o$ is required to make this term small. This really is not so, because for a fixed beam-width $x^2$ also increases with an increase in range.

For a stationary Gaussian distributed random surface $\zeta(x)$ of variance $\sigma^2$ and surface correlation coefficeint $\rho(x-x')$, the average of the magnitude square of the scattered field is

$$\langle E_s E_s^* \rangle = - \frac{k^2 a_o^2}{(4\pi R_o^2)^2} \int\int_{-\infty}^{\infty}$$

$$\cdot\, e^{jk(x^2-x'^2)/R_o - jk_x(x-x') - k_z^2\sigma^2[1-p(x-x')]}$$

$$\cdot\, e^{-2g_o^2\cos^2\theta_o(x^2+x'^2)} dx\, dx' \qquad (4a)$$

By letting $q = x - x'$, $v = 1/2(x + x')$ and after integrating v, $\langle E_s E_s^* \rangle$ reduces to

$$\langle E_s E_s^* \rangle = - \frac{k^2 a_o^2}{(4\pi R_o^2)^2} \int\int_{-\infty}^{\infty}$$

$$\cdot\ e^{-jk_x u - u^2\left(g_o^2\cos^2\theta_o + \frac{k^2}{4R_o^2 g_o^2\cos^2\theta_o}\right)}$$

$$\cdot\ e^{-k_z^2\sigma^2} \sum_{n=0}^{\infty} [k_z^2\sigma^2\rho(u)]^n/n!du \qquad (4b)$$

Hence, the incoherent backscattering power from a surface with the Gaussian surface height correlation $\rho(u) = \exp(-u^2/l^2)$ is given after u-integration by

$$\langle E_s E_s^* \rangle = c\ e{-k_z^2\sigma^2} \sum_{n=1}^{\infty} \frac{(k_z^2\sigma^2)^n}{n!} \sqrt{\frac{\pi}{\beta}} \exp(-kx^2/4\beta) \quad , \qquad (5)$$

where $c = k^2a_o^2A/(4\pi R_o^2)$

$$\beta = \frac{n}{\quad} \left( 1 + \frac{l^2\cos^2\theta_o}{nR_o^2\beta_o^2} + \frac{k^2l^2\beta_o^2}{4n\cos^2\theta_o} \right)$$

l = surface height correlation length.

Note that the effects of antenna beamwidth and sphericity are reflected in $\beta$ in terms of $l^2\cos^2\theta_o/(nR_o^2\beta_o^2)$ and $k^2l^2\beta_o^2/(4n\cos^2\theta_o)$, respectively.

The rough estimates for the relative importance of antenna beamwidth and sphericity can be obtained for two special cases shown as follows:

i) In case of a high altitude (air borne or space borne) radar, the condition $R_oB_o >> 1$ may be easily met. Hence, $\beta$ becomes $\beta = \frac{n}{l^2}(1 + k^2l^2\beta_o^2/4n\cos\theta)$. It is apparent from this expression of $\beta$ that the radar backscattering power is independent of radar altitude, but is strongly dependent on antenna beamwidth $\beta_o$ if $l\beta_o/\lambda >> 1$.

ii) In case of the plane wave assumption, $\beta$ simplifies to $\beta = \frac{n}{l^2}$ which cross-checks with the result in [4].

Hence from the previous two special case considerations, if the radar operates at a high altitude and the surface height correlation is small enough to satisfy the condition $l << \lambda/\beta_o$, then the assumption of uniform plane waves for incident and scattered waves should be valid.

In the next section, the theoretical behavior of backscattered power, given in Eq.(5), is shown and their effects of antenna beamwidth and sphericity are examined.

## III. THEORETICAL BEHAVIOR OF BACKSCATTERED POWER

For the purpose of theoretical illustration, an L-band airborne SAR-imaging radar with the following radar parameters is considered: $\lambda$ = 20 cm, $\beta_o$ = 0.17 rad, and $R_o$ = 1 km, the rough surface is chosen to have $k\sigma$ = 0/5 and kl = 25, which are on the order of typical surface roughness structures. It is very informative to notice from the above chosen parameters that the effect of wave sphericity $(k^2 l^2 \beta_o^2/(4\eta\cos^2\theta_o))$ is far more important than that of the beamwidth $(l^2\cos^2\theta_o/\eta R_o^2\beta_o^2)$.

Fig. 2 shows the comparison of incoherent radar backscattered power versus incidence angles between plane wave and spherical wave incidences. It is seen that due to the large beamwidth associated with the antenna, the actual radar backscattered power of a spherical wave exhibits a much broader angular backscattering pattern as compared to the plane wave assumption. In order to examine the effects of surface roughness, the surface correlation length is reduced to kl = 10 from 25. (Hence, the surface considered in Fig. 3 becomes rougher than the one in Fig. 2) and its effects are shown in Fig. 3. The effects of sphericity as well as antenna beamwidth become less appreciable compared to Fig. 2.

## 4. CONCLUSION

The effects of antenna beamwidth and wave sphericity associated with the spherical wave are investigated using the one-dimensional Kirchhoff surface. It is found that the backscattering coefficient depends strongly on the beamwidth due to the effects of wave sphericity, in particular, in SAR imaging problems. It has also been shown that unless the surface correlation length l is much smaller than $\lambda/\beta_o$, the plane wave assumption is invalid for scattering computation for randomly rough surfaces. The theoretical study indicates that the effects of antenna beamwidth must be accounted for in radar scattering computation of the L band airborne/spaceborne SAR imaging radar for systems operating at larger wavelength.

## ACKNOWLEDGEMENTS

This research was supported by the US Navy, Office of Naval Research, Contract Nos. N00014-80-C-0773 and N00014-90-J-1405 and by the US Army, Army Research Office under Contract No. DAAK-03-89-K-0116. The skillful typing of the manuscript and the preparation of the figures by Richard W. Foster is greatfully acknowledged.

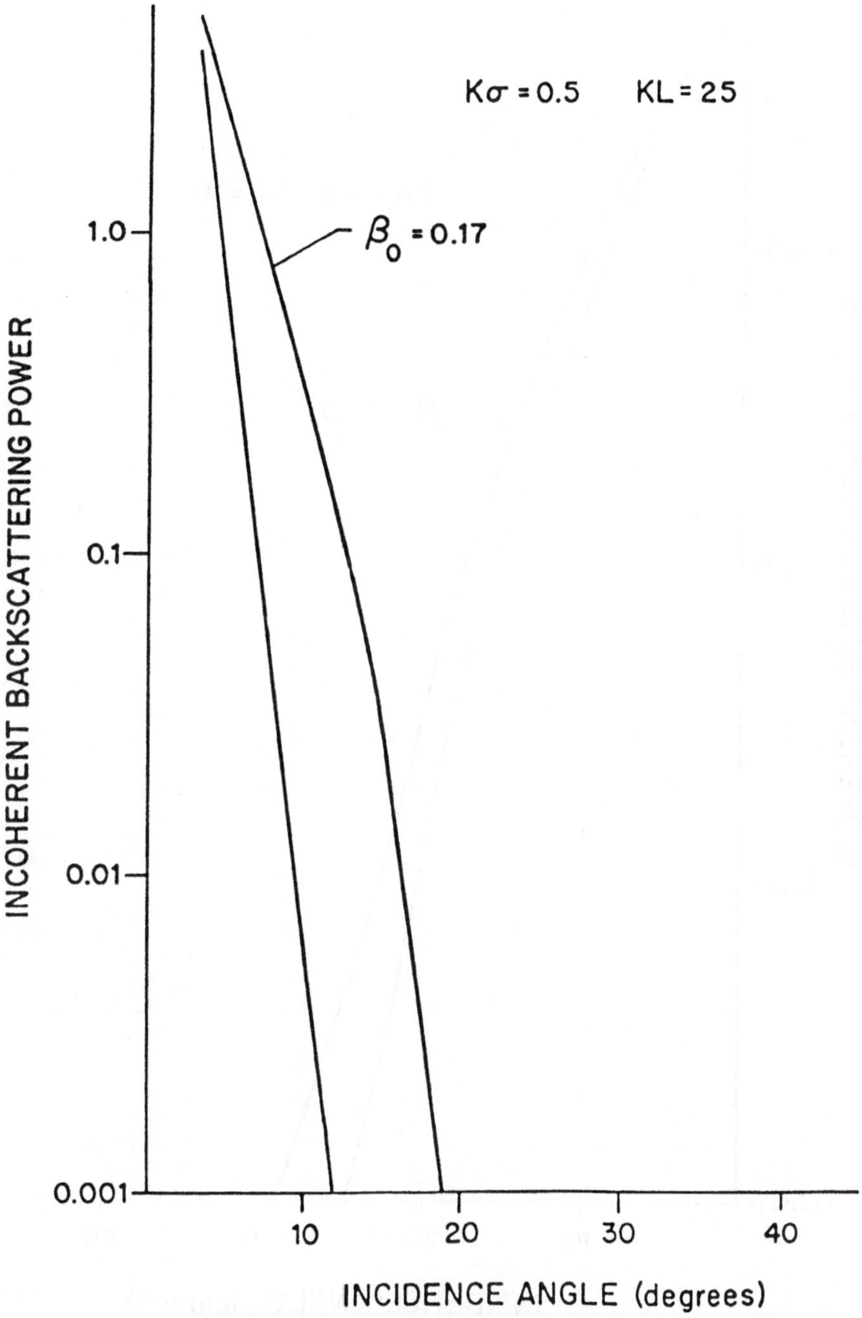


Fig. 2 Angular Behavior of Incoherent Radar Backscatter Between Spherical ($\beta_o$ = 0.17) and Plane Wave Incidences.

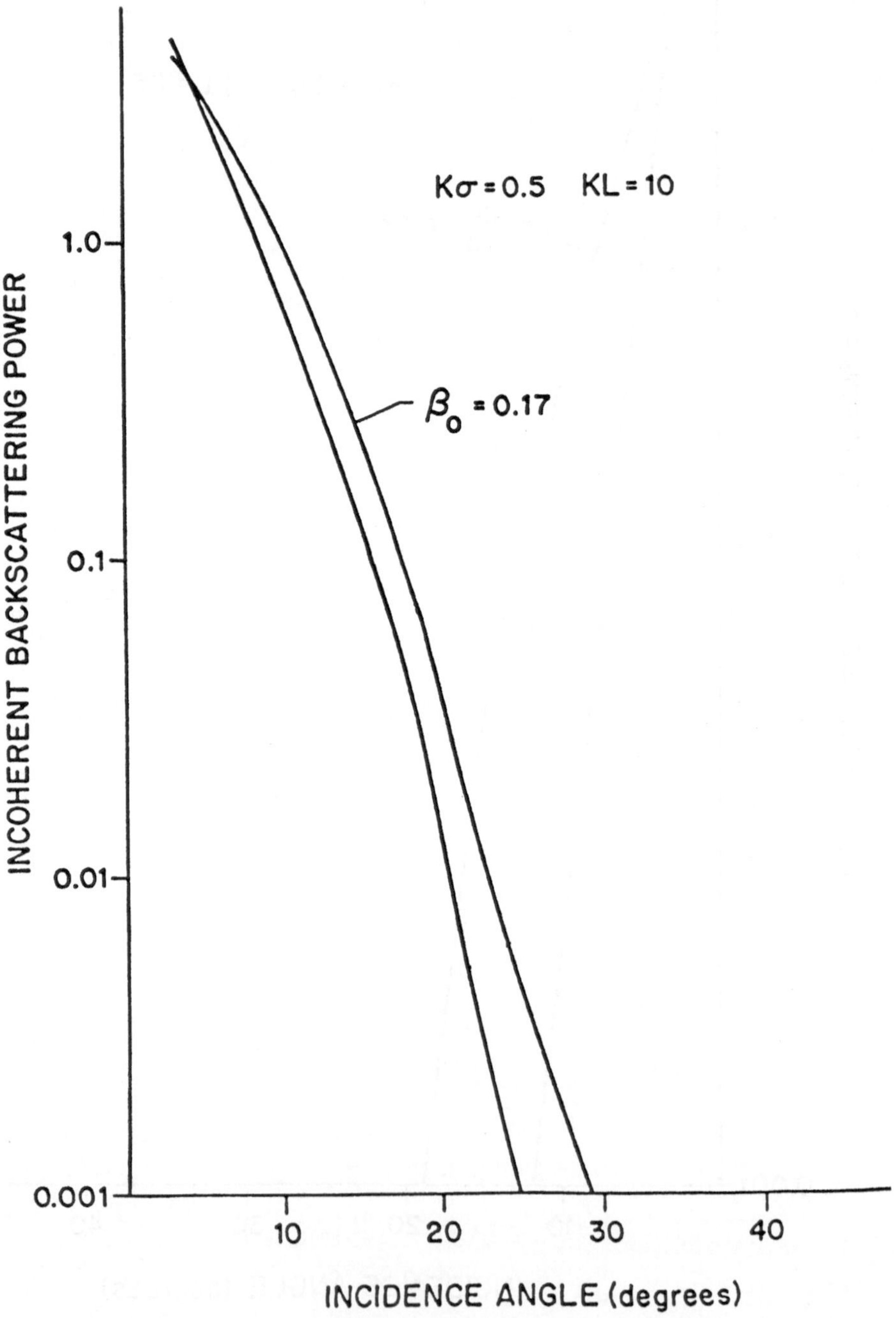


Fig. 3 Angular Behavior of Incoherent Radar Backscatter Between Spherical ($\beta_o$ - 0.17) and Plane Wave Incidences.

REFERENCES

[1] H.J. Eom and W-M. Boerner, "A Re-Examination of Radar Terrain Backscattering at Nadir", IEEE Trans. on Geoscience and Remote Sensing, Vol. TE-24, No. 2, pp. 232-234, March 1986

[2] T. Kojima, "Scattering of Hermite-Gaussian Beams from an Irregular Surface", J. Appl. Phys., Vol. 65, No. 4, pp. 1424-1428, Feb. 1989

[3] T. Kojima, "Scattering of Hermite-Gaussian Beams from Random Rough Surface-3D Scalar Analysis", Proc. of Int. Symp. on Ant. and Prop., Vol. 4, pp. 969-972, Aug. 1989

[4] A.K. Fung and H.J. Eom, "Coherent Scattering of Spherical Wave from an Irregular Surface", IEEE Trans. on Antenna and Propagation, Vol. AP31, No. 1, pp. 68-72, Jan. 1983

[5] J.T. Ulaby, R.K. Moore, and A.K. Fung, "Microwave Remote Sensing", Vol. II, Reading MA, Addison-Wesley, 1984

# POLARIMETRIC SIGNAL PROCESSING TECHNIQUES

D. GIULI, M. GHERARDELLI
*Dept. of Electronics Engineering*
*University of Florence*
*Via di S.Marta*, 3
50139 *Firenze - Italy*

ABSTRACT. Recent developments on polarimetric processing techniques are presented, referring to the objective of improving target detection performance in low resolution radars equipped with dual polarization receiving channels. To this end the polarization behaviour of different radar objects is recalled with reference to several measurement results obtained with a dual polarization radar. The detection of targets in the cases of presence of unpolarized background white noise, polarized jamming white noise and ground clutter are considered. Based on a statistical detection theory approach several structures of optimum receivers for detection of target in the clear are presented and discussed. Several adaptive polarization cancellers for improving the signal-to-clutter ratio are described and examined. The performance of some adaptive polarization cancellers is evaluated in the presence of polarized jamming white noise or ground clutter. This is made through the processing of experimental dual-polarization data.

## 1. Introduction

In conventional radar systems the received backscattered wave is converted to a scalar signal; in this way the wave polarization is not recovered.

To acquire the entire information contents of the backscattered wave, the wave polarization information has to be retained by a vector measurement process. This can allow for improving radar performance in the presence of disturbance, through dual-polarization signal processing [1-3].

This operation requires that the radar be capable of decomposing the received wave into two ortogonally polarized components, which independently feed two identical and coherent reception channels. In other words, for the above purpose a dual-polarization receiver is needed.

When polarization diversity is used not only on reception but also in transmission, the object's scattering properties are determined completely, but the system complexity increases.

Polarimetric signal processing techniques are subject of current studies. Their objectives are to improve radar performance in terms of target detection, discrimination, classification, and resolution [1].

In this paper some recent developments on polarimetric processing techniques are systematically presented, referring to the objective of improving target detection performance in



*W.-M. Boerner et al. (eds.), Direct and Inverse Methods in Radar Polarimetry, Part 1*, 909–938.

low resolution radars equipped with dual polarization receiving channels. In particular, the detection of targets in the cases of presence of unpolarized background white noise, polarized jamming white noise and ground clutter is considered.

To this end, the basic polarization analysis tools (Sect.2) and the polarization behaviour of different radar objects (Sect.3) are recalled with reference to several measurement results obtained with a dual polarization radar. Based on the extension of a statistical detection theory approach introduced in [4] several structures of optimum receivers, derived in [5],[6] for detection of target in the clear are presented and discussed. In Sect.5 several adaptive polarization cancellers recently proposed for improving the signal-to-clutter ratio are described and examined. Their performance is evaluated in the presence of polarized jamming white noise or ground clutter. This is made through the processing of experimental dual-polarization data.

Throughout this paper emphasis is posed on the different adaptive vector signal processing schemes that can be used to improve target detection performance in low resolution radars equipped with dual polarization receiving channels, but transmitting with fixed polarization. It is pointed out that even with quite affordable polarimetric signal techniques applied to low-resolution radars, valuable target detection improvements can be expected, while further improvements, also in the field of polarimetric target classification, can be expected with high resolution radars. These aspects are briefly discussed in Sect. 6.

## 2. Polarization Analysis

In this section some fundamental analytical tools are briefly recalled in order to describe the signal processing techniques presented in the following sections.

### 2.1 PARTIALLY AND COMPLETELY POLARIZED WAVES

In a right handed cartesian x-y-z coordinate system, the e.m. field vector of a plane, harmonic wave propagating along the z-axis (positive sense) can be represented by a complex vector given by:

$$\underline{E}(z,t) = [\, E_H(z,t),\, E_V(z,t)\,]^T = \underline{h}(t)\, e^{j(\omega t - kz)} \tag{1}$$

where the labels H and V denote the horizontal and vertical electric field components respectively, k is the propagation constant, the 'T' upper label means transposition of the vector, and $\underline{h}(t)$ is a time-varying vector which can be expressed as:

$$\underline{h}(t) = [\, h_H(t)\, ,\, h_V(t)\,]^T \tag{2}$$

where $h_H(t)$ and $h_V(t)$ are the complex envelopes of the horizontal and vertical field components, considered as outcomes of stationary and ergodic random processes, with phase uniformly distributed and independent of moduli. In this condition, the propagating wave is called "partially polarized".

When the factorization $\underline{h}(t) = a(t)\,[h_H,h_V]^T$ holds, the wave is said completely polarized, and its time invariant polarization is specified by the constant vector $\underline{h} = [h_H,h_V]^T$. In this case the wave polarization is usually represented through the polarization ellipse (Fig. 1), which is the locus described in a transverse plane by the tip of the field vector $\underline{E}(z,t)$.

The wave polarization is thus related to the shape and orientation of the ellipse together with the rotation sense of the field vector, when looking along the direction of propagation. Therefore the following ellipse parameters define the polarization state:

- the ellipticity angle $\tau$, with $|\tau| \leq 45°$;
- the orientation angle $\phi$, with $0° \leq \phi \leq 180°$;
- the polarization sense, given by the sign of $\tau$ (the positive value applies for right handed polarizations).

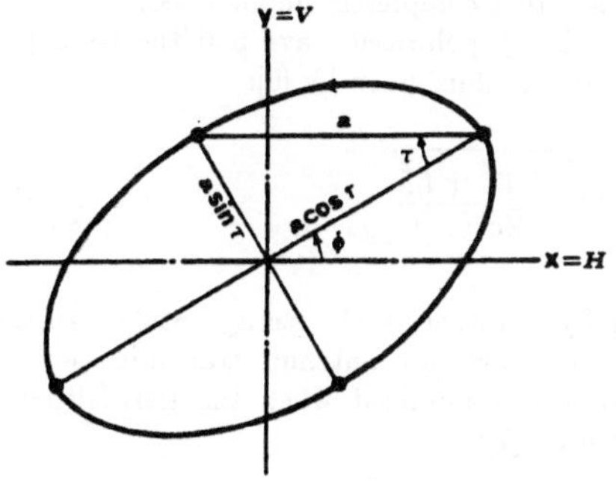


Fig. 1 - Polarization ellipse

Notice that according to the above definition the wave polarization is invariant with respect to any scalar complex factor multiplying the polarization vector $\underline{h}$. Therefore it is generally convenient to refer to a canonical polarization vector, subject to the normality constraint: $|h_H|^2+|h_V|^2=1$.

Disregarding a complex factor, the following relation holds between the constant vector $\underline{h}$ and the polarization ellipse parameters:

$$\underline{h} = \begin{bmatrix} \cos\phi & -\sin\phi \\ \sin\phi & \cos\phi \end{bmatrix} \begin{bmatrix} \cos\tau \\ j\,\sin\tau \end{bmatrix} \tag{3}$$

The polarization of a partially polarized wave can also be described through the average Stokes vector $\underline{\bar{g}}$ whose components are defined as follows [7]:

$$\underline{\bar{g}} = \begin{bmatrix} \bar{g}_0 \\ \bar{g}_1 \\ \bar{g}_2 \\ \bar{g}_3 \end{bmatrix} \hat{=} \begin{bmatrix} <|h_H(t)|^2>+<|h_V(t)|^2> \\ 2\,<\mathrm{Im}[h_H^*(t)\; h_V(t)]> \\ <|h_H(t)|^2>-<|h_V(t)|^2> \\ 2\,<\mathrm{Re}[h_H^*(t)\; h_V(t)]> \end{bmatrix} \tag{4}$$

where the symbol "$\hat{=}$" means definition of the parameters on the left side and <(.)> denotes time average. When the wave is completely polarized, the associated Stokes vector $\underline{g}$ is defined as in (4), where no time-averaging is performed.

The total average power associated with the wave, given by $\bar{g}_0$, meets the following condition:

$$\bar{g}_0^2 \geq \bar{g}_1^2 + \bar{g}_2^2 + \bar{g}_3^2 \tag{5}$$

In (5) the equality holds when the wave is completely polarized, while $\bar{g}_1=\bar{g}_2=\bar{g}_3=0$ when the wave is completely unpolarized.

In terms of the Stokes vectors, a partially polarized wave can be uniquely decomposed as [3]:

$$\underline{\bar{g}} = [\, g_{op},\, \bar{g}_1,\, \bar{g}_2,\, \bar{g}_3 \,]^T + [\, \bar{g}_0 - g_{op},\, 0,\, 0,\, 0 \,]^T \tag{6}$$

where $g_{op} = (\bar{g}_1^2 + \bar{g}_2^2 + \bar{g}_3^2)^{1/2}$ and $(\bar{g}_0 - g_{op})$ are the average powers pertaining to two independent waves, one completely polarized and the other completely unpolarized.

The ratio between the average power of the completely polarized wave and the total power is called the 'degree of polarization" p of the wave, and it is thus given by [8]:

$$p \triangleq (g_{op}/\bar{g}_0) = \frac{\sqrt{\bar{g}_1^2 + \bar{g}_2^2 + \bar{g}_3^2}}{\bar{g}_0} \tag{7}$$

When a partially polarized wave backscattered by a target, with average Stokes vector $\underline{\bar{g}}$, is received by an antenna with polarization $\underline{h}_r$ (i.e.: such an antenna transmits waves with polarization vector $\underline{h}_r$), the average received power is minimized when the normalized Stokes vector representing the antenna polarization is given by [7]:

$$\underline{f}(\underline{h}_r) = [\, f_0,\, f_1,\, f_2,\, f_3]^T = [\, 1,\, -\bar{g}_1/g_{op},\, -\bar{g}_2/g_{op},\, -\bar{g}_3/g_{op} \,]^T \tag{8}$$

while the orthogonal polarization yields the maximum average power.

The minimum and the maximum received powers are respectively given by [7]:

$$\bar{P}_{min} = \bar{g}_0\,(1-p)\,/2 \quad ; \quad \bar{P}_{max} = \bar{g}_0\,(1+p)\,/2 \tag{9}$$

A useful description of partially polarized waves is also given by the complex correlation coefficient $\mu$ between the voltages $v_r$ and $v_{r_\perp}$, received from two orthogonally polarized antennas. This is defined as [9]:

$$\mu \triangleq \frac{<v_r \cdot v_{r_\perp}^*>}{\sqrt{<|v_r|^2>\,<|v_{r_\perp}|^2>}} \tag{10}$$

with $0 \leq |\mu| \leq 1$.

It is also convenient to define the transformation of a polarization vector $\underline{h}$ from the polarization basis A-B to the vector $\underline{h}$' represented in a new polarization basis A'-B'. The latter is obtained through a linear transformation [10]:

$$\underline{h}' \triangleq [\, h_{A'}\,,\, h_{B'}\,]^T = [T]\,\underline{h} = [T]\,[\, h_A\,,\, h_B\,]^T \tag{11}$$

where the matrix [T] can be expressed as [10]:

$$[T] = \begin{bmatrix} 1 & -\rho^* \\ \rho & 1 \end{bmatrix} \frac{1}{\sqrt{1+|\rho|^2}} \tag{12a}$$

$$\rho \triangleq h_{A'}(\underline{h}_A)\,/\,h_{B'}(\underline{h}_A) \tag{12b}$$

where $h_{A'}(\underline{h}_A)$ and $h_{B'}(\underline{h}_A)$ are the components of the versor $\underline{h}_A$ in the A'-B' basis, while $\rho$ represents the complex polarization ratio of the polarization A expressed in the new basis.

## 2.2 POLAR REPRESENTATION OF POLARIZATION

Several polar representations of the polarization state can be usefully applied for analyzing the polarization properties of signals [11].

In the Poincaré sphere representation [12], any polarization state of a completely polarized wave can be represented by a point P whose cartesian coordinates are expressed through Stokes parameters as follows:

$$P = ( g_2/g_0, g_3/g_0, g_1/g_0) \tag{13}$$

where the involved parameters are not time averaged because the wave is supposed completely polarized.

Polarizations are thus mapped onto the surface of a sphere, said the Poincaré sphere (see Fig. 2).

In such a representation the sphere poles correspond to the circular polarizations; the equatorial circle is the locus of the linear polarizations.

The extrema of each diameter correspond to a pair of orthogonal polarizations. Symmetrical points with respect to the equatorial plane change polarization for the rotation sense only.

Once the average Stokes parameters are used in (13), the Poincaré sphere representation is extended to the general case of partially polarized waves [13]. A partially polarized wave is thus represented by a point inside the Poincaré sphere. In fact from the eqs. (7) and (13) it follows that the degree of polarization (p<1) of a partially polarized wave equals the distance of the representative point P from the centre of the sphere.

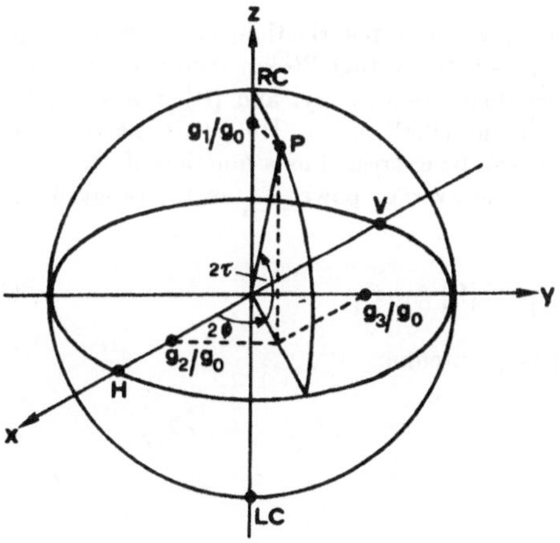


Fig. 2 - Poincaré sphere

## 2.3 TARGET SCATTERING MATRIX

The polarization state of radar echoes strongly depends on the polarization transformation induced by the e.m. backscattering from the illuminated radar object. If $\underline{h}_T$ is the polarization vector of the incident wave, and $\underline{h}_S$ is that of the backscattered wave, both defined in polarization basis $\underline{i}_1$-$\underline{i}_2$, the following relationship holds [10],[13]:

$$\underline{h}_S = [S]\, \underline{h}_T \tag{14}$$

where [ S ] is the object's scattering matrix defined as:

$$[S] = [S(t)] = \begin{bmatrix} s_{11}(t) & s_{12}(t) \\ s_{21}(t) & s_{22}(t) \end{bmatrix} \tag{15}$$

The complex element $s_{ij}(t)$ represents the echo signal received when transmitting with polarization $\underline{i}_i$ and receiving with polarization $\underline{i}_j$, with i=1,2 and j=1,2 as pedices.
If $\underline{h}_R$ is the polarization vector of the receiving antenna, the complex amplitude of the received signal is given by:

$$V = [S]\, \underline{h}_T \cdot \underline{h}_S \tag{16}$$

The absolute phase of [ S ] does not affect the wave polarization, so the matrix is defined by only five independent parameters [14], which depend on:
- target electrical and geometrical characteristics;
- the transmission frequency;
- the target aspect with respect to the radar line of sight;
- the radar transmitted waveform.

Since the wave backscattered from the object is generally partially polarized, it can also usefully be expressed in terms of the average Stokes vectors, as follows:

$$\bar{\underline{g}}_S = [R]\, \underline{g}_T = \underline{g}_{SP} + \bar{\underline{g}}_{SU} \tag{17}$$

where $\underline{g}_T$ is the Stokes vector of the transmitting antenna polarization, $\bar{\underline{g}}_S$ is the average Stokes vector of the backscattered field, $\bar{\underline{g}}_{SU}$ and $\underline{g}_{SP}$ are the average Stokes vectors of the unpolarized and polarized components of the backscattered field respectively, and [R] is a 4x4 real valued symmetric matrix named "reflection Stokes" or "Mueller" matrix [14]. This matrix is generally specified by nine independent parameters and it can be expressed as a function of the second-order statistical moments of the [S] matrix elements. The average power $\bar{P}_R$ of the received signal can thus be expressed as [14]:

$$\bar{P}_R = \bar{\underline{g}}_S \cdot \underline{g}_R = [R]\, \underline{g}_T \cdot \underline{g}_R \tag{18}$$

where $\underline{g}_R$ is the polarization vector of the receiving antenna.

## 3. Polarization Behaviour

### 3.1 SOME MEASUREMENT RESULTS

The actual improvement of radar performance which can be obtained through polarization processing techniques can be evaluated if the needed information on the actual behaviour of radar-signal polarization is made available. In this concern, the essential polarization behaviour of different radar objects is described in this section, basing upon some measurement results obtained with an operational S band Air-Traffic-Control (ATC) radar suitably modified for dual polarization reception [15], [16].

The experimental results have been obtained by using right-circular polarization in transmission. The same circular polarization and the orthogonal one were made available through two reception channels.

The pulse duration was 1.5$\mu$s, while the pulse repetition frequency 1KHz. The antenna scan period and dwell time on target were 4s and 20ms respectively.

The data have been acquired through an acquisition system [15] which allows sampling and acquisition of dual-polarization signals within a radar coverage window, selectable by software both in size and position. The data acquisition has been performed during successive, even if not sequential, scans.

Some experimental results were obtained which pertain to polarized barrage jamming, aircraft echoes and ground clutter. The data which refer to jammer noise have been received during a pulse repetition period, for all resolution cells in a sweep; while those data which refer to the other types of radar signals have been received during a dwell time and within a specified resolution cell.

The results have been illustrated by resorting to the polar representation on the polarization chart, namely the projection of the Poincaré surface (see Sect. 2) on the equatorial plane, where the polarization plots associated with each target echo have been reported for all the echo samples received from the same object at a specified antenna scan. Different symbols have been used to distinguish orthogonal polarizations with different sense of rotation.

The polar representations of signals produced by a horizontally-polarized white-noise jammer source have been considered referring to two different acquisition modes: in the first case the source radiation was received through the radar mainlobe, while in the second case a radar sidelobe was pointing at the same source. As a consequence of the different antenna effect, a slight signal depolarization has been observed.

The polarization behaviour of an aircraft target has been analysed for two different scans with a time interval of 20s from each other: the target was tracked while taking off Fiumicino Airport. The related representations show that the target echo polarization does not change significantly during dwell time, while its average polarization changes considerably from one scan to another, still keeping itself nearly linear.

Some examples of ground clutter polarization behaviour have also been considered by resorting to the polarization chart representation of sequences of samples collected during dwell time from one resolution cell of different types of ground clutter: extended ground clutter phenomena and clustered strong ground clutter. The scanned ground surface was characterized by the low relieves of the suburban area in the city of Rome.

Two main features can be thus observed: the first (Figs. 3 and 4) is that the polarization state of samples lines up along some preferred direction; the other is a quite clustered distribution of the polarization state of the samples around their short-term average polarization on the Poincaré sphere. In both cases the mean polarization behaviour of the sequences doesn't change considerably from scan to scan, as illustrated in Figs. 3 and 4 which pertain to ground clutter acquired during two successive scans with 8 s time delay between them. In order to better evaluate such features, a polarization basis transformation for the representation on the polarization chart of the echo sequences was performed, so as to map the mean polarization of each sequence, collected during the dwell-time, on the circularly left handed polarization point (centre point of the polarization chart).

By a statistical analysis thus purposely performed [16], it was pointed out that, in the case of space-distributed ground clutter, the window sample alignment on the Poincaré sphere, along the prevalent direction, increases as the clutter-to-noise ratio (C/N) increases. Moreover it can be observed that the values of the polarization degree, evaluated for each acquired sample sequence during dwell time approach to the maximum value ($p \sim 1$) as the C/N ratio increases. A parallel investigation concerning the scan to scan stationarity of the polarization state of each analyzed resolution cell has been accomplished: the observed ground clutter data indicate frequent high stationarity of polarization on the same window from one scan to another, for high C/N ratios.

This result, together with the spatially narrow patterns on the polarization domain and a high polarization degree during dwell-time points out that good performance of polarization based techniques for clutter rejection can be expected in the presence of strong ground clutter.

This capability becomes more evident when aircraft targets have to be detected in the presence of screening obstacles giving rise to strong clustered ground clutter: because of high polarization degree of both phenomena during dwell time, and because of fast changement of the target mean polarization from scan to scan, in contrast with the stationary behaviour of clutter mean polarization, polarization based discrimination of such phenomena can be successfully applied, so as the signal-to-clutter ratio at the input of the receiver can be considerably improved (see Sect. 5).

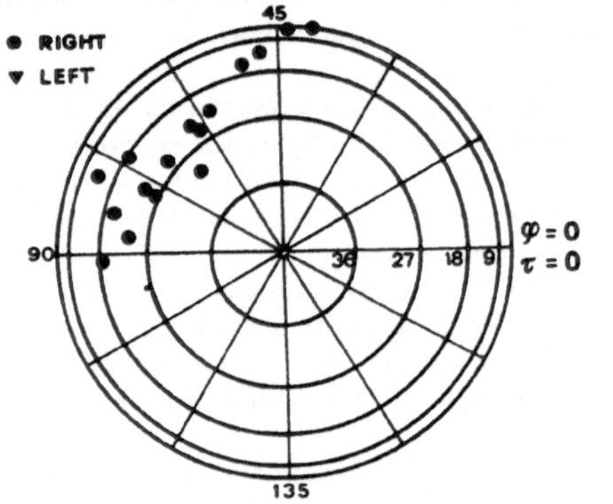


Fig. 3 - Polar representation of space-distributed ground clutter.

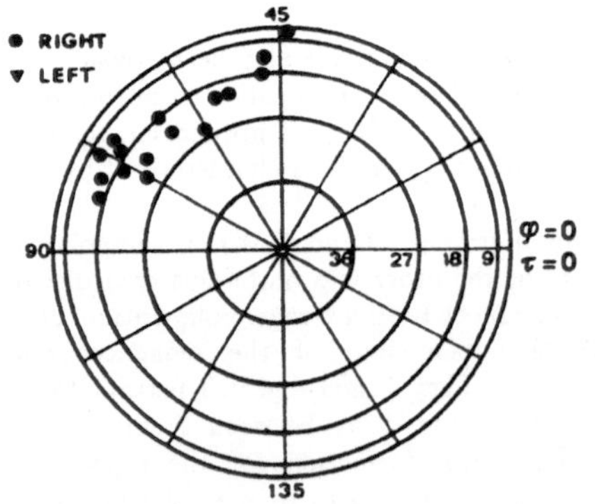


Fig. 4 - Polar representation of space-distributed ground clutter (8 s after).

### 4. Optimum Radar Receivers for Target Detection in the Clear

Several techniques have been proposed to improve the target detection capability by means of dual polarization radars, relying on the fact that the overall energy contained in the target backscattered wave can be collected by the two channel receiver of such radars, while only the e.m. field component parallel to the radar antenna polarization is collected with single polarization radars. An improvement can thus be expected, especially when a significant mismatch exists between the polarization of the incident wave and the radar antenna characteristic polarization.

The subject in this section is the description of some dual channel receiver structures, derived through a statistical theory of detection approach. They refer to the case of targets in the clear (that is, no clutter is present and the disturbance consists of a white additive noise only): the optimality criterion is to extract from the received vector signal a sufficient statistic for a decision test such that the probability of detection is maximized for given values of the probability of false alarm. This general approach allows a joint optimization of time - and polarization - domain processing.

It is assumed that a fixed polarization is used in transmission, while two orthogonally polarized channels are available on reception.

For more details on the subject and results discussed in this section the reader is referred to [5] and [6].

Few attempts have been made in devising optimum schemes for target detection with dual-polarization radars in a systematic way (ref. to [6], [17] and [4]). In [5] and [6], the statistical detection theory approach followed in [4] has recently been extended in order to determine the structure of optimum dual-polarization radar receivers for the detection of targets in the clear, while relating such structures to various statistical behaviours of targets in the time-polarization domain. The vector signal statistical approach allows indeed joint optimization of signal processing in the time and polarization domain. Such structures are optimum in that they maximize the probability of detection $P_d$ for given values of the probability of false alarm $P_{fa}$ and for given statistical properties of the received signal. The latter is recognized to have a random nature which is not limited to a mere complex multiplying random factor (as it happens in the single channel case) but arising also from the random way the signal power parts between the two orthogonally polarized channels.

To this end the total received signal (i.e., target return plus background noise) is properly modelled and its statistical behaviour is described in terms of its complex covariance matrix in the time polarization-domain. The additive noise is furtherly supposed to be unpolarized and statistically independent of the target return. Both signal and noise are assumed to have Gaussian statistics. It is also assumed that the radar pulse duration is so short that amplitude, phase and polarization fluctuation may be neglected within the pulse itself.

Based on such models, through a simple generalization of the single channel optimum receiver for random signal detection [18, sect.2.1], an optimum dual-channel receiver structure of general validity has been derived. Such a structure, based on the estimator-correlator principle (ref. [18], sect.2.1.1), is one of canonical implementations of the likelihood-ratio test and was the starting point for the derivation of the structures valid for the specific problems of single hit and pulse train detection. This derivation has been carried out in [6] by introducing specific hypotheses on the target echo which are briefly recalled below.

In the single hit detection case the observation period of the received waveform is limited to the radar pulse duration; therefore the polarization vector (including absolute phase and amplitude terms), although random, is constant throughout the observation period. The expression of the sufficient statistic and, consequently, the structure of the receiver, that implements it, are greatly simplified.

In the pulse train detection case a decision test is performed after an observation period equal to the dwell-time of the antenna on the target. The signal polarization can be no longer considered constant; conversely, a partially polarized target return must be considered, which can be conveniently decomposed in two statistically independent terms, according to the wave decomposition theorem by Born and Wolf [8].

The wave decomposition of the target return by Born and Wolf [8] was extended to a target decomposition by Huynen [14]; the result of the latter work was the association of the two wave components with two targets, a so called "mean" or "effective" target, associated with the completely polarized wave component, and a "noise" target associated with the uncorrelated wave component. Such a decomposition of targets was later used for the modelling of partially fluctuating targets in [19] and [23].

The short discussion above grounds the made assumption that in the pulse train detection case the target echo signal may be generally decomposed in two statistically independent terms:

a) a "mean target" vector signal contribution resulting from the completely polarized wave component; such a signal has a constant polarization vector during dwell-time, although a priori unknown, and in the time-domain will consist of a coherent or an incoherent pulse train, depending on the particular case considered.

b) a "noise target" vector signal contribution, resulting from the completely unpolarized wave component; such a signal has a polarization vector that fluctuates in a random, uncorrelated way from pulse to pulse and in the time-domain will be represented by an incoherent pulse train in all particular cases; furthermore the two orthogonal polarization signal components will be mutually uncorrelated and have the same average power.

The above model jointly accounts for:

- limited a priori knowledge of the mean target return polarization, represented by the randomness of the contribution a);
- target amplitude and polarization fluctuation during dwell-time, represented by the contribution b).

The following specific cases of pulse train detection can thus be considered:

i) Non-fluctuating target: this is the case where only the target return contribution of type a) is present.

ii) Fluctuating target: this is the case where only the target return contribution of type b) is present.

iii) Partially fluctuating target: this is the case where both the target contributions of type a) and b) are present.

Referring to such cases the optimum receiver structures have been derived in [5] and [6], for both coherent and non-coherent pulse train transmissions.

### 4.1 SOME OPTIMUM RECEIVER STRUCTURES

Through the statistical approach above outlined it can be demonstrated that in the most relevant cases the optimum receiver structures for detection of targets in the clear can be implemented through the following operations:

- time-domain pulse matched filtering independently performed on both the dual polarization channels;
- optimum polarization basis transformation;
- coherent or non-coherent pulse integration (with pulse train transmission);
- optimum two-channel power linear combination.

This is made clearer through the following examples of optimum dual polarization receiver structures pertaining to some particular cases of target detection.

4.1.1 *Case* 1: *Single-pulse detection.* An optimum receiver structure for this case is reported in Fig. 5. We can note that matched filtering is performed for each orthogonally polarized component of the dual polarization vector received signal $\underline{r}(t)$. The vector sample $\underline{x}$ thus obtained is transformed through the 2x2 matrix [P']. This matrix transformation is equivalent to a polarization basis transformation, as that expressed by eq. (12a), which is the optimum one for target detection. An optimum linear combination of the square amplitudes of the vector components thus obtained provides the logarithmic likelihood ratio $l_r$, i.e. the sufficient statistic, which is the quantity to be compared with a set threshold for the target detection decision test.

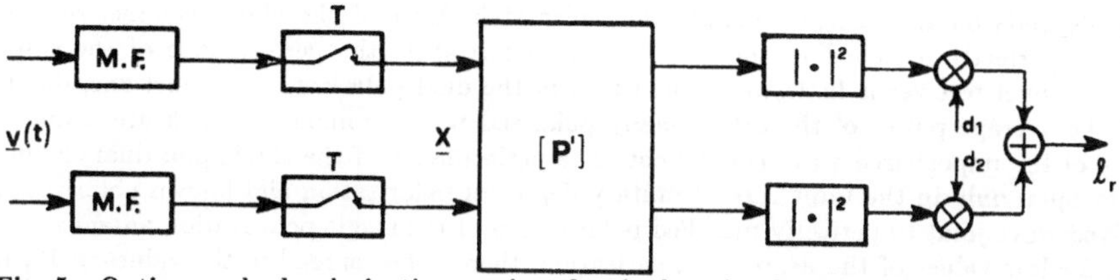


Fig. 5 - Optimum dual polarization receiver for single pulse detection of target in the clear.

It can be demonstrated that the optimum transformation [P'] corresponds to setting the optimum polarizations on receptions, i.e. that couple of polarizations providing the maximum and the minimum average received signal power, as well as uncorrelated orthogonally polarized signals. The latter property justify the successive linear power combination.

The analytical expressions of the matrix [P'] and weights $d_1$ and $d_2$ depend on the polarization state of the received wave as well as on the signal-to-noise ratio.

4.1.2. *Case 2: Non fluctuating target with non-coherent pulse train transmission.* An optimum dual polarization receiver structure for this case is reported in Fig. 6. The receiver scheme is quite analogous to that of Fig. 5. The changement is just the final non-coherent pulse sample integration which is now carried out, under the hypothesis that M pulses of duration $\tau$ and interpulse period $T_r$ are transmitted.

Other specific cases are dealt with in [5] and [6].

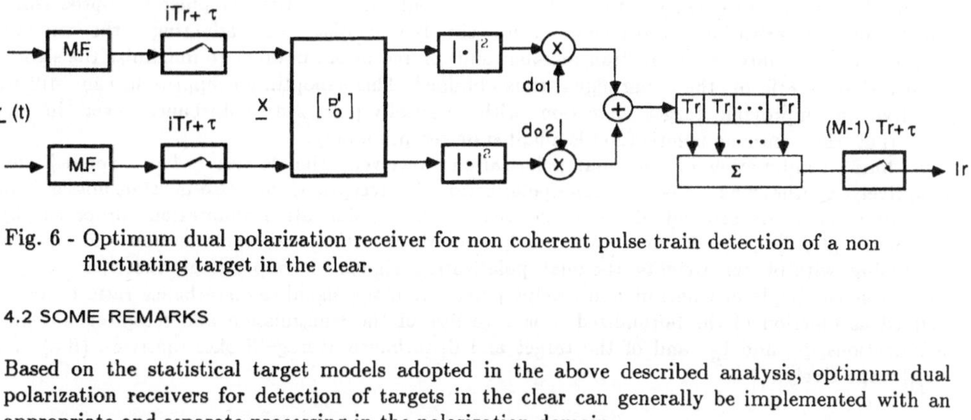


Fig. 6 - Optimum dual polarization receiver for non coherent pulse train detection of a non fluctuating target in the clear.

## 4.2 SOME REMARKS

Based on the statistical target models adopted in the above described analysis, optimum dual polarization receivers for detection of targets in the clear can generally be implemented with an appropriate and separate processing in the polarization domain.

This is made by just a polarization basis transformation which compensates for the antenna polarization basis mismatch with respect to the one providing mutually uncorrelated dual-polarization signals with maximum and minimum average power.

The application of the devised receivers is related to the target behaviour. Such a behaviour has to be statistically known, through the covariance matrix of the dual-polarization receiver signals.

The target detection performance of the described optimum dual-polarization receivers is of particular interest, particularly when compared with that of the corresponding optimum single polarization receivers. This problem is dealt with in [5] where the comparison is made in terms of lower bound of detection probability $P_d$.

The results thus obtained show that for high values of degree of polarization (nearly constant polarization of the echo from an observation to another) there is a large spread of the values of $P_d$ in the case of single channel receiver; this is due to the fact that the power received in such

case highly depends on the antenna polarization mismatch. Even if the single channel receiver polarization is perfectly matched to the mean target polarization the performance of the single polarization channel receiver is little worse than that of the dual polarization channel one, due to the loss of the average power of the orthogonally polarized wave component which arises due to the presence of the unpolarized wave component. The performance of the single and dual channel receivers are equal only in the limit case of unitary degree of polarization and known polarization of the received wave joint to perfectly matched polarization of the single polarization antenna.
Conversely, for low values of the degree of polarization, there is less spread in the values of $P_d$ in single channel case, such values being always smaller than the corresponding value in the dual polarization case. In the limit case of unpolarized target echo the same value of $P_d$ would be obtained with the dual-channel receiver with half as the signal power necessary in the single polarization receiver case, whatever is the polarization on reception of latter.

## 5. Adaptive Polarization Cancellation of Partially Polarized Disturbance

### 5.1 IMPROVING SIGNAL/DISTURBANCE RATIO THROUGH POLARIZATION ADAPTATION

The statistical approach, described in the preceding section, for optimum detection of target in the presence of background unpolarized white-noise, could be extended to other types of disturbance, such as clutter. In this case we can expect that the optimum dual polarization receiver structure would become more complex, as well as the vector processing procedure would not necessarily imply a separate polarization processing in the polarization domain.

However, in order to reduce receiver complexity polarization processing can be carried out separately and independently from the conventional time/Doppler processing. Furthermore polarization processing can simply consist of adapting the antenna polarization, possibly both in transmission and on reception, in order to maximize the signal-to-disturbance ratio of the scalar signal thus obtained. This suboptimum approach, can still be effective in improving target detection with partially polarized disturbance, even in the case that the antenna polarization is adapted on reception only.

A linear combination of the dual polarization receiver signals can be applied for adaptively synthesizing any antenna polarization in reception, but also in transmission. In the latter case we can indeed report to the virtual polarization adaptation proposed by Poelman [21,22], based on alternately radiating orthogonal polarization and linearly combining with proper weights the dual polarization signals correspondingly received.

For a given couple of transmit and receipt polarization the signal-to-disturbance ratio r can be defined as function of the normalized Stokes vectors of the transmission and reception antenna polarizations, $\underline{f}_T$ and $\underline{f}_R$, and of the target and disturbance average Stokes matrices, $[R_T]$ and $[R_D]$ respectively:

$$r = \frac{\underline{f}'_R\,[R_T]\,\underline{f}_T}{\underline{f}'_R\,[R_D]\,\underline{f}_D} \tag{19}$$

where the apostrophe means transposition and conjugation of the vector, $\underline{f}_D$ is the polarization Stokes vector of the disturbance source.

If target and disturbance features are statistically known, the optimum adaptation of the polarization both in transmission and on reception, or on reception only, requires the maximization of r, given by (19), with respect to $\underline{f}_T$ and/or $\underline{f}_R$

5.1.1. *Target in the clear.* In the simplest detection problem, such that of a target in the presence of background unpolarized white Gaussian noise, the maximization of the signal-to-disturbance ratio is not to be accounted for anymore. Actually, in this case, the additive noise signals present on the reception channels are statistically independent and have equal power, so that no polarization adaptation can change their contributed average power.

Therefore the mere maximization of the target return power defined as $\bar{P}_T = \underline{f}'_R\,[R_T]\,\underline{f}_T$ is equivalent to the maximization of r. Note that this operation is performed by the optimum receivers illustrated in the preceding section (see Figs. 5 and 6) when only the channel with the maximum average power is used after the optimum polarization basis transformation. This confirms that a detection loss is incurred when resorting to a simple antenna polarization adaptation.

When polarization is under control on reception only, the antenna polarization basis has a strong influence on the optimum detection process. In the case that transmission and reception polarizations are linear, target detection based on the co-polar return processing provides a quite optimum performance, because weak power is generally expected through the cross polarized channel.

In the case that a circular polarization basis is utilized the optimum solution to detection problem is preferably to be based on the processing of two receiving channels because of the high wave depolarization produced by the interaction between target and incident wave (see sect. 3).

Since the signals received from the target on the orthogonally circularly polarized channels have nearly equal power and are weakly correlated, a simple detection scheme is preferable which combines the detection statistics derived from each channel by simply summing up the sample statistic powers.

5.1.2. *Target in the presence of partially polarized disturbance.* When the target has to be detected in the presence of a partially polarized disturbance, the maximization of the expression (19) would be profitable. But a noticeable simplification can still be achieved when the disturbance is characterized by a high degree of polarization. Actually the most common phenomena, which characterize the radar environment, such as ground clutter, atmospheric clutter and jamming, often show a stationary polarization behaviour. At the same time very little a-priori information on an aircraft target can be exploited because of its time/space non stationarity which does not allow the adaptive estimation of the $[R_T]$ matrix.

Thus a suboptimum procedure for the detection problem solution can be applied by only minimizing the disturbance signal power defined as:

$$\bar{P}_D = \underline{f}'_R\,[R_D]\,\underline{f}_T \tag{20}$$

This expression requires the estimation of the $[R_D]$ matrix, which can be performed when disturbance is space-time distributed and stationary. Moreover, when the degree of polarization of the disturbance is high, the minimization of $\bar{P}_D$ allows a considerable improvement of r (see expression (19)) with a decreasing performance loss as that polarization degree increases.

The minimization of the disturbance power given by expression (20) can be thus considered as a convenient approach for the optimal choice of antenna polarization both in transmission and on reception, especially under the hypothesis of high polarization degree of disturbance.

Two steps are thus required to perform the disturbance rejection: the first one is devoted to minimize the average unpolarized component of the received signal by suitably choosing the transmitting antenna polarization (i.e.: the Stokes vector $\underline{f}_T$); the second step is devoted to determine the Stokes vector $\underline{f}_R$ by estimating the Stokes reflection features of the disturbance so as $\bar{P}_D$ in (20) is minimized.

When the above described procedure is optimally fulfilled, the disturbance power at the receiver output equals $\bar{P}_{min}$ in expression (9). This is due to the fact that the polarization-based disturbance cancellation is obviously limited by the partially polarized nature of such a phenomenon: the totally polarized component of the related wave is completely suppressed, while the unpolarized component is present as residual at the output of the receiver.

These considerations point out the possibility of facing the detection problem as a problem of adaptive polarization cancellation of disturbance. Actually this technique is still profitable when polarization control is possible only on reception. In such a case the first optimization step cannot be applied and the procedure reduces to perform the second step, but it can still achieve significant disturbance cancellation if the polarization degree approaches unity.

Because disturbance cancellation is a function of the received wave polarization degree, the performance of a polarization based disturbance canceller, which applies optimum selection of the antenna polarization on reception can be evaluated through the disturbance cancellation ratio defined (in decibels) as [20]:

$$Cr_{max} \triangleq 10\ \log_{10}[\bar{P}/\bar{P}_{min}] \tag{21}$$

where $\bar{P}$ is the average power present on the least-powered between the two orthogonally polarized channels of the receiver, and $\bar{P}_{min}$ is the average power obtained when the optimum polarization would be selected on reception.

If $\bar{P}_1$ and $\bar{P}_2$ are the average powers of the two orthogonally polarized signals, while $\bar{g}_0$ and p are the total average power and the polarization degree of the incoming wave, it can be easily demonstrated that [20]:

$$\bar{P}_1 = \bar{g}_0\ \frac{(1-\alpha p)}{2} \qquad\qquad \bar{P}_2 = \bar{g}_0\ \frac{(1+\alpha p)}{2} \tag{22}$$

where $\alpha$ expresses the mismatch between the actual and the optimum polarization bases: $|\alpha|=1$ means optimum polarization basis on reception, while $\alpha=0$ means maximum polarization mismatching on reception.

Thus $Cr_{max}$ can be expressed as [20]:

$$Cr_{max} = 10\ \log_{10}\left[\frac{1-|\alpha|p}{1-p}\right] \tag{23}$$

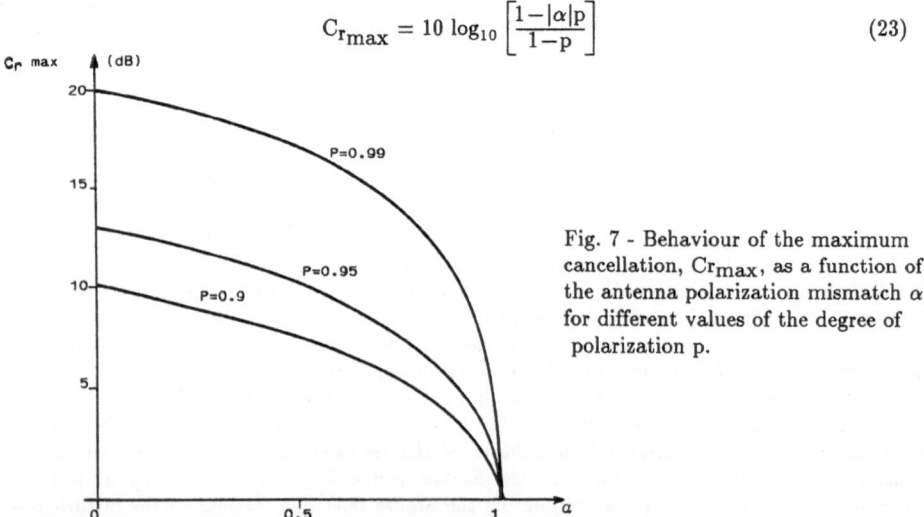

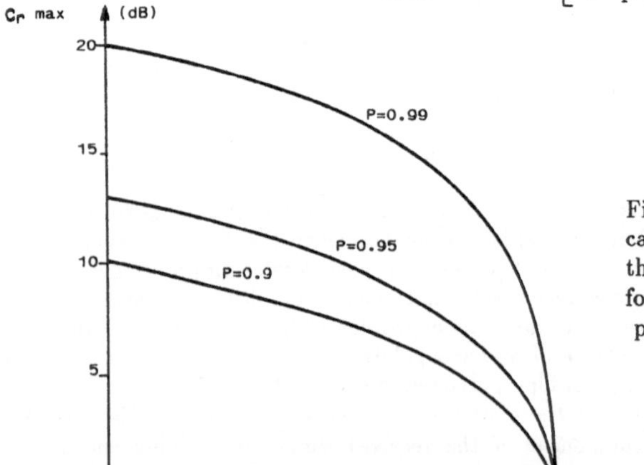


Fig. 7 - Behaviour of the maximum cancellation, $Cr_{max}$, as a function of the antenna polarization mismatch $\alpha$, for different values of the degree of polarization p.

In Fig. 7 the behaviour of $Cr_{max}$ is reported as a function of $\alpha>0$ for set values of p. The behaviour of $Cr_{max}$ indicates that a cancellation improvement can be gained when the optimum polarization is synthesized on reception. Such an improvement increases as the pair polarizations, characterizing the two receiving channels, are more mismatched with respect to the optimum polarization basis ($\alpha \rightarrow 0$). This improvement becomes larger as the degree of polarization increases.

### 5.2 POLARIZATION ADAPTATION FOR DISTURBANCE CANCELLATION

In this subsection some alternative adaptive polarization cancellers are described with reference to the results obtained in [20].

5.2.1 *The optimum polarization adaptation procedure.* The optimum selection of the receiver polarization, and then the optimum disturbance rejection by means of the filtering process in the polarization domain of the vector signal received through two orthogonally polarized channels, can be performed through the following optimum estimation procedure, which is still based on the minimization of expression (20):

1) Estimate the elements of the average Stokes vector $\underline{\bar{g}}$ of the disturbance.

2) Calculate the optimum antenna Stokes vector $\underline{f_r} = [f_0, f_1, f_2, f_3]^T$ through equation (8):

$$\begin{cases} g_{op} = \sqrt{\bar{g}_1^2 + \bar{g}_2^2 + \bar{g}_3^2} \\ f_0 = 1 \\ f_i = - \bar{g}_i / g_{op} \qquad i = 1, 2, 3 \end{cases} \tag{24}$$

The elements of the average Stokes vector of the disturbance are easily obtained by resorting to the estimation of three parameters: the average cross-correlation $\bar{M}_{12}$, and the average powers $\bar{P}_1$ and $\bar{P}_2$ of the orthogonally polarized components of the disturbance. This is achieved by performing averages on signal samples of the observed radar signals, within a range/cross-range window where stationary behaviour of the disturbance is expected.

Denoted the generic vector sample components on the two receiving channels as $s_1(i)$ and $s_2(i)$, the above mentioned parameters, defined within the said window of the radar coverage [20], can be estimated through the following expressions:

$$\hat{M}_{12} = N^{-1} \sum_{i=1}^{N} \left[ s_1(i)\, s_2^*(i) \right] \tag{25a}$$

$$\hat{P}_1 = N^{-1} \sum_{i=1}^{N} |s_1(i)|^2 \tag{25b}$$

$$\hat{P}_2 = N^{-1} \sum_{i=1}^{N} |s_2(i)|^2 \tag{25c}$$

where N is the number of vector samples.

Through the above optimum procedure, it is possible to perform a linear filtering in the polarization domain of the receiving vector signal. This linear filtering process is applied by linearly and coherently combining the orthogonally polarized signals present on the two reception channels: the received wave is thus differently attenuated according to the angular distance of its polarization from the antenna polarization synthesized on reception, when these are represented on the Poincaré sphere. If adaptation is optimum, the polarized component of the wave is totally rejected.

The weights of the linear combination can easily be obtained from the receiving antenna Stokes vector $\underline{f}_r=[f_0, f_1, f_2, f_3]^T$.

In fact it can be verified that, if $h_{r1}$ and $h_{r2}$ are the complex weights in subject [20]:

$$|h_{r1}| = \sqrt{\frac{f_0 + f_2}{2}}$$

$$|h_{r2}| = \sqrt{\frac{f_0 - f_2}{2}} \qquad (26)$$

$$\arg(h_{r1}) - \arg(h_{r2}) = \arg(h_{r1}\, h_{r2}^*) = \tan^{-1}(f_1/f_3)$$

where $\arg(h_{r1})$ or $\arg(h_{r2})$ can be chosen arbitrarily.

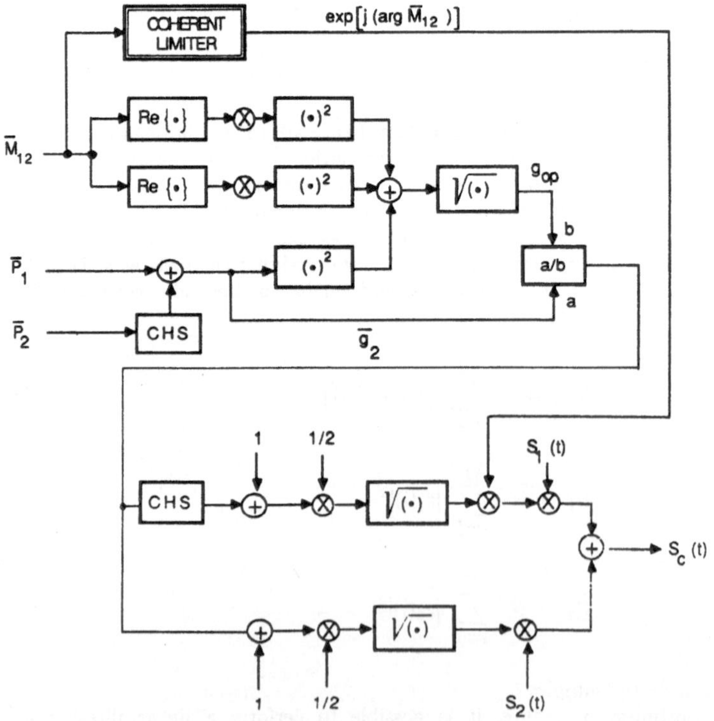


Fig. 8 - Optimum adaptive polarization canceller [20].

Once $h_{r1}$ and $h_{r2}$ are obtained substituting in (26) the $\underline{f}_r$ components estimated by means of the optimum adaptive procedure, then the optimum antenna polarization is adaptively set through the following linear operation:

$$s_c(t) = h_{r1}\, s_1(t) + h_{r2}\, s_2(t) \tag{27}$$

where $s_c(t)$ is the residual output of the receiver, while $s_1(t)$ and $s_2(t)$ are the signals pertaining to the orthogonally polarized channels.The scheme reported in Fig. 8 represents an optimum receiver for polarization-based disturbance cancellation: it implements the above described optimum procedure. In such a scheme CHS means change sign, a/b means division, and the double frame box is used to indicate an operation giving rise to a complex output.

5.2.2 *Some suboptimum procedures.* Some suboptimum procedures can be suggested to perform the antenna polarization adaptation on reception [20]. It can be verified that their performance are characterized by a limited loss of cancellation ratio with respect to the optimum procedure when the hypotheses made for this technique still hold.

The first suboptimum adaptation procedure performs the following linear combination of the dual polarization signals:

$$s_{c1}(t) = s_1(t) - w_1\, s_2(t) \tag{28a}$$

where $s_{c1}(t)$ is the receiver output, while $w_1$ is defined through the estimates of the parameters $\bar{M}_{12}$, $\bar{P}_1$ and $\bar{P}_2$:

$$w_1 = \frac{\bar{M}_{12}}{\bar{P}_2} = \sqrt{\frac{\bar{P}_1}{\bar{P}_2}}\, \mu \tag{28b}$$

$\mu$ being defined in eq. (10). It can be dually operated as follows:

$$s_{c2}(t) = s_2(t) - w_2\, s_1(t) \tag{29a}$$

with

$$w_2 = \frac{\bar{M}^*_{12}}{\bar{P}_1} = \sqrt{\frac{\bar{P}_2}{\bar{P}_1}}\, \mu^* \tag{29b}$$

These operations allow to cancel the component of the unweighted signal which is correlated with the other signal.

The performance evaluation of such a procedure can be carried out once the related cancellation ratio is calculated. To this end the canonical polarization vectors $\underline{h}_{r1}$ and $\underline{h}_{r2}$, having unitary modulus, and associated with operations (28) and (29) respectively, have to be determined. They are indeed given by:

$$\begin{aligned} \underline{h}_{r1} &= \frac{1}{\sqrt{1+|w_1|^2}}\,[1,\ -w_1]^T \\ \underline{h}_{r2} &= \frac{1}{\sqrt{1+|w_2|^2}}\,[-w_2,\ 1]^T \end{aligned} \tag{30}$$

Thus the cancellation ratios have to be evaluated with respect to the following normalized output signals:

$$s_{c1n} = \frac{s_{c1}(t)}{\sqrt{1+|w_1|^2}} \tag{31a}$$

$$s_{c2n} = \frac{s_{c2}(t)}{\sqrt{1+|w_2|^2}} \tag{31b}$$

The corresponding cancellation ratios are given by [20]:

$$C_{r1} = C_{rmax} - C_\delta(\, p,\, \alpha) \tag{32a}$$

$$C_{r2} = C_{rmax} - C_\delta(\, p, -\alpha) \tag{32b}$$

with $0 \leq \alpha \leq 1$ and $C_\delta$ given by [20]:

$$C_\delta(p,\alpha) \triangleq 10 \log_{10}\left[\frac{(1+p)\,(1+\alpha p)}{p^2+2\alpha p+1}\right] \tag{33}$$

which is defined for $-1 \leq \alpha \leq 1$.

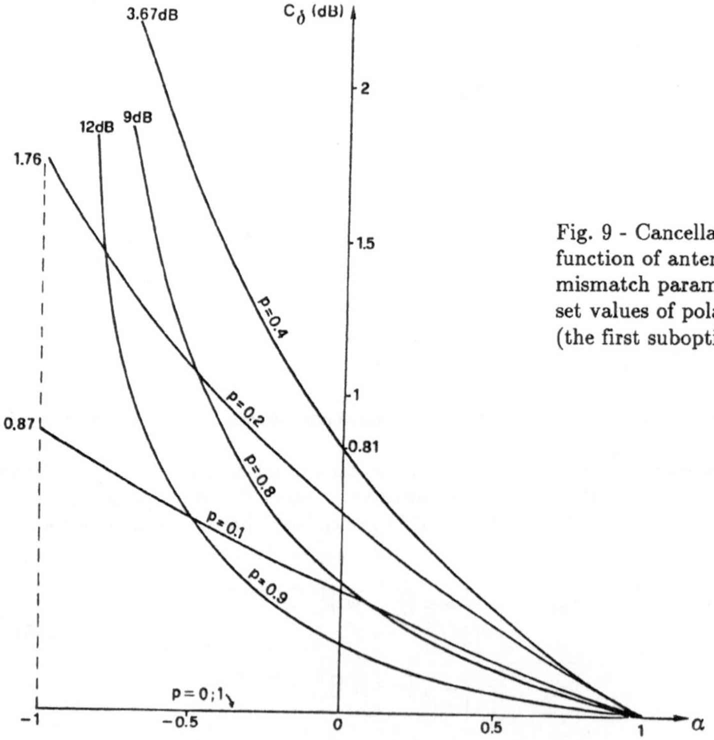


Fig. 9 - Cancellation loss $C_\delta$ as a function of antenna polarization basis mismatch parameter $\alpha$, for different set values of polarization degree p (the first suboptimum procedure)

The function $C_\delta$ expresses the cancellation loss of this procedure with respect to the optimum polarization adaptation procedure: Fig. 9 shows the behaviour of $C_\delta(p,\alpha)$ as a function of $\alpha$ for different values of p.

It can be pointed out that an optimum performance of this procedure is expected when p approaches unity. Moreover the cancellation loss vanishes as the polarization mismatch decreases ($\alpha \rightarrow 1$), whatever p is.

The best performance is achieved when the least-powered orthogonally polarized channel is chosen and the linear combination (28) is continuously applied: this fact can be argued by observation of Fig. 9, because $C_\delta$ takes on higher values when $\alpha<0$.

The results, sinthesized in Fig. 9, show that disturbance cancellation by means of the proposed polarization adaptation is quite effective.

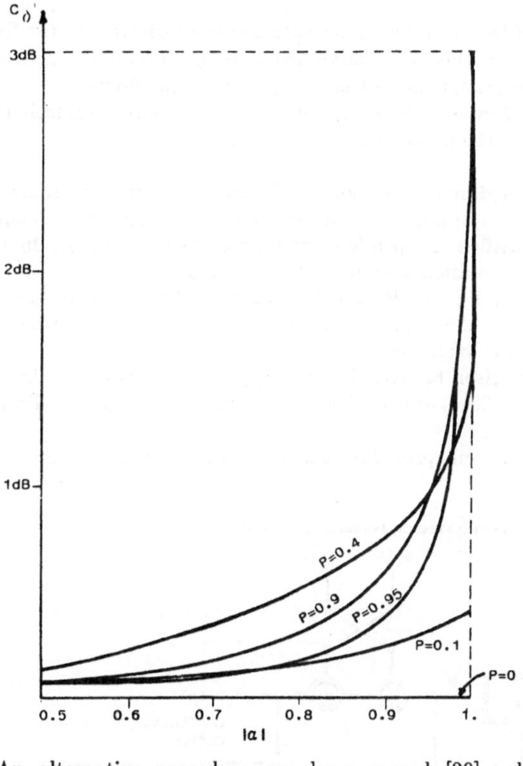


Fig. 10 - Cancellation loss as a function of antenna polarization basis mismatch $|\alpha|$, for different values of polarization degree p (second suboptimum procedure).

An alternative procedure can be proposed [20], which reduces the cancellation loss as $|\alpha|$ approaches zero (large antenna polarization mismatch). In this procedure the following linear combination are now taken into account :

$$s'_{c1} = s_1(t) - w'_1\, s_2(t) \tag{34a}$$

$$s'_{c2} = s_2(t) - w'_2\, s_1(t) \tag{34b}$$

where:

$$-w'_1 = -\frac{1}{w'_2} \triangleq \sqrt{\frac{\overline{P}_1}{\overline{P}_2}}\; e^{j\,\arg(w_1)} \tag{35}$$

The cancellation ratio now becomes [20]:

$$C'_r = C_{rmax} - C_{\delta'}(p,\alpha) \tag{36}$$

with

$$C_{\delta'}(p,\alpha) = 10 \log_{10}\left[\frac{1-\alpha^2p^2}{1-p}\left(1-p\sqrt{\frac{1-\alpha^2}{1-\alpha^2p^2}}\right)\right] \qquad p\neq 0,1$$

while (37)

$$C_{\delta'}(p,\alpha) = 0 \qquad p = 1$$

which is meaningful when $-1\leq\alpha\leq 1$.

In Fig. 10 $C_{\delta'}$ is reported as a function of $|\alpha|$ for different set values of p. In contrast to the first suboptimum procedure, the performance of this alternative suboptimum procedure is near-optimum when $|\alpha|\rightarrow 0$, i.e. when the antenna polarization basis mismatch is maximum.

A proper choice of the suboptimum procedure is to be based on both the a-priori information of the antenna polarization basis mismatch and the implementation aspects.

5.2.3 *Implementation schemes of the suboptimum procedures.* Different adaptive polarization schemes based on the just described suboptimum procedures can be conceived to cancel disturbance on reception: they can be classified in open-loop and closed-loop cancellers. In the latter case the cross correlation parameter estimation is embodied in a loop operation.

In open-loop schemes the parameters $\hat{M}_{12}$, $\hat{P}_1$ and $\hat{P}_2$ are the input of the canceller; Fig. 11 shows the computation scheme which refers to the implementation of the first procedure, when no information on the ratio between $\hat{P}_1$ and $\hat{P}_2$ is available.

Such a canceller performs firstly the comparison between $\hat{P}_1$ and $\hat{P}_2$ powers, afterward it chooses the linear combination, between (28a) and (28b), which gives rise to the minimum cancellation loss.

The implementation of the second suboptimum procedure can be realized through a canceller, which block diagram is reported in Fig. 12.

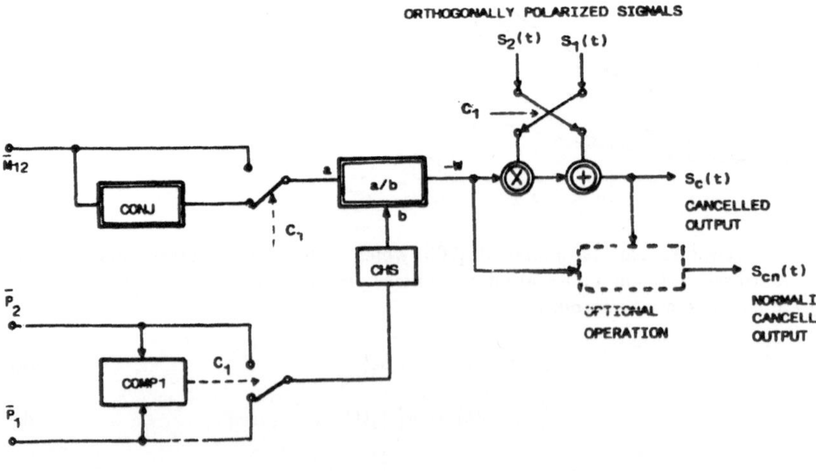


Fig. 11 - Computations involved in the first suboptimum procedure.

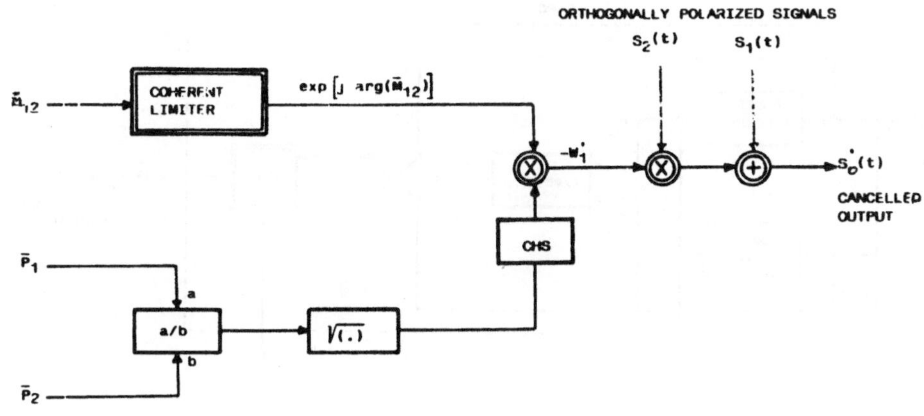


Fig. 12 - Computations involved in the second suboptimum procedure [20]

A closed-loop implementation of the first sub-optimum procedure is given by the scheme of Fig. 13 [24], called adaptive polarization canceller (APC). It can be observed that such a scheme embodies the estimation process of the cross correlation parameter: this is obtained by recursively processing the dual-polarization signals along the sweep time, when a stationary behaviour of the disturbance in the range domain is hypothesized.

When a significant antenna polarization mismatch can occur the adaptation scheme of Fig. 14 is more profitable [16]. It is called symmetric adaptive polarization canceller (SAPC) and can be classified as a closed loop implementation of the scheme reported in Fig. 11.

The closed loop implementation of the second suboptimum procedure does not appear profitable because it requires a more complex implementation than in the previously described realizations. However, in terms of performance the second procedure achieves a quite limited loss in disturbance cancellation, even in the worst case of significant antenna polarization mismatch.

Therefore, based on the steady-state analysis of the adaptive cancellers, such considerations suggest the use of the adaptation schemes of Figs. 13 and 14.

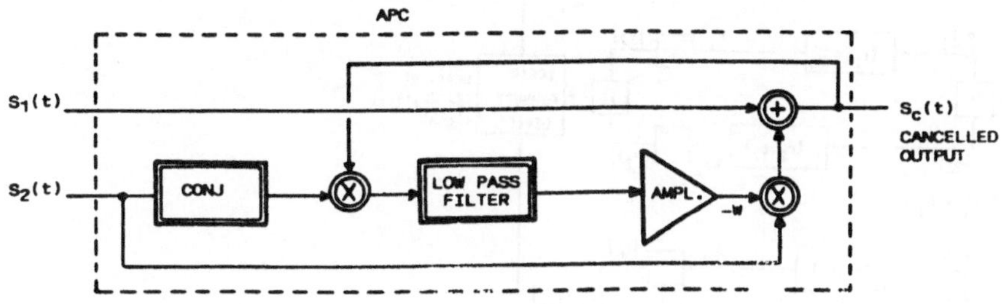


Fig. 13 - Adaptive Polarization Canceller [APC].

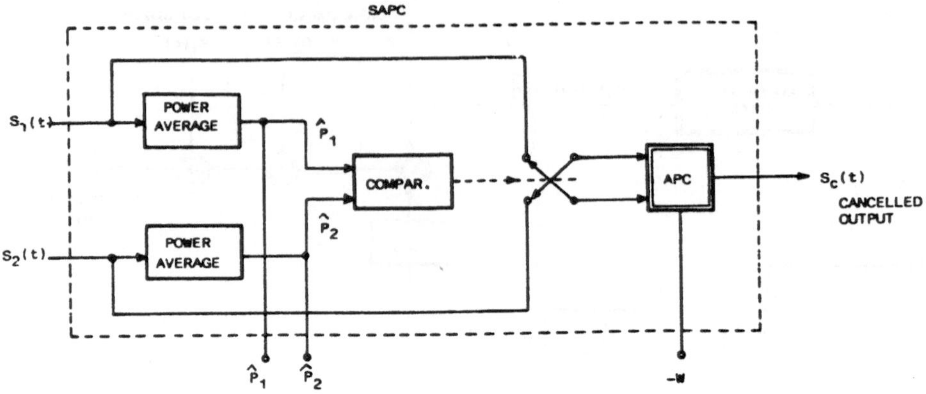


Fig. 14 - Symmetric Adaptive Polarization Canceller [SAPC].

5.2.4 *Other Adaptive Polarization Cancellers.* The disturbance rejection can be operated by means of non-linear polarization filters too. The non-linear polarization filtering process is performed by transforming at any time the orthogonally polarized components of the vector signal present on reception into a scalar signal, according to a law which is a function of the incoming field polarization at the same time.

A non linear polarization filter is usually used to improve the rejection of partially polarized disturbance. Actually the non linear polarization filtering process can attenuate the signals present in a suitably selectable area of the Poincare' sphere, whose dimension is even predeterminable.

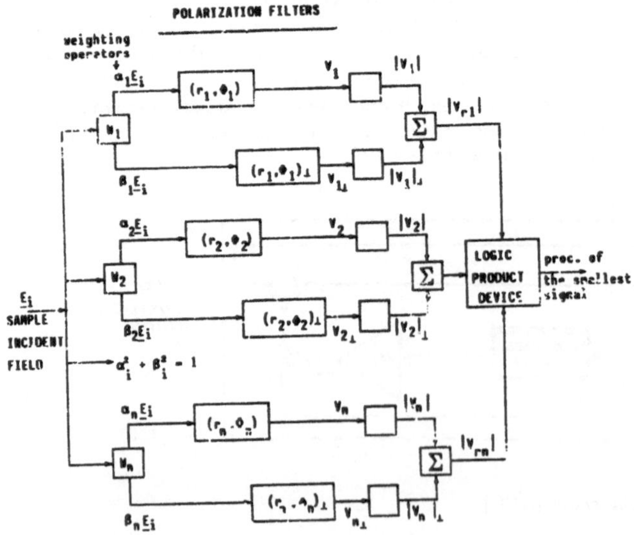


Fig. 15 - Multinotch Logic Product MLP polarization filter [23].

An example of this kind of filters is the Multinotch Logic Product polarization filter (MLP) proposed by Poelman [22, 25]. Its scheme is reported in Fig. 15: it is constituted by a bank of LP filters with different polarization settings. The output powers of the channels pertaing to each LP filter are weighted and summed up. The output of the MLP filter at any time coincides with the weakest output at that time among those of the LP filters in the bank.

The MLP filter can be utilized together with other adaptive devices which add the capability of suitably selecting the suppression area in the polarization domain [26, 27]. A canceller was indeed proposed [26] which is also an adaptive polarization filter. In Fig. 16 is reported the block scheme of such a canceller. It is called MLP-SAPC filter and it allows to exploit the SAPC capability of tracking the average disturbance polarization. In fact at any time the estimation of the weight w can be utilized to perform a translation of the polarization samples, present at the input of the MLP-SAPC filter, over the Poincare' sphere through the PVT transformation, firstly introduced by Poelman [27] . The processed samples are then filtered by the MLP polarization filter.

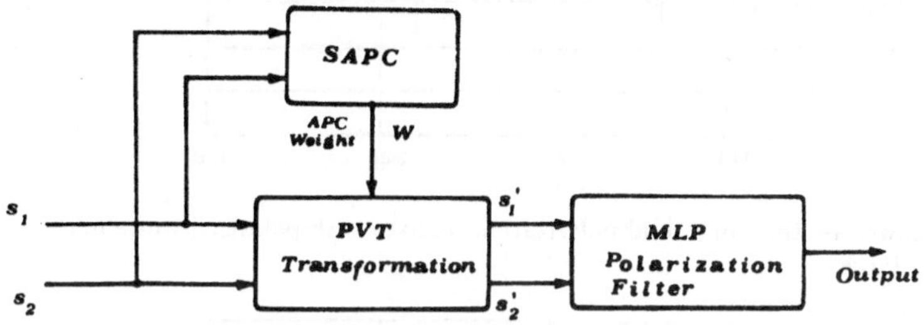


Fig. 16 - MLP-SAPC polarization filter [24].

### 5.3 RESULTS ON ADAPTIVE POLARIZATION CANCELLATION OF PARTIALLY POLARIZED DISTURBANCE

The performance of some adaptive polarization cancellers are now considered with reference to partially polarized disturbance: barrage jamming signals and ground clutter data, both acquired during measurements carried out with the modified ATC radar system previously described (Sect. 3).

5.3.1 *Adaptive cancellation of barrage jamming.* Data sequences have been utilized in this analysis, which have separately been acquired and produced by a white-noise jammer and by an aircraft target.

Specifically three sequences obtained during three successive scans by an aircraft target have been summed up to the sequence obtained by a horizontally polarized jammer.

Three dual-polarization experimental radar data sequences have thus been obtained which have been applied to test the performance of the SAPC filter and of the MLP-SAPC filter.

The curve in Fig. 17 represents the total power of the incoming field as a function of time (range): it refers to one of the described sequences. The target signal power to jamming power ratio is about 0 dB.

In Figs. 18 the output of both the SAPC and MLP-SAPC cancellers is represented for the above input dual polarization signals.

In Table I the improvements in the target/jamming power ratio are reported for the analyzed filtering configurations, for different target echo sequences, when MTI prefiltering at each channel is optionally considered. It can be noticed that the best improvements are obtained when the MLP-SAPC filter is used together with the MTI preprocessing. This is related to the presence of low-intensity clutter in the experimental data.

Moreover the MLP-SAPC operation appears more profitable than the SAPC operation.

Each column in the Table corresponds to a different target experimental data sequence at the input of the analyzed filter.

It can be noticed that the best performance is obtained in the column marked as "Target 2", i.e. when the target echoes and the jamming signals in the considered sequence are characterized by polar representations on the polarization chart with the largest distance between their average polarizations.

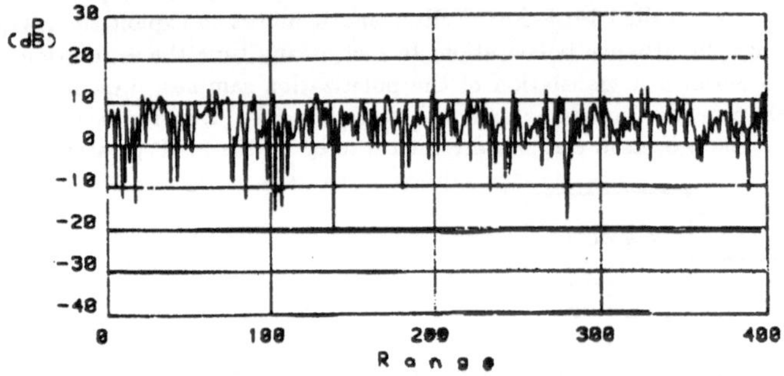


Fig. 17 - Total input power vs. time in a dual polarization receiver with polarized white noise jamming and target.

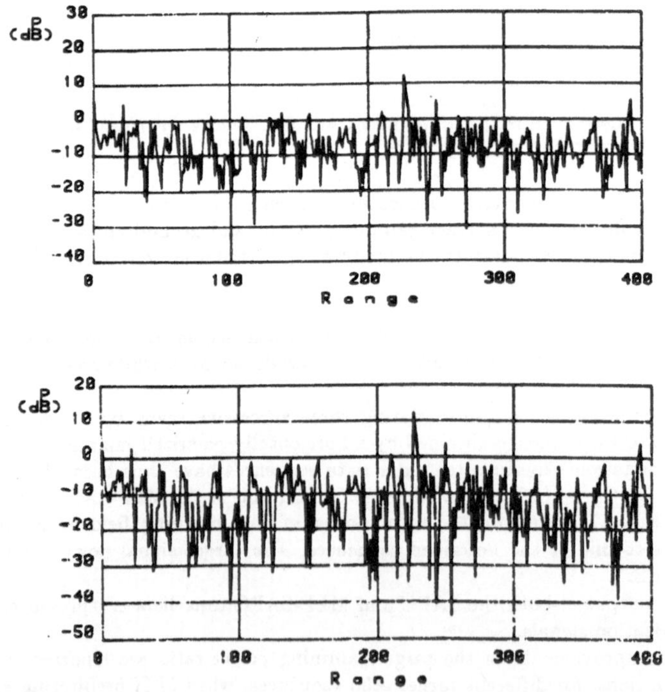


Fig. 18 - Cancelled output in the presence of polarization white noise jamming and target: a) SAPC; b) MLP-SAP.

TABLE 1. Signal/Jamming Ratio Improvements (dB)

| | TARGET 1 | TARGET 2 | TARGET 3 |
|---|---|---|---|
| APC | 7.5 | 11.5 | 10. |
| with MTI | 12. | 16. | 15.6 |
| MLP-APC | 10.24 | 15. | 12.12 |
| with MTI | 15. | 20. | 18. |

5.3.2 *Adaptive rejection of ground clutter.* As pointed out in Sect. 3, based on experimental results the polarization behaviour of ground clutter and aircraft targets can be summarized as follows:

- targets often present a noticeable degree of polarization during dwell time, but a high non-stationarity from scan to scan;
- ground clutter in urban area often presents a highly stationary polarization from scan to scan and its degree of polarization can take on high values especially with strong ground clutter.

This behaviour makes adaptive polarization filtering on reception profitable to improve the signal to ground clutter ratio, expecially in urban area. This improvement can be achieved by using a polarization map which is dynamically updated from one scan to another. The stored polarization is then used to adapt the polarization on reception at the successive scans, in order to filter out ground clutter at each specified coverage cell.

Through the analysis of experimental data, the experienced improvement of signal to clutter ratio was up 15 dB with strong ground clutter.

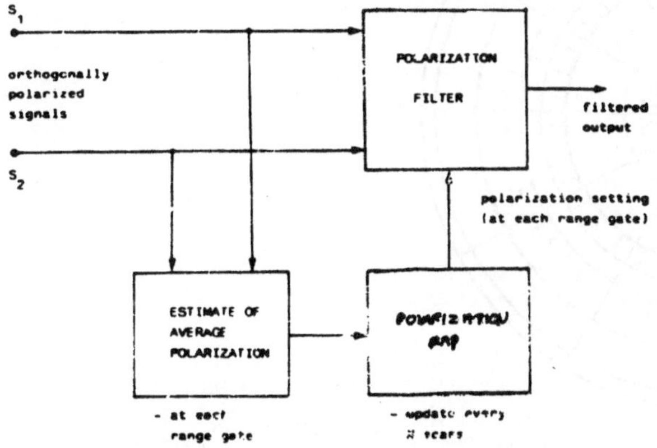


Fig. 19 - Ground clutter canceller based on a polarization map.

In Fig. 19 the block diagram of a ground clutter canceller applying the polarization map estimation is shown. Its performance is related to the use of a circular polarization in transmission. Actually, due to the scattering properties of the different radar objects, better polarization discrimination on reception can be expected when circularly polarized radiation is used rather than the linear polarization.

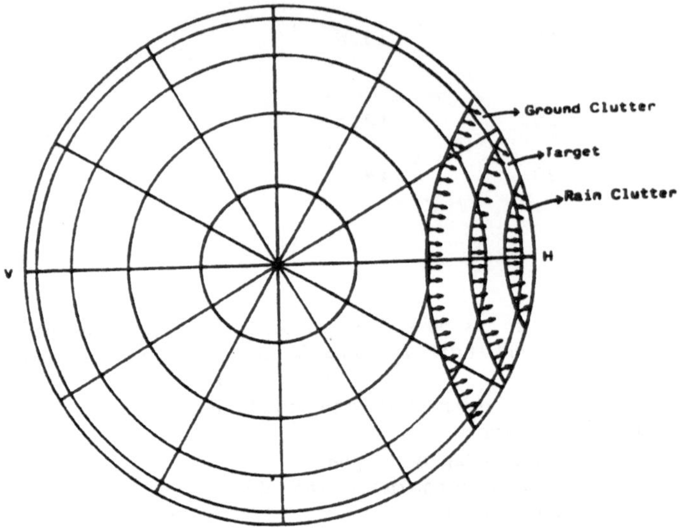


Fig. 20 - Polar distribution of different types of radar signals with linear polarization at trasmit.

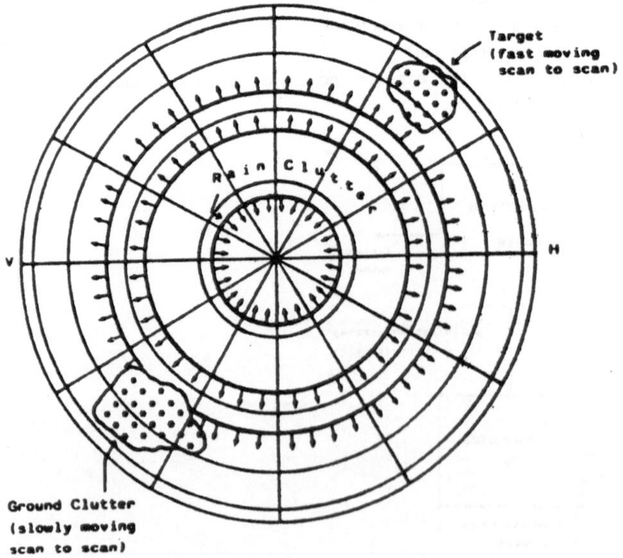


Fig. 21 - Polar distribution of different types of radar signals (circular polarization at transmit).

In Figs. 20 and 21 two polarization charts are shown where the general polarization backscattering of different radar objects is represented when they are illuminated by a linearly polarized wave and by a circularly polarized wave, respectively. Such charts show roughly the expected distribution of polarization of different types of returns with linearly or circularly polarized antennas. They point out the better discrimination capability which is to be expected with circular polarization. Another effective type of ground clutter polarization filtering can be applied by resorting to a non-linear polarization filter, which can estimate the ground clutter parameters at one scan, such as: average polarization, polarization state preferred direction of sample alignment (see sect. 3.1), polarization sample distribution around the mean direction. Such a filter has to adapt its parameters for each vector signal sample, so as the input signals at successive scans can be attenuated when their suppression polarization area is placed optimally on the Poincare' sphere, according to the ground clutter behaviour estimated through the polarization map.

A polarization filter which is able to operate such a processing has been simulated and applied to experimental data sequences. Its block diagram is constituted by three blocks: the estimation section, the non-linear-PVT transformation block, the so called non linear polarization filter. The second block is realized by means of a "PVT transformation" operator [27] , whose parameters have been modified so as the input data samples can be properly processed: the corresponding output samples exhibit modified polarization, specifically their polarizations are moved toward or away from a characteristic mean polarization, which is the null polarization of the linear polarization filter.

In Figs. 22 and 23 the performance of such a filter is illustrated: Fig. 22 refers to the signal present on the two input channels, while Fig. 23 refers to the output signal after adaptive non-linear polarization filtering. The signal representation is three-dimensional: on the vertical axis the signal amplitude is reported at each cell of the processed range/cross-range window.

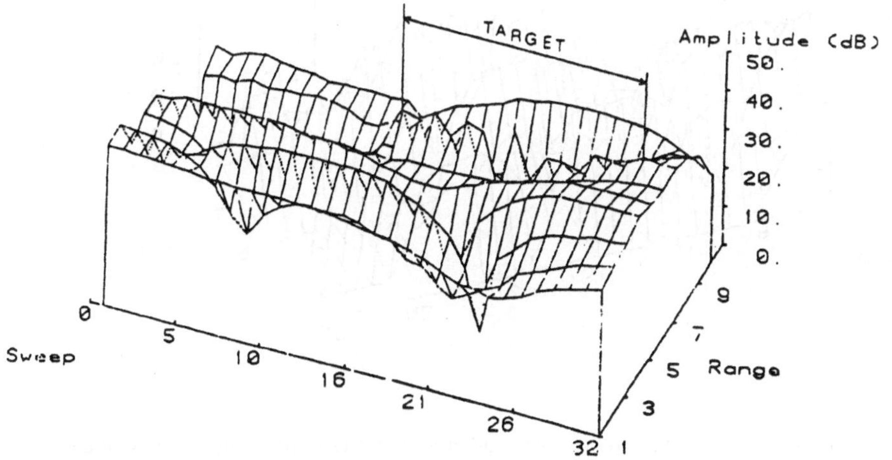


Fig. 21 a) - Input power distribution vs. range and cross range (sweep) in the presence of space-distributed ground clutter and target: copolar reception channel.

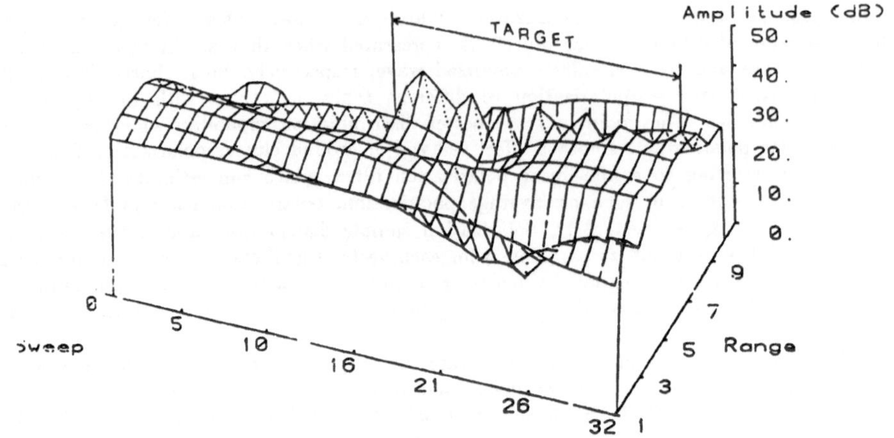


Fig. 21 b) - Input power distribution vs. range and cross range (sweep) in the presence of space-distributed ground clutter and target: cross-polar reception channel.

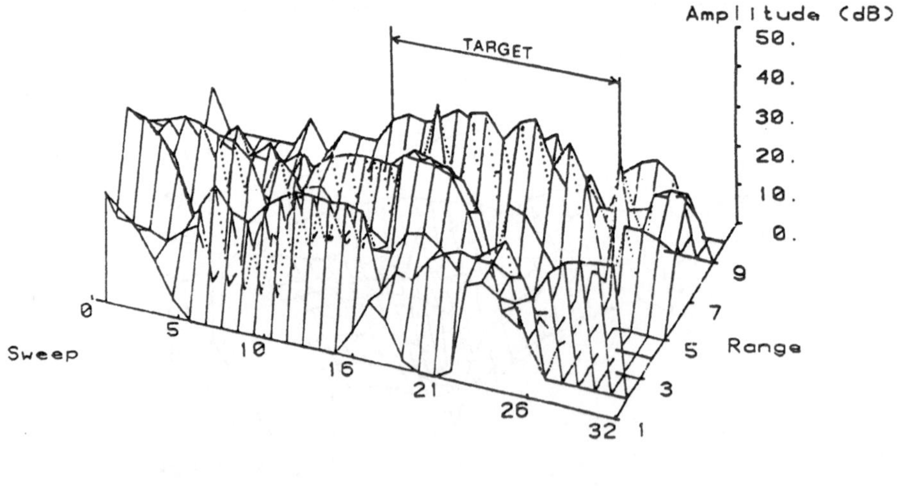


Fig. 22 - Output power distribution vs. range and cross range (sweep) in the presence of space-distributed ground clutter and target.

## 6. Conclusions and Perspectives

As shown in this paper, polarimetric signal processing can profitably be applied to improve target detection capabilities, even when a low resolution radar with fixed polarization at transmit is used. Several dual polarization receiver structures have been purposely illustrated. Through a simple approach based on improving signal-to-disturbance ratio through adaptive polarization cancellation of partially polarized disturbance, good performance can be obtained. This has also been confirmed by some experimental results which refer to target in the presence of polarized jamming and ground clutter. Antenna polarization adaptation implemented through adaptive linear combination of dual polarization signals is a simple, viable, but suboptimum approach. This has been pointed out in the simplest case of target in clear, where a more complex vector processing procedure has to be applied to implement an optimum receiver devised accordingly to a statistical decision theory approach.

MTI prefiltering independently performed for each orthogonally polarized received signal is compatible with polarimetric signal processing, but also can enhance the cancellation of partially polarized disturbance which is contributed by polarimetric techniques based on adaptive estimation of disturbance polarization parameters.

Further improvements can be expected with future advances in field of polarimetric signal processing. Particularly promising is their application in high resolution radars. In this field of applications performance improvements can be expected not only in terms of target detection but also of target classification and identification [28]. This requires further advances in research areas such as broadband polarimetric synthesis and processing of radar signals, polarization coding, target and clutter modelling, analysis of experimental polarimetric radar data.

*This work has been supported by Selenia Co. and the Italian Ministry of Education.*

## References

[1] Giuli, D. (1986) 'Polarization diversity in radars', Proc. IEEE, n.74, pp. 245-269.

[2] Boerner, W.M. (1982) 'Polarization control in radar meteorology', Proc. of Conf. on Multiple-Parameter Radar Measurements of Precipitation, Bournemouth (U.K.).

[3] Poelman, A.J. and Guy, J.R.F. (1985) 'Polarisation information utilization in primary radar', in W.M. Boerner et. al. (Eds.), *Inverse methods in electromagnetic imaging*, Part I, D. Reidel, pp 521-572.

[4] Vannicola, V.C. and Lis, S. (1985) 'Polarization vector signal processing for Radar clutter suppression', in W.M. Boerner et al. (Eds.), *Inverse Methods in Electromagnetic Imaging*, Part II, D. Reidel, pp 721-737.

[5] Giuli, D. and Rossettini, A. (1987) 'Analysis of Radar Receivers for Dual Polarization Target Detection', IEE Int. Conf. RADAR 87, pp. 170-174.

[6] Giuli, D. and Rossettini, A. 'Dual polarization Radar Detection of Targets in Clear: Optimum Receiver Structures', Submitted for publication to IEEE Trans. on AES.

[7] Poelman, A.J. (1976) 'Reconsideration of the target detection criteria based on adaptive antenna polarizations', Agard Conf. Proc. on New devices, techniques and systems in radar, The Hague (The Netherlands).

[8] Born, M. and Wolf, E. (1965) *Principles of optics*, Pergamon Press, New York.

[9] Poelman, A.J. (1976) 'Cross correlation of orthogonally polarized backscatter components', IEEE Trans. on AES, AES-12, pp. 674-682.

[10] Kennaugh, E.M. (1952) 'Polarization properties of radar reflections' M. Sc. thesis,

Department of Electrical Engineering, Ohio State University, Columbus (OH, USA).

[11] Huang. B.X. (1985) 'Mercator conformal, and Lambert, Mollweide, Aitoff-Hammer equal area projection of the polarization sphere and their applications to radar polarimetry'. M. Sc. thesis, Graduate College, of University of Illinois at Chicago, Chicago (IL, USA).

[12] Poincare',H. (1892) *Theorie mathematique de la lumiere*, Georges Carr Ed. Paris.

[13] Deschamps, G.A. and Mast P.E. (1973) 'Poincare' sphere representation of partially polarized fields', IEEE Trans. on AP, AP-21, pp. 474-478.

[14] Huynen, J.R. (1978) 'Phenomenological theory of radar target', in *Electromagnetic Scattering*, P.L.E. Uslenghi (Ed.), New York, ch. 11.

[15] Biffi Gentili,G. Fossi,M. Gherardelli,M. Giuli,D. Giaccari,E. (1982) 'A system for dual-polarization radar measurement and analysis of precipitation echoes', in Proc. Conf. on Multiple-Parameter Radar Measurements of Precipitation, Bournemouth (U.K.), pp. 61-65.

[16] Fossi,M. Gherardelli,M. Giuli,D. Pirri,F. Ponziani,G. (1987) ' Experimental results on a double polarization radar', in Proc. of Colloque International sur le Radar, Versailles (France), pp. 419-424.

[17] Poelman,A.J. (1975) 'On using orthogonally polarized noncoherent receving channels to detect target echoes in Gaussian noise', IEEE Trans. on AES, vol. AES-11, pp. 660-663.

[18] H.L. Van Trees,H.L. (1971) *Detection Estimation, Modulation Theory: Part III*, Wiley Ed., New York.

[19] Huynen, Mc. Nolty, Hansen, (1975) 'Component distributions for fluctuating radar targets', IEEE Trans. on AES, vol. AES-11, n. 6, pp. 1316-1331.

[20] Gherardelli,M. Giuli,D. Fossi,M. (1988) ' Suboptimum adaptive polarisation cancellers for dual-polarisation radars', IEE Proc., Pt. F, vol.135(1), pp. 60-72.

[21] Poelman, A.J. (1980): 'Study of controllable polarization applied to radar', in Military Microwaves '80 Conf. Rec., London (U.K.), pp. 389-404.

[22] Poelman, A.J. (1981) 'Virtual polarization adaptation: A method of increasing the detection capability of a radar system polarization-vector processing', IEE Proc., pt. F, vol 128, pp. 261-270.

[23] Dalle Mese,E. Giuli,D. (1984) 'Detection probability of a partially fluctuating target', IEE Proc., Pt. F, vol. 131(2), pp. 179-182.

[24] Nathanson,F.E. (1975) 'Adaptive circular polarization'. Proceedings of IEEE International Radar Conference, Arlington (VA, USA), pp. 221-225.

[25] Poelman,A.J. and Guy,J.R.F. (1984) 'Multinotch Logic-Product polarization suppression filters. A typical design example and its performance in rain clutter environment', IEE Proc., Pt. F, vol. 131(4), pp. 383-396.

[26] Giuli,D. Fossi,M. and Gherardelli,M. (1985) 'A technique for adaptive polarization filtering in radars', Proc. of IEEE 1985 International Radar Conference, Arlington (VA, USA), pp. 213-219.

[27] Poelman,A.J. and Guy,J.R.F. (1984) 'Nonlinear polarisation-vector translation in radar systems: a promising concept for real-time polarisation-vectr signal processing via a single-notch polarisation suppression filter', IEE Proc., Pt. F, vol 131(5), pp. 451-465.

[28] Giuli,D. Gherardelli,M. and Fossi,M. (1986) 'Using polarization discriminants for target classification and identification', Proc. of CIE 1986 Int. Conf. on Radar, Nanjing (China), pp. 889-898.

# POLARIZATION RADAR SIGNAL DESCRIPTION: MODELLING AND PROCESSING OF POLARIMETRIC PULSE RADAR DATA

G. Wanielik
AEG Research Institute
7900 Ulm
West Germany

## Abstract

Many signal processing algorithms are based on signal models. For such signal processing purposes we develop coherent polarimetric reflecting models. Because of the stimulating effect of visual data interpretations we introduce a tangential vector description for spinors and an elementary matrix decomposition of the scattering matrix (SM), which offers new physically relevant features of the scattering mechanism.

## 1. Introduction

Pulse radars transmit a sequence of pulses at the same carrier frequency and receive a transformed sequence of reflected signals. The reflected sequence is then used in the signal processing part of the radar. The description and modelling of such reflected sequences is well known for a non-polarimetric, say conventional, radar. When we engage in signal processing it seems to be necessary to use such models to get a better understanding of the reflection behaviour. In contrast to the conventional radar there is a lack of signal models, if we want to process the multichannel data of a polarimetric radar. Therefore, before starting with polarimetric signal processing we have to deal with the modelling of polarimetric data. Because visual representations very often have a stimulating effect on the creation of new polarimetric signal processing algorithms it is an aim of this work to obtain such representations for the signals themselves as well as for the description of the reflecting object.

## 2. Polarimetric signal description

Let us start with a signal description for a pulse radar, which can be graphically interpreted. Due to the narrowband character of the pulse radar considered here we can use a monochromatic electromagnetic wave (MEW) to describe the state of the radar signal with regard to one single pulse. Considered from the analytical point of view we can therefore use



*W.-M. Boerner et al. (eds.), Direct and Inverse Methods in Radar Polarimetry, Part 1*, 939–959.

the known basic form of the spinor $\underline{E}$ in its horizontal-vertical polarization basis:

$$\underline{E}(a_x,\delta_x,a_y,\delta_y) = \begin{pmatrix} a_x\, e^{j\delta_x} \\ a_y\, e^{j\delta_y} \end{pmatrix} = \begin{pmatrix} E_x \\ E_y \end{pmatrix} \tag{1}$$

In (1) $a_x, a_y$ are the amplitudes and $\delta_x, \delta_y$ are the phases of the known phasor representation of the harmonic RF-signal. From the graphical point of view we know a visualization of a MEW in the three-dimensional real Poincaré space due to the partition $\underline{g}$ of the Stokes vector $\underline{G} = (I,\underline{g})^T$ of the wave, /1/ ( see equation (5) ).

We can interpret $\underline{g}$ as a vector in this space. However, there is a lack of information when we use $\underline{g}$ instead of $\underline{E}$. $\underline{g}$ is an incoherent description of a MEW because it does not contain the so-called "zero phase" $\alpha=\delta_x$ of the spinor $\underline{E}$. However, we consider coherent processing here, and therefore we need a coherent signal description. As a consequence of this we have to look for a visualization of the spinor $\underline{E}$, which for further work will be represented in a new "polarization vector form", /2/:

$$\underline{E}(I,\alpha,\underline{p}) = I^{1/2}\cdot e^{j\alpha}\cdot 2^{-1/2}\cdot \begin{pmatrix} (1+p_1)^{1/2} \\ (1+p_1)^{-1/2}\cdot (p_2+jp_3) \end{pmatrix} \tag{2}$$

The parameters of $\underline{E}(a,\alpha,\underline{p})$ are the wave intensity I, the zero phase $\alpha$ and the polarization vector $\underline{p}=(p_1,p_2,p_3)^T$:

$$I = E_x E_x^* + E_y E_y^* \tag{3}$$

$$\alpha = \delta_x \quad \varepsilon\ (0,2\pi) \tag{4}$$

$$\underline{p} = \frac{1}{I}\cdot \begin{pmatrix} E_x E_x^* - E_y E_y^* \\ E_x E_y^* + E_y E_x^* \\ j\cdot(E_x E_y^* - E_y E_x^*) \end{pmatrix} \tag{5}$$

Because of the relationship between the new wave description and the so-called "null matrix" $\underline{\boldsymbol{Q}}$, discussed later, we first calculate the known coherency matrix $\underline{\boldsymbol{J}}$.

$$\underline{\boldsymbol{J}} = \underline{E}\cdot\underline{E}^{*T} = \begin{pmatrix} E_x E_x^* & E_x E_y^* \\ E_y E_x^* & E_y E_y^* \end{pmatrix} \tag{6}$$

In connection with equation (2) and its Pauli decomposition we get, /2/:

$$\underline{J}(I,\underline{p}) = \frac{1}{2}\cdot I\cdot\left( \begin{pmatrix} 1 & 0 \\ 0 & 1 \end{pmatrix} + p_1\begin{pmatrix} 1 & 0 \\ 0 & -1 \end{pmatrix} + p_2\begin{pmatrix} 0 & 1 \\ 1 & 0 \end{pmatrix} + p_3\begin{pmatrix} 0 & -j \\ j & 0 \end{pmatrix} \right)$$

$$\underline{J} = \frac{1}{2}\cdot I\cdot\left( \underline{\sigma}_0 + \sum_{i=1}^{3} p_i\cdot\underline{\sigma}_i \right) = \frac{1}{2}\sum_{i=0}^{3} g_i\cdot\underline{\sigma}_i \qquad (7)$$

In this formula $(g_0, g_1, g_2, g_3)^T$ is the Stokes vector $\underline{G}$,

$$\underline{G} = \begin{pmatrix} g_0 \\ \hdashline g_1 \\ g_2 \\ g_3 \end{pmatrix} = \begin{pmatrix} g_0 \\ \hdashline \underline{g} \end{pmatrix} = I\cdot\begin{pmatrix} 1 \\ \hdashline \underline{p} \end{pmatrix} \qquad (8)$$

$\underline{\sigma}_0$ is the identity matrix and $\underline{\sigma}_1, \underline{\sigma}_2, \underline{\sigma}_3$ are the Pauli matrices:

$$\underline{\sigma}_0 = \begin{pmatrix} 1 & 0 \\ 0 & 1 \end{pmatrix};\ \underline{\sigma}_1 = \begin{pmatrix} 1 & 0 \\ 0 & -1 \end{pmatrix};\ \underline{\sigma}_2 = \begin{pmatrix} 0 & 1 \\ 1 & 0 \end{pmatrix};\ \underline{\sigma}_3 = \begin{pmatrix} 0 & -j \\ j & 0 \end{pmatrix} \qquad (9)$$

The null matrix $\underline{Q}$ is defined analogous to $\underline{J}$. Instead of taking $\underline{E}$ for both, the first and the second vector in the definition of $\underline{J}$, we use the conjugate spinor $\underline{E}_C$ to define $\underline{Q}$, /3/:

$$\underline{E}_C = \begin{pmatrix} -E_y^* \\ E_x^* \end{pmatrix} \qquad (10)$$

$$\underline{Q} = \underline{E}\cdot\underline{E}_C^{*T} = \begin{pmatrix} -E_xE_y & E_xE_x \\ -E_yE_y & E_xE_y \end{pmatrix} = \begin{pmatrix} Q_{11} & Q_{12} \\ Q_{21} & Q_{22} \end{pmatrix} \qquad (11)$$

To see, which parameters of the spinor $\underline{E}(I,\alpha,\underline{p})$ are elements of the null matrix $\underline{Q}$, we use equation (2) for the calculation:

$$\underline{Q}(I,\alpha,\underline{n}) = I\cdot e^{j2\alpha}\cdot\frac{1}{2}\cdot\begin{pmatrix} -(p_2+jp_3) & (1+p_1) \\ -(p_2+jp_3)^2\cdot(1+p_1)^{-1} & (p_2+jp_3) \end{pmatrix} \qquad (12)$$

Comparing the parameters used to describe $\underline{J}$ and $\underline{Q}$ in equation (7) and (12) we can see that the null matrix $\underline{Q}$ contains the zero phase $\alpha$, which is not present in $\underline{J}$. This is the reason why $\underline{Q}$ can be used for the coherent signal description. Knowing $\underline{Q}$ from equation (12) we can now calculate the spinor $\underline{E}$ up to a sign:

$$\underline{E} = \pm \begin{pmatrix} Q_{12}^{1/2} \\ -Q_{21}^{1/2} \end{pmatrix} \tag{13}$$

The interpretation of equation (13) shows that $\underline{E}(I,\alpha,\underline{p})$ and $\underline{E}(I,(\alpha+\pi),\underline{p})$, which differ only in a sign, lead to the same null matrix, which is due to the argument $2\alpha$ in equation (12).

Next, we want to discuss the visualization of the null matrix $\underline{Q}$ in the Poincaré space. Analogous to the coherency matrix $\underline{J}$, which can be visualized by its Pauli vector $\underline{g}=(g_1,g_2,g_3)^T$, we can interpret $\underline{Q}$ using its own Pauli vector $\underline{q}=(q_1,q_2,q_3)^T$, which is called the "null vector".

$$\underline{Q} = \frac{1}{2} \cdot \sum_{i=1}^{3} q_i \cdot \underline{\sigma}_i \tag{14}$$

We define the null vector $\underline{q}$ from the $q_i$ as follows:

$$\underline{q} = \begin{pmatrix} q_1 \\ q_2 \\ q_3 \end{pmatrix} = \underline{q}_R + j \cdot \underline{q}_I \quad \varepsilon\ \mathbb{C}^3 \tag{15}$$

The real valued vectors $\underline{q}_R$ and $\underline{q}_I$ used can be calculated from the parameters $I,\alpha,\underline{p}$ of $\underline{E}(I,\alpha,\underline{p})$:

$$\underline{q}_R = I \cdot \begin{pmatrix} -\ p_2 \cdot \cos(2\alpha) + p_3 \cdot \sin(2\alpha) \\ \dfrac{p_1(1+p_1) + p_3^2}{1+p_1} \cos(2\alpha) + \dfrac{p_2 \cdot p_3}{1+p_1} \sin(2\alpha) \\ -\dfrac{p_2 \cdot p_3}{1+p_1} \cos(2\alpha) + \dfrac{p_1(1+p_1) + p_2^2}{1+p_1} \sin(2\alpha) \end{pmatrix} \tag{16}$$

$$\underline{q}_I = I \cdot \begin{pmatrix} -\ p_2 \cdot \sin(2\alpha) - p_3 \cdot \cos(2\alpha) \\ \dfrac{p_1(1+p_1) + p_3^2}{1+p_1} \sin(2\alpha) - \dfrac{p_2 \cdot p_3}{1+p_1} \cos(2\alpha) \\ -\dfrac{p_2 \cdot p_3}{1+p_1} \sin(2\alpha) + \dfrac{p_1(1+p_1) + p_2^2}{1+p_1} \cos(2\alpha) \end{pmatrix} \tag{17}$$

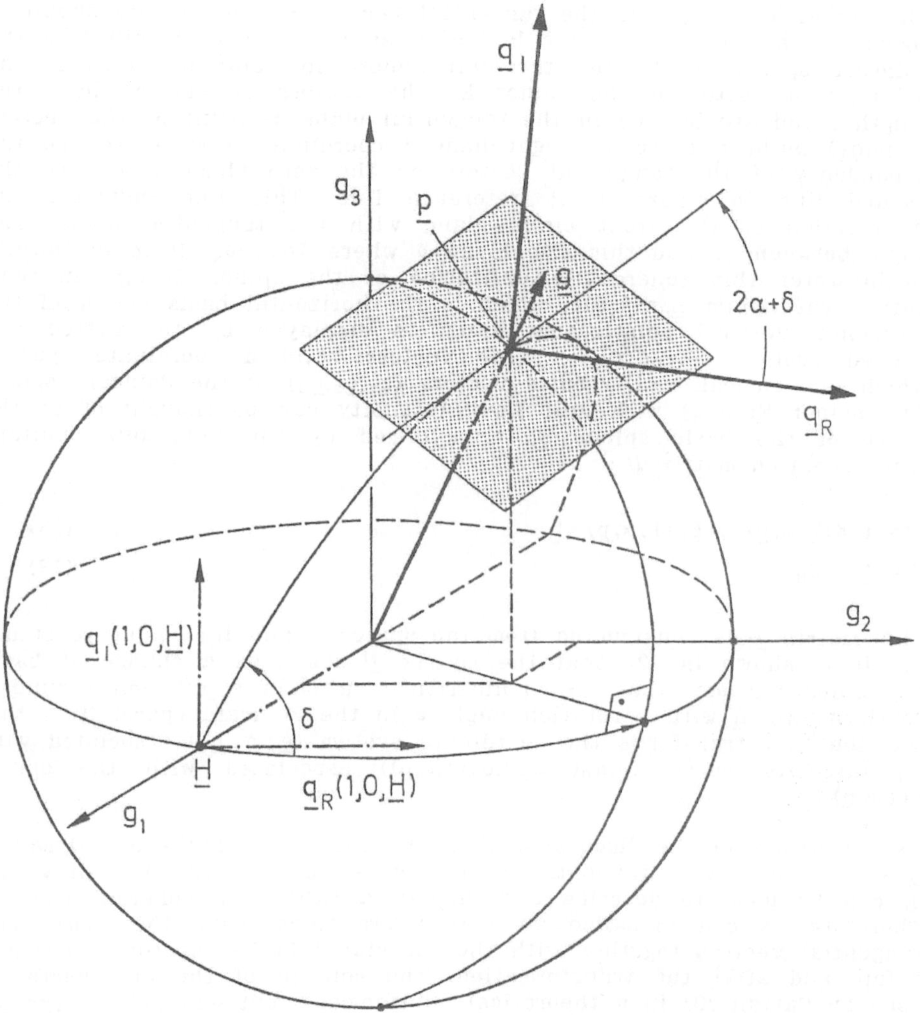


<u>Figure 1:</u> **Representation of a MEW by its tangential vectors**

The Stokes vector $\underline{g}$ and the tangential vectors $\underline{q}_R$ and $\underline{q}_I$ are shown in figure 1. The Stokes vector $\underline{g}$ is known as a vector of length I in the Poincaré space, penetrating the unit sphere in point $\underline{p}$, which is the polarization vector of the spinor $\underline{E}$. The vectors $\underline{q}_R$ and $\underline{q}_I$ have the length I and are located in the tangential plane of point $\underline{p}$. The vectors $(\underline{g},\underline{q}_R,\underline{q}_I)$ build a cartesian right-handed coordinate system. To see the dependency of the tangential vectors on the zero phase $\alpha$ we use the dashed line in figure 1 as reference line. This line indicates the intersection of the great circle plane with the tangential plane. The angle between $\underline{q}_R$ and this line is $2\alpha+\delta$ where $\delta=\delta_y-\delta_x$. It is worthwhile to interpret this coherent visualization of the spinor $\underline{E}(I,\alpha,\underline{p})$ in more detail. The spinor $\underline{E}_H=(1,0)^T$, which is the horizontal basis vector of the horizontal-vertical polarization basis, is displayed by the dotted and dashed vectors $(\underline{H},\underline{q}_{RH},\underline{q}_{IH})$. These vectors build a coordinate system, which is identical to the basis vectors $(\underline{e}_1,\underline{e}_2,\underline{e}_3)$ of the Poincaré space. Any spinor $\underline{E}(1,\alpha,\underline{p})$ with unit wave intensity can be thought of as the image of the basic spinor $\underline{E}_H$ transformed by the unimodular unitary transformation matrix $\underline{U}$ :

$$\underline{U} = ( \underline{E}(1,\alpha,\underline{p}) \ ; \ \underline{E}_C(1,\alpha,\underline{p}) ) \tag{18}$$

$$\underline{E} = \underline{U} \cdot \underline{E}_H \tag{19}$$

This matrix $\underline{U}$ is constructed from the spinor $\underline{E}$ and its conjugate spinor $\underline{E}_C$. It is shown in /2/ that the matrix $\underline{U}$ describes a change of basis transformation and that its application results in a rotation around a rotation axis $\underline{n}$ with a rotation angle $\theta$ in the Poincaré space. It is this rotation that transforms the coordinate system $(\underline{e}_1,\underline{e}_2,\underline{e}_3)$ associated with $\underline{E}_H$ into its image $(\underline{p},\underline{q}_R(1,\alpha,\underline{p}),\underline{q}_I(1,\alpha,\underline{p}))$ associated with the spinor $\underline{E}(1,\alpha,\underline{p})$.

As a summary of the discussion above we can state that the null matrix $\underline{Q}$ and its associated tangential vectors $\underline{q}_R,\underline{q}_I$, which form the null vector $\underline{q}$, can be used to describe and display a MEW in a coherent manner. Therefore we can visualize wave transformations using the geometrical tangential vectors together with the associated Stokes vector partition $\underline{g}$ before and after the transformation. The concept of the null vector is used by Cartan /3/ in a theoretical environment, but was not interpreted by him in any way. The name results from the fact that the sum of the squares of the vector components $q_i$ is zero.

It can be shown that the coherency matrix $\underline{J}$ and therefore the Stokes vector partition $\underline{g}$ as well can be generated only with the knowledge of $\underline{Q}$.

$$\underline{J} = \frac{1}{2}\cdot(\det(\underline{Q}\cdot\underline{Q}^{*T} - \underline{Q}^{*T}\cdot\underline{Q}))^{1/4}\cdot\left[ \begin{pmatrix} 1 & 0 \\ 0 & 1 \end{pmatrix} + \frac{(\underline{Q}\cdot\underline{Q}^{*T} - \underline{Q}^{*T}\cdot\underline{Q})}{(\det(\underline{Q}\cdot\underline{Q}^{*T} - \underline{Q}^{*T}\cdot\underline{Q}))^{1/2}} \right]$$

$$= \frac{1}{2}\cdot\sum_{i=0}^{3} g_i \ \underline{\sigma}_i \tag{20}$$

This is not surprising because of the geometry discussed in figure 1. We can also generate the Stokes vector $\underline{g}$ in a direct way, when we use the vector product of the tangential vectors $\underline{q}_R$ and $\underline{q}_I$.

$$\underline{g} = \frac{\underline{q}_R \times \underline{q}_I}{(\|\underline{q}_R\| \cdot \|\underline{q}_I\|)^{1/2}} = \frac{\underline{q}_R \times \underline{q}_I}{I} \tag{21}$$

However, the reverse is not true because $\underline{J}$ does not contain the zero phase $\alpha$. In the following section we discuss the question how these tangential vectors geometrically behave during the transformation by a radar target.

## 3. Polarimetric Radar Target Description

Stimulated by the result of the new wave description in chapter 2 we are concerned in this chapter with the answer to the following question: Can we find a description of the radar target, which gives both, an analytical description as well as a visual interpretation? We start with the well-known transformation description of the radar target through its scattering matrix $\underline{S}$. The spinor of the scattered wave $\underline{E}_s$ is gained from the spinor $\underline{E}_i$ of the incident wave by the following transformation:

$$\underline{E}_s = \underline{S} \cdot \underline{E}_i \tag{22}$$

Using this result and the matrix $\underline{C}$

$$\underline{C} = j \cdot \underline{\sigma}_1 = \begin{pmatrix} 1 & 0 \\ 0 & -1 \end{pmatrix} \tag{23}$$

we can calculate the null matrix $\underline{Q}_s$, which describes the scattered wave.

$$\begin{aligned} \underline{Q}_s &= \underline{E}_s \cdot \underline{E}_{sC}^{*T} = \underline{E}_s \cdot (\underline{E}_s^T \cdot \underline{C}) \\ &= \underline{S} \cdot \underline{E}_i \cdot (\underline{E}_i^T \cdot \underline{S}^T \cdot \underline{C}) = \underline{S} \cdot (\underline{E}_i \cdot \underline{E}_i^T \cdot \underline{C}) \cdot \underline{C}^T \cdot \underline{S}^T \cdot \underline{C} \\ &= \underline{S} \cdot \underline{Q}_i \cdot (\underline{C}^T \cdot \underline{S}^T \cdot \underline{C}) \end{aligned} \tag{24}$$

This result shows the transformation law for the null matrices of the incident and the scattered waves, describing the alteration caused by a radar target.
Instead of this transformation law we can use another one, which shows how the associated null vectors are transformed:

$$\underline{q}_s = (\ \underline{U}_3 \cdot (\ \underline{S} \otimes (\underline{C}^T \cdot \underline{S}^T \cdot \underline{C})) \cdot \underline{U}_3^{-1}\ ) \cdot \underline{q}_i = \underline{T} \cdot \underline{q}_i \tag{25}$$

$$= (\ \underline{U}_3 \cdot \left( \begin{array}{c:c} S_{11} \cdot \begin{pmatrix} S_{22} & -S_{21} \\ -S_{12} & S_{11} \end{pmatrix} & S_{12} \cdot \begin{pmatrix} S_{22} & -S_{21} \\ -S_{12} & S_{11} \end{pmatrix} \\ \hdashline S_{21} \cdot \begin{pmatrix} S_{22} & -S_{21} \\ -S_{12} & S_{11} \end{pmatrix} & S_{22} \cdot \begin{pmatrix} S_{22} & -S_{21} \\ -S_{12} & S_{11} \end{pmatrix} \end{array} \right) \cdot \underline{U}_3^{-1}\ ) \cdot \underline{q}_i$$

$$\underline{U}_3 = \begin{pmatrix} 1 & 0 & 0 & -1 \\ 0 & 1 & 1 & 0 \\ 0 & j & -j & 0 \end{pmatrix} \tag{26}$$

We notice that the analogous transformation description of the coherency matrix $\underline{J}$ and the Stokes vector $\underline{G}=(g_o,\underline{g})^T$ is:

$$\underline{J}_s = \underline{S} \cdot \underline{J}_i \cdot \underline{S}^{*T} \tag{27}$$

$$\underline{G}_s = (\ \underline{U}_4 \cdot (\ \underline{S} \otimes \cdot \underline{S}^*)\ \cdot \underline{U}_4^{-1}\ ) \cdot \underline{G}_i = \underline{M} \cdot \underline{G}_i \tag{28}$$

$$= (\ \underline{U}_4 \cdot \left( \begin{array}{c:c} S_{11} \cdot \begin{pmatrix} S_{11}^* & S_{12}^* \\ S_{21}^* & S_{22}^* \end{pmatrix} & S_{12} \cdot \begin{pmatrix} S_{11}^* & S_{12}^* \\ S_{21}^* & S_{22}^* \end{pmatrix} \\ \hdashline S_{11} \cdot \begin{pmatrix} S_{11}^* & S_{12}^* \\ S_{21}^* & S_{22}^* \end{pmatrix} & S_{12} \cdot \begin{pmatrix} S_{11}^* & S_{12}^* \\ S_{21}^* & S_{22}^* \end{pmatrix} \end{array} \right) \cdot \underline{U}_4^{-1}\ )\ \cdot \underline{G}_i$$

$$\underline{U}_4 = \begin{pmatrix} 1 \quad 0 \quad 0 \quad 1 \\ \cdots\cdots\cdots \\ \underline{U}_3 \end{pmatrix} ; \qquad \underline{G} = \begin{pmatrix} g_o \\ \cdots \\ \underline{g} \end{pmatrix} \tag{29}$$

We can now use the tangential vectors, which are the real and imaginary parts of the null vectors $\underline{q}_i$ and $\underline{q}_s$ to study the different transformation behaviour of different radar targets. In the polarimetric literature this study is usually done using examples of different SM, e.q. /4/. Because this method cannot describe the effect of all possible SM, we looked for a new way to do this systematically.
As a result we found that we only have to study the transformation behaviour of four so called "elementary matrices" $\underline{K}$, $\underline{U}$, $\underline{H}$ and $\underline{P}$ by using

the in- and output tangential vectors before and after the elementary SM transformation. Knowing this implies that we know the full transformation variability of all possible SM because each SM $\underline{S}$ can be written in the following form, /2/:

$$\underline{S} = \underline{U} \cdot \underline{\hat{H}} \cdot \underline{K} \tag{30}$$

$$\underline{\hat{H}} = \begin{cases} \underline{H} & \text{if } \underline{S} \text{ is regular} \\ \underline{P} & \text{if } \underline{S} \text{ is irregular} \end{cases} \tag{31}$$

which can be interpreted as a cascade of three elementary matrices as shown in figure 2.

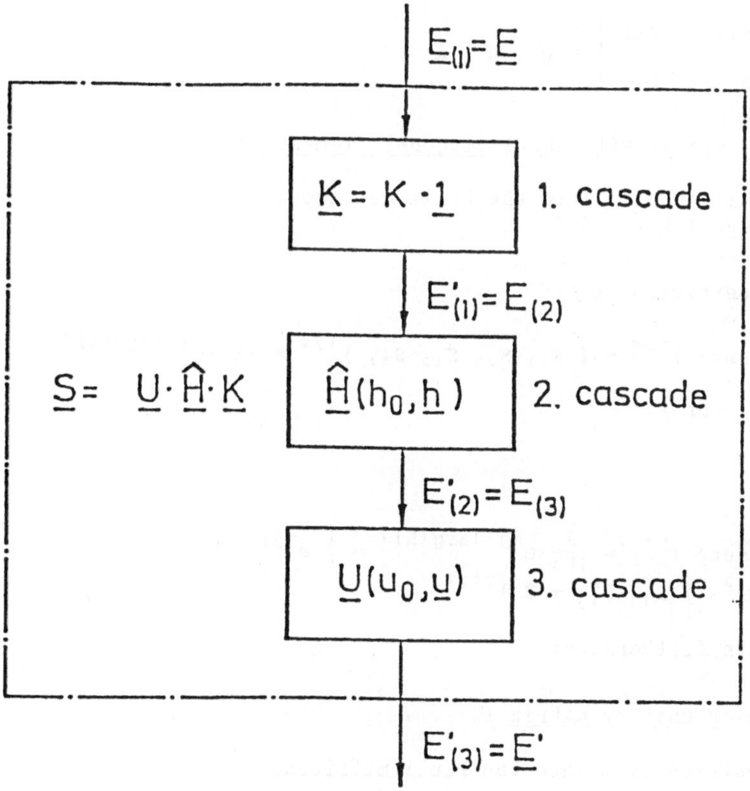


Figure 2: Scattering matrix as a cascade of three elementary matrices

$$\underline{E}_s = \underline{S} \cdot \underline{E}_i = (\underline{U} \cdot \underline{\hat{H}} \cdot \underline{K}) \cdot \underline{E} \tag{32}$$

$$\left.\begin{array}{l}\underline{E}_{(1)} = \underline{E} \\ \underline{E}'_{(1)} = \underline{K} \cdot \underline{E}_{(1)}\end{array}\right\} \longrightarrow \left.\begin{array}{l}\underline{E}_{(2)} = \underline{E}'_{(1)} \\ \underline{E}'_{(2)} = \underline{\hat{H}} \cdot \underline{E}_{(2)}\end{array}\right\} \longrightarrow \begin{array}{l}\underline{E}_{(3)} = \underline{E}'_{(2)} \\ \underline{E}'_{(3)} = \underline{U} \cdot \underline{E}_{(3)}\end{array}$$

$$\underline{E}_s = \underline{E}_{(3)}$$

Before we discuss the visually interpretable result of this cascaded description of $\underline{S}$ we want to calculate the elementary matrices of each cascade analytically from a given SM $\underline{S}$. Again, we use the parameters of the Pauli decomposition of the elementary matrices to describe those because of their physical and geometrical meaning. Depending on whether the SM is regular ( det $\underline{S} \neq 0$ ) or not we get the following results:

**I. Decomposition of a <u>regular</u> SM $\underline{S}$ in elementary matrices:**

$$\underline{S} = \begin{bmatrix} S_{11} & S_{12} \\ S_{21} & S_{22} \end{bmatrix} = \underline{U} \cdot \underline{H} \cdot \underline{K} \tag{33}$$

<u>diagonal matrix with equal diagonal elements</u> $\underline{K}$:

decomposition of $\underline{K}$ into the Pauli matrices:

$$\underline{K} = k_0 \cdot \underline{\sigma}_0 + \underline{k} \cdot \underline{\sigma} = K \cdot \underline{1} \tag{34}$$

Pauli coefficient $k_0$:

$$k_0 = K \tag{35}$$

$$K = (\det \underline{S})^{1/2} = (S_{11} \cdot S_{22} - S_{12} \cdot S_{21})^{1/2} = |K| \cdot e^{j \cdot |\arg(K)|} \tag{36}$$

Pauli vector $\underline{k}$:

$$\underline{k} = \underline{0} \tag{37}$$

with

$$\frac{1}{K} = (\det \underline{S})^{-1/2} = \frac{1}{|K|} \cdot e^{-j \cdot |\arg(K)|} = \frac{1}{\kappa} e^{-j\xi} \tag{38}$$

$$f = \Big( \kappa^2 \cdot \sum_{i,j} S_{ij} \cdot S^*_{ij} + 2 \Big)^{-1/2} \tag{39}$$

we obtain furthermore:

<u>unimodular unitary matrix</u> $\underline{U}$:

decomposition of $\underline{U}$ into the Pauli matrices:

$$\underline{U} = u_0 \cdot \underline{\sigma}_0 + \underline{u} \cdot \underline{\sigma} \tag{40}$$

Pauli coefficient $u_0$:

$$u_0 = \frac{1}{2}(U_{11} + U_{22}) = f \cdot \kappa \cdot \mathrm{Re}[e^{j\xi} S_{11} + e^{-j\xi} S^*_{22}] \tag{41}$$

Pauli vector $\underline{u}=(u_1,u_2,u_3)^T$:

$$u_1 = \frac{1}{2}(U_{11}-U_{22}) \quad = j\cdot(f\cdot\kappa\cdot Im[e^{j\xi}S_{11}+ e^{-j\xi}S^*_{22}])$$
$$u_2 = \frac{1}{2}(U_{12}+U_{21}) \quad = j\cdot(f\cdot\kappa\cdot Im[e^{j\xi}S_{12}- e^{-j\xi}S^*_{21}]) \qquad (42)$$
$$u_3 = \frac{1}{2}\cdot j\cdot(U_{12}-U_{21})= j\cdot(f\cdot\kappa\cdot Re[e^{j\xi}S_{12}- e^{-j\xi}S^*_{21}])$$

positive definite hermitian matrix $\underline{H}$:

decomposition of $\underline{H}$ into the Pauli matrices:

$$\underline{H} = h_0\cdot\underline{\sigma}_0 + \underline{h}\cdot\underline{\sigma} \qquad (43)$$

Pauli coefficient $h_0$: (44)

$$h_0 = \frac{1}{2}(H_{11}+H_{22}) = \frac{1}{2}\cdot f\cdot\kappa^2\cdot(S_{11}\cdot S^*_{11}+ S_{21}\cdot S^*_{21} + S_{12}\cdot S^*_{12}+ S_{22}\cdot S^*_{22}+ \frac{2}{\kappa^2})$$

Pauli vector $\underline{h}=(h_1,h_2,h_3)^T$:

$$h_1 = \frac{1}{2}(H_{11}-H_{22}) = \frac{1}{2}\cdot f\cdot\kappa^2\cdot(S_{11}\cdot S^*_{11}+ S_{21}\cdot S^*_{21} - (S_{12}\cdot S^*_{12}+ S_{22}\cdot S^*_{22}) )$$
$$h_2 = \frac{1}{2}(H_{12}+H_{21}) = f\cdot\kappa^2\cdot Re[S_{11}\cdot S^*_{12}+ S_{21}\cdot S^*_{22}] \qquad (45)$$
$$h_3 = \frac{1}{2}\cdot j\cdot(H_{12}-H_{21})= f\cdot\kappa^2\cdot Im[S_{11}\cdot S^*_{12}+ S_{21}\cdot S^*_{22}]$$

**II. Decomposition of an irregular SM $\underline{S}$ in elementary matrices:**

$$\underline{S} = \begin{pmatrix} S_{11} & S_{12} \\ S_{21} & S_{22} \end{pmatrix} = \underline{U}\cdot\underline{P}\cdot\underline{K} \qquad (46)$$

diagonal matrix with equal diagonal elements $\underline{K}$:

decomposition of $\underline{K}$ into the Pauli matrices:

$$\underline{K} = k_0\cdot\underline{\sigma}_0 + \underline{k}\cdot\underline{\sigma} \qquad (47)$$

Pauli coefficient $k_0$:

$$k_0 = F^{-1} \qquad (48)$$
$$F = (\sum_{i,j} S_{ij}\cdot S^*_{ij})^{-1/2} \in \mathbb{R} \qquad (49)$$

Pauli decomposition vector $\underline{k}$:

$$\underline{k} = \underline{0} \qquad (50)$$

<u>unimodular unitary matrix</u> $\underline{U}$:

decomposition of $\underline{U}$ into the Pauli matrices:

$$\underline{U} = u_0 \cdot \underline{\sigma}_0 + \underline{u} \cdot \underline{\sigma} \tag{51}$$

Pauli coefficient $u_0$:

$$u_0 = \frac{1}{2}(U_{11}+U_{22}) \quad = F \cdot \mathrm{Re}[S_{11} + S_{22}^*] \tag{52}$$

Pauli vector $\underline{u}=(u_1,u_2,u_3)^T$:

$$u_1 = \frac{1}{2}(U_{11}-U_{22}) \quad = j \cdot (F \cdot \mathrm{Im}[S_{11} + S_{22}^*])$$

$$u_2 = \frac{1}{2}(U_{12}+U_{21}) \quad = j \cdot (F \cdot \mathrm{Im}[S_{12} - S_{21}^*]) \tag{53}$$

$$u_3 = \frac{1}{2} \cdot j \cdot (U_{12}-U_{21}) = j \cdot (F \cdot \mathrm{Re}[S_{12} - S_{21}^*])$$

<u>positive semidefinite hermitian matrix</u> $\underline{P}$:

decomposition of $\underline{P}$ into the Pauli matrices:

$$\underline{P} = h_0 \cdot \underline{\sigma}_0 + \underline{h} \cdot \underline{\sigma} \tag{54}$$

Pauli coefficient $h_0$: (55)

$$h_0 = \frac{1}{2}(P_{11}+P_{22}) = \frac{1}{2} \cdot F^2 \cdot (S_{11} \cdot S_{11}^* + S_{21} \cdot S_{21}^* + S_{12} \cdot S_{12}^* + S_{22} \cdot S_{22}^*)$$

Pauli vector $\underline{h}=(h_1,h_2,h_3)^T$:

$$h_1 = \frac{1}{2}(P_{11}-P_{22}) \quad = \frac{1}{2} \cdot F^2 \cdot (S_{11} \cdot S_{11}^* + S_{21} \cdot S_{21}^* - (S_{12} \cdot S_{12}^* + S_{22} \cdot S_{22}^*))$$

$$h_2 = \frac{1}{2}(P_{12}+P_{21}) \quad = F^2 \cdot \mathrm{Re}[S_{11} \cdot S_{12}^* + S_{21} \cdot S_{22}^*] \tag{56}$$

$$h_3 = \frac{1}{2} \cdot j \cdot (P_{12}-P_{21}) = F^2 \cdot \mathrm{Im}[S_{11} \cdot S_{12}^* + S_{21} \cdot S_{22}^*]$$

A deeper insight in this problem is given in the authors dissertation /2/. Let us now interpret this matrix cascade using as an example the intensity transformation behaviour of a regular SM. Figure 3 shows the transformation behaviour at each stage of the cascade (32). We start our discussion with the locus of all possible incident waves with unit intensity, which is the shell of the unit sphere marked with a zero in figure 3. The first cascade $\underline{K}$ transforms this shell onto the shell of a new sphere marked with one. We see that this cascade amplifies the input wave independently of its polarization state.

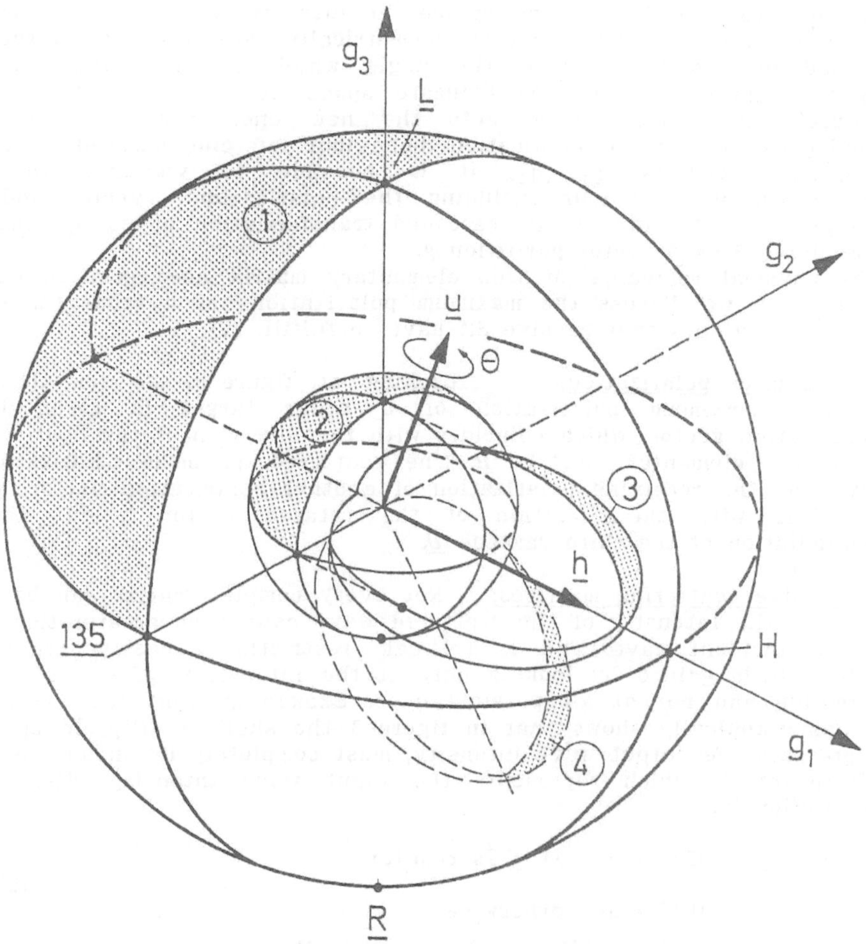


**Figure 3:** **Intensity transformation behaviour of a regular SM**

**The second cascade *H* however is a polarization dependent amplifier. *H* transforms the shell *1* onto the shell of an ellipsoid of revolution marked with *2*. The direction of the axis of revolution is given by the unit vector of the Pauli vector h of cascade *H*, the form of this ellipsoid is described by the length of h, /2/.**

The last cascade $\underline{U}$ does not change the intensity of its input wave. Its transformation behaviour can be geometrically interpreted as a rotation around an axis $\underline{u}$ with a rotation angle, which is functionally described by the length of $\underline{u}$ in the Poincaré space. As a result the shell of ellipsoid *2* is transformed onto the new one, namely, ellipsoid *3*. Analogous to the transformation behaviour of one parameter of the tangential vectors $\underline{q}_R$, $\underline{q}_I$, it is possible to visualize the full transformation behaviour including the polarization vector and the zerophase by looking at the cascaded transformation of $\underline{q}_R$, $\underline{q}_I$ and the associated Stokes vector paritition $\underline{g}$.
The physical relevance of this elementary matrix decomposition can be seen when we discuss the maximum polarizations of a radar target /5/ and requirements that passive SM have to fulfill, /2/:

1. Maximum polarizations: Looking at figure 3, we see that the incident maximum polarization of a radar target is given by a polarization vector, which coincides with the direction of the Pauli vector $\underline{h}$ of the elementary matrix $\underline{H}$. The scattered maximum polarization is given by the receiving polarization of another polarization vector, which coincides with the direction of the rotated vector $\underline{h}$ due to the manipulation of the third cascade $\underline{U}$.

2. Passive scattering matrices: Not every complex matrix can be a SM because the intensity of the scattered wave cannot be greater than that of the incident wave. This is a "weak" restriction a passive SM has to fulfill. It is gained by looking only at the intensities of one scattering direction and not at all possible ones. Examining what this restriction means graphically shows that in figure 3 the shell of ellipsoid *3*, which represents the output wave intensity, must completely lie inside the shell of sphere *1*, which represents the input wave intensity. The formal restriction is:

$$\begin{cases} (h_0+|\underline{h}|)^2\cdot|K|^2 \leq 1 & \text{if } \underline{S} \text{ is regular} \\ |K|^2 \leq 1 & \text{otherwise} \end{cases} \qquad (57)$$

Until now we have discussed geometrically interpretable description methods of waves and radar targets, which can be used to model the reflection when we transmit and analyse only one narrowband pulse. In the following chapter we present a model of the radar reflection, which describes the situation when we transmit and analyse a whole sequence of pulses, which is characteristic for a pulse radar.

## 4. Time variant modelling of the radar scattering

Figure 4 shows the time variant "total" model of the radar reflection. "Total" means that we describe the transmit and receive antennas and channels by the same matrix description method as usually used for the description of the radar target by its SM. "Time variant" means that we do not only describe the reflecting scenario for one pulse, which is the usual way, but that we also describe the system state for the i-th pulse and therefore model the time dependence of the radar reflection. One advantage of this modelling is that at each stage of the reflecting path the same analytical methods can be used to calculate the signals.

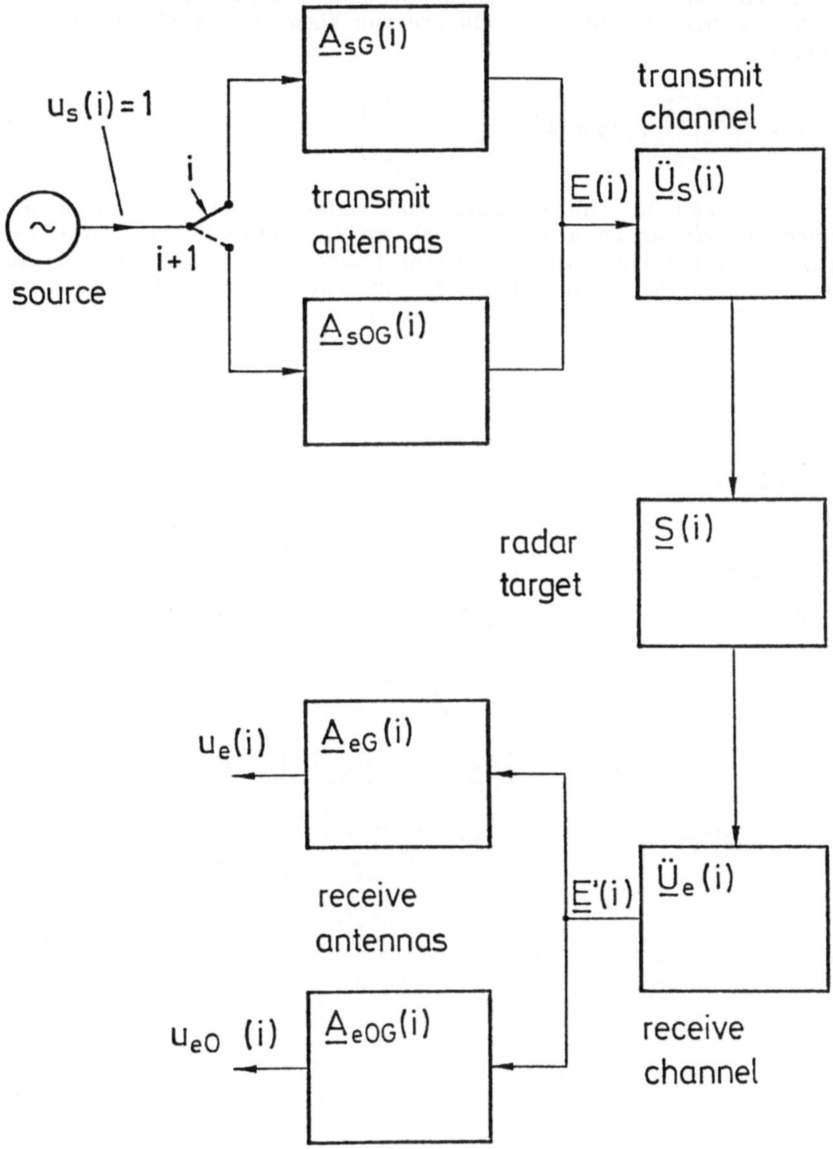


Figure 4: Time variant "total" model of the radar reflection

Another advantage is that we can use ideal and non-ideal cascades to describe, for example, the behaviour of the antennas and compare the results. And a third advantage of this modelling is the flexible addition of time variant parts to special cascades of the model. Let us explain this last point in a little more detail looking at the matrix description of the transmit and receive channels in a monostatic enviroment. In this case the matrices of the channels are the same. One model matrix for the channel is:

$$\underline{K}_{\ddot{u}}(r) = e^{-j2\pi f_0 \cdot \frac{r}{c}} \cdot (4\pi r^2)^{1/2} \cdot \begin{pmatrix} 1 & 0 \\ 0 & 1 \end{pmatrix} \qquad (58)$$

This matrix can be chosen when the radar target is at rest and its distance to the antenna is r as shown in figure 5. The first term in equation (58) describes the change of phase of a spinor due to the free space wave propagation where $f_0$ is the carrier frequency of the

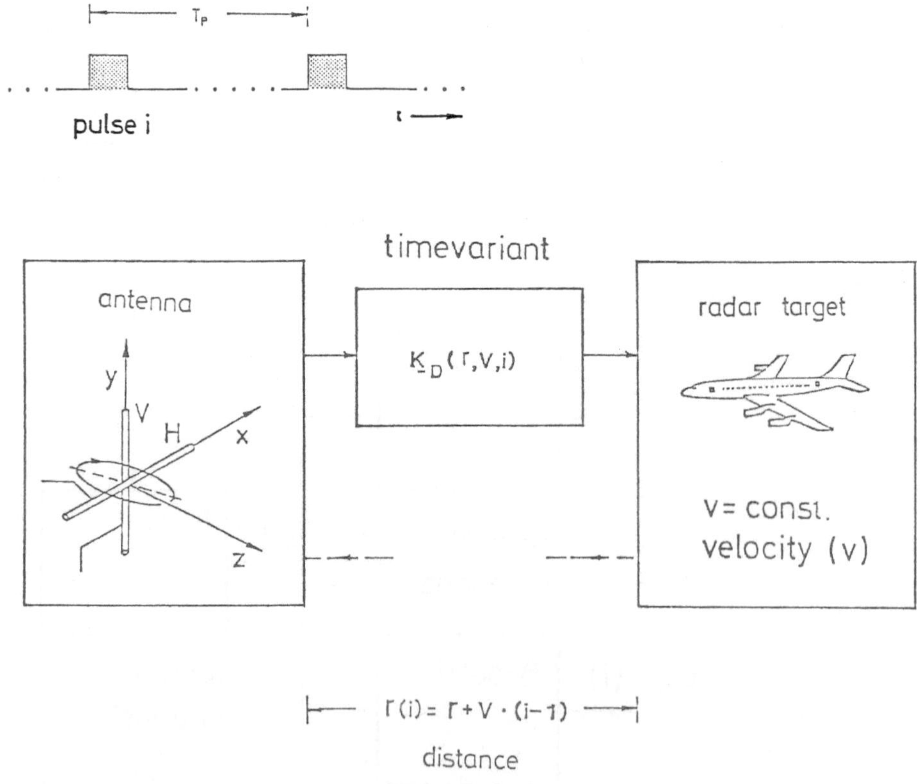


Figure 5: Time variant description of the transmit and receive channel

RF-signal and c is the velocity of light in vacuum. The second term in equation (58) shows the damping of the spinor due to the free space wave propagation. We have to use another matrix description of the channel when the radar target moves with constant velocity $\pm v$ in the r-direction ( positive sign of v when the target moves in positive r-direction away from the antenna). Then the time variant matrix

$$\underline{K}_D(r,v,i) = \underline{K}_{\ddot{u}}(r) \cdot e^{-j2\pi f_0 \cdot \frac{v \cdot T_P \cdot (i-1)}{c}} \cdot \begin{pmatrix} 1 & 0 \\ 0 & 1 \end{pmatrix} \tag{59}$$

describes the state of the channel for the i-th pulse. Due to the movement of the radar target, the i-th pulse "sees" the radar target at the time variant place r(t).

$$r(t) = r \pm v \cdot (i-1) \cdot T_p \tag{60}$$

This leads to the time variant phase term in equation (59), the argument of which is proportional to the well known Doppler frequency of the pulse radar. In equation (59) $T_p$ is the pulse repetition time and i is the time index, which varies from i=1 to i=N and where N is the number of transmitted pulses. The matrix $\underline{K}_{\ddot{u}}(r)$ in equation (59) describes the state for the first pulse.
Under the assumption that the SM of the target $\underline{S}(i)$ remains constant during the time interval of N pulses, the transmitted signal and the signal at the input of the receiving antennas can be visualized by the tangential vector $\underline{q}_R(i)$ shown in figure 6. We see, that the tangential vector $\underline{q}_R(i)$ at the input of the receiving antennas does not vary in its polarization state and wave intensity but only in its zero phase $\alpha'(i)$. The zero phase varies from pulse to pulse by the constant increment $\Delta/2$, where $\Delta$ is the angle between two successive tangential vectors. The second tangential vector $\underline{q}_I$ is not displayed because of its known relation to $\underline{q}_R$. This reflecting model is called the non-fluctuating model because of the time invariant SM of the radar target during the observation time.
Contrary to this model is the fluctuating target model, in which the SM $\underline{S}(i)$ of the radar target is time variant and the locus r(i) varies also with time due to another equation than equation (60). In this case the tangential vector $\underline{q}_R(i)$ varies in the way shown in figure 7. We see the variation of all spinor parameters $\underline{E}'(i)$ at the input of the receiving antennas. Depending on the nature of the radar reflection one of the two models is more appropriate for the signal and system description. For example, the non-fluctuating model should be used to model the reflection of a jet when observed by a polarimetric radar with low pulse repetition frequency, whereas the fluctuating model should be used to model the reflection from a helicopter with the same radar.

The discussed time variant total reflecting model of figure 3 offers considerable insight in the dynamic scattering process, which can be used to describe other dynamic phenomena, for example the necessary corrections of the measured data to obtain the SM when the radar target is moving, /2/.

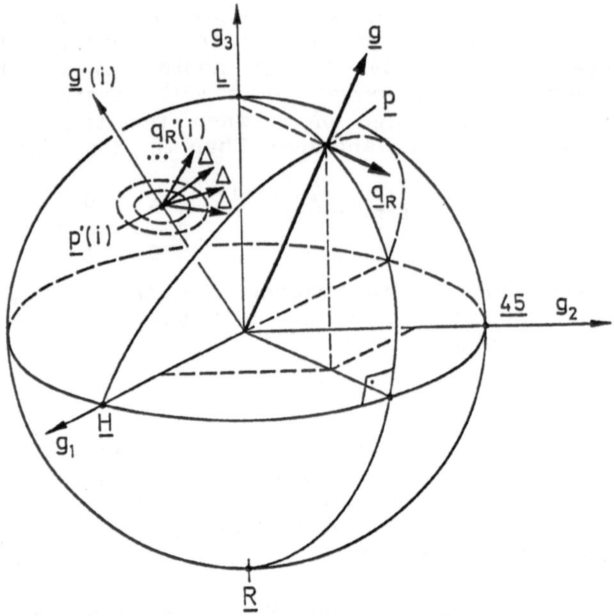


Figure 6: Tangential vector variation in the non-fluctuating model

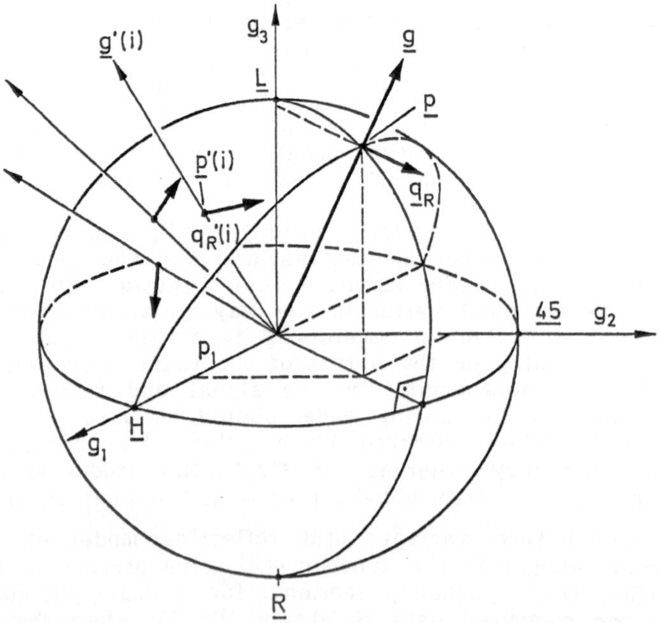


Figure 7: Tangential vector variation in the fluctuating model

## 5. Coherent processing of polarimetric vector sequences

In this chapter we want to deal with coherent signal processing possibilities, which make use of the information of the zero phases of the observed spinors. This is contrary to other signal processing algorithms discussed, for example, in /6/ where the zero phase is not of interest. Figures 6 and 7 show that polarimetric pulse radars use as input data a sequence of complex vectors:

$$\underline{v}(i) = \begin{pmatrix} v_1(i) \\ \vdots \\ v_M(i) \end{pmatrix} \Bigg|_{i=1,\ldots,N} \tag{61}$$

The parameter N describes the number of observed vectors $\underline{v}(i)$ using the same pulse repetition time $T_p$. The parameter M ( which equals 2 or 4 ) describes whether we measure one row of the SM or both. One coherent method of analysing such a vector sequence is to estimate the power spectral density matrix of the vector sequence, /7/. There exist a number of spectral estimators with different behaviour. In the following we discuss two of them, beginning with the multichannel correlogram method /7/. We use the following estimates of the multichannel correlation matrices for time lag k whith $k \geq 0$:

$$\hat{\underline{R}}_{VV}(k) = \frac{1}{N} \cdot \sum_{n=0}^{N-k-1} \underline{v}(n+k) \cdot \underline{v}^{*T}(n) \tag{62}$$

With

$$\hat{\underline{R}}_{VV}(k) = \hat{\underline{R}}_{VV}^{*T}(-k) \tag{63}$$

we yield an estimate

$$\hat{\underline{P}}_1(f) = T_p \cdot \sum_{k=-L}^{L} w(k) \cdot \hat{\underline{R}}_{VV}(k) \cdot e^{-j2\pi fkT_p} \tag{64}$$

of the hermitian power spectral density matrix of the sequence $\underline{v}(i)$. The parameter L in eq. (64) is the maximum time lag, for which we want to analyse the dependencies of the vectorial data. The values w(k) describe a lag window similar to a Hamming or Hanning window /8/. Instead of this "windowed" correlogram method we can use other techniques to estimate the power spectral density matrix of the process $\underline{v}$. One such method is to use an autoregressive spectral estimator. The sequence $\underline{v}(i)$ is thought to be an output process of a vectorial filter with coefficient matrices $\underline{A}(k)$, which is driven by the multichannel white noise input process $\underline{u}(i)$ with the identity matrix $\underline{1}$ as correlation matrix:

$$\underline{\boldsymbol{R}}_{uu}(k) = \begin{cases} \underline{\boldsymbol{1}} & \text{if } k=0 \\ \underline{\boldsymbol{0}} & \text{else} \end{cases} \tag{65}$$

$$\underline{v}(i) = \underline{u}(i) - \sum_{k=1}^{L} \underline{\boldsymbol{A}}(k) \cdot \underline{v}(i-k) \tag{66}$$

The coefficient matrices $\underline{\boldsymbol{A}}(k)$ can be evaluated using the matrix form of the Yule-Walker equation and the estimated correlation matrices of the process $\underline{v}$:

$$\begin{bmatrix} \hat{\underline{\boldsymbol{R}}}_{VV}(0) & \hat{\underline{\boldsymbol{R}}}_{VV}(-1) & \cdots & \hat{\underline{\boldsymbol{R}}}_{VV}(-(L-1)) \\ \hat{\underline{\boldsymbol{R}}}_{VV}(1) & \hat{\underline{\boldsymbol{R}}}_{VV}(0) & \cdots & \hat{\underline{\boldsymbol{R}}}_{VV}(-(L-2)) \\ \vdots & \vdots & \cdots & \vdots \\ \hat{\underline{\boldsymbol{R}}}_{VV}(L-1) & \hat{\underline{\boldsymbol{R}}}_{VV}(L-2) & \cdots & \hat{\underline{\boldsymbol{R}}}_{VV}(0) \end{bmatrix} \cdot \begin{bmatrix} \underline{\boldsymbol{A}}^{T}(1) \\ \underline{\boldsymbol{A}}^{T}(2) \\ \vdots \\ \underline{\boldsymbol{A}}^{T}(L) \end{bmatrix} = \begin{bmatrix} \hat{\underline{\boldsymbol{R}}}_{VV}(1) \\ \hat{\underline{\boldsymbol{R}}}_{VV}(2) \\ \vdots \\ \hat{\underline{\boldsymbol{R}}}_{VV}(L) \end{bmatrix} \tag{67}$$

Using the multichannel system function

$$\underline{\boldsymbol{A}}_{s}(f) = \underline{\boldsymbol{1}} + \sum_{k=1}^{L} \underline{\boldsymbol{A}}(k) \cdot e^{-j2\pi f k T_p} \tag{68}$$

we can calculate the autoregressive estimate of the power spectral density matrix:

$$\hat{\underline{\boldsymbol{P}}}_{2}(f) = T_p \cdot (\underline{\boldsymbol{A}}_s(f))^{-1} \cdot (\underline{\boldsymbol{A}}_s(f)^{T})^{-1} \tag{69}$$

The use of the different spectral estimators is data dependent so that no a priori choice can be made without looking at the data.

## 6. Conclusions

We have developed coherent polarimetric reflecting models, which are usefull in the design of signal processing algorithms and in the physical interpretation of the scattering mechanism. To make the different transformation behaviour of different radar targets more transparent we used the tangential vector description for spinors and the elementary matrix decomposition of the scattering matrix. Both descriptions have visual interpretations in the Poincaré space and are therefore very usefull in the handling of measured polarimetric data. Analogous to the transformation law of the Stokes vector by the Mueller matrix we have shown the transformation law for the null vector associated with the tangential vectors. In contrast to the Stokes vector the null vector of the wave contains the zero phase and can therefore be used for the coherent signal description. Furthermore we have shown some possibilities for the coherent processing of the measured polarimetric vector sequences.

## 7. Notation

A - scalar ( normal )

$\underline{A}$ - column vector ( underlined )

$\underline{\boldsymbol{A}}$ - matrix ( underlined, bold and cursiv)

$\|\underline{A}\|$ - length of the real vector $\underline{A}$

$\underline{A}\cdot\underline{B}$ - scalar product

$\underline{A}\mathrm{x}\underline{B}$ - vector product

$\underline{A}\otimes\underline{B}$ - Kronecker product

$^{T}, ^{*}$ - transposed, conjugate complex

## 8. Literature


/1/ M. Born, E. Wolf
Principles of optics
Pergamon Press, New York, 1986

/2/ G. Wanielik
Signaturuntersuchungen an einem polarimetrischen Pulsradar
Diss. Universität Karlsruhe ( Prof. W. Wiesbeck ), 1988

/3/ E. Cartan
The Theory of spinors
Dover Publications Inc., New York, 1981

/4/ A.P. Agrawal
A Polarimetric Rain Backscatter Model developed for Coherent Polarization Diversity Radar Applications
Diss. UIC-EECS University of Illinois at Chicago ( Prof. W.M. Boerner ), 1986

/5/ W.M. Boerner, M. Davidovitz
Extension of Kennaugh's Optimal Polarization Concept to the Asymmetric Scattering Matrix Case
IEEE AP-34, April 1986

/6/ G. Wanielik, D.J.R. Stock
A proposed Polarimetric CFAR-Detector and an Analysis of its Operation ( this issue )

/7/ L.M. Koopmans
The Spectral Analysis of Time Series
Academic Press, New York, 1974

/8/ A.V. Oppenheim, R.W. Schafer
Digital Signal Processing
Prentice-Hall Inc., Englewood Cliffs, 1975

# THE FOUR-DIMENSIONAL SPHERE APPLICATION TO THE REPRESENTATION AND ANALYSIS OF PARTIALLY POLARIZED ELECTROMAGNETIC WAVES

**Lev A. Zhivotovsky**
**V.I. Ul'yanov (Lenin)**
**Electronic Technical Institute**
**LETI, Popova Street 5**
**USSR 197-022 LENINGRAD**

**ABSTRACT:** The possibility of representing the state of a partially polarized wave at one point on a four-dimensional sphere is presented based on the author's modified interpretation of the Stokes vector and his approach of visualizing the multi-dimensional sphere and of mapping pertinent sub-sphere projections. Because it is cumbersome to use the classical Stokes vector formulation for the analysis of partially polarized waves due to the fact that the modulus of the Stokes vector changes with a change of the degree of polarization at unchanged wave intensity, a modification of the Stokes vector is introduced which allows for retaining the modulus constant although the degree of polarization changes. This is accomplished by utilizing certain properties of the four-dimensional Sphere (4-Sphere) which turns out not only to be useful for the presentation of partially polarized waves but also for the analysis of 2x2 Hermitian matrices of arbitrary determinants.

Especially, it is shown that various radar polarization problems for the partially coherent case can be drastically simplified; for example, the summation of non-coherent waves or the unique decomposition of a wave into non-coherent component waves. The relation between a partially polarized wave and the wave accepted by the receiving antenna is defined as the geodesic distance between two correspondent points on the 4-sphere, which should be of considerable interest in developing proper mathematical tools in fundamentals of polarization antenna theory as applied to the reception of partially polarized waves.

## 1. INTRODUCTION

The polarization sphere - the Poincaré sphere - is widely used to represent and analyze the polarization parameters of plane electromagnetic waves [1-4]. Each point on the Poincaré sphere corresponds to one particular polarization ellipse. If, on the other hand, we wish to represent a partially polarized wave on the Poincaré sphere then, using Stokes' well-known theorem, the wave can be associ- ated with two diametrically opposite (antipodal) points [4-6]. Such a representation of partially polarized waves is not always convenient: it does not enable one to estimate accurately and conveniently the "distance" between the partially polarized wave and the receiving antenna and the distance between partially polarized waves, as well as the parameters of the



*W.-M. Boerner et al. (eds.), Direct and Inverse Methods in Radar Polarimetry, Part 1*, 961–975.

resulting electromagnetic wave when incoherent waves are superimposed, etc., [3].

It is obvious that these problems would be greatly clarified if we could represent the partially polarized wave by one point in some parametric space. To this end, in this paper we introduce a four-dimensional polarization sphere [5]. In so doing, it turns out that the proposed geometric apparatus is useful not only for solving the problems of "polarization transformation of partially polarized waves", but also in the more general case of the analysis of 2 x 2 Hermitian matrices.

To elucidate the possibility of constructing a convenient representation of a four-dimensional sphere (four-sphere), we shall first examine the method for visualizing n-dimensional spheres, the n-spheres.

## 2. A METHOD FOR VISUALIZING AN N-SPHERE

It is well known that an n-dimensional Elucidian space $\varepsilon^n$ can be represented as a direct sum of orthogonal subspaces $[P_j(\kappa_j)]$ [7]:

$$\varepsilon^n = [P_0(\kappa_0)]\oplus[P_i(\kappa_i)]\oplus \; . \; . \; . \; \oplus[P_m(\kappa_m)] \quad , \tag{1}$$

where $\kappa_j$ is the dimensionality of the j-th subspace, and $1 \leq \kappa_j \leq n$. It is obvious that $m \leq (n-1)$, and

$$\sum_{j=0}^{m} \kappa_j = n \quad .$$

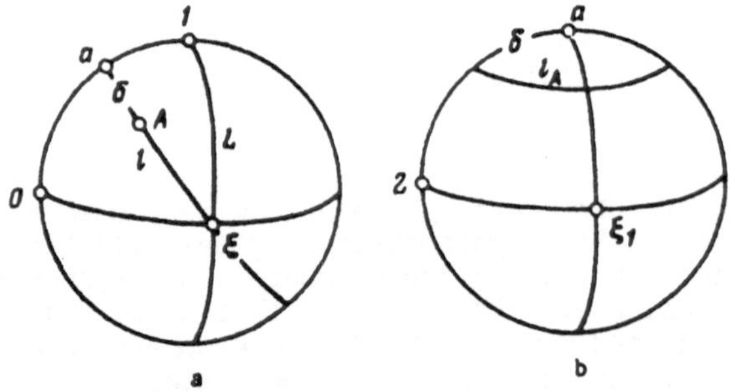


Fig. 1 Visualization of a multi-dimensional sphere:
a) n-sphere; b) (n-1)-sphere.

Let $m = 2$ and $\kappa_0 = \kappa_1$, $\kappa_2 = n-2$, then any vector $\vec{A}$ in $\varepsilon^n$ can be defined as

$$\vec{A} = a_0\hat{e}_0 + a_1\hat{e}_1 + a_2\hat{e}_2 \quad , \tag{2}$$

where $a_j$ is the projection of $\vec{A}$ on $P_j$, and $\hat{e}_j$ is the unit vector in $P_j$, $j = \overline{0,2}$.

If the analysis is restricted to the set of normalized vectors $\vec{A} = \{\vec{A} \in \varepsilon^n, |\vec{A}| = 1\}$ then, by (2), each vector can be associated with some point on a unit sphere (Fig. 1a). In this case the points 0 and 1 will correspond to fixed vectors $\hat{e}_0$ and $\hat{e}_1$ in $P_0(1)$ and $P_1(1)$, while the point $\xi$ will correspond to some vector $\hat{e}_2$ in $P_2(n-2)$. Each point on the sphere, not lying on some circle passing through the points 1 and 0, for example the point $\vec{A}$, represents the set of vectors $[A_j]$ with the same projections on $\hat{e}_0$ and $\hat{e}_1$ and, in general, different projections on the orthonormal vectors $\hat{e}_j$, $j = \overline{2, n-1}$, selected in the subspace $P_2(n-2)$, $n > 3$. [7] Any great circle, passing through the point $\xi$, for example the circle 1 in Fig. 1a, defines some hyper-plane, for which it is possible to construct an (n-1)-sphere (Fig. 1b), using as orthogonal subspaces $P_a(1)$, $P_2(1)$, $P_3(n-3)$, where $P_a(1)$ coincides with the vector $\hat{e}_a$ in the $(\hat{e}_0, \hat{e}_1)$ plane: $P_2(n-2)$, while $P(n-3)$ is a normalized subspace in $P_2(n-2)$, orthogonal to $\hat{e}_2$. At the same time the point $\vec{A}$ in Fig. 1a corresponds to the circle $1_A$ in Fig. 1b.

Thus one or another feature of different sets of n-dimensional vectors can be revealed by successively reducing the dimensionality of the subspace [7].

If $n = 4$, then the point $\xi$ in Fig. 1a corresponds to a circle, while the great circle L passing through the points 1 and $\xi$ correspond to the ordinary sphere (3-sphere).

## 3. POLARIZATION 4-SPHERE PRESENTATION OF PARTIALLY POLARIZED WAVES

Let $\vec{s}(t)$ be a partially polarized plane wave. We shall briefly examine the well-known methods for representating and analyzing the polarization state of the wave $\vec{s}(t)$. The electric field vector $\vec{E}(t)$ and the coherence matrix $[\Phi]$ in the orthonormal basis $(\hat{e}_x, \hat{e}_y)$, in the plane of the wavefront of the wave $\vec{s}(t)$, can be defined [8] as

$$\vec{E}(t) = \begin{bmatrix} E_1(t) \\ E_2(t) \end{bmatrix} \quad , \qquad [\Phi] = \langle[\vec{E}(t)\cdot\vec{E}^*(t)]\rangle \quad ,$$

where $^*$ indicates Hermitian conjugation and the brackets < > indicate time averaging. We shall represent the coherence matrix in the form

$$[\Phi] = [U]^*[\Phi_p][U] + [\Phi_u] \ , \ [\Phi_p] = \begin{bmatrix} \lambda_p & 0 \\ 0 & 0 \end{bmatrix} \ , \quad [\Phi_u] = \begin{bmatrix} \lambda_u & 0 \\ 0 & \lambda_u \end{bmatrix} .$$

Here [U] is a unitary matrix which diagonalizes $[\Phi]$ :

$$[\Phi_d] = \begin{bmatrix} \lambda_1 & 0 \\ 0 & \lambda_2 \end{bmatrix} \ , \quad \lambda_1 = \lambda_p + \lambda_u \ , \quad \lambda_2 = \lambda_u \ . \tag{3}$$

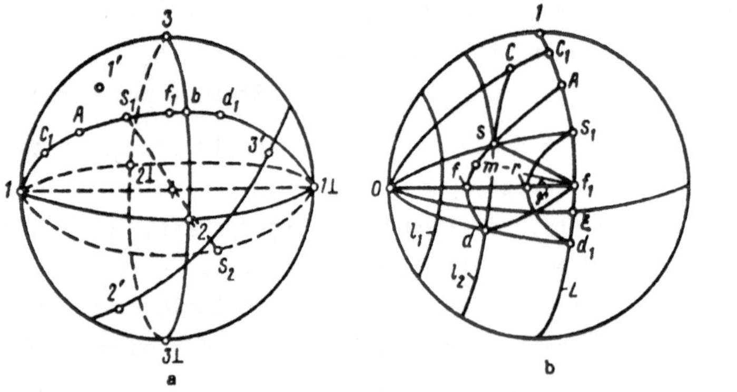


Fig. 2 (a) The Poincaré sphere; (b) the polarization 4-sphere.

The parameters $\lambda_p$ and $2\lambda_u$ characterize the intensity of the polarized and unpolarized components of the electromagnetic wave ($J_p$ and $J_u$), respectively. In this case, the degree of polarization is defined by the expression

$$p = \frac{J_p}{J} = \frac{\lambda_p}{2\lambda_u + \lambda_p} = \frac{\lambda_1 - \lambda_2}{\lambda_1 + \lambda_2} \ , \tag{4}$$

where $J = J_p + J_u$ is the total intensity of the electromagnetic wave.

On the Poincaré sphere (Fig. 2a), the wave $\vec{s}(t)$ can be associated either with the single point $S_1$, corresponding to the polarized component of the wave, and the "weight" $\lambda_p$ assigned to it, or with two points $S_1$ and $S_2$ with the weights $\lambda_1$ and $\lambda_2$; we assume at the same time that the basis vectors $\hat{e}_x$ and $\hat{e}_y$ correspond to the points 1 and $1_\perp$.

The polarization state of an electromagnetic wave is often evaluated with the aid of the Stokes parameters [1]:

$$J = \langle|\dot{U}_1(t)|^2\rangle + \langle|\dot{U}_{1_\perp}(t)|^2\rangle \ ,$$

$$q_1 = \langle|\dot{U}_1(t)|^2\rangle - \langle|\dot{U}_{1_\perp}(t)|^2\rangle \ ,$$

$$q_2 = \langle|\dot{U}_2(t)|^2\rangle - \langle|\dot{U}_{2_\perp}(t)|^2\rangle \ , \quad (5)$$

$$q_3 = \langle|\dot{U}_3(t)|^2\rangle - \langle|\dot{U}_{3_\perp}(t)|^2\rangle \ ,$$

where $\langle|\dot{U}_1(t)|^2\rangle$ and $\langle|\dot{U}_{1_\perp}(t)|^2\rangle$ are the average intensities of the oscillations received in the loads of the receiving antennas with the corresponding (i-th and $i_\perp$-th) polarizations. In particular, if the points 1 and $1_\perp$ correspond to the vertical and horizontal linear polarizations, then the points 2 and $2_\perp$ correspond to linear polarizations tilted at an angle of 45° and -45° to the horizontal, while the points 3 and $3_\perp$ correspond to circular polarizations with clockwise and anti-clockwise rotation of the electric field vector.

For a completely polarized wave the equality $J^2 = \sum_{i=1}^{3} q_i^2$ holds. In the case of a partially polarized wave $J^2 > \left[\sum_{i=1}^{3} q_1^2\right]$ . In addition,

$$J_p = \left[\sum_{i=1} q_i^2\right]^{1/2} \ , \quad J_u = J - \left[\sum_{i=1}^{3} q_i^2\right]^{1/2} \ .$$

The parameters (5) can be interpreted as the coordinates of some vector $\vec{Q}$: Stokes' vector in $\varepsilon^4$. It is not difficult to see that

$$|\vec{Q}|^2 = J^2 + J_p^2 \ .$$

Therefore, for a constant wave intensity J and a variable degree of polarization p, the modulus of Stokes' vector varies in the range J - $\sqrt{2}\cdot J$ . **This in our opinion, is the reason for the inconvenience in the use of the vector $\vec{Q}$ for analyzing partially polarized waves.** We replace Stokes' first parameter (J) by the parameter

$$q_0 = \sqrt{J^2 - \sum_{i=1}^{3} q_i^2} \ ,$$

and the vector $\vec{Q}$ by the vector $\vec{Q}_1$ : $\vec{Q}_1^{\,T} = (q_0, q_1, q_2, q_3)$. It is easy to see that $|\vec{Q}_1| = J$. In what follows, we can set $J = 1$, and then the vector $\vec{Q}_1$ can be associated with some points on a 4-sphere with the basis vectors $\hat{e}_0$ , $\hat{e}_1$ , $\hat{e}_2$ , $\hat{e}_3$; in addition, $q_i = (\vec{Q}_1 \cdot \hat{e}_j)$ , $j = \overline{0,3}$.

As the degree of polarization is varied $(o \leq p \leq 1)$, the value of $q_0$ varies from 0 (completely polarized wave) to 1 (unpolarized wave). The parameters $q_1$ , $q_2$ , $q_3$ determine the position of the projection of the vector $\vec{Q}_1$ on the Poincaré sphere: the image of the polarized component.

On the basis of what was said above, the 4-dimensional polarization sphere can be represented in the manner shown in Fig.2b, where the point 0 corresponds to the vector $\hat{e}_0$ , point 1 corresponds to the vector $\hat{e}_1$ , and the point $\xi$ corresponds to the great circle passing through the points 2 and 3 on the Poincaré sphere, shown in Fig.2a. Thus the Poincaré sphere is represented on the 4-sphere of Fig.2b by the circle L passing through the points 1 and $\xi$.

Remark 1. It is evident that in each specific case the point $\xi$ can be associated with some specific vector, for example, the vector $\vec{b}$ (the point b in Fig. 2a); in this case, the sphere of Fig.2b transforms into the ordinary 3-sphere.

Remark 2. The basis vectors $\hat{e}_1$ , $\hat{e}_2$ , $\hat{e}_3$, are not "rigidly fixed". Depending on the problem under study, it may be useful to rotate the coordinate system; in this case, the points 1 and $\xi$ in Fig.2b will correspond, for example, to the point 1′ and the circle passing through the points 2′ and 3′ in Fig.2a.

Let us assume that the partially polarized wave $\vec{s}(t)$ corresponds to the vector $\vec{Q}_1$, which on the sphere of Fig.2b corresponds to the point S with the projection $S_1$ (Figs. 2a and b) on the Poincaré sphere, while the receiving antenna has the polarization $\vec{A}$ (the point A in Figs.2a and b). We assume that the poin $\xi$ on the 4-sphere corresponds to the point b on the Poincaré sphere. Let $AS_1 = \delta$, $SS_1 = \gamma$, $SA = \beta$. From the spherical triangle $SAS_1$ we obtain

$$\cos^2\frac{\beta}{2} = \frac{1}{2}(1 - \cos\gamma) + \cos\gamma\cos^2\frac{\delta}{2} \quad . \tag{6}$$

On the other hand, it is well-known that the normalized value of the intensity of the partially polarized wave received in the load of the receiving antenna is equal to [7]

$$W = \frac{1}{2}(1 - p) + p\cos^2\frac{\delta}{2} \quad . \tag{7}$$

It is evident from the expressions (6) and (7) that

$$W = \cos^2\frac{\beta}{2}\,, \qquad p = \cos\gamma \quad . \tag{8}$$

Therefore, on the polarization 4-sphere of Fig. 2b the partially polarized waves with the same degrees of polarization correspond to a set of points located at equal distances from the point 0, i.e., the points on the small circles $l_1$ , $l_2$ , etc. In this case, since $J = 1$,

$$J_p = \cos\gamma_i\,, \qquad J_u = 1 - \cos\gamma_i \tag{9}$$

where $\gamma_i$ is the angular distance between $l_i$ and L.

Thus the "distance" between the partially polarized electromagnetic wave $\vec{s}(t)$ and the receiving antenna $\vec{A}$, just as in the case of completely polarized waves and Poincaré's sphere, can be characterized by the value

$$\rho_{S,A} = \cos\left(\frac{\beta(S,A)}{2}\right) \quad ,$$

where $\beta(S,A)$ is the angular distance between the corresponding points on the sphere.

We define analogously the distance between two partially polarized waves:

$$\rho_{S,C} = \cos\left(\frac{\beta(S,C)}{2}\right) \quad ,$$

where S and C are the images of the partially polarized waves $\vec{s}(t)$ and $\vec{c}(t)$ on the sphere in Fig.2b. Let $CC_1 = \gamma_1$ , $S_1C_1 = \delta_1$ (Fig.2b), then from the spherical triangle OCS we obtain

$$\rho_{S,C}^2 = 0.5(1 + \cos[\beta(S,C)]) = 0.5[\sin\gamma\sin\gamma_1 + \cos\gamma\cos\gamma_1\cos\delta_1 + 1]. \tag{10}$$

On the other hand, the mean-square modulus of the scalar product of the vectors $\vec{s}(t)$ and $\vec{c}(t)$ is equal to

$$R^2_{S,C} = \langle|(\vec{s}(t)\cdot\vec{c}(t)|^2\rangle = 0.5(1 + \cos\gamma_1\cos\gamma\cos\delta_1) \quad . \tag{11}$$

It is evident from Eqs.(10) and (11) that the values of $\rho_{S,C}$ and $R_{S,C}$ are identical, if the angle $\gamma$ or $\gamma_1$ is equal to 0, i.e., if one of the waves is completely polarized. In general, $(p,\ p_1 < 1)\rho_{S,C} > R_{S,C}$. For $p = p_1 = 0$, $\rho_{S,C} = 1$, $R_{S,C} = 2^{-0.5}$ . This is explained by the fact that according to the polarization parameters (the coherence matrix, Stokes' parameters) unpolarized waves are identical.

## 4. 4-POL-SPHERE PRESENTATIONS OF THE SUPERPOSITION OF PARTIALLY POLARIZED WAVES

Let us find the image of the resulting partially polarized wave, formed by the superposition of two incoherent waves

$$\vec{f}(t) = \vec{s}(t) + \vec{d}(t)$$

on the 4-sphere. We will assume that the indicated waves have the same intensities and degrees of polarization. Let the waves $\vec{s}(t)$ and $\vec{d}(t)$ correspond to the points S and d on the sphere in Fig.2b. The points $S_1$ and $d_1$ on the Poincaré sphere correspond to the polarized components of the indicated waves. We write $S_1d_1 = \alpha$. In accordance with the results of [6], the image of the polarized component of the resulting wave on the Poincaré sphere lies on the great circle passing through the points $S_1$ and $d_1$, and in this case divides the arc in half (the point $f_1$ in Fig.2b). In this case, $Sf_1 = df_1 = \beta_1$ and $\cos\beta_1 = \cos\gamma\cos(\alpha/2)$.

It is evident that in the basis $(\vec{f}_1, \vec{f}_{1\perp})$, where $\vec{f}_{1\perp}$ is the polarization orthogonal to $\vec{f}_1$, the coherence matrix of the resulting wave is equal to

$$[\Phi_f] = 2 \begin{bmatrix} \cos^2\frac{\beta_1}{2} & 0 \\ 0 & \sin^2\frac{\beta_1}{2} \end{bmatrix} .$$

Therefore, the degree of polarization of the resulting wave is given by

$$p_f = \cos^2\frac{\beta_1}{2} - \sin^2\frac{\beta_1}{2} = \cos\beta_1 \tag{12}$$

Thus the image of the resulting wave on the polarization 4-sphere (the point f) lies at the center of the arc of a small circle, passing through the image of the summed waves, centered at the image point of its polarized component. The degree of polarization of the resulting wave is equal to the cosine of the angular size of the radius of this circle. In this case, when the summed waves are completely polarized (the points $S_1$ and $d_1(S_1d_1 = \alpha)$, the degree of polarization of the resulting wave (the point r) becomes $p_r = \cos(\alpha/2)$.

Therefore, when two completely polarized waves with equal intensity are summed incoherently, the degree of polarization of the resulting wave is numerically equal to the cosine of half the angular distance between them on the Poincaré sphere. It is somewhat more difficult to determine the parameters of the resulting wave, if the summed waves have different intensities. If, on the other hand, the degrees of polarization of the summed waves are the same ($p_i = p$) while the intensities of the polarized components are different, it can be shown that the image of the resulting wave lies on the same arc of the small circle as in the case of equal intensities, however, not at the center, but rather displaced toward the more intense wave (the point m in Fig.2b).

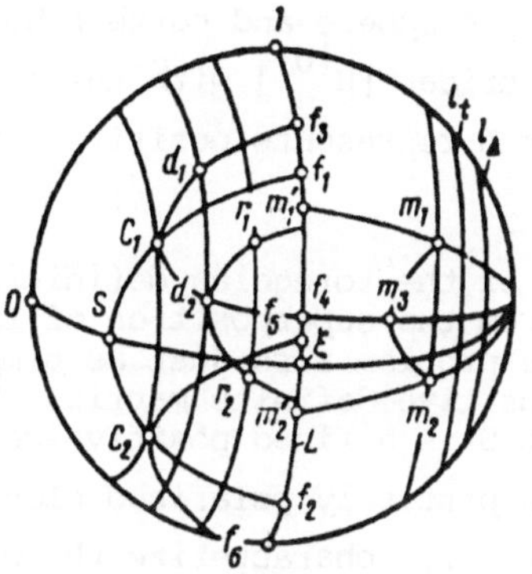


Fig.3. Representation of Hermitian matrices on a 4-sphere.

In this case the angle $\psi$, characterizing this displacement, is equal to

$$\psi = \arcsin\left( \left(\frac{1-k}{1+k}\right) \frac{p \sin\frac{\alpha}{2}}{\sqrt{1 - p^2\cos^2\frac{\alpha}{2}}} \right) \qquad (13)$$

where k is the ratio of the intensities of the summed waves. For p = 1

$$\psi = \arcsin\left( \frac{1-k}{1+k} \right) .$$

Since the coherence matrix of real electromagnetic waves is a positive-definite Hermitian matrix with $J_u > 0$, we used above only the left hemisphere of the 4-sphere in Fig.2b. In some (polarization) problems, however, "fictitious" waves with negative-definite coherence matrices appear [9]. In this context, it is deisrable to study the possibility of using both the left and right hemispheres of the 4-sphere to represent and analyze physical objects (without going into their nature), characterized by 2x2 Hermitian matrices with arbitrary, both positive and negative, definiteness.

We shall represent, as done previously, the diagonal form of an Hermitian matrix [H] in the form of a sum

$$[H_d] = [H_1] + [H_2]; \quad [H_1] = \begin{bmatrix} \lambda_p & 0 \\ 0 & 0 \end{bmatrix}; \quad [H_2] = \begin{bmatrix} \lambda_u & 0 \\ 0 & \lambda_u \end{bmatrix}, \; \lambda_p \geq 0 ;$$

$$[H_d] = \begin{bmatrix} \lambda_1 & 0 \\ 0 & \lambda_2 \end{bmatrix}; \quad \lambda_1 = \lambda_p + \lambda_u ; \quad \lambda_2 = \lambda_u .$$

We write $J = J_p + |J_u|$ , $J_p = \lambda_p$ , $J_u = 2\lambda_u$ , and let $J = \text{const}$. It is evident that the circle **L** of the 4-sphere and correspondingly the sphere in Fig.2a represent singular matrices $[H^{(0)}]$ with the determinant $\Delta = 0$, and the left hemisphere without **L** represents positive-definite matrices $[H^{(+)}]$ with $\lambda_1$ , $\lambda_2 > 0$.

The simplicity and convenience of the foregoing definition of the parameters of the wave resulting from the superposition of incoherent waves enable us to solve the inverse problem also just as simply: namely to find the different sets of a positive-definite matrix. Figure 3 illustrates this problem: the point S is a fixed positive-definite Hermitian matrix (in other words, a fixed partially polarized electromagnetic wave $\vec{s}(t)$),and the points $c_i$, $d_i$, $r_i$, $f_i$ characterize the waves into which $\vec{s}(t)$ can be "decomposed." This can be written conditionally as $S \to (c_1 + c_2) \to (c_2 + d_1 + d_2) \to (c_2 + d_1 + r_1 + r_2) \to (f_1 + f_2) \to (f_3 + f_4 + f_5 + f_6)$ , etc. Here, the values of $P_i$ of the indicated waves and the images of the polarized components are easily determined from Fig.3 with the help of simple formulas from spherical trigonometry, and the values of $J_i$ are pairwise equal and must sum to J. It is natural to suppose that the right hemisphere can be used to represent negative-definite matrices $[H^{(-)}]$ with $\lambda_j < 1$, $J_u < 0$.

The invariants of the Hermitian matrix are the determinant $\Delta$ and the trace tr = J, since $\Delta = \lambda_1\lambda_2$ and $\text{tr} = \lambda_1 + \lambda_2$ , where

$$\lambda_1 = \frac{1}{2}[\text{tr} + \sqrt{(\text{tr})^2 - 4\Delta}\,] , \quad \lambda_2 = \frac{1}{2}[\text{tr} - \sqrt{(\text{tr})^2 - 4\Delta}\,] .$$

We write $\lambda_1/\lambda_2 = h$. It is not difficult to see that for points on the left hemisphere

$$\lambda_1 = J_p + 0.5J_u = J[\cos\gamma + 0.5(1 - \cos\gamma)] = J\cos^2\frac{\gamma}{2} ;$$

$$\lambda_2 = 0.5J_u = 0.5(1 - \cos\gamma) = J\sin^2\frac{\gamma}{2} ; \tag{14}$$

$$h = \text{ctg}^2\frac{\gamma}{2} .$$

For points on the right hemisphere

$$\lambda_1 = J\cos\gamma - 0.5(1 - \cos\gamma) = J\,\frac{3\cos\gamma - 1}{2}\ ;$$
$$\lambda_2 = -0.5J(1 - \cos\gamma) = -J\sin^2\frac{\gamma}{2}\ ; \qquad (15)$$
$$h = 3 - \sec^2\frac{\gamma}{2}\ .$$

In (14) and (15), $\gamma$ is the angular distance on the 4-sphere between the corresponding point and the circle **L**.

We point out some characteristic points on the right hemisphere. From (15), it is evident that for $\gamma = 60°$ (the circle $l_t$ in Fig.3) $h = -1$, and therefore tr = 0; for $\gamma = 70.5°$ (the circle $l_\Delta$) $h = 0$, and therefore $\Delta = 0$. Thus the circle $l_\Delta$, like the great circle **L**, represents singular matrices, but unlike **L** matrices with negative eigenvalues.

Let us examine the problem of determining the matrix [H] obtained by summing two matrices $[H_1]$ and $[H_2]$ with equal values of J and h on the right hemisphere. These matrices, by (9), correspond to points lying on the same small circle, for example, the points $m_1$ and $m_2$ in Fig.3. Let the angular distance between the points $m_1'$ and $m_2'$ be equal to $\delta$ and let the distance from $m_1$, $m_2$ to the circle **L** be equal to $\gamma$. We recall that $\delta/2$ is the angle between the characteristic bases of the matrices $[H_1]$ and $[H_2]$. It is evident that

$$J_{p1} = J_{p2} = J\cos\gamma\ , \quad J_{u1} = J_{u2} = J(\cos\gamma - 1)\ . \qquad (16)$$

If the diagonal form of the matrix [H] is written in the form

$$[H_d] = \begin{bmatrix} \lambda_{p\Sigma} & 0 \\ 0 & 0 \end{bmatrix} + \begin{bmatrix} \lambda_{u\Sigma} & 0 \\ 0 & \lambda_{u\Sigma} \end{bmatrix}$$

and $J_{p\Sigma} = \lambda_{p\Sigma}$, $J_{u\Sigma} = 2\lambda_{u\Sigma}$, $J_\Sigma = J_{p\Sigma} + |J_{u\Sigma}|$, then by (16)

$$J_{p\Sigma} = 2J\cos\gamma\cos\frac{\delta}{2}\ ; \quad J_{u\Sigma} = J_u^{+} + J_u^{-}\ , \quad J_u^{+} = 2J\cos\gamma\left(1 - \cos\frac{\delta}{2}\right)\ ;$$
$$J_u^{-} = 2J(\cos\gamma - 1)\ ; \quad J_{u\Sigma} = 2J\left(2\cos\gamma - \cos\gamma\cos\frac{\delta}{2} - 1\right)\ . \qquad (17)$$

We will assume that the resulting matrix [H] corresponds on the right hemisphere to the point $m_3$, located at some angle $\beta$ from **L**. Therefore, we can construct the system

$$2J\cos\gamma\cos\frac{\delta}{2} = J_{\Sigma}\cos\beta \quad ;$$

$$2J\left(2\cos\gamma - \cos\gamma\cos\frac{\delta}{2} - 1\right) = J_{\Sigma}(\cos\beta - 1) \quad .$$

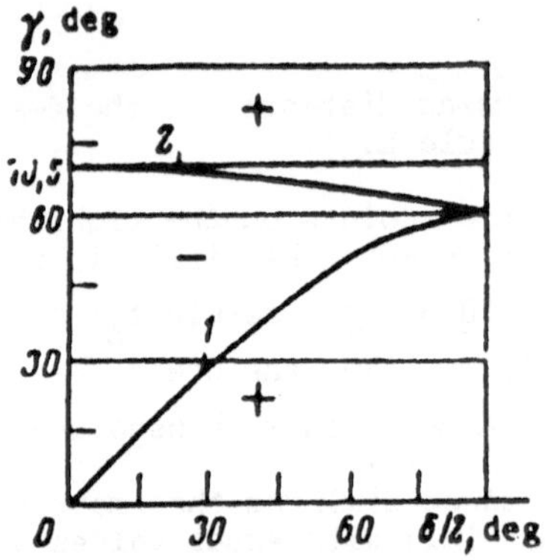


Fig. 4 Curves of $\gamma(\delta)$ determining the singularity of the matrix $[H] = [H_1] + [H_2]$ .

Solving (18), we obtain

$$\cos\gamma = \frac{\cos\beta\cos\frac{\delta}{2}}{2\cos\gamma\cos\frac{\delta}{2} - 2\cos\gamma + 1} \quad . \tag{19}$$

It is evident from (19) that for $\gamma = 60°$ and any value of $\delta$, $\beta = 60°$, which is to be expected, since $tr\{[H]\} = tr\{[H_1]\} + tr\{[H_2]\}$ ; and, in the case $\gamma = 60°$, we will find that $tr\{[H_{1(2)}]\} = 0$. Let us determine the conditions for which $\beta = 0$, i.e., $\Delta\{[H]\} = 0$, $\lambda_1 > 0$, $\lambda_2 = 0$. From (19), we obtain

$$\cos\gamma = \frac{1}{2 - \cos\frac{\delta}{2}} \quad . \tag{20}$$

We determine the conditions for which $\beta = 70.5°$ , i.e., $\Delta\{[(H)]\} = 0$, $\lambda_1 = 0$, $\lambda_2 < 0$:

$$\cos\gamma = \frac{1}{2 + \cos\frac{\delta}{2}} \quad . \tag{21}$$

The graphs of $\gamma(\delta)$, which enable us to determine the relation between the parameters $\gamma$ and $\delta$, characterizing the matrices $[H_1]$ and $[H_2]$ which are responsible for the singularity of the matrix $[H] = [H_1] + [H_2]$, are constructed in Fig.4 (curves 1 and 2, respectively) from the formulas (20) and (21). Here the region under the curve 1 corresponds to positive-definite matrices $[H]^{(+)}$ with

$$\lambda_1 , \lambda_2 > 0 \text{ and } \Delta > 0; \tag{22}$$

the region between the curves 1 and 2 corresponds to matrices with indefinite signs with $\lambda_1 > 0$ , $\lambda_2 < 0$ and $\Delta < 0$; and, the region above curve 2 corresponds to negative-definite matrices $H^{(-)}$ with $\lambda_1$ , $\lambda_2 < 0$ and $\Delta > 0$. The sign of the determinant of the matrix corresponding to each region is shown in this figure. Specific values of $\beta$ are easily found from formulas (17), (19).

## 5. CONCLUSIONS

In this paper we have examined the possibility of using the convenient apparatus of spherical trigonometry to represent and analyze partially polarized waves. To this end, we introduced the 4-dimensional polarization sphere (4-pol-sphere) and proposed a method for visualizing it. It was shown that a real partially polarized wave can be associated with a point on the left hemisphere; and that the distance between a partially polarized wave and the receiving antenna and also that between different partially polarized waves are easily determined. We examined the problem of determining the parameters of a partially polarized wave formed by the sum of incoherent waves; and, in analyzing this problem we examined the possibility of using both the left and right hemispheres to represent and analyze 2x2 Hermitian matrices with arbitrary definiteness. This 4-pol-sphere description of partially polarized waves was effectively applied to the analysis of radar targets imbedded in non-coherent clutter [10-12], and more recently to a systematic approach of dealing with the analysis of polarization properties of fluctuating targets in polarization antenna space [13]. In this latter paper several important new theorems in antenna theory for the transmission and reception of partially polarized waves are derived which deserve serious studies. The reader is referred to either the Journals "Radiotekhnika i elektronika" (Russian) or its translated versions in the concurrent issues of the "Soviet Journal of Communications Technology and Electronics (formerly: Radio Engineering and Electronic Physics)", published with Scripta Technical Inc., (Wiley), for forthcoming papers on continuing advances made by the author.

## 6. ACKNOWLEDGEMENT

The author wishes to express his sincere gratitude to Professor Boernerov, Wolfgang-Martinovich for assisting in the knowledge transfer of the author's papers, for accomodating this paper in these Proceedings of the NATO- ARW-DIMRP'88 as a post-workshop contribution, and for assiting on the updating of an earlier version of this paper [5]. Here, also, the author wishes to thank him for enabling him to meet and interact with Dr. Jean Richard Huynen to whose original work the last referenced paper [13] was dedicated.

## 7. REFERENCES

[1] Shurcliff, W.A. Polarized Light (Russian translation), Moscow: Mir Press, 1965 (English: ibid, Cambridge, MA: Harvard Press, 1962).

[2] Kanareykin, D.B., N.F. Pavlov and V.A. Potekhin, Polyarizatsiya radiolokatsionnykh signalov (Polarization of Radar Signals). Moscow: Sovetskoye Radio Press, 1966 (English: Radar Polarization Effects, New York: CCM Inf. Corp., G. Collier & McMillan, 900 Third Ave., New York, NY, 10023, 1972).

[3] Zhivotovskiy, L.A. Radiotekhnika i Elektronika, 17, No.10, pp.2184- , 1972 (English: Radio Eng. and Electronic Physics, Vol.17, No. 10, pp. 1972).

[4] Zhivotovskiy, L.A., Optimum Polarization of Radar Signal, Izv. VUZ., Radioelectronika, Vol.16, No. 12, pp. 48-50, 1973 (English: Radio Engineering and Electronic Physics, Vol.18, No.4, pp.630-632, 1973).

[5] Zhivotovskiy, L.A., "A modification of the polarization sphere for representing partially polarized electromagnetic waves", Radiotek-nika i Elektronika, Vol.30, No.8, pp. 1497-1504, 1985 (English: Soviet Journal of Communications Technology & Electronics (formerly: Radio Engineering & Electronic Physics), Vol.30, No.12, pp.20-28, 1985).

[6] Chandrasekhar, S., Radiant Energy Transfer (Russian translation), Moscow, IIL Press, 1953 (English: ibid, Osford, England: Oxford Press (1950): New York: Dover 1960).

[7] Efimov, N.V. and E.R. Rozendorn, Lineynaya algebra i mnogomernaya geometriya (Linear Algebra and Multidimensional Geometry), Moscow, Fizmatgiz Press, 1970.

[8] Ko, H.C., "On the reception of quasi-monochromatic, partially polarized radio waves", Proc. IRE, 50, No.9, pp. 1950-1957, 1962.

[9] Zhivotovskiy, L.A., "Calculation of the depolarizing properties of radar targets", Radiotekhnika, Vol.31, Part 2, pp.46-48, 1976 (English translation: Radio Egnineering & Electronic Physics, Part 2, Vol.31, pp.46-48, 1976).

[10] Zhivotovskiy, L.A., "Dependence of the power flux density of echo signals on radar signal polarization", (English translation: Radio Engineering, Part 2, Vol.31, pp.49-52).

[11] Zhivotovskiy, L.A., "Some features of the depolarizing properties of radar targets", (English translation: Soviet Journal of Communications TEchnology and Electronics (formerly Radio Engineering & Electronic Physics), Vol.33/34, No.10, pp.110-112, 1980).

[12] Zhivotovskiy, L.A., "Intrinsic bases and polarization protraits of stable and fluctuating radar targets", Radioteknika i Elektronika, Vol.33, No.2, pp.326-330, 1988, (English translation: Soviet Journal of Communications Technology and Electronics (formerly: Radio Engineering and Electronic Physics), Vol.33, No.4, pp.128-132, 1988).

[13] Zhivotovskiy, L.A., "Decomposition of fluctuating radar targets in antenna space", Radioteknika i Elektronika, Vol.33, No.6, pp.1186-1191, 1988, (English: Soviet Journal of Communications Technology and Electronics, Vol.34, No.2, pp.38-43, 1989).

# WEIBULL DISTRIBUTION IN RADAR POLARIMETRY

Matsuo SEKINE
Department of Applied Electronics
Tokyo Institute of Technology
4259, Nagatsuta, Midori-ku, Yokohama,
Japan

ABSTRACT. Most of the clutter received by a pulse radar obeys Weibull distribution with various values of the shape parameters. Rice crop field was measured using an FM-CW radar with a frequency of 9.6 GHz, frequency excursion 200 MHz, modulation frequency 850 Hz, horizontal and vertical polarizations, transmitter power 13 dBm and horn antenna. We obtained the following polarization effects of the backscattered amplitude from rice crop field:

Pol. VV: Weibull distribution
Pol. VH: Weibull distribution
Pol. HH: Log-normal distribution
Pol. HV: Weibull distribution.

Our results are the same as Olin's results on sea clutter for VV and HH polarizations.

## 1. INTRODUCTION

Recently there has been a continuing great interest in various Weibull-distributed clutter, such as ground, sea and weather clutter. In regard to ground clutter, Weibull distribution was reported for various radar resolutions, wavebands and land types(1). For example, ground clutter amplitude from the Rocky mountains(2-3) obeys a Weibull distribution with the shape parameter of 0.512 for S-band, a 2.0 μs pulsewidth, and a $1.5^{\circ}$ beamwidth. The data were taken for range less 32 km from an unmasked location. From grassland, low rolling wooded hills, wooded mountains, and some man-made structures(4), a Weibull distribution with the shape parameter of o.626 was observed from L-band radar having a pulsewidth of 3.0 μs, a beamwidth of $1.7^{\circ}$, and a depression angle of about $0.5^{\circ}$. The data were taken near Huntsville, Alabama and the measurements were made out to a maximum range of approximately 40 km.



W.-M. Boerner et al. (eds.), Direct and Inverse Methods in Radar Polarimetry, Part 1, 977–988.

Data taken at X-band by Linell(5) at the Research Institute of National Defence in Sweden, also obeys a Weibull distribution with the shape parameter of 0.506 to 0.531 for forest at a depression angle of 0.7°, and 0.606 to 2.0 for cultivated land at depression angles of 1.25°, 2.5° and 5.0°. The resolution cell is defined by a pulsewidth of 0.17 μs and a beamwidth of 1.4°. The data were taken with a rotating antenna atop a water tower with the height of approximately 30 m in different seasons, that is, March, April, May, August and November. The shape parameters were slightly different for different seasons. It was also concluded that as the depression angle increases, the shape parameters of the Weibull distribution increase and approach the Rayleigh distribution.

Recently ground clutter data of cultivated land has been measured by the present author, using a high-powered low-resolution L-band long-range air-route surveillance radar(ARSR) having a 3.0 μs pulsewidth and a 1.23° beamwidth at very low grazing angles between 0.21° and 0.32° (6). It was discovered that the ground clutter amplitudes obey the Weibull distribution with the shape parameters of 1.5 to 2.0, corresponding to the Rayleigh distribution.

Miller(7) studied the hill clutter and town clutter with a non-coherent radar. The hill clutter consists of part of the range of Cotswold Hills to the south of Malvern, England. The town clutter approximately covers the center of Worcester, which lies to the north west of Malvern. The distribution for the hill clutter is roughly Weibull, though the tail is more log-normal. For the town clutter, the entire distribution is approximately log-normal, though the fit is not good. Other, large samples of clutter have shown the tails to be consistently log-normal distributed, with the distribution itself tendind to be closen to the Weibull distribution.

Savchenko et al.(8) measured the Weibull-distributed grass and forest clutter using a coherent X-band radar with a wavelength of 3.2 cm, antenna beamwidth 1.2°, vertical beamwidth 20° , pulsewidths 50 ns and 200 ns, pulse repetition frequencies 4.4 kHz and 2.1 kHz, and horizontal linear polarization. For grass clutter with the wind velocity of 5 m/s, they obtained the large shape parameters, that is, 10.7 for pulsewidth 50 ns and 10.9 for 200 ns. Thus the variation of the pulsewidth changed slightly the shape parameter. On the other hand, for a strong wind of 7 m/s, the larger shape parameter of 14.96 was obtained for grass clutter using a pulsewidth of 200 ns. This means that the stronger wind broadens the histogram of the Weibull distribution, because of the comparatively greater contribution of the fluctuating part of the signal. Also they obtained the shape parameter of 4.6 for forest clutter with the wind velocity of 7 m/s using a pulsewidth of 200 ns.

Sea clutter is somewhat different from ground clutter. In general, the backscatter coefficient of sea waves is smaller than that of ground terrain. It increases with grazing angle, radar frequency and sea state. It is also greater for vertical polarization than horizontal polarization and is a maximum value for upwind, minimum value for downwind and intermediate value for crosswind.

It has been long believed that sea clutter amplitude statistics obey a Rayleigh distribution. However, in the last decade, because of rapid advances in radar technology, non-Rayleigh sea clutter has been observed with relatively high resolution radars. For example, Schleher proposed the Weibull clutter model to describe sea clutter data measured at the Applied Physics Laboratory(APL) of the John Hopkins University using a $K_u$-band airborne radar of horizontal polarization, pulsewidth 0.1 μs and grazing angles between $1^\circ$ and $30^\circ$(9). The increasing shape parameter from 1.16 to 1.78 with increasing grazing angle from $1^\circ$ to $30^\circ$ indicates that the distribution is approaching a Rayleigh distribution.

Weibull-distributed sea clutter was also reported by Clarke and Peters(10) and Fay, Clarke and Peters(11). The data were taken by Bishop at the Royal Signals and Radar Establishment(RSRE) at X-band, pulsewidths 0.07 and 0.27μs (either vertical or horizontal polarization), azimuth beamwidths of $0.6^\circ$ and $1.2^\circ$ and depression angles of $1^\circ$ and $1.5^\circ$(12). The instrumentation is sited on a cliff overlooking the ocean.

Weibull-distributed sea clutter data also has been measured by the present author(13) using a fixed antenna of an L-band ARSR at low grazing angles between $0.50^\circ$ and $0.72^\circ$. We used the horizontal linear polarization. Radar echoes were taken from sea state 3 in a range interval of 23-28.6 n miles at a fixed azimuth angle. It was shown that the sea clutter amplitude statistics obey a Weibull distribution with a shape parameter of 1.585.

The author(14) has also reported the log-Weibull distributed sea clutter using a rotating antenna of an L-band long-range ARSR at very low grazing angles between $0.13^\circ$ and $0.25^\circ$.

Weibull-distributed weather clutter was also observed using an L-band ARSR(15-17) in a range interval of 42 to 47.6 n miles and 60 to 65.6 n miles. It was concluded that Weibull-distributed weather clutter from rain clouds also obey a Weibull distribution with the shape parameter of 1.2 to 2.0.

Thus, Weibull-distributed ground, sea and weather clutter was observed using a pulse radar. In the following, we will see that the amplitude statistics of rice crop field using an FM-CW radar obey a Weibull distribution for VV, VH and HV polarizations, and a log-normal distribution for HH polarization.

## 2. OBSERVATIONS OF RICE CROP FIELD

For purposes of vegetation inventory, the measurement of the radar backscatter coefficient is very useful. However, the backscatter properties of vegetation vary with wavelength, grazing angle and polarization. For example, Bush and Ulaby(18) investigated the fading characteristics of backscattered radar signals from four agricultural targets, such as row crops(corn and soybeans), continuous canopy(alfalfa) and bare ground. They used an FM-CW radar with the frequency of 9 GHz. Comparison with theory of Rayleigh fading backscatter showed adequate agreement with the experimental results provided crop type, soil moisture condition and incident angle are correctly incorporated in the target model. In order that the distribution of the fading amplitude of radar backscatter from vegetation can be closely approximated by a Rayleigh distribution, the target must consist of many random and independent scatters. However, for agricultural vegetation, such as corn, soybeans and bare cultivated fields, the distribution deviates from the Rayleigh assumption, since they are normally prepared and planted in a very ordely fashion by the farmer(18).

Kurosu et al. (19-22) measured ordely planted rice crop field using an FM-CW radar having a frequency of 9.6 GHz, frequency excursion 200 MHz, modulation frequency 850 Hz, horizontal and vertical polarizations, transmitter power 13 dBm, and horn antenna(transmit and recive). The rice crop intervals between two adjacent range bins, and between two adjacent range sweeps in the azimuth direction, were 25 cm and 21 cm, respectively. An antenna was mounted at the height of 3.8 m above a head of rice crops. This horn antenna with the beamwidth $11^{\circ}$ was moved with the velocity of approximately 7.1 m/s during the measurement. In this case, the radar illuminates the rice crop field over the range interval of 1.5 m with an azimuth interval of 1.0 m. The rice crop field data also includes bare ground.

Data was recorded on magnetic tape after passing through antenna, high pass filter with cutoff frequency of 6 kHz, band pass filter with bandwidth 3 kHz, envelop detection and logarithmic amplifier. Data was converted from analog to digital with sampling frequency of 100 Hz. To determine the probability density function, the number of data points of 2995 was used. Kurosu et al. (22) showed the backscattered power data for VV, VH, HH and HV polarizations. From these data, we converted the rice crop field data from the power to the amplitude. Next we will show that the backscattered amplitude statistics obey a Weibull distribution for VV, VH and HV polarizations, and a log-normal distribution for HH polarization.

## 3. WEIBULL AND LOG-NORMAL DISTRIBUTIONS

The Weibull probability density function is written as follows:

$$p_c(x) = \begin{cases} \frac{c}{b}\left(\frac{x}{b}\right)^{c-1} e^{-\left(\frac{x}{b}\right)^c} & \text{for } x>0,\ b>0 \text{ and } c>0 \\ 0 & \text{otherwise.} \end{cases} \tag{1}$$

Here x is the amplitude of the return signals, b is a scale parameter and c is a shape parameter. For c=2.0, the Weibull distribution is identical to the Rayleigh distribution. Eq. (1) is easily integrated to obtain

$$Y = cX - c\ \ln(b) \tag{2}$$

where

$$\left.\begin{aligned} Y &= \ln\left[-\ln\{1 - \int_0^x p_c(x)\,dx\}\right] \\ &\text{and} \\ X &= \ln(x). \end{aligned}\right\} \tag{3}$$

From Eq. (2), the shape parameter c is easily estimated from a plot of Y against X. One example of HV polarization is shown in Fig. 1.

A straight line was fitted to the values of Y and X by the least squares method. If the data follow a Weibull distribution, they lie on a straight line in this representation, and the slope gives the shape parameter. The root mean square error(R.M.S.E) is the deviation of the data points from the straight line drawn by the least squares methods.

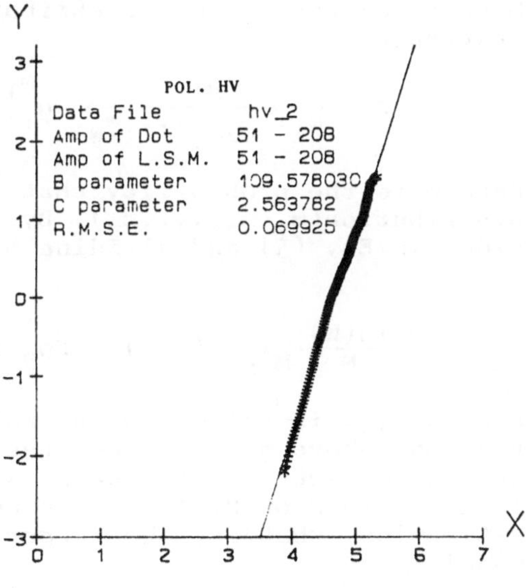


Fig. 1. Determination of the shape parameter c from rice crop field data for HV polarization. c=2.56 and R.M.S.E=0.07.

However, if the amplitude obeys a log-normal distribution, the data deviate slightly from this straight line. The log-normal distribution is defined as

$$p(x) = \frac{1}{\sqrt{2\pi}\ \sigma x}\ e^{-\frac{1}{2\sigma^2}\left(\ln\frac{x}{x_m}\right)^2} \tag{4}$$

where x is the amplitude of the radar return signals, $x_m$ is the median value of x and σ is the standard deviation of $\ln(x/x_m)$.

For rigorous fit of the distribution to the data, next we will consider the Akaike Information Criterion.

## 4. AKAIKE INFORMATION CRITERION

By considering two or more hypothetical probability distribution models, we can have a criterion which the model gives an optimum fit to the data. This is called Akaike Information Criterion which is abbreviated by AIC(23).

The AIC is explained as follows:
First we will assume that the true probability distribution $(p_1,p_2,...,p_N)$ is known. Here $p_N$ is the probability which the Nth event occurs. Next we shall consider sufficiently large number of trials. Then the Nth event will occur approximately $m_N=Mp_N$ times. As a model, we assume the probability distribution$(q_1,q_2,...,q_N)$. By observing the M samples obeying this distribution, the probability W is written as

$$W = \frac{M!}{m_1!,m_2!,...,m_N!}\ q_1^{m_1}\ q_2^{m_2}\cdots\ q_N^{m_N}. \tag{5}$$

Here W is the probability that we obtain the probability distribution$(m_1,m_2,...,m_N)$. By taking a logarithm of both sides in Eq. (5) and dividing by M, we obtain

$$\frac{\ln(W)}{M}\ \underset{M\to\infty}{\to}\ B(p,q) = \Sigma p_N\ \ln(\frac{q_N}{p_N}). \tag{6}$$

Here B(p,q) is called as the Kullback-Leibter's entropy(24). From the above discussions, the probability that the predicted distribution realizes becomes large with the larger values of B. In this sence, B is used as a model estimation, that is, the larger values of B mean a good model.

The Kullback-Leibler entropy is rewritten as

$$B = \Sigma p_N \ln(q_N) - \Sigma p_N \ln(p_N). \tag{7}$$

The second term on the right side depends only on a true distribution. Therefore, only the first term plays an important role in estimating the model. This term is interpreted as an expected value of $\ln(q_N)$. Therefore, the first term is estimated from the M numbers of the observed values $x_1, x_2, \ldots, x_M$. Then the logarithmic likelihood L is defined as

$$L = \Sigma \ln \{f(x_N)\}, \quad f(x_k) = q_N \quad \text{for } x_k = N. \tag{8}$$

Here a function f(x) is a probability that the observed values are x and depends on the model. The larger L is the better model. Now we assume that the probability density function f(x) has parameters θ. Then we can write the probability density function model as f(x:θ) for a stochastic variable x and parameters θ. Ifθ have k numbers of parameter, then θ are k-dimensional vectors. In this case, the logarithmic likelihood L(θ) is defined as

$$L(\theta) = \Sigma \ln\{f(x:\theta)\}. \tag{9}$$

Equation (8) is determined from the observed values. However, if $f^*(x)$ is a true probability density function, then the true logarithmic likelihood is written as

$$L^*(\theta) = M \int f^*(x) \ln \{f(x:\theta)\} dx \tag{10}$$

Usually, $L^*(\theta)$ cannot be calculated, as long as the true probability density function is not known. However, it is well-known that $L(\theta_0) - k$ is an unbiased estimation of the logarithmic likelihood $L^*(\theta_0)$. Here $\theta_0$ is the maximum likelihood estimation to obtain the largest $L^*(\theta)$. Therefore, finally, the AIC for a given model is defined as

$$\begin{aligned} AIC = -2[&\{\text{maximum logarithmic likekihood}\} \\ &-\{\text{the number of parameters included in the model}\}] = -2\{L(\theta_0) - k\} \end{aligned} \tag{11}$$

The model which yields the smallest AIC is regarded as the best one.

## 5. DETERMINATION OF THE OPTIMUM PROBABILITY FUNCTION FOR THE OBSERVED DATA

The results are shwon in Figs. 2, 3, 4 and 5.

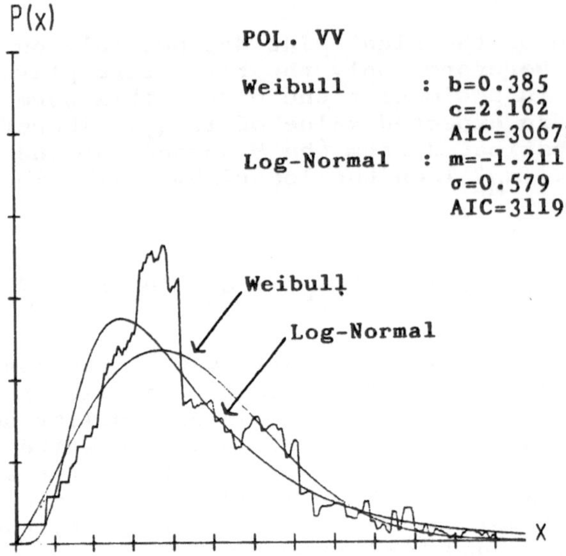


Fig.2. Weibull is an optimum distribution for VV polarization.

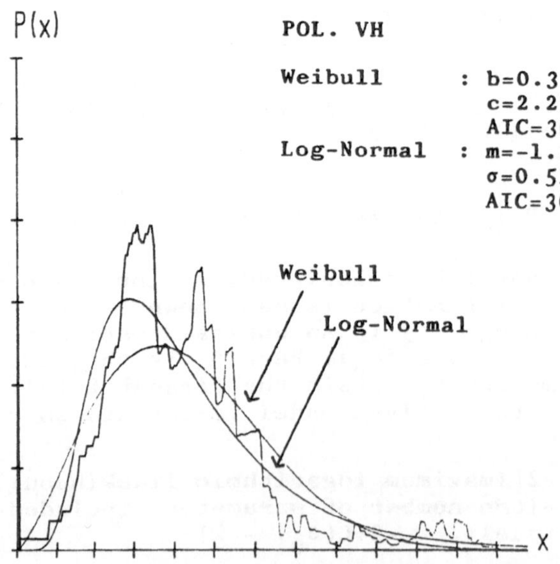


Fig.3. Weibull ia an optimum distribution for VH polarization.

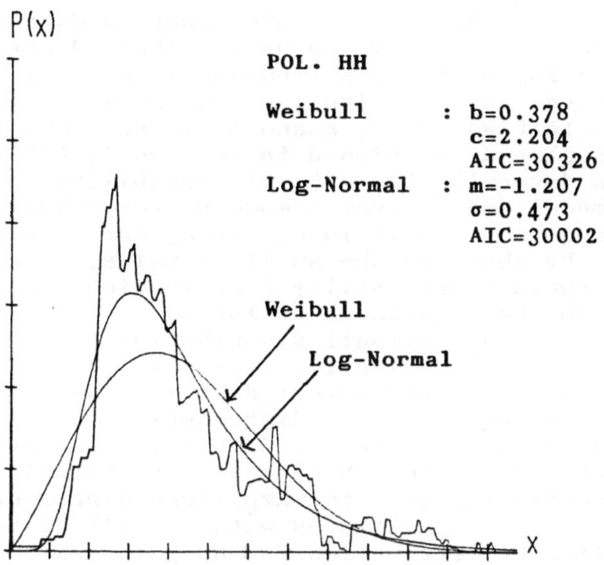


Fig.4. Log-normal is an optimum distribution for HH polarization.

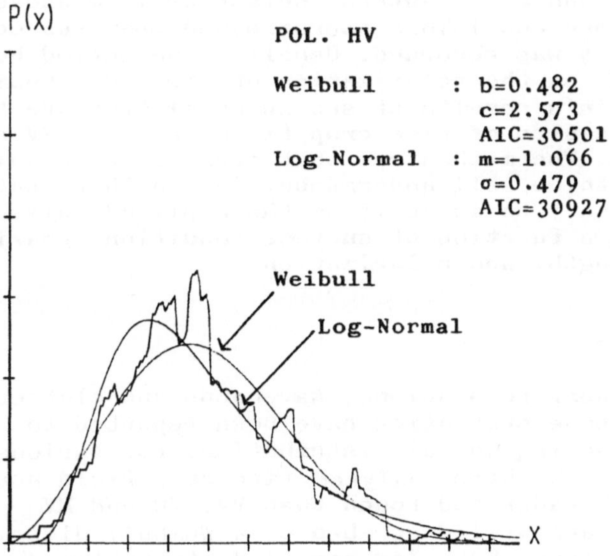


Fig.5. Weibull is an optimum distribution for HV polarization.

As shown in Figs. 2, 3, 4 and 5, the observed data of rice crop field was approximated by a Weibull distribution in Eq. (1) or a log-normal distribution in Eq. (4), since many radar clutter obey a Weibull or log-normal distributions. In Figs. 2, 3, 4 and 5, m means $m = \ln x_m$ in Eq. (4). As already mensioned in Section 4, AIC is very useful in determing the optimal probability distribution model among several models, for example, Weibull or log-normal models in our case, for fitting observed data. By choosing the smallest value, we can determine the optimal probability distribution model. Now we can summarize the results as follows:

VV. Pol: Weibull distribution
VH. Pol: Weibull distribution
HH. Pol: Log-normal distribution
HV. Pol: Weibull distribution

Olin(25) analysed the sea clutter data collected by Hansen and Cavaleri at Naval Research Laboratory(NRL)(26) and obtained the conclusion that the amplitude distribution for VV polarization can be fitted with Weibull distribution, while the amplitude distribution for HH polarization can be fitted with log-normal distribution. They measured sea spike clutter using a high resolution(40 ns x $1^{\circ}$) X-band radar. The radar was installed on a platform site about 16 km offshore from Panama city, Florida at a water depth of about 30 m and at an antenna height of 15 m above mean sea level. These conditions approximated open sea conditions. The sea state 5 was recorded. Usually, the period between spikes depends on the polarization of the radar beam.

Thus, Olin's results of sea spike clutter are the same as our results of rice crop field data for VV and HH polarizations. The mechanisms that cause thses results continue not to be well understood. For further study, it will be necessary to investigate the amplitude distributions of clutter as a function of surface condition, grazing angle, wavelength, and polarization.

## 6. CONCLUSIONS

For various radar resolutions, wavebands and clutter, clutter amplitude statistics have been reported to obey the Weibull and log-normal distributions for various grazing angles. We have analysed rice crop field statistics from an FM-CW radar and found that VV, VH and HV polarizations are well described by a Weibull distribution and HH polarization by a log-normal distribution.

## REFERENCES


(1) Boothe, R: "The Weibull distribution applied to the ground clutter backscatter coefficient", U.S.Army Missile Command, Technical Report RE-TR-69-15, ASTIA Document AD691109, June 1969.

(2) Barton, D. K: "Target detection in land clutter", Raytheon Company, Wayland, MA, Internal Memorandum, Nov. 30, 1966.

(3) Nathanson, F. E: "Sea and land backscatter", Radar training program notes, The John Hopkins University Applied Physics Laboratory, ch. 7.

(4) Holliday, E. M., Wood, W. E., Powell, D. E., and Basham, C. E: "L-band clutter measurements", U.S.Army Missile Command Report RE-TR-65-1, Nov. 3, 1964.

(5) Linell, T: "An experimental investigation of the amplitude distribution of radar terrain return", Institute of National Defence, Stockholm, Sweden, Report D 313<sup>-</sup>-62, Oct. 1966.

(6) Sekine, M.et al.: "Weibull-distributed ground clutter", IEEE Transaction on Aerospace and Electronic Systems, <u>AES-17</u>, No. 6, pp. 596-598, July 1981.

(7) Miller, R: "Characterisation of noncoherent ground clutter", Proceedings of the 1984 International Symposium on Noise and Clutter Rejection in Radars and Imaging Sensors, October 22-24, Tokyo, Japan, pp. 59-64, 1964.

(8) Savchenko, A. K., Haimov, S. J. and Kulemin, G. P: "On the experimental study of radar backscattering from land ", Proceeding of the 18th European Microwave Conference, Stockholm, Sweden, 12-16 September, 1988, pp. 705-709, 1988.

(9) Schleher, D. C: "Radar detection in Weibull clutter", IEEE Transaction on Aerospace and Electronic Systems, <u>AES-12</u>, pp. 736-743, 1976.

(10) Clarke, J. and Peters, R.S.: "The effect of pulse length changes on Weibull clutter", Royal Establishment Memorandum 3033, 1976.

(11) Fay, F. A., Clarke, J., and Peters, R. S.: "Weibull distribution applied to sea clutter" in "Radar - 77", IEE Conf. Pub. 105, pp. 101-104, 1977.

(12) Bishop, G.: "Amplitude distribution characteristics of X-band radar sea clutter and small surface targets", Royal Radar Establishment Memorandum 2348, 1976.

(13) Sekine, M. : "Weibull-distributed sea clutter", IEE Proceedings, <u>130</u>, Pt. F, No. 5, p. 476, August 1983.

(14) Sekine, M. et al. : "Log-Weibull distributed sea clutter", IEE Proceedings, <u>127</u>, Pt. F, No. 3, pp. 225-228, June 1980.

(15) Sekine, M. et al.: "Suppression of Weibull-distributed weather clutter", IEEE International Radar Conference, pp. 294-298, 1980.

(16) Sekine, M. et al.: "Non-Rayleigh weather clutter", IEE Proceedings, 127, Pt. F, No. 6, pp. 471-474, December 1980.

(17) Sekine, M. et al.: "Suppression of ground and weather clutter", IEE Proceedings, 128, Pt. F, No. 3, pp. 175-178, June 1981.

(18) Bush, T. F. and Ulaby, F. T.: "Fading characteristics of panchromatic radar backscatter from selected agricultural targets", IEEE Trans. Geosci. Electron., GE-13, 4, pp. 149-157, 1975.

(19) Kurosu, T., Suitz, T., Kozu, T. and Umehara, T.: "An analysis of microwave fading data backscattered by rice crop in the X band FM-CW radar measurement", SANE 87-35, pp. 1-8, 1987.

(20) Suitz, T., Yoshikado, S., Kurosu, T., Kozu, T. and Umehara, T.: "Backscattering coefficient of rice crops and rice fields by an X-band scatterometer", 21st Int. Symp. Remote Sensing of Environ., Ann Arbor, Michigan, pp. 767-773, 1987.

(21) Kurosu, T., Suitz, T. and Umehara, T.: "Measurement of agricultural plant's size by analysis of fadings of microwave backscatter data", SANE 88-8, pp. 23-30, 1988.

(22) Kurosu, T., Suitz, T., Kozu, T. and Umehara, T.: "A method of analysis of fading of microwave backscatter data from rice crop", The Transactions of the Institute of Electronics, Information and Communication Engineers, Vol. J71-B, No.7, pp. 882-890, July 1988.

(23) Akaike, H.: "Information theory and an extension of the maximum likelihood principle", 2nd International Symposium on Information Theory edited by B.N.Petrov and F.Csaki, Akademiai Kiado, Budapest, pp. 26-281, 1973.

(24) Kullback, S. and Leibler, R. A.: "On information and sufficiency", Ann. Math. Statist., vol. 22, pp. 79-86, 1953.

(25) Olin, I. D.: "Amplitude and temporal statistics of sea spike clutter", Radar-82, IEE Conference Publication Number 216, pp. 198-202, 1982.

(26) Hansen, J. P. and Cavaleri, V. F.: "High-resolution radar sea scatter experimental observations and discriminants", NRL Report 8557, 1982.

# ON ADAPTIVE RADAR POLARIMETRY: CONCEPTS OF IMPLEMENTATION

LEONID B. PREISER
Department of Electrical Engineering
Northrop University
5800 W. Arbor Vitae Street
Los Angeles, California 90045
U.S.A.

ABSTRACT. Basic fundamentals of adaptive signal processing applicable to radar polarimetry when either the scattering properties of the target are unknown or they change in a real time have been examined. Suggested algorithms for a reference signal to be used in a process of adaptive evaluation are based on a generalized analysis of polarization states in a nonorthogonal basis. A calibration procedure of the transmit–receive radar system involving a one-to-one mapping from a sample transmit space to a sample receive space, as well as a real-time estimation technique and adaptive refinement in radar polarimetry are discussed.

## 1. INTRODUCTION

There are numerous applications of scattering matrix analysis in radar polarimetry, centered around backscatter modeling, target discrimination, contrast optimization, polarization detection, optimum polarization of radar signals, and based on an assumption of known, measured, predicted or approximated structure of scattering matrix [1]-[5].

On some occasions the scattering signature of the target either is not known or it is synthesized from the particular experimental setup or theoretical prediction and, being convenient for interesting and diversified practical and academic analysis, precludes, however, from accounting for "coherent" perturbations caused by the effects of propagation, reflection and scattering subsequent to scattering by the target of interest. Yet in other cases, the target signature might change in a real time.

Seems reasonable to assume that implementation of some adaptive methods in radar polarimetry could possibly bridge this and similar gaps and supplement the existing techniques. The main purpose of this paper is an introduction to the concepts and ways of implementation of this approach.

Section 2 introduces a generalized analysis of polarization states. In particular, the nonorthogonal basis appears to be more flexible and convenient for the purposes of adaptive estimation. A new polarization descriptor is introduced in section 3. Section 4 examines some fundamentals of adaptive signal processing applicable to our task and analyses the relevant search methods, adaptive algorithms and the quadratic performance surface of an adaptive finite impulse response filter.

The techniques of a real-time adaptive estimation and ways of its implementation in radar polarimetry are discussed in Section 5. Section 6 examines an adaptive refinement (interference canceling) in radar polarimetry.

## 2. GENERALIZED ANALYSIS OF POLARIZATION STATES (NONORTHOGONAL BASIS)

By polarization of a monochromatic plane wave, we mean, as it is commonly accepted, the curve traced out at a fixed point in space, as time $t$ varies, by the tip of the instantaneous electric field intensity $\mathbf{F}(t)$ in the form

$$\mathbf{F}(t) = \mathbf{F}e^{j(\omega t - \mathbf{k}\cdot\mathbf{r})} = \mathbf{E}e^{j\omega t} \tag{1}$$

where the complex vector $\mathbf{E}$ depends on the position vector $\mathbf{r}$

$$\mathbf{E} = \mathbf{E}_0 e^{j\mathbf{k}\cdot\mathbf{r}} \tag{2}$$

In (2), $\mathbf{E}_0$ is a complex-constant amplitude vector, independent of $\mathbf{r}$, and $\mathbf{k}$ is a wave vector. Rewriting the complex vector $\mathbf{E}_0$ as

$$\mathbf{E}_0 = \mathbf{E}_1 + j\mathbf{E}_2 \tag{3}$$



*W.-M. Boerner et al. (eds.), Direct and Inverse Methods in Radar Polarimetry, Part 1*, 989–998.

where $\mathbf{E}_1$ and $\mathbf{E}_2$ are the real vectors, and substituting (3) into (1), we obtain the instantaneous electric field intensity (IEFI)

$$\mathbf{F}(t) = (cos\varphi \mathbf{E}_1 - sin\varphi \mathbf{E}_2) + j(cos\varphi \mathbf{E}_2 + sin\varphi \mathbf{E}_1) \tag{4}$$
$$\varphi = wt - \mathbf{k}\cdot\mathbf{r}$$

In what follows, a real part of (4) will be considered as an IEFI:

$$\mathbf{F}(t) = cos\varphi \mathbf{E}_1 - sin\varphi \mathbf{E}_2 \tag{4a}$$

From (4a), the square of the amplitude (by using the identities $cos^2\varphi = (1 + cos2\varphi)/2$, $sin^2\varphi = (1 - cos2\varphi)/2$

$$|\mathbf{F}(t)|^2 = \frac{1}{2}(\mathbf{E}_1^2 + \mathbf{E}_2^2) + \frac{1}{2}(\mathbf{E}_1^2 - \mathbf{E}_2^2)cos2\varphi - \mathbf{E}_1 \cdot \mathbf{E}_2 sin2\varphi \tag{5}$$

As it is shown in [6]-[7], the equation for the curve obtained from (4a)

$$(\mathbf{F} \times \mathbf{E}_1)^2 + (\mathbf{F} \times \mathbf{E}_2)^2 = (\mathbf{E}_1 \times \mathbf{E}_2)^2 \tag{6}$$

represents a polarization ellipse.
Actually, since the vector $\mathbf{F}$ lies in the plane formed by $\mathbf{E}_1$ and $\mathbf{E}_2$, it can be represented as a linear combination of these vectors

$$\mathbf{F} = \alpha \mathbf{E}_1 + \beta \mathbf{E}_2 \tag{7}$$

where $\alpha$ and $\beta$ are two arbitrary constants.
Substituting (7) into (6), we also obtain a limiting condition

$$\alpha^2 + \beta^2 = 1. \tag{8}$$

By employing the method of the Lagrange multiplier [8] in a problem of determining the extrema of the function

$$f(\alpha,\beta) = F^2 - \lambda(\alpha^2 + \beta^2 - 1) = \alpha^2\mathbf{E}_1^2 + \beta^2\mathbf{E}_2^2 + 2\alpha\beta\mathbf{E}_1 \cdot \mathbf{E}_2 - \lambda(\alpha^2 + \beta^2 - 1) \tag{8a}$$

where $\lambda$ is a parameter to be determined, and applying the conditions

$$\frac{\partial f}{\partial \alpha} = \frac{\partial f}{\partial \beta} = 0 \tag{8b}$$

in a search for the extrema of $F^2$ (subject to the constraint (8)), it can be shown that the major and minor axes of the polarization ellipse are

$$F_{max}^2 = \frac{1}{2}\left[(\mathbf{E}_1^2 + \mathbf{E}_2^2) + \sqrt{(\mathbf{E}_1^2 + \mathbf{E}_2^2) - 4(\mathbf{E}_1 \times \mathbf{E}_2)^2}\right] \tag{9}$$

and

$$F_{min}^2 = \frac{1}{2}\left[(\mathbf{E}_1^2 + \mathbf{E}_2^2) - \sqrt{(\mathbf{E}_1^2 + \mathbf{E}_2^2) - 4(\mathbf{E}_1 \times \mathbf{E}_2)^2}\right] \tag{10}$$

In particular, as it follows from (9) and (10),

$$F_{max}^2 + F_{min}^2 = \mathbf{E}_1^2 + \mathbf{E}_2^2 \tag{11}$$
$$F_{max}^2 F_{min}^2 = (\mathbf{E}_1 \times \mathbf{E}_2)^2 \tag{12}$$

On the other hand, two linear equations in $\alpha$ and $\beta$ resulting from applying conditions (8b) to (8a), and a linear presentation (7) yield the directions of the major and minor axes of the polarization ellipse:

$$\mathbf{F}_{max} \parallel [\mathbf{E}_2 \times (\mathbf{E}_1 \times \mathbf{E}_2) - F_{max}^2 \mathbf{E}_1] \tag{13}$$
$$\mathbf{F}_{min} \parallel [\mathbf{E}_1 \times (\mathbf{E}_2 \times \mathbf{E}_1) - F_{min}^2 \mathbf{E}_2] \tag{14}$$

At this point, it should be emphasized once more that 1) the directions of oscillations of two linearly polarized waves $\mathbf{E}_1$ and $\mathbf{E}_2$ may be arbitrary (not necessarily mutually perpendicular), and 2) wave of any (elliptical, circular, or linear) polarization may be considered as a sum of these two waves.

## 3. POLARIZATION DESCRIPTOR

It seems reasonable and justifiable to introduce the general (not necessarily orthogonal) basis for the polarization descriptor.
The rationale and justification of this approach would include the following.
1) Better match between the set of nonorthogonal eigenvectors of the scattering matrix and the transmitting (incident on the target) field.
2) In general case of bistatic scattering when the receiving antenna is different from the transmitting one, bilinear form

$$\mathbf{Y}^T C\mathbf{X}$$

does not feature one of very important properties of the quadratic form

$$\mathbf{X}^T C\mathbf{X} = \mathbf{X}^T(S+K)\mathbf{X} = \mathbf{X}^T S\mathbf{X} + \mathbf{X}^T K\mathbf{X}$$
$$\mathbf{X}^T K\mathbf{X} = 0;$$

thus, only the <u>symmetric</u> (or <u>Hermitian</u>) component of C contributes to the value of the quadratic form.
The new polarization descriptor would be a two-dimensional column vector

$$\mathbf{P} = \begin{pmatrix}\mathbf{A}_1\\ \mathbf{A}_2\end{pmatrix},$$

where dependence on $t$ and $\mathbf{r}$ is suppressed. This vector represents a decomposition into two linearly polarized components with arbitrary directions of oscillations and with conjugation of one of them. In particular,

$$\mathbf{A}_1 = \mathbf{E}_1$$
$$\mathbf{A}_2 = \mathbf{E}_2 e^{j\pi/2}\frac{(\mathbf{E}_1\times\mathbf{E}_2)\cdot\hat{\mathbf{k}}}{|\mathbf{E}_1|\cdot|\mathbf{E}_2|}$$

where $\hat{\mathbf{k}}$ is a unit vector in the direction of propagation (transmitting mode), with sign minus in receiving mode. As it can easily be shown, for any polarization (elliptical, circular, linear), the right-handed polarization wave would be described by the inequality

$$\hat{\mathbf{k}}\cdot(\mathbf{E}_1\times\mathbf{E}_2) > 0$$

whereas the left-handed polarization by

$$\hat{\mathbf{k}}\cdot(\mathbf{E}_1\times\mathbf{E}_2) < 0.$$

## 4. SOME FUNDAMENTALS OF ADAPTIVE SIGNAL PROCESSING APPLICABLE TO RADAR POLARIMETRY

The general configuration of an adaptive system applicable to the radar polarimetry would include the transmitting and receiving systems.
The transmitting antenna system would consist of two linearly polarized sources with a total of five adjustable parameters: amplitudes $A_{ti}$, phases $\phi_{ti}, i = 1,2$ and a spatial angle $\psi_t$ between them. Similarly, at the receiving end a pair of two identical antennas with an adjustable angle $\psi_r$ betweeen them would form the signals $x_i(t), i = 1,2$.

The signal form from each element of this 2-element array, $x_i(t)$, has its magnitude and phase adjusted in the box labeled $c_i$ and is then added to produce the array output $S(t)$. The adjustments $c_i$ are controlled by a feedback system, which adjusts the $c_i$ to minimize the mean-square value of the error signal $\varepsilon(t)$, i.e. the difference between a reference signal $d(t)$ and the array output $S(t)$.

Let us assume, for example, that the basic purpose of the bistatic or monostatic adaptive radar system is to maximize target response, thus eliminating a combination of clutter, cross-polarization, post-scattering propagation effects, background noise, etc.; we will call this combination an interference signal.

Suppose a desired signal is being received, along with interference and background noise. If the reference signal $d(t)$ is made to equal the desired signal component of the array output, the error signal $\varepsilon(t)$ will then consist of the undesired components of the array output $S(t)$. Minimizing the error signal then corresponds to minimizing the antenna response to these undesired components.

At this point it is important to address the subject of reference signal generation. From section 2, the equations (9)-(14) relate the magnitudes of extrema as well as their spatial orientations to the parameters of the vectors $\mathbf{E}_1$ and $\mathbf{E}_2$. Assume for a moment that the preliminary calibration of the transmit-receive system has been completed, i.e. a one-to-one mapping of different combinations $A_{ti}, \phi_{ti}, \psi_t$ on the transmit side (sample transmit space STS) into a set of polarimetric reference signal structures $d_{\psi r}$ where parameter $\psi_r$ is a spatial angle between two linearly polarized receiving antennas (a sample receive space SRS).

It becomes evident that, based on such $STS-SRS$ mapping, different algorithms for a reference signal, to be used in a process of adaptive evaluation, can be generated.

One of possible ways to implement the amplitude and phase adjustments denoted by $c_i$ would be by splitting the signal from each element, $x_i(t)$ into two quadrature components with a quadrature hybrid:

$$x_i(t) = x_{ir}(t) + jx_{ij}(t)$$

The two outputs from the hybrid would be separately weighted with real coefficients $w_{ir}, w_{ij}$, whose values are under the control of the feedback loop.

Thus, the total array output is

$$S(t) = \sum_{i=1}^{2}[w_{ir}x_{ir}(t) + w_{ij}x_{ij}(t)] \tag{15}$$

The error signal is

$$\varepsilon(t) = d(t) - \sum_{i=1}^{2}[w_{ir}x_{ir}(t) + w_{ij}x_{ij}(t)] \tag{16}$$

If we let $x_1(t)$ and $x_2(t)$ represent the in-phase and quadrature components of the signal from the receiving antenna 1, $x_3(t)$ and $x_4(t)$ represent the corresponding components from antenna 2, then the input signal vector $\mathbf{X}(t)$ is defined as

$$\mathbf{X}(t) = [x_1(t)x_2(t)x_3(t)x_4(t)]^T \tag{17}$$

where $T$ denotes transpose.

Accordingly, the weighting coefficients or multiplying factors $w_i, i = 1-4$, which are adjustable, form the weight vector

$$\mathbf{W} = [w_1 w_2 w_3 w_4]^T \tag{18}$$

The output $S(t)$ is equal to the inner product of $\mathbf{X}(t)$ and $\mathbf{W}$:

$$S(t) = \mathbf{X}^T(t)\mathbf{W} = \mathbf{W}^T\mathbf{X}(t) \tag{19}$$

The error $\varepsilon(t)$ is defined as the difference

$$\varepsilon(t) = d(t) - \mathbf{X}^T(t)\mathbf{W} = d(t) - \mathbf{W}^T\mathbf{X}(t) \tag{20}$$

Our next objective is to see how the adaptive process adjusts the weights to minimize the mean square of the error $\varepsilon(t)$.

We assume for now that all input signals $x_i(t), i = 1-4$, are statistically stationary and have finite correlation functions (during adaptation, the weight factors $w_i$ vary, so that even with stationary inputs the output $S(t)$ and error $\varepsilon(t)$ will generally be nonstationary).

Assume also that the adaptive process progresses recursively or by iterative cycles, so that at the $j-th$ iteration the weight vector is $\mathbf{W}_j$. Squaring and expanding (20), letting $\mathbf{W} = \mathbf{W}_j$ and statistically averaging (with expected value operator $E$) yields

$$E[\varepsilon^2(t)]_{\mathbf{W}=\mathbf{W}_j} = E[d^2(t)] - 2E[d(t)\mathbf{X}^T(t)]\mathbf{W}_j + \mathbf{W}_j^T E[\mathbf{X}(t)\mathbf{X}^T(t)]\mathbf{W}_j \tag{21}$$

The mean-square error (MSE) (21) can also be expressed as [9]

$$E[\varepsilon^2(t)]_{\mathbf{W}=\mathbf{W}_j} = E[d^2(t)] - 2\mathbf{R}_{dx}\mathbf{W}_j + \mathbf{W}_j^T\mathbf{R}_{xx}\mathbf{W}_j \tag{22}$$

where the vector $\mathbf{R}_{dx}$ is a cross correlation between a reference signal (a scalar) and the input vector $\mathbf{X}$

$$\mathbf{R}_{dx} = E[d(t)\mathbf{X}^T(t)] = E[d(t)x_1(t)\ d(t)x_2(t)\dots d(t)x_4(t)]^T \tag{23}$$

and the input correlation matrix $\mathbf{R}_{xx}$ is

$$\mathbf{R}_{xx} = E[\mathbf{X}(t)\mathbf{X}^T(t)] = E\begin{bmatrix} x_1(t)x_1(t) & \dots & x_1(t)x_4(t) \\ x_2(t)x_1(t) & \dots & x_2(t)x_4(t) \\ x_3(t)x_1(t) & \dots & x_3(t)x_4(t) \\ x_4(t)x_1(t) & \dots & x_4(t)x_4(t) \end{bmatrix} \tag{24}$$

As it was mentioned before, we assume that the adaptive processor is a linear filter with a transfer function $H(z)$, which is adapted by adjusting weights. Assuming further that $H(z)$ represents a causal FIR filter with, say, $L$ adjustable weights $W_i$ [10],

$$H(z) = \sum_{i=0}^{L-1} w_i z^{-i} \tag{25}$$

Thus the dynamics of the adaptive process which progresses recursively or by iterative cycles are closely related to the dimension $L$ of the $H(z)$.
In other words, the number $L$ would give an indication on how many iterations (or how many weight vectors $\vec{W}_j, j = 1-L$) would be necessary to arrive to the MSE minimum.

Referring to the input correlation matrix $\mathbf{R}_{xx}$ (24) which is real, symmetric and positive definite [11], it appears that the MSE is a quadratic function of the weights and therefore, (22) is the general expression for the performance surface of a causal FIR adaptive filter.
Since the MSE is always positive, the performance surface, being quadratic, must be parabolic and "concave upward", that is, extending toward a positively increasing MSE. Again, with stationary signals, the performance surface is thus a parabolic "bowl" in $(L+1)$ dimensional space. As to some typical applications with nonstationary, slowly varying signal statistics, one would think of the bowl drifting in space as the signal properties change slowly with time, and of adaptation as the process of searching for and continuously tracking the bottom of the bowl.
When searching for the bottom of the bowl one finds generally that a knowledge or estimate of the gradient is useful [8]. The gradient vector is the column vector obtained by differentiating Eq.(22) with respect to the weight vector $\mathbf{W}$:

$$\begin{aligned} \nabla &= \left[\frac{\partial E[\varepsilon^2(t)]}{\partial w_0} \frac{\partial E[\varepsilon^2(t)]}{\partial w_1} \dots \frac{\partial E[\varepsilon^2(t)]}{\partial w_{L-1}}\right]^T \\ &= 2\mathbf{R}_{xx}\mathbf{R}_{dx} \end{aligned} \tag{26}$$

Since the bowl is quadratic, the global minimum MSE is obtained where $\nabla = 0$. Setting $\nabla = 0$ in (26) gives the optimum weight vector, $\mathbf{W}_{opt}$, as

$$\mathbf{W}_{opt} = \mathbf{R}_{xx}^{-1}\mathbf{R}_{dx} \tag{27}$$

which is the Wiener solution for the optimum weight vector [12] for the adaptive linear combiner.

Thus the corresponding minimum MSE is found by substituting Eq. (27) into Eq. (22):

$$\begin{aligned}(MSE)_{min} &= \mathbf{R}_{dd} + (\mathbf{R}_{xx}^{-1}\mathbf{R}_{dx})^T\mathbf{R}_{xx}\mathbf{W}_{opt} - 2\mathbf{R}_{dx}^T\mathbf{W}_{opt} \\ &= \mathbf{R}_{dd} - \mathbf{R}_{dx}^T\mathbf{W}_{opt}\end{aligned} \tag{28}$$

## 5. REAL-TIME ADAPTIVE ESTIMATION AND WAYS OF ITS IMPLEMENTATION IN RADAR POLARIMETRY

We will now put another limitation on statistical features of the signals, $x_i(t)$ i.e. we will assume that they are ergodic, so that ensemble (statistical) averages (expectation operator) can be replaced by time averages in evaluating the mean function, autocorrelation function, or any function of interest. For example, for an ergodic random process an alternative way to obtain the mean function of the random process $Y(t)$ is by averaging $Y(t)$ over an infinitely long time interval as

$$M_y = \lim_{T\to\infty}\frac{1}{2T}\int_{-T}^{T} Y(t)dt \tag{29}$$

and the autocorrelation function can be obtained by averaging the value of $Y(t)Y(t+\tau)dt$

$$R_y(\tau) = \lim_{T\to\infty}\frac{1}{2T}\int_{-T}^{T} Y(t)Y(t+\tau)dt \tag{30}$$

There should not be any doubts why in (30) we write $R_y(\tau)$ instead of $R_y(t_1, t_2)$: if a random process is ergodic it must be stationary. The converse of this is not necessarily true (that is why an assumption of ergodicity is wider sense than stationarity). Thus all correlation functions involved in our adaptive system can be measured from actual sample functions. Combining now (15), (16), (18) and replacing the statistical averaging (22) with the time-average value of $\varepsilon^2(t)$ yields

$$\overline{\varepsilon^2}(t) = \overline{d^2(t)} - 2\sum_{i-1}^{4} w_i\overline{x_i(t)d(t)} + \sum_{i-1}^{4}\sum_{j=1}^{4} w_i w_j \cdot \overline{x_i(t)x_j(t)}. \tag{31}$$

$\overline{\varepsilon^2(t)}$ is a quadratic function of the weights, so there is a unique optimum value for each weight $w_i$ where $\overline{\varepsilon^2(t)}$ is minimum. As it can be shown, according to the LMS (least mean-square) algorithm (also referred to as a gradient technique in the adaptive control literature), to minimize $\overline{\varepsilon^2(t)}$ each $w_i$ is adjusted according to the following feedback rule [13], [14]:

$$\frac{dw_i}{dt} = -u\frac{\partial[\overline{\varepsilon^2(t)}]}{\partial w_i}$$

where a constant $u, 0 < u < 1$, the so called convergence factor, would determine the convergence time (for example, with $u > \frac{1}{2}$ the convergence is oscillatory, with a weight vector $\mathbf{W}$ jumping "back and forth" across the bowl and converging toward $\mathbf{W}_{opt}$).

Differentiating (31) with respect to $w_i$ yields

$$\frac{\partial[\overline{\varepsilon^2(t)}]}{\partial w_i} = -2\overline{x_i(t)d(t)} + 2\sum_{j=1}^{4} w_j \cdot \overline{x_i(t)x_j(t)} = -2\overline{x_i(t)\varepsilon(t)} \tag{33}$$

where, according to (15) and our conventions thereafter ($x_{1r} = x_1, x_{ij} = x_2, x_{2r} = x_3, x_{2j} = x_4, w_{1r} = w_1, w_{1j} = w_2, w_{2r} = w_3, w_{2j} = w_4$)

$$\varepsilon(t) = d(t) - \sum_{i-1}^{4} w_i x_i(t) \tag{15a}$$

Thus, the feedback algorithm (32) becomes [15]

$$\frac{dw_i}{dt} = 2u\overline{x_i(t)\varepsilon(t)} \tag{34}$$

Integration of (34) yields

$$w_i = w_{io} + 2u \int_0^t \overline{x_i(t')\varepsilon(t')}dt' \tag{35}$$

The suggested instrumentation structure of this real-time adaptive system would have total number of four adaptive loops. Each adaptive loop includes two multipliers, one to multiply the error signal $\varepsilon(t)$ by the signal $x_i(t)$ and the other to multiply the weight $w_i$ by the signal $x_i(t)$, and an integrator between them. Also, it should be noted that $w_{i0}$ represent starting values for each loop.

Two remaining issues to be addressed are: 1) measurement technique concepts and 2) the reference signal algorithm.

1. We will first consider the worst scenario when there is either no information available regarding scattering properties of the target or these properties change with time.

Providing that the calibration ("mapping"), described in Section 4, has been accomplished, two consecutive steps in adaptation procedure should be exercised.

The first step, a "target adaptation procedures" (TAP) would be a "discrete screening", by checking different pairs of elements in STS (sample transmit space) and SRS (sample receive space). For example, let us assume, that the first substep of this TAP would include an "element" of STS like vertical linear polarization (and accordingly, and "element" of SRS, i.e. some particular form of the complex reference signal $d_{\psi r}$). The adaptation procedure then would follow, and, as a result, the output adaptively filtered signal $S_1(t)$ would be recorded (or observed). Subsequently, several other substeps of the TAP (say, LHP, RHCP, LHCP, and two or three elliptical polarizations) would provide the operator with sufficient information $(S_1(t), S_2(t) \ldots, S_N(t))$ for matching the transmit side with the target.

The second step would actually be an adaptive radar polarimetry when the target's "fine analysis" is exercised by "sliding" through different reference algorithms.

Certainly, an ambiguity of the first step could be reduced by, say, modulating a CW transmitting signal (and, accordingly, an adequate reference signal) as it is used in spread spectrum communication systems.

2. Our next concern is with some polarization states relations that would be instrumental for a reference signal generation.

As a follow-up of the discussion in Section 2, here are some applicable relationships derived from (3)-(14).

a) Eq. (5) can be written in the form

$$|\mathbf{F}(t)|^2 = a + b cos(2\varphi + \theta) \tag{36a}$$

where

$$a = \frac{1}{2}(\mathbf{E}_1^2 + \mathbf{E}_2^2) \tag{36b}$$

$$b = \sqrt{\frac{1}{4}(\mathbf{E}_1^2 + \mathbf{E}_2^2)^2 - \mathbf{E}_1^2\mathbf{E}_2^2 sin\xi} \tag{36c}$$

$$tan\theta = \frac{2|\mathbf{E}_1||\mathbf{E}_2|cos\xi}{\mathbf{E}_1^2 - \mathbf{E}_2^2} \tag{36d}$$

and $\xi$ is the space angle between vectors $\mathbf{E}_1$ and $\mathbf{E}_2$ (as it can be seen, such an explicit dependence on $\xi$ could substantially diversify the "screening" procedures through different polarization states).
a) For a circularly polarized wave

$$|\mathbf{E}_0|^2 = |\mathbf{E}_0 \times \mathbf{E}_0^*|$$

and for a linearly polarized wave

$$|\mathbf{E}_0|^2 = |\mathbf{E}_0^2|$$

where $\mathbf{E}_0$ is a complex-constant amplitude vector ((2), (3)).
c) Because in applications it is more convenient to describe the polarization of a wave in terms of the complex amplitude vector $\mathbf{E}_0 = \mathbf{E}_1 + j\mathbf{E}_2$ (and not in terms of $\mathbf{E}_1$ and $\mathbf{E}_2$), the following formulas for a major semiaxis $\mathbf{A}_0$ and the minor semiaxis $\mathbf{B}_0$ of the polarization ellipse (6) may prove to be instrumental:

$$\mathbf{A}_0 = \frac{1}{2}\left[\sqrt{|\mathbf{E}_0^2|/\mathbf{E}_0^2}\mathbf{E}_0 + \sqrt{|\mathbf{E}_0^2|/\mathbf{E}_0^{*2}}\mathbf{E}_0^*\right],$$
$$\mathbf{B}_0 = \frac{1}{2j}\left[\sqrt{|\mathbf{E}_0^2|/\mathbf{E}_0^2}\mathbf{E}_0 - \sqrt{|\mathbf{E}_0^2|/\mathbf{E}_0^{*2}}\mathbf{E}_0^*\right] \quad (37)$$

## 6. ADAPTIVE REFINEMENT (INTERFERENCE CANCELING) IN RADAR POLARIMETRY

Yet another development to demonstrate a versatile nature of an adaptive approach in radar polarimetry would be to show its applicability not only as a self-contained but also as a refining one to supplement some standard methods.
Our approach here follows that of a three-stage optimization procedure [1] when the transmit and receive antennas have already been adjusted to match the known scattering matrix of the target, and perturbations caused by the effects of propagation, reflection, or scattering subsequent to scattering by the target are to be removed.
To provide an example, we take the simple case. We might, for instance, be trying to use this system to cancel the "perturbation" interference, a Gaussian random process $M(t)$ from the resulting signal $S(t)$ at the output of the receiving antenna where

$$S(t) = M(t) + 5cos(wt + \theta) \quad (38)$$

with 5 and $w$ constants, $\theta$ a random variable uniformly distributed from 0 to $2\pi$, and $\theta$ and $M(t)$ statistically independent for all $t$.
To keep up with the terminology in Sec. 4 and Sec. 5, the signal $S(t)$ will represent a "desired" signal, and random process $N(t)$ (correlated with an interference $M(t)$), will represent an "input" signal.
We assume that

$$\overline{M(t)} = \overline{N(t)} = 0$$
$$R_{MM}(\tau) = e^{-3|\tau|}$$
$$R_{NN}(\tau) = e^{-5|\tau|} \quad (39)$$
$$R_{MN}(\tau) = 0.5\tau$$

where $R_{MM}(\tau), R_{NN}(\tau)$ are the autocorrelation functions of random processes $M(t)$ and $N(t)$, and $R_{MN}(\tau)$ is a cross-correlation function.
From equations (21), (22), (25), (31), one of possible forms for MSE can be presented as

$$MSE = \overline{\varepsilon^2(t)} = R_{dd}(0) + \sum_{i=0}^{L-1}\sum_{m=0}^{L-1} w_i w_m R_{xx}(i-m) - 2\sum_{i=0}^{L-1} w_i R_{xd}(i) \quad (40)$$

In (40), $L$ is a number of weights of the recursive process (associated with a transfer function (25) of an adaptive filter), $R_{dd}$ is an autocorrelation function of the "desired" signal $S(t), R_{xx}$ is an autocorrelation function of an "input" signal $N(t)$, and $R_{xd}$ is a cross-correlation function of signals $S(t)$ and $N(t)$.
With $L = 2$ (only two iterations), we have the MSE (40) in this case as follows:

$$MSE = R_{SS}(0) + R_{NN}(0)(w_0^2 + w_1^2) + 2w_0w_1R_{NN}(1) - \\ - 2w_0R_{SN}(0) - 2w_1R_{SN}(1) \tag{41}$$

First, we should evaluate functions $R_{SS}(0), R_{NN}(0), R_{NN}(1), R_{SN}(0)$ and $R_{SN}(1)$.

$$\begin{aligned}
R_{SS}(\tau) &= R_{MM}(\tau) + (\frac{25}{2})cosw\tau \\
R_{SS}(0) &= R_{MM}(0) + \frac{25}{2} \\
R_{MM}(0) &= 1 \\
R_{SS}(0) &= 1 + \frac{25}{2} = 13.5 \\
R_{NN}(0) &= 1 \\
R_{NN}(1) &= e^{-5} = 6.73 \times 10^{-3} \\
R_{SN}(\tau) &= E\{[M(t) + 5cos(wt + \theta)][N(t + \tau)]\} \\
&= E[M(t)N(t + \tau)] + E\{[5cos(wt + \theta)][N(t + \tau)]\} \\
&= R_{MN}(\tau) + E[5cos(wt + \theta)]E[N(t)] = 0.5\tau + 0 \\
R_{SN}(0) &= 0 \\
R_{SN}(1) &= 0.5
\end{aligned}$$

Thus, MSE (41) is a quadratic surface, and one can easily visualize a plot of this parabolic "bowl". From equation (27) we can see that the optimum weight values in this example are

$$\begin{bmatrix} w_0^{opt} \\ w_1^{opt} \end{bmatrix} = \begin{bmatrix} R_{NN}(0) & R_{NN}(1) \\ R_{NN}(1) & R_{NN}(0) \end{bmatrix}^{-1} \begin{bmatrix} R_{SN}(0) \\ R_{SN}(1) \end{bmatrix} \tag{42}$$

Or

$$\begin{bmatrix} w_0^{opt} \\ w_1^{opt} \end{bmatrix} = \begin{bmatrix} 1 & 6.73 \times 10^{-3} \\ 6.73 \times 10^{-3} & 1 \end{bmatrix}^{-1} \begin{bmatrix} 0 \\ 0.5 \end{bmatrix} \tag{42a}$$

Evaluation of (42a) gives the weight values delivering the minimum of MSE:

$$\begin{aligned} w_0^{opt} &= 3.36 \times 10^{-3} \\ w_1^{opt} &= 0.5 \end{aligned} \tag{43}$$

Thus, when the weights are optimized, $p(t)$ cancels $M(t)$ as we might have anticipated, and the output, $\varepsilon(t) = 5cos(wt + \theta)$, is free of interference $M(t)$.

## 7. CONCLUSION

A generalized analysis of polarization states in a nonorthogonal basis has been undertaken, and it was demonstrated how different relations between two linearly polarized waves $\mathbf{E}_1$ and (the quadrature of) $\mathbf{E}_2$ could be used for the reference signal $d(t)$ generation in an adaptive system ((9)-(14), (36), (37)). A new polarization descriptor has been introduced so that both transmit and receive adjustments in a process of adaptation would gain from the generally nonorthogonal basis.

Basic fundamentals of adaptive signal processing applicable to radar polarimetry have been examined. In particular, suggested calibration procedure of the transmit- receive radar system involving a one-to-one mapping from a sample transmit space to a sample receive space would permit functioning of the system in situations when either the scattering properties of the target are unknown or they change with time.

Some physical, mathematical and geometrical aspects of optimization procedure involving the performance surface for a causal finite impulse response adaptive filter, and detailed analysis of the suggested real-time adaptive estimation procedure and ways of its implementation (algorithms, instrumentation) demonstrate conceptual feasibility of adaptive radar polarimetry as a supplement to or enhancement of already developed techniques.

In particular, a versatile nature of an adaptive approach in radar polarimetry is demonstrated in a problem of so called adaptive refinement when it supplements a three-stage optimization procedure with known scattering matrix by removing (canceling) the "perturbation" interference.

REFERENCES


[1] A.B. Kostinski and W.-M. Boerner, 'On foundations of radar polarimetry', *IEEE Trans. Antennas Propagat.*, **AP-34**, pp.1395-1404, Dec. 1986.

[2] C.D. Graves, 'Radar polarization power scattering matrix', *Proc. IRE*, **44**, pp. 248-252, Feb. 1956.

[3] G.A. Ioannidis and D.E. Hammers, 'Optimum antenna polarizations for target discrimination in clutter', *IEEE Trans. Antennas Propagat.*, **AP-27**, pp. 357-363, May 1979.

[4] A.B. Kostinski and W.-M. Boerner, 'On the polarimetric contrast optimization', *IEEE Trans. Antennas Propagat.*, **AP-35**, pp. 988-991, August 1987.

[5] M. Davidovitz and W.-M. Boerner, 'Extension of Kennaugh's optimal polarization concept to the asymmetric scattering matrix case', *IEEE Trans. Antennas Propagat.*, **AP-34**, pp. 569-574, April 1986.

[6] M.L. Kales, 'Elliptically polarized waves and antennas', *Proc. of IRE*, pp. 544-549, May 1951.

[7] H.C. Chen, *Theory of electromagnetic waves*, McGraw-Hill, 1983.

[8] J.G. Reid, *Linear system fundamentals*, McGraw-Hill, 1983.

[9] J.J. Komo, *Random signal analysis in engineering systems*, Academic Press, 1987.

[10] N. Wiener, *Extrapolation, interpolation, and smoothing of stationary times series with engineering applications*, Wiley, New York, 1949.

[11] S.H. Friedberg, A.J. Insel, and L.E. Spence, *Linear algebra*, Prentice-Hall, New Jersey, 1979.

[12] N. Levinson, 'The Wiener RMS error criterion in filter design and prediction', *J. Math and Physics*, **25**, pp. 209- 278, 1946.

[13] B. Widrow, P.E. Mantley, L.J. Griffiths, and B.B. Goode, 'Adaptive antenna systems', *Proc. IEEE*, **55, no. 12**, p. 2143, Dec. 1967.

[14] R.L. Riegler and R.T. Compton, Jr., 'An adaptive array for interference rejection', *Proc. IEEE*, **61, no. 6**, pp. 748-758, June 1973.

[15] R.T. Compton, Jr., 'An experimental four-element adaptive array', *IEEE Trans. Antenna Propagat.*, **AP-24, no. 5**, pp. 697-706, Sept. 1976.

# A PROPOSED POLARIMETRIC CFAR-DETECTOR AND AN ANALYSIS OF ITS OPERATION

G. Wanielik
AEG Research Institute

D.J.R. Stock
AEG Radar Department

D-7900 Ulm
West Germany

**Abstract**

In contrast to the conventional CFAR-detector which uses scalar amplitudes for a decision, we propose here a polarimetric CFAR-detector which can render a decision based on the polarization difference between clutter and target. This CFAR-system allows a target detection for low signal/clutter values where the conventional system fails.

## 1. Introduction

Polarimetric radars are able to measure the scattering matrix (SM) of a reflecting radar object and obtain complete information in the far-field of the object. This polarimetric information is used in the Polarimetric Signal Processing, which deals with multivariate signals instead of processing only one channel information as most of the existing radars do. All signal processing problems which are known from non-polarimetric radars as the detection, discrimination and classification have to be discussed again in the multiple channel case, /1, 2, 3/. In the following we deal with the polarimetric detection problem, especially with a constant false alarm rate (CFAR) processor.

The conventional CFAR-circuit uses scalar amplitudes of the return signal of a radar to establish if a target is present or not. Each resolution cell of a set of selected cells is tested to see if the amplitude in the cell under test exceeds the average amplitudes of the neighboring cells. If this is the case, the system decides for a target. Emphasis here is that the process uses only scalar information. Furthermore, the CFAR is adaptive in that it uses local clutter information for a decision without having to know specific properties of the clutter in advance.

A polarimetric CFAR, on the other hand, uses polarimetric information to decide on the presence of a target. This implies that in regions where a conventional CFAR cannot decide for a target because the clutter amplitude equals the target amplitude the polarimetric-CFAR can render a decision based on the difference in polarization between clutter and target.



*W.-M. Boerner et al. (eds.), Direct and Inverse Methods in Radar Polarimetry, Part 1*, 999–1010.

## 2. Theory of the Polarimetric CFAR-Detector

The described detector uses the complex measurement vector sequence

$$[\underline{s}(i)] = [\begin{pmatrix} S_{11}(i) \\ S_{21}(i) \end{pmatrix}] = [\begin{pmatrix} s_1(i) \\ s_2(i) \end{pmatrix}] ; \quad i = 1,\dots,2N+1 \tag{1}$$

of a dual channel receiver. That is, the radar is polarimetric only on receive. The same processing concept discussed now is also usable for a full polarimetric radar where $[\underline{s}(i)]$ contains the full scattering matrix $\underline{S}$ in vectorform. Because of the parallelism we discuss only the case for the receiving polarimetric radar.

The vector-data are selected from a small part of the range-azimuth raster picture shown in figure 1. Considering CFAR-window 1, one has to decide whether the signal vector $\underline{s}(N+1)$ of the cell under test is a target or is clutter like the surrounding data

$$[\underline{s}(i)] = [\begin{pmatrix} S_{11}(i) \\ S_{21}(i) \end{pmatrix}] = [\begin{pmatrix} s_1(i) \\ s_2(i) \end{pmatrix}] ; \quad i \neq N+1 \tag{2}$$

The clutter process is modelled as a two dimensional complex normal distribution with expectation value zero and covariance matrix $\underline{J}$, which is the well known coherency matrix /4/ of the clutter-process. The density function is given by $d_c$:

$$\underline{J} = \langle \underline{s} \cdot \underline{s}^{*T} \rangle = \begin{pmatrix} J_{11} & J_{12} \\ J_{21} & J_{22} \end{pmatrix}, \quad \underline{s}(i);\ i \neq N+1 \longrightarrow \text{clutter} \tag{3}$$

The $\langle\ .\ \rangle$ denotes "expected value".

$$d_c = \frac{1}{\pi^2 \cdot \det(\underline{J})} \cdot e^{-\underline{s}^{*T} \cdot \underline{J}^{-1} \cdot \underline{s}} \tag{4}$$

In order to see the operation of the polarimetric CFAR in detail, one first considers the regions of constant density of the complex normal distribution representing the clutter process. Assuming $\det(\underline{J}) \neq 0$ and $\underline{J}$ is constant in the region, one may collect the constant terms of (4) on the left, giving:

$$d_c \cdot \pi^2 \cdot \det(\underline{J}) = e^{-Q} \tag{5}$$

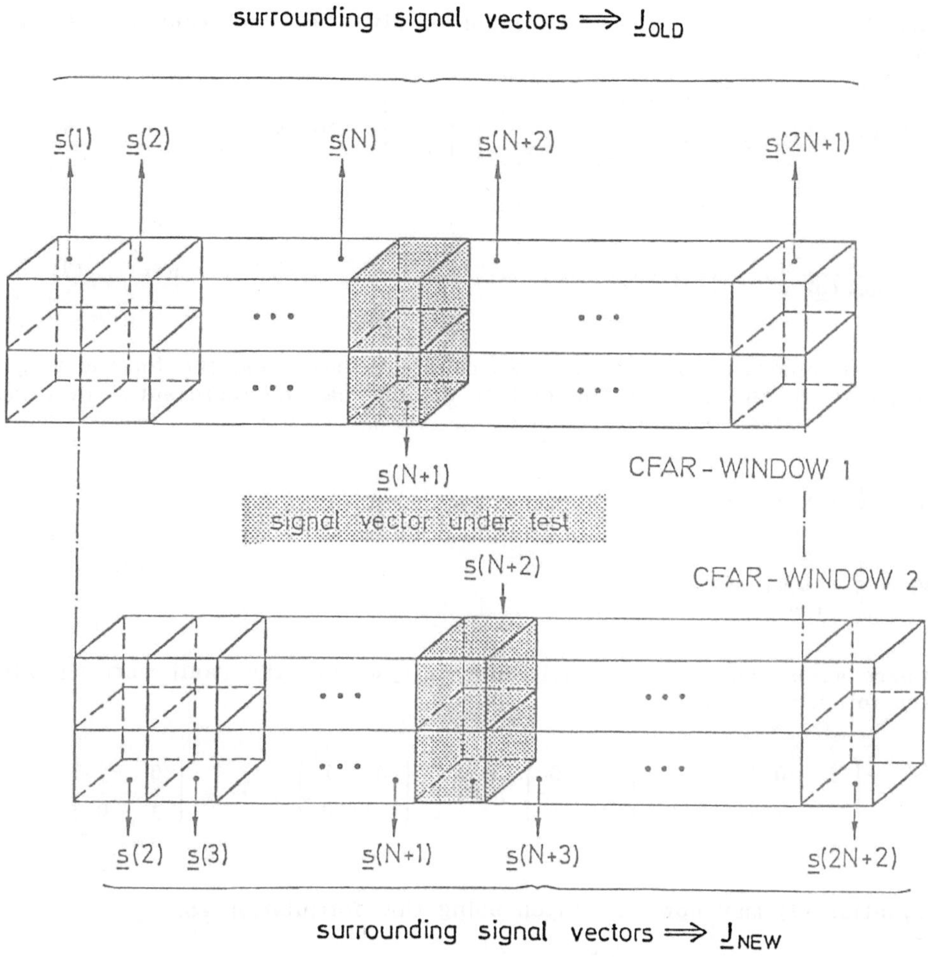


**Figure 1:** **Polarimetric CFAR for two successive CFAR-windows**

One looks for the value of (5) for a given $d_c$ and constant $\underline{J}$. This implies

$$Q = (s_1^*, s_2^*) \cdot \frac{1}{\det(\underline{J})} \cdot \begin{pmatrix} J_{22} & -J_{12} \\ -J_{21} & J_{11} \end{pmatrix} \cdot \begin{pmatrix} s_1 \\ s_2 \end{pmatrix} = \text{const.} \qquad (6)$$

or

$$Q = \frac{1}{\det(\underline{J})} \cdot (J_{22} \cdot s_1 s_1^* - J_{12} \cdot s_2 s_1^* - J_{21} \cdot s_1 s_2^* + J_{11} \cdot s_2 s_2^*) = \text{const.} \qquad (7)$$

In CFAR-operation one obtains two coherency matrices, one for the target response $\underline{J}_t$ and one for the clutter $\underline{J}$. Both can be represented in terms of their Stokes vector components

$$\underline{J} = \frac{1}{2} \cdot \sum_{i=0}^{3} g_{ci} \, \underline{\sigma}_i$$
$$\underline{J}_t = \frac{1}{2} \cdot \sum_{i=0}^{3} g_{ti} \, \underline{\sigma}_i \qquad (8)$$

where $\underline{\sigma}_0$ is the identity matrix and $\underline{\sigma}_1, \underline{\sigma}_2, \underline{\sigma}_3$ are the Pauli matrices with the following notation:

$$\underline{\sigma}_0 = \begin{pmatrix} 1 & 0 \\ 0 & 1 \end{pmatrix} ; \; \underline{\sigma}_1 = \begin{pmatrix} 1 & 0 \\ 0 & -1 \end{pmatrix} ; \; \underline{\sigma}_2 = \begin{pmatrix} 0 & 1 \\ 1 & 0 \end{pmatrix} ; \; \underline{\sigma}_3 = \begin{pmatrix} 0 & -j \\ j & 0 \end{pmatrix}$$

Equation (7) may now be written using this formulation as:

$$Q \cdot \det(\underline{J}) = \frac{1}{2} \cdot \left( g_{co} \cdot g_{to} - \sum_{i=1}^{3} g_{ci} \cdot g_{ti} \right) \qquad (9)$$

where $\quad g_{to} = (g_{t1}^2 + g_{t2}^2 + g_{t3}^2)^{1/2}$

After some manipulation one obtains

$$\sum_{i=1}^{3} g_{ti}^2 = \frac{1}{g_{co}^2} \cdot \left( 2 \cdot Q \cdot \det(\underline{J}) + \sum_{i=1}^{3} g_{ci} \cdot g_{ti} \right)^2 \qquad (10)$$

which is a quadratic form in the variables $g_{ti}$ and has a geometric interpretation.

## 3. Example of a Polarimetric CFAR-Detector

It is preferable for clarity of the presentation to select a simple example to derive the three-dimensional geometric form represented by Q=const. Defining $p_c$ as the degree of polarization of the clutter, one writes the Stokes vector of the clutter process:

$$\underline{G}_c = \begin{pmatrix} g_{c0} \\ g_{c1} \\ g_{c2} \\ g_{c3} \end{pmatrix} = \begin{pmatrix} g_{c0} \\ \underline{g}_c \end{pmatrix} = \begin{pmatrix} 1 \\ p_c \begin{pmatrix} 1 \\ 0 \\ 0 \end{pmatrix} \end{pmatrix} \tag{11}$$

Then (10) simplifies to

$$\sum_{i=1}^{3} g_{ti}^2 = ( 2 \cdot Q \cdot \det(\underline{J}) + p_c \cdot g_{t1} )^2$$

Using

$$p_c = ( 1 - \frac{4 \cdot \det(\underline{J})}{g_{c0}^2} )^{1/2} \tag{12}$$

one obtains finally

$$\frac{g_{t3}^2}{a^2} + \frac{g_{t2}^2}{a^2} + \frac{(g_{t1}-A)^2}{c^2} = 1 \tag{13}$$

where

$$\begin{aligned} c &= 2Q \cdot (1-p_c^2)^{-1} \det(\underline{J}) = \frac{1}{2} g_{c0}^2 \cdot Q \\ A &= p_c \cdot 2Q \cdot (1-p_c^2)^{-1} \det(\underline{J}) = \frac{1}{2} p_c \cdot g_{c0}^2 \cdot Q = p_c \cdot c \leq c \\ a &= c(1-p_c^2)^{1/2} = \frac{1}{2} g_{c0}^2 \cdot Q \cdot (1-p_c^2)^{1/2} \end{aligned} \tag{14}$$

Equation (13) is the equation of an ellipsoid of revolution as shown in figure 2. A section through a family of these ellipsoids is shown in figure 3. The rotation axis $g_1$ is the axis of the Stokes vector of eq. (11).

This example shows that the locus of constant density $d_c$=constant is the shell of an ellipsoid in the polarization vectorspace shown in figure 3. Analog to the scalar detector which has to guarantee a constant false-alarm rate, we have to decide for a target, when the associated Stokes vector of the cell under test does not lie inside the shell of this ellipsoid, that is, points which lie outside of a chosen ellipsoid $d_c$=const (in the shaded area of figure 3) are to be considered as targets.

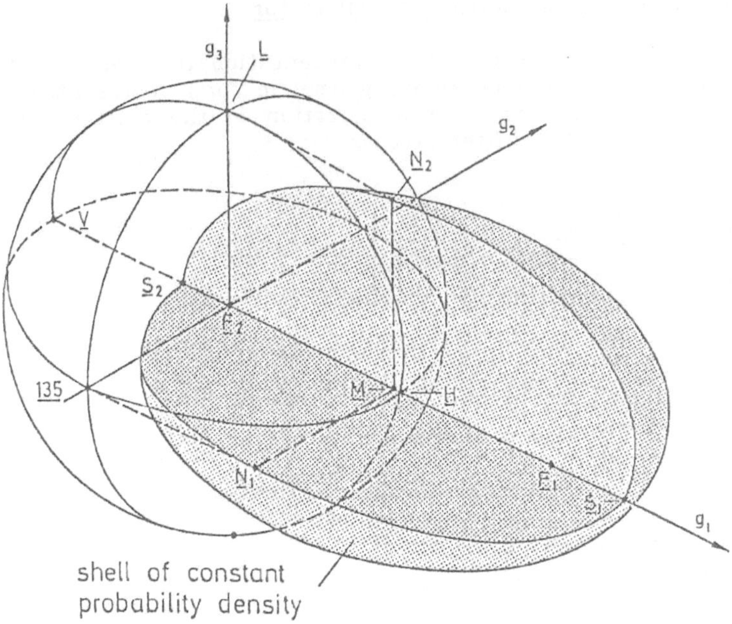


Figure 2: Shell of constant probability density in the Poincaré space

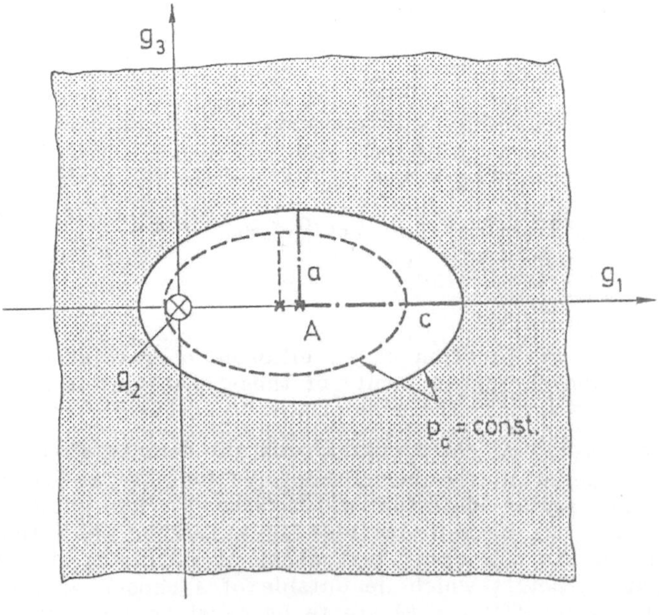


Figure 3: Section through a family of ellipsoids of revolution

To prove this, one does not have to use the Stokes vector approach, but can go directly from the quadratic form Q. One decides for a target if:

$$Q = (s_1^*, s_2^*) \cdot \frac{1}{\det(\underline{J})} \cdot \begin{pmatrix} J_{22} & -J_{12} \\ -J_{21} & J_{11} \end{pmatrix} \cdot \begin{pmatrix} s_1 \\ s_2 \end{pmatrix} \geq T \tag{15}$$

where T is chosen in such a way that a given false-alarm rate can be guaranteed.

## 4. Interpretation of the operation of a Polarimetric CFAR-Detector using the Stokes-Vector

A very clear interpretation of the CFAR detection process in terms of the Stokes vectors may be obtained by starting with equation (9), which determines Q.

$$Q = \frac{g_{co} \cdot g_{to}}{2\det(\underline{J})} \cdot \left( 1 - \sum_{i=1}^{3} \frac{g_{ci}}{g_{co}} \cdot \frac{g_{ti}}{g_{to}} \right) \geq T \tag{16}$$

Using the Stokes vector from eq. (12) one defines the polarization vector $\underline{p}_c$ for the clutter and in an analog way the polarization vector $\underline{p}_t$ for the target

$$\underline{p}_c = \frac{1}{p_c \cdot g_{co}} \cdot \begin{pmatrix} g_{c1} \\ g_{c2} \\ g_{c3} \end{pmatrix}$$

$$\underline{p}_t = \frac{1}{g_{to}} \cdot \begin{pmatrix} g_{t1} \\ g_{t2} \\ g_{t3} \end{pmatrix} \tag{17}$$

One may now write eq. (16) as:

$$Q = \frac{2 \cdot g_{co} \cdot g_{to}}{(1-p_c^2) g_{co}^2} \cdot \left( 1 - p_c(\underline{p}_t \cdot \underline{p}_c) \right) \geq T \tag{18}$$

where $\underline{p}_t \cdot \underline{p}_c$ denotes the scalar product of the two vectors.
With the definition of the signal to clutter ratio:

$$(SCR) = \frac{g_{to}}{g_{co}} \tag{19}$$

one obtains

$$Q = (SCR) \cdot \left[ \frac{2}{(1-p_c^2)} \cdot (1 - p_c(\underline{p}_t \cdot \underline{p}_c)) \right] = (SCR) \cdot k \geq T \tag{20}$$

where k is a scalar factor within the following bounds:

$$\frac{2}{1+p_c} \leq k \leq \frac{2}{1-p_c}$$

Assuming that k=constant, one sees that Q is dependent on (SCR). That is, the system functions as a conventional CFAR.

One now assumes (SCR)=const. and considers the polarization-dependent term k. Recalling that the scalar product may be written as

$$\underline{p}_c \cdot \underline{p}_t = \cos(\alpha)$$

where $\alpha$ is the angle between the two polarization vectors one has:

$$k = \left[\frac{2}{(1-p_c^2)} \cdot (1- p_c \cdot \cos(\alpha))\right] \tag{21}$$

When $\alpha = 0$, that is, when the clutter polarization vector lies parallel to the target polarization vector, the value of k is a minimum. When, however, the two vectors are orthogonal, the value of k is a maximum. In other words the value of k increases as the angle between the two vectors increases.

One may now interpret eq. (20) in the following way:

Considering the degree of polarization of the clutter $p_c$ as a weighting function for the scalar product, when $p_c$ is small, the scalar product as a distance measure of the two polarizations has little influence on Q, and the (SCR)-term predominates. When $p_c$ is large, the term k has the greater effect on Q.

As was seen above, this means that even though (SCR) may be too small to override the fixed threshold T, a large value of k can cause Q > T. That is, a detection based on polarization may be possible even though the conventional (SCR)-term may be too small.

Our CFAR-detector uses the covariance matrix $\underline{J}$ of the clutter process. For the actual CFAR-window $\underline{J}$ can be estimated from the measurement data of the 2N cells representing the clutter-process which surrounds the cell under test. A preferred method of calculation which allows more efficient use of the signal processor is to calculate the covariance matrix $\underline{J}_{new}$ of the CFAR-window 2 by subtracting from the previously calculated matrix $\underline{J}_{old}$ of CFAR-window 1 the signal vector of the 1st cell and adding the signal vectors of the (N+2)th cell. In addition, the cell under test in window 1 must be added as clutter in window 2, and the signal in cell (N+2), which is now the cell unter test, must be subtracted:

$$\underline{J}_{new} = \underline{J}_{old} + \frac{1}{2N} \sum_i \underline{s}_i \cdot \underline{s}_i^{*T} - \frac{1}{2N} \sum_j \underline{s}_j \cdot \underline{s}_j^{*T} \tag{22}$$

i=N+1, 2N+2; j=1, N+2

## 7. Determination of the probability of false alarms and detection

In section 6 the operation of the CFAR was described using the Stokes vector approach, which resulted in an intuitively satisfying description of the process. It is now necessary to consider the calculations which lead to numerical values for the probability of false alarms $P_{fa}$ and the probability of detection $P_d$.

It should be noted here that the calculation can proceed in a similar manner to that for the conventional CFAR, since one is concerned here with a scalar function Q and asks again about the conditions necessary for $Q > T$.

Q is a hermitian form whose statistical distribution is dependent on the statistics of the cell under test, assuming that $\underline{J}^{-1}$ is known. Assuming that the cell under test has a complex normal distribution with covariance matrix $\underline{J}_s$, the density function for Q can be calculated as follows, using results from /5/:

$$d_s(Q) = \frac{e^{-\frac{\mu\cdot Q}{2}}}{\det(\underline{J}_s\cdot\underline{J}^{-1})\cdot\frac{\nu}{2}}\cdot\sinh\left(\frac{\nu\cdot Q}{2}\right) \tag{23}$$

$$\mu = \operatorname{tr}(\underline{J}_s\cdot\underline{J}^{-1})^{-1}$$

$$\nu^2 = \mu^2 - 4\det(\underline{J}_s\cdot\underline{J}^{-1})^{-1}$$

Considering first that the cell under test contains only clutter, then we get:

$$\underline{J}_s = \underline{J}$$

and from eq. (23)

$$d_c(Q) = \begin{cases} Q\cdot e^{-Q} & 0 \leq Q \leq \infty \\ 0 & \text{elsewhere} \end{cases} \tag{24}$$

The distribution can now be determined:

$$D_c(Q) = \int_{-\infty}^{Q} d_c(x)\cdot dx = 1-(Q+1)e^{-Q} \tag{25}$$

Using $D_c(Q)$ one may now evaluate the false alarm rate:

$$P_{fa}(Q) = 1 - D_c(Q) = (Q+1)e^{-Q} \tag{26}$$

One now considers the case where the test-cell contains a target:

This process will have a complex normal distribution with expectation value zero and covariance matrix $\underline{J}_{tc}$. The density function can be obtained again from eq. (23) with

$$\underline{J}_s = \underline{J}_{tc}$$

where $\underline{J}_{tc}$ is the covariance matrix for the target in clutter:

$$d_t(Q) = \frac{1}{\nu \cdot \eta} \left( e^{-\frac{(\mu-\nu)\cdot Q}{2}} - e^{-\frac{(\mu+\nu)\cdot Q}{2}} \right) \tag{27}$$

with

$$\begin{aligned} \eta &= \det(\underline{J}_s \cdot \underline{J}^{1}) \\ \mu &= tr(\underline{J}_s \cdot \underline{J}^{1})^{-1} \\ \nu^2 &= \mu^2 - 4 \cdot \eta^{-1} \end{aligned} \tag{28}$$

Similarly to the above operations for the clutter process, one can calculate for the target process:

$$D_t = \frac{2}{\nu \cdot \eta} \left( \frac{1-e^{-\frac{(\mu-\nu)\cdot Q}{2}}}{\mu - \nu} - \frac{1-e^{-\frac{(\mu+\nu)\cdot Q}{2}}}{\mu + \nu} \right) \tag{29}$$

The detection probability can then be evaluated as:

$$P_d(Q) = 1 - D_t(Q) \tag{30}$$

In this section a procedure for the calculation of the $P_d$, $P_{fa}$ curves for the polarimetric detector as a function of Q has been given. One notes however that the parameters necessary for the calculation, $\mu$, $\nu$ and $\eta$, are given only as a function of $(\underline{J}_{tc} \cdot \underline{J}^{1})^{-1}$. These parameters can be given more explicitly when the clutter and target processes are more closely defined, as will be done in the next section.

## 8. Determination of the process parameters $\mu$, $\nu$, $\eta$ for a geometrically simple case which preserves generality

Without loss of generality, one considers the clutter process to be centered on the point of the Poincaré sphere representing left circular polarization, while the target process is centered on an arbitrary point on the sphere. On normalizing the average power of the clutter process to unity, one obtains:

$$\underline{J} = \frac{1}{2}\begin{bmatrix} 1+p_c & 0 \\ 0 & 1-p_c \end{bmatrix} \tag{31}$$

where $p_c$ is the degree of polarization of the clutter, previously introduced for eq. (11).

Define $t_1$, $t_2$, $t_3$ as the polarization vector components of the arbitrary target point on the Poincaré sphere. The polarization degree of the target is represented by $p_t$. Due to the above normalization of the average power of the clutter process, one may write the signal to clutter ratio (SCR) as equal to the average power of the target process. One writes now $\underline{J}_t$ as the covariance matrix of the target process alone (without clutter), giving:

$$\underline{J}_t = \frac{1}{2}\cdot(\text{SCR})\cdot\begin{bmatrix} 1+p_t t_3 & p_t(t_1-jt_2) \\ p_t(t_1+jt_2) & 1-p_t t_3 \end{bmatrix} \tag{32}$$

Assuming that the clutter process and the target process are statistically independent, one may superimpose the two covariance matrices giving /6/:

$$\underline{J}_{tc} = \underline{J} + \underline{J}_t \tag{33}$$

where $\underline{J}_{tc}$ is as defined following eq. (26). The parameters of eq. (28) may then be calculated to be:

$$\eta = 2\left(1 + \frac{(\text{SCR})}{1-p_c^2}(1-p_c p_t t_3)\right) - \left(1 - (\text{SCR})^2\frac{1-p_t^2}{1-p_c^2}\right)$$

$$\mu = \frac{2}{\eta}\left(1 + \frac{(\text{SCR})}{1-p_c^2}(1-p_c p_t t_3)\right) \tag{34}$$

$$\nu = \frac{2}{\eta}\left(\frac{(\text{SCR})}{1-p_c^2}\left((1-p_c p_t t_3)^2 - (1-p_t^2)(1-p_c^2)\right)^{1/2}\right)$$

It is important to realize here that $t_3$ represents the cosine of the angle between the target centerpoint and the clutter centerpoint. Recalling the discussion following eq. (20), one identifies

$$t_3 = \cos(\alpha) \tag{35}$$

The above representation is simple, but without loss of generality. Its advantage is that the process parameters are obtained in terms of interpretable and measurable clutter and target parameters (SCR), $p_c$, $p_t$, and the separation angle $\alpha$ between clutter and target on the Poincaré sphere.

## 9. Summary

We have shown here the form for a new polarimetric CFAR-detector and we have given a simple interpretation of its function in Poincaré space. The analytic expressions for $P_d$, $P_{fa}$ have been derived. This work will form the basis for future considerations of the multiple target problem.

## 10. Notation

| | |
|---|---|
| A | - scalar ( normal ) |
| $\underline{A}$ | - column vector ( underlined ) |
| $\underline{\boldsymbol{A}}$ | - matrix ( underlined, bold and cursiv ) |
| det($\underline{\boldsymbol{A}}$) | - determinant of the matrix $\underline{\boldsymbol{A}}$ |
| tr($\underline{\boldsymbol{A}}$) | - trace of the matrix $\underline{\boldsymbol{A}}$ |
| $\underline{A} \cdot \underline{B}$ | - scalar product |
| $^T$, $^*$ | - transposed, conjugate complex |

## 11. Literature

/1/ G. Wanielik
Polarization Weighting Functions as a useful Tool in Clutter Supression, Target Classification and Electromagnetic Wave Simulation
IEEE Intern. Geoscience and Remote Sensing Symposium Oct. 1985, Amherst Mass.

/2/ G. Wanielik, D.J.R. Stock
Classification and Cluster Algorithms based on Features of the Scattering Matrix
U.R.S.I. Int. Symp. on Microwave Signatures in Remote Sensing, Jan. 87, Göteborg

/3/ G. Wanielik
Signal Processing Possibilities for Pulse Radars using Polarimetric Information
International Conference "Radar 87", Oct. 87, London

/4/ M. Born, E. Wolf
Principles of Optics
Pergamon Press, New York, 1986

/5/ G.L. Turin
The Characteristic Function of Hermitian Quadratic Forms in Complex Normal Variables
Biometrica 47, 1960, p. 199-201

/6/ G. Wanielik, D.J.R. Stock
Radar Polarization Jamming using the Superposition of two fully polarized Waves
International Conference "Radar 87", Oct. 87, London

# POLARIMETRIC RADAR STATIONARY TARGET/CLUTTER DISCRIMINATION TECHNIQUES

Dr. William A. Holm
Georgia Tech Research Institute
Georgia Institute of Technology
Atlanta, Georgia 30332
U. S. A.

ABSTRACT. The discrimination of stationary ground targets from ground clutter with an airborne radar sensor is discussed. Discrimination techniques are categorized as either scalar, vector, or matrix techniques based upon the amount of polarimetric information that they utilize. Specific discrimination techniques from these three categories are then discussed.

## 1. INTRODUCTION

Airborne radar detection of stationary man-made ground vehicles (targets) in a ground clutter background is a difficult task. Achieving the desired high probabilities of detection ($P_d$) using traditional amplitude thresholding techniques [e.g., constant false alarm rate (CFAR) processing] usually requires that thresholds be set so low that an unacceptable number of false-target alarms due to clutter is obtained. Raising these thresholds to reduce the number of false alarms to an acceptable level, however, results in an unacceptably low detection probability. This problem occurs because, for a given transmit and receive polarization, many clutter objects have an effective radar cross section (RCS) that is approximately equal to or greater than the RCS of the targets of interest. Therefore, those radar signal processing techniques that rely on (single-polarimetric-channel, single-radar-resolution-cell) amplitude data to make a target/false-target decision do not, in general, provide the level of detection performance desired. Some other target attribute (other than, or in addition to, RCS or amplitude) must, therefore, be exploited to discriminate stationary man-made targets from clutter false targets. For moving targets, this other attribute is target radial velocity, i.e., Doppler frequency shift. Doppler frequency processing techniques have been shown to be very effective at discriminating moving targets from clutter, but these techniques are not, in general, applicable for stationary



*W.-M. Boerner et al. (eds.), Direct and Inverse Methods in Radar Polarimetry, Part 1*, 1011–1019.

targets. For stationary targets, *geometrical* attributes, or features, must be exploited.

Target geometrical features can be sensed with a radar through the effective use of the polarization of the radar's transmitted and received signals, or through the use of (one- or two-dimensional) high resolution imaging techniques. Polarization is the description of the vector nature of the electromagnetic (EM) wave that is transmitted and received by a radar sensor. Since the radar signal is a vector quantity, it is sensitive to the geometrical characteristics of the target from which it scatters. For example, if the electric field vector of a plane EM wave transmitted from a radar is rotating in a counterclockwise sense (at a fixed point in space as viewed by someone toward whom the EM wave is propagating -- "right-handed" sense polarization by the convention adopted by the IEEE), then this EM wave's electric field vector will, upon reflecting from a flat plate, rotate in a clockwise sense (at a point in space as viewed by someone toward whom the EM wave is propagating -- "left-handed" sense polarization by the IEEE convention). If this same EM wave were to reflect off of a dihedral corner reflector (aimed at the monostatic radar), then the sense of polarization would remain unchanged -- it would remain "right handed." Thus, the geometry of the target affects the polarization state of the reflected EM wave. If the target and the clutter have different geometrical characteristics, then with the proper polarization signal processing the radar return from these objects might be distinguishable. In this paper, various radar polarization processing techniques for stationary target/clutter discrimination are discussed. This is preceded by a discussion of non-polarimetric (scalar) discrimination techniques.

In Section 2, some definitions that are commonly used in radar stationary target identification are presented. This is followed by a discussion of radar target/clutter discrimination techniques in Sections 3, 4, and 5. A summary is given in Section 6.

## 2. DEFINITIONS

Radar stationary target identification consists of three phases: *detection*, *discrimination*, and *recognition* (see Table I). In the *detection* phase, noise interference signals and the radar return from benign clutter are separated from the remaining radar return signals. What is meant here by "benign" clutter is that type of clutter whose RCS is low enough that its radar return can be eliminated in the radar's signal processor through the use of standard amplitude thresholding techniques, e.g., CFAR processing. What remains following the detection phase are signals from targets of potential interest and from strong (high-RCS), target-like clutter, i.e., "false-targets." In the *discrimination* phase, the signals from the targets of potential interest are separated from the false-target signals. Finally, in the *recognition* phase, signals from targets of potential interest are classified or identified; classification being the determination of a generic target class (e.g., truck, tank, etc., or "tracked" versus "wheeled") to which a target belongs, and identification being the determination of the specific targets within the target classes (e.g., M-48, T-62, etc., within the tank class). In this paper, only discrimination techniques are discussed.

TABLE I. PHASES OF STATIONARY TARGET IDENTIFICATION

| PHASE | PURPOSE |
|---|---|
| Detection | Separation of targets from noise and benign clutter |
| Discrimination | Elimination of strong, target-like clutter returns |
| Recognition | Recognition of targets |
| Classification | Determination of generic target classes |
| Identification | Specification of targets within a generic class |

Stationary target/clutter discrimination signal processing techniques fall into three broad categories: *scalar*, *vector*, and *matrix* (see Table II). *Scalar* discriminants do not use any aspect of polarization to separate targets from false-targets. *Vector* discriminants use some limited aspect of polarization, while *matrix* discriminants require the radar to measure the polarization scattering matrix (PSM).

## 3. SCALAR DISCRIMINATION TECHNIQUES

Scalar discrimination techniques do not use any aspect of polarization to effect target/clutter discrimination. Examples of scalar discrimination techniques include *clutter decorrelation techniques*, *spatial feature techniques*, and *high resolution techniques*.

The basic idea behind *clutter decorrelation techniques* is that, if it were possible to make the returns from clutter statistically uncorrelated on a sample-to-sample basis, then the discrimination performance of the radar would be enhanced. The amplitude of a radar return is a random variable; and, as such, it must be described using statistical descriptors, e.g., probability density functions (PDFs). For example, the single-cell-averaged RCS of ground clutter as a function of resolution cell (location on the ground) for spatially-varying clutter will tend to be lognormally distributed for millimeter-wave radars. This means that the PDF for spatially-varying clutter as a function of averaged RCS is described mathematically by lognormal curves. These PDF curves must be obtained for both clutter cells and target-plus-clutter cells to determine the discrimination performance for a radar that measures RCS only. The degree of overlap of these PDF curves, determined by their means and standard deviations, will dictate the discrimination performance of the radar. The larger the separation of the means of the PDF curves and the smaller their standard deviations, the better the discrimination performance. If the RCS samples used to produce the averaged-RCS values are made statistically uncorrelated, then the standard deviations of the resulting PDF curves will be smaller; thus, improving radar discrimination performance. This can be accomplished by transmitting a frequency-agile waveform (e.g., a stepped-frequency waveform for a pulsed radar or a frequency modulated waveform for a continuous wave radar). If the clutter return is sampled at different frequencies, the samples will tend to be more statistically uncorrelated than if they were sampled at a constant frequency.

TABLE II. CLASSES AND EXAMPLES OF TARGET/CLUTTER DISCRIMINATION ALGORITHMS

SCALAR ALGORITHMS
(Do not use any aspect of polarization)

- Clutter decorrelation techniques
- Spatial feature algorithms
- High resolution techniques

VECTOR ALGORITHMS
(Use some limited aspect of polarization)

- Depolarization techniques
- Depolarization techniques with frequency agility

MATRIX ALGORITHMS
(Based on measuring the PSM)

- Null-polarization techniques
- Matrix-parameter techniques
- Matrix decomposition techniques

When *spatial feature techniques* are employed, target/false-target decisions are based on the returns from many radar resolution cells. Two examples of spatial feature algorithms are spatial extent filtering and contextual feature processing. In spatial extent filtering, detections from N contiguous range cells are rejected when $N\delta R$ ($\delta R$ being the radar's range resolution) is much greater than the target length. Here, the assumption is made that these detections are from extended clutter, e.g., tree lines. In contextual feature processing, discrimination is accomplished by looking for spatial patterns of targets in, for example, a range versus cross-range radar map; for example, a convoy of tanks traversing a road. Isolated detections are dismissed as being from discrete clutter.

Targets and clutter objects are imaged through the use of *high resolution techniques*. These images can be one-dimensional in the range dimension, or two-dimensional in the range and cross-range dimensions through the use of synthetic aperture radar techniques. The range resolution used for one-dimensional high-resolution techniques results in a range bin whose range dimension is much less than the minimum target length. Therefore, targets will appear in many contiguous range bins, i.e., a high range resolution profile (HRRP) of the target will be obtained. Isolated detections, e.g., single range bin detections, can thus be rejected as discrete clutter returns. Various features can then be extracted from candidate target HRRPs and matched with the features extracted from known target HRRPs to effect target discrimination and recognition.

## 4. VECTOR DISCRIMINATION TECHNIQUES

Vector discrimination techniques use some limited aspect of polarization to effect target/clutter discrimination. These techniques are often implemented using a frequency-agile waveform to enhance discrimination performance. Many vector discrimination techniques have been developed and tested over the years with varying results. These discriminants are usually based on some simple polarization-dependent scattering property that differs for a target of interest and clutter. For example, probably the simplest vector discrimination used today is the one that discriminates between a complex target and rain clutter through the use of circular polarization. Since the shapes of raindrops can usually be well approximated by spheriods, an incident circularly polarized EM wave will have the opposite sense of circular polarization, i.e., it will be cross-polarized, when backscattered from rain. The backscatter from complex targets, however, will have both copolarized and cross-polarized components. Therefore, if the radar transmits a given sense of circular polarization and receives that same polarization sense (i.e., receives only the copolarized component of the backscattered signal), then the rain clutter backscatter will have been effectively eliminated, albeit at the expense of some of the target backscatter signal.

While the rain clutter suppression technique described above only requires one radar polarimetric channel, most vector discrimination techniques require two polarimetric channels. Typically, in the implementation of these discriminants, a signal with a given polarization is transmitted, and both the copolarized and cross-polarized components are received. For example, in the *polarization ratio discriminant*,[1] the ratio of the copolarized receive channel to the cross-polarized receive channel is formed. The value of this ratio can vary significantly for a target return and a clutter return.

Many other vector discriminants exist (e.g., pseudocoherent detection[2,3]), but they all have one thing in common; they all require the implementing radar to be in a specific polarization transmit/receive configuration. Therefore, in general, a radar configured to support one vector discriminant may not be configured to support another. This can be a severe limitation since no one vector discriminant exploits all the polarization information that is available to the radar sensor. Therefore, multiple vector discriminants may be necessary to achieve the discrimination performance level desired. This is especially true in dynamic target/clutter scenarios. These limitations of vector algorithms are overcome with the implementation of matrix algorithms.

## 5. MATRIX DISCRIMINATION TECHNIQUES

Matrix discrimination techniques require that the implementing radar measure the full PSM. These techniques are potentially the most powerful polarimetric discriminants, since they can use all of the polarimetric information available in the received signals to effect discrimination. This information includes target geometrical information such as maximum RCS, target orientation, target symmetry, target odd/even bounce characteristics, and target polarizability. In addition, this backscatter is independent of the radar's transmit and receive antenna polarizations, effectively decoupling radar-induced polarization effects (other than radar system imperfections) from the data.

Several approaches have been taken in an attempt to achieve discrimination using matrix techniques. One approach is to use the *null-polarization technique*.[4] Every object has at least one copolarization null (most have two). The measured RCS of an object will be zero (a null) when the transmit/receive antenna polarization is one of these copolarization nulls. Therefore, if a copolarization null of clutter is known, then by simply transmitting and receiving this polarization, the clutter return would be eliminated. Unfortunately, the radar-clutter interaction is usually non-stationary, i.e., the clutter return is partially polarized. Thus, the clutter null-polarizations are time and space dependent and random. Still, if the null polarizations of the clutter remain fairly well "behaved" during the observation time (i.e., do not vary greatly), then an optimum transmitting and receiving polarization can be found to maximize the target-to-clutter ratio.[5]

Another matrix discrimination technique is the *matrix-parameter technique.* This technique involves using the elements of the PSM or extracting the five Huynen parameters[6] from the PSM and using these elements/parameters (or functions thereof) to effect target discrimination. The elements of the Mueller matrix may also be used for this purpose. These parameters are then used as features in a multi-dimensional feature space to which pattern recognition algorithms are applied to separate the target and clutter signatures.[7] Discrimination results using this technique have been mixed. As mentioned above, the radar return from targets and clutter is usually partially polarized. Therefore, matrix parameters tend to be "noisy" and discrimination performance in some cases can actually be reduced using this technique.

The last class of matrix discrimination techniques discussed here are the *matrix decomposition techniques.* As was mentioned above, the radar returns from targets and clutter are usually partially polarized. This partial polarization can be caused by random target motions, radar platform motions, or changes in the radar waveform. When this happens, the elements of the PSM are complex (usually zero-mean) random variables. Therefore, an ensemble of PSMs must be used to represent the radar return. Before taking ensemble averages, the PSMs are usually transformed into Mueller matrices or density matrices[8,9] to obtain meaningful (non-zero) results upon averaging. The resulting averaged Mueller or density matrix can then be decomposed into two or more matrix components, one of which represents a target (or object) that gives rise to a fully-polarized radar return. The hope is that this matrix component will contain the relevant geometrical attributes of the target (or clutter) to which the polarization of the radar signal was sensitive. Therefore, this matrix component (and the other residue matrix components) can be analyzed for features to be used in the discrimination processing.

There are an infinite number of ways to decompose a Mueller or density matrix into two or more constituent matrix components. One way is to use the Huynen decomposition method[6] and another way is to apply the characteristic decomposition.[8,9] The Huynen decomposition method, which can be applied to both Mueller and density matrices, is a physically realizable, unique decomposition which results in two matrix components. One component represents a fully-polarized radar return, and the remaining component represents a partially-polarized radar return. The Huynen decomposition has physical merit and has been shown to be effective in target/clutter discrimination.[8] The characteristic decomposition is obtained by solving the characteristic (or eigenvalue) equation for the density matrix. It results in three matrix components; one representing a fully-polarized radar return, one representing a partially-polarized radar return, and one representing a fully unpolarized radar return. The characteristic decomposition has also shown promise as an effective target/clutter discrimination technique.[8]

## 6. SUMMARY

Target geometrical features must be exploited to achieve airborne radar detection/discrimination of stationary man-made ground vehicles in a ground clutter background. These geometrical features are sensed by the radar either through the use of high resolution techniques or polarization techniques. After radar detection, in which noise interference and benign clutter returns are eliminated, the returns from stationary targets of potential interest are sorted from the clutter false-target returns in the discrimination processor. Discrimination processing techniques fall into three broad categories: scalar, vector, and matrix techniques. The latter two involve the use of the radar signal's polarization, with the matrix techniques requiring the measurement of the full PSM. These techniques have shown promise for being effective in stationary target/clutter discrimination.

## 7. REFERENCES


1. Hayes, R. D. and Eaves, J. L., 'Study of Polarization Techniques for Target Enhancement,' AF 33(615)-2523, AD-316270, Report A-871, Georgia Institute of Technology, Atlanta, Georgia, USA, August 1966.

2. Holm, W. A., 'Polarimetric Fundamentals and Techniques', Principles of Modern Radar, Van Nostrand Reinhold, New York, 1987, Chapter 20.

3. Echard, J. D., et al., 'Discrimination Between Targets and Clutter by Radar,' DAAG29-78-C-0044, Report A-2230, Georgia Institute of Technology, Atlanta, Georgia, USA, December 1981.

4. Weisbrod, S. and Morgan, L. A., 'RCS Matrix Studies of Sea Clutter,' Teledyne-Micronetics Report No. R2-79, N00019-77-C-0494, AD B036684, January 1979.

5. Ioannidis, G. A., 'Optimum Antenna Polarizations for Target Discrimination in Clutter,' IEEE Transactions on Antennas and Propagation, Vol. AP-27(3), May 1979, pp. 357-363.

6. Huynen, J. R., Phenomenological Theory of Radar Targets, Ph.D. Dissertation, Drukkerij Bronder-Offset N.V., Rotterdam, 1970.

7. Holm, W. A., 'Polarization Scattering Matrix Approach to Stationary Target/Clutter Discrimination,' Colloque International sur le Radar, May 1984, pp. 461-465.

8. Holm, W. A., 'MMW Radar Signal Processing Techniques,' Principles and Applications of Millimeter-Wave Radar, Currie and Brown, editors, Artech House, Dedham, Massachusetts, USA, 1987, Chapter 6.

9. Holm, W. A. and Barnes, R. M., 'On Radar Polarization Mixed Target State Decomposition Techniques,' Proceedings of the 1988 IEEE National Radar Conference, April 1988, pp. 248-254.

# TARGET DETECTION AND CLASSIFICATION WITH POLARIMETRIC HIGH RANGE RESOLUTION RADAR

A. Farina, F. Scannapieco, and F. Vinelli
Radar Dept, Selenia SpA
Via Tiburtina Km. 12,400
00131 Rome, I

ABSTRACT. This paper describes the signal processing theory of a polarimetric radar having very high resolution along range. The radar is meant to be used for surveillance to detect and classify targets against a clutter background. A very high range resolution radar (VHR[3]) is characterised by a very short pulse which cuts the target into many pieces from which separate echoes can be collected. This radar concept allows to obtain better detection and classification performance with respect to conventional low resolution radar transmitting a pulse which encompasses the whole target extension.

## 1. INTRODUCTION

A high resolution radar is commonly regarded as a sensor which detects and measures details of a target of interest; this contrasts to the low resolution radar in which the whole target, when detected, is represented by just one plot. High resolution may occur in the angular and range domains. High resolution in the angular domains (i.e. azimuth and elevation) is obtained by resorting to SAR (synthetic aperture radar) techniques such as the spotlight SAR and the ISAR (inverse SAR). Nonlinear spectral estimation, i.e. the Maximum Entropy Method and the like, is another approach to tackle the problem. High resolution in down range is achieved by using a short duration pulse. Such short pulse can be obtained by coding a long pulse or by direct generation and transmission of the same short pulse. In these cases the radar bandwidth is available instantaneously; however, other techniques are conceived in which the required bandwidth is made available over a longer time interval [1, Ch.5]. The corresponding systems, referred also to as synthetic high range resolution radar, play a key role in the



*W.-M. Boerner et al. (eds.), Direct and Inverse Methods in Radar Polarimetry, Part 1*, 1021–1041.

instrumentation radars while they presently are of limited use in the surveillance. This paper considers surveillance radars having very high resolution along range.

The technology of pulse coding is well established; practical demonstrations of the capability to achieve range resolution of few cms. are available today [2]. The generation and transmission of a short pulse is generally related to the use of a high carrier frequency (e.g.X, millimetric band) so that the required instantaneous bandwidth is few percents of the carrier frequency. However, the digital processing of the base-band signal still remains a formidable problem to solve. For completeness, it is worth mentioning the carrier free radar (CFR) [3] system concept which refers to the transmission of a so short pulse which cannot contain any carrier in it. The capabilities of such a system for surveillance purposes has not been demonstrated and many theoretical and practical issues remain without satisfactory answers. Nevertheless, achievements in the technology of CFR have been obtained [4].

In a more comprehensive way, high resolution techniques should be regarded as means to obtain improved performance in the whole suite of functions of a modern radar. The suite includes detection, tracking, classification, and imaging. Such challenging statement will be partially proved in this paper. More in detail, this paper considers a very high range resolution radar ($VHR^3$), that is a sensor having a so large instantaneous bandwidth to provide a down range resolution considerably lesser than the target extent. Consequently, the radar receives more than one echo from the target in reply to a single transmitted short pulse. The echoes, when properly processed, allow to obtain better detection and classification performance when compared to a conventional low resolution radar. It is also explored the benefit of having a $VHR^3$ with polarimetric features, i.e. a radar having a large bandwidth and, contemporary, two receiving channels with orthogonal polarizations. Numerical results are provided showing the detection and classification capabilities of such advanced radar.

Apparently, limited research has been done in the area of detection with very large bandwidth radars. References [5], [6], and [7] represent the limited effort undertaken in this promising topic. The situation does not change for the polarimetric $VHR^3$. A limited number of papers([8] thru [11]) describe experimental results obtained with high resolution polarimetric set-up. A theory of detection for such polarimetric radar has not been developed up to now, even though remarkable references (e.g. [12]) describe the concept of vector signal processing for low resolution polarimetric radar.

The aim of this paper is to shed some light in the intriguing theories of detection and classification for $VHR^3$ with polarimetric features. This is in reply to the recommendations of the first NATO-ARW-DIMRP-1985, see [13]; consequently, there is an ideal continuation between the previous and the present Workshop.

The ensuing material is organised as follows. Chapter 2 affords the theory of detection for a $VHR^3$. Specifically, Sect. 2.1 considers processing schemes and corresponding detection performance when just one pulse is transmitted towards the target. Subsequently, the case of more than one pulse is described in Sect. 2.2. The polarimetric radar is tackled in Sect. 2.3; again, schemes and corresponding performance are provided. Chapter 3 is devoted to the classification. In particular, Sect. 3.1 illustrates the classification based on the range profile of the target, i.e. the scattered signal amplitude in down range. Sect. 3.2 deals with the classification according to the Doppler frequency shift. Finally, the classification based on the range profiles received in the two channels of a polarimetric radar is described in Sect. 3.3.

## 2. DETECTION WITH A VERY HIGH RANGE RESOLUTION RADAR

The Chapter briefly outlines a novel theory of detection for $VHR^3$ sensors. The Chapter describes a number of detection schemes able to process the echoes from N cells along range and M cells along azimuth. The information concerning the polarization states of the target and the clutter is also incorporated into the schemes. The detection performance of these schemes are also evaluated and compared with the performance offered by a conventional low resolution radar. A number of interesting results are obtained, such as: (i) the optimum number $\hat{N}$ of target slices along range, and (ii) the operational conditions in which the polarization information is useful as compared to the case of processing just the NM echoes from the target.

### 2.1 Detection for one transmitted pulse

Assume to have received N echoes scattered by a target hit by one transmitted pulse. The problem is to process in an optimum way the N echoes to detect a target against clutter background and thermal noise.

If the range profile of the target scattered energy would be known a-priori, an optimum filter (i.e. which maximizes the signal-to-noise ratio, SNR, at the output) along range could be applied. In practice, sub-optimum approaches have to be used. Here, the detection strategies 1 and 2, shown in fig. 1, have been considered. Strategy 1 is of the type "K out of N" with K = 1 and where N is the number of range cells encompassing the target extension along range. Strategy 2 corresponds to a non-coherent integration of the scattered signal along range. For comparison, the same figure illustrates the conventional detection scheme (strategy 3) adopted for low-range resolution radar. Throughout the paper it is assumed that the target and clutter signals are uncorrelated in down range.

First of all, the performance of the three detection schemes is evaluated when the disturbance is mainly due to clutter (e.g. from ground) In this situation of high C/N (clutter-to-noise-ratio) value, there is a great benefit in using high range resolution radar. As a matter of fact, the signal-to-clutter ratio (S/C) is the same for the three strategies, however the high resolution strategies have more opportunities to detect the target. It can be shown that the detection performance of strategy 1 depends on the distribution of target energy along range. More specifically, strategy 1 gives the best performance for target energy concentrated in one or few range cells. The performance of strategy 2 is independent of the energy distribution of target. Strategy 2 should be preferred over strategy 1 when the target energy is evenly distributed along range. Fig. 2 illustrates the receiving operating characteristics (ROCs) of strategy 2 against strategy 3 having N as parameter. It is seen that the SNR gain of the $VHR^3$ increases with N; in particular, assuming a probability of false alarm ($P_{FA}$) of $10^{-6}$ and a probability of detection ($P_D$) of 0.8, the gain is 8dB for N = 10 and raises to 14,5dB for N = 100. Fig. 3 illustrates the ROCs for strategy 1 and for different distributions of target energy along range. The SNR gain is larger for a target having scattering centers concentrated in down range as compared to the case of scattering centers evenly distributed along the target extent. In summary, figures 2 and 3 show the considerable benefit of using the $VHR^3$ concept when C/N >>1.

This type of system concept has limited performance against thermal noise. In fact, as the range resolution increases, the receiver bandwidth increases as well and the target is embedded in more noise power. In this operational situation, the use of very low noise receiver technologies is mandatory.

Let now consider the more interesting and practical case in which the target is embedded in both clutter and noise. In this situation, there are two contrasting effects to consider. As the number N of target slicing increases there is a detection benefit against clutter, accompanied by a detection reduction against thermal noise. Consequently, it is interesting to evaluate the optimum number $\hat{N}$ of target slicing as a function of the C/N value, and the detection strategy adopted: fig. 4 shows the key result. It follows that the best operational bandwidth of the radar, to be used to obtain the maximum detection performance, is a function of the target length, the C/N value and the detection strategy. Also in the case of clutter and noise contemporary present, the ROCs have been evaluated and fig.5 gives an explanatory example. Here the strategy 1 has been considered; the C/N is 30dB and the SNR is 39dB while the detection probability is drawn against the number N of target slicing. It is seen that $P_D$ is maximum around the value N predicted by fig. 4.

## 2.2 Detection for more than one transmitted pulse.

Assume to transmit M pulses T sec. apart, where T is the pulse repetition time (PRT). The number of received echoes is NM. This case is important because the clutter can be better suppressed and the required transmitting power reduced. These benefits require the solution of the following "range migration" problem. The target radial velocity $\dot{R}$ induces, in addition to the usual Doppler frequency shift, a range migration of the received radar echoes from contiguous azimuthal pulses. In other words, a same target scatterer appears at different ranges in response to different azimuthal pulses. Figure 6 illustrates the problem for a specific example: the short transmitted pulse has a time duration $\tau$ = 1 nsec., the pulse repetition time is T = 1 msec. If the target radial velocity $\dot{R}$ is 150 m/s, the range migration is one cell each PRT. This situation prevents the use of the optimum coherent processing which allows contemporary the coherent target integration and clutter cancellation.

Two procedures are conceived to tackle the problem. The first refers to the use of an estimate of R to compensate for the Doppler shift and the range migration of the target. In the second case, here considered in more detail, the clutter, which does not migrate, is removed by coherent cancellation and the target is enhanced by non-coherent integration. Figure 7 illustrates the corresponding processing scheme. For each i-th range cell a stream $\underline{z}i$ of M azimuthal samples is available. A coherent cancellation of clutter is achieved by means of a finite impulse response (FIR) filter having a weight $\underline{W}i$. The filter weight $\underline{W}$ is calculated by eigen analysis of the clutter covariance matrix, which is assumed to be a-priori known in this context. The output "yi" of the filter is envelope detected; the whole set of envelopes from the N range cells are non coherently integrated by means of either strategy 1 or strategy 2. A sample of detection performance for the strategy 2 is shown in fig. 8. This refers to a number M = 2 of azimuthal pulses and several values of the slicing factor N. The one-lag autocorrelation coefficient of clutter is $\rho = 0.9$, the mean Doppler frequency of the target is $F_s = 0.5 * PRF$. The thermal noise is assumed negligible. The figure compares the detection performance in presence of target migration of one range cell and in absence of range migration. It is demonstrated that the probability of detection increases with M because of the better clutter cancellation and the non-coherent integration of target. The performance improvement with N is independent of the number M of azimuthal pulses. The range migration reduces the detection performances.

## 2.3 Detection for polarimetric radar.

The last topic covered in this Chapter is concerned with the use of polarization in $VHR^3$. It is assumed to have a system being able to transmit a wave with a selectable polarization and to receive the backscattered energy by means of two orthogonally polarized channels. This study is motivated by the need to assess the benefit of polarization in terms of target vs clutter discrimination. For sake of simplicity, the polarization states of target and clutter are taken deterministic, stationary and a-priori known. This means that target and clutter are represented by points onto the Poincarè sphere. The concepts of Stokes vector and Stokes reflection matrix are not used in this context. Here, the 2MN radar echoes are processed to remove the clutter and enhance the target by the contemporary exploitation of the polarization domain, the high resolution in range and the Doppler analysis in azimuth. For each range cell it is advisable to apply the same coherent processing which can be adopted in a low resolution radar. After such kind of coherent filtering each filter output is envelope detected and the full stream of envelopes are processed thru either strategy 1 or strategy 2. The detection scheme is that shown in Fig. 9.

Let consider in some more detail the processing applied to the i-th range cell. Indicate with $\underline{Z}_{Hi}$ and $\underline{Z}_{Vi}$ the set of M samples each received by the horizontally and vertically polarized channel, respectively. Indicate with $\underline{S}_{Hi}$ and $\underline{S}_{Vi}$ the a-priori known values of the target signal. The disturbance (clutter and noise) are assumed to have a zero-mean Gaussian probability density function with the following covariance matrix, [14]:

$$\underline{M} = \underline{M}_p \oplus \underline{M}_t + \underline{I}, \qquad (1)$$

where the operator $\oplus$ indicates the Kronecher matrix product, $\underline{M}_p$ is a (2,2) matrix incorporating the state polarization of clutter, $\underline{M}_t$ is a (M,M) matrix in the time domain. Briefly speaking, $\underline{M}_p$ contains parameters related to the orientation and shape of clutter polarization ellipse; $\underline{M}_t$ contains the power and the time correlation properties of the clutter. Finally, the identity matrix $\underline{I}$ in equn. (1) represents the thermal noise with unity power. The processing for each range cell is as follows:

$$y_i = [\underline{Z}^T_{Hi} \; \underline{Z}^T_{Vi}] \; \underline{M}^{-1} \begin{bmatrix} \underline{S}_{Hi} \\ \underline{S}_{Vi} \end{bmatrix}^* \qquad (2)$$

A sample of detection performance is shown in fig. 10. It is assumed that the target and the clutter have different polarization states on the Poincarè sphere: the angular distance between the two states is $\theta°$. The C/N value is 20dB, the one-lag clutter autocorrelation coefficient $\rho$ is 0.99, the target energy is evenly distributed along range, the $P_{FA}$ value has been set to 10**-6 and the $P_D$ value is maintained to 80%. The curves show the SNR vs the slicing factor N having as parameter the polarization distance $\theta$ and the number M of azimuthal pulses. It is seen that the polarization greatly helps when M = 1; when more azimuthal pulses are available, the relevance of the polarization decreases because the clutter has already been cancelled by the Doppler processing in azimuth.

## 3. CLASSIFICATION WITH A VERY HIGH RANGE RESOLUTION RADAR

The second part of the paper is devoted to the problem of target classification. A number of processing schemes are illustrated and the corresponding classification performance are evaluated. The processing schemes are conceived by resorting to the theory of multiple hypotheses testing [14]. Each target class is represented by a hypothesis; the processor selects that hypothesis which best fits the received echo sequence in spite of clutter and receiver noise which overwhelm the scattered signals from the target.

To set up the mathematical model of the multiple hypotheses testing (MHT), consider the simple case of two classes of targets; extensions to more classes is almost straightforward. The MHT model for this case is as follows:

$$\underline{Z} = \underline{S}_1 + \underline{C} + \underline{N} \; ; \; H_1$$
$$\underline{Z} = \underline{S}_2 + \underline{C} + \underline{N} \; ; \; H_2$$

$$\underline{Z} = \underline{C} + \underline{N} \; ; \; H_0 \qquad (3)$$

where $\underline{Z}$ is a vector containing the set of received samples, $\underline{S}_1$ and $\underline{S}_2$ represent the samples expected by target class 1 and class 2 respectively; $\underline{C}$ is the clutter vector and $\underline{N}$ is the thermal noise vector. In the more general case, the common dimension of the vectors is 2MN when the radar has two orthogonally polarized receiving channels and process M azimuthal samples and N samples in down range.

Examples of classification performance are worked out for two target classes having different amplitude and polarization profiles in range and different Doppler frequency signatures.

3.1 Classification based on the target range profile

Assume to have just N samples to process (i.e. the radar is not polarimetric and M = 1); the two target classes can be distinguished on the basis of their different profiles of signal amplitude in down range. The corresponding processing scheme is that shown in fig. 11. After envelope detection, the set $\underline{X}$ of envelopes is processed in parallel by two filters: namely, $FIR_1$ and $FIR_2$. The filter weights, $\underline{W}_1$ and $\underline{W}_2$, are matched to the expected amplitude profiles of the two targets. The maximum value, $Y_{MAX}$, between the two filter outputs, $Y_1$ and $Y_2$, is compared with a suitable threshold $\lambda$ for detection. Finally, a classification logic selects among the three hypotheses, i.e. $H_0$, $H_1$, and $H_2$. It should be noted that the scheme of Fig. 11 is a generalization of strategy 2 illustrated in Fig. 1-b.

Fig. 12 is an explanatory example of the performance pertaining to the classifier of Fig. 11. The joint probability of correct classification and target detection, referred also to as $P_{cc,d}$, is drawn versus the signal-to-clutter ratio (thermal noise is negligible). The parameter of the curves is the number N of target slicing. The figure refers to two target classes: the first is characterised by an evenly distributed amplitude profile, while the second profile is concentrated in just one range cell. It is noted that $P_{cc,d}$ increases with N, thus showing the benefit of having a VHR[3] also for classification purposes. Of course, the classification performance degrades if the "distance" between the two target classes reduces.

3.2 Classification based on the target mean Doppler shift

This Section considers the case of a non polarimetric radar able to process NM echoes. The two target classes can be distinguished on the basis of their range profiles and their mean Doppler frequency values. This means that more than three hypotheses have to be tested. Consider, for instance, two mean Doppler values and two amplitude profiles: the hypotheses to be tested are five. The corresponding processing scheme is quite involved; here it is preferred to show the scheme pertaining to the classification based just on the mean Doppler frequency value (see fig. 13). For each range cell, the M samples are processed in parallel by two coherent FIR filters indicated by $FIR_{D1}$ and $FIR_{D2}$. They are filters matched to the target Doppler frequency $f_{D1}$ and $f_{D2}$ respectively. They also allow to suppress the clutter. The filter outputs are envelope detected and not coherently integrated; the corresponding two signals $y_1$ and $y_2$ are directly used for detection and classification.

An example of classification performance is shown in fig. 14. The example refers to M = 2 azimuthal pulses and a number N of target slices ranging from 1 to 10. The two target classes have the same uniform amplitude range profile, but two different mean Doppler frequency values, namely: $f_{D1} = 0.5$ PFR and $f_{D2} = 0.3$ PFR. The interfering clutter has one lag autocorrelation coefficient $\rho = 0.9$; thermal noise is assumed to be negligible. Once again, it is noted the advantage of the $VHR^3$ system concept over the conventional low resolution radar.

## 3.3 Classification with a polarimetric $VHR^3$

In the more general case, we should process 2MN echoes in order to distinguish target classes on the basis of their amplitude range profiles received in the two polarization channels and on the basis of their mean Doppler frequency values. For simplicity, the following hypotheses are considered, namely: (i) one azimuthal sample, M = 1; (ii) polarization states of target and clutter, evenly distributed along range, and a-priori known; and (iii) classification not based on the target range profile.

The processing scheme is that shown in fig. 15. The 2N echoes are coherently processed in parallel by two FIR filters $F_1$ and $F_2$ which are matched to the polarization states of the two target classes and, at the same time, provide clutter filtering in the polarization domain. Downstream, the usual non-coherent processing is implemented for detection and classification.

Fig.16 gives an example of the classification performance.

## 4. CONCLUSIONS

In this paper several novel processing schemes have been

presented. They are conceived for detection and classification purposes to be used with a polarimetric VHR[3]. The corresponding performance demonstrate the benefit of using such kind of advanced system concept. Of course, the complexity of the processing schemes is related to the complexity of the corresponding mathematical models taken for target and clutter. In this paper, simple models were worked out to proof the working principle of the radar. Future work will be mainly devoted to conjugate more realistic target and clutter models with adequate processing schemes.

## REFERENCES

[1] WEHNER, D.R.
"High Range-Resolution Radar", Artech House Inc, Dedham, 1987.

[2] LOSQUADRO, G., et al.
"The Radar Altimeter for ERS-1 Satellite", Proc. of the Intl Radar Symposium India, IRSI-83, Bangalore, October 1983, pp. 393-398.

[3] HARMUTH, H.F.
"Nonsinusoidal Waves for Radar and Radio Communications", Academic Press, N.Y. 1981.

[4] VAN ETTEN, P.
"The Present Technology of Impulse Radars", International Conference Radar-77, London (UK), pp. 535-539.

[5] HUGHES, P.K.
"A High-Resolution Radar Detection Strategy", IEEE Trans on Aerospace and Electronic Systems, Vol. AES-19, N.5, September 1983, pp. 663-667.

[6] VANNICOLA, V.C. HILLMAN, K.G., and W.L. SIMKINS
"Detection Performance for over Resolved Targets", Agard Conf. Proc. N. 381, Multifunction Radar for Airborne Applications, 1985, p. 22.

[7] GNISS, H. KRUCKER, K., MAGURA, K., and D. PERKUHN
"Problems of Signal Processing in a High Resolution Radar-Synthetic Aperture Imaging of Rotating Targets with Narrow band and Broadband Signals", Proc. of the Intl. Radar Conf., Paris, 1978, pp. 242-250.

[8] MANSON, A.C., and W.M. BOERNER

"Interpretation of High-Resolution Polarimetric Radar Target Down-Range Signatures Using Kennaugh's and Huynen's Target Characteristic Operator Theories", Inverse Methods in Electromagnetic Imaging, Part 2, pp. 695-720, W.M. Boerner et al. (eds), 1985, D. Reydel Publishing Company.

[9] RACKSON, M.M., WEI, P.S.P., and T.H. MEYER "High Resolution Polarimetric Radar Precision Limits", Proc. of the 1984 IEEE National Radar Conference, Atlanta (Georgia), pp. 8-10.

[10] RUSSEL, R.F., SEDENQUIST, F.W., and D.P. GAINES "Frequency Agile/Polarimetric Radars. Simulation and Testing", Proc. of the 1984 IEEE National Radar Conference, Atlanta (Georgia), pp. 58-62.

[11] WEI, P.S.P., RACKSON, M.M., BRADLEY, T.C. MEYER, T.H., and J.D. KELLY "Study of Two Scatterer Interference with a Polarimetric FM/CW Radar", Inverse Methods in Electromagnetic Imaging, Part. 2, pp. 673-682, W.M. Boerner et al. (eds), 1985, D. Reidel Publishing Company.

[12] VANNICOLA, V.C., and, L. STANLEY "Polarization Vector Signal Processing for Radar Clutter Suppression", Inverse Methods in Electromagnetic Imaging, Part 2, pp. 739-770. W.M. Boerner et al. (eds), 1985, D. Reidel Publishing Company.

[13] WRIGHT, J. "VI.3 Final Report of Working Discussion Group W-C on Polarization Utilization in High Resolution Imaging", Inverse Methods in Electromagnetic Imaging, Part 2, pp. 1277-1280, W.M. Boerner et al. (eds), 1985, D. Reidel Publishing Company.

[14] FARINA, A. and A. VISCONTI "Classification of Radar Targets by means of Multiple Hypotheses Testing", Proc. of the 1987 IEE Intl Radar Conference, London (UK), pp. 73-78.

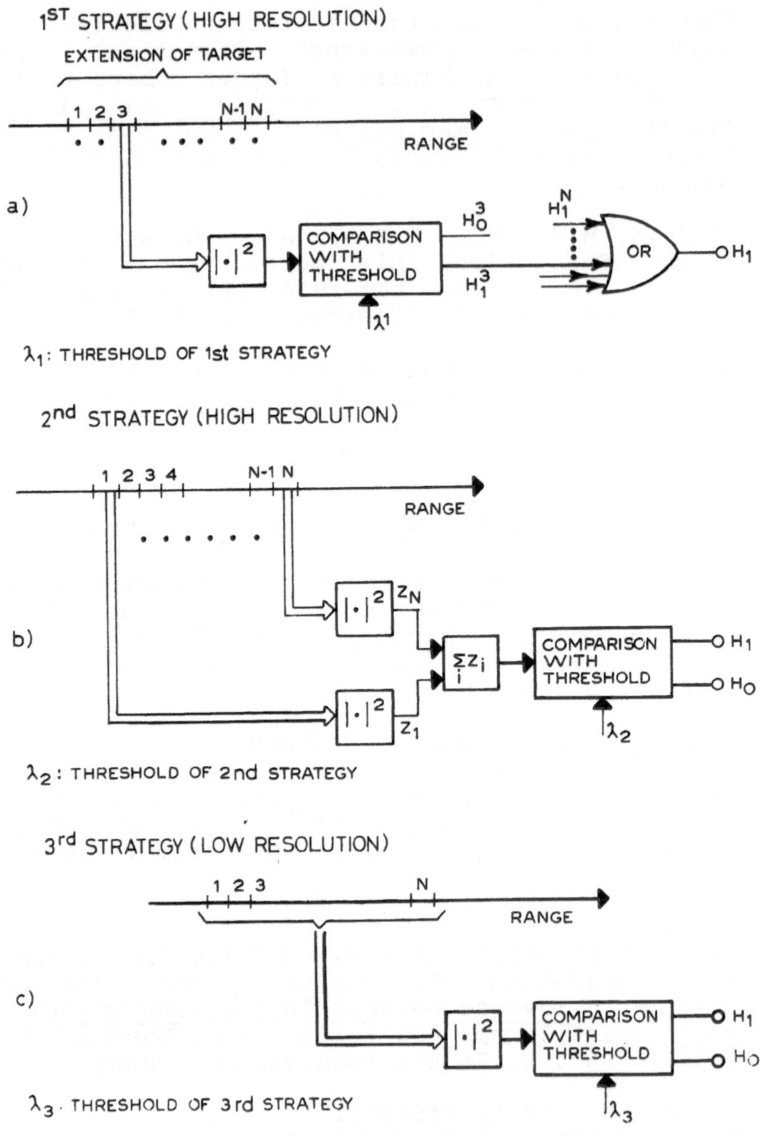


FIG. 1 - DETECTION SCHEMES

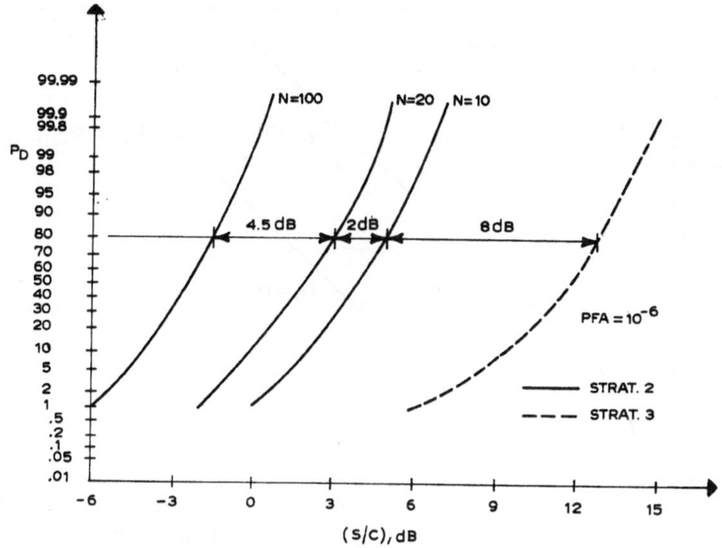


FIG. 2 - COMPARISON OF DETECTION PERFORMANCE BETWEEN STRATEGY 2 AND STRATEGY 3 IN CLUTTER

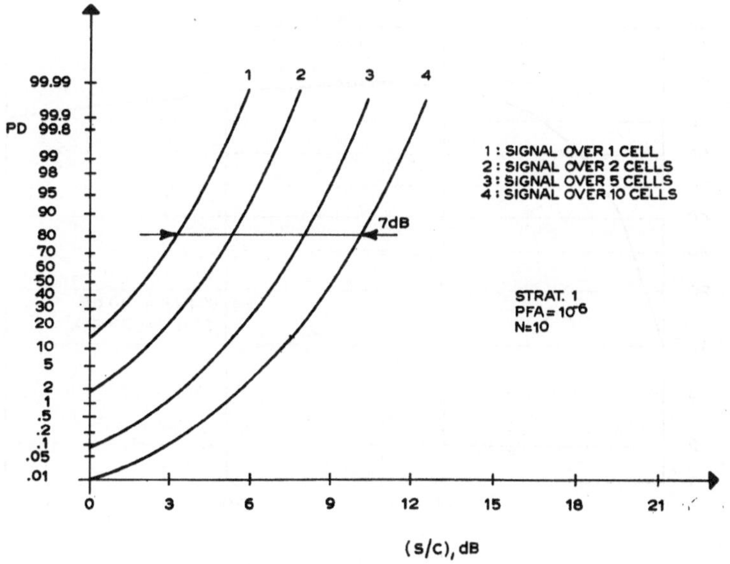


FIG. 3 - DETECTION PERFORMANCE FOR STRATEGY 1 IN CLUTTER

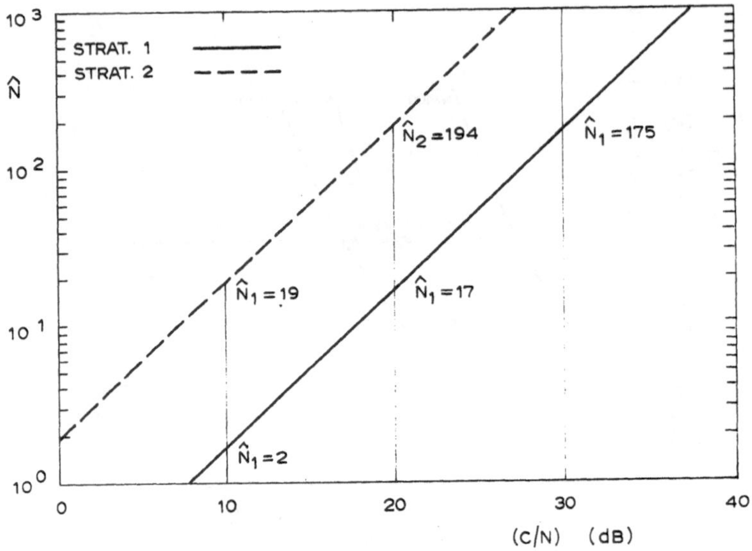


FIG. 4 - THE OPTIMUM VALUE N OF THE SLICING FACTOR ALONG RANGE FOR STRATEGIES 1 AND 2

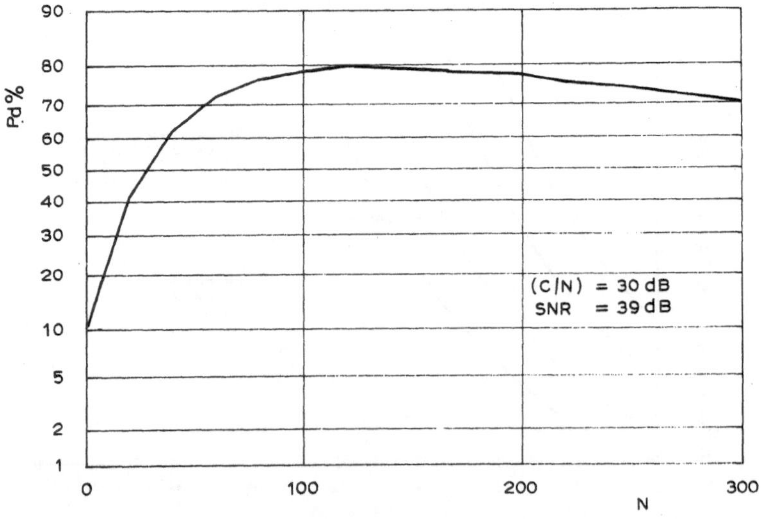


FIG. 5 - DETECTION PERFORMANCE OF STRATEGY 1 IN CLUTTER AND NOISE

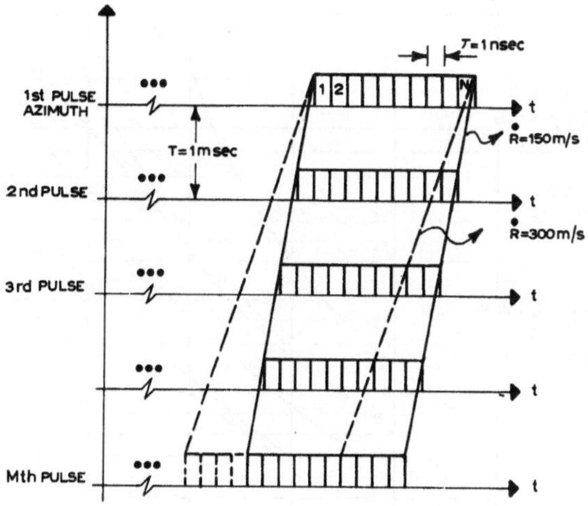


FIG. 6 - THE RANGE MIGRATION EFFECT

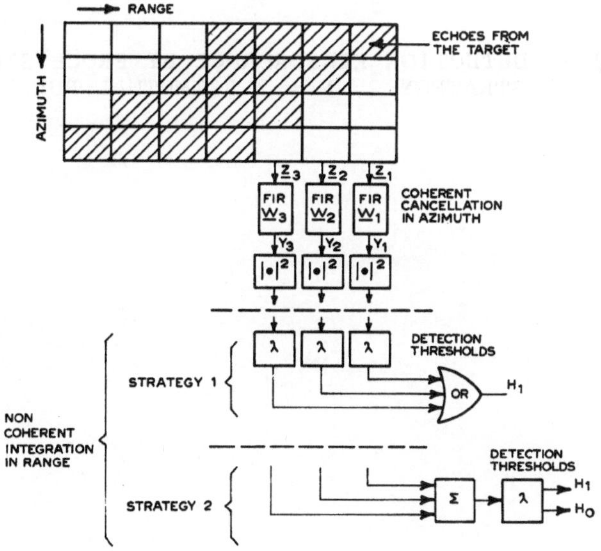


FIG. 7 - PROCESSING SCHEME FOR MORE THAN ONE AZIMUTHAL PULSE

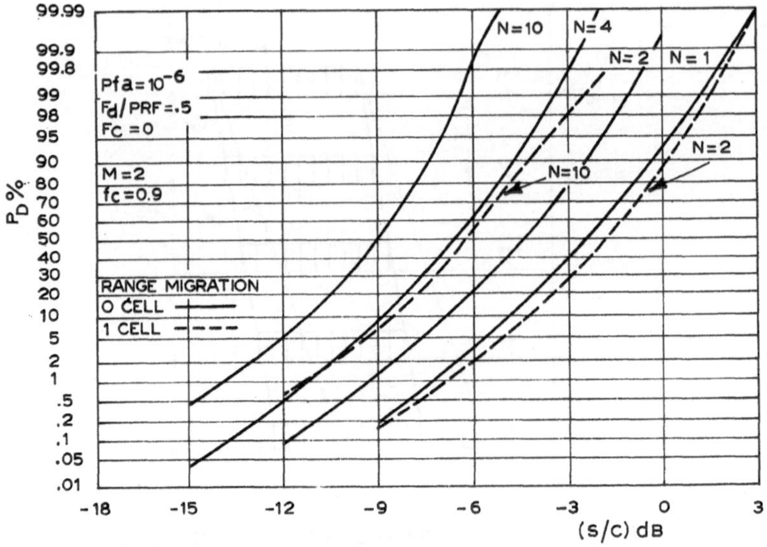


FIG. 8 - DETECTION PERFORMANCE BY PROCESSING, WITH STRATEGY 2, M = 2 AZIMUTHAL PULSES

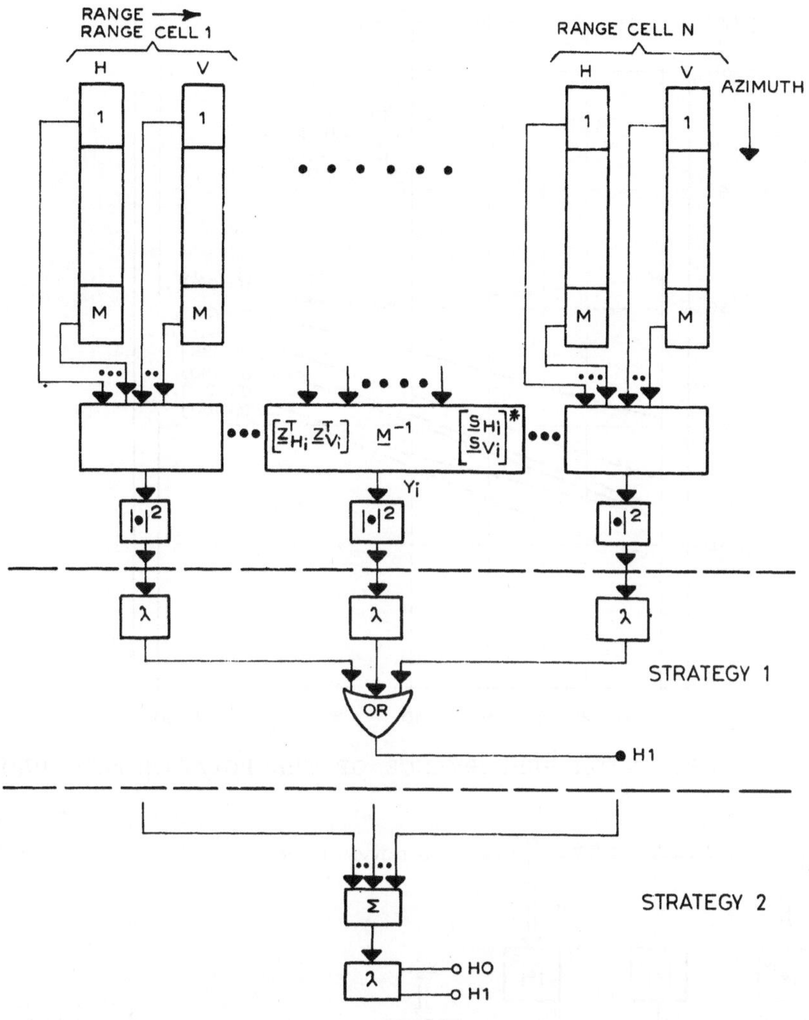


FIG. 9 - DETECTION SCHEME FOR A POLARIMETRIC HIGH RANGE RESOLUTION RADAR

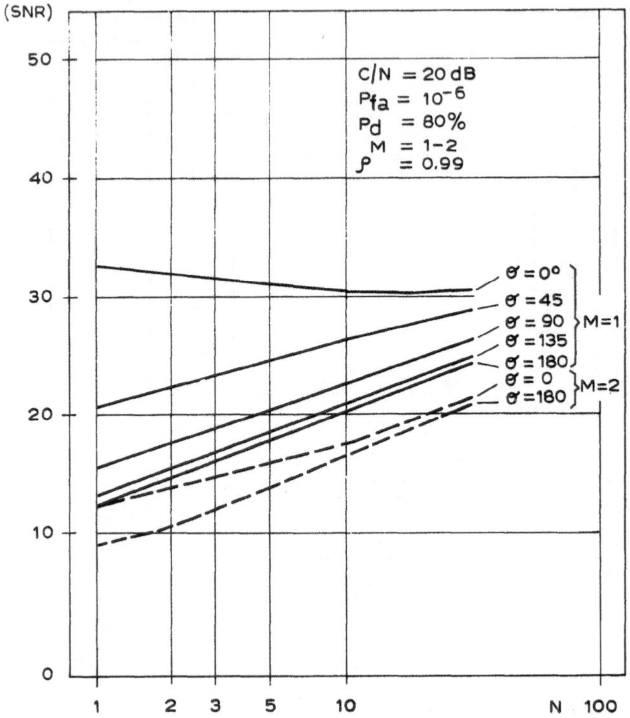


FIG.10 - DETECTION PERFORMANCE OF THE POLARIMETRIC RADAR

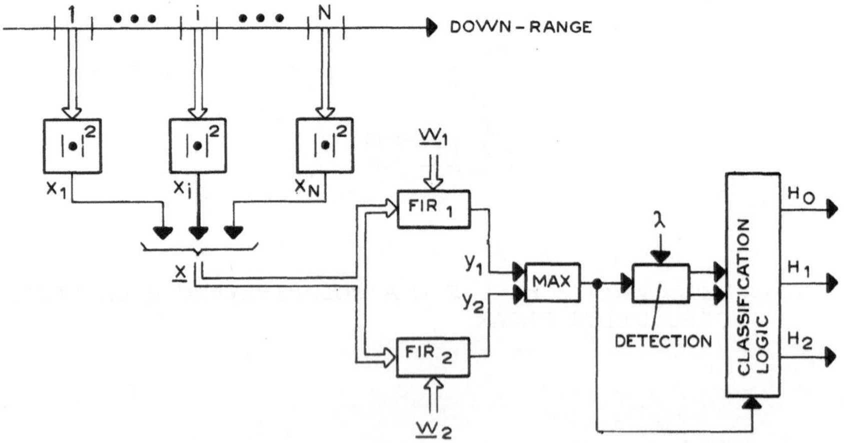


FIG.11 - CLASSIFIER BASED ON TARGET RANGE PROFILE

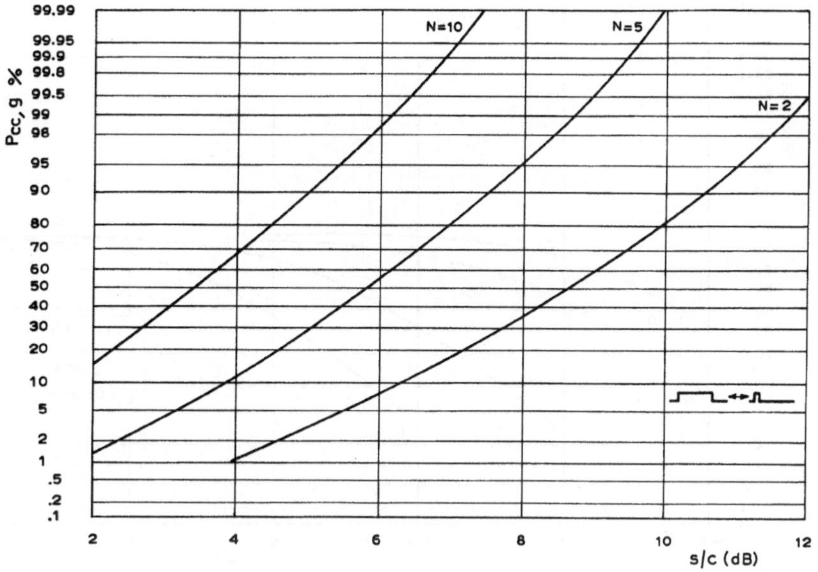


FIG.12 - PERFORMANCE OF A CLASSIFIER BASED ON AMPLITUDE PROFILE IN DOWN RANGE OF TARGET

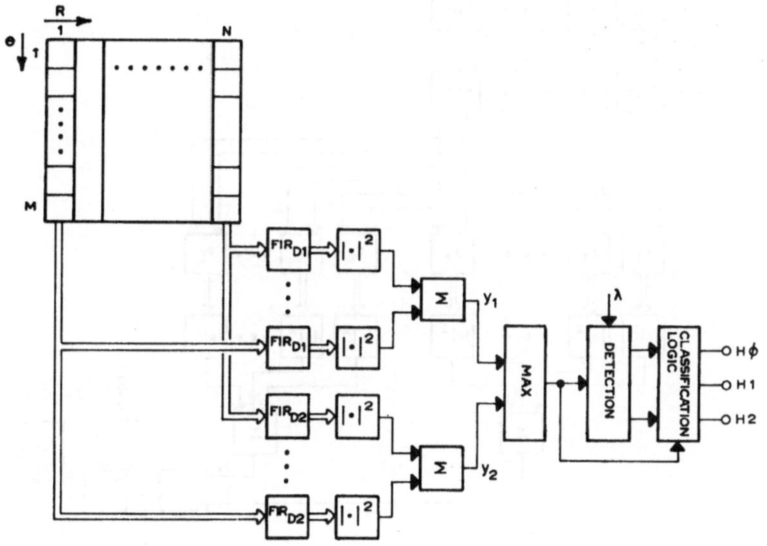


FIG.13 - CLASSIFIER BASED ON MEAN DOPPLER FREQUENCY VALUE

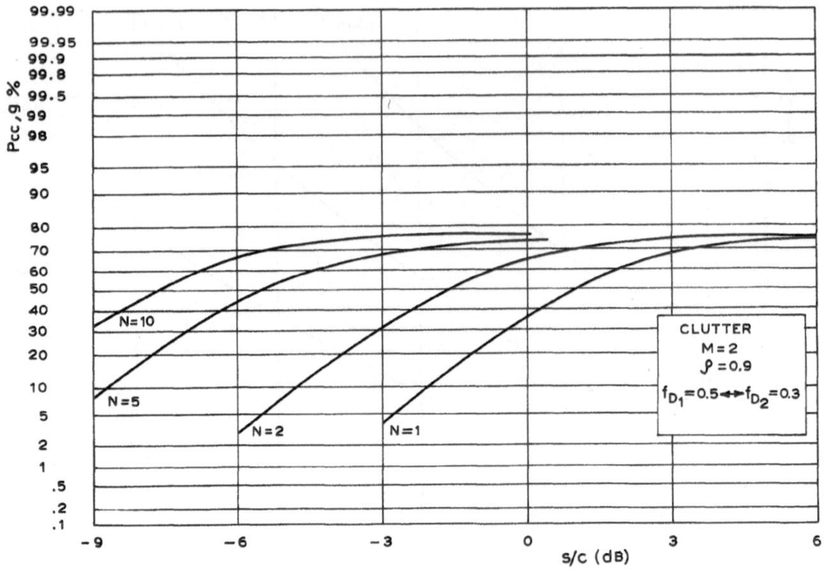


FIG.14 - PERFORMANCE OF A CLASSIFIER BASED ON MEAN DOPPLER FREQUENCY OF TARGET

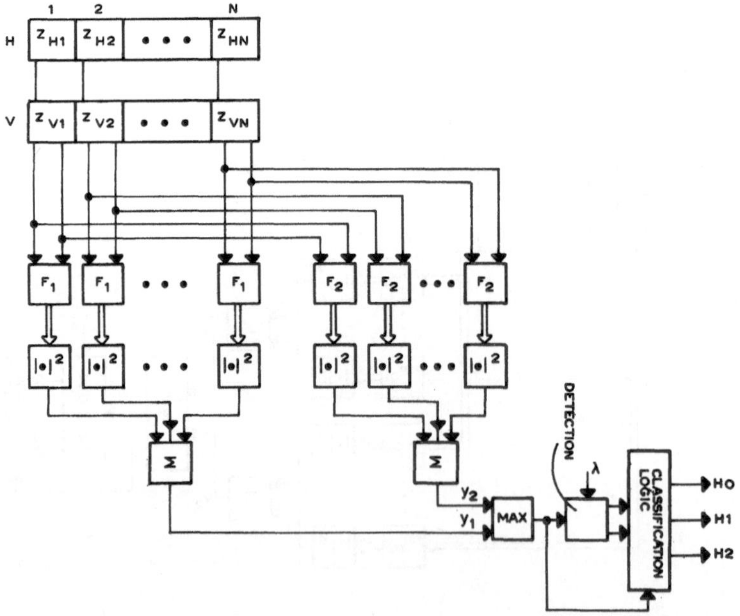


FIG.15 - CLASSIFIER BASED ON TARGET POLARIZATION IN DOWN RANGE

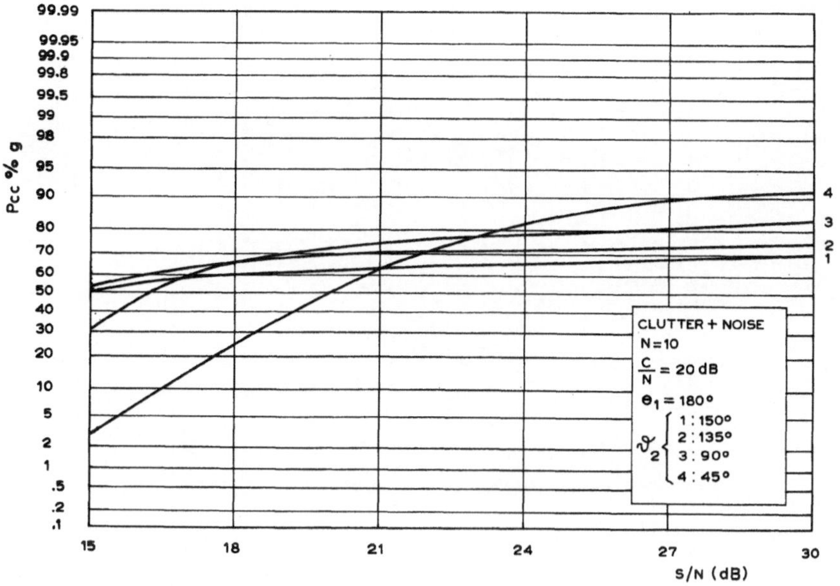


FIG.16 - PERFORMANCE OF A CLASSIFIER BASED ON TARGET POLARIZATION IN DOWN RANGE

# PLASMA RESONANCE EFFECTS IN RADAR BACKSCATTERING FROM METEOR TRAILS AS STUDIED BY THE SCATTERING MATRIX METHOD

P. S. P. Wei
Boeing Aerospace, Mail Stop 82-35
P. O. Box 3999
Seattle, Washington 98124, U.S.A.

**ABSTRACT.** The meteor trail is an ionized column of cylindrical symmetry. In the monostatic radar-target S matrix representation, the line-of-sight orientation angle dependence is separable from the rest of the target parameters.

Using polarimetric radio echoes measured from meteor trails in the 1950's, we analyze the time-dependent scattering contributions from the specular as well as the plasma resonance effects.

## 1. INTRODUCTION

Meteor bursts formed at heights of 85-115 km may provide a means for point-to-point communications (called MBC) beyond line-of-sight.[1] In a document issued by DCA in 1987,[2] the state-of-the-art on polarization[3-5] and other criteria for MBC was described. From a comprehensive review[3], we find that most of the theoretical and experimental research was done in the 1950's. With renewed interest on MBC, it is timely to reevaluate the physical mechanisms affecting the polarization, namely: Thomson scattering, Faraday rotation, and plasma resonance. To our best knowledge, the last effect has not been included in MBC modeling and simulation.

For underdense meteor trails (with electron line-density $\leq 10^{14}$/m) which occur more frequently, there is a plasma resonance effect associated with the polarization transverse to the axis of the trail so that the reflection coefficient may be several times higher than that for the longitudinal direction in the first half of lifetime for a trail. For meteor trails with preferred directions, one may need to take into account the resonance effect in order to correctly predict the usable hotspot distribution for a given combination of the transmit/receive antenna beam patterns and polarizations. Another consequence of the resonance effect is that once an underdense trail is excited it survives for a finite time (>0.1 sec) independent of the electron line-density. Thus, using the smaller yet more abundant meteor trails may be an



W.-M. Boerner et al. (eds.), *Direct and Inverse Methods in Radar Polarimetry, Part 1*, 1043–1056.

advantage for communications with short messages.

In this paper we analyze the available polarimetric radio echoes from the meteor trails[3-9] in terms of the radar-target scattering matrix S.[10] The work was initiated in the hope of better understanding the plasma resonance effect and obtaining an improved representation for radio wave scattering from the underdense meteor bursts. For an ionized column of cylindrical symmetry, it turns out that the S matrix may be written in a simple form containing 3 contributions, namely: the specular, the resonance, and the orientation. In what follows we will first review the meteor burst phenomena, give a brief description of the theoretical method, and then present the results on the time dependence of S. Possible implications are discussed.

## 2. RADIO WAVE ECHOES FROM METEOR BURSTS

The upper atmosphere of the earth is bombarded daily by billions of small meteors of speeds from 11 to 72 km/sec in the solar system. On collisions with the air molecules, the meteors are heated to evaporation and ionization with their trails, called meteor bursts, detectable by radio echoes due to Thomson scattering by the free electrons.[11] Although the trails may be of lengths up to a few ten's of km, it is the first half of a Fresnel zone (square root of "one-half wavelength times range", typically on the order of 1 km and formed in about 20 msec) on each side of the normal which contributes mainly to the backscattered signal. The higher order Fresnel zones cancel each other. As the ionized column expands, destructive interference from the front and back sides leads to an exponential decay of the echoes in a few tenths of a second. The distributions of underdense meteor bursts are given in Table 1.

Table 1. Abundance of Underdense Meteor Bursts.

| Meteor Mass (mg) | Radius (mm) | Number (per day) | Electron line-Density (per m) |
|---|---|---|---|
| 1 | 0.4 | $10^{8}$ | $10^{14}$ |
| 0.1 | 0.2 | $10^{9}$ | $10^{13}$ |
| 0.01 | 0.08 | $10^{10}$ | $10^{12}$ |

Meteors of a larger mass occur less frequently but do produce a higher line-density, called overdense, which is also observable by optical means. In addition to the sporadic meteors, there are also concentrated meteor streams or showers which move in comet-like elliptical orbits around the sun in varying periods of several years. It has been estimated that the earth would reduce the number of solid debris by sweeping them in its orbit around the sun to a factor of 1/e in some $10^8$ years. Therefore, we may consider the meteor distributions to be qualitatively similar from year to year. Some typical properties about the meteor burst environment are summarized in Table 2.

Table 2. Properties of Meteor Burst Environment.

| Height (km) | Mean Free Path (cm) | Diffusion Coef. D ($m^2$/sec) | Radius at 1-sec (m) |
|---|---|---|---|
| 85 | 1 | 1 | 2 |
| 115 | 70 | 140 | 20 |

For bistatic or forward scattering, the center of the principal Fresnel zone is the point of tangency of the meteor-trail axis with a prolate ellipsoid having the transmitter and receiver as foci.[3,12] For the underdense case, the ratio of power received over transmitted as a function of the scattering geometry and time t is given by:[2,3,12]

$$\frac{P_{rec}}{P_{trans}} = \frac{K_1 \sin^2\alpha}{(1 - \cos^2\beta \sin^2\phi)} \exp[-(K_2 + K_3\ t)\cos^2\phi], \quad \ldots(1)$$

where $\alpha$ is the angle between the incident electric vector at the meteor trail to the direction of scattering, $\phi$ is the half-angle subtended at the meteor trail by the transmitter and the receiver, and $\beta$ is the angle between the trail and the plane of scattering.

The factor $\sin^2\alpha$ is due to Thomson scattering.[11] For a linear horizontal polarized wave from the transmitter directed above the midpoint along a great circle path towards the receiver, the factor $\sin^2\alpha$ is always unity. However, for the vertically polarized wave, which is polarized in the plane of scattering, $\sin^2\alpha$ will be less than 1.

In eq.(1), the parameters $K_i$'s ($i$ = 1,2,3) are functions of range, antenna gain, wavelength, electron line-density, and other constants. In particular, the time for the signal amplitude to fall to 1/e of its initial value is given by:[3,12]

$$\tau = 2/(K_3 \cos^2\phi) = (\lambda \sec\phi)^2 / 16\pi^2 D, \quad \ldots(2)$$

where $\lambda$ is the wavelength and D is the diffusion coefficient. We see that $\tau$ is a minimum when $\phi = 0$ for the backscattering case. Note that the trail lifetime does not depend on the electron line-density.

Reflections of radio waves from electrical discharge columns had been studied in laboratories.[13,14] The amplitude of reflection, while depending on the discharge current and geometry, was much higher when the incident wave polarization was transverse rather than parallel to the column. From the literature, we can find several papers on the measurements of the polarization dependence of radio echoes from meteor bursts.[5-8] Their experimental parameters are summarized in Table 3.

Table 3. Conditions for The Backscattering Experiments.

| Reference | f (MHz) | λ (m) | Pulse (μsec) | Rate (Hz) | Power (kW) | Antenna Polarization |
|---|---|---|---|---|---|---|
| 6. Clegg & Closs (1951) | 75 | 4 | 8 | 600 | 40 | Linear Cross-dipoles |
| 7. Closs, Clegg, & Kaiser (1953) | 72 | 4.17 | 8 | 600 | 40 | Linear |
| 5. Billam & Browne (1956) | 55 | 5.4 | 8 | 600 | 50 | Linear |
| 8. Van Valkenburg (1952, 1954) | 22.9 | 13.1 | 350 | 240 | 3 | Circular L/R helices |

Some typical results are reproduced in Figure 1. Clegg and Closs[6] at the University of Manchester were the first to use cross-dipole antennas of parallel/transverse ( $\parallel$ , $\perp$ ) linear polarizations to study the radar echoes from meteor trails during meteor showers of known orientation in the 100-250 km range. They found that the ratios of amplitudes $g_\perp$ to $g_\parallel$ , measured alternately, reached a value of about 3 for 48 long duration echoes (>0.15 sec, such as that of the upper frame), and that for 170 short duration echoes (<0.15 sec, lower frame) the distribution of ratios could range between 1 and 12. The polarization dependence of the radio echoes, especially large in the transition region between short and long duration echoes with electron line-densities near $10^{14}$/m, was further studied by Closs et al[7] and by Billam and Browne[5], which confirmed the findings and provided more data.

Van Valkenburg from Stanford University used helical antennas, transmit right (R) and receive both R and left (L) circular, to study the meteor trails of arbitray orientations which occurred above his test station in the 80-120 km range at the quiet times after midnight.[8] He observed that the reflected wave polarization was linear at onset ($T_0$ to $T_1$), changed to elliptical through midway ($T_1$ to $T_2$), further changed to

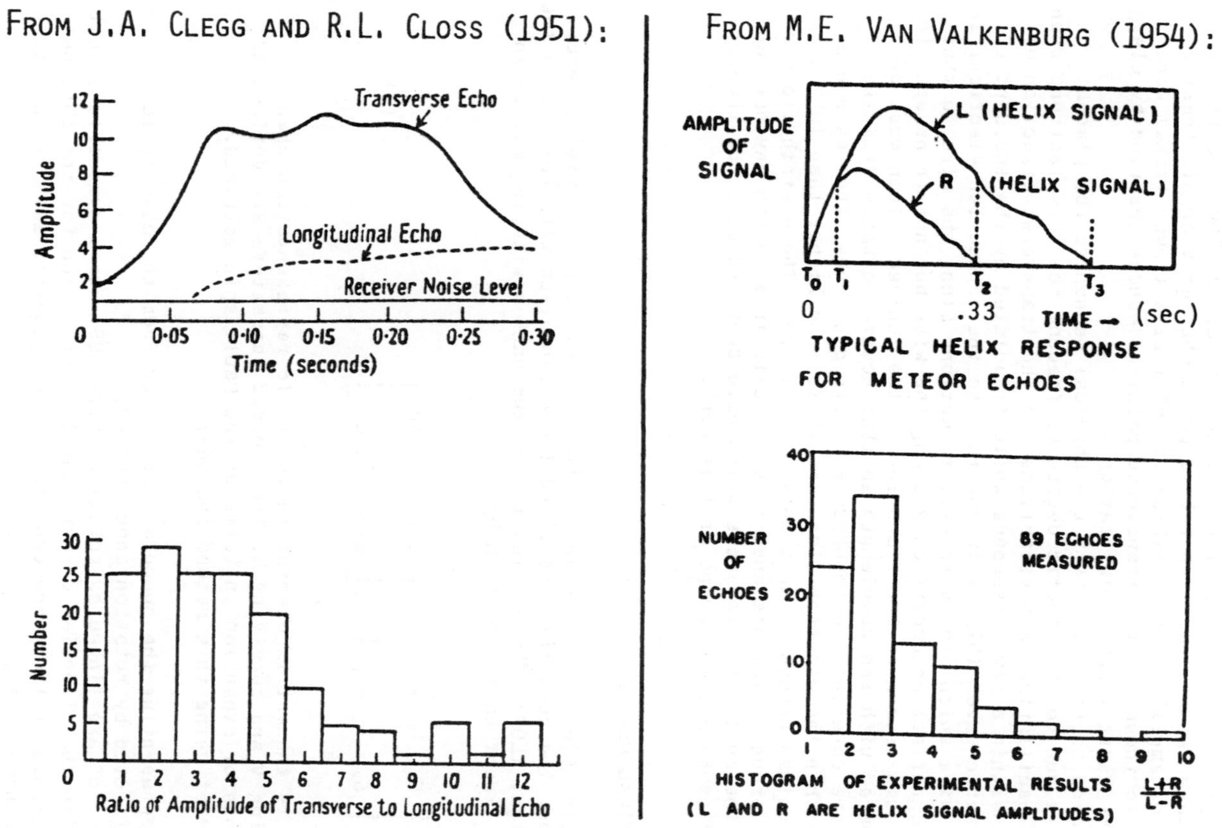


Figure 1. Typical radar echo amplitudes from a meteor trail measured by Clegg and Closs using linear polarizations,[6] and those by Van Valkenburg using circular polarizations.[8,9] The lower frames show the distribution of ratios for many underdense trails compiled by them.

L circular towards the end of the trail lifetime($T_2$ to $T_3$), and that L received amplitude was always larger than R received. The time from $T_0$ to $T_2$ was taken as 1/3 sec with reference to Chapter V of his thesis.[9] The ratios of amplitudes (R+L)/(R-L) for 89 trails (lower frame) were consistent to those for the transverse/parallel echoes observed by Clegg and Closs[4] using linear polarizations.

According to theory,[15-18] the physical mechanism for plasma resonance is due to the kinetic degree of freedom for the electrons when collective oscillations are established in the transverse direction of the column within a few $\mu$seconds while irradiated by the incident electromagnetic wave. This is in contrast to the conduction electrons in a metallic wire which are restricted to motions along the wire so that appreciable reflection occurs only along the wire but not transverse to it. For plasma resonance the dimension of the ionized column must be small compared with the wavelength so that scattered waves from the electrons are coherent and in phase at large distances. The resonance diminishes when the electron line-density is much higher than $10^{14}$/m, at which the column behaves like a metallic cylinder. The distributions of electrons versus radial distance from the axis in a column may have an effect on the multiple resonances which have been subjects for theoretical studies beyond the scope of this work.

## 3. THEORETICAL METHODS

For a single object, measurement of the complete S matrix yields information about the geometric shape and its orientation with respect to the line-of-sight (LOS). The complex two-by-two back scattering matrix in the (H,V) linear basis is given by:

$$S_{\substack{H\\V}} = \begin{bmatrix} S_{HH} & S_{HV} \\ S_{VH} & S_{VV} \end{bmatrix} = \begin{bmatrix} a+b & c-jd \\ c+jd & a-b \end{bmatrix} \qquad \ldots(3)$$

where $S_{HV}$ is the matrix element in the field representation when transmitting V and receiving H. The a,b,c,d parameters are defined as coefficients of expansion in terms of the Pauli spin matrices. Advantages of using this method include:

"a", representing the specular reflection from the object, is not affected by rotation about the LOS.

"b" and "c" are linked by rotation about LOS. For a symmetric object with the direction of horizontal polarization for the antenna aligned in the plane containing the symmetric axis of object, we have $c = 0$.

"d", representing the off-diagonal asymmetric contribution, is not affected by any change of orthonormal polarization basis. For a monostatic radar in a reciprocal condition, $d = 0$.

The radar cross section is given by:

$$RCS = Tr(S^{*T} \cdot S) = 2(a^*a + b^*b + c^*c + d^*d). \qquad \ldots(4)$$

We have studied the transformation of polarization basis for the monostatic S matrix.[10] In the left-right (L,R) circularly polarized basis, we have:

$$S_{\substack{L\\R}} = \begin{bmatrix} S_{LL} & S_{LR} \\ S_{RL} & S_{RR} \end{bmatrix} = \begin{bmatrix} (b-jc)e^{j2\psi} & -j(a+d) \\ -j(a-d) & -(b+jc)e^{-j2\psi} \end{bmatrix} \qquad \ldots(5)$$

where the a,b,c,d coefficients are the same ones as in (3) except that they are combined differently. In the circular basis,[19] the LOS orientation angle $\psi$ appears as phase factors for the diagonal terms.

The ionized meteor trail may be regarded as an object of cylindrical symmetry. For the purpose of analysis, we may define the longitudinal and transverse direction of the trail as our H and V, respectively. The measured backscattering amplitudes may then be readily identified as follows:

| | | |
|---|---|---|
| In linear basis: | Parallel echoes | $= S_{HH} = a+b$, |
| (Clegg, Closs, et al)[5-7] | Transverse echoes | $= S_{VV} = a-b$; |
| In circular basis: | Right receive | $= S_{RR} = b+jc = b$, |
| (Van Valkenburg)[8,9] | Left receive | $= S_{LR} = a+d = a$. |

For the present case, we have $c = 0$ and $d = 0$. Thus, from the published radar echoes[5-9] we may read off points as a function of time from the figures and compute the a and b coefficients, which may be referred to as the specular and the resonance contributions, respectively.

If the meteor trail, specified in the (H,V) frame, occurs at an LOS angle $\psi$ counter-clockwise with respect to the antenna reference frame (h,v), the measured S matrix may be written as:

$$S_{\substack{h\\v}} = \mathbf{R}(-\psi) \cdot S_{\substack{H\\V}} \cdot \mathbf{R}(\psi) = \begin{bmatrix} a + b\cos 2\psi & b\sin 2\psi \\ b\sin 2\psi & a - b\cos 2\psi \end{bmatrix} \qquad \ldots(6)$$

$$\text{where } \mathbf{R}(\psi) = \begin{bmatrix} \cos\psi & \sin\psi \\ -\sin\psi & \cos\psi \end{bmatrix} \text{ rotates the object counter-clockwise}$$

by $\psi$ , which is equivalent to rotating the antenna reference frame by $-\psi$ . Therefore, a given trail at any LOS angle may be represented by the S matrix of three coherent contributions, namely: the specular a, the resonance b, and the LOS orientation $\psi$ . The relative phase between a and b is taken to be zero, unless further studies may show it to be otherwise, or even as a function of time.

The above simplification holds for monostatic radar target analysis only. In the bistatic case, a rotation about the LOS with respect to the transmitter leads to changes in both the aspect angle and the LOS rotation angle with respect to the receiver. The LOS angle can be separated from the rest of the target parameters only when the bistatic angle is zero, namely, for the monostatic case. Except the recent paper by Cannon(4) on Faraday rotation losses in MBC systems, well defined bistatic experiments on the full polarimetric effects are yet to be found. Note that Faraday rotation by the D-region (around 60 km height) at daytime may be appreciable at 40 MHz but not as severe at 60 MHz.

## 4. ANALYTICAL RESULTS

From the literature,(5-9) we are glad to find fourteen sets of good measurements on the polarization dependence of radio echoes, all done with monostatic radars as summarized in Table 3. Though not exhaustive, we believe that these data are fairly representative of the physical phenomena occurring in meteor burst scattering. Fortunately, the available data allow us to derive the time-dependent behaviors of the trails.

For the purpose of easy comparison, we have plotted the normalized $a/a_o$ and $b/a$ as a function of time in Figures 2 to 4 using the method outlined above. Individual references or run numbers are as marked. The maximum peak heights $a_o$ (in units of mm, as measured from the published figures) are given on top of the respective peaks.

In Figure 2, normalized a and b coefficients from six sets of radio echoes are shown. Of particular interest are the two sets of curves computed with data (referring to the upper parts of Figure 1) taken from Clegg and Closs(6) (in dashes) and from the model curves of Van Valkenburg(8,9) (in dots). Although the radio echoes were measured by two groups of workers separated by a large distance on earth using different methods and at different times, the normalized a and b coefficients are in very good agreement with each other. Within about 60 msec from the onset, the fact that $b/a = -1$ means that the meteor trail responds like a transverse line-scatterer. The negative sign is due to the definition of longitudinal direction of the trail as H of the coordinates.

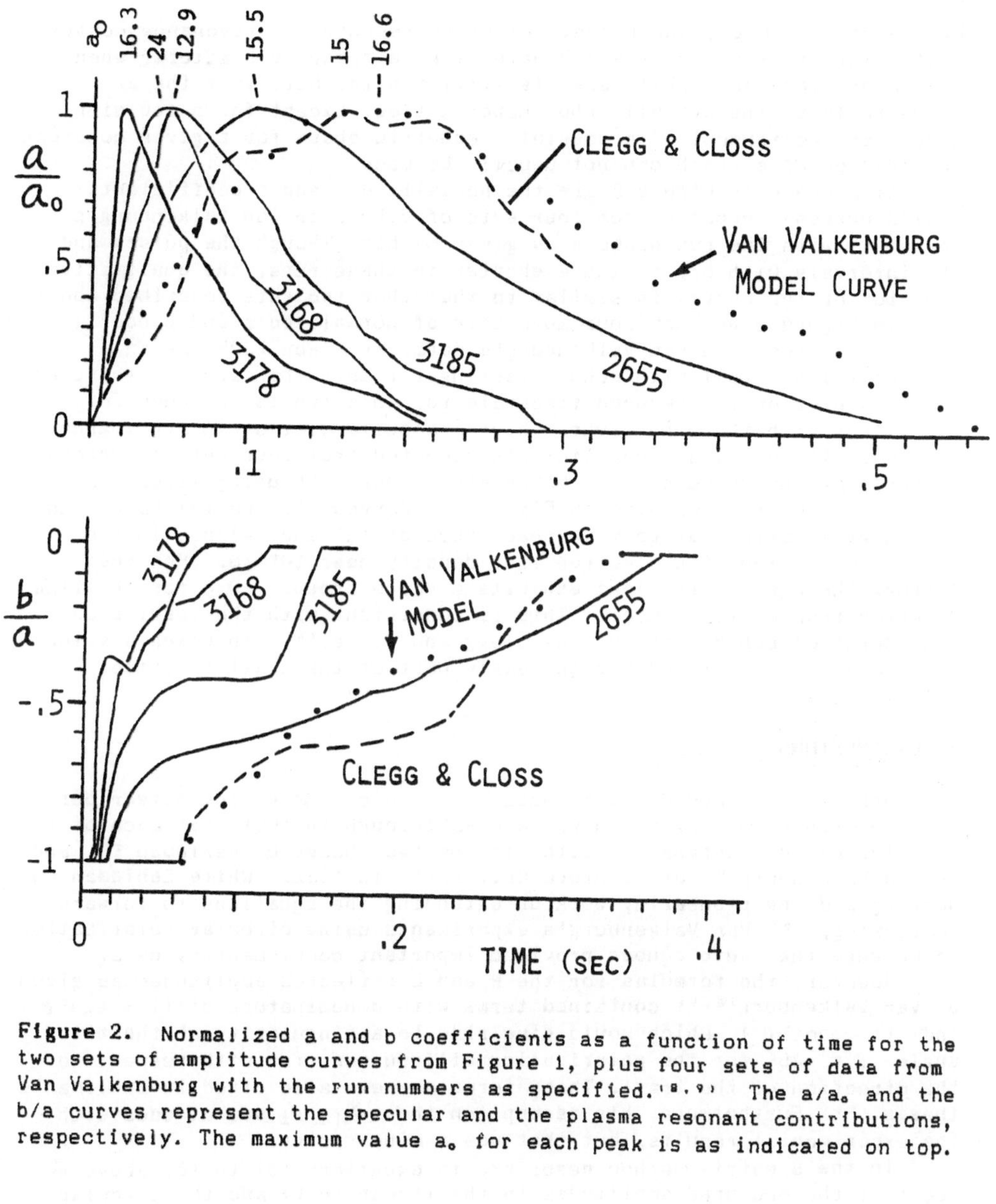


**Figure 2.** Normalized a and b coefficients as a function of time for the two sets of amplitude curves from Figure 1, plus four sets of data from Van Valkenburg with the run numbers as specified.[8,9] The $a/a_0$ and the b/a curves represent the specular and the plasma resonant contributions, respectively. The maximum value $a_0$ for each peak is as indicated on top.

From 60 to 300 msec, the b coefficient represents the resonance contribution. Afterwards, the trail behaves like a specular scatterer when b/a approaches zero while $a/a_o$ is still finite. Note that the $a/a_o$ curve reflects the overall echo-shape in time. Except for a few simple cases, the assignment of a possible geometric shape for a given coherent combination of a and b can not usually be made.

Also shown in Figure 2 are the normalized a and b coefficients (solid curves) computed from four sets of echoes in Van Valkenburg's work[8,9] with the run numbers as given by him. Though the pulses and the intervals with b/a = -1 are shorter in these runs, the qualitative behavior of the curves is similar to the other two sets described above.

In Figure 3 we show four more sets of normalized a and b coefficients computed from Van Valkenburg's data.[8,9] Run 2305 was an unusually long trail with echoes lasting for more than 1.4 sec (circled dots). A plot on 1/2 reduced timescale for this run is included for comparison with the other runs, which are quite similar in appearance.

Normalized a and b coefficients computed from four sets of echoes measured by the University of Manchester group[7,3] using linear polarizations are presented in Figure 4. Curves (1) and (2) belong to the short duration category whereas those of (3) and (4) are in the transition region with electron line-density near $10^{14}$/m. For the latter, the b/a coefficient exhibits a value around -0.8 for the time duration from 40 to 80 msec. This is consistent with the results on long-duration echoes observed by Clegg and Closs,[4] in which a value of b/a of -1 is found for the early part of the trail lifetime.

## 5. DISCUSSIONS

The work by Clegg and Closs[4] measuring radio echoes from meteor bursts with the linear polarization was a breakthrough in 1951 that each of the two figures were separately cited in the two theses by Eshleman[20] and by Van Valkenburg[9] of Stanford University in 1952. While Eshleman and Manning did the pioneering work on extending the equations to forward scattering,[12] Van Valkenburg's experiments using circular polarization to measure the radio echoes provided important complementary data.

However, the formulas for the R and L reflected amplitudes as given by Van Valkenburg[8,9] contained terms with denominators of $(1 - \tan^2\theta)$ and $(1 - \cot^2\theta)$ which would give rise to a singularity at the tilt angle $\theta = \pm 45°$ for the polarization ellipse measured with respect to the direction of the trail. These formulas were also cited in Eshleman's thesis.[20] Furthermore, his assumption that $\rho_{||} \geq \rho_{\perp}$ was at odds with the experimental results, but that was a minor point.

In the S matrix method described in equations (3) to (6) above we see that the measured amplitudes in the linear (H,V) and the circular (L,R) bases are simply related and that no singularity is encountered.

In eq.(1) for forward scattering from meteor bursts, there is a $\cos^2\beta$ dependence in the denominator. If it is mainly the transverse

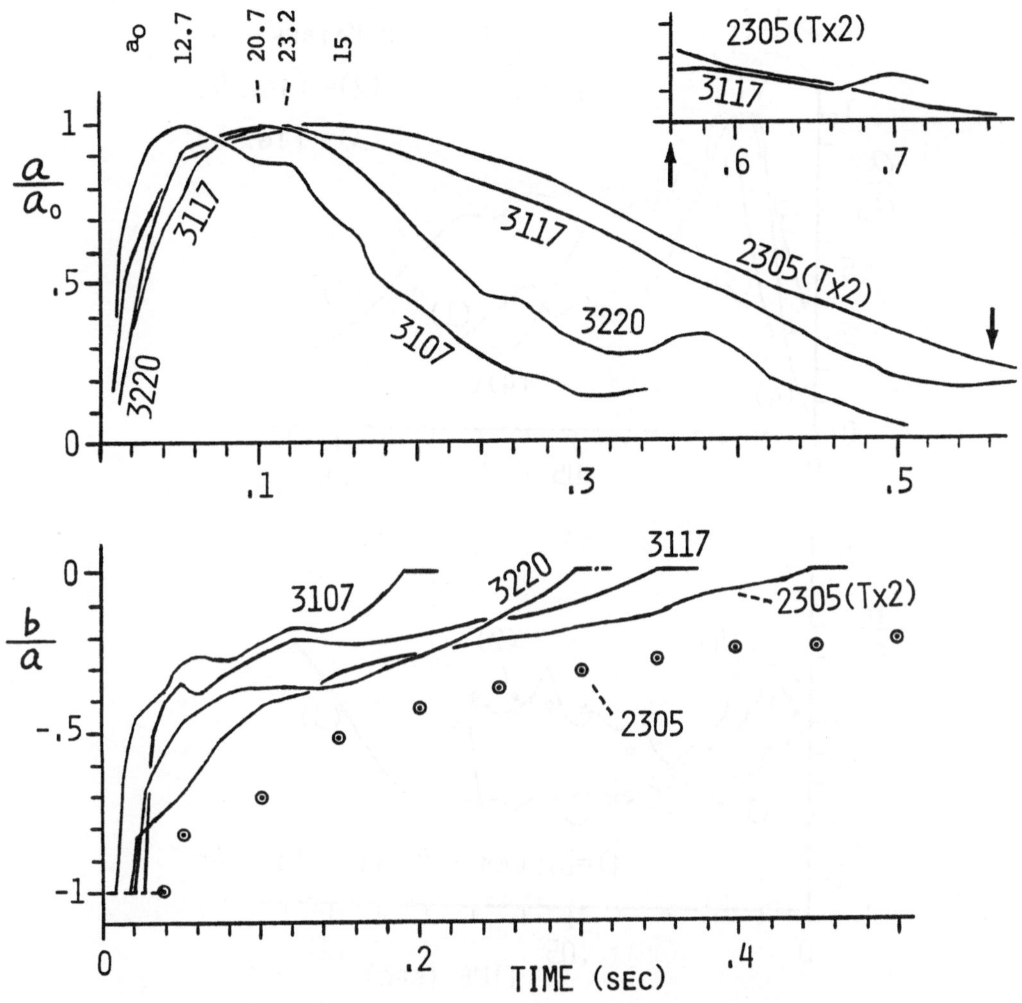


**Figure 3.** Normalized a and b coefficients as a function of time for four more sets of radar echoes measured by Van Valkenburg using circular polarizations with the run numbers as specified.[8,9] (See captions of Fig. 2 for other descriptions.)

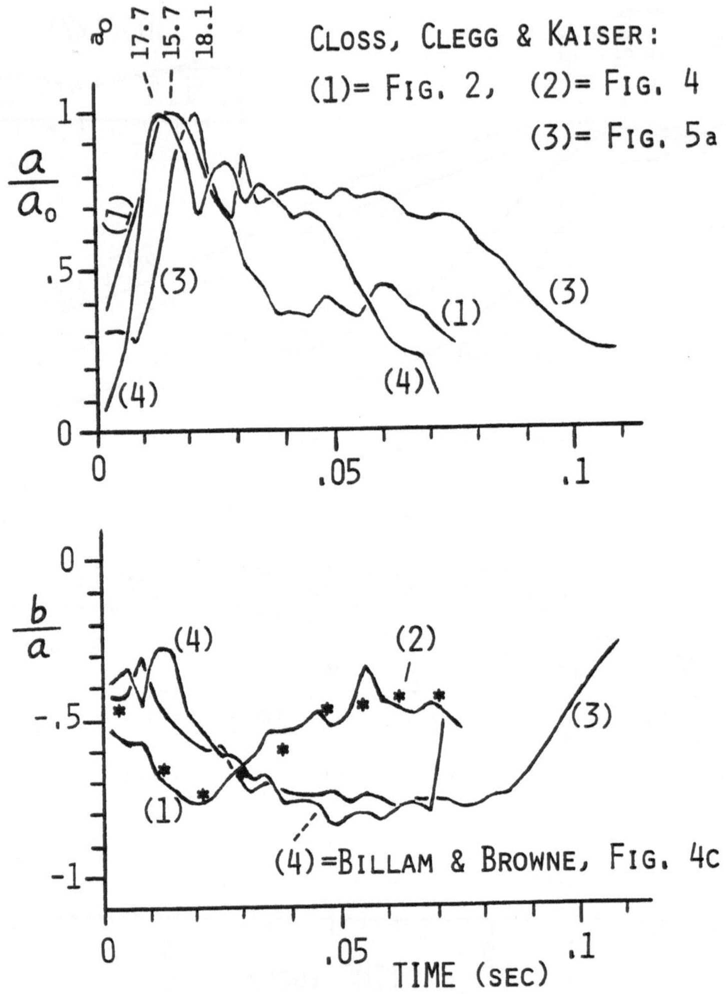


**Figure 4.** Normalized a and b coefficients as a function of time for data measured by the University of Manchester group using linear polarizations.[7,5] (See captions of Fig. 2 for other descriptions.)

direction which contributes to scattering, then the $\beta$-dependence for a given trail may have to be modified as $\sin^2\beta$. Otherwise, the calculated scattered power would not be correct.

If for each meteor trail there is another occurring at 90° to it in the same tangential plane to the prolate ellipsoid, then the statistical properties of the ensemble will not be sensitive to the angle $\beta$. This would not be true, however, if the meteor trails do occur with a preferred orientation rather than random, such as those occurring in morning and evening hours or during a meteor shower.

Due to the plasma resonance effect, once an underdense meteor trail occurs it is going to survive for a finite time ($\tau > 0.1$ sec) independent of the electron line-density. Thus, although the occurrance of meteor masses may follow certain (Poisson) distributions, the meteor trail lifetimes do not. This is clearly demonstrated by figure 12 in the paper by Kaiser and Closs.[17] The smaller yet more abundant meteor trails may be used to our advantage for designing an MBC system with short messages.

## 6. CONCLUSIONS

From a study of theory and experiments in the literature we conclude that the radio echoes from underdense meteor bursts are stronger in the transverse direction than in the longitudinal during the first half of its lifetime due to the plasma resonance effect. Using the S matrix method, we have separated the backscattering contribution into three parts, namely, the specular, the resonance, and the LOS orientation. In Figures 2, 3 and 4, normalized a and b coefficients are plotted to illustrate the temporal behaviors of the specular and the resonance contributions, respectively. Our results may be useful for an improved modeling on vector scattering from meteor bursts.

## 7. ACKNOWLEDGEMENTS

The author is grateful to Henry A. Gratrix for his interest and support in this work. Helpful discussions with William M. Leavens on meteor burst physics and with J. Richard Huynen on the S matrix format are appreciated.

**REFERENCES**


1. E. J. Morgan, "The resurgence of meteor burst", SIGNAL magazine, January, pp.69-73 (1983).
2. "A proposed standard for Meteor Burst Communications interoperability", U.S. Defense Communications Agency draft document, Part I pp. 4-1 to 4-19 (24 August 1987).
3. G. R. Sugar, "Radio propagation by reflection from meteor trails", Proc. IEEE 52, 116-136 (1964).
4. P. S. Cannon, "Polarization rotation in meteor burst communication systems", Radio Sci., 21, 501-510 (1986).
5. E. R. Billam and I. C. Browne, "Characteristics of radio echoes from meteor trails, IV: polarization Effects", Proc. Phys. Soc. B 69, 98-113 (1956).
6. J. A. Clegg and R. L. Closs, "Plasma oscillations in meteor trails", Proc. Phys. Soc. B 64, 718-719 (1951).
7. R. L. Closs, J. A. Clegg & T. R. Kaiser, "An experimental study of radio reflections from meteor trails", Phil. Mag.44,313-324 (1953).
8. M. E. Van Valkenburg, "The two-helix method for polarization measurement of meteoric radio echoes", J. Geophys. Res. 59,359-364 (1954).
9. M. E. Van Valkenburg, "Polarization and fading studies of meteoric radio echoes", PhD dissertation, Stanford University (Feb., 1952).
10. P. S. P. Wei, J. R. Huynen and T. C. Bradley, "Transformation of polarisation bases for radar target scattering matrix", Electr. Lett., 22, pp.13-14 (1986).
11. A. H. Compton and S. K. Allison, "X-Rays in theory and experiment", (Van Nostrand Co., N. Y., 2nd ed., 1935) pp. 116-119.
12. V. R. Eshleman and L. A. Manning, "Radio communication by scattering from meteoric ionization", Proc. IRE 42, 530-536 (1954).
13. L. Tonks, "Plasma-electron resonance, plasma resonance, and plasma shape", Phys. Rev. 38, 1219-1223 (1931).
14. D. Romell, "Radio reflexions from a column of ionized gas", Nature, 167, 243 (1951).
15. N. Herlofson, "Plasma resonance in ionospheric irregularities", Arkiv Fysik, 3, 247-297 (1951).
16. J. Feinstein, "The interpretation of radio echoes from meteor trails", J. Geophys. Res. 56, 37-51 (1951).
17. T. R. Kaiser and R. L. Closs, "Theory of radio reflections from meteor trails", Phil. Mag. 43, 1-32 (1952).
18. W. M. Leavens, "Scattering resonances of a cylindrical plasma", Radio Sci. 69D, 1321-1333 (1965).
19. J. T. deBettencourt, "The polarization of downcoming ionospheric radio waves", ScD thesis, Harvard University (May, 1949).
20. V. R. Eshleman, "The mechanism of radio reflections from meteoric ionization", PhD dissertation, Stanford University (July, 1952).

# FORMULATION OF SCATTERING MATRIX FOR ELECTROMAGNETIC WAVES UNDER NON-LINEAR TRANSFORMATION: NON-LINEAR OXIDIZED METAL JUNCTION HIGHER HARMONICS RERADIATION EFFECTS

**Aleksander I. Logvin and Anatolij I. Kozlov**

Moscow Civil Aviation Engineering Institute
MIIGA, Kronshtadsky Boulevard 20
USSR 125-493 MOSCOW

**ABSTRACT:** The phenomenon of the strongly polarization-dependent, non-linear metal junction higher harmonics reradiating radar effect is re-visited and approached using a rather generalized novel m x n scattering matrix formulation. For this purpose the concept of the "effective polarization cross-section" (EPC), $\sigma_n$, is introduced, a formalistic exposition on its derivation is developed, and a final expression for $\sigma_n$ is presented.

## 1. INTRODUCTION

Improving the quality of radar systems is based on updating existing methods and developing new methods of forming radar signals and studying reflected ones. The methods of studying radar signals based only on the use of energy characteristics without paying due consideration to a target's polarization state transformation properties have been exhausted to a considerable extent. They make it impossible to use many specific features of radar targets inherent in the sophisticated structure of reflected signals.

The necessity for detecting and singling out objects from a group of local objects aroused interest in studying non-linear polarization-dependent effects resulting from electromagnetic waves reflected from the targets. The need for the detection and the selection of small-dimensional metallic objects in a heterogeneous complex backscattering environment aroused interest for studying non-linear effects arising when electromagnetic waves are reflected from such objects.

In irradiating a radar target having "non-linear" polarization-dependent features by transmitting signals at f frequency and receiving reflected signals at 2f and 3f harmonic frequencies, a combination of useful signals (at 2f or 3f harmonics) and the receiver's thermal noise will be formed at the receiving device output [1]. Hence it is not the task of differentiating between signals reflected from a target and clutter but the task of detecting a useful signal in the noise background that will be solved here. In this case an increase in sensing signal energy results in an increase in the energy of the reflected one, and with it in an increase of the $J_c/N_o$ ratio (where $J_c$ is the energy of a signal



*W.-M. Boerner et al. (eds.), Direct and Inverse Methods in Radar Polarimetry, Part 1,* 1057–1064.

an increase of the $J_c/N_o$ ratio (where $J_c$ is the energy of a signal received, and $N_o$ is noise spectral density) especially under consideration of proper transmit/receive polarization state selection. In solving the task of differentiation, the useful signal-to-noise ratio remains constant whatever the value of sensing signal energy, which radically limits the possibilities for signal accumulation. As for the task of detecting it, the problem becomes quite the contrary; namely, then signal intensity accumulation of optimally polarized waves, received at higher order harmonics, is the main instrument for detecting "non-linear" targets.

Non-linear effects are caused by the presence of all kinds of oxides on the object surface which form contacts of different types, which at the same time display marked polarization state sensitivity toward the transmitted and received waves [1]. Non-linear contacts of the semiconductor-semiconductor type are characterized by a transfer function of an exponential type [2]. As a result, a variety of harmonics with maximum level at the second harmonic (Fig 1a [2]) is formed. For non-linear contacts of metal-oxide-metal type the spectrum of a reflected signal has an explicit maximum on the third harmonic (Fig. 1b [2]).

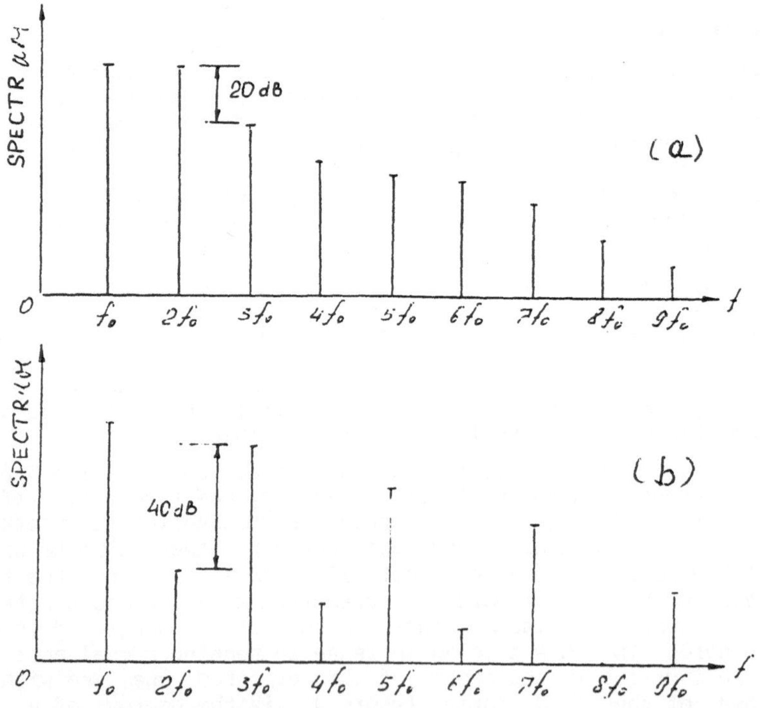


Fig. 1 Reflected signal spectrum for "non-linear" object with contacts semiconductor (a) and metal-oxide-metal (b) types (according to [2]).

## 2. FORMULATION OF THE PROBLEM

In non-linear reflection of polarized waves from oxidized metal junctions, it is useful to introduce the concept of the effective polarization cross-section (EPC) on the nth harmonic - $\sigma_n$, defined by the following equation:

$$\sigma_n = 4\pi R^2 E^2_{ref,n}/E^2_{inc} \quad , \tag{1}$$

where $E_{ref}$, $E_{inc}$ are the moduli of the scattered and incident wave electric vectors, and R - is the distance from the object to the receiver. Dimensions of the effective polarization-cross-section (EPC) are unusual: $M^{2n}\ B^{(1-n)} \triangleq (\text{length})^{2n}\ (\text{effective power})^{(1-n)}$ [3].

It may be shown [3] that the maximum range of detection, $R_{max}$, will be determined by the following equation:

$$R_{max} = \sqrt[2(n+1)]{\frac{G_n\lambda_n^2\sigma_n}{30(4\pi)^3}\ \frac{[(30P_\Sigma G_1)^n]_m}{P_{Amin}}} \quad , \tag{2}$$

where $G_1$ is the gain ratio of a transmitting antenna on the first harmonic, $G_n$ is the same for a receiving antenna on the nth harmonic, $\lambda_n$ is the wavelength corresponding to n-th frequency, index m implies mean value, $P_{Amin}$ - minimum power, fixed by a receiver.

Equation (2) results in a very important conclusion, namely that the maximum range depends on the form of the sensing signal. That is why the expressions of $R_{max}$ will be different for continuous and impulsive modes of operation.

For continuous mode:

$$R_{max} = \sqrt[2(n+1)]{\frac{G_n\lambda_n^2\sigma_n}{30(4\pi)^3}\ \frac{(30P_\Sigma G_1)^n 2t_o}{g_{on}\ N_o}} \tag{3}$$

For impulsive mode:

$$R_{max} = \sqrt[2(n+1)]{\frac{2P_m\ P_i^{n-1}\ t_o\ G_1^n G_n \lambda_n^2 \sigma_n}{(4\pi)^3\ g_{on}^3 N_o}} \tag{4}$$

where $g_{on} = \sqrt{2\ \ J_{min}/N_o}$, $J_{min}$ is the minimum power value necessary for object detection, $t_0$ is the time of signal observation, $P_m$ is the mean signal power, $P_i$ is the impulsive signal power; and $N_O$ represents the product of the system noise figure, system noise temperature, and Boltzmann's constant [3].

It is clear from inspection of the equation given that in order to increase $R_{max}$, besides the technical design aspects associated with the characteristics of antennas $G_1, G_n$, transmitter and receiver $P_1$, $P_m$, $q_{on}$; there exists the possibility of increasing the effective polarization cross-section (EPC) of a non-linear object (target) by changing the incident electromagnetic wave polarization state, and possibly also that of the receiving antenna (requiring a separate transmit versus receive, coherent, dual polarization, multi-spectral radar system), as is discussed elsewhere.

Classical radiolocation is based on presenting coherent radiowave reflection in the form of some linear process, the scattering matrix of radar objects being their essential characteristic. The scattering matrix for the coherent linear case is a square 2x2 matrix, the three complex elements of which are independent of each other for reciprocal monostatic operation. However, when the reflection process is of a non-linear character, i.e., described by a power series, the scattering matrix in the above sense becomes meaningless and to describe the reflective properties of a non-linear object in interacting with polarized waves, a radically new mathematical mechanism based on some generalized concept of a "scattering matrix" is required. The task of obtaining it and analyzing these matrix properties in different polarization bases is not trivial and require a special approach. As an example, consider the solution to this problem for a particular simplified case of "square-law" reflection.

In non-linear scattering matrices, columns characterizing the incident $\vec{E}_{inc} = (E_x\ E_y)^T_{inc}$ and the reflected $\vec{E}_{ref} = (E_x\ E_y)^T_{ref}$ fields are related to each other through some non-linear transformation (T means a transposing operation):

$$E_{x,y_{ref}} = \sum_k C_{K_{x,y}} \; E^k_{x,y_{inc}} \tag{5}$$

The physical cause of the existence of the correlation (5) is, as has already been mentioned, the emerging of spectral frequency components different from those of an incident wave spectrum. In this case of a purely square-law non-linearity all $C_k$, except at $k = 2$, are equal to 0.

In order for developing a special mathematical apparatus and to introduce a generalization of the scattering matrix concept, it is necessary to find such a matrix transformation between the elements of matrix-columns $\vec{E}_{inc}$ and $\vec{E}_{ref}$ which would be invariant to different decompositions of the complex vector describing an incident wave. In other words, the desired transformation should retain its form (but not the coefficients) in all of the decompositions mentioned above. It can be shown that the invariance needed is provided by the following transformation relating the square-law forms of the incident wave components with the reflected wave components:

$$(E_x \; E_y)^T_{ref} = \begin{bmatrix} \varepsilon_{11} & \varepsilon_{12} & \varepsilon_{21} & \varepsilon_{22} \\ q_{11} & q_{12} & q_{21} & q_{22} \end{bmatrix} (|E_x|^2 \;\; E_x E_y^* \;\; E_x^* E_y \;\; |E_y|^2)^T_{inc} \tag{6}$$

where * denotes complex conjugation, and T transposition.

Equation (6) can be presented in the compact symbolic form:

$$\vec{E}_{ref} = [S_2] \, |\vec{E}_{inc}|^2 \tag{7}$$

The meaning of the symbolism introduced is quite clear. It is seen that the non-linear reflector is described by a rectangular 2×4 matrix comprising 8 members. The $[S_2]$ matrix which is an analogue to the classic scattering matrix is a complete characteristic feature of the square law reflector. It should be pointed out that out of the 16 real members describing $[S_2]$, 12 matrix elements are independent ones. Deducing equation (7) was not based on relating it with some polarization basis (PB). Specially chosen X,Y should be related not only to the linear (Cartesian) basis but to any orthogonal elliptic polarization basis. When the polarization basis changes, the $[S_2]$ matrix elements change too. Let us consider this phenomenon in detail.

As it is known [3,4], when the polarization basis changes, the matrix columns of the electric vector of the reflected wave in the old $(\vec{E}_0)_{ref}$ and new $(\vec{E}_n)_{ref}$ are related to each other via a linear transformation by means of an unitary matrix $[Q_1]$:

$$(\vec{E}_n)_{ref} = [Q_1](\vec{E}_0)_{ref}$$

$$[Q_1] = \begin{bmatrix} e^{i\eta}\cos\alpha & e^{-i\xi}\sin\alpha \\ -e^{i\xi}\sin\alpha & e^{-i\eta}\cos\alpha \end{bmatrix} , \tag{8}$$

where the parameters $\alpha$, $\eta$, $\xi$ determine parameters of the new polarization basis.

As for the transformation of the square-law matrix of the incident wave $[\vec{E}]^2_{inc}$, when the polarization basis changes, the following transformation is valid:

$$(\vec{E}_0)^2_{inc} = [Q_2](\vec{E}_N)^2_{inc}\ , \tag{9}$$

where, the 4 x 4 transformatio matrix $[Q_2]$ assumes the form:

$$[Q_2] = \begin{pmatrix} \cos^2\alpha & 0.5e^{i(\eta+\xi)}\sin2\alpha & 0.5e^{-i(\eta+\xi)}\sin2\alpha & \sin^2\alpha \\ -0.5e^{i(\eta-\xi)}\sin2\alpha & e^{2i\eta}\cos^2\alpha & -e^{-2i\xi}\sin^2\alpha & 0.5e^{i(\eta-\xi)}\sin2\alpha \\ -0.5e^{-i(\eta-\xi)}\sin2\alpha & -e^{2i\xi}\sin^2\alpha & e^{-2i\eta}\cos^2\alpha & 0.5e^{-i(\eta-\xi)}\sin2\alpha \\ \sin^2\alpha & -0.5e^{i(\eta+\xi)}\sin2\alpha & -0.5e^{-i(\eta+\xi)}\sin2\alpha & \cos^2\alpha \end{pmatrix} \tag{10}$$

Direct verification shows that the matrix $[Q_2]$ is also unitary. It's interesting to note that the matrix $[Q_2]$ is the same as the matrix for transforming the covariance matrix in linear scattering [3,4].

In order to find out how the matrix $[S_2]$ changes when the polarization basis changes, it is necessary to insert equations (7) and (9) into equation (6) and to take the complex conjugated value as its right-hand part:

$$(\vec{E}_n)_{ref} = [Q_1]^T[S_2]_o\ [Q_2](\vec{E}_n)^2_{inc} \tag{11}$$

Thus, in changing the polarization basis, the matrix $[S_2]$ undergoes the following transformation:

$$[S_2]_n = [Q_1]^T[S_2]_o[Q_2], \tag{12}$$

where $[Q_1]$ and $[Q_2]$ are unitary matrices.

To determine the power of a wave reflected from a target, the following transformation must be carried out:

$$\begin{aligned} \Pi = \vec{E}^T_{ref}\cdot\vec{E}^*_{ref} &= \{[Q_1]^T[S_2]_o[Q_2](\vec{E}_n)^2_{inc}\}^T\{[Q_1]^{T*}[S_2]^*_o\times \\ \times [Q_2]^*(\vec{E}^*)^2{}_{inc}\} &= (\vec{E}^T_n)^2_{inc}[Q_2]^T[S_2]^T_o[Q_1][Q_1]^{T*}[S_2]^*_o\times \\ \times [Q_2]^*(\vec{E}^*_n)^2_{inc} &= (\vec{E}^T_n)^2_{inc}[Q_2]^T\ [G_o][Q_2]^*(\vec{E}^*_n)^2_{inc}\ , \end{aligned} \tag{13}$$

where the unitary matrix $[Q_1]$ is used, i.e., $[Q_1]\ [Q_1]^{T*} = [I]$ and the designation $[G_0] = [S_2]^T_0[S_2]^*_0$ is introduced. The matrix $[G_0]$ is a square conjugated 4x4 matrix which is an analogue to Graves power matrix. It is seen from equation (13) that in transforming the polarization basis, the matrix $[G_0]$ undergoes a similarity transformation:

$$[G] = [Q_2]^T [G_0] [Q_2]^* \tag{14}$$

The explicit form of matrix $[G_0]$ then becomes:

$$[G_0] = \begin{pmatrix} |\varepsilon_{11}|^2+|q_{11}|^2 & \varepsilon_{11}\varepsilon_{12}^*+q_{11}q_{12}^* & \varepsilon_{11}\varepsilon_{21}^*+q_{11}q_{21}^* & \varepsilon_{11}\varepsilon_{22}^*+q_{11}q_{22}^* \\ \varepsilon_{11}^*\varepsilon_{12}+q_{11}^*q_{12} & |\varepsilon_{12}|^2+|q_{12}|^2 & \varepsilon_{12}\varepsilon_{21}^*+q_{12}q_{21}^* & \varepsilon_{12}\varepsilon_{22}^*+q_{12}q_{22}^* \\ \varepsilon_{11}^*\varepsilon_{21}+q_{11}^*q_{21} & \varepsilon_{12}^*\varepsilon_{21}+q_{12}^*q_{21} & |\varepsilon_{21}|^2+|q_{21}|^2 & \varepsilon_{21}\varepsilon_{22}^*+q_{21}q_{22}^* \\ \varepsilon_{11}^*\varepsilon_{22}+q_{11}^*q_{22} & \varepsilon_{12}^*\varepsilon_{22}+q_{12}^*q_{22} & \varepsilon_{21}^*\varepsilon_{22}+q_{21}^*q_{22} & |\varepsilon_{22}|^2+|q_{22}|^2 \end{pmatrix} \tag{15}$$

Equations (6,15) and presentation (16) make it possible to give recommendations on the experimental determination of the elements of the matrix $[S_2]$. In order to determine the latter, it is necessary to carry out measurements on 4 polarizations: for instance, horizontal $\vec{E}^{(1)}_{inc} = (1\ \ 0)^T$ ,vertical $\vec{E}^{(2)}_{inc} = (0\ 1)^T$,linear with $\pi/4$ angle $\vec{E}^{(3)}_{inc} = 0.5\sqrt{2}(1\ \ 1)^T$, and circular $\vec{E}^{(4)}_{inc} = 0.5\sqrt{2}(1\ i)^T$ .

The knowledge of the reflected wave amplitudes and phases of the above polarizations makes it possible to determine $\varepsilon_{mn}$, $q_{mn}$.

If in equation (5) all $k \neq 0$, then instead of equation (7), we'll have:

$$\vec{E}_{ref} = \sum_{k=1}^{\infty} [S_k]\ (\vec{E})^{(k)}_{inc} \tag{16}$$

It can be shown that in changing the polarization basis, the matrices $[S_k]$ will be transformed in the following way

$$[S] = \sum_{k=1}^{\infty} [Q_k]^T [S_k][Q_k] \tag{17}$$

where $[Q_k]$ is a unitary m x m matrix which is formed in a way similar to $[Q_2]$, being determined by equation (10).

In order to determine the dependence of the non-linear effective polarization cross section (EPC) $\sigma_2$ on polarization type, it is possible to use equations (1) to (11) which, after strongly stirring the equations, yields

$$\sigma_2 = \frac{(\vec{E}_n^T)^2_{inc}[Q_2]^T[G_o][Q_2]^*(\vec{E}_n^*)^2_{inc}}{\{(\vec{E}_n^T)_{inc}\ (\vec{E}_n^*)_{inc}\}^2} \tag{18}$$

To obtain the explicit dependence $\sigma_2$ on polarization type, it is necessary to know the values of all of the complex elements of the matrix $[G_o]$ which cannot be made available with existing dual polarization radar systems at the present time. Instead, we badly require more advanced coherent dual orthogonal polarization, polarization state adaptive radar systems with appropriate supra-fast switching capabilities of polarization state adaptive (dual polarization, purely coherent) antenna systems.

## CONCLUSION

An important problem of detecting man-made metallic objects (targets) with oxidized metal junctions in a severe heterogeneous background environmental (clutter) was considered. The novel innovative approach of dealing with non-linear polarimetric wave reflection and scattering effects should be of special interest to the instantaneous detection, discrimination and identification of hostile metallic targets in a heterogeneous vegetative and/or rocky natural environment for which standard linear detection schemes fail.

## ACKNOWLEDGEMENTS

The authors wish to acknowledge the generous assistance of the workshop director and proceedings editor in accommodating and in the re-editing of this paper.

## REFERENCES

1. Elsner, R.F., ed., 'Vehicular Variable Parameter **MET**al Reradiating **RA**dar (METRRA) Systems', Final Report, IITRI, Techn-Rept. No. E6224 (AD 782214), May (1974) (Illinois Institute of Technology-Research Institute)

2. Flemming, M., Mullins, F., Watson, W., 'Harmonic radar system', Radar-77, Int. Conf., London, 25-28 Oct, (1977), p.552-554

3. Bogorodtsky V.V., Kanareikin D.B., Kozlov A.I., Scattering and Proper Radioradiation Polarization of Earth Surfaces, Leningrad, Gidrometeoizdat, (1980)

4. Kozlov, A.I., 'Several properties of scattering matrix element parameters of radiolocation (radar) targets', Radio-Electronika, (1979), Vol. 22